ZOOLOGY

FOR DEGREE STUDENTS

Comparative Anatomy of Vertebrates | Physiology II | Biochemistry II

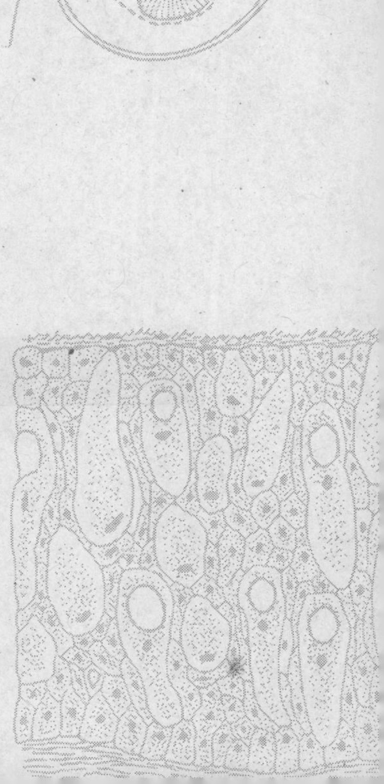

About the Author

Dr V K Agarwal, MSc; PhD, has a rich teaching experience of more than four decades at Meerut College, Meerut. He has taught subjects such as Cell Biology, Genetics, Molecular Biology, Chordate Embryology (Developmental Biology), Invertebrate Zoology, Chordate Zoology, Evolution, Ecology and Animal Behaviour (Ethology) to undergraduate and postgraduate students.

Dr Agarwal has to his credit the authorship of several widely acclaimed textbooks such as *Cytology*, *Cell Biology*, *Genetics*, *Chordate Embryology* (*Developmental Biology*), *Ecology* (*Environmental Biology*), *Ethology* (*Animal Behaviour*), *Zoology for Degree Students–BSc First Year* (includes *Invertebrate Zoology*), *Zoology for Degree Students – Semesters I, II, III* and *V*, *Molecular Biology, Genetic Engineering, CBSE–IX Biology,* and a multicoloured edition of *Cell Biology, Genetics, Molecular Biology, Evolution and Ecology.* He has also published a few research papers on Insect Physiology and Toxicology.

ZOOLOGY

FOR DEGREE STUDENTS

Comparative Anatomy of Vertebrates |

Physiology II | Biochemistry II

BSc (Hons) Semester IV

As per UGC Choice Based Credit System (CBCS)

V K Agarwal
MSc, PhD
Former Head of the Zoology Department
Meerut College, Meerut, Uttar Pradesh

S. CHAND
PUBLISHING

S Chand And Company Limited

(ISO 9001 Certified Company)

S Chand And Company Limited

(ISO 9001 Certified Company)

Head Office: D-92, Sector–2, Noida – 201301, U.P. (India), Ph. 91-120-4682700

Registered Office: A-27, 2nd Floor, Mohan Co-operative Industrial Estate, New Delhi – 110 044, Phone: 011-49731800

www.**schandpublishing.com**; e-mail: **info@schandpublishing.com**

Marketing Offices:

Chennai	:	Ph: 23632120; chennai@schandpublishing.com
Guwahati	:	Ph: 2738811, 2735640; guwahati@schandpublishing.com
Hyderabad	:	Ph: 40186018; hyderabad@schandpublishing.com
Jalandhar	:	Ph: 4645630; jalandhar@schandpublishing.com
Kolkata	:	Ph: 23357458, 23353914; kolkata@schandpublishing.com
Lucknow	:	Ph: 4003633; lucknow@schandpublishing.com
Mumbai	:	Ph: 25000297; mumbai@schandpublishing.com
Patna	:	Ph: 2260011; patna@schandpublishing.com

First Edition 2020
Reprints 2020 (Twice), 2021

Reprint 2022 (Thrice)

ISBN: 978-93-525-3410-4 **Product Code:** H6ZOO68ZOOL10ENAA20O

PRINTED IN INDIA

By Vikas Publishing House Private Limited, Plot 20/4, Site-IV, Industrial Area Sahibabad, Ghaziabad – 201 010 and Published by S Chand And Company Limited, A-27, 2nd Floor, Mohan Co-operative Industrial Estate, New Delhi – 110 044.

Preface

It gives me immense pleasure and satisfaction to present this textbook *Zoology for Degree Students – Semester IV* covering Comparative Anatomy of Vertebrates, Animal Physiology and Biochemistry. This comprehensively written textbook provides up-to-date information on the subject in an informative and systematic manner. It is an outcome of my long teaching experience of over four decades at both undergraduate and postgraduate levels.

This textbook has been designed as per the UGC Choice Based Credit System (CBCS) for the students of BSc (Hons) Zoology. It includes the following three core courses (CC):

1. CC VIII: Comparative Anatomy of Vertebrates
2. CC IX: Animal Physiology–Life Sustaining Systems
3. CC X: Biochemistry of Metabolic Processes

The organization of the *Comparative Anatomy of Vertebrates* is intended to support and encourage Zoology students to be actively engaged in the material and to think critically about the functional and evolutionary significance of the anatomy introduced and explored.

Comparative anatomy of vertebrates has become an integrative field bringing together physiology, embryology, biomechanics, behaviour, evolution and ecology as well as deploying modern methods of systematics and the new material arriving from palaeontology. As an integrative discipline, vertebrate anatomy (morphology) is well-suited to show the working of science itself.

Comparative vertebrate anatomy clearly explains differences and similarities between organisms. The concepts of homology, analogy and homoplasy help us understand the basis of these comparative features. Many of the concepts were birthed in the nineteenth century and have grown into the guiding themes of biology today. Evolution, descent with modification through time, is one of the foundation concepts in biology.

Comparative vertebrate anatomy provides compelling examples of how an integrated organism might evolve. For example, a remarkable fossil record documents an undeniable change in jaw articulation within synapsids (a group of extinct reptiles), seeing the two participating bones (articular, quadrate) of basal synapsids replaced by two different bones in derived groups including mammals. Fossil intermediates between the two conditions mark the anatomical changes, but they also suggest how functional changes, which must accompany evolving systems, also change without disrupting performance.

Further, within many vertebrate systems, the close coupling of form and function with lifestyle are illustrated. Built on a basic vertebrate plan, the tetrapod locomotor system illustrates the close relationship between limbs and axial skeleton, and the type of locomotion—flight, cursorial, burrowing. Likewise, the cardiovascular system, especially in organisms that exploit water and air, illustrates the close relationship between vascular anatomy and physiological flexibility that permits. The basic concepts of form, function and adaptive evolution parade before us as we move from system to system in vertebrate morphology (or anatomy).

Evolution proceeds most often by remodelling, modification of a basic underlying plan, not by all new construction. This is illustrated in the skeletal system as well as within the cardiovascular (aortic arches) system.

Physiology is a synthesizing science that applies physical and chemical methods to biology. It is the analysis of functions in living organisms. Physiologists attempt to understand the mechanisms that operate in living organisms at all levels, ranging from the subcellular to the integrated whole organism.

Physiology is firmly rooted in the laws and concepts of chemistry and physics. The principles of evolution, natural selection and speciation underlie animal physiology, just as they do in any other study of life on the Earth. Thus, the history of animal physiology is rich in studies of the adaptations by individual species to particular environmental constraints and demands. Physiology provides an intellectual bridge between molecular biology and whole animal biology.

Animal physiology has made innumerable contributions in the understanding of human physiology. It is the cornerstone of scientific medical practice. Understanding of the functioning and malfunctioning of living tissues provided the foundation for developing effective, scientifically sound treatment for human diseases. A physician and a medical researcher with the knowledge of animal physiology is better equipped to make intelligent and preceptive use of information from animal models.

Animal physiology has provided a viewpoint that adds to our understanding of human nature. It combines viewpoints of genetics and cultural heritage. It provides tools for diagnosis and therapy. Every kind of animal that has been successful during its period in Earth's history has been adopted to its physical and biotic environments. Human beings are currently successful because they excel in cognition, communication and social organization via the understanding of physiological processes.

Biochemistry is perhaps the most fascinating subject as it deals with the chemical language of life, be it human, animal, plant or microorganism. No other science subject has as much application as biochemistry to the disciplines of biology, life science, medicine, health, veterinary, agriculture bioengineering and technology.

Biochemistry deals with the chemistry of life and living processes (e.g. metabolism). There is no exaggeration in the statement, "The scope of biochemistry is as vast as life itself". Every aspect of life involves biochemistry. For that matter, every movement of life is packed with hundreds of biochemical reactions. The discipline of biochemistry serves as a torchlight to trace the intricate complexities of biology besides unravelling the chemical mysteries of life.

The following are the salient features of the textbook:

- Provides an extensive coverage of comparative anatomy of vertebrates with an emphasis on evolutionary origin, embryological development, phylogeny and functional utility of various organ systems.
- Gives a detailed discussion on physiology of digestion, respiration, excretion, osmoregulation, blood and heart. Also deals with foregut and hindgut fermentation.
- Focuses on metabolism of carbohydrates, lipids, proteins and oxidative phosphorylation (energy metabolism).
- Profusely illustrated with a number of well-labelled diagrams and tables to help reinforce students' understanding of the concepts discussed.
- Contains several boxed items throughout the text to provide information on intricate terms or topics.

- Incorporates chapter-end questions containing long, short and very short answer questions, fill in the blanks and multiple choice questions to test students' knowledge of the subject matter.

I am indebted to the patient, able and ever-supportive editorial people—Surabhi Khare, Shalini Negi, K.K. Chaturvedi, Priyanka Sharan, Arun Kumar, Deepak Kashyap and Sanjeev Arora at S. Chand And Company Limited—for the timely release of the textbook.

It is again pleasure to work with an artist as accomplished and knowledgeable as Trilok Jain who contributed figures to this textbook. My heartiest thanks to him.

To friends and family, I remain grateful and thank them for their support during preparation of the manuscript. My grandchildren–Prafful, Muskan and Khushi deserve great appreciation; they understand me and kept me on the right track of hope and enthusiasm. Indeed, my wife Usha Gupta and three lovely daughters Dr. Surabhi, Dr. Shubhra, Er. Anubha and their spouses made vital contributions to do something challenging.

Suggestions and feedback from the readers will be gratefully acknowledged.

V K Agarwal
Villa-20, Green Wood City
Godwin Colony
Haridwar Bypass Road
Meerut–250 005
Uttar Pradesh (India)

Syllabus

BSc (Hons) Zoology–Semester IV

CORE COURSE VIII

COMPARATIVE ANATOMY OF VERTEBRATES

Unit 1: Integumentary System **8**
Structure, functions and derivatives of integument.

Unit 2: Skeletal System **8**
Overview of axial and appendicular skeleton, Jaw suspensorium, Visceral arches.

Unit 3: Digestive System **8**
Alimentary canal and associated glands, Dentition.

Unit 4: Respiratory System **8**
Skin, gills, lungs and air sacs; Accessory respiratory organs.

Unit 5: Circulatory System
General plan of circulation, Evolution of heart and aortic arches.

Unit 6: Urinogenital System **6**
Succession of kidney, Evolution of urinogenital ducts, Types of mammalian uteri.

Unit 7: Nervous System **8**
Comparative account of brain, Autonomic nervous system, Spinal cord, Cranial nerves in mammals.

Unit 8: Sense Organs **6**
Classification of receptors, Brief account of visual and auditory receptors in man.

CORE COURSE IX

ANIMAL PHYSIOLOGY: LIFE SUSTAINING SYSTEMS

Unit 1: Physiology of Digestion **14**
Structural organization and functions of gastrointestinal tract and associated glands; Mechanical and chemical digestion of food; Absorptions of carbohydrates, lipids, proteins, water, minerals and vitamins; Hormonal control of secretion of enzymes in gastrointestinal tract.

Unit 2: Physiology of Respiration **12**
Histology of trachea and lung; Mechanism of respiration, Pulmonary ventilation; Respiratory volumes and capacities; Transport of oxygen and carbon dioxide in blood; Respiratory pigments, Dissociation curves and the factors influencing it; Carbon monoxide poisoning; Control of respiration.

Unit 3: Renal Physiology **8**

Structure of kidney and its functional unit; Mechanism of urine formation; Regulation of water balance; Regulation of acid-base balance.

Unit 4: Blood **14**

Components of blood and their functions; Structure and functions of haemoglobin, Haemostasis: Blood clotting system, Kallikrein-Kinninogen system, Complement system and Fibrinolytic system, Haemopoiesis, Blood groups: Rh factor, ABO and MN.

Unit 5: Physiology of Heart **12**

Structure of mammalian heart; Coronary circulation; Structure and working of conducting myocardial fibers, Origin and conduction of cardiac impulses, Cardiac cycle; Cardiac output and its regulation, Frank-Starling Law of the heart, nervous and chemical regulation of heart rate, Electrocardiogram, Blood pressure and its regulation.

CORE COURSE X

BIOCHEMISTRY OF METABOLIC PROCESSES

Unit 1: Overview of Metabolism **10**

Catabolism vs Anabolism, Stages of catabolism, Compartmentalization of metabolic pathways, Shuttle systems and membrane transporters; ATP as "Energy Currency of Cell"; Coupled reactions; Use of reducing equivalents and cofactors; Intermediary metabolism and regulatory mechanisms.

Unit 2: Carbohydrate Metabolism **16**

Sequence of reactions and regulation of glycolysis, Citric acid cycle, Phosphate pentose pathway, Gluconeogenesis, Glycogenolysis and Glycogenesis.

Unit 3: Lipid Metabolism **14**

β-oxidation and omega-oxidation of saturated fatty acids with even and odd number of carbon atoms; Biosynthesis of palmitic acid; Ketogenesis.

Unit 4: Protein Metabolism **10**

Catabolism of amino acids: Transamination, Deamination, Urea cycle; Fate of C-skeleton of Glucogenic and Ketogenic amino acids.

Unit 5: Oxidative Phosphorylation **10**

Redox systems; Review of mitochondrial respiratory chain, Inhibitors and un-couplers of Electron Transport System.

Contents

CC VIII: Comparative Anatomy of Vertebrates

CC IX: Animal Physiology–Life Sustaining Systems

CC X: Biochemistry of Metabolic Processes

CC VIII

Comparative Anatomy of Vertebrates

1

Definition, Historical Background and Concepts of Comparative Vertebrate Anatomy

1.1 INTRODUCTION

Comparative vertebrate anatomy is the study of the structure of vertebrates (descriptive morphology) and of the functional significance of structure (functional morphology) (Box 1.1). Because structure involves development of the individual (**ontogenesis**), and individuals have an ancestral history (**phylogenesis**), the discipline embraces these areas of inquiry as well. Every area of biological inquiry provides relevant data to the discipline. Ecology, embryology, genetics, molecular biology, serology, biochemistry and paleo biology are all sources of valuable data.

Box 1.1 Comparison

The use of "comparison" in comparative anatomy/morphology is not just convenience. It is a tool. Comparison of structure throws similarities and differences into better assistance. Comparison emphasizes the functional evolutionary themes the vertebrates carry within their structures. Comparison also help formulate the questions we might ask of structure (**Kardong** 2006).

For example, different fishes have different tail shapes. In the **homocercal tail**, both lobes are equal in size, making the tail symmetrical (Fig. 1.1 A). In the **heterocercal tail**, found in sharks and a few other groups, the upper lobe is elongated (Fig. 1.1 B). Why this difference?

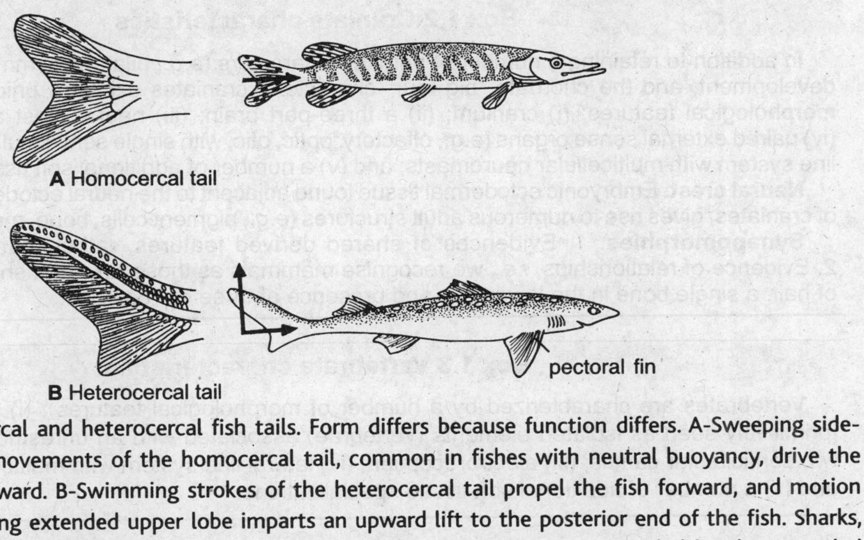

Fig. 1.1. Homocercal and heterocercal fish tails. Form differs because function differs. A-Sweeping side-to-side movements of the homocercal tail, common in fishes with neutral buoyancy, drive the body forward. B-Swimming strokes of the heterocercal tail propel the fish forward, and motion of the long extended upper lobe imparts an upward lift to the posterior end of the fish. Sharks, which are a good deal denser than water, need the upward forces provided by the extended lobe of the tail to counteract a tendency to sink (after Kardong 2006).

The homocercal tail is found in the teleost fishes – salmon, tuna, trout, and the like. These fishes have a swim bladder, an air-filled sac that gives their dense bodies neutral buoyancy. They neither sink to the bottom nor bob to the surface, so they need not struggle to keep their vertical position in the water. Sharks, however, lack swim bladder, and so tend to sink. The extended lobe of their heterocercal tail provides lift during swimming to help counteract this sinking tendency. So, the differences in structure, homocercal versus heterocercal are related to differences in function.

Geology is an indispensable source. In this book, the thrust of the chapters will be the organs and systems, their roles in survival, their embryogenesis, and their historical background in geological time. The geological (or palaeontological) treatment involves consideration of the phylogenesis of vertebrates in general. In studying vertebrates, it is important to consider related organisms, so the craniates (hagfish or cyclostomes and vertebrates) will be compared to protochordates.

The comparative anatomy is concerned with phylogenesis (*i.e.*, origin of new taxa). In fact, it is a study of history of chordates and of animals that no longer inhabit the earth (*i.e.*, extinct animals) and are known to us only by a fossil record. It is interested in the survival value of structure (an adaptation), of the struggle for compatibility with ever-changing environment, of the invasion of new territory by those most effectively equipped for survival, and of the extinction of species. The history of vertebrates is a fascinating story from which is developing a genealogy (gen-origin) based on data just described. Comparative anatomy addresses curiousity about the origin of species, including our own. The generalizations and conclusions arrived at in the discipline add to the enlightenment of the human mind.

We should note that vertebrates are a subgroup of a larger taxon that includes hagfish or cyclostomes (*i.e.*, Craniata). Many traditional Linnean classifications equate craniates with vertebrates (they include hagfish despite the absence of vertebrae; see **Kent** and **Carr** 2001).

Thus, **chordates** are animals that have a notochord in the embryo stage at least. **Craniates** are chordates with a neurocranium (braincase; Box 1.2). **Vertebrates** are chordates with vertebrae (Box 1.3). Vertebrae appear during embryonic development after the notochord has formed. Subsequently, they reinforce the notochord or replace it functionally.

Box 1.2 Craniate characteristics

In addition to retaining a number of primitive characters (*e.g.*, bilateral symmetry, deuterostome development, and the chordate "big four" characters, craniates exhibit a unique combination of morphological features: (i) cranium; (ii) a three-part brain; (iii) neural crest and its derivatives; (iv) paired external sense organs (e.g., olfactory, optic, otic, with single semicircular canal, and lateral line system with multicellular neuromasts; and (v) a number of additional soft tissue specializations.

Neural crest. Embryonic ectodermal tissue found adjacent to the neural ectoderm; synapomorphy of craniates; gives rise to numerous adult structures (*e.g.*, pigment cells, bone, muscle and neurons).

Synapomorphies. 1. Evidences of shared derived features, rather than overall similarity. 2. Evidence of relationships, *i.e.*, we recognise mammals as those taxa that share the possession of hair, a single bone in the lower jaw, and presence of three ear ossicles.

Box 1.3 Vertebrate characteristics

Vertebrates are characterized by a number of morphological features : (i) a vertebral column [primitively seen as isolated elements (vertebrae) associated with an unrestricted notochord]; (ii) two semicircular canals; (iii) electroreception; (iv) lateral line system with multicellular neuromasts; and (v) a number of additional soft tissue specializations.

1.2 GENERAL PLAN OF CRANIATE BODY

All craniates confirm to a generalized pattern and anatomic structure representing a collection of primitive (**plesiomorphic**) and unique (**derived**) anatomical features. This is revealed by dissection and is the result of the expression of similar DNA molecules inherited during the course of evolution. Craniates also exhibit similar but not identical, patterns of embryonic development. This, too, is the result of common ancestry. Both morphology and developmental processes have been altered during the passage of time, which, as it lengthens, provides increasing opportunities for genetic changes that result in anatomic diversity. Yet, despite these changes, innumerable primitive structural and developmental similarities still exist. These similarities and diversities will be examined in detail in later chapters. In this chapter, will be discussed the structural highlights that characterize craniates in general and history of comparative vertebrate morphology.

1. Regional Differentiation of Body of Craniates

The typical craniate body consists of three regional components: head, trunk and postanal tail.

***(i)* Head.** Concentrated on or within the head are (1) special sense organs for monitoring the external environment; (2) a brain that is at least large enough to receive and process essential

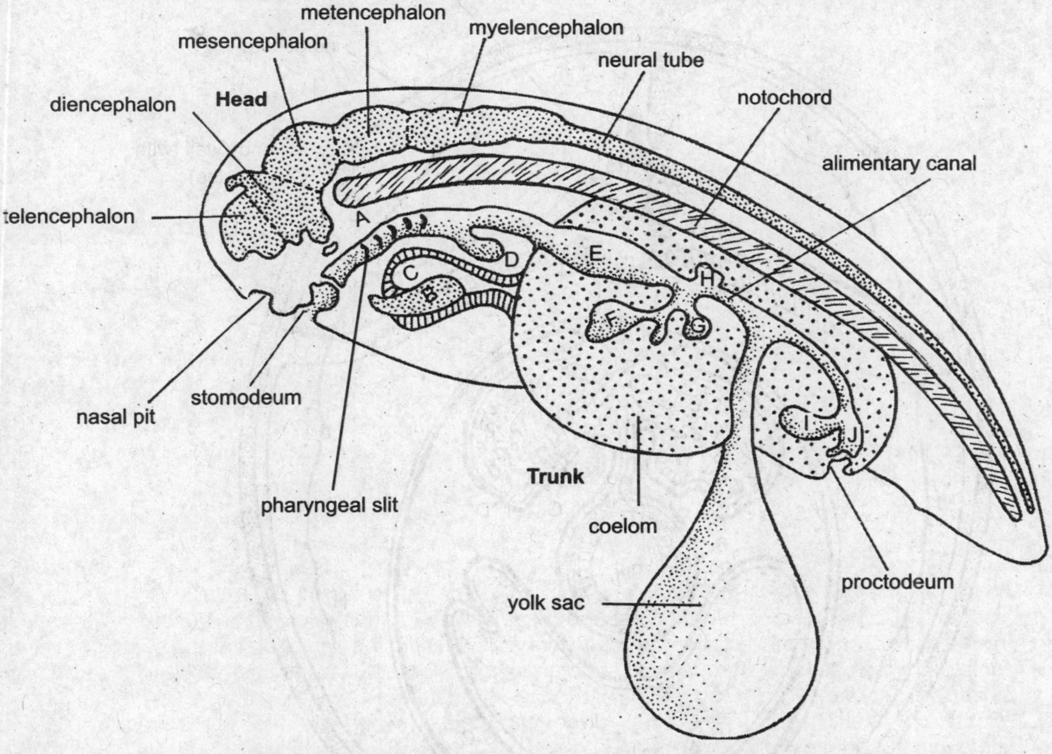

Fig. 1.2. Sagittal section of a generalized craniate embryo. **A**, pharynx with pharyngeal slits; **B**, and **C**, ventricle and atrium of heart; **D**, diverticulum that gives rise lungs in tetrapods and swim bladder in fishes; **E**, stomach, **F**, liver bud and associated gall bladder; **G**, ventral pancreatic bud; **H**, dorsal pancreatic duct; **I**, urinary bladder of tetrapods; **J**, cloaca separated from the proctodaeum by a cloacal membrane. The stomodaeum is separated from pharynx by a thin oral plate. The differentiated brain has major five major subdivisions: telencephalon and diencephalon (forebrain), mesencephalon (midbrain) and metencephalon and myelencephalon (hind brain) (after Kent and Carr 2001).

incoming information and to provide appropriate stimuli to the body musculature; (3) jaws in some species for acquiring, retaining and macerating food; and (4) in fishes, gills for respiration. Expansion of brain over hundreds of millions of years has resulted in larger braincases that have become increasingly movable independent of the trunk during that time. **Cephalization** has developed to a greater degree in craniates than in any other group of animals.

***(ii)* Trunk.** The trunk contains a cavity, the **coelom** which houses most of the viscera (Fig. 1.2 and Fig. 1.3). Surrounding the coelom is the body wall consisting of chiefly of muscle, vertebral column and ribs. The body wall has to be opened to expose the viscera. In many, but not all, craniates paired pectoral and pelvic appendages (*i.e.*, fins and limbs) are associated with the trunk. The **neck** is a narrow extension of the trunk of amphibians, reptiles, birds and mammals. Neck lacks the coelom. It consists primarily of vertebrae, muscles, spinal cord, nerves, and elongated tubes—esophagus, blood vessels, lymphatics, trachea—that connect structures of the head with those of trunk.

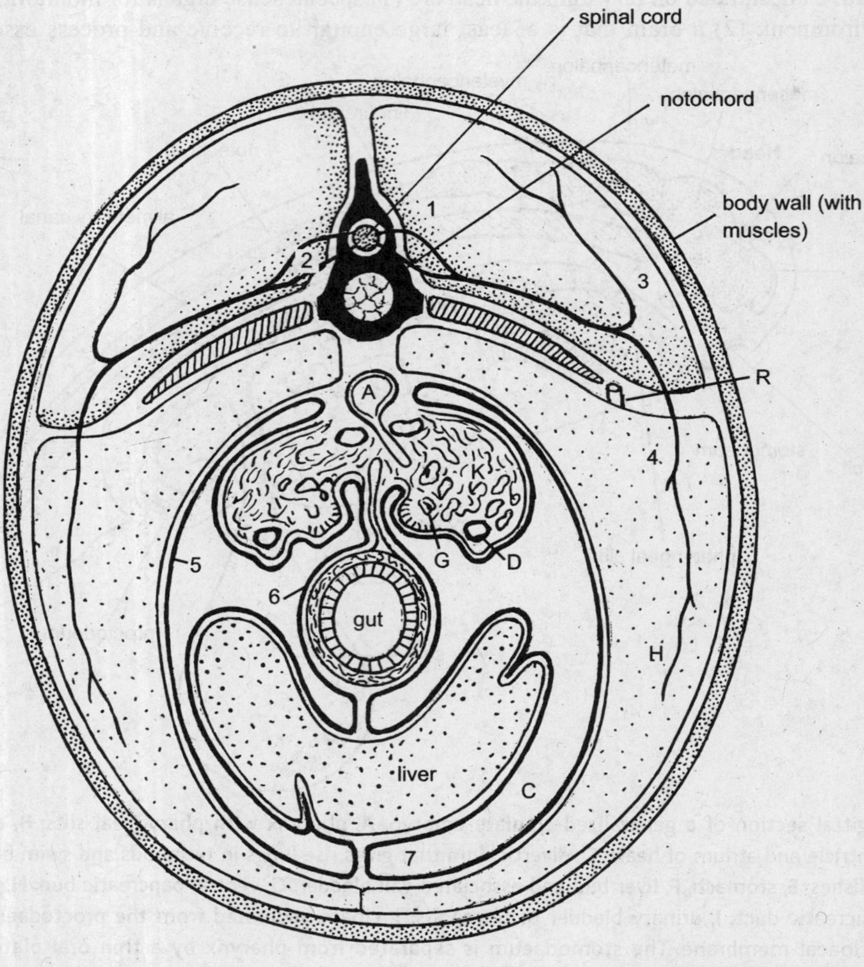

Fig. 1.3. Cross section of a generalized craniate trunk. A, dorsal aorta, giving off renal artery to kidney; C, coelom; D, kidney duct; E, epaxial muscle; G, future gonad (genital ridge); H, hypaxial muscle in body wall; K, kidney; R, rib projecting into a horizontal skeletogenous septum from the transverse process of a vertebra. 1. dorsal root of spinal nerve; 2. ventral root; 3. dorsal ramus of spinal nerve; 4. ventral ramus ; 5. parietal peritoneum; 6. visceral peritoneum; 7. ventral mesentery (after Kent and Carr 2001).

***(iii)* Tail.** The tail begins at the anus or vent; hence, it is postanal. It consists almost exclusively of a caudal continuation of body wall muscles, axial skeleton, nerves and blood vessels. Some adult craniates lack a postanal tail, although it is present in all embryos. The swimming larvae of frogs, toads, and worm-like amphibians (caecilians) have a tail, but it is resorbed at metamorphosis. Modern birds have tails reduced to a nubbin (small and stunted structure), but the first birds had long tails. Human beings have a vestigial postanal tail early in embryonic life. Remnanats remain in adult human beings as the "tail bone" (coccyx).

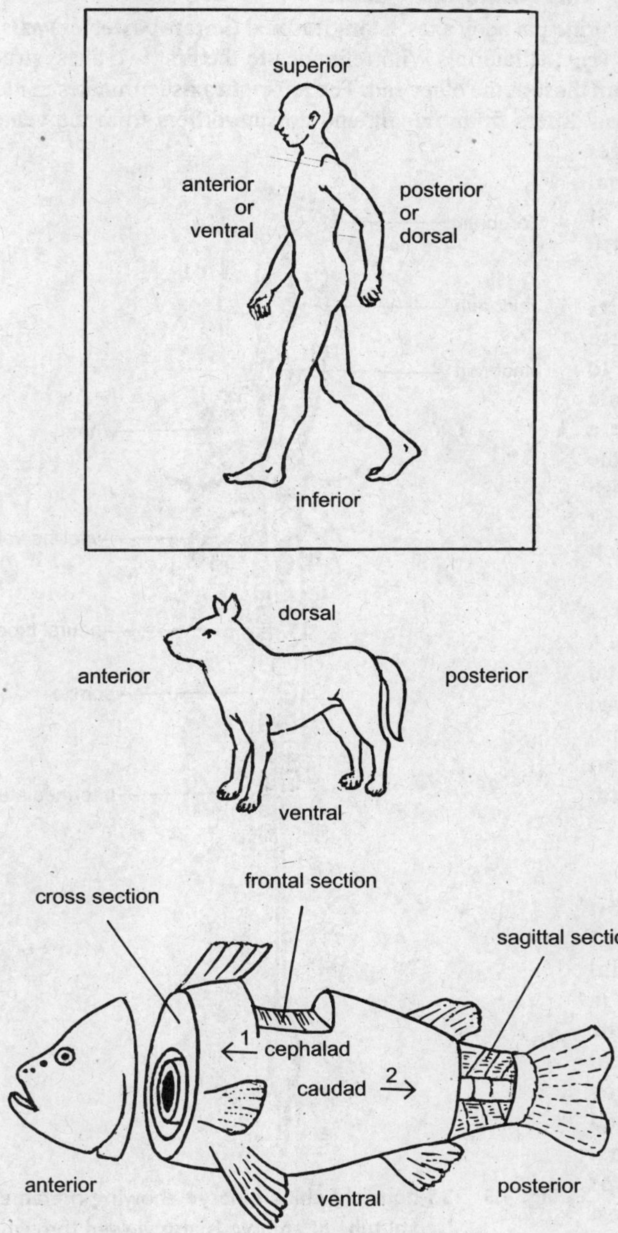

Fig. 1.4. Terms of direction and position and planes of sectioning of the craniates body. Terms employed in human anatomy are shown in box (after Kent and Carr 2001).

Appendages (fins or limbs). Two pairs of **appendages**—pectoral and pelvic, supported by an internal skeleton and operated by contributions from the trunk musculature—are characteristic of most vertebrates. However, they are sometimes vestigial or have been completely lost. The earliest known craniates lacked one or both pairs of these appendages, with the latter condition seen in living jawless craniates (agnathans or cyclostomes). The pelvic appendages appeared first in gnathostomes (craniates with jaws).

2. Bilateral Symmetry and Anatomic Planes

Craniates have three principal body axes, a **longitudinal (anteroposterior) axis,** a **dorsoventral axis** and a **left-right axis** (*e.g.*, Bilateria). With reference to the first two axes, structure at one end of the axis are different from those at the other end. The left-right axis terminates in identical structures at each end. Thus, the head differs from the tail, and dorsum differs from the venter, but right and left sides are mirror images of each other. An animal with this arrangement of body parts exhibits **bilateral symmetry**.

More often parts of the craniates body are discussed with reference to three **principal anatomic planes**. Two axes define a plane. The left-right and the dorsoventral axes establish the **transverse plane**. A cut in this plane is a **cross section** (Fig. 1.4, Fish). The left-right and longitudinal axes establish the **frontal plane.** A cut in this plane is a **frontal section**. The longitudinal and dorsoventral axes establish the **sagittal plane**. A cut in this plane is a **sagittal section.**

3. Metamerism

Craniates exhibit the primitive feature of **metamerism**, the serial repetition of structures in a longitudinal axis of the body. It is clearly evident in craniate embryos (Fig. 1.5) and is retained in many adult systems. No external evidence of metamerism is seen in most adult reptiles, birds and mammals because the skin

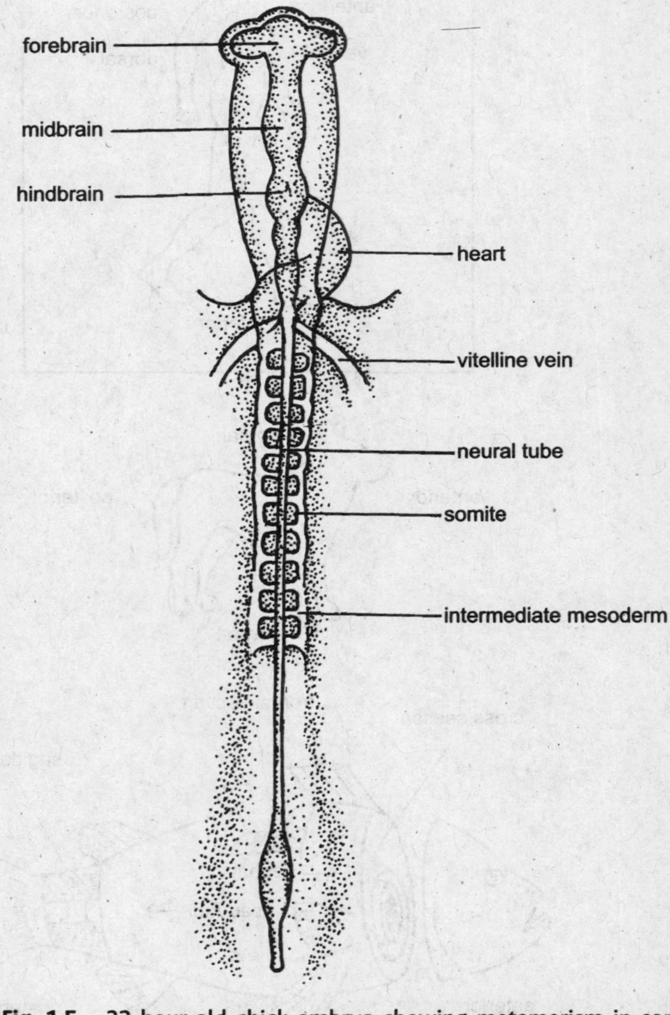

Fig. 1.5. 33-hour-old chick embryo showing metamerism in somites. Neural tube of embryo is also viewed through tissues rendered transparent (after Kent and Carr 2001).

is not metameric. If, however, the skin is stripped from the trunk or tail of fishes, most amphibians and some reptiles, one sees a series of muscle segments that are reflections of embryonic metamerism (Fig. 1.6). In addition, the serial arrangement of vertebrae, ribs, spinal nerves, embryonic kidney tubules, and segmental arteries and veins are further expression of the fundamental metameric nature of craniates.

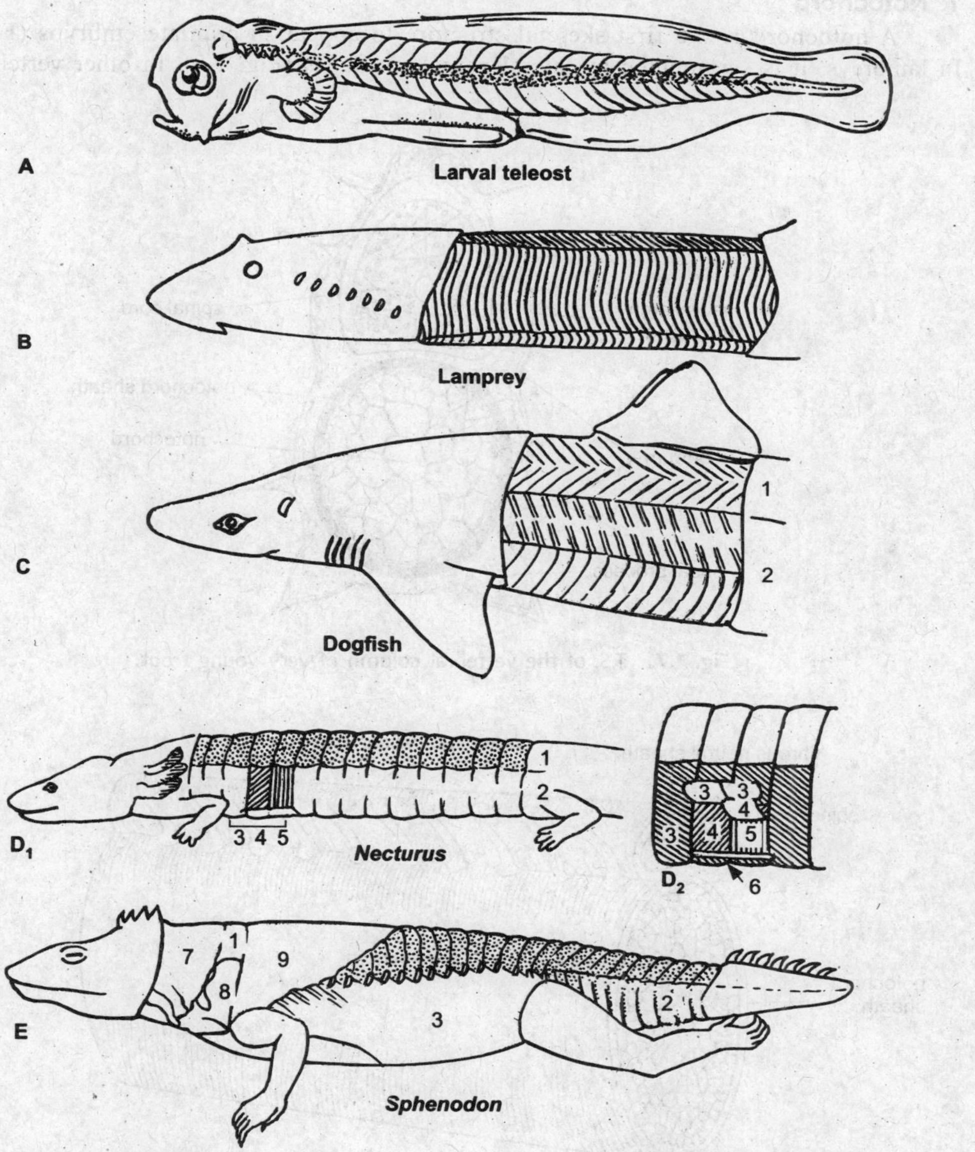

Fig. 1.6. Embryonic metamerism in trunk muscles of selected vertebrates. 1, epaxial; 2, hypaxial; 3, external oblique; 4, internal oblique; 5, transverse muscle of abdomen; 6, rectus abdominis. In (E) the appendicular muscles are 7, trapezius; 8, dorsalis scapulae; 9, latissimus dorsi. In (A) stippel indicates position of notochord by the horizontal skeletogenous septum; (D_2) a close-up of the bracketed myomeres in (D_1) showing the layered hypaxial muscles. In (D) and (E), the epaxialis are the dorsalis trunci muscles (after Kent and Carr 2001).

1.3 COMMON STRUCTURES OF CRANIATES

Primitive and unique structures of craniates include— 1. Notochord and vertebral column; 2. Pharynx; 3. Pharyngeal pouches and slits; 4. Pharyngeal arches; 5. Dorsal hollow central nervous system.

1. Notochord

A notochord is the first skeletal structure to appear in craniate embryos (Fig. 1.7). In lampreys, it is surmounted by lateral neural cartilages (Fig. 1.8). In other vertebrates, it

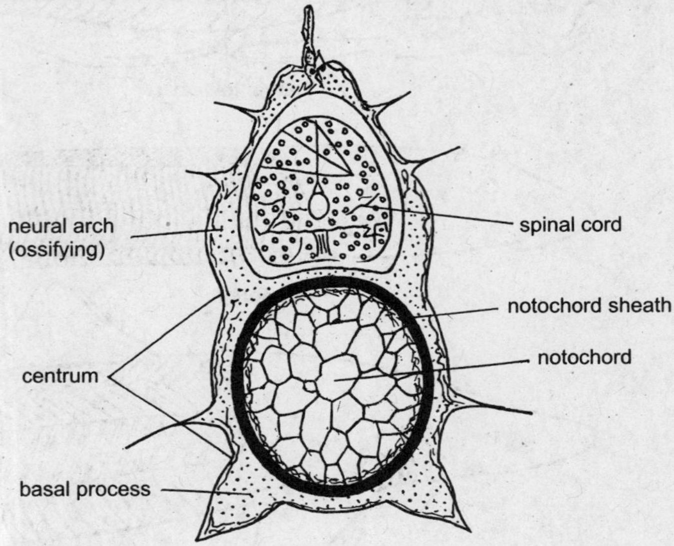

Fig. 1.7. T.S. of the vertebral column of very young trout.

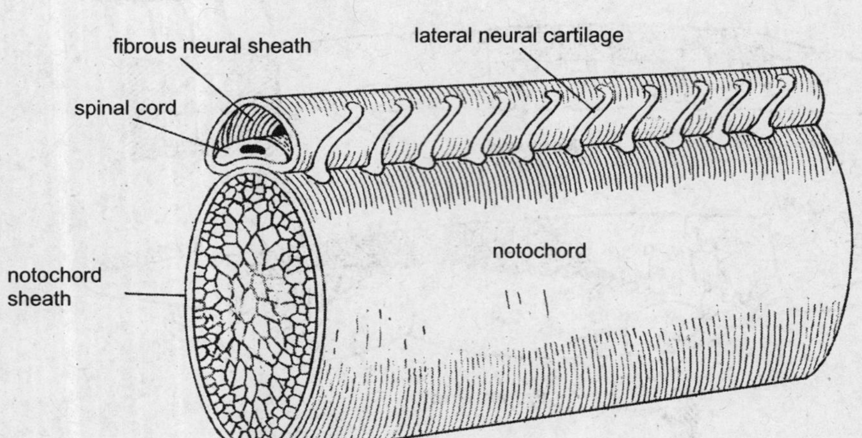

Fig. 1.8. Lateral neural cartilage and notochord of a lamprey (after Kent and Carr, 2001).

becomes surrounded by cartilaginous or bony vertebrae after which the notochord may be reduced to remanants or disappear.

2. Pharynx

The pharynx is a vital part of the craniate embryo. It produces the gills of fishes, the lungs of tetrapods, the skeleton and musculature of the jaws, the endocrine glands that regulate metabolic rates in every cell of the body and that maintain appropriate calcium level of bones, other tissues and the circulating blood. It gives rise to the middle ear cavity of tetrapods. Very importantly, pharynx provides, in humans, during fetal life and for a short time thereafter, the initial cells of the immune system, which if destroyed, leads to death.

3. Pharyngeal Pouches

Pharyngeal pouches (Fig. 1.9) tend to rupture to the exterior to form temporary or permanent pharyngeal slits (Fig. 1.10). These persist as gill slits in fishes and larval amphibians. They close permanently in other vertebrates.

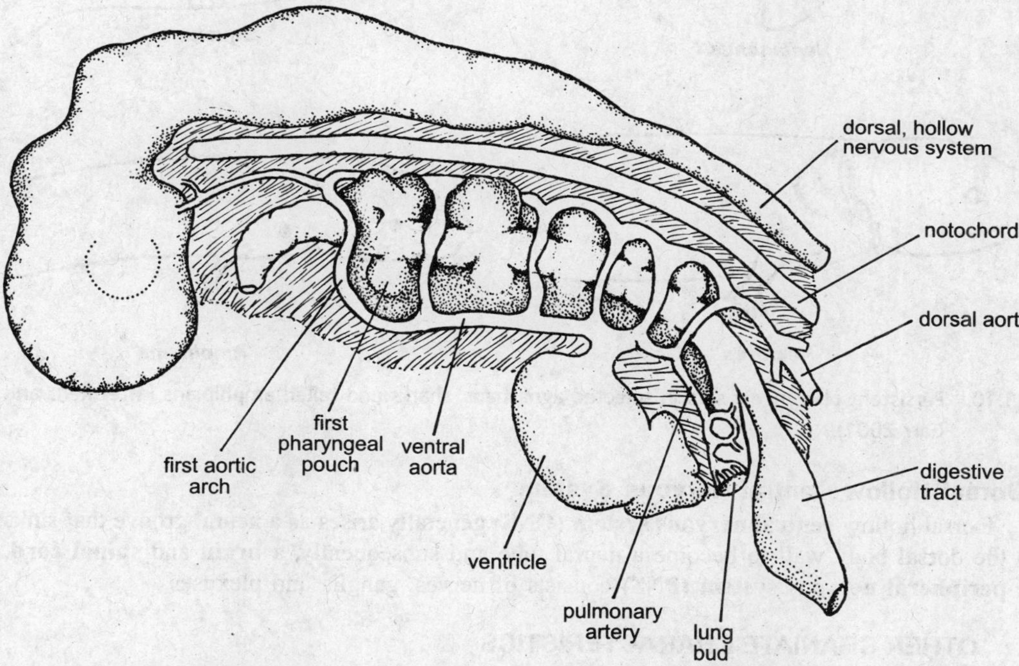

Fig. 1.9. Basic pharyngeal architecture exemplified by a composite craniate embryo. The dorsal aorta is paired in the head. The ventral aorta is unpaired. A series of pharyngeal pouches has evaginated from the lateral walls of the digestive tract. Six aortic arches connect the heart and ventral aorta with the dorsal aorta. (Typically, the first aortic arch disappears before the sixth one has formed in gnathostomes). Although a lung does not form in all craniates, it is an ancient structure and is represented by a swim bladder in many fishes. The anterior end of the dorsal hollow nervous system is enlarging to form the brain. The notochord commences at the level or the midbrain. All reptiles, birds and mammal embryos exhibit a cephalic flexure at the level of the midbrain (after Kent and Carr 2001).

4. Pharyngeal Arch

Pharyngeal arch lies between each pharyngeal slit, anterior to the first and immediately caudal to the last. The pharyngeal arches support gills when the latter are present. The first two are **mandibular** and **hyoid** arches.

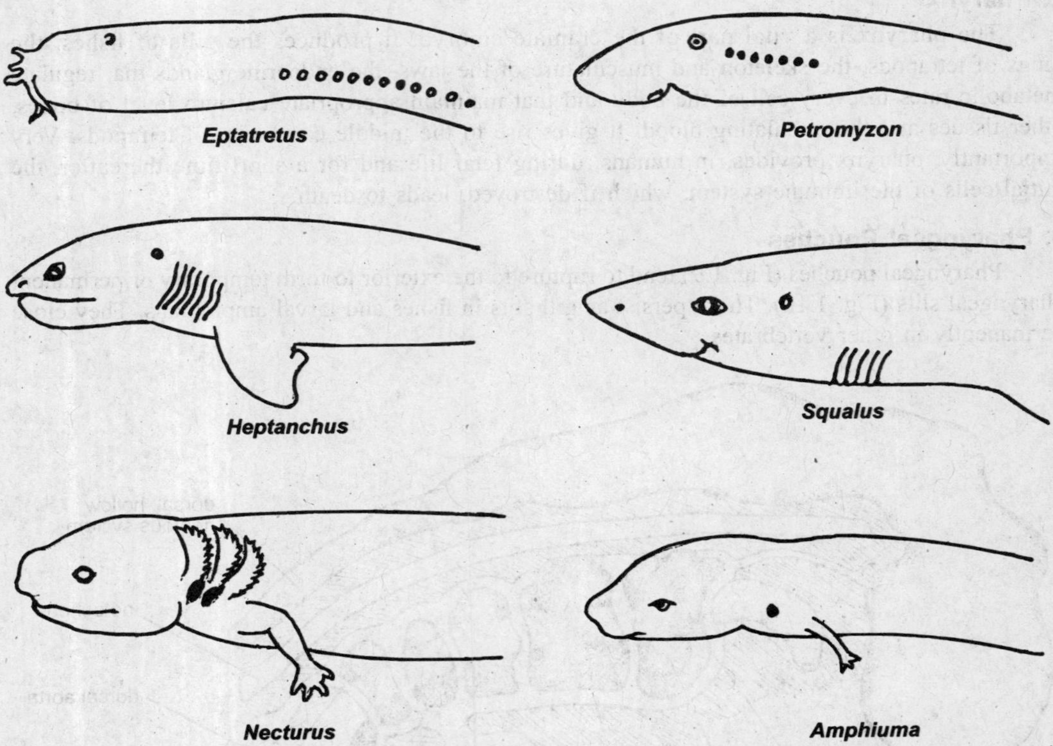

Fig. 1.10. Persistent pharyngeal slits in selected agnathans, sharks and tailed amphibians (after Kent and Carr 2001).

5. Dorsal Hollow Central Nervous System

Dorsal hollow **central nervous system (CNS)** generally arises as a neural groove that sinks into the dorsal body wall to become a neural tube and subsequently, a **brain** and **spinal cord**. The **peripheral nervous system (PNS)** consists of nerves, ganglia and plexuses.

1.4 OTHER CRANIATE CHARACTERISTICS

These include **integument, respiratory mechanisms** (lungs, gills, skin, etc.), **coelom, digestive system, urinogenital organs, circulatory system, skeleton, muscles** and **sense organs.** Here, we will elaborate only two characteristics : coelom and receptors (or sense organs).

1. Coelom

Like the body of many invertebrates, the craniate trunk is built like a tube within a tube, the outer tube (body wall) and inner tube (gut) being separated by a cavity, the **coelom** (Fig. 1.2). In fishes, amphibians and some reptiles, the coelom is partitioned into a **pericardial cavity** housing the heart, and a **pleuroperitoneal cavity** housing most of the other viscera, including lungs in tetrapods (Fig. 1.11, fish and amphibian). The pericardial and pleuroperitoneal cavities in these craniates are separated by a fibrous **transverse septum**. In other reptiles and mammals, the heart occupies a separate **pleural cavity**, and the digestive tract caudal to the oesophagus occupies an **abdominal (peritoneal) cavity** (Fig. 1.11; mammal). In most male mammals, a pair of **scrotal cavities** houses the testes. The scrotal cavities arise as evaginations from the abdominal cavity.

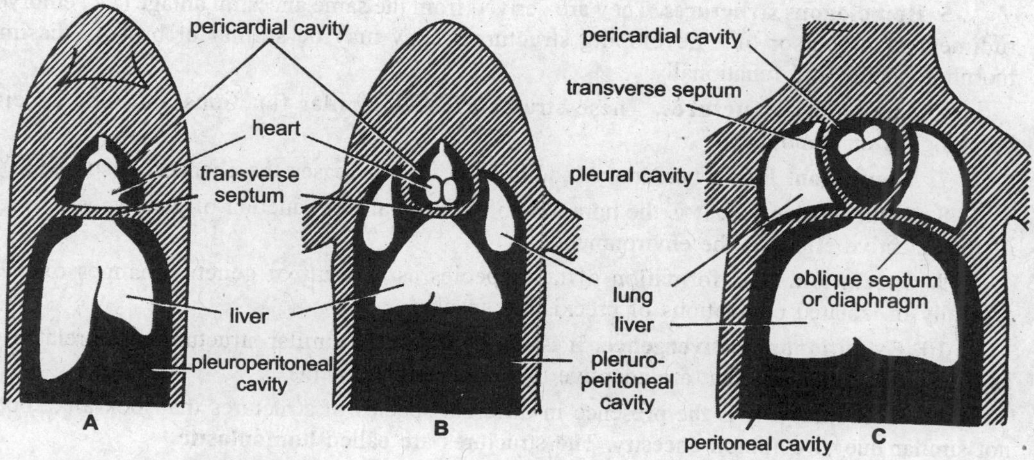

Fig. 1.11. Chief subdivisions of coelom of craniates (black shade): A-Fishes; B-Amphibians; C-Mammals.

The coelom is totally enclosed by a **peritoneal membrane** that lines the body wall and invests the coelomic viscera (**parietal** and **visceral peritoneum,** 5 and 6 in Fig. 1.3). These membranes are continuous via **dorsal mesenteries** and when present, via **ventral mesenteries.** The few visceral organs that do not develop mesentries–chiefly the kidneys–lie against the dorsal body wall just external to the peritoneum. Therefore, they are retroperitoneal (in some species such as *Necturus*, the kidneys may be suspended in the coelom).

2. Sense Organs or Receptors

A wide variety of sense organs or receptors monitor the external and internal environment. **Exteroceptors** monitor the external environment; they include **mechanoreceptors, chemoreceptors, electroreceptors** and **thermoreceptors,** and **receptors for radiation** in the visible and infrared spectra. **Mechanoreceptors** are stimulated chiefly by vibrations of varying amplitudes in the water or air, by barometric pressure, or by direct contact of another object with the skin (touch). With the exception of receptors for radiation, exteroceptors are distributed widely on the head, except those for touch and temperature. **Interoceptors** gather information and respond to general sensations of organs within the internal environment. **Proprioceptors** (a type of interoceptors) monitor the activity of muscles, joints and tendons. **Visceral receptors** monitor the rest of the internal environment. The information the receptors acquire is employed by the central nervous system to regulate skeletal muscle and visceral muscle and gland activity in a manner conducive to survival.

1.5 CONCEPTS OF COMPARATIVE ANATOMY

As stated earlier, one aspect of the discipline of comparative vertebrate anatomy is the study of the structure of living and extinct species. The former reveals similarities and differences among animals of today. The latter reveals what vertebrates were like in the past. Assembling the data in a geological time frame reveals the panorama of changes that have taken place from the distant past to the immediate present. That panorama has led to formulation of certain concepts that we will now examine.

1. Ontogeny. It is developmental history of an individual.

2. Phylogeny. It is evolutionary history of a taxon.

3. Homology. It is manifestation of homologous structures in different species.

4. Analogy. Two structures that have the same function.

5. Homologous structures. They are derived from the same ancestral anlage (*i.e.*, embryonic rudiment or precursor of a developing structure). They may be similar or broadly dissimilar morphologically and functionally.

6. Analogous structures. These structures have similar functions and may be either homologous or homoplastic.

7. Adaptation. It is a hereditary modification that increases the chances of survival.

8. Natural selection. It is the nonrandom differential reproduction of genotypes as a result of the selective effect of the environment.

9. Speciation. It is formation of new species as a result of genetic changes over time typically in isolated populations of preexisting species.

10. Evolutionary convergence. It is the evolution of similar structures in unrelated taxa as a result or mutations that are adaptive to similar environments.

11. Homoplasy. It is the presence in different species of structures that look alike but are not similar due to common ancestry. The structures are called **homoplastic.**

12. Heterochrony. It is the temporary or permanent retention of new or larval traits in sexually mature individuals, populations, or species. These traits are due to changes in the rate and/or timing of development events.

13. Paedomorphosis. It is an evolutionary process wherein certain larval or immature features of the immediate ancestor become the end product of metamorphosis in the descendant species.

14. Peramorphosis. It is contrast to paedomorphosis, results in a descendant that exceeds ancestral development.

15. Systematics. It is the discipline and practice of ordering organisms into hierarchies that reflect their morphological similarities and phylogenetic history.

16. Taxon. Taxonomic unit such as phylum or species.

17. Taxonomy. It is the process and rules by which we apply names to the groups determined in a systematic analysis. An international code provides the rules for binomial nomenclature.

18. Organic evolution. It involves changes that have taken place in animals and plants through historical time.

19. Teleology. It describes natural phenomena to purposeful design rather than to natural laws.

20. Darwin. His major contribution was the Theory of Natural Selection, an alternative to Lamarck's Doctrine of Acquired characteristics and their inheritance.

21. Georges Cuvier. He was the founder of the disciplines of comparative anatomy.

22. Vesalius, Andreas (1514-1564). He was an early anatomist who published on the human body.

23. Theology. The study of God and religious belief, *e.g.*, Christian theology.

24. Scholasticism. The system of theology and philosophy taught in medieval universities.

1.6 HISTORICAL BACKGROUND OF COMPARATIVE VERTEBRATE ANATOMY

Cave dwellers had some knowledge of the internal organs of mammals, as did the Babylonians and ancient Egyptians, who practiced surgery and embalming. **Embalming**, which involved removal of most of the internal organs, was an advanced art. Aside from a small number of Egyptian medical papyri dated to about 3000 B.C., oldest anatomical work in western civilization were written during the last 400 years B.C., by Greek philosophers and physicians. These works were incomplete, mostly superficial and often imaginative. Anatomy of the classical era culminated in the works of **Galen**, a Greek philosopher-physician who practiced in Rome between A.D. 165 and 200. Galen assembled all available Greek anatomical writings, supplemented them with his own dissections of apes from the Barbary coast (human dissection was as that time prevented by public opinion and superstition), and in addition, wrote more than 100 treaties on medicine and human anatomy. Shortly thereafter, scholasticism took over, and during the next 1300 years Galen's descriptions were considered infallible and dissent was punishable.

It was not until the fifteenth century that **Leonardo da Vinci** (1452-1519) and other Italian artists began to make anatomical observations of their own (at the risk of ex-communication from their church), and a renaissance in human and animal anatomy began. In 1543, **Vesalius** published *De humani corporis fabrica* (On the Structure of Human body), ushering in a renewed interest in anatomy of animals in general.

Eight years after publication of Vesalius's anatomy, **Pierre Belon** (1517-1564), a French naturalist and physician, published an important work on the dissection of cetaceans and other marine animals, invertebrates and vertebrates, including fishes. Belon was also interested in birds, saying he never missed an opportunity to dissect one and had examined the internal parts of 200 different species of birds. In 1555, Belon published his classic illustration of a human and bird skeleton side by side, showing that the skeletons correspond almost bone for bone. He published the illustration because he had made the observation and thought the general public would be enlightened. Like other public figures of his day, he was a teleologist. He attributed the similarity of the skeletons, like the convergent evolutionary similarities that he observed between fishes and cetaceans, to the manifestation of a basic architectural plan, or **archetype**, in the mind of Creator. Today, the term **basic architectural plan** refers to the generalized pattern of development and morphology that has been inherited from the earliest craniate ancestors.

Georges Cuvier (1769-1832), a French naturalist is regarded as **founder of comparative anatomy**. Cuvier was a prolific author in the field of natural history. His works include, among many others, nine volumes entitled *Lecons d'anatomie comparee* (*Paris*, 1801-1805), four volumes of palaeontological observations on fossil skeletons, a geological description of the environs of Paris, and a work that is now classic (**classic** is a work of art/literature which is generally agreed to be of high quality), a four-volume 1816 abridgement (*i.e.*, a shortened version of a larger work) of his course in comparative anatomy at the Ecole Centrale due Pantheon in Paris, which he titled *The animal kingdom arranged in accordance with its structure for the purpose of serving as a basis for the natural history of animals and as an introduction of comparative anatomy*. In this volume, **Cuvier** summed up his lifetime of research on the morphology of fossil and living animals, which he divided in the role of taxonomist, into four taxa, **Radiata**, **Mollusca**, **Articulata**, and **Vertebrata**, based on differences in structure of the skeleton and organs. Cuvier's final contribution was a 22-volume masterpiece, *Natural history of fishes*, published over a 4-year span ending in 1832. He died in 1832 in Paris after an energetic life dedicated to natural history and teaching including chancellorship at the University of Paris. Throughout life, Cuvier constantly maintained the absolute immutability of species (Box 1.4), objecting vehemently to the mutability ideas of **Lamarck** (*i.e.*, Doctrine of Acquired Characteristics and their Inheritance), of **Geoffroy Saint-Hillaire** (1772-1844), with whom he had collaborated in several published works, and of others who began to envision the possibilities of changes in species with time.

Box 1.4 Cuvier's idea of Evolution

According to **Cuvier**, parts of organisms, like parts of a machine, serve some purpose. Consequently, for the entire organism (or machine) to perform properly, the parts must **harmonize**. If one part is altered, function of connected parts is disrupted and performance fails. From this,

Cuvier reasoned that *evolution simply could not occur because if an animal were altered beyond a certain limit, harmony among the parts would be destroyed, function would fail, and animal no longer be viable.* Change (evolution) would cease before it began. Cuvier's functional morphology put him in intellectual company with **Linnaeus** but in opposition to **Lamarck's** evolutionary ideas (see **Kardong**, 2006).

Cuvier would have been a leading opponent of Darwin in the great uproar in scientific circles, in academia, among theologians, and consequently, in the press, that was evoked by the publication of **Charles Darwin's** masterpiece in 1859, "*On the origin of species by means of natural selection or in the preservation of favoured race in the struggle of life*" (**Kent** and **Carr** 2001).

English anatomist **Richard Owen** (1804-1892) believed like Cuvier that species were immutable, but unlike Cuvier, he felt that the correspondence between parts (homologies) could not be left without explanation. Virtually the same bones and pattern are present in the flipper of a dugong, the forelimb of a mole, and the wing of a bat (Fig. 1.12). Each possesses the same bones. Why ?

Owen with Cuvier's emphasis on adaptation, however, he felt some explanation for homologies was required and therefore introduced the idea of **archetype** (Fig. 1.12).

From our 20th century perspective, the answer is clear. Out of common ancestry, evolution passes along similar structures to perform new adaptive functions. But Owen, opposed to evolutionary ideas, was determined to find an alternative explanation. His

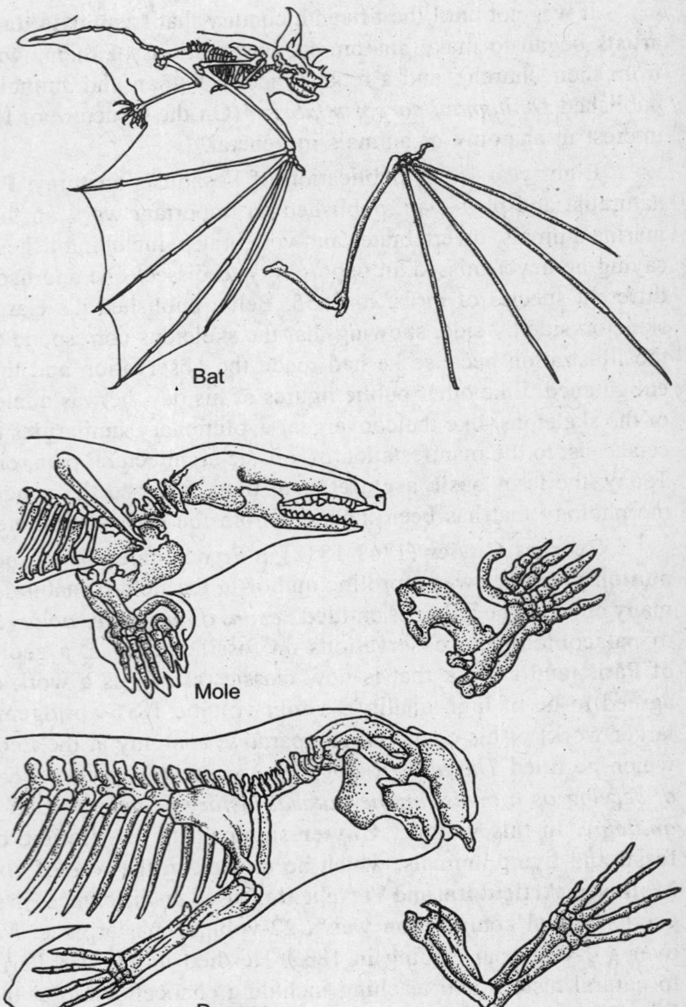

Fig. 1.12. Forelimbs of bat, mole and dugong. Owen noted that each limb performs a different function—flight, digging, and swimming, respectively— and each is superficially different, but he could trace all three to an underlying common plan he called the archetype. Today we recognize that common ancestry account for these underlying similarities, although we would join Owen crediting adaptation for the superficial differences among these homologous parts (From Kardong 2006).

answer centered around **archetypes**. *An archetype was a kind of biological blueprint, a supposed underlying plan upon which an organism was built.* All parts arose from it. Members of each major animal group were constructed from the same essential, basic plan. All vertebrates, for instance, were thought to share the same archetype, which explained why all possessed the same fundamental parts. Specific differences were forced on this underlying plan by particular functional needs. Owen was unclear about why he ruled out an evolutionary explanation, but he was vigorous in promoting his idea of archetypes.

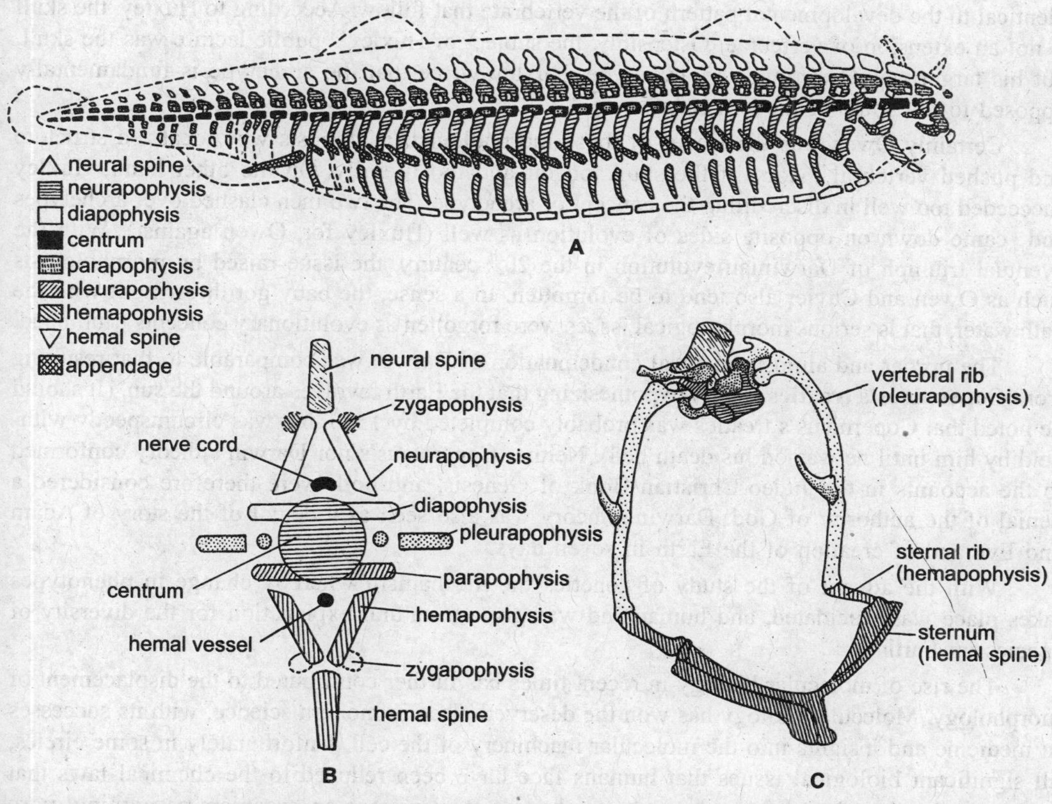

Fig. 1.13. Vertebrate archetype and Owen's vertebral theory of skull. A-Richard Owen saw the underlying pattern of the vertebrate body as a repeating series of vertebral units, collectively the vertebrate archetype. Owen supported the view that these vertebral units, carried forward into the head, even produced basic elements of the skull. B-Ideal vertebra. Each vertebra potentially included numerous elements, although not all were expressed in each segment. C-An actual section from a bird's skeleton indicates how this underlying plan might be realized (after Kardong, 2006).

Owen even carried this idea to repeated parts within the same individual (Fig. 1.13A). For example, he envisioned that the vertebrate skeleton consisted of a series of idealized segments he termed vertebrae (Fig. 1.13 B). Not all available parts of these serially repeated vertebrae were expressed at each segment, but all were available if demanded. Taken together, this idealized series of vertebrae constituted the archetype of the vertebrate skeleton. **Johann Wolfgang von Goethe** (1749-1832), although perhaps best remembered as a German poet, also dabbled (*i.e.*, take part in an activity in a casual way) in morphology and was the first to suggest that the **vertebrate skull** was created from modified and fused vertebrae. His idea was expanded by others, such as **Lorenz Oken** (1779-1851), so by Owen's time, the concept was well known. Owen considered the skull to be formed of vertebrae extended forward into the head. He held that all four vertebrae contributed, and even went so far as to derive human hands and arms from parts of the fourth contributed vertebra, "the occipital segment of the skull".

T.H. Huxley, in a public lecture (published in 1857-1859) took the task the **"vertebral theory of the skull"**, as it had become known. Bone by bone, Huxley traced homologies and developmental appearances of each skull component. He reached two major conclusions. First, all vertebrate skulls are constructed on the same plan. Second this developmental plan is not identical to the developmental pattern of the vertebrate that follow. According to Huxley, the skull is not an extension of vertebrae. Ostensibly, the subject of Huxley's public lecture was the skull, but his target was Owen and the archetype. Huxley wrote that the archetype is fundamentally opposed to the spirit of modern science.

Certainly Owen was the leader of those morphologists/anatomists who idealized structure and pushed vertebral theory of the skull too far and too literally. On the other hand, Huxley succeeded too well in discrediting the concept of archetype. The two men clashed over archetypes and came down on opposite sides of evolution as well (Huxley for, Owen against). With the eventual triumph of Darwinian evolution in the 20th century, the issue raised by morphologists such as Owen and Cuvier also tend to be forgotten. In a sense, the baby got thrown out with the bath water, that is serious morphological issues were forgotten as evolutionary concepts triumphed.

The uproar and almost universal condemnation of Darwin was comparable to that resulting from **Copernicus's treaties** (1543) hypothesizing that the Earth revolves around the sun. (It should be noted that Copernicus's treaties was probably completed by 1530, but was circumspectly withheld by him until he was on his death bed). Neither Copernicus's nor Darwin's theory conformed to the accounts in the Judeo-Christian Book of Genesis, and both were therefore considered a denial of the authority of God. Darwin's theory was also seen as a denial of the story of Adam and Eve and of creation of the Earth in seven days.

With the advent of the study of genetics the mechanism whereby change in phenotypes takes place was elucidated, and humankind was given a natural explanation for the diversity of species on Earth.

The rise of molecular biology in recent times has further contributed to the displacement of morphology. Molecular biology has won the deserved place in modern science, with its successes in medicine and insights into the molecular machinery of the cell. Unfortunately in some circles, all significant biological issues that humans face have been reduced to the chemical laws that govern molecules. In its extreme, such a reductionist view sees an organism as nothing more than the simple sum of its parts—know the molecules to know the person (see **Kardong** 2006).

1.7 EXPLANATIONS OF SOME TERMS

1. Primitive. It is a relative term. It refers to a beginning or origin. A primitive trait is one that appears in a stem ancestor from which arose an array of subsequent species, some of which may retain the trait. The notochord is primitive because it occurred in the first chordates. However, one cannot always be certain that a given structure is primitive. For example, the lateral neural cartilages of lampreys (Fig. 1.8) are primitive only if they reveal an original condition from which typical vertebrae later evolved. They could represent a secondary reduction from some unknown vertebral condition.

The term primitive should not be applied to organisms (cyclostomes). Although lampreys may retain a number of primitive characters, they are highly **specialized parasities**. Their invasion of the Great Lakes in North America devastated the local commercial fishing industry. (How did a "primitive" form outcome so many "advanced" taxa? (see **Kent** and **Carr** 2001).

2. Generalized. It refers to structural complexes that, at least in some of the descendants, have undergone subsequent adaptation to a variety of conditions. The hand of an insectivore (ancestor of mammals) was and remains a generalized mammalian hand. It was competent to evolve into the wing of a bat, the hoof of a horse, the flipper of a seal, and the hand of a primate. A generalized group of animals has demonstrated that it was genetically suitable for divergent evolution (that is, evolution in many directions). Labyrinthodonts were generalized tetrapods. The term generalized connotes (*i.e.*, suggests in addition to its primary meaning) a state of potential adaptability.

3. Specialized. A specialized condition is one that represents an adaptive modification. Craniate wings are specializations of anterior limbs, and beaks are specialized upper and lower jaws. Beaks (Fig. 1.14) may be needle-like and tubular for extracting nectar from flowers (humming birds), chisel-like for

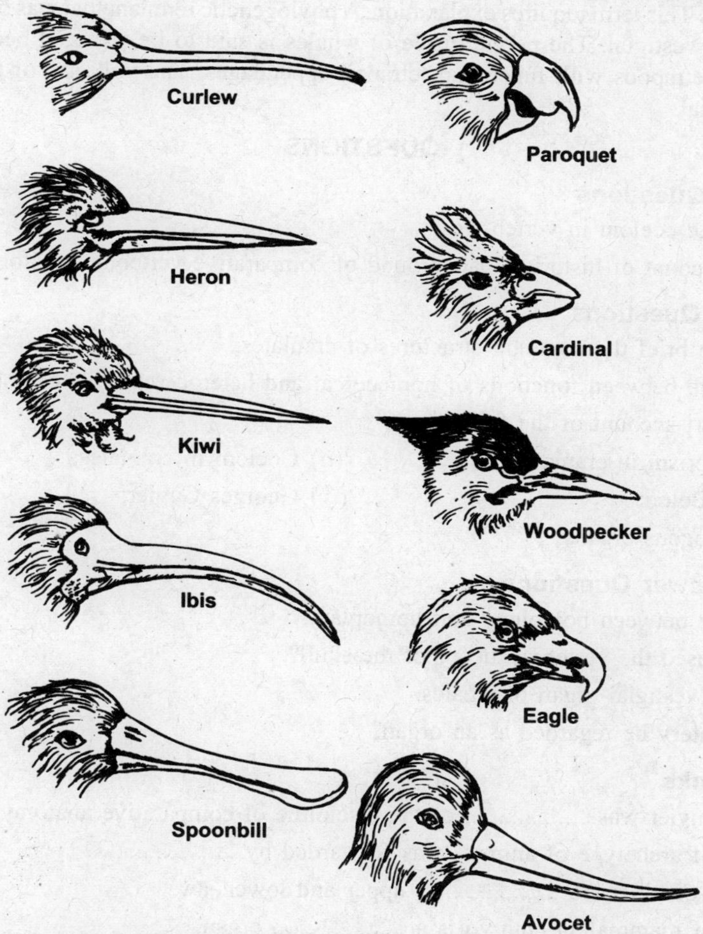

Fig. 1.14. Adaptations of the beak for feeding in selected birds.

drilling (woodpeckers); hooked for piercing and tearing prey (predatory birds, such as hawks and eagles); sturdy and spear-like, for capturing small fish and other aquatic animals in marshes (herons); long, spear like, but less sturdy, used for nocturnal foraging for grubs and worms (kiwi); or long and upcurved for sweeping through shallow waters and mud in search for small crustaceans, molluscs, and other marsh inhabitants (avocets). Increased specialization connotes increased adaptation. The greater the specialization, the less may be the potential for further adaptive changes.

4. Derived or modified. Both of these terms connotes any state of change from an ancestral condition, a mutated state. If the presence of bone is a primitive trait, wholly cartilaginous skeletons are modifications of the condition. The modification (*i.e.*, loss of the potential to form bone) was a specialization if it adaptively modified the animal. Perhaps not all modifications are adaptive but if they are not, they portend (*i.e.*, be a sign or warning that something important or unpleasant is likely to happen) the demise of the species because any change is statistically more likely to make the animal less competitive.

5. Higher and lower. These terms are used to express the relative position of major taxa on a conventional phylogenetic scale. They are less informative when applied below the class level. Birds and mammals evolved from a common ancestor and hence are said to be higher than the ancestral species.

6. Vestigial. This term requires explanation. A phylogenetic remnant that was better developed in an ancestor is vestigial. The pelvic girdle of whales is said to be vestigial because ancestors of whales were tetrapods with functional tetrapod appendages. The yolk sac of the mammalian embryo is vestigial.

QUESTIONS

Long Answer Questions

1. Describe the coelom in vertebrates.
2. Give an account of historical background of comparative vertebrate anatomy.

Short Answer Questions

1. Describe in brief the common structures of craniates.
2. Differentiate between functions of homocercal and heterocercal tails of fish.
3. Give a short account of the following :
   (*i*) Metamerism in craniates; (*ii*) Coelom in craniates;
   (*iii*) Pierre Belon; (*iv*) Georges Cuvier;
   (*v*) Homologous organs.

Very Short Answer Questions

1. Distinguish between homology and homoplasy.
2. Who proposed the vertebral theory of the skull?
3. Name any vestigial organ of whales.
4. Can mesentery be regarded as an organ?

Fill in the Blanks

1. Georges Cuvier was.................... of the discipline of comparative anatomy.
2. The idea of archetype of animals was forwarded by ....
3. The beaks of birds are .... upper and lower jaw.
4. Yolk sac of mammalian embryo is a .... organ.

Multiple Choice Questions

1. Homology is exhibited by
   (*a*) wings of butterfly, birds and bats
   (*b*) paddle of whale, forelimbs of horse and fore arms of human beings
   (*c*) tail of monkeys and birds
   (*d*) sting of scorpion and *Apis*

2. The study of palaeontology is associated with

(*a*) primates (*b*) birds

(*c*) fossils (*d*) bones

3. The term ontogeny means

(*a*) development of individual organism (*b*) place of an animal in classification system

(*c*) puberty period of an individual (*d*) development history of a group

4. Characters that are of common origin and hence indicative of common ancestry in descent are

(*a*) homologous characters (*b*) analogous characters

(*c*) adaptive characters (*d*) radioactive characters

5. Analogy is an example of

(*a*) speedy evolution (*b*) slow evolution

(*c*) convergent evolution (*d*) divergent evolution

6. Divergent evolution is due to

(*a*) analogy (*b*) reproductive isolation

(*c*) homology (*d*) vestigial organs

7. In the history of animal science, the person who for the first time, recognized the importance of the fossils as evidence of evolution was

(*a*) Leonardo da Vinci (1452-1519) (*b*) George Cuvier (1769-1832)

(*c*) Charles Darwin (1809-1882) (*d*) Lamarck (1744-1829)

ANSWERS

Very Short Answer Questions

1. Homology is similarity due to common ancestry. Homoplasy is similarity due to any cause other than common ancestry.

2. Richard Owen.

3. Pelvic girdle.

4. Mesentery is one or more vertebrate membranes that consists of a double fold of the peritoneum and invest the intestines (*e.g.*, jejunum and ileum in mammals) and their appendages and connect them with the dorsal and posterior wall of the abdominal cavity. **Dr. J. Calvin Coffey**, Professor at University of Limerick in Ireland (2017) provided a status of complete organ to mesentery (*i.e.*, it is regarded as 79th organ of human body, see **Carolyn Gregoire**, 2017). Coffey found mesentery to be a continuous structure.

Fill in the Blanks

1. Founder. **2.** Richard Owen.

3. Specialized. **4.** Vestigial.

Multiple Choice Questions

1. (*b*) **2.** (*c*) **3.** (*a*) **4.** (*a*) **5.** (*c*) **6.** (*c*)

7. (*a*).

2

Integumentary System

The integument is the outer covering of the body, a protective wrapping that includes the **skin** and all structures derived from or associated with the skin, such as **hair**, **scales**, **feathers** and **horns**. Some other hard derivatives of integument include **claws, hooves** and **nails**. In most animals integument is tough and pliable, providing mechanical protection against abrasion and puncture and forming an effective barrier against the invasion of bacteria. It may provide moisture proofing against fluid loss or gain.

Integument covers all exposed surfaces of the vertebrate body, such as exposed portion of the eyeball (where it is covered by conjunctiva and is usually transparent) and external surface of ear drum. Besides this integument is also directly continuous with the lining of all passageways opening to the exterior (*e.g.*, mouth, anus, etc.).

2.1 EMBRYONIC ORIGIN OF INTEGUMENT

The integument of vertebrates is a composite organ having its origin from the embryonic layers such as **ectoderm** and **mesoderm**. The epidermis of skin is thin, outer stratified layer and it is derived from the ectoderm that remains on the surface of embryo after development of neural tube. Next to epidermis, the dermis is inner thicker layer and is derived from mesoderm, *i.e.*, the dermatome of the epimere and somatic mesoderm of the somatopleure (Fig. 2.1). The dermis of skin contains in it connective tissue, the blood vessels and nerve which grow throughout the dermis. The pigment cells migrate into dermis from the neural crest.

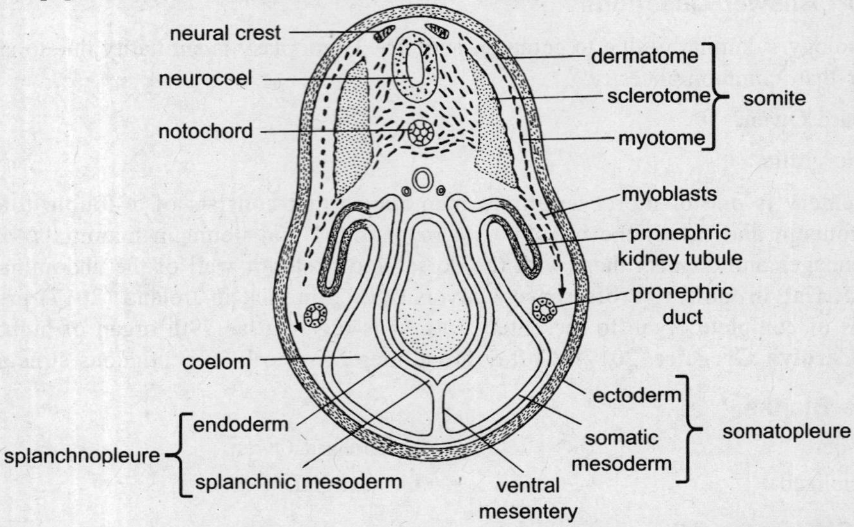

Fig. 2.1. T.S. vertebrate embryo (showing fate of the mesoderm). From dorsal mesoderm of the somite, sclerotome cells move toward notochord and form vertebra; myoblasts (arrows) form myotomal muscle and dermatome forms dermis of skin.

2.2 STRUCTURE OF INTEGUMENT

Fundamentally, the integument of chordates is composed of two layers, **epidermis** and **dermis**, both are separated by the **basement membrane**.

1. Epidermis. Various layers of epithelial layers of epithelial cells which compose the epidermis, arise in the deepest layer, the **stratum germinativum** (or **stratum Malpighii** or **stratum basale**). The cuboidal or columnar cells of stratum germinativum rest on a fine, fibrous basement membrane and undergo mitotic divisions throughout life. With each mitotic division, one of the two daughter cells stays in contact with the basement membrane, but the other cell is pushed outwards and differentiate according to species and body region (Fig. 2.2).

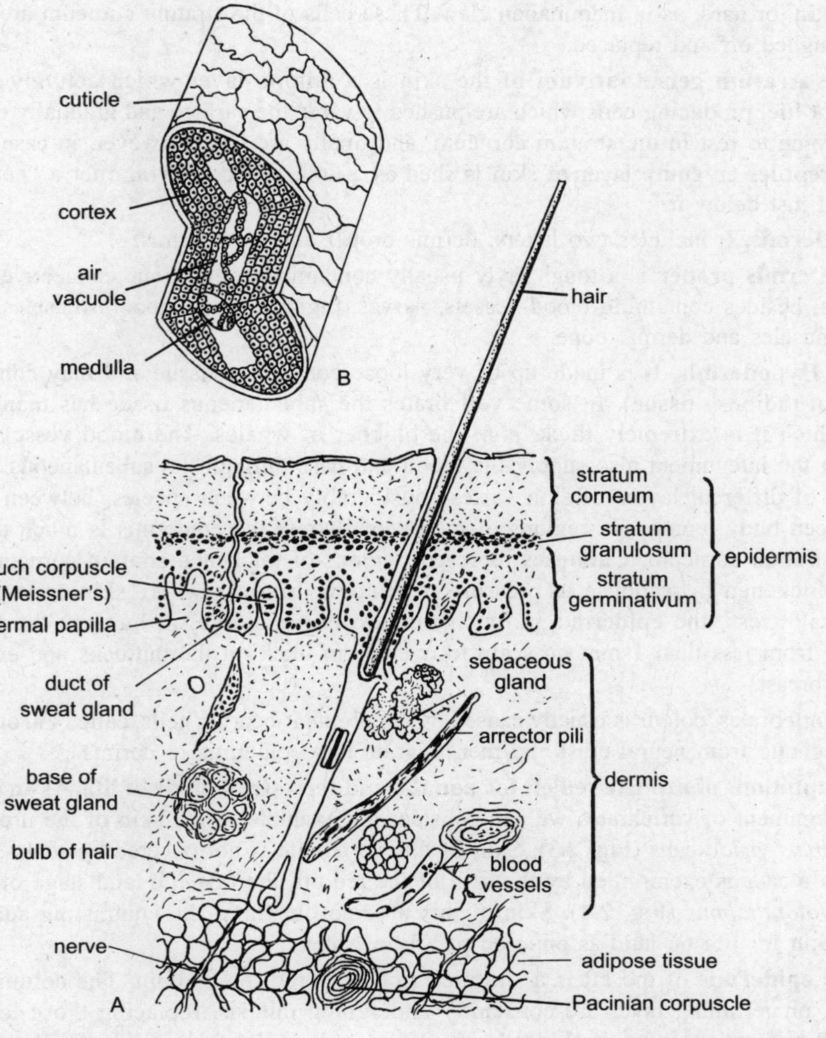

Fig. 2.2. A – Mammalian skin. Supporting the constituents of the dermis are bundles of collagenous connective tissue. Sweat glands are tortuous; hence, they are seen only in cross section. They are longer than the dermis is deep. The touch corpuscle occupies a dermal papilla. The pacinian corpuscle (kinesthetic receptor for the sensation of touch and deep pressure) is in the subcutaneous connective tissue. B – Section of the hair (Kent and Carr 2001).

The cells superficial to the stratum germinativum usually are connected to adjacent cells by desmosomes, which give the epidermis physical coherence. The cytoplasm of epidermal cells located peripherally may contain granules of the structural protein (scleroprotein) **keratin**. Some cells of the stratum germinativum can differentiate into glands or other integumentary derivatives.

The **stratum corneum** is the most superficial layer of the skin. It consists of dead cells which contain in it the keratin. In amniotes, the keratin is formed massively and exists in two molecular forms: **α-keratin** which contains filaments 70-80 Å wide, and **β-keratin** which contains filament size about 30 Å wide. The β-keratin layer is invariably hard, for example, the surface of avian and crocodilian scales. In contrast, α-keratin layers may be either soft as in mammalian flexible skin, or hard, as in mammalian claw. These cells of the stratum corneum are continually being sloughed off and replaced.

The **stratum germinativum** of the skin is a mitotic layer which actively proliferates throughout life, producing cells which are pushed towards the surface and gradually differentiate, die and come to rest in the stratum corneum, and finally are lost. However, in case of **Lepidosaurian reptiles** an entire layer of skin is shed as a unit, called **exuvia**, after a "new skin" has developed just below it.

2. Dermis. It includes two layers: dermis proper and hypodermis.

(*i*) Dermis proper is a tough layer usually containing collagenous connective tissue cells and fibers, besides containing blood-vessels, nerves, pigment cells, smooth muscles, occasional skeletal muscles and dermal bone.

(*ii*) Hypodermis. It is made up of very loose connective tissue and may contain a great deal of fat (adipose tissue). In some vertebrates the subcutaneous tissue has many fat layers due to which it is extremely thick, *e.g.*, the blubber of whales. The blood vessels and nerve present in the integument also supply the blood and pass through the subcutaneous tissues. The thickness of different layers of skin varies considerably between species, between individuals and between body regions. In regions of mechanical stress the epidermis is much thicker, with a heavy stratum corneum. **Callouses** represent regions of thickened stratum corneum that have become thickened in response to pressure or friction. In areas where skin copes with lesser mechanical stress, the epidermis is much thinner. For example, in humans the hypodermis may vary from less than 1 mm on scalp to over 3 cm thick on the buttocks and even more in women's breast.

In vertebrates, colour is usually caused by the pigment-bearing cells, called **chromatophores** which originate from neural crest and then migrate into epidermis or dermis.

Adaptations of urodele red eft for aquatic and terrestrial mode of life. As an introduction to the integument of vertebrates we can consider the example of the skin of the urodele red eft *Notopthalmus viridescens* (Fig. 2.3). Skin of this vertebrate is not covered by scales, feathers or hair. Such a case is exemplified by the skin of the red eft, the juvenile land stage of the aquatic urodele *Notopthalmus* (Fig. 2.4). Skin of this urodele illustrates the contrasting adaptations of craniate skin for life on land as opposed to life in water.

The **epidermis** of the eft is a stratified (multilayered) epithelium. The columnar cells in the basal, or germinal, layer are constantly undergoing mitosis, replacing those lost from the surface. Proliferation from the basal layer causes older cells to be pushed outward. As they approach the surface, they synthesize **keratin** and they become flattened (squamous). They are then said to be keratinized or cornified. Keratinization causes the cells to die. In efts, the cornified layer is thin in contrast to the thicker cornified stratum of terrestrial craniates (*e.g.*, human beings).

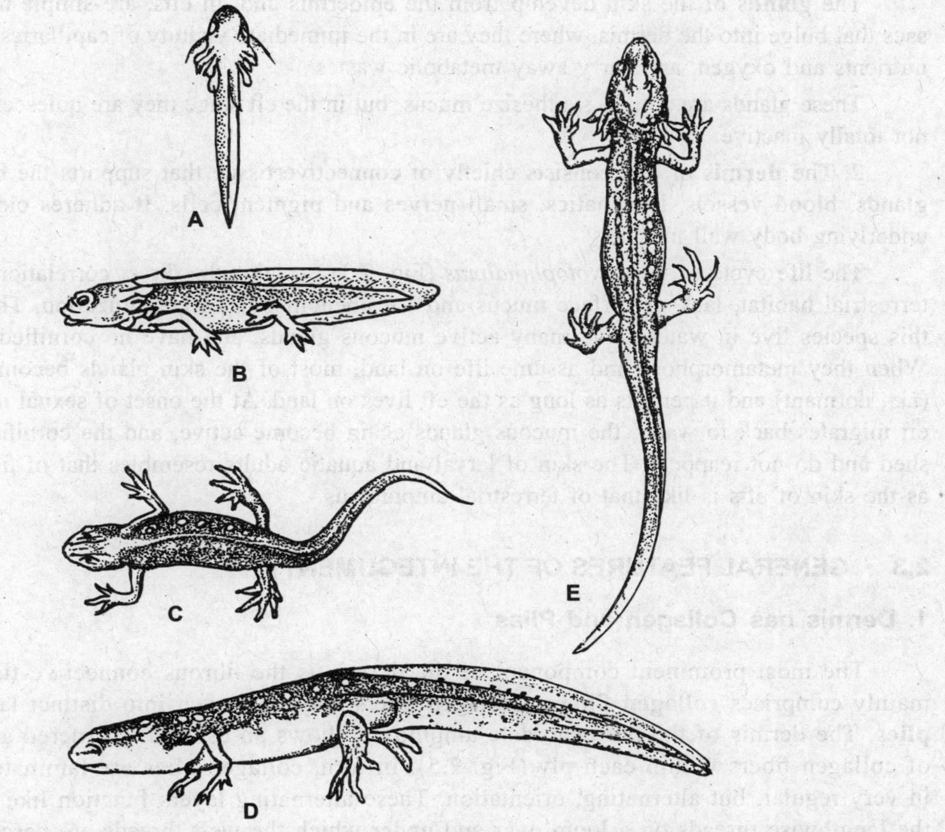

Fig. 2.3. Life history of newt *Notopthalmus viridescens*; A – Newly hatched larva (7 mm), balancers behind the eyes, gills only slightly branched, forelimbs are blunt buds, hind limbs not formed, keeled back and tail; B – Late larva; C – Terrestrial eft; D – Adult male (95 mm); E – Heterochronic adult (after Kent and Carr 2001).

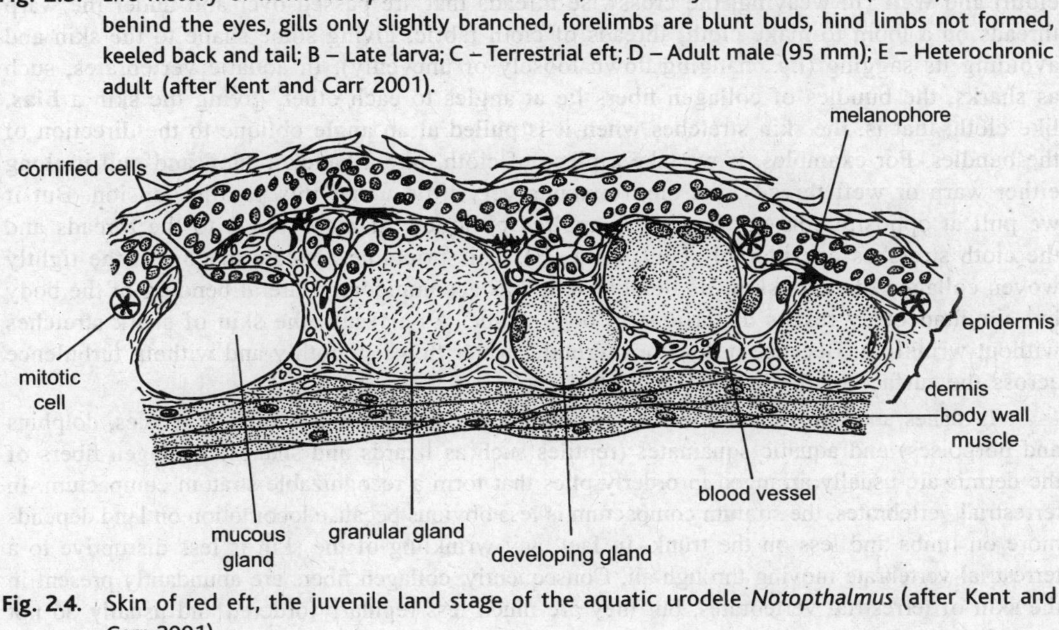

Fig. 2.4. Skin of red eft, the juvenile land stage of the aquatic urodele *Notopthalmus* (after Kent and Carr 2001).

The **glands** of the skin develop from the epidermis and, in efts, are simple multicellular sacs that bulge into the dermis, where they are in the immediate vicinity of capillaries that supply nutrients and oxygen, and carry away metabolic wastes.

These glands are able to synthesize mucus, but in the eft stage they are quiescent, although not totally inactive.

2. The **dermis** of efts consists chiefly of connective tissue that supports the bases of the glands, blood vessels, lymphatics, small nerves and pigment cells. It adheres closely to the underlying body wall muscle.

The life cycle of newt *Notophthalmus* (Fig. 2.3) explains the direct correlation between a terrestrial habitat, lack of surface mucus and the phenomenon of keratinization. The larvae of this species live in water, have many active mucous glands, and have no cornified epidermis. When they metamorphose and assume life on land, most of the skin glands become quiescent (*i.e.*, dormant) and it persists as long as the eft lives on land. At the onset of sexual maturity, the eft migrates back to water, the mucous glands again become active, and the cornified cells are shed and do not reappear. The skin of larval and aquatic adults resembles that of fishes, where as the skin of efts is like that of terrestrial amphibians.

2.3 GENERAL FEATURES OF THE INTEGUMENT

1. Dermis has Collagen and Plies

The most prominent component of the dermis is the fibrous connective tissue which mainly comprises **collagen fibers**. Collagen fibers may be woven into distinct layers called **plies**. The dermis of the protochordate amphioxus shows an especially ordered arrangement of collagen fibers within each ply (Fig. 2.5). In turn, collagen plies are laminated together in very regular, but alternating, orientation. These alternating layers function like **warp** (*i.e.*, the lengthwise threads on a loom over and under which the weft threads are passed to make cloth) and **weft** (in weaving, the crosswise threads that are passed over and under the warp threads on a loom to make cloth) threads of cloth fabric, giving some shape to the skin and avoiding its sagging (*i.e.*, hanging down loosely or unevenly). In aquatic vertebrates, such as sharks, the bundles of collagen fibers lie at angles to each other, giving the skin a **bias**, like cloth; that is, the skin stretches when it is pulled at an angle oblique to the direction of the bundles. For examples, if we take a piece of cloth, like a handkerchief, and pull it along either warp or weft threads, the cloth extends very little under this parallel tension. But if we pull at opposite corners, tension is applied obliquely at a 45° angle to the threads and the cloth stretches considerably (Fig. 2.6A, B). This principle appears to govern the tightly woven collagen of shark skin. Its flexible skin bias accommodates lateral bending of the body but simultaneously resists distortions in body shape. As a result, the skin of shark stretches without wrinkling. Because it does not wrinkle, water flows smoothly and without turbulence across the surface of the body (Fig. 2.6D).

In fishes and aquatic vertebrates, including cetaceans (mammals such as whales, dolphins and porpoises) and aquatic squamates (reptiles such as lizards and snakes), collagen fibers of the dermis are usually arranged in orderly plies that form a recognizable stratum compactum. In terrestrial vertebrates, the stratum compactum is less obvious because locomotion on land depends more on limbs and less on the trunk. In fact, any wrinkling of the skin is less disruptive to a terrestrial vertebrate moving through air. Consequently, collagen fibers are abundantly present in the skin of terrestrial vertebrates, but they are much less regularly ordered and usually do not form distinct plies.

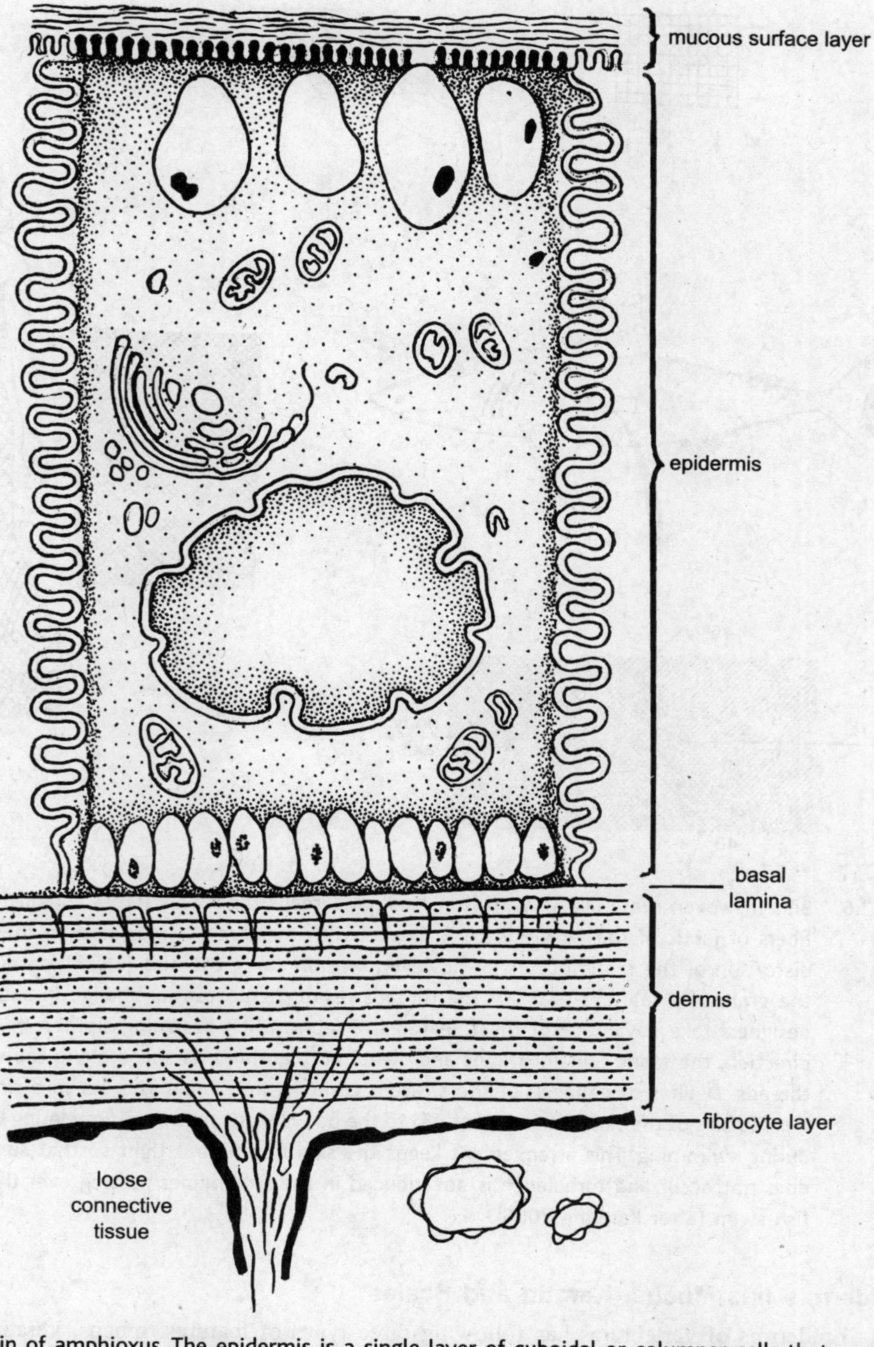

Fig. 2.5. Skin of amphioxus. The epidermis is a single layer of cuboidal or columnar cells that secrete a mucus that coats the surface and rests upon a basal lamina. The dermis consists of very highly ordered collagen fibers arranged in alternating plies (layers) to form a "fabric" that brings structural support as well as flexibility to the outer body wall. Pigment is secreted by the epidermal cells (after Kardong 2006).

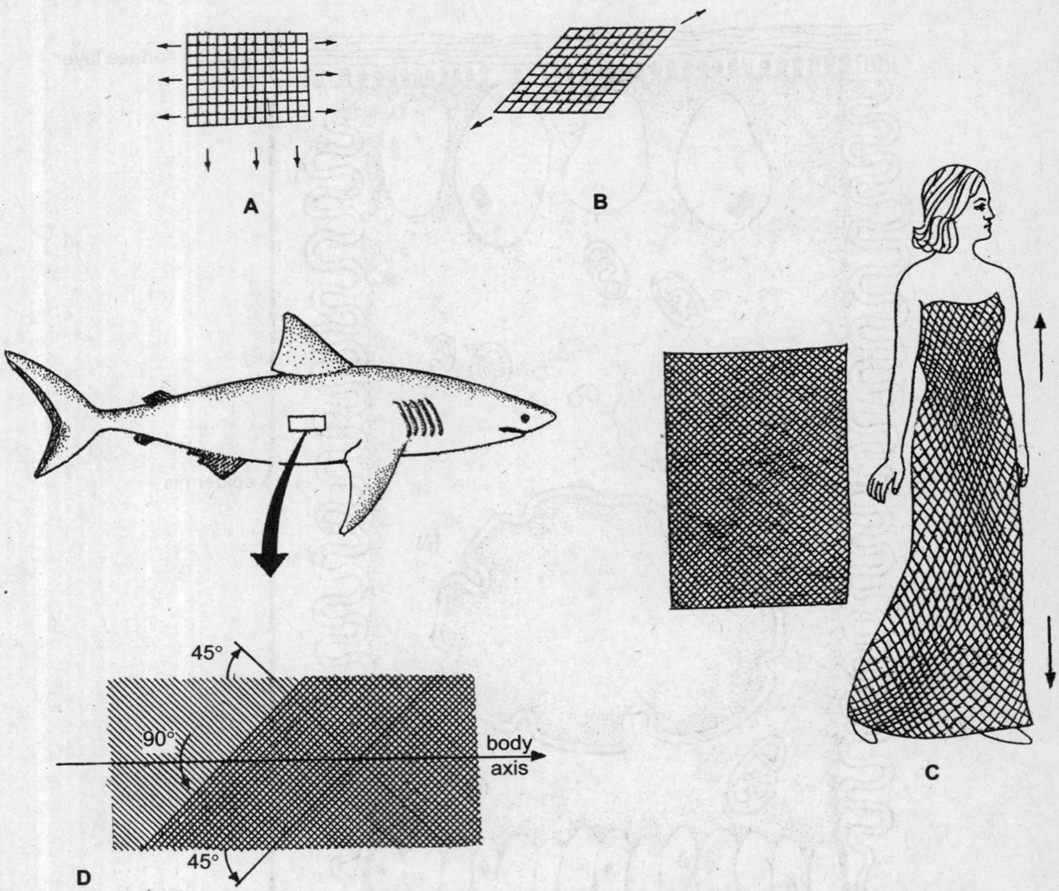

Fig. 2.6. Bias in woven material. A–Warp (longitudinal threads) and weft (cross threads) compose the fibers of fabric. If the tensile force is parallel with the threads (indicated by the arrows), little distortion of the fabric occurs. B–However, tension along the bias (a slanting direction across the grain of a fabric) at 45° to the threads result in substantial change in shape. C–Fashion designers take advantage of these features of fabric when designing clothes. In the loose bias direction, the fabric falls into folds and pleats but can hold its shape along the warp-and-weft threads. D–Plies of collagen of the stratum compactum of fish skin act in a similar way. The flexible bias of the skin is oriented at 45° to the body length thus accommodating lateral bending during swimming. This arrangement keeps the skin flexible but tight so that surface wrinkling does not occur and turbulence is not induced in the streamlines passing over the body as the fish swim (after Kardong 2006).

Epidermis has Mucus, Keratin and Scales

Epidermis of vertebrates has following three types of features: mucus, keratin and scales.

1. Mucus. The epidermis of many vertebrates produces mucus to moisten the surface of the skin. In fishes, mucus seems to afford some protection from bacterial infection and helps ensure the laminar flow of water across the body surface. In amphibians, mucus tends to serve similar functions and it also keeps the skin from drying during animal's visits on land.

2. Keratin. The epidermis of terrestrial vertebrates forms a protective layer of **karatinized** or **cornified** cells, called **stratum corneum**. New epidermal cells are formed by mitotic division, primarily in the deep **stratum basale**. These new epidermal cells push more superficial cells toward the surface, where they tend to self-destruct in an orderly fashion. During their death process, various protein products accumulate and collectively form **keratin** in the process of **keratinization**. Keratin is a scleroprotein that is insoluble in water. Thus, keratin is a class of protein produced during keratinization, and the specific epidermal cells that participate are keratinocytes. The resulting superficial stratum corneum of epidermis of flattened squamous cells is a nonliving layer that serves to reduce water loss through the skin in dry terrestrial environments.

Keratinization and formation of a stratum corneum also occur where friction or direct mechanical abrasion injure the epithelium. For example, the epidermis in the oral cavity of aquatic and terrestrial vertebrates often exhibits a keratinized layer, especially if the food eaten is unusually sharp or abrasive. In areas of the body where friction is common, such as the soles of the feet or palms of the hands, the cornified layer may form a thick protective layer, or **callus** to prevent mechanical damage (Fig. 2.7). The stratum corneum may be differentiated into hair, hooves, horn sheathes, or other specialized cornified structures. The term keratinizing system refers to the elaborate interaction of epidermis and dermis that produces the orderly transformation of keratinocytes into such cornified structures.

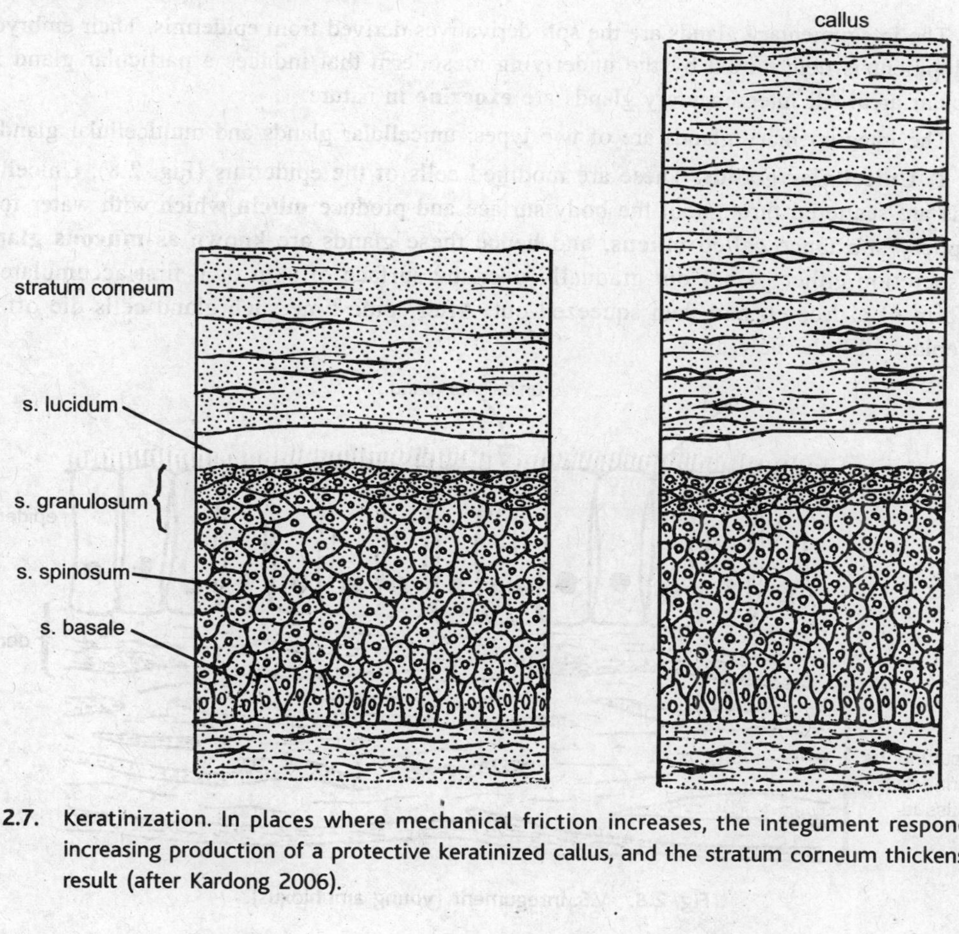

Fig. 2.7. Keratinization. In places where mechanical friction increases, the integument responds by increasing production of a protective keratinized callus, and the stratum corneum thickens as a result (after Kardong 2006).

3. Scales. Scales are formed within the integument of many aquatic and terrestrial vertebrates. **Scales** are basically folds in the integument. If dermal contributions predominate, especially in the form of ossified dermal bone, the fold is termed a **dermal scale**. An epidermal fold, especially in the form of a thickened keratinized layer, produces a **epidermal scale**.

Most fishes synthesize little or no keratin in their epidermis, but some aquatic urodeles, especially those that venture on land, have a thin desiccation–impeding stratum of cornified cells. In lampreys, conical cornified epidermal spines and "teeth" develop in the buccal funnel and on the rasping lingual cartilage. Hagfishes also possess cornified "teeth". Anuran tadpoles have rows of horny toothlike structures surrounding the entrance to the mouth that enable them to feed by scraping vegetation during their herbivorous larval stage. Tadpoles shed these horny structures at metamorphosis. Callus-like caps develop on the toes of aquatic urodeles (**Kent** and **Carr** 2001).

2.4 DERIVATIVES OF INTEGUMENT

Following two types of structures are derived from the integument: soft derivatives and hard derivatives.

1. Soft Derivatives of Integument

The integumentary glands are the soft derivatives derived from epidermis. Their embryonic development is determined by the underlying mesoderm that induces a particular gland in a particular area. All integumentary glands are **exocrine** in nature.

The integumentary glands are of two types: unicellular glands and multicellular glands.

1. Unicellular glands. These are modified cells of the epidermis (Fig. 2.8). Unicellular glands are scattered throughout the body surface and produce **mucin** which with water forms a slippery substance called **mucus**, and hence these glands are known as **mucous glands**. The secretion either passes out gradually over the surface of skin or it first accumulates in the form of a droplet, and then squeezed out. After some time such gland cells die off and are replaced by the new ones.

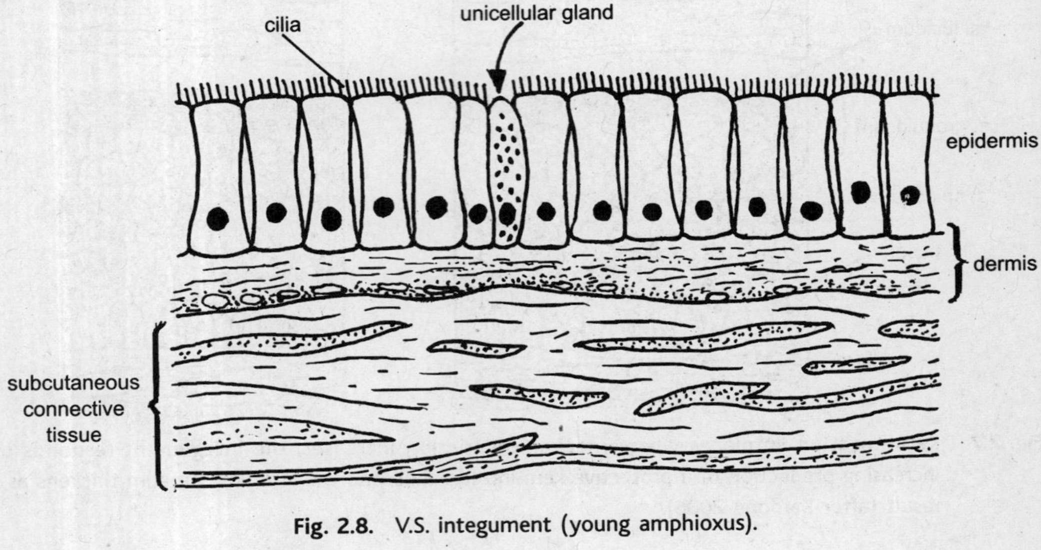

Fig. 2.8. V.S. integument (young amphioxus).

2. Multicellular glands. Usually an invagination from the stratum germinativum extends down into the dermis to differentiate into multicellular glands. Multicellular glands may be classified into **tubular** and **alveolar** (or **saccular**) and each of these may again be subdivided further (Fig. 2.9).

Tubular glands are deeper, some even extend into the **hypodermis**. They are lined by a single layer of epithelial cells; usually those in the deeper part of the gland do secretion. If the tube is straight; it is called a **simple tubular gland**; and if the tube coils at the end, it is called **coiled tubular gland**; and if after entering the dermis or hypodermis the tube branches into smaller tubes; it is called a **branched tubular gland**. Some time each branch coils in the hypodermis, forming a branched, coiled tubular gland.

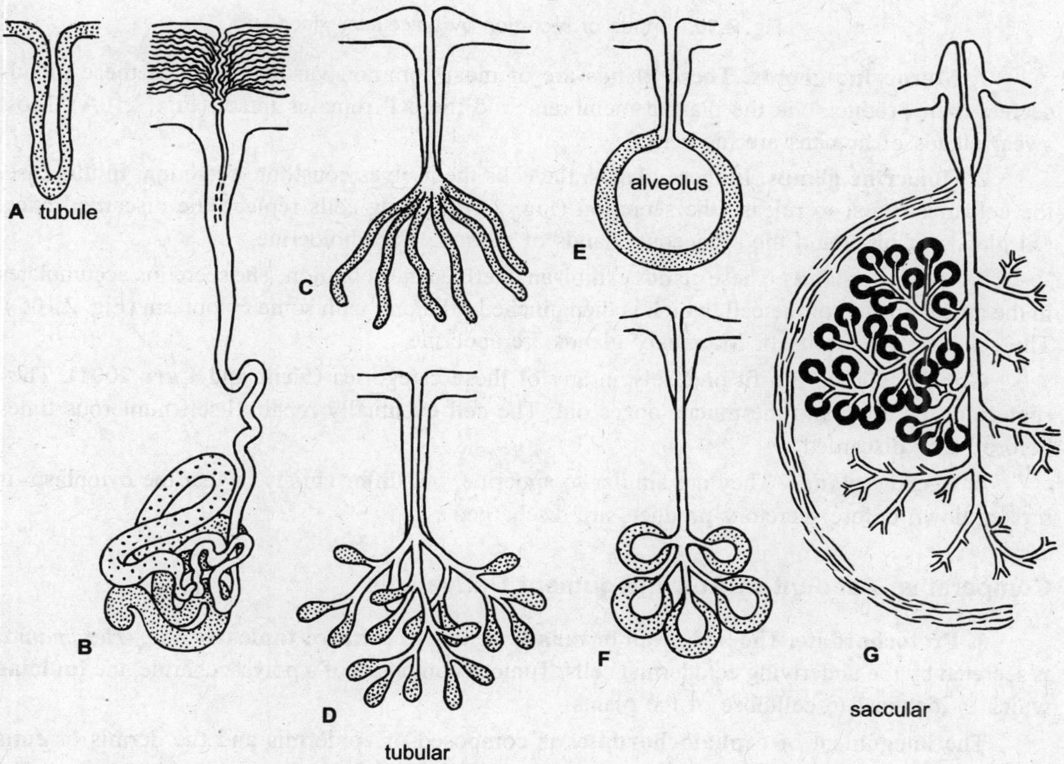

Fig. 2.9. Major morphologic varieties of multicellular glands. A – Simple tubular; B – Coiled tubular; C – Simple branched tubular; D – Compound tubular; E – Alveolus of a simple saccular gland; F – Simple branched saccular; G – Compound saccular (compound alveolar). Shaded part shows secretory epithelia (after Kent and Carr 2001).

Alveolar glands seldom go deep into the dermis (Fig. 2.9E) and only rarely enter the hypodermis (Fig. 2.9B). However, they have more bulk, being several cells thick with a small lumen. If arranged as one large clump of cells, the gland is called a **simple alveolar gland**; if in several clumps, it is called a **compound alveolar gland**.

Glands vary in the way their cells liberate the substances they synthesize. There are four basic categories (Fig. 2.10).

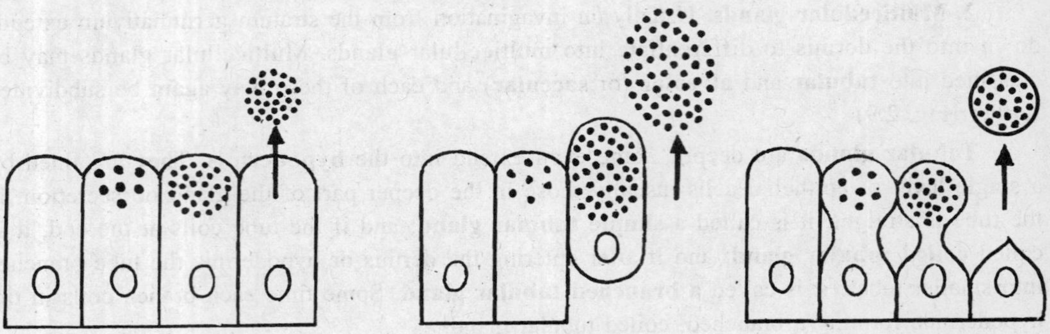

Fig. 2.10. Mode of secretion by unicellular gland.

1. Merocrine glands. These glands are of most common variety. Cells of these glands secrete their products via the plasma membrane and the cell remains intact (Fig. 2.10A). Most sweat glands of humans are merocrine.

2. Holocrine glands. In these glands, the cells themselves constitute secretion. In this case, the cell must burst to release the secretion (Fig. 2.10B), new cells replace the discarded ones. Oil glands of birds and the sebaceous glands of mammals are holocrine.

3. Apocrine glands. These glands exhibit an intermediate condition. The secretion accumulates in the apical portion of the cell, which is then pinched off along with some cytoplasm (Fig. 2.10C). The cell than repairs itself. Mammary glands are apocrine.

Goblet cells do not fit precisely in any of these categories (**Kent** and **Carr** 2001). They rupture at the apex and the mucus oozes out. The cell eventually repairs itself numerous times before being discarded.

4. Eccrine glands. They are similar to apocrine, but differ chiefly in that the cytoplasm is broken down before secretory products are discharged.

Comparative Account of Soft Integument Derivatives

1. Protochordate. The body wall or **mantle** or **test** or **tunic** of **tunicates** (*e.g., Herdmania*) is secreted by the underlying ectodermal cells. Tunic is composed of a polysaccharide, the **tunicine** which is identical to cellulose of the plants.

The integument of **cephalochordates** is composed of epidermis and the dermis or cutis and is followed by muscles. Epidermis of the amphioxus (*Branchiostoma*) is very thin and is composed of a single layer of **cubical epithelial cells** (Fig. 2.11). The nucleated cells in young or immature individual bear cilia (Fig. 2.8). Later these cells shed cilia and secrete a thin cuticle on outer border. Unicellular gland cells and chemoreceptor cells are scattered between the epidermal cells.

The integument of **hemichordates** (*e.g., Balanoglossus, Saccoglossus*) is composed of ciliated and secretory cells (Fig. 2.12). The ciliated cells which cover the body surface are distinct and recognized by the presence of cilia and associated bundle of rootlet fiber. The secretory cells may be goblet shaped and contain a pear shaped secretory product or long **granular** or **alveolated cells**: they secrete mucus that spread over the body wall. The **mulberry cells** are also present in the integument; they secrete amylase.

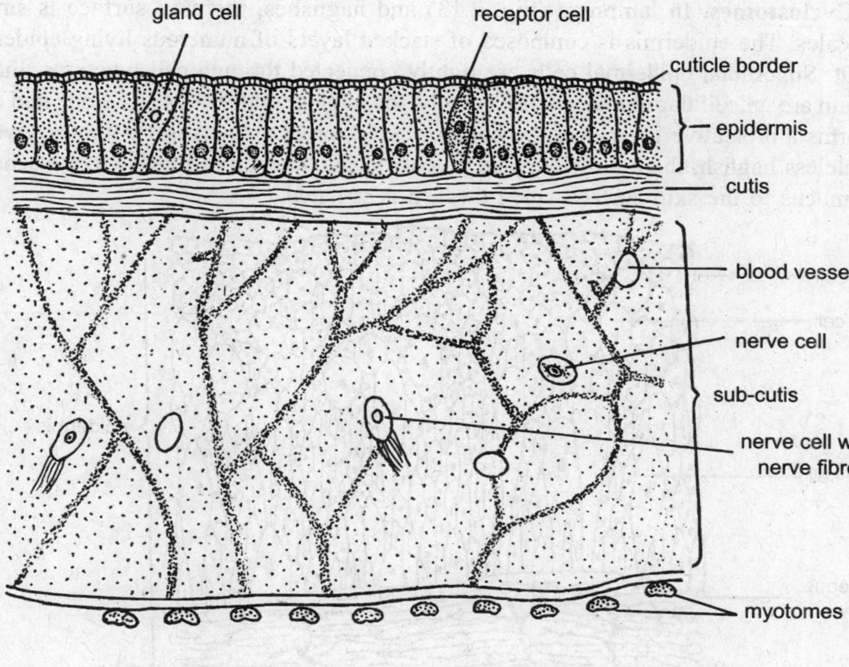

Fig. 2.11. V.L.S. of integument of *Branchiostoma* (amphioxus).

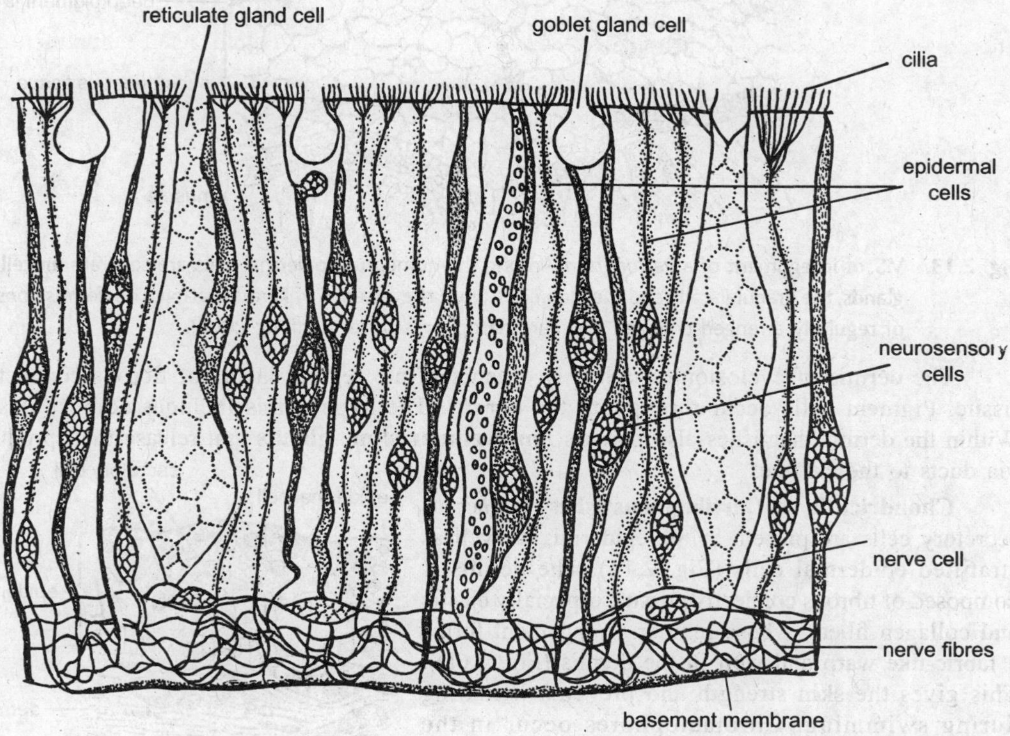

Fig. 2.12. T.S. integument of *Saccoglossus*.

2. Cyclostomes. In lampreys (Fig. 2.13) and hagfishes, the skin surface is smooth and without scales. The epidermis is composed of stacked layers of numerous living epidermal cells throughout. Superficial epidermal cells are tightly connected through cell junctions. Interspersed among them are unicellular mucus glands, namely, the large granular cells and elongated club cells. Mucus forms a protective layer over the body of cyclostomes. Mucus is a complex glycoprotein. In the scaleless hagfish, there are elongated mucous cells, "the **thread cells**" that discharge thick cords of mucus to the skin surface when the fish is irritated.

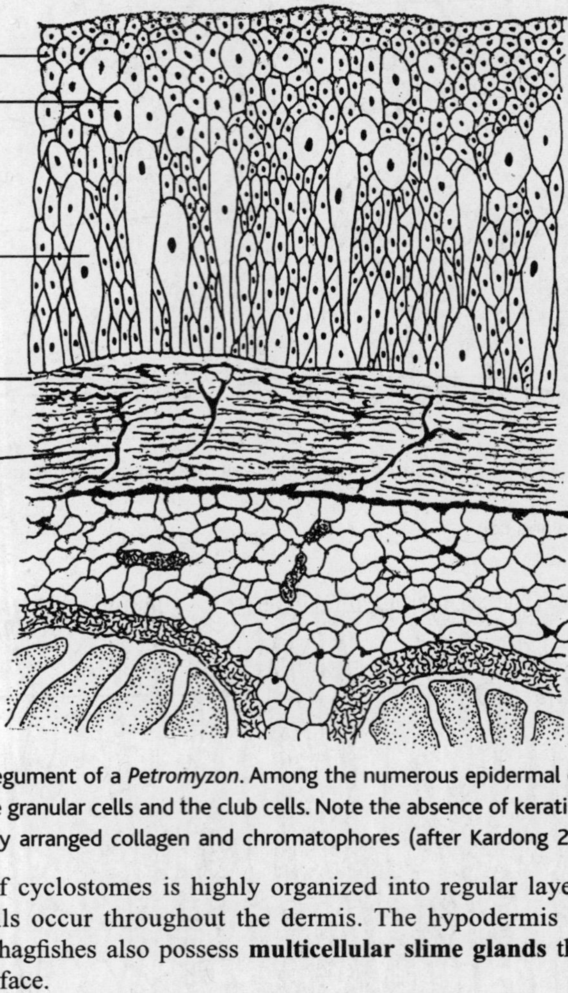

Fig. 2.13. V.S. of integument of a *Petromyzon*. Among the numerous epidermal cells are separate unicellular glands, the granular cells and the club cells. Note the absence of keratinization. The dermis consists of regularly arranged collagen and chromatophores (after Kardong 2006).

The dermis of cyclostomes is highly organized into regular layers of fibrous connective tissue. Pigment cells occur throughout the dermis. The hypodermis includes adipose tissue. Within the dermis, hagfishes also possess **multicellular slime glands** that release their products via ducts to the surface.

Chondrichthyes (Cartilaginous fishes). Numerous secretory cells are present in the epidermis as well as stratified epidermal cells (Fig. 2.14). The dermis is composed of fibrous connective tissue, especially elastic and collagen fibers, whose regular arrangement forms a fabric-like warp and weft in the dermis (Fig. 2.6D). This gives the skin strength and prevents wrinkling during swimming. Chromatophores occur in the lower part of the epidermis and upper region of the dermis.

Fig. 2.14. V.S. of skin of shark.

Osteichthyes (Bony fishes). The dermis of bony fishes is subdivided into a superficial layer of loose connective tissue. Chromatophores are found within the dermis. The epidermal covering

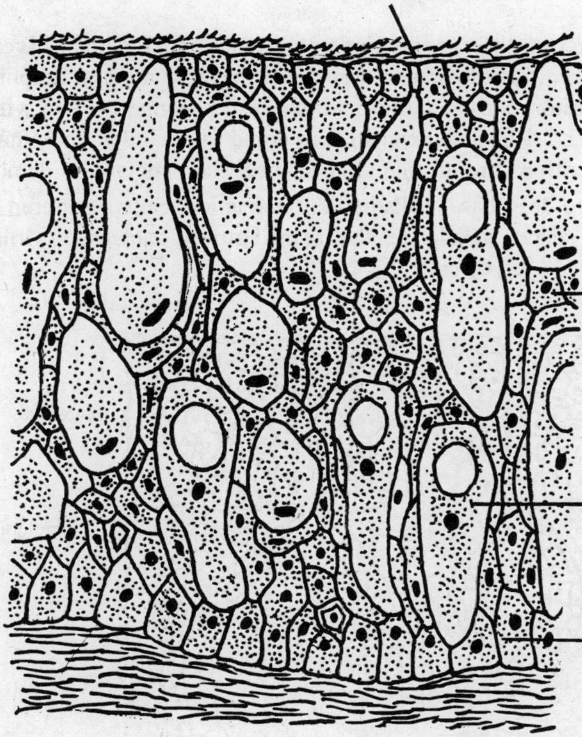

Fig. 2.15. Bony fish skin. Enlargement of epidermis to show club-cells and mucous cuticle (after Kardong 2006).

includes a basal layer of cells. Above this layer are stratified epidermal cells. As they move toward the surface, epidermal cells undergo cytoplasmic transformation, but they do not become keratinized. Within these layered epidermal cells occur single unicellular glands, the secretory and club cells (Fig. 2.15). These unicellular glands, along with epidermal cells, are the source of the **mucous cuticle**, or surface "slime".

The electron microscope has helped distinguish an additional type of unicellular gland in the epidermis of bony fishes, the **sacciform cell**. It holds a large, membrane-bound secretory product that seems to function as a **repellent** or **toxin** against enemies once it is released (**Kardong** 2006).

Integument of Tetrapods

While keratinization occurs in fishes, among terrestrial vertebrates it becomes a key feature of the integument. Vast keratinization produces a prominent outer cornified layer, the stratum corneum, that resists mechanical abrasion. Lipids are often added during the process of keratinization or spread across the surface from specialized glands. The cornified layer along with these lipids increases the resistance of the tetrapod skin to desiccation.

Multicellular glands are more common in the skin of tetrapods than in the skin of fishes. In fishes, the mucous cuticle and secretion of the unicellular glands at or near the surface of the skin coat it. In contrast, among tetrapods, multicellular glands usually reside in the dermis and reach the surface through common ducts that pierce the cornified layer (*e.g.*, sebaceous glands and sweat or sudoriferous glands). Thus, the stratum corneum that protects skin and prevent desiccation also controls the release of secretions directly to the surface. Due to these openings in the stratum corneum, the surface of the skin remains coated or lubricated by the secretions of skin glands.

Amphibia. Amphibians are of special interest because during their lives they usually metamorphoses from an aquatic form to a terrestrial form. In most modern amphibians, the skin is also specialized as a respiratory surface across which gas exchange occurs, with the capillary beds in the lower epidermis and deeper dermis. In fact, some salamanders lack lungs and depend entirely on **cutaneous respiration** through the skin to meet their metabolic demands.

Within the epidermis are deep basal cells and surface apical cells. Scattered throughout are large **Leydig cells** thought to secrete substances that resist entry of bacteria or viruses (Fig. 2.16).

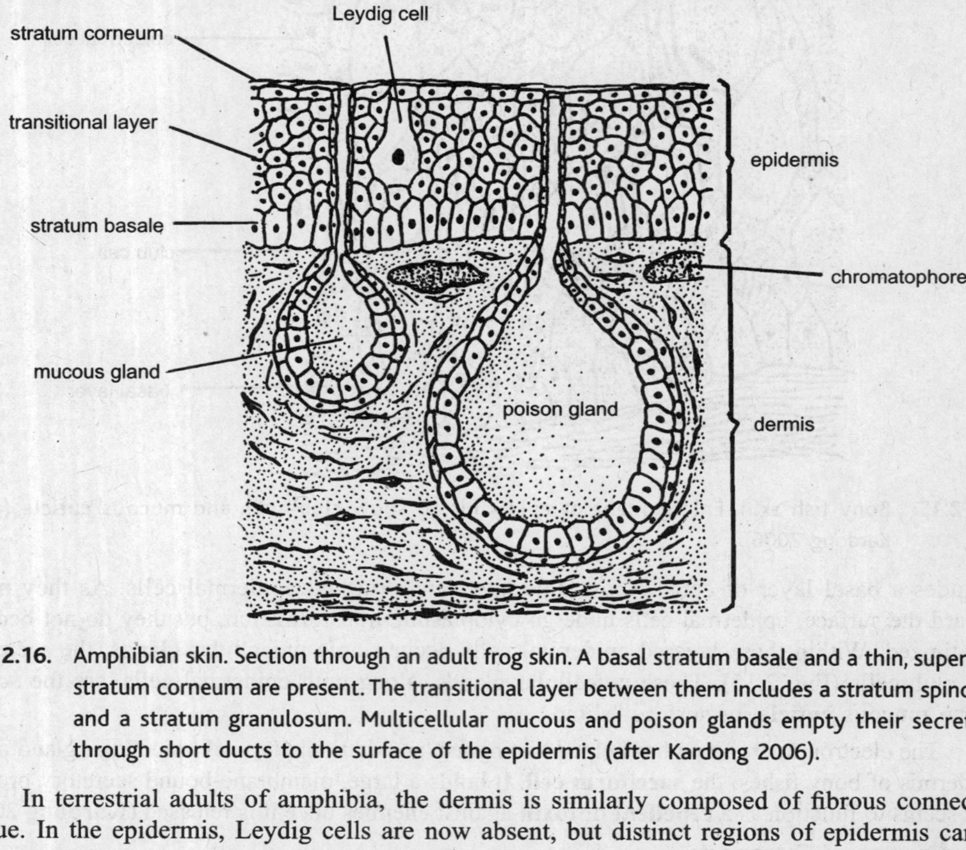

Fig. 2.16. Amphibian skin. Section through an adult frog skin. A basal stratum basale and a thin, superficial stratum corneum are present. The transitional layer between them includes a stratum spinosum and a stratum granulosum. Multicellular mucous and poison glands empty their secretions through short ducts to the surface of the epidermis (after Kardong 2006).

In terrestrial adults of amphibia, the dermis is similarly composed of fibrous connective tissue. In the epidermis, Leydig cells are now absent, but distinct regions of epidermis can be recognized, such as the strata basale, spinosum, granulosum and corneum. Presence of a thin stratum corneum affords some protection from mechanical abrasion and retards off cutaneous exchange. During the breeding season, **nuptial pads** may form on digits or limbs of male salamenders or frogs. Nuptial pads are raised **calluses** of cornified epidermis that help the male hold the female during mating.

The skin of amphibians contains two types of multicellular glands: mucous and poison glands. Both are located in the dermis and open to the surface tnrough connecting ducts. The **mucous glands** tend to be smaller, each of them is made up of a little cluster of cells that release their product into common duct. The **poison glands** (or granular glands) tend to be larger and often contain stored secretion within the lumen of each gland. Secretions of poison glands tend to be distasteful or even toxic to predators (Box 2.1).

Box 2.1 Poison dart frog

The bright colours of poison dart frog advertise toxic skin secretions that are poisonous to most predators. The skin of most amphibians contains glands that secrete products that are distasteful or even toxic to predators. Often these skin toxins are not manufactured by the frogs themselves, but instead are copted from toxins in their prey, such as from ants or other arthropods. Some ant specialists suggest that the insects, in turn, may get the toxins or toxin precursors from their diet. In tropical regions of New World lives a group of frogs, the *poison dart frogs*, with specially toxic skin secretions. Native people of the region often gather these frogs, hold them on sticks over a fire to stimulate release of these secretions, and then collect the secretions on the tips of their darts. Game shot with these toxin-laced darts is quickly tranquilized or killed. Cooking denatures the toxins, making the game safe for humans to eat (see **Kardong**, 2006).

Chromatophores may occasionally be found in the amphibian epidermis, but most reside in the dermis. Capillary beds, restricted to the dermis in most vertebrates reach into the lower part of the epidermis in amphibians, a feature serving cutaneous respiration.

Reptiles. The skin of reptiles is adapted to life in arid and ofter hostile environment. Keratinization is much more extensive, and skin glands are fewer than in amphibians. **Scales** are present, but these are fundamentally different from the dermal scales of fishes, which are built around bone of dermal origin (Fig. 2.18). The reptilian scale usually lacks the bony under support or any significant contribution from dermis. Instead it is a fold in the surface epidermis, hence, an **epidermal scale**. The junction between adjacent epidermal scale is the flexible **hinge** (Fig. 2.17A). If the epidermal scale is large and platelike, it is called **scute**. Additionally epidermal scales of reptiles may be modified into **crests**, **spines** or **hornlike processes**.

Dermal bone is present in many reptiles. The **gastralia**, a collection of bone in the abdominal area, are examples of gastralia. Where dermal bones support the epidermis, they are called **osteoderms**, plates of dermal bone located under the epidermal scales. Osteoderms are found in crocodilians, some lizards and extinct reptiles. Some bones of the turtle shell are probably modified osteoderms.

The **dermis** of reptilian skin is composed of fibrous connective tissue. The epidermis is generally delineated into three regions: **stratum basale**, **stratum granulosum** and **stratum corneum**. However, stratum corneum changes prior of molting in those reptiles that slough large pieces of the cornified skin layer. In turtles and crocodiles, sloughing of skin is moderate, comparable to birds and mammals, in whom small flakes fall of at irregular intervals. But in lizards and especially in snakes, shedding of the cornified layer, termed **molting** or **ecdysis**, results

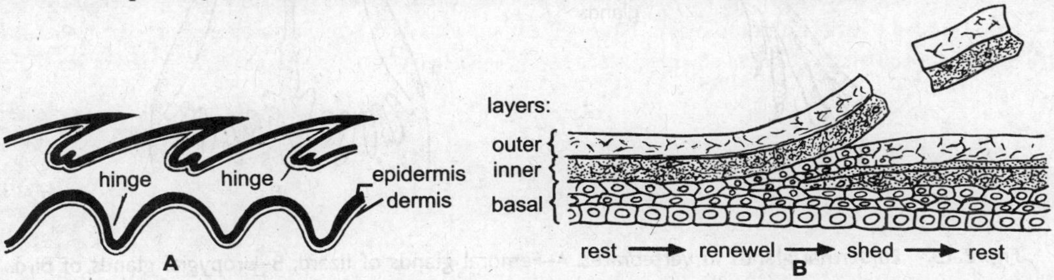

Fig. 2.17. Reptile skin. A–Extent of projection and overlap of epidermal scales varies among reptiles and even along the body of the same individual. Snake body scales (top) the tubercular scales of many lizards (bottom) are illustrated. Between scales is a thinned area of epidermis, a "hinge" allowing skin flexibility. B–Skin shedding. Just before the old outer layer of epidermis is shed, the basal cells produce an inner epidermal generation. White blood cells collect in the splitting zone to promote separation of new from old outer epidermis (after Kardong 2006).

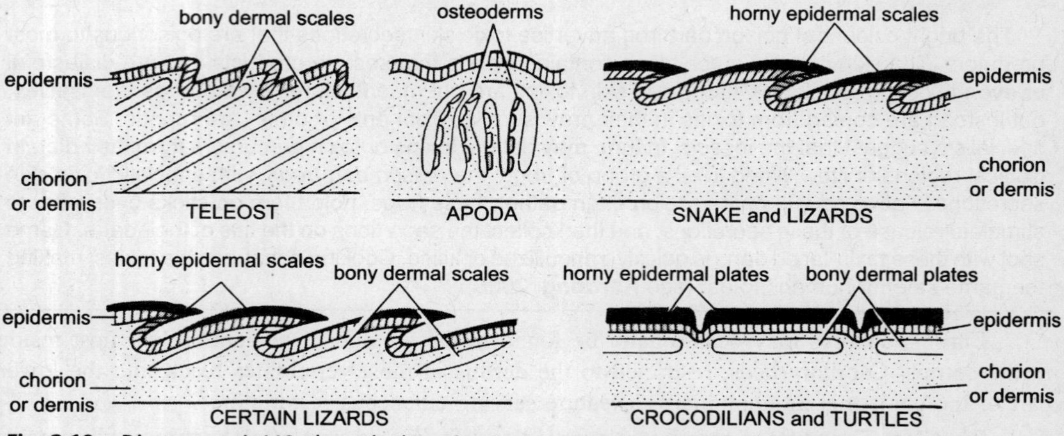

Fig. 2.18. Diagrammatic V.S. through skin of various vertebrates showing relationship of various types of scales.

in removal of extensive sections of superficial epidermis. As molting begins, the stratum basale, which has given rise to the strata granulosum (inner) and corneum (outer), duplicates the deeper layers of granulosum and corneum, pushing up under the old layers. White blood cells invade the **stratum intermedium**, a temporary layer between old and new skin (Fig. 2.17B). These white blood cells are thought to promote the separation and loss of the old superficial layer of the skin.

In reptiles, integumental glands are usually restricted to certain areas of the body. Many lizards have rows of femoral glands along the underside of the hind limb in the thigh region (Fig. 2.19A). Crocodiles and some turtles have **scent glands**. In alligators of both sexes, one pair

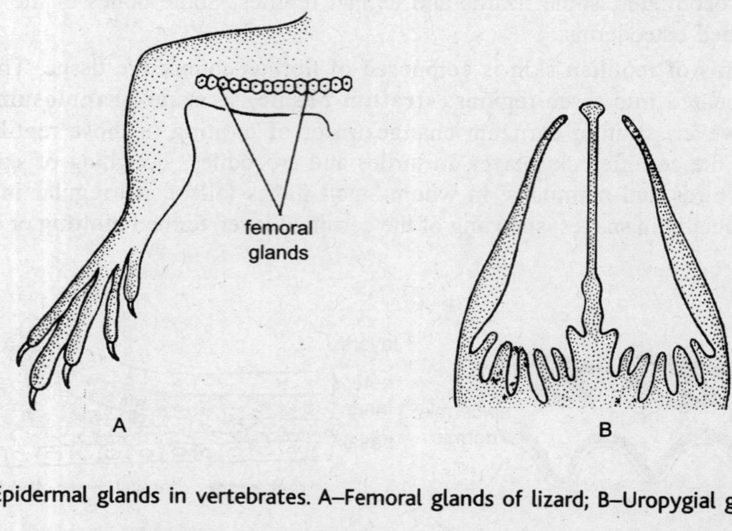

Fig. 2.19. Epidermal glands in vertebrates. A–Femoral glands of lizard; B–Uropygial glands of bird.

of scent glands opens into the cloaca, another pair opens on the margins of the lower jaw. In some turtles, scent glands can produce quite pungent odors, especially when the animal is alarmed by handling. Most integumental glands of reptiles are thought to play a role in reproductive behaviour or to discourage predators (see **Kardong** 2006).

Birds. Birds are thin-skinned animals. The epidermis and dermis constitute a delicate membrane that is loosely applied to the underlying muscle. Only on the feet and head is the skin relatively thick and intimately attached. In these regions are the horny beak, epidermal scales, spurs, and claws. Elsewhere the body of a bird is clothed in feathers.

The feathers of birds have been called nothing more than elaborate reptilian scales. Certainly the presence of epidermal scales along the legs and feet of birds testifies to their debt to reptiles. Feathers are most complex of all derivatives of the craniate stratum corneum.

The dermis of bird skin, especially near the feather follicles, is richly supplied with blood vessels, sensory nerve endings, and smooth muscles. During the brooding season, the dermis in the breast of some birds becomes increasingly vascularised, forming a **brood patch** in which warm blood can come into close association with incubated eggs.

The epidermis of bird skin comprises the **stratum basale** (old name stratum germinativum) and the **stratum corneum**. Between them is the **transitional layer** of cells transformed into the keratinized surface of the corneum (Fig. 2.20).

The exceptional thinness, and the mobility of the skin as a whole, may be correlated with the thermoregulatory role of feathers, which can be fluffed to provide insulation for the body. Birds have no osteoderms. Some, however, such as male game cocks, develop **spur** of dermal bone on each ankle, fused to the tarsometatarsus (bone). In some other species, both sexes have a spur on the carpometacarpus of the wrist.

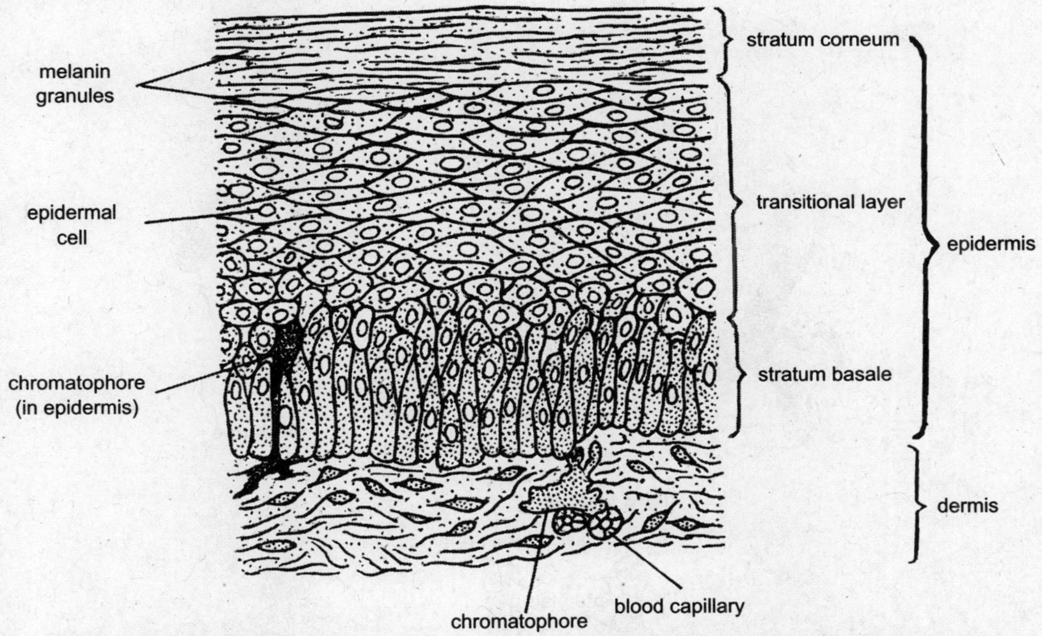

Fig. 2.20. Bird skin. Section of skin of bird showing the stratum basale (*i.e.*, stratum germinativum) and the keratinized surface layer, the stratum corneum. Cells moving out of the basal layer spend time first in the transitional layer before reaching the surface. This middle transitional layer is equivalent to the spinosum and granulosum layers of mammals (after Kardong, 2006).

Bird skin has few glands. The **uropygial gland**, located at the base of the tail (Fig. 2.19B and Fig. 2.21), secretes a lipid and protein product that bird collect on the sides of their beak and then smear it on their feathers. **Preening** coats the feathers with this secretion, making them water repellent, and probably conditions the keratin of which they are composed. The other gland, located on the heads of some birds, is the **salt gland**. Salt gland is well developed in marine birds. Salt glands excrete excess salt obtained when these birds ingest marine foods and seawater.

Feathers

The feathers and skin of the colourful hooded pitohui from New Guinea (which is about the size of blue jay) are laced with a potent toxic secretion, surprisingly similar chemically to toxins of the poison of dart frog. Apparently, the bird's poison works to repel snake or hawk attacks, which might also account for the bright colouration announcing its toxicity to predators.

Feathers distinguish birds from all other vertebrates. They can be structurally elaborate and come in a variety of forms. Feathers are nonvascular and nonnervous products of the skin, principally of the epidermis and keratinizing system. They are laid out along distinctive tracts termed **pterylae** (Fig. 2.21C). Feathers are replaced each year via one or many molts. They develop embryologically from **feather follicles** (Fig. 2.22), invaginations of the epidermis that dip into the underlying dermis. The root of the feather follicle, in association with the dermal pulp cavity, begins to form the feather. The feather itself grows outward in a sheathed case (Fig. 2.22A). Within the sheath, the central axis is divided into a distal **rachis** that bears **barbs** with interlocking connections, termed **barbules**, and a proximal **calamus** that attaches to the body (Fig. 2.22B).

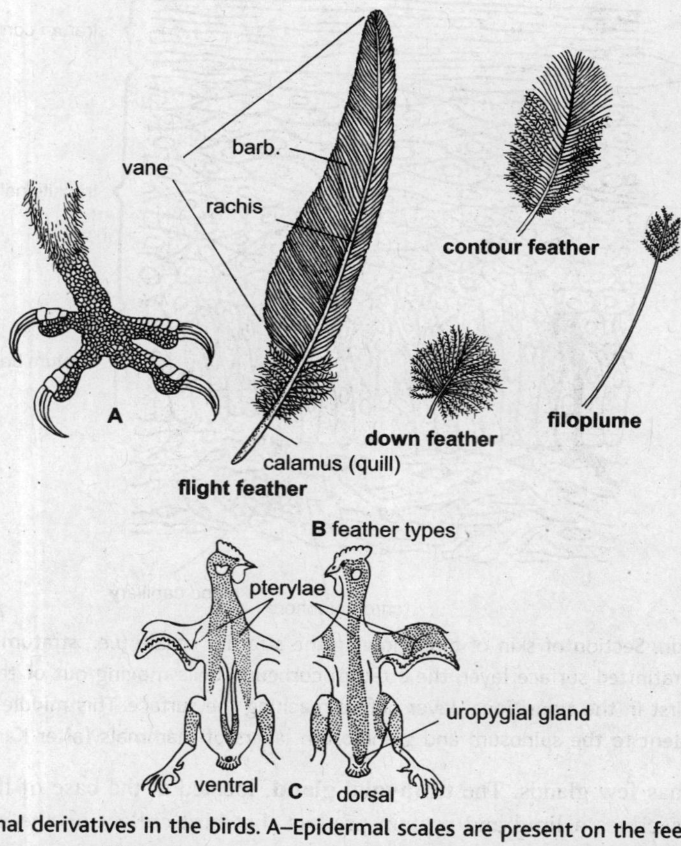

Fig. 2.21. Epidermal derivatives in the birds. A–Epidermal scales are present on the feet and legs of birds. B–Feather types. Flight feathers constitute the major locomotor surfaces. Contour feathers on the body aerodynamically shape the surface of a bird. Filoplumes are often specialized for display. Down feathers lie close to the skin as thermal insulation, of feather tracts. C–Pterylae (after Kardong 2006).

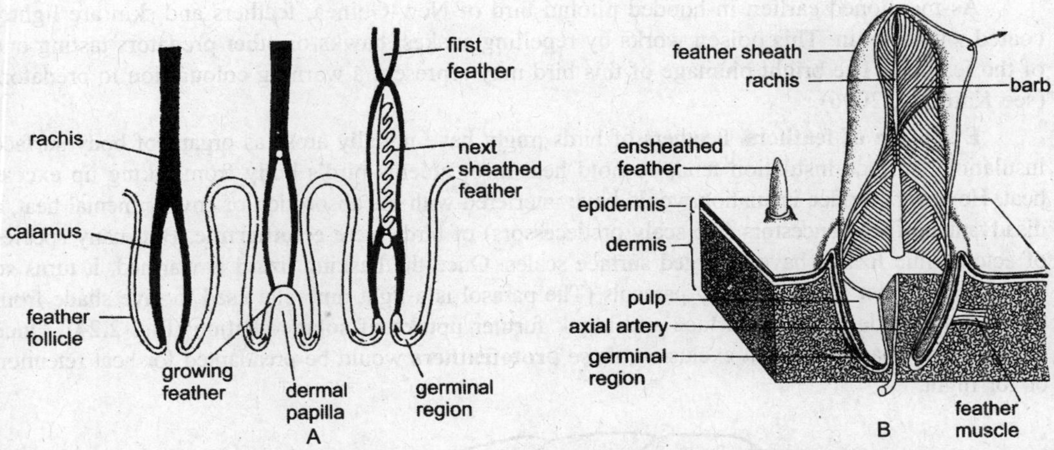

Fig. 2.22. Growth of feather. A–Successive stages in the growth of new feathers. During molting of old feathers, new ones arise in the feather follicle to replace those shed. B–Growth of a feather follicle. The feather itself forms within a sheath that, like the feather, is a keratinized derivative of the epidermis (after Kardong, 2006).

Functions of feathers. There are several types of feathers in birds (Fig. 2.21B) and each type has a distinct role in the life of birds. *Contour feathers* aerodynamically shape the surface of the bird. *Down feathers* lie close to the skin as thermal insulation. *Filoplumes* are often specialized for display, and *flight feathers* constitute major aerodynamic surfaces (Fig. 2.23). Flight feathers of the wings are a type of contour feathers. They are characterized by a long rachis and prominent vane. These feathers have some value as insulation but their primary function is locomotion. Most feathers receive sensory stimuli and carry colours for display or courtship. Chromatophores occur within the epidermis, and their pigments are carried into the feathers to give them colour. However, light refraction on the feather barbs and barbules also create some of the iridescent colours that feather display.

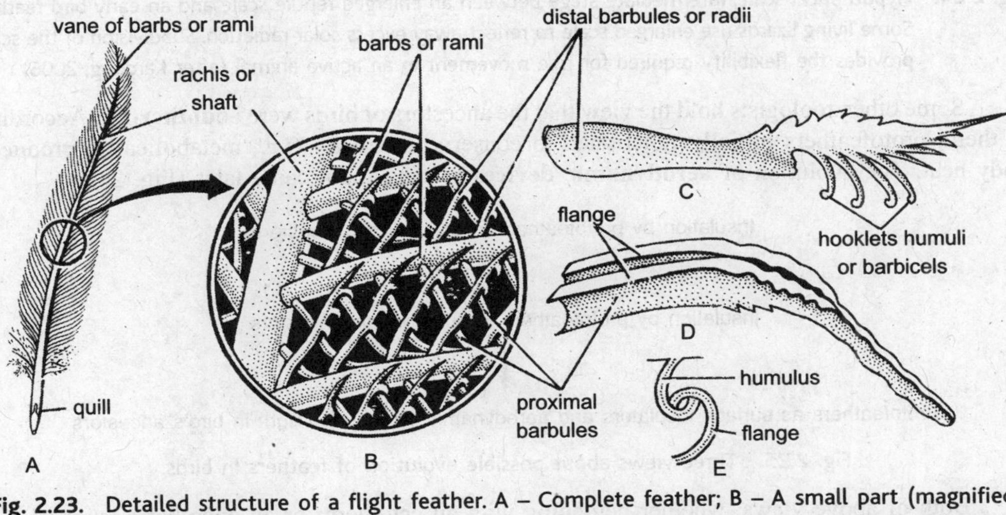

Fig. 2.23. Detailed structure of a flight feather. A – Complete feather; B – A small part (magnified to show the interlocking mechanism); C – A distal barbule; D – Two proximal barbules; E – The humulus (hooklet) of a distant barbule interlocking with the flange of a proximal barbule.

As mentioned earlier, in hooded pitohui bird of New Guinea, feathers and skin are lightly coated with a toxin. This poison works by repelling snakes, hawks or other predators tasting one of the feathers. The bright plumage of this bird may represent a warning colouration to predators (see Kardong, 2006).

Evolution of feathers. Feathers of birds might have initially arose as organs of body surface insulation. Surface insulation tends to hold heat in or shields bird's body from taking up excess heat. However, surface insulation would have interfered with the absorption of environmental heat, a disadvantage if the ancestors (the scaly predecessors) of birds were **ectothermic**. But many species of ectothermic lizards have enlarged surface scales. Once the basking lizard is warmed, it turns so that the scales act like many tiny parasols (The parasol is a light umbrella used to give shade from the sun) to shade the skin surface and block further uptake of solar radiation (Fig. 2.24). Once enlarged and shaped for heat exclusion, these **protofeathers** would be preadapted for heat retention or for flight.

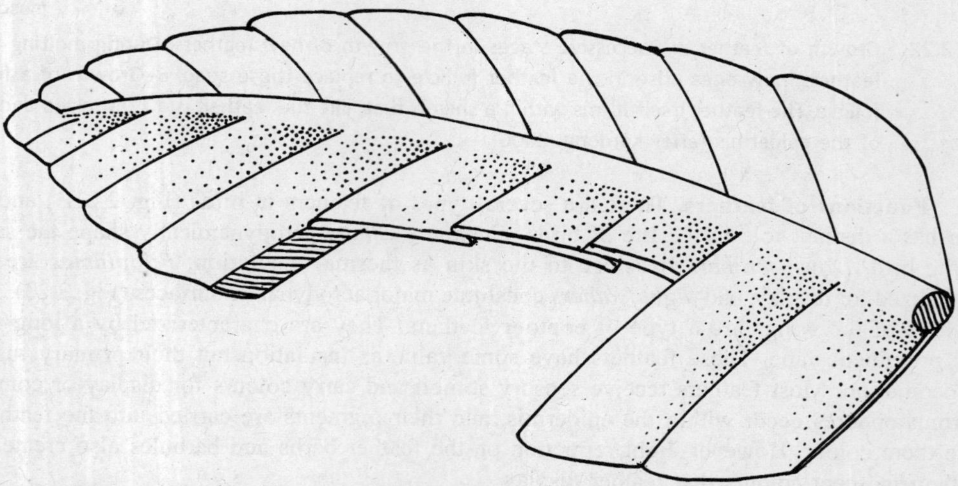

Fig. 2.24. Hypothetical scale, intermediate stage between an enlarged reptile scale and an early bird feather. Some living lizards use enlarged scale to reflect away excess solar radiation. Subdivision of the scale provides the flexibility required for free movement in an active animal (after Kardong, 2006).

Some other zoologists hold the view that the ancestors of birds were **endothermic**. According to them, protofeathers initially functioned to conserve internally (*i.e.*, metabolically) produced body heat. The evolution of **aerodynamic devices** serving flight came later (Fig. 2.25).

Insulation by protofeathers in ectothermic ancestors

↓

Insulation by protofeathers in endothermic ancestors

↓

Protofeathers as surface insulators and aerodynamic devices for flight in bird's ancestors

Fig. 2.25. Three views about possible evolution of feathers in birds.

Both of above views, whether beginning with an ectotherm or an endotherm, suggest that feathers played a role in surface insulation when they first appeared and became secondarily co-opted for a role in flight.

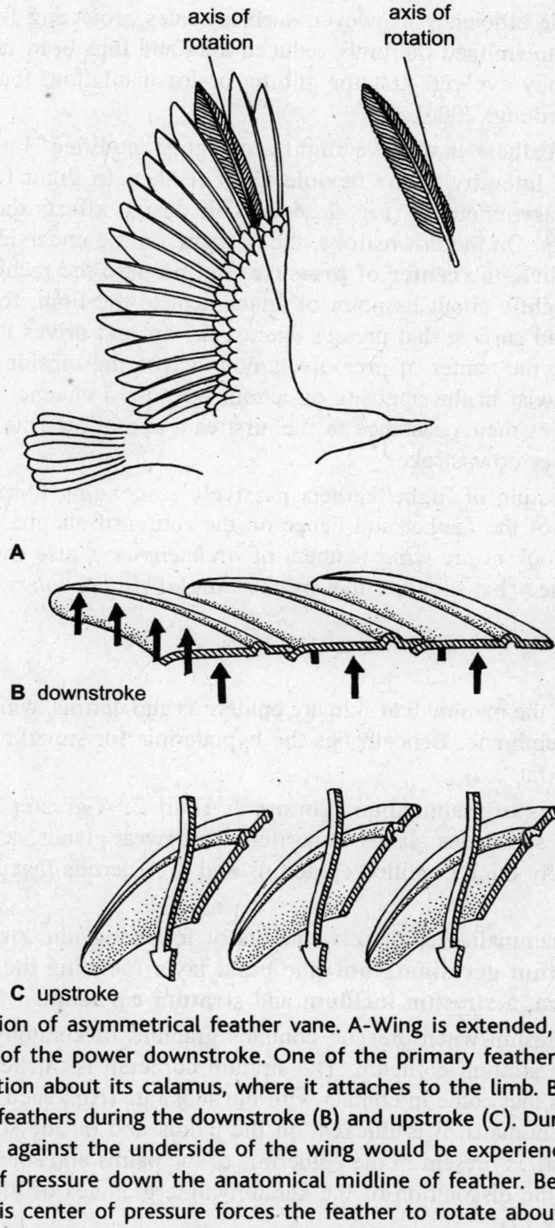

Fig. 2.26. Flight function of asymmetrical feather vane. A-Wing is extended, as it might appear during the middle of the power downstroke. One of the primary feathers is removed to show the axis of rotation about its calamus, where it attaches to the limb. B, C-Cross sections through three flight feathers during the downstroke (B) and upstroke (C). During the power downstroke, air pressure against the underside of the wing would be experienced by each feather along its center of pressure down the anatomical midline of feather. Because the rachis is offset, however, this center of pressure forces the feather to rotate about its axis, and the primary feathers temporarily form a closed uniform surface. During the recovery upstroke, air pressure against the back of the wing forces rotation in the opposite direction, spaces open between feathers, the air slips between the resulting slots, thus reducing resistance to wing recovery (after Kardong, 2006).

There is also a totally different view about the origin of feathers. According to this view, feathers evolved initially as aids to **gliding** and then to **flight**. Feathers were selected because of their favourable effect on the airstream passing over the body or limbs of a gliding animal. If the protoavian limbs were not streamlined, then pressure drag would result and turbulence would have reduced aerodynamic efficiency. However, surface scales projecting from the trailing edge of the limb would have streamlined the limb, reduced drag and thus been favoured by selection. Regardless of whether they evolved first for gliding or for insulation, feathers were modified from reptilian scales (Kardong, 2006).

In modern birds, feathers that serve flight are highly modified. Interlocking barbs and barbules give strucutural integrity to the flexible flight feather. In flight feathers, the rachis is offset, making the vane asymmetrical (Fig. 2.26A). This design affects the action of the flight feathers during wing beats. On the downstroke, the pressure on the underside of each feather act along its anatomical midline, the **center of pressure**. But because the rachis is offset, the result is to twist the feather slightly about its point of attachment to the limb, forcing feathers of the wing together into a broad surface that presses against the air and drives the bird forward (Fig. 2.26B). On the upstroke, the center of pressure is now across the topside of the asymmetrical feather and forces it to twist in the opposite direction, opening a channel between the feathers (Fig. 2.26 C). This reduces their resistance to the airstream and allows the wing to recover and prepare for the next power downstroke.

This controlled twisting of flight feathers passively responding to wing beat depends on the asymmetrical design of the feather and hence on the action of air pressure against it during powered flight. A close look at the wing feathers of *Archaeopteryx* also reveals an offset rachis and an asymmetrical vane. This suggests that by the time of *Archaeopteryx*, powered flight had already evolved.

Mammals

Two main layers of the mammalian skin are epidermis and dermis, which joins and interface through the basement membrane. Beneath lies the hypodermis (or superficial fascia), composed of connective tissue and fat.

The notable features of mammalian skin are 1. Hair; 2. A greater functional variety of epidermal glands such as sebaceous glands, sudoriferous or sweat glands, scent glands, mammary glands, etc.; 3. A relatively thick cornified epidermis and 4. A dermis that is many times thicker than the epidermis.

Epidermis. The mammalian epidermis consists of following four strata or cellular layers: an actively mitotic **stratum germinativum** (the basal layer touching the dermis), a relatively thin **stratum granulosum**, a **stratum lucidum** and **stratum corneum.**

The stratum granulosum when present, contains granules of keratohyaline, a precursor of the **keratin** found in the stratum corneum. The stratum corneum is thickest on the parts of the forelimbs and hind limbs that come in contact with the substrate (*e.g.*, such as hoofs, apical pads, knee pads, and tori). In humans, it is thickest on the palms and on the soles. A fourth cellular layer, the stratum lucidum, is present in the epidermis of the palms and soles. It is a layer of cells rendered translucent by the dissolution of the keratohyaline granules of the stratum granulosum in those locations (Fig. 2.27).

The epithelial cells of the epidermis of skin of mammals are **keratinocytes** and belong to the keratinizing system that forms the dead, superficial cornified layer of skin. The surface keratinized cells are continually exfoliated and replaced by cells arising primarily from the deepest layer of the epidermis, the stratum basale (Fig. 2.27). Cells within the basale divide mitotically, producing some that remain to maintain the population of **stem cells** and others that are pushed outward. As they are displaced to higher levels, they pass through keratinization stages exhibited as distinct, successive layers towards the surface, *i.e.*, stratum spinosum, stratum granulosum, often a stratum lucidum and a stratum corneum (Fig. 2.27).

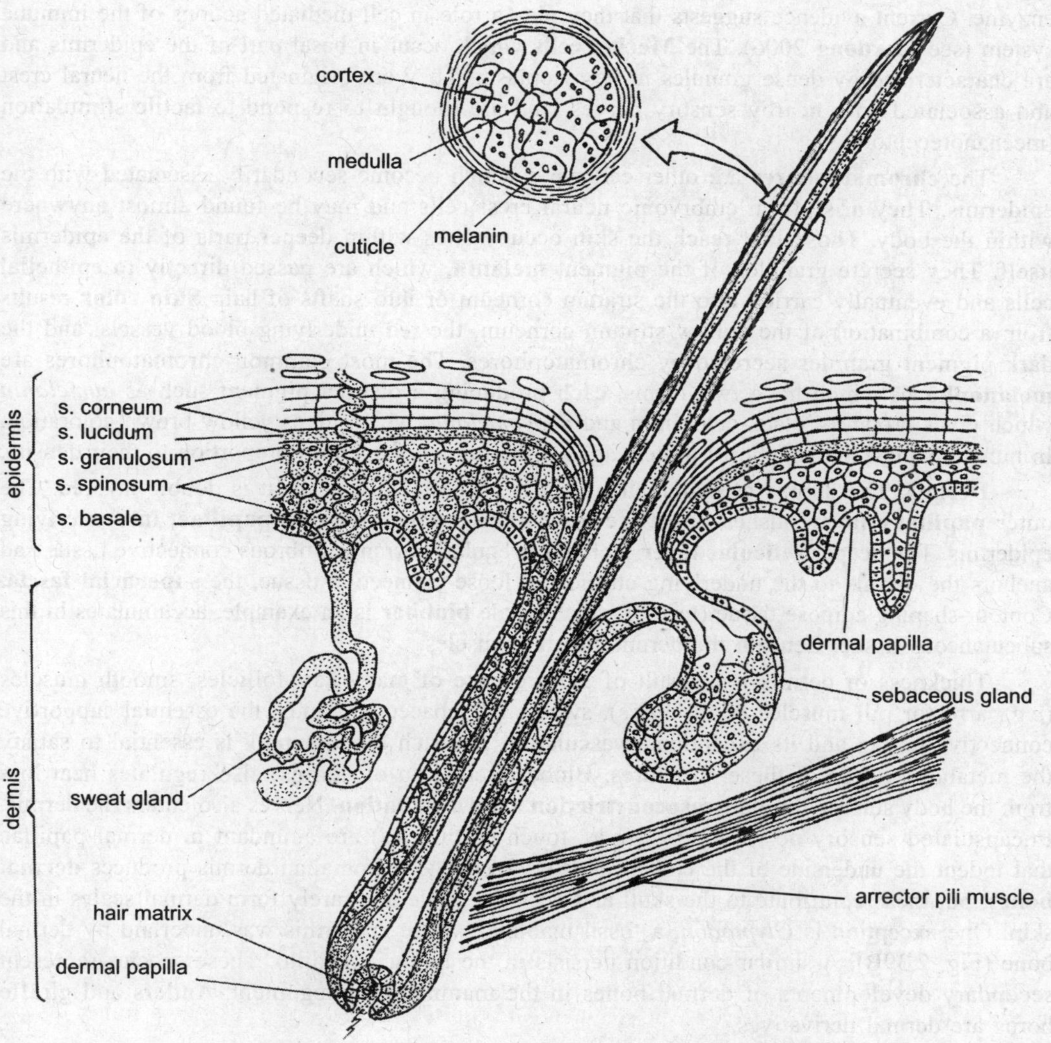

Fig. 2.27. Mammalian skin. The epidermis is differentiated into distinct layers. As in other vertebrates, the deepest is the stratum basale, which through mitotic division produces cells that as they age become successively part of the stratum spinosum, the stratum granulosum, often the stratum lucidum, and finally the surface stratum corneum. The dermis pokes up dermal papillae that gives overlying epidermis an undulating appearance. Sweat glands, hair follicles, and sensory receptors lie within the dermis. Sweat ducts tends to pass through the overlying epidermis to release their watery secretion on the surface of the skin (after Kardong 2006).

Cell types of Mammalian epidermis. Above described keratinocytes are the most prominent cell types of the epidermis. Other types are recognized, although their functions are less clearly known. The **Langerhans cells** are dendritic or stellate cells dispersed singly throughout the upper parts of the stratum spinosum. These cells resemble the melanocytes but lack tyrosinase enzyme. Current evidence suggests that they play a role in cell-mediated actions of the immune system (see **Kardong** 2006). The **Merkel cells** which occur in basal part of the epidermis and are characterized by dense granules in its cytoplasm. They are originated from the neural crest and associated with nearby sensory nerves, and are thought to respond to tactile stimulation (mechanoreceptors).

The **chromatophores** are other cell types which become secondarily associated with the epidermis. They arise from embryonic neural crest cells and may be found almost anywhere within the body. Those that reach the skin occupy sites within deeper parts of the epidermis itself. They secrete granules of the pigment **melanin**, which are passed directly to epithelial cells and eventually carried into the stratum corneum or into shafts of hair. **Skin color** results from a combination of the yellow stratum corneum, the red underlying blood vessels, and the dark pigment granules secreted by chromatophores. The most common chromatophores are **melanophores** which have two forms, each producing a distinct pigment such as *eumelanin* which gives brown or black coloration and *phaeomelanin* which gives yellow brown coloration. In human beings, melanophores protect against the rays of the sun in proportion to their density.

Dermis. The dermis of mammalian skin is exceptionally thick. It is double layered. The outer **papillary layer** pushes fingerlike projections, termed **dermal papillae**, into overlying epidermis. The deeper **reticular layer** includes irregularly arranged fibrous connective tissue and anchors the dermis to the underlying cushion of loose connective tissue, the **superficial fascia**. Contour-shaping adipose tissue (fat), of which whale **blubber** is an example, accumulates in this subcutaneous tissue. Beneath the dermis lies the muscles.

Thickness of dermis is a result of the presence of many hair follicles, smooth muscles (*e.g.*, arrector pili muscles of the hairs), sweat and sebaceous glands, the essential supportive connective tissue, and its exceptional vascularity. The rich blood supply is essential to satisfy the metabolic needs of these structures. Blood circulation of dermis, also regulates heat loss from the body surface through **vasoconstriction** and **vasodilation**. Nerves also enters the dermis. Encapsulated sensory nerve endings (*e.g.*, touch corpuscles) are abundant in dermal papillae that indent the underside of the epidermis (Fig. 2.2). The mammalian dermis produces **dermal bones**, but these contribute to the skull and pectoral girdle and rarely form dermal scales in the skin. One exception is *Glyptodon*, a fossil mammal whose epidermis was underlaid by dermal bone (Fig. 2.39B). A similar condition persists in the living armadillo. These species represent secondary developments of dermal bones in the mammalian integument. Antlers and giraffe horns are dermal derivatives.

Hair follicles and glands project inward from the epidermis (Fig. 2.2). The dermis is usually composed of irregularly arranged fibrous connective tissue that is often impregnated with elastic fibers to give it some **stretch** but return it to its original shape. As a person ages, this elasticity is lost and the skin sags.

The skin of most mammals is hidden by fur; dermal pigments other than those in hairs play little role in skin colouration. The vividly coloured ischial callosities of baboons are among the exceptions.

Hair

Hairs are slender, keratinous filaments. The base of a hair is the **root.** Its remaining length form the **shaft**. The outer surface of the shaft often forms a scaly **cuticle** (*i.e.*, has overlapping microscopic scales). Cuticle of hair is composed of thin, transparent, cornified squamous cells (scales) arranged like shingles (shingle is a mass of small rounded pebbles especially on a seashore) with the edges directed away from the root. Beneath the cuticle is the **hair cortex**, and its core is the **hair medulla**. Medulla is formed of rounded cells. In larger hairs, it contains air spaces (Fig. 2.2).

The hair shaft projects above the surface of the skin but it is produced within an epidermal **hair follicle** rooted in the dermis. The surface of the epidermis dips down into the dermis to form the hair follicle. At its expanded base, the follicle receives a small tuft of the dermis, the **hair papilla**. This papilla seems to be involved in *stimulating activity* of the **matrix cells** of the epidermis, but does not itself directly contribute to the hair shaft. The tiny clump of living matrix cells, like the rest of the stratum basale, is the germinal region that starts the process of keratinization to produce hair within the follicle. Unlike keratinization within the epidermis, which is general and continuous, keratinization within the hair follicle is localized and intermittent.

Chromatophores in the follicle contribute pigment granules to the hair shaft to give it further colour. The arrector pili muscle, a thin band of smooth muscle anchored to the follicle and makes the hair strand erect in response to cold, fear, or anger. Thus, when the arrector pilorum contract, the hairs are elevated and the skin is pulled into tiny mounds resembling goose-flesh (Goose-flesh or goose-bumps is a pimply state or roughness of the skin with the hairs erect, produced by cold, fright or fear or sudden feeling of excitement). Elevation of the hairs gives carnivores a ferocious appearance. It is also a device for thermoregulation, increasing insulating effect of the fur.

The scales of pangolins, and perhaps the horns of rhinoceros, may be evolutionary products of agglutinated hairs. Other modifications of hair includes bristles, the spines of spiny anteaters and porcupine quills.

The thick covering of hair, **fur** or **pelage** is generally composed of guard hairs and underfur (Box 2.2). The **guard hairs**, the larger, coarse hairs, are the most apparent on the outer surface of the fur. The **underfur** is located beneath the guard hairs and is usually much finer and shorter. Both guard hairs and underfur function mainly as insulators. In most marine mammals, the underfur is reduced or lost entirely, and only a few guard hairs are evident. For example, some whales have only one or two bristles on the upper lip. Hair has **grain**, that is, it is laid out in a particular direction. Strokes against the grain are restricted (Fig. 2.28). An exception occurs in moles, which lack turnaround space and must back up in their tunnels. The hair of moles can be combed forward or backward without much difference in grain.

Box 2.2 Types of pelages

Just as birds in their lifetime have several plumages, mammals have more than one pelage.

1. The first is the **lanugo**-the soft downy hair that covers the fetus of some mammals. Lanugo hair is frequently developed and usually also lost in the uterus. Hence, most mammals are borne quite naked except for vibrissae.
2. Next is **juvenile pelage**, often not of the same colour as the adult pelage. At about puberty, the juvenile pelage is replaced by the adult pelage.
3. **Adult pelage** appears often in an orderly moult which starts at one end of the animal's body and progresses to the other, and is marked along its way by a distinct line between the new and the old hair.

Some hairs are specialized. Sensitive nerves are associated with the roots of **vibrissae** or **whiskers**, around the snout of many mammals. Whiskers are common in nocturnal mammals that live in burrows with limited light. The **quills** of porcupines are stiff coarse hairs specialized for defense (safeguard). Thus, hairs have an insulating as well as sensitive tactile effect. The roots of each hair is surrounded by a basket-like network of sensory nerve endings, and displacement of the root initiates a train of sensory impulses to the brain.

Hairs are not permanent. After an interval of time—one or many months, and often seasonally, depending on the species-proliferation of hair cells in the follicle ceases temporarily and the hair becomes loose in the follicle. Subsequently, when formation of hair in the follicle restarts, the old hair is shed if it has not already fallen out. When all follicles in an area permanently cease activity, the area becomes bald.

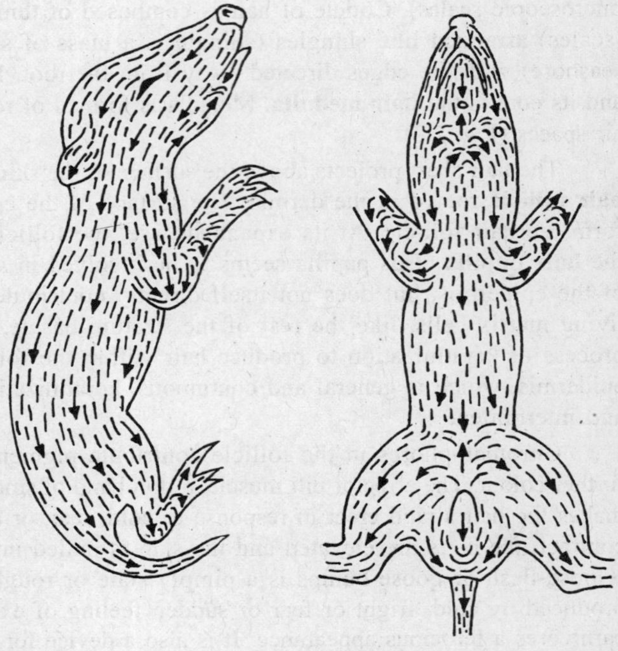

Fig. 2.28. Hair tracts. Hair grows with a grain, a particular direction in which it slants. Notice the various growth directions (arrows) in which the hairs of the marsupial bandicoot lie (after Kardong 2006)

Origin and evolution of hair. Phylogenetic origin of hair remains hypothetical. One view holds that hair arose initially as surface insulation, retaining body heat in primitive endothermic mammals (Box 2.3). An alternative view is that hair evolved first as tiny projecting rods in the hinges between scales and served as tactile devices. These hairs are called **protohairs** and they could help monitor surface sensory data when an animal was hiding from an enemy or retreating from the weather. If such a role increased in importance, it would have favored longer shafts and perhaps the evolution of structure resembling vibrissae. This sensory protohair might then have evolved secondarily into an inclusive pelage as mammals became endothermic. Although insulative in modern mammals, hair still retains a sensory function.

Hairs are soft and decomposable, so they do not leave a reliable trace in the fossil record. Some therapsids (*i.e.*, cynodonts), ancestors to mammals, have tiny pits in the facial region of their skulls. These pits resemble pits on the skulls associated with sensory vibrissae in modern mammals. Some have interpreted these **indirect evidence** of hair in therapsids. However, the skulls of some modern lizards with scales have similar pits with a tactile bristle, called **protothrix**, and of course, lizards have no hair. Thus, such pits on the skulls are not conclusive evidence of the presence of hair. Further, one especially well-preserved skin impression of *Estermmenosuchus*, a therapsid from the Upper Permian, shows no evidence of hair. The epidermis was smooth, without scales, and undifferentiated, although it was supplied with glands. In life, skin of this ancestral therapsid was probably soft and pliable. Thus, we still do not know when hair first arose in primitive mammals or in their therapsid ancestors.

Box 2.3

The interscale distribution of hairs (*e.g.*, *Armadillo*, Fig. 2.29) may reflect the amniote origin of mammals, but tactile hairs and tactile protothrixes seem to be products of convergent evolution. Feathers and hairs may be **neomorphs** (structures that have no ancestral precursor; see **Kent** and **Carr** 2001).

Nails, Claws and Hooves

Nails, claws and hooves (or hoofs) are modifications of the stratum corneum at the ends of digits. Nails are plates of tightly compacted, cornified epithelial cells on the surface of fingers and toes. Only primates have nails (Fig. 2.30). **Claws** or **talons** are curved, laterally compressed keratinized projections from the tips of the digits. They are seen in some amphibians and in most birds, reptiles and mammals. **Hooves** are enlarged keratinized plates on the tips of the ungulate digits.

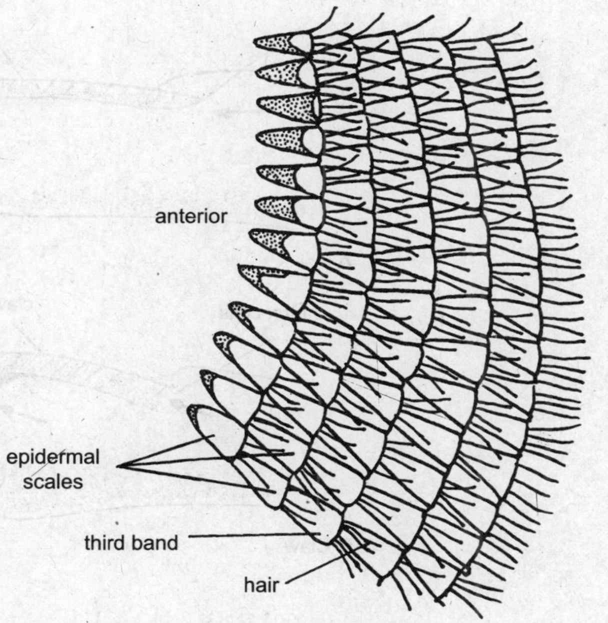

Fig. 2.29. Epidermal scales and interspersed hairs of armadillo skin.

Nails. Nails are products of the keratinizing system of the skin. The **nail matrix** forms new nail at the nail base by pushing the existing nail forward to replace that which is worn or broken at the free edge. Nails protect the tips of digits from inadvertent mechanical injury. Nails also help stabilize the skin at the tips of the fingers and toes, so that on the opposite side the skin can establish a secure friction grip on objects grasped.

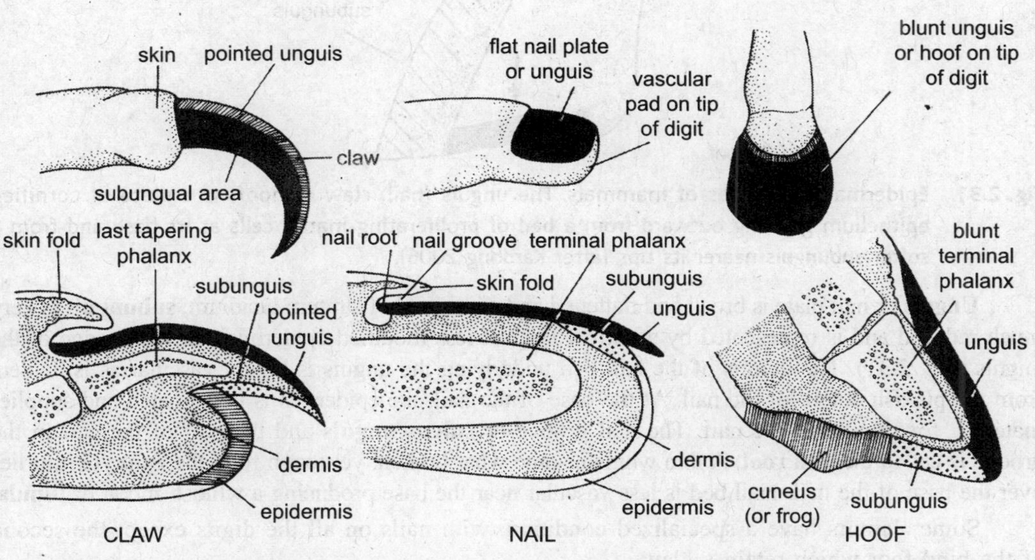

Fig. 2.30. Relation between claw (eagle), nail (human) and hoof (horse). Digital tips show complete above and in sagittal section below.

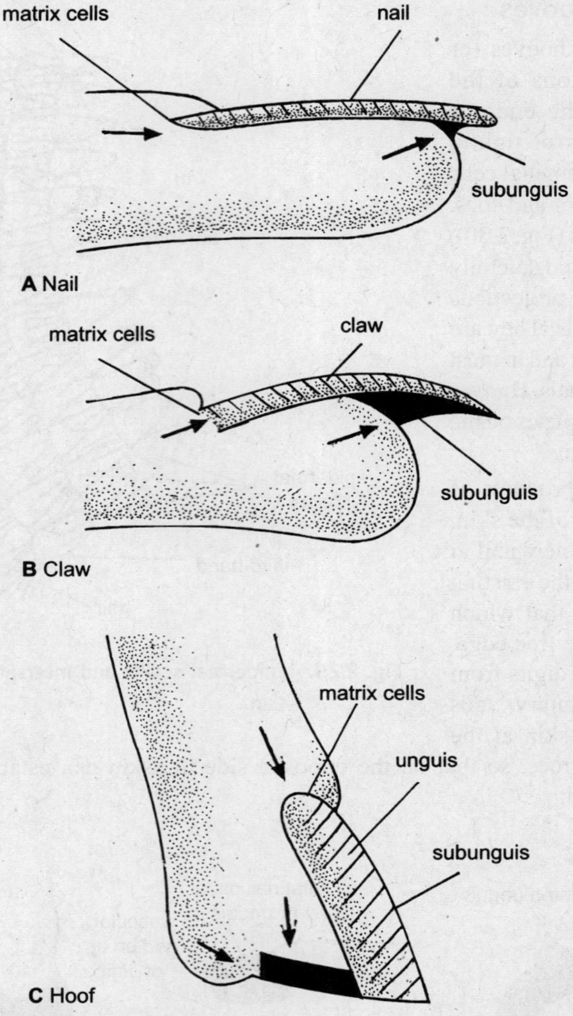

Fig. 2.31. Epidermal derivatives of mammals. The unguis (nail, claw or hoof) is a plate of cornified epithelium growing outward from a bed of proliferating matrix cells at its base and from a softer subunguis nearer its tips (after Kardong 2006).

Unguis or nail plate is broad and flattened and composed of stratum lucidium; **subunguis** is very much reduced and is represented by a narrow strip of less modified epidermis below the edge of the unguis (Fig. 2.31). The dermis of the nail bed underlying the unguis is highly vascular as is evident from the pinkish colour of the nail. At the base of the nail, the epidermis is invaginated and supplies material for growth of the nail. The nail groove or **sulcus unguis** and the part of the nail in the groove is called the **nail root**. A thin whitish, some what broken, yet tough fold, the **eponychium** lies over the base of the nail. Nail bed is less vascular near the base producing a whitish moon or **lunula**.

Some lemurs have a specialized condition, with nails on all the digits except the second of the hind foot which retain a claw.

Hooves. These are characteristics of ungulate mammals such as Perissodactyla (*e.g.*, horse, donkey, zebra, rhinoceros, etc.) and Artiodactyla (*e.g.*, pig, camel, deer, giraffe, cow, buffalo, sheep, goat, yak, etc.). No part of their digit touches the ground. Unguis is large (or heavy) and the animal actually walk over it, and the sub-unguis is carried beneath it. **Fleshy pad** or **torus** of the digit lie behind the hoof and is called the **frog.** Frog or cuneus is composed of hard keratin, unguis wear away slowly.

Horse hoof. The horse hoof consists of the hoof wall, sole and the frog (Fig. 2.32A). The **hoof wall** is U-shaped and open at the heel, a derivative of integument. It consists of a keratinized **stratum externa** (or tectorium), a thin shiny surface layer, the **stratum medium**, thicker and keratinized and permeated with coiled, tubular channels; and an inner **stratum internum** (or lamellatum), a highly and regularly laminated, infolded layer that interdigitates with the dermis (5 corium beneath). The hoof wall grows out from its base, the germinal region (matrix cells), not from underlyaing dermis, at about 6 mm per month, taking 9 to 12 months overall for the toe to renew.

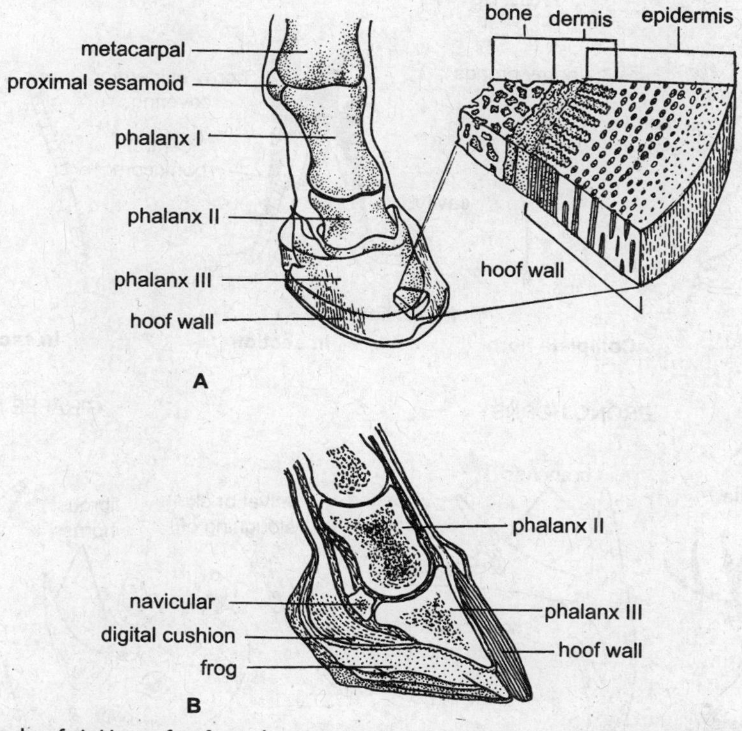

Fig. 2.32. Horse hoof. A-Horse forefoot showing internal bones and hoof wall. Enlarged cross section of hoof showing layers from outer wall to inner bone. B-Longitudinal section (L.S.) through horse forefoot. (after Kardong, 2006).

The bottom of the hoof, the ground surface includes the wedge-shaped frog; a mostly keratinized derivative of the integument that fills the opening in the heel of the hoof wall. The **sole** fills the ground surface space between the wall and triangular (Fig. 2.32B). It consists of epidermis and thickened dermis, the corium of the sole. Deep to the sole is the fatty **digital cushion**, or **pad**, a derivative of the **hypodermis.**

Horns and Antlers

Horns are found in ungulate mammals only. Mammals have following five types of horns (Fig. 2.33).

1. True horns of the hollow type are found in pronghorn, cattle, antelope, bison, sheep, goat and wildebeests (Fig. 2.34). They consist of an inner core of bone which is an outgrowth

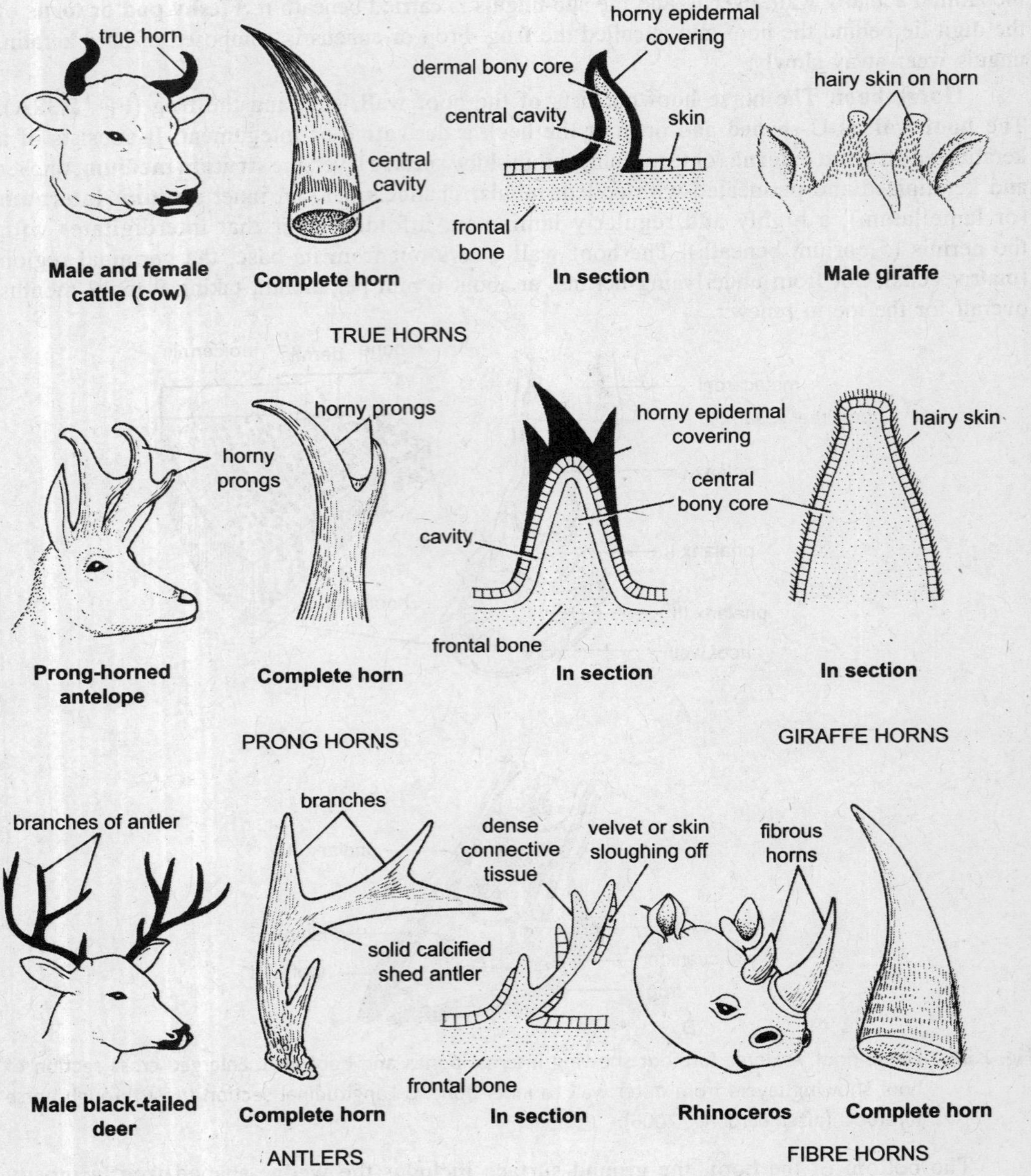

Fig. 2.33. Types of mammalian horns and antlers.

from the frontal bone. It is encased in a keratinized, epidermal covering. True horns are unbranched. They continuously grow throughout life and are not shed. Horns of the males are designed to withstand the forces encountered during head-butting combat.

2. Rhinoceros or keratin fiber horn. This horn has no skeleton (bone) and is exclusively a product of the integument. It is made by keratinized cells of the epidermis and consists of matted keratin fibers bound together, but its fibers are not true hair. There is one horn in the Indian rhinoceros and two in the African species (Fig. 2.33).

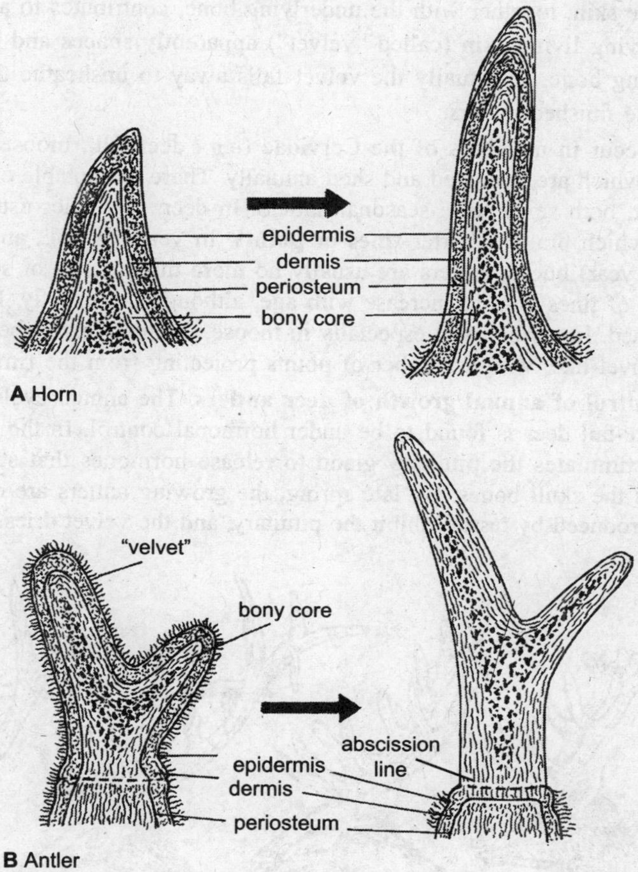

Fig. 2.34. Horns and antlers. A-Horns appear as outgrowths of skull beneath the integument which forms a keratinized sheath. Horns occur in bovids of both sexes and are usually retained year round. B-Antlers also appear as outgrowths of the skull beneath the overlying integument, which is referred to as "velvet" because of its appearance. Eventually this overlying velvet dries and falls away, leaving the bony antlers. Antlers are restricted to members of the deer family and except for caribou (and reindeer), they are present only in males. Antlers are shed and replaced annually (after Kardong, 2006).

3. Pronghorn. It is a true horn and is found in Russian antelope *Antilocapra* (Fig.2.33). It consists of a permanent projection of the frontal bone covered by a hard horny, epidermal sheath (Fig. 2.34). The sheath is forked bearing one to three prongs made only of horny sheath. The horn sheath is shed annually and is replaced by another which grows from the skin that surrounds the core. Female pronghorns also have horns whose keratinized sheath is replaced annually, but these are much smaller and only slightly forked.

4. Giraffe horns. They develop from cartilaginous protrusions which are present at birth. They ossify and fuse at the top of the skull, where they appear as knobs permanently covered with living skin and hair. Giraffe possesses three of these knobs, one is median and anterior to the other two. These horns are short, unbranched and permanent, and are present in both sexes.

5. Antlers. The skin, together with the underlying bone, contributes to antlers (Fig. 2.34). In antlers, the overlying living skin (called "velvet") apparently shapes and provides vascular supply to the growing bone. Eventually the velvet falls away to unsheathe the bare bone, the actual material of the finished antlers.

True antlers occur in members of the Cervidae (*e.g.*, deer, elk, moose). Typically, only males have antlers, which are branched and shed annually. There are notable exceptions. Among caribou and reindeer, both sexes have seasonal antlers. In deer, the antler usually consists of a **main beam**, from which branch shorter **tines** or **points**. In yearling (*i.e.*, an animal of a year old or in its second year) bucks, antlers are usually no more than prongs or spikes that may be forked. The number of tines tend to increase with age, although not exactly. In old age, antlers may even be deformed. In caribou and especially in moose, the main antler beam is compressed and **palmate**, or shovel-like, with a number of points projecting from the rim.

Hormonal control of annual growth of deer antlers. The annual cycle of antler growth and loss in the white-tail deer is found to be under hormonal control. In the spring, increasing length of daylight stimulates the pituitary gland to release hormones that stimulate antlers to sprout from sites on the skull bones. By late spring, the growing antlers are covered by velvet. By fall, hormones produced by testes inhibit the pituitary, and the velvet dries. By thrashing and

Fig. 2.35. Annual growth of elk antlers. A-B-New antlers begin to grow in April. C-By May, antlers are nearly fully formed even though they are still covered by the living integument (velvet). D-By the summer, the velvet has begun to dry and peel off. E-Fully formed bony antlers are in place (after Kardong, 2006).

rubbing the deer wipes the velvet off to expose the fully formed, but now dead, bone of the antlers (Fig. 2.35). Males use their antlers during clashes with other males to maintain access to reproductively receptive females. Following this brief mating season, further hormone changes lead to a weakening of the antlers at the base where it attaches to the living bone of the skull. The antlers break off, and for a short time during winter, deer are without antlers.

Modification of the stratum corneum

Due to environmental stress and strain mammals exhibit various **callus-like** pads that develop at sites where friction occur. The **ischial callosities** of monkeys are thick, cornified pads of epidermis, covering the ischial portion of the hip bone, on which the monkey sits. Camels have **knee pads**, which absorb some of the shock when the animal kneels preparatory to lying down.

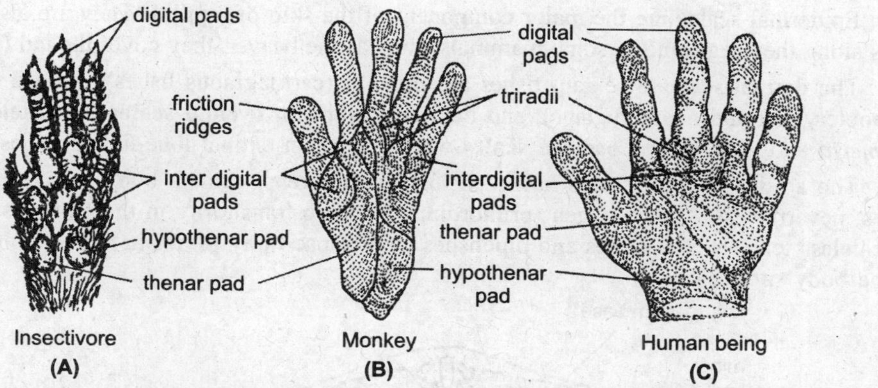

Fig. 2.36. Tori and friction ridges on the hands. A – Insectivore; B – Monkey; C – Human being.

Tori. These are epidermal pads on the tips of the digits (apical or digital pads) and on the palms and soles of numerous mammalian species (Fig. 2.36). The cat is able to "pussy foot" by retracting the claws and walking silently upon the apical and interdigital pads. Tori are prominent on the human hands (palms) but they disappear before birth. **Corns** and **callouses** (or callosities) are thickenings of the stratum corneum that develop where the integument is subjected to continued friction.

Baleen. The integument within the mouths of mysticete whales (*e.g.*, right whale, blue wale, common fin whale, etc.) forms plates or **baleen** that act as strainers to extract krill (a component of plankton) from water gulped in the distended mouth (Fig. 2.37). Although baleen is sometimes referred to as "whale bone", it contains no bone. Baleen is a series of keratinized plates that arise from the integument. During its formation, groups of dermal papillae extend and lengthen outward, carrying the overlying epidermis. The epidermis forms a cornified layer over the surface of these projecting papillae. Collectively, these papillae and their covering of epidermis constitute the plates of frilled baleen (Fig. 2.38).

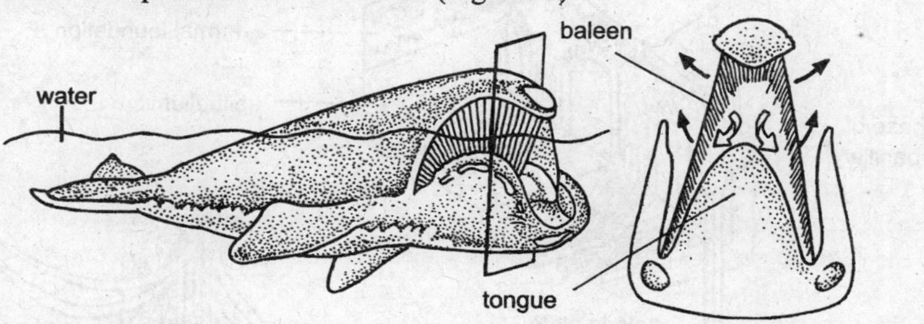

Fig. 2.37. Whale feeding. The right whale has long plates of baleen suspended from the upper jaw, and it feeds by swimming through plankton with its mouth parted. Water enters, passes along the sides of the tongue and then departs through the curtain of baleen, leaving the plankton entangled in the baleen (after Kardong, 2006).

Scales. Scales have many functions. Both epidermal and dermal scales are hard, so when they receive mechanical insult and surface abrasion, they prevent damage to soft tissues beneath. The density of scales also makes them a barrier against invasion of foreign pathogens, and they retard water loss from the body. In sharks and other fishes, scales dampen (*i.e.,* make less intense) the boundary layer turbulence to increase swimming efficiency. Some reptiles regulate the amount of surface heat they absorb by turning their bodies toward or away from the sun. This determines whether the sun rays are deflected off the full face of the scale or shine under the lifted posterior edge of the scale to reach the thin epidermis beneath.

Epidermal scales are the major component of the skin of reptiles. They are also present in birds along their legs, and in some mammals, such as the beaver, they cover the tail (Fig. 2.39A).

The dermal scales of extant fishes are **placoid** (cartilaginous fishes), **ganoid** (some basal actinopterygian and neopterygians), and flexible **ctenoid** or **cycloid scales** (most teleosts, *Amia*, *Latimeria* and dipnoans). **Cosmoid scales** were present in extinct lobe-finned fishes.

The ability to form scales has been lost by lampreys, catfish, bony eels and some other fishes; nevertheless, scale anlagen (primordium) develop transitorily in the embryos. In contrast to scaleless teleosts, sea horses and pipefishes have dense **bony plates** in their dermis over most of the body surface.

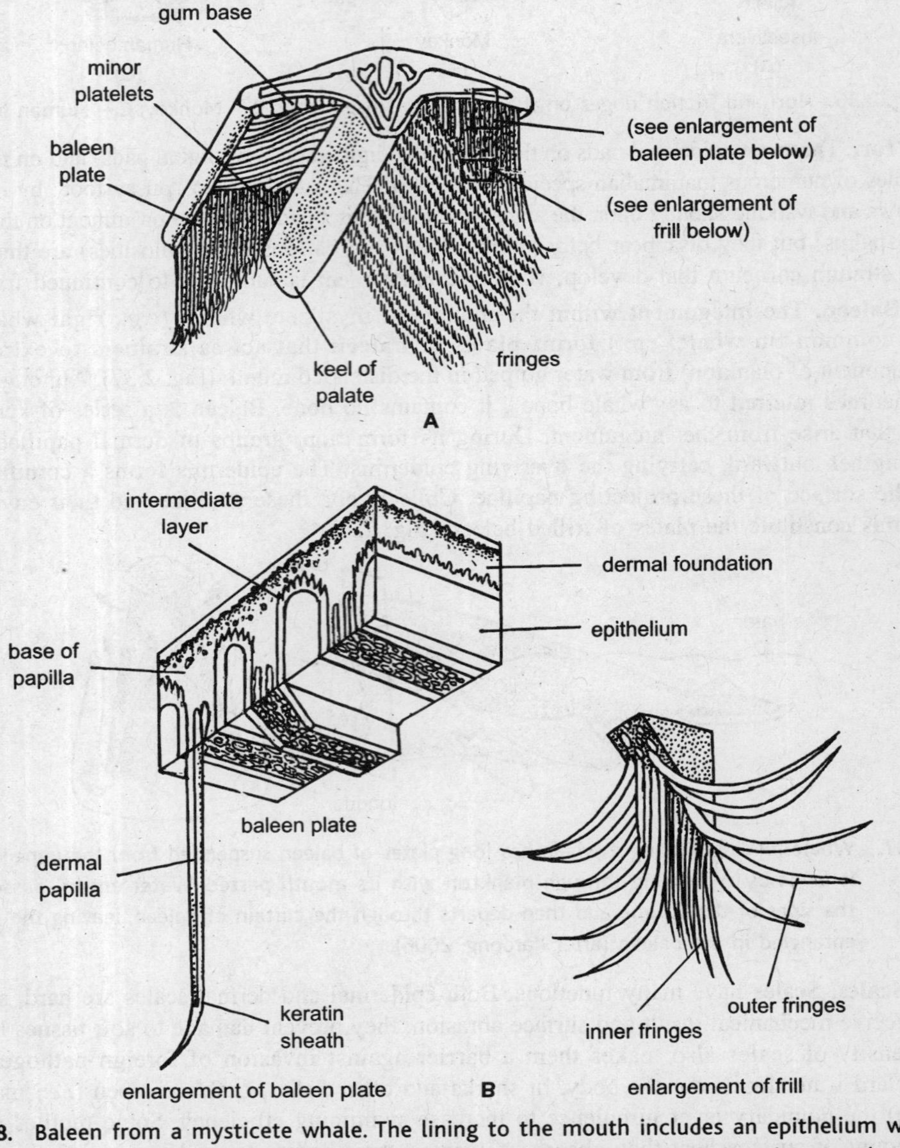

Fig. 2.38. Baleen from a mysticete whale. The lining to the mouth includes an epithelium with the ability to form keratinized structures. A-Groups of overgrowing epithelium becomes keratinized and frilly to form the baleen. B-Longitudinal sections of the keratinized layer and the frill are enlarged (after Kardong, 2006).

Dermal Armor

Dermal bone forms the armor of ostracoderm and placoderm fishes. Being a product of the dermis, dermal bones find their way into alliances with a great variety of structures. Dermal bone supports the scales of bony fishes but tends to lost in tetrapods (Box 2.4). It is absent in the

Box 2.4

Some labyrinthodonts (fossil amphibians descended from crossopterygian fishes) and even some basal amniotes (reptiles, aves and mammals collectively), had large bony plates in the dermis; others had minute bony scales, called **osteoderms** in tetrapods. Among amphibians, caecilians and some tropical toads still have osteoderms. Crocodilians (*e.g.*, alligator) have oval osteoderms, especially along the back.

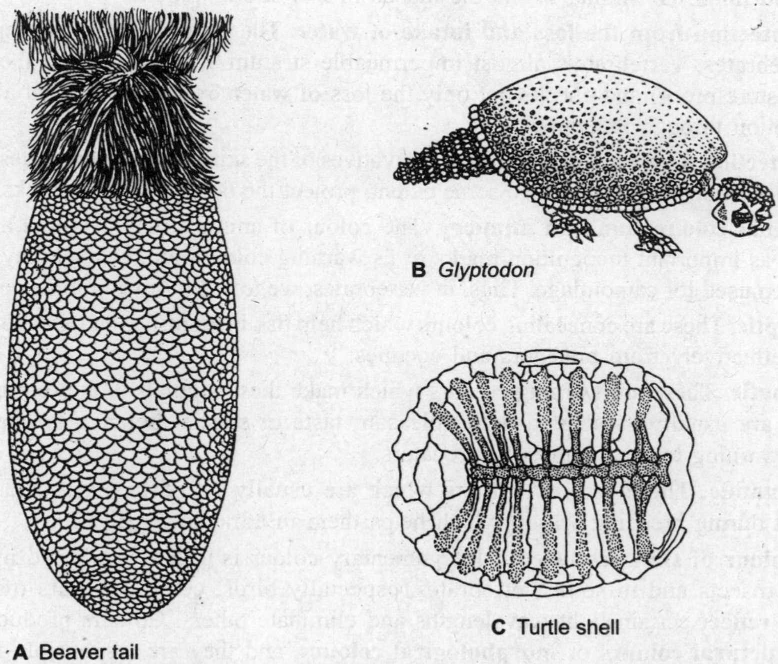

Fig. 2.39. Epidermal and dermal derivatives. A-Epidermal scales occur in some mammalian structures, such as this tail of a beaver (dorsal view). B-Although dermal scales are rare in mammals they fused to form armor of *Glyptodon*, a fossil mammal. C-Turtle shell. The encasing shell (carapace) of a turtle is derived from three sources—the ribs and the vertebrae of the endoskeleton (shaded) and the dermal bone arising in the integument (white). The surface of this bony shell is covered by large, thin epidermal scales (not shown) (after Kardong 2006).

skin of birds and most mammals. Exceptions have been noted earlier, namely, in the fossil mammal *Glyptodon* (Fig. 2.39B) and in the skin of the living armadillo. However, selected dermal bones take up residence in the fish skull and pectoral girdle and have persisted into modern groups of vertebrates. Most dermal bones of the skull and shoulder girdle all began phylogenetically in the skin and later sank inward to become parts of the skeleton. Thus, the dermis of vertebrates has ancient and persistent potential to form bone.

The shell of turtles is a composite structure. The dorsal half of the shell is the **carapace**, formed by fusion of dermal bone with expanded ribs and vertebrae (Fig. 2.39C). Ventrally, the

plastron represents fused dermal bones along the belly. On the surface of both carapace and plastron, keratinized plates of epidermis cover this underlying bone.

2.5 FUNCTIONS OF INTEGUMENT

Integument or skin of vertebrates have the following functions :

1. Protection. Skin provides protection of the animal from injurious external influences.

(*i*) From mechanical injuries. Skin and its derivatives such as scales, feathers and hair, and sub-dermal fats offer protection from pressure, friction and blow.

(*ii*) From the access of bacteria and fungi. Connective tissues of the dermis contain a variety of cells that remove foreign bodies, fight infection, take part in healing of the wounds and the replacement of damaged tissues. Saline sweat in mammals does not favour the survival of bacteria and fungi. Of similar use is the mucus in fish and amphibia.

(*iii*) Protection from the loss and intake of water. The water is of prime importance for the land vertebrates. Vertebrate's almost impermeable stratum corneum makes the skin water proof (or moisture proof) and checks not only the loss of water from the body but also does not permit absorption through integument.

(*iv*) Protection from enemies. The hard derivatives of the skin such as osteoscutes, scales, hair and their modifications and feathers, to some extent, protect the body from the attacks of enemies.

(*v*) Animal colouration and mimicry. The colour of animals may be vivid and dramatic when serving as important recognition marks or as warning colouration, or they may be subdued or cryptic when used for camouflage. Thus, in vertebrates, we found following three color patterns:

(*a*) Cryptic. These are concealing colours which help fish in assimilation with the background and conceal effectively from predators and enemies.

(*b*) Sematic. These are warning colours which make these animals look prominent because such colours are usually associated with upleasant taste or special defence mechanism. Such colours give warning to the enemies or predators.

(*c*) Epigamic. These are sex colours which are usually found in males and make them more brilliant during breeding season which helps them in attractic females.

How colour of skin produced ? Integumentary colour is usually produced by **pigments**, but in many insects and in some vertebrates, especially birds, certain colours of the surface tissue, which **reflect** certain light wavelengths and eliminate others. Colours produced this way are called **structural colours** or **morphological colours**, and they are responsible for the most beautifully iridescent and metallic hues to be found in the animal kingdom. A few fishes and birds are earth's most resplendent (*i.e.*, bright and colourful in an impressive way) animals. Certain structural colours of feathers are caused by minute, air-filled spaces or pores that reflect white light (white feathers) or some portion of the spectrum (*e.g.*, Tyndall blue colouration produced by scattering of light). Iridescent colours that change hue as the animal's angle shifts with respect to the observer are produced when light is reflected from several layers of thin, transparent film. By **phase interference**, light waves reinforce, weaken, or eliminate each other to produce some of the purest and most brilliant colours we know.

Pigments. Pigments or **biochromes** form an extremely varied group of large molecules that reflect light rays. In ectothermic vertebrates (fishes, amphibians and reptiles) these pigments are contained in large cells with branching processes, called **chromatophores**. The pigment may concentrate in the center of the cell in an aggregate too small to be visible, or it may disperse throughout the cell and its processes, providing maximum display. The most common chromatophores are **melanophores** which have two forms, *i.e.*, eumelanin which gives brown or black colouration and **phaeomelanin** (gives yellow brown colouration).

The most widespread of animal pigments are the **melanins**, a group of black or brown polymers that are responsible for the various earth-colored shades that most animals wear. Yellow and red colours are often caused by **carotenoid** pigments which are frequently contained within special pigment cells called **xanthophores**. Most vertebrates are incapable of synthesizing their own carotenoid pigments but must obtain them directly or indirectly from plants. Two entirely different classes of pigments called **ommochromes** and **pteridines** are usually responsible for the yellow pigments of molluscs and arthropods. Green colours are rare, when they occur, they are usually produced by yellow pigment overlying blue structural colour. **Iridophores**, a third type of chromatophores, contain crystals of guanine or some other purine, rather than pigment. They produce a silvery or metallic effect by reflecting light.

By vertebrate standards, mammals are a somber-coloured group. Most mammals are more or less **colour blind**, a deficiency that is doubtless connected with the lack of bright colours in the group. Exceptions are the brilliantly coloured skin patches of some baboons and mandrills. Significantly, primates have colour vision and thus can appreciate such eye-catching ornaments. The subdued colours of mammals are caused by melanin, which is deposited in growing hair by dermal melanophores.

***(vi)* Exteroception**. Free nerve-endings and special cutaneous receptors in the integument give warnings of heat, cold, contact or damage. Mucus membranes in the nostrils and over the tongue gives a variety of informations.

***(vii)* Respiration**. Nearly all amphibia respire through skin, a thin film of water is retained over the surface of skin with the help of mucus; exchange of gases takes place through the water film. In certain fish, such as eels and periopthalamus, skin is an additional (accessory) organ of respiration.

***(viii)* Regulation of body temperature**. Skin is vitally concerned with temperature regulation, since most of the body's heat is lost through the skin; it contains mechanisms that cool the body when it is too hot and slow heat loss when the body is too cold. In cold-blooded (poikilothermous) animals such as frogs, evaporation of water from the surface helps keeping the body temperature from rising above the tolerable limit. Cold makes the frogs dormant and forces them to hibernate. The same is true of reptiles.

To keep themselves active warm blooded animals, the birds and the mammals, produce and conserve heat and prevent its loss through various means. Hair and feathers tend to trap a layer of air which insulates the body, reducing heat exchanges with the environment. Interlocking mechanism developed in feathers and in hair such as those of sheep and some rabbits increases the heat-retaining capacity.

A layer of nonconducting fat in the dermis serves the same purpose (*i.e.*, conservation of heat) in mammals (*e.g.*, whale, sea cow, seal, polar bear, etc.), which have a few hair or no hair. Equally important are the means and ways for producing heat. Heat is chiefly produced by muscular activity. Liver and other organs give heat as they work. Excess heat is lost with the help of the following means.

1. The **sweat glands**. When the blood supply is increased, evaporation of sweat from the surface involves heat loss.

2. The **lungs**. Heat is lost with the expired air, it is for this reason that dogs which have few sweat glands breathe faster to get rid of excess heat from the wet surface as tongue and upper respiratory tract.

3. Special areas. When it is hot, flow of blood is increased in the ear pinnae of elephant which are then raised and flapped. An elephant increases its surface area which may be nearly one-sixth.

Thermoregulatory action of **insectivorous bat** is different as they have no mechanism of control of the amount of heat produced. Every time a bat hangs up and goes to sleep its temperature falls and within an hour it reaches that of surroundings.

4. **Hibernation** is practiced by several mammals. Ground squirrels, chipmunks, and marmots undergo hibernation and their heart beats are slowed and the body temperature falls considerably.

(*ix*) Integument stores food. Large amounts of fat is deposited in the sub-dermal parts of the skin. Subcutaneous fat of the whale is called **blubber** and it is about 45 cm thick. Besides insulating it helps in propulsion or locomotion (*e.g.*, polar bear).

(*x*) Integument contributes to sexual selection. Bright colour and bold patterns are related to sexual habits (or behavior) and apparently attracts mates and perhaps establish territories.

(*xi*) Integument helps in locomotion. Wing membranes of bats, flying lizards, flying frogs, flying squirrels and lemurs; webs in the feet of frogs and ducks, are folds of integument that help in propulsion or locomotion.

(*xii*) Skin secretions can make the animal sexually attractive or repugnant or provide olfactory cues that influence behavioural interactions between individuals.

(*xiii*) Integument also gives rise to various modifications such as the several forms of **claws, hooves, horns, antlers** and **rattle** (of rattale snake), **antlers** all confer advantages in struggle for the existence of the animals.

QUESTIONS

Long Answer Questions

1. What is integument? Describe the integument and its derivatives in vertebrates.
2. Give an account of integument of the mammals. Explain the functions of integument.
3. Give a comparative account of integument of reptiles, birds and mammals.

Short Answer Questions

1. Write short notes on the following:

   (*i*) Baleen; (*ii*) Horse hoof;
   (*iii*) Horns; (*iv*) Antlers;
   (*v*) Keratinization; (*vi*) Epidermal glands.

2. Describe the process of annual growth of deer antlers.

Very Short Answer Questions

1. What is vibrissa?
2. What is baleen.
3. Define the integument.

Fill in the Blanks

1. .... of the right whale is called whalebone but it does not contain any ....
2. .... is synthesized from 7-dehydrocholesterol in the skin of primates.

Multiple Choice Questions

1. Outermost protective covering of animals

   (*a*) scales (*b*) integument
   (*c*) feathers (*d*) hairs

2. Which is not a function of skin?

   (*a*) protection (*b*) locomotion
   (*c*) secretion (*d*) digestion

3. Dermis of vertebrate integument is derived from
(*a*) ectoderm (*b*) endoderm
(*c*) mesoderm (*d*) ecto-mesoderm

4. Typical cornification are modification of
(*a*) stratum corneum (*b*) stratum lucidum
(*c*) stratum germinativum (*d*) dermis

5. Integumentary gland secreting tears is
(*a*) sebaceous gland (*b*) mammary gland
(*c*) lacrimal gland (*d*) uropygial gland

6. In hippopotamus, sweat glands are restricted in
(*a*) lips (*b*) ears
(*c*) muzzle (*d*) soles of feet

7. Melanocytes are located in
(*a*) stratum corneum (*b*) stratum germinativum
(*c*) stratum lucidum (*d*) dermis

8. In presence of sunlight vitamin D is synthesized in mammalian skin from
(*a*) sweat (*b*) earwax
(*c*) 7-dehydrocholestrol (*d*) tears

9. Reflecting pigment cells in luminescent glands contain
(*a*) guanine crystal (*b*) adenine crystal
(*c*) cytosine crystal (*d*) thymine crystal

10. Poison secreted by parotid glands of amphibians is
(*a*) alkaloid (*b*) fatty acid
(*c*) alcohol (*d*) carbohydrate

11. In birds, uropygial gland is present just above the
(*a*) beak (*b*) eye
(*c*) ear (*d*) tail

ANSWERS

Very Short Answer Questions

1. Vibrissa are stiff tactile hairs of facial region of mammals, *e.g.*, cat's whiskers.
2. Baleen is a hard keratinous structure hanging from the roof of the mouth in toothless whales (mysticete whales); functionally baleen act as a plankton filter or strainer.
3. Integument is the skin covering the animal body.

Fill in the Blanks

1. Baleen, bone. 2. Vitamin D.

Multiple Choice Questions

1. (*b*) 2. (*d*) 3. (*c*) 4. (*a*) 5. (*c*) 6. (*b*)
7. (*d*) 8. (*c*) 9. (*a*) 10. (*d*) 11. (*d*).

3

Skeletal System

3.1 INTRODUCTION

The term skeleton refers to the framework of the animal body. In vertebrates, it is composed of **cartilage**, **bone** or a combination of the two (Box 3.1). The skeleton gives the vertebrate body shape, supports its weight, offers a system of levers that together with muscles produces movement, and protects the soft parts such as nerves, blood vessels and other viscera.

Box 3.1

Because skeleton is hard, bits of the skeleton often survive fossilization better than does soft tissue anatomy; so our most direct contact with long-extinct animals is often their skeletons. The story of vertebrate function and evolution is written in the architecture of the skeleton.

The skeleton system is composed of an exoskeleton and endoskeleton (Fig. 3.1 A). The **exoskeleton** is formed from or within the integument, the dermis giving rise to bone and the epidermis keratin. The **endoskeleton** forms deep within the body from **mesoderm** and other sources, not directly from the integument. Tissues contributing to the endoskeleton include fibrous connective tissue, bone and cartilage.

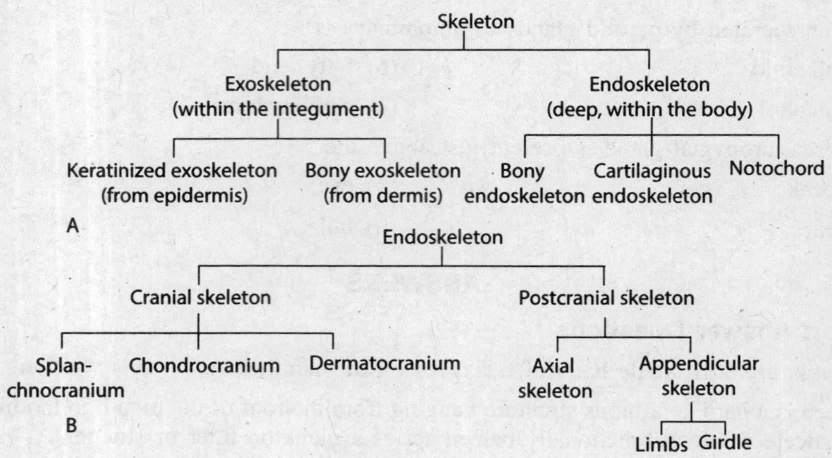

Fig. 3.1. Organization of skeletal tissue in vertebrates. Components of the skeletal system function together as a unit but as a convenience, they can be divided into manageable parts for closer analysis. A–As a protective and supportive system, the skeleton can be divided into structures on the outside (exoskeleton) and inside (endoskeleton) of the body. B–On the basis of position, the skeleton can be treated as two separate components, the cranial skeleton (skull) and the postcranial skeleton. The postcranial skeleton includes the axial and appendicular skeletons (after Kardong, 2006).

During the course of vertebrate evolution, most bones of the exoskeleton stay within the integument and protect surface structures. Dermal armor of ostracoderms are bony scales of fishes are examples. Other bones have sunk inward, merging with deeper bones and cartilaginous elements of the endoskeleton separatly. Parts of one often found in company with the other. Instead, we select *composite structural units* and follow their evolution. This way of dividing the skeleton for study gives us two units: the **skull**, or **cranial skeleton**, and the **postcranial skeleton** (Fig. 3.1 B). The cranial skeleton includes **splanchnocranium**, **chondrocranium** and **dermatocranium**. The postcranial skeleton includes the **vertebral column**, **limbs**, **girdles**, and associated structures such as **ribs** and **shells**.

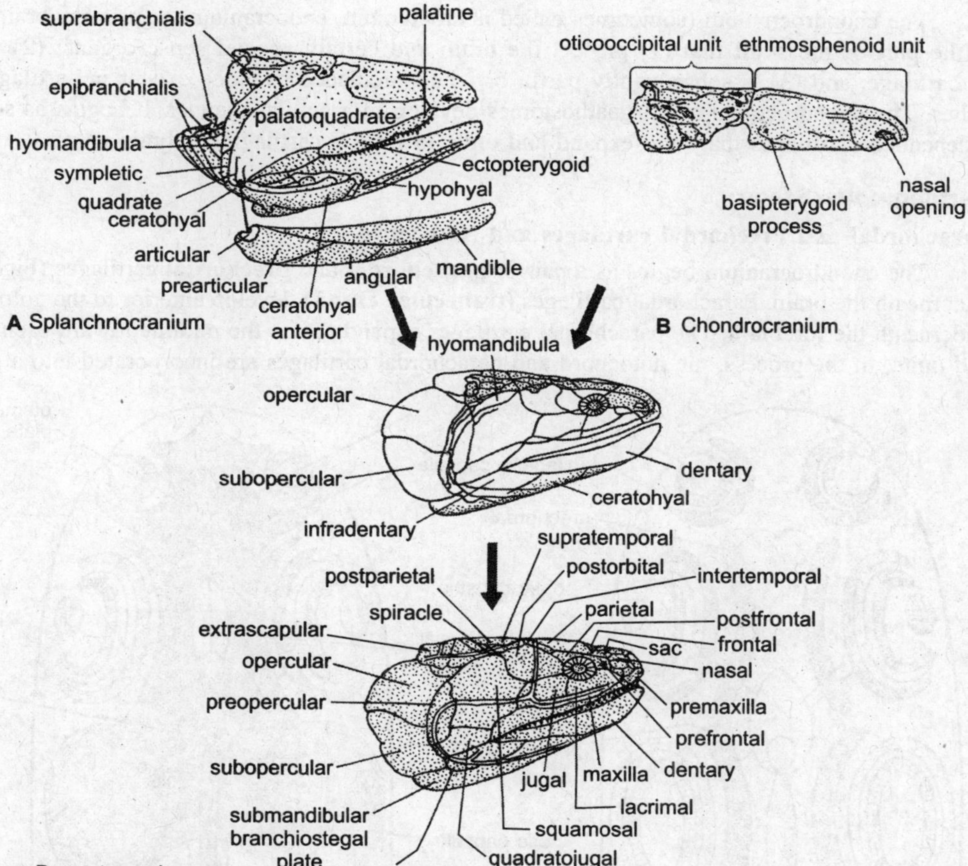

Fig. 3.2. Composite skull. The skull is a mosaic composed of three primary contributing parts: the chondrocranium, the splanchnocranium, and the dermocranium. Each has a separate evolutionary background. The skull of *Eusthenopteron*, a Devonian rhipidistian fish, illustrates how parts of all three phylogenetic sources contribute to the unit. A – The splanchnocranium arose first and is shown in association with the chondrocranium and parts of the dermatocranium. The right mandible is lowered from its point of articulation better to reveal depper bones. B – The chondrocranium in *Eusthenopteron* is formed by the union between the anterior ethmosphenoid and the posterior oticooccipital units. C – The superficial wall of bones composes the dermatocranium. The central figure depicts the relative position of each contributing set of bones brought together in the composite skull (after Kardong 2006).

3.2 CRANIAL SKELETON

Although merged into a harmonious unit, the vertebrate skull, or **cranium**, is actually a composite structure formed of three distinct parts. Each part of the skull arises from a separate phylogenetic source. The most ancient part is the **splanchnocranium (visceral cranium)**, which first arose to support pharyngeal slits in protochordates (Fig. 3.2 A). The second part, the **chondrocranium**, underlies and supports the brain and is formed of endochondral bone or of cartilage, or both (Fig. 3.2 B). The third part of the skull is the **dermatocranium**, a contribution that in later vertebrates forms most of the outer casing of the skull. As its name suggests, the dermatocranium is composed of dermal bones (Fig. 3.2 C).

A. Chondrocranium

The chondrocranium (sometimes called neurocranium, endocranium, or primary braincase) is the part of the skull that (1) protect the brain and certain special sense organs; (2) arises as cartilage, and (3) is subsequently partly or wholly replaced by bone except in cartilaginous fishes. The chondrocrania of all gnathostomes develop in a similar manner. It begins as several independent cartilages that later expand and unite to form a cartilaginous brain case.

Cartilaginous Stage

Parachordal and Prechordal cartilages and Notochord

The chondrocranium begins as a pair of **parachordal** and **prechordal** cartilages (Fig. 3.3), underneath the brain. Parachordal cartilages (**trabeculae cranii**) develop anterior to the notochord underneath the forebrain. The parachordal cartilages expand across the midline toward each other and unite. In the process, the notochord and parachordal cartilages are incorporated into a single

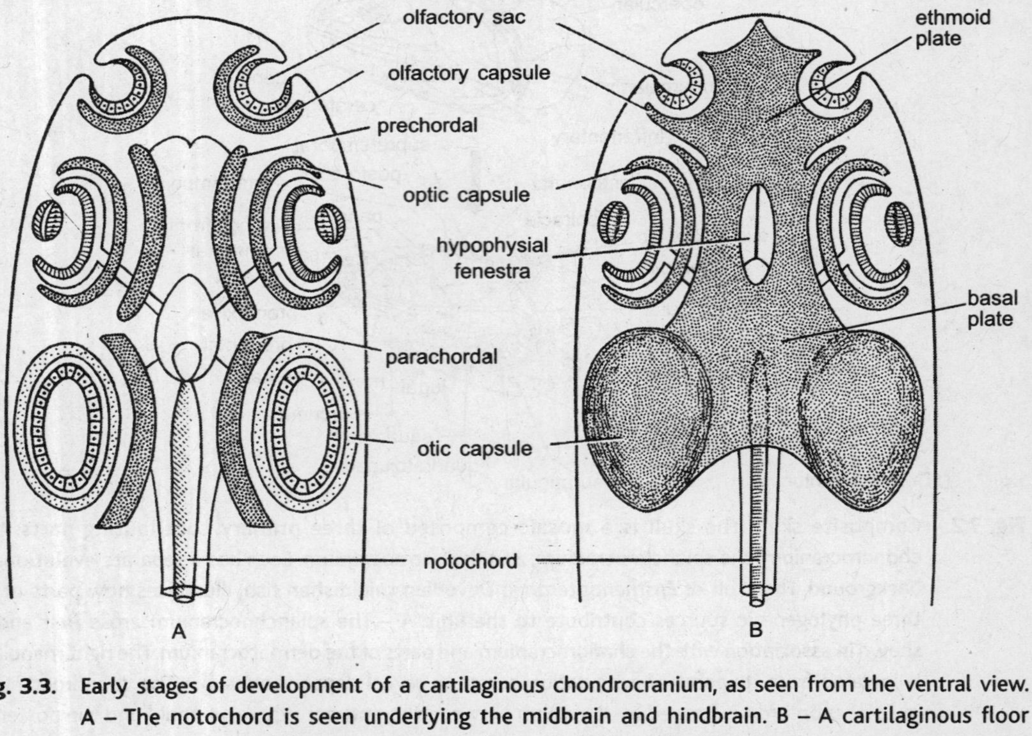

Fig. 3.3. Early stages of development of a cartilaginous chondrocranium, as seen from the ventral view. A – The notochord is seen underlying the midbrain and hindbrain. B – A cartilaginous floor has been completed beneath the entire brain and a hypophysial fenestra remains between the ethmoid and basal plates. The optic capsule becomes the sclerotic coat of the eye ball.

broad, cartilaginous **basal plate**. The prechordal cartilages likewise expand and unite across the midline at their anterior ends to form an **ethmoid plate**.

Sense Capsules

While parachordal and prechordal cartilages are forming, cartilage also appears in two other locations: (1) as an **olfactory** or **nasal capsule** partially surrounding the olfactory epithelium and (2) as an **otic capsule** completely surrounding the otocyst, which is the developing inner ear (Fig. 3.3). The olfactory capsules are incomplete anteriorly, because water (in fishes) or air (in tetrapods) must have entrance to the olfactory epithelium. The walls of the olfactory and otic capsules are perforated by foramina that transmit nerves and vascular channels.

An **optic capsule** forms around the retina of eye, but it is not the orbit, or skeleton socket, in which the eyeball lies. It is the **sclerotic coat** of the eyeball. Although this capsule is fibrous in mammals, cartilaginous or bony plates often forms within the sclerotic coat in reptiles and birds (Fig. 3.4). This is a primitive condition, having been present in some ostracoderms, placoderms, and basal osteichthyans. Because the optic capsule does not fuse with the rest of the chondrocranium, the eyeball is free to move independent of the skull. Therefore the sclerotic coat is not conventionally considered part of the chondrocranium.

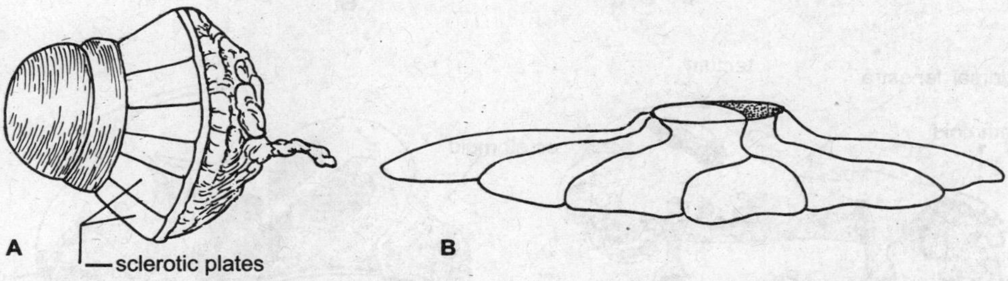

Fig. 3.4. Ossicles of the sclerotic coat of the eye. A – Owl's eye, showing sclerotic plates in place. B – Scleral ring of overlapping ossicles from a lizard's eye (after Kent and Carr 2001).

Completion of the Floor, Walls and Roof of Skull

The expanding ethmoid plate unites anteriorly with the olfactory capsules, and the expanding basal plate unites with the otic capsules, which are lateral to the hindbrain. The ethmoid and basal plates also expand toward one another until they meet to form a floor upon which the brain thereafter rests (Fig. 3.3 B). In the latter process, there remains in the midline between the two plates a **hypophyseal fenestra** accomodating the hypophysis and internal carotid arteries which are being en route to the brain. Later the fenestra is usually reduced to a pair of foramina transmitting the arteries. Further development of the chondrocranium primarily involves construction of cartilaginous walls around the brain and as a primitive condition in some craniates, a cartilaginous roof (called **tectum**) above the brain with one or two prominent fenestrae (Fig. 3.5 C). The cranial nerves and blood vessels are already present by this time, and cartilage is deposited in such a manner as to leave foramina for these structures. The largest is the foramen magnum in the rear wall of the chondrocranium.

The mesenchyme that gives rise to the chondrocranium comes from at least two sources. The prechordal cartilages are formed from **neural crest** ectoderm that streams ventrad infront of developing eyestalks (optic stalks) that connect the brain with the developing eyeball.

The mesenchyme that forms the parachordal cartilages, encapsulates the inner ear, in part, comes from **sclerotome** (epimeric mesoderm). The source of the mesenchyme that forms the remaining bulk of the chondrocranium is derived from neural crest.

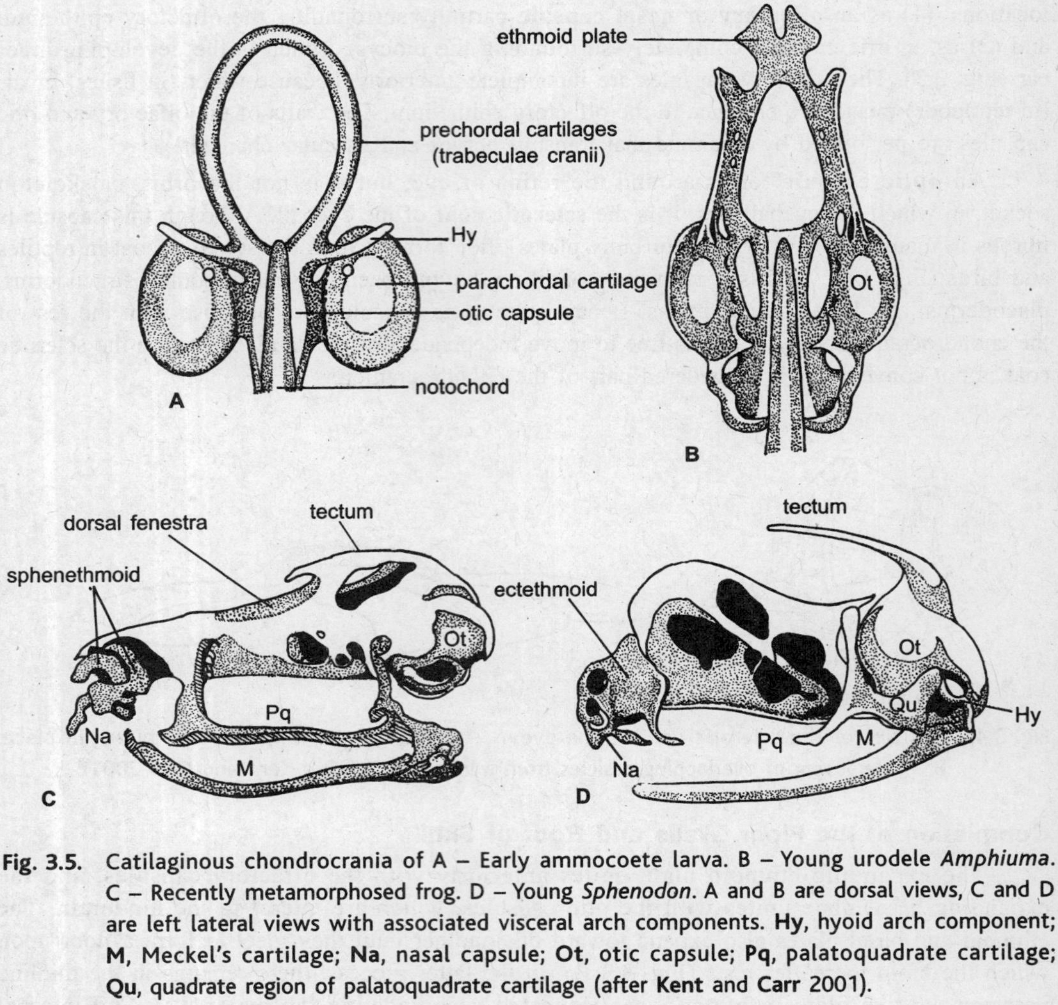

Fig. 3.5. Catilaginous chondrocrania of A – Early ammocoete larva. B – Young urodele *Amphiuma*. C – Recently metamorphosed frog. D – Young *Sphenodon*. A and B are dorsal views, C and D are left lateral views with associated visceral arch components. **Hy**, hyoid arch component; **M**, Meckel's cartilage; **Na**, nasal capsule; **Ot**, otic capsule; **Pq**, palatoquadrate cartilage; **Qu**, quadrate region of palatoquadrate cartilage (after **Kent** and **Carr** 2001).

Cartilaginous Chondrocrania of Adult Craniates

1. Living Agnathans. Various cartilaginous components of the embryonic chondrocranium remain more or less independent throughout life (Fig. 3.6). An olfactory capsule, two otic capsules and a basal plate, and a notochord (not fused with the basal plate) are identifiable and several other cartilages of unknown homology are also present. The roof above the brain remains unchondrified and therefore fibrous.

2. Cartilaginous Fishes and Bony Fishes. Cartilaginous fish retain a cartilaginous chondrocranium throughout life. The chondrocranium in these fishes completely encloses the brain (Fig. 3.7), and the otic and olfactory capsules are fused into it along with the notochord (Fig. 3.8). Dorsally, an endolymphatic fossa is perforated by endolymphatic and perilymphatic

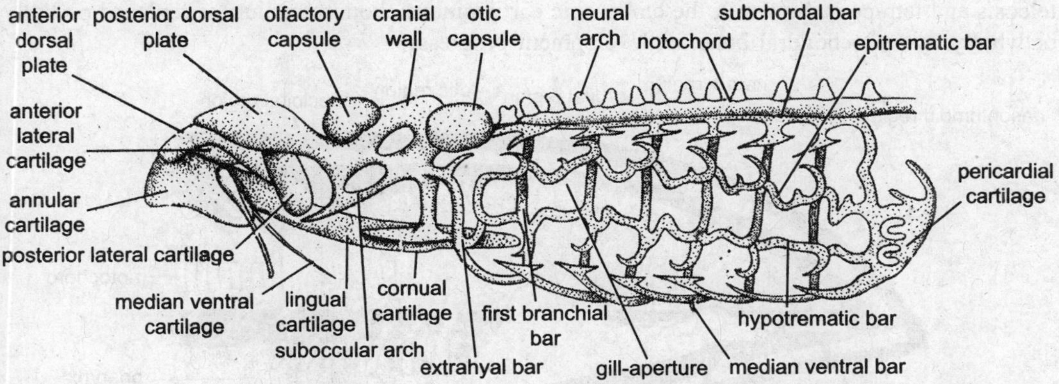

Fig. 3.6. *Petromyzon* (lamprey). Chondrocranium branchial apparatus (basket) and feeding apparatus. The olfactory capsule is a midline structure; otic capsules are paired. The lingual cartilage is also referred to as tongue.

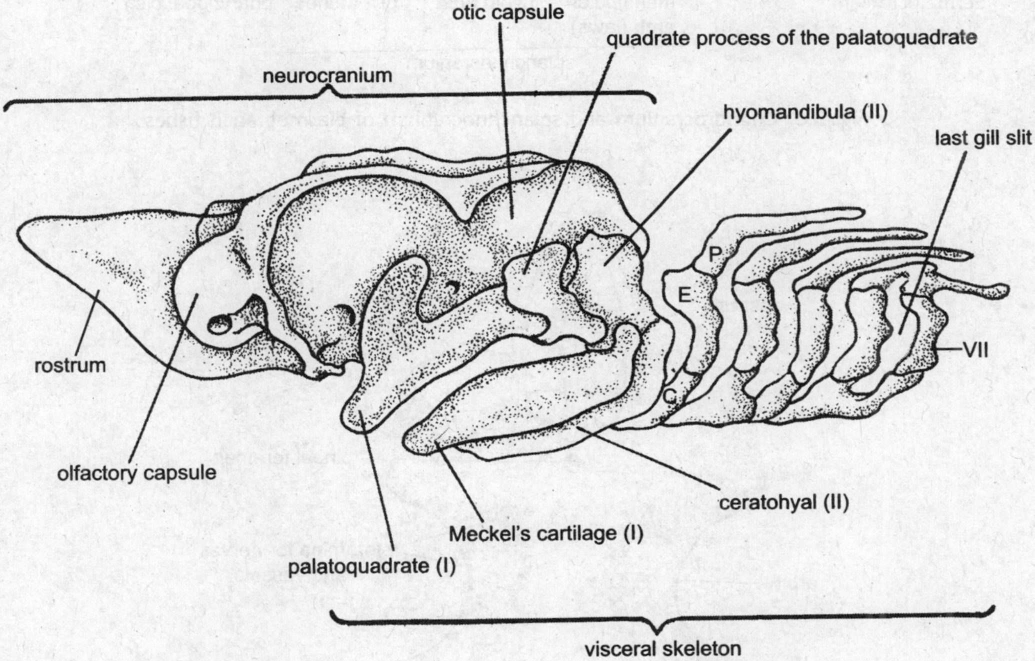

Fig. 3.7. Skull and visceral skeleton of shark *Squalus acanthias*. **I**, **II**, and **VII**, skeleton of first, second and seventh pharyngeal arches; **C**, ceratobranchial; **E**, epibranchial; **P**, pharyngobranchial cartilages of the second gill arch. Labial cartilages, gill rakers and gill rays are omitted. The hyomandibula suspends the entire visceral skeleton from the skull at the otic capsule (after Kent and Carr, 2001).

ducts and there is an opening, the pineal foramen, that is occupied by the pineal body (Fig. 3.9). Among bony fishes and lung fishes and most ganoids retain a highly cartilaginous chondrocranium throughout life. Such chondrocranium is covered by the membrane or dermal bones (Fig. 3.10). In teleosts and tetrapods, however, the embryonic cartilaginous chondrocranium is replaced partially or wholly by endochondral bone as development progresses.

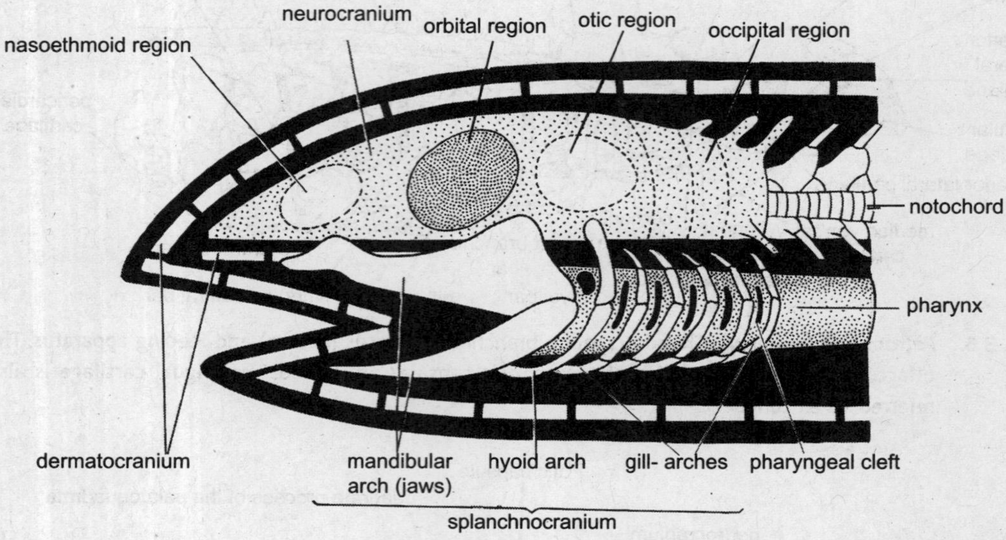

Fig. 3.8. Chondrocranium and splanchnocranium of elasmobranch fishes.

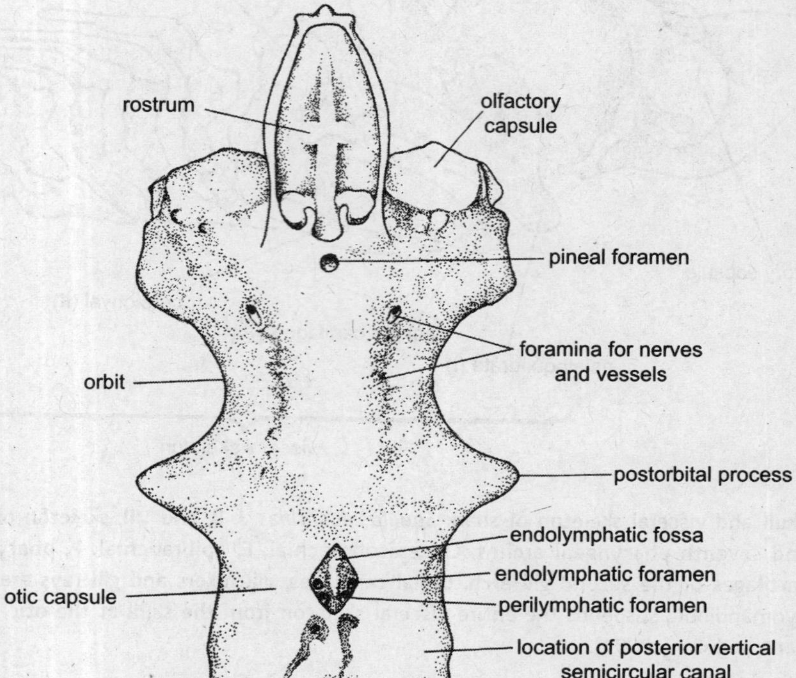

Fig. 3.9. Chondrocranium of *Squalus* (Dorsal view).

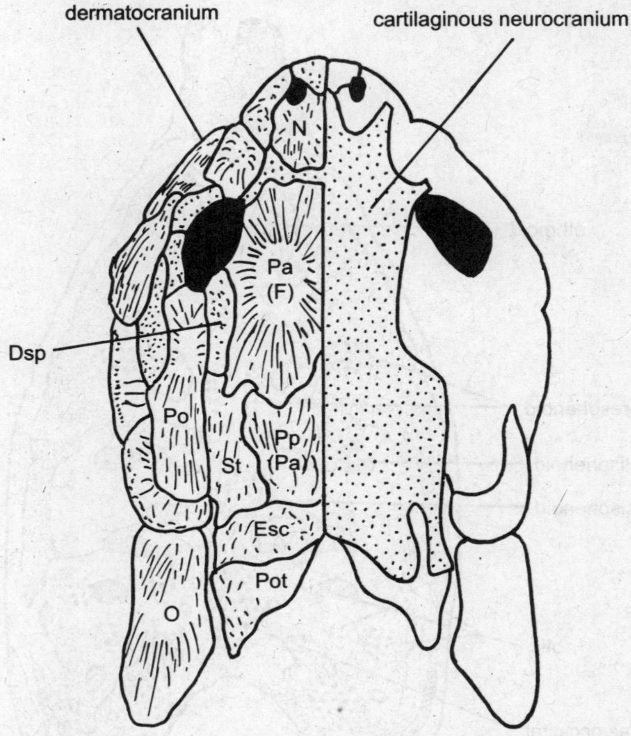

Fig. 3.10. Skull of *Amia calva*, dorsal view. Membrane/dermal bones have been removed on the right side to reveal underlying cartilaginous chondrocranium (or neurocranium). **Dsp**, dermosphenotic (*i.e.*, infraorbital), **Esc**, extrascapular; **F**, frontal; **N**, nasal; **O**, opercular; **Pa**, parietal; **Po**, postorbital; **Pp**, postparietal; **Pot**, post-temporal; and **St**, supra-temporal. The bones anterior to the nasals are ethamoids. Paramaxillae are not visible in this view of the skull (after Kent and Carr, 2001).

Centers of Ossification in Chondrocranium

The process of endochondral ossification within the chondrocranium occurs more or less simultaneously at numerous separate ossification centers. The specific number of such centers varies among species, but four regional groups are occipital, sphenoid, ethmoid and otic (Fig. 3.11 and Fig. 3.12).

Occipital centers. The cartilage surrounding the foramen magnum may be replaced by as many as four bones. One or more endochondral ossification centers ventral to the foramen magnum produce a **basioccipital bone** underlying the hindbrain (Fig. 3.11 and Fig. 3.13 A). Ossification centers in the lateral walls of the foramen magnum produce two **exoccipital bones**. Above the foramen, a **supraoccipital bone** may develop. In mammals, all four occipital elements usually fuse to form a single **occipital bone**. In modern amphibians, one or more of these may remain cartilaginous, although they were bony in stem amphibians.

The chondrocranium of tetrapods articulates with the first vertebra via one or two **occipital condyles**. Stem amphibians had a single condyle bone chiefly on the basioccipital bone. Living reptiles and birds still have single condyle. Modern amphibians and mammals diverged from the early tetrapod condition, gradually shifting the occipital condyles from the median basioccipital to the two exoccipitals.

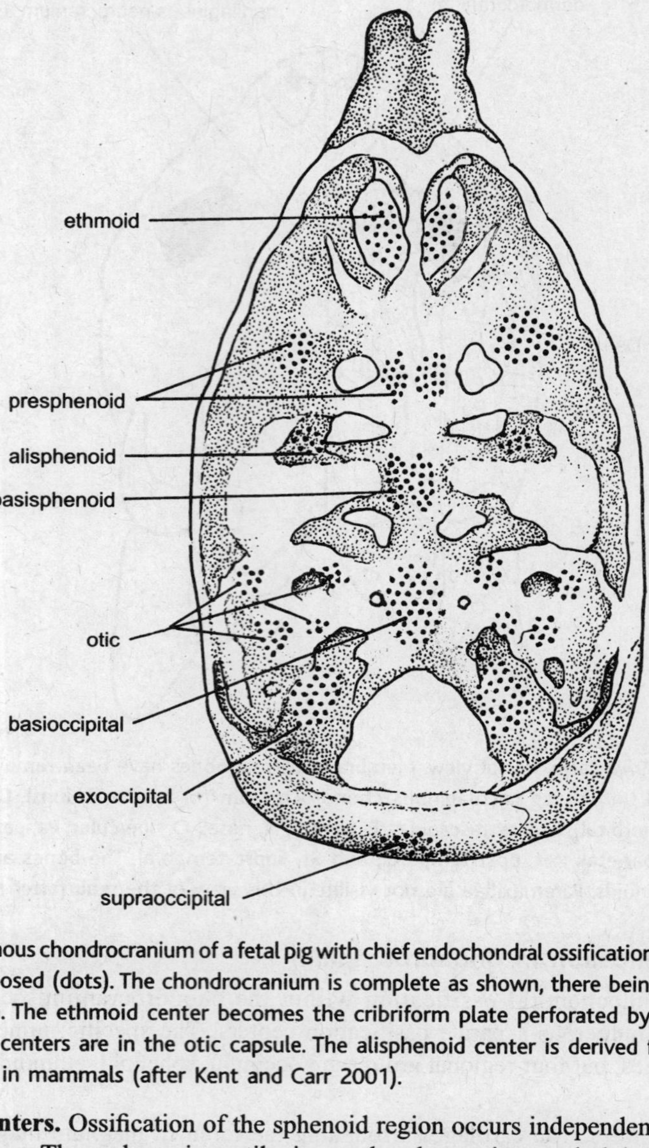

Fig. 3.11. Cartilaginous chondrocranium of a fetal pig with chief endochondral ossification centers of mammals superimposed (dots). The chondrocranium is complete as shown, there being no cartilage above the brain. The ethmoid center becomes the cribriform plate perforated by olfactory foramina. The otic centers are in the otic capsule. The alisphenoid center is derived from palatoquadrate cartilage in mammals (after Kent and Carr 2001).

Sphenoid centers. Ossification of the sphenoid region occurs independently in the synapsid and reptilian lineages. The embryonic cartilaginous chondrocranium underlying the midbrain and pituitary gland ossifies to form a **basisphenoid bone** (anterior to the basioccipital). In mammals, a **presphenoid bone** (Fig. 3.11) ossifies anterior to the basisphenoid. Thus, a bony plateform consisting of the basioccipital and the sphenoid bones underlies the brain. The side wall in the sphenoid region above the basisphenoid is formed by an additional presphenoid ossification in mammals. In archosaurs (crocodiles and dinosaurs), an independent **laterosphenoid bone** forms the lateral ossification of the sphenoid region. A separate interorbital septum forms in archosaurs as the **orbitosphenoid bone** (Fig. 3.14 A). The **alisphenoid bone**, of at least some mammals, helps to form the lateral wall, but it is derived from the palatoquadrate cartilage rather than the chondrocranium. The sphenoid elements in mammals (*i.e.*, basisphenoid, presphenoid and alisphenoid) may remain separate or unite to form a single **sphenoid bone** with "wings" (Fig. 3.12). The pituitary rests in the sella turcica of the basisphenoid region. No replacement bones develop above the brain.

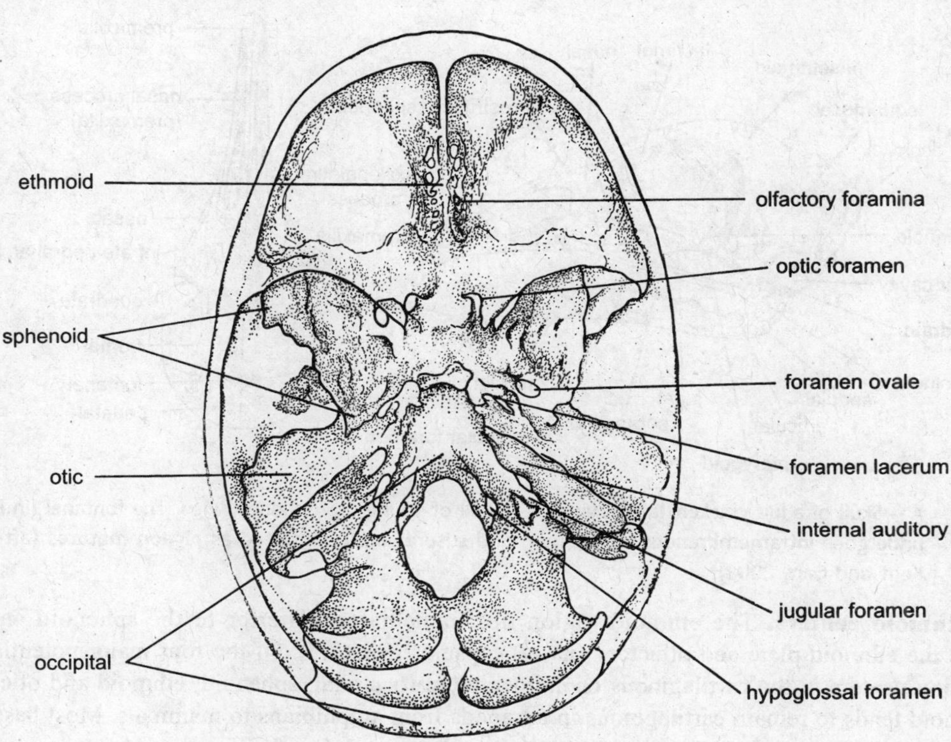

Fig. 3.12. Bony chondrocranium of human skull. The calvarium (roof) of the skull has been sawed off, and the view is looking down into the skull from above. Major endochondrial ossification centers are labeled at left. The sphenoid bone exhibits an anterior (lesser) wing enclosing the optic foramina and a posterior (greater) wing between the two label lines for the sphenoid bone. The olfactory foramina are in the cribriform plate of the ethmoid (after Kent and Carr 2001).

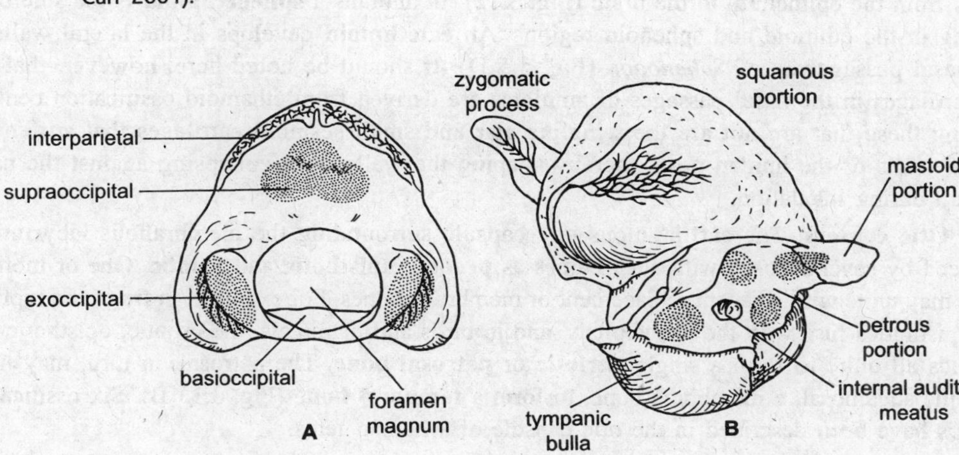

Fig. 3.13. Typical mammalian endochondral ossification centers (dots) and intramembranous centers (reticulum) in two cat skull bones. A – Occipital bone (caudal view). B – Right temporal (medial view). The petrous region houses the inner ear (after Kent and Carr 2001).

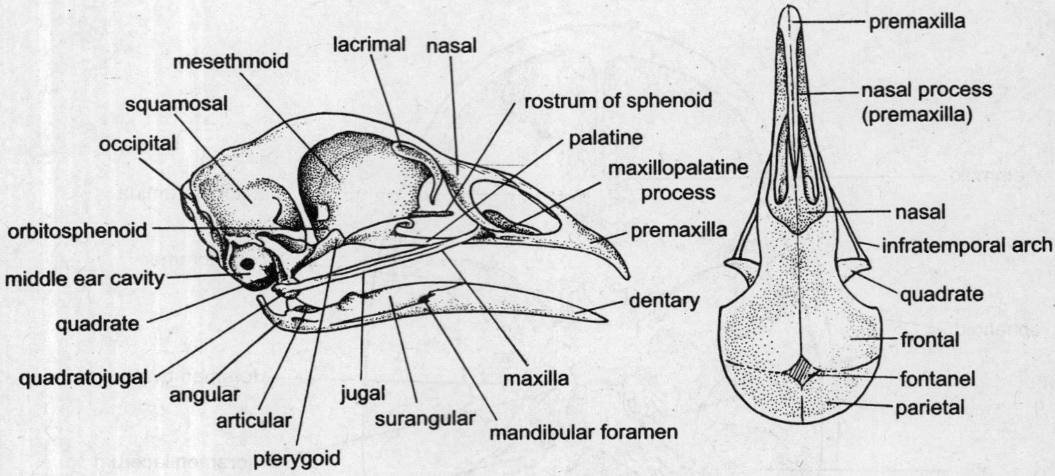

Fig. 3.14. A – Skull of adult chicken, lateral view. B – Skull of young pigeon, dorsal view. The fontanel (in B) undergoes intramembranous ossification and sutures are obliterated as pigeon matures (after Kent and Carr, 2001).

Ethmoid centers. The ethmoid region lies immediately anterior to the sphenoid and includes the ethmoid plate and olfactory (nasal) capsules (Fig. 3.5). Of the four major potential ossification centers in the cartilaginous chondrocranium (occipital, sphenoid, ethmoid and otic), the ethmoid tends to remain cartilaginous in tetrapods from amphibians to mammals. Most basal tetrapods develop no ethmoidal ossification centers whatsoever. Ossification centers in amniotes are chiefly **mesethmoid bones**, which contribute to (1) the otherwise cartilaginous nasal septum of birds and mammals, and (2) in birds, to the anterior portion of the interorbital septum; (3) one or more scroll-like **turbinal bones** (**conchae**) in the walls of the nasal passageways of most reptiles, birds, and mammals; (4) in mammals, a **cribiform plate** (*cribiform* means perforated such as a sieve) perforated by **olfactory foramina** that transmit bundles of olfactory nerve fibers from the epithelium to the brain (Fig. 3.12). In anurans a **sphenethmoid** is the sole bone arising in the ethmoid and sphenoid regions. An **ectethmoid** develops in the lateral walls of the nasal passageway of *Sphenodon* (Fig. 3.5 D). It should be noted here, however, that not all cartilages in the nasal passages of amniotes are derived from ethamoid ossification centers. Among these that are not are the winglike alar and small sesmoid cartilages that stiffen the lateral walls of the human nose, thereby keeping the walls from collapsing against the nasal septum during inhalation.

Otic centers. The cartilaginous otic capsule surrounding the membranous labyrinth is replaced by several bones with such names as **prootic**, **opisthotic** and **epiotic**. One or more of these may unite with adjacent replacement or membrane bones. For example, in frogs and reptiles, the opisthotics fuse with the exoccipitals, and in birds and mammals, the prootic, opisthotic and epiotics all unite to form a single **periotic** or **petrosal bone**. The petrosal, in turn, may unite with the squamosal, a membrane bone, to form a **temporal bone** (Fig. 3.13 B). Six ossification centers have been described in the otic capsule of a human fetus.

B. DERMATOCRANIUM

The membrane bones of the skull collectively constitute the dermatocranium. In modern vertebrates from fish to humans, the membrane bones of the head no longer ossify from mesenchyme in the dermis, but from subdermal mesenchyme. The inductors that convert this mesenchyme to fibroblasts and scleroblasts may be the same ones that induced ossification in the dermis of earlier generations. *It is these membrane bones that constitute the modern dermatocranium from fish to mammals*. On going research on mesenchymal cell migration and inductor substances support the concept that the dermatocranium is derived from the dermal armor of the earliest vertebrates.

Dermal Bone Series of Dermatocranium

The dermal skull of amniotes may contain a considerable series of bones (about six) joined firmly at sutures in order to box in the brain and other skull elements (Table 3.1; Fig. 3.15; Fig. 3.16):

1. Facial series. It encircles the external naris and collectively form the snout. The **maxilla** and **premaxilla** (also called **incisive bone** since it is relating to, or situated near the incisor teeth) define the margins of the snout and usually bear teeth. The **nasal** lies medial to the naris. The **septomaxilla** is a small dermal bone of the facial series that is often absent. When present, it is usually sunken below the surface bones and aids in forming the nasal cavity.

Table 3.1 The six series of bones of dermatocranium (Source. Kardong 2006)

BRAIN CASE					***MANDIBLE***
Facial series	***Orbital series***	***Temporal series***	***Vault series***	***Palatal series***	***Mandibular series***
1. Premaxilla	1. Lacrimal	1. Intertemporal	1. Frontal	1. Vomer	**A. Lateral bones:**
2. Maxilla	2. Prefrontal	2. Supratemporal	2. Parietal	2. Palatine	1. Dentary (teeth)
3. Nasals (septomaxilla)	3. Postfrontal	3. Tabular	3. Postparietal	3. Ectopterygoid	2. Splenials (2)
	4. Postorbial	4. Squamosal		4. Pterygoid	3. Angular
	5. Jugal	5. Quadratojugal		5. Parasphenoid (unpaired)	4. Surangular
					B, Median bones:
					1. Prearticular
					2. Coronoids

2. Orbital series. The dermal bone encircle the eye to define the orbit superficially. The **lacrimal** takes its name from the nasolacrimal (tear) duct of tetrapods that passes through or near this bone.

The **Prefrontal, postfrontal,** and **postorbital** continue the ring of bones above and behind the orbit.

3. Temporal series The temporal series lies behind the orbit, completing the posterior wall of the braincase. In many primitive tetrapods, this series is indented posteriorly by a **temporal notch**. Once thought in life to suspend an eardrum, this notch was named accordingly an **otic notch**. This now seems unlikely, and instead the notch perhaps accommodated a spiracle, a respiratory tube. Openings called **fenestrae** (singular fenestra) arise within this region of the outer braincase in many tetrapods in association with the jaw musculature. A row of bones, the **intertemporal, supratemporal,** and **tabular**, make up the medial part of the temporal series. This row is reduced in early tetrapods and usually lost in later species. Laterally, the **squamosal** and **quadratojugal** complete the temporal series and form the cheek.

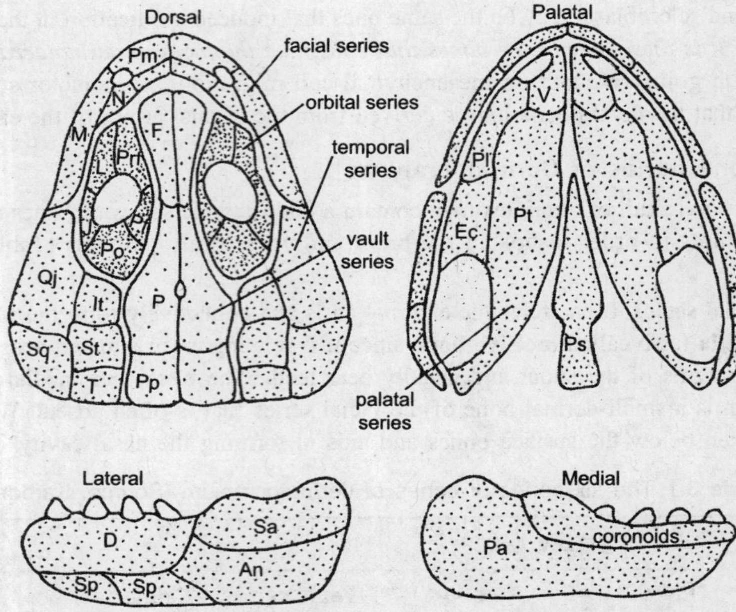

Fig. 3.15. Major bones of the dermatocranium. Sets of dermal bones form the facial series surrounding the nostril. The orbital series encircles the eye, and the temporal series composes the lateral wall behind the eye. The vault series, the roofing bones, run across the top of the skull above the brain. Covering the top of the mouth is the palatal series of bones. Meckel's cartilage (not shown) is encased in the mandibular series of the lower jaw. Abbreviations: angular (An), dentary (D), ectopterygoid (Ec), frontal (F), intertemporal (It), Jugal (J), lacrimal (L), maxilla (M), nasal (N), parietal (P), prearticular (Pa), palatine (Pl), premaxilla (Pm), postorbital (Po), postparietal (Pp), prefrontal (Prf), parasphenoid (Ps), pterygoid (Pt), quadratojugal (Qj), surangular (Sa), splenial (Sp), squamosal (Sq), supratemporal (St), tubular (T), vomer (V) (after Kardong 2006).

4. Vault series. The **vault** or **roofing bones**, run across the top of the skull and cover the brain beneath. These include the **frontal** anteriorly and the **postparietal** (interparietal) posteriorly. Between them is the large **parietal**, occupying the center of the roof of skull and defining the small **parietal foramen** if it is present. The parietal foramen is a tiny skylight an endocrine gland to direct sunlight.

5. Palatal series. The dermal bones of the **primary palate** cover much of the roof of the mouth. The largest and most medial is the **pterygoid**. Lateral to it are the **vomer**, **palatine** and **ectopterygoid**. Teeth may be present on any or all four of these palatal bones. In fishes and lower tetrapods, there also is an unpaired medial dermal bone, the **parasphenoid**.

6. Mandibular series. Meckel's cartilage (the cartilaginous axis of the mandibular arch in the embryo and fetus) is usually encased in dermal bones of the mandibular series. Laterally, the wall of this series includes the tooth-bearing **dentary** and one or two **splenials**, the **angular** at the posterior corner of the mandible and the **surangular** above. Many of these bones wrap around the medial side of the mandible and meet the **prearticular** and one or several **coronoids** to complete the medial mandibular wall. Left and right mandibles usually meet anteriorly at the midline in a **mandibular symphysis**. If firm, the mandibular symphysis unites them into an arched unit. However, in snakes, the mandibular symphysis is composed of soft tissues, permitting independent movement of each mandible.

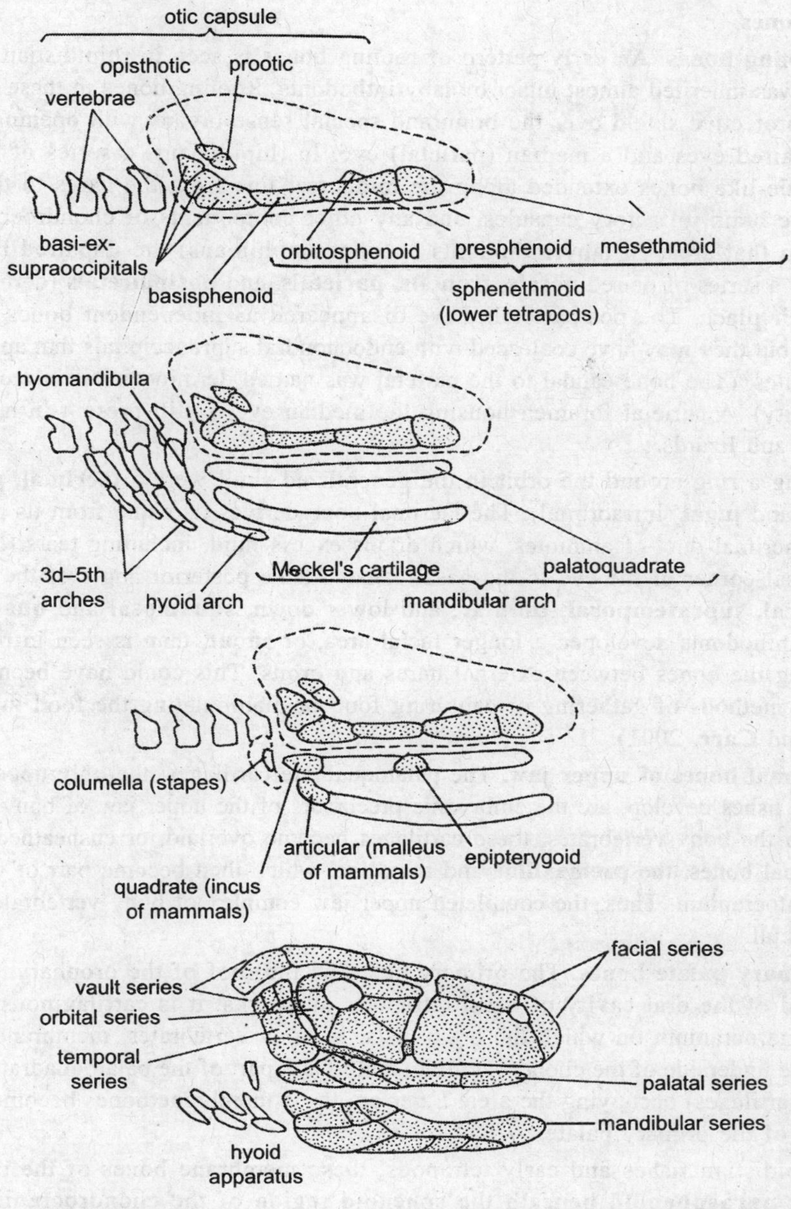

Fig. 3.16. Contributions to the skull. The chondrocranium establishes a supportive plateform that is joined by contributions from the splanchnocranium, in particular the epiterygoid. Other parts of the splanchnocranium give rise to the articular, quadrate, and hyomandibula, as well as to the hyoid apparatus. The dermatocranium encases most of the chondrocranium together with contributions from the splanchnocranium (after Kardong 2006).

The generalized dermatocranium can be divided into four types of bones: 1. bones that form above and along side of the brain and chondrocranium (**roofing bones**); 2. dermal bones of the upper jaw (**marginal bones**); 3. dermal bones of the **primary palate**; and 4. **opercular bones**.

1. Roofing bones. An early pattern of roofing bones is seen in rhipidistian fishes and this pattern was inherited almost intact by labyrinthodonts. Roofing bones in these vertebrates provided a protective shield over the brain and special sense organs with openings only for the nares, paired eyes and a median (parietal) eye. In rhipidisians, a series of paired and unpaired scale-like bones extended along the middorsal line from the nares to the occiput, overlying the brain, olfactory capsules, and any other components of chondrocranium that developed in that area. In labyrinthodonts (extinct amphibians) the unpaired bones were missing, and a series of paired **nasals**, **frontals**, **parietals**, and **postpairetals** (dermoccipitals) too take their place. The postparietals have disappeared as independent bones in modern amphibians, but they may have coalesced with endochondral supraoccipitals that appeared first in early reptiles. (The bone caudal to the parietal was named dermioccipital in recognition of this possibility). A parietal foramen housing the median eye is still present in many fishes, amphibians, and lizards.

Forming a ring around the orbit in the generalized skull were a **lacrimal**, **prefrontal**, **postfrontal** and **jugal** (infraorbital). The lacrimal bone derives its name from its relationship to the nasolacrimal duct of amniotes, which drains excess fluid, including tears (lachrymae), from the nasal corner of the eye to the nasal canal. At the posterior angle of the skull were **intertemporal**, **supratemporal**, **tubular**, and lower down, **squamosal** and **quadratojugal** bones. Labrinthodonts developed a longer facial area, or **snout**, than is seen in rhipidistians by elongating the bones between external nares and orbits. This could have been correlated with altered methods of gathering or capturing food or manipulating the food in the mouth (see **Kent** and **Carr**, 2001).

2. Dermal bones of upper jaw. The palatoquadrate cartilages, the only upper jaws that cartilaginous fishes develop, are the embryonic precursors of the upper jaw of bony vertebrates (Fig. 3.7). In the bony vertebrates, these cartilages become overlaid, or ensheathed, by tooth-bearing dermal bones, the **premaxillae** and **maxillae**, which then become part of the margins of the dermatocranium. Thus, the completed upper jaw complex of bony vertebrates becomes part of the skull.

3. Primary palate bones. The primary palate is the roof of the oropharyngeal cavity of fishes and of the oral cavity of basal tetrapods. In sharks, it is cartilaginous, being the floor of chondrocranium on which the brain rests. In bony vertebrates, membrane bones are applied to the underside of the chondrocranium and to any part of the palatoquadrate cartilages (upper jaw cartilages) occupying the area. Later on, these membrane bones become the major components of the primary palate.

In rhipidistian fishes and early tetrapods, these membrane bones of the palate were an unpaired **parasphenoid** beneath the sphenoid region of the chondrocranium, paired **vomers** beneath the ethmoid region, and paired **palatines**, **ectopterygoids**, and **pterygoids** (Fig. 3.17). The last three membrane bones have invested the palatoquadrate cartilage to one degree or another. Primitively, teeth formed on all these bones, and several of them still bear teeth in extant basal vertebrates. Internal nares pierced the palate anterolaterally. A primary palate is still present with modifications in all tetrapods (Fig. 3.21), but in those that also develop a secondary palate (e.g., mammals) the primary palate remains in the roof of the nasal passageways (Fig. 3.29).

4. Opercular bones. The operculum is a flap of tissue that arises as an outgrowth of the hyoid arch and extends caudad the gill slits. It is membranous in holocephalans and absent in elasmobranchs. In bony fishes, operculum is stiffened by squamous plates of dermal bones. Most constant membrane bones are a large **opercular**, smaller **preoperculars** overlying the site of articulation of the upper and lower jaws, **suboperculars** and **interoperculars**. In some of the more basal bony fishes (*e.g.*, *Polypterus*), one or more **gular bones** lie in the opercular membrane in the floor of the opercular chamber. They are represented in more specialized ray-finned fishes and dipnoans by **branchiostegal rays** located in caudally directed flaps, the **branchiostegal membranes** of the operculum. No opercular or gular plates or vestiges of them remain in tetrads.

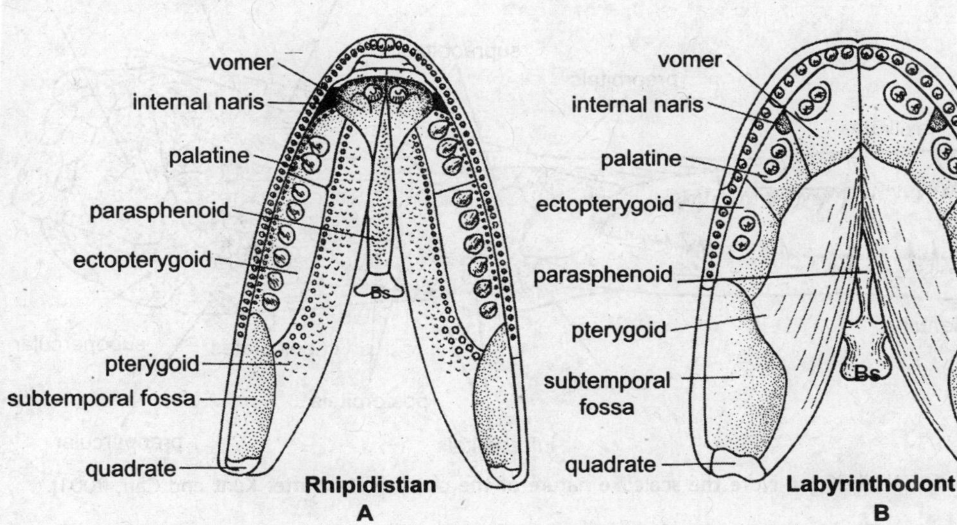

Fig. 3.17. Primary palate (shaded part) of a rhipidistian (A) and a late Paleozoic labyrinthodont (B). Small denticles cover the parasphenoid and pterygoid bones in A. Bs, basisphenoid (after Kent and Carr, 2001).

Chondrocranial – Dermatocranial Complex of Bony Fishes

Basal actinopterygians. Sturgeons (*Acipenser*) and spoonbills or paddlefishes (*Polyodon*) have a skull that justifies the name chondrostei. The chondrocranium remains almost completely cartilaginous throughout life. The only traces of endochondral ossification in the chondrocranium of sturgeons are isolated sites in the otic capsules and in the sphenoid bone where it contributes to the walls of the orbit. Elsewhere, the chondrocranium remains cartilaginous. The spoonbill is not much different. Its unusual bill is an extension of the cartilaginous rostrum. Numerous overlying dermal bones hide the cartilaginous components of the upper jaws and hyoid arch.

Polypterus (ganoid fish) retains more of the features of Paleozoic fishes than do sturgeons and spoonbills. The chondrocranium is well ossified and is overlaid in the middorsal line by paired nasals, parietals and postparietals of dermal origin. Lateral to these is a linear series of small, mostly unnamed bony dermal plates (bones). Premaxillae and maxillae overlie the endochondral palatoquadrate bones of the first visceral arch, and a preopercular bone overlies the site of articulation of the upper and lower jaws. Other opercular bones complete the dermatocranium bones of dermal origin lie in the opercular membranes.

Basal neopterygians. The skulls of *Amia* (bow-fin) and gars (*Lepidosteus*) bear little resemblance to the skulls of most modern fishes (Fig. 3.10 and Fig. 3.18). The dermal bones are grooved and pitted (*i.e.*, having hollows in the surface), exhibiting the sculpturing effects of the overlying skin. They can be charactrerized equally well as skull bones or dermal scales. The cheek plates of gars are especially compelling evidence that the dermatocranium may have been derived from earlier dermal scales. The chondrocranium of *Amia* remains highly cartilaginous (**Jarvik**, 1980); in gars it is bony.

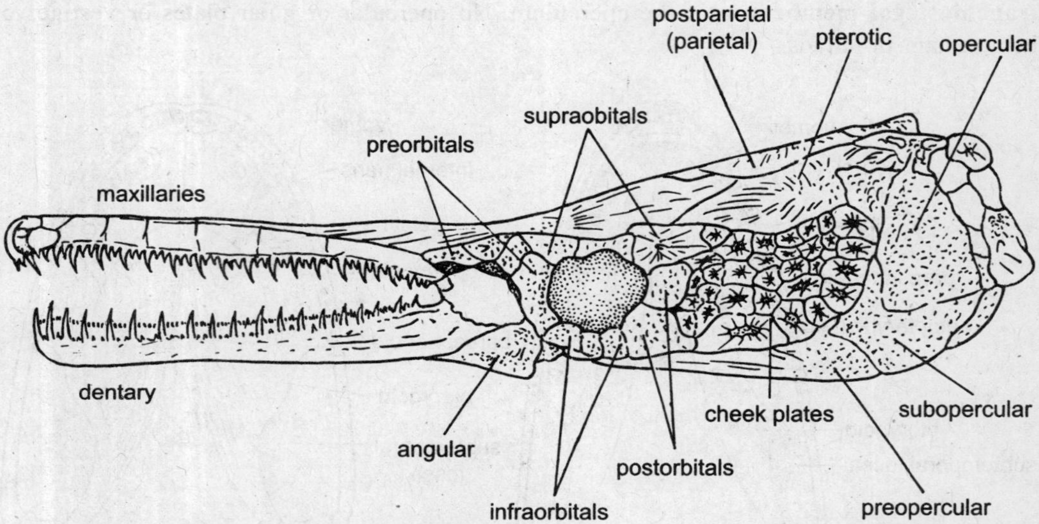

Fig. 3.18. Skull of gar. Note the scalelike nature of the cheek plates (after Kent and Carr, 2001).

Teleosts. The skull of most teleosts are highly specialized and architecturally diverse. Such a diversity in the skull of teleosts is correlated with the diverse feeding habits of the group, which includes species that harvest plant tissues or capture animal life in every conceivable niche of the streams, lakes and oceans of the world. The anatomic diversity of skull of bony fishes is due to a combination of highly maneuverable jaws and palates, coupled with the largest number of bones in any modern vertebrate skull.

The skull of carp, like many teleost skulls, is compressed laterally and vaulted (arched) dorsally (Fig. 3.19). The maxillae, premaxillae, dentary, articular, quadrate and symplectic are associated with the visceral skeleton. The posttemporal is the dorsalmost segment of the pectoral girdle. It has been secondarily incorporated into the skull. The mesethmoid, epiotic and pterotic often incorporate some membrane bone along with the usual replacement bone. Branchiostegal rays in paired branchiostegal membranes beneath the gill chamber take the place of the more primitive gular bones of early ray-finned fishes.

The chondrocranium in most teleosts is fully ossified except for the olfactory capsules. However, in cyprinids (carps), chondrocranial ossification is less than complete, and the islands of cartilage seen on the surface of the carp skull are unossified regions of the chondrocranium, or at some sites, new cartilage.

Despite the kinetic nature of many teleost palates and the architectural disruptions that this has entailed, the dermal bone complement of the palate reflects that of ancestral bony fishes. The vomer, parasphenoid, and pterygoid bones are common components.

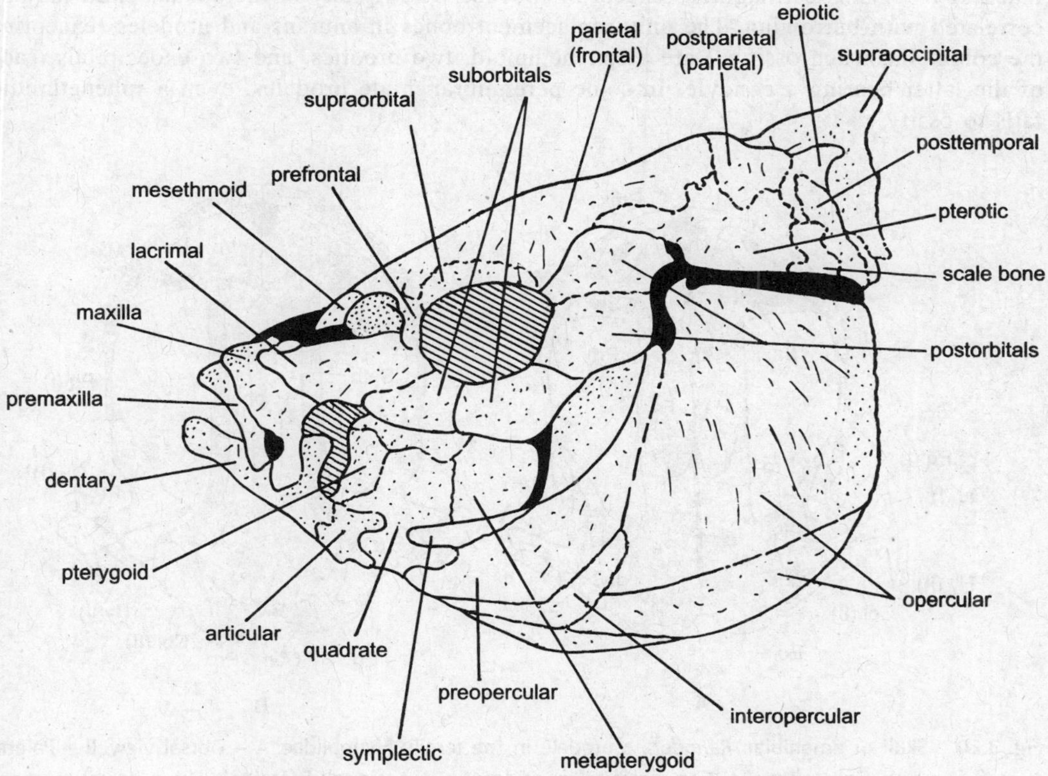

Fig. 3.19. Skull of teleost (carp). Dark and hatched areas show unossified cartilage.

Further, names given to bones of the skull many years ago have been assigned on the basis of location (palatine), shape (pterygoid), or some other characteristic (squamosal) rather than on the basis of whether they are endochondral or intramembranous in origin. A bone of the same name in two species may be of endochondral origin in one speices and of dermal origin in another (**Kent** and **Carr** 2001).

Dipnoans. After their first appearance in the early Devonian, dipnoans (*i.e.*, lung fishes such as *Neoceratodus*, *Protopterus* and *Lepidosiren* (see **Agarwal**, 2016) did not experience the burst of speciation subsequently undergone by teleosts. Therefore, changes in the architecture of dipnoan skulls have been more moderate. With reference to the dermatocranium, time has taken lung fishes in the direction quite different from that of teleosts, whose *dermal bones increased in number and decreased in size*. The dermatocranium of lungfishes has evolved from a large number of scale-like dermal bones in fossils to a relatively few broad bony plates in today's species. The chondrocranium is largely cartilaginous like that of some chondrosteans and basal neopterygians, rather than bony, as in teleosts. The palate accommodates openings of the nasal canal into the oral cavity just behind the mouth, as in other rhipidistians.

Chondrocranial – Dermatocranial Complex of Modern Tetrapods

1. Amphibians. The skulls of modern amphibians are considerably modified from those of labyrinthodonts, although they are still flattened (**platybasic**) compared with the vaulted skulls of amniotes. The chondrocranium is incomplete dorsally (*Sphenodon*, Fig. 3.5 D), and much of it remains cartilaginous except in apodans. The rigidity of the apodan skull may be correlated with burrowing. The only replacement bones in anurans and urodeles (excepting the columella, when ossified) are a sphenethmoid, two prootics, and two exoccipitals, each of the latter bearing a condyle. In some perennibranchiate urodeles, even a sphenethmoid fails to ossify.

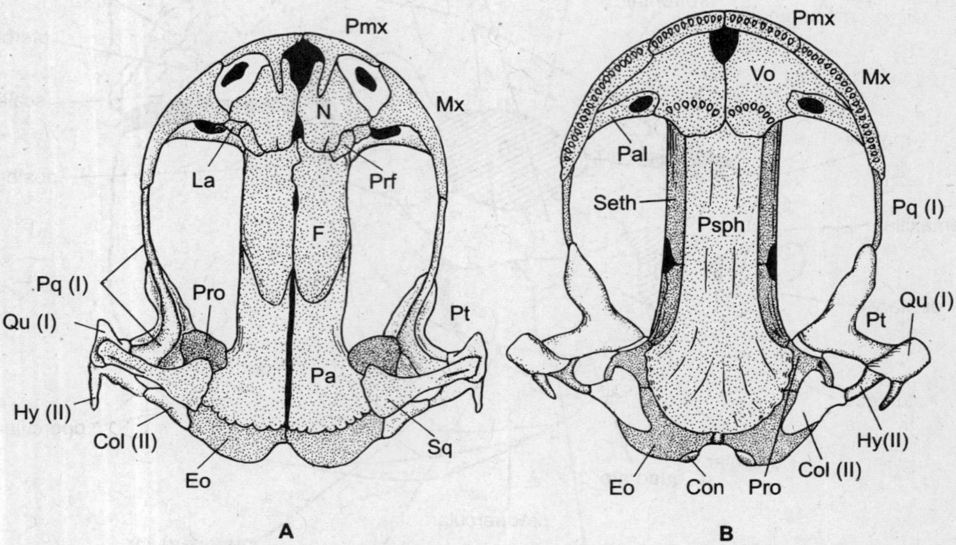

Fig. 3.20. Skull of amphibian *Ranodon*, a urodele in the family Hynobiidae. A – Dorsal view. B – Palatal view. **Col**, columella; **Con**, occipital condyles; **Eo**, exoccipital; **F**, frontal; ;**Hy**, a dorsal segment of the hyoid skeleton; **La**, lacrimal; **Mx**, maxilla; **N**, nasal; **Pa**, parietal; **Pmx**, premaxilla; **Pq**, palatoquadrate cartilage; **Prf**, prefrontal; **Pro**, prootic; **Psph**, parasphenoid; Pt, pterygoid; **Qu**, quadrate; **Seth**, sphenethmoid; **Sq**, squamosal; **Vo**, vomer; **I** and **II** indicate origin from the first or second visceral arch (after Kent and Carr, 2001).

In amphibians and other tetrapods, abutting against the otic capsule is a skeletal rod, the **columella** (stapes), a middle ear ossicle that conducts sound waves from an eardrum to the capsule.

The dermatocranium of amphibians is quite incomplete. The bones that surrounded the orbit of early amphibians have been lost except for the lacrimals and prefrontals of basal urodeles (Fig. 3.20 A). Also missing are the primitive bones of the temporal region from the orbits cauded (*i.e.*, intertemporals, supratemporals, tubulars, and postparietal). This loss of ensheathing bones leaves the optic capsule (prootic) exposed dorsally and laterally. Only the squamosal and sometimes a quadratojugal remain in this region. Premaxillae and maxillae are usually present to ensheath the upper jaw cartilages at the margin of the dermatocranium. However, in perennibranchiates even maxillae may fail to develop.

In amphibians, the primary palate has been altered. In anurans and a few urodeles, large palatal vacuities ("empty places") have evolved beneath the orbit, reducing the palatines to transverse splinters that brace the upper jaw against the palate anteriorly and reducing the pterygoids to a pair of bipartite (*i.e.*, consisting of two parts) bones that brace the upper jaw with the braincase posteriorly (Fig. 3.20 B and Fig. 3.21, *Rana*). Because of the palatal vacuities, the large eye balls of anurans, which protrude just above the waterline to monitor the environment in an otherwise submerged frog, can be retracted until the membranous floor of the orbit bulges into the oral cavity. In urodeles, the parasphenoid has become exceptionally broad (Fig. 3.20 B and Fig. 3.21, *Necturus*) and palatal vacuities are not present in some amphibian orders.

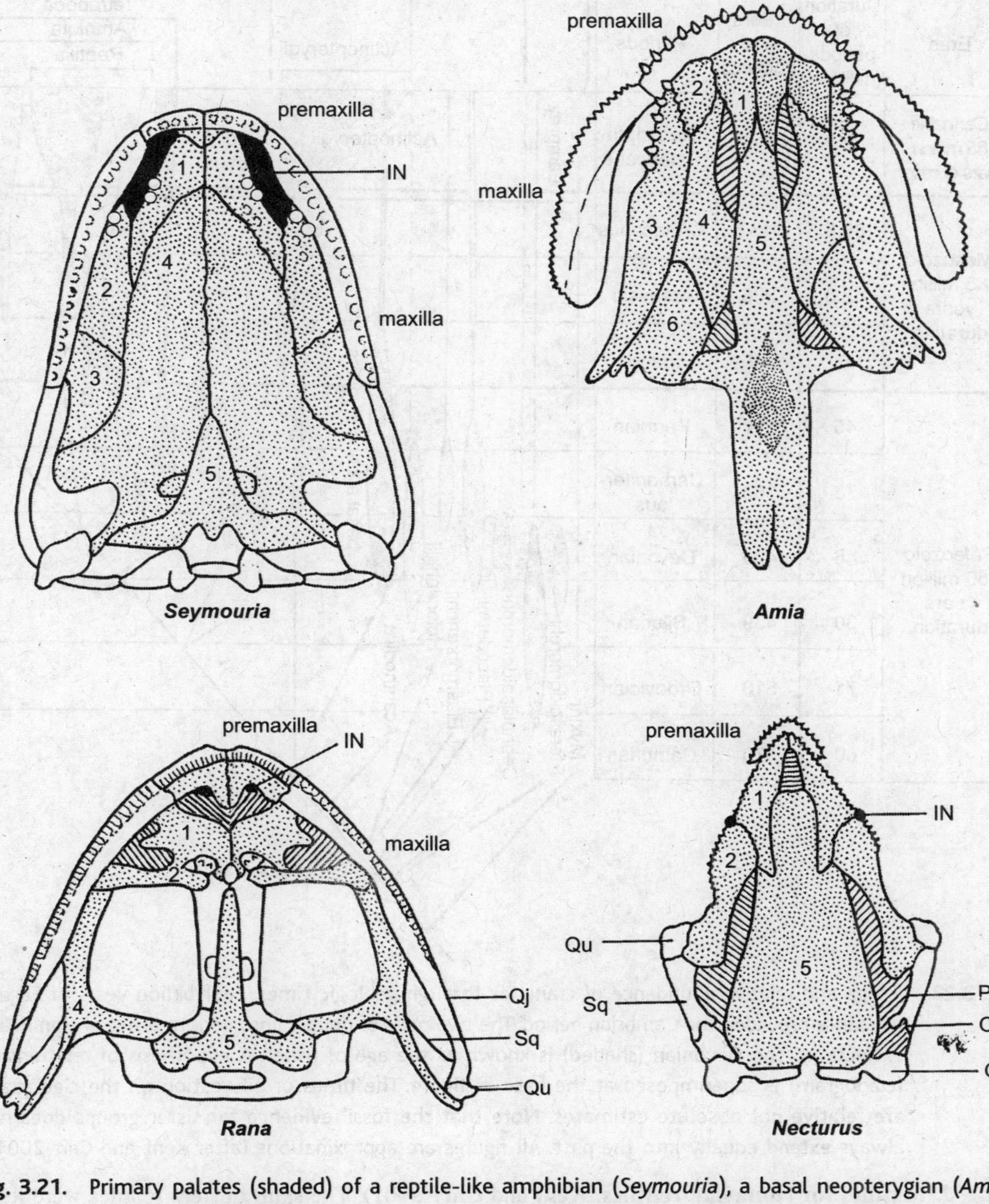

Fig. 3.21. Primary palates (shaded) of a reptile-like amphibian (*Seymouria*), a basal neopterygian (*Amia*) and two amphibians (*Rana* and *Necturus*). Cartilage is indicated by diagonal lines; internal nares are black. **1**, vomer; **2**, palatine (called palatopterygoid in *Necturus*); **3**, ectopterygoid; **4**, pterygoid; **5**, parasphenoid; **6**, epipterygoid; **Oc**, cartilaginous portion of otic capsule; **Op**, opisthotics; **Qu**, quadrate; **Qj**, quadratojugal; **Sq**, squamosal; **IN**, internal nares (after Kent and Carr, 2001).

Eras	Duration of periods (millions)	Years ago (millions)	Periods
Cenozoic 65 million years duration	65	65	Recent to Paleocene
Mesozoic 185 million years duration	80.5	145.5	Cretaceous
	62.5	208	Jurassic
	37	245	Triassic
Paleozoic 350 million years duration	45	290	Permian
	73	363	Carboniferous
	46	409	Devonian
	30	439	Silurian
	71	510	Ordovician
	60	570	Cambrian

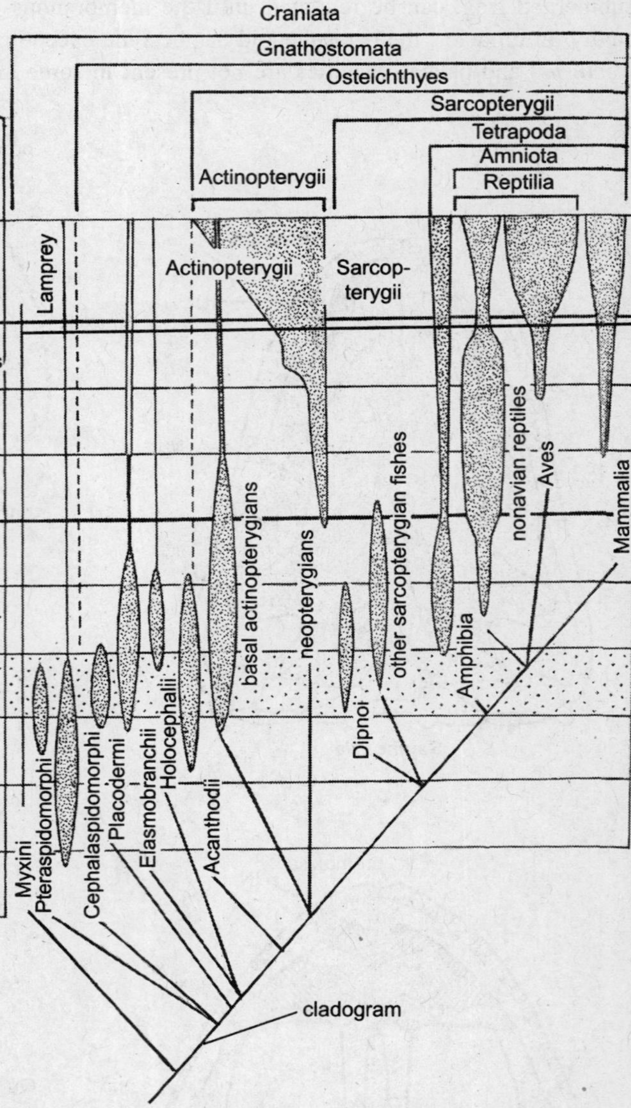

Fig. 3.22. Range and relative abundance of craniates through geologic times. Four billion years of Earth's existence preceded the Cambrian period. The two pairs of double lines denote times of significant extinctions. The Devonian (shaded) is known as the age of fishes. A hypothesis of relationship (cladogram) is superimposed at the base of figure. The times of bifurcation on the cladogram are relative not absolute estimates. Note that the fossil evidence for sister groups does not always extend equally into the past. All figures are approxinations (after Kent and Carr 2001).

2. Reptiles (or Nonavian reptiles, Kent and **Carr** 2001). The skull of stem reptiles were little changed from those of labyrinthodonts and some primitive features are still present in living reptiles. Among these are a well-ossified chondrocranium, a single occipital condyle, a large complement of membrane bones, and in rhynchocephalians and many lizards, a parietal foremen housing median eye, *e.g.*, *Sphenodon* (Fig. 3.23 C) and iguana lizard. Major changes from stem reptile skulls included the appearance of temporal fossae and development of a partial or complete secondary palate. Temporal fossae and secondary palate were inherited by birds (Fig. 3.22).

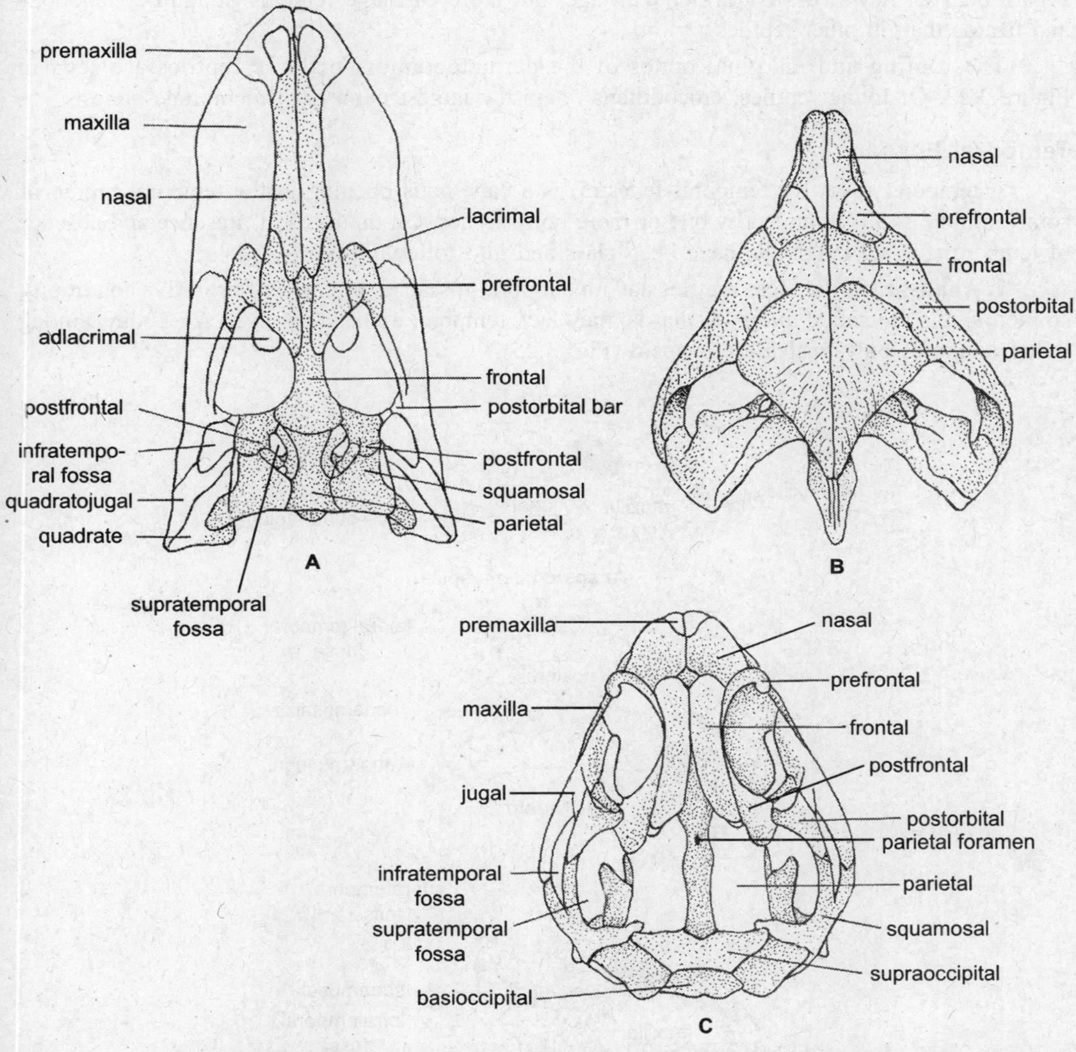

Fig. 3.23. Roofing bones (light shade) and marginal bones of the skulls of three reptiles. A – Diapsid skull of alligator; B – Derived anapsid skull of the snapping turtule, *Marcroclemys temminckii*. C – Diapsid skull of juvenile *Sphenodon* (Rhyncocephalia). *Dark shade*, unossified regions of chondrocranium. The supraoccipital and basioccipital are ossified parts of the chondrocranium (after Kent and Carr 2001).

The fully ossified chondrocranium of an adult alligator consists of two exoccipital bones, a supraoccipital, a basioccipital that bears a single condyle for articulation with the atlas vertebra, laterosphenoids, ethmoids, and several otics that arise from separate ossification centers in the otic capsules. None of the otic bones are on the surface, being overlaid by the squamosal, a dermatocranial bone (Fig. 3.23 A). Some of the otic bones are coalesced with adjacent occipital. The chiefly cartilaginous chondrocranium of a juvenile *Sphenodon* is seen in situ in Figure 3.23 C. It will ossify further with age, but more cartilage remains in adult sphenodons and lizards than in other reptiles.

The roofing and marginal bones of the dermatocranium of three reptiles are seen in Figure 3.23. Of living reptiles, crocodilians retain the largest number of membrane bones.

Temporal Fossae

A temporal fossa (or temporal fenestra) is a cavernous opening in the temporal region of some amniote skulls bounded by one or more bony arches. On the basis of presence and absence of temporal fossae, the skulls have been classified into following types:

1. Anapsid skulls. Stem reptiles had no temporal fossae (a retention of primitive condition), so their skulls are called **anapsid**, that is, they lack temporal arches (Fig. 3.24 A). Today, among living amniotes only turtles are anapsid (Fig. 3.25).

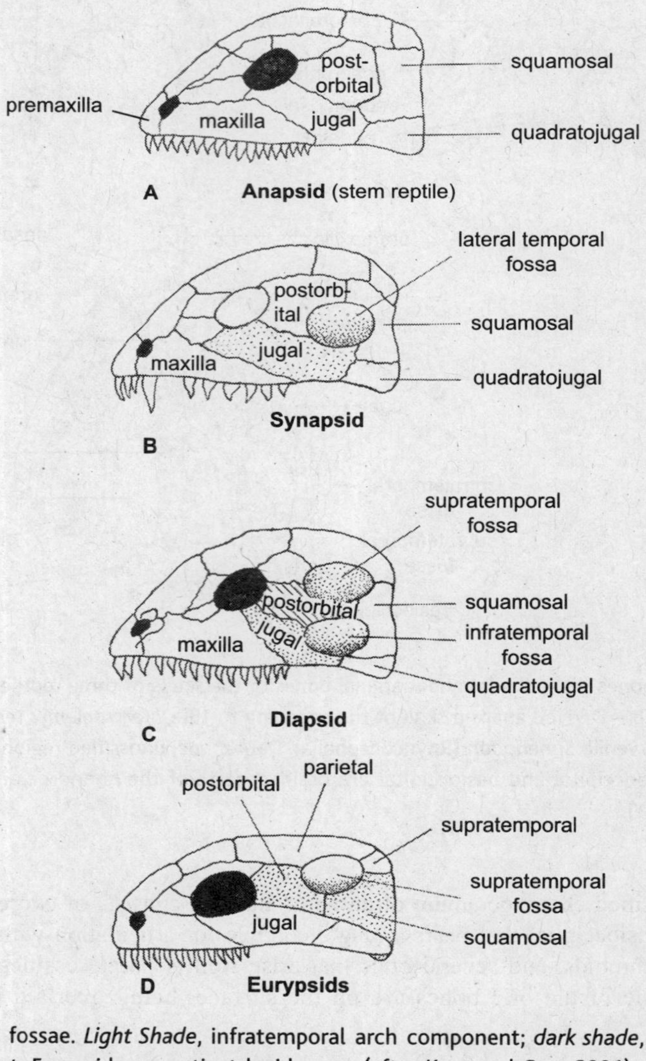

Fig. 3.24. Temporal fossae. *Light Shade*, infratemporal arch component; *dark shade*, supratemporal arch component. Eurypsids are extinct lepidosaurs (after Kent and Carr, 2001).

2. Synapsid skull. Synapsida developed a single **lateral temporal fossa** surrounded by postorbital, squamosal and jugal bones (Fig. 3.24 B). The last two bones form and underlying **infratemporal** (zygomatic) **arch**. This synapsid skull was transmitted to mammals. In humans, the cheek-bones are the zygomatic arches.

3. Diapsid skull. Archosaurs and predecessors of *Sphenodon*, lizards, and snakes had superior and inferior temporal fossae (Fig. 3.24 C). When two fossae are present, the skull is said to be **diapsid**. The lower arch corresponds to the zygomatic arch of mammals. The upper arch is called **supratemporal** and it exists between the two fossae. It consists of parts of the postorbital and squamosal bones. Crocodilians and sphenodon still have diapsid skulls (Fig. 3.23 and 3.26 A), but modern lizards have lost part or all of the lower arch, and snakes have lost both arches (Fig. 3.26 B, C). This loss has left a cavernous void in the posterolateral walls of squamate skulls. These are the **modified diapsid skulls**. Reduction or loss of the arches along with gaining of intracranial joints facilitated **cranial kinesis**, the movement of one section of skull independent of others.

Ichthyosaurs and plesiosaurs (Fig. 3.27) had one temporal fossa, dorsally located, that resembled the dorsal fossa of diapsids in its anatomic relationship (compare Fig. 3.24 C and 3.24 D). However, it is likely that this **euryapsid** condition is an example of evolutionary convergence in lineages distinct from that leading to other reptiles. All euryapsids reptiles are extinct (Fig. 3.27).

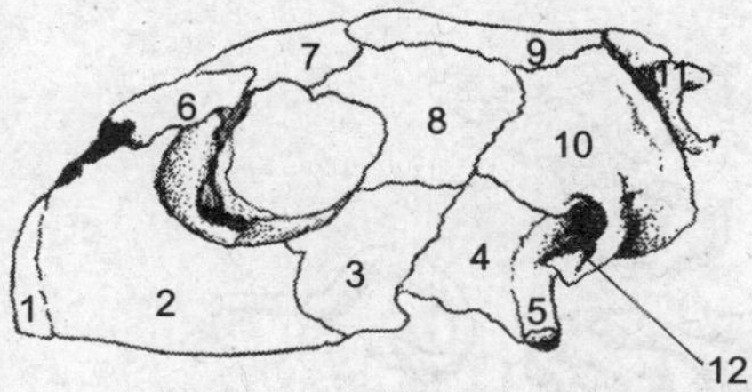

Fig. 3.25. Anapsid skull of small sea turtle, lateral view, lower jaw removed. **1**, premaxilla; **2**, maxilla; **3**, jugal; **4**, quadratojugal; **5**, quadrate; **6**, prefrontal; **7**, frontal; **8**, postorbital; **9**, parietal; **10**, squamosal; **11**, supraoccipital; **12**, middle ear cavity (after Kent and Carry 2001).

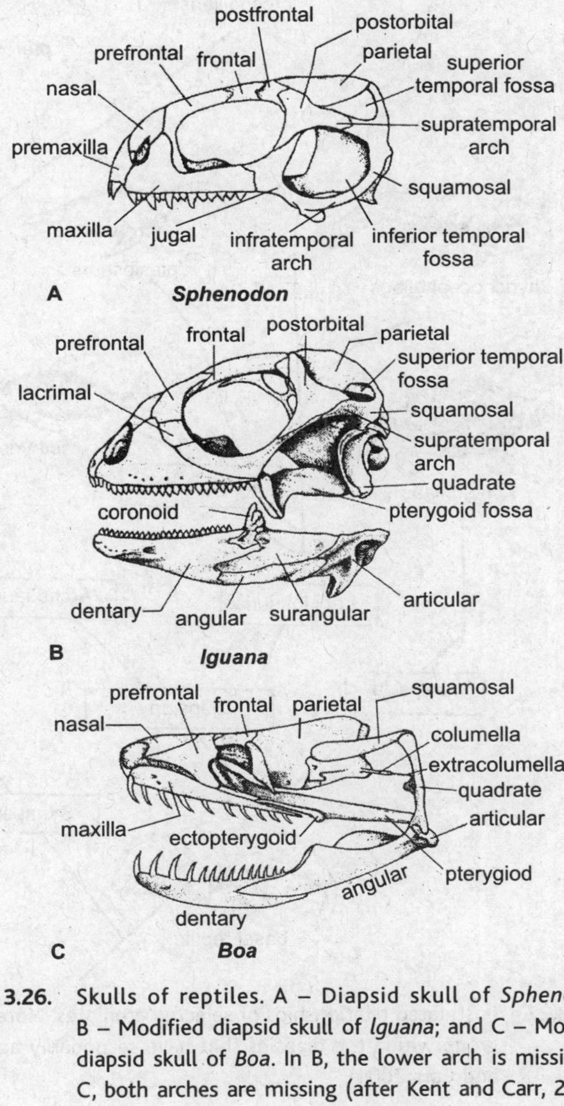

Fig. 3.26. Skulls of reptiles. A – Diapsid skull of *Sphenodon*; B – Modified diapsid skull of *Iguana*; and C – Modified diapsid skull of *Boa*. In B, the lower arch is missing. In C, both arches are missing (after Kent and Carr, 2001).

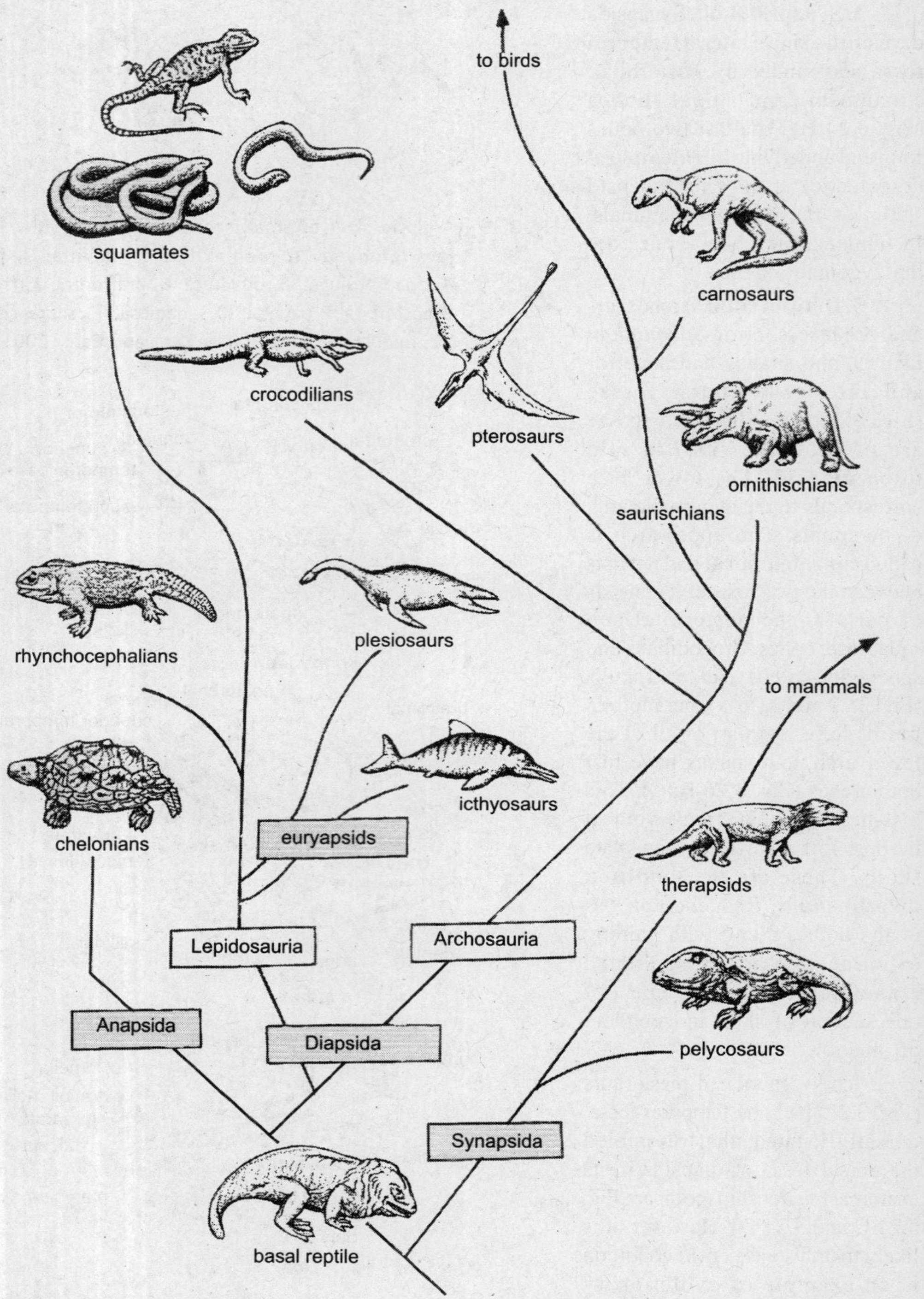

Fig. 3.27. Postulated relationships of selected amniotes. Note that the euryapsids represent a paraphyletic group within the diapsids that have secondarily acquired a single temporal fenestra (after Kent and Carr 2001).

The temporal region of a turtle skull is an riddle. Absence of temporal fossae suggests a primitive condition. However, there has been extensive loss of dermal bones and excavation of the temporal region dorsal to the otic capsule (Fig. 3.23 B), a condition more distinct in some families than in others. Supratemporal, tabular and postparietal bones are missing, postorbitals and postfrontals have united, and parietals give the impression of having regressed from the rear, leaving a wide vacuity that is best seen from a caudal or ventral view.

Advantages of temporal fossae. Temporal fossae provide space and surfaces in functionally advantageous positions for accommadating the powerful adductor muscles needed to operate the lower jaws of amniotes. In labyrinthodonts and extinct basal reptiles, the **adductor mandibulae** (chief levator muscle of the lower jaw) was confined to cramped quarters internal to the temporal region of the dermatocranium. (In cartilaginous fishes, it lies just under the skin of the head because there is no dermatocranium). The adductor mandibulae muscle enabled bony fishes and early tetrapods to seize food, bite off pieces, and close the mouth to prevent return of the food to the environment. The food was swallowed whole. Temporal fossae provided space for an expanding amniote adductor muscle to shorten and thicken during contraction; they provided access to the surface of the dermatocranium so that one portion of the adductor, called **temporalis muscle**, could acquire more leverage by spreading upward on to the temporal region of the skull; and they provided a zygomatic arch to which another portion of the adductor, called **masseter**, could acquire an anchorage. Figure 3.28 shows the temporalis and masseter muscles in relation to the mandibular and hyoid arches, make possible the complex side-to-side, forward-backward, and rotary chewing movements seen in herbivorous mammals that grind grasses or chew their cud and in carnivores that macerate flesh and crush bones.

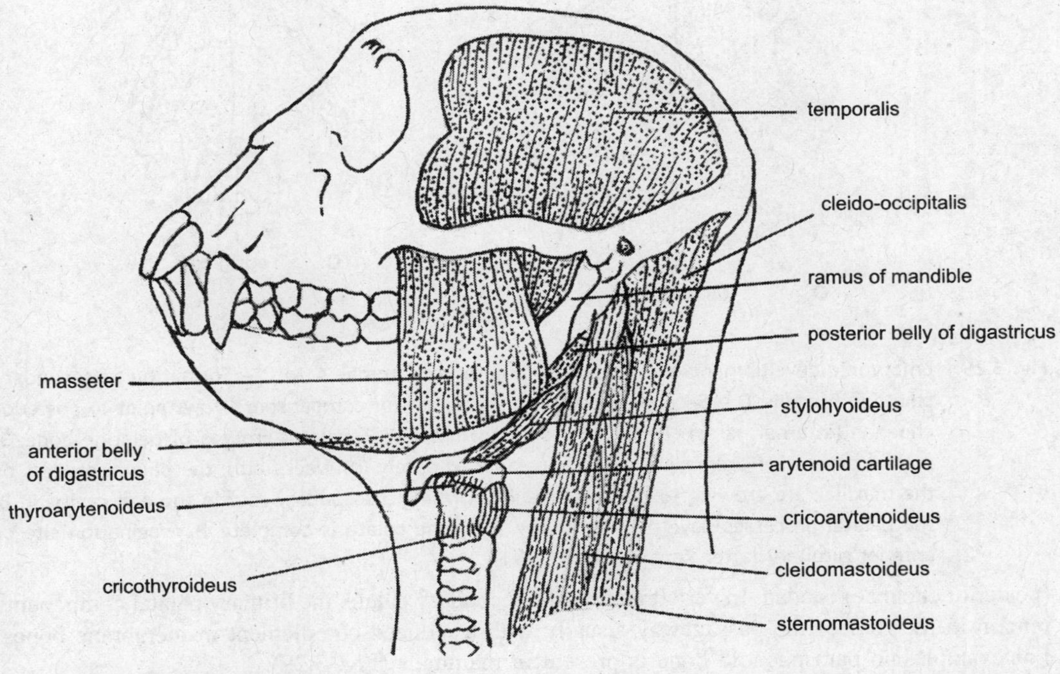

Fig. 3.28. Branchiomeric muscles of male rhesus monkey. The temporalis inserts on the coronoid process of the ramus of the mandible. The pterygoideus muscle can be seen only after the mandibular symphysis has been separated and dentary bone retracted laterad.

Secondary Palate

A **secondary palate** is a horizontal partition that partially or completely divides the primitive oral cavity into separate oral and nasal passageways, thereby displacing the **internal nares**

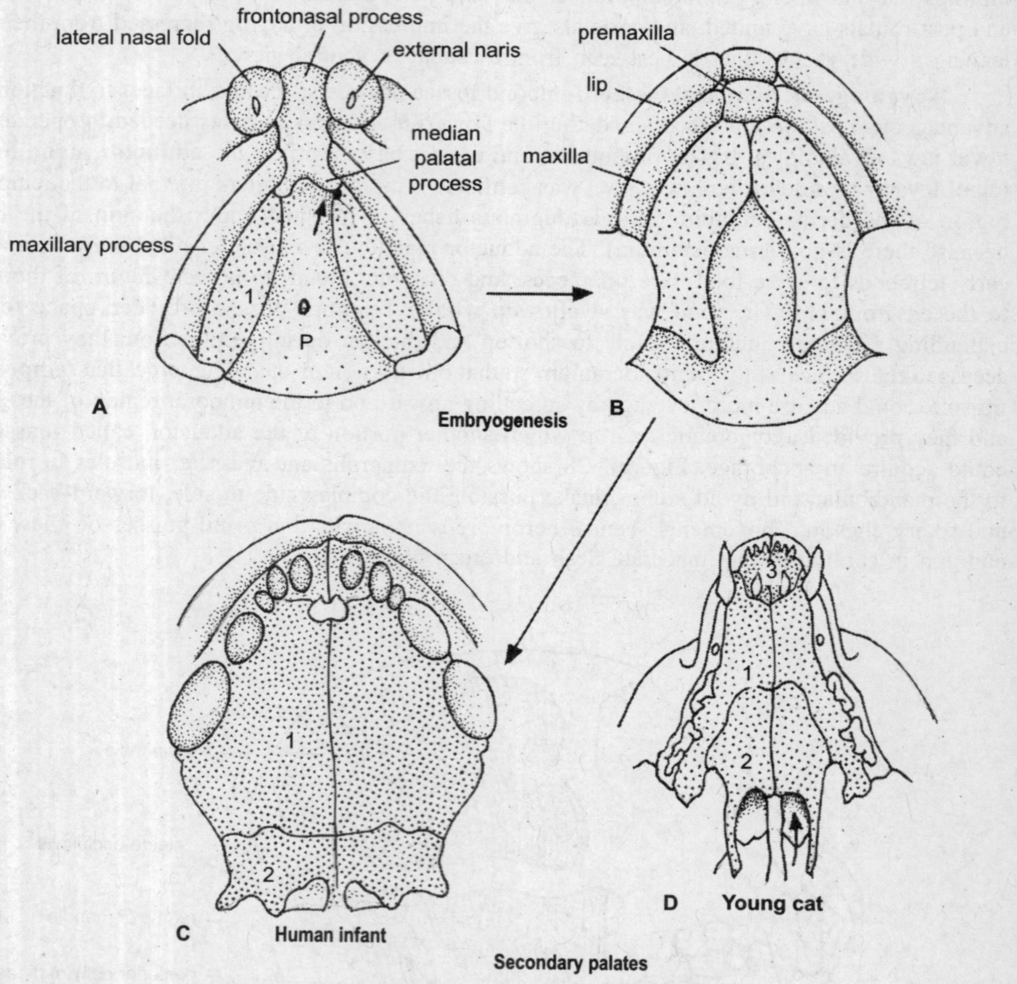

Fig. 3.29. Embryonic development of a secondary palate in mammal. A to C – Formation of secondary palate in humans. D – Secondary palate of young cat for comparison. *Arrows* point to posterior choanae (internal nares). **1**, palatal process of maxilla; **2**, palatal process of palatine bone; **3**, palatal process of premaxilla. In A (fetus approximately 18 weeks old), the palatal process of the maxillae are growing toward the midline, forming a secondary roof in the oral cavity. In B, the palatal processes have met anteriorly. In C, the palate is complete. **P**, evagination site of anterior pituitary (after Kent and Carr 2001).

(posterior choanae) caudad. In vertebrates with a secondary palate, the primary palatal components remain in the roof of the passageway, usually with a reduced complement in membrane bones. For example, no parasphenoid bone is present in mammals (Fig. 3.29).

In crocodilians, medially directed, shelflike **palatal processes** of the premaxillae, maxillae, palatine, and pterygoid bones meet the midline to form a long bony secondary palate with internal nares far to the rear. In other reptiles, not all the palatal processes reach the midline, so the

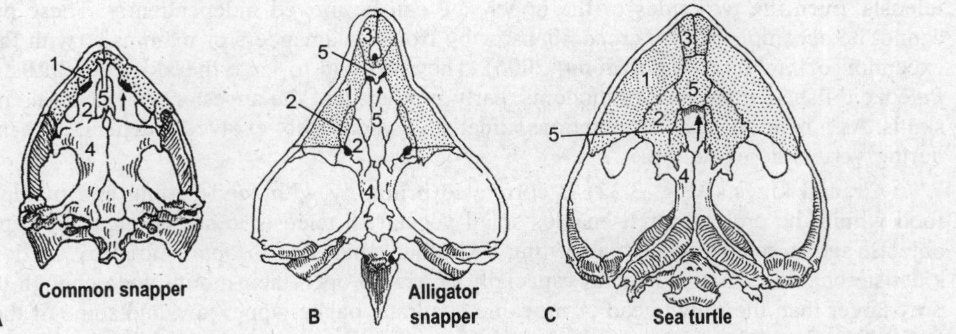

Fig. 3.30. Species differences in the secondary palates (dot shaded) of turtles. A – *Chelydra serpentina*. B – *Macroclemys temminckii*. C – *Lepidochelys olivacea*, temporal and supraoccipital regions omitted. **1**, maxilla; **2**, palatine; **3**, premaxilla; **4**, pterygoid; **5**, vomer. In A, only the maxilla has the palatal processes, and these are rudimentary. *Arrows* point to an internal naris (after Kent and Carr, 2001).

secondary palate is incomplete. Varying degrees of completion can be seen by examining several species of turtle (Fig. 3.30). When the secondary palate is incomplete, the respiratory air stream is channeled in a fairly deep longitudinal groove, the **palatal fissure**, in the roof of oral cavity (Fig. 3.30 B, C). The borders of the fissure are fleshy **palatal folds**, seen also in birds. In mammals, the secondary palate extends all the way to the pharynx, but the caudal portion – the soft palate lacks bone.

Cranial kinesis. Kinesis means movement. **Cranial kinesis** or **kinetism**, is the movement of a functional component of a skull independent of another component. It is made possible by the presence of movable intracranial joints between the two components.

Teleosts, lizards, snakes and birds are especially well endowed with such joints. They are said to have **kinetic skulls**. Animals having kinetic skulls raise the upper jaw and palate as a unit independent of the chondrocranium when they open their mouth (Fig. 3.31). In some

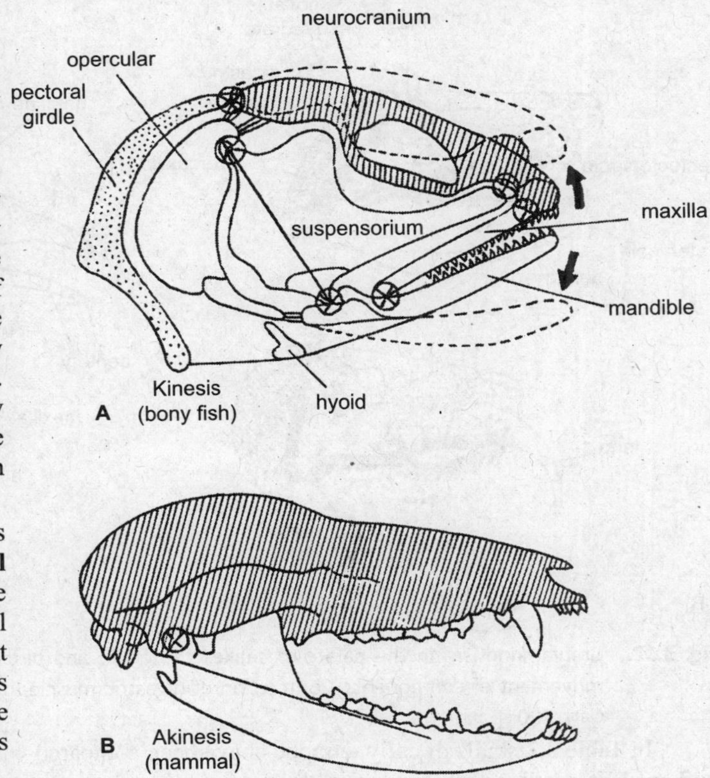

Fig. 3.31. Mobility of skull bones. A – The fish skull is kinetic. The upper jaw and other lateral skull bones rotate upon each other in a linked series, resulting in displacements of these bones (dashed outline) during feeding. Circles represent points of relative rotation between articulated elements. B – The mammal skull is akinetic because no relative movement occurs between the upper jaw and the braincase. There are no hinge joints through the braincase nor any movable linkages of lateral skull bones (after Kardong, 2006).

teleosts, even the two sides of the upper jaw can be moved independently. These movements cannot be accomplished by crocodilians or by frogs, salamanders or mammals (with the possible exception of rabbits, see **Kardong**, 2006). They are said to have the **akinetic skulls**. However, Paleozoic fishes, many labyrinthodonts, early reptiles, and the ancestors of mammals had kinetic skulls. As a result of chance mutations, kinetism has probably evolved independently many times during vertebrate evolution.

Cranial kinesis (Fig. 3.32) is correlated primarily with food-getting and manipulation of food within the oral cavity. It enables small vegetarian teleosts to scrape algae from underwater objects, and it enables plankton-feeding and other small fish to obtain food by suction. Cranial kinetism enables some squamates especially vipers, to open their mouth wide enough to swallow prey larger than their own head. A movable quadrate, palate, upper jaw, and some of the bones in front of or above the orbit contribute to the process, along with a hinge joint between the kinetic unit and rest of the braincase. The palate and temporal region of squamates and some species of birds have become so modified for feeding that primitive dermatocranial – splanchnocranial complex has become greatly disrupted (Fig. 3.33).

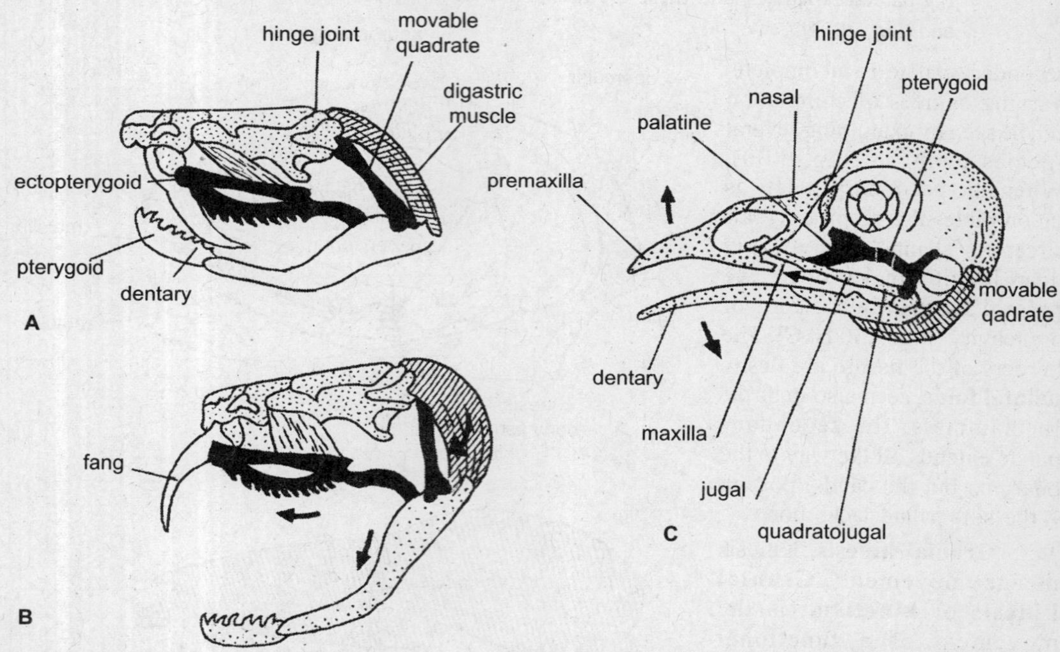

Fig. 3.32. Cranial kinetism in the palate of snakes (A and B) and birds (C). Arrows indicate direction of movement of components. Contraction of digastric muscle lowers the mandible (after Kent and Carr 2001).

In Table 3.2 skulls of early tetrapods have been compared with skulls of modern amphibians and modern reptiles with respect to the palates and other traits.

Table 3.2 Skulls of early tetrapods contrasted with those of modern amphibians and reptiles with reference to a few selected characteristics (Source. Kent and Carr 2001).

S. No.	*Characteristics*	*Early tetrapods*	*Modern reptiles*	*Modern amphibians*
1.	Chondrocranium	Well ossified One condyle	Well ossified One condyle	Mostly cartilage Two condyle
2.	Primary palate	Complete complement of dermal bones	Relatively complete	Fewer

S. No.	Characteristics	Early tetrapods	Modern reptiles	Modern amphibians
		Parasphenoid small	Small	Large in urodeles
		Vacuity small	Small	Large in anurans
3.	Secondary palate	None	Partial or complete	None
4.	Temporal fossae	None	Present except turtles	None
5.	Dermal roofing bones	Complete complement	Some reduction	Extensive reduction
6.	Parietal foramen	Present	Present in some	Confined to larvae
7.	Marginal bones	Complete complement	Usually complete	Fewer
8.	Bones ensheathing Meckel's cartilage	Numerous	Numerous	Fewer

3. Skull of Birds

The skull of a bird is basically reptilian but exhibits modifications correlated with flight, altered feeding habits, and a much larger brain. In avian skulls, some of the roofing bones have been lost, and the remaining dermal bones are much thinner; in carinate skulls, most of the sutures have been obliterated. Therefore, to identify individual bones, one have to use the skull of a ratite bird (the flightless bird such as emu, cassowary, kiwi, or ostrich) or a nestling.

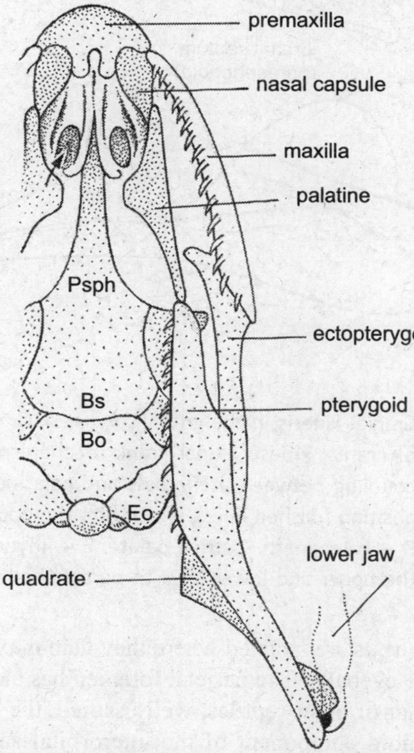

Fig. 3.33. Kinetic palatal complex of the snake *Tropidonotus*. **Bo**, basioccipital; **Bs** and **Psph**, ankylosed basisphenoid and parasphenoids of primary palate; **Eo**, exoccipital. Vomer bones, not visible, lie anterior to the parasphenoid. The left and right palatine and pterygoid bound a medial palatal fissure. *Arrow* indicates an internal naris. The posterior part of the lower jaw is drawn as being transparent to show the site of articulation with the quadrate (after Kent and Carr 2001).

One might think of the avian skull as comprising two functional regions (Fig. 3.34): (1) at the rear, a solid bony box (chondrocranium and dermatocranium) that houses the brain, olfactory organs, eye ball, and equilibratory and hearing complex, and therefore protects all structures needed for input and processing of information; and (2) at the front, a food procuring and handling area, the elongated beak and palate.

The posterior part of avian skull is highly vaulted, bulging outward and arching upward alongside the greatly enlarged

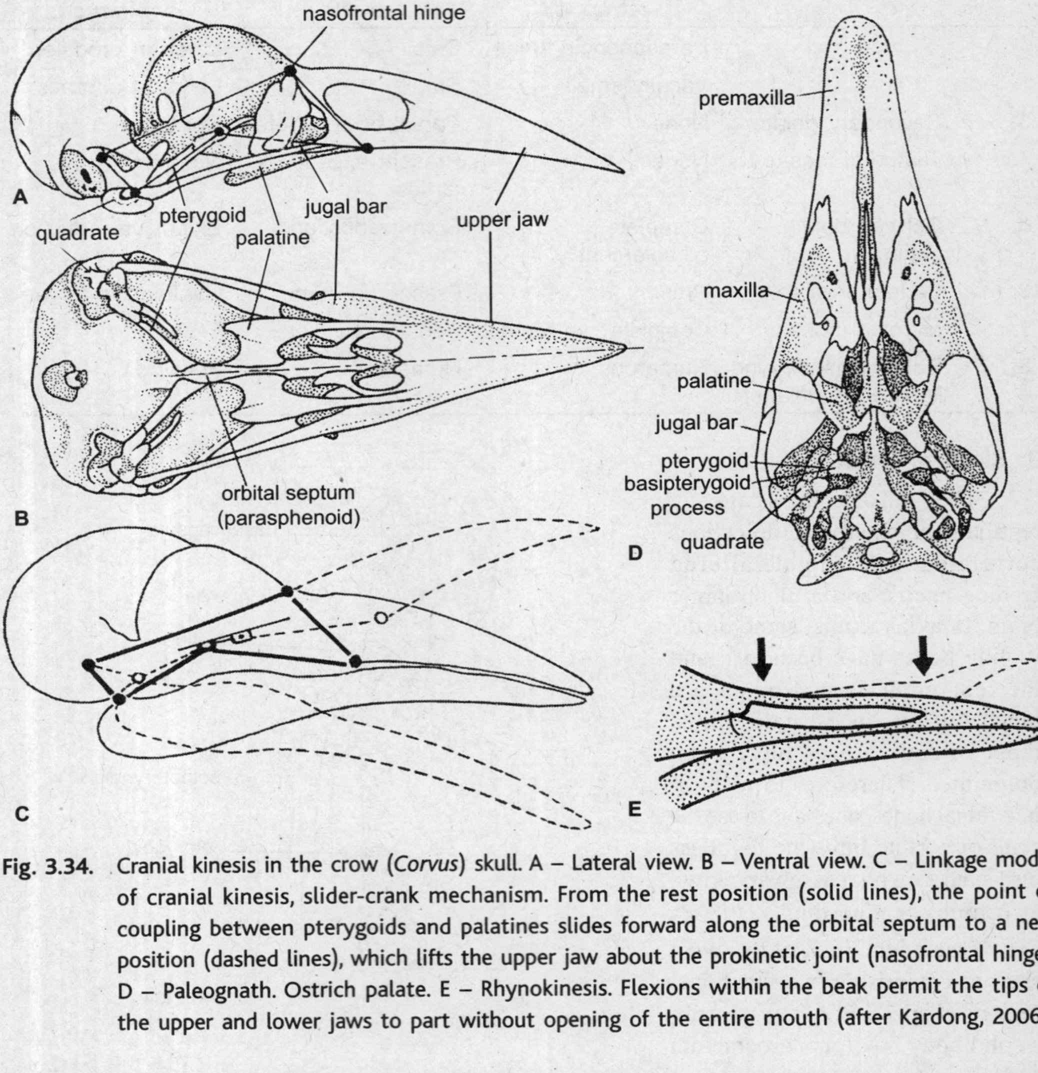

Fig. 3.34. Cranial kinesis in the crow (*Corvus*) skull. A – Lateral view. B – Ventral view. C – Linkage model of cranial kinesis, slider-crank mechanism. From the rest position (solid lines), the point of coupling between pterygoids and palatines slides forward along the orbital septum to a new position (dashed lines), which lifts the upper jaw about the prokinetic joint (nasofrontal hinge). D – Paleognath. Ostrich palate. E – Rhynokinesis. Flexions within the beak permit the tips of the upper and lower jaws to part without opening of the entire mouth (after Kardong, 2006).

brain. The orbits are situated where they gain maximal arcs of visibility, and they reflect the size of the massive eyeballs. The parietal foramen has closed. The chondrocranium is incomplete dorsally and, like that of other reptiles, well ossified, the only cartilage being in the olfactory capsule and in mesethmoid component of the interorbital septum. As in other reptiles, the chondrocranium bears a single occipital condyle.

Like turtles and some dinosaurs, birds are toothless, and their jaws are covered by keratinized sheath. Birds that feed on slippery prey, such as fish-eating shorebirds, have beaks whose keratinized margins are serrated to improve friction grip. The jaws are drawn out into a beak. The upper temporal bar is absent, and lower temporal bar is a slender rod, called the **jugal bar** (*i.e.*, quadratojugal-jugal bar), which extends from the beak posteriorly to the side of the movable (streptostylic) quadrate. The skull is prokinetic. A strong **postorbital ligament** extends from behind the eye to the lower jaw. In neognathous birds, the palate is functionally divided at

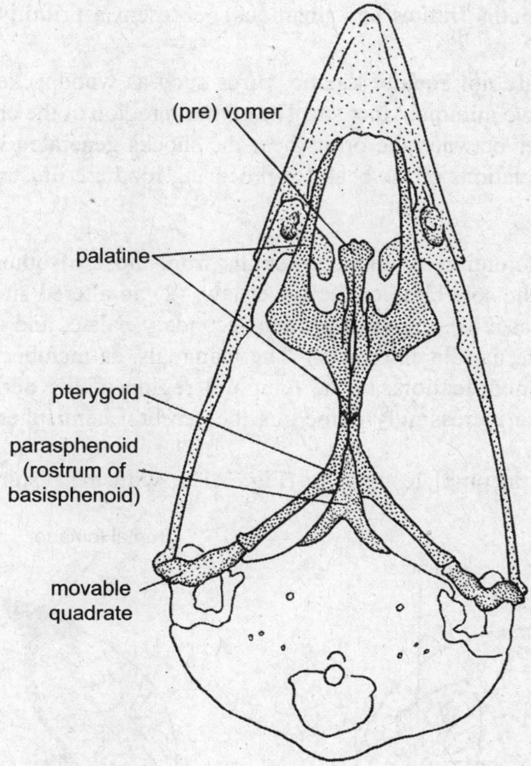

Fig. 3.35. Palatal complex of a kinetic bird, the cotinga, ventral view. The parasphenoid is immovable. The homology of the prevomer is hypothetical (after Kent and Carr 2001).

the pterygopalatine joint (Fig. 3.34 A, B; Fig. 3.35). The converging paired pterygoids meet or nearly meet at the midline to couple these with the palatine (Fig. 3.34 B). At this coupling, these bones form a joint that slides along the ventral edge of the orbital septum. The nasofrontal and palato-maxilla joints are thin, flexible, bony joints, not synovial, but they can be represented as hinges about which rotation occurs. The mechanically important bones are modeled as a linkage system on each side (Fig. 3.34 C). When inserting muscles pull the quadrate and palate forward, the divided palate slides along the septum pushing the palatines forward, which in turn push against the base of the beak, rotating it about the nasofrontal joint, and raising the beak. The coupled pair of gliding linkages is a **slider-crank mechanism** (*i.e.*, muscles acting directly on the lower jaw activates its opening). Jaw-closing muscles act in the opposite fashion to move the beak downward to grasp food, and return the kinetic machinery to the rest position. The thin jugal bar usually bows outward during jaw opening, but does not contribute significantly to the mechanism producing upper jaw elevation.

Many birds use their beak like a probe to reach buried grubs or insects embedded in tree barks or soft soil. Such birds often use a form of rhynchokinesis, the lifting of the beak tips about points of rotation within the jaws (Fig. 3.34 E). The jaws need not be parted far to seize the food. Other birds have beaks that open tough seeds and short, stout jaws that concentrate closing forces at the base of their beaks.

In paleognathus birds, the pterygoids do not meet at the midline, but slide on projecting struts, the basipterygoid process (Fig. 3.34 D). So structurally distinct is the palate that it has been used to argue that all paleognaths (ratites and tinamous) represent a primitive condition and a monophyletic group.

The skulls of all birds are not equally kinetic. Birds such as woodpeckers that subject the beak to rough usage usually have minimal kinetism. The ethmoid region of the chondrocranium and a narrow arch of roofing bones between the orbits bear the shocks generated when woodpeckers drill in solid wood. Some adaptations of the beak for procuring food are illustrated in figure 1.14.

4. Skull of Mammals

Major features that differentiate mammalian skulls from those of other amniotes are (1) emergence of the dentary as the sole bone of the lower jaw; (2) an altered site of articulation of the lower jaw with the braincase; (3) alterations in the secondary palate; and (4) the presence of three bones (ear ossicles in the middle ear cavity). The mammals, as members of the synapsida, differ from reptiles in their modifications in the temporal region of the dermatocranium. The skull of mammals has become increasingly domed as the cerebral hemispheres have ballooned dorsally, laterally and caudad.

The chondrocranium is incomplete dorsally (Fig. 3.11), with the result that membranous

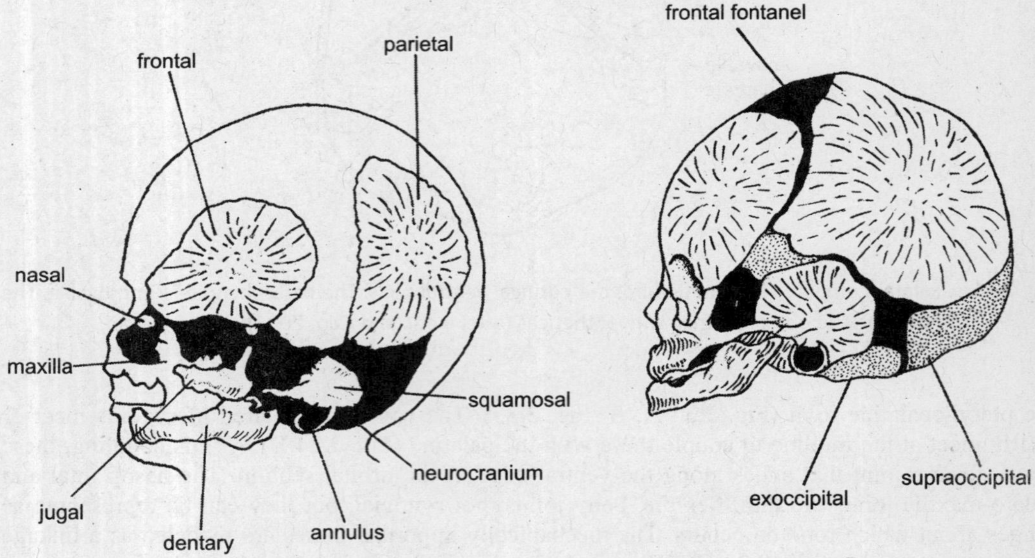

Fig. 3.36. Two stages in the development of a human skull. A – Intramembranous ossification is under way. The cartilaginous chondrocranium (solid black) is incomplete lateral to and above the brain. B – Intramembranous ossification has progressed, but "soft spots" (fontanels) remain where there is no membrane bone (after Kent and Carr 2001).

soft spots or **fontanels**, can be felt in heads of newborn human infants until ossification of the roofing bone is complete (Fig. 3.36). Fontanels enable the fetal skull to be molded as necessary during delivery through the narrow birth canal. One or more small **bregmatic bones** may ossify in the frontal fontanel in some species, and a single bregmatic bone may develop as an anomaly in human skulls. **Paracelsus** called this bone the "**antiepileptic bone**" because he believed that this bone served as a pop-up valve for relieving pressure within the cranial cavity.

Ossification centers in the chondrocranium of a fetal pig have been shown in figure 3.11. The basioccipital, basisphenoid and presphenoid bones form a floor on which the brain rests, exoccipitals, alisphenoid, and lateral extensions of the presphenoids form partial side walls of skull; and a supraoccipital completes, along with the basioccipital and exoccipitals, a bony ring that resembles a vertebra. Ethmoid cartilages and bones house the olfactory epithelium, underlie the olfactory bulbs of the brain, constitute much of the nasal septum (mesethmoid) and are perforated by foramina for olfactory nerve bundles (see Figs., 3.68, 3.71 and 3.41). Ossification centers in the otic capsules unite solidly to form petrosal (periotic) bones beneath the overgrown temporal lobes of the mammalian cerebral hemispheres of brain. They are often incorporated into a temporal bone complex. Each exoccipital bone bears a condyle inherited from therapsid ancestors. The chondrocranial derivatives of a generalized mammalian skull are diagrammed in figure 3.37.

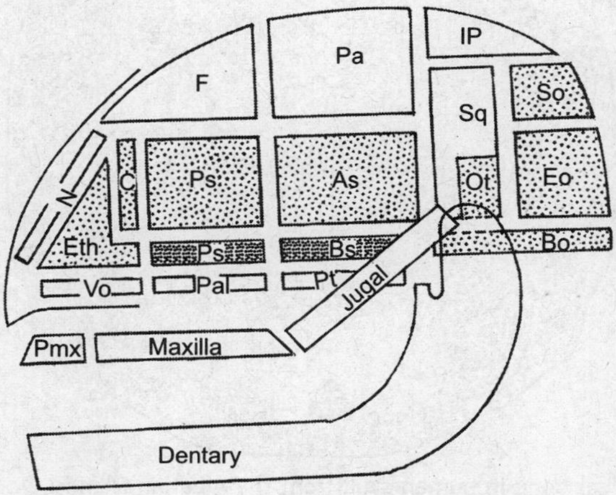

Fig. 3.37. Main endochondral bones (shaded) of a generalized mammalian skull. The vomer, palatine and pterygoid are parts of the primary palate. The premaxilla, maxilla, and palatine processes form the secondary palate. **As**, alisphenoid (pleurosphenoid), **Bo**, basioccipital; **Bs**, basisphenoid; **C**, cribriform plate of ethmoid; **Eth**, ethmoid, perpendicular plate; **Eo**, exoccipital; **F**, frontal; **Ip**, interparietal; **N**, nasal; **Ot**, otic (petrous); **Pa**, parietal; **Pal**, palatine; **Pmx**, premaxilla; **Ps**, presphenoid; **Pt**, pterygoid; **So**, supraoccipital; **Sq**, squamosal; **Vo**, vomer (after Kent and Carr 2001).

The sequence in mammals of centrum-like basioccipital, basisphenoid and presphenoid bones and their dorsal wings (the exoccipitals, alisphenoids and lateral aspects of the presphenoids) resembles a series of three successive vertebrae that are incomplete dorsally. It insipired poet-biologist **Goethe's Vertebral Theory** of origin of the skull (**Kent** and **Carr** 2001).

In mammals, the dermatocranium is represented by paired premaxillae, maxillae, jugals (malars), nasals, lacrimals, and squamosals; and by paired or unpaired frontals and parietals; and by an unpaired interparietal, all of which are paired in embryos and in some species, in neonates. A postparietal (dermoccipital) was present in *Homo erectus* (a likely immediate ancestor of humans) and is still present in some human populations, chiefly Mongolians. It is sometimes called an **inca bone** (Fig. 3.38) because it was common among Inca Indians (**Kent** and **Carr** 2001). Premaxillae are not identifiable in adult human skulls because they unite with the maxillae early in embryonic life, a discovery made by **Goethe**. The zygomatic arch varies from massive to slender, depending on the magnitude of the force exerted on it by the masseter muscle. This arch becomes extremely delicate, or even incomplete, in some insectivores.

The temporal complex of mammals consists of numerous components of intramembranous and endochondral origin (Fig. 3.39). The **squamous portion** is the squamosal of lower tetrapods. A **tympanic bulla** (Fig. 3.40) consists of two parts, a **tympanic** and **entotympanic.** The tympanic bone (or annulus tympanicus) surrounds the tympanic membrane (eardrum) as a bony ring. It is derived, according to evidence from embryonic opossums; from the angular bone of nonmammalian amniotes (The annulus of anurans is not a homologue. It is thought to be derived

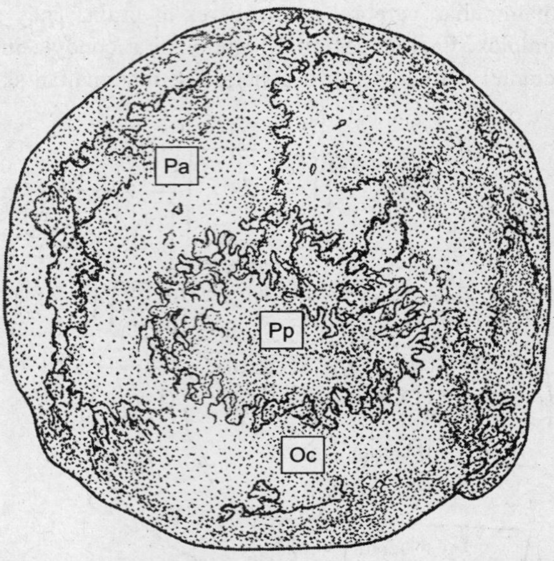

Fig. 3.38. Postparietal (Inca) bone in human skull from the Aleutian Islands. **Oc**, occipital, **Pa**, parietal, **Pp**, postparietal (after Kent and Carr 2001).

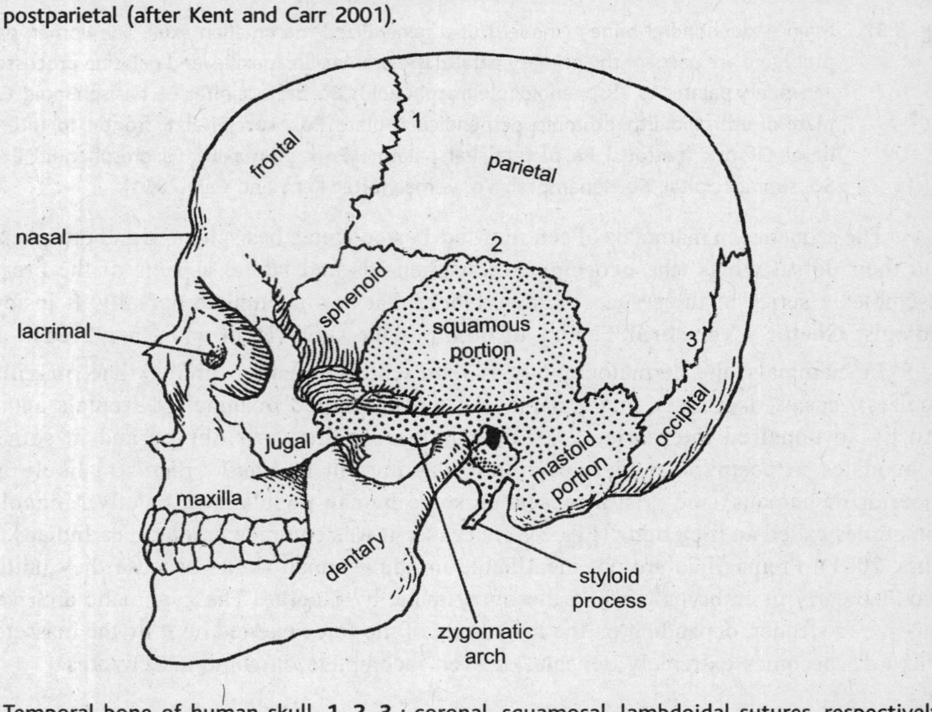

Fig. 3.39. Temporal bone of human skull. **1, 2, 3** : coronal, squamosal, lambdoidal sutures, respectively (Kent and Carr 2001).

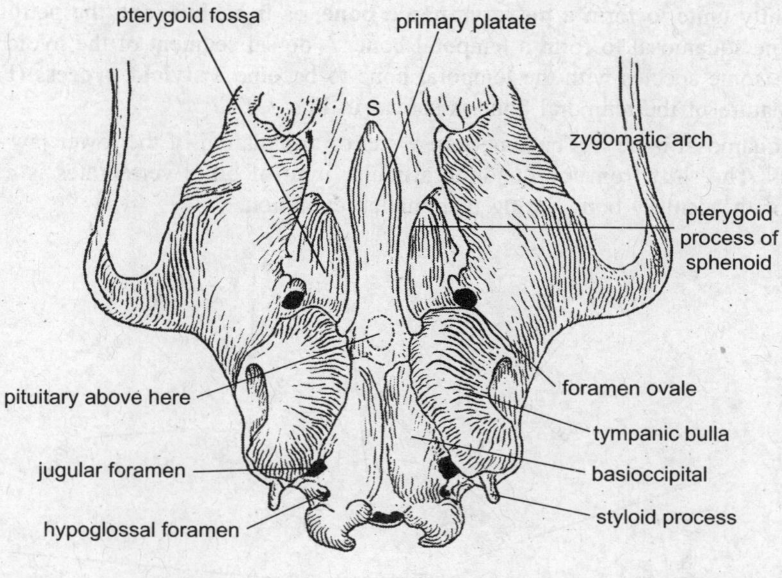

Fig. 3.40. Hamester skull, caudal part, ventral view. **S**, secondary palate. The primary palate is the roof of the nasopharynx (after Kent and Carr 2001).

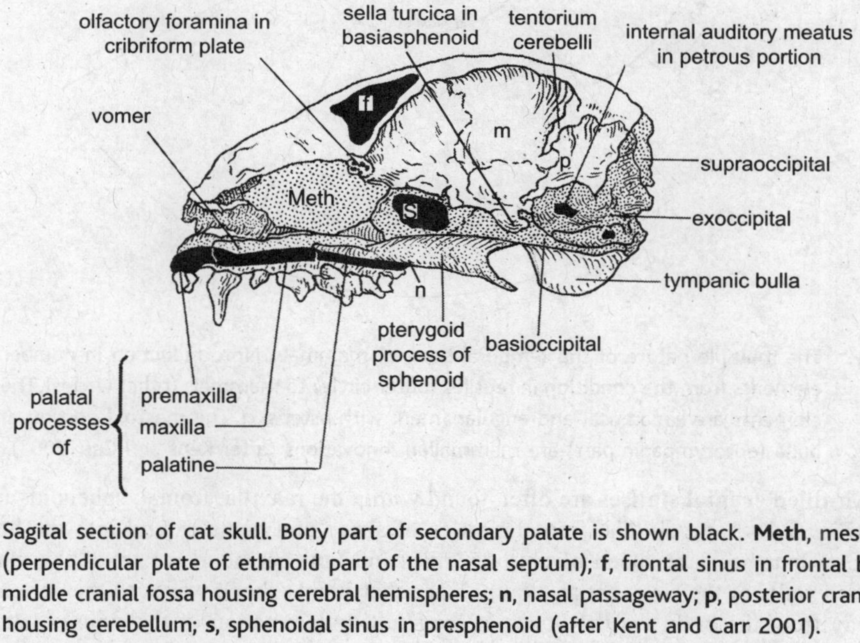

Fig. 3.41. Sagital section of cat skull. Bony part of secondary palate is shown black. **Meth**, mesethmoid (perpendicular plate of ethmoid part of the nasal septum); **f**, frontal sinus in frontal bone; **m**, middle cranial fossa housing cerebral hemispheres; **n**, nasal passageway; **p**, posterior cranial fossa housing cerebellum; **s**, sphenoidal sinus in presphenoid (after Kent and Carr 2001).

from the palatoquadrate cartilage). A large portion of the **bulla**, when present, is represented by **entotympanic**, a cartilage replacement ossification, which is a new structure in some mammals. The **petrous portion** (*i.e.*, fused prootic, opisthotic and epiotic) is ossified otic capsule (Fig. 3.41). As a primarily internal structure, it may still be seen externally as the **mastoid process** of the mastoid portion. The bulk of the **mastoid portion** of endochondral origin is new in mammals

(Fig. 3.39). Although the typanic and petrous portions are separate bones in some mammals, they frequently unite to form a **petrotympanic bone**, as in rabbits, and the petrotympanic may unite with the squamosal to form a temporal bone. A dorsal segment of the hyoid arch skeleton coalesces in some species with the temporal bone to become a **styloid process** (Fig. 3.39). The composite nature of the temporal bone is shown in Fig. 3.42.

The squamosal bone has become a new site of articulation of the lower jaw with the skull in mammals. The shift from the quadrate articular joint of other vertebrates is correlated with expansion of the dentary bone during mammalian evolution.

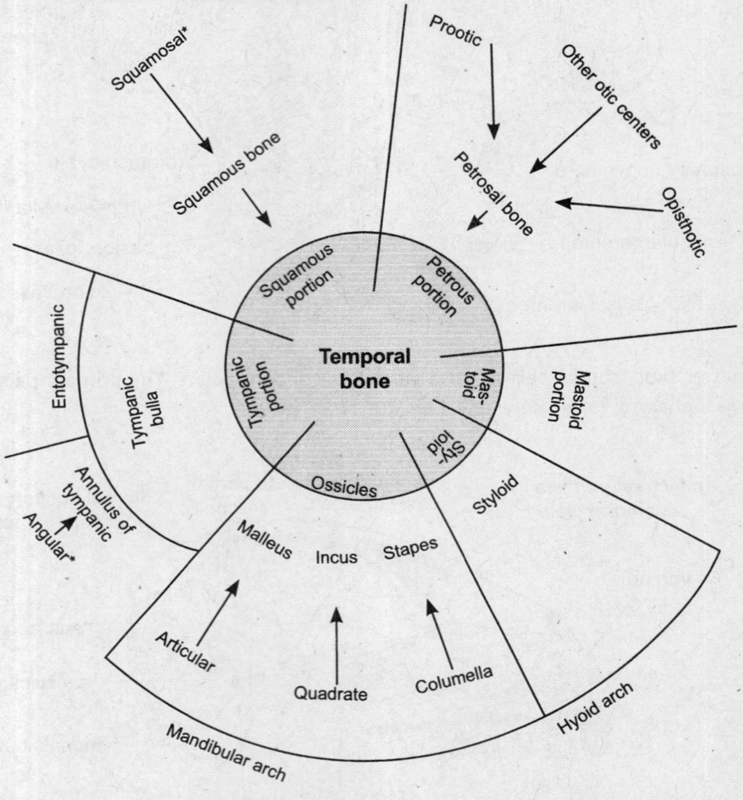

Fig. 3.42. The multiple nature of the temporal bone of mammals. Note reduction in number of separate elements from the condition in reptiles (outer circle) to mammals (other circles). The two dermal elements are squamosal and angular (mark with asterisks). The mastoid portion and tympanic bulla (entotympanic part) are mammalian innovations (after Kent and Carr 2001).

Air-filled **cranial sinuses** are often found within the maxilla, frontal, sphenoid, and ethmoid bones (Fig. 3.41 f and s). The frontal sinuses of sheep and goats extend into the horns. When male goats butt heads at speed of upto 35 miles (60 km) per hour as part of the mating ritual, the arched walls of the sinus act as a bony brace that shunts shock waves to the vertebral column and away from the brain via the bones of the skull. Inflammation of the sinuses (**sinusitis**) is a common ailment in human beings.

Of the inherited primary palate, an unpaired vomer lies at the base of the nasal septum (Fig. 3.41). The septum consists of a masethmoid bone and more or less cartilage. A **nasal process** of the **palatine** develops in the lateral wall of the nasopharynx where it contributes to the wall of the orbit, and a palate process of the palatine bone contributes to the secondary palate. The pterygoids are reduced to small, winglike, **pterygoid processes** of the sphenoid complex.

The parasphenoid and ectopterygoids have been lost. The bony part (or hard palate) of secondary palate consists of palatae processes of premaxilla, maxilla and palatine bones. Caudal to the bony part is a membranous soft palate.

C. Splanchnocranium or Visceral Cranium

The visceral skeleton, or splanchnocranium, is the skeleton of the pharyngeal arches. In fishes, therefore, it is the skeleton of the jaws and gill arches. In tetrapods, it includes jaws and hyoid apparatus and in them visceral skeleton has become modified to perform new functions on land.

The splanchnocranium is an ancient chordate strucutre. In amphioxus, the splanchnocranium, or at least its forerunner, is associated with filter-feeding surfaces.

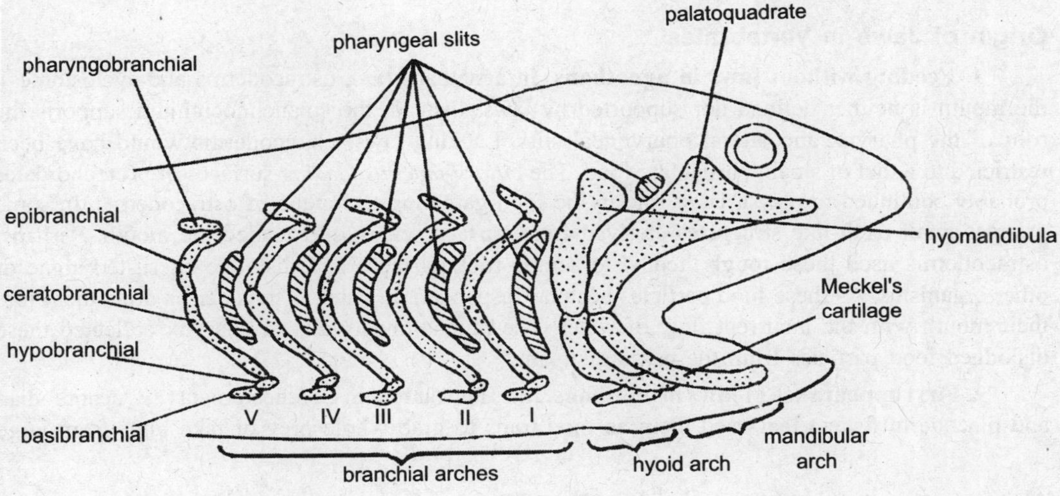

Fig. 3.43. Primitive splanchnocranium. Seven branchial arches are shown. Up to five elements form an arch on each side, beginning with pharyngobranchial dorsally and in sequence to the basibranchials most ventrally. The first two complete branchial arches are named mandibular arch for the first and hyoid arch for the second that supports it. The characteristic five-arch elements are reduced to just two in the mandibular arch: the palatoquadrate and Meckel's cartilage. The large hyomandibula, derived from an epibranchial element, is the most distinct component of the next arch, the hyoid arch. Behind the hyoid arch are variable numbers of branchial arches I, II and so on. Labial cartilages are not included (after Kardong 2006).

Embryology of visceral skeleton. The mistaken view that the splanchnocranium develop from the same embryonic source as the walls of the digestive tract inspired the name "visceral" cranium, a name that unfortunately has stuck despite being a misnomer (see **Kardong** 2006). Embryologically, the splanchnocranium arises from **neural crest cells**, not from lateral plate mesoderm like the smooth muscle in the walls of the digestive tract. In protochordates (*e.g.*, Amphioxus) neural crest cells are absent. Pharyngeal bars, composed of fibrous connective tissue, but never bone or cartilage, arise from mesoderm and form the unjointed branchial basket, the phylogenetic predecessor of the vertebrate splanchnocranium. In vertebrates, cells of neural crest depart from the sides of the neural tube and migrate into the walls of the pharynx between successive slits to differentiate into the respective pharyngeal arches. Pharyngeal arches of aquatic vertebrates usually are associated with their respiratory gill system. Because of this association, they are called **branchial arches, visceral arches** or **gill arches**.

Each gill arch can be composed of a series of up to five articulated elements per side beginning with the **pharygobranchial** element dorsally and then, in descending order, the **epibranchial**, **ceratobranchial**, **hypobranchial**, and **basibranchial** elements (Fig. 3.43). One or more of these anterior branchial arches may come to border the mouth, support soft tissue and bear teeth. Branchial arches that support the mouth are called **jaws**, and each contributing branchial arch is numbered sequentially or named. The first fully functional arch of the jaw is the **mandibular arch**, the largest and most anterior of the modified series of branchial arches. The mandibular arch is composed of the **palatoquadrate** dorsally and **Meckel's cartilage** (or mandibular cartilage) ventrally. The **hyoid arch**, whose most distinct element is the **hyomandibula**, follows the mandibular arch. A varying number of branchial arches, often designated with roman numerals, follow the hyoid arch (Fig. 3.43).

Origin of Jaws in Vertebrates

1. Feeding without jaws in agnathans. In agnathans (*i.e.*, ostracoderms and cyclostomes), the mouth is neither defined nor supported by jaws. Instead, the splanchnocranium supports the roof of the pharynx and lateral pharyngeal slits. Lacking jaws, ostracoderms would have been restricted to a diet of small, particulate food. The *ciliary-mucous feeding* surfaces of protochordates probably continued to play a large role in the food-gathering technique of ostracoderms. In some groups, small teeth-like structures, derived from surface scales, surrounded the mouth. Perhaps, ostracoderms used these rough "teeth" to scrape rock surfaces and dislodge encrusted algae or other organisms. As these food particles became suspended in water, ostracoderms drew them into their mouth with the incurrent flow of water. The mucus-lined walls of pharynx collected these dislodged food particles from the passing stream.

2. First appearance of jaws in gnathans. Jaws appear first in Gnathostomata (*i.e.*, acanthodian and placoderm fishes) that used them as food traps to grab whole prey or take bites from large

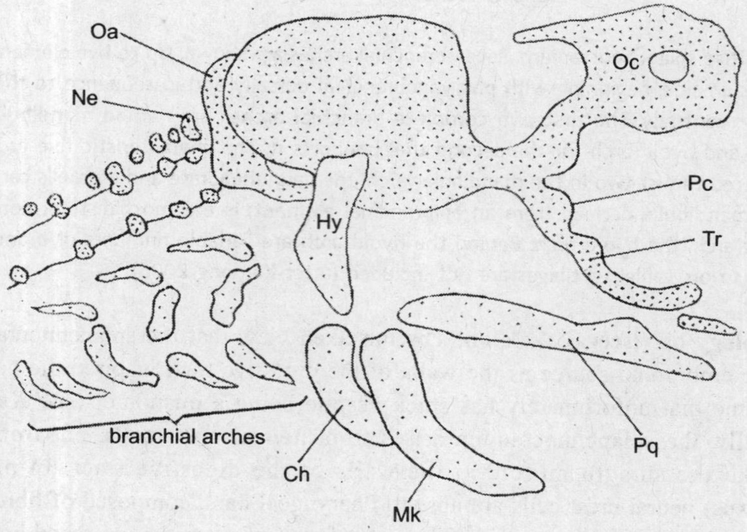

Fig. 3.44. Shark embryo, the dogfish *Scyllium*. Jaws appear to be in series with the branchial arches. The mandibular arch is first, followed by the hyoid and then several branchial arches. Such a position of the jaws, in series with the arches, is taken as evidence that the jaws derive from the most anterior branchial arch. Abbreviations: **Ch**, ceratohyal; **Hy** hyomandibula; **Mk**, Meckel's cartilage; **Ne**, neural arch; **Oa**, orbital arch; **Oc**, orbital cartilage; **Pc**, polar cartilage, **Pq**, palatoquadrate, **Tr**, trabecula. Labial cartilages are not included (after Kardong 2006).

prey. Within some groups, jaws also served as crushing or chewing devices to process food in the mouth. With the advent of jaws, these fishes became more free-ranging predators of open waters.

Jaws arise from one of the anterior pair of gill arches. Evidence supporting this fact came from several sources. 1. The embryology of sharks suggests that jaws and branchial arches develop similarly in series (Fig. 3.44), and both arise from neural crest. The spiracle appears to have once been a full-sized gill slit, but in modern sharks it is crowded and reduced by the enlarged hyoid arch next in series. 2. Nerves and blood vessels are distributed in a pattern similar of branchial arches and jaws. The musculature of the jaws appears to be transformed and modified from branchial arch musculature.

Evidently, the branchial arches phylogenetically gave rise to jaws. But it has been debated whether jaws represent derivatives of first, second or third, or even fourth branchial arches of primitive ancestors. Following two theories have been forwarded to explain this riddle: serial theory and composite theory.

(1) Serial theory. It is simplest view and holds that the first or perhaps second ancient branchial arch gave rise exclusively to the mandibular arch, the next branchial arch exclusively to the hyoid arch, and the rest of the arches to the branchial arches of gnathostomes (Fig. 3.45 A).

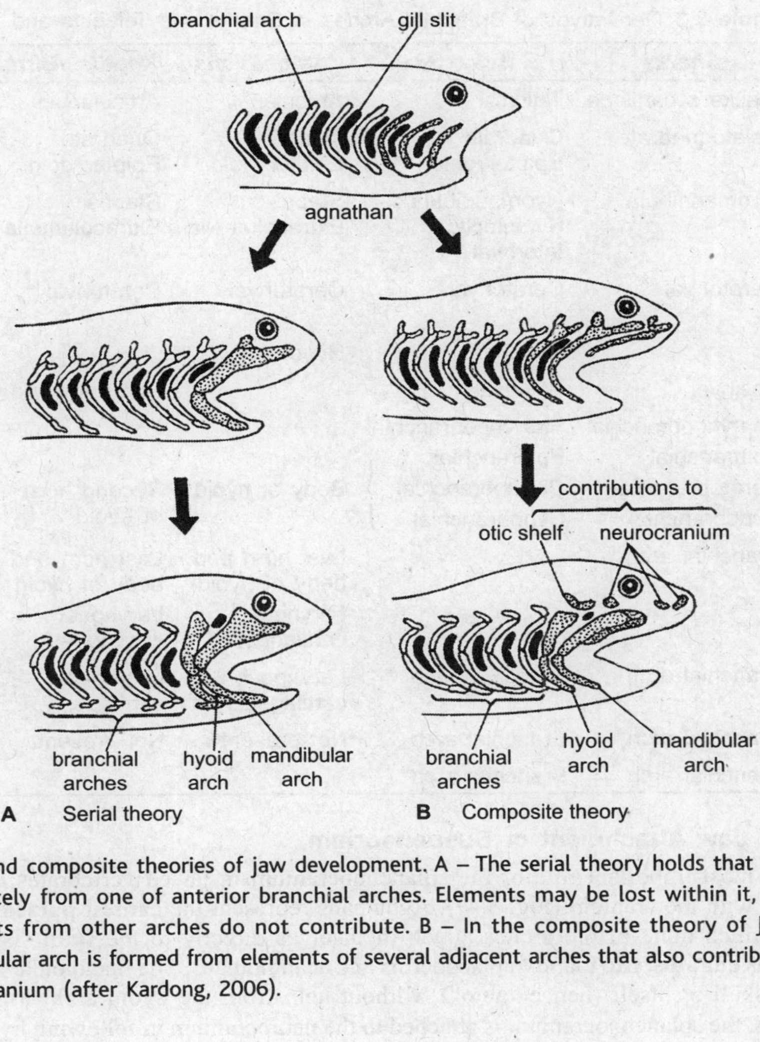

Fig. 3.45. Serial and composite theories of jaw development. A – The serial theory holds that jaws arise completely from one of anterior branchial arches. Elements may be lost within it, but other elements from other arches do not contribute. B – In the composite theory of Jarvik, the mandibular arch is formed from elements of several adjacent arches that also contribute to the neurocranium (after Kardong, 2006).

(2) Composite theory. A Swedish palaeontologist, **Erik Jarvick**, in 1980 proposed the **composite theory** which is a more complex view based on his examination of fossil fish skulls and embryology of living forms (Fig. 3.45 B). Jarvic hypothesized that ten branchial arches were present in primitive species, the first and following arches being named **terminal premandibular**, **hyoid**, and six **branchial arches**. Rather than the "*one arch, one mandible*" view, Jarvik envisioned a complex series of losses or fusions between selective parts of several arches that come together to produce the single composite mandible.

According to composite theory, the **mandibular arch** of gnathostomes is formed by fusion of parts of the premandibular arch and parts of the mandibular arch of jawless ancestors. The **palatoquadrate** forms from the fusion of the epibranchial of the premandibular arch with the epibranchial and one pharyngobranchial of the mandibular arch. **Meckel's cartilage** arises from the expanded ceratobranchial element. Likewise, the **hyoid arch** arises phylogenetically from the epibranchial, ceratobranchial and hypobranchial elements of the third primitive gill arch. The remaining branchial arches persist in serial order. The other elements of the primitive arches are lost or fused to the neurocranium.

Finally, it has been concluded that jaws of vertebrates are derivatives of ancient gill arches (Table 3.3).

Table 3.3 Derivatives of Branchial Arches in Sharks and Teleosts and Tetrapods

Arch	*Sharks*	*Teleosts*	*Amphibians*	*Reptiles/Birds*	*Mammals*
I	Meckel's cartilage	Articular	Articular	Articular	Malleus
	Palatoquadrate	Quadrate Epipterygoid	Quadrate Epipterygoid	Quadrate Epipterygoid	Incus Alisphenoid
II	Hyomandibula	Hyomandibula Symplectic Interhyal	Stapes Extracolumella	Stapes Extracolumella	Stapes
	Ceratohyal	Ceratohyal	Ceratohyal	Ceratohyal	Anterior horn hyoid
		Hypohyal	Hypohyal		
	Basihyal	Basihyal		Body of hyoid	Body of hyoid
III	Pharyngobranchial	Pharyngobranchial			
	Epibranchial Ceratobranchial Hypobranchial	Epibranchial Ceratobranchial Hypobranchial	Body of hyoid	Second horn of hyoid	Second horn of hyoid
IV	Branchial arch		Last horn and body of hyoid Laryngeal cartilages(?)	Last horn and body of hyoid Laryngeal cartilages(?)	Thyroid cartilages(?)
V	Branchial arch	Branchial arch	Laryngeal cartilages(?)	Laryngeal cartilages(?)	Laryngeal cartilages
VI	Branchial arch	Branchial arch	Not present	Not present	Not present
VII	Branchial arch	Branchial arch			

Types of Jaw Attachment or Suspensorium

The most important role of the splanchnocranium in jawed vertebrates is to suspend the lower jaw with the cranium (Fig. 3.47). Agnathans represent the earliest **palaeostylic stage** (Fig. 3.46) in which none of the arches attach themselves directly to the skull. The earliest jawed condition is **euautostylic** found in placoderms and acanthodians. The mandibular arch is suspended from the skull by itself (hence "auto"), without help from the hyoid arch (Fig. 3.47). In other vertebrates, the splanchnocranium is attached to the neurocranium in following five principal ways:

1. Autodiastylic type. This type of suspensorium is found in the earliest gnathostomes Acanthodians. In this case, the jaw is supported by direct attachment to the cranium by ligaments both at its hind as well as the front end. The hyoid arch is a typical branchial arch, not modified in any way. The spiracle is a fully functional gill slit.

2. Amphistylic type (Fig. 3.46). This type of jaw suspension occurs in early elasmobranchs such as *Hexanchus*. In this case both the pterygoquadrate articulate with the neurocranium at two points: (i) basal process at the level of the posterior end of the embryonic trabecula and (ii) otic process articulating with the side of the otic capsule (Fig. 3.47).

3. Hyostylic type. In this type of jaw suspension, the upper jaw is not fixed with the cranium but is slung from the latter by the hyomandibular and a prespiracular ligament (Fig. 3.47). Thus, hyomandibular, a part of the hyoid arch, acts as suspensorium. This naturally allows a wide gap for swallowing large prey. This type of jaw suspension occurs in dog fish.

4. Autostylic type. In this case, the upper jaw is completely fused with the cranium (Fig. 3.47). It is the lower jaw which is articulated with cranium and this articulation is effected in different ways and is found in dipnoans, amphibians, reptiles and birds and is of various types:

***(a)* Holostylic**. This is a modified form of hylostyly (Fig. 3.46) where the upper jaw articulates with the neurocranium through two or more pairs of processes without the intervention of the hyomandibular. These articulations are usually the immovable type. The hinge occurs between the articular of lower jaw and the quadrate of the upper jaw. Jaws of *Chimaera* are also similar but in this case the upper jaw fuses with cranium rather completely and the hyoid arch is free.

***(b)* Autosystylic**. It is found in lungfish and tetrapods except mammals. Here the hyoid arch breaks up and is no longer in the service of jaws, hyomandibular becomes reduced and forms the ear ossicles, the stapes. The lower jaw is articulated with the quadrate part of the upper jaw, thus quadrate becomes the suspensorium. it is of following two subtypes:

***(i)* Monimostylic.** In this case the upper jaw is fused with the cranium and thus is immovable. The lower jaw articulates with the quadrate. it is mostly found in amphibians.

***(ii)* Streptostylic.** In this type of jaw suspension, the quadrate is movable at both the ends and not firmly fixed with the skull. With this arrangement, the lower jaw has two movable joints. Here the articulation, in fact, takes place between the quadrate of the cranium and the articular of lower jaw. Quadrate is not firmly fixed. Streptostylic suspension is found in snakes, lizards and birds.

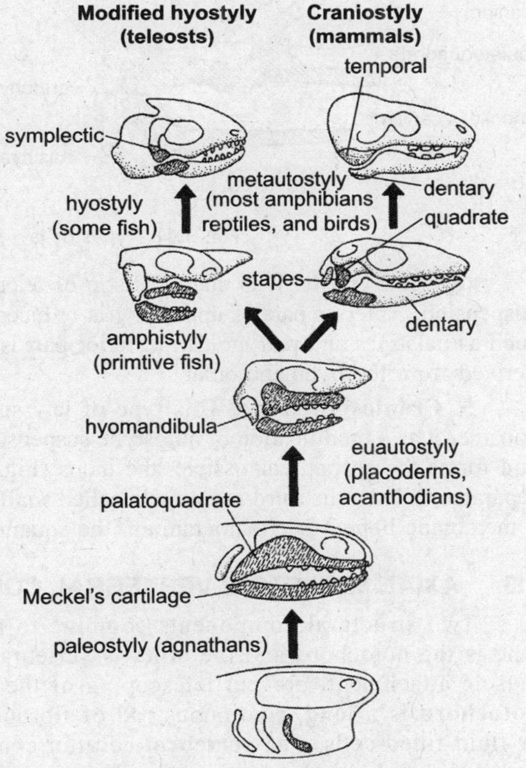

Fig. 3.46. Jaw suspension. The points at which the jaws attach to the rest of the skull defines the type of jaw suspension. Note the mandibular arches (crosshatched areas) and hyoid arches. The dermal bone (white areas) of the lower jaw is the dentary (after Kardong 2006).

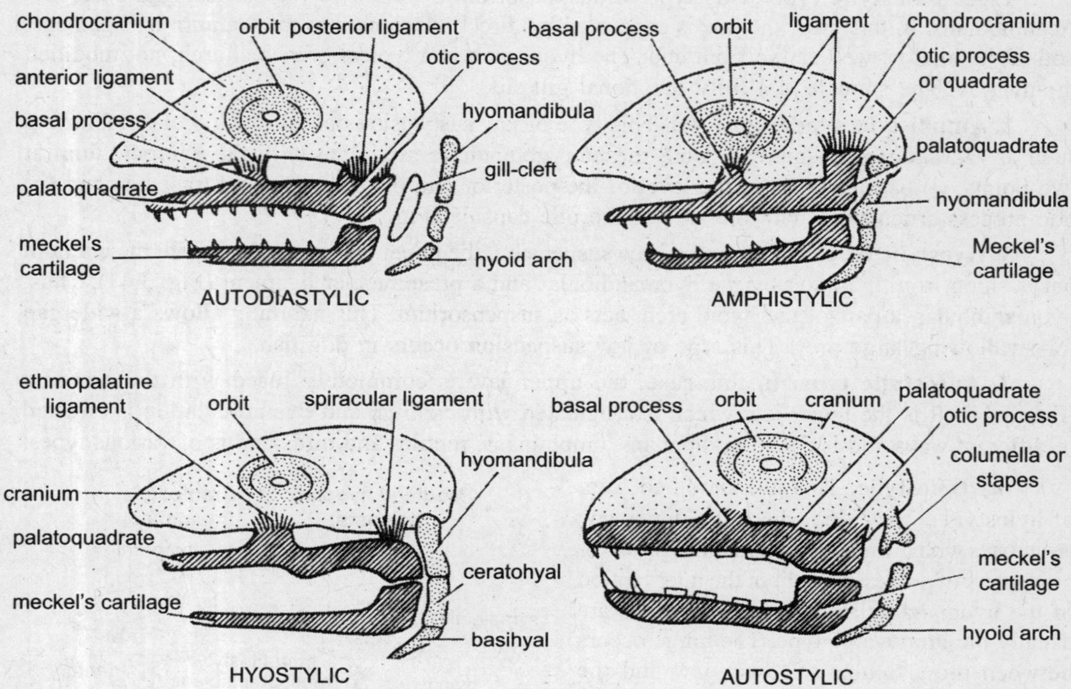

Fig. 3.47. Types of jaw suspension in vertebrates.

(*c*) **Methystylic.** It is characteristic of teleosts (bony fish) and is a variation of autostylic suspension. Anterior part of upper jaw is reduced and its place is taken up by membrane bones such as palatines and pterygoids, posterior part is replaced by metapterygoid and quadrate bones derived from the hyomandibular.

5. Craniostylic type. This type of jaw suspension is characterisitc of mammals. Some consider it as a modification of autostylic suspensorium where quadrate of the upper jaw separates and forms the second ear ossicle, the incus (Fig. 3.48). A part of the lower jaw, the **articular**, separates to form the third ear ossicle called **malleus**, and the lower jaw articulated directly with a membrane bone of dermatocranium, the squamosal.

3.3 AXIAL SKELETON: VERTEBRAL COLUMN AND ASSOCIATED STRUCTURES

Two structural components combine to define the long axis of the vertebrate body. One is the notochord and the other is vertebral column. The axial skeleton offers sites for muscle attachment, prevent telescoping of the body, and support much of the weight. The **notochord** is a long, continuous rod of fibrous connective tissue wrapping a core of fluid or fluid-filled cells. The **vertebral column** consists of a discrete (*i.e.*, individually separate and distinct) but repeating series of cartilaginous or bony elements (the **vertebrae**). The notochord is phylogenetically the oldest of the two structural components, but it tends to give way to the vertebral column, which assumes the role of body support in most later vertebrates.

The vertebral column debuts (*i.e.*, its first appearance in a role) early with evidence of segmental blocks along a notochord in the fossils of *Haikouella* and *Haikouichthys* (Fig. 3.49). Living hagfishes possess a notochord but lack vertebrae. In lampreys a few, small cartilaginous elements, neural arches and spines, ride upon a prominent notochord, but the main bodies of vertebrae are absent. In ostracoderms, there is a few vertebrae with even a hint of a vertebral

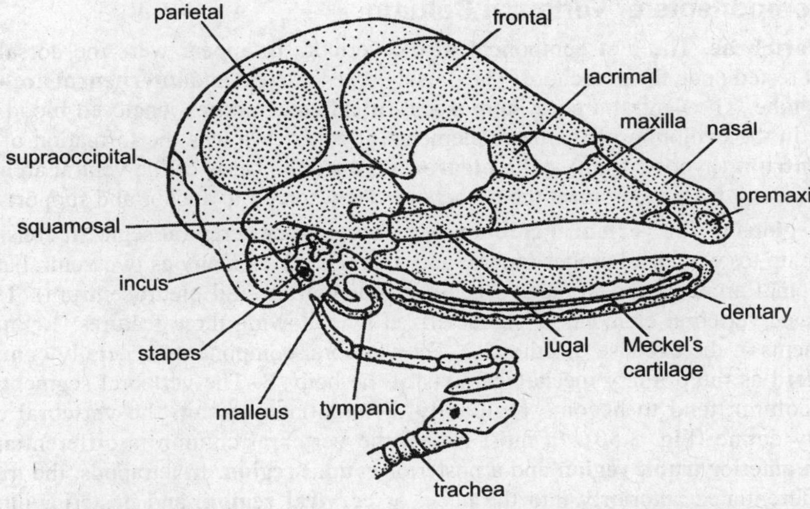

Fig. 3.48. Skull of armadillo embryo (mammal). During embryonic formation of the three middle ear ossicles (incus, stapes, malleus), the incus and stapes arise from the mandibular arch, testifying to the phylogenetic derivation of these bones from this arch. The dermal dentary is cut away to show Meckel's cartilage which ossifies at its posterior end to form the malleus (after Kardong 2006).

column. This may be because vertebrae were present but unossified or poorly preserved. Or, more probably, vertebrae were uncommon (**kardong** 2006). The subsequent evolution of the vertebral column in fishes and tetrapods is also complicated, partly because some components became enlarged, others were lost, and some evolved independently several times.

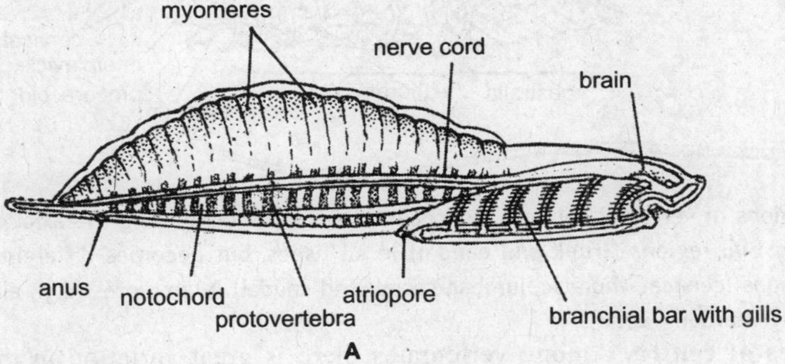

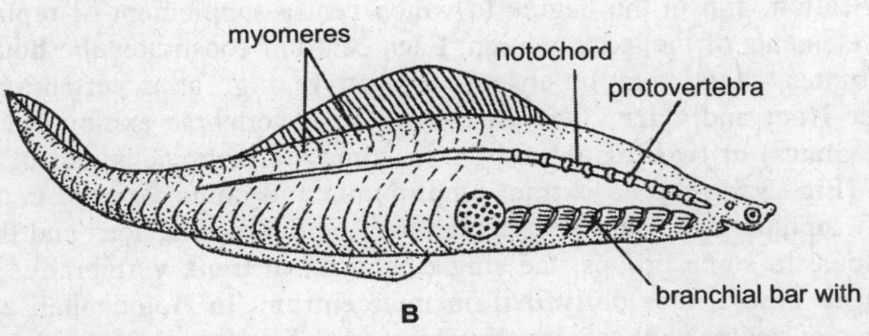

Fig. 3.49. Early vertebrate fossils. A – *Haikouella*, from the early Cambrain. Note the gills, brain, and other vertebrate features; 2.75 cm in length. B – *Haikouichthys* from early Cambrian, 2.75 cm in length. Both fossils were obtained from China (Chengjiang) and placed in Cephalochordata (after Kardong 2006).

Basic Components of Vertebral Column

1. Vertebrae. The first component of the vertebra to appear were the dorsal and ventral arches that rested upon the notochord. The dorsal arches, **neural** and **interneural arches**, protected the neural tube. The ventral arches, **hemal** and **interhemal arches**, enclosed blood vessels. The next stage in the evolution of the basic elements of a vertebra was the formation of two **centra**: an **intercentrum** (hypocentrum) and a **pleurocentrum**. The bases of the ventral arches expand to form these centra where they meet the notochord. They served to anchor and support these arches.

2. Regions of the vertebral column. Each vertebra or vertebral segment consists of arches and centra: up to two dorsal arches (neural and interneural), as many as two ventral arches (hemal and interhemal arches), and often two centra (intercentrum and pleurocentrum). The evolution of these basic vertebral elements is characterized by following three features: 1. Enlargement of some elements at the expense of others. 2. The vertebral components generally come to displace the notochord as the primary mechanical axis of the body. 3. The vertebral segments composing the axial column tend to become regionally differentiated within the vertebral column they collectively define (Fig. 3.50). In most fishes, the vertebral column is differentiated into two regions: an anterior **trunk region** and a posterior **caudal region**. In tetrapods, the trunk becomes further differentiated anteriorly into the neck, or **cervical region**, and posterior into the **hip** or **sacral region**. In early tetrapods, a **postsacral region** of up to five or six vertebrae, bearing the fused, proximal bases of their respective ribs, typically continues behind the sacral region. In some tetrapods, there is further differentiation of the trunk into the chest, or the **thoracic region**, and into the area between the thorax and the hips, the **lumber region.**

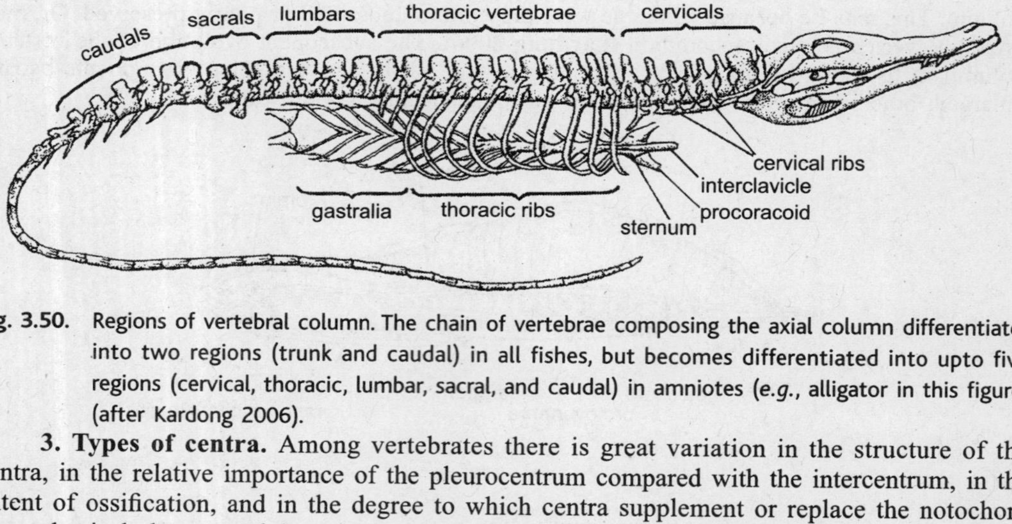

Fig. 3.50. Regions of vertebral column. The chain of vertebrae composing the axial column differentiates into two regions (trunk and caudal) in all fishes, but becomes differentiated into upto five regions (cervical, thoracic, lumbar, sacral, and caudal) in amniotes (*e.g.*, alligator in this figure) (after Kardong 2006).

3. Types of centra. Among vertebrates there is great variation in the structure of the centra, in the relative importance of the pleurocentrum compared with the intercentrum, in the extent of ossification, and in the degree to which centra supplement or replace the notochord as mechanical elements of the axial column. Each centrum constitutes the **body** of the vertebra. In some vertebrates, centra may be absent (**aspondyly**, *e.g.*, atlas vertebra of frog and other amphibians, see **Kent** and **Carr** 2001; Box 3.2). Other vertebrae exhibit one (**monospondyly**, *e.g.*, most vertebrates) or two [**dispondyly**, *e.g.*, some elasmobranchs, tail of some bony fishes such as *Amia* (Fig. 3.51), some extinct amphibians and some lizards] centra per segment. Indeed in many amniotes, the caudal vertebrae may be diplospondylous and the trunk vertebrae monosphondylous. In some groups, the single centrum in trunk vertebrae is derived from the pleurocentrum; in others, it is derived from intercentrum. In Holocephali and Dipnoi (lung-fish) the number of centra may secondarily increase to five to six per segment (**polyspondyly**, Fig. 3.52). In aminotes, the pleurocentrum predominates and becomes the body of each vertebral segment. Remnanats of the intercentrum contribute to cervical vertebrae and perhaps in a minor way to other vertebral elements, but for the most part, the intercentrum becomes **intervertebral cartilage** (disc) of the amniote vertebral column.

Box 3.2

The first vertebra of frog and other amphibians is the cervical vertebra and is called **atlas**. it is known as the atlas, because it supports the skull on it. This name is given after the giant "Atlas" of Greek legend, who is believed to carry the Earth on his shoulders.

In tetrapods, two main types of vertebrae occur (Fig. 3.53): 1. **Aspidospondylous vertebra**, in which all elements are separate. Specifically, the three arch elements (*i.e.*, intercentrum, pleurocentrum, and neural arch), either paired or single, remain as separate ossified elements. For example, **rachitomous vertebra** which is found in some rhipidistians (Osteolepiformes – extinct crossopterygians) and some early tetrapods, is a specialized type of aspidospondylous vertebra. The term rhachitomous means "cut-up spine", a reference to the numerous separate parts that constitute each vertebral segment (Fig. 3.53 A). Both large intercentrum and small pleurocentrum are usually paired, and a separate neural arch is associated with them. Aspidospondylous vertebrae are of many types, but following two types are important: **(i) Embolomerous vertebra** in which separate but approximately equal-sized centra are present (*e.g.*, extinct amphibians). **(ii) Stereospondylous vertebra** which consists of a single body derived entirely from the centrum (*e.g.*, modern amphibians).

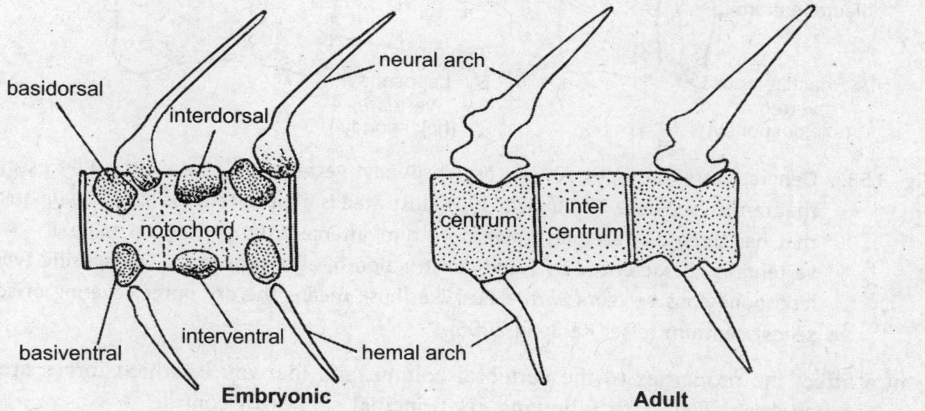

Fig. 3.51. Tail vertebrae of *Amia*. A centrum and intercentrum develop in each body segment (a diplospondylous condition). Intercentrum does not bear arches. Diplospondyly may provide the tail with increased flexibility for locomotion. (after Kent and Carr 2002).

2. Holospondyl vertebra. In this type of vertebra, all vertebral elements in a segment are fused into a single piece. Typically, the fused centrum derives from a pleurocentrum. The

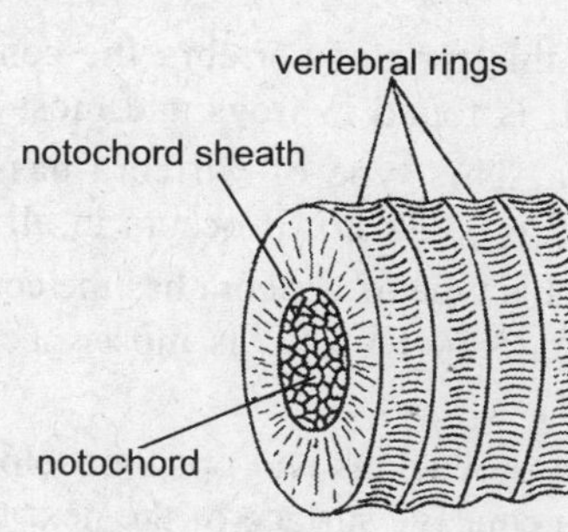

Fig. 3.52. Vertebral rings from an extinct holocephalan. Neural arches have been removed (after Kent and Carr 2002).

intercentrum, if present remains as an unossified contribution to the intervertebral cartilage between vertebrae. For example, the **lepospondylous vertebra** (*e.g.*, extinct aquatic amphibians) (Fig. 3.53 B) is a specialized type of holospondylous vertebra in which the centrum of the solid vertebra is husk shaped and usually perforated by a notochordal canal.

Types of vertebra based on shape of centra. The centra of vertebrae are linked successively into a chain of vertebrae, called **axial column**. The shapes of surfaces at the articular ends of the

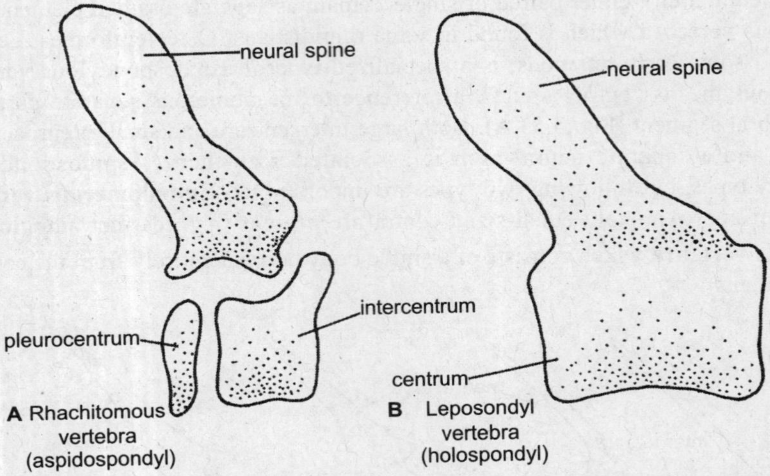

Fig. 3.53. General vertebral types. A – An aspidospondyl vertebrae is characterized by ossified elements that remain separate. The specific type illustrated is a rhachitomous (*i.e.*, "cut-up-spine") vertebra that has three discrete parts: pleurocentrum, intercentrum and neural spine. B – A holospondyl vertebra is characterized by fused construction of all components. The specific type shown is a lepospondylous vertebra with a husklike (husk means the dry outer covering of some fruits or seeds) centrum (after Kardong 2006).

centra affect the properties of the vertebral column and the way in which forces are distributed between vertebrae. There are following six principal shapes of centra.

1. Acoelous or amphiplatyan vertebra. It is found in mammals and has centrum which is flat at its both ends (Fig. 3.54 A). Since centra seem especially suited to receive and distribute compressive forces within the vertebral column.

2. Amphicoelous vertebra. This vertebra has its centrum concave at both ends (Fig. 3.54 B). This is the most primitive type and is found nearly in all fishes. Such centra seem to allow limited motion in most directions.

3. Procoelous vertebra. In this type of vertebra the centrum is concave anteriorly and convex posteriorly (Fig. 3.54 C). It is found in frogs and most reptiles.

4. Opisthocoelous vertebra. This type of vertebra has a reverse shape. It is concave anteriorly and convex posteriorly (Fig. 3.54 D). It occurs in all classes of vertebrates.

5. Heterocoelous vertebra. This type of vertebra has the centrum which bears saddle-shaped articular surfaces at both ends (Fig. 3.54 E). This is most specialized vertebra and is found in neck region of birds.

Functions of centra. In both procoelous and opsithocoelous vertebra, the convex articular surface of one centrum fits into the concave surface of the next to form a kind of ball-and-socket joint, permitting extensive motion in most directions without stretching the nerve cord that their neural arches protect.

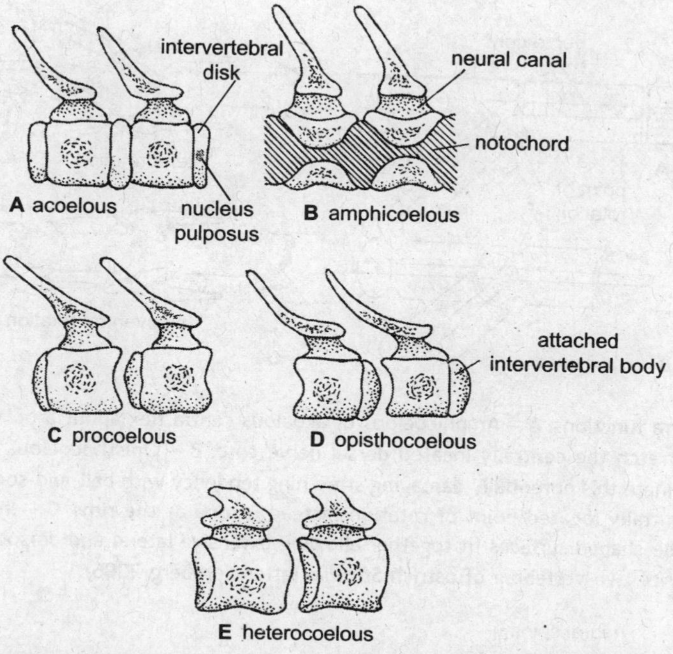

Fig. 3.54. General shapes of vertebra. The shapes of articulating centra ends, as viewed in sagittal section, define specific anatomical types. A – Acoelous, both ends are flat. B – Amphicoelous, both ends are concave. C – Procoelous, anterior end is concave; D – Opisthocoelous, posterior end is concave. E – Heterocoelous, saddle-like articulating ends: Anterior to the right (after Kardong 2006).

By comparison, if acoelous or amphicoelous vertebral series are flexed, adjacent centra hinge about their edges. If rotation is extensive, then like opening a door, the space will tend to widen between centra and stetch the central nerve cord running dorsally above them (Fig. 3.55 A). However, in procoelus and opisthocoelus centra, with ball-and-socket articulation, the point of rotation is not at the edge but at the center of the convex surface of the centrum. Flexion of the vertebral series does not open a space between them, and the central nerve cord is not unduly stretched (Fig. 3.55 B). Heterocoelus centra allow great lateral and vertical flexion, but they prevent wringing or rotation of the vertebral column about its long axis (Fig. 3.55 C). Heterocoelous centra are most common in turtles that retract their neck and in cervical vertebrae of the birds.

Intervertebral disk. Beside the centra, soft tissues are often associated with the vertebral column and they are usually extremely important in affecting function. The notochord, or its adult derivatives, often runs through and fills the concavities at the articular ends of centra, which are capped by cartilaginous pads. The term **intervertebral disk** has been used broadly to designate any pad of tissues between articular surfaces of centra. In fact, an intervertebral disk in the adult is a pad of fibrocartilage whose gel-like core, the **nucleus pulposus**, is derived from the embryonic notochord. By this strict definition, intervertebral disks are found only in mammals, in whom they reside between successive surfaces of adjacent centra. In other groups, the pad between centra is called an **intervertebral cartilage** or **body**, joining the rims of adjacent centra is the **intervertebral ligament**, which is important in controlling the stiffness of the vertebral column when it flexes.

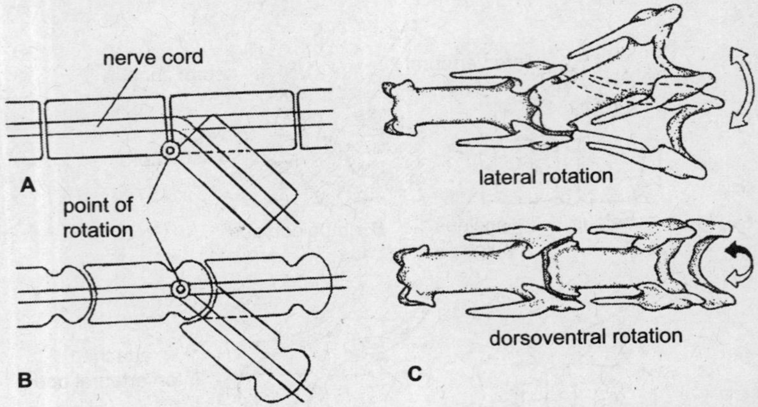

Fig. 3.55. Centra functions. A – Amphicoelous or acoelous centra flex about a point on their rim tending to stretch the centrally located dorsal nerve cord. B – Opisthocoelous and procoelous centra eliminate this potentially damaging stretching tendency with ball-and-socket ends that establish a centrally located point of rotation instead of one at the rims. C – In heterocoelous centra, saddle-shaped surfaces fit together allowing extensive lateral and dorsoventral rotation. Ventral view of two vertebrae of ostrich *Struthio* (after Kornberg 2006).

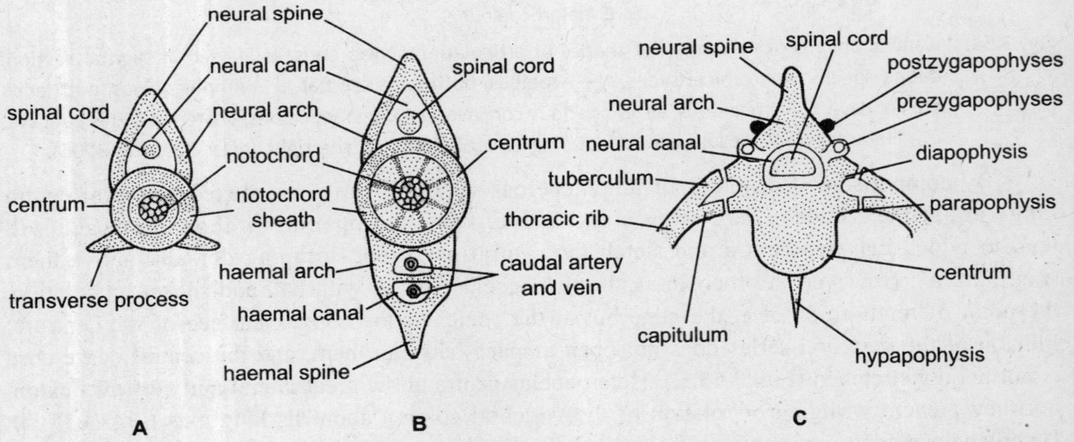

Fig. 3.56. Structure of a vertebra showing process (apophyses) in cephalic view A – Trunk vertebra of shark; B – Caudal vertebra of shark; C – Typical tetrapod vertebra.

Types of processes of vertebra. Various kinds of processes, called **apophyses**, project from centra and from their arches (Fig. 3.56 C) **(i) Zygapophyses** present over anterior and posterior faces of neural arch are articulations between successive vertebrae. **Prezygapophyses** are anterior projections facing upwards and **postzygapophyses** are posterior projections facing downward. **(ii) Transverse processes.** Lateral transverse processes arise from the neural arch or centrum and serve for attachement of the upper head (tuberculum) of a two-headed (bicipital) rib; **(iii) parapophyses** arise laterally from the centrum for attachment of the ventral head (capitulum) of a bicipital rib; **(iv) basapophyses** are ventro-lateral processes of the centrum or haemal arch; **(v) pleurapophyses** are lateral projections to which short ribs are fused; **(vi) hypapophysis** is a mid-ventral projection from the centrum.

4. Ribs

Ribs are struts that sometimes fuse with vertebrae or articulate with them. Ribs provide sites for secure muscle attachment, help suspend the body, form a protective case around viscera (rib cage), and sometimes serve as accessory breathing devices. Embryologically, ribs preform in cartilage within **myosepta** (myocommata), that is, within the dorsoventral sheets of connective tissue that partition successive blocks of segmental body musculature (Fig. 3.57 A). In tetrapods, one of these sets of ribs is lost and the other, apparently the dorsal ribs, persists to become the **trunk ribs** of terrestrial vertebrates. Ribs of primitive tetrapods are **bicipital** having two heads that articulate with the vertebrae. The ventral rib head, or **capitulum**, articulates with the parapohysis, a ventral process on the intercentrum. The dorsal head, or **tuberculum**, articulates with the diapophysis, a process on the neural arch (Fig. 3.58). If these vertebral processes fail to develop, the articular surface persists, forming a small concavity, the **facet**, to receive the rib. In amniotes, the intercentrum is lost or incorporated into other elements, so the capitulum must shift its articulation to the pleurocentrum (in most reptiles and birds) or between centra (in mammals).

Classification of tetrapod ribs is based on the type of association they establish with the sternum. Ribs that meet ventrally with the sternum are **true ribs**. Those ribs that articulate with each other but not with the sternum are **false ribs**. Those false ribs articulating with nothing ventrally are **floating ribs**. True ribs consists of two jointed segments, the **vertebral (costal) rib**, a proximal segment articulated with the vertebrae, and the **sternal rib**, a distal segment that is usually cartilaginous and meets the sternum. The joint between vertebral and sternal segments accommodates changes in chest shape during respiratory expansion and compression.

In birds, cervical ribs are reduced and fused to the vertebrae. In the thoracic region, the first several ribs are floating ribs, followed by true ribs that articulate with the sternum. Some floating and most true ribs bear **uncinate processes** (Fig. 3.58 D) projections that extend posteriorly from proximal rib segments. Uncinate processes, like the rib cage generally, offer sites of attachment for respiratory and shoulder muscles. In birds, these processes act primarily as lever arms for inhalatory muscles that flare (*i.e.*, gradually become wider at one end) the rib cage. Similar rib projections are also found in some living and fossil reptiles as well as in the early labyrinthodonts, *Acanthostega*, and *Ichthyostega*, where they project posteriorly to overlap with the next adjacent rib. This

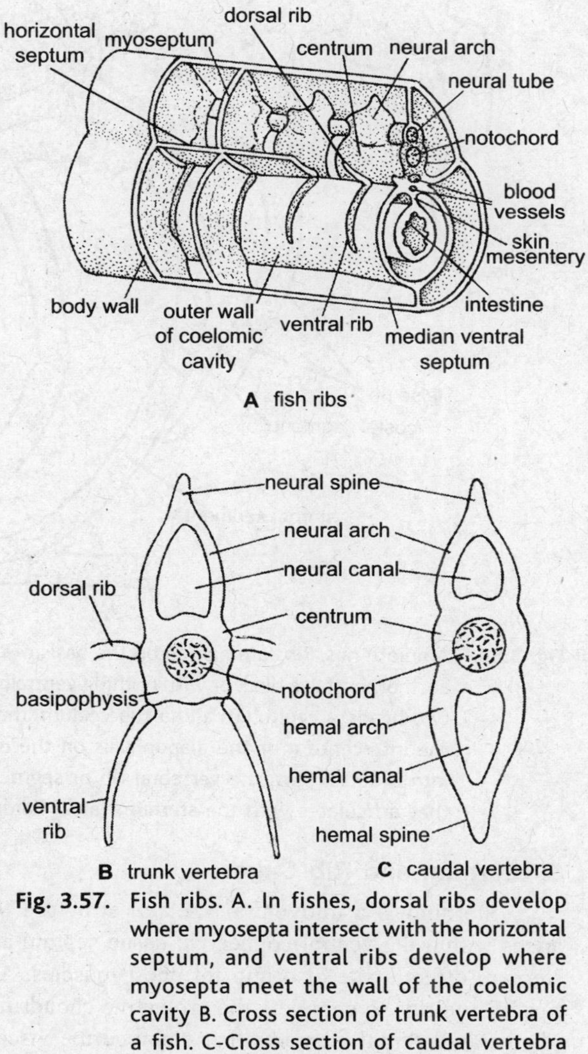

Fig. 3.57. Fish ribs. A. In fishes, dorsal ribs develop where myosepta intersect with the horizontal septum, and ventral ribs develop where myosepta meet the wall of the coelomic cavity B. Cross section of trunk vertebra of a fish. C-Cross section of caudal vertebra of a fish. Ventral trunk ribs are serially homologous with caudal hemal arches (after Kardong 2006).

overlap between successive ribs may introduce some overall firmness into the thoracic ribs, giving them the functional integrity to act as a unit during lung ventilation and locomotion.

In mammals, ribs are present on all thoracic vertebrae and define this region. Some are floating and others are false. Most are true ribs, and they meet the sternum through cartilaginous sternal rib segments. Within cervical and lumber regions, ribs exist only as remnants fused with transverse processes, forming what should properly be termed **pleurapophyses** (transverse process plus rib remnant).

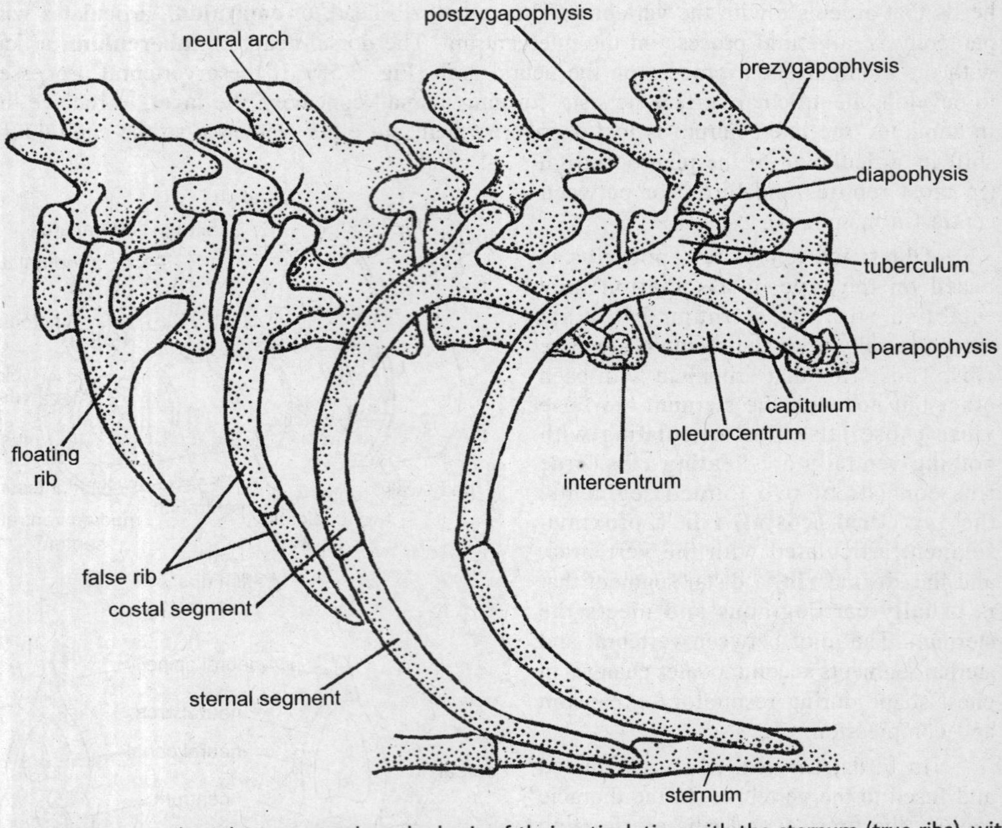

Fig. 3.58. Amniote ribs. Ribs are named on the basis of their articulation with the sternum (true ribs), with each other (false ribs), or with nothing ventrally (floating ribs). Primitively, ribs are bicipital, having two heads, a capitulum and a tuberculum, that articulate respectively with the parapophysis on the intercentum or the diapophysis on the neural arch. The body of the rib may differentiate into a dorsal part, the vertebral rib or segment, and a ventral part, the sternal rib or segment that articulates with the sternum (after Kardong 2006).

5. Sternum and Rib Cage

Sternum is a midventral skeletal structure that is endochordal in embryonic origin and arises within the ventral connective tissue septum and adjacent myosepta (Fig. 3.59 A to F). The sternum offers a site of origin for chest muscles. As described, sternum also secures the ventral tips of true ribs to complete the protective chondrified or ossified rib cage. The **rib cage** consists of ribs and sternal elements that embrace the viscera. Changes in the size and shape in the rib cage also act to compress or expand the lungs, promoting ventilation. The sternum may consists of a single bony plate or several elements in series.

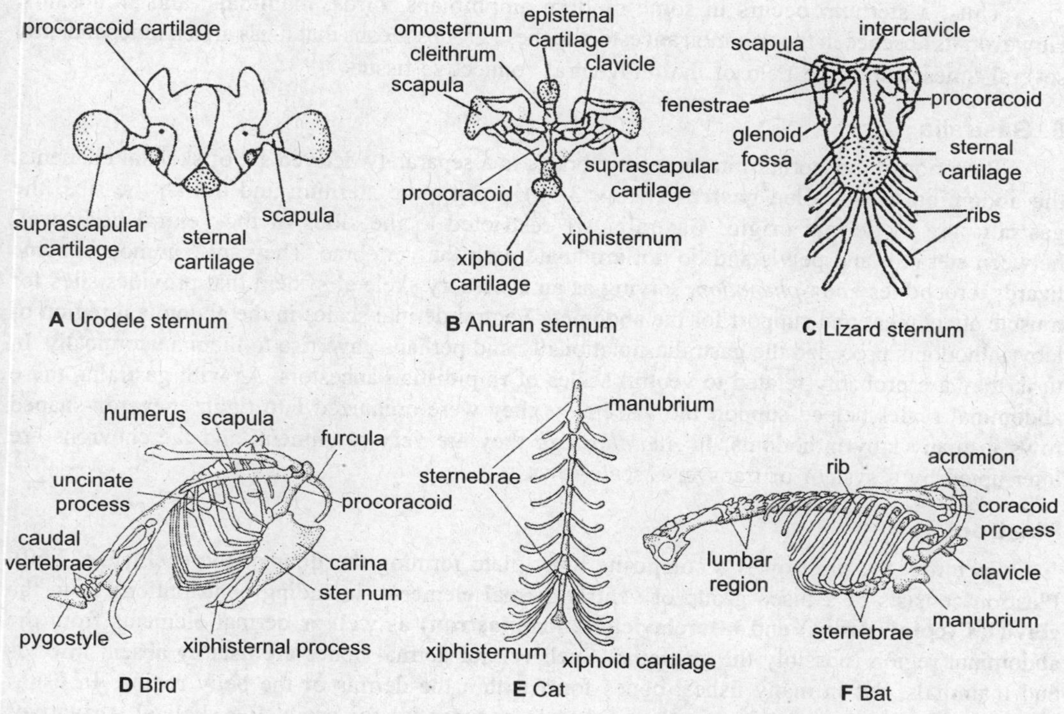

Fig. 3.59. Tetrapod sterna. A – Urodele, ventral view. B – Anuran, ventral view; C – Lizard sternum, ventral view. D – Bird sternum, lateral view. In birds, the sternum is deeply keeled, forming a carnia that offers increased attachment area for enlarged flight muscles within the axial column, the tail is short, ending in a specialized pygostyle that supports a fan of tail feathers; the pelvic bones and many of the vertebrae are fused; and shoulder is braced by the large procoracoid. E – Mammal (cat) sternum, ventral view. F – Bat sternum, lateral view. In bats, the sternebrae are robust and fused. Within the axial column, the lumber region and neural spines are short; the ribs are broad; and the coracoid process and clavicle are large, reflecting enhanced roles in flight (after Kardong 2006).

Fishes lack a sternum. When it first appears in tetrapods, the sternum is apparently not a phylogenetic derivative of either the ribs or the pectroal girdle, although in many groups it has secondarily associated with each. A sternum is absent in the first fossil tetrapods, but it is present in modern amphibians. In many urodeles, the sternum is a single midventral **sternal plate** grooved along its anterior borders to receive the ventral elements of the shoulder girdle, the **procoracoid plate** (Fig. 3.59 A). In anurans, a single element, the **xiphisternum**, often tipped with the **xiphoid cartilage**, lies posterior to the pectoral girdle; and in some, a second element, the **omosternum**, capped by the **episternal cartilage**, lies anterior to the girdle (Fig. 3.59 B). The sternum is absent in turtles, snakes and many limbless lizards, but is common in other reptiles, where it consits of a single, midventral element associated with the shoulder (pectoral) girdle (Fig. 3.59 C). During locomotion, the reptilian sternum confers stability on weight-bearing girdle elements. In flying birds, the massive flight muscles arise from a large sternum that bears a prominent ventral keel, the **carina**. The carnia provides additional surface for muscle attachment (Fig. 3.59 D). In most mammals, the sternum consists of a chain of ossified elements in series, the **sternebrae** (Fig. 3.59 F). The first and last of these are often modified and are called the **manubrium** and **xiphisternum**, respectively.

Thus, a sternum occurs in some modern amphibians, birds, mammals, and archosaurs. However, its absence in the common ancestors to these groups means that it has arisen independently several times within the field of the midventral connective tissue.

6. Gastralia

Posterior to the sternum in some vertebrates is a separately derived set of skeletal elements, the abdominal ribs, called **gastralia** (Fig. 3.50). Unlike the sternum and unlike the ribs, the gastralia are of dermal origin. Gastralia are restricted to the sides of the ventral body wall between sternum and pelvis and do not articulate with the vertebrae. They are common in some lizards, crocodiles and *Sphenodon*, serving as an accessory skeletal system that provides sites for muscle attachment and support for the abdomen. Ventral dermal scales in the abdominal region of labyrinthodonts preceded the gastralia functionally and perhaps gave rise to them anatomically. In turn, they are probably related to ventral scales of rhipidistian ancestors. As with gastralia, these abdominal scales helped support the viscera, as they were organized into tight, cheveron-shaped rows in many labyrinthodonts. In *Acanthostega* they are very prominent, and the chevrons are interrupted by a system of transverse scale rows.

Plastron

In turtle, the plastron is a composite bony plate forming the floor of the shell (Fig. 3.60). Plastron consists of a fused group of ventral dermal elements, including contributions from the **clavicles (epiplastrons)** and **interclavicle (entroplastron)** as well as dermal elements from the abdominal region (possibly the gastralia). Such ventral dermal bones are usually absent in birds and mammals, but in many fishes, bones form within the dermis of the belly region. In fishes and other vertebrates, the dermis shows the ready potential for producing skeletal derivatives like the gastralia independently in different phylogenetic lineages.

Median or Unpaired Fins

In discussing the axial skeleton, certain unpaired, median appendages have to be taken into consideration. Only aquatic vertebrates possess median appendages, which are always fin like in nature. Unpaired fins are found in cyclostomes, fish, larval amphibians, some adult urodele amphibians and cetaceans. Typically they are of three types: **dorsal fin, anal fin** and **caudal fin**. In some forms the median or unpaired fins are continuous (Fig. 3.61), but most frequently the continuity is interrupted and the various fins are separated by gaps. In some of the newts the dorsal fin of male becomes developed only during the breeding season. At other times, it is rather inconspicuous. Fin development and regression in these forms *(i,e.,* urodele amphibians) are controlled by the endocrine activity of the testes. In cyclostomes and fishes the median fins are supported by skeletal structures (fin rays) and special muscles, but in higher forms they are merely elaborations of the integument. Dorsal and anal fins are used chiefly in directing the body during locomotion. The caudal fins also help in this respect, in addition to being the main organ of propulsion.

The skeleton supporting the unpaired dorsal and anal fins consists of radial cartilaginous or bony **pterygiophores** supporting slender fin rays at their distal ends (Fig, 3.65A). Union of several pterygiophores may take place at the base of fin to form one or more **basipterygia** or **basalia**. These are ossified in bony fishes. Pterygiophores and basalia may form secondary connections with the neural spines of the vertebral column. There may or may not be segmental connection between the skeletal elements of the fin and the vertebrae. Distal to the basalia the pterygiophores continue as cartilaginous or bony radialia. From the radialia, extend numerous fin rays of dermal origin, supporting greater part of the fin. Frequently, the fin rays connect directly

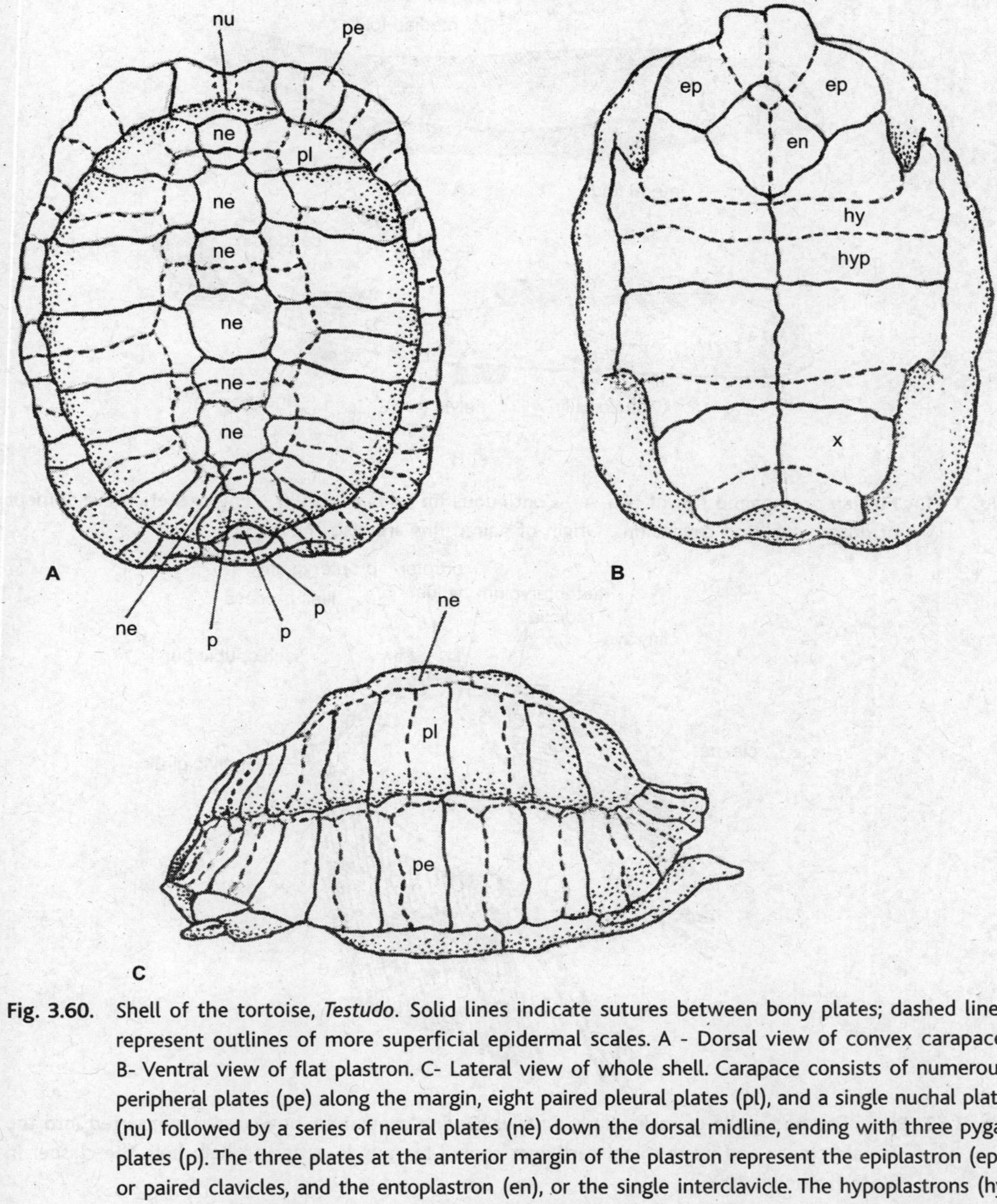

Fig. 3.60. Shell of the tortoise, *Testudo*. Solid lines indicate sutures between bony plates; dashed lines represent outlines of more superficial epidermal scales. A - Dorsal view of convex carapace. B- Ventral view of flat plastron. C- Lateral view of whole shell. Carapace consists of numerous peripheral plates (pe) along the margin, eight paired pleural plates (pl), and a single nuchal plate (nu) followed by a series of neural plates (ne) down the dorsal midline, ending with three pygal plates (p). The three plates at the anterior margin of the plastron represent the epiplastron (ep), or paired clavicles, and the entoplastron (en), or the single interclavicle. The hypoplastrons (hy, hyp) and the xiphiplastron (x) are the remaining plastral plates to be incorporated into the shell (after Kardong 2006).

with the basalia, the radialia being absent. Various types of joints permit the fin to be moved in a complex undulating manner or merely to be raised or lowered, as the case may be. In the males of many fish (elasmobranchs and holcephalians) the anal fin and its skeletal elements are modified to form a gonopodium (Fig. 3.62), which serves in the transport of sperm from male to female in internal fertilization. Development of gonopodium has been shown to be under the control of testosterone hormone.

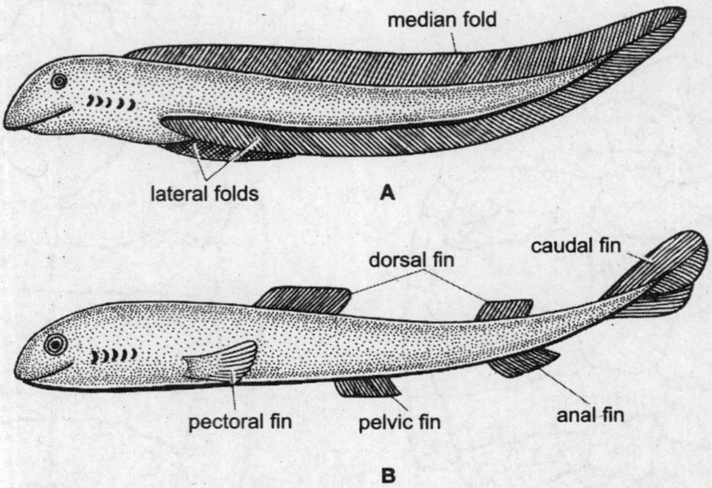

Fig. 3.61. Median or unpaired fins of fish. A – Continuous fin folds of fish. B – Phylogenetic development of dorsal, anal and caudal fins. Origin of paired fins are also shown.

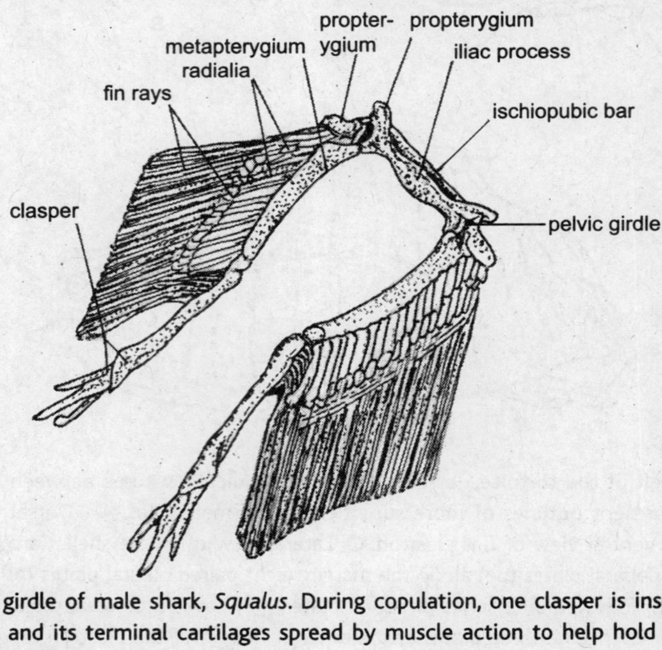

Fig. 3.62. Pelvic fins and girdle of male shark, *Squalus*. During copulation, one clasper is inserted into the female cloaca and its terminal cartilages spread by muscle action to help hold the clasper in place (kardong, 2006).

Caudal fin or tail fin. The caudal fin is well developed in most fishes because it plays the most important role in forward propulsion during swimming. The caudal fin is absent in *Hippocampus* (sea horse) and some of the eels. In *Hippocampus* the tail is prehensile. In bottom dwelling rays, the caudal fin tends to become reduced and in the sting rays the caudal fin is lacking. The caudal fin exhibits different shapes in different groups of fishes correlated with their habits, so that it is of great importance in classification of fishes. Three main types of caudal fins are found in fishes. These are diphycercal (protocercal), heterocercal and homocercal fins (Fig. 3.63).

1. Diphycercal (protocercal) caudal fin. This type of caudal fin of fishes is regarded to be the most primitive type. It is not exhibited by many living fish. The vertebral column extends up to the tip of the tail and divides the caudal fin into two equal halves. The dorsal half is called **epichordal lobe** and the ventral half is known as **hypochordal lobe**. The epichordal and hypochordal parts of the caudal fin are symmetrical and equal in size. Diphycercal caudal fin occurs in modern cyclostomes, primitive sharks, Holocephali (*Chimaera*), living Dipnoi (lung-fishes), living Crossoptergii (*Latimaria*), many larval teleosts and deep sea fishes. In *Chimaera* and some deep sea fishes, the caudal fin is called **isocercal** which is very much elongated and symmetrical. *Latimeria* and extinct coelocanths have unique symmetrical three- lobed tail with a marked median lobe. In living dipnoans (lung fishes), the caudal fin is of **protocercal type**.

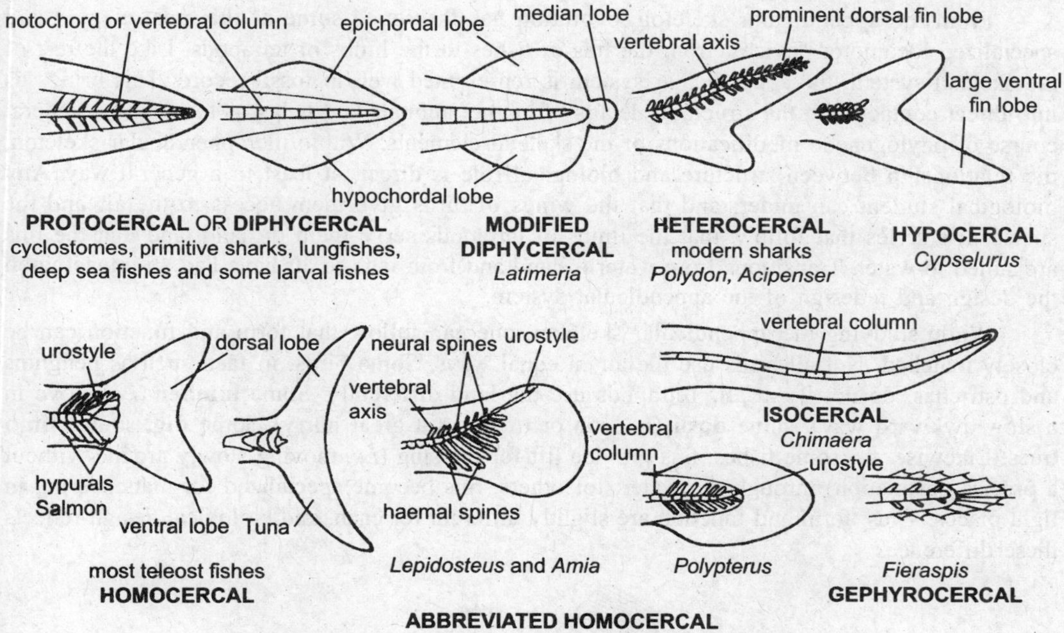

Fig. 3.63. Different types of caudal fin and fish tails.

2. Heterocercal caudal fin. It is the intermediate type in which vertebral column bends inwards and reaches up to the tip of the more prominent dorsal lobe, thus, making the caudal fin strongly asymmetrical (Gr., *heteros* = different, other). This type of caudal fin is found in modern elasmobranchs, extinct dipnoans (*Dipterus*) and living holosteans (*Acipenser, Polyodon*). Heterocercal caudal fin is characteristic of bottom feeders (fishes with ventral mouth and without swim-bladder or air-bladder). The strokes of larger dorsal lobe in swimming serve to direct fish towards bottom of water.

The opposite of heterocercal condition, in which the vertebral column terminates into a larger ventral lobe, is known as **hypocercal** (Gr., *hypos* = under, beneath) type. Hypocercal caudal fin is peculiar to flying fish, some primitive fishes and fossil agnatha (*i.e.*, ostracoderms). In flying fish (*Cypselurus*), the larger ventral lobe of caudal fin enables the fish to attain maximum speed for gliding as it leaves the water.

3. Homocercal caudal fin. This is the advanced and most common type of caudal fin of fishes. The homocercal (Gr., *homos* = common, alike) caudal fin is the characteristic of the higher bony fishes (teleosts). It is symmetrical externally but internally it is asymmetrical. The posterior end of vertebral column is turned upwards and becomes greatly reduced. The tip of vertebral column do not reach the posterior limit of fin. There is no apparent dorsal lobe but the ventral lobe is greatly enlarged and divide into two equal superficial lobes. Most of the teleosts retain the typical homocercal condition, but in some forms this condition is slightly modified.

For example, in cod and tuna, homocercal condition is modified. The upward terminal portion of vertebral column is withdrawn. In deep sea fishes, the terminal part of vertebral column is straight and greatly elongated to form **isocercal** caudal fin. In *Fieraspis* and *Orthagoriscus*, the caudal fin completely disappears forming a **gephyrocercal** caudal fin. In chondrosteans (*Polypterus Amia, Lepidosteus*) the vertebral column is a bit elongated and upturned, also with or without an upturned fleshy lobe. This is called **abbreviated homocercal** caudal fin. Most of these variations represent an intermediate stage between heterocercal and homocercal types.

3.4 APPENDICULAR SKELETON

From the appendicular skeleton, evolution has fashioned some of the most elegant and specialized locomotor devices, from the fins of fishes to the limbs of tetrapods. Like the rest of the skeletal system, the appendicular system is represented well in fossil record. This brings us into direct contact with the structural details of extinct animals and helps us to track the general course of phylogenetic modifications of the skeletal elements. Within the appendicular skeleton, the relationship between structure and biological role is direct, at least in a general way. Any biological student can understand that the wings of birds give them access to the air and the special life styles that follow, that the limbs of tetrapods serve them on land, and that the fins are suited to water. Transitions from water to land and from land to air have had an impact upon the design and redesign of the appendicular system.

While studying the appendicular skeleton, one can follow that form and function can be closely matched. Not all birds use the air in equal ways. Some birds, in fact, such as penguins and ostriches, do not fly at all. Tetrapods use the land differently. Some **lumber** (*i.e*, move in a slow awkward way), some **dash** (*i.e*, run or travel in a great hurry), some **dig**, some **climb** trees. Likewise, for some fishes, fins provide lift for cruising (*i.e*, moving slowly around without a precise destination) through the water; for others, fins become specialized for maneuvering in tight places. Thus, form and function are slightly different for each, and biological design reflects these differences.

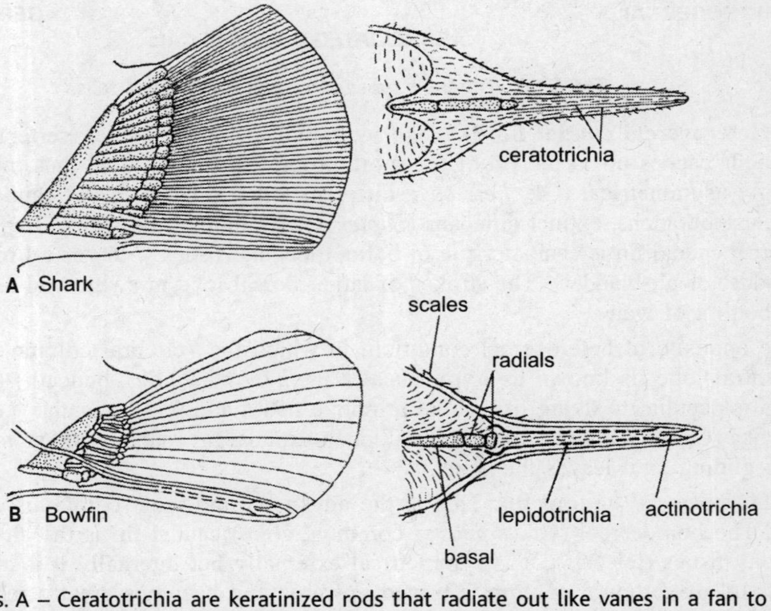

Fig. 3.64. Fin rays. A – Ceratotrichia are keratinized rods that radiate out like vanes in a fan to internally support the fins of elasmobrach or chondricthian (or chondrichthyan) fishes. B – lepidotrichia are cartilaginous or ossified supports within fins of bony or teleostian fishes (after Kardong 2006).

Basic Components of Appendicular Skeleton

The appendicular skeleton of vertebrates includes the **paired fins** or **limbs** and the **girdles**, the braces within the body that support them. The anterior girdle is the **shoulder** or **pectoral girdle**, to which dermal and endochondral skeletal elements contribute, and that support a **pectoral fin** or **limb**. The posterior girdle is the **hip** or **pelvic girdle**, consisting of endochondral skeletal elements that support the **pelvic fin** or **limb**.

Paired Fins

The body of primitive fishes is apt to carry projecting spines, lobes, or processes. Unlike these projections, **fins** are membranous or webbed processes internally strengthened by radiating and thin **fin rays**. These rays form initially at the interface between dermis and epidermis, like scales, but then sink into the dermis, and so are sometimes called **dermal fin rays**. Dermal fin rays are of two types:

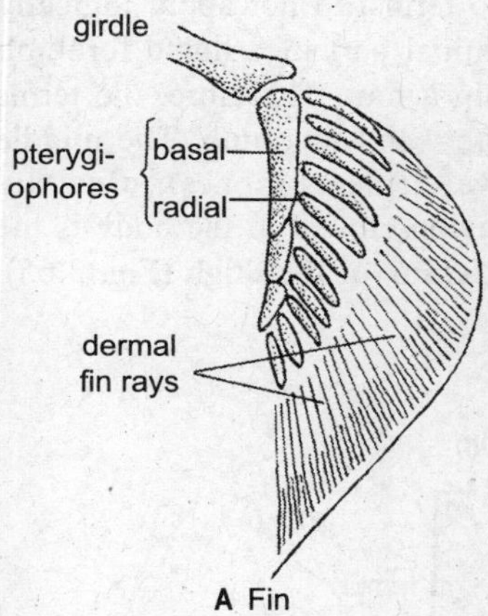

1. Ceratotrichia. These dermal fin rays occur in elasmobranchs and they are slender keratinized rods (Fig. 3.64 A).

2. Lepidotrichia. These dermal fin rays occur in bony fishes and are usually ossified or chondrified series of tiny elements that strengthen this web (Fig 3.64B). In some bony fishes, the tip of the fin may be further stiffened by additional keratinized rods, called **actinotrichia**. The proximal part of the fin close to the body is supported by **pterygiophores** of two general types: the enlarged **basals** within the proximal part of the fin, and the slender **radials** that extend support from the basals into the middle region of the fin (Fig. 3.65A).

As already described, fins occur singly, except for a pair near the head and a second pair posterior to this, the **pectoral** and **pelvic fins**, respectively. The basal pterygiophores of these projecting paired fins articulate with and are braced by girdles inside the body wall. These paired fins are the phylogenetic source of the tetrapod limbs.

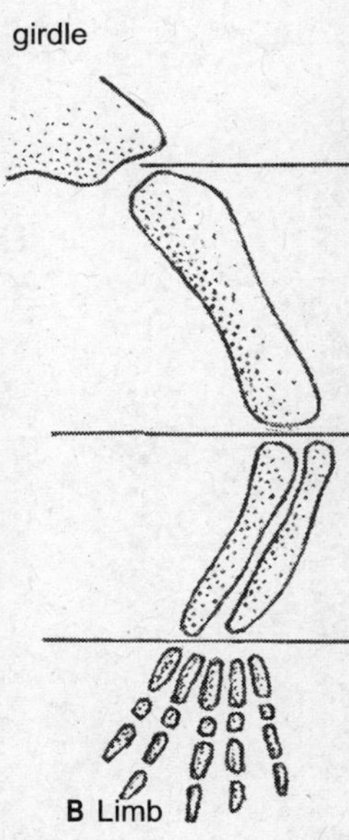

Morphological term	Forelimb	Hindilimb
Stylopodium	Upper arm	Thigh
Zeugopodium	Forearm	Shank (crus)
Autopodium	Manus (wrist-palm-fingers)	Pes (ankle-sole-toes)

Fig. 3.65. Basic components of the fin and the limbs A – The fin is composed of pterygiophores, basals and radials and dermal fin rays. Fin rays are called lepidotrichia in bony fishes and ceratotrichia in elasmobranchs. B – The limb, either fore - or hindlimb, includes three regions: stylopodium (upper arm/ thigh), Zeugopodium (forearm/shank), and autopodium (manus/pes) (after Kardong 2006).

Limbs

The limbs of all tetrapods have evolved from a common basic plan of a **pentadactyl limb** though they now show a wide variation in adaptation to their use—walking, running, leaping, climbing, flying, swimming and burrowing. The name **pentadactly** refers to typical condition of having five fingers or toes (G. *pente* = five + *dactylos* = finger). The pentadactyl limbs typically occur in two pairs: **forelimbs** and **hindlimbs**. Both the pair of limbs have the same segments, which are supported by skeletal elements arranged on the same general plan (Fig. 3.66).

Formally, a limb is termed a **chiridium**, a muscular appendage with well-defined joints bearing digits (fingers and toes) not a fin, at its ends. The chiridia of tetrapods forelimbs and hindlimbs, are all built, on the same pattern, composed of three recognized regions. The **autopodium**, the distal end of the limb, consists of numerous elements composing the **wrist** and **ankle**, which in turn support their respective digits (Fig. 3.65B). The special term **hand** implies a structure modified for grasping, and **foot** suggests a role in standing. But these two terms do not apply logically across the tetrapods (see **Kardong** 2006). For instance, the terminal part of a horse forelimb is not a hand, nor is the terminal part of the hindlimb of a dolpin a foot. Therefore, the terms **manus** and **pes**, respectively, are used for the autopodium of fore - and hindlimb. The middle limb region is called **zeugopodium**, with two internal supportive elements (bones): **ulna** and **radius** of the forearm, **tibia** and **fibula** of the shank. The limb region closest to the body is the **stylopodium**, with a single element: **humerus** of the upper arm, **femur** of the thigh (Fig. 3.65).

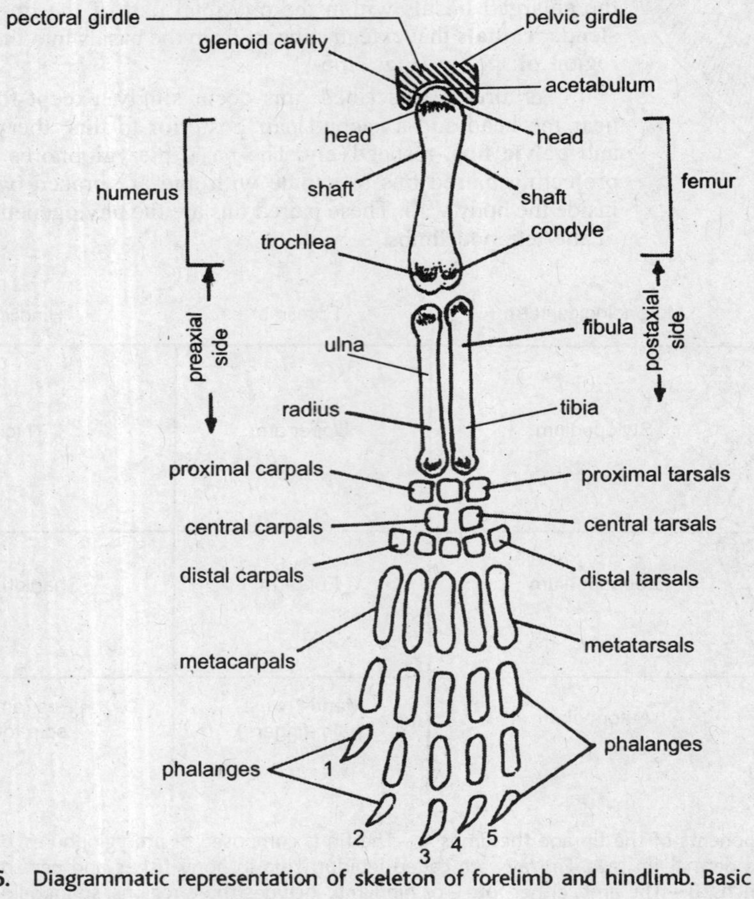

Fig. 3.66. Diagrammatic representation of skeleton of forelimb and hindlimb. Basic resemblance between the two is noteworthy.

A depression within the pectoral girdle, the **glenoid fossa**, articulates with the humerus. A deep socket in the pelvis (pelvic girdle), the **acetabulum**, receives the femur.

Origin of Paired Fins

The body of a fish is liable to deflections from its line of travel in three-dimensional space about its center of mass. It may swing from side to side (called **yaw**), rock about its long axis (called **roll**), or buck (jump) forward and back (called **pitch**, Fig. 3.67A). **Wind tunnel tests** using shark models with selected fins removed have helped clarify how fins bring stability to a streamlined body. It is worked out that perhaps dorsal and lateral fins control the body by resisting perturbations of the body about its center of mass. Separate tests of pectoral fins show

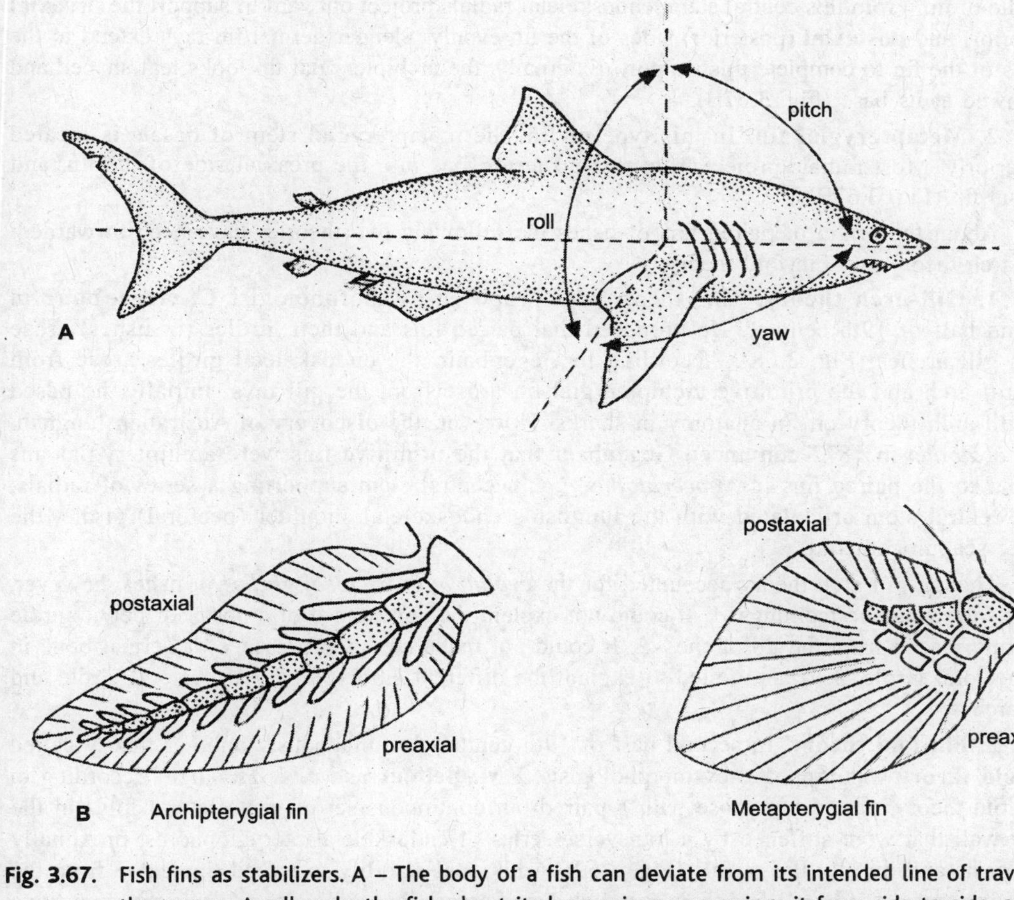

Fig. 3.67. Fish fins as stabilizers. A – The body of a fish can deviate from its intended line of travel in three ways. A roll rocks the fish about its long axis, a yaw swings it from side to side, and a pitch bucks it up and down about its center of mass. B – There are two types of fin structure, the archipterygial fin with a symmetrical central axis (left) and metapterygial fin with the asymmetrical axis displaced toward the postaxial side (right). The metaphygial axis or stem present in both is shaded (after Kardong 2006).

that these fins do not produce significant lift as a fixed-winged air craft. Instead, pectoral fins are used for close maneuvering within enclosed vegetation (in some teleosts) and during steady, horizontal swimming (sharks) for initiating rising or sinking turns.

As early fishes became more active, they would have experienced instability while in motion. Presumably, just such conditions favoured any body projection that resisted pitch, roll or yaw, and led to the **evolution of the first paired fins.** The associated girdles stabilized the fins, served as sites for muscle attachment and transmitted propulsive force to the fish body.

In gnathostome fishes following *two* basic types of fins have been evolved due to two different arrangements of the **metapterygial stem** or **axis**, a chain of endoskeletal basals: Archipterygial fin and Metapterygial fin.

1. Archipterygial fin. In this type of fin, the axis or metapterygial stem run down the middle of fin. From this central stem, endoskeletal radials project outward to support the **preaxial** (anterior) and **postaxial** (posterior) sides of the fin evenly. Slender dermal fin rays extend to the edges of the fin to complete this support. Externally, the archipterygial fin looks leaf-shaped and narrowed at its base (Fig. 3.67B).

2. Metapterygial fin. In this type of fin, the metapterygial stem of basals is located posteriorly. Most radials project from this posterior axis into the preaxial side of the fin, and dermal fin (Fig. 3.67B).

About the origin of paired fins of fishes the following two theories have been forwarded: Gill-arch theory and Fin-fold theory.

1. Gill-arch theory. This theory was proposed by morphologist **C. Gegenbaur** in second half of 19th century. He proposed that paired fins and their girdles (in fishes) arose from gill arches (Fig. 3.68). According to Gegenbaur, the endoskeletal girdles arose from the gill arch and the primitive archipterygial fin arose from the gill rays. Initially, he based his gill-arch theory on fin anatomy in sharks. However, the discovery of Australian lungfish, *Neoceratodus* in 1872 convinced Gegenbaur that the primitive fins were archipterygial fins similar to the paired fins in *Neoceratodus*, *i.e.*, a central stem supporting a series of radials. This central stem articulated with the lungfish's endoskeletal shoulder (pectoral) girdle, the future scapulocoracoid.

Though gill arch theory accounted for the *evolution of pectoral girdles* in fishes, however, it has following shortcomings: 1. It could not explain the appearance of a posterior pelvic girdle distantly placed from the gill arches. 2. It could not make clear the presence of dermal bone in the pectoral girdle. 3. It also failed to explain the different embryologies of pectoral girdle and gill arches.

2. Fin-fold theory. In second half of 19th century simultaneously another theory called **fin-fold theory** was proposed by morphologists, **F.M. Belfour** and **J.K. Thacher**. According to fin -fold theory, paired fins arose with a paired but continuous set of ventrolateral folds in the body wall that were stiffened by a transverse series of endoskeletal pterygiophores, proximally the basals and distally the radials (Fig. 3.69). Additional stability to paired fins came from the inward extension of basals and their eventual fusion across the midline to produce the supportive girdles. Dermal bone, a contribution of the overlying bony armour, was later added to the pectoral girdle to strengthen the paired fins further.

The fin-fold theory has been supported by several indirect evidences. The pregnathostomes with the first fin-folds were *Myllokunmingia* and *Haikouichthys* (Fig. 3.49). Both of these early Cambrian agnathans were equipped with fin-folds. Further, many surviving fossils of early fishes carry hints of presumed remnants of these earlier fin-folds. For instance, some primitive ostracoderms possessed lateral continuous folds along the ventral body wall. Acanthodians possessed a paired row of spines to mark where a paired fin-fold presumably resided in their

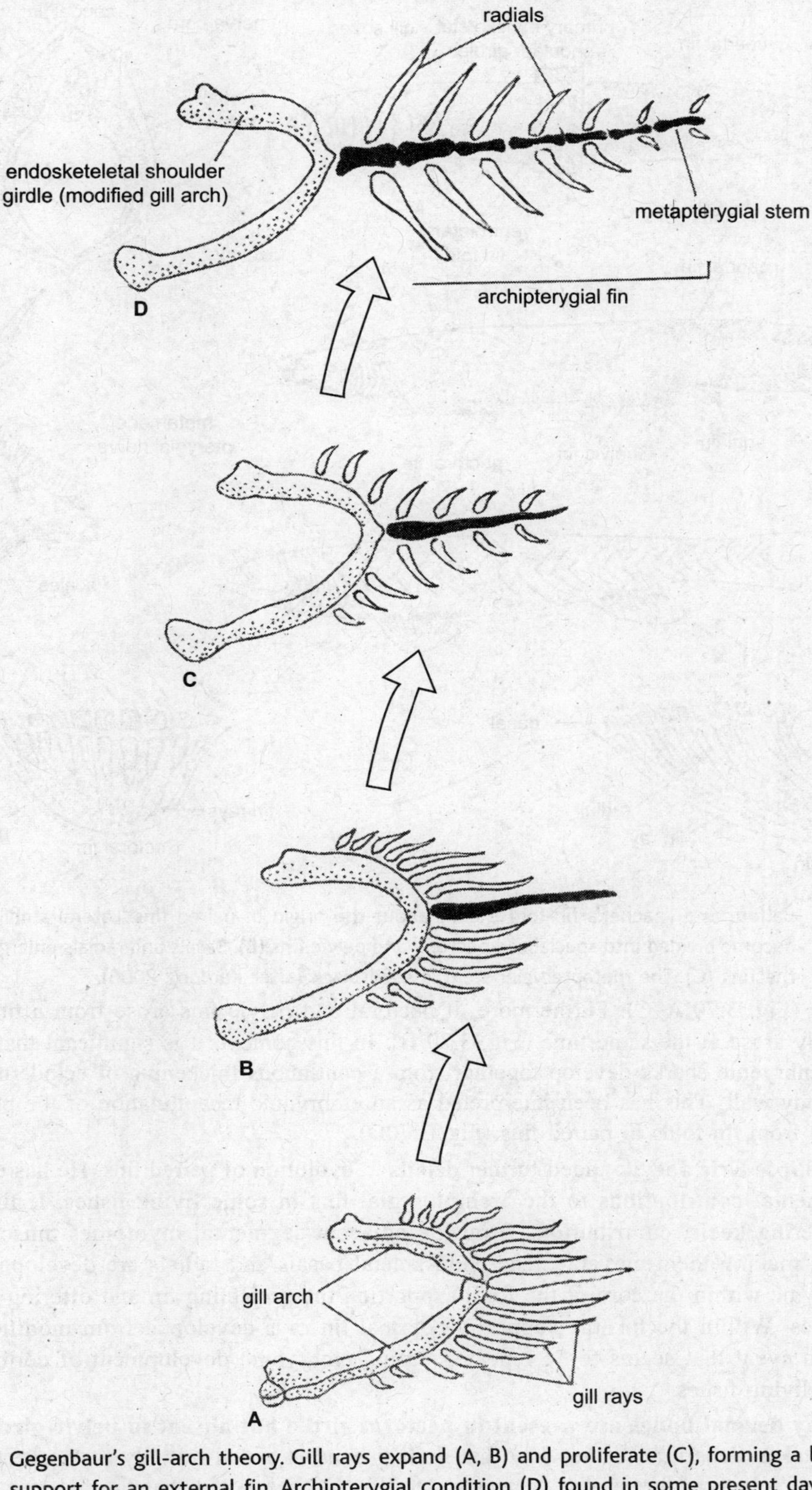

Fig. 3.68. Gegenbaur's gill-arch theory. Gill rays expand (A, B) and proliferate (C), forming a long central support for an external fin. Archipterygial condition (D) found in some present day lungfishes (after Kardong 2006).

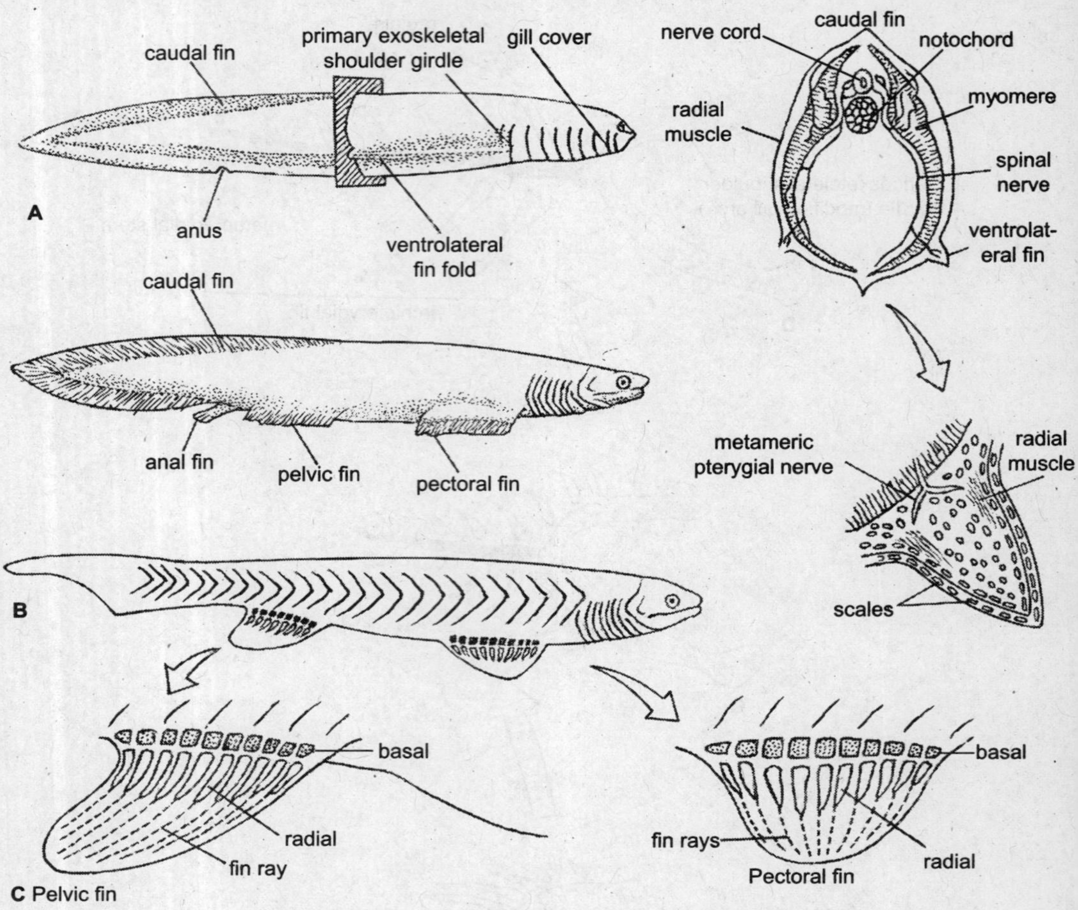

Fig. 3.69. Balfour and Thacher's fin-fold theory about the origin of paired fins. Lateral stabilizer fins (A) become divided into specialized pectoral and pelvic fins (B). Basals and radials enlarge to support the fins (C). The metapterygial stem is solid black (after Kardong 2006).

ancestors (Fig. 3.70 A–C). Furthermore, if pectoral and pelvic fins arose from a fin-fold, then they likely arose at the same time (Fig. 3.70 B). In this context, it is significant that the paired fins of embryonic sharks develop together from a continuous thickening of ectoderm along the lateral body wall. This has been interpreted as an embryonic recapitulation of the phylogenetic transition from fin-folds to paired fins (Fig. 3.70D).

In 1965, **Eric Jarvik** added further details to evolution of paired fins. He has emphasized the segmental contributions to the archipterygial fins in some living fishes. If fins initially were steering keels, contributions from the adjacent segmental myotomes muscularize the fin folds, making them movable. The endoskeletal basals and radials are developed from the mesenchyme within the core of the fold, supporting the projecting fin and offering attachment to muscles. Within the fin, the supportive dermal fin rays developed from modified rows of scales, an event that seems to be repeated during embryonic development of dermal fin rays in many living fishes.

Why dermal bones are present in pectoral girdle but absent in pelvic girdle? Jarvik has also taken departure to the view that dermal bone is added initially to the pectoral girdle because of selection forces favouring fin stability. He has noted that the fish pectoral girdle lies at the **transition** from trunk to head. At this point, the axial musculature is interrupted by the pharyngeal slits. Fusion of small dermal skin bones into a composite dermal pectoral girdle may have been initially advantageous because it offered an anterior site for attachment of the interrupted axial musculature at this point of transition. This dermal pectoral girdle would also form the posterior wall of the buccal cavity, protect the heart, and be the attachment site of some sets of muscles of jaws and gill arches. For one or all these reasons, an anteriorly placed (pectoral) girdle may have arisen and only secondarily joined with endoskeletal elements in support of fin.

Indeed, no similar selection force would be acting posteriorly, where the axial musculature runs uninterrupted from the trunk to the tip of the tail. This would help account for the presence of dermal contributions to the pectoral girdle and their absence from the pelvic girdle (see **Kardong** 2006).

Appendicular Skeleton in Tetrapods

The first tetrapods quickly displayed changes in the appendicular skeleton correlated with locomotion on land and exploitation of the terrestrial environment. The pectoral girdle tends to lose its attachment to the skull, a feature allowing increased cranial mobility and perhaps reduced jarring of the head. Girdles and limbs in tetrapods became stronger, more robust and more completely ossified (Fig. 3.71). One of the earliest tetrapod was *Ichthyostega*. Its pelvic girdle was composed of a single bone with three parts: pubis, ischium and ilium (Fig. 3.72).

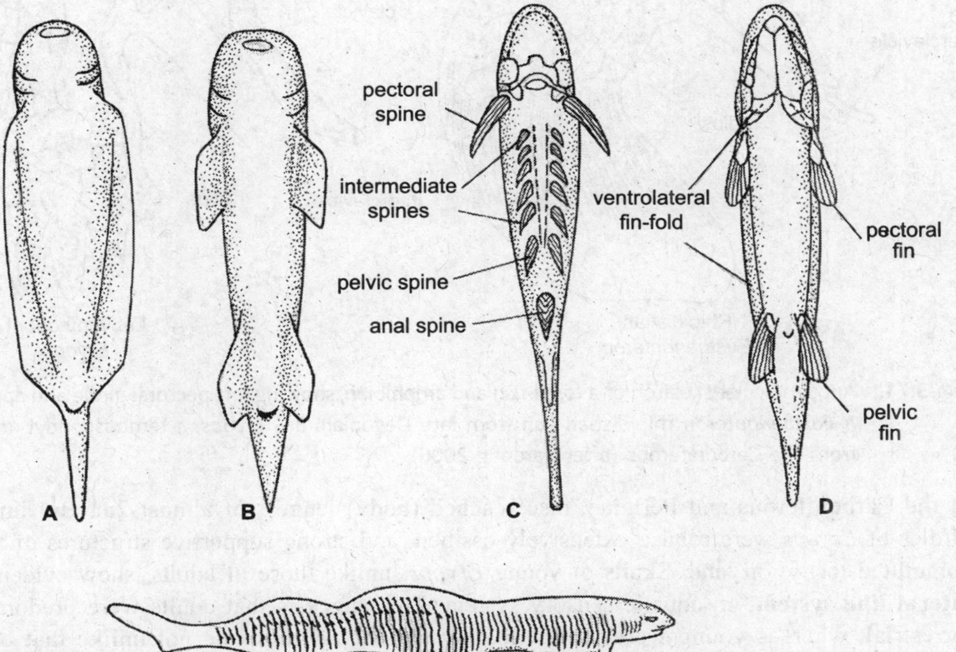

Fig. 3.70. Evidences of fin-fold theory. Indirect evidence pointing to the fin-fold theory can be seen from the position of rows of ventral spines in acanthodians (C), presumably remnants of the lost fin-fold in pregnathosome ancestors (A, B). Ventral views (A- C) can be compared with the ventral view of a bony fish (D). Further evidence can be drawn from shark embryology (E), in which the discrete paired fins develop from a continuous thickening along the lateral body wall of the shark, a developmental event taken to be reminiscent of phylogenetic events (after Kardong 2006).

Through the ilium, the pelvic girdle became attached to the vertebral column and so defined the sacral region (Fig. 3.72). The pectoral girdle lost its attachment to the skull. The fins of fish ancestors were replaced by digits.

Eogyrinus, a primitive labyrinthodont of the Carboniferous, reached over 2 m in length. The limbs and girdles, although strengthened, were relatively small for its size and mostly cartilaginous, so they were unlikely to be used to romp(sport) across the land. Instead, like modern salamanders and some fish, *Eogyrinus* must have used its limbs as pivot around which it moved. Limbs and girdles of early tetrapods became progressively larger and strengthened, reflecting an increased terrestrial habit and exploitation of the land. *Eryops* (Fig. 3.71B), a temnospondyl

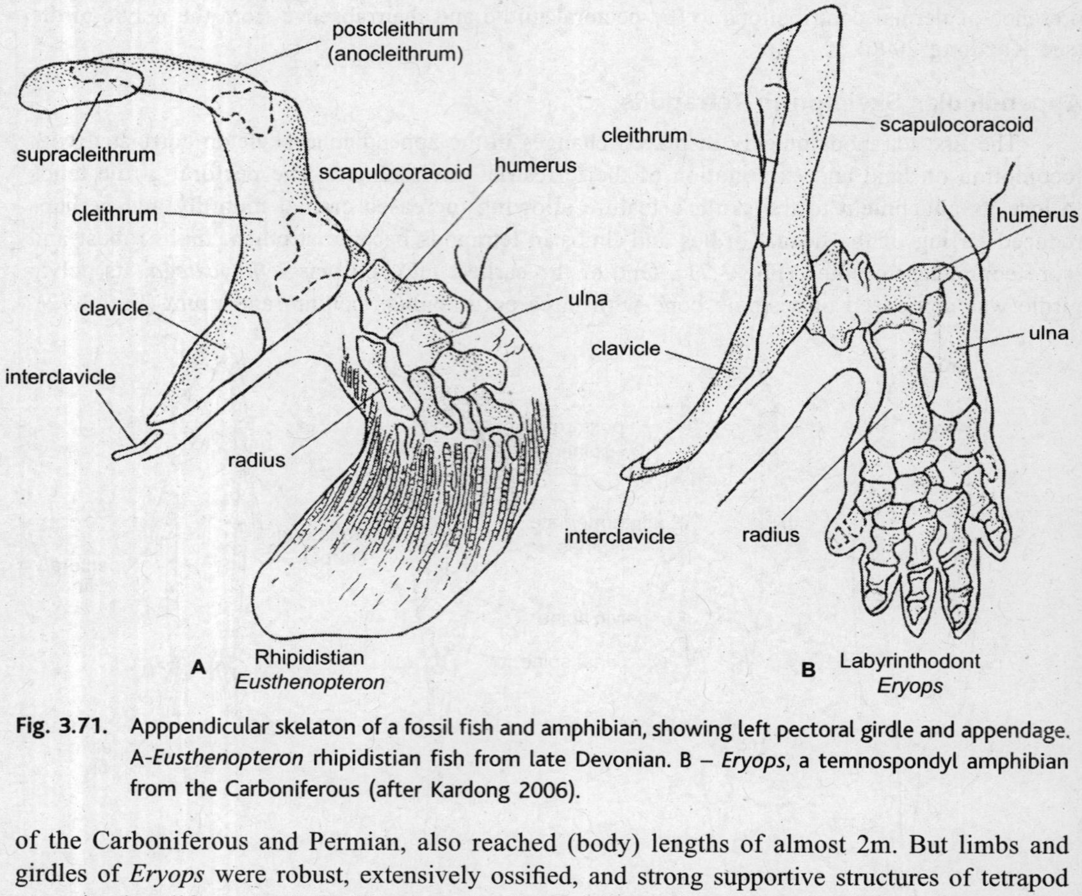

Fig. 3.71. Apppendicular skelaton of a fossil fish and amphibian, showing left pectoral girdle and appendage. A-*Eusthenopteron* rhipidistian fish from late Devonian. B – *Eryops*, a temnospondyl amphibian from the Carboniferous (after Kardong 2006).

of the Carboniferous and Permian, also reached (body) lengths of almost 2m. But limbs and girdles of *Eryops* were robust, extensively ossified, and strong supportive structures of tetrapod committed to life on land. Skulls of young *Eryops*, unlike those of adults, show evidence of a **lateral line system**, an aquatic sensory system. This suggests that adults were predominantly terrestrial, whereas young were predominantly aquatic, a life cycle not unlike that of many modern amphibians.

Pectoral Girdle of Tetrapods

Tetrapods carry over from rhipidistians a pectoral (or shoulder) girdle consisting of dermal and endoskeletal elements(bones), however, unlike their fish ancestors, tetrapods have a pectoral girdle that is structurally and functionally detached from the skull. Because the pectoral girdle is no longer connected to the back of the skull, the dorsal series of dermal bones, which previously were involved in establishing this connection in fish, are lost as well. Thus, in early tetrapods, the connecting skull bone, the **posttemporal**, and adjoining shoulder bones, **supracleithrum** and **postcleithrum** are absent, leaving a dermal pectoral girdle composed of remaining ventral elements: the paired **cleithrum** and **clavicle**, and an unpaired midventral **interclavicle** that joins both halves of the girdle across the midline (Fig. 3.73). In modern amphibians, the dermal bone are usually lost entirely, as in salamanders, or reduced in prominence, as in anurans. The endoskeletal scapuloprocoracoid becomes the predominant girdle element, yet retains its fidelity to the cleithrum (Fig. 3.73). In fishes, the scapulocoracoid, as its composite name suggests, tends to be a single element. However, in early tetrapods, it actually arises from two distinct embryonic centers of endochondral ossification and produces two distinct bones, the **scapula** and the **procoracoid** (Fig. 3.74).

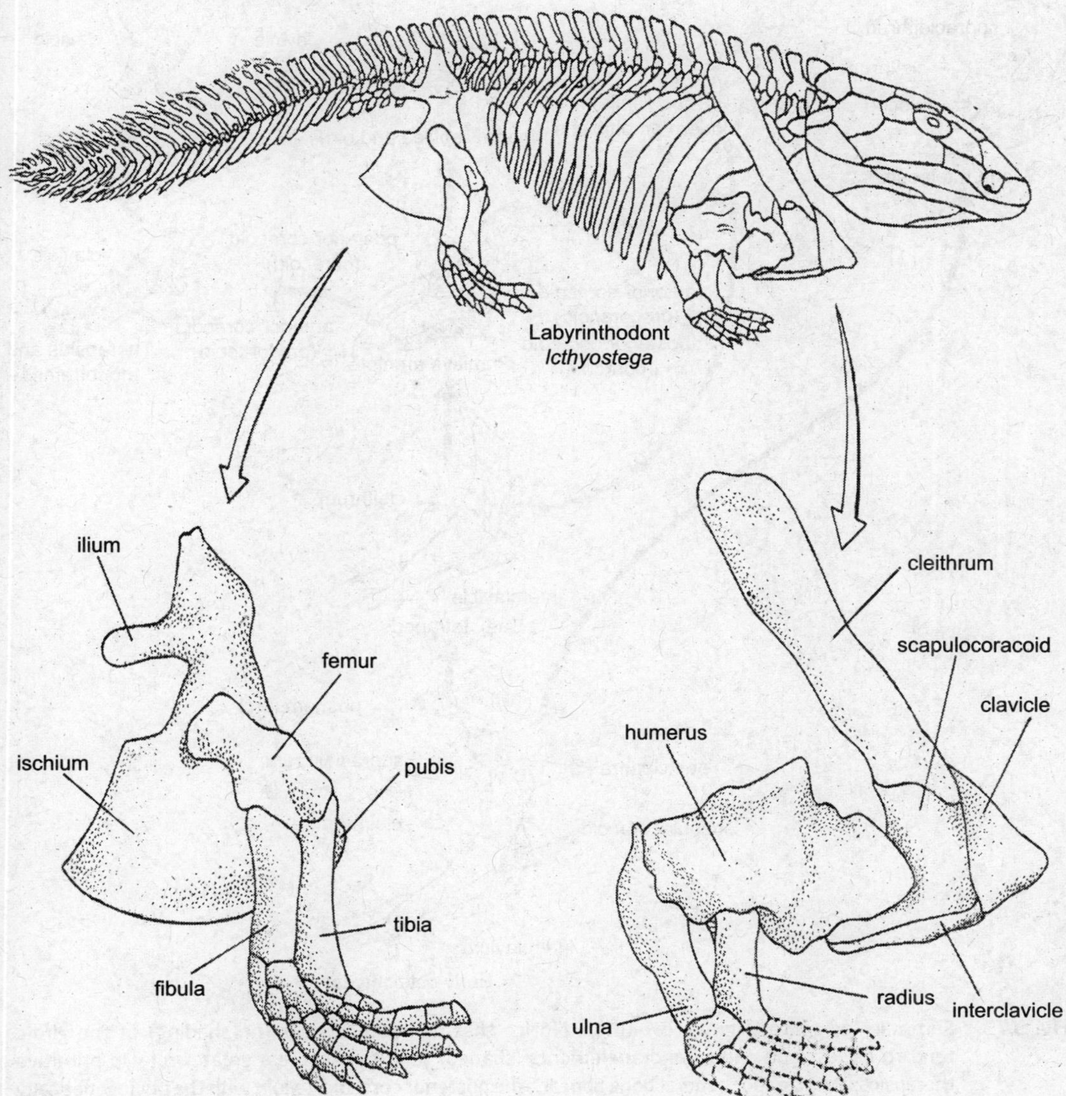

Fig. 3.72. Primitive labyrinthodont *Ichthyostega* showing components of appendicular skeleton The number of digits carried by the forelimbs is not known. Seven digits were present on the hindlimb (after Kardong 2006).

In primary amniotes, the **clavicle** and **interclavicle** persist, but the cleithrum is usually absent. The clavicle is lost in some modern reptiles, but it is retained in many. In turtles it is incorporated into the plastron as the **entoplastron**. In birds, the paired clavicle usually fuses with the unpaired interclavicle, producing the composite wishbone or **furcula**. Both scapula and procoracoid of the endochondral girdle persist. In fact, they now become a more prominent part of the pectoral girdle in birds and modern reptiles.

As mentioned earlier, endochondral elements of the early tetrapod shoulder develop from two centers of ossification, giving rise to a scapula and a "coracoid". However, in primitive

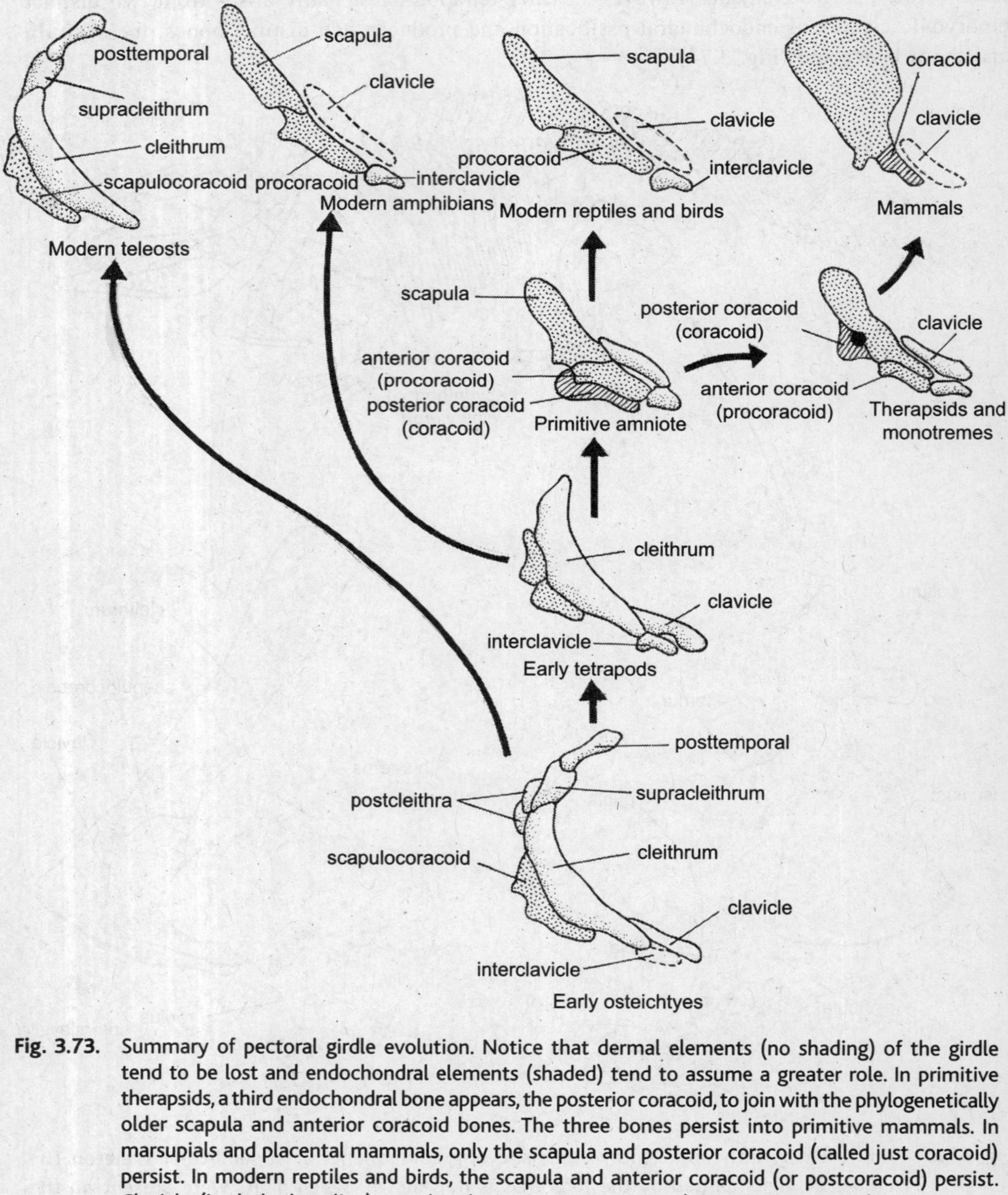

Fig. 3.73. Summary of pectoral girdle evolution. Notice that dermal elements (no shading) of the girdle tend to be lost and endochondral elements (shaded) tend to assume a greater role. In primitive therapsids, a third endochondral bone appears, the posterior coracoid, to join with the phylogenetically older scapula and anterior coracoid bones. The three bones persist into primitive mammals. In marsupials and placental mammals, only the scapula and posterior coracoid (called just coracoid) persist. In modern reptiles and birds, the scapula and anterior coracoid (or postcoracoid) persist. Clavicle (in dashed outline) may be absent in some species (after Kardong 2006).

synapsid reptiles, three centers of ossification develop. The dorsal center gives rise to the **scapula**, and two ventral centers produce two coracoids. To avoid confusion and to track the fate of each coracoid, separate names are given to these two coracoids. The anterior of these synapsid coracoid is homologous to the coracoid of fishes, amphibians and other reptiles. This **anterior coracoid** is more often called the **procoracoid** (precoracoid). The posterior of these synapsid coracoid is a new center of ossification, the **posterior coracoid**; it is more often called just the **coracoid**.

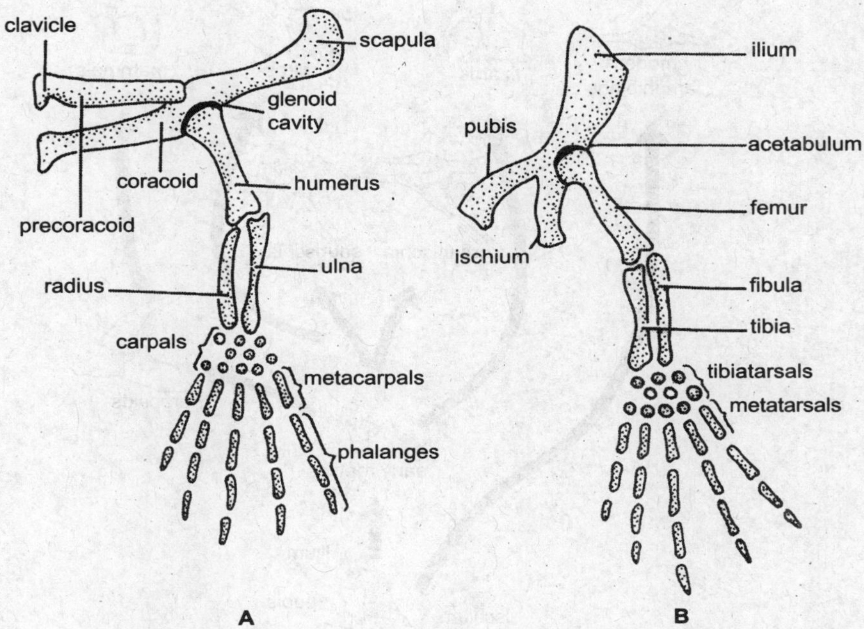

Fig. 3.74. Girdles and limb-bones of tetrapods. A – Pectoral girdle and forelimb. B – Pelvic girdle and hindlimb.

Both coracoids are present in pelycosaurs, therapsids and monotreme mammals, but only the coracoid (*i.e.*, posterior coracoid) persists in marsupial and placentals; the interclavicle is absent, the clavicle often is reduced in size and the scapula becomes the predominant shoulder element. The coracoid (*i.e.*, posterior coracoid) is reduced and fused to the scapula as the **coracoid process** (*e.g.*, opossum, *Didelphis*).

Pelvic Girdle of Tetrapods

The pelvic girdle is never joined by contributions of dermal bone. From its first appearance in placoderms, the pelvic girdle is exclusively endoskeletal. It arose from pterygiophores, perhaps several times, in support of the fin. In most fishes and in very early tetrapods, it is formed of single element (bone), but in tetrapods, three endochondral bones contribute: **ilium**, **ischium** and **pubis** (Fig. 3.75). Through the ilium, the pelvic girdle is attached to the vertebral column first in amphibians, establishing and therefore defining the sacral region. Throughout later amniotes, these three bones of the pelvic girdle persist, although their general pattern varies. For example, two distinctive patterns, the **saurichian** and **ornithischian pelvic girdles**, define two respective groups of dinosaurs. In birds, all three bones appear embryologically as distinct centers of ossification, but then they fuse to form the composite **innominate bone**, usually with no trace of sutures between them. Further, fusion between the composite **innominate** and composite **synsacrum** introduces considerable firmness in the posterior avian skeleton.

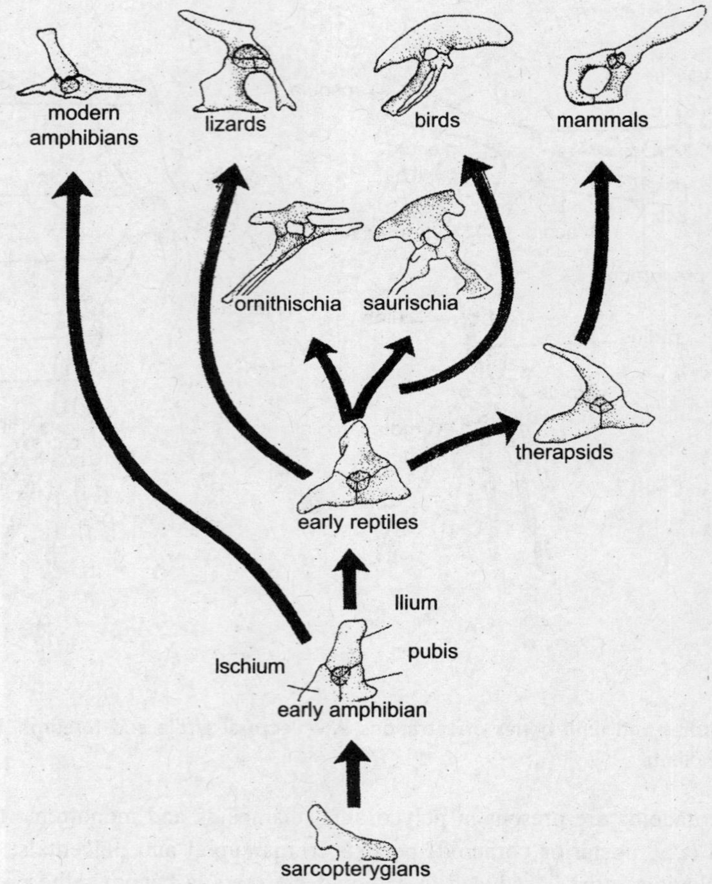

Fig. 3.75. Summary of pelvic girdle evolution. Three endochondral elements – ilium, ischium and pubis – characterize the pelvic girdle in early tetrapods. This basic pattern persists into later tetrapods (after Kardong 2006).

Manus and Pes

The autopodium at the end of each tetrapod limb has undergone a complex evolution of its own. Tracking this evolution has been difficult, largely because of the numerous elements that participate. There are several **digits**, each beginning proximally with a **metapodial element** (bone) (*i.e.*, **metacarpals** on the forelimb, **metatarsals** on the hindlimb) followed by a chain of **phalanges** (singular phalanx). The digits rest upon several separate bones, collectively known as the **carpals** in the wrist and the **tarsals** in the ankle. In some marine vertebrates (ichthyosaurs,

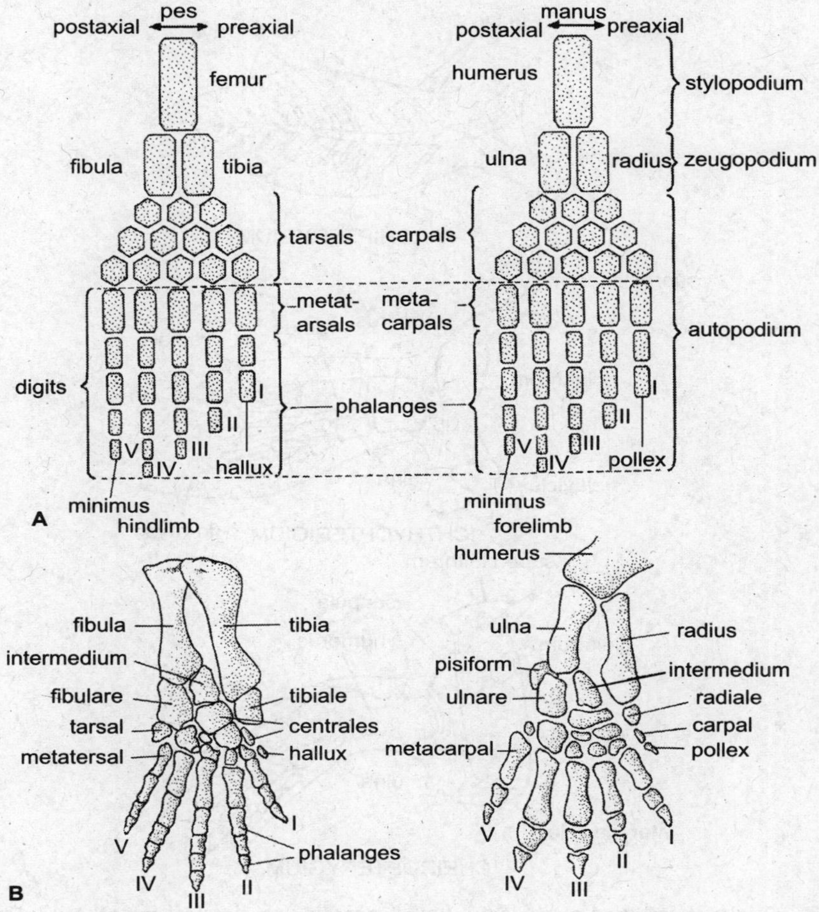

Fig. 3.76. Basic organization of the fore- and hindlimbs. A – Manus and pes have five digits; each digit includes its metacarpal or metatarsal and chain of phalanges. These digits in turn articulate with various wrist and ankle bones. B – Fore - and hindlimbs of primitive tetrapod (after Kardong 2006).

plesiosaurs, cetaceans, sirenians and marine carnivores, for example), the major trend has been toward **polyphalangy**, a proliferation in the number of phalanges. It is very unusual to species with more than five digits, a condition known as **polydactyly**. But in many groups, such as ungulates and some terrestrial carnivores, the opposite trend has occurred, namely, toward reduction in number of phalanges and loss or fusion of associated carpals and tarsals.

Usually, the basic tetrapod limb was thought to consist of five digits that were named and numbered (roman numerals) by their **pentadactylus** pattern (Fig. 3.76). Some were lost in specialized lines, but this underlying five-digit pattern is a logical hypothesis because the limbs of most primitive amniotes commonly carry five digits. New discoveries of early tetrapod specimens test this five-digit hypothesis. The hindlimb of *Ichthyostega* had seven digits (its manus remains unknown still, Fig. 3.78), the manus of *Acanthostega* included eight digits (the pes number is known; **Coats** and **Clark** 1990), and the fore- and hindlimbs of *Tulerpeton* had six digits. These late Devonian fossils are the earliest tetrapod remains available. Collectively, they indicate that the primitive tetrapods pattern was polydactylous and five-digit pattern is a later stabilization.

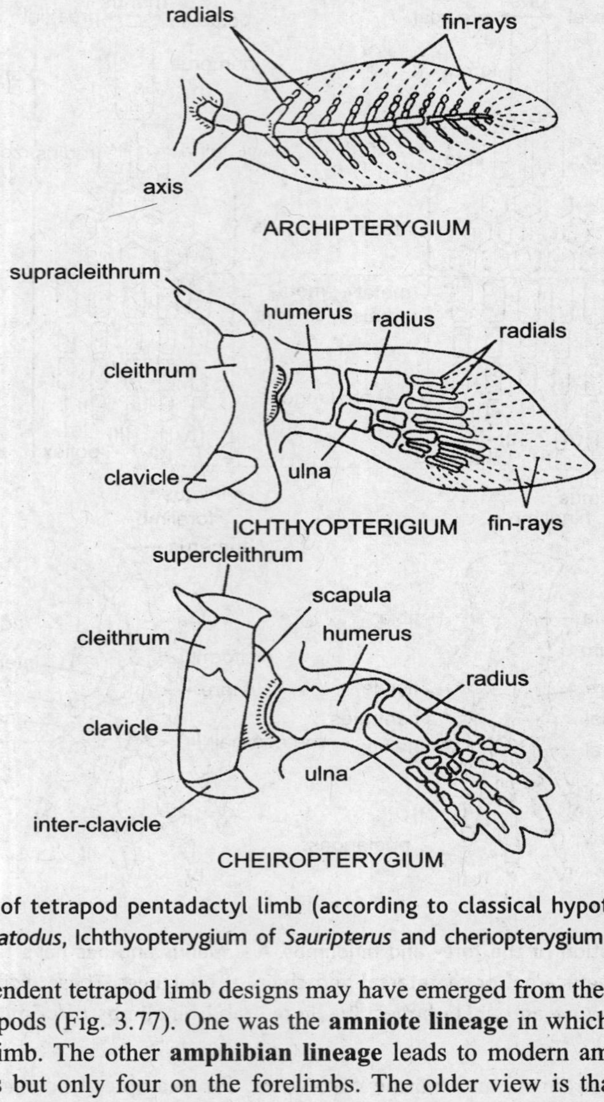

Fig. 3.77 Origin of tetrapod pentadactyl limb (according to classical hypothesis). Archipterygium of *Neoceratodus*, Ichthyopterygium of *Sauripterus* and cheriopterygium of *Eryops*.

Two independent tetrapod limb designs may have emerged from the polydactylous condition in primitive tetrapods (Fig. 3.77). One was the **amniote lineage** in which digit number stabilized at five on each limb. The other **amphibian lineage** leads to modern amphibians, with five toes on the hindlimbs but only four on the forelimbs. The older view is that this four-digit pattern is derived from a five-digit ancestor. If we trace modern amphibian ancestry back to these earliest tetrapods, then amphibian independently derived their reduced digit number directly from polydactylous ancestors. In fact, the five-digit pattern represents a useful basis for discussing limb evolution and change in functional design.

In the manus, a digit consists of several phalanges with a metacarpal at its base. In turn, each of the five metacarpals articulates with a carpel (Fig. 3.78). The wrist bones that articulate with the radius and ulna are, respectively, the **radiale** and the **ulnare**. The **intermedium** lies between these two wrist bones. Within the middle of the wrist are one of three **centrales**. In the pes, the primitive number of digits is also taken as five, each with a metatarsal at its base. In turn, each digit articulates proximally with the following sequence of bones: **tarsal**, **centrale**, **tribiale**, **intermedium** and **fibulare**, the last three meeting the **tibia** and **fibula** of the shank (Fig. 3.76).

While this grand pattern of expected manus and pes elements (bones) gives us a starting point when we look at distal limb anatomy, the actual morphology is often considerably modified by **fusions**, **elongations** and **additions** of apparently new elements to this pattern (Fig. 3.79

and Fig. 3.80). For example, the **pisiform** is a **sesmoid bone** that can lie to the outside of the carpus, espeically in reptiles and mammals. In birds, fusion of the forelimb elements produce

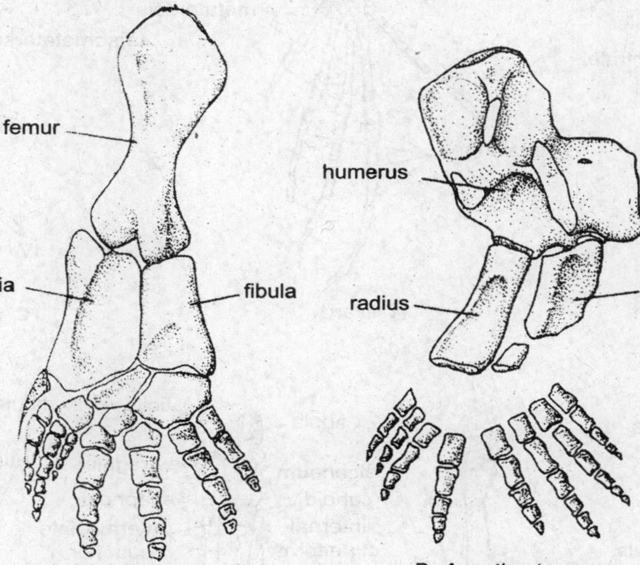

Fig. 3.78. Polydactylous limbs of the earliest tetrapods (dorsal views). A – Hindlimb of *Ichthyostega* with seven digits. B – Forelimb of *Acanthostega* (an ichthyostegalian) with eight digits (after Kardong, 2006).

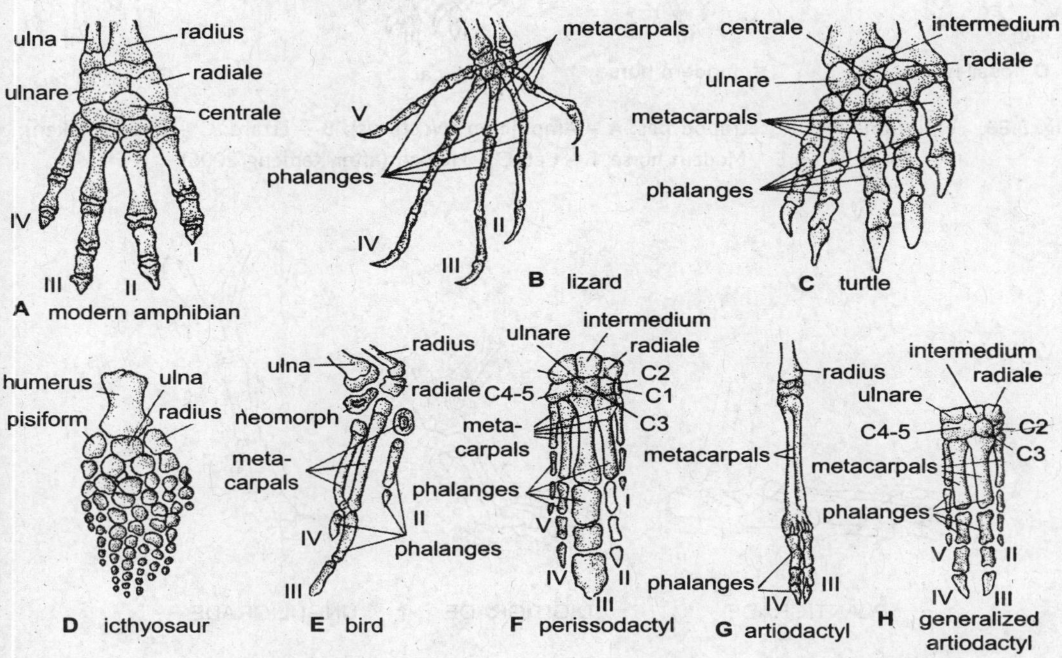

Fig. 3.79. Variations of the tetrapod manus. A – Amphibian (*Necturus*); B – Lizard; C – Turtle (*Pseudemys*); D – *Ichthyosaur*; E – Bird; F – Hypothetical primitive pressodactyl; G – Artiodactyl; H – Generalized artiodactyl (after Kardong 2006).

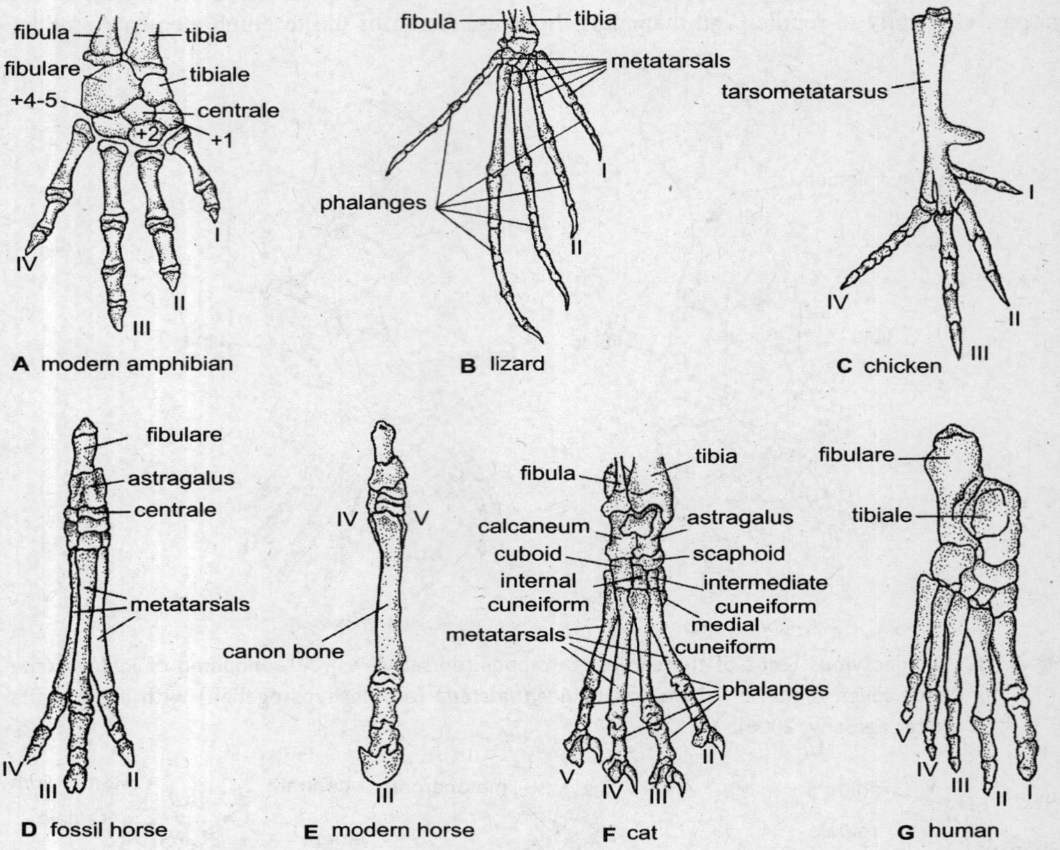

Fig. 3.80. Variations of the tetrapod pes. A – Amphibian (*Necturus*); B – Lizard; C – Bird (chicken); D – Fossil horse; E – Modern horse; F – Cat; G – Human (after Kardong 2006).

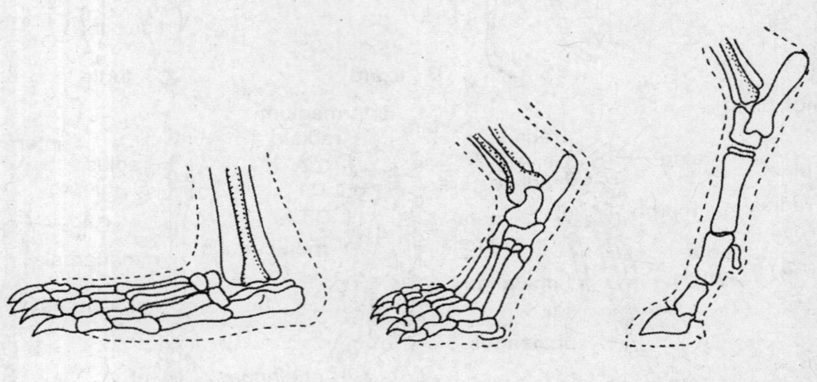

Fig. 3.81. Types of foot posture in mammals.

an autopodium with three digits (II, III, IV). The ulnare regresses during ontogeny, and in its space a neomorph arises from a new embryonic condensation (Fig. 3.79 E). This new wrist bone has not been named (see **Kardong** 2006). Some still call it an ulnare; others say "ulnare" in quotation marks. Some number it; others call it a neomorph.

In the hindlimb, lateral digits may be lost in cursorial mammals, and the medial metatarsals (III and IV) fuse into a composite ankle bone commonly called the **cannon bone** (Fig. 3.80 E). In mammals, the fibulare is the specific tarsal that articulates with the fibula, but it is more commonly called the **calcaneum**. The **astragalus**, a fusion of the tibiale and intermedium, articulates with the tibia (Fig. 3.80 F). Although originating in mammals, these names have been carried over to ankles of reptiles and birds; however, differences in embryonic contributions to these two bones in reptiles, birds and mammals perhaps reveal differences in homology and, strictly speaking, call for different names. But some uncertainties regarding embryonic contributions remain, and the convenience of the terms calcaneum and astragalus for tarsal bones of similar function justifies using these terms for all groups at least for the present.

Postures of limbs. In terrestrial mammals there are three kinds of postures of hands and feet in locomotion (Fig. 3.81).

1. Plantigrade. It is most primitive. In it the entire hand and foot are in contact with the ground, as in bears, rabbit, some insectivores and human beings.

2. Digitigrade. It is used for increased speed. In it only the digits are placed on the ground. The wrist and ankle are lifted above, as in dogs, cats and wolves.

3. Unguligrade. It is found only in the most swiftly running mammals. It is an extreme condition of locomotion in which only the tips of the digits are placed on the ground and the tips are covered with hoofs. This is the most specialized condition, as in deer, horse and cattle.

The appendicular skeleton in linked closely to the locomotor demand placed upon this system, especially in tetrapods. Consequently, it is a good system to illustrate the matching of form and function to changing environmental and, consequently, changing biochemical demands. The transition from water to land was accompanied by modifications in girdles and limbs, which became weight-bearing and important in developing the propulsive forces moving the tetrapod over land. Generally within terrestrial locomotion, the appendicular system was brought into the service of specialized biological roles - cursorial, aerial, fossorial and other modes of locomotion. These specialized modes of locomotion are reflected in the specific adaptations of form and function of appendicular system.

3.5 OSTEOGENESIS: DEVELOPMENT OF BONE

Three distinct embryonic lineages generate the skeleton. The somites generate the axial (vertebral) skeleton, the lateral plate mesoderm generates the limb skeleton, and the cranial neural crest gives rise to the pharyngeal arch and craniofacial bones and cartilage. There are two major modes of bone formation, or **osteogenesis**, and both involves the transformation of preexisting **mesenchymal tissue** into bone tissue. The direct conversion of mesenchymal tissue into bone is called **intramembranous ossification** (Box 3.3). In other cases, the mesenchymal cells differentiate into cartilage, which is later replaced by bone. The process by which a cartilage intermediate is formed and then replaced by bone cells is called **endochondral ossification** (**Horton**, 1990). Endochondral ossification is seen predominantly in the vertebral column, ribs, pelvis, and limbs (**Gilbert** 2010).

Box 3.3

There are three main types of intramembranous bone: **sesamoid bone** (such as patella), which forms as a result of mechanical stress; **periostal bone** which adds thickness to long bones; and

dermal bone, which forms in the dermis of skin. The first two types of intramembranous ossification are characteristics of mesodermal cells. Many of the dermal bones of the skull and face, however originate from cranial neural crest-derived mesenchymal cells.

The vertebrate skull, or **cranium**, is composed of the **neurocranium** (skull vault and base) and the **viscerocranium** (jaws and other pharyngeal arch derivatives). Skull bones are derived from both the neural crest and the head mesoderm (**Evans** and **Noden** 2006).

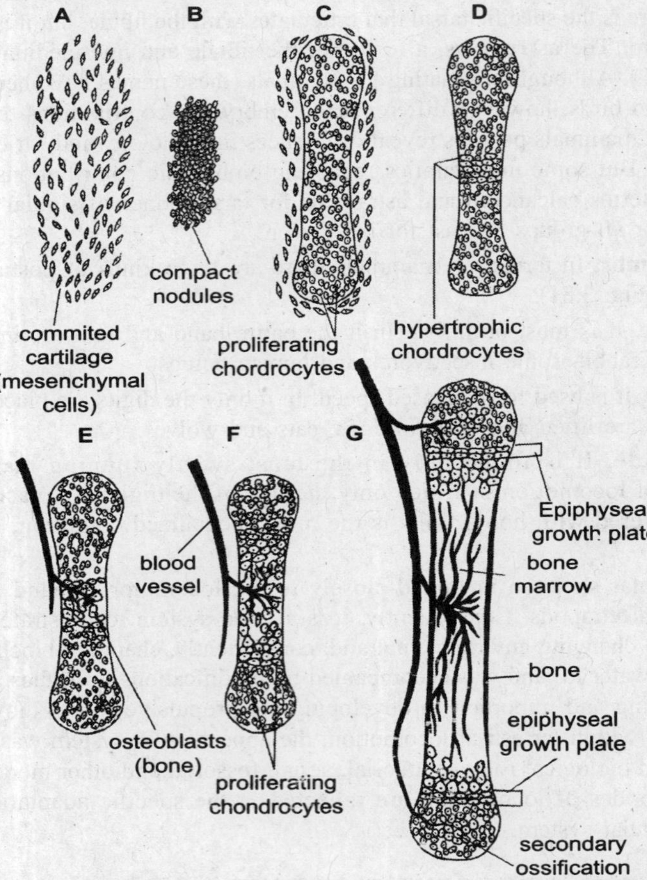

Fig. 3.82. Process of endochondral ossification (Horton 1990). A – Mesenchymal cells commit to become cartilage cells (chondrocytes). B – Committed mesenchyme condenses into compact nodules. C – Nodules differentiate into chondrocytes and proliferate to form the cartilage model of bone. D – Chondrocytes undergo hypertrophy and apoptosis while they change and mineralize their extracellular matrix. E – Apoptosis of chondrocytes allows blood vessels to enter. F – Blood vessels bring in osteoblasts, which bind to the degenerating cartilaginous matrix and deposit bone matrix. G – Bone formation and growth consist of ordered array of proliferating, hypertrophic, and mineralizing chondrocytes. Secondary ossification centers also form as blood vessels enter near the tips of the bone (after Gilbert 2010).

Endochondral Ossification

The process of endochondral ossification involves the following five stages (Fig. 3.83):

1. The mesenchymal cells commit to becoming cartilage cells.

2. The committed mesenchyme cells condense into compact nodules and differentiate into chondrocytes, or cartilage cells.

3. The chondrocytes proliferate rapidly to form the cartilage model for the bone (Fig. 3.82 C). As they divide, the chondroctyes secrete a cartilage-specific extracellular matrix.

4. The chondrocytes stop dividing and increasing their volume dramatically, becoming **hypertrophic chondrocytes** (Fig. 3.82 D; Fig. 3.83). These large chondrocytes alter the matrix they produce (by adding collagen x and more fibronectin) to enable it to become mineralized (calcified) by calcium phosphate. They also secrete the angiogenesis factor VEGF, which can transform mesodermal mesenchymal cells into blood vessels. Hypertrophic chondrocytes secrete numerous small, membrane-bound vesicles into extracellular matrix. These vesicles contain enzymes that are active in the generation of calcium carbonate crystals, called **hydroxyapatite**, which mineralize the cartilaginous matrix (**Wu** *et al.*, 1997). The hypertrophic chondrocytes, with their metabolism and mitochondrial membranes altered, then die by apoptosis (**Rajpurohit** *et al.*, 1999).

5. The blood vessels induced by VEGF invade the cartilage model (Fig. 3.82 E-G). As the hypertrophic chondrocytes die, the cells that surround the cartilage model differentiate into **osteoblasts**. Later on the blood vessels bring in both osteoblasts and **chondroclasts** (which eat the debris of the apoptotic chondrocytes). The osteoblasts begin forming bone matrix on the partially degraded cartilage matrix and construct a **bone collar** around the calcified cartilage matrix. New bone material is added peripherally from the internal surface of the **periosteum**, a

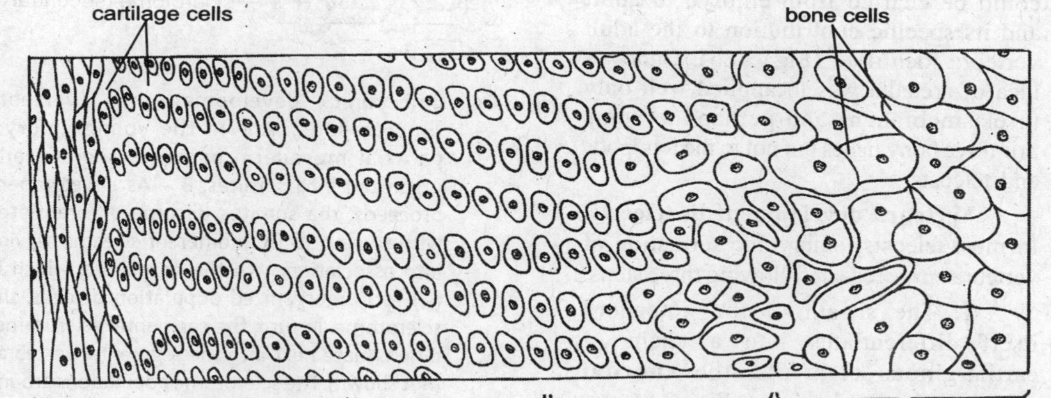

Fig. 3.83. Endochondral ossification. Long bone undergoing endochondral ossification. Diagram showing transition zone wherein the chondrocytes (cartilage cells) divide, enlarge, die and are replaced by osteocytes (bone cells) (after Gilbert 2010).

fibrous sheath containing connective tissue, capillaries, and bone progenitor cells and that covers the developing bone. At the same time, there is a hollowing out of the internal region of the bone to form the **bone marrow cavity**. This destruction of bone tissue is carried out by **osteoclasts**, multinucleated cells that enter the bone through the blood vessels (**Manolags** and **Jilka** 1995). Osteoclasts are derived from lateral plate mesoderm.

Vertebrae Formation

For the most part, vertebrae arise embryologically from mesenchyme. Early in embryonic development, the paraxial mesoderm becomes segmentally arranged into somites. As they differentiate, somites form vertebrae (and ribs), skeletal muscles and the dermis of the dorsal skin. Commitment of cells within the somites to these respective fates occurs relatively late, after the somite forms. Soon after formation, lateral cells nearest the ectoderm differentiate into a dermatome (dermis) and myotome (skeletal muscles). Medial cells depart and migrate toward the nearby notochord, forming streams of mesenchymal cells that settle segmentally along the notochord in cell groups called sclerotomes (Fig. 3.84). These sclerotomes in turn contribute to the vertebrae and ribs.

Vertebra development in elasmobranch fishes. Among elasmobranchs and many primitive fishes, the inward streams of cells arriving from the sclerotomes first congregate in discrete clusters and differentiate into paired cartilages rather than into ossified vertebrae directly. Up to four pairs of cartilages are formed per segment. The embryologist **Hans Gadow** called these paired cartilages **arcualia**. The developmental fate of each arculium could be charted from embryo to adult and its specific contribution to the adult vertebra identified (Fig. 3.85). Gadow's idea of arcualia was to applied well only to elasmobranchs and perhaps to some primitive bony fishes but not to the tetrapods and teleosts.

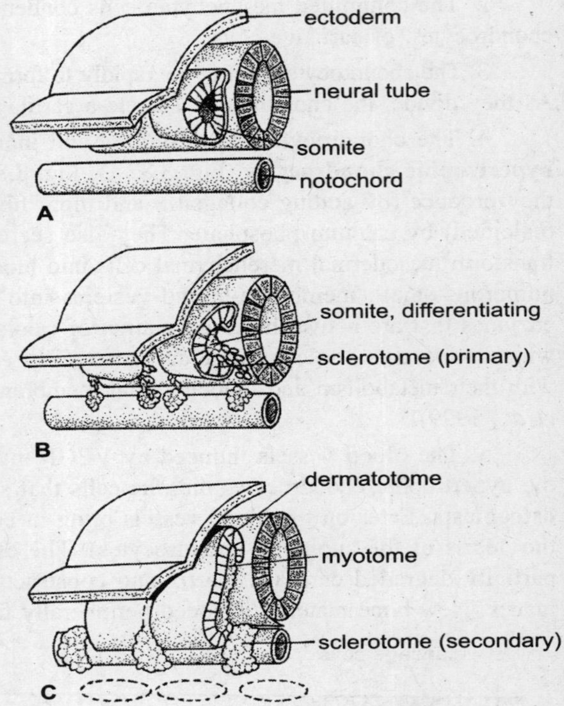

Fig. 3.84. Embryonic development of vertebra (in tetrapods). A – In the young embryo, paraxial mesoderm becomes segmentally organized into somites. B – As development proceeds, the somites become differentiated with some cells from different streams moving into association with the notochord, which in (C) form reorganized population of cells, the sclerotome. In turn, the segmental sclerotomes differentiate next into the segmental vertebrae (not shown). The sclerotome populations around the notochord represent the merger of cell streams from adjacent somites and hence, each is a reorganized or secondary sclerotome. This reorganization places them between axial muscle masses, derived from the myotome, which will next take up residence in the regions indicated by the dashed oval areas (after Kardong 2006).

Vertebra development in teleosts. In most teleosts, embryonic formation of vertebrae proceeds in following three steps:

1. The sheath of the notochord itself differentiates into a chain of cartila-ginous elements, the **chordal centers** (or chordacentra; Fig. 3.86 A). Between successive chordal centers, the undifferentiated notochordal sheath is destined to become the intervertebral ligament between vertebrae of the adult.

2. Local mesenchyme condenses at the level of myosepta. These condensations become cartilaginous anlagen called **arch centers** (arcualia by some workers) that give rise to the dorsal and ventral arches.

3. Cells of sclerotomal origin condense on the surface of the notochordal sheath, forming the perichordal tube that becomes ossified without first passing through a cartilaginous stage (Fig. 3.86 B). As vertebra formation proceeds, the deep chordal centers fuse with their respective perichordal centers on their surfaces. The arch centers often, but not always, fuse with the perichordal tube and ossify (Fig. 3.86 C).

Vertebra development in tetrapods. In the trunk and tail of tetrapods, each of the segmental chains of somites becomes internally subdivided into discrete layers of cells. Laterally, somites form the dermatome; beneath it, the myotome; and median to both, the sclerotome (Fig. 3.84).

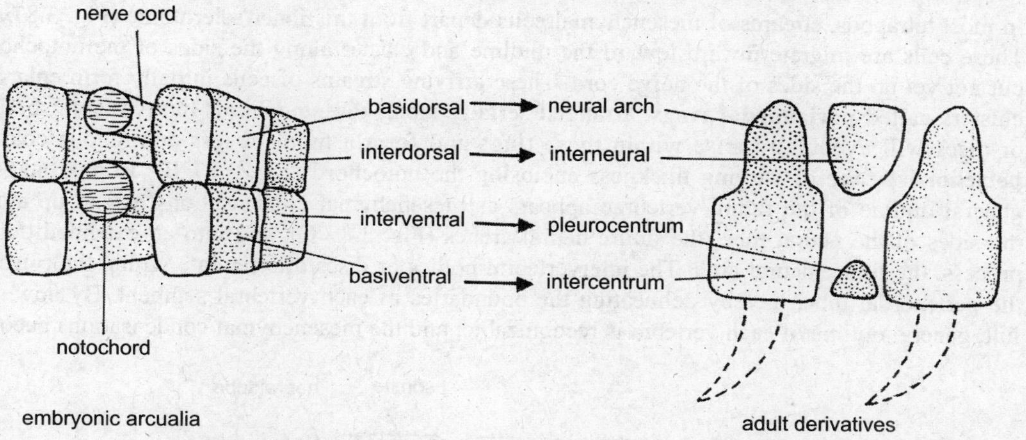

Fig. 3.85. Arcualia. During embryonic development in some primitive fishes, mesenchymal cells that gather around the notochord first form discrete blocks of cartilage, upto four pairs per segment. These blocks are called arcualia. In such fishes, each arcualium can be followed through its subsequent embryology to the part of vertebra it forms in the adult (after Kardong, 2006).

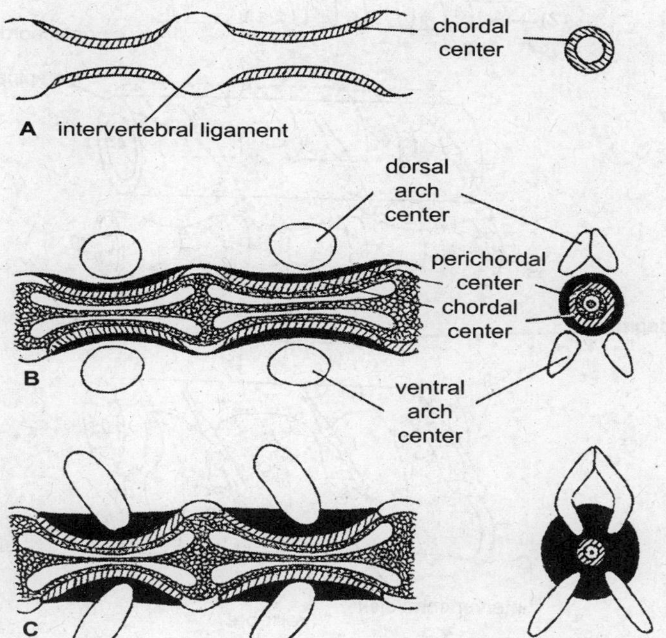

Fig. 3.86. Embryonic formation of teleost vertebrae. Successive stages in embryonic development are shown in lateral views (left of each figure) and corss-sectional views (right of each figure) through the middle of a forming centrum. A – A chordal center (chordacentrum) forms within the notochordal sheath. B – Paired cartilaginous rudiments, or anlagen, of dorsal and ventral arches form at myosepta from condensations of mesenchyme. Within the perichordal tube, itself formed from sclerotomal cells, perichordal centers (autocentra) of ossification appear. C – The chordal center becomes incorporated within the ossified perichordal center, forming the centrum. Arches often, but not always, fuse with the centrum to ossify along with it. The notochord may persist as intervertebral cartilaginous pads, as intervertebral ligaments, and as a constricted notochord running through the centers of the centra (after Kardong 2006).

In most tetrapods, streams of mesenchymal cells depart from this inner sclerotome (Fig. 3.87 A). These cells are migrate inward toward the midline and cluster along the sides of the notochord, but not yet up the sides of the nerve cord. These arriving streams of cells initially form enlarged clusters called **perichordal rings**, arranged serially along the notochord. Intervertebral bodies or discs will eventually arise within these rings and form a more or less continuous coat or **perichordal tube** of varying thickness enclosing the notochord (Fig. 3.87 B). In next stage a general outline of the future vertebrae appears as mesenchymal condensations extend up along the sides of the neural tube, the future neural arches (Fig. 3.87 C), and into an extended dorsal process, the future neural arch. The intervertebral bodies or discs differentiate within the rings of the perichordal tube, thereby delineating the boundaries of each vertebral segment. By now, the full, general outline of each vertebra is recognizable, and the mesenchymal condensations become

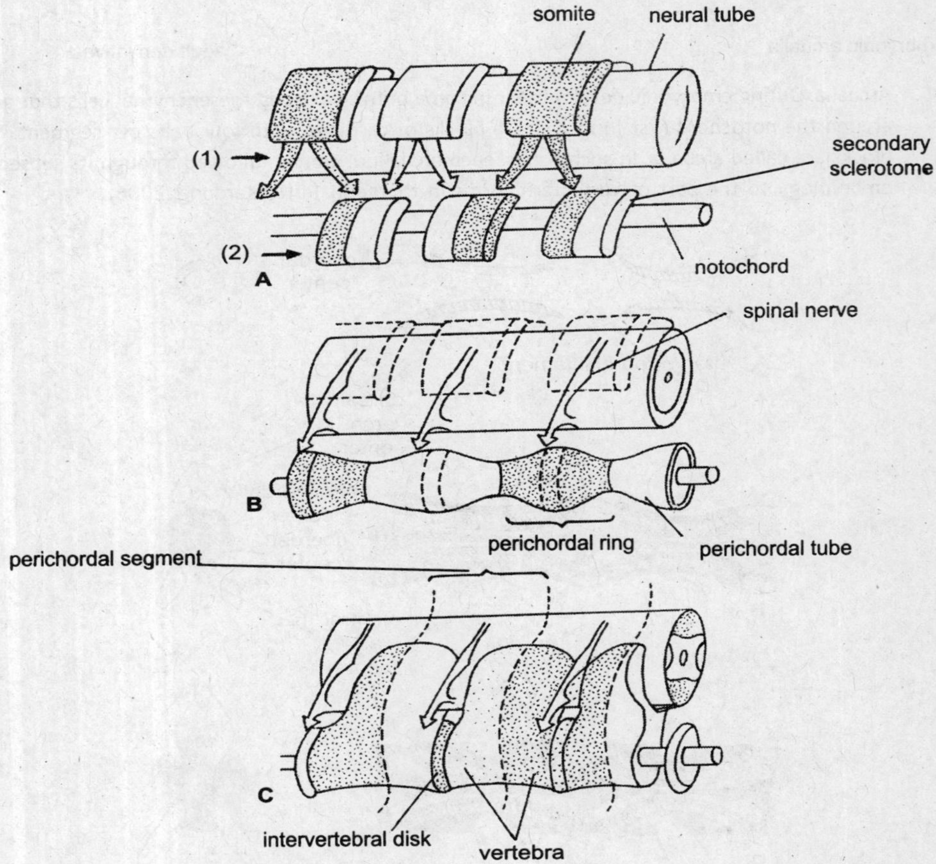

Fig. 3.87. Development of vertebrae in a generalized mammal. A – The primary sclerotome (not exposed) forms from cells on the medial side of the somite that depart and stream downward toward the notochord (arrows). These arriving streams of cells form secondary sclerotomes, represented here as blocks, half derived from one somite, and the other half from the next adjacent somite. B – The arriving sclerotomal cells soon amass a repeating perichordal rings that grow into contact, forming a continuous perichordal tube. C – The amassed cells of the perichordal tube grow upward, around the nerve cord, then above it, forming the outline of neural arches and spines. Chondrification, followed usually by ossification, produces the adult bony vertebrae. The intervertebral discs differentiate between vertebrae within the former perichordal ring (after Kardong 2006).

condrified. Bone formation is endochondral, so that in most tetrapods, subsequent ossification replaces these cartilaginous precursors, producing the bony vertebrae of the adult.

It is important to note that the bony vertebra is not formed directly or exclusively from its adjacent sclerotome, one-to-one. Instead, halves of adjacent sclerotomes fuse, forming resegmented sclerotomes that produce the basic vertebra (Fig. 3.87 C and 3.88). As cells stream out of the segmental **primary sclerotomes** on their way to form the perichordal tube, they first regroup, or regroup along the way (Fig. 3.87 A). This cellular regrouping is accomplished by the caudal

Box 3.4

The notochord appears to induce its surrounding mesenchymal cells to secrete **epimorphin**, and this epimorphin attracts sclerotome cells to the region around the notochord and neural tube, where they begin to condense and differentiate into cartilage. However, before the sclerotome cells form a vertebra, they must split into a rostral (anterior or cranial) and a caudal segment (Fig. 3.88). As the motor neurons from the neural tube grow laterally to innervate the newly forming muscles, the rostral segment of each sclerotome recombines with the caudal segment of the next anterior sclerotome to form the vertebral rudiment. Such **resegmentation** enables the muscles to coordinate the movement of the skeleton, permitting the body to move laterally. The bending and twisting movements of the spine are permitted by the intervertebral (synovial) joints that form the arthrotome region of the sclerotome.

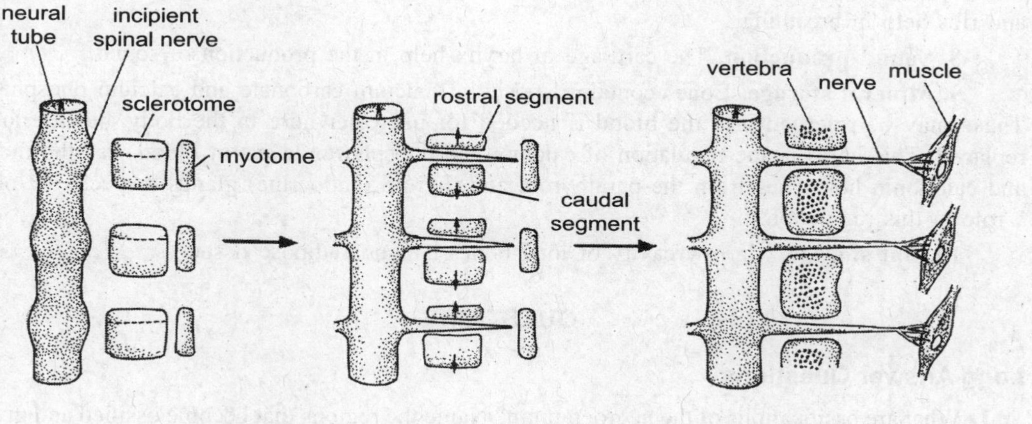

Fig. 3.88. Respecification of the sclerotome to form each vertebra. Each sclerotome splits into a rostral and caudal segment. As the spinal neurons grow outward to innvervate the muscles from the myotome, the rostral segment of each sclerotome combines with the caudal segment of the next anterior sclerotome to form a vertebral rudiment (after Gilbert 2010).

half of one sclerotome fusing with the cranial half of the next to form resegmented blocks of cells, the **secondary sclerotomes** (Box 3.4). These regrouped cells continue to the notochord to form the perichordal tube (Fig. 3.87 B). This provides a developmental mechanism by which the sclerotome and its mytome, initially in register with one another, become offset or staggered before differentiating into respective vertebrae and muscles (Fig. 3.84). Thus, musculature forms across adjacent vertebrae, rather than within the same vertebra. In this way, muscles act on adjacent vertebrae in a suitable functional position. If a muscle were attached to only one vertebra, the muscle would, of course, have no functionally significant role. Spinal nerves sprout and grow out between successive vertebrae.

3.6 FUNCTIONS OF SKELETON

The main function of the skeleton in the vertebrates are as follows:

1. Support. The skeleton forms a rigid framework, which gives support to the body against gravity. This role of the skeleton (*i.e.*, endoskeleton) depends on the deposition of calcium phosphate and calcium carbonate in the bones.

2. Shape. The skeletal framework gives and maintains the shape of the body. Without skeletal support, the animal body would be flexible like that of jellyfish.

3. Protection. The skeleton encloses the more vital organs of the body to protect them from injury. For example, brain and sense organs are enclosed in the cranium and sense capsules of the **skull**, spinal cord in the vertebral column, and the heart and lungs in a bony cage formed of backbone, sternum and ribs.

4. Movements. The skeleton provides a hard surface for the attachment of muscles to make them effective in movements. Both the ends of a skeletal muscle are attached to bones by means of tendons. The skeleton has joints to make movement possible. The various parts of the skeleton are so arranged that they can absorb the shocks generated in locomotion.

5. Formation of blood cells/corpuscles (Haemopoiesis). Blood cells (red and granulocytes) and platelets are produced in the red bone marrow present at the ends of the long bones; in the sternum and scapula, and in the centra of the vertebrae.

6. Hearing. Certain bones (ear ossicles present in the middle ear) help in hearing.

7. Breathing. Cartilages of the larynx, cartilaginous/bony rings of the trachea and bronchi, and ribs help in breathing.

8. Sound production. The cartilage of larynx help in the production of sound.

9. Mineral storage. Bones contain deposits of calcium carbonate and calcium phosphate. These may be released into the blood if needed for use elsewhere in the body, but are later replaced. This helps in the regulation of calcium and phosphorus levels of blood. Parathormone and calcitonin hormones from the parathyroid and thyroid (endocrine) glands respectively play a role in this regulation.

10. Fat storage. Marrow cavity of long bones contains adipose tissue.

QUESTIONS

Long Answer Questions

1. What are ossifications of the neurocranium? Name the regions that become ossified and gives at least one example for each region.
2. What are phylogenetic patterns for temporal fossae? Include the primitive condition.
3. Crocodila and Mammalia independently evolve a secondary palate. What functional role may be related to the origin of these structure.
4. Describe a typical branchial arch skeleton. What are the serial homologues of the components of a branchial arch in the mandibular and hyoid arches?
5. Why the origin of a sternum associated with tetrapods?
6. What are the structural components of the pectoral and pelvic fins?
7. Compare various hypotheses for the origin of the paired fins.
8. Give two mechanistic hypotheses for digit-number reduction.
9. Give an account of different types of jaw suspensorium.
10. Describe the skeleton of a tetrapod limb.

11. Describe the development of bone tissue.
12. Discuss the functions of skeleton.

Short Answer Questions

1. Give a labelled figure of the skeletal elements of a tetrapod limb.
2. Mention the bones of stylopodium.
3. How are replacing bones developed in vertebrates?
4. Write short notes on the following:
   (*i*) Development of chondrocranium.
   (*ii*) Jaw suspensorium.
   (*iii*) Pectoral girdle.
   (*iv*) Pelvic girdle.

Very Short Answer Questions

1. Which visceral arch forms the jaws?
2. Name the bones which form the lower jaw in *Labeo* (a bony fish)?
3. Name the components of the suspensorium in frog's skull.
4. Name the bones that form pelvic girdle.
5. Which bones contribute to the glenoid cavity?
6. Mention two vertebrates that have a double articulation for the lower jaw?
7. What is ossification?
8. Which minerals are stored in the bones?
9. A tetrapod limb consists of three main segments. Name these.
10. Mention the terms used for the first digit in the manus and the pes.
11. What are podials? Name their two types.
12. List the elements that are included in axial skeleton?
13. Are the visceral arches of fishes included in the axial or appendicular skeleton?
14. Why sternum is included in axial skeleton?
15. What is cannon bone?
16. Define composite theory.
17. What is centrum?
18. Define ceratotrichia.
19. What is lepidotrichia?

Fill in the Blanks

1. Fin rays are ceratotrichia in .... and lepidotrichia in ....
2. Meckel's cartilage is a part of .... of frog.
3. Study of bones is called ....
4. The tissues that form the endoskeleton are .... and ....
5. .... bones develop in the dermis of the skin.
6. Ribs are a part of .... skeleton.

7. The two heads of a rib are .... and ....
8. In .... turbinates occur in the olfactory capsule.
9. .... bones are absent in pelvic girdle.

Multiple Choice Questions

1. Cranium is a constituent of
   (*a*) splanchnocranium (*b*) chondrocranium
   (*c*) neurocranium (*d*) visceral skeleton
2. Jaw suspensorium in elasmobranchs is
   (*a*) autostylic (*b*) autodiastylic
   (*c*) hyostylic (*d*) amphistylic
3. Cartilage bone is
   (*a*) parietal (*b*) prefrontal
   (*c*) quadrate (*d*) lacrimal
4. Which of the following is a pelvic bone?
   (*a*) ilium (*b*) scapula
   (*c*) coracoid (*d*) clavicle
5. Skull of ___________ is kinetic
   (*a*) *Rana* (*b*) *Varanus*
   (*c*) *Oryctolagus* (*d*) *Labeo*
6. Carpometacarpus is characteristic of __________ hand.
   (*a*) amphibian (*b*) reptilian
   (*c*) mammalian (*d*) avian
7. Lower jaw consists of a single bone in
   (*a*) frog (*b*) lizard
   (*c*) fowl (*d*) rabbit
8. Visceral arches are a component of
   (*a*) vertebral column (*b*) skull
   (*c*) girdles (*d*) none of these
9. In modern amphibians, the columella of middle ear is a modified
   (*a*) mandibular (*b*) hyomandibular
   (*c*) sphenoid (*d*) basisphenoid

ANSWERS

1. Very Short Answer Questions

1. Mandibular arch.
2. Articular, dentary and angular.
3. Quadrate cartilage, quadratojugal bone and squamosal bone.
4. Ilium, ischium and pubis.
5. Scapula and coracoid.

6. Fowl and *Varanus*.
7. Ossification is a process of bone formation in cartilage, dermis or tendons.
8. Calcium carbonate and calcium phosphate.
9. Stylopodium, zeugopodium and autopodium.
10. Pollex, hallux.
11. Podials are the small bones in the wrist and ankle. Carpals and tarsals.
12. Skull, vertebral column, sternum and ribs.
13. Axial skeleton as a part of skull.
14. Sternum is included in the axial skeleton because of its position. Its embryonic development from lateral plate mesoderm shows it to be a part of appendicular skeleton.
15. It is a hindlimb bone resulting from fusion of metatarsals III and V, *e.g.*, horse.
16. The hypothesis that jaws of vertebrates evolved from several fused anterior branchial arches.
17. Centrum is body or base or vertebra.
18. Ceratotrichia is a fan-shaped array of keratinized rods internally supporting the fin of the elasmobranch fishes.
19. Lepidotrichia is a fan-shaped array of ossified or chondrified dermal rods that internally supports the fin of bony fishes.

Fill in the Blanks

1. *Scoliodon*, *Labeo*.
2. Lower jaw.
3. Osteology.
4. Cartilage, bone.
5. Investing or membrane or dermal.
6. Axial.
7. Capitulum, tuberculum.
8. Rabbit.
9. Dermal.

Multiple Choice Questions

1. (*c*)	2. (*c*)	3. (*c*)	4. (*a*)	5. (*b*)
6. (*d*)	7. (*d*)	8. (*b*)	9. (*b*).	

4

Digestive System (Alimentary Canal, Associated Glands and Dentition)

4.1 ALIMENTARY CANAL

The digestive tract or alimentary canal, is a tube, seldom straight and often tortuously coiled, that begins at the mouth and empties either into a vented cloaca or directly to the exterior via an anus (Fig. 4.1). Alimentary canal functions in the digestion and absorption of foodstuffs and in elimination of undigested wastes (Box. 4.1). Food is propelled from the pharynx to the vent or anus by smooth muscle tissue in the walls of the tract. This process is called **peristalsis**.

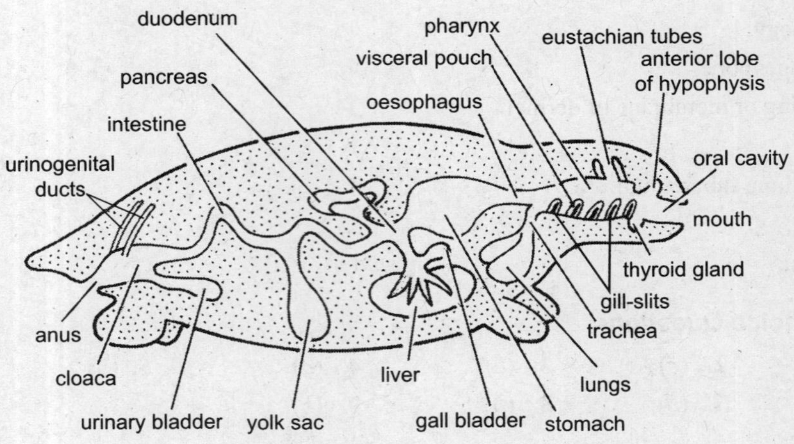

Fig. 4.1. Alimentary canal and its chief derivatives in a vertebrate.

Box 4.1 Necessity and Function of Digestive system

In order to draw energy from any food material (an indiffusible form), it has to be broken down into a form that is diffusible from the alimentary canal. This breakdown of indiffusible food material into its diffusible form is the necessity of an organism for furtherance of the absorbance of products.

The functions can be ascertained in the following points:

- Ingestion of food,
- Digestion of food,
- Secretion of digestive juices,
- Elimination of assorted infectious agents that enter digestive system,
- Mixing of food with the digestive juices facilitated by the movement of gastro-intestinal tract, and
- Defecation.

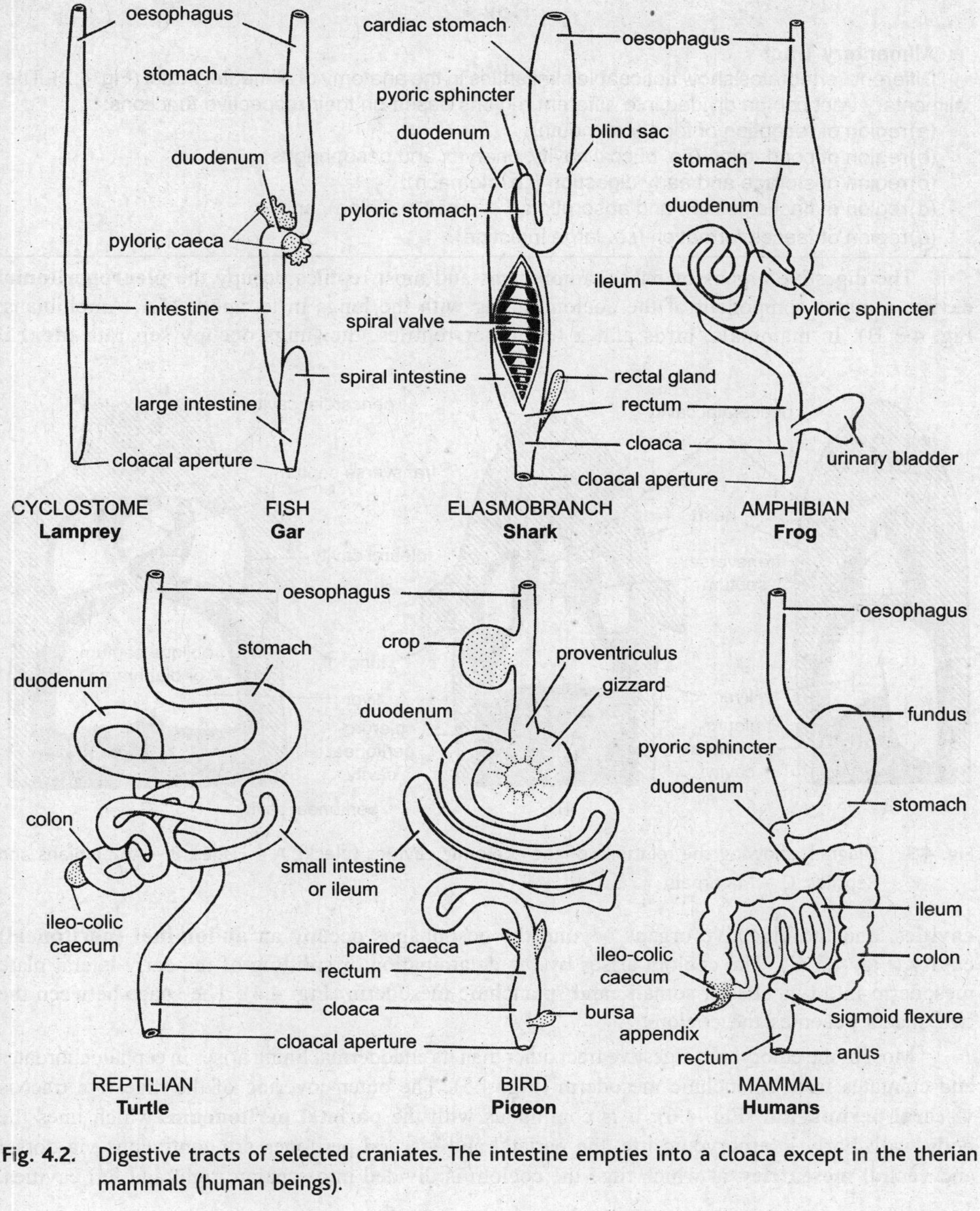

Fig. 4.2. Digestive tracts of selected craniates. The intestine empties into a cloaca except in the therian mammals (human beings).

Major subdivisions of the alimentary canal are in oral cavity and pharynx (oropharyngeal cavity in fishes), oesophagus, stomach, and intestine (Box 4.2). The intestine is divided into small and large intestine in tetrapods. Emptying into the alimentary canal are ducts from accessory organs (or glands) principally pancreas, liver, and gall bladder. These organs and their ducts arise as evaginations from the embryonic digestive tract. Other evaginations are known as caeca, having one function or another are common (Fig. 4.1). The digestive tract or alimentary canal and accessory organs (glands) constitute the **digestive system**.

Box 4.2

Alimentary Tract

Different vertebrates show noticeable similarities in the anatomy of alimentary tract (Fig. 4.2). The alimentary tract can be divided into different regions based on their respective functions:

(a) region of reception of food (*i.e.* mouth),
(b) region of conduction (*i.e.* buccal cavity, pharynx and oesophagus),
(c) region of storage and early digestion (*i.e.* stomach),
(d) region of final digestion and absorption (*i.e.* small intestine), and
(e) region of faeces formation (*i.e.* large intestine).

The digestive organs of fishes, amphibians and most reptiles occupy the **pleuroperitonial cavity**, a major component of the coelom, along with the lungs in tetrapods (*i.e.*, Amphibians; Fig. 4.3 B). In mammals, birds and a few other reptiles, the lungs occupy separate **pleural** **cavities**, and the digestive organs beyond the oesophagus occupy an **abdominal (peritoneal) cavity** (Fig. 4.3 C). The coelom arises by the delamination or splitting of the early lateral plate mesoderm into two sheets, somatic and splanchnic mesoderm (Fig. 4.4). The space between the two sheets becomes the coelom.

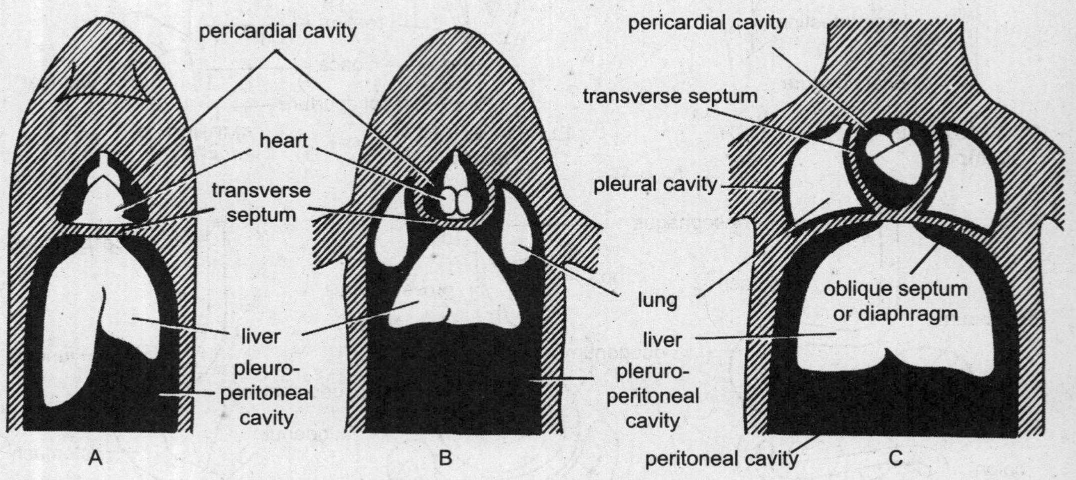

Fig. 4.3. Diagram showing the relations of the coelomic cavities (black). A – Fishes; B – Amphibians and Reptiles; C – Mammals.

Most of the embryonic digestive tract other than its endodermal lining arises in cephalochordates and craniates from splanchnic mesoderm (Fig. 4.5). The outer covering of the digestive tract is **visceral peritoneum** (Fig. 4.6). It is continuous with the **parietal peritoneum**, which lines the body wall. Early in embryonic life, the parietal and visceral peritonea are continuous via dorsal and ventral **mesentries**, at which time the coelom is divided into separate right and left cavities.

The dorsal mesentery remains essentially intact throughout life, conducting blood vessels and nerves from the roof of the coelom to the digestive organs. The ventral mesentery eventually disappears except at the level of the liver and urinary bladder.

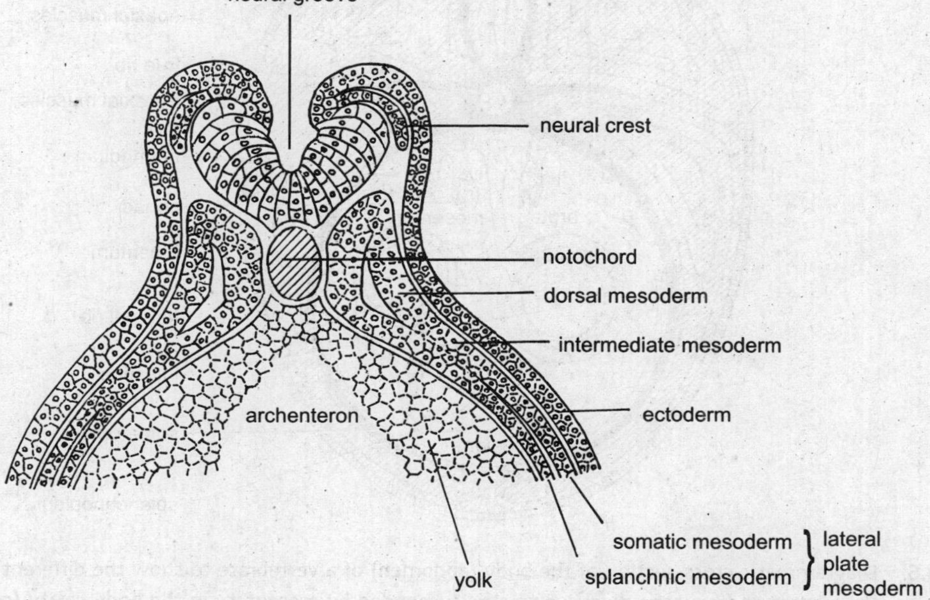

Fig. 4.4. Cross section of neurula of frog at level of future midgut. The coelom is the space between the somatic and splanchnic mesoderm (of lateral plate mesoderm) (after Kent and Carr 2001).

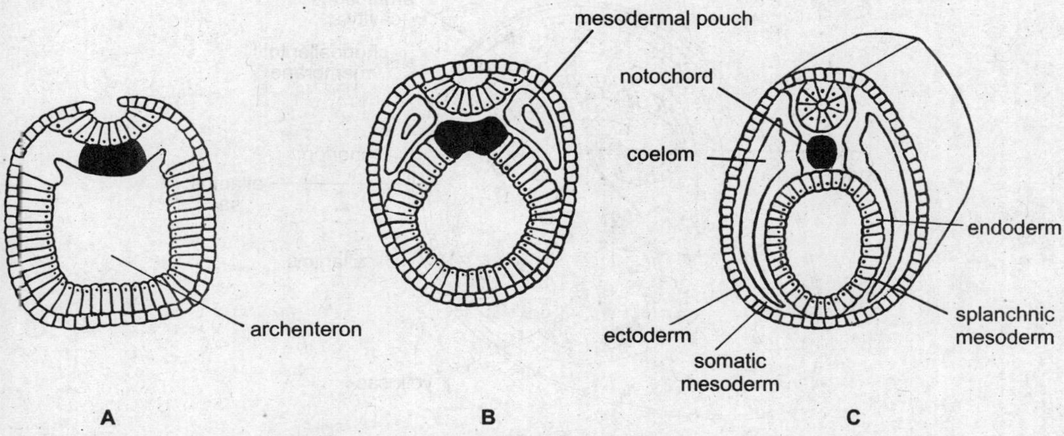

Fig. 4.5. Three steps of coelom formation in *Amphioxus* (after Kent and Carr 2001).

The embryonic digestive tract (or alimentary canal) consists of three regions. The part containing the yolk, when present, or to which the yolk sac is attached, the **midgut** (Fig. 4.7 and Fig. 4.8). Caudal to the midgut is the **hindgut**, and anterior to the midgut is the **foregut**, The foregut elongates to form part of the oral cavity, the pharynx, oesophagus, stomach and much of the small intestine. The hindgut becomes the remainder of the intestine and the cloaca. Little of midgut remains in the adults.

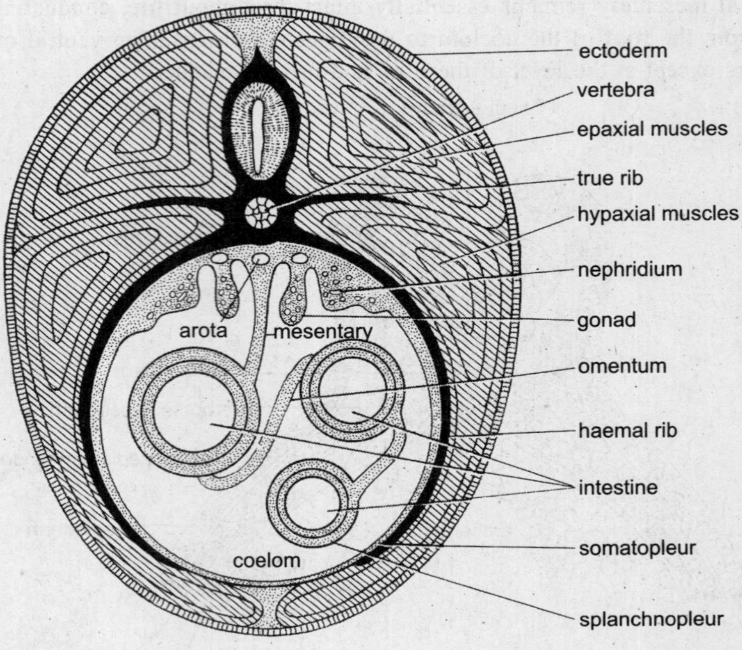

Fig. 4.6. Diagrammatic cross-section of the body (abdomen) of a vertebrate to show the different part of the viscera and how the digestive tract is suspended by mesentary in the body cavity (coelom).

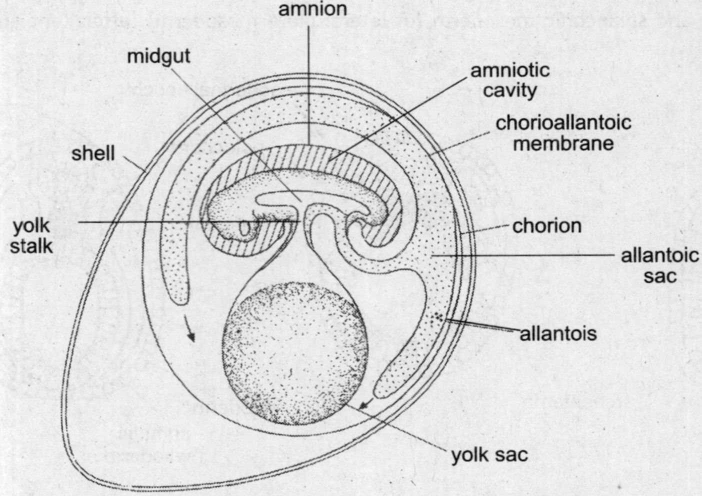

Fig. 4.7. Development of midgut in chick embryo in relation to yolk stalk of yolk sac (after Kent and Carr 2001).

The anterior portion of the oral cavity, or of the oropharynx of fishes, arises as a midventral invagination of the ectoderm of the head, the **stomodeum** (Fig. 1.2 and Fig. 4.8). A thin membrane, the **oral plate**, temporarily separates the early embryonic foregut from the exterior. It soon ruptures to provide an anterior entrance to the digestive tract. A similar invagination, the **proctodeum**, provides an exit from the hindgut when the **cloacal plate** ruptures (Fig. 1.1). Like all chordates, therefore, craniates, including humans, are deuterostomous.

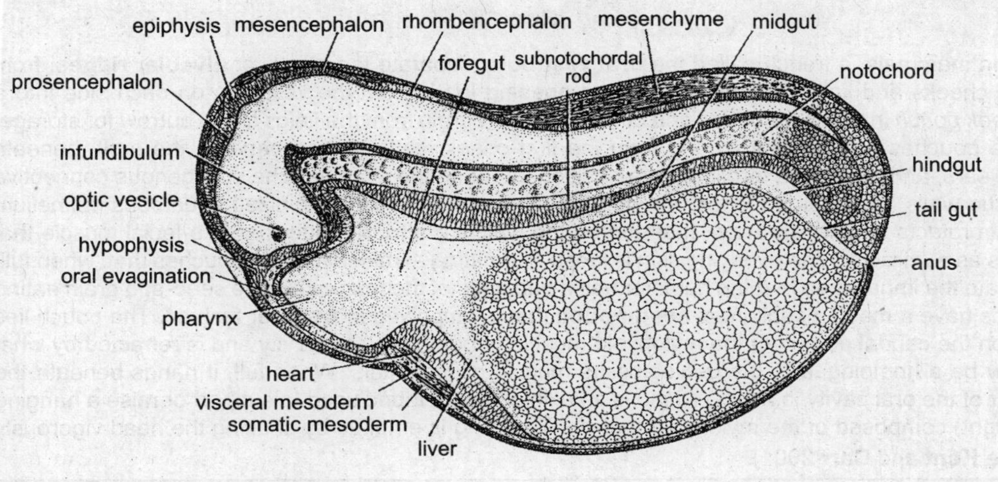

Fig. 4.8. Median longitudinal section of an early tadpole larva (8 mm). Midgut is near yolk-filled cell mass.

Differences in the anatomy of digestive tracts caudal to the pharynx are correlated not so much whether an animal (*e.g.*, craniate) lives on land or in water as with the nature and abundance of food. Is it readily absorbable when ingested, as in vampire bats, or does it require extensive enzymatic activity or mechanical maceration, as in carnivores? Is the food supply constant, so that whenever an animal is hungry the food is likely to be there, or does it have to be stalked? If the latter, the food is probably bulky, and there has to be room for it in an appropriately expandable stomach until it can be digested. And what is the shape of the animal's body? If it is long, like that of lamprey or snake, the digestive tract will probably be straight. If the trunk is short, as in turtles and anurans, the intestine must be coiled in order to provide, sufficient absorptive area. Some fishes have a special membrane that increases the absorptive area (Kent and Carr 2001).

4.2 MOUTH AND ORAL CAVITY

The mouth is the entrance of the digestive tract. In gnathostome fishes, it opens into an **oropharyngeal cavity** that exhibits teeth and walls perforated by gill slits. The oropharyngeal cavity terminates at a short oesophagus. In tetrapods, the mouth opens into an **oral cavity** or **buccal cavity** housing the teeth and tongue. The oral or buccal cavity leads to pharynx.

The buccal or oral cavity of vertebrates contains the teeth, tongue and palate. Oral glands empty into it. Salivary glands help moisten food and secrete enzymes to begin chemical digestion. In some species, mastication begins the mechanical breakdown (digestion) of food.

Boundaries of the Buccal Cavity

The **oral opening** (*i.e.*, mouth), the margins of which are the **lips**, creates the buccal cavity entrance. Generally upper and lower lips follow the line of the tooth rows to meet posteriorly at the angle of the jaws. In mammals, upper and lower lips meet well forward of the jaw angle near the front of the mouth, thus creating a skin-covered **cheek** region. Cheeks prevent loss of food out sides of the mouth during chewing. In some rodents and old word monkeys, they expand into **cheek pouches**, small compartments in which gathered food can be temporarily held until it is chewed or carried to caches (*cache* is a hidden store of things; Box 4.3). Lips are usually pliable, although birds, turtles, some dinosaurs and a few mammals have rigid beaks with firm margins. In most mammals, lips are fleshy, a feature helping an infant form a seal around the nipple during nursing. Human lips help from the vocalization of speech.

Box 4.3

In mammals, a trench called the **oral vestibule**, separate the gums, or **alveolar ridges**, from the cheeks and lips. In many rodents, an opening leads from the vestibule on each side into a *cheek pouch* in which hamsters and others transport grain from the fields to the burrow for storage. The pouches of hamsters extend from the first cheek teeth cauded beneath the orbit, beneath the ears, and above the forelimb, to a position lateral to the scapula. The collagenous connective tissue walls (of cheek pouches) are lined by a moderately low kertainized squamous epithelium that protects against abrasion. Inserted on the wall is a **slip** of a buccinator, a facial muscle that acts as a retractor. The overlying skin is loose, permitting expansion of the pouches that, when full, create the impression that the hamster has a bad case of the mumps! Some seed-and grain eating birds have a median sublignual **seed pouch** that is empolyed in a similar fashion. The pouch lies upon the caudal portion of the mylohyoid muscle beneath the oral cavity and is retracted by what may be a homologue of the genioglossus muscle of mammals. When full, it hangs beneath the rear of the oral cavity in a sling (*Sling* is a flexible loop of fabric used to support or raise a hanging weight) composed of the mylohyoid muscle. The sling is empited by shaking the head vigorously (see **Kent** and **Carr** 2001).

The lips define the anterior border of mouth, and the palatoglossal arch in a fold that marks the posterior border of mouth and lies between the mouth and pharynx (Fig. 4.9 A). However, if this or other anatomical markers are absent, the mouth and pharynx form a collective chamber called the **oropharyngeal cavity** (Fig. 4.9 B).

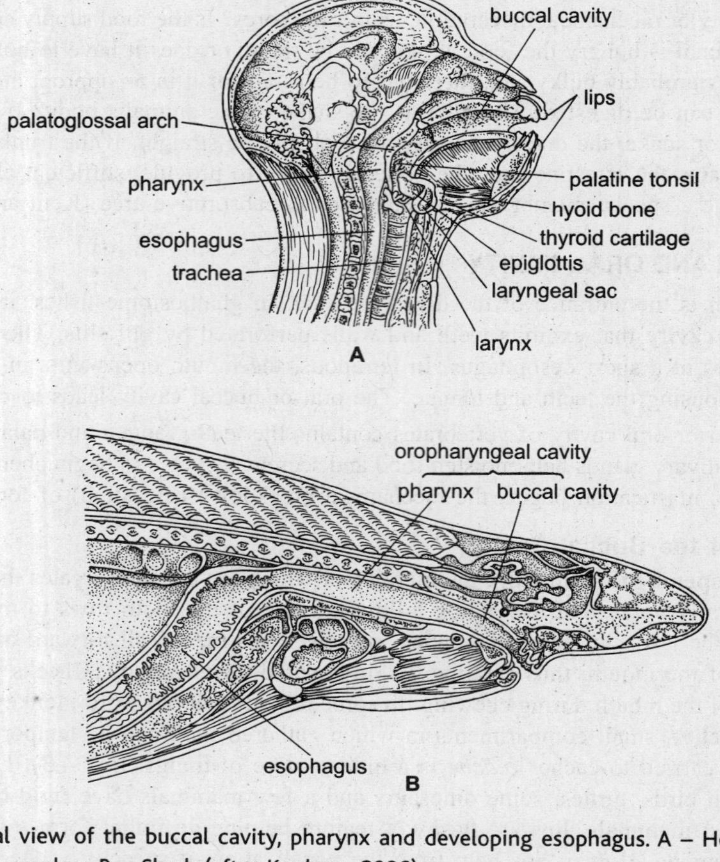

Fig. 4.9. Sagittal view of the buccal cavity, pharynx and developing esophagus. A – Head and neck of *Rhesus* monkey. B – Shark (after Kardong, 2006).

The stomodeum not only forms the buccal cavity, but it also contributes to surface features of the head in some vertebrates. Two good embryonic landmarks present within the stomodeum, the **hypophyseal pouch** (or adenohypophyseal placode) and the **nasal placode**, make this evident (Fig. 4.10). In cyclostomes, only the posterior part of the stomodaeum turns inward contributing the mouth. The anterior part turns outward, contributing to the outer surface of the head. In sharks and bony fishes, the hypophyseal pouch lies within the mouth, the nasal placode differentiates on the outside of the head, and the margins of the mouth form between these anatomical landmarks. In rhipidistians and tetrapods, both landmarks (*i.e.*, nasal placode and hpophyseal canal), lie within the buccal cavity (Fig. 4.10).

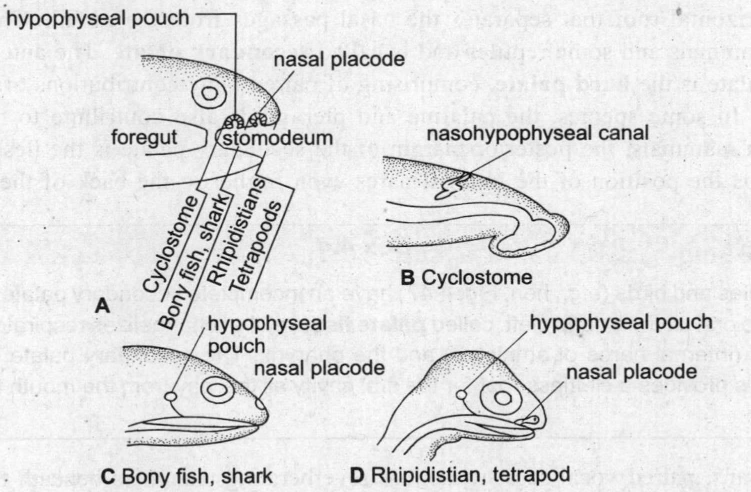

Fig. 4.10. Boundaries of the buccal cavity in vertebrates. The extent to which the embryonic stomodeum contributes to the mouth can be followed by two markers: the nasal placode and hypophyseal pouch. A – Comparative positions of the anterior margin of the mouth in various groups of vertebrates. B – to D – Diagrammatic view of each group. These two markers remain outside the mouth in cyclostomes. In sharks and bony fishes, the hypophyseal pouch is pinched off from the mouth. In rhipidistians and tetrapods, both nasal placode and hypophyseal pouch opens to or are derived from the mouth (after Kardong, 2006).

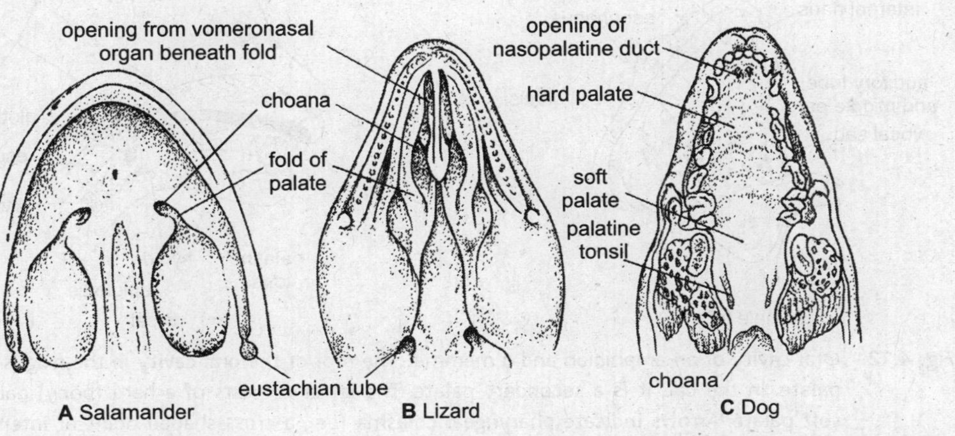

Fig. 4.11. Roof or palate of the mouth in tetrapods. A – Salamander; B – Lizard; C – Mammal (dog). Notice the point of entry of the choana or internal naris, in each animal (after Kardong, 2006).

Palate

The roof of the buccal cavity is called **palate**. It is formed from the fusion of ventral skull bones overlying the mouth. The primitive, or **primary palate** includes a medial series of bones (such as vomers, pterygoids and parasphenoid) and also a lateral series of bones such as palatines and ectopterygoids. In most fishes, the primary palate is low vault (*vault* is the roof in the form of an arch or a series of arches) which has no opening. In rhipidistians (members of osteolepiformes - the lobefin (fishes) and tetrapods, the nasal passages reach the mouth through paired openings in the primary palate, called the **internal nares** or **choanae** (Fig. 4.11). The **palatal folds** are inward growths of lateral bones that meet the midline and form a second horizontal roof that separates the nasal passages from the mouth. This new roof is present in mammals and some reptiles and is called **secondary palate**. The anterior part of the secondary palate is the **hard palate**, comprising of paired bony contributions of the premaxilla and maxilla. In some species, the palatine and pterygoids also contribute to the hard palate (Box 4.4). In mammals, the posterior margin of the secondary palate is the fleshy **soft palate**, which extends the position of the internal nares even farther to the back of the buccal cavity.

Box 4.4

Most reptiles and birds (*e.g.*, hen, Fig. 4.47) have an incomplete secondary palate. This leaves in the roof of the oral cavity a deep cleft, called **palate fissure**, that channelizes respiratory air between the choanae (internal nares of amniotes) and the pharynx. The secondary palate of crocodilians and mammals provides a cleftless roof for the oral cavity all the way from the mouth to the pharynx. (Fig. 4.13).

In anurans, paired vocal sacs, which are reverberating chambers beneath the floor of the pharynx, open into the oral cavity near the angles of the jaw (Fig. 4.12). In all tetrapods a variety of oral glands, or their ducts, empties into the oral cavity.

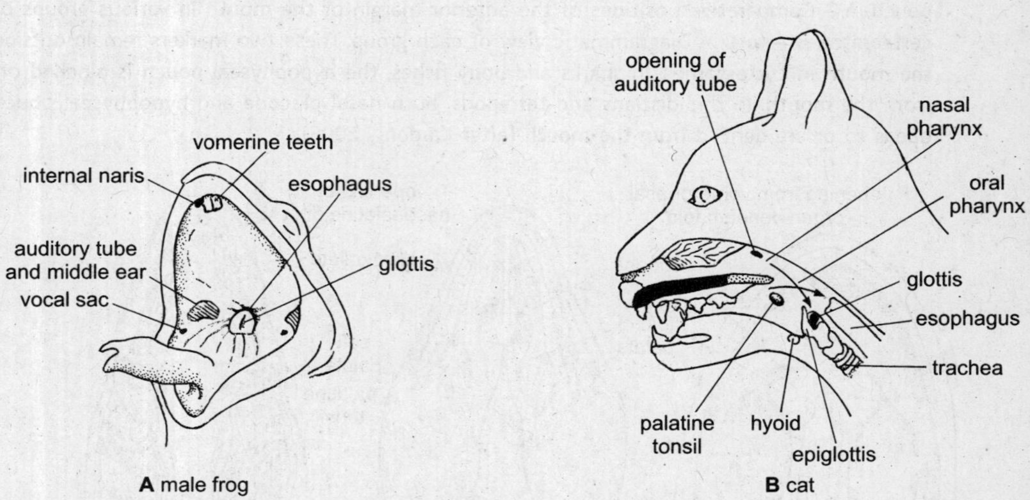

Fig. 4.12. Oral cavity of an amphibian and a mammal. The roof of the oral cavity in the frog is a primary palate. In the cat, it is a secondary palate. The latter consists of a hard (bony) palate and a soft palate. Arrows indicate pharyngeal chiasma (*i.e.*, a cross-shaped point of intersection of two parts) where food and airstreams cross (after Kent and Carr, 2001).

4.3 TEETH

Bony teeth are found among jawed fishes, amphibians, reptiles and mammals with few exceptions, and they were present in the earliest birds. In mammals, teeth have achieved a peak in regional specialization. Lacking teeth are sturgeons, numerous teleosts including sea horses, a few amphibians, all turtles, and modern birds; and among mammals, whalebone whales, South American and scaly anteaters and the monotreme, *Echidna* (spiny ant-eater). But even among tootless speices, there are many that develop an embryonic set that simply does not erupt or after erupting, disappears.

Teeth are unique among vertebrate animals. They are usually capped with **enamel**, a mineralized coat found only in vertebrates. Embryologically, neural crest cells contribute to the parts of the forming tooth. Teeth are generally thought to have arisen phylogenetically from the bony armour of primitive fishes, probably from surface structures together with contributions from the underlying dermis. The fact that neural crest cells also contribute to the dermis supports the concept of a close relationship between teeth and bony armour.

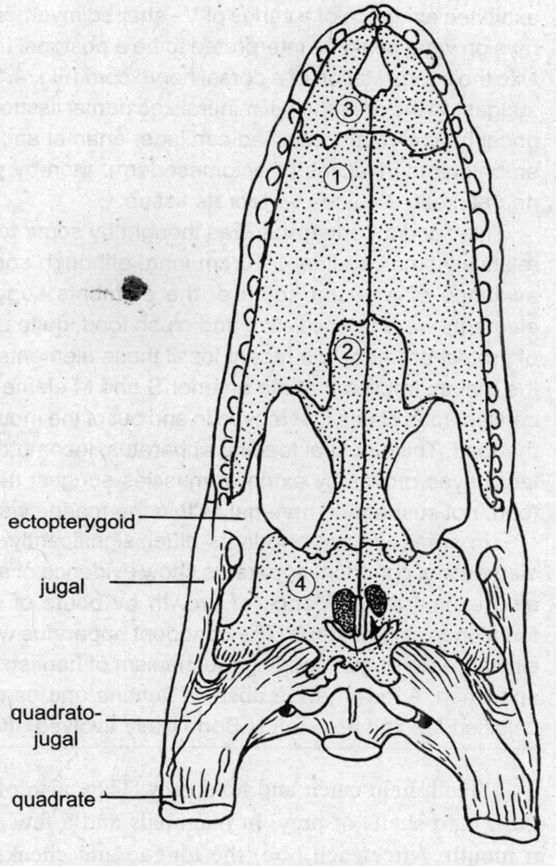

Fig. 4.13. The long rigid secondary palate of an alligator (a reptile). 1, 2 and 3, palatal processes of maxilla, palatine and premaxilla bones, respectively; 4, pterygoid bones. Arrow indicates an internal naris (after Kent and Carr 2001).

Teeth were first appeared in condonts, the very early vertebrates (Box 4.5). The presence of teeth, but not dermal armor, in condonts complicates this picture. Indeed the teeth of condonts were mineralized, but it is still not clear if this mineral was the enamel. If codonts prove to be very early vertebrates, then teeth apparently preceded exterior armor rather than the other way around. However, at the moment, it seems most likely that *teeth evolved from parts of the dermal armor* (see **Kardong** 2006; **Agarwal** 2016).

Box 4.5 Story of Condonts

For almost a century and a half, **toothlike microfossils** known as condont elements have been important index fossils in many geological studies (see V.K. Agarwal's Zoology for Semester III - B.Sc. Hons.) Although extremely common in rocks from the late Cambrian to the end of the Triassic, the organism owing this assortment of pointed and comblike phosphatic elements was not known, leading to speculation that it might be a mollusc, or chordate, or even part of aquatic plants. This mystery was solved in the early 1980s with the discovery of fossil impressions of laterally compressed, slender, soft-bodied animal bearing a complete set of condont elements in its mouth. But there was much more than this. These fossils bore evidence that condonts were in fact vertebrates. The trunk exhibited evidence of a series of V - shaped myomeres, a notochord down the midline and caudal fin rays on what could be interpreted to be a postanal tail. Above the notochord was a streak consistent with the interpretation of a dorsal nerve cord (Fig. 4.14). Histological examination of codont elements suggested the presence of mineralized dental tissues known from vertebrates, cellular bone, calcium phosphate crystals, calcified cartilage, enamel and dentine. Dentine is laid down by odontoblasts, embryonic derivatives of ectomesoderm, thereby providing indirect evidence for the presence of neural crest, a unique vertebrate tissue.

This condont apparatus was thought by some to be a filter-feeding system serving animals with relatively small bodies, 3-10 cm long, although some reached perhaps 30 cm or more. However, evidence of wear on some of the elements suggests instead that the blade-like posterior (P) elements were used to slice and crush food, quite unlike the endostyle-based, filter-feeding system of protochordates. In a recent fossil these elements were preserved in their natural positions within the mouth, implies that the anterior S and M elements (Fig. 4.14C) were attached to tongue-like or cartilaginous plates that moved in and out of the mouth catching and delivering, respectively, speared the food. This unusual feeding apparatus, locomotor system (notochord, myotomes), and relatively large eyes moves by extrinsic muscles suggest further that conodonts selected and fed on large food, not suspended material, within the marine waters where they lived.

In some ways, conodonts differ significantly from vertebrates and remain puzzling. Dental elements of conodont apparatus show evidence of replacement, but worn or broken dental elements also exhibit the evidence of growth by bouts of renewed layering (= appositional growth) and functioned like the teeth. The conodont apparatus was a very specialized structure, it would function similar to the lingual feeding mechanism of hagfish. Pharyngeal slits are present within at least one specimen. A lower jaw is absent. Dentine and dentine plus enamel were present in the conodont calcified feeding apparatus. Bone likely evolved after conodonts.

Teeth help catch and hold prey. They also offer strong opposing surfaces that jaws work to crush hard shells of prey. In mammals and a few other vertebrates, **mechanical digestion** begins in mouth. After each bite, the tongue and cheeks collect food and place it between the upper and lower teeth rows, and the teeth break down the bolus mechanically, reducing it to smaller chunks to make swallowing easier. By breaking the large bolus into many smaller pieces, chewing also increases the surface exposed to chemical digestion. Even in vertebrates that do not chew their food, sharp teeth puncture the surface of the prey, creating sites through which digestive enzymes penetrate when food reaches the alimentary canal. For vertebrates that feed on insects and other arthropods, punctures through the chitinous exoskeleton are especially important in giving proteolytic enzymes approach to the digestible tissues within.

Anatomy of teeth. The part of tooth projecting above the gum line, or **gingiva**, the **crown**; the region below is the **base**. If the base fits into a hole, or **socket (alveolus)** within the jaw bone, the base is referred to as a **root**. Within the crown, the **pulp cavity** narrows when it enters the root, forming the **root canal**, and opens at the tip of the root as the **apical foramen**. Mucous connective tissue, or **pulp**, fills the pulp cavity and root canal to support blood vessels and nerves that enter the tooth via the apical foramen. The **occlusal surface** of the crown of the teeth makes contact with opposing teeth. The **cusps** are tiny, raised peaks or ridges on the occlusal surface (Fig. 4.15). The tooth is built of following three tissues: enamel, dentine and cementum.

***(i)* Enamel**. It is the hardest substance in body and forms the surface of the tooth crown. Concentric rings seen under microscopic examination are believed to result from pulses of calcium salt deposits before tooth eruption. However, with only a few exceptions among mammals, no further enamel is deposited on the crown after the tooth erupts.

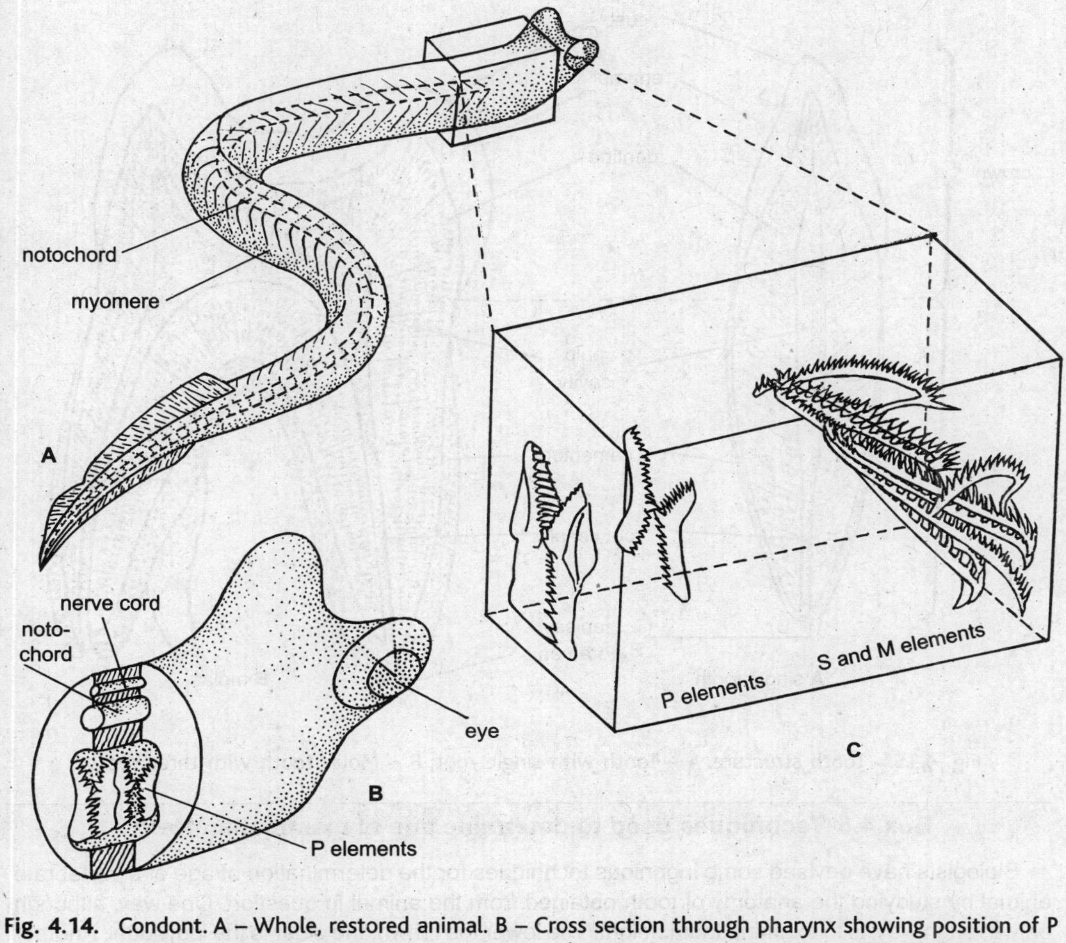

Fig. 4.14. Condont. A – Whole, restored animal. B – Cross section through pharynx showing position of P elements. C – Isolated condont apparatus, showing P, S, and M elements (after Kardong 2006).

***(ii)* Dentine**. It resembles with bone in chemical composition but it is harder. Dentine lies beneath the surface enamel and cementum and forms the walls of the pulp cavity. Even after tooth eruption, new dentine is laid down slowly throughout the life of an individual. Growth of dentine occurs by daily deposition along the walls of the pulp cavity, so that in very old animals, dentine may almost fill the entire cavity. The daily layers of dentinal growth are called the *incremental lines of von Ebner* (Box. 4.6).

***(iii)* Cementum.** It, like the bone, has cellular and acellular regions. Cementum rests upon the dentine and grows in layers on the surface of the roots. In many herbivores, cementum tends to extend up along the crown to between the enamel folds and actually contribute to the occlusal surface. Cells within the cementum, termed **cementocytes**, elaborate the matrix but in seasonally related pulses, so that cementum increases irregularly with age. The result is the production of **cemental annuli**, concentric rings that characterise the cementum layer. The appearance of these annuli changes predictably with mechanical properties of food (hard), nutritional state (lean times) and season (winter). In fossils, cementum appearance can thereby answer question about diet and even the season of the animal's death.

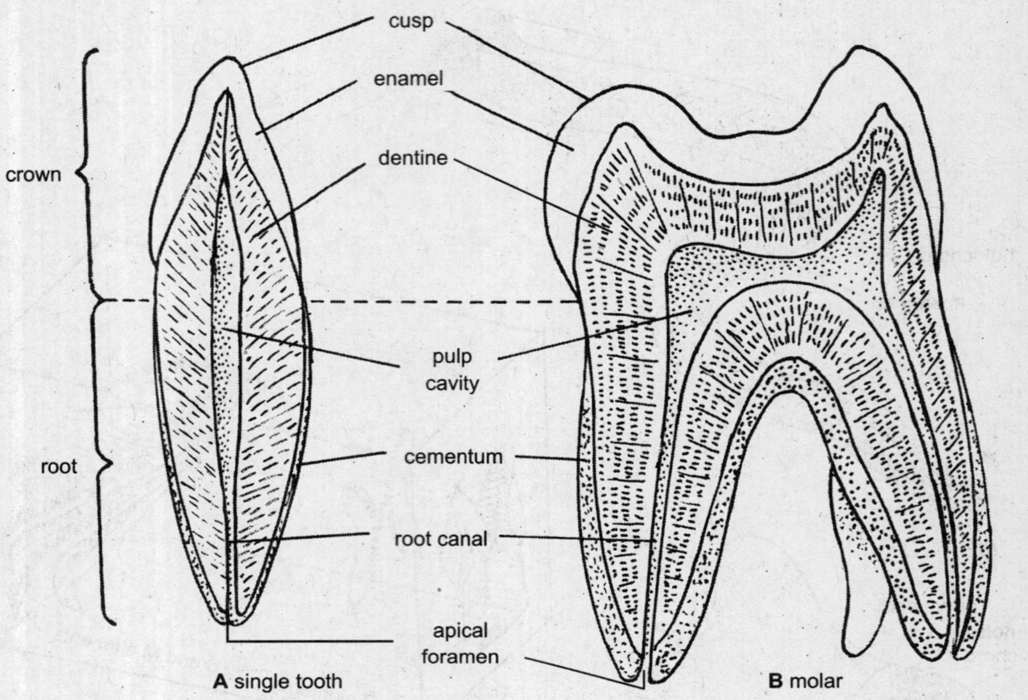

Fig. 4.15. Tooth structure. A – Tooth with single root; B – Molar tooth with three roots.

Box 4.6 Techniques used to determine age of mammalian teeth

Biologists have devised some ingenious techniques for the determination of age of a vertebrate animal by studying the anatomy of tooth obtained from the animal in question. One way, although only a rough way, is to examine tooth height. The lower the crown, the older is the individual. Another way is to examine the width of the pulp canal. Inside the tooth, odontoblasts persist and continue to add layers of detine slowly to the inner walls of the pulp cavity throughout the life of the individual. Therefore, the pulp cavity progressively narrows with advancing age. On the outer surface of the root, additional cementum is added usually on a seasonal basis to produce annuli, or rings, within the cementum (Fig. 4.16 A, B). These three features of teeth — wear, narrowing of the pulp cavity and deposition of cemental annuli — are roughly proportional to age. Techniques for estimating the ages of individuals are most reliable when age estimates are compared against similar characteristics from a sample containing individuals of known age. In lower vertebrates, where new teeth replace worn teeth, these techniques do not work, of course. But in mammals, who have only one set of permanent teeth to last a lifetime, these three dental features are rough indicators of age.

The determination of age of members of a population is important to wildlife biologists. A scarcity of young individuals may imply a decline in potential new breeding members of the population. On the other hand, if there are low members of older individuals, there may be too few young produced to breed and sustain the population. Management decisions are based on such information.

The old adage, "*Don' t look a gift horse in the mouth*", means that if you are kindly given a free horse, don't insult the donor by looking for defects in the gift. It is based on the fact that as horse grind away at fibrous food, their teeth wear down. Eventually, even hypsodont teeth are reduced to small stubs. Because wear increases progressively with age, tooth height is important to age.

The **peridontal membrane** (or peridontal ligament) consists of thick bundles of collagenous fibers (Sharpey's fibers) that connect the cementum-covered root to the bone of socket.

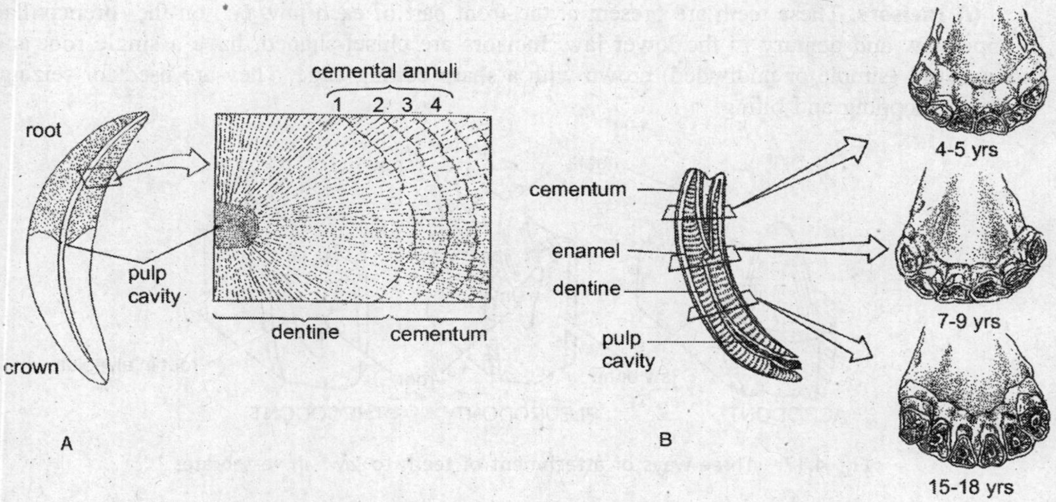

Fig. 4.16. Techniques used to determine age of mammalian teeth. A – upper canine tooth from a carnivore. A cross section through the root reveals cemental annuli in the cementum superficial to the dentine. The rate of rhythmic formation of these annuli can be caliberated if biologists first study canines from known age groups in the same species. Then, by counting annuli in teeth from free-ranging individuals, they can determine the age of individuals in the population and estimate the age structure of the population. B – Horse teeth. Progressive use with age wears horse teeth down through successive levels, exposing different layers within the teeth. By examining these teeth and comparing them with teeth of known age, biologists can identify the distinctive pattern on the crowns and determine a approximate age of horse (after kardong, 2006).

Characteristic Of Mammalian Dentition

1. Thecodont. Mammalian teeth are lodged in sockets or alveoli in the jaw bones (Fig. 4.17). (There are two more types of teeth attachment in vertebrates. **(i) acrodont type** which occurs in most vertebrates (shark, frog) and involves attachment of teeth with crest of jaw bones. **(ii) pleurodont type** which occurs in urodeles and lizards and involves attachment of teeth to the shelf-like indentation at inner margins of the jaw bones.)

2. Diphyodont. The mammalian teeth grow in tow successive sets. The teeth of the first set are called the **milk teeth** or **deciduous teeth**. These appear early in life and are, after some time, pushed out by the second set of teeth termed the **permanent teeth** which last throughout the life of the animal. The permanent teeth are often more than the milk teeth. In tubulidentates (aardvark), however, the milk teeth outnumber the permanent teeth. In bats and guinea pigs the milk teeth are shed before birth. Insectivores, platypus, toothed whales, sloths and sirenians have **monophyodont dentition**, *i.e.,* they cut only milk teeth. Milk dentition is reduced to the replacement of third premolar in marsupials.

3. Homodont. In dolphins and porpoises all teeth are similar as in fishes, amphibians and most reptiles.

4. Heterodont. Teeth of most mammals are differentiated in form and function to suite their food habits. Such teeth are said to form a **heterodont dentition**. Mammalian teeth are of following *four* types:

(i) **Incisors.** These teeth are present at the front part of each jaw, *i.e.*, on the premaxillae in upper jaw and dentary in the lower jaw. Incisors are chisel-shaped, have a single root and **monocuspid** (simple or undivided) crown with a sharp cutting edge. They are used for seizing, snipping, cropping and biting.

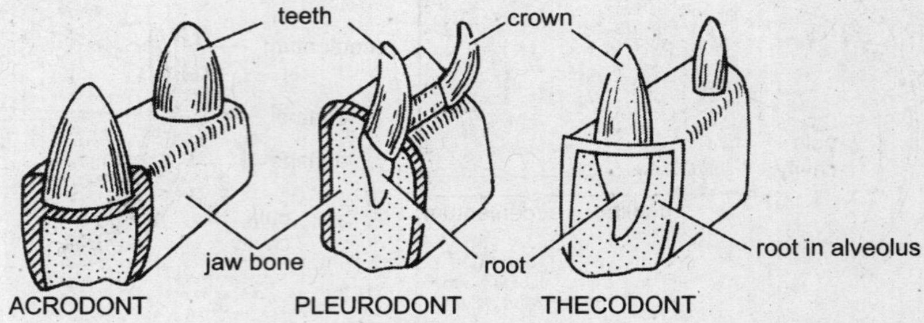

Fig. 4.17. Three ways of attachment of teeth to jaws in vertebrates.

(ii) **Canines.** The canines occur next to the incisors, and there is a single such tooth in each half of each jaw. These are present on maxillae in the upper jaw and dentary in the lower jaw, between premolars and incisors. These used for piering and tearing the prey, or in defence. Canines have single root and long, conical, monocuspid sharp crown.

(iii) **Premolars.** The premolars follow the canines. They are cheek teeth and born on the maxillae. Premolars have a double root and compressed crown with one or two cusps. Premolars are used for shearing and slicing.

(iv) **Molars.** These are also cheek teeth, borne on the maxillae but occupy the posterior parts of jaws, lying next to the premolars. Molars have more than two roots and cusps. Molars always develop in the permanent set, having no milk predecessors. They are used for crushing and mastication.

Depending upon the shape, size and number of cusps, the cheek teeth (premolars and molars) are distinguised into the following types (Fig. 4.18):

1. Monocuspid teeth are with one cusp or tubercle.

2. Bicuspid teeth are with two cusps.

3. Tricuspid teeth are with three cusps. The tricuspid teeth might be **triconodont**, having all three cusps arranged in one plane or **tritubercular** with three cusps forming a triangle.

4. Bunodont teeth with a number of distinct blunt pointed cusps, *e.g.,* human beings and monkeys.

5. Lophodont teeth are with cusps fused in the form of ridges, *e.g.*, horse and elephant.

6. Solenodont teeth are where cusps wear away, exposing the underlying dentine, which gets surrounded by crescent-shaped bands of enamel, *e.g.*, ruminats.

7. Secodont teeth are where cusps having sharp cutting surfaces, *e.g.*, dog and other carnivorous mammals.

8. Hypsodont teeth are with long crown and deeply placed neck.

9. Brachydont teeth are with short crown and neck at the surface of gum.

Diastema. The form of teeth is related to the food habits of the mammals. The insectivores (mole, bat) have narrow teeth; the carnivores have well developed canines, but poorly developed molars, except the first; the herbivores have short canines, but broad molars with enamel ridges for grinding. The cetaceans have sharp, conical, curved teeth for capturing fish.

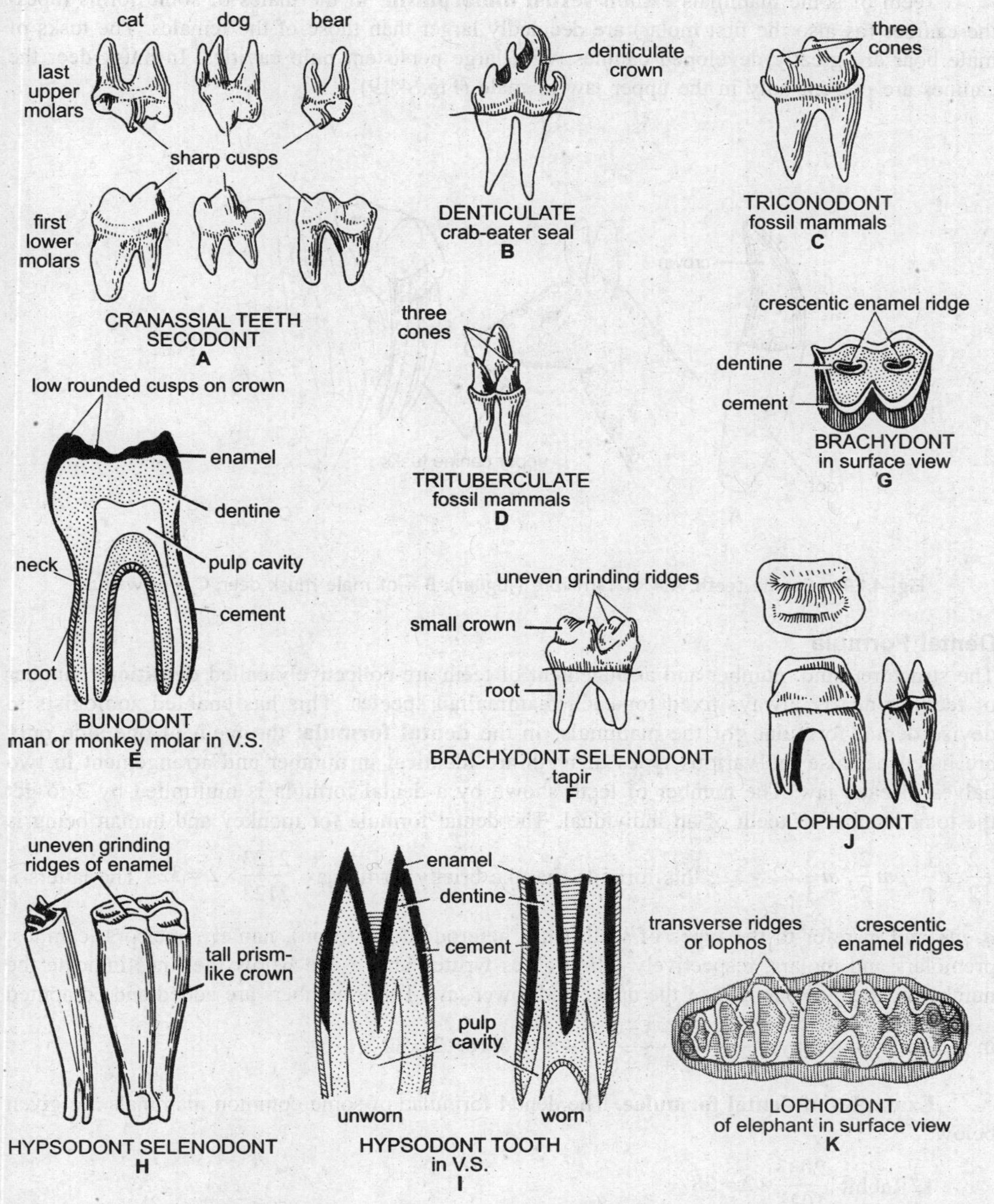

Fig. 4.18. Modification of cheek teeth (premolars and molars). A – Carnassial teeth (secodot); B – Denticulate molar; C – Triconodont tooth; D – Trituberculate tooth; E – Bunodont molar in V.S, F–Brachydont selenodont molar; G – Surface view of crown of brachydont molar; H – Hypsodont selenodont molars; I – Hypsodont teeth in V.S.; J – Lophodont molar; K – Lophodont in surface view.

One or more types of teeth are absent in certain cases, *e.g.*, canines in rabbit and squirrel, both canines and premolars in rat and elephant. This leaves a gap between the teeth. Such a natural gap is known as the **diastema**.

Teeth of some mammals exhibit **sexual dimorphism**. In the males of some forms (apes) the canines (as also the first molar) are decidedly larger than those of the females. The tusks of male boar are greatly developed canines, with large persistent pulp cavities. In musk deer the canines are present only in the upper jaw of male (Fig. 4.19)

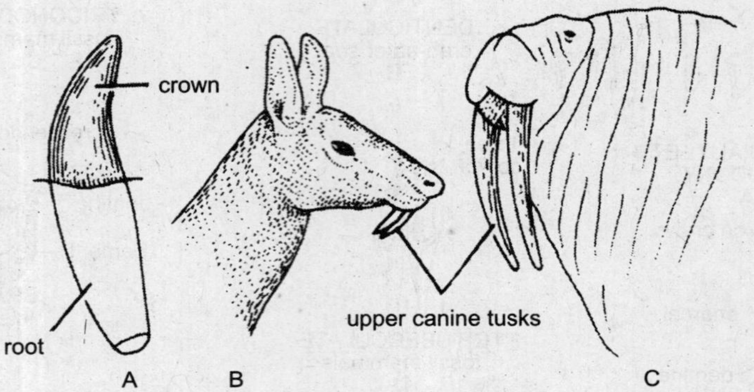

Fig. 4.19. Canine teeth. A – of carnivore (Juguar); B – of male musk deer; C – of walrus.

Dental Formula

The structure, kind, number and arrangement of teeth are collectively called **dentition**. Number of teeth is nearly always fixed for each mammalian species. This has enabled zoologists to devise dental formulae for the mammals. In the **dental formula**, the teeth of one side only are listed because the various types of teeth are identical in number and arrangement in two halves of each jaw. The number of teeth shown by a dental formula is multiplied by 2 to get the total number of teeth of an individual. The dental formula for monkey and human being is $i\frac{2}{2}, c\frac{1}{1}, pm\frac{2}{2}, m\frac{3}{3} \times 2 = 32$; this formula may be briefly written as $\frac{2123}{2123} \times 2 = 32$. The latters *i, c, pm* and *m* refer to the types of teeth (in a heterodont dentition), namely, incisors, canines, premolars and molars, respectively. The figures written above and below the lines indicate the number of teeth in one half of the upper and lower jaw. These numbers are added and computed in the following way: $\frac{2+1+2+3=8}{2+1+2+3=8}; 8+8=16; 16\times 2=32$.

Examples of dental formulae. The dental formulae of some common mammals are given below:

1. Rabbit $\frac{2033}{1023} \times 2 = 28$.

2. Rat $\frac{1003}{1003} \times 2 = 16$.

3. Dog $\frac{3142}{3143} \times 2 = 42$.

4. Cat $\frac{3131}{3121} \times 2 = 30$.

5. Cow, Goat and Sheep $\frac{0033}{3133} \times 2 = 32$.

6. Horse, Pig and Mole $\frac{3143}{3143} \times 2 = 44$.

7. Elephant $\frac{1003}{0003} \times 2 = 14$.

8. Kangaroo $\frac{3124}{1024} \times 2 = 34$.

9. Squirrel $\frac{1023}{1023} \times 2 = 24$.

The dentition of pig and horse is considered as typical in Eutheria, and that of others as modified. The toothed cetaceans (dolphins, porpoises) are peculiar among mammals in having indefinite number of teeth.

Development of Teeth

The tooth develops partly from the epidermis and partly from the mesenchyme of dermis. Its enamel is derived from the epidermis while the remaining parts from the mesenchyme. Mammalian teeth develop in two sets. The **milk** or **deciduous set** develops first in the buccal epithelium of the gums and is replaced later by the **permanent set**.

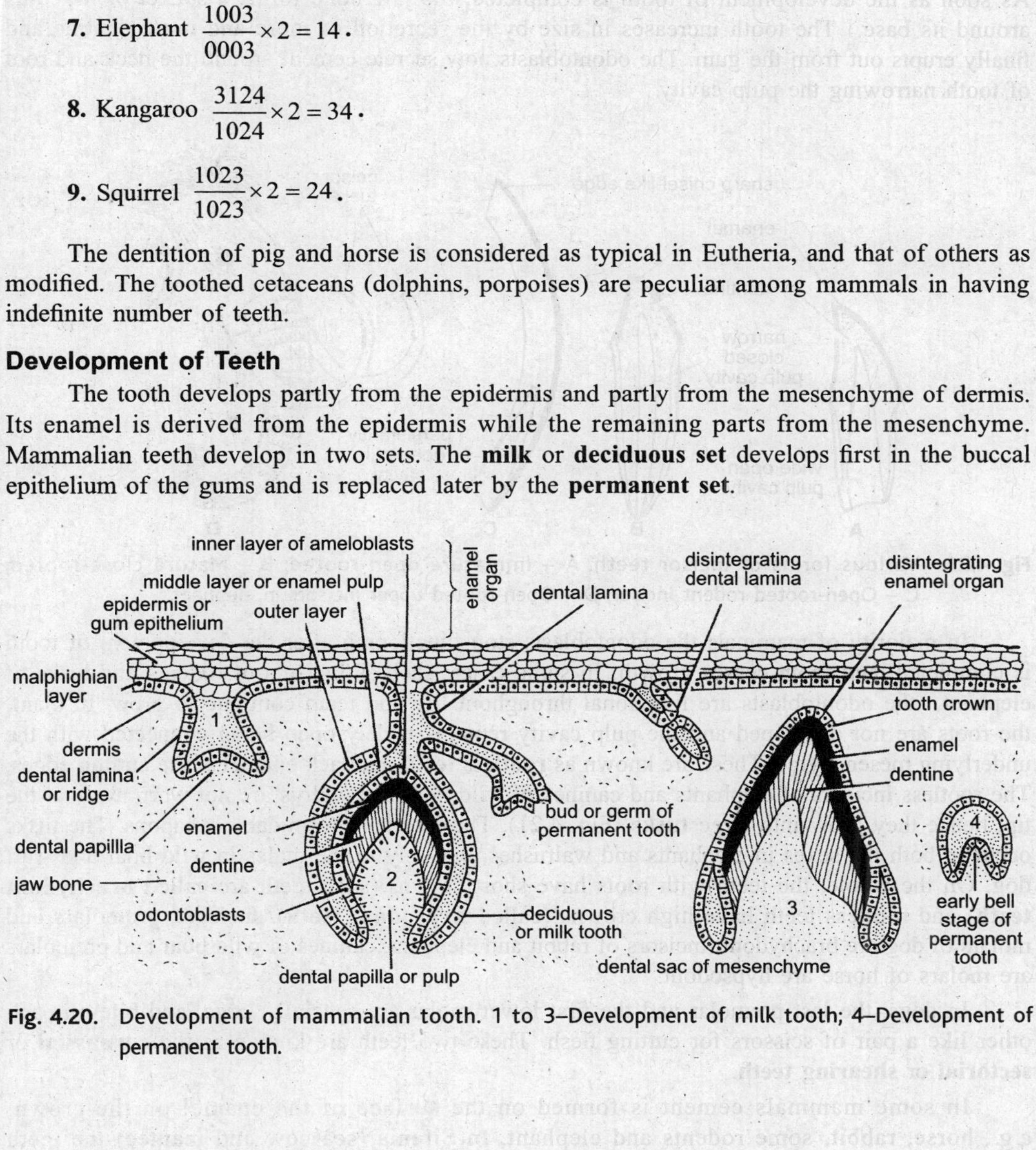

Fig. 4.20. Development of mammalian tooth. 1 to 3–Development of milk tooth; 4–Development of permanent tooth.

In the beginning the buccal epithelium of the gums sinks down forming the **dental groove** or **dental ridge** (Fig. 4.20). Simultaneously, the cells of Malpighian layer of epidermis grows down into the dermis forming the **dental lamina.** Along the dental lamia, the mesenchymal cells multiply and condense into a **dental** or **tooth bud.** This bud grows upwards and pushes the dermal lamina, which assumes the form of an inverted cup. The cells of Malpighian layer become markedly columnar and are known as **ameloblasts**. The mesenchymal cells below it differentiate into a layer of **odontoblast cells**. The tooth bud is now know as **enamel organ**. The ameloblast cells secrete enamel towards their inner side, while the odontoblasts secrete dentine. The enamel surrounds the upper part of dentine like a cap. The remaining mesenchymal cells of the dental papilla form the **pulp** and the cavity of the enamel changes into the **pulp cavity**. As soon as the development of tooth is completed, the jaw bone forms a socket or **alveolus** around its base. The tooth increases in size by the secretion of more and more dentine and finally erupts out from the gum. The odontoblasts now secrete cement around the neck and roof of tooth narrowing the pulp cavity.

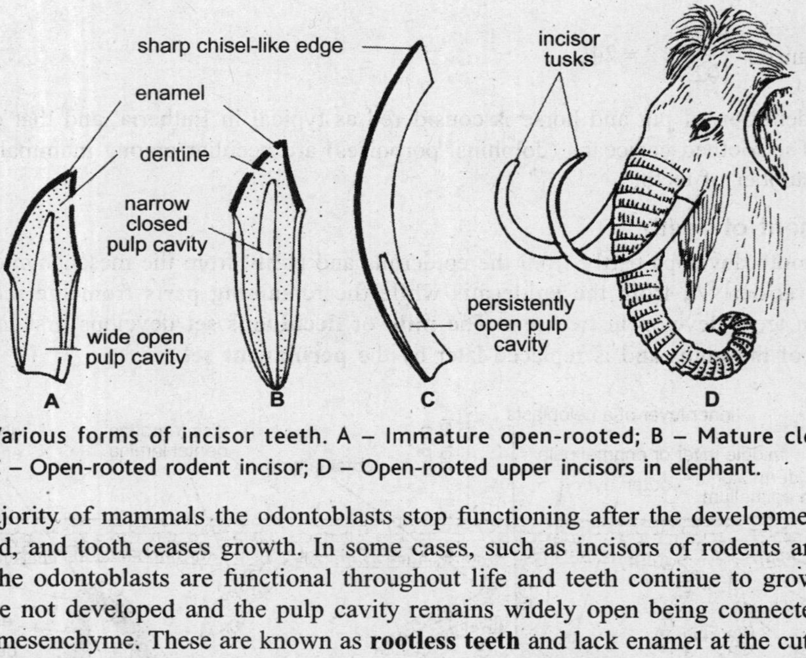

Fig. 4.21. Various forms of incisor teeth. A – Immature open-rooted; B – Mature close-rooted; C – Open-rooted rodent incisor; D – Open-rooted upper incisors in elephant.

In majority of mammals the odontoblasts stop functioning after the development of tooth is completed, and tooth ceases growth. In some cases, such as incisors of rodents and tusks of elephant, the odontoblasts are functional throughout life and teeth continue to grow. In them, the roots are not developed and the pulp cavity remains widely open being connected with the underlying mesenchyme. These are known as **rootless teeth** and lack enamel at the cutting edges. The rootless incisors of elephants and canines of wild boar and walrus are not worn away at the tip, hence they grow into large **tusks** (Fig. 4.21). The latter are formidable weapons. The tusks occur in both the sexes in elephants and walrushes, but only in the males in wild boar and wart dog. On the whole, the teeth with roots have short crowns (such teeth are called **brachydont teeth**) and rootless teeth have high crowns (called **hypsodont teeth**). Incisors, premolars and molars of dog are brachydont; incisors of rabbit and elephant, canines of wild boar and premolars are molars of horse are hypsodont.

In dogs, the last premolar and the first lower molar are specially large, and bite on each other like a pair of scissors for cutting flesh. These two teeth are known as the **carnassial** or **sectorial** or **shearing teeth**.

In some mammals cement is formed on the surface of the enamel on the crown, *e.g.*, horse, rabbit, some rodents and elephant. In Sirenia (sea cow and mantee) the teeth lack enamel.

Horny Teeth

Agnathans or cyclostomes lack true teeth but the buccal funnel and tongue are provided with horny epidermal teeth for rasping flesh. Horny teeth arise from the stratum corneum. The lips of the suctorial mouth of some species of frog tadpoles develop horny teeth. In turtles, sphenodon, crocodiles, birds and monotremes (prototherian mammals) a horny **egg tooth** for cracking the shell of cleidoic egg develops on the upper jaw at the tip of the beak prior to hatching. In lizards and snakes embryos that develop within an egg shell, the egg tooth is transitory genuine tooth of dentine. The adult duck-billed platypus (*Ornithorhynchus*) has only horny teeth. Horny beaks in some turtles and birds have serrations that simulates teeth.

4.4 TONGUE

Cyclostomes posses a tongue derived not from hypobranchial musculature, but from the floor of the pharynx. During feeding, cyclostomes protrude their soft tongue, bearing ridges of rasping keratinized "teeth". The tongue of elasmobranchs, bony fishes and perennibranchiate amphibians is mere crescentic or angular elevation of the floor of the oro-pharyngeal cavity shaped by the underlying basihyal and ceratohyal cartilages that constitute a skeleton. This lean hyoid elevation is a **primary tongue** (Fig. 4.22). A primary tongue has no musculature; hence, it cannot be independently manipulated. It may assists the jaws in holding prey within the oropharyngeal cavity until it can be swallowed, especially if the surface of the tongue has teeth-like denticles. It is, however, the forerunner of the tongues of tetrapods.

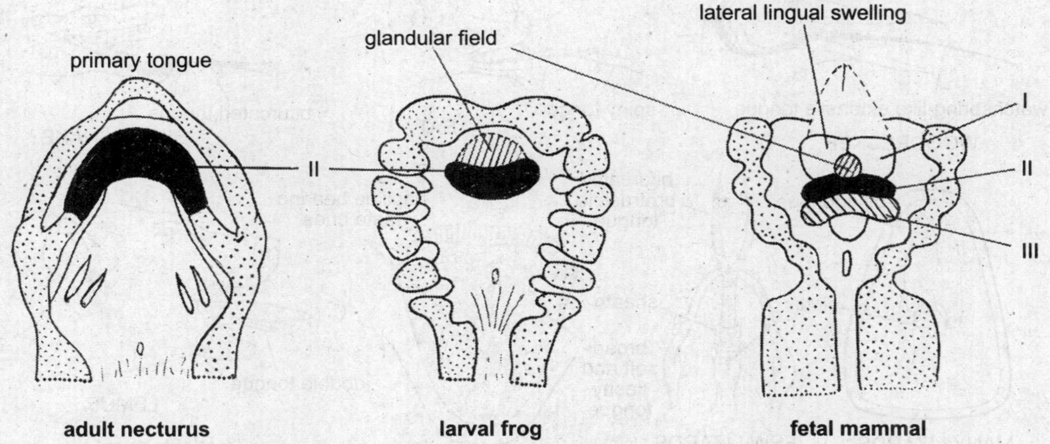

Fig. 4.22. Floor of the oral cavity depicting stages in evolution of the tongue. I, II and III, derivatives of pharyngeal arches 1,2 and 3. Dotted outline in mammal show final extent of lateral lingual swelling (after Kent and Carr, 2001).

A **mobile tongue** develops first in tetrapods from the hypobranchial musculature attached to resting on the underlying **hyoid apparatus**, a skeletal derivative of modified lower ends of the hyoid arch and adjacent branchial arches.

The tongue of terrestrial urodeles and of anurans consists of a primary tongue and of an extension that can be flipped out of the mouth. In both groups, the primary tongue develops from hyoid arch mesenchyme, and the extension develops from an embryonic **glandular field** in the pharyngeal floor anterior to the hyoid arch (Fig. 4.22; larval frog). The glandular field is so named because this region of the adult tongue secretes a sticky mucus that entangles insects when the tongue is suddenly thrust out of the mouth. The tip of the tongue of insectivorous

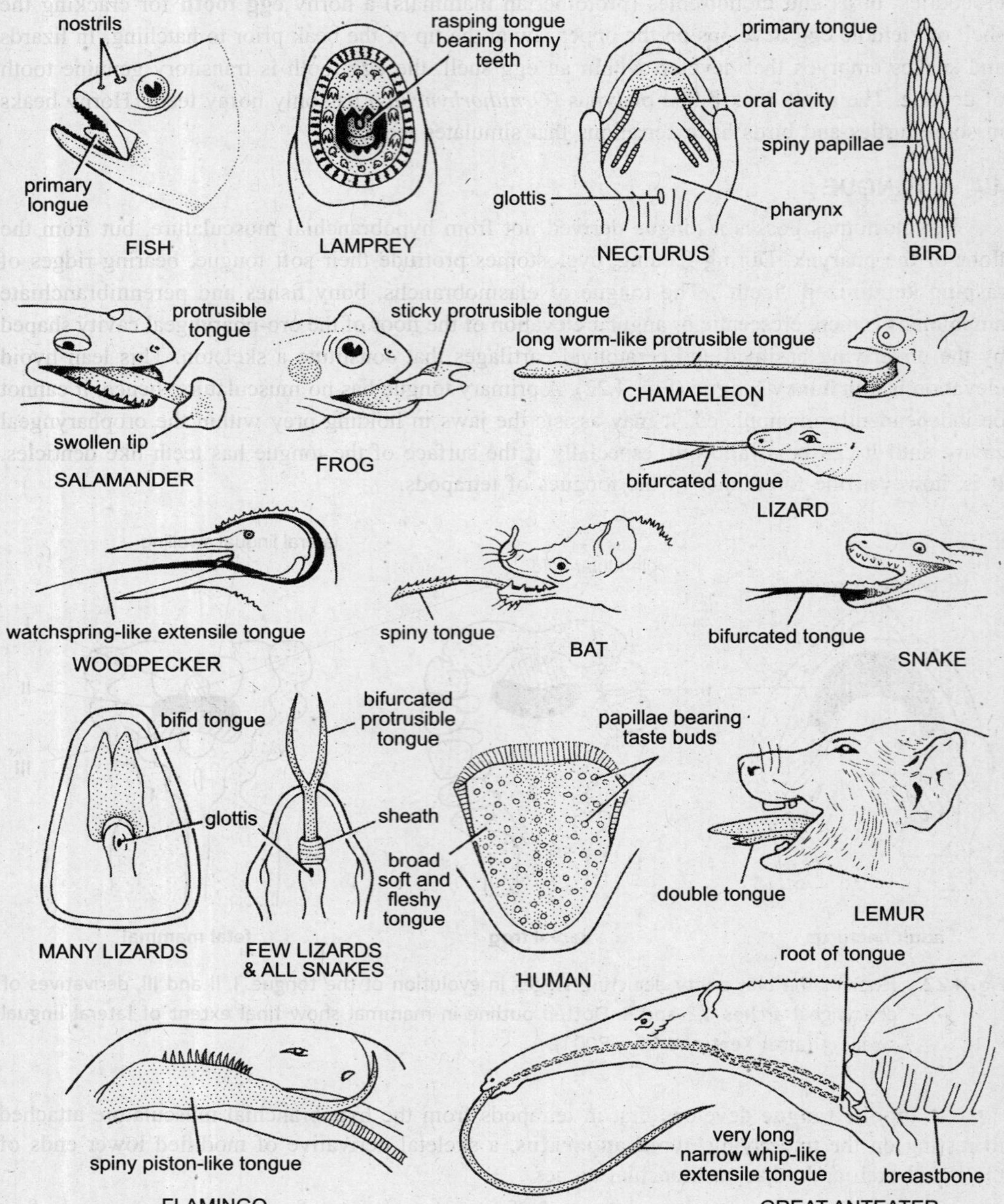

Fig. 4.23. Types of tongues in some vertebrates.

amphibians usually terminates in a broad, sticky expansion that increases the probability that the prey will be caught and then delivered into the oral cavity. The root of the tongue of terrestrial urodeles becomes anchored to the basihyal and ceratohyal cartilages in the floor of the pharynx; that of anurans, unlike other tetrapods, becomes anchored to the floor of the oral cavity immediately behind the mandibular symphysis. In the toad family *papidae*, no tongue develops.

The tongue of many tetrapods holds the **taste buds**, sensory organs responsive to chemicals entering the mouth. Lizards and snakes project their foretongue out of the mouth to retrieve air and/or substrate chemicals that their sensory **vomeronasal organ** (Jacobson's organ) evaluates (Box 4.7). In many mammalian carnivores the surface of the tongue is roughened, like a file, with numerous keratinous spiny projections, or **filiform papillae**, that help rasp flesh from the bones.

Box 4.7 Vomeronasal Area

Named for the bones that usually house this chemosensory organ, the **vomeronasal organ** is known only in some tetrapods. It is absent in most turtles, crocodiles, birds, some bats and aquatic mammals. In amphibians, it is in a recessed area off the main nasal cavity. In reptiles, the vomeronasal organ is a separate pit to which the tongue and oral membranes deliver chemicals. In mammals possessing this organ, it is an isolated area of olfactory membrane within the nasal cavity that is usually connected the mouth via the **nasopalatine duct**.

The vomeronasal organ is an accessory olfactory system. It includes basal, sustentacular and bipolar sensory cells similar to those of the olfactory epithelium. Sensory receptor cells of the vomeronasal organ project into the lumen by means of microvilli, whereas olfactory sensory cells have cilia. The neural circuitry of the vomeronasal system runs parallel to but remains entirely separate from the main olfactory system. Like the main olfactory system, the vomeronasal system can be traced through the limbic system to the hypothalamus and thalamus.

In many vertebrates with a vomeronasal organ, the respiratory airflow carries particles to the organ. But the vomeronasal organ may also establish an association with the mouth, leading some biologists to suggest that it may sense the chemical composition of food in the buccal cavity. However, the vomeronasal organ also seems to be especially sensitive to chemical important in social or reproductive behaviour. Theoretically, individuals lacking this organ could increase their respiratory ventilation rate to draw more frequent samples of air across the olfactory epithelum, but this would be energetically more expensive and might result in problems with regulation of blood pH. In order for us to recognize the anatomical and functional distinctiveness of the nasal epithelium and the vomeronasal organ, **Kardong** (2006) has referred to the chemical detected as **odours** and **vomodours** and to the processes as **olfaction** and **vomerolfaction**, respectively.

Snakes and lizards extend their tongue from the mouth, sweep air in front of their snout to collect vomodours, and then retract their tongue into the mouth to deliver these chemicals to the vomeronasal organ (Fig. 4.24). In snakes and possibly in lizards, the returning tongue does not directly enter the lumen of the vomeronasal organ. Instead, the reptile wipes its tongue against the entry ducts to the organ and against small ridges in the lower part of the mouth. These ridges are lifted up to the ducts, apparently adding vomodours they have gathered to those the tongue delivered directly to the entry ducts. In snakes, the removal or inactivation of the vomeronasal organ leads to deficits in courtship, pheromonal trailing and prey detection (see **Kardong**, 2006).

Many tetrapods use their tongue in **lingual feeding**. They protract their tongue out of the mouth at prey. Its sticky surface holds the catch until jaws advance or tongue retraction brings it back into the mouth. Many terrestrial salamanders and lizards use this technique. In fact, some anatomists argue that development of just such a mobile and projectile tongue represents

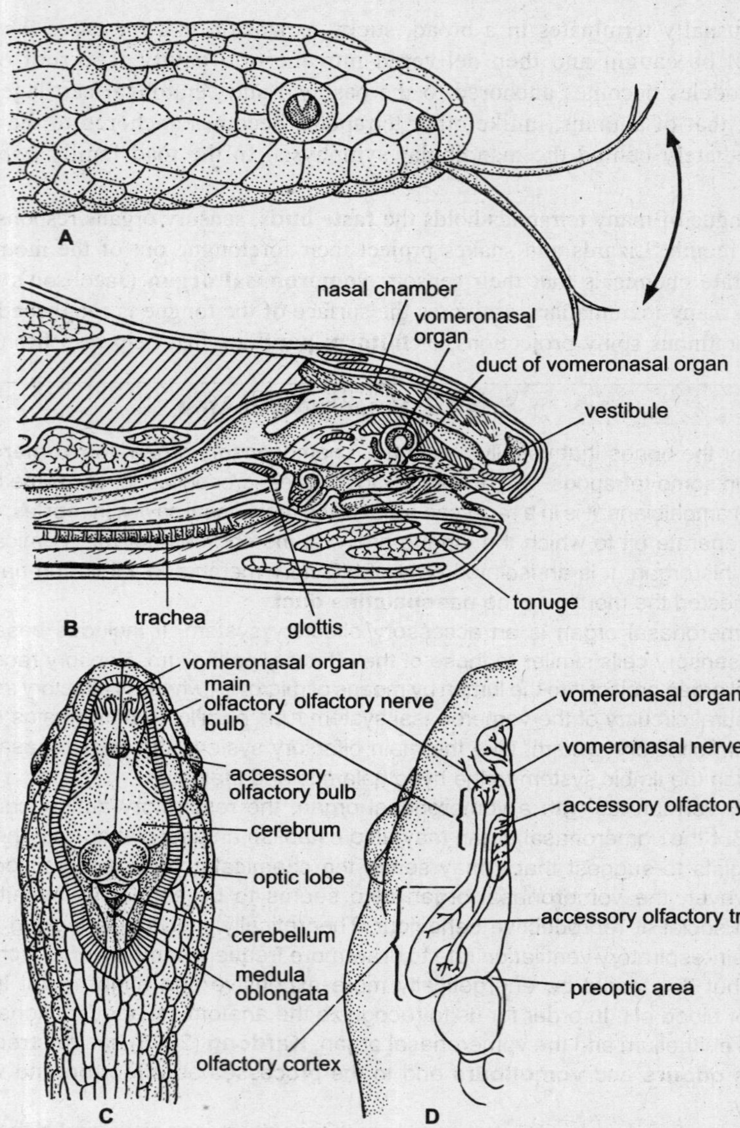

Fig. 4.24. Tongue flicking in snakes. A – Snakes, like lizards, extend their tongues to sweep air in front of them. The tongue collects and then transports airborne particles into the mouth. Probably along with other oral membranes, the tongue wipes these particles onto the vomeronasal organ on the roof of the mouth. B – Sagittal section of the head of a constrictor (*Constrictor* is a snake that kills by squeezing and choking its prey). The vomeronasal organ is a blind pocket with a lumen that opens directly into the mouth via a duct. The tip of the retracted tongue projects from its sheath beneath the trachea. C – Skull and overlying tissue have been cut away to reveal dorsal view of the snake brain. D – Neuroanatomy of a snake's olfactory organs. The main olfactory bulb receives inputs from the olfactory epithelium. The accessory olfactory bulb, via a separate tract, receives information from the vomeronasal organ. Vomeronasal organ and olfactory systems are separate chemoreceptive organs whose fibers travel separately within the olfactory tract. Thereafter, sensory information tends to be brought together in the olfactory cortex of the telencephalon (after Kardong, 2006).

a major **feeding innovation** in the transition of early amphibians to life on land woodpeckers (*e.g., Picus*) use their long specialized tongues like a probe to obtain insects between cracks in tree barks or in holes they create (Fig. 4.25).

The tongue of tetrapods can also transport captured prey, that is, move prey through the buccal cavity to the back of the pharynx where it is swallowed (Fig. 4.26 A). This process is called **intraoral transport** and proceeds in several steps. First, the jaw part slowly, and the tongue is advanced forward and beneath the food to fit partially around it (Fig. 4.26 B). Second, the jaws open more rapidly, and the tongue draws the adhering food backward in the mouth (Fig. 4.26 C). Third, the jaws close slowly and grip the most posteriorly positioned food, and the tongue comes away from its adhesive hold on the food. The cycle is repeated, with the tongue working in synchrony with the jaws to move the food in increments to the back of the buccal cavity and into the pharynx (Fig. 4.26 D).

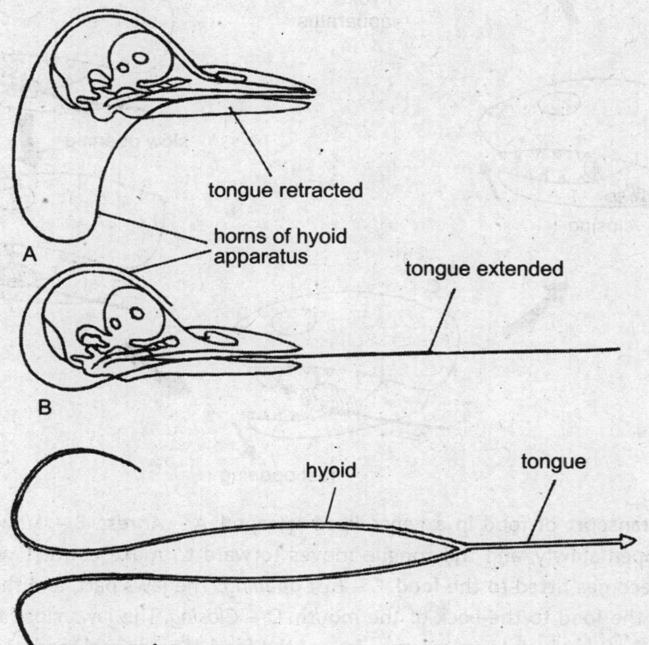

Fig. 4.25. Tongue protrusion in a woodpecker. A – The flexible and thin hyoid apparatus supports the fleshy tongue. B – When the woodpecker protrudes its tongue, the hyoid apparatus slips forward and the tongue extends from it. C – Ventral view of the hyoid apparatus from the woodpecker *Picus* (after Kardong, 2006).

The grip of tongue on food during intraoral transport depends partially on its surface irregularities that interlock or physically engage the prey. The ability of the fleshy tongue to shape itself to the food may help it gain a physical grip on the prey. This tongue adhesion also depends on a "wet" adhesion, the sticky effects created by surface tension in air and by capillary action. Where tetrapods have returned to aquatic feeding, these physical phenomena are less effective. This may account for the much smiplified tongue devoid of intrinsic musculature found, for example, in crocodilians. Many aquatic tetrapods such as aquatic feeding turtles resort to modified suction feeding in which the tongue has little role to play. In fact, if the tongue of aquatic tetrapods were a large fleshy structure occupying the floor of the buccal cavity, it could interfere with the sudden expansion of the buccal cavity necessary for suction feeding.

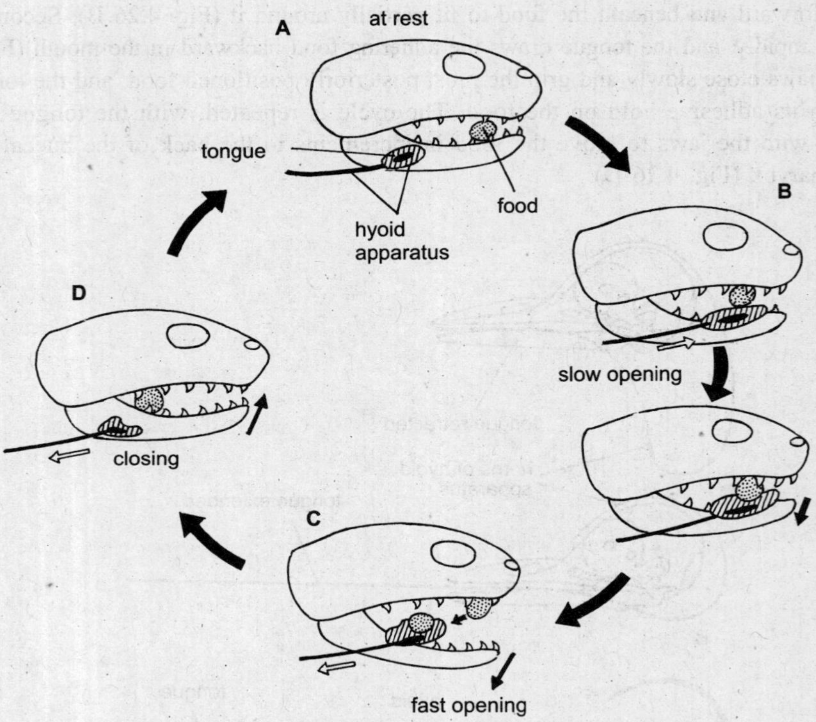

Fig. 4.26. Intraoral transport of food in a generalized tetrapod. A – *At rest*. B – *Slow opening*. The jaws begin to open slowly, and the tongue moves forward to make contact with the food, after which it becomes fitted to the food. C – *Fast opening*. The jaws part, and the retracting tongue transports the food to the back of the mouth. D – *Closing*. The jaws close to establish a grasp on the food, after which tongue comes free of the food. If an animal repeats this sequence, food is moved successively to the back of the mouth and into the pharynx, where it is swallowed into the oesophagus (after Kardong 2006).

Certain other adaptations of tetrapod tongue. The tongue is widely used for capturing or gathering food. Long, sticky tongues of insectivorous or ant eaters dart in and out at lightening speed. The sticky tongue of the 4-foot-long great anteater is 2 feet in length (Fig. 4.23). When the mouth of a toad is closed, the tongue lies folded back over itself so that the tip is directed toward oesophagus. Flipping the tongue out of the mouth occurs when the long fibers of the **genioglossus medialis** muscle, which extend to the tip of the tongue, stiffen to form a complex of intrinsic rods, and the **genioglossus basalis muscle**, which forms a wedge beneath the anchored anterior end of the tongue, suddenly swells. The combination flips the rigid tongue over the mandibular symphysis. It returns to mouth by contraction of the **hypoglossal muscles** (**Gans** and **Gorniak**, 1982).

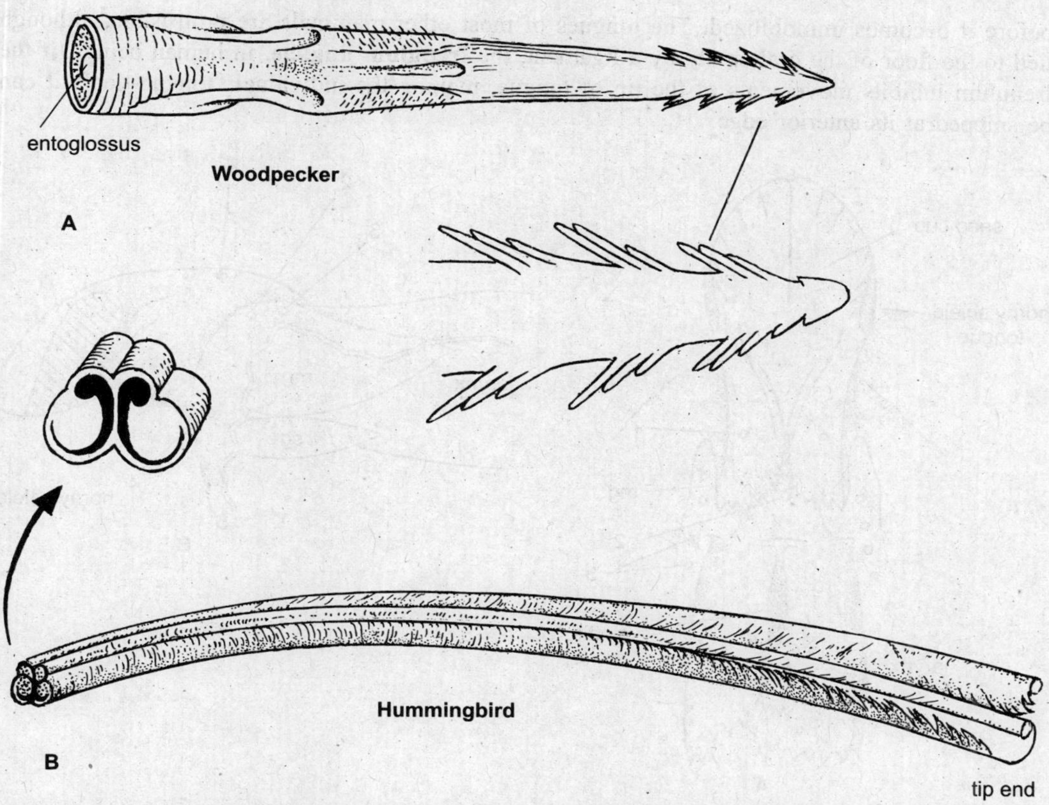

Fig. 4.27. Adaptations of the tongues of two birds with different feeding habits. A – Ventral view of tongue of a grub-eater (woodpecker). B – Lateral view of a nectar-feeder (after Kent and Carr, 2001).

Woodpeckers have a barbed tongue that shoots like an arrow into dark crevices in tree trunks, spearing grubs and carrying them to the mouth (Fig. 4.27 A). The tiny tongue of humming birds darts rapidly back and forth between flower and mouth, collecting droplets of nectar at the hollow frayed tip (Fig. 4.27 B). A parrot's tongue is armed with two flexible hory shields in the wall of a seed cup that is used for shucking (*shuck* means to remove the husk or pods of corn, peas, etc.) hard-coated seeds and preparing grain for swallowing (Fig. 4.28). Like fingernails, the shields of tongue are composed of keratinized epithelial cells that grow forward from nail-like bed halfway back on the tongue. The anterior edge is constantly being worn away and replaced. Embedded within the tongue of birds and lizards is an entoglossus bone, an anteriorly directed process of the hyoid (Fig. 4.28 A). In many birds, a pair of long paraglossal bones are attached to the entoglossus near the tip of the tongue and extend caudal embedded in its edges.

The huge immobilized muscular tongue of some baleen whales directs tons of incoming sea-water into the huge reservoirs under the throat and chest. When these are being emptied by compression of their walls, the water is strained back into the sea through frayed sheets of baleen, and organisms strained out of the water accumulate on the deeply furrowed tongue and are then swallowed. Newborn baleen whales have only rudiments of baleen, so there is room in the neonatal oral cavity for a mobile tongue to be manipulated for suckling for about six months before it becomes immobilized. The tongues of most other mammals are protrusible, although tied to the floor of the oral cavity by a ligament, the **frenulum linguae**. In human beings, if the frenulum inhibits movements of the tip of tongue, making the individual "tongue-tied", it can be snipped at its anterior edge.

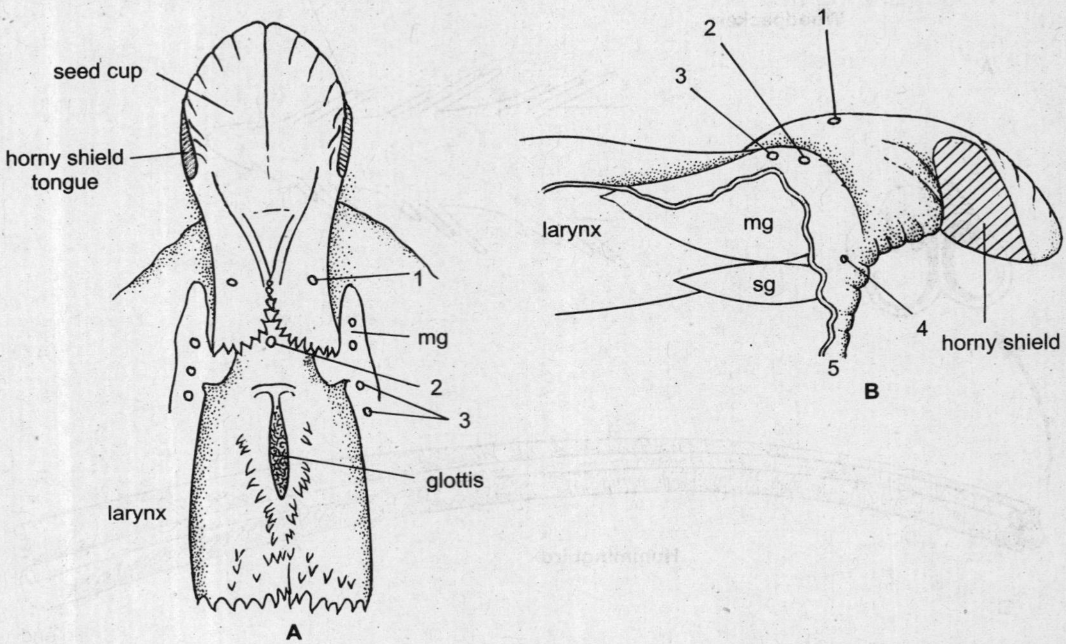

Fig. 4.28. Tongue of the parrot *Psittacula roseata*. A – Dorsal view; B – Right lateral view, the mucosa of the oral cavity cut away to reveal glands. mg = mandibular gland; sg = sub-lingual gland. 1-4, opening of following glands: **1**, paired lingual; **2**, median lingual; **3**, mandibular; **4**, sublingual. In B, **5**, is the site of the mandibular symphysis (after Kent and Carr, 2001).

The mucosa of the tongue contains receptors (*i.e.*, taste buds) not only for taste but for other stimuli, including, in amniotes, **stereognosis** (perception of the shape, weight, and texture of a solid body by feeling, handling, or lifting it). These encapsulated nerve endings on the tip of the tongue enable insectivores to locate food by exploring dark interiors with the tongue. Seed-eating birds use stereognostic information in manipulating a seed that is being husked in the seed cup.

Tongues which are not used to procure food or water, manipulate fluids and solids in the oral cavity, and the tongues of most tetrapods participate in swallowing. Long tongues function in overheated mammals as a site for cooling the blood by evaporation of saliva while **panting**. Lizards clean their spectacles (transparent eyelids) with them. Spiny papillae on the tongue's surface are used by carnivores (Tigers, lions, etc.) for rasping bones and by many mammals for grooming. The latter accounts for hair-balls that often form in the stomach of domestic cats. And, of course, human speech as we know it would not be possible without a tongue.

Oral Glands

Terrestrial tetrapods have a variety of multicellular oral glands that secrete watery or viscous fluids into the oral cavity. The chief ingredient is usually mucus of varying viscocities and chemical composition. Mucus moistens the food as necessary to produce a bolus that can be manipulated by the tongue, when present and lubricates dry food for passage through the pharynx and down the oesophagus. Moisture is also essential for taste buds to function because the stimulant for taste must be in solution to evoke a gustatory response. Other secretions include, in one species or another, serous fluids, toxins, venoms, and an enzyme (amylase) that digests starch, although the latter is rare outside of mammals. Viscous secretions keep the tongue sticky in species where this is essential, and venom tranquilizes prey or dispatches them promptly.

As a rule, oral glands are named according to their location. **Labial glands** open into the oral vestibule at the base of lips, **molar glands** are in the floor of the orbit, **palatal glands** open onto the palate, and **sublingual** and **sub-mandibular glands** open via common papillae under the tongue. **Intermaxillary (internasal) glands** lie near the premaxillary bone. Frogs have upto 25 small intermaxillary glands, each with its own duct that delivers a sticky secretion onto the palate. The large **venom glands** of the four families of venomous snakes are palatal glands. Their ducts open at the base of a maxillary tooth, and the venom exudes into a groove or tube in the tooth (fang; Fig. 4.29). In *Heloderma*, the only poisonous lizard, sublingual glands secrete the toxin.

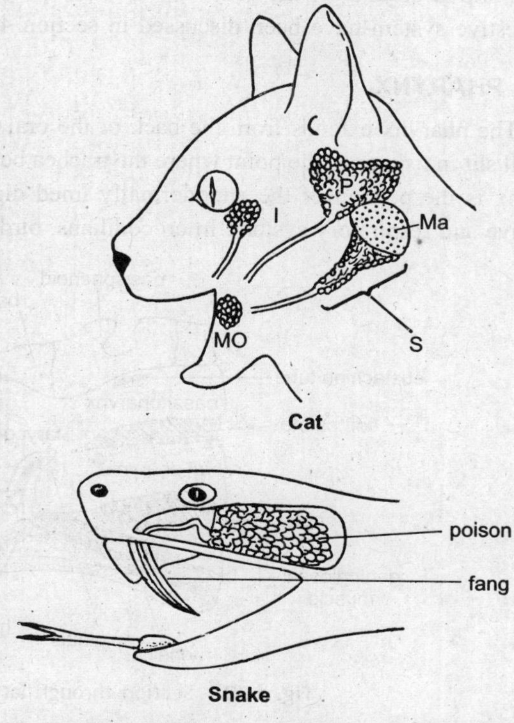

Fig. 4.29. Oral glands in a cat and a snake. **P**, parotid; **Ma**, mandibular; **S**, sublingual; **Mo**, molar; **I**, infraorbital. A proximal portion of the mandibular duct (dotted) is surrounded by the sublingual gland. Ducts for molar and sublingual glands are small and cannot be seen. A duct from the poison gland drains into a grooved maxillary tooth (Fang) (after Kent and Carr, 2001).

The **saliva** is a mixture of oral secretions, but the term is usually reserved for oral secretions of the mammals. The mammalian **parotid** is the largest tetrapod salivary gland and **ptyalin** (amylase), an enzyme that initiates digestion of starch, is always one of its secretions. Mammals have several salivary glands, but not all of them produce ptyalin. The specific mix of mucus, serous secretions, and ptyalin produced by any mammalian salivary gland varies with the species and is correlated with dietary habits. The duct of parotid gland crosses the masseter muscle to open into the oral vestibule near one of the upper molar teeth (Fig. 4.29; cat).

The poison glands of reptiles resemble the parotid gland histologically. Birds do not have a copious supply of saliva, but the openings of some of the oral glands of birds are shown in Fig. 4.28 and 4.47).

Aquatic craniates have adequate moisture in the oral cavity for taste buds to function, but glands that secrete digestive enzymes would be a waste of energy. Any enzymatic secretion would be diluted and washed away. Mucus-producing **goblet cells** are about the only source of oral secretions, and these are common. Mucus lubricates the esophagus. Goblet cells perform a special function in the male population of a few species of catfishes. Male catfishes carry fertilized eggs in temporary folds, or crypts, called **brood pouches**, in the mucosa of the palate and retain the hatchlings in the crypts for about two months during which time the male does not feed. The goblet cells in the brood pouches produce a copious nutritious secretion that maintains a suitable environment for development of the eggs and later nourishes the hatchlings. After the hatchlings leave, the brood pouches atrophy in response to altered hormonal ratios. A rare instance of multicellular oral glands in aquatic craniates are the **anticoagulant glands** of lampreys, which open into the oral cavity beneath the tongue just internal to the buccal funnel. Their secretion inhibits blood clotting of the prey while the lamprey is feeding. (Note. Other associated glands of digestive system have been discussed in section 4.7).

4.5 PHARYNX

The pharynx extends from the back of the oral cavity caudally to the posterior limit of the last gill slit in fish, or to the point where the trachea begins in tetrapods (Fig. 4.30). In other words, pharynx is the portion of the endodermally lined digestive tract which is common to both the digestive and respiratory system. In crocodilians, birds and mammals it can be divided into three

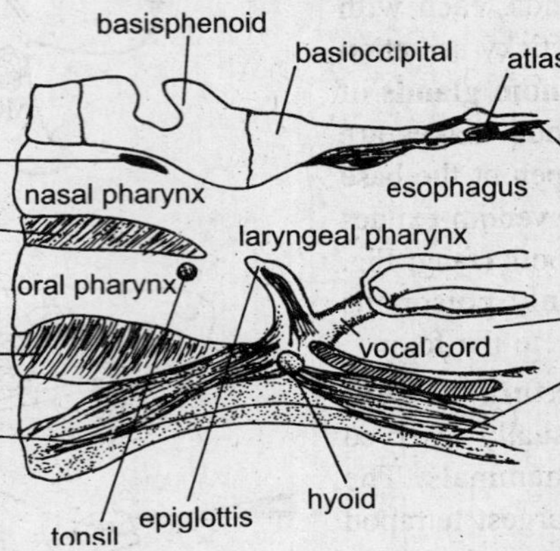

Fig. 4.30. Section through larynx of a mammal.

regions. **Nasal pharynx** above the secondary palate; **oral pharynx** below nasal pharynx and **laryngeal pharynx** caudally opening into both esophagus and larynx (Fig. 4.32).

In fish, food and water passes through the pharynx. In them, most of the water is then passed out through the gill slits while the food is pushed into the esophagus.

In **tetrapods**, air may reach the pharynx from either the mouth or the nostril. In either case the air enters the pharynx dorsally and must pass ventrally through the pharynx to reach the larynx and trachea. At the same point the food must pass dorsally to enter the oesophagus (Fig. 4.34).

This area of laryngeal pharynx where the air current crosses the food current, called the **pharyngeal chiasma** (Fig. 4.32). Only in whales there is a different and much refined system, the most anterior part of the larynx, the

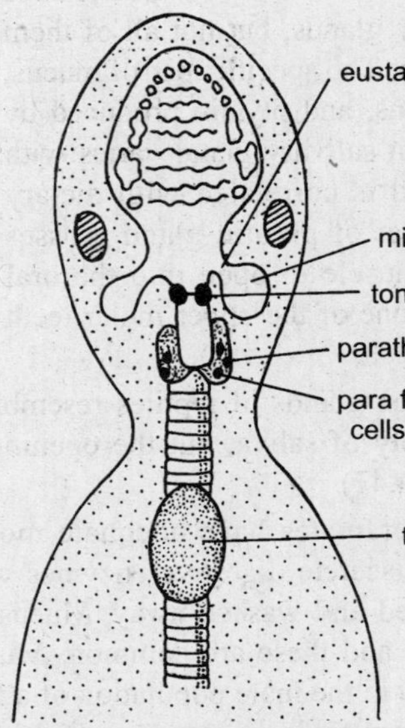

Fig. 4.31. The derivatives of pharyngeal pouches in mammals. Roman numbers indicate pouch numbers.

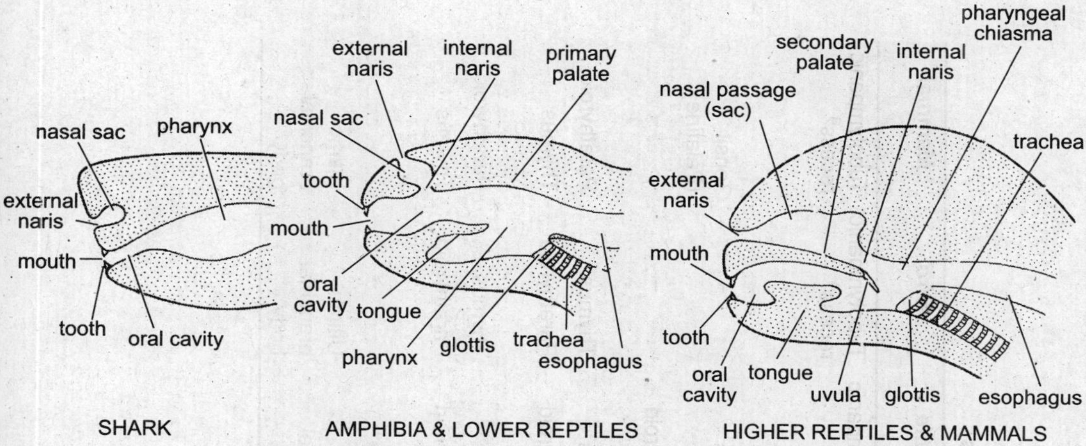

Fig. 4.32. Relation of nasal passages to buccal or oral cavities in different vertebrates.

epiglottis wraps around the pharynx and inserts into the internal nostrils eliminating pharyngeal chiasma. In all vertebrates, of course, six pairs of pharyngeal pouches develop in the pharyngeal region (except for cyclostomes, which have many more). In fish, the pharyngeal pouches (gill pouches) unite with ecotodermal invaginations to form gill slits which break through the surface. In tetrapods the endodermal lining of pharyngeal pouches forms several different structures (Fig. 4.31). Each of **first pair of pouches** forms the **middle ear** cavity and with eustachian tube it retains its continuity with pharynx.

Second pair of pouches form **palatine tonsil** – the lymphatic tissue on each side of the pharynx that retain continuity with the pharynx. **Third**, **fourth**, **fifth** and **sixth** pairs of pouches lose their continuity with pharynx and the endoderm migrate caudally and ventrally. Third and fourth pairs of pouches contribute to thymus (Box 4.8) and parathyroid gland. Fifth the sixth pairs of pouches contribute to ultimobranchial and parathyroid. The ultimobranchial bodies are separate

Box 4.8

All pouches contribute to the thymus in fishes, variable numbers contribute in amphibians and pharyngeal puches III and IV contribute in mammals.

glands in fishes, amphibians, reptiles and the birds, but in mammals, they become part of the thyroid gland and apparently form its internal population of C cells. The C cells are involved in controlling blood calcium levels (**Kardong**, 2006). The thyroid gland is formed from mid ventral part of the anterior pharynx. The only prominent glands in the pharynx itself are the lymphoid tissue of palatine tonsils and adenoids. The roof of pharynx gives rise to the pharyngeal tonsil, the floor gives to the thyroid, part of the tongue, lingual tonsil and lung primordium (Table 4.1).

• **Swallowing.** It is also called **deglutition** and involves the forceful movement of the bolus from the mouth and pharynx into the esophagus and then the stomach. Most vertebrates bolt (*i.e.*, eat the food quickly) their food, swallow it whole, and expand their esophaguses to accomodate its size. Sea-birds catch fishes in their bills and toss them to the back of the throat. The esophagus becomes distended as food enters it. Contractions of muscles within its walls sequeeze the fish along and into the stomach. Snakes work their flexibly articulated jaw bones over the prey, drawing the sides of their mouth over the prey to engulf it. When food enters the esophagus, waves of contraction in its walls and general neck movements force the bolus along to the stomach (Fig. 4.33).

Table 4.1 Pharyngeal Pouch Derivatives in Vertebrates

Pharyngeal pouch	Position	Lamprey	Elasmo-branch	Urodele	Anuran	Reptile	Bird	Mammal
1	Dorsal	Thymus	Spiracle	Tubotympanic recess[a]	Tubotympanic recess[a]	Tubotympanic recess[a]	Tubotympanic recess[a]	Tubotympanic recess[a]
	Ventral	Branchial pouch	–	–				
2	Dorsal	Thymus	Thymus	–	Thymus	Thymus[1]	–	Tonsil (Palatine)
	Ventral	Branchial pouch	–	–	–	Parathyroid	–	–
3	Dorsal	Thymus	Thymus	Thymus	–	Thymus[1,2]	Thymus	Parathyroid
	Ventral	Branchial pouch	–	Parathyroid	Parathyroid	Parathyroid	Parathyroid	Thymus
4	Dorsal	Thymus	Thymus	Thymus	–	Thymus[2,3]	Thymus	Parathyroid
	Ventral	Branchial pouch	–	Parathyroid	Parathyroid	Parathyroid	Parathyroid	Thymus
5	Dorsal	Thymus	Thymus	Thymus	–	Thymus[3]	–	–
	Ventral	Branchial pouch	Ultimo-branchial body	Ultimo-branchial body	Ultimo-branchial body	Ultimo-branchial body	Ultimo-branchial body	Ultimo-branchial body

[a]Auditory cavity of middle ear and eustachian tube
[1]Lizard
[2]Turtle
[3]Snake

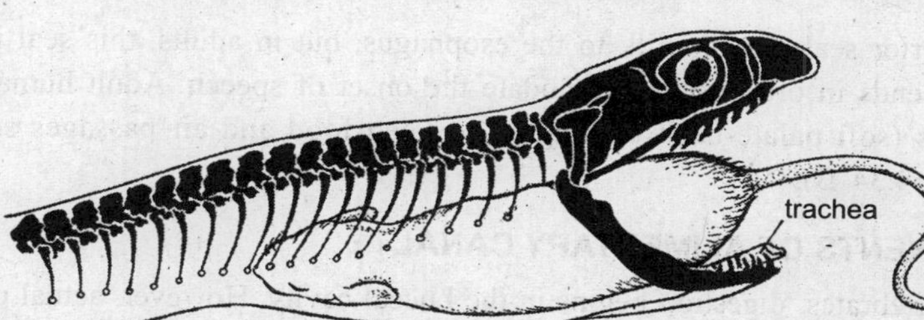

Fig. 4.33. Swallowing of prey by the snake. A snake steps its jaws alternately left and right along its prey, "walking" them along its surface until the prey reaches the back of the snake's throat. Smooth muscles in the wall of esophagus aided by striated muscles in the lateral body wall move the prey toward the stomach. While the prey is in the snake's mouth, the trachea slips beneath it and forward to maintain an open route through which the snake breathes even while swallowing (after Kardong 2006).

Three temporary seals form as most mammals chew and swallow their food. The anterior oral seal is formed by the lips. The middle oral seal develops between the soft palate and back of the tongue. The third posterior oral seal occurs between the soft palate and the **epiglottis**. As an animal chews, food tends to gather temporarily in the **vallecula**, the space infront of the epiglottis, and the **pyriform recess**, the passageways around either side of the **larynx.** The epidglottis sits above the larynx, and the trachea sits below it. When the animal swallows, the back and sides of the tongue expand against the soft palate, forcing food out of the vallecula, through the pyriform recess, and into the esophagus. The **glottis** is a muscular slit that closes momentrily across the larynx to prevent inadvertent aim of food into the trachea and lungs. In most mammals, the posterior seal (soft palate-epiglottis) is in place during swallowing to ensure passage of food into the esophagus without blocking the air passage (Fig. 4.34). In human

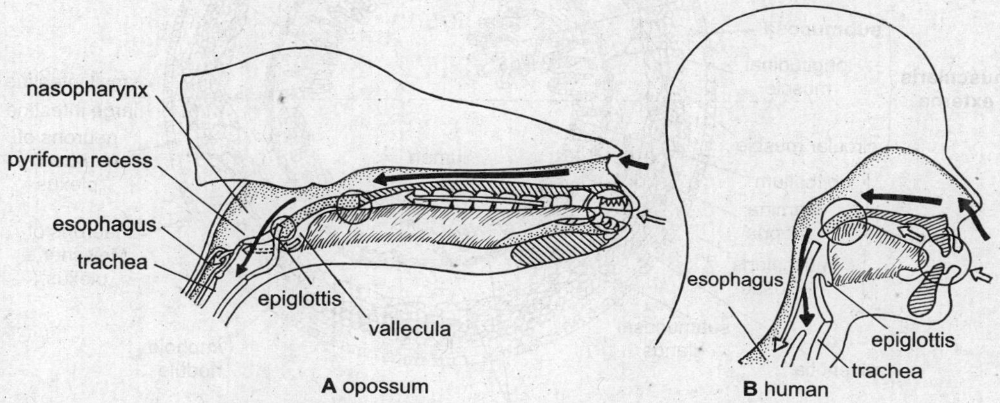

Fig. 4.34. Food and air passages. A – Sagittal view of an opossum head showing three oral seals: anterior (lips), middle (soft palate and tongue), and posterior (back of tongue and epiglottis). Air (solid arrows) flows directly from the nasal passages into the trachea. Food (open arrows) passes around the sides of the larynx to reach the esophagus. B – Sagittal view of a human head. The posterior oral seal is absent because the pharynx drops lower in the neck to accommodate sound production for speech. Thus, food and air potentially cross in the lengthened human pharynx (after Kardong, 2006).

infants, the posterior seal directs milk to the esophagus, but in adults, this seal is lost because the pharynx descends in order to accommodate the onset of speech. Adult humans rely on the middle oral seals (soft palate-back of tongue) to keep food and air passages separate during swallowing (Fig. 4.34 B).

4.6 COMPONENTS OF ALIMENTARY CANAL

In some vertebrates, digestion begins in the buccal cavity. However, actual process of food processing starts in the alimentary canal. The alimentary canal covers further *breakdown* of the bolus, *absorption* of its available constituents and *elimination* of indigestible ramnants. The design of the alimentary canal suits the diet of the organism. Because diet can differ even between related groups, alimentary canals can differ significantly between phylogenetically related vertebrates. Most vertebrates have an alimentary canal made up of an **esophagus**, **stomach**, **intestines** and **cloaca** (Fig. 4.2). These parts of the alimentary canal are distinctive, but all of them share an underlying unity of design (Fig. 4.35).

Each region or part of the alimentary canal is built on a common plan of organization, namely a hollow tube with walls composed of four layers. The innermost layer is the **mucosa**, which includes the *epithelium* that lines the lumen, the thin smooth muscle fibers of the *muscularis mucosae*, and the region of loose connective tissue, the *lamina propria* between the epithelial

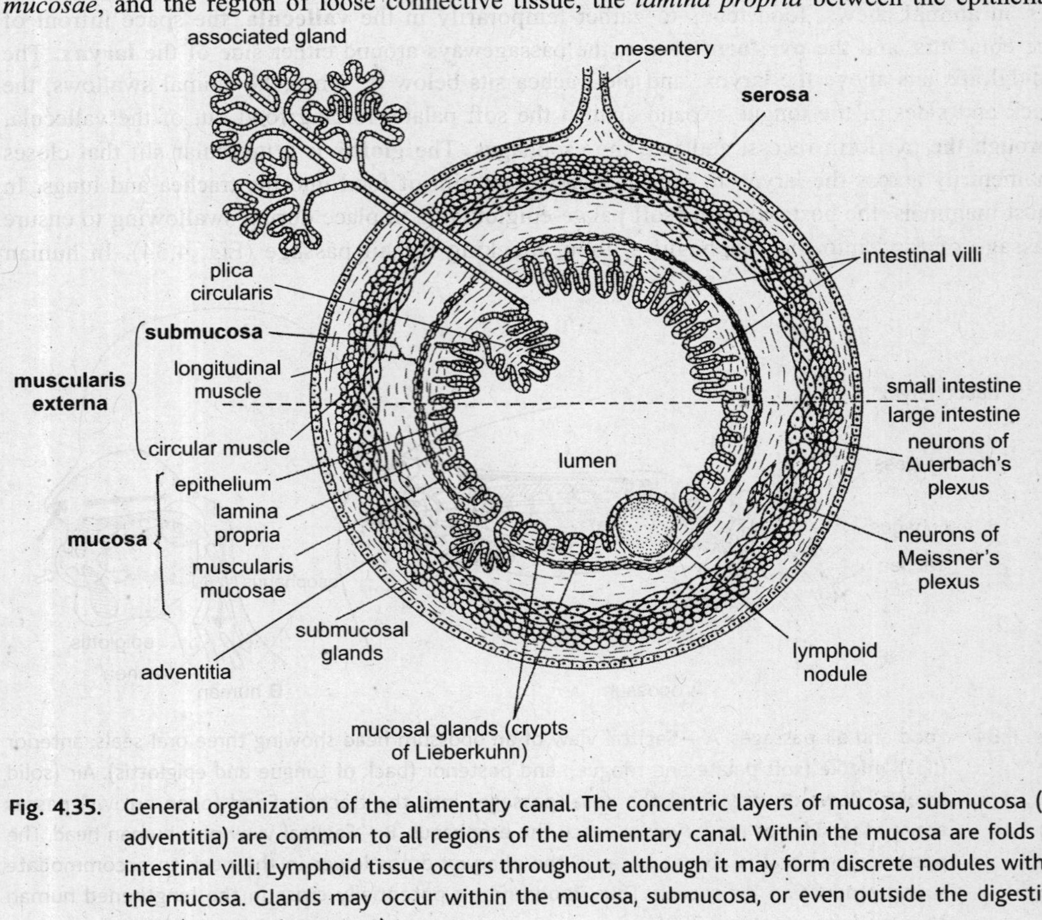

Fig. 4.35. General organization of the alimentary canal. The concentric layers of mucosa, submucosa (or adventitia) are common to all regions of the alimentary canal. Within the mucosa are folds or intestinal villi. Lymphoid tissue occurs throughout, although it may form discrete nodules within the mucosa. Glands may occur within the mucosa, submucosa, or even outside the digestive tube (after Kardong 2006).

lining and the muscularis mucosae. The **submucosa**, consisting of loose connective tissue and nerve plexuses of the autonomic nervous system, forms the second layer of the digestive tract. Outside this layer lies the **muscularis externa**, composed of circular and longitudinal sheets of smooth muscle. The surface layer is the **adventitia**, consisting of fibrous connective tissue. If a mesentary envelops the alimentary canal, this outer layer of connective tissue plus mesentery is called the **serosa**.

Embryonic differentiation of gut. During embryonic development, the endoderm gives rise to the lining of the gut, and the surrounding mesoderm forms smooth muscles, connective tissue, and blood vessels. In most tetrapods, a series of positional changes transform the relatively straight gut of the embryo into the coiled digestive tube of the fetus (Fig. 4.36). First, the *primary gut loop* forms, a large bend in what was upto this point a straight gut. Next, rapid elongation of the gut twists this loop into the *first major coil*. Thereafter, continued elongation, and coiling produce the compact digestive tube, and distinctive regions within it become delineated. The end result of **embryonic twisting**, **growth** and **looping** of the gut is to solve a packing problem – stuffing the long, active gut into a confined space, the body cavity. Remarkably, this long tube is packed into a small space in such a way as to avoid kinking (twisting sharply) the coiled gut that, like kinking of garden hose, would interrupt flow through the tube. When food arrives, the section of the gut becomes active mechanically (churning, peristalsis) and chemically (digestive secretions) as it is suspended by and controlled in its activity by the mesenteries.

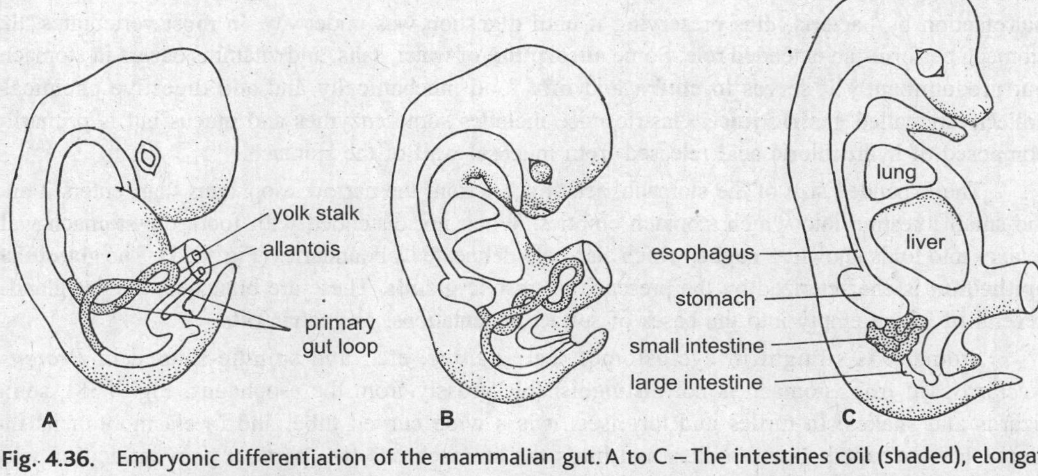

Fig. 4.36. Embryonic differentiation of the mammalian gut. A to C – The intestines coil (shaded), elongate and become tucked behind the stomach within the body cavity (after Kardong 2006).

Esophagus

The esophagus connects the pharynx with the stomach. It is short in most fishes and amphibians because they lack neck, but it is longer in amniotes due to presence of neck. In mammals, long esophagus passes through the diaphragm, the portion below the diaphragm is covered with visceral peritoneum which is lacking from the upper part.

Esophagus is a slender tube that easily becomes distended to accommodate even a large bolus of food. Its lining has longtiudinal folds or finger-like fleshy papillae (elasmobranchs) or horny papillae in marine turtles. Mucus, to aid in the passage of food, is often secreted, but the esophagus seldom produces enzymes that contribute to chemical digestion. In some vertebrates, the esophageal mucosa is lined with ciliated cells that control the flow or lubricating mucus around the food. The ciliated epithelium may also help gather small crumbs from the meal and move these along to the stomach. In others, the mucosa is stratified epithelium; this may even be keratinized in animals ingesting rough or abrasive foods. Anteriorly, muscle coats tend to be composed of striated muscle that becomes replaced posteriorly by smooth muscle.

In most birds (*e.g.*, granivorous and carnivorous birds) the posterior part of the esophagus forms a sac, called the **crop** which is particularly large in seed eaters. The crop acts primarily as a storage depot; food is eaten rapidly and then passed on gradually for digestion. In pigeons and few other birds the crop solughs of its squamous epithelium when there are nestlings. This cast off material forms a highly nutritious **pigeon milk** or **crop milk** which is regurgitated and fed to the young bird. Both males and females produce crop milk, apparently at the mere sight of the nestlings.

Stomach

The esophagus delivers the bolus of food to the stomach. It is an expanded region of the alimentary canal. Stomachs are absent in protochordates, except for some urochordates in which a stomach is present to receive **mucus** that is laden with collected food particles from the branchial basket. Vertebrates that take in large quantities of food on an irregular basis, such as many carnivores (tiger, loin, cat, snakes, etc.) have stomachs that serve as **storage compartments** until the processes of mechanical and chemical digestion catch up. Such food storage may have been an initial function of the stomach when early vertebrates evolved from suspension feeding to feeding on larger chunks of food. **Hydrochloric acid (HCl)** produced by the stomach may have functioned to retard food putrefaction by bacteria, thus preserving it until digestion was underway. In most vertebrates, the stomach performs an extended role. Some **absorption** of water, salts, and vitamins occurs in stomach, but predominantly it serves to **churn** and **mix** food mechanically and add digestive chemicals collectively called **gastric juice**. Gastric juice includes some enzymes and mucus but is primarily composed of hydrochloric acid released from mucosal wall of the stomach.

The expanded size of the stomach, sets it apart from the narrow esophagus that enters it and the small intestine into which stomach empties. When not distended with food, the stomach wall relaxes into folds known as **rugae**, which also help delineate its boundaries (Fig. 4.37). The **glandular epithelium** is characterized by the presence of **gastric glands**. These are branched, tubular glands, several of which empty into the bases of surface indentations, or **gastric pits.**

Stomach is straight in cyclostomes, gar, *Belone*, etc., and spindle-shaped in *Proteus*, *Necturus*, (in frog, stomach is not distinguishable grossly from the esophagus; Fig. 4.38), some lizards and snakes. In turtles and tortoises, it is a wide curved tube, and in elasmobranchs the stomach is J-shaped. In crocodiles and birds the stomach has two parts, a **proventriculus** with gastric glands, and a highly muscular **gizzard**, which represents the pyloric region and has a hard, cornified lining for grinding food. In mammals the stomach lies transversely. There are three divisions of the stomach–cardia, fundus and pylorus–based on the relative position and types of gastric glands (Fig. 4.37 and Fig. 4.39). The **cardia** is a very narrow region found only in mammals, and it marks the transition between esophagus and the stomach. Its gastric glands, called **cardiac glands**, are composed mainly of mucus-sectreting cells. The **fundus** is usually the largest region of the stomach and contains its most important gastric glands, the **fundic glands**. Fundic glands have two types of mucous cells: 1. **parietal cells** which secrete hydrochloric acid 2. **chief cells** which secrete various types of enzymes. Before emptying into the intestine, the stomach usually narrows into a **pylorus**, whose mucosal walls hold distinct glands called **pyloric glands**. The pyloric glands are predominantly composed of mucous cells

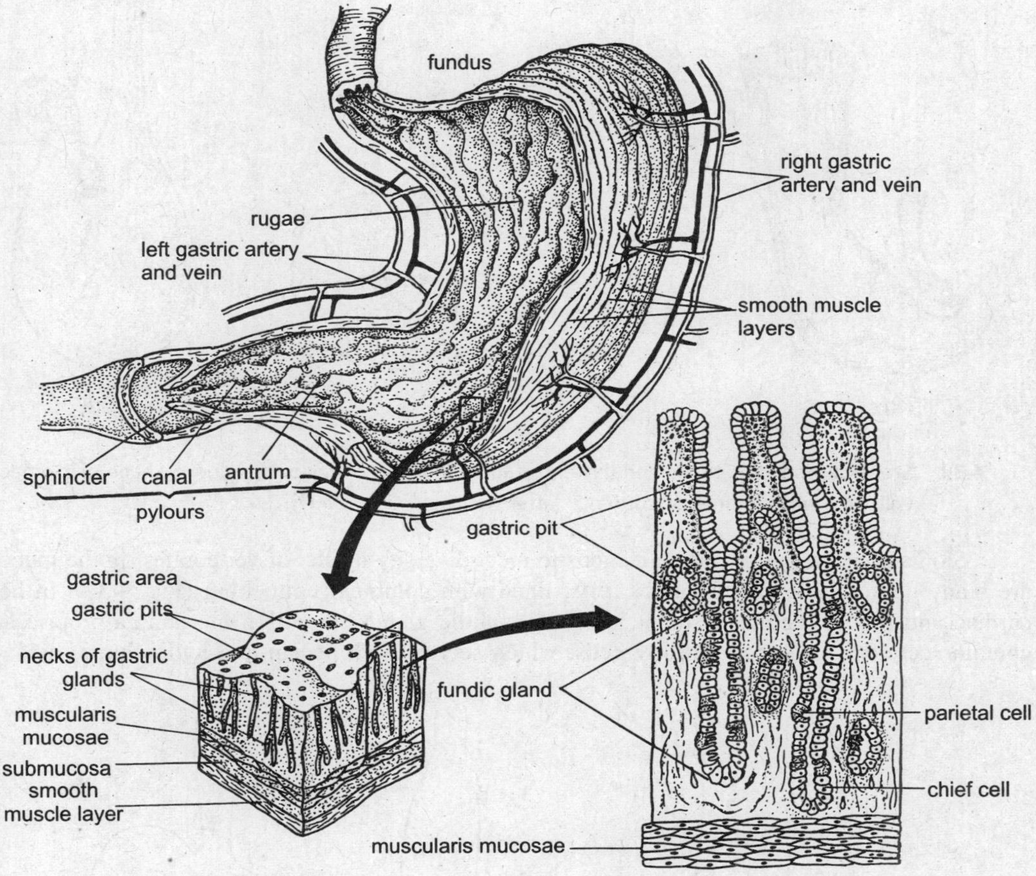

Fig. 4.37. Anatomy and histology of the stomach. Up to three regions are usually discernible in the stomach of vertebrates. The largest of these is the fundic region. Gastric pits open into fundic glands having parietal and chief cells at their bases. The two other regions of the stomach are the cardia and the plyorus, which contain cardiac and pyloric glands, respectively, at the bases of their gastric pits. Various types of mucous cells predominate in these glands (after Kardong, 2006).

whose secretions help to neutralize the acidic chyme as it moves next to intestine. Thus, most of the chemical and mechanical processes of gastric digestion occur in the fundus. The cardia and pylorus add mucus. Smooth muscle bands in their walls act as sphincters to prevent the retrograde transfer of food (Fig. 4.37).

In addition to a region of glandular epithelium, the stomach of some vertebrates also has a second region characterized by **nonglandular epithelium**, devoid of gastric glands. As in some herbivores, the nonglandular region may develop from the base of esophagus. In other species, such as rodents, loss of gastric glands in the mucosa leaves a nonglandular epithelial stomach in which smooth muscle contractions knead and mix digesta. This nonglandular epithelium in rodents also can be keratinized, perhaps as a result of mechanical abrasion from rough foods such as seeds, grasses and insect chitinous exoskeletons. Chemical insults from digestive enzymes added in the mouth may also cause a keratinized nonglandular epithelium.

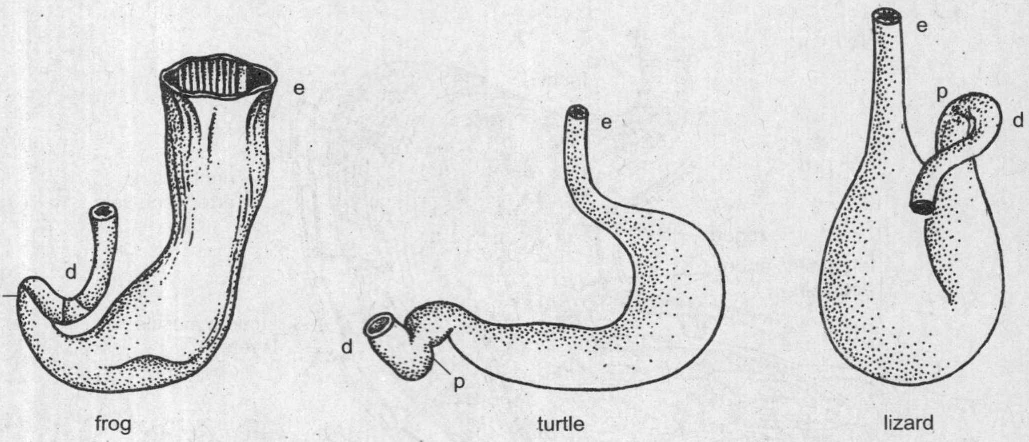

Fig. 4.38. Stomach of a frog, turtle and lizard. **d**, duodenum, **e**, esophagus, **p**, pylorus. Shape of stomachs will vary with volume of contents (after Kent and Carr, 2001).

Stomach has acquired specializations to meet diversity of diet of vertebrates. In the mucosa are many depressions, called **gastric pits**, lined with columnar epithelium (Fig. 4.37). In both cardiac and fundic region of stomach, tubular fundic glands lie within the lamina propria and contain secretory cells, called **body cells**, which secrete both pepsin and hydrochloric acid. In

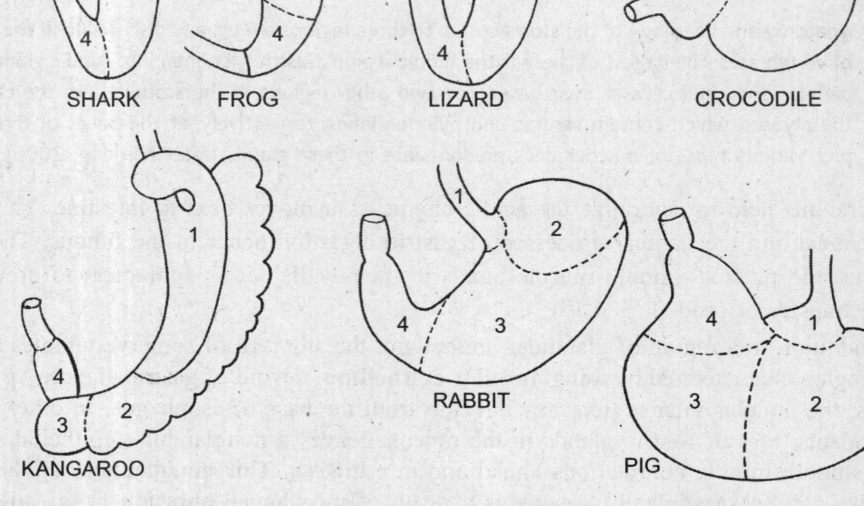

Fig. 4.39. Stomachs of some vertebrates showing different glandular regions. 1, esophagus; 2, cardiac stomach; 3, fundic region; 4, pyloric region.

the pyloric region instead of fundic glands there are mucus secreting pyloric glands also opening into the base of the gastric pits.

Box 4.9

Flexure is a sharp bend of the anterior part of the primary axis of the vertebrate embryo. In mammals, *e.g.*, mouse, a constriction in the middle part of the stomach mark off a cardiac chamber from a pyloric chamber. In human beings the stomach bears resemblance to this condition found in mice and is called "hour glass" stomach.

The stomach is straight when it first develops in the embryo and may remain so throughout life in lower vertebrates. More often, flexon (Box. 4.9) develops producing a J-shaped or U-shaped stomach (Fig. 4.40). As a result, the stomach may exhibit a concave border (lesser curvature) and a convex border (greater curvature). The stomach also undergoes torsion in higher forms so that it may finally lie across the long axis of the trunk. As flexion and torsion gets pronounced during development of mammalian stomachs; the dorsal mesentery of the stomach (mesogaster) becomes twisted and finally get suspended from the greater curvature. The part of the dorsal mesentery attached to the greater curvature is then called the **greater omentum**. The stomach of some vertebrates, especially of fish, exhibit one or more blind

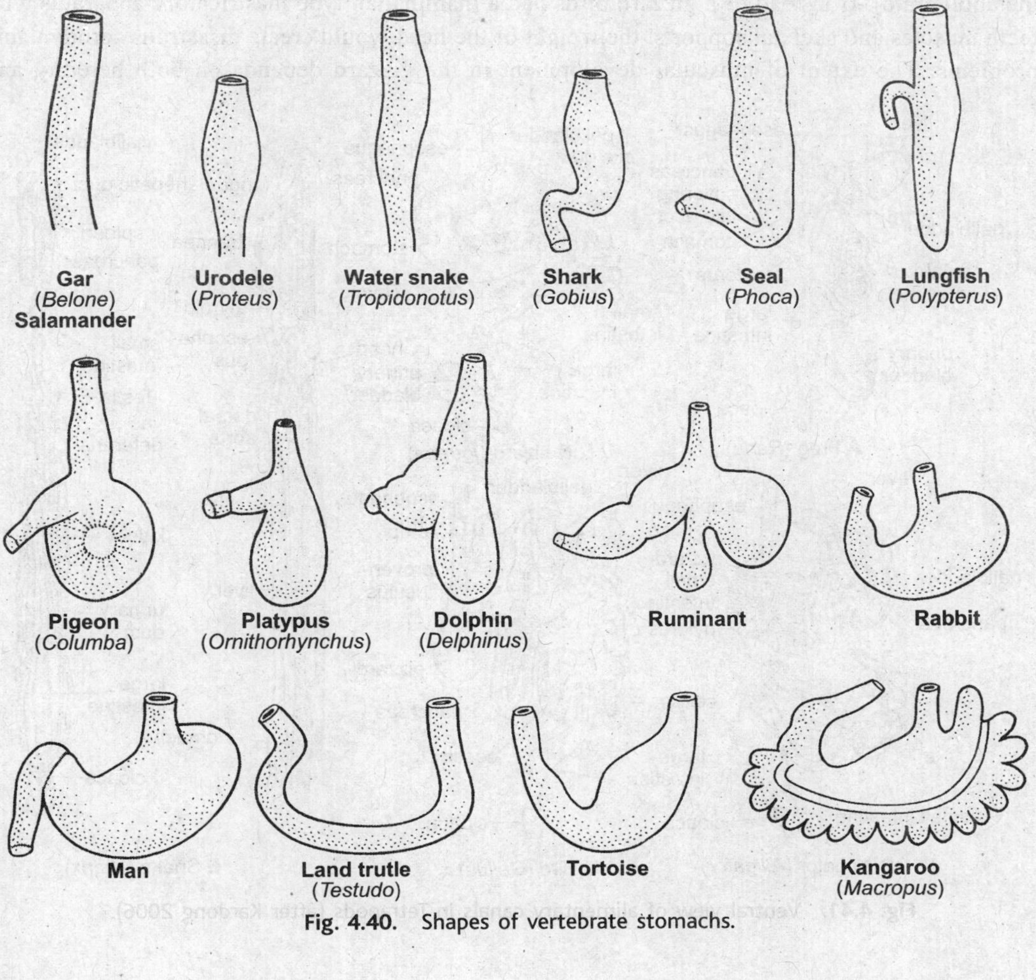

Fig. 4.40. Shapes of vertebrate stomachs.

pouches or caeca. The stomach of some marine teleosts is lined with a horny membrane that may exhibit spiny projections.

The stomach of amphibians and reptiles, when examine grossly are relatively simple dilations of the alimentary canal. On detailed microscopic examination they can be divided into regions by types of glands present. In the stomach of crocodilians the smooth musculature is thickened and extremely powerful, it function analogously to the avian gizzard, violently and efficiently macerating the food. The gizzard of crocodiles and alligators grinds food against indested hard objects, usually small stones deliberately swallowed into the stomach (Fig. 4.41).

The stomach of the birds is differentiated into two sections: (*i*) the glandular portion, the **proventriculus** and (*ii*) masticatory portion, the **gizzard**. The proventriculus secretes gastric juice to help digest the bolus, and the gizzard, together with selected pieces of hard grit and pebbles, grinds large food into smaller pieces. In birds, the food is stored in the crop located at the distal end of the esophagus and is periodically released in the stomach at the junction between gizzard and proventriculus. The avian gizzard has extremely thick powerful muscular walls and a glandular mucosa that produces a very hard, keratin like protein called **koilin**. The koilin is analogous to teeth, when the gizzard muscles contract, the food is crushed. The gizzard is another means of accomplishing mechanical digestion (essential for an animal with a high metabolic rate). If instead of a gizzard birds had a mammalian type mastricatory apparatus with teeth muscles and skeletal supports, the weight of the head would create disastrous aerodynamic problems. The extent of muscular development in the gizzard depends on both heredity and

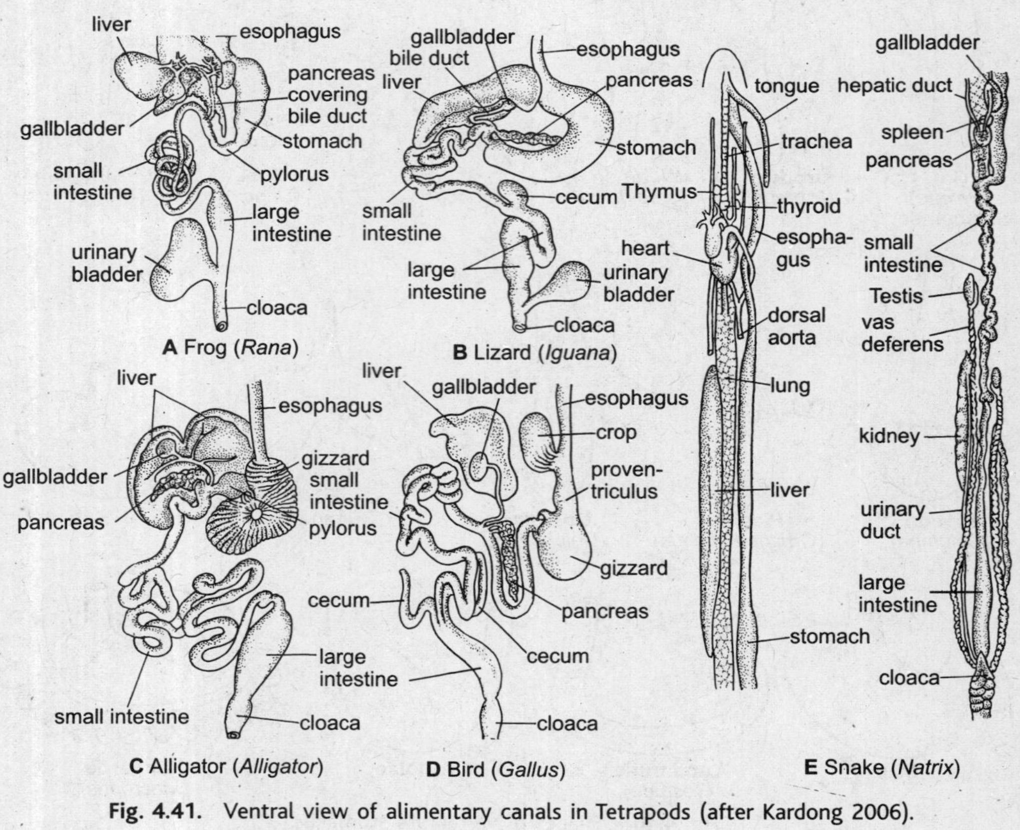

Fig. 4.41. Ventral view of alimentary canals in Tetrapods (after Kardong 2006).

diet. For example, experimentally if herring gulls are fed their usual soft fish diet their gizzard remain thin walled and relatively inconspicuous. If they are forced to eat grain, however, the gizzard wall soon proliferate and strengthen.

In ruminants (the rumen containing mammals) and in camels, the stomach is highly specialized. It has four chambers, although the first three, namely **rumen**, **reticulum** and **omasum**, arise from the esophagus and only the fourth, the **abomasum**, is an actual derivative of the stomach (Fig. 4.42A). The large rumen receives the food after it is clipped by teeth and swallowed. The reticulum is a small accessory chamber with a honeycombed texture (Fig. 4.42, sheep). Like the

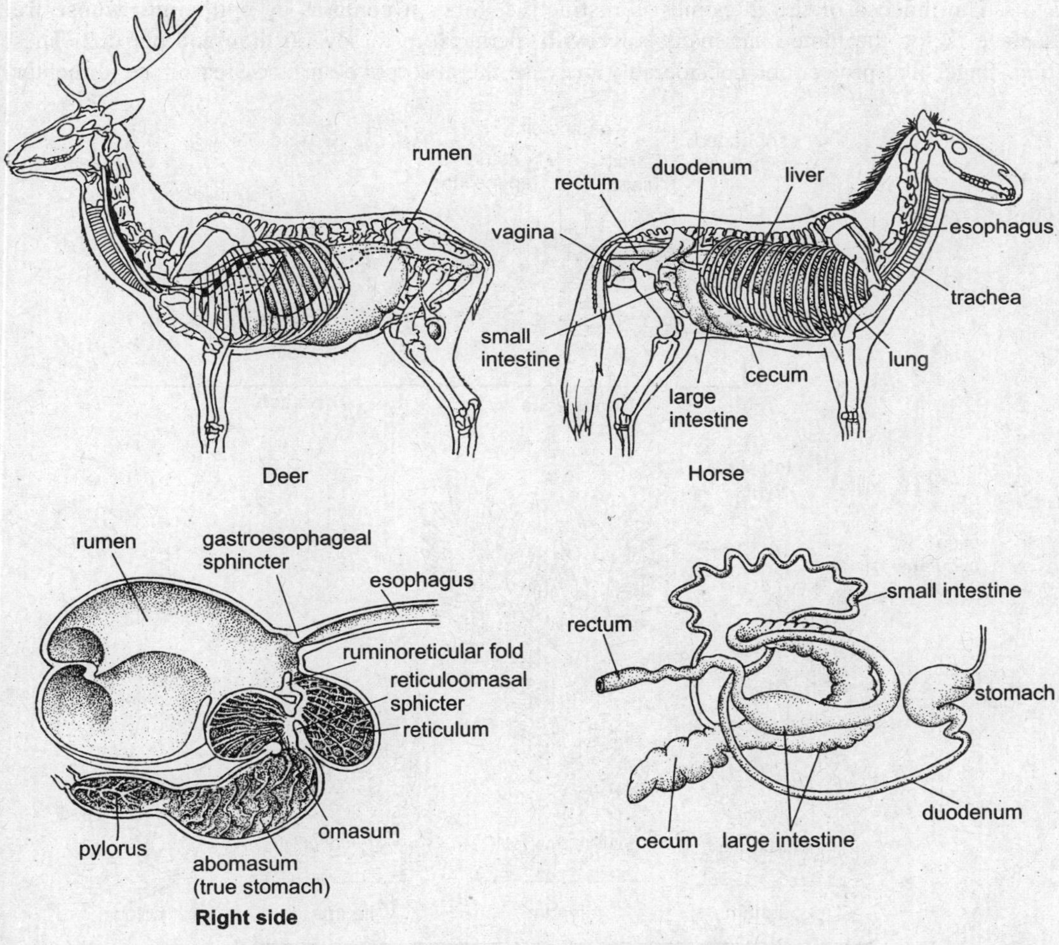

Fig. 4.42. Alimentary canal of foregut and hindgut fermenter mammals. A – Ruminants ferment food in their foregut. Note the position of the alimentary canal (top) in the deer. The sagittal section of a sheep's "stomach" is illustrated below the deer. Note the series of four chambers of stomach of sheep. Rumen, reticulum and omasum are derivatives of the esophagus. The fourth chamber, the abomasum is the actual or true stomach. B – Hindgut fermenters. Position of the alimentary canal (top) in a horse. Isolated view (bottom) of the large caecum near the juncture of small and large intestines. There is no four-chambered stomach. In hindgut fermenters (*e.g.*, horse), the caecum (cecum) and/or large intestine are the major sites of fermentation (after Kardong, 2006).

first two chambers of the stomach, the omasum is lined with esophageal epithelium, although it is folded into overlapping leaves. The three types of mucosa distinctive of the mammalian stomach (cardia, fundus, pylorus) are found only in the abomasum, most probably the "true" stomach.

In many herbivores (*e.g.,* horse), digestion of plant cellulose is enhanced by a caecum situated between small and large intestines (Fig. 4.42 B). The caecum contains additional microorganisms effective in cellulose digestion and provides an expanded region prolonging the time available for digestion.

Intestine

The mucosa of the intestines is distinctive. First, it contains an epithelium whose free surface facing the lumen has many **microvilli**, perhaps up to several thousand per cell. These tiny, finger-like projections considerably increase the absorptive surface area of the alimentary

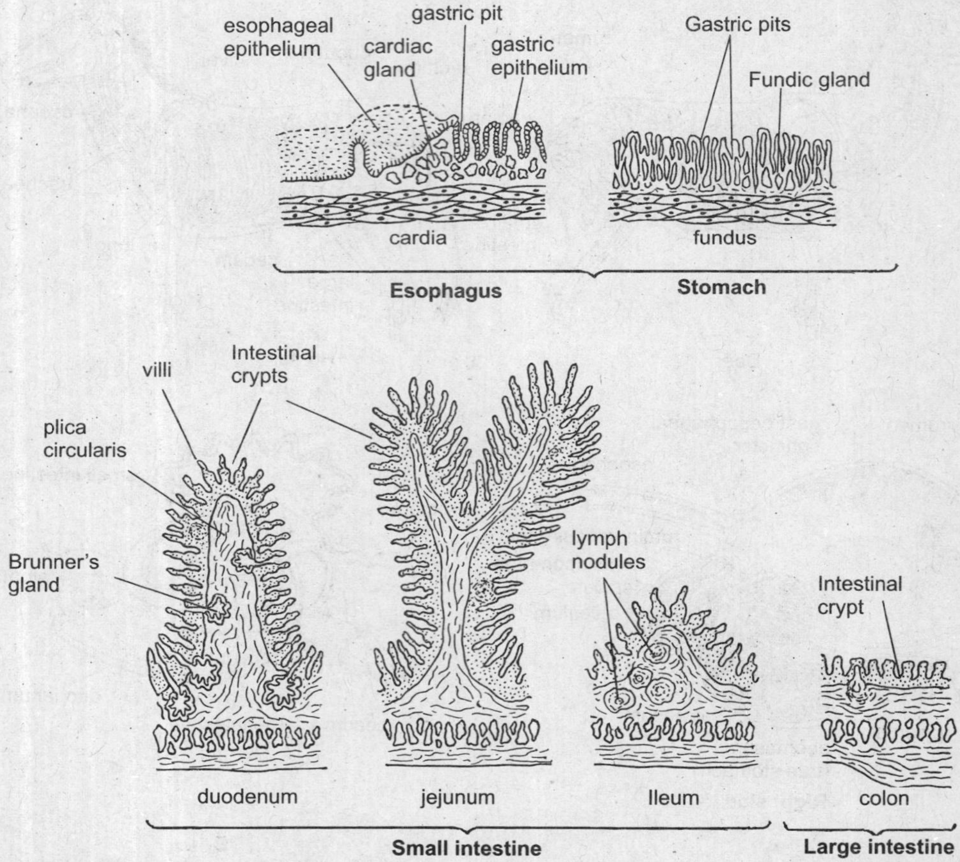

Fig. 4.43. Comparative histology of the mucosa along the alimentary canal of a mammal. Note that the large intestine, like the small intestine, contains intestinal glands, but villi are absent (after Kardong, 2006).

canal. Their surfaces also seem to harbor a microenvironment away from the large central lumen, where digestive enzymes can more favourably act upon food. Second, the intestinal mucosa also includes **intestinal glands** (crypts of Lieberkuhn), the source of many of the digestive enzymes released into the intestines (Fig. 4.43).

Generally, there are two major regions of the intestines, the **small** and **large intestines**. The small intestine may be quite lengthy, but it is usually smaller in diameter than the large intestine. It possesses **villi**, small surface projections that increase the surface area of the mucosa (not to be confused with the much smaller microvilli, tiny projections of individual cells). There can be three successive parts of the small intestine: **duodenum**, **jejunum** and **ileum**. The duodenum receives chyme from the stomach and exocrine secretions primarily from the liver and pancreas. The jejunum and ileum are best delineated in mammals on the basis of histological features of the mucosal wall (Fig. 4.43). The ileocolic valve (ileocecal valve) is a sphincter between the ileum of the small intestine and the large intestine. This valve regulates the movement of food into the large intestine.

The large intestine, so named for its large diameter, is usually a straight tube passing to the colaca or anus. Its mucosa lacks villi. It may be pushed to one side of the body cavity or, as in many mammals, form a large, gentle loop called the **colon**. The large intestine often straightens, forming a section called the **rectum**. The rectum narrows into the anal canal in which a transition occurs from simple columnar epithelium to stratified epithelium within its mucosal wall. A smooth muscle sphincter within the muscularies of the anal canal controls the release of waste products from the digestive tract.

The intestines serve several functions. 1. Peristalsis within the intestinal walls moves food along the digestive tract. 2. The intestines add secretions to food being digested. Mucous secretions protect the epithelial lining from digestive enzymes and lubricate it to facilitate passage of food. The **intestinal juice** produced by the intestinal glands include enzymes for digestion of proteins, carbohydrates and lipids. Accessory glands add some secretions as well. For example, the **duodenal glands** (Brunner's glands), located in the submucosa, empty their secretions into the duodenum to help neutralize the acidity of the chyme entering from the stomach. The pancreas releases its proteolytic enzymes into the duodenum also. 3. The intestines selectively absorb the final products of digestion – amino acids, carbohydrates, and fatty acids, water is also absorbed, especially in the large intestine.

Cloaca

The proctodeum at the end of the embryonic gut gives rise to the **cloaca**, a common chamber that receives products from the intestines and urogenital tracts. In some fishes and most mammals, the cloaca is absent. Instead, the intestine opens to the outside through the anus, which is a separate opening from that of the urogenital system.

Specializations of the Alimentary Canal

Structural modifications accommodating specialized diets occur in several ways within the digestive tract. *First*, the path that food travels can be lengthened according to the time required for digestion. A **spiral valve** within the lumen of the alimentary canal is one way to increase the length of the route through the digestive tract (*e.g.*, sharks; Fig. 4.44B). This valve creates a helical partition that forces passing food to wind through a spiral channel, thereby increasing the amount of time the bolus spends in the intestines and prolonging digestion. Another way to increase the path traveled by food is by increasing the length of the alimentary canal. Carnivores have relatively short intestines, but herbivores that must extract nutrients from resistant plant cells usually have long intestines.

In larval lampreys, a prominent longitudinal fold, the **typhlosole** projects from one wall of intestine into the lumen. The typhlosole is a site of haemopoiesis, blood formation, in the larva, but it also increases the surface area available for absorption of digestive end products. In the adult lampreys, the typhlosole is generally lost and is replaced by numerous **longitudinal folds** (Fig. 4.44A) in the mucosal wall, which dramatically increase the intestinal absorptive area.

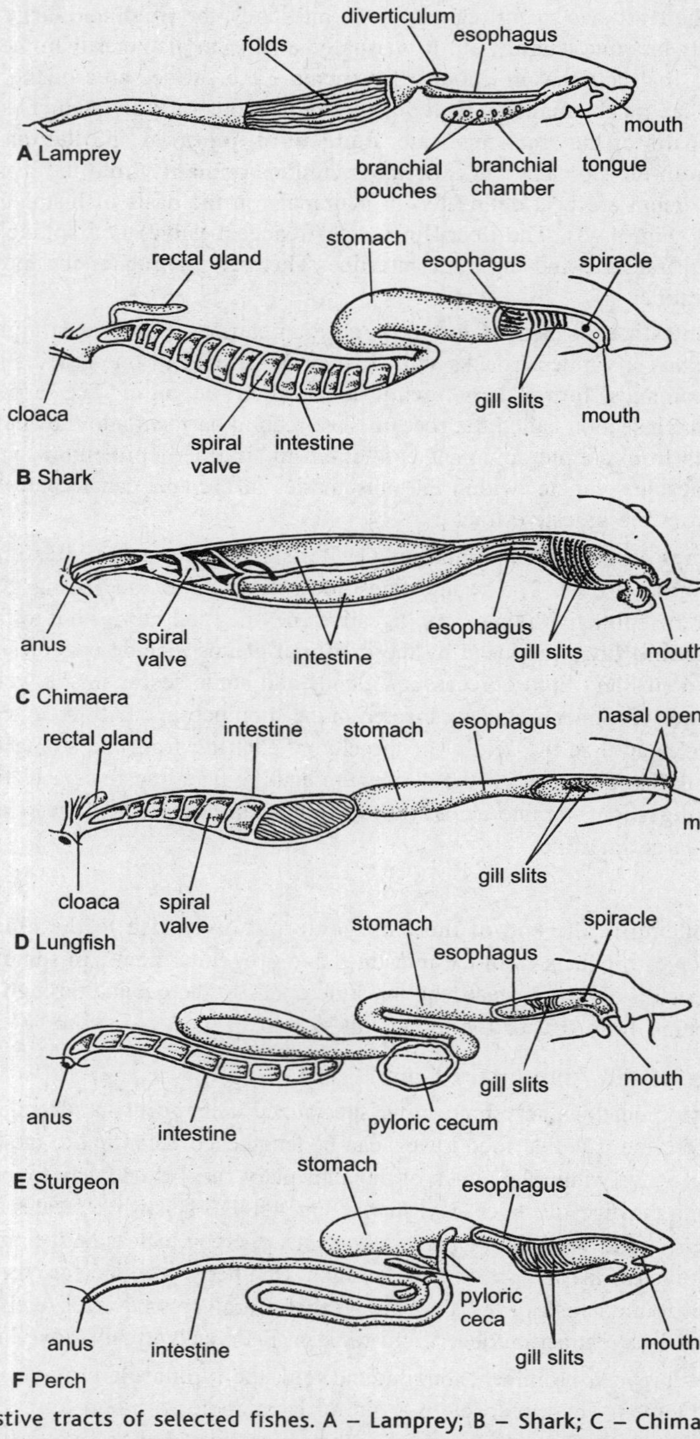

Fig. 4.44. Digestive tracts of selected fishes. A – Lamprey; B – Shark; C – Chimaera; D – Lungfish; E – Sturgeon; F – Perch. When a spiral valve is absent, the intestine is often lengthened, as in the perch (after Kardong, 2006).

Second, expansions or extensions of the alimentary canal also may develop to accommodate specialized diets. A **crop** is a baglike expansion of the esophagus often used to store food temporarily during processing. One of the most common extensions is a **caecum**, a blind-ended outpocketing from the intestines through which food circulates as part of the digestive process.

4.7 ASSOCIATED GLANDS OF DIGESTIVE SYSTEM

The epithelium lining the buccal or oral cavity contains a rich source of cells that secrete mucus and serous (*i.e.*, serum-like thin watery) fluid. When these secretory cells are gathered together and emptied by a common duct, they constitute an **oral gland**. Such discrete (*i.e.*, individually separate and distinct) glands are rare in fishes.

In tetrapods, oral glands are more common perheps reflecting the absence of the watery medium to moisten food. The most common are the **salivary glands**, a term loosely applied to most major oral glands in tetrapods. These glands are not all homologous, however. In salamanders, **mucous glands** are found on the tongue, and a large **intermaxillary gland** is located within the palate. Reptiles also possess oral glands. Usually, strips of glandular tissue, called **supralabial** and **infralabial glands**, are present along the upper and lower lips. In addition, glands may occur within the tongue (**lingual gland**), or below it (**sublingual glands**),

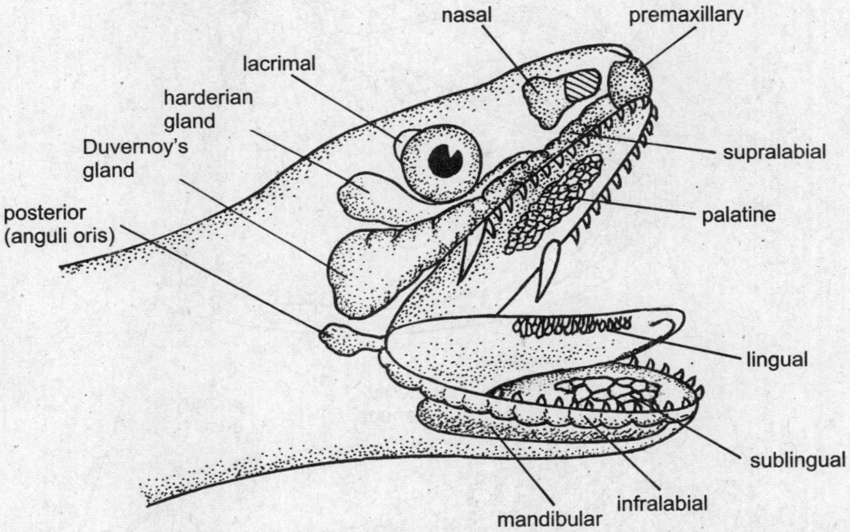

Fig. 4.45. Oral glands of reptiles. Not all oral glands are present in all reptilian species. The venom gland of advanced snakes is a phylogenetic derivative of Duvernoy's gland and is located within the temporal region behind the eye, a position similar to that of Duvernoy's gland (after Kardong, 2006).

in association with the snout (**premaxillary** and **nasal glands**), and along the roof of the mouth (**palatine glands**). These glands release mucus to lubricate the prey during intraoral and esophageal transport. The **lacrimal** and **Harderian glands** release secretions that bathe the eye and vomeronasal organ. **Duvernoy's gland**, situated along the posterior upper lip, is found in many nonvenomous snakes, and releases its serous secretions via a duct adjacent to the posterior maxillary teeth (Fig. 4.45).

Secretions from one or most snake's oral glands lubricate the food and may also help maintain healthy oral membranes, neutralize the toxins carried by the prey, and perhaps initiate the chemical stages of digestion. In poisonous snakes, the **venom gland**, a homologue of Duvernoy's gland, secretes a cocktail of different chemicals with various functions — some toxic, some digestive (Fig. 4.46). Thus, the injected secretion of the poison gland not only functions to convey prey quickly, but it also contains a suite of digestive enzymes introduced along with the toxins during the strike. These digestive enzymes are injected deep into the prey to initiate the breakdown of the prey's tissues from within. Gastric and intestinal juices released later from the walls of the alimentary canal act on the surface of the prey. The overall result is to provide a very speedy digestion from within (venom) and from without (alimentary tract digestive juices). This efficient system apparently allows the snake to swallow large prey and digest it before putrefaction sets in. Rapid digestion may also allow venomous snakes to take advantage of seasonally abundant prey or to clear the digestive tract of bulky prey that might otherwise interfere with lateral bands of the body during locomotion.

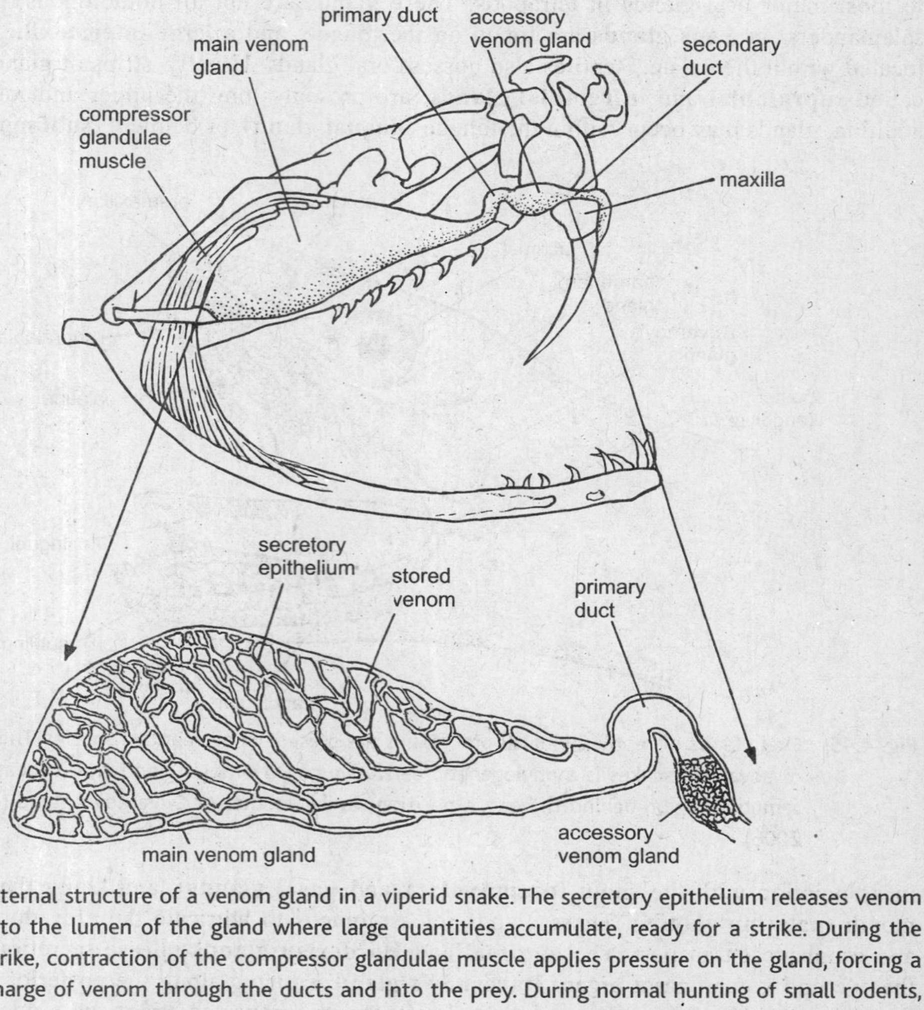

Fig. 4.46. Internal structure of a venom gland in a viperid snake. The secretory epithelium releases venom into the lumen of the gland where large quantities accumulate, ready for a strike. During the strike, contraction of the compressor glandulae muscle applies pressure on the gland, forcing a charge of venom through the ducts and into the prey. During normal hunting of small rodents, the snake usually does not expend all of its venom reserves within the lumen in a single strike. If a snake is artificially forced to expend all its reserves of venom, full replacement of the venom takes about two days (after Kardong 2006).

Most birds, particularly those feeding in water, lack oral glands, but there are exceptions (Fig. 4.47). Some passerine birds (*e.g.*, songbirds such as bulbuls, sparrows, weavers, etc.) use mucus from oral secretions to help bind together materials composing their nests.

The most common oral glands in mammals are the salivary glands. There are usually three primary pairs of salivary glands, named for their approximate positions: **mandibular** (submandibular or submaxillary), **sublingual** and **parotid**. They form the **saliva**, which is added to food in the mouth. These three pairs of glands lie at the angle of the jaws, usually at about the juncture between the head and the neck, but they are positioned superficial to neck musculature. Ducts from the mandibular and sublingual glands run anteriorly and release secretions into the floor of the buccal cavity. The duct from the parotid gland opens into the roof of the buccal cavity. In some species, additional salivary glands may be present. In dogs and cats, but in no other domestic animals, a **zygomatic** (orbital) **gland** is present, usually beneath the zygomatic arch (Fig. 4.48). Like most digestive secretions, saliva contains mucus, salts, proteins, and a few enzymes, most notably **amylase**, which initiates starch digestion. Saliva also aids swallowing by lubricating food.

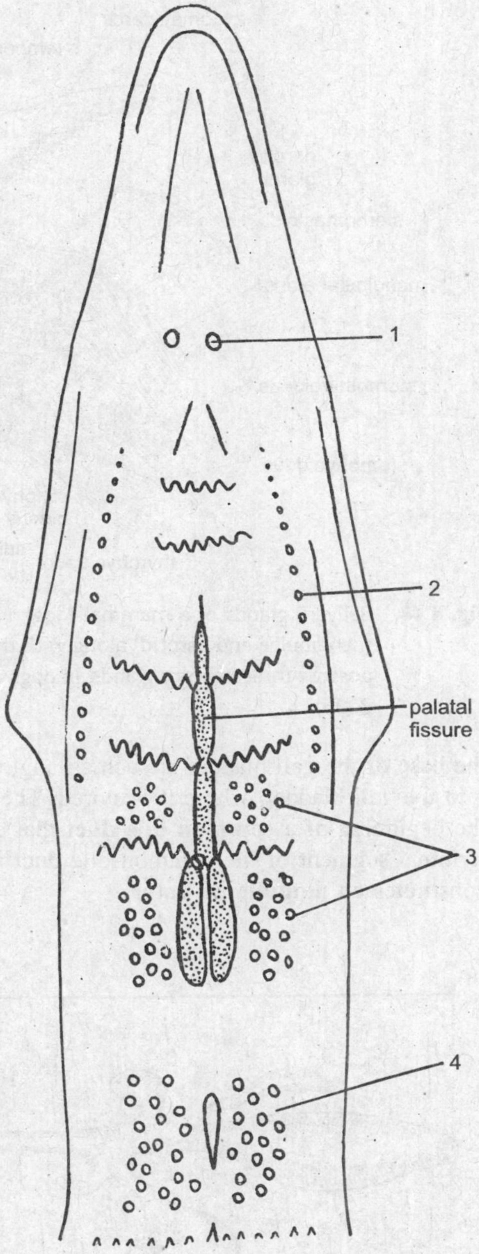

Fig. 4.47. Palate of domestic hen. **1** to **4**, gland openings as follows: **1**, paried maxillaries, **2**. lateral palatines; **3**, medial palatines, **4**, sphenopterygoids. Internal nares (choanae) are above palatal folds at anterior end of palatal fissure (after Kent and Carr, 2002).

Liver and Gall Bladder

The liver is the second largest organ in humans, exceeded in size only by the skin. The liver arises from the midventral aspect of the midgut as a hollow caecumlike diverticulum, the **liver bud** (Fig. 4.49). The bud grows cephalad in the ventral mesentery of the stomach. The growing tip of the bud gives rise to numerous sprouts that become the liver and the gallbladder. The cephalic pole of the liver finally becomes anchored to the embryonic transverse septum by a **coronary ligament**.

The shape of the adult liver conforms to the space available in the coelom. In forms with an elongated trunk the liver is elongated. In animals with short trunks (*e.g.*, humans), it is broad and flattened. It is variously subdivided in different species but usually exhibits median and lateral lobes.

The lobes of adult liver are drained by several **hepatic ducts**, and the **gall bladder** is drained by a **cystic duct**. These ducts converge very close to

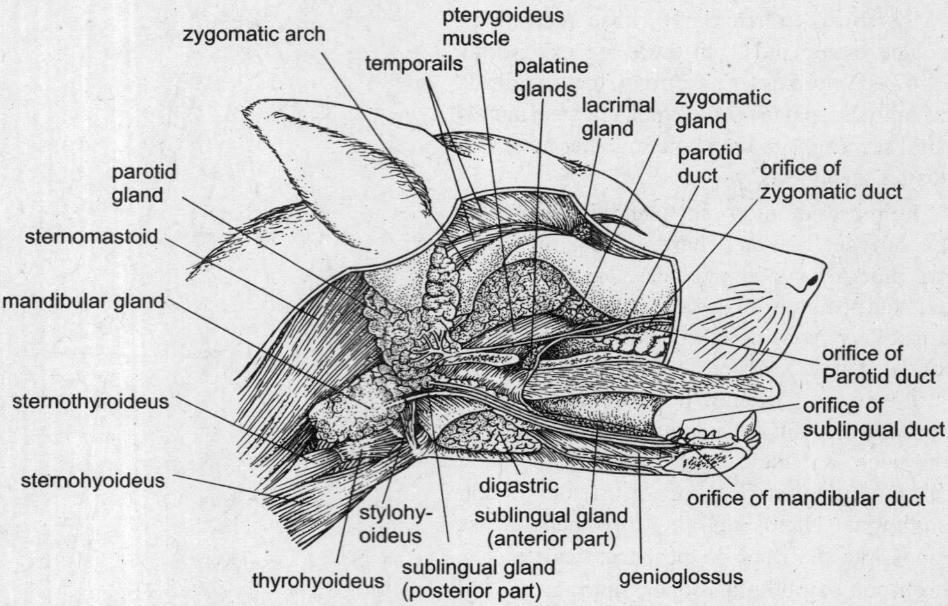

Fig. 4.48. Salivary glands of a mammal (dog). Note the locations of the main salivary glands (sublingual, mandibular, and parotid) along with their ducts leading to the oral or buccal cavity. All mammals possess three salivary glands. In dogs and cats, a zygomatic gland is also present (after Kardong, 2006).

the base of the gall bladder at such an angle that bile from the liver can pass via the cystic duct into the gall bladder, where it is stored. The site where the hepatic and cystic ducts converge is the beginning of a **common bile duct** that empties into the duodenum (8 in Fig. 4.49). A short terminal segment of the common bile duct is embedded in the wall of the duodenum, where it constitutes an **ampulla of Vater**.

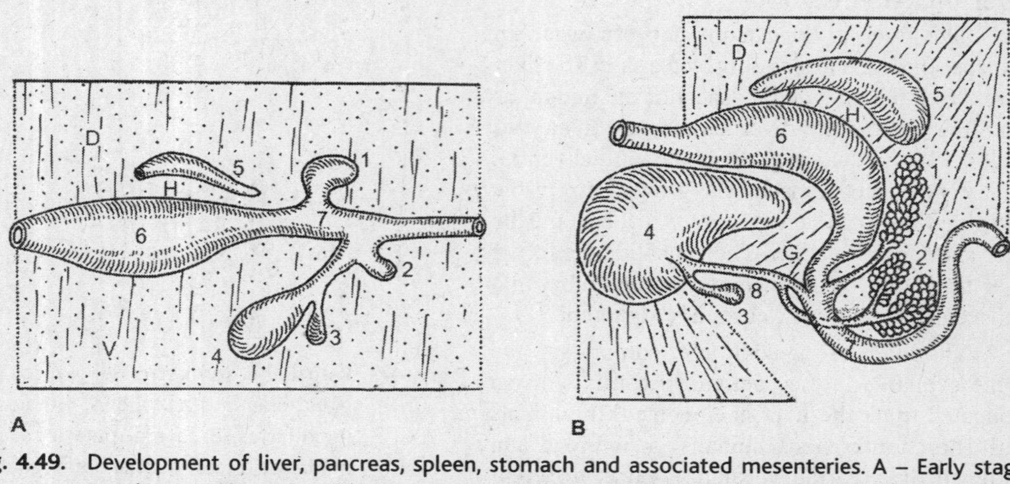

Fig. 4.49. Development of liver, pancreas, spleen, stomach and associated mesenteries. A – Early stage; B–A later stage. **1**, dorsal pancreatic bud from duodenum; **2**, ventral pancreatic duct; **3**, gall-bladder; **4**, liver bud in (A) and liver in (B); **5**, spleen; **6**, stomach; **7**, duodenum; **8**, common bile duct; **D**, dorsal mysentry; **G**, lesser omentum; **H**, gastrosplenic ligament; **V**, ventral mesentry in (A), falciform ligament in (B) (after Kent and Carr 2001).

Although most of the embryonic ventral mesentery disappears during subsequent development, the mesentery ventral to the duodenum and stomach that was invaded by the liver but remains as the **hepatoduodenal ligament** connecting the duodenum and liver, and as the **gastrohepatic ligament** that connects the pyloric stomach and the liver. These two ligamentous mesenteries constitute the **lesser omentum**. It serves as a bridge that conducts the common bile duct to the duodenum and the hepatic artery and hepatic portal vein to the liver (Box 4.10). The embryonic mesentery ventral to the liver remains in adults as the **falciform ligament**.

Box 4.10

The liver is one of the most heavily vascularized organs of the vertebrate body, being supplied with arterial blood via the hepatic artery. However, unlike most organs, it is also supplied with venous blood via the hepatic portal vein that runs directly from the intestines and spleen to the liver, delivering absorbed products of digestion.

The liver has many roles. It produces **bile**, an alkaline fluid containing bile salts that emulsify lipids (*i.e.*, break them up into small droplets) in the small intestine and impart to them an alkalinity necessary for digestion. Without emulsification only a small proportion of dietary fat would be digested. Lipids are stored in liver cells and in adipose tissue throughout the body.

Some liver cells phagocytose aging red blood cells, splitting the haemoglobin molecule and freeing the iron. The remainder of the molecule is converted into red and green pigments (**bilirubin** and **biliverdin**) that are excreted as part of the bile (Box 4.11) (*Note*. Red blood cells are phagocytosed also in the spleen and red bone marrow. The pigments from those sites are extracted from the bloodstream in the liver and contribute to the bile). The fetal liver is a source of red blood cells, but hat role is taken by other tissues.

Box 4.11

The black and blue disfiguration of bruised human skin is caused by the pigments released from traumatized red blood cells being phagocytosed at the site of trauma (physical injury).

The liver removes from the circulation glucose in excess of immediate tissue needs. The glucose is converted to glucose-6-phosphate and then stored in the liver as glycogen. The latter is converted back to glucose as needed to maintain appropriate level in circulation.

Throughout life, liver detoxifies and removes toxic substances from the blood. It removes dietary amino acids from the hepatic portal system and deaminates them. The byproducts of deamination include ammonia, uric acid and urea. These are eventually excreted in the urine in one form or another, depending on the species.

Several blood proteins are manufactured in the liver, including **fibrinogen** and **prothrombin,** which are essential for blood clotting.

A gall bladder develops in most craniates. None develops in cyclostomes (lampreys), a few teleosts, most birds, perissodactyles or odd-toed ungulate mammals (*e.g.*, horses, zebras, asses, etc.), whales, and some rodents, including rats. Craniates lacking a gall bladder are among those that have little fat in their diet. Human beings can live without a gall bladder, but they have to remain careful about the amount of lipids they consume.

Exocrine Pancreas

The pancreas consists of two histologically distinct and functionally independent components. An **exocrine** portion produces alkaline digestive enzymes in alveoli (acini). The enzymes (*e.g.*, trypsin, amylase, and lipases) are transported via **pancreatic ducts** to the duodenum. An **endocrine** portion bears **pancreatic islets (islands of Langerhans)**, lacks ducts and secretes its hormonal products, insulin and glucagon, into the bloodstream. These two components are not always part of the same organ.

The pancreas varies from diffuse to compact. When diffuse, as in teleosts and many other craniates, pancreatic tissue is distributed along the blood vessels in the ventral mesentery of the stomach and duodenum. When compact, it may consist of several discrete lobes. In living agnathans, there is no definitive pancreas, the exocrine and endocrine components are spatially separated, and many exocrine cells remain in the intestinal epithelium.

The pancreas arises typically as one or two ventral **pancreatic ducts** off the liver bud and as a single dorsal pancreatic bud from the foregut immediately caudal to the stomach (Fig. 4.49 A). The ventral buds invade the ventral masentery of the duodenum and stomach and generally coalesce to form a ventral lobe (called **body**) of the pancreas (Fig. 4.49 B). The dorsal bud becomes the dorsal lobe (called **tail**) of the pancreas. This pattern varies in different animals. In sharks, the entire pancreas develops from the dorsal bud, and in most mammals it develops from one ventral and one dorsal bud.

Craniates may have as many pancreatic ducts as there were embryonic pancreatic buds, but more often one or more of the ducts loses its connection with the bile duct or gut from which it invaginated, and the pancreas is drained by the remaining duct or ducts. For example, in sheep the dorsal lobe loses its connection with the gut, and the entire pancreas drains into the common bile duct. In pigs and oxen the duct of the ventral lobe loses its connection with the common bile duct, and the entire pancreas drains directly, into the duodenum. In still other mammals, both ducts remain when one duct is larger, as in cats and humans, the other is demoted to the status to **accessory pancreatic duct**.

4.8 CELLULOSE DIGESTION, FERMENTATION AND COPROPHAGY

Cellulose is a carbohydrate which is structural component of all plants. It is insoluble and extremely resistant to chemical attack. Many herbivores depend on it as a major energy source, yet surprisingly, no vertebrate is able to manufacture **cellulases**, the enzymes that can digest cellulose. Symbiotic microorganisms, bacteria and protozoans, that live in the digestive tract of the host vertebrate produce cellulases to breakdown cellulose from ingested plants. The process of breaking down cellulose is known as **fermentation**, which yields organic acids that are absorbed and utilized in oxidative metabolism. Carbon dioxide and methane (CH_4) gases are unsuable by-products released by belching.

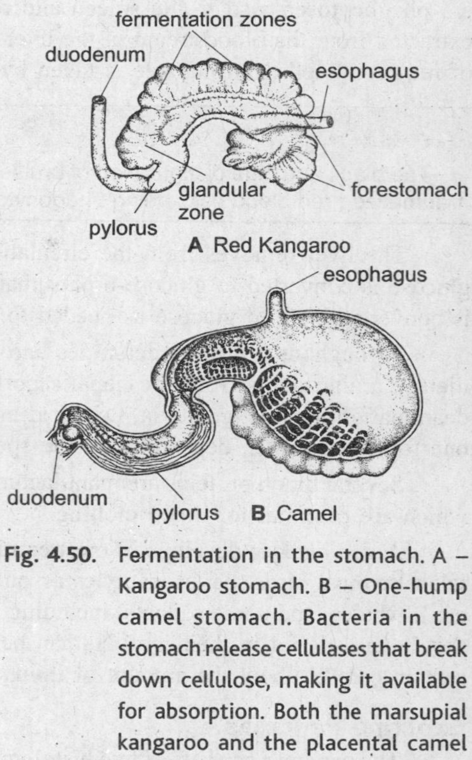

Fig. 4.50. Fermentation in the stomach. A – Kangaroo stomach. B – One-hump camel stomach. Bacteria in the stomach release cellulases that break down cellulose, making it available for absorption. Both the marsupial kangaroo and the placental camel independently evolved stomachs that harbor bacteria to ferment cellulose of the fibrous plants that constitute a large part of their diet (after Kardong, 2006).

In many vertebrate herbivores, parts of the digestive tract are specialized as fermentation chambers in which symbiotic microorganisms eventually digest cellulose. Because microbial fermentation is relatively slow and cellulose relatively resistant, these chambers are quite extensive and lengthy. Microbial fermentation may occur in specialized stomachs or pouches that open off the intestine, and is known, respectively, as gastric and intestinal fermentation.

Gastric Fermentation

When the digestion of cellulose is centered in or near a specialized stomach, it is called **gastric fermentation**. More generally, animals in which microbial fermentation arisen within the esophagus as well as stomach are categorized as **foregut fermenters** (Fig. 4.50). In birds, foregut fermentation is known only in the Hoatzin (A South American bird of genus *Opisthocomus* having a pair or hooked claws on each wing) where it occurs in an enlarged crop and wide lower esophagus. Among mammals, many employ such fermentation but ruminants are specialists and take their taxonomic name from the process. When a ruminant feeds, food initially collects in the sac-like **rumen**, the first of four chambers. The rumen is thin walled and lined with numerous projecting papillae that increase its absorptive surface area. It serves as a large holding and fermentation vat. Later, food in the rumen is *regurgitated* back into the mouth, *remasticated* and *swallowed* again. This process is repeated until there has been thorough mechanical breakdown of plant material (mastication) and chemical attack on cellulose (fermentation).

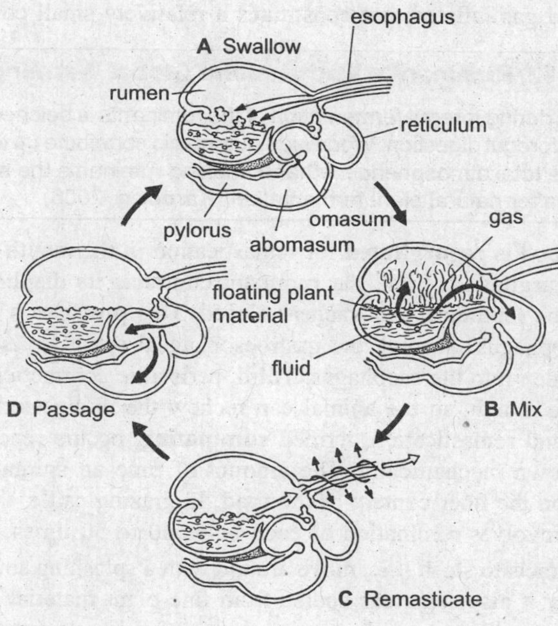

Fig. 4.51. Foregut fermentation in the bovine stomach (based on domestic cow). A – In ruminants, food is clipped, rolled into a bolus, mixed with saliva, and swallowed. B – Contractions spread through rumen and reticulum in cycles that circulate and mix the digesta. Contents separate into fluid and particulate material. Floating, fibrous plant material and a pocket of gas form during fermentation. C – poorly masticated boli of plant material are regurgitated and rechewed later to break down fibrous cell walls mechanically and exposes further plant tissue to cellulase enzymes. Respiratory inhalation, without opening the trachae, produces negative pressure around the esophagus to draw some of this poorly masticated material into the esophagus through the gastroesophageal sphincter. Peristaltic waves moving forward in the wall of the esophagus carry the bolus into the mouth for rechewing. D – THe omasum transports reduced digesta from the reticulum to the abomasum in two phases. First, the relaxation of omasal walls produces negative pressure that draws fine particulate material from the reticulum into its own lumen. Next, contraction of omasum forces these particulates into the abomasum, the stomach region rich in gastric glands. Thus, the abomasum is the first true part of the stomach (after Kardong, 2006).

Process of rumination. Rumination involves complicated waves of contraction that sweep through the rumen and that are synchronized with remastication and with passage of food along the digestive tract. Initially, ruminant animals clip plant material, mix it with saliva, roll it into boli, and swallow it into the rumen (Fig. 4.51A). Cycles of contraction pass through the rumen and reticulum and mix the ingested food with microorganisms. In domestic cows that eat hay (Hay is the grass that has been mown and dried for use as fodder), this mixing also results in physical separation of coarse and fine food particles. Small particles sink into the fluid that accumulates ventrally within the rumen. Large, undigested plant fibers float on top of this fluid. **Methane** gas that forms during fermentation collects above this fluid and plant fiber (Fig. 4.51 B; Box 4.12). In wild ruminants, the reticulum collects much finer particles than does the rumen. However, when eating natural forage, which is usually of poor quality, no such physical separation of coarse and fine food particles occurs within the rumen as it does in cows. Instead, in wild ruminants, muscle contractions in the rumen are very strong, less fluid is present, food particles are generally small, and gas produced by fermentation is quickly eliminated. Consequently, the rumen of wild ruminants is not large and gas inflated, but constitutes a relatively small compartment.

Box 4.12 Ruminant's Methane and Global Warming!

Methane gas produced during foregut fermentation of the ruminants is belched up, and constitutes a substantial byproduct of foregut digestion. Worldwide, ruminants contribute up to 60 tons of methane per year, about 15% of the total atmospheric methane, making ruminants the second major source of atmospheric methane, after natural plant fermentation (**Kardong**, 2006).

Poorly masticated food is **regurgitated** for remastication in the mouth (Fig. 4.51 C). Three steps are involved in regurgitation. First, the ruminant contracts its diaphragm as if taking in breath but keeps the glottis (entrance into trachea) closed. This produces a negative pressure in the thorax around the esophagus. Second, the gastroesophageal sphincter is relaxed and digesta are aspirated from the rumen into the esophagus. Third, peristaltic contractions sweep the digesta up the esophagus into the mouth, so the animal can rechew the undigested plant material. The process of regurgitation and remastication, termed **ruminating**, occurs repeatedly until most of the material is broken down mechanically. The amount of time an animal spends ruminating depends proportionately on the fiber content of the food. In grazing cattle, this may occupy upto eight hours per day and involves rumination of each bolus 40 to 50 times.

The **reticulum** contracts to slosh (*i.e.*, move around with a splashing sound) digesta between itself and rumen. Possibly it also separates coarse from fine plant material available for further transit. The **omasum** operates like a two-phase pump to transfer digesta from the reticulum to the abomasum (Fig. 4.51D). First, relaxation of the muscular walls of the omasum aspirates fluid and fine particles from the reticulum into the lumen of the omasum. Second, the omasum contracts to force this digesta into the abomasum. The abomasum is the fundic part of the stomach in which further digestion occurs before digesta pass to the intestines.

In animals that feed upon fibrous plants, the combination of remastication and the fermentation process is very efficient. In cattle, *organic acids* produced in the rumen alone make up 70 percent of their total energy requirements. Eventually, rechewed food travels through the reticlulum into the omasum. The omasum absorbs volatile **fatty acids** and **ammonia** and water and at the same time separates the fermenting contents of the rumen and reticulum from the highly acidic contents of the abomasum. The omasum moves smaller food particles into the abomasum, the true stomach in which enzymatic and acidic hydrolysis takes place. Finally, digesta enter the intestine (Fig. 4.51D).

In the suckling ruminant neonate, the abomasum and intestine digest milk, so fermentation in the rumen is unnecessary. Milk bypass the rumen of a neonate through a **reticular groove** that reflexively closes when the neonate swallows milk. Thus, milk passes directly from the esophagus to the abomasum via the reticular groove.

Foregut fermentation has arisen independently in groups other than ruminants. Only slightly less elaborate in the multichambered stomach of some nonruminants, including leaf-eating sloths, langur monkey, peccaries (*i.e.*, pig-like mammals found from the south-western US to Praguay), hippopotamuses, many rodents, and rock hyraxes. Among marsupials, wallabies and kangaroos depend on microbial fermentation in a specialized region of the stomach to digest plants on which they feed (Fig. 4.50 A). However, only ruminants, camels and some marsupials regurgitate contents and remasticate food, such remastication together with microbial fermentation is called **rumination** in ruminants and **merycism** in nonruminants.

Intestinal Fermentation

Microbial digestion of cellulose centered in intestine is called **intestinal fermentation**. Extensive elongation of the intestine and large ceca extend the volume available for fermentation. More generally, herbivores in which fermentation occurs primarily in or associated with the intestine and its caeca are **hindgut fermenters**, or **caecal fermenters**. Rabbits, pigs, horses, and koalas are examples of hindgut fermenters (Fig. 4.52), as are many birds. Herbivorous and omnivorous birds tend to have large caeca; fish- and grain-eating birds have small caeca.

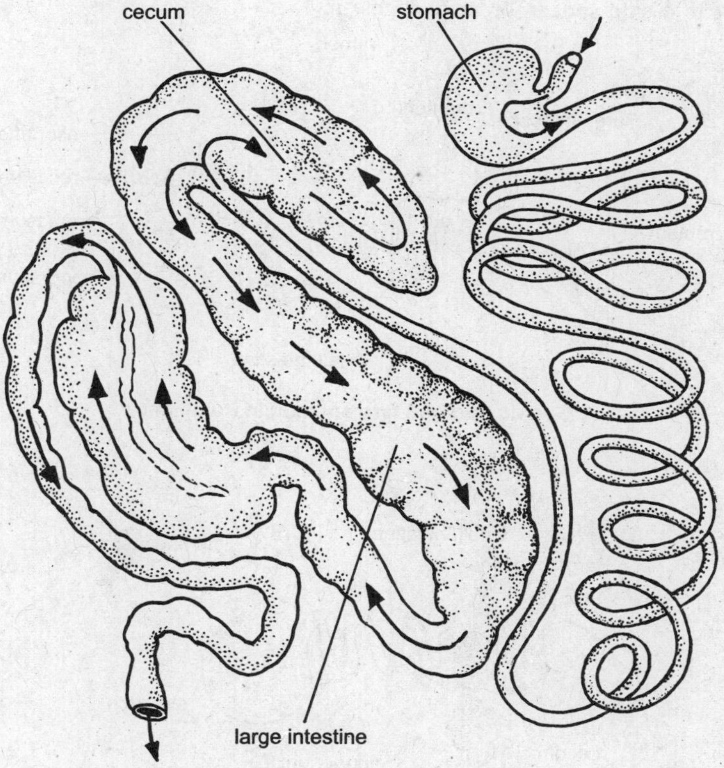

Fig. 4.52. Hindgut fermentation in the alimentary canal of a horse. The caecum (or cecum) acts as a diverticulum at the junction between small and large intestines. In the caecum, microbial fermentation of the digesta helps break down plant cell walls. The large intestine continues the fermentation process (after Kardong, 2006).

Compaıison of Gastric and Intestinal Fermenters

In both gastric and intestinal fermenters, microorganisms of the digestive tract release enzymes that digest the plant cellulose (Fig. 4.53). However, the physiological advantages of such fermentation differs between gastric (foregut) and intestinal (hindgut) ferments. At first glance, it might seem that gastric ferments, such as ruminants, camels, kangaroos, and wallabies, enjoy all the advantages when it comes to efficient digestion. First, fermentation takes place in the anterior part of alimentary canal, yielding end products of digestion early in the digestive process so they are ready for uptake next in the intestine (Fig. 4.53). Second, the ruminant system allows rechewing and more complete mechanical breakdown of the cells walls. By shuttling food between mouth and rumen via the esophagus, the ruminant can keep grinding away at plant fibers. The distant caeca of intestinal fermenters make such shuttling impossible. Third, the ruminant system turns much of the nitrogen, which in most vertebrates is a waste product, into the resource. This is particularly useful in ruminants that consume low-protein diets. Initially, the ruminant system converts nitrogen into waste products, such as ammonia is a by-product of fermentation of protein in the rumen (and of deamination of amino acids in liver). Microorganisms take up ammonia and use it to make their own cell proteins. Periodically, the rumen contracts, flushing these microorganisms into the abomasum and intestine, where, like any food, the microorganisms themselves are digested and their high-quality proteins are absorbed. Finally, the large storage capacity of the rumen means that the animal can quickly gather a large meal in an exposed site, and then retreat to a safe spot in order to digest it.

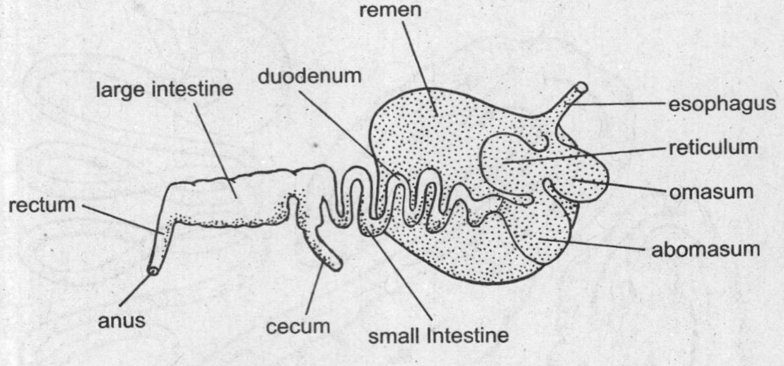

A Gastric (foregut) fermentation in Ruminant

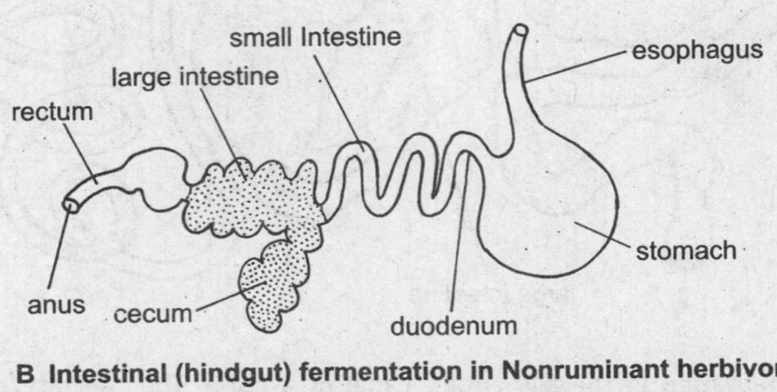

B Intestinal (hindgut) fermentation in Nonruminant herbivore

Fig. 4.53. Gastric (foregut) versus intestinal (hindgut) fermentation. A – Ruminants and other foregut fermenters depend upon microbial activity early in the digestive process as the ruminant "stomachs" chemically attacks cellulose in plant cell walls. B – In intestinal fermenters, such as horses fermentation takes place in the long intestines and extended caeca (after Kardong, 2006).

Gastric fermenters are also able to turn urea, another waste product, into a resource. For example, a camel fed nearly protein-free foods excretes almost no urea in its urine. Certainly urea forms during metabolism, but it reenters the rumen, partly by direct transfer across the rumen wall and partly in camel's saliva. In the rumen, the urea is broken down into carbon dioxide and ammonia. Microorganisms take up this ammonia as well and combine it with other organic carbon compounds to make cell protein. Finally, in addition to utilizing ammonia, microorganisms also aid in the breakdown of cellulose, a carbohydrate in the cell walls of plants. Cellulose fermentation produces carbon dioxide, water and volatile fatty acids.

Gastric fermentation is especially efficient at extracting the most even from food of poor quality. Ruminant and ruminant-like animals have been especially successful in habitats where the forage is scarce, fibrous and poor for at least part of the year, such as in alpine regions (goats), deserts (camels) and harsh winter areas (bison).

However, intestinal or hindgut fermentation has some advantages. For the intestinal fermenter, the bolus passes through the major absorptive regions of the alimentary canal before reaching the principal sites of fermentation, usually the caeca (Fig. 4.53 B). Soluble nutrients such as carbohydrates, glucose and proteins can be safely absorbed before fermentation begins. By contrast, among foregut fermenters, fermentation occurs early, and many of these necessary nutrients are disposed of before they can be absorbed. To compensate for premature digestion, foregut fermenters must rely on occasional flushing of microorganisms into the intestine where they are digested to replace nutrients lost in fermentation.

Further, although foregut fermentation is thorough, it is also slow. To ferment fibrous plants, food must occupy the rumen for extended cellulose processing. Where forage is abundant, the intestinal fermenter can move large quantities of food through the digestive tract, process the most easily digestible component of the forage, excrete the low quality component, and replace that which has been excreted with fresh forage. Horses are hindgut fermenters that depend on a rapid transit and a high food intake to meet their nutritional needs.

Coprophagy

In some animals, feces are a resource. Tinamous, a family of neotropical birds, and rabbits, hare, many rodents, and even gorillas eat their feces, a behaviour termed **coprophagy**, or **refection** (**Kardong**, 2006). But the feces eaten usually come only from the caecum, not from the main intestine. The caecum is emptied in the early morning, and only those droppings are consumed. Within the alimentary canal, there is also a selection process at the caecal-intestinal junction. Liquids and fine particles are diverted into the caecum for extended fermentation and coarser fibers are excluded. Thus, the coarse fiber that bypasses the caecum is not reingested, and only a small percentage of the digesta from the caecum is consumed a second time.

Coprophagy allows reingestion, an additional opportunity for the full length of the alimentary canal to capture the products of fermentation, namely, vitamins (K and all B vitamins), amino acids, and volatile fatty acids. If normal coprophagy is prevented, the animal may require vitamin supplements to remain healthy. Coprophagy has been reported in captive apes, cervids, and some other animals, but it is not known if such behaviour is importnat in the wild among these groups.

For the marsupial koala of Australia, feces are eaten by the growing neonate as a transitional food between milk and leaves. The koala mother feeds her own droppings to the six-month-old neonate to begin the process of weaning it form milk to feces to eucalyptus leaves.

The odour from feces can alert a predator to the presence of vulnerable young. Among many herbivores that hide from predators, the young animal does not pass feces until it is licked by its mother. Licking stimulates elimination of feces, which the mother eats so that feces does not collect at sites where young hide and leave a telltale odour. Many young birds bundle their feces. As feces move into the cloaca, its wall secrete a mucous bag that holds the digesta. Parents carry off these bundles of dirty feces, both contributing to good **housekeeping** (*i.e.*, the nest is not fouled) and removing any smelly feces that might attract the attention of a predator.

Digesting Toxins

Many animals have digestive tracts involved in the detoxification of potentially poisonous chemicals in the food. The koala exists on a diet exclusively of eucalyptus leaves. There are over 500 species of eucalyptus in Australia, and the koala favours perhaps a dozen of these as food. Because plants cannot run from herbivorous animals, many produce chemical defenses to make their tissues unpalatable or toxic to herbivores. Examples include tannins in grapes, caffeine in coffee plants, and cannabis in marijuana leaves. These chemicals are bitter or distasteful, causing animals to avoid them. Some plants produce altered states of alertness, leading perhaps an easy mark for a predator not on drugs. Such unpalatable or toxic antiherbivore compounds are called **secondary plant compounds** because they are not part of the plant's primary metabolic activity. Similarly, many prey animals have evolved toxins of their own, analogous to secondary plant toxins, that discourage predators. The toxic skin glands of most amphibians are examples.

Eucalyptus produce volatile oils that are toxic to most animals if eaten. Thanks to these secondary compounds, few herbivores can safely eat eucalyptus. The koala is an exception. Its digestive system can detoxify the harmful oils and exploit a resource that is, for the most part, unavailable to other herbivore competitors.

For many ruminants that browse on foods high in tannins, the ruminant saliva plays a central role in neutrializing in deterimental effects. In addition to being toxic, tannins bind with proteins to reduce their absorption, thus reducing the deigestibility of plants. However, ruminants that consume plants rich in tannins produce salivary proteins that bind firmly with tannins when food first enters the mouth to reduce the toxicity of tannins immediately. But salivary proteins further neutralize such secondary compounds to reduce their deterimental effects upon digestibility. Not surprisingly, the salivary glands of such ruminants are usually enlarged. In particular, their parotid gland is three times the size of that of ruminants not grazing on tannin-laden plants.

QUESTIONS

Long Answer Questions

1. Give a brief account of teeth.
2. In procuring and processing food what systems other than the digestive system are employed?
3. Give a comparative account of stomach in different vertebrates.
4. The mammalian air and food passages cross over. When breathing through your nose and chewing, how is food delivered to the correct passage when swallowing?
5. Outline the mechanical and chemical digestive processes in the digestive tract of a typical mammal.
6. Describe the cross section of the gut wall. What are the primary embryological sources for gut?
7. What taxa develop a complete secondary palate? What biological roles do the secondary palate serve in these taxa?
8. What modifications of the digestive tract occur in the cow?
9. In the rabbit, the site of bacterial digestion occurs distal to the small intestine (site of absorption for nutrients). Given a unidirectional flow through the digestive tract can you explain this observation.

Short Answer Questions

1. Write notes on the following:
   (*i*) Gizzard; (*ii*) Ileum;
   (*iii*) Colon; (*iv*) Ruminant stomach;
   (*v*) Digestive glands responsible for digestion;
   (*vi*) Palate.

Very Short Answer Questions

1. What is fermentation?
2. Define gastric fermentation.
3. What is gizzard.
4. Define intestinal fermentation.
5. Define coprophagy.
6. What is palate?
7. What is merycism?
8. Define rumination.
9. What is advantage of coprophagy?
10. What is mesentery? Can it be regarded as an organ.

Fill in the Blanks

1. .... defines the anterior border of mouth.
2. .... prevent loss of food out the sides of the mouth during chewing.
3. Vertebrate pharynx contains .... skeleton in its wall.
4. Mammalian stomach has three divisions ...., fundus and ....
5. The masticatory portion of stomach of birds is ....
6. Gastric fermentation is also called ....
7. Intestinal fermentation is also called .... or .... fermentation.

Multiple Choice Questions

1. True muscular lips are found in
   (*a*) cyclostomes (*b*) fishes
   (*c*) birds (*d*) mammals
2. Largest oral glands are found in
   (*a*) amphibians (*b*) reptiles
   (*c*) birds (*d*) mammals
3. In frogs tongue is
   (*a*) immobile tongue (*b*) primary tongue
   (*c*) definite tongue (*d*) horny tongue
4. Oesophagus in adult vertebrates in derived from
   (*a*) primitive gut (*b*) foregut
   (*c*) midgut (*d*) hindgut
5. Large oesophagus is found in
   (*a*) dogfish (*b*) frog
   (*c*) sparrow (*d*) giraffe

6. Epithelial lining in a true stomach contains

(*a*) salivary glands
(*b*) goblet glands
(*c*) mucous glands
(*d*) gastric glands

7. In ruminants, true stomach is represented by

(*a*) abomasum
(*b*) omasum
(*c*) rumen
(*d*) reticulum

ANSWERS

Very Short Answer Questions

1. Fermentation is a process in which microorganisms anaerobically extract energy from food in vertebrates by releasing cellulase enzyme that breaks down plant material.

2. Gastric fermentation is a process in which microorganisms digest food in a specialized stomach, *e.g.*, ruminants such as cow.

3. Gizzard is an especially well-muscularized region of the stomach used to grind hard foods.

4. Intestinal fermentation is a process in which microorganisms digest food in the intestine, *e.g.*, rabbits, pigs, horses, koalas.

5. Coprophagy is eating of feces (especially which released from caecum and that released during early morning), a behaviour performed usually to process undigested material again, *e.g.*, rabbit, koala, gorilla.

6. Palate is the roof of buccal cavity. It is formed from fusion of ventral skull bones overlying the mouth.

7. Merycism is remastication together with microbial fermentation of food in nonruminants.

8. Rumination is remasticiation together with microbial fermentation in ruminants.

9. Coprophagy allows reingestion, an additional opportunity for the full length of the alimentary canal to capture the products of fermentation, namely vitamins (K and all B vitamins), amino acids, and volatile fatty acids.

10. Mesentery is a thin double-walled sheet of peritoneum (and connective tissue) that supports organs (*e.g.*, intestine) in the abdominal cavity of vertebrates. Currently it is regarded as an organ (**J. Calvin Coffey**, 2017). It is 79th organ of human body according to Gray's Anatomy (see **C. Gregoire** 2018).

Fill in the Blanks

1. Lips.

2. Cheeks.

3. Visceral.

4. Cardia, pylorus.

5. Gizzard.

6. Foregut.

7. Hindgut, cecal.

Multiple Choice Questions

1. (*d*) **2.** (*d*) **3.** (*c*) **4.** (*b*) **5.** (*d*) **6.** (*d*) **7.** (*a*)

5

Respiratory System

5.1 PASSIVE DIFFUSION AND 'PUMPS' OF RESPIRATION

To metabolise effectively and survive, cells within the body of a vertebrate (or craniate) must replenish the oxygen used and rid themselves of the byproducts (*e.g.*, carbon dioxide) accumulated during metabolism . These tasks fall primarily to two **transport systems**, the circulatory and respiratory systems. The **circulatory system** (see Chapter 6) basically connects cells deep within the body with the environment. The **respiratory system** involves gas exchange between the surface of an organism and its environment. Basically, these two systems aid in the process of **passive diffusion**, the random movement of molecules from an area of high to an area of low partial pressure (Fig. 5.1A). Oxygen usually at high partial pressure in the environment and tends to diffuse into the organism. Carbon dioxide collects in tissues and tend to diffuse out.

However, if unaided, passive diffusion alone is incapable to meet the needs of large multicellular organisms (animals). For example, a hypothetical spherical aquatic organism could have a radius no larger than 0.5 mm for its core tissues to receive adequate oxygen by passive diffusion alone, even if the surrounding water were saturated with oxygen. For oxygen to diffuse passively from human lungs to his/her extremities, the journey would require several years! Certainly, this would not do. In large multicellular animals, transport systems assist passive diffusion. The circulatory and respiratory systems speed the process.

Factors Affecting Passive Diffusion of Respiratory Gases

During evolution of animals, major modifications in the design of respiratory organs have occurred to optimize the diffusion of important gases. The rate of passive diffusion between an organism and its environment depends on several factors.

1. One is **surface area**. The greater is the available surface area, the greater is the scope for molecules to move across an epithelial surface (Fig. 5.1B). For instance, the gas exchange organs of vertebrates (*e.g.*, gills, lungs) are highly subdivided to increase the surface available to transfer gases between air and blood.

2. The second factor is **distance**. The greater the distance, the longer it will take for molecules to reach their goals (Fig. 5.1C). Thick tissues slow diffusion, and thin barriers help in the process of diffusion. The thin walls of the respiratory organs reduce the distance between the environment and the blood.

3. A third factor is the **resistance** to diffusion by the tissue barrier itself. The moist skin of living amphibians facilitate gas exchange. In contrast to this situation, the skin of most mammals is cornified and thick, a feature that slows gas diffusion with the environment.

One of the most important factors affecting diffusion is *the difference in partial pressure across the exchange surface*. The gills of most fishes face a high partial pressure of oxygen relative to the blood; therefore, oxygen diffuses across the gills into the blood. Sometimes, fishes living in warm, stagnant water may encounter a partial pressure of oxygen in the water below that of

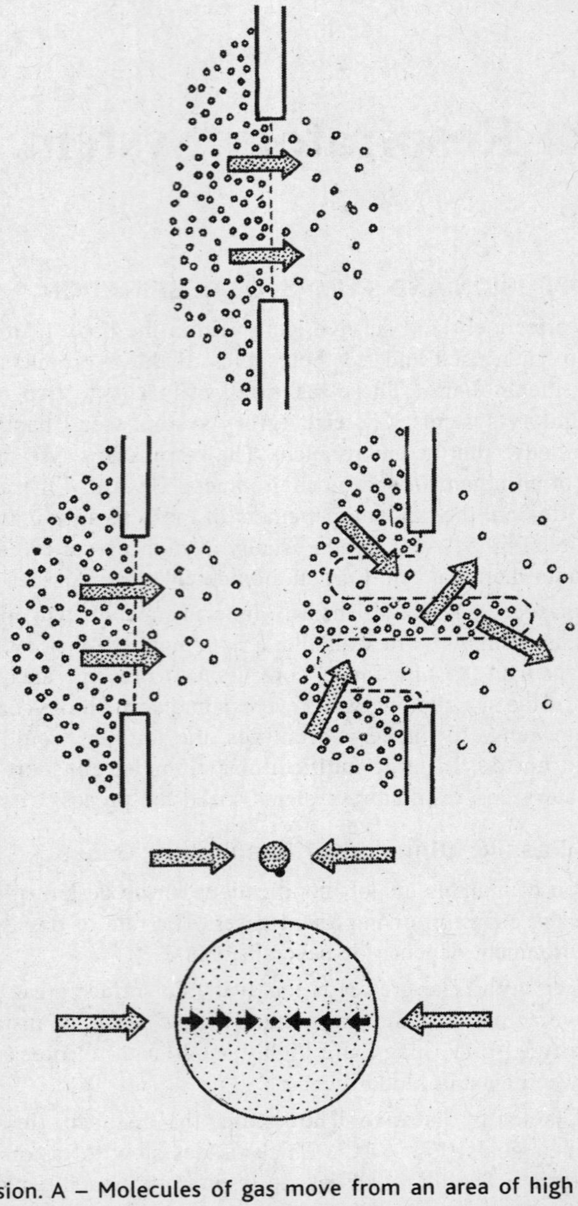

Fig. 5.1. Passive diffusion. A – Molecules of gas move from an area of high partial pressure to an area of low partial pressure. Eventually an equilibrium is reached when the concentration of molecules becomes equal on both sides of the surface across which diffusion occurs. B – The rate of movement of diffusing molecules depends upon available surface area. Increasing the surface area, increases the rate at which diffusion occurs, although the final equilibrium concentration will eventually be the same regardless of the surface area available. C – The time it takes for molecules to reach deep tissues depends upon the distance they must travel. Molecules moving into cells at the center of the small circle reach the core much sooner than molecules that must traverse thicker tissues to reach those cells within the center (after Kardong 2006).

its blood. Under these unusual conditions, oxygen may actually diffuse in the reverse direction, and the fish is in danger of losing oxygen to the water (see **Kardong** 2006).

Pumps Aiding Respiration and Ventilation

Both respiratory and circulatory systems have "pumps" that move fluids, such as air or water (respiration) or blood (circulatory). The heart is a pump that circulates blood. In fishes, the predominant **respiratory pump** is the branchial apparatus that drives water across the gills (Fig. 5.2 A). In tetrapods, one familiar pump is the **rib cage**, sometimes assisted by a **diaphragm**, that moves air through the lungs (Fig. 5.2B). Also, many types of supplementary pumping devices in vertebrates are also part of the respiratory mechanism. By moving fluids that contain gases these pumps function to maintain high partial pressure gradients across exchange surface.

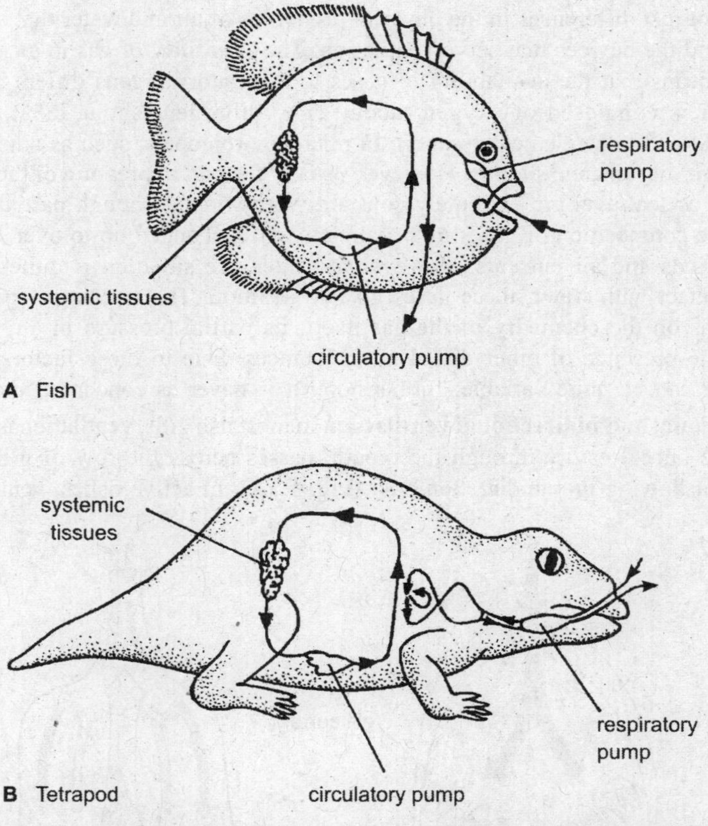

Fig. 5.2. Respiratory and circulatory systems cooperate to deliver oxygen to deep tissues and carry away carbon dioxide. During external respiration, air or water is inhaled and transported to exchange capillaries of the blood. Thereafter, blood circulates oxygen to all systemic (body) tissues, represented here by a small patch of tissue, where internal respiration occurs. Oxygen is delivered to these tissues and carbon dioxide is carried away. A – In fishes, the respiratory pump usually includes the branchial arches and their musculature. External respiration occurs in the gill capillaries. The heart, being the primary circulatory pump, drives blood through the gills and then to systemic tissues. B – In tetrapods, this respiratory pump can include the buccal cavity, which forces air into elastic lungs against resistance, and a rib cage around the lungs. External respiration occurs in lungs. The circulatory pump, or heart, drives blood though vessels. Internal respiratory exchange occurs between blood and systemic tissues (after Kardong 2006).

During external respiration, gases diffuse between the environment and the organism-oxygen enters, carbon dioxide departs. **Ventilation** or **breathing** is the active process of moving the respiratory medium, water or air, across the exchange surface. Ceasing the movement of the respiratory medium is **apnea**, or breath holding. Pumping of blood through an organ via capillaries is known as **perfusion**. The **respiratory organs** specialize in ventilation to deliver oxygen and remove carbon dioxide accumulated during perfusion. The demands on the respiratory organ vary, depending on whether the medium is water or air. This is partly due to differences in density. Water, being denser than air, requires more energy to set in motion. Other things being equal, ventilation that involves moving water is energetically more expensive than ventilation that involves moving the air. In addition, because water is more dense, structures are more byoyant (*i.e.*, able to keep afloat) in water than in air. Gills supported by water tend to collapse in air and therefore fail as respiratory organ on land. Lungs are structurally strengthened to work better in air.

But it is not just differences in the physical properties of air and water that affect ventilation (or breathing) and the devices that serve ventilation. The **solubility** of gas in air differs from that in water. This means that the availability of gases to respiratory organs differs in air and water. Atmospheric air is composed of oxygen (about 21%), nitrogen (about 78%), carbon dioxide (less than 0.03%) and trace elements (gases). In microenvironments, such as animal burrows, the composition of air may change slightly. However, usually the partial pressure of the physiologically important gases at sea level are extremely constant word-wide. Although partial pressure varies with altitude, the composition of gases in air is relatively unchanged up to over 1000 km, thanks to mixing by winds and air currents. However, in water, the situation is quite different. When brought into contact with water, these gases go into solution. The amount of gas that dissolves in water depends on the chemistry of the gas itself, its partial pressure in air, the temperature of water, and the presence of other dissolved substances. Due to these factors, the amount of oxygen in water can be quite variable; furthermore, it is never as concentrated as it is in air.

Unidirectional and bidirectional ventilation. In most fish gills, ventilation is **unidirectional**. Water enters the buccal cavity through the mouth, passes across the row of gills known as **gill curtain**, and exit flowing in one direction only (Fig. 5.3A). In active fishes, ventilation is almost

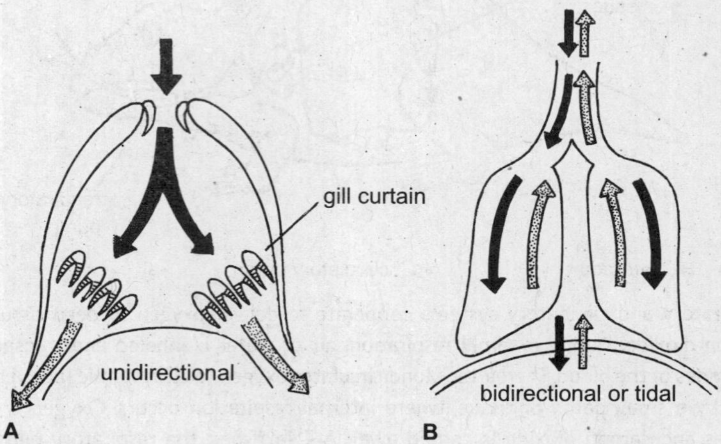

Fig. 5.3. Unidirectional and bidirectional flow. A – In fishes and many aquatic amphibians, water movement is unidirectional because water flows through the mouth, across the gill curtain, and out the lateral gill chamber. B – In many air – breathing vertebrates, air flows into the respiratory organ and then reverses its direction to exit along the same route, creating a bidirectional or tidal flow (after Kardong 2006).

continuous to keep a more or less steady stream of new water washing across the exchange surfaces of the gills. Lung ventilation, however, is usually **bidirectional** (tidal), with air entering and exiting through the same channels (Fig. 5.3B). A fresh breath of air that is **inhaled** into the lungs mixes with spent air and is **exhaled**. The exchange capillaries of the lung are replenished discontinuously, with air.

Vertebrates that live in aqueous environment most often encounter too little oxygen, called **hypoxia**, partly because water is already low in dissolved oxygen. For this reason, most organs that supplement respiration are found among aquatic rather than strictly terrestrial animals. Indeed, one of the major transitions in vertebrates evolution was the change from water breathing to air breathing.

5.2 RESPIRATORY ORGANS

The chief organs of external respiration in adult craniates (or vertebrates) are external and internal gills, the oropharyngeal mucosa, air sacs or lungs, and the skin. Less common respiratory devices include bushy or filamentous outgrowths of the pectoral fins, as in male *Lepidosiren*, or of the posterior trunk region and thigh, as in the African hairy frog; the cloacal, rectal, or anal lining, and the lining of the esophagus, stomach, or even of the intestine in one species or another. Embryos employ a variety of respiratory surfaces including extraembryonic membranes (**Kent** and **Carr** 2001).

I. Gills

A. Internal Gills

Vertebrate gills are designed for water breathing. Specifically, they are dense capillary beds in the branchial region that serve external respiration. They are supported by skeletal elements, the branchial arches.

1. Gills in Agnathans

Hagfishes in the genus *Myxine* have five or six pairs of well-vascularized gill pouches, rarely seven, associated with the pharynx (Fig. 5.4A). Various species of *Eptatretus*, another hagfish genus, have five to 15 pairs of gill pouches. **Afferent branchial ducts** conduct respiratory water from the pharynx to the pouches, and **efferent ducts** lead from the pouches to the exterior. Each efferent duct usually has its own external aperture (*e.g.*, *Eptatretus*), but in myxinoids the efferent ducts unite to open via a single common external slit on each side (Fig. 5.4A).

In hagfishes, respiratory water enters the median unpaired **naris** and passes via a **nasopharyngeal duct** to a **velar chamber** at the anterior end of the pharynx (Fig. 5.4A). The wall of the chamber is a pulsating muscle, or velum, which pumps water form the velar chamber caudad into the pharynx and according to some observers, into the gill pouches, thereby creating a vacuum in the velar chamber. The vacuum draws additional water through the naris and into the velar chamber. Muscle tissue in the walls of gill pouches expel the water to the exterior (Fig. 5.5A). The velum is supported by pharyngeal cartilages and pulsates 50 to 100 times a minute in sleeping animals. On the left side only, a **pharyngocutaneous duct** of greater diameter than the afferent branchial ducts connects the pharynx with the last efferent branchial duct (*Myxine*), or , in some hagfishes, with the exterior. Periodically, debris or particles too large to enter the afferent branchial ducts are forcefully ejected through the pharyngocutaneous duct in a manner somewhat analogous to coughing. The pharygocutaneous duct appears to be modified last gill pouch.

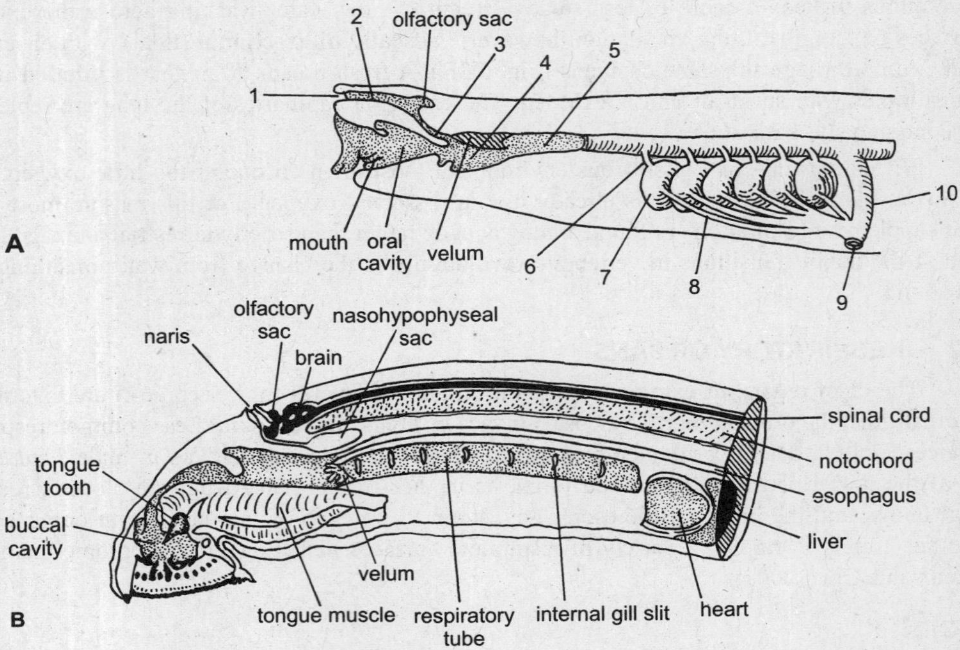

Fig. 5.4. Head and pharyngeal region of hagfish and lampreys. A – The hagfish *Myxine glutinosa*. **1**, naris; **2**, **3**, nasopharyngeal duct; **4**, velar chamber; **5**, pharynx; **6**, afferent branchial duct; **7**, gill pouch; **8**, efferent branchial duct; **9**, common external gill aperture (present on both sides); **10**, pharyngocutaneous duct (present on left side only). B – The lamprey, *Petromyzon*, sagittal section. Connecting naris and nasohypophyseal sac is the nasohypophyseal duct (after Kent and Carr 2001).

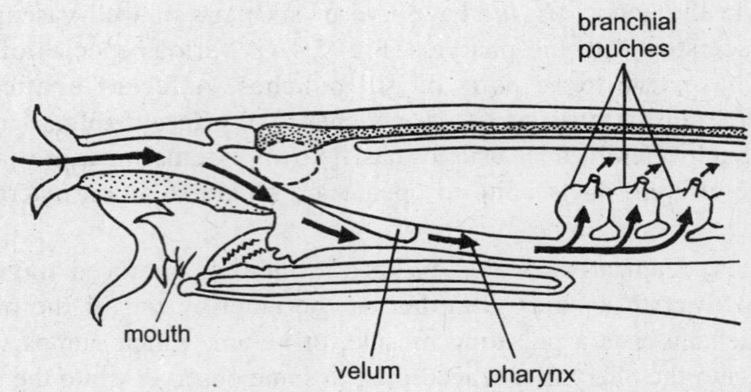

Fig. 5.5. Ventrilation in hagfish. Longitudinal section. Water (indicated by arrows) enters via the nostril, not the mouth, to reach the pharynx. The scroll-shaped velum rolls up and down as the branchial pouches contract to drive this current across the gills and out the gill pores (after Kardong 2006).

The mechanism of respiration differs considerably in lampreys in comparison with hagfishes. The buccal cavity of lampreys is pressed tightly into the flesh of a host, and the duct from the unpaired naris **(nasohypophyseal duct)** does not open into the pharynx but ends blindly in a nasohypohyseal sac (Fig. 5.4B). Therefore neither the nasal nor buccal route is available for the

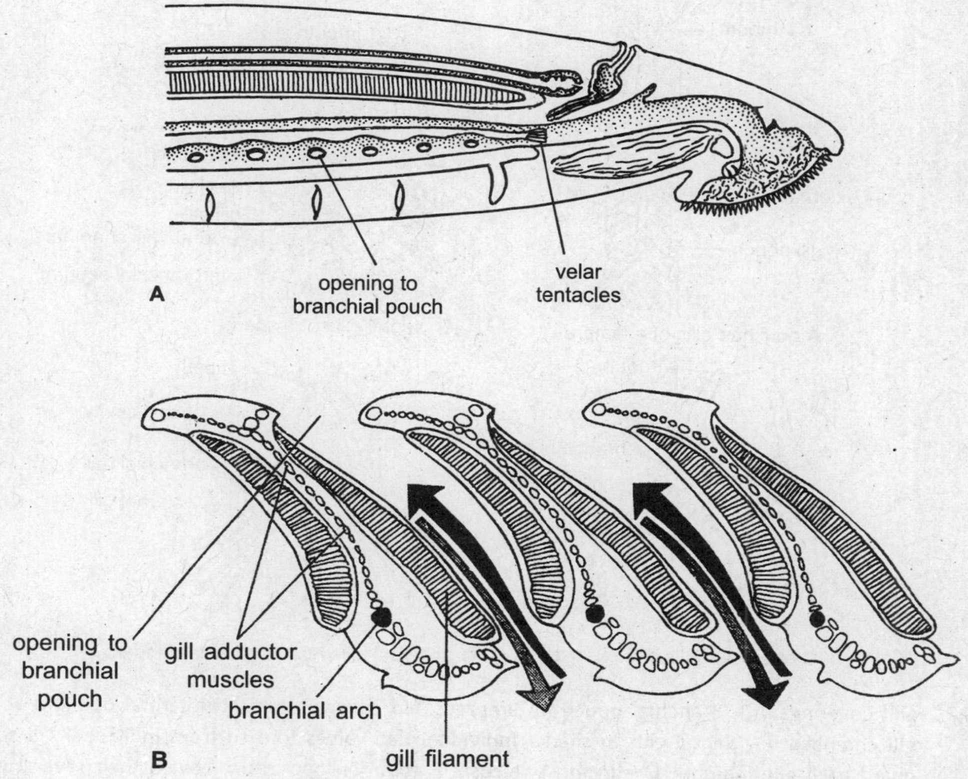

Fig. 5.6. Tidal ventilation in the adult lamprey. A – Longitudinal section. Because the adult lamprey's mouth often is attached to prey, water must alternatively enter as well as exit via pharyngeal slits. Thus, unlike the most fishes, gill ventilation in the lamprey is tidal. B – Frontal section of three gill arches. Double arrows indicate tidal flow of water: black inflow; gray, outflow (Kardong 2006).

passage of respiratory water to the gills. As a **parasitic adaptation**, respiratory water *enters and exits* the pharyngeal pouches via the external gill slits as a result of **pulsations** of the pharyngeal and gill pouch musculature (this represents tidal ventilation; Fig. 5.6).

The external apertures are guarded by thin flaps of skin that serve as two-way valves (Fig. 5.7A). As a further adaptation, the lamprey pharynx becomes subdivided longitudinally at metamorphosis into an esophagus that continues to the stomach and a ventral respiratory tube that ends blindly (Fig. 5.4B). The entrance to the adult respiratory tube is guarded by a valve-like velum that prevents nutrients rasped from a host from entering the respiratory tube, becoming diluted by respiratory water, and then being lost through the external gill slits.

The seven pairs of voluminous gill pouches are lined with gill lamellae and communicate directly with the respiratory tube and the exterior with no intervening afferent and efferent branchial ducts (Fig. 5.7A). The respiratory systems of hagfishes and lampreys have little in common with those of jawed fishes structurally (Fig. 5.7B,C). The former represent two variant adaptations to parasitism.

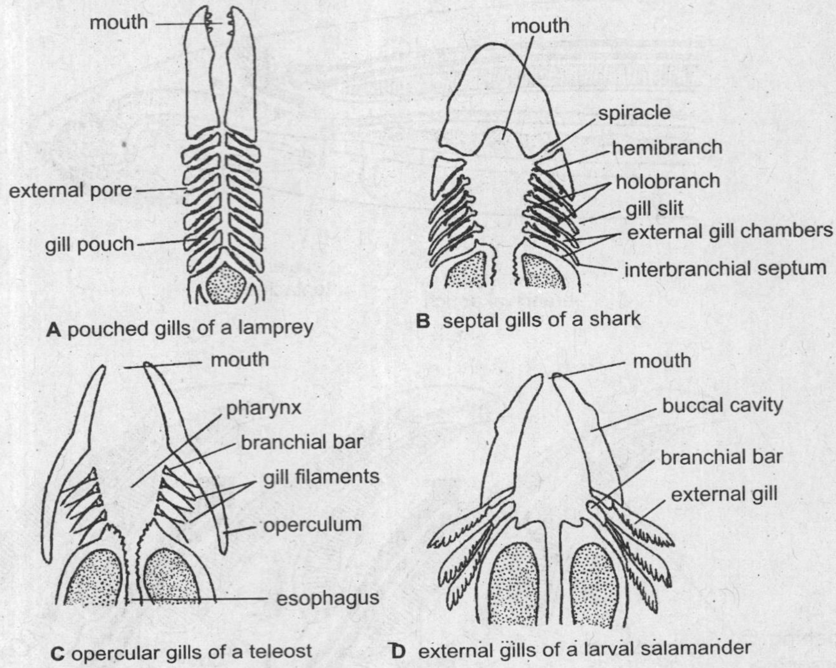

Fig. 5.7. Gill coverings. A – Branchial pouch in lampreys. No cover protects the lateral opening of the gill chamber. B – Septal gills in sharks. Individual flap valves formed from individual gill septa guard each gill chamber. C – In most teleosts a common operculum covers their several gills. D – In larval salamanders, the branchial arches support vascular external gills that project into the surrounding water (after Kardong 2006).

2. Gills in Elasmobranchs

Most elasmobranchs – sharks, skates and rays – are **pentanchid**, having five pairs of gill pouches, each with an internal and external gill slit and a pair of functional **spiracles** located dorsally immediately in front of the hyomandibular cartilages (Fig. 5.8A). In dogfish, *Scoliodon*, the spiracles are vestigial pits in the pharynx with no pseudobranchs and external branchial apertures (see **Jordan** and **Verma** 2013). Certain other sharks are also exceptional, *e.g.*, *Hexanchus* has a spiracle and six gill pouches and *Heptanchus* has a spiracle and seven gill pouches (this being the largest number of gill pouches in any gnathostome). No gill surface develops in the posterior wall of the last pouch. The external gill slits in sharks are visible on the side of the head; those in Rajiformes (skates) are on the underside. The slits are said to be "naked" because, unlike in other gnathostomes, there is no operculum. A miniature gill-like structure, or **pseudobranch**, consisting of a vascular rete (network) develops in its anterior wall. The rete may regulate blood pressure in the eyeball (Box 5.1).

The spiracel has a one-way intake valve and is the exclusive incurrent aperture for respiratory water in Rajiformes and for much of the respiratory water in sharks except some fast-swimming voracious ones in which the spiracle becomes secondarily closed by a membrane of skin (*e.g.*, *Scoliodon*). Dorsal position of spiracle in rays minimizes the entrance of muddy debris-laden water in the phyarynx of these bottom dwelling species.

As in all gnathostome gills, elasmobranch gills lie lateral to the branchial arch. Each gill consists of a central partition, the interbranchial septum, covered on each face by primary lamellae

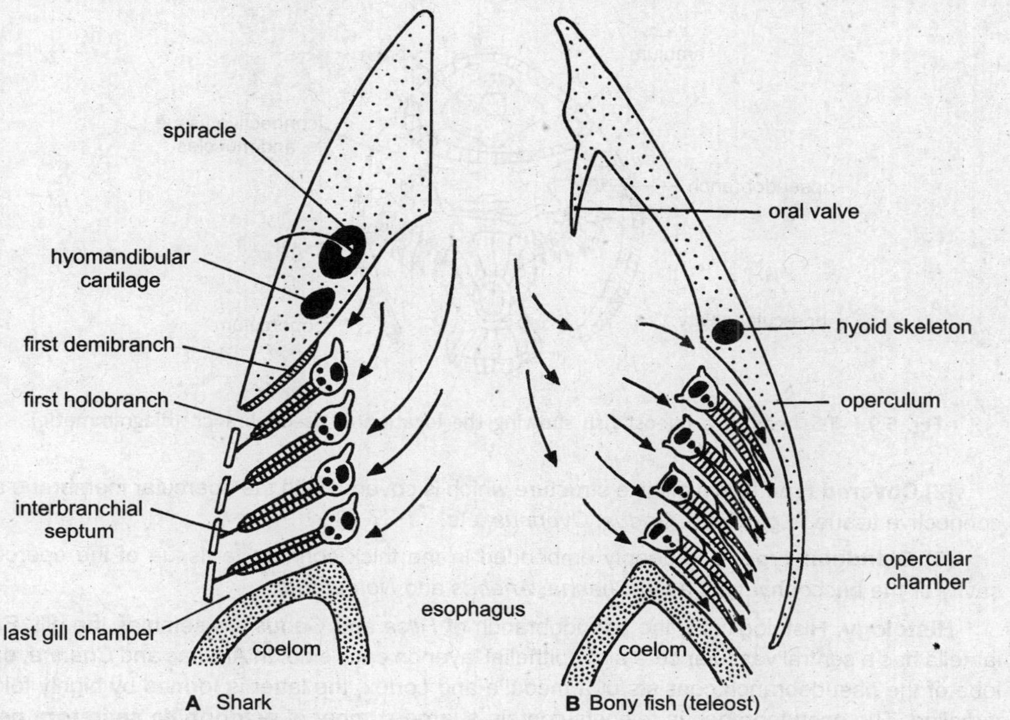

Fig. 5.8. Gills of a pentanchid shark and teleost, frontal section, schematic. Arrows show routes of respiratory water (after Kent and Carr 2001).

(gill filaments). The primary lamellae are composed of standing rows of secondary lamellae. Across their sides flows water that irrigates the gills. Like ribs of a fan, **gill rays** within the septum give it support. The term **holobranch** refers to a branchial arch and (gill) lamellae on both anterior and posterior faces of its septum.

Box 5.1 Pseudobranch

The term pseudobranch was used for a hemibranch of fishes which has lost its respiratory function, and receives already oxygenated blood from the efferent branchial artery. A pseudobranch is present in several actinopterygian fishes. It may be free or covered with a layer of mucous membrane and consists of a series of gill filaments. Pseudobranch develops early in the embryo and may be respiratory in embryonic stage but not in the adult. The pseudobranch in teleosts receives oxygenated blood directly from the dorsal aorta or an efferent branchial artery and has a vascular connection with the internal carotid artery. It is present in a large number of teleosts and is free or glandular; but is absent in cat fishes (*e.g.*, *Wallago*, *Mystus*), the eels, and *Gymnarchus* and *Cobitis*. A teleostean pseudobranch is reported to represent the posterior mandibular hemibranch (**Goodrich**, 1930). It sharks, rays, chondrosteans, and holosteans, the pseudobranch may be derived from the posterior mandibular hemibranch, but in the teleosts it is derived from the posterior hyoidean hemibranch.

Structure. Pseudobranch varies in shape, size and location in various teleosts, but its blood supply and innervation suggest a common plan. It is either attached to the operculum (Fig. 5.9), as in *Hilsa ilisha* or is deeply embedded in the connective tissue of the roof of the buccal cavity, as in *Channa*. According to **Bertin** (1958), pseudobranch may be of following three types:

(1) Free type. It contains a row of filaments and secondary lamellae, *e.g.*, clupeidae (*Hilsa*, *Gadusia*).

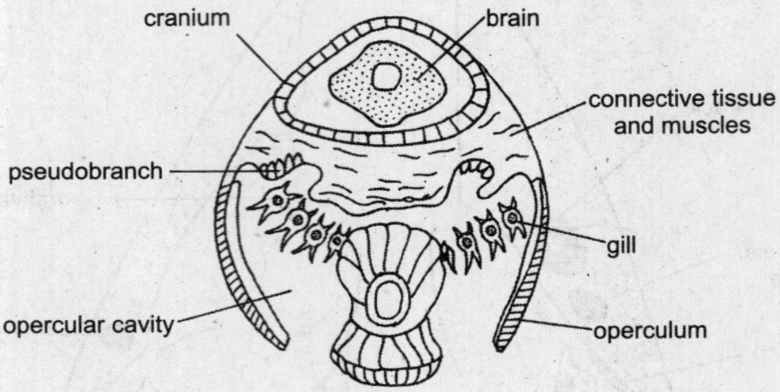

Fig. 5.9. T.S. head of a teleost fish showing the location of pseudobranch (diagrammatic).

(2) Covered type. It is a gill-like structure which is covered with the opercular membrane and connective tissue, *e.g.*, *Catla*, *Gadus*, *Cyprinus*, etc.

(3) Glandular type. It is deeply embedded in the thick connective tissue of the opercular cavity or the buccophyarynx as in *Channa*, *Anabas* and *Notopterus*.

Histology. Histologically, the pseudobranch of *Hilsa* and *Gadusia* resembles the gill. Each lamella has a central vascular core and epithelial layer on each side. In *Anabas* and *Channa*, each lobe of the pseudobranch consists of a medulla and cortex, the latter is formed by highly folded lamellae. The pseudobranch in teleosts contain a large number of **acidophilic secretory cells**, which replace the respiratory cells of the gills. Typical **chloride cells** and mucous cells are also present in the free pseudobranch of both fresh water and marine speices.

Functions. Pseudobranch has the following functions (see **Khanna** and **Singh** 2003):

1. Pseudobranch having chloride cells performs osmoregulatory function and is possibily involved in regulation of uptake and secretion of sodium (Na^+) and chloride (Cl^-) ions from water.

2. Pseudobranchial cells may produce carbonic anhydrase enzyme which may help in concentration of oxygen in the blood going to retina and in regulation of intraocular pressure.

3. Pseudobranch has also the sensory functions: to sense O_2/CO_2 tension in blood, monitoring blood pressure in the arteries and detection of pH level. Sensory function is performed with the help of nerve endings and receptors.

4. The pseudobranch of *Channa* and *Anabas* performs the function of an oxygen multiplier (**Munshi** *et al.*, 1994). The blood flow in the lamellae is built on counter current principle due to folding. The direction of blood flow in the lamellae of the first half is opposite to that in the second half, acting as a multiplier of oxygen.

5. **Srivastava** and **Gopesh** (1981) have suggested a role in neurosecretion for the pseudobranch.

A gill arch with lamellae on only one face is a **hemibranch**. Facing plates of lamellae on adjacent gills constitute a **respiratory unit** (Fig. 5.10A,B).

Ventilation in sharks is based on a dual **pump mechanism** that creates alternative negative (suction) and positive pressures to draw water in and then drive it across the gill curtain. Pressures recorded on either side of the gill curtain within the **buccal** and **parabranchial cavities** (compartments) reveal the efficiency of this dual pump. Although pressures rise and fall in each cavity, the pressure is always relatively lower in the parabranchial cavity, located lateral to the gills, than in the buccal cavity, located medial to the gills. The dual pump mechanism does two

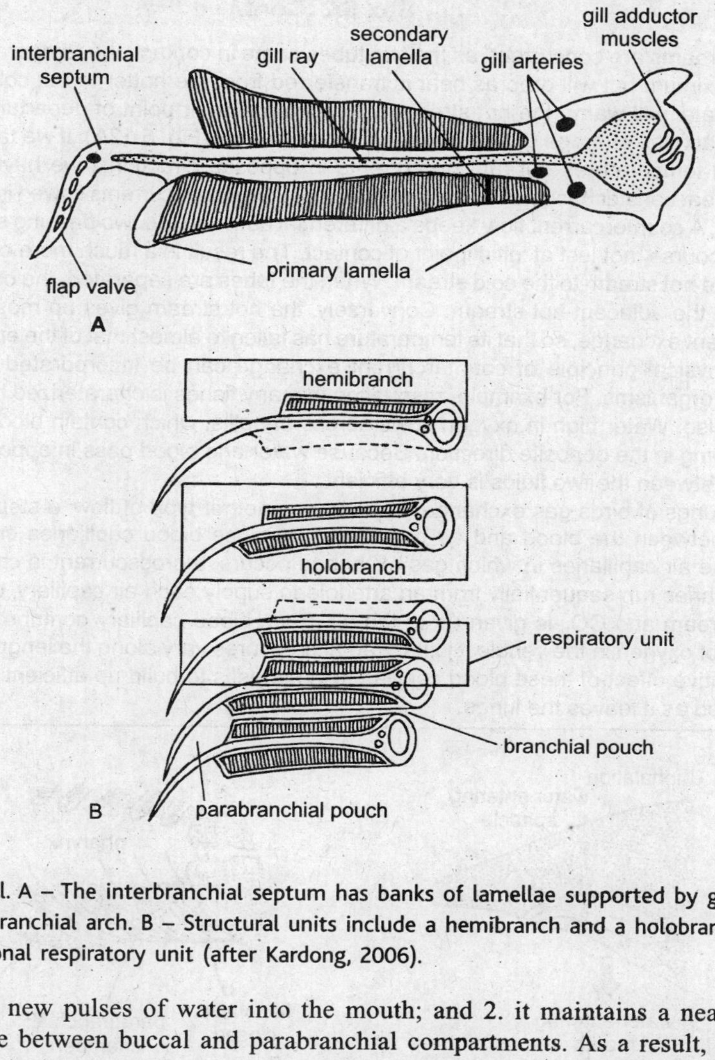

Fig. 5.10. Shark gill. A – The interbranchial septum has banks of lamellae supported by gill rays and a medial branchial arch. B – Structural units include a hemibranch and a holobranch as well as a functional respiratory unit (after Kardong, 2006).

tasks: 1. it brings new pulses of water into the mouth; and 2. it maintains a nearly constant pressure difference between buccal and parabranchial compartments. As a result, the pressure ossilations of the dual pump are converted into a smoother, almost continuous unidirectional irrigation of the gills (Fig. 5.11A, B). Blood coursing within the capillaries of the secondary gill lamellae sets up a **countercurrent** or perhaps a **crosscurrent** pattern (Box 5.2), promoting efficient gas exchange.

Box 5.2 Countercurrent, Concurrent and Crosscurrent Exchange

Exchange is a large part of life. The process of exchange, whether it involves gases (O_2, CO_2), or heat, or ions is sometimes supplemented by air or water currents passing one another. Efficiency of exchange depends on whether the currents pass in opposite or equivalent directions.

Imagine two parallel, but separate, identical tubes through which streams of water flow at the same speed. Water entering one tube is hot, and water entering the other is cold. If the tubes are made of conducting material and contact each other heat will pass from one to the other (Fig. 5.12 A,B). Water flow may be in the same direction, as in **concurrent exchange** or in opposite directions, as in **countercurrent exchange.** The efficiency of heat exchange between the tubes is affected by the directions of flow.

Box 5.2 (Cont.)

If the streams are concurrent, as the two tubes come in contact, the temperature difference will be at its maximum but will drop as heat is transferred from the hotter to the colder tube. The cold stream of water will warm, the hot stream will cool; so at their point of departure, both streams of water approach the average of their two initial temperatures (Fig. 5.12A). If we take the same tubes and starting temperatures but run the currents in opposite directions, we have a countercurrent exchange; heat transfer becomes much more efficient than if both currents flowed in the same direction (Fig. 5.12B). A countercurrent flow keeps a differential between the two passing streams throughout their entire course, not just at initial point of contact. The result is a much more complete transfer of heat from the hot stream to the cold stream. When the tubes are separated, the cold stream is nearly as warm as the adjacent hot stream. Conversely, the hot stream gives up most of its heat in this countercurrent exchange, so that its temperature has fallen to almost that of the entering cold stream.

This physical principle of countercurrent exchange can be incorporated into the design of many living organisms. For example, respiration in many fishes is charaterized by a countercurrent exchange also. Water high in oxygen flows across the gills, which contain blood capillaries low in oxygen flowing in the opposite direction. Because water and blood pass in opposite directions, gas exchange between the two fluids is very efficient.

In the lungs of birds gas exchange is based on another type of flow, a stepwise cross-current exchange between the blood and air capillaries. Because blood capillaries cross at nearly right angles to the air capillaries in which gas exchange occurs, a crosscurrent is created (Fig. 5.12C). Blood capillaries run sequentially from an arteriole to supply each air capillary, oxygen passes into the bloodstream and CO_2 is given up to the air. Each blood capillary contributes stepwise to the rising level of oxygen in the venule it joins. Partial pressures vary along the length of an air capillary, but the additive effect of these blood capillaries in series is to build up efficient levels of oxygen in venous blood as it leaves the lungs.

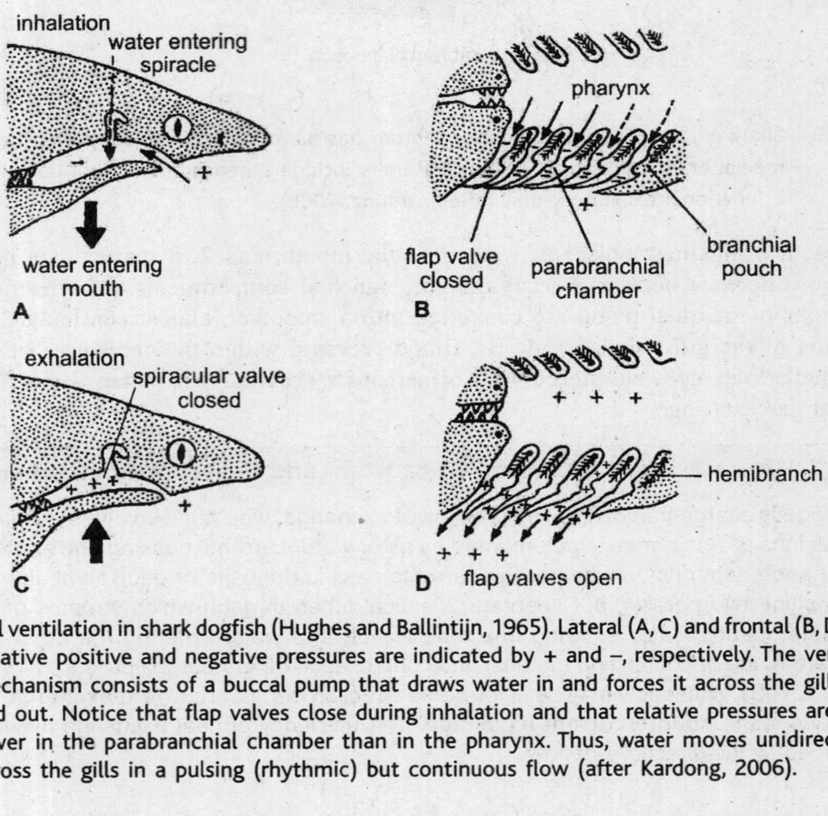

Fig. 5.11. Gill ventilation in shark dogfish (Hughes and Ballintijn, 1965). Lateral (A, C) and frontal (B, D) views. Relative positive and negative pressures are indicated by + and –, respectively. The ventilation mechanism consists of a buccal pump that draws water in and forces it across the gill curtain and out. Notice that flap valves close during inhalation and that relative pressures are always lower in the parabranchial chamber than in the pharynx. Thus, water moves unidirectionally across the gills in a pulsing (rhythmic) but continuous flow (after Kardong, 2006).

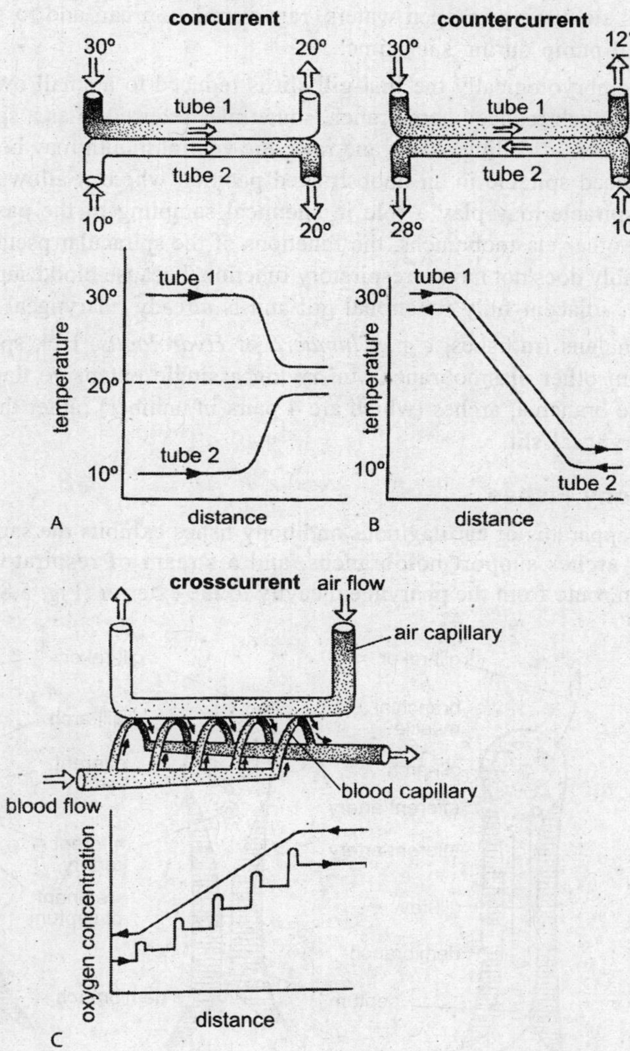

Fig. 5.12. Systems of exchange. Direction and design of exchange tubes affect the efficiency of transfer regardless of whether the exchange involves heat, gases, ions, or other substances. The first two examples (A and B) illustrate heat transfer. The third example (C) depicts gas exchange. A – Concurrent exchange describes the condition in which separated fluids flow in the same direction. Because the temperature gradient between the fluids is high when they enter the tubes and rather low when they exit, the average difference in heat exchanged, between the two fluids is relatively high. The fluid in tube 2 is at 10° when it enters and at 20° when it exits. B – In countercurrent exchange, the fluids pass in opposite directions within the two tubes so that temperature difference between them remains relatively low all along their lengths. The fluid in tube 2 is at 10°C when it enters and at 28°C when it exits. Thus, more heat is transferred with countercurrent exchange then with concurrent exchange. C – In a stepwise crosscurrent exchanger, each blood capillary branch passes across an air capillary at about right angles to it and picks up oxygen. The levels of oxygen rise serially in the departing blood. Arrows indicate the direction of flow (after Kardong 2006).

In sharks swimming in open waters, ram ventilation can add to gill irrigation and nearly replace the dual pump during such time.

What is embryologically the first gill slit is reduced to a small oval opening, the **spiracle**, which carries a much reduced hemibranch, sometimes referred to as a **spiracular pseudobranch** (Box 5.1). In bottom-dwelling skates and rays, the ventral mouth may be partially buried, leaving the dorsally placed spiracle in an unobstructed position where it allows water for gill irrigation to enter. The spiracle may play a role in chemical sampling of the passing stream of water as well. For most other elasmobranchs, the functions of the spiracular pseudobranch is unknown. In sharks, it probably does not have a respiratory function, because blood supplying the pseudobranch comes from an adjacent fully functional gill and is already pharyngeal slit.

Holocephalans (ratfishes, *e.g.*, *Chimaera* or *Hydrolagus*) lack spiracles, altogether. They also differ from other elasmobranchs in having a single extensive flap of skin or operculum, covering all the branchial arches (which are 4 pairs in number) rather than individual flap valves over each pharyngeal slit.

3. Gills in Bony Fishes

The gill apparatus of cartilaginous and bony fishes exhibits the same basic pattern. A series of pharyngeal arches support holobranchs, and a stream of respiratory water flows over the demibranchs en route from the pharyngeal cavity to the exterior (Fig. 5.8B). The chief differences

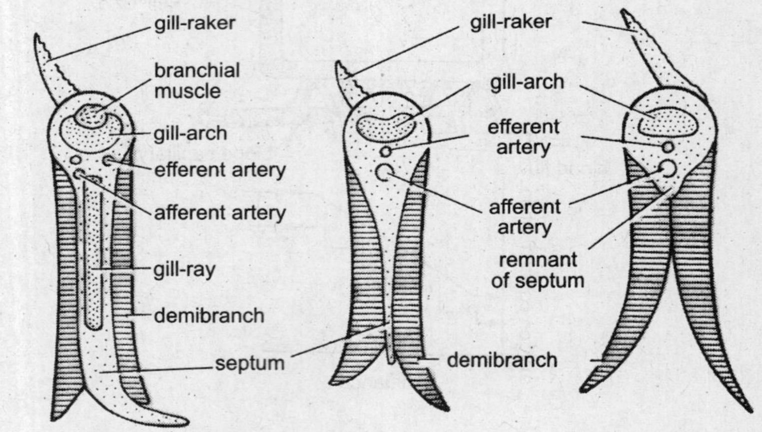

Fig. 5.13. Types of gills in fishes in section. Note the septum.

lie in the presence of an operculum and an opercular chamber and in the shorter length of the interbranchial septa in bony fishes (Fig. 5.13). The interbranchial septa are very short, with the result that the demibranchs of each holobranch are unattached distally, allowing water freer access to their capillaries (Fig. 5.14). The **operculum** is a bony flap that commences at the hyoid arch and extends caudal, covering the gill chambers on each side. The operculum of osteichthyans is bony or cartilaginous. It provides a protective cover over the branchial arches and gills they support. In addition, the operculum is part of the dual pump used to ventilate the gills.

In cross section, each gill is V-shaped and composed of primary lamellae (gill filaments) that are subdivided into secondary lamellae and supported on a branchial arch. Tiny adductor muscles cross between filaments to control the arrangement of adjacent gills that govern the flow of water across the secondary lamellae (Fig. 5.15A). As in most other fish gills, the blood in the secondary lamellae flows in one direction and water flows in the opposite direction to establish a countercurrent exchange (Fig. 5.15B; Box 5.2).

4. Aerial Respiration in Bony Fishes

Craniates arose in an aqueous environment, and water was the earliest source of oxygen. During the Devonian, however, if not earlier, air became a source of oxygen for many fishes. One compelling factor may have been that the water at that time was warm and swampy and, as such, low in dissolved oxygen. Air, on the contrary, contains 20 times the oxygen that oxygen-saturated water can hold. Little wonder, then, that many Devonian fishes obtained some or all of their oxygen from above the surface of water. Today's air-breathing fishes including dipnoans, basal actinopterygians, gars, *Amia*, and many teleosts. The fishes snatch bubbles of air from just above the water surface and the air then comes in contact with the oropharyngeal lining or, as we will see shortly, with the lining of **swim bladders** in those few fishes in which these swim bladders serve as lungs. A few teleosts swallow the bubble and extract oxygen in the stomach or intestine. Although oxygen is acquired from the atmosphere most of the excess carbon dioxide is eliminated through the gills into the water.

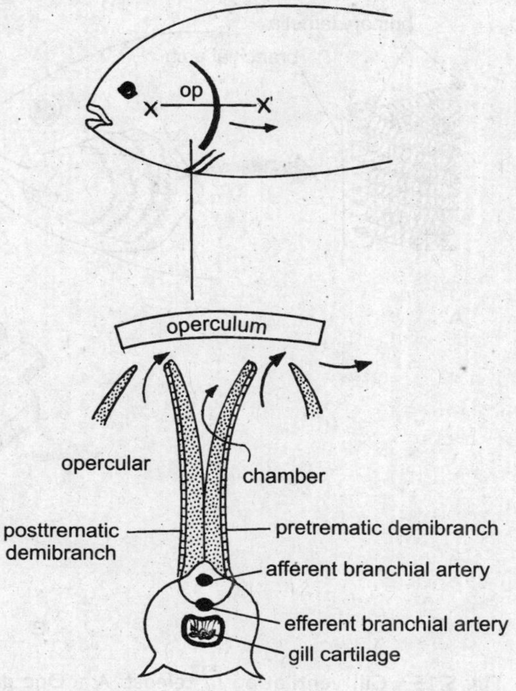

Fig. 5.14. Operculum (op) and holobranch of a teleost. The holobranch, bottom, is seen in frontal section (X – X′). Arrows indicate direction of efferent water flow (after Kent and Carr 2001).

Fishes that ventilate a **gas bladder** do so by gulping and forcing fresh air through the pneumatic duct. Usually, a fish expels spent air as it approaches the water's surface, capture and swallows a new gulp of fresh air, and descends again. In the jeju, a freshwater fish of the Amazonian region, the anterior muscular compartment of the gas bladder is connected to a posterior compartment through a sphincter. As the jeju breaks the water surface, the fresh air it gulped into the buccal cavity is forced along the pneumatic duct and preferentially enters the anterior chamber of the gas bladder (Fig. 5.16A, B). The sphincter closes, and spent air in the posterior chamber exits. Finally the sphincter opens, and the muscular walls of the anterior chamber contract, forcing the new air into the vascularized posterior chamber (Fig. 5.16C,D).

5. Larval Gills

The temporary gills of larva of fishes and amphibians are called **larval gills**. Larval gills are of three kinds:

(i) External gills. These are outgrowths from the external surface of one or more gill arches.

(ii) Filamentous extensions. These are filament extensions of external gills that project through gill slits to the exterior.

(iii) Internal gills. These are gills of late anuran tadpoles, which are hidden behind the larval operculum.

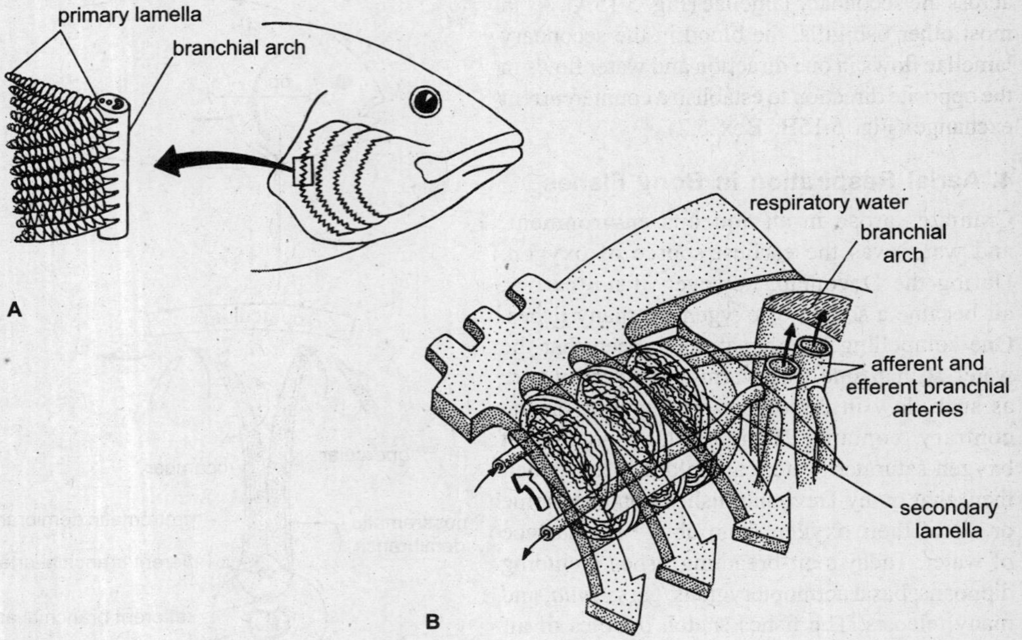

Fig. 5.15. Gill ventilation in teleost. A – One gill bar is removed showing the stack of gill lamellae. B – Water flow is directed across the secondary lamellae opposite to that of blood flowing within each secondary lamella, establishing a countercurrent exchange between them (after Kardong 2006).

External gills usually develop before gills slits open and before any opercular fold has started to develop. They can be waved about by branchial muscles and can often be retracted. They develop in the embryonic or larval stages of most dipnoans, amphibians and in a few ray-finned fishes such as sturgeons and lung fish, *Polypterus* which has one pair of external gills. (Fig. 5.17). Although external gills are larval structures, there are perennibranchiate urodeles (*Amphiuma* Fig. 5.17) and perennibranchiate African lungfish, *Protopterus*. *Protopterus* has four pairs of larval gills and retains three pairs in a reduced state throughout life (Fig. 5.19). *Lepidosiren* (Dipnoi) also has four pairs of filamentous larval external gills attached to the outer edges of branchial arches (Fig. 5.20).

External gills develop in anuran tadpoles on pharyngeal arches III to V. Later, when pharyngeal pouches II to V rupture to the exterior, the pouch linings become folded to form a set of external gills (Fig. 5.21). Thereupon a fleshy, operculum grows backward from the hyoid arch over the gill region, enclosing external gills and internal gills in an opercular chamber that retains only a single excurrent pore, this being located at the posterior edge of the left operculum. Thereafter, the external gills gradually atrophy, and the internal gills function until metamorphosis. At that time, the internal gills and operculum are absorbed along with the larval tail. All these structures are, of course, very small in a metamorphosing anuran.

Elasmobranchs develop either in a yolk-laden egg case attached to underwater vegetation, or they develop in the maternal uterus. During early development, **filamentous outgrowths** of the developing internal gills project through external gill slits and into the fluids within the egg case or the uterine chamber, where they serve temporarily as respiratory organs (Fig. 5.18C and D).

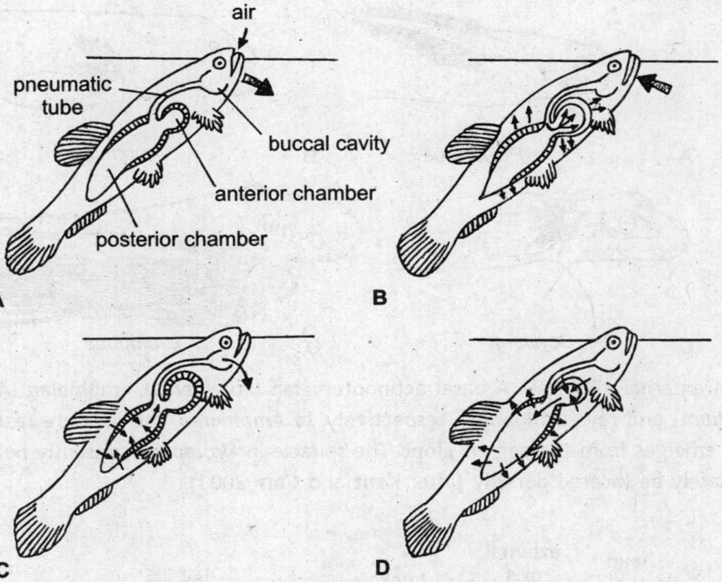

Fig. 5.16. Air-breathing fishes. Most air-breathing fishes use a buccal pump to fill their air bladders or lungs, which are able to separate spent and incoming air during ventilation. The mouth breaks the surface of water (A) so that air drawn in along the pneumatic tube preferentially enters the anterior air chamber (B). Spent air in the posterior chamber is forced out through the pneumatic tube and exits under the operculum (C). The sphincter between the anterior and posterior chambers opens, allowing air to replenish the posterior chamber as well (D) (after Kardong 2006).

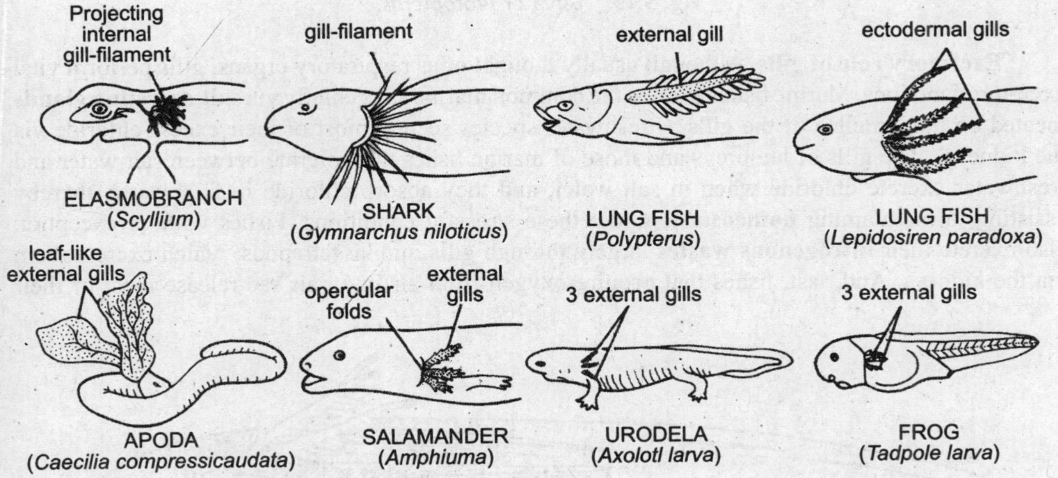

Fig. 5.17. Larval external gills of fishes and amphibians.

In viviparous species, they also absorb nutrients from the uterine fluid. A few viviparous chondrostean and teleost larvae utilize filamentous external gills in a similar manner. Absorption of nutrients from uterine or other maternal fluids is known as **histotrophic nutrition**.

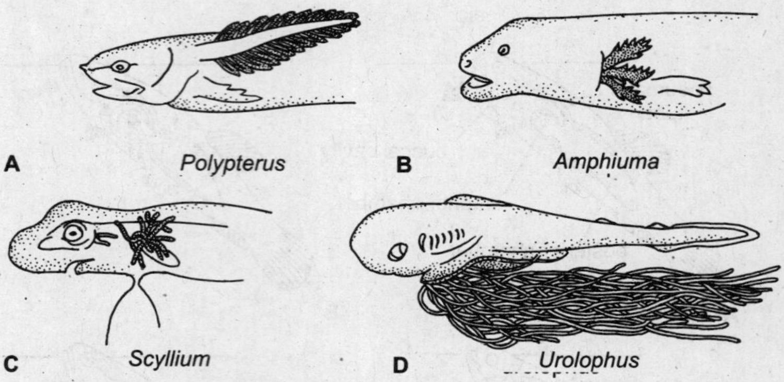

Fig. 5.18. Larval external gills. A–D. A basal actinopterygian (*Polypterus*), amphibian (*Amphiuma*), shark (*Scyllium*) and ray (*Urolophus*), respectively. In *Amphiuma*, the gills are resorbed before the larva emerges from the egg envelope. The spiracle in *Urolophus*, presently behind the eye, will ultimately be located dorsally (after **Kent** and **Carr** 2001).

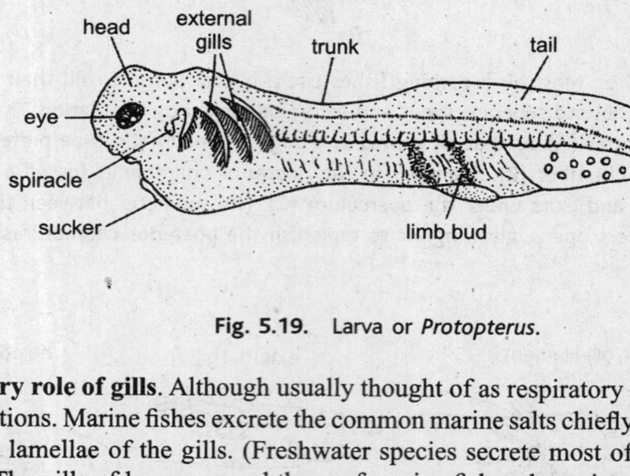

Fig. 5.19. Larva or *Protopterus*.

Excretory role of gills. Although usually thought of as respiratory organs, gills perform vital excretory functions. Marine fishes excrete the common marine salts chiefly via **salt-secreting glands** located on the lamellae of the gills. (Freshwater species secrete most of their excess chloride via the kidneys). The gills of lampreys and those of marine fishes that migrate between salt water and freshwater excrete chloride when in salt water, and they absorb chloride in freshwater, thereby assisting in maintaining **homeostasis** under these stressful conditions. Fishes without exception also excrete their **nitrogenous wastes** largely through gills, unlike tetrapods, which excrete them via the kidneys. And, last, fishes that acquire oxygen from air in an air sac release most of their

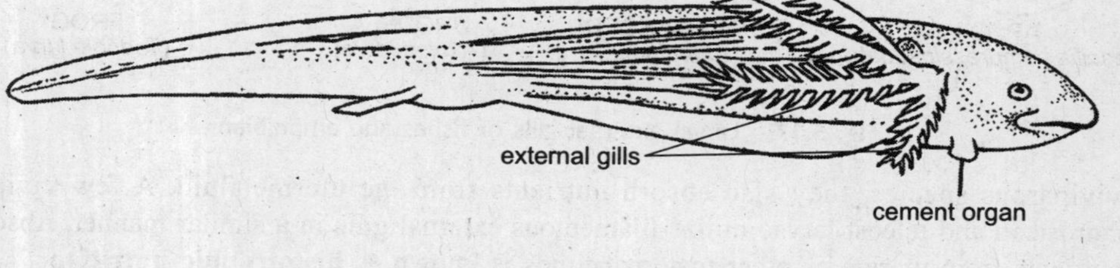

Fig. 5.20. Larva of *Lepidosiren*.

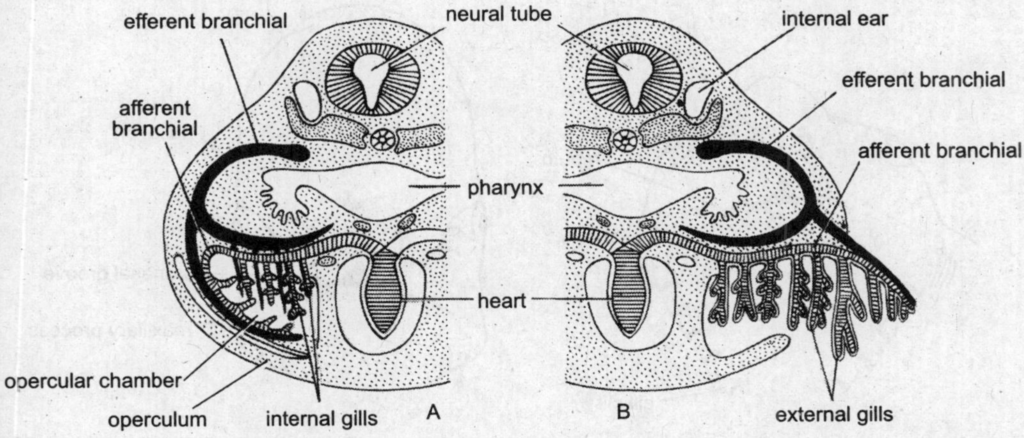

Fig. 5.21. T.S. of head of frog tadpole showing blood supply of gills. A – Through internal gills; B – Through external gills.

carbon dioxide into the water flowing over the gills. Carbon dioxide dissolves more readily in water than in air; hence, the partial pressure of CO_2 in the water passing over the gills is low enough to assure uninterrupted release of CO_2 from the circulation.

6. NARES AND NASAL CANALS

The paired **nostrils** or **external nares** of cartilaginous and ray-finned fishes (Actinopterygii) open directly into blind olfactory sacs that contain the sensory epithelia for smell. Each nostril is divided into a forward-directed **incurrent aperture** through which water is driven as a fish swims forward, and a laterally or ventrally directed **excurrent aperture** through which water exit after having bathed the olfactory epithelium. In lobe-finned fishes (*e.g.*, *Latimeria*), the nostrils are connected with the oropharyngeal cavity by a pair of nasal canals that conduct respiratory water. Their openings into the oropharyngeal cavity are **internal nares**. As we learned earlier, the nasal canals of agnathans are unpaired, and only in myxinoids do they extend to the pharynx. In no living fish, including those that obtain their oxygen from air, are nostrils used for breathing. They are part of a sensory system for monitoring one constituent of the environment — chemicals in solution in the surrounding water.

If the rhipidistian ancestors of labyrinthodonts were not using their nostrils for breathing and whether they were is not known. Labyrinthodonts were opportunists because they commenced to use notrils for drawing air into their oral cavity. With subsequent development of a secondary palate in amniotes, the internal nares opened farther caudad, and, as we have seen, the longer the secondary palate, the farther caudad the nares are located. In mammals, internal nares open into the nasopharynx above the soft palate (see Fig. 4.12; cat).

Nasal canals are also called **choanae**. They arise from paired **nasal pits** and **oronasal grooves**, the dorsolateral walls of which roll together to form a tube (Fig. 5.22B). Although elasmobranchs lack nasal canals, oronasal grooves form and remain throughout life (Fig. 5.22A). In mammals, the **olfactory epithelium** is restricted to the upper chambers of the nasal passageways. The lower chambers have a ciliated glandular **nasal epithelium**. Hairs at the entrance to mammalian nasal canals trap insects and coarse particulate matter in the incoming air; venous plexuses in the submucosa overlying the turbinal bones heat cold air; and air sinuses that open into the upper reaches of the nasal passageways of mammals serve partly as resonating chambers for vocalization.

A fleshy, partly cartilaginous, proboscis (**nose**) develops in some mammals and carries the nares to positions characteristic of the species. Compare the location of the nostrils of a catarrhine

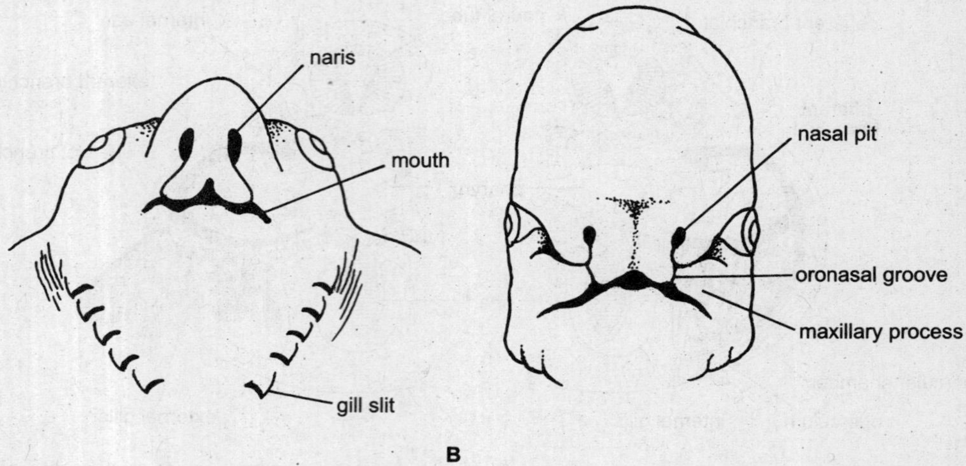

Fig. 5.22. Oronasal relationships in (A), an adult skate, and (B), a 6-week (12 mm) human fetus (after Kent and Carr 2001).

(orangutan, anthropoid) with that of an elephant. Whales have no nose. The nostrils are far back on the top of the head and are known as **blowholes**. The blowholes are equipped with valves that close when the whales dives. In some whales, the two nares unite during development to become a median blowhole.

II. Lungs

Vertebrate lungs are desinged for air breathing. Lungs are elastic bags that lie within the body. Their volume expands when air is inhaled and decreases when air is exhaled. Embryologically, lungs arise as endodermal outpocketings from the pharynx. In primitive fishes and most tetrapods, the lungs of adults are usually paired. They lie ventral to the digestive tract and are connected to the outside environment through the **trachea**. Entrance into the trachea is gained through the **glottis**, which is guarded by tiny sets of muscles that open and close it. Usually the trachea branches into two **bronchi**, one to each lung (Fig. 5.23). In some species, each bronchus branches into successive smaller **bronchioles** that eventually supply air to the respiratory surfaces within the lung. In tetrapods with slender bodies, one lung may be reduced in size (Fig. 5.24); and in some amphisbaenids and most advanced snakes, only a single lung is present. (Fig. 5.24).

The trachea, bronchi, and bronchioles can hold a significant volume of air. Although exhalation forces most of the spent air from the lungs, some remains in these passageways. Upon inhalation, this "spent air" is drawn back into the lungs before fresh air from outside reaches the lungs to mix with the used air. This volume of used air within the respiratory passageways is called the **dead space**. The total volume inhaled in a single breath is referred to as the **tidal volume**. In a chicken, the dead space may represent upto 34% of the total tidal volume. Normal tidal volume of a human at rest is about 500 ml. Because the dead space is about 150 ml (30%), 350 ml (500–150 ml) of fresh air actually reaches the lungs.

III. Cutaneous Respiratory Organs

Although lungs and gills are the primary respiratory organs, the skin can supplement breathing. Respiration through the skin of animals is called **cutaneous respiration** and it can take place in air, in water, or in both. In the European eel and plaice (an edible flatfish with orange spots), oxygen uptake through the skin may account for up to 30% of total gas exchange (Fig. 5.25).

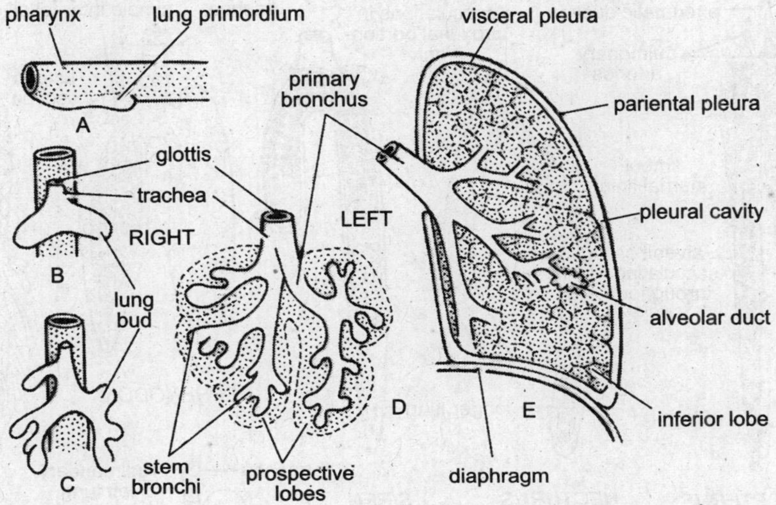

Fig. 5.23. Development of a vertebrate lung in embryo. A – Lung primordium; B – and C – Lung buds; D – Embryonic lung; D – Lung at birth.

Amphibians rely heavily on cutaneous respiration, often developing accessory skin structures to increase the surface area available for gas exchange. In fact, in salamanders of the family Plethodontidae, adults lack lungs and gills and depend entirely on cutaneous respiration to meet their metabolic needs. Like most mammals, humans respire very little cutaneously, although our skin is permeable to some chemicals applied topically (spread on the surface). In fact, many medicinal ointments are absorbed through the skin. Bats take advantage of cutaneous respiration across their well-vascularized wing membranes to eliminate as much as 12% of their total carbon dioxide waste, but take up only 1% or 2% of their total oxygen requirement through this cutaneous route (Fig. 5.25). Feathers and poorly vascularized skin of birds preclude cutaneous respiration. Similarly, in reptiles, the surface coverings of scales limit cutaneous respiration. However, in areas between scales (at the hinges of scales) and in areas with reduced scales (*e.g.*, around the cloaca), the skin is heavily vascularized to allow some cutaneous respiration. Sea snakes can supplement up to 30% of their oxygen intake via cutaneous respiration across the skin on their sides and back. Many turtles pass the cold winter in hibernation safely at the bottom of ponds where the limited respiration around their cloaca is sufficient to meet their reduced metabolic needs.

The newly hatched larva of the teleost fish *Monopterus albus*, an inhabitant of south-east Asia, uses predominantly cutaneous respiration during its early life. At hatching, the large and heavily vascularized pectoral fins beat in such fashion as to drive a stream of water backward across the surface of the larva and its yolk sac. Blood in superficial skin vessels course forward. This establishes a countercurrent exchange between water and blood to increase the efficiency of cutaneous respiration in this larva (Fig. 5.26A). Such a respiratory organ allows the larva to inhabit the thin layer of surface water into which nearby oxygen from the air has dissolved. Similarly, in many amphibians, increased surface area allows for increased cutaneous gas exchange (Fig. 5.26B, C).

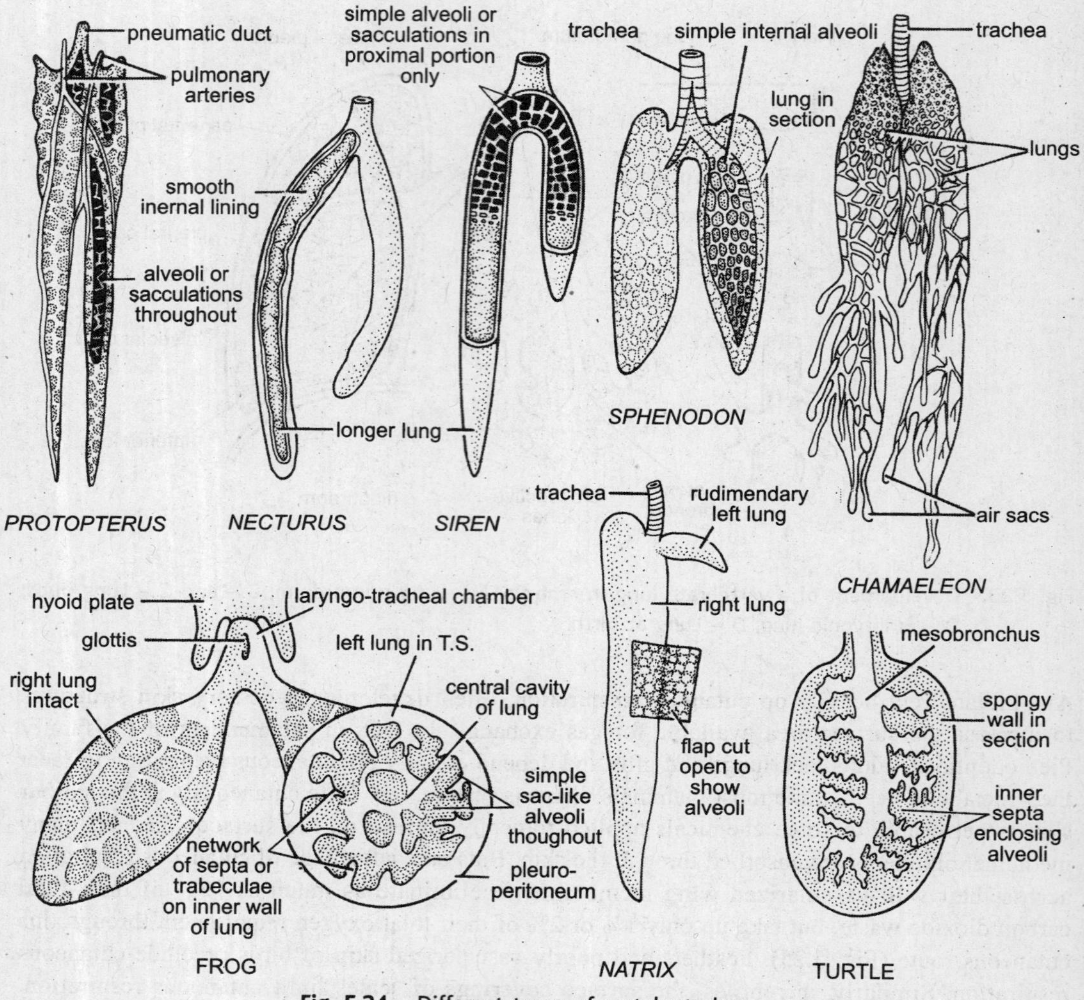

Fig. 5.24. Different types of vertebrate lungs.

5.3 ACCESSORY RESPIRATORY ORGANS OF FISHES

Majority of the teleostean fishes have well developed gills for respiration which are very efficient organs for gaseous exchange in the aquatic environment. But some tropical fishes have evolved special structures called the additional or **accessory respiratory organs**. These organs supplement or replace gill respiration and enable the fishes to live in oxygen deficient water to aestivate in dry summer and to take short excursions on land. Oxygen deficient water occurs in the swamps and muddy ponds infested with weeds. It has been shown that the absorption of oxygen in the accessory breathing organs is much greater than the excretion of carbon dioxide. Most of this carbon dioxide is ultimately excreted in the gills.

Generally, some existing respiratory and other organs of fishes have become highly vascular and modified for aerial respiratory organs. The organs so modified are skin, gills, branchial cavity, air bladder or the intestine. However, in a few species of fishes, new structures have developed for accessory respiration and these are called **neo-morphic organs**. Such accessory organs are found in fish species living in shallow, stagnant water and swamps where deoxygenated water

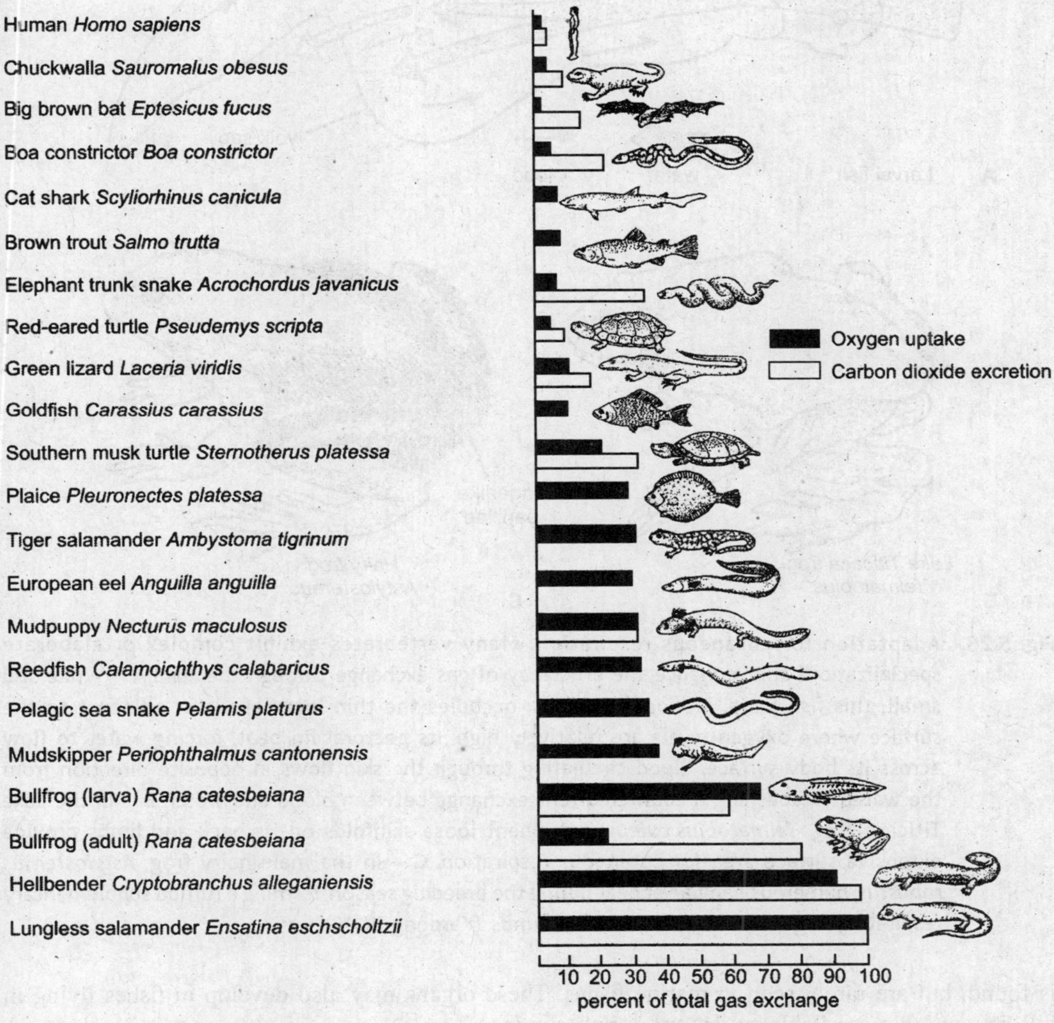

Fig. 5.25. Cutaneous respiration among vertebrates. Most amphibians depend largely on cutaneous respiration to meet their metabolic needs, and some, such as the lungless family of salamanders (plethodontids), use it exclusively. Other vertebrates also are known to supplement gills or lungs with cutanoeous respiration. Gas exchange through the skin involves oxygen uptake from, and release of carbon dioxide into, the environment, but these exchanges are not necessarily of equal magnitude. For example, loss of carbon dioxide through the wing membranes of bats accounts for about 12% of the total gas exchange, but oxygen uptake is considerably less. Cutaneous excretion of carbon dioxide (gray bars) and uptake of oxygen (black bars) are indicated as the percent of total gas exchange (after Kardong 2006).

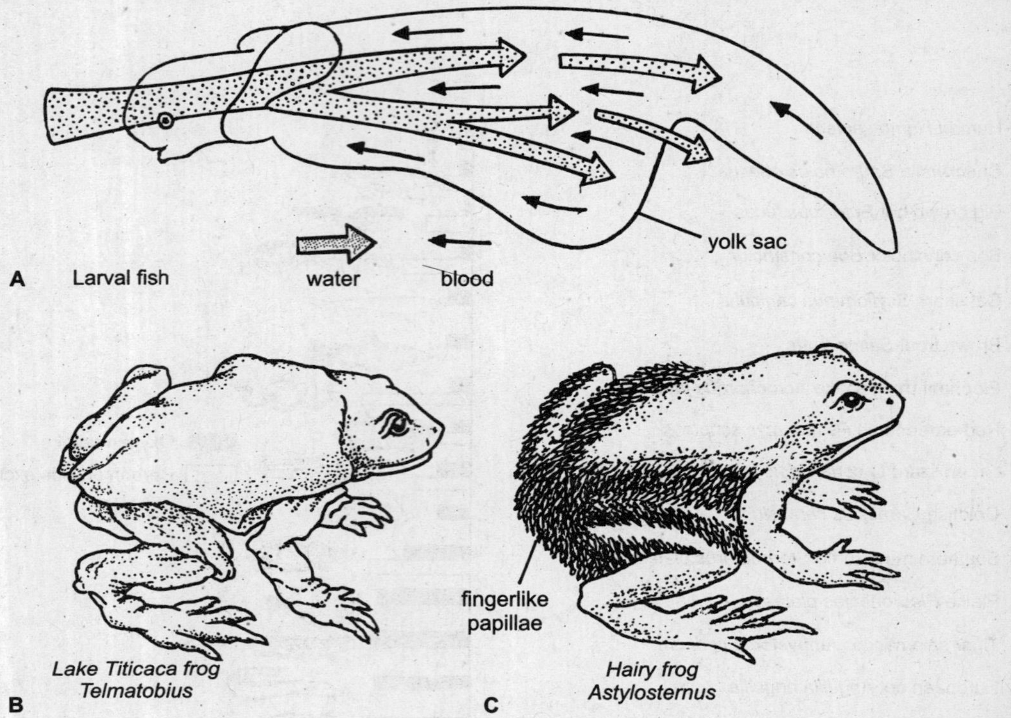

Fig. 5.26. Adaptation for cutaneous respiration. Many vertebrates exhibit complex or elaborate specializations that enhance the efficiency of gas exchange through the skin. A – While still small, this fish larva, *Monopterus albus*, occupies the thin layer of water adjacent to the surface where oxygen levels are relatively high. Its pectoral fin beat, forcing water to flow across its body surface. Blood circulating through the skin flows in opposite direction from the water, establishing a countercurrent exchange between blood and water. B – In the Lake Titicaca frog, *Telmatobius culeus*, prominent loose skinfolds on its back and limbs provide extensive surface area for cutaneous respiration. C – In the male hairy frog, *Astylosternus robustus*, numerous papillae appear during the breeding season, forming a ruffled supplementary respiratory organ on its sides and hindlimbs (Kardong 2006).

is found, but are rarely seen in marine fishes. These organs may also develop in fishes living in hill streams that are liable to dry up during summer.

1. Skin

In several species of fishes, the skin becomes highly vascular and is kept moist by mucus, so that exchange of gases takes place between blood and air.

(i) *Periophthalmus*, the mudskipper or goggle-eyed fish of Indian Ocean often skips mud and walk about through damp areas round the roots of mangrove trees. In mudskippers, the gill covers can completely seal the gill cavities and a small amount of sea water retained inside enable the fish to remain outside water for a long time. These fish love to bask in the sun at ebb tide (**Chhapgar** 2008). The old idea that the mudskippers use the vascular tail as the respiratory organ is not supported by recent ichthyologists (see **Jordan** and **Verma**, 2013).

(ii) *Anguilla anguilla*, the common eel, can crawl on damp grass for some distance, respiring through the skin during the venture. When on land, the eel keeps its small branchial aperture tightly closed so that water retained in the branchial chambers enables it to continue gill respiration also.

(iii) Median fin folds of many fishes are supplied with numerous blood vessels that help in cutaneous respiration.

(iv) ***Lepidosiren***, the South American lungfish, has delicate outgrowths of the pelvic fins. These outgrowths develop in males during breeding season and serve as cutaneous respiratory organs during parented care.

(v) The highly vascular opercular folds of *Sturgeons* and many cat fishes serve as accessory respiratory organs.

2. Bucco-pharyngeal epithelium

The buccal cavity of the South American fish, *Symbranchus,* has a highly vascular lining capable of respiration. When water runs short of oxygen, the fish rises to the surface, fills its buccal cavity with air, closes the mouth and uses the trapped air in respiration.

In most of the fishes, the epithelial lining of buccal cavity and pharynx is usually highly vascular and permeable to gases in water. It may remain simple or may develop folds, pleats or tongues projecting into the buccal cavity and pharynx to make it an efficient respiratory organ. However, in mudskippers (*Periophthalmus* and *Boleophthalmus*) the highly vascularised bucco-pharyngeal epithelium help in absorbing oxygen directly from the atmosphere.

3. Gut epithelium

In some species, a part of the alimentary canal (such as stomach or intestine) becomes modified for aerial respiration. Thus, in several fishes epithelial lining of gut becomes thin and vascular and modified to serve as a respiratory organ. Such a respiratory epithelium may be present just behind the stomach (*Misgurus fossilis*) or intestine (*Hoplosternum, Lepidocephalus guntea, Gobitus* (giant loach of Europe) or rectum (callichthyes (catfish), *Hypostomus* and Doras). Fresh air is drawn through mouth or anus and after gaseous exchange the air is voided through the anus.

4. Pharyngeal diverticula

In *Channa* (= *Ophiocephalus*) spp. suprabranchial cavities are formed in the roof of the pharynx (Fig. 5.27). These cavities are lined with highly vascular respiratory membrane raised into folds, pleats and tongues, and contains respiratory islets. The suprabranchial cavities have alveoli also, and open antero-ventrally into the opercular cavity. All the gill arches bear gill filaments but they are reduced in size and those on the fourth arch are greatly reduced.

In *Channa,* processes of cartilage develop from the parasphenoid, hyomandibular and the epibranchial of the first arch and project into the supra-branchial cavity. These projections are covered by a thin, highly vascular respiratory membrane. The epibranchial of the first gill arch becomes flattened and the respiratory membrane covering it becomes folded and studded with a large number of nodules or papillae forming a dendritic labyrinthine organ. Each of these nodules is believed to be a modified gill filament. In *Channa* the roof of the buccal cavity is also covered by the vascular respiratory epithelium (or membrane), which may be over the tongue also. Thus, the entire buccopharyngeal cavity along with the supra-branchial chamber serves for aerial respiration and is filled with air for this purpose.

In cuchia eel, *Amphipnous cuchia* the accessory respiratory organs consists of a pair of vascular sac-like diverticulae from the pharynx above the gills at both lateral sides of the head. These sacs develop as diverticulae of the pharynx and are lined with highly vascular epithelium raised into folds. The epithelium consists of vascular areas called 'islets' bearing

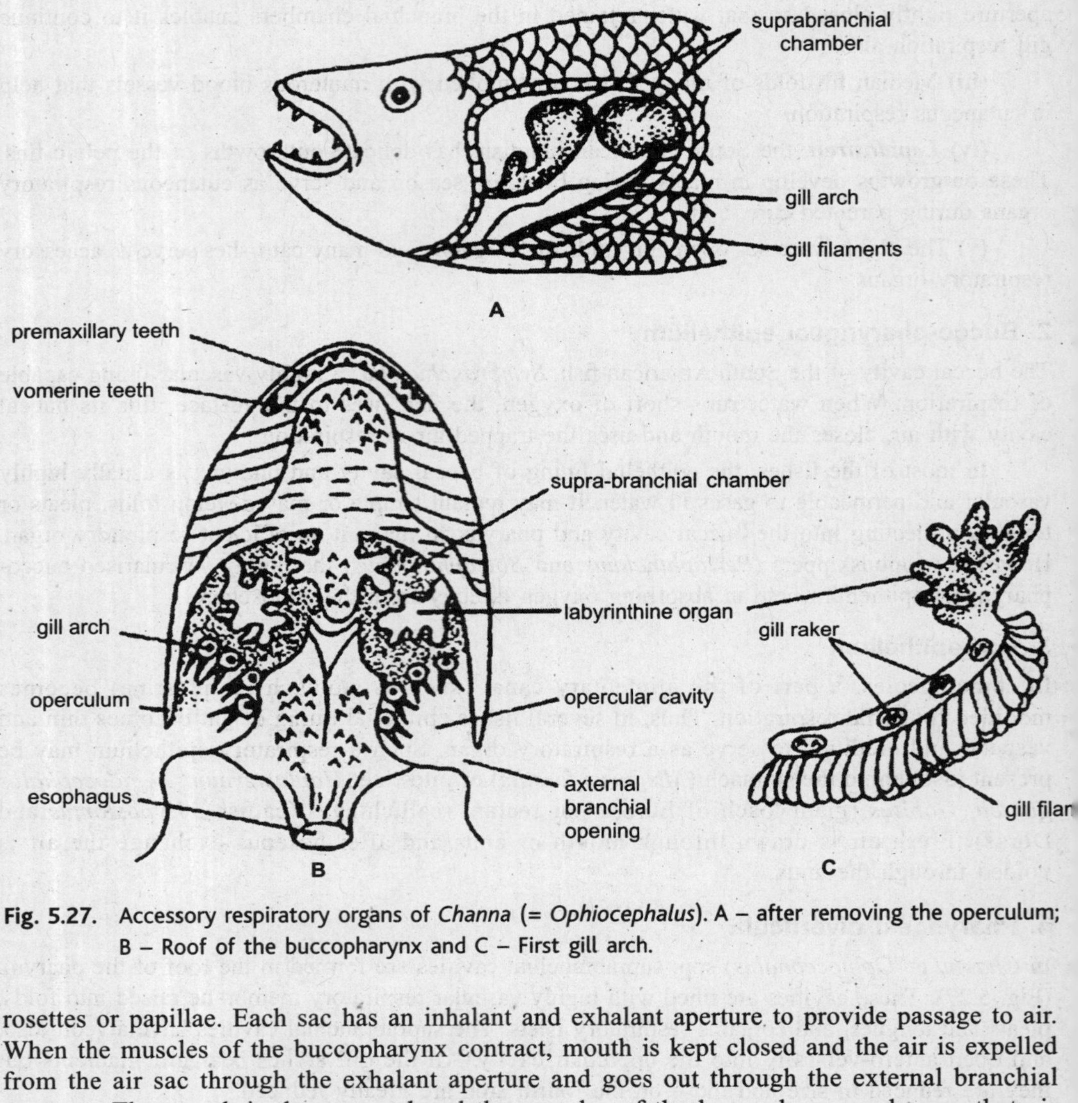

Fig. 5.27. Accessory respiratory organs of *Channa* (= *Ophiocephalus*). A – after removing the operculum; B – Roof of the buccopharynx and C – First gill arch.

rosettes or papillae. Each sac has an inhalant and exhalant aperture to provide passage to air. When the muscles of the buccopharynx contract, mouth is kept closed and the air is expelled from the air sac through the exhalant aperture and goes out through the external branchial aperture. The mouth is then opened and the muscles of the buccopharynx relax, so that air enters to fill the empty space.

In cuchia eel, the gills are reduced and only the second gill arch bears filaments, while the third arch has a fleshy vascular membrane.

In *Periophthalmus* also, a small, shallow pharyngeal diverticulum lined with respiratory epithelium (vascular epithelium) is present on each side of the roof of the pharynx.

5. Opercular chamber modified for aerial respiration

In *Periophthalmus* and *Boleophthalmus* the opercular chamber serves for aerial respiration. For this purpose the opercular cavity bulges and to form two little balloon-like structures in the hinder region of the head and inhaled air is passed through the gill slits and is stored for sometime in the cavity of balloons. The membrane lining the opercular cavity is thin and highly vascular. After exchange of gases, the air is passed out through the external branchial aperture.

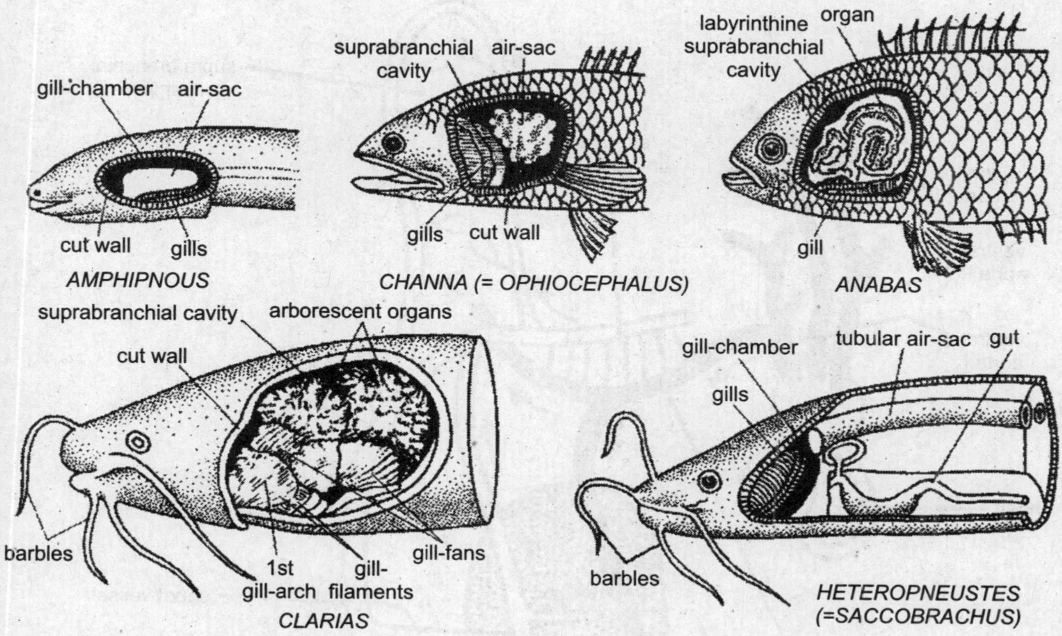

Fig. 5.28. Head of some air-breathing teleost fishes dissected to show the accessory respiratory organs.

6. Branchial diverticula

The outgrowths from gill-chambers form more complicated aerial respiratory organs than the simpler pharyngeal outgrowths in other fishes. Such air breathing organs are present in *Heteropneustes, Clarias, Anabas* etc.

(i) *Heteropneustes fossilis* (- *Saccobrachus*). The accessory air breathing organs of Indian catfish, *Heteropneustes fossilis,* are highly specialised and consists of (a) two air sacs, (b) fans and (c) respiratory membrane.

(a) Air sacs. A pair of simple sac-like structures extend posteriorly, one on each side, from the suprabranchial chamber of the opercular cavity. These sacs are thin walled, tubular structures having highly vascular walls and lie embedded between the myotomes of the body, extending upto the middle of the caudal region (Fig. 5.28). Each air sac receives blood from the fourth **afferent branchial vessel** of its own side, through a thick branch called the **afferent air sac vessel** (Fig. 5.29). This branch gives off several capillaries to supply blood to the respiratory islets of the sacs. The blood is collected from the islets by efferent capillaries which open into an **efferent air sac vessel**, which in its turn carries blood to the fourth efferent branchial vessel of its own side **(Dutta Munshi** 1993).

(b) Fans. The gills of the Indian cat fish bear **fans** (Fig. 5.31.), Gill filaments of all the gills are reduced in size and those of the upper limb of the gill arch are fused to form fans, one on each gill arch.

(c) Respiratory membrane. It lines the air sacs of Indian cat fish and is composed of vascular and non-vascular areas, and is thrown into folds and ridges to increase its surface. The vascular areas consist of a number of islets having lamellae arranged biserially are the sites of gaseous exchange. The nonvascular areas consist of lanes and possess mucous glands. The air sacs and the fans are the result of fusion and modification of gill filaments and reduction of the lamellae. During respiration,

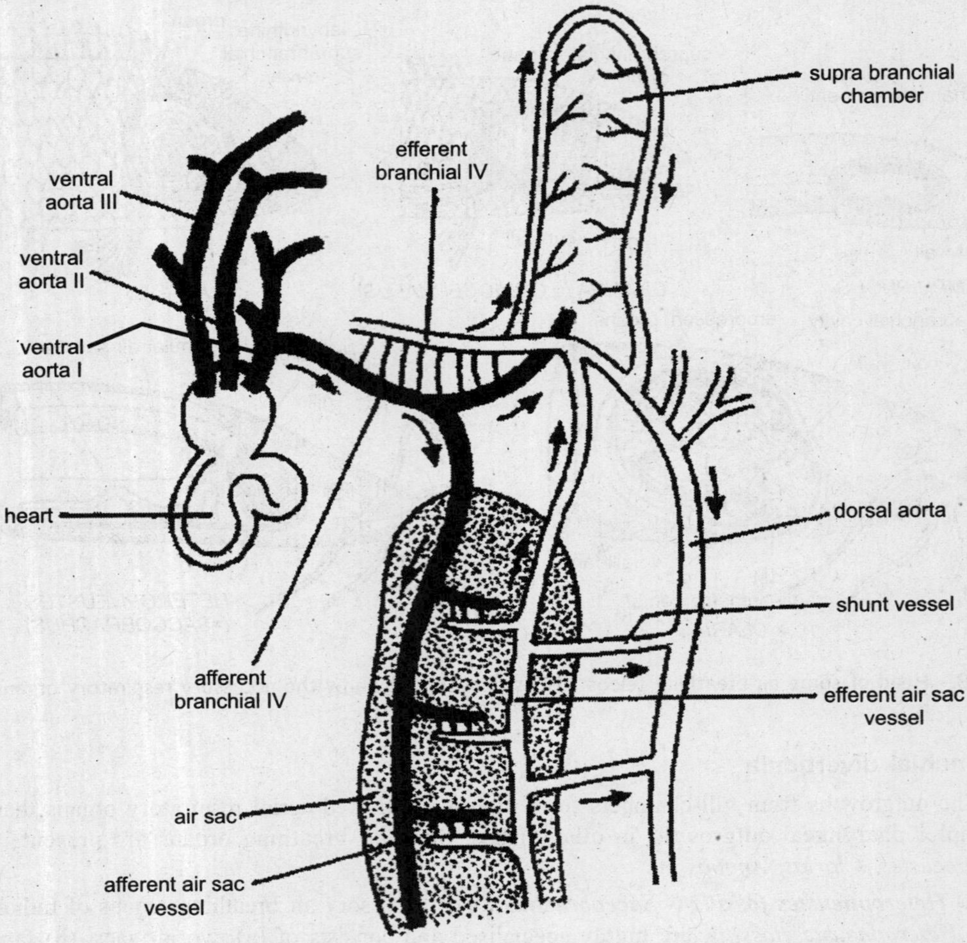

Fig. 5.29. Blood supply of the air sac and suprabranchial chamber in *Heteropneustes fossilis*.

H. fossilis inhales air which enters the suprabranchial chamber through the inhalant aperture and is passed into the tubular sac. After the exchange of gases, used air is expelled through the external branchial opening, and the fans help in this process.

(ii) *Anabas*. Air breathing organs of Indian climbing perch, *Anabas testudineus,* consist of a specious air chamber and a labyrinthine organ on each side (Fig. 5.28). The air chamber lies between the first gill arch and the hyomandibular and opens into the bucco-pharyngeal cavity as well as the opercular cavity. The respiratory epithelium which lines the air chamber is highly vascular and may be folded to increase the respiratory surface. A well developed **labyrinthine organ** is present on the first gill arch. This organ develops from the first epibranchial and consists of three concentric bony plates, the first plate is fused with the respiratory epithelium of the air chamber and divides it incompletely into two. The second and the third plates are greatly folded due to which the labyrinthine organ becomes a complex structure. The labyrinthine organ is also covered by a vascular respiratory epithelium and serves to increase the respiratory area for exchange of gases.

(iii) *Trichogaster*. Air breathing organs of *Trichogester* includes a suprabranchial chamber, a labyrinthine organ and the respiratory membrane (Fig. 5.30). As in *Anabas*, the

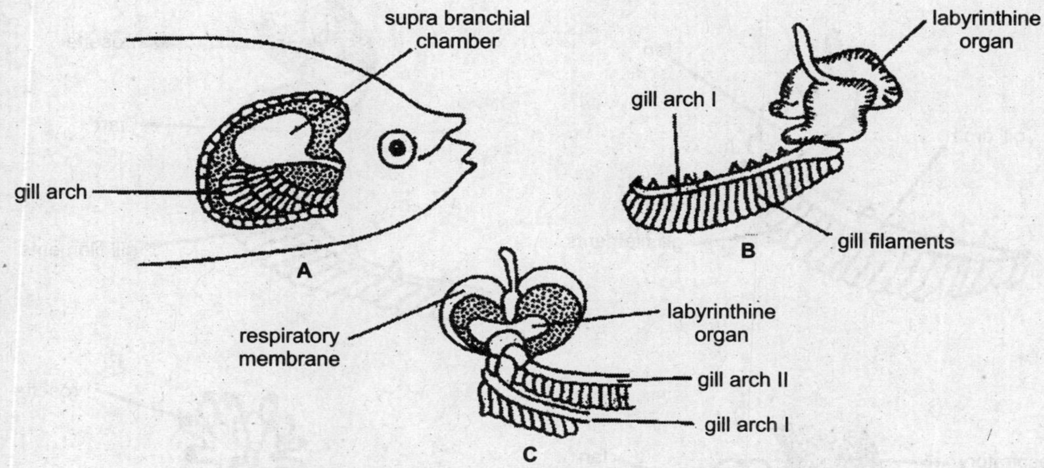

Fig. 5.30. *Trichogaster*: Air breathing organs.

suprabranchial chamber lies on either side above the gills, and opens into the pharynx by means of an inhalant aperture, and into the opercular chamber by means of an exhalant aperture. A labyrinthine organ has two leaf-like expansions and develops on the epibranchial of the first gill arch. It looks like a spiral organ and is simpler in organization than that of *Anabas*. The respiratory membrane covers the suprabranchial chamber and the labyrinthine organ and it consists of vascular and non-vescular areas. The vascular area contains a number of respiratory islets, which have numerous parallel blood capillaries, and are believed to be derived from the lamellae of the gill filaments.

(iv) *Clarias batrachus*. The air breathing organs of Indian cat fish *Clarias batrachus* are highly evolved and complex. These include following types of organs:

(a) a suprabranchial chamber on each side of head;

(b) two air trees, rosettes, dendritic organs or absorescent organs on each side;

(c) the fans; and

(d) the respiratory membrane.

The suprabranchial chamber is situated above the gills. It is lined by a highly vascular respiratory membrane and is divided into two cup-like compartments. A rosette or arborescent organ is present in each compartment. The anterior rosette is smaller in size and is borne on the epibranchial of the second gill arch, while the posterior one is larger and is present on epibranchial of the fourth gill arch. Each rosette is a highly branched tree-like structure, each branch ending in a knob. Each knob has a core of cartilage and is covered over by a vascular epithelium having eight folds. Each knob is believed to represent eight shortened and fused gill filaments. In addition to the rosettes some of the gill filaments of each gill arch are fused to form a "fan", so that there are four fans, one on each gill arch. The remaining filaments are reduced in size.

The suprabranchial chamber has inhalant and exhalent apertures. The respiratory membrane lining the suprabranchial chamber and covering the rosette, is composed of vascular and non-vasular areas. The vascular areas have numerous islets which serve as the sites of gaseous exchange.

The fish (*C. batrachus*) frequently rises to the surface of water and gulps in air, which enters the suprabranchial chamber through the inhalant aperture which is a narrow slit between the second and third arch. The air entering the opercular cavity is directed into the suprabranchial

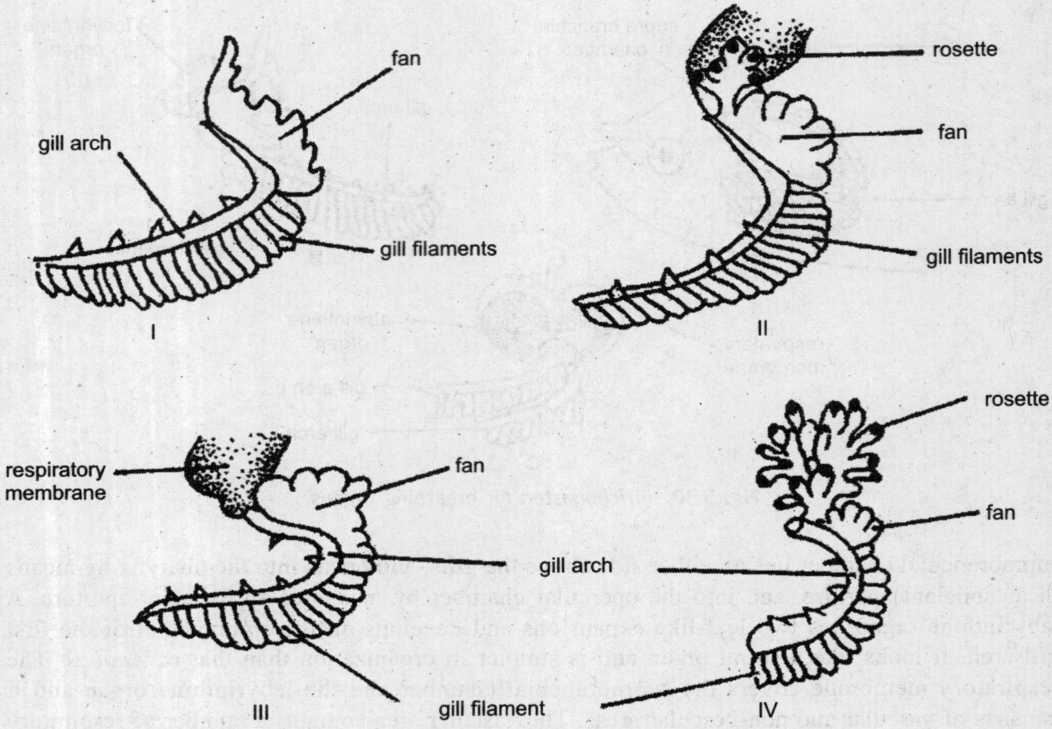

Fig. 5.31. *Clarias batrachus (Indian cat fish).* I, II, III and IV gill arch.

chamber by the action of the fans. The slit between the third and the fourth gill arches is used as the exhalant aperture, and the air, after the gaseous exchange is passed into the opercular cavity and then to the exterior.

7. Air Bladders

The air bladder is an outgrowth from the esophagus in bony fishes. It may retain or lose communication with the esophagus. The fishes which retain this communication are said to be **physostomous**. For example, teleost fishes of orders such as Cypriniformes, Anguilliformes and Esociformes are physostomous. The physostomous fishes frequently rise to the water surface and gulp air into air bladder where exchange of gases takes place. Lung fishes store air in the air bladder for use in respiration during aestivation in burrows.

Diurnal fluctuation in aerial respiration in fishes. Air breathing fishes exhibit fluctuation in their breathing pattern. These fishes are nocturnal in habit and are usually inactive restive during the day, and are able to meet their oxygen requirement through aquatic respiration by gills. During night these fishes become active and need more oxygen. Which is in short supply in water during night. Hence, there fishes frequently rise to the surface to gulp in air. According to **Dutta Munshi** and **Ghosh** (1994), the two abiotic ecological factors (*i.e.,* dissolved oxygen and free carbon dioxide) act as the **steering wheel** controlling the metabolic activities of the living biota in the swamps.

Breathing in Embryos of Cleidoic Eggs

In birds and most reptiles, the embryo is wrapped in extraembryonic membranes and enclosed in a shell. One of these membranes, the **choriollantois**, lies directly beneath the shell and acts as the respiratory organ. The porous shell allows oxygen to be picked up by the blood circulating within the chorioallantois and carbon dioxide to be eliminated from it. The chorioallantois sustains the respiratory needs of the chicken embryo for most of its time in the egg (Fig. 5.32A,B). About six hours before allows its lungs to fill for the first time and begin to participate, along with the chorioallantois, in air breathing. When the chick breaks through the outer shell several hours later, its lungs breathe atmospheric air directly and the chorioallantois quickly shuts down (Fig. 5.32C).

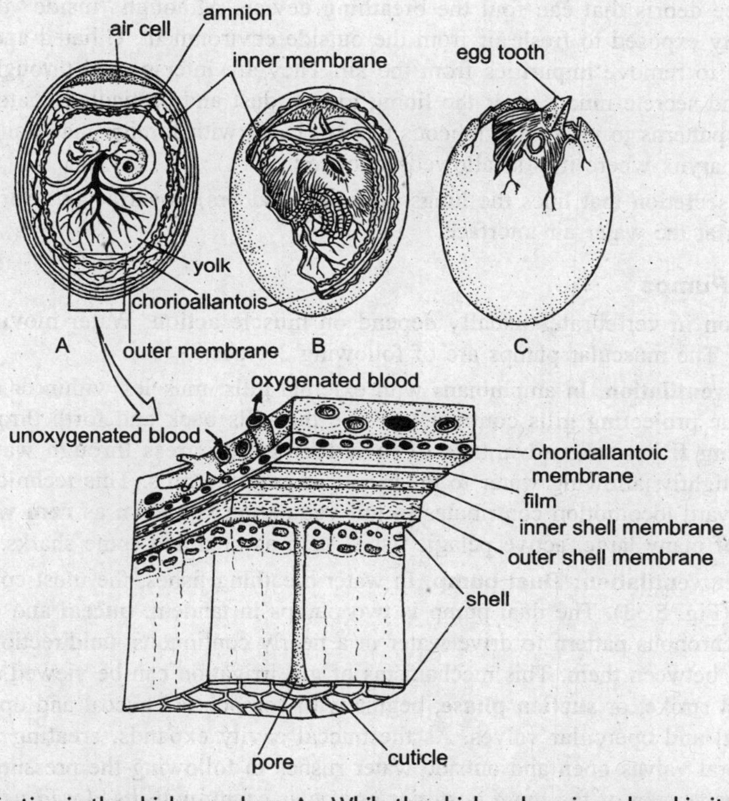

Fig. 5.32. Respiration in the chicken embryo. A – While the chick embryo is enclosed in its shell, it respires through this porous shell. The chorioallantois carries blood to the inner surface of the shell to exchange gases at this interface. The shell proper is made up of calcite crystals pierced by tiny pores. Inner and outer shell membranes separate the shell from the vascualrized chorioallantois. The chick embryo meet all its respiratory needs, up to day 19 of incubation, as air passes through the porous shell and exchanges gases with blood in the chorioallantois. B – On day 19, the embryo pokes its beak through the inner shell membrane into the air space between both membranes. Its lungs inflate and the chick breathes air in addition to continued respiration via the chorioallantois. C – Six hours later, the chick pecks through the shell proper, a process called **pipping**, to breathe atmospheric air directly. Therefore, chorioallantoic respiration declines and the chick further cracks the shell and soon steps out (after Kardong 2006).

5.4 VENTILATORY MECHANISMS

What are the exchange organs-lungs, gills, skin or accessory structures—water or air moves actively across the respiratory surfaces to increase the rate of diffusion. Ventilation mechanisms are of two main types: ciliary and muscular pumps.

1. Ciliary Pump

If an animal is small (*e.g.*, protochordate) and its metabolic demands are modest, microscopic cilia are sufficient to move across the respiratory surfaces and support the exchange of gases between tissues and the environment. Cilia line the routes along which the water current flows. Their coordinated sweeps drive water, a relatively viscous medium through the pharynx and across the gills. Cilia, like oars, are ineffective against a relatively thin medium such as air.

In large vertebrates, the respiratory channels often retain cilia, but they are involved in clearing surface debris that can foul the breathing device. Although "inside" the body, lungs are continuously exposed to fresh air from the outside environment. Ciliated and mucous cells are specialized to remove impurities from the air. They are interspersed throughout the lining of the lungs and secrete mucus over the lining to trap dust and particulate material. Cilia beat in coordinated patterns to move this mucous blanket laden with foreign material up the airways and into the pharynx where it is swallowed unnoticed.

Another secretion that lines the lungs (and gas bladders) is surfactant. Surfactant reduces surface tension at the water air interface.

2. Muscular Pumps

Ventilation in vertebrates usually depend on muscle action. Water moving across gills ventilate them. The muscular pumps are of following 3 types:

(i) Ram ventilation. In amphibians with external gills, muscles within or associated with the bases of the projecting gills contract to wave the gills back and forth through the water. Some swimming fishes take advantage of their forward progress through water. They open their mouths slightly, allowing water to enter and irrigate the gills. This technique by which a fish's own forward locomotion contributes to gill ventilation is known as **ram ventilation**. It is characteristic of many large, active pelagic fishes such as tuna and some sharks.

(ii) Water ventilation: Dual pump. In water-breathing fishes, the most common pump is a **dual pump** (Fig. 5.33). The dual pump is two pumps in tandem, **buccal** and **opercular**, that work in a synchronous pattern to drive water in a nearly continuous, unidirectional flow across the gill curtain between them. This mechanisms of gill irrigation can be viewed as a two-stroke pump. The first stroke, or **suction phase**, begins with compressed buccal and opercular cavities and closed oral and opercular valves. As the buccal cavity expands, creating a low intraoral pressure, the oral valves open and outside water rushes in following the pressure gradient. The simultaneous expansion of the more posterior opercular cavity with its closed valve also creates a pressure that is even lower than in adjoining buccal cavity. Consequently, water that first enters the buccal cavity is encouraged by the pressure differential to continue on across gill curtain and into the opercular cavity.

During the second stroke, or **force phase**, the oral valves close and the opercular valves open. Simultaneous muscle compression of the buccal and opercular cavities raises pressure in both, but because of the open opercular valve, pressure in the opercular cavity is slightly lower. Consequently, water flows from the buccal cavity across the gill curtain and exits via the open opercular valve. The timing of the suction and force phases, together with the pressure differential between them, results in a unidirectional, nearly continuous flow of new water across the gills.

(iii) Air ventilation: Buccal pump. Air breathing fishes and amphibians utilize a buccal pump to ventilate their lungs. The **buccal pump** (or pulse pump) employs the mouth cavity; which expands to fill with fresh air and then compresses to pump this air into the lungs. Spent gases from the lung exit in synchrony under the buccal forces. In the **two-stroke buccal pump** (Fig. 5.34), initial expansion of the buccal cavity brings fresh air and expired gas from the lungs

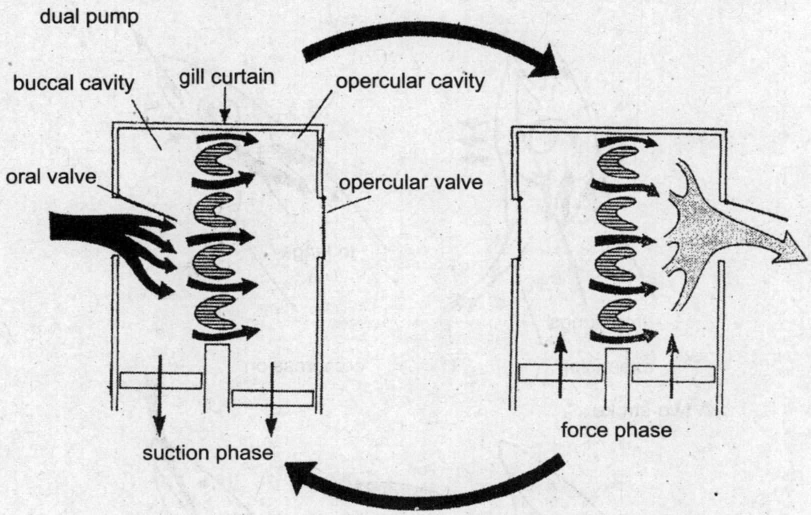

Fig. 5.33. Dual pump of water-breathing fish. In most fishes, the buccal and opercular cavities form dual pumps on opposite sides of gill curtain. Muscle action expands both cavities, represented by the falling pistons (downward dark arrow, left) in the suction phase. During the force phase, muscles contract to compress the cavities, represented by the rising pistons (upward dark arrows, right). As pressure within each cavity falls and rises, more water (suction phase), is drawn in and expelled (force phase). Because of the slight difference in pressure between buccal and opercular cavities, water is almost continuously moving from buccal to opercular cavity. The valves of the mouth and operculum prevent reverse flow of water. Thus, a one-way and more or less continuous flow of water across the gills is established (after Kardong 2006).

into the mouth where they mix during the first stroke. In the second stroke, buccal compression forces these mixed buccal gases into the lungs, with the excess gas expelled via the nares or mouth. Exhalation and inhalation of gases may also be based on a four-stroke mechanism. The **four-stroke buccal pump** (Fig. 5.34B) begins with buccal expansion that brings gas from the lungs into the mouth, so upon the second stroke, buccal compression, this gas forced out the nares. In the third stroke, buccal expansion draws fresh air into the mouth, via the nares, so that in the fourth stroke, buccal compression forces this air into the lungs. Tracheal sphincters and narial valves are synchronized with buccal displacements to help control gas movement.

(a) Air-breathing fishes. Fishes that occasionally gulp atmospheric air, such as lungfishes, are no different from other fishes when they are actively water breathing. They use the same dual pump mechanism to irrigate their gills. However, when the lungfish breathes air, the dual pump is modified into a buccal pump to move air in and out of the lungs. The *four-stroke buccal pump* can be summarized as an exhalation phase and an inhalation phase. The **exhalation phase** begins with the **transfer** (expansion 1) of spent air from the lungs into the buccal cavity. In some fishes, relaxation of sphincter around the glottis permits this transfer from the lungs to the buccal cavity. Exhalation concludes with **expulsion** (compression 2) of air from the buccal cavity to the outside either trough the mouth or under the operculum. As the fish rises and its snout breaks the surface of water, its mouth opens to **intake** (expansion 3) atmospheric air, the first step in the **inhalation phase**. Inhalation concludes with *compression* (4), which forces a bubble of fresh air from the buccal cavity into the lungs (Fig. 5.35).

(b) Dual buccal pump in amphibians. Like air-breathing fishes amphibians use a pulse-pump to ventilate their lungs. Airflow is bidirectional. The two-stroke buccal pump method

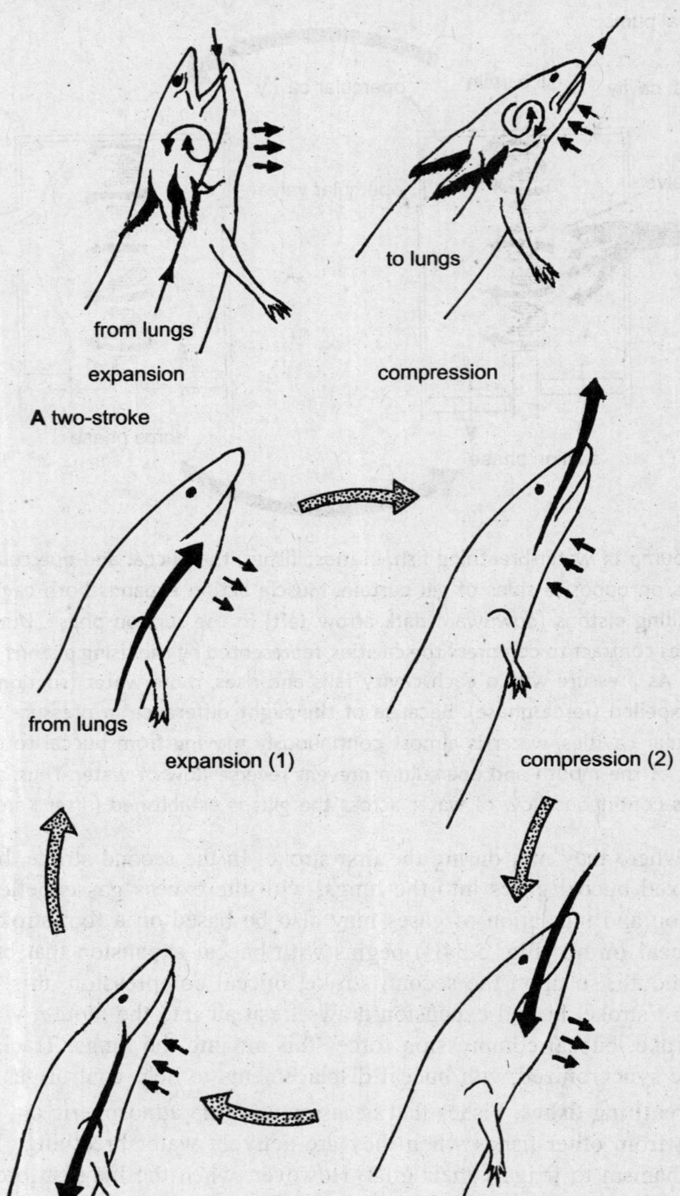

Fig. 5.34. Buccal pump: two-stroke and four-stroke. Air-breathing fishes and amphibians use an expansion and compression of the buccal cavity to move gases to and from the lungs. A – Two-stroke buccal pump (*e.g., Ambystoma tigrinum*). Expansion mixes gas from the lungs and fresh air in the mouth, upon compression, this mixture is forced into the lungs, and excess expelled through the nares. B – Four-stroke buccal pump (*e.g., Amphiuma tridactylum*). Initial buccal expansion (1) draws air into the mouth to be expelled next during buccal compression (2); now buccal expansion (3) draws in fresh air; to be forced into the lungs upon buccal compression (4) (after Kardong 2006).

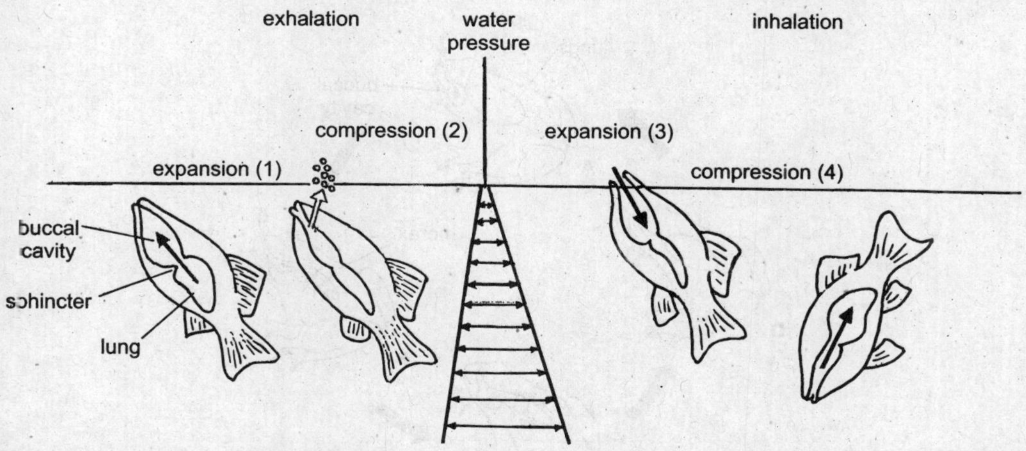

Fig. 5.35. Air breathing fishes: buccal pump. The four-stroke pump occurs in two phases—exhalation and inhalation. During exhalation, surrounding water pressure on the body walls forces air from the lungs and out the open mouth. During inhalation, the fish's head breaks the water surface and its buccal cavity expand, drawing in air. Muscle compression of the buccal cavity forces the valve in the mouth closed, and positive pressure moves air into the lungs. As sphincter between the mouth and lungs closes to prevent escape of air from the lungs (after Kardong 2006).

of ventilation is primitive for amphibians and is found in most. It has been reported in some lungfishes (*e.g.*, African lungfish). A few aquatic amphibians use the four-stroke buccal pump.

The four stage of lung ventilation in frog are best understood. In the first stage, the buccal cavity expands to draw fresh air in through the open nares (Fig. 5.36A). In the second stage, the glottis opens rapidly, releasing spent air form the elastic lungs. This air streams across the buccal cavity with little mixing and is vented through the open nares (Fig. 5.36B). In the third stage, the nares close, and the floor of the buccal cavity rises, forcing the fresh air held in this cavity into the lung through the open glottis (Fig. 5.36C). In the fourth stage, the glottis closes, retaining the air that has just filled the lungs, and the nares open again. Between cycles, the buccal cavity may oscillate repeatedly (Fig. 5.36D). This rapid oscillation was once thought to trun the lining of the mouth temporarily into an accessory breathing organ. However, experimental evidence refutes this (see **Kardong** 2006). The capillaries lining the mouth do not serve in gas exchange. Instead, such buccal oscillations between lung fillings serve mainly to flush the buccal cavity of any stray residue to expired air in the mouth following each ventilatory cycle.

(iv) Air ventilation: aspiration pump. The **aspiration pump** is a third type, after dual and buccal pumps, that does not push air into the lung against a resisting force. Rather, air is sucked in, or aspirated, by the low pressure created around the lungs (Fig. 5.37).

The lungs are located *within* the pump so that the force required to ventilate them is applied directly. The "pump" includes the rib cage and often muscular diaphragm. A movable diaphragm in the thorax causes pressure changes rather than the action of the buccal cavity. The diaphragm, like a plunger, alters the pressure on the lungs to favour entry or exit of air.

The aspiration pumps is bidirectional and moves air tidally. It is found in terrestrial amniotes reptiles, mammals and birds. In birds especially, the aspiration pump is highly modified. The buccal cavity is no longer part of the amniote pumping mechanism. Unlike the buccal pump, feeding and ventilation are decoupled in vertebrates using the aspiration pump.

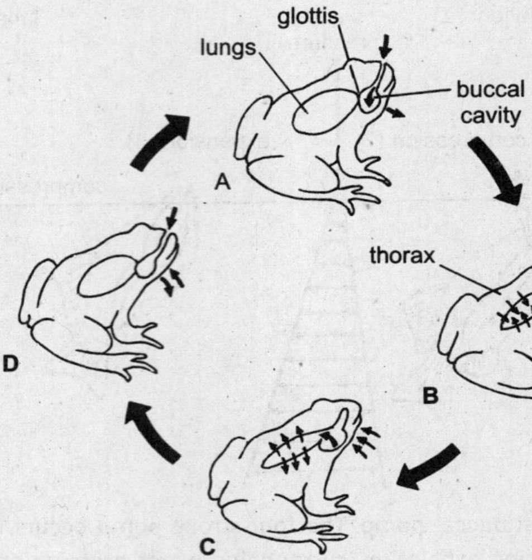

Fig. 5.36. Lung ventilation in the forg. A – The frog's throat drops to replenish the air in the buccal cavity. B – As the glottis is opens, the thorax is compressed, forcing spent air from the lungs past that held in the buccal cavity and expelling it (open arrows). C – Elevation of the throat and closure of the nares forces fresh air from the bcucal cavity into the lungs. D – Repeated pumping of the throat (multiple arrows) flushes the buccal cavity (after Kardong 2006).

Phylogeny of Lungs in Terrestrial Vertebrates

1. Amphibians. The respiratory surface within the lungs is usually developed best anteriorly, and it decreases posteriorly along the inner walls. This surface is **septal**, meaning that partitions form and subdivide to increase the surface area exposed to incoming air. The inter-connecting septa divide the internal wall into compartments, called **faveoli**, that open into the central chamber within each lung. Faveoli differ from the alveoli of mammalian lungs in that they are not found at the end of a highly branched tracheal system. Faveoli are internal subdivision of the lung wall that open into a common central chamber. Inspired air travels along the trachea into the central lumen of the lung and from here it diffuses into the surrounding faveoli. Capillaries located within the thin septal wall of the faveoli take up oxygen and give carbon dioxide.

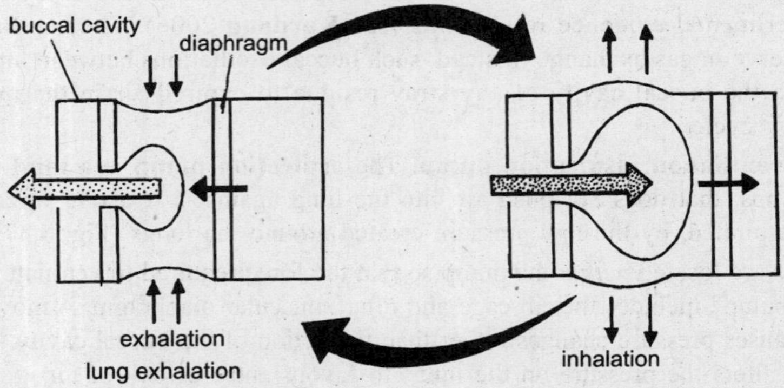

Fig. 5.37. Air-breathing amniotes: aspiration pump. In most amniotes, the buccal cavity has little to do with forcing air in and out of the lungs. Instead, a rib cage expands and compresses and/or a diaphragm moves forward and back within the body cavity to create a positive pressure that expels air or negative pressure that draws air into the lungs (after Kardong 2006).

2. **Reptiles.** The lungs of snakes and most lizards typically include a single central air chamber into which faveoli open (Fig. 5.38A,B). Like purse strings, cords of smooth muscle define and encircle the opening into each faveolus. The thin walls of each faveoli carry capillary beds and may be subdivided by even smaller internal septa. Sometimes the faveoli are reduced in the posterior part of the lung, leaving it as a non-exchange region. In monitor lizards, turtles and crocodiles, the single central air chamber itself is subdivided into numerous internal chambers that receive air from the trachea. These internal chambers are ventilated by respiratory movements, whereas the exchange of gas between the faveoli and these chambers appears to occur by diffusion.

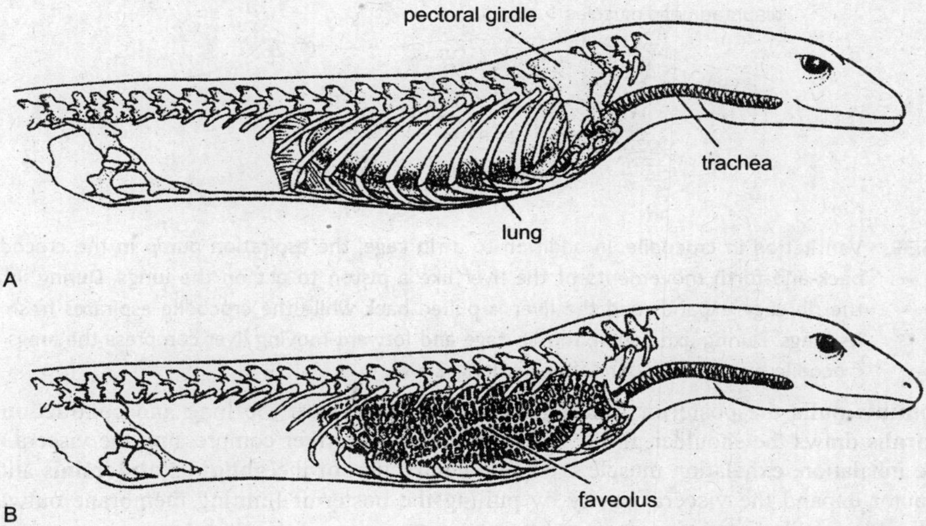

Fig. 5.38. Lung ventilation in a lizard. A – The lungs are located in the thorax, surrounded by ribs and connected to the trachea. Compression and expansion of the rib cage forces air in or out of the lungs. B – Cutaway view of the internal lining of the lungs showing numerous faveoli that collectively give the lining a honeycomb appearance. The internal faveoli of the lungs increase their respiratory surface area and function in gas exchange with capillaries lining their walls (after Kardong 2006).

Filling of the lungs in all reptiles is based on an **aspiration pump mechanism,** although the anatomical parts that actually participate may differ. Ventilation in turtles represents a special problem in design. The shell around the lungs prevents change in shape and precludes (*i.e.,* prevents from happening) aspiration pumping using the ribs. In soft-shelled turtles, movements of the hyoid apparatus draw water in and out of the pharynx. Oxygen is absorbed in the pharynx to sustain the turtle while it is submerged. In snapping turtles, the plastron is reduced, permitting deformations of the body wall that contribute to lung ventilation. More commonly in-and-out movements of the limbs alter pressure on the lungs, and special sheets of muscles within the shell change pulmonary pressure (Fig. 5.40A). Turtle lungs and other viscera resides in a single fixed cavity, so any changes in volume alters pressure on lungs. A limb extended from or pulled into the shell affects pressure in this cavity and aids the aspiration pump (Fig. 5.40B). In addition, the posterior visceral cavity is closed by a **limiting membrane**, connective tissue to which the **transversus abdominis** and **obliquus abdominis muscles** are attached. Contraction and relaxation of these muscles alters the volume of the cavity within the shell and contributes to the inhalation or exhalation of air (Fig. 5.40C). Thus, during active exhalation, contraction of the **transverse**

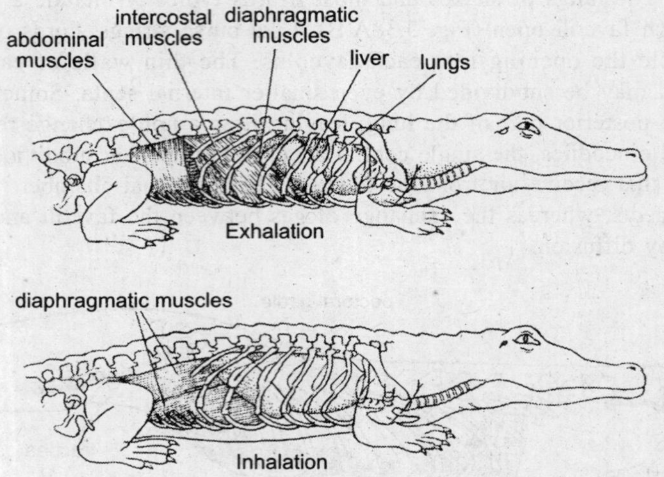

Fig. 5.39. Ventilation in crocodile. In addition to a rib cage, the aspiration pump in the crocodile uses back-and-forth movements of the liver like a piston to act on the lungs. During inhalation, the rib cage expands and the liver is pulled back while the crocodile aspirates fresh air into its lungs. During exhalation, the rib cage and forward-moving liver compress the lungs and the crocodile expels spent air (after Kardong 2006).

abdominis pulls the posterior limiting membrane up against the lung and contraction of the **pectoralis** draws the shoulder girdle back into the shell, further compressing the viscera. During active inhalation, exhalation muscle relax and contraction of the **obliquus abdominis** and girdle protractor expand the visceral cavity by pulling the posterior limiting membrane outward and the shoulder girdle forward, respectively.

The diaphragmaticus muscle, although absent in tortoises, is present in most other turtles. The diaphragmaticus together with the transversus abdominis compress the visceral cavity to act as exhalation muscles. The glottis opens and the obliquus abdominis expands the visceral cavity to the act as an inhalation muscle.

3. Mammals. An aspiration pump ventilates the lung of mammals. Changes in shape of the rib cage and piston-like action of a muscularized **diaphragm** contribute to this pumping mechanism. The diaphragm consists of **crural, costal and sternal** parts, all of which converge on a **central tendon.** Unlike the diaphragmatic muscles of crocodiles (Fig. 5.39), which are located posterior to the liver, the diaphragm of mammals lie anterior to the liver, and acts directly on the **pleural cavities** in which the lungs reside (Fig. 5.41A, B). Intercostal muscles run between ribs. The transversus abdominis, serratus on the ribs and originate outside the rib cage, all aid in mammalian lung ventilation.

Mammalian ventilation is **bidirectional** and involves the rib cage and diaphragm. The mammalian diaphragm lies immediately posterior to the lungs and separates the thoracic cavity containing the lungs from the abdominal cavity containing the lungs from the abdominal cavity.

As we have seen in lower tetrapods, faveoli along the interior walls of the lungs form the respiratory exchange surface. Air is drawn into the centre core of the lung and diffuses outward into the faveoli. However, in mammals the sites of respiratory exchange are reached, via a different route. The respiratory passageway (including trachea, bronchi, bronchioles) repeatedly divides, producing smaller and smaller branches until they finally terminate in blind-ended compartments, the alveoli, which characterize the respiratory bronchioles and air sacs (Fig. 5.43A, C). The trachea, bronchi and terminal bronchioles that transport gas to and from the alveoli are called

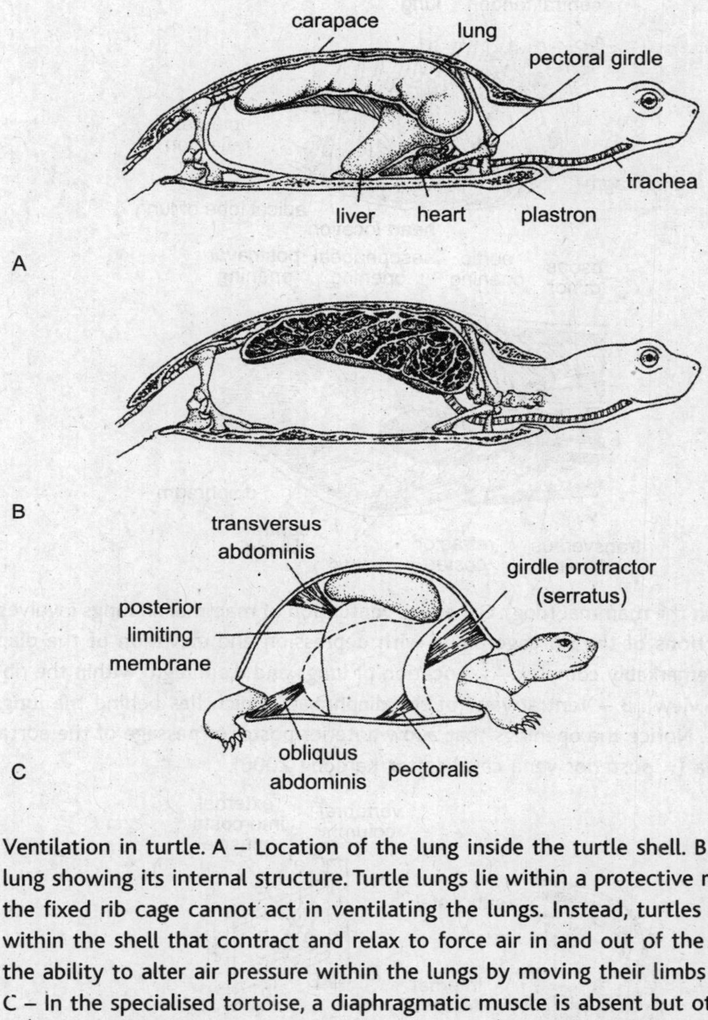

Fig. 5.40. Ventilation in turtle. A – Location of the lung inside the turtle shell. B – Cutaway view of the lung showing its internal structure. Turtle lungs lie within a protective rigid shell. Consequently, the fixed rib cage cannot act in ventilating the lungs. Instead, turtles have sheets of muscles within the shell that contract and relax to force air in and out of the lungs. Turtles also have the ability to alter air pressure within the lungs by moving their limbs in and out of the shell. C – In the specialised tortoise, a diaphragmatic muscle is absent but other respiratory muscles take its place. Within the rigid shell, the viscera are enclosed by limiting membranes that under muscle action after their position during exhalation (solid line) and inhalation (dashed line) (after Kardong 2006).

the **respiratory tree** in recognition of their branching pattern. *No gas exchange* occurs along the conducting passageway of the respiratory tree until air reaches the respiratory bronchioles and alveoli. In mammals, the total alveolar area is extensive, perhaps over ten times that of amphibians of similar mass. Such a large exchange area is essential in mammals to sustain the high rate of oxygen uptake required by an active endotherm. The nasal passages not only form part of this conducting system but serve to warm and moisten the entering air.

4. Birds. Respiration by skin is quite meagre in birds. The almost exclusive respiratory organ in birds is lungs. Like mammals, birds have two lungs connected to a trachea and ventilated by an aspiration pump. Beyond this, however, the structural similarities are few. For example, there are no blind-ended alveoli in and out of which air moves. Instead, the conducting passages branch repeatedly and eventually form numerous tiny, one-way passageways, the **parabronchi**, that permit air to flow through the lungs. Small **air capillaries** open off the walls of each

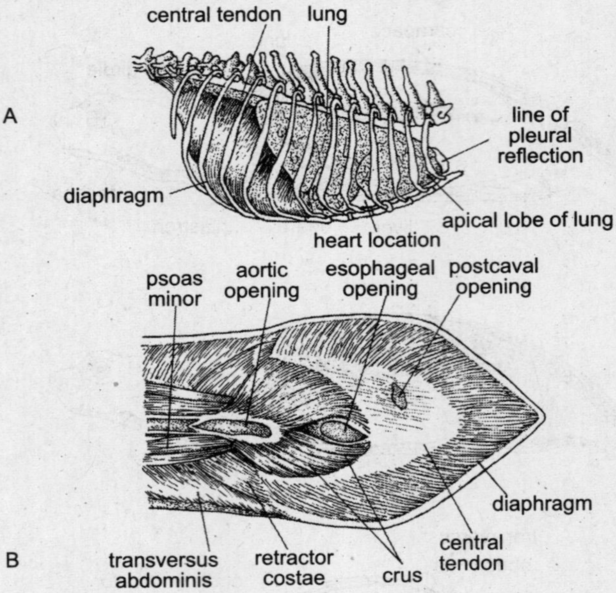

Fig. 5.41. Ventilation in the mammal (dog). Generally, ventilation of mammalian lungs involves expansions and contractions of the rib cage along with depression and elevation of the diaphragm. The details are remarkably complex. A – Location of lungs and diaphragm within the rib cage of the dog (lateral view). B – Ventral view of the diaphragm, which lies behind the lungs and has a dome shape. Notice the openings that allow anterior-posterior passage of the aorta, esophagus and postcava (= posterior vena cava) (after Kardong 2006).

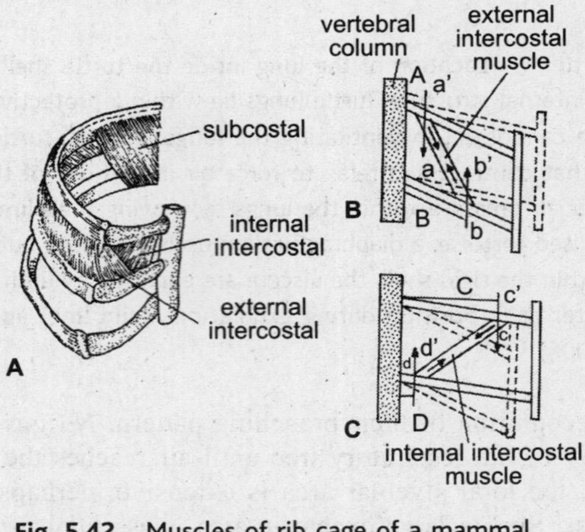

Fig. 5.42. Muscles of rib cage of a mammal.

parabronchus, and gas exchange with the blood actually occurs in the air capillaries. Further, nine avascular **air sacs** are connected to the lungs, although they are tucked in among the viscera and extend into the cores of most large bones (Fig. 5.44A,B). Thus, the bones of birds contain air, not marrow. From 6 (house sparrow) to 12 (shorebirds) air sacs may be present. Generally, the anterior air sacs include the single **interclavicular sac** and the paired **cervical** and **anterior thoracic air sacs.** The **posterior air sacs** include the paired **posterior thoracic** and paired **abdominal air sacs** (Fig. 5.44A).

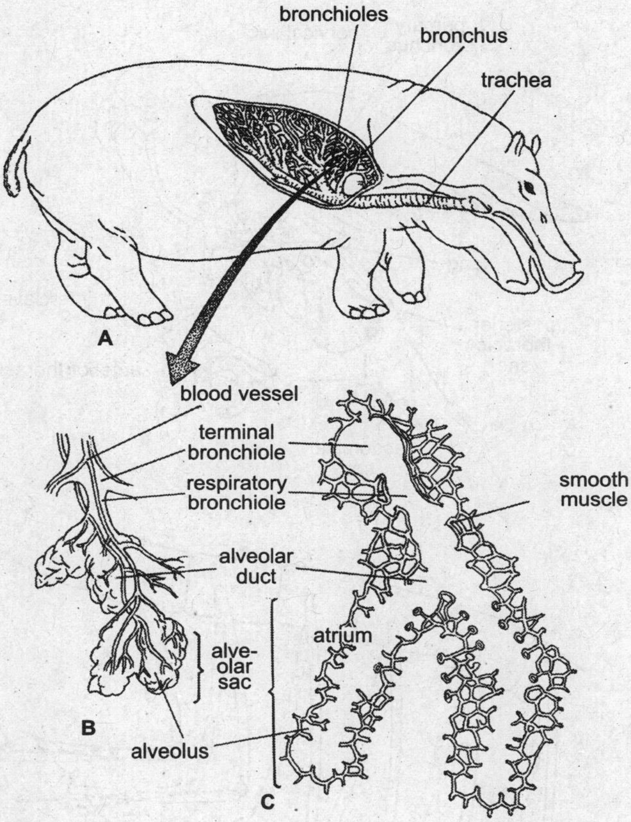

Fig. 5.43. Mammalian lung. The lungs of mammals (e.g., Hippopotamus) are blind-ended, terminating in small alveoli. A – The trachea leads to the pleural cavities and branches into bronchi to supply left and right lungs. Repeated bronchial branchings produce smaller and smaller bronchioles that eventually lead to alveolar sacs. B – Enlarged alveolar sac. Arteries and veins supply the alveoli to accommodate gas exchange within them. C – Internal subdivisions of the alveolar sacs are shown. Each small compartment is an alveolus where actual respiratory exchange between blood and air occurs. Note the smooth muscle bands at the openings to the alveolar sacs (after Kardong 2006).

The trachea is divided into two **primary bronchi** (= mesobranchi) that do not enter the lung but extend posteriorly to reach the posterior air sacs. Along the way, the primary bronchi give rise to numerous branches, the most prominent of which include **latero-, ventro-,** and **dorso-bronchi** as well as **secondary bronchi.** These lead to the **parabronchi** (Fig. 5.45A–C). During passage through the parabronchus, gases diffuse between the lumen of the parabronchus and the connecting, blind-ended air capillaries into the adjacent blood capillaries that give up carbon dioxide to the air capillaries. Thus, the walls of air and blood capillaries constitute the sites of gas exchange.

Within this vast system of connecting passageways, there are no values to suggest what the pattern of airflow might be. This has led to much speculation about the roles played by the different parts of the respiratory system. Recent research suggests that air sacs act as bellows. If we follow a single breath, its passage through the sacs and lungs include two complete cycles of inhalation and exhalation (Fig. 5. 46A,B). During the first inhalation, air enters the trachea, passes along the primary bronchi, and then is divided: some air passes directly to the lungs

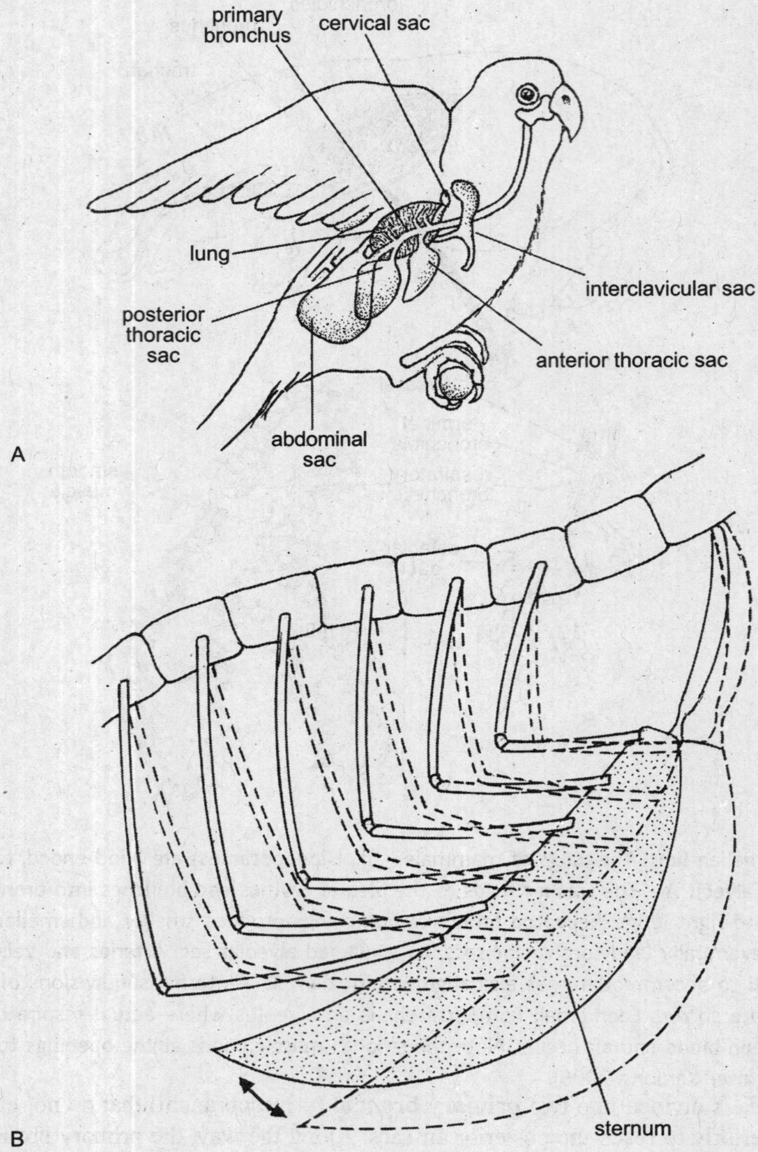

Fig. 5.44. Avian respiratory system. A – The respiratory system of birds consists of paired lungs located in the dorsal wall of the thoracic cavity. Air sacs that lie among the viscera and extend into the cores of adjacent bones are attached to the lungs. Apparently the lungs themselves do not change shape with rib cage motion. Rather, compression and expansion of the rib cage acts on the air sacs, drawing air through them and then into the lung. B – Ventilation of the avian lung. Ribs are hinged to each other and to the sternum in such a way that lowering of the sternum results in expansion of the rib cage and inhalation. Elevation of the sternum compresses the air sacs and air is expelled (after Kardong 2006).

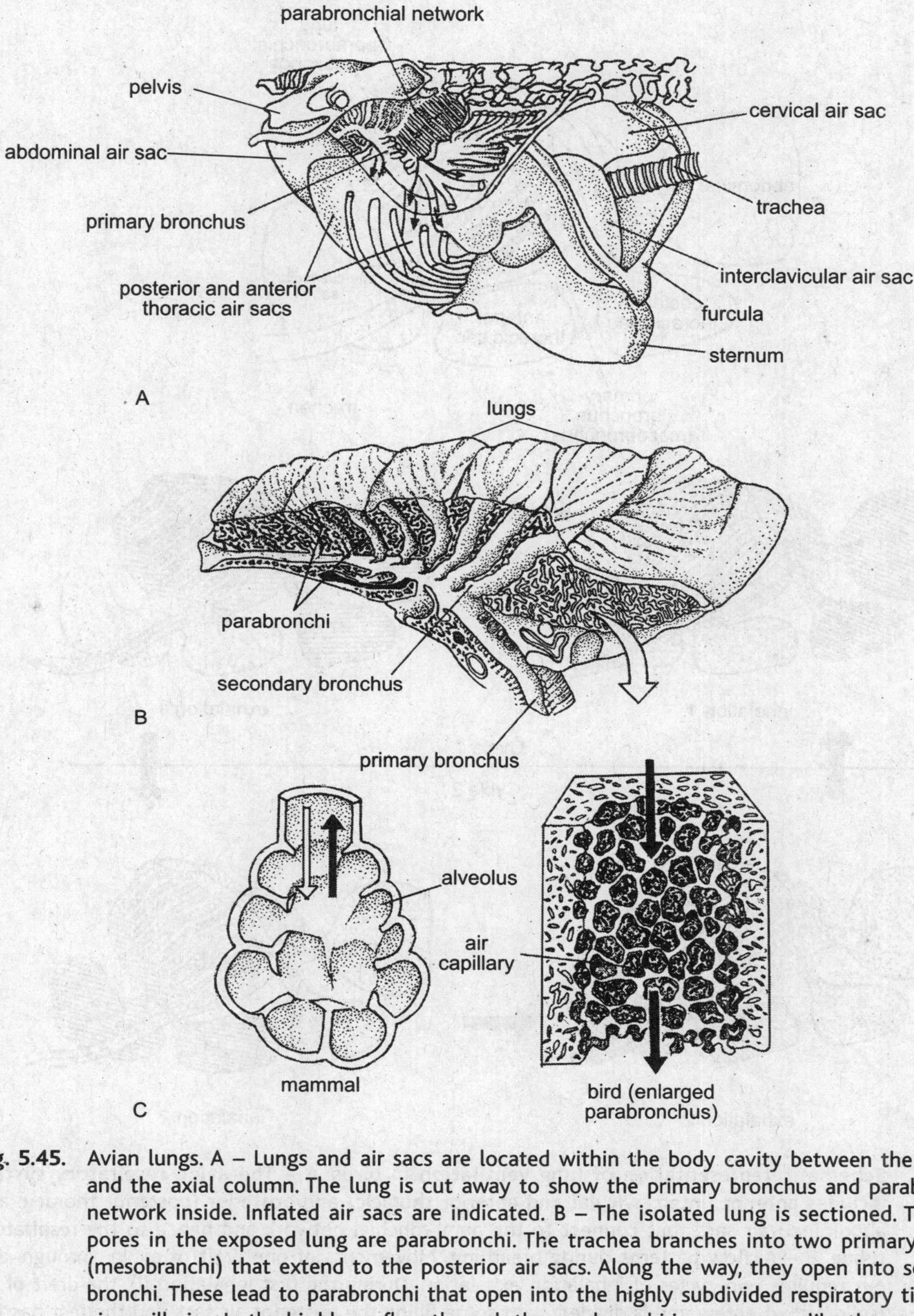

Fig. 5.45. Avian lungs. A – Lungs and air sacs are located within the body cavity between the sternum and the axial column. The lung is cut away to show the primary bronchus and parabronchial network inside. Inflated air sacs are indicated. B – The isolated lung is sectioned. The small pores in the exposed lung are parabronchi. The trachea branches into two primary bronchi (mesobranchi) that extend to the posterior air sacs. Along the way, they open into secondary bronchi. These lead to parabronchi that open into the highly subdivided respiratory tissue, the air capillaries. In the bird lung, flow through the parabronchi is one way, unlike the mammalian airflow that ends in blind alveoli. C – Comparison of avian and mammalian respiratory surfaces. In the avian lung, air passes one way (solid arrows) through the parabronchi replenishing the air capillaries that surround and open into the parabronchi. In the mammalian lung, the alveoli are blind-ended. For gas exchange to take place, the air must move tidally (open and solid arrows) (after Kardong 2006).

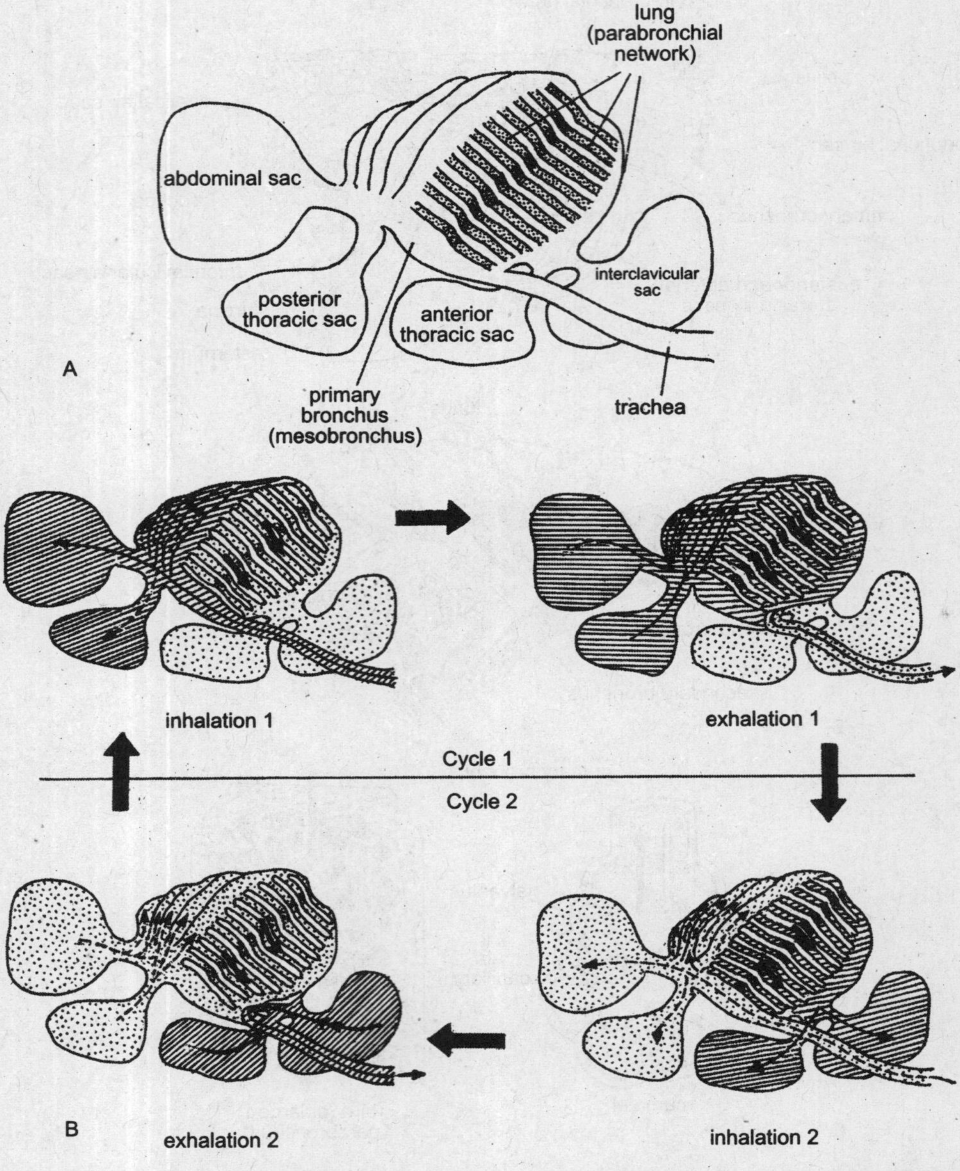

Fig. 5.46. Schematic representation of lung ventilation in birds. A – The avian respiratory system includes anterior (interclavicular and anterior thoracic) and posterior (posterior thoracic and abdominal) air sacs that connect to the parabronchial network and hence to the respiratory tissue. B – Airflow patterns during breathing. Movements of one draft of air in, through, and out requires two cycles of inhalation/exhalation. During the first inhalation (I), the draft of air (solid arrow) enters and is divided, with some filling the posterior air sacs and the rest passing to the parabronchial network. Upon exhalation (I), fresh air from the posterior air sacs passes into the lung, pushing spent air out. The second breath (dashed arrow) of inhalation (2) enters and is divided, with some refilling posterior air sacs and the rest passing to the parabronchial network, pushing the remainder of the air from the first breath into the anterior air sacs. During exhalation (2), spent air is pushed out of the lungs and exits with air from the anterior air sacs (after Kardong 2006).

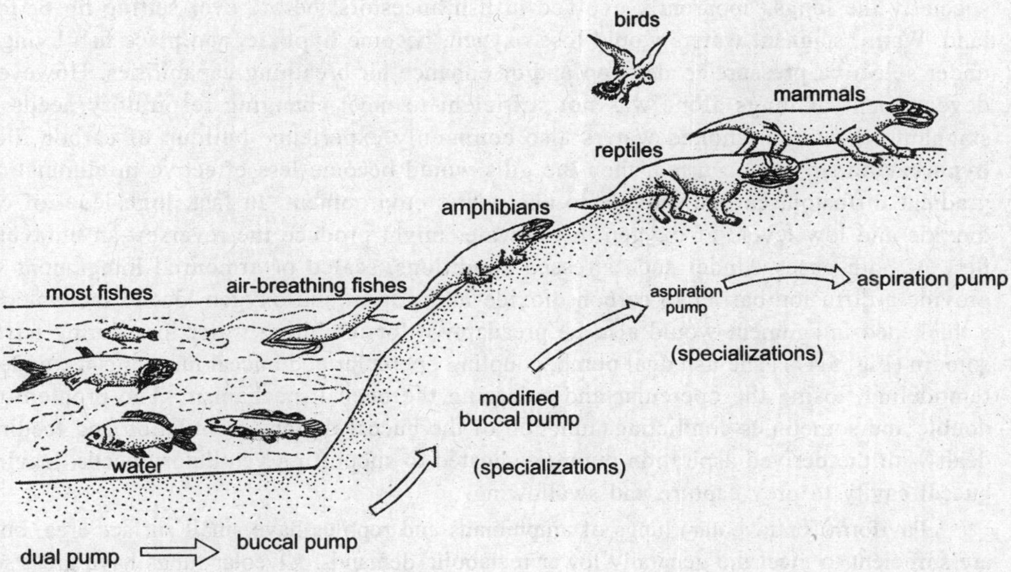

Fig. 5.47. Evolution of ventilatory mechanisms. Water-breathing fishes irrigate their gills with a dual pump mechanism in which buccal and opercular cavities operate in tandem. Air-breathing fishes use a buccal pump mechanism, a modification of the dual pump in which the buccal cavity is the major mechanical component. In adult amphibians, lung ventilation is based on a modified buccal pump in which the opercular pump is lost entirely. However, the amphibian skull design is compromised because the buccal cavity must function in feeding and in lung ventilation. One solution is found in plethodontid salamanders. Gas exchange is taken over entirely by cutaneous respiration, and lungs are lost; therefore, the buccal cavity serves only feeding. In amniotes, the aspiration pump completely separates feeding from lung ventilation, uncoupling demands on the jaws for service in both activities. The basic dual and buccal pumps are specialized in many fishes, Bird respiration represents a specialization of the aspiration pump (after Kardong 2006).

and the rest fills the posterior air sacs (*i.e.,* posterior thoracic and abdominal air sacs). Upon the first exhalation, air from the posterior air sacs now flows through the lungs, displacing the spent air that exits via the trachea. As the second inhalation begins, the entering air again divides, some refilling the posterior air sacs and the rest flowing through the lungs, pushing the remainder of the spent air of the previous cycle out and temporarily into the anterior air sacs (anterior thoracic and interclavicular air sacs). With the second exhalation, air in these anterior sacs now exits along with air from the lungs, replaced by air from the posterior air sacs, which now flows through the lungs. Thus, this pattern of ventilation produces a nearly continuous **unidirectional flow** of fresh air across the lungs. Speculating further, such a unidirectional flow may also establish a **crosscurrent exchange** within the lungs, with air flowing from posterior to anterior air sacs as circulating blood flows next to it in the opposite direction.

Evolution of Ventilatory Mechanisms in Vertebrates

The vertebrate transition from fish to tetrapod included a transition from a water-breathing to an air-breathing respiratory system. However, much of the air-breathing organs, specially the lungs, apparently evolved in fish ancestors, before ever setting fin or foot on land. Warm, stagnant waters would lose oxygen, become **hypoxic**, and place fish living there under selective pressure to develop and/or enhance air-breathing capabilities. However, the development of lungs alone was not sufficient to meet changing respiratory needs. Such stagnant, vegetation-choked waters also commonly experience buildup of carbon dioxide, **hypercarbia**. If this occurred, then the gills would become less effective in eliminating, by gradient diffusion, carbon dioxide to an aquatic environment. In fact, high level of carbon dioxide and low levels of oxygen in the water might produce the reverse—an unfavourable flow of both gases. Under such stressed conditions, scaled or armoured integument would provide a diffusion barrier to carbon dioxide influx from and oxygen loss to the water. Such a thickened integument would also be preadaptive for inhibiting water loss during terrestrial sojourn (Fig. 5.47). The fish dual pump, coupling opercular and buccal mechanisms, underwent remodeling, losing the opercular and enlarging the buccal mechanism. The problem of the double and sometimes conflicting, function of the buccal region—ventilation and feeding—is dealt with the derived aspiration pump, dedicated to supporting ventilatory needs, leaving the buccal cavity to prey capture and swallowing.

Faviform (or faveolar) lungs of amphibians and reptiles have small surface area, but they are sufficient to meet the generally lower metabolic demands. Alveolar lungs have great surface area and support the great metabolic requirements of mammals. Alveolar lungs are elastic and ventilated by an aspiration pump, permitting ventilation to occur at low metabolic cost. In birds, the air sac system and the crosscurrent gas exchange within the parabronchial lungs support a high metabolic scope and uptake of oxygen at high altitudes.

QUESTIONS

Long Answer Questions

1. How are the respiratory and circulatory system related?
2. Lampreys occlude their oral cavity during feeding. How can they respire given this observation.
3. Compare holobranchs of a typical teleost and shark and their external openings.
4. Give an account of ventilation systems in tetrapods.
5. Compare the lungs of birds and mammals.

Short Answer Questions

1. Compare aspiration pump of mammals with buccal pump of amphibians.
2. Write short notes on the following:
   - **(i)** Spiracle;
   - **(ii)** Larval gills;
   - **(iii)** Pseudobranch;
   - **(iv)** Swim bladder;
   - **(v)** Air bladder;
   - **(vi)** Alveolus;
   - **(vii)** Faveolus.

Very Short Answer Questions

1. What is diffusion?
2. What is faveolus?

3. Define the alveolus?
4. What is air capillary?
5. Which vertebrate solely depends on cutaneous respiration?
6. Which are the parts of respiration tree of a mammal?
7. Define the air bladder?
8. What is aspiration?
9. Define cleidoic egg.
10. What is concurrent?
11. Define countercurrent?
12. Define ventilation?

Fill in the Blanks

1. In living fishes, the nostrils are _______ used for breathing.
2. _____ have no nose, however, they are mammals.
3. In mammalian respiratory track, _______ in the submucosa overlying the turbinal bones heat cold air.
4. ______ salamanders depend entirely on the cutaneous respiration.
5. Bones of birds have _______, they do not contain marrow.
6. Air-breathing fishes have a ______ -stroke ______ pump.
7. ________ of avian lungs are _______ passageways of the air.

Multiple Choice Question

1. A complete gill is called.
   (a) holobranch (b) demibranch
   (c) hemibranch (d) pseudobranch
2. Gills and lungs of vertebrates are the derivatives of embryonic
   (a) archenteron (b) pharynx
   (c) kidney (d) heart
3. In bony fishes operculum arises from which arch bone
   (a) mandibular (b) hyomandibular
   (c) hyoid (d) sphenethmoid
4. Lungs of reptiles are located in
   (a) neck (b) thorax
   (c) abdomen (d) tail
5. During submergence in aquatic turtles the accessory respiratory organs are
   (a) skin (b) gills
   (b) air sacs (d) cloacal bladder
6. Respiratory organs of embryonic vertebrates is
   (a) amnion (b) chorion
   (c) egg membranes (d) yolk sac

ANSWERS

Very Short Answer Questions

1. Diffusion is movement of molecules from an area of high concentration to an area of low concentration; if the movement is random and unaided, it is known as passive diffusion.
2. Faveolus is a tiny respiratory compartment within the lung that opens to a central air chamber and results from the subdivisions of the lung lining, *e.g.,* lungs of amphibians and reptiles.
3. Alveolus is smallest division of respiratory tissue in mammalian lungs located at the ends of the branching respiratory tree.
4. Air capillary is a small compartment within a bird lung in which gas exchange occurs.
5. Lungless salamander, *Ensatina eschscholtzii.*
6. Trachea, bronchi and terminal bronchioles of mammalian lungs.
7. Air bladder is a gas bladder for respiratory exchange or buoyancy control.
8. Aspiration is drawing in by suction.
9. The cleidoic egg is the shelled container in which the fetus is laid, as in reptiles, birds and primitive mammals.
10. Concurrent is flow of adjacent currents in the same direction.
11. Countercurrent is flow of adjacent currents in opposite direction.
12. Ventilation is active movement of water or air across respiratory exchange surface.

Fill in the Blanks

1. Not. **2.** Whales. **3.** Venous plexuses. **4.** Lungless. **5.** Air. **6.** Four, buccal. **7.** Parabronchi, one-way.

Multiple Choice Questions

1. (a) **2.** (b) **3.** (c) **4.** (c) **5.** (d) **6.** (b).

6

Circulatory System

The circulatory system of craniates consists of the heart, arteries, veins or venus sinuses, capillaries or sinusoids, and blood (**blood vascular system** or **cardio-vascular system**) and of lymph channel and lymph (**lymphatic system**). The blood carries *oxygen* from respiratory organs; nutrients from extra-embryonic membranes, digestive tract, and storage sites (liver, adipose tissue, bones); *hormones* and other substances associated with homeostasis and immunity to diseases, and *waste products* of metabolism to the excretory organs. Blood also conducts *heat* to and from the skin and other surfaces where heat is exchanged, thereby regulating and equalizing internal temperatures. Lymph channels collect interstitial tissue fluids not taken up by the bloodstream, and they collect lipids from the villi of the small intestine. Lymph vessels ultimately terminate in venous channels.

6.1 ARTERIES, VEINS AND BLOOD CAPILLARIES

Arteries carry blood away from the heart. They have muscular and elastic cells capable of distention with each intrusion of blood. The smallest arteries, 0.3 mm or less in diameter, are **arterioles**. They dilate and constrict reflexly and thereby assist in regulating blood pressure. They terminate in **blood capillaries**. **Veins** begin in capillaries (other than the respiratory capillaries of the gills) and carry blood toward the heart. They have proportionately less muscle and elastic tissue and more fibrous tissue than arteries and are therefore capable of less distention or constriction. The smallest are **venules**, which provide a direct connection to the capillaries. Arteries and veins, like most other organs, have a loose connective tissue covering called **adventitia**.

Capillaries (Fig. 6.1) generally consist of endothelium alone, although certain capillaries are accompanied by a delicate investment of **mesenchyme** and a scattering of smooth muscle fibers. Their lumen is just large enough to accommodate red blood cells in single file. In fact, the cells must often "squeeze through" becoming temporarily deformed until they "pop out" into the smallest venule. A capillary "bed" or network represents all the individual capillaries served by a single arteriole. At the site where a capillary emerges from an arteriole, a short section of the capillary wall, the **pre capillary sphincter**, has smooth muscle fibers, given an adequate neural or hormonal stimulus, close off the entrance to the capillary bed at that site. Relaxation of the constrictor muscle readmits blood to the capillary. Blanching or blushing of human skin is a result of the constrictor and dilatory effects of these sphincters. The smallest arterioles and venules are connected directly by short vascular **capillary shunts** that assure uninterrupted circulation between the arterial and venous sides of the capillary beds when other capillaries are constricted.

A **portal system** is a system of veins bounded by capillary beds (Fig. 6.2). In most craniates blood from the capillaries of the tail passes via a **renal portal system** to capillaries of the kidneys before continuing to the heart. Blood from the digestive tract, pancreas, and spleen passes via a

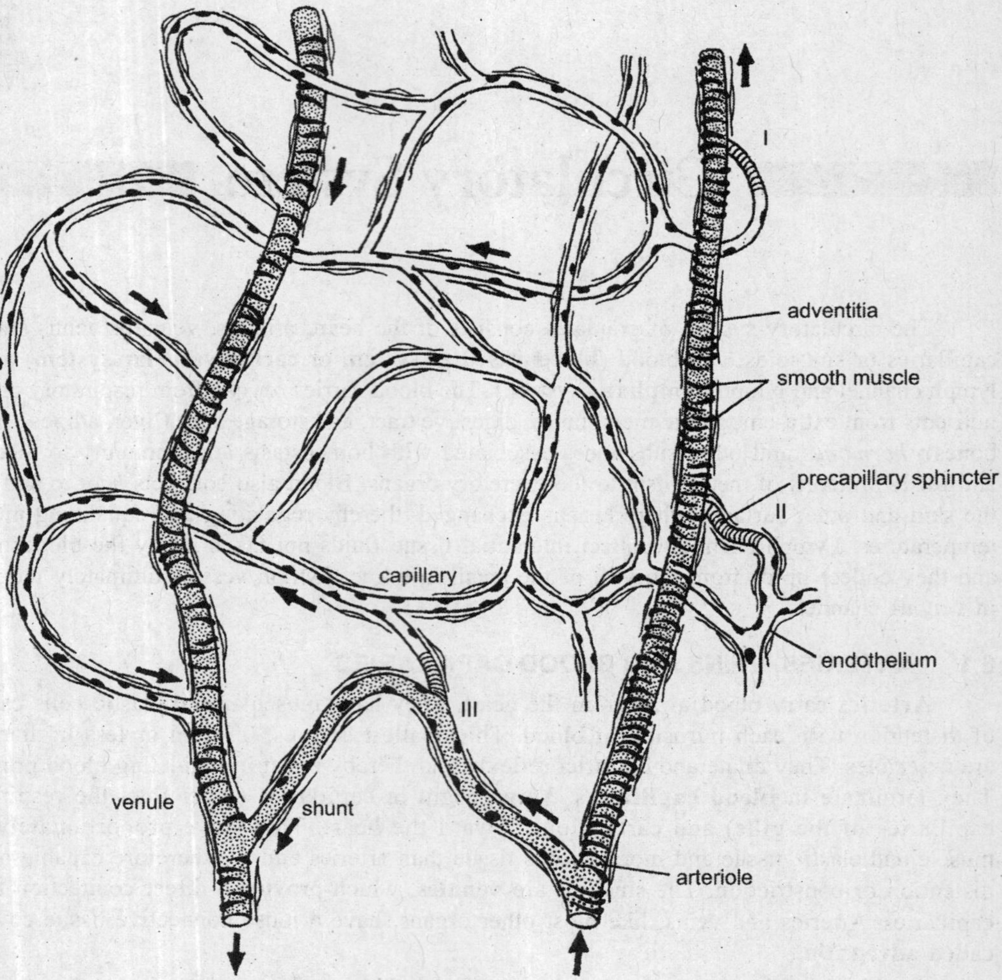

Fig. 6.1. A capillary bed and its connections. Three crosshatched segments (I, II and III) are sites of pre-capillary sphincters. Arrows indicate direction of flow (after Kent and Carr 2007).

hepatic portal system to the capillaries of liver before continuing to the heart. Blood containing pituitary regulating hormones from the hypothalamus passes via a **hypophyseal portal system** to the adenohypophysis before continuing to the heart.

6.2 BLOOD

Cells produced by hemopoietic tissues usually enter the circulation to become the **peripheral** or **circulating blood**. Circulating blood comprises plasma and formed elements. The **plasma** is the fluid component and can be thought of as the ground substance of blood, a special connective tissue. The **formed elements** are the cellular components of blood. **Red blood cells**, or **erythrocytes**, are one cell type of the formed elements. All erythrocytes have nuclei, except those in mammals. Mature red blood cells in mammals lack nuclei. **Haemoglobin**, the major oxygen transport molecule, is excreted by the kidneys if it is left free in the

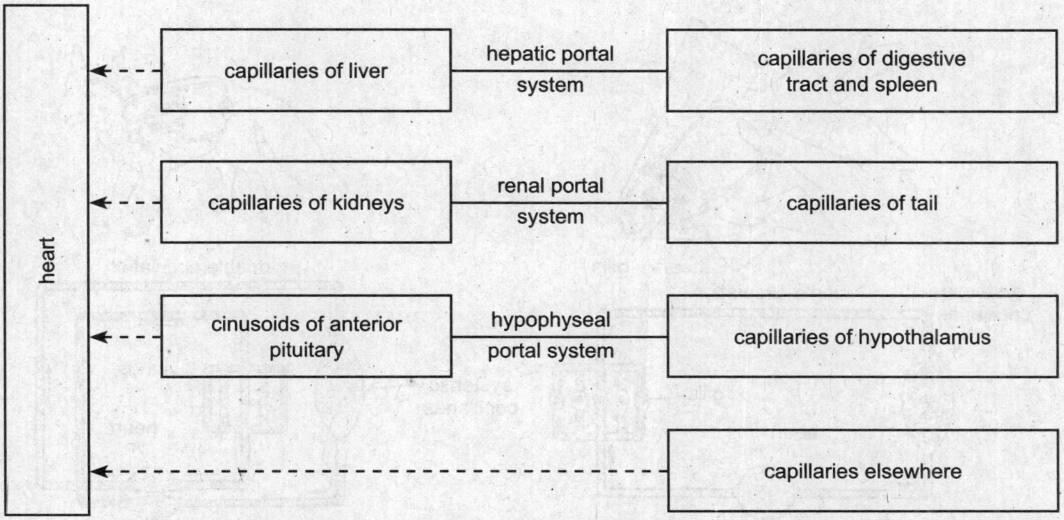

Fig 6.2. Chief portal systems of craniates. The renal portal system is lacking in typical mammals, and the hypophyseal portal system has not been demonstrated in most bony fishes. Channels indicated by broken lines are not part of a portal system (after Kent and Carr 2001).

plasma. Thus, erythrocytes function as containers for haemoglobin, preventing its eliminating. Erythrocytes vary in individual size, from 8 μm in humans, to 9 μm in elephants, to 80 μm in some salamanders. Most red blood cells live three to four months in the circulating blood before being broken down and replaced.

Leucocytes or **white blood cells (WBCs)**, are a second major cellular constituents of the formed elements. They (WBCs) defend the body from infection and disease. The **platelets** are a third formed element in the blood. They release factors that produce a cascade of chemical events leading to the formation of a **clot**, or **thrombus**, at sites of tissue damage.

Plasma and formed elements give blood a wide variety of roles in body processes. In addition to the functioning in respiration and disease protection, blood also plays a part in nutrition (carries carbohydrates, fats, proteins), excretion (carries spent metabolites), regulation of body temperature (carries and distributes heat), maintenance of water balance, and transport of hormones.

6.3 SINGLE AND DOUBLE CIRCULATION

Blood travels in one of two general patterns. Most fishes have a **single circulation** pattern in which blood passes only once through the heart during each complete circuit. With this design, blood moves from the heart to the gills to the systemic tissues and back to the heart (Fig. 6.3A). Amniotes have a **double circulation** pattern in which blood passes through the heart twice during each circuit, traveling from the heart to the lungs, back to the heart a second time (Fig. 6.3 B). The rise of this double circulation involving the addition of a pulmonary circuit was a major evolutionary event. Between those vertebrates with a single and those with double circulation stand functional intermediates include lungfishes, amphibians, and reptiles. They suggest the adaptive advantages of transitional forms venturing onto land and highlights the evolution of the circulatory system design.

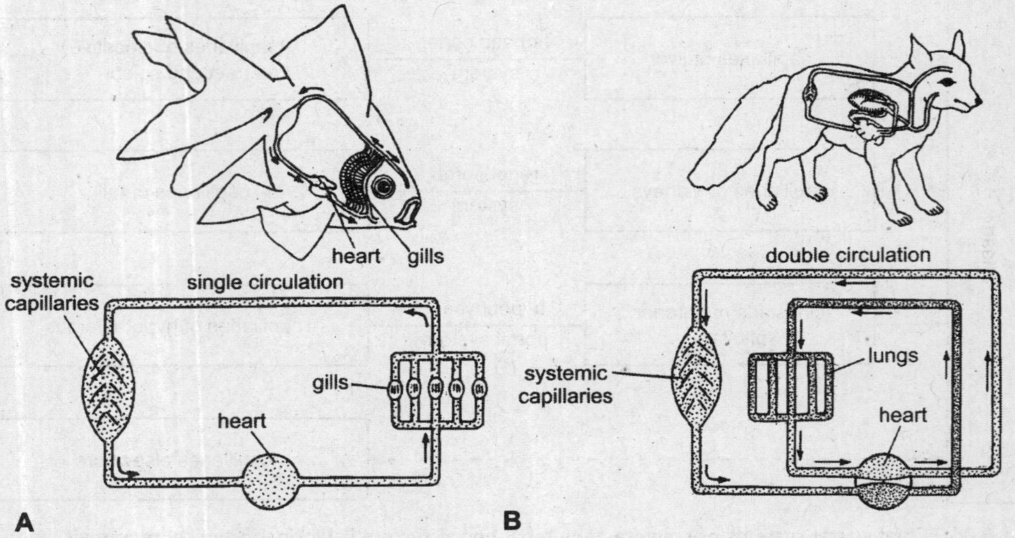

Fig. 6.3. Single and double blood circulation. A – The single circulation of fishes includes heart, gills and systemic capillaries in series with one another (arrows indicate path of the blood flow). B – The double circulation of most amniotes includes heart, lungs, and systemic capillaries. Blood passes twice through the heart before completing one route. This places lungs and systemic tissues in separate circuits that parallel each other (after Kardong 2006).

6.4 EMBRYONIC DEVELOPMENT OF THE CARDIOVASCULAR SYSTEM

Most blood vessels arise within embryonic mesoderm (or from mesenchyme) almost as soon as this germ layer becomes established in the vertebrate embryo. Small clusters of mesodermal cells called **blood islands** mark the embryonic debut of the cardiovasculary system (Fig. 6.4). Embryonic blood islands yield both blood vessels and blood cells, so they are involved in **angiogenesis** (blood vessel formation) and **hemopoiesis** (blood cell formation). Blood islands merge, forming a connected vascular network that eventually links parts of the embryo to each other and connects it to its nutrient supply and respiratory organs. The embryonic heart is tubular. Early on, it has autonomic rhythm, *i.e.*, autonomous rhythmic beats that drive blood through the developing vascular network. In the early embryo, such heart beats help primarily to promote new blood vessel formation. As in the adult the cardiovascular system of the mid-to late-term embryo also assumes an active and essential role in respiration, metabolism, excretion, and growth.

When first formed, the embryonic vertebrate heart is already contractile and includes four major adjoining chambers. The **sinus venosus** is the first chamber to receive returning blood. Blood flows next into the **atrium**, then into the **ventricle**, and finally in the fourth chamber, the **bulbus cordis**. From the bulbus cordis, blood leaves the heart to enter arteries, departing for the body of the embryo. In most tetrapods, splanchnic mesoderm forms the basic four-chambered, tubular heart. Development of the heart begins when cells leave the splanchnic mesoderm to form a medial pair of **endocardial tubes** (Fig. 6.5). Cells remaining in the splanchnic mesoderm proliferate, producing a thickened lateral region, the paired **epimyocardium**. Cells of the endocardial tube and epimyocardium grow toward the midline and fuse into the single, centrally located tubular heart. Specifically, the fused endocardial tubes form the endothelial

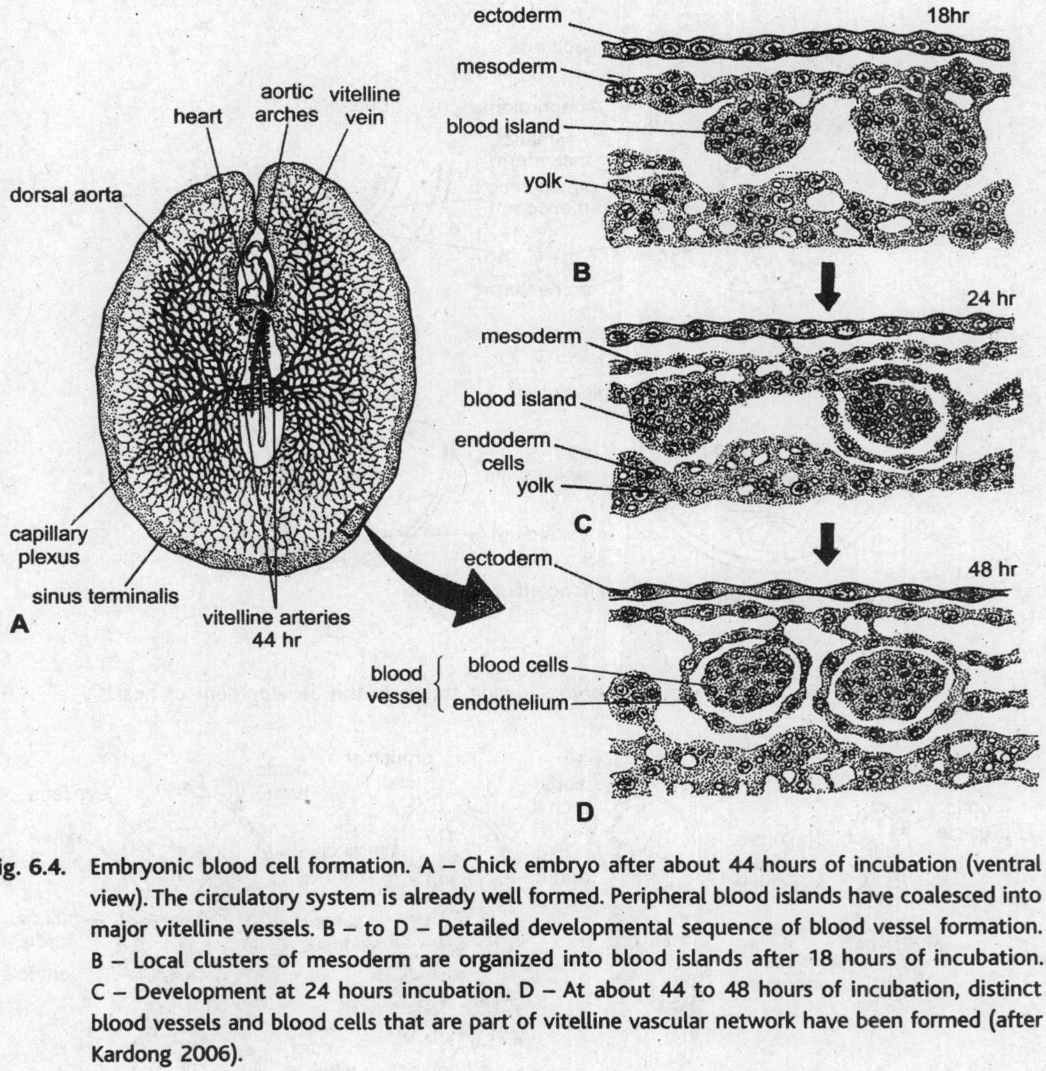

Fig. 6.4. Embryonic blood cell formation. A – Chick embryo after about 44 hours of incubation (ventral view). The circulatory system is already well formed. Peripheral blood islands have coalesced into major vitelline vessels. B – to D – Detailed developmental sequence of blood vessel formation. B – Local clusters of mesoderm are organized into blood islands after 18 hours of incubation. C – Development at 24 hours incubation. D – At about 44 to 48 hours of incubation, distinct blood vessels and blood cells that are part of vitelline vascular network have been formed (after Kardong 2006).

lining of the heart, called the **endocardium**, and the epimyocardium gives rise to extensive cardiac muscle of the heart wall, the **myocardium**, together with the thin visceral peritoneum covering the heart's surface. With these fusions, the basic four-chambered embryonic heart is established.

Subsequent folding of the tubular heart into different configurations, but the internal sequence of blood flow remains the same. In most fishes, adult retain this basic four-chambered embryonic heart. However, in lungfishes and tetrapods, varying degrees of internal subdivisions cordon off additional compartments within the heart, and some of the original chambers may become reduced or appropriated by other parts of the adult vascular system.

Subsequent folding of the tubular heart twists the heart into different configurations, but the internal sequence of blood flow remains the same (Fig. 6.6). In most fishes, adults retain this basic four-chambered embryonic heart. However, in lungfishes and tetrapods, varying degrees of internal subdivisions cordon off additional compartments within the heart, and

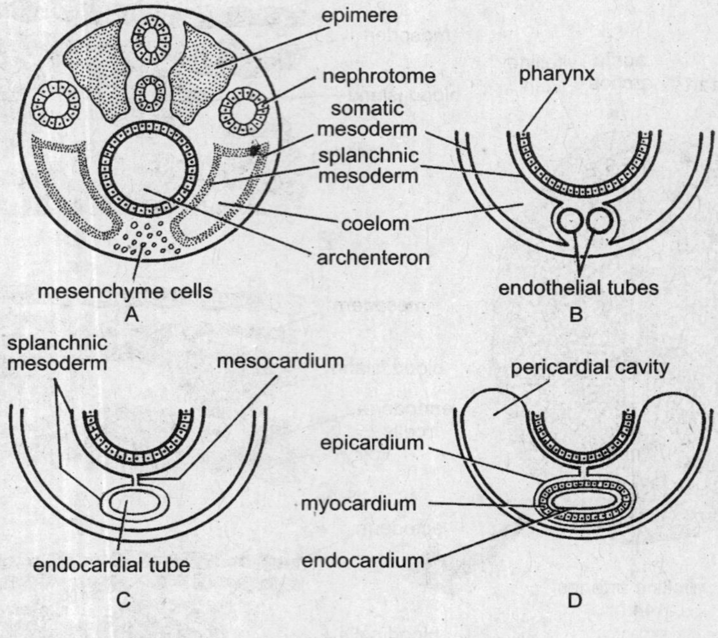

Fig 6.5. T.S. of a vertebrate embryo showing stages in the development of heart.

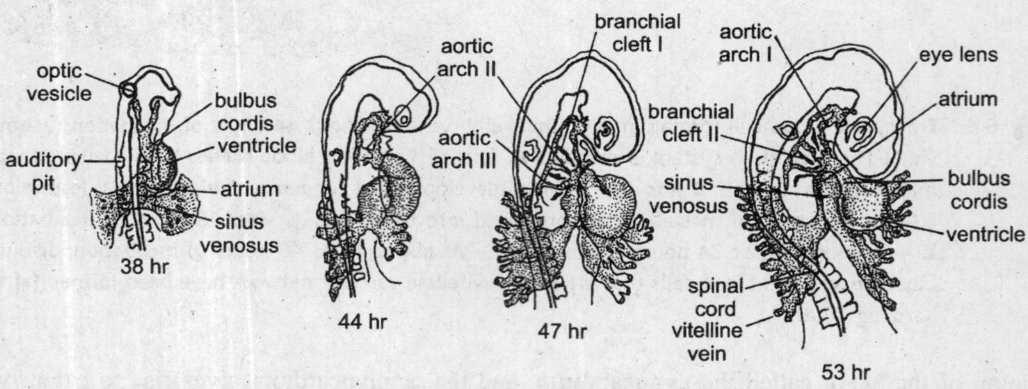

Fig. 6.6. Growth of chick heart. The four-chambered heart consists of sinus venosus, atrium, ventricles and bulbus cordis. Once heart is formed, subsequent folding and enlargement shift the relative positions of these chambers. This process does not alter the route of blood flow through the functioning embryonic heart (Kardong 2006).

some of the original chambers may become reduced or taken over by other parts of the adult vascular system.

6.5 PHYLOGENY OF THE CARDIOVASCULAR SYSTEM OF VERTEBRATES

The vessels of the cardiovascular system are as varied as the diverse organs they supply. However, these variations are based on modifications of a fundamental plan of organization common to vertebrates. Because it is usually highly modified in advanced forms, this fundamental organization of the cardiovascular system is most evident in primitive vertebrates. Blood leaving the heart first enters an unpaired **ventral aorta** and courses forward below the pharynx. Anteriorly, the ventral aorta divides into the **external carotids**, which carry blood into the ventral region of the head. Before producing these external carotids, however, the ventral aorta gives off a series of **aortic arches**, which pass dorsally within the branchial arches between pharyngeal slits. Above the pharynx, these aortic arches meet a paired **dorsal aorta**. Sprouting from the anterior end of the dorsal aorta are the **internal carotids**, which carry blood forward into the head and usually penetrate the braincase to supply the brain. The dorsal aorta itself, however, carries blood posteriorly (Fig. 6.7).

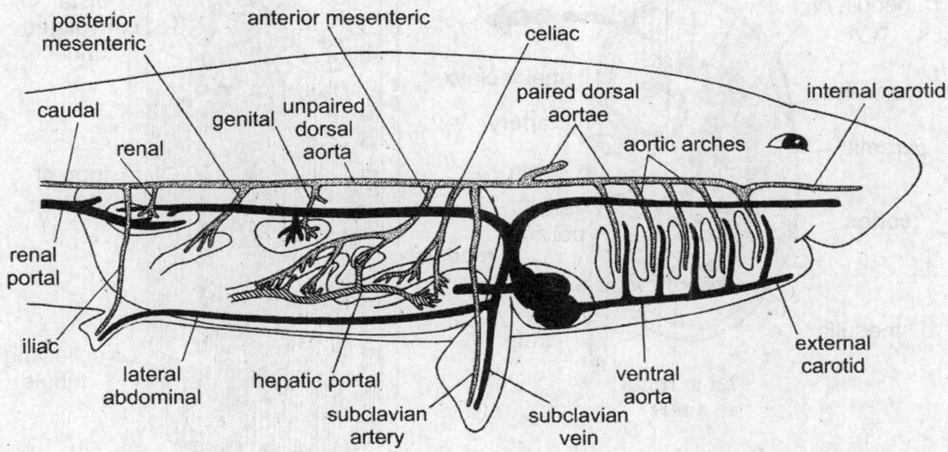

Fig 6.7. Basic vertebrate circulatory pattern illustrated in a shark. The heart pumps blood to the ventral aorta, from which it is distributed to the paired aortic arches and then to the single dorsal aorta. From the dorsal aorta, blood flows forward to the head and posteriorly to the body, where major branches carry it to visceral and somatic tissues (after Kardong 2006).

At about the level of the liver, the paired vessels of the dorsal aorta unite to form the unpaired **aorta**, which distributes blood to the posterior part of the body and eventually extends into the tail as the **caudal artery**. Along the way, the dorsal aorta gives off numerous small **parietal arteries** to the local body wall, as well as several major arteries, usually paired, to somatic tissues. Paired **subclavian arteries** supply the anterior appendages (fins or limbs) and usually branch from the dorsal aorta, as do the caudal **iliac arteries**, which supply the posterior appendages. The gonads receive blood from paired **genital arteries** (ovarian or spermatic). Paired **renal arteries** to the kidneys are large, major branches from the dorsal aorta. This ensures that the kidneys receive blood early in the arterial circuit while blood pressure is still relatively high, a feature of the hemodynamics that aids renal filtration (Box 6.1).

Box 6.1 Glomerulus

The renal artery, one of the major branches from the dorsal aorta, delivers blood to the kidneys. Through a series of subsequent branches, it eventually forms tiny capillary beds known as **glomeruli**, each being associated with a **renal capsule** (Bowman's capsule) constituting the first part of the nephron. Collectively, the glomerulus and renal capsule form the **renal corpuscle**. An ultrafiltrate without blood cells and proteins is forced through the capillary walls and collects in the renal capsule before it passes through the **proximal** convoluted tubule, **intermediate** tubule, and

distal convoluted tubule of the nephron, eventually entering the collecting tubules. During transit, the composition of the fluid is altered and water is removed. After circulating through the glomerulus, blood flows through an extended capillary network entwined about the rest of the uriniferous tubule (Fig 6.8). Thereafter, blood is collected in progressively larger veins that join the common **renal veins** leaving the kidneys.

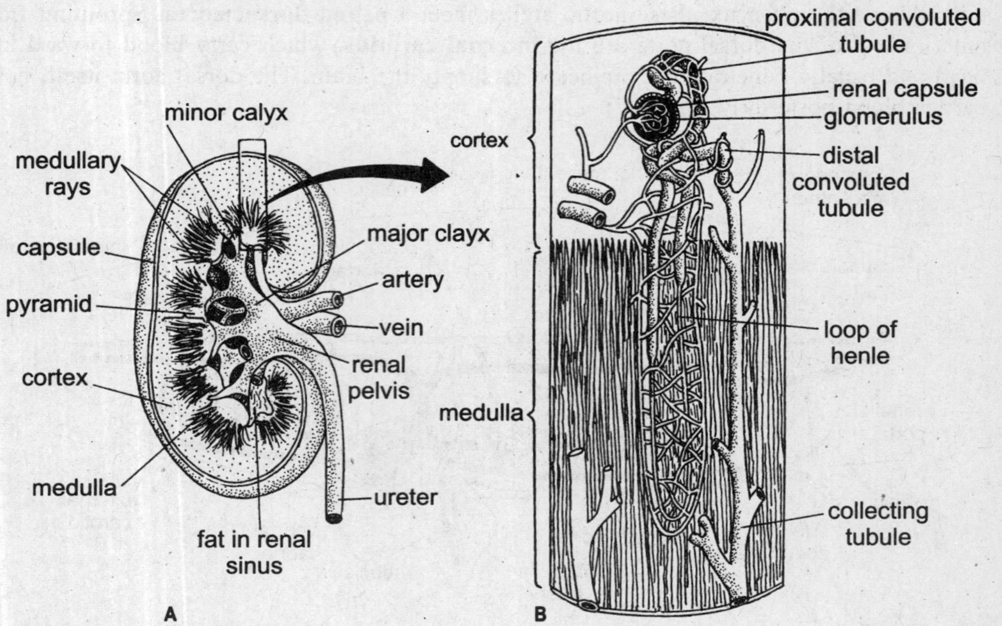

Fig 6.8. Blood supply to a mammalian kidney (after Kardong 2006).

In primitive vertebrates, blood return to the heart includes several prominent veins. The **common cardinal vein**, or **sinus**, is the major vein that receives blood returning from the **anterior cardinal vein** (precardinal) and the **posterior cardinal vein** (postcardinal). These drain the anterior and posterior regions of the body, respectively. Tributaries from the appendage usually empty into the common cardinal vein via the **subclavian vein**. Veins from the lateral body wall and posterior appendage also empty into the common cardinal vein via the **lateral abdominal vein** (Fig 6.7).

A **portal system** is a vascular pathway that begins in one set of capillaries and runs to another without coursing through the heart in between. There are two major portal systems of the venous circulation in vertebrates. The **hepatic portal vein** to the liver, where it empties into the capillaries and blood sinuses of the liver. This hepatic portal vein transports absorbed nutrients directly from the digestive tract to the liver for storage or processing of many end products of digestion. The **renal portal system** transports blood returning from capillary beds within the tail or hindlimbs through the paired **renal portal veins**, which empty into capillaries within the kidney.

The function of the renal portal system is not well understood. Because it carries caudal blood to the kidneys, some have suggested that it provides a direct route for delivering metabolic byproducts to the kidneys that result from active locomotion involving the caudal musculature. Others suggest that it may represent a way of improving kidney filtration. Arterial blood entering the renal arteries directly from dorsal aorta has high pressure, venous blood of the renal portal system has low pressure. Kidney filtration depends in part on an initial high pressure to move fluid out of the blood and into the kidney tubules, but low pressure in the renal portal veins aids in the recovery of water and other usable solutes, returning these fluids to the general circulation. The renal portal system is present in all classes of vertebrates except mammals. Even though mammals lack a renal portal system, the mammalian kidney nevertheless has a low-pressure vascular network that may be its counterpart and functions similarly to recover fluids from the forming urine.

Phylogenetic modifications within this basic pattern of arteries and veins are largely correlated with functional changes. In the transition from water to land, gills gave way to lungs, accompanied by the establishment of a pulmonary circulation. In some fishes and certainly in tetrapods, the cardinal veins become less involved in blood return. Instead, the composite prominent **post cava (posterior vena cava)** arose to drain the posterior part of the body and the **pre cava (anterior vena cava)** developed to drain the anterior part of the body.

6.6 EVOLUTION OF HEART

The craniate heart is chiefly a muscular pump that occupies the pericardial cavity. The walls consist of an **endocardium, myocardium** and **epicardium**, which correspond, respectively, to the intima, media, and adventitia of the arteries. However, the myocardium is cardiac muscle, rather than smooth muscle as in arteries. It is especially thick in the ventricular wall. Lying on the epicardium is the **visceral pericardium**, the equivalent of the visceral peritoneum or pleuroperitoneum of the main coelom. The parietal and visceral pericardia are continuous with one another, being reflected over the blood vessels that enter and leave the heart. The space between these pericardia is the **pericardial cavity**.

The tissues of the heart are supplied with arterial blood by **coronary arteries** and drained by **coronary veins**. The heart muscle pulsates in response to specific electrolytes in the blood that perfuses it. The rhythmicity of the beat, is imposed by the autonomic nervous system in vertebrates, although all craniates possess an intrinsic control.

The heart as a pump moves blood through vessels both by **pushing** blood through the circulatory system, but also by **aspiration**, creating a negative pressure that sucks blood into the heart. In a slow-swimming dogfish shark, the heart can move 7.5 liters of blood per hour; in a resting hen, 24 liters per hour; in a human, 280 liters per hour. In a giraffe, almost 1,200 liters blood can circulate throughout the body per hour. If heart rate increases, a response known as **tachycardia**, these values can increase fivefold. If heart rate decreases, **bradycardia** results and these values can fall precipitously. For example, when a turtle dives, its cardiac output can drop to less than one-fiftieth (1/50) of its prior output. In addition to functioning as a pump, the heart also serves to channel deoxygenated and oxygenated blood to appropriate parts of the circulation, thereby preventing their mixing.

Basic Vertebrate Heart

Phylogenetically, the heart probably began as a contractile vessel, much like those found within the circulatory system of amphioxus. In most fishes, the heart is part of a *single circulation*. Vessels serving gas exchange in the gills and systemic capillary beds are in series, with each other. The embryonic fish heart consists of **four chambers**, which are also in series, so that blood flows in sequence from the sinus venosus, to the atrium, to the ventricle, and finally to fourth and most anterior heart chamber, the bulbus cordis, before entering the ventral aorta. Differences in structure, doubts about homology, and loose use of terms have led to confusion about nomenclature for this fourth chamber. **Kardong** (2006) have used the term **bulbus cordis** for this chamber in embryos. In adults, Kardong used the term **conus arteriosus** for this anterior chamber of heart if it is composed of cardiac muscle, is contractile, and contains various numbers of **conal values** internally.

A conus arteriosus is generally present in chondrichthyans, holosteans and dipnoans. Although absent in adult tetrapods, during embryonic development its forerunner, the **bulbus cordis**, divides into the bases of the major arteries leaving the heart. In some fishes, most notably in teleosts, this anterior chamber is thin walled with smooth muscle and elastic fibers, but it lacks both cardiac muscle and conal valves, so it usually is called the **bulbus arteriosus.** The adult bulbus arteriosus, like the conus arteriosus, arises generally from the embryonic bulbus cordis, but in some fishes the adult bulbus arteriosus may incorporate part of the adjoining ventral aorta as well (Fig. 6.9). Another term often used ambiguously in the older literature is **truncus arteriosus**, which should apply only to the ventral aorta or its immediate derivatives but not to any part of the heart proper (**Kardong** 2006). In tetrapods, the ventral aorta often becomes reduced, sometimes persisting only as a small section of vessel at the base of major departing aortic arches. In these cases, the term **truncus arteriosus** is most suitable.

The heart muscles are quite active ones, so requires gas exchange (oxygen, carbon dioxide) to support its metabolism. In many fishes and primitive tetrapods, this demand is met by direct gas exchange between the myocardium and blood passing through its lumen. The inner wall of the myocardium, especially of the ventricle, often forms projecting cones of muscle termed **trabeculae** that are set off by deep recesses. The resulting texture, when viewed from the lumen,

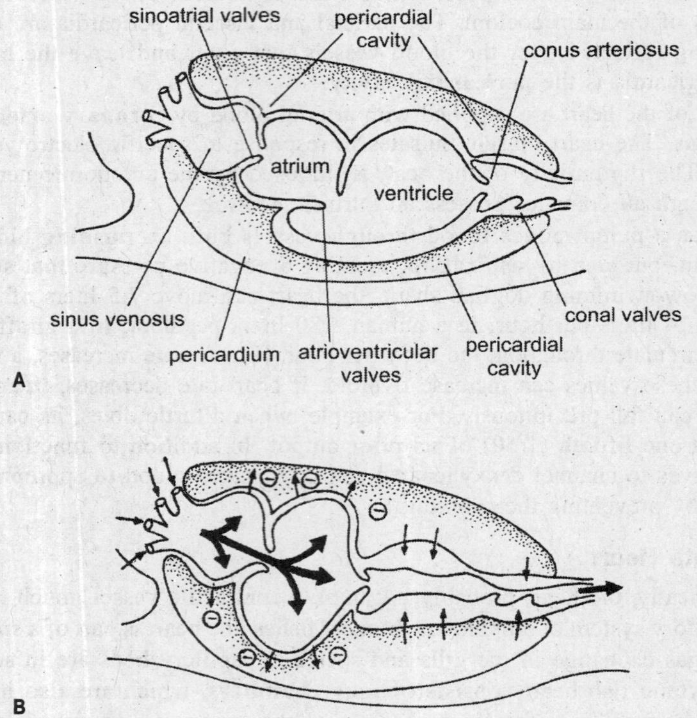

Fig. 6.9. Basic heart structure and aspiration filling. A – The four chambers of the fish heart are enclosed within the pericardial cavity. One-way valves between each chamber prevent reverse flow of blood as successive chambers contract. B – When the ventricle contracts, the volume occupied by the ventricle within the pericardial cavity is momentarily reduced. Reduced ventricular volume creates a negative pressure around the other chambers. Because the walls of the sinus venosus and atrium are thin, this low surrounding pressure causes them to expand, creating within their lumina a negative pressure that sucks in or aspirates blood from the returning veins (after Kardong 2006).

looks spongy and is referred to as **trabeculate**. The **coronary vessels** perfuse the heart wall, usually only the outer part of the myocardium. They are especially well developed in elasmobranchs, crocodiles, birds, and mammals, in which they supply most of the myocardium. In fishes, the coronary arteries are derived from the efferent arches or collecting loops of the gills, which carry oxygenated blood. The coronary veins enter the sinus venosus.

The intrinsic property of cardiac muscle is contraction. Individual cells even show rhythmic contractions if they are isolated outside the body in a suitable culture medium. Cardiac cells tend to beat in synchrony. Contraction of entire heart usually begins within a restricted region in the sinus venosus called the **pacemaker** or **sinoatrial (SA) node**, and then spreads through a conducting system of fibers into the ventricle and other contracting regions of the heart. In mammals, the contracting system includes, in addition to the SA node, a second node, the **atrioventricular (AV) node** in the wall of the heart. The AV node consists of **Purkinje fibers**, neuron-like fibers that are modified cardiac muscle cells. The Purkinje fibers depart from the AV node, divide into left and right bundles traveling within the interventricular septum to the apex of the heart, then turn and sweep around the respective sides of the ventricles. The rate at which heart beats are initiated is under the influence of the nervous and endocrine systems. Heart rate also responds to the rate of venous filling (the Frank-Starling reflex).

Birds and mammals have four-chambered hearts, but of the original four fish chambers, only two persist as major receiving compartments, the atrium and the ventricle, both of which are divided into left and right compartments to produce four chambers. Although the hearts of birds and mammals are both derived from reptiles, they arose independently from different reptilian ancestors.

Evolution of Heart in Different Classes of Vertebrates

1. Heart of Cyclostomes

Hagfishes are primitive fishes. They live a life of slow, prowling (*i.e.*, moving about in a stealthy or restless way) scavengers. Their cardiovascular system holds some unique features. Like all vertebrate hearts, the hagfish heart, called **branchial** or **venous heart**, lies within the anterior trunk region and is composed of cardiac muscle, and receives blood returning from the general systemic circulation (Fig. 6.10). Branchial heart of hagfish includes three chambers in series: the sinus venosus, the atrium, and the ventricle (Fig. 6.10 B). Blood returning from the two common cardinals and liver first enters the sinus venosus. It flows through the atrium and then the ventricles, finally it is pumped directly into the ventral aortic and hence to the gills. One-way valves between heart chambers prevent reverse blood flow. No major nerves innervate the hagfish heart to stimulate contraction. Instead, filling of the sinus venosus by returning blood elicits the Frank-Starling reflex, which stimulates stronger contractions that originate in the pacemaker and then spread sequentially to the other chambers.

Besides the branchial heart, there occur two types of unique accessory blood pumps or hearts, namely **portal heart** and **caudal heart** (Fig. 6.10 C, D). These hearts contract but usually lack the cardiac muscle of true branchial heart. In most fishes, systemic tissues drain via discrete venules and veins. However, in the hagfish, venous drainage of some regions, such as the head and the subcutaneous caudal regions, is provided with large open **sinuses**. A probable consequence is that venous blood pressure is especially low. Accessory hearts on the venous side of circulation are an apparent answer to the problem of returning low-pressure venous blood.

The **cardinal hearts** lying within the anterior cardinal veins are like sacs whose pumping action is initiated by skeletal muscles around their outer walls. The paired **caudal hearts**, which are located in the tail, represent a unique blood-pumping mechanism among vertebrates. These hearts are composed of a central **cartilaginous rod**, skeletal muscles on the sides and veins

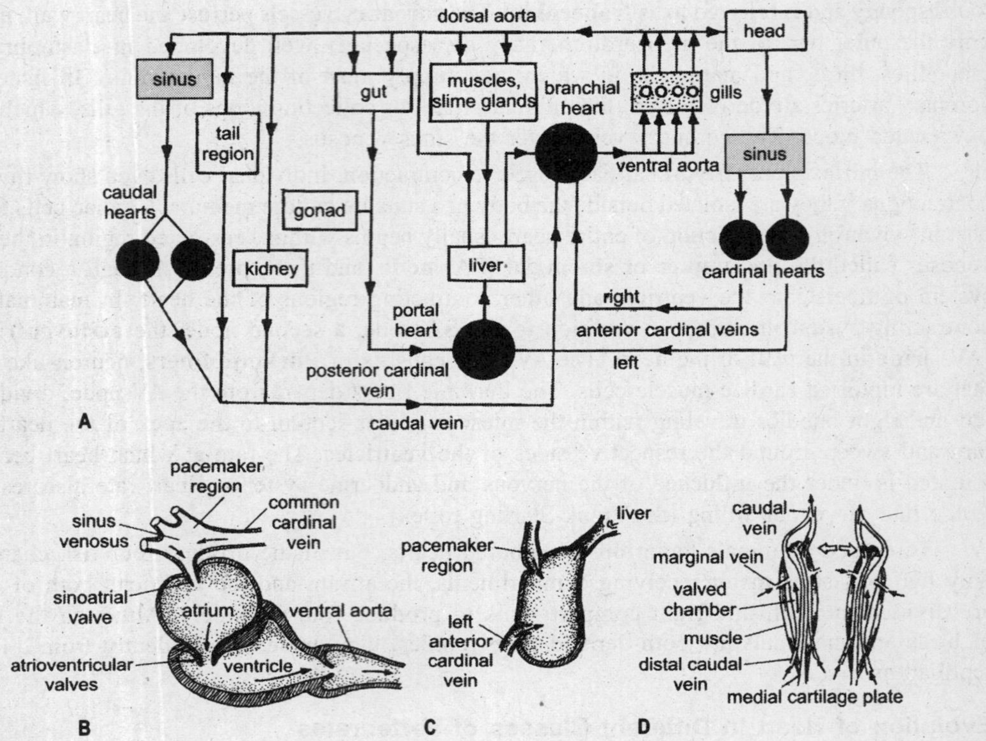

Fig. 6.10. Heart and blood circulation in hagfishes. A – Diagram of cardiovascular system. B – Branchial heart illustrating the three basic chambers. C – Accessory heart, the portal heart pulsates to help drive blood. D – Another types of accessory hearts, the caudal hearts have paired striated muscles and central flexible support. Caudal hearts lie almost at the end of the tail of hagfishes. Contraction of the left muscle bends the medial cartilage plate expelling blood from the compressed right chamber and allowing the left to fill. Contraction of the right muscle has the opposite effect. Alternating contractions of the caudal heart musculature enlarge and compress the veins, causing them to fill and then empty (after Kardong 2006).

between. Alternative contraction of these muscles bends the rod back and forth, which presses on the walls of the vessels and pumps blood into the caudal vein (Fig. 6.10 D).

The **portal heart** is a single, expanded vascular sac that receives venous blood from one anterior and one posterior cardinal vein, and then it contracts to drive the blood through the liver (Fig. 6.10 C). Only hagfishes have such an accessory heart in the course of the hepatic portal vein, which **elevate blood pressure** prior to the blood's entry into the liver sinusoids. Furthermore, the portal heart is the only accessory heart of true branchial hearts.

The lamprey heart is the branchial heart which is more like the typical fish heart, with four basic compartments (Fig. 6.11). Blood flows sequentially into the sinus venosus, the atrium, the ventricle, and the conus arteriosus. One-way valves separate compartments of the heart. The sinoatrial and atrioventricular valves prevent retrograde blood flow. Like the other heart chambers, the conus arteriosus is muscular, although this is thick, smooth, and not cardiac musculature. Luminal walls of the conus arteriosus are thrown into leaflets, the conal valves, collectively forming the **semilunar valves**, which prevent reverse (backward) blood flow and possibly aid in distributing blood-to the aortic arches. From the arches, blood flows to the delicate gills capillaries next in line in the circulation.

2. Heart of Fishes

Like the lamprey, the hearts of chondrichthyans and bony fishes consist of four basic chambers–sinus venosus, atrium, ventricle, and conus arteriosus (or bulbus arteriosus)– with one-way valves stationed between compartments (Fig. 6.11). Like the other chambers, the muscular conus arteriosus, acting as an auxiliary pump to help maintain blood flow into the ventral aorta after the onset of ventricular relaxation. Contraction of conus arteriosus also brings together the conal valves located on its opposing walls. When these valves meet, they prevent the backflow of departing blood. In teleosts, this fourth chamber is an elastic, noncontractile bulbus arteriosus. A single pair of **bulbar valves** at the juncture of the bulbus arteriosus and the ventricle prevents retrograde flow (Fig. 6.12). When receiving blood following ventricular contraction, the bulbus arteriosus stretches and then gently undergoes elastic recoil to maintain blood flow into ventral aorta. The result is a **depulsation**. This has been proposed as a means of protecting the delicate

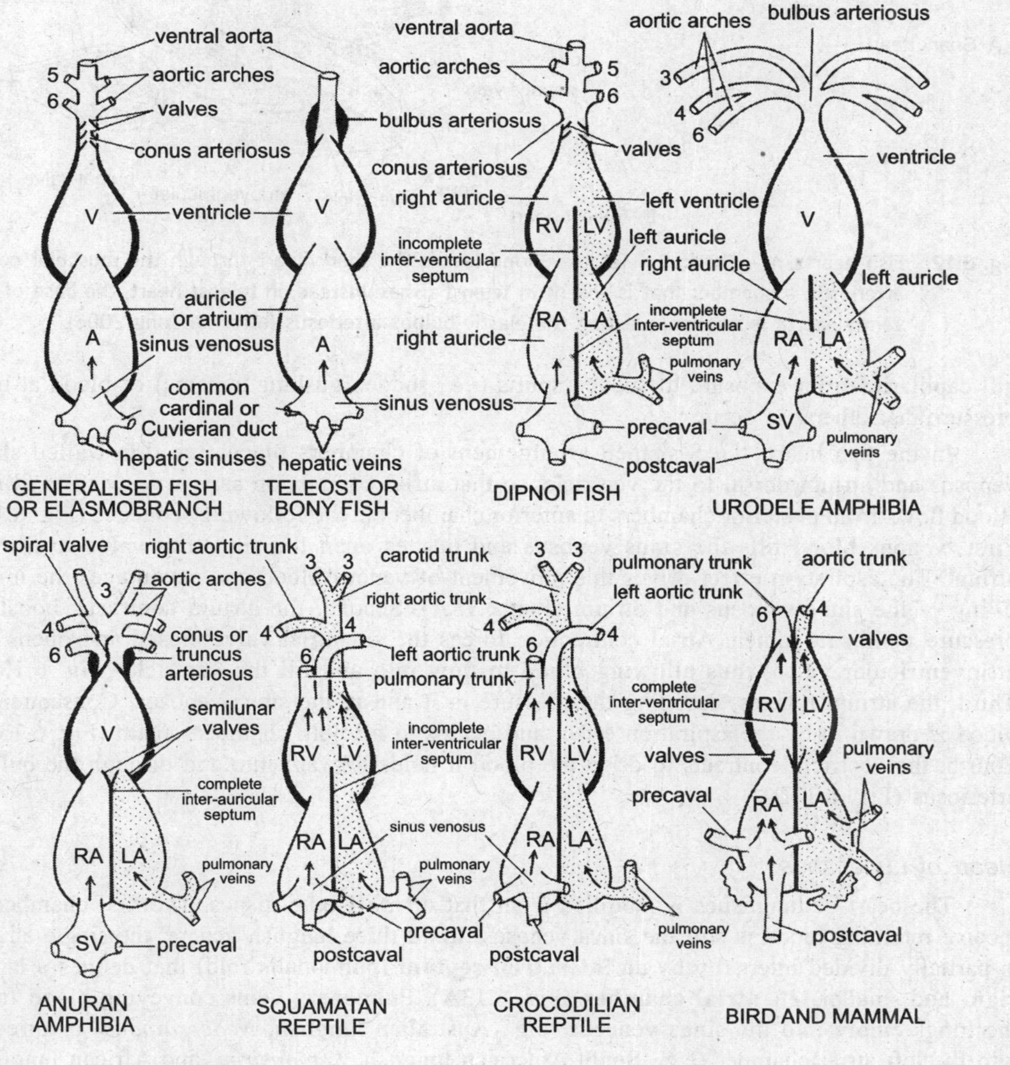

Fig. 6.11. Evolution of heart in different classes of vertebrates.

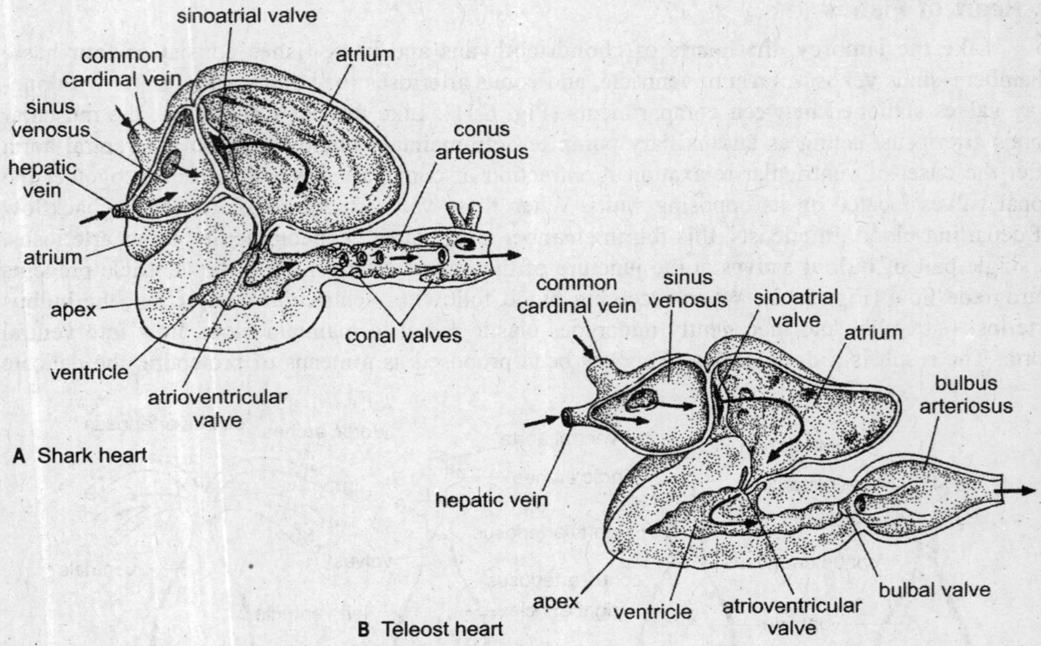

Fig. 6.12. Fish hearts. A – Shark; B – Teleost. Blood leaves the shark heart through the muscular conus arteriosus, a chamber that is absent in teleost fishes. Instead, in teleost heart, the base of the ventral aorta is swollen, creating the elastic bulbus arteriosus (after Kardong 2006).

gill capillaries from exposure to sudden spurts (*i.e.*, sudden gushing streams) of blood at high pressure that otherwise occur.

In the fish heart, the S-shaped arrangement of chambers places the thin-walled sinus venosus and atrium dorsal to the ventricle, so that atrial contraction assists ventricular filling. Blood flows from posterior chambers to anterior chambers in the following sequence (Fig. 6.13). First, venous blood fills the sinus venosus and pushes open the sinoatrial valve to fill the atrium. The aspiration effect drives this movement of venous blood and encourages the initial filling of the sinus venosus and atrium (Fig. 6.13A). Second, the atrium contracts, boosting pressure within its lumen. Atrial contraction forces the sinoatrial valve closed and opens the atrioventricular valve, thus allowing blood to flow into and fill the ventricle (Fig. 6.13B). Third, the atrium relaxes, dropping the pressure in it and in the sinus venosus. Consequently, blood is drawn in by the aspiration effect and begins to fill both chambers again (Fig. 6.13C). Fourth, the ventricle contracts to drive the blood it holds forward into and through the bulbus arteriosus (Fig. 6.13D).

Heart of Lungfishes

The heart of lungfishes is modified from that of other bony fishes. The first chamber to receive returning blood is still the sinus venosus. In all three lungfish genera, the single atrium is partially divided internally by an **interatrial septum** (pulmonalis fold) that defines a larger **right** and smaller **left atrial chamber** (Fig. 6.13A). Pulmonary veins conveying blood from the lungs empty into the sinus venosus (e.g., Australian lungfish, *Neoceratodus*) or directly into the left atrial chamber (e.g., South American lungfish, *Lepidosiren*, and African lungfish, *Protopterus*). The sinus venosus conveying systemic venous blood opens into the right atrial

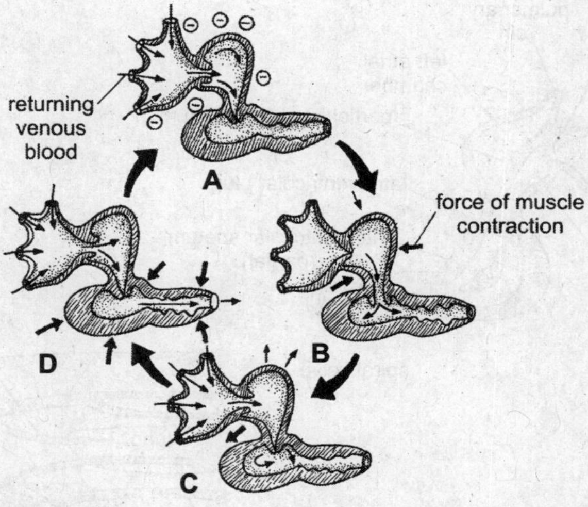

Fig. 6.13. Contraction cycle of the teleost heart. A – Relaxation of the sinus venosus and atrium draws in blood via the hepatic and common cardinal veins. B – Contraction of the atrium closes the sinoatrial valve and forces blood into the ventricle. C – As the atrial walls relax again, blood enters the atrium. D – Contraction of the ventricle forces blood through the bulbus arteriosus and distributes it to the aortic arches. The aspiration effect (negative signs) contributes to and completes the refilling of the sinus venosus and atrium to start the cycle again (after Kardong 2006).

chamber (Fig. 6.13A). In place of the atrioventricular valves is the **atrioventricular plug**, a raised cushion in the wall of the ventricle. It moves into and out of the opening from the atrium, like the AV valves, to prevent retrograde flow of blood into the atrium. The ventricle of lungfish is also divided internally, but only partially by an **interventricular septum**. Within dipnoans, the South American lungfish shows the greatest degree of both ventricular and atrial subdivision. The Australian lungfish shows the least. Alignment of the interventricular septum, atrioventricular plug, and interatrial septum establishes internal channels within and through the heart. When the lungfish breathes air, the left channel tends to receive oxygenated blood returning from the lungs. The right channel tends to carry deoxygenated blood (Fig. 6.13B). Thus, despite the anatomical incomplete internal septation of the lungfish heart, blood entering from the sinus venosus does not tend to mix with blood returning from the lungs.

The **spiral valve** within the conus arteriosus aids in separating oxygenated and deoxygenated blood (Fig. 6.14; Fig. 6.15). Apparently derived from conal valves, the spiral valve consists of two endocardial folds whose opposing free edges touch but do not fuse.

3. Heart of Amphibians

Amphibians depend on cutaneous gas exchange (plethodontid salamanders, which lack lungs), on gills (many larval forms), on lungs (most toads and frogs) or on all three modes (most amphibians). Because the sources of oxygenated and deoxygenated blood vary, structure of amphibian heart varies as well. Generally, in amphibians with functional lungs, the heart includes a sinus venosus, right and left atria divided by an anatomically complete interatrial septum, a ventricle lacking any internal subdivision, and a conus arteriosus with a spiral valve (Fig. 6.11). Except for salamanders of the genus. *Siren*, which have a partial interventricular septum, amphibians are unique among air - breathing vertebrates in lacking any internal division within the ventricle.

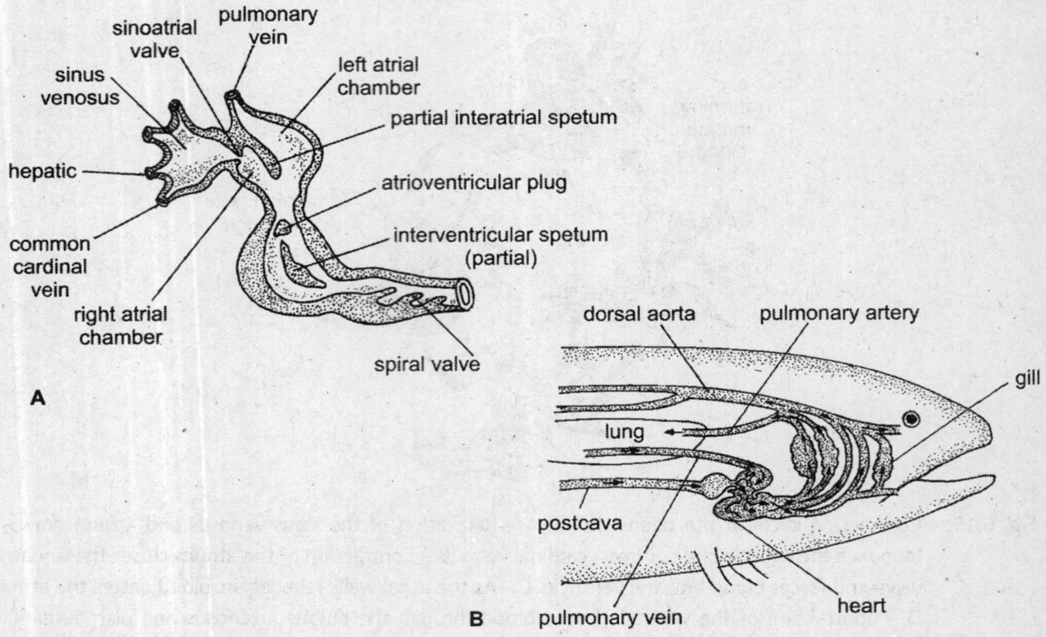

Fig. 6.14. Heart of African lungfish *Protopterus*. A – Internal structure of the heart. B – Path of blood. When the lungfish breathes air, venous blood returing from systemic tissues flows through the heart and tends to be directed to the last aortic arch. The pulmonary artery carries most of the deoxygenated blood to the lung. Blood high in oxygen returning from the lung passes through the heart and then tends to enter the aortic arches without gills. In this manner, blood is shunted directly to the general circulation. Thus, when lung fishes breathe air, they display the beginnings of a double circulation system. The five aortic arches phylogenetically represent the second through the sixth (roman numerals). The first (II) and last two (V, VI) of these carry gills (after Kardong 2006).

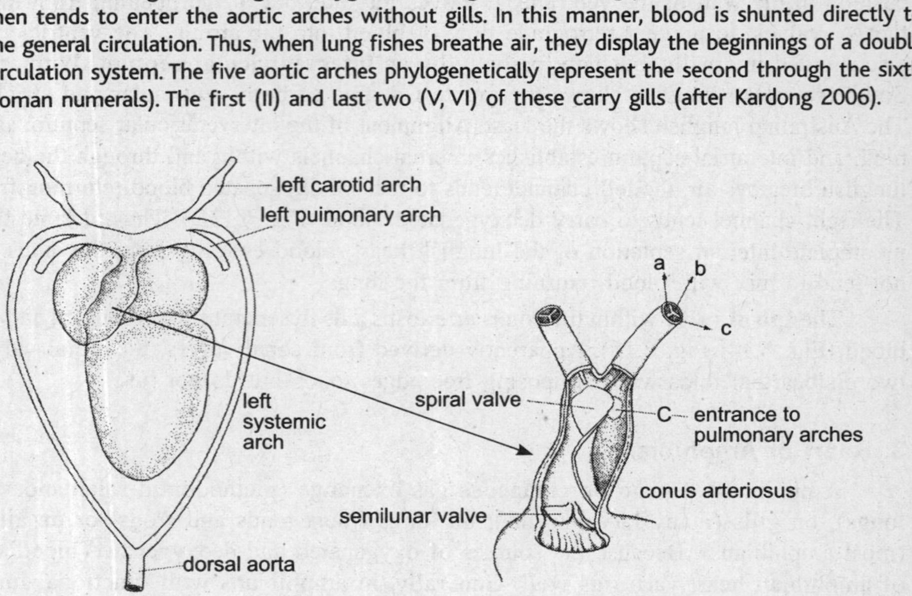

Fig. 6.15. A – Heart and aortic arches of frog, ventral view. B – Conus arteriosus opened to show passageways to left carotid arch (arrow a – a), and left systemic arch (arrow b – b), c, arrow enters common passageway to left and right pulmonary arches, then turns to enter passageway to left pulmonary arch (after Kent and Carr, 2001).

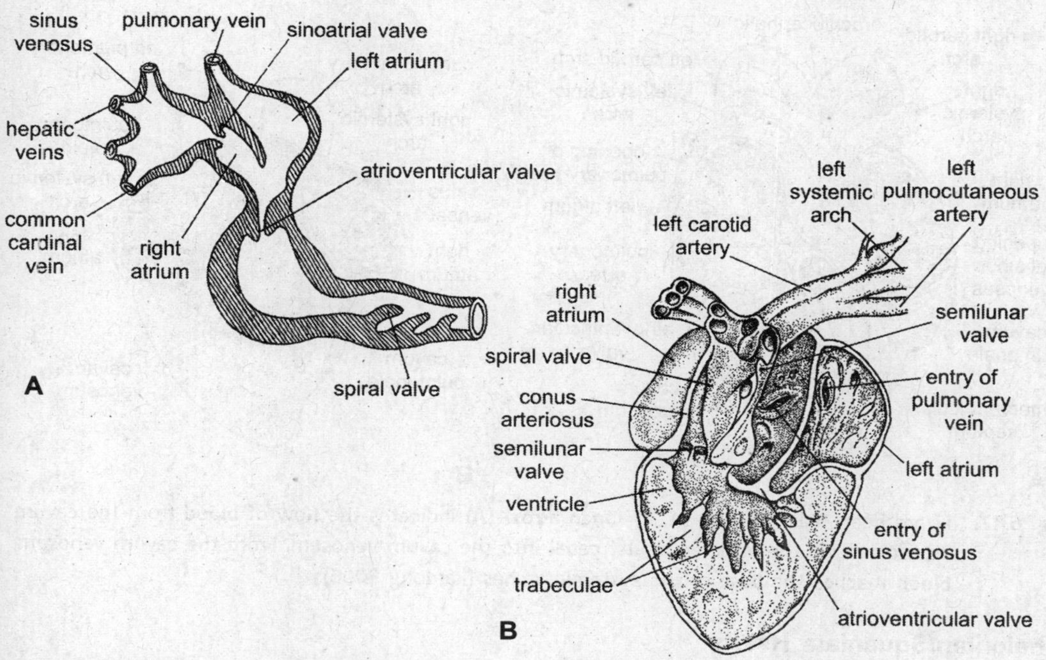

Fig 6.16. Amphibian hearts. A – Diagram of typical amphibian heart. Notice that the atrium is divided into left and right chambers but the ventricle lacks an internal septum. B – Bullfrog (*Rana catesbeiana*) heart. Although lacking internal septa, the wall of the ventricle folds into numerous trabeculae. The small compartments between these trabeculae are thought to aid in separating bloodstreams that pass through the heart (after Kardong 2006).

The cardiovascular system is perhaps best studied in frogs (Fig. 6.15). The conus arteriosus of the frog heart arises from a single trabeculate ventricle (Fig. 6.16B). Semilunar valve lie at the base of the conus arteriosus and prevent retrograde flow of blood back into the ventricle. Internally, a spiral valve twisting through nearly a complete rotation establishes within the conus, each of which guides blood to specific sets of systemic and pulmocutaneous arches. The systemic and pulmocutaneous arches both arise from the truncus arteriosus, a remnant of the ventral aorta, but the two sets of arches receive blood from different sides of the spiral valve.

In lungless salamanders or those with reduced lung function, the interatrial septum and spiral valve may be much reduced or absent entirely. Unlike frogs, in which the pulmocutaneous artery branches give rise to the cutaneous artery, salamander's lack a cutaneous artery. Instead, branches from vessels supplying the systemic circulation carry blood to the salamander skin. The pulmonary artery and the systemic arches in salamanders arise from the truncus arteriosus.

4. Heart of Reptiles

Reptiles have entered more fully terrestrial environments and adopted more active lifestyles than the amphibians that preceded them. The cardiovascular system of reptiles supports accompanying higher metabolic rates and elevated levels of oxygen and carbon dioxide transport. It is capable of generating elevated blood pressure, higher cardiac output, and efficient separation of oxygenated and deoxygenated bloodstreams. The diversity of hearts and heart function in reptiles is becoming better understood. It is clear that no single reptile heart can represent all others. In general, two basic reptilian heart patterns are recognised. One is found in chelonians and squamates and the other in crocodilians.

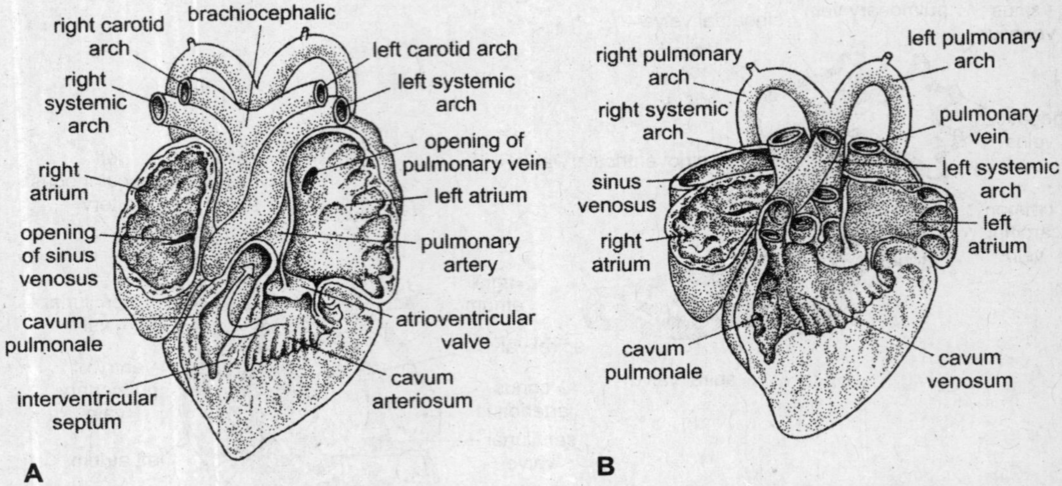

Fig. 6.17. Lizard heart, ventral view (A, B). Open arrow (A) indicates the flow of blood from the cavum arteriosum via the interventricular canal into the cavum venosum. From the cavum venosum, blood reaches the base of the systemic arches (kardong 2006).

Chelonian/Squamata Hearts

In these reptiles, the *sinus venosus* is reduced in comparison to amphibians but it retains the same functions. It is still the first chamber to receive venous blood and contains the pacemaker. The *atrium* is completely divided into right and left atria. Prominent atrioventricular valves guard the entrance to the ventricles. The *conus arteriosus* (or bulbus cordis) appears during early embryonic development but becomes divided in the adult to form the bases (trunks) of three large arteries leaving the ventricle: the **pulmonary trunk** and the **right** and **left aortic trunks**. In snakes, a valved interaortic foramen connects the bases of adjacent aortae. But the shunting of blood made possible by this foramen has not been explored. Usually, the **brachiocephalic artery**, delivering blood to the subclavians and carotids, comes out directly from the ventricle, crowded in with the trunks of the three aortic arches (Fig. 6.17A, B). The conus also gives rise to a band of contractile tissue at the base of pulmonary trunk to control the resistance blood meets as it flows to the lungs. The *ventricle* is a single chamber functioning as a single fluid pump to drive blood into the major arteries leaving the heart. Internally, however, ventricle has *three* interconnected compartments: the **cavum venosum** and the **cavum pulmonale** separated from each other by a **muscular ridge**, and the **cavum arteriosum** connected to the cavum venosum via an **interventricular canal**. The cavum arteriosum fills with blood from the left atrium but has no direct arterial output. During systole, the blood it receives flows through the interventricular canal to the aortic arches. The cavum pulmonale does not receive blood directly from the atria. Instead, blood from the cavum venosum moving across the muscular ridge fills the cavum pulmonale. In turn, much of the blood filling the cavum venosum is deoxygenated blood from the right atrium. Thus, the heart has five chambers, composed of two atria and three compartments of the ventricle, or six chambers if we count the sinus venosus.

The *pattern of blood flow* through the hearts of Chelonia and Squamata differs depending on whether they breathe air or hold their breath. For example, in an air-breathing turtle on land, most deoxygenated blood returning from systemic tissues is directed to the lungs, and most oxygenated blood from the lungs is directed to the systemic tissues via the aortic trunks.

The right atrium receives deoxygenated blood returning from the body from the sinus venosus. The left atrium receives oxygenated blood returning from the lungs (via pulmonary vein). When the atria contract, deoxygenated blood in the right atrium flows into the cavum venosum and then across the muscular ridge to the cavum pulmonale. Additionally, when the right AV (= atrioventricular) valve opens, they lie across the opening to the interventricular canal and temporarily close it. Oxygenated blood in the left atrium enters the cavum arteriosus and temporarily remains there while the AV valves occlude the interventricular canal. While the ventricle contracts, the muscular ridge is compressed against the opposite wall to separate the cavum venosum from the cavum pulmonale. The AV valves close to prevent retrograde backflow into atria, but in so doing, the right AV valve opens the interventricular canal and allows blood to flow through it. Thus, blood leaves the ventricle via the most accessible routes; 1. Deoxygenated blood in the cavum pulmonale exits primarily through the pulmonary artery to the lung, although some also squirts across the muscular ridge to enter the left aortic arch; 2. Oxygenated blood in the cavum arteriosum moves through the interventricular canal to reach the bases of the aortic trunks through which it then exits (Fig. 6.18A,B).

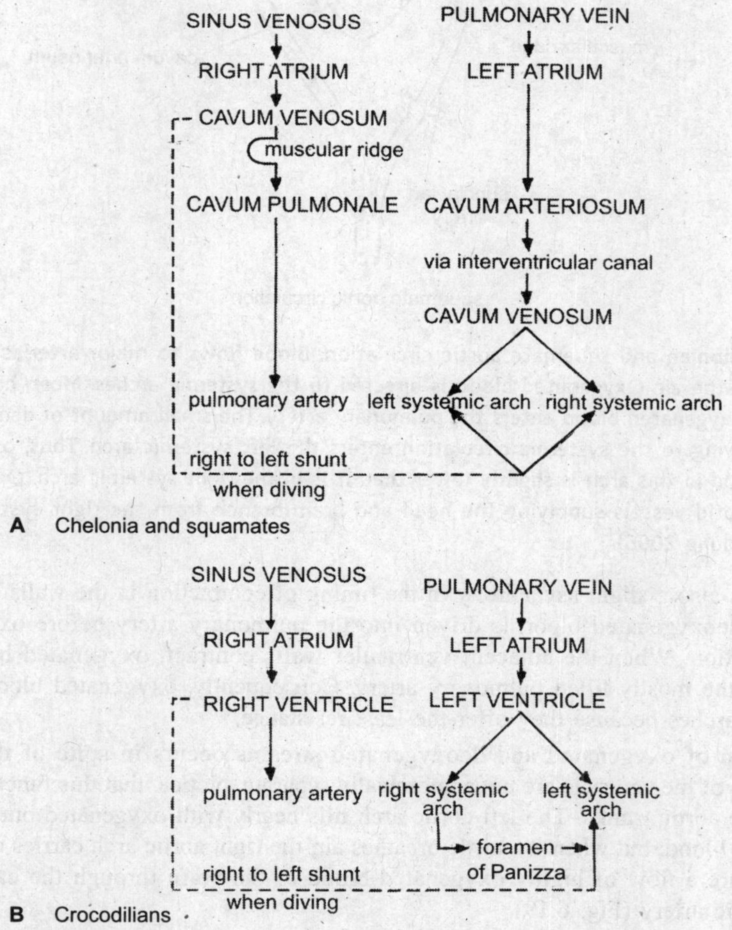

Fig. 6.18. Reptilian hearts. This flowchart compares the path of blood in the hearts of Chelonia and Squamata (A) and crocodilians (B). Dashed lines indicate the cardiac diving shunts that divert blood from the pulmonary circuit directly to the systemic circuits (after Kardong 2006).

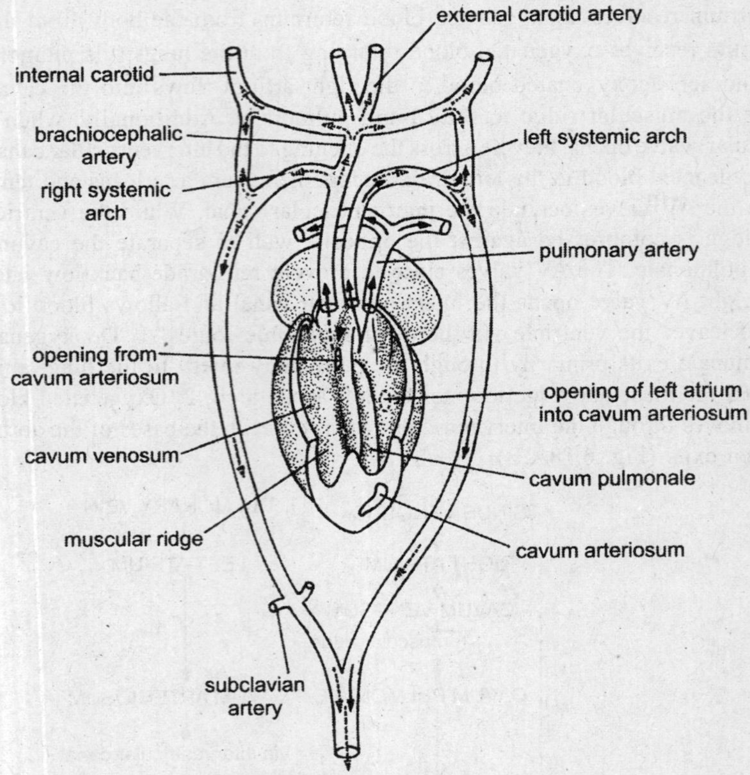

Fig. 6.19. Chelonian and squamate aortic circulation. Blood flows to major arteries when squamate breathe air. Oxygenated blood is directed to the systemic arches. Most, but not all, of the deoxygenated blood enters the pulmonary artery. The small amount of deoxygenated blood flowing to the systemic circulation enters the left systemic arch. Thus, oxygen tension of blood in this arch is slightly lower than it is in the right systemic arch. Quite interestingly, carotid vessels supplying the head and brain branch from the right systemic arch (after Kardong 2006).

There is also a slight asynchrony in the timing of contraction in the walls of the ventricle. As a result, deoxygenated blood is driven into the pulmonary artery before oxygenated blood is sent in motion. When the adjacent ventricular walls contract, oxygenated blood finds high resistance in the mostly filled pulmonary artery. Consequently, oxygenated blood exits through the systemic arches because they offer the least resistance.

Isolation of oxygenated and deoxygenated streams occurs in spite of the fact that the compartments of the ventricle are not anatomically separate. Notice, that this functional separation extends to the aortic trunks. The left aortic arch fills nearly with oxygenated but also with some deoxygenated blood; but when the turtle breathes air, the right aortic arch carries only oxygenated blood to ensure a flow of highly oxygenated blood to the brain through the carotids from the brachiocephalic artery (Fig. 6.19).

When a turtle dives, blood that would pass to the lungs during air breathing is shunted instead through the aortic arches to the systemic circuit. Thus, a diving turtle makes the best of a difficult situation. The air held in its lungs is soon depleted of oxygen, so there is little advantage in circulating large quantities of blood to it. The energy used to pump blood to the lungs also saves energetic costs. Instead, by diverting blood to the systemic circuit, this increases the blood volume that can remove metabolites or gather oxygen stored in tissues.

Crocodilian Hearts

The hearts of alligators and crocodiles are structurally similar (in many respects) to those of other reptiles. Thus, the *conus arteriosus* (bulbus cordis) produces the bases of the trunks of the three departing arteries — pulmonary and left and right aortic trunks. One way **lunar valves** at the bases of each trunk permit blood to enter the conus but halt reverse backflow into the ventricle. The *sinus venosus* is reduced but still functions as the receiving chamber for returning systemic blood. The *atrium* is completely subdivided into two distinct left and right chambers, and the sinus venosus empties into the right atrium. The pulmonary vein enters the left atrium in adults, but it does not open into left atrium during embryonic development. What are initially separate pulmonary veins, one from each lung, unite as a single stem, the pulmonary vein, that enters sinus venosus. However, as embryonic development proceeds, this part of the sinus venosus together with the associated pulmonary vein become incorporated into developing left atrium (Fig. 6.20).

In other respects, the crocodilian heart is quite different. The ventricle is divided into by an anatomically complete interventricular septum into distinct left and right chambers. The pulmonary trunk and *left* aortic arch open off the thick-walled *right* ventricle. The *right* aortic arch open off the *left* ventricle. A narrow channel called the **foramen of Panizza** connects the left and right aortic arches shortly after they depart from the ventricle.

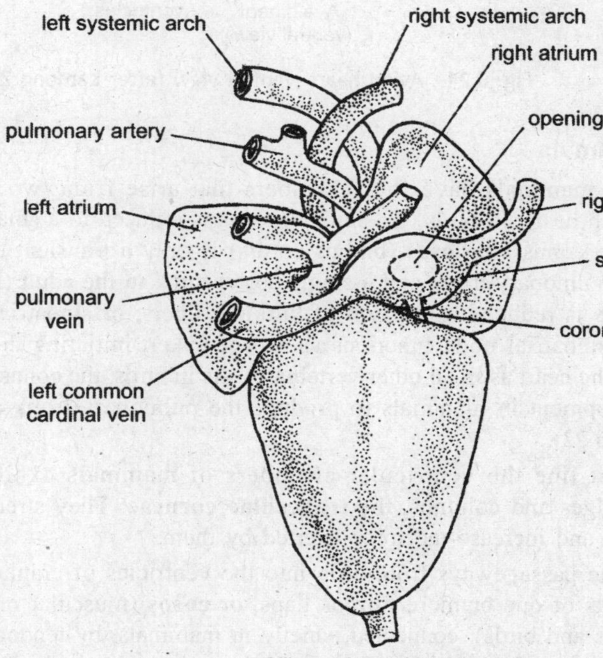

Fig. 6.20. Aligator heart in dorsal view. The sinus venosus is although reduced but is the first chamber of the heart to receive returning venous blood via the common cardinals. The coronary vein also flows into the sinus venosus (after Kardong 2006).

In a crocodile breathing air, right and left atria fill with deoxygenated systemic and oxygenated pulmonary blood, respectively. Contraction of the atria delivers blood to the respective ventricles. When the ventricles contract, blood flows through the nearest portals of least resistance. At the moment of systole, pressure is greatest in the left ventricle. The oxygenated blood it holds enters the base of the right aortic arch, but because of its high pressure, it also enters the left aortic arch via the foramen of Panizza. High pressure in the left aortic arch keeps the lunar valves at its base closed, leaving only the pulmonary route to exit for blood in the right ventricle. As a result, both aortic arches carry oxygenated blood to systemic tissues, and the pulmonary artery carries deoxygenated blood to the lungs.

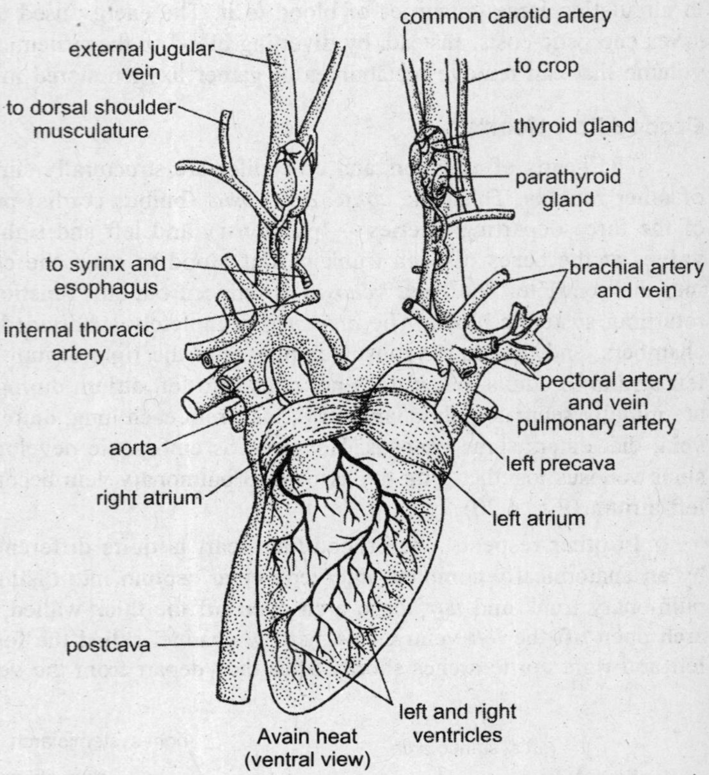

Fig. 6.21. Avian heart, ventral view (after Kardong 2006).

5. Heart of Birds and Mammals

The hearts of birds and mammals have four chambers that arise from two chambers (atrium and ventricle) of the fish heart. In birds, the sinus venosus is reduced to a small but still anatomically discrete area. The conus arteriosus (bulbus cordis) is only a transient embryonic chamber that gives rise to the pulmonary trunk and a single aortic trunk in the adult (Fig. 6.21). In mammals, the sinus venosus is reduced to a patch of Purkinje fibers, or sinoatrial node, in the wall of right atrium. The sinoatrial node functions as a pacemaker, initiating the wave of contraction that spreads across the heart as in all other vertebrates. As in birds, the conus arteriosus splits during embryonic development in mammals to produce the pulmonary trunk and single aortic trunk of the adult (Fig. 6.22).

The muscular walls that line the ventricular chambers of mammals exhibit sturdy interanastomosing muscular ridges and columns, the **trabeculae corneae**. They strengthen the walls of these powerful pumps and increase the force exerted by them.

One-way valves guard the passageways from atria into the ventricles of craniate's heart. In amniotes, each valve consists of one or more fibrous flaps, or **cusps** (muscular on the right side of the heart in crocodilians and birds), connected, chiefly in mammals, by tendonous cords (**chordae tendinae**) to **papillary muscles** that project from the ventricular walls (Fig. 6.22). During relaxation of the ventricle (**diastole**), blood from the atria flows freely past the cusps and into the ventricles. During ventricular contraction (systole), the cusps are forced forward and upward into the atrioventricular passageway, preventing reflux of blood into the atria. Each valve

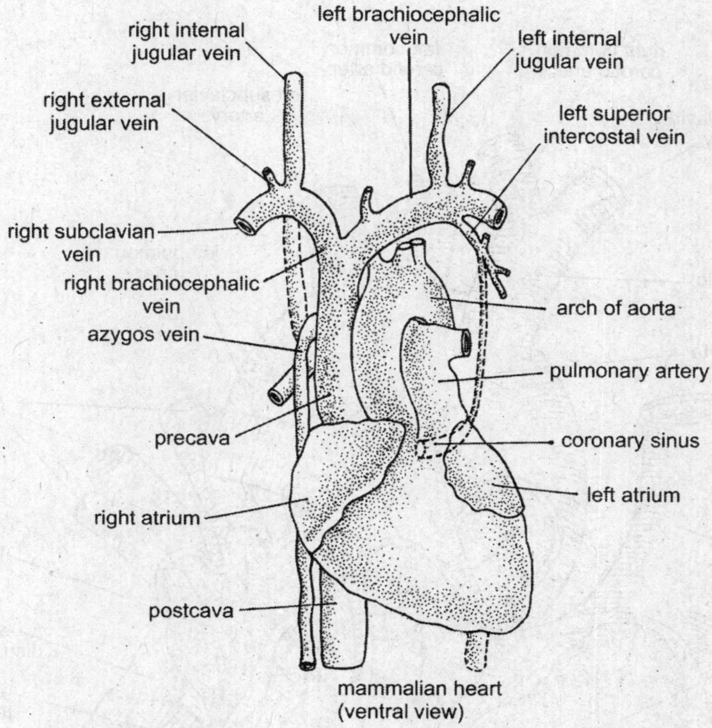

Fig. 6.22. Mammalian heart, ventral view (after Kardong 2006).

has one or two cusps in reptiles. In most mammals, the left valve has two cusps (**bicuspid or mitral valve**), and the right valve has three (**tricuspid valve**). **Semilunar valve** at the exits of the ventricles into the pulmonary and aortic trunks prevent backflow into the ventricles as the latter relax (Fig. 6.23).

Although structurally similar, bird and mammal hearts arose independently from different groups of reptilian ancestors. This difference is reflected in their embryonic development. Appearance of the interventricular and interatrial septa that form the paired chambers occurs quite differently in the two groups. Both consist of parallel pumps with double circulation circuits. The right side of the heart gathers deoxygenated blood from systemic tissues and pumps it into the pulmonary circuit. The left side of the heart pumps oxygenated blood from the lungs through the systemic circuit. The hearts of birds and mammals are anatomically divided into left and right compartments; thus, there is no cardiac shunting with changing ventilation rates. Therefore, unlike in amphibians and reptiles, a cardiac shunt cannot be used in birds and mammals to decouple perfusion of the lung and systemic tissues. Although the reasons are not well understood, some workers propose that endothermic animals (birds and mammals) may require complete anatomical separation of the cardiac chambers.

6.7 ARTERIAL CHANNELS AND THEIR MODIFICATIONS

Arterial channels supply most organs with oxygenated blood, although they carry deoxygenated blood to respiratory organs. In the primitive pattern of gnathostomes, the major arterial channels consist of (1) a **ventral aorta** (paired above the pharynx); (2) a **dorsal aorta**, paired above the

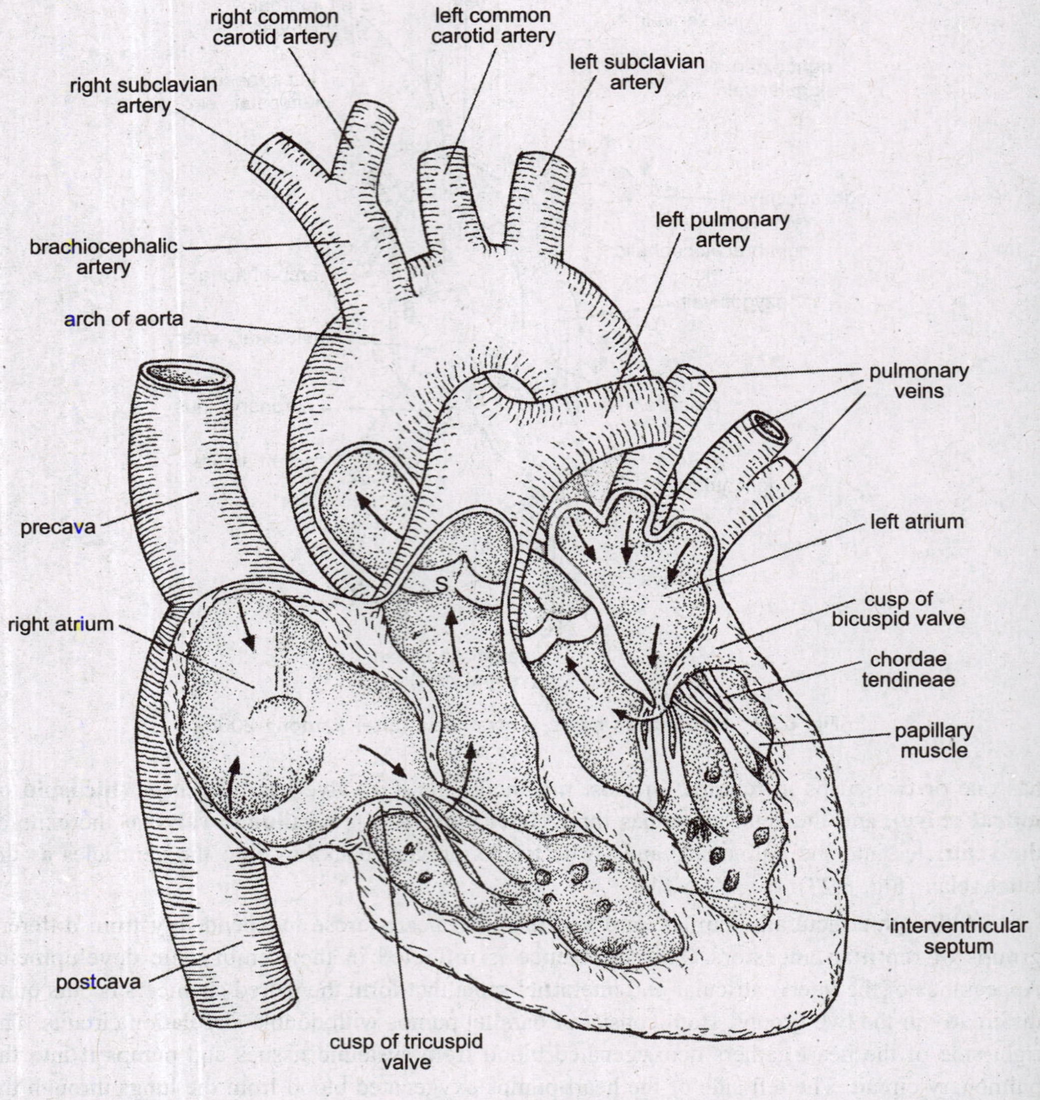

Fig. 6.23. Human heart, auricles removed. **S**, semilunar valve at entrance to pulmonary trunk. A similar valve is seen at entrance of arch of the aorta. The third cusp of the tricuspid valve has been removed. The sinus venosus has been incorporated into the wall of the right atrium (after Kent and Carr 2001).

pharynx only; and extending caudal in roof of the coelom; and (3) six pairs of **aortic arches** connecting the **ventral aorta** with the dorsal aorta (the number of arches varies in living agnathans with the number of branchial arches) (Fig. 6.24). Branches of these major channels supply all parts of the body. Modifications of this inherited basic pattern during embryonic development involve, most prominently, the aortic arches, which become modified for respiration by whatever means is employed during the life history of the species (Fig. 6.28).

1. Aortic Arches in Cyclostomes. The number of primitive aortic arches and branchial arches through which they run are still debated (see **Kardong** 2006). Some ostracoderms had as many as ten pairs of branchial arches and presumably ten pairs of aortic arches. The number of pairs of aortic arches varies in living forms. Lampreys have 8 (Fig. 6.25), hagfishes 15.

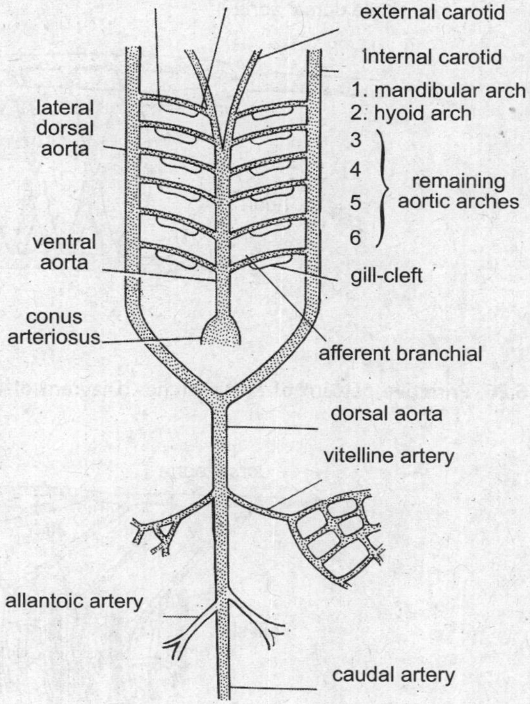

Fig. 6.24. Embryonic aortic arches and arteries in vertebrates.

2. Aortic Arches in Fishes. Some species of sharks possess 10 or 12 pairs of aortic arches. Nevertheless, only up to six pairs of aortic arches customarily appear during embryonic development. Accordingly, six is the number of aortic arches usually taken as the basic embryonic pattern, and they are designated by roman numerals (I-VI, Fig. 6.26).

In fishes, soon after branching from the ventral aorta, the aortic arches divide into capillary beds within the gills. The ventral section of the aortic arch delivering blood to the gills is the **afferent artery**, and the dorsal section carrying blood away is the **efferent artery**. The capillary bed between them partially or completely encircle the gills and empty first to the **collecting loop**, which joins the efferent artery.

In chondrichthyans (cartilaginous fishes) the first pharyngeal slit becomes reduced but not lost, forming the small **spiracle**. During embryonic development, the ventral section of the first aortic arch, does not appear. Instead, a vascular sprout from the adjacent collector loop grows to the spiracle, feeding a small capillary bed in its wall. This vessel constitutes the **afferent spiracular**

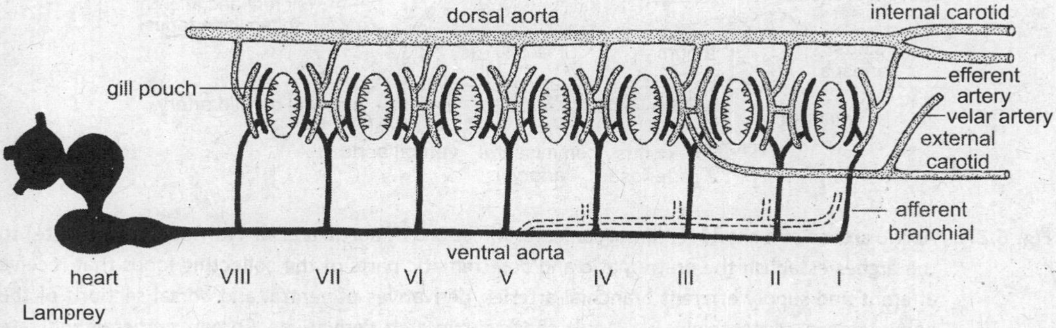

Fig. 6.25. Aortic arches, gills and anterior arteries of lamprey (after Kardong 2006).

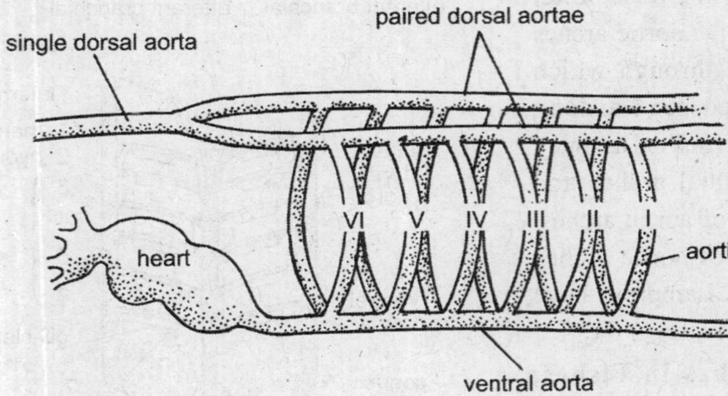

Fig. 6.26. Primitive pattern of aortic arches. Diagram of the basic six-arch pattern (after Kardong 2006).

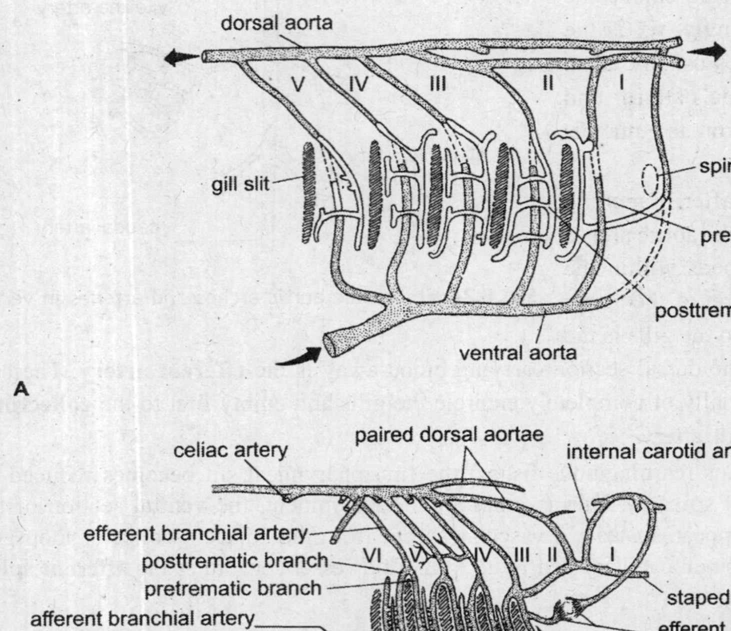

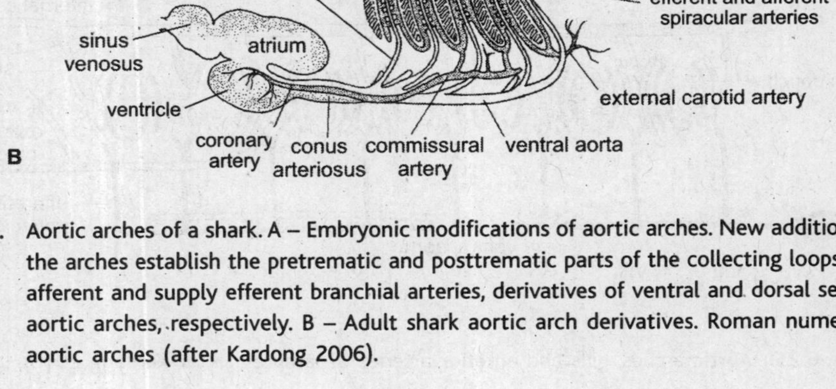

Fig. 6.27. Aortic arches of a shark. A – Embryonic modifications of aortic arches. New additions (white) to the arches establish the pretrematic and posttrematic parts of the collecting loops that receive afferent and supply efferent branchial arteries, derivatives of ventral and dorsal sections of the aortic arches, respectively. B – Adult shark aortic arch derivatives. Roman numerals indicate aortic arches (after Kardong 2006).

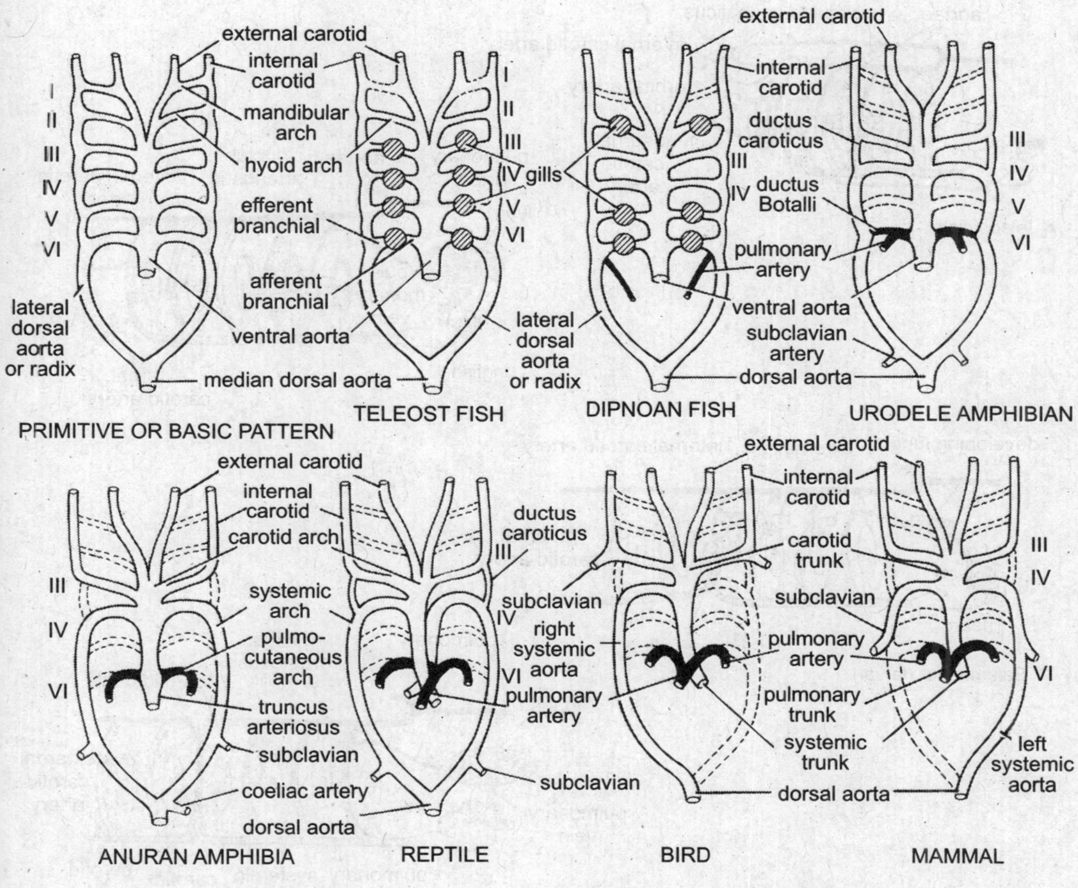

Fig. 6.28. Modification of aortic arches in representative vertebrates.

artery. The dorsal section of the first arch forms the **efferent spiracular artery**, which drains this small capillary bed. Because of its small size and the fact that it receives oxygenated blood via the afferent spiracular artery; the spiracle is thought to play little role in respiratory exchange. Instead it may develop as part of a secretory or sensory organ.

The remaining aortic arches (II–VI) form small sprouts halfway along their lengths. These merge and cross connect as the collecting loops serving the vascular capillary beds within the gills that form adjacent to the enlarged pharyngeal slits. The anterior and posterior halves of each collecting loop are its **pretrematic** and **post-trematic branches**, respectively (Fig. 6.27A). Although the external carotid arises embryologically from the anterior end of the ventral aorta, it becomes associated with the collecting loop, an understandable change if it is to carry oxygenated blood to the lower jaw. The internal carotid artery supplying the brain receives oxygenated blood from the first fully functional collecting loop (pharyngeal slit II) via the efferent branchial artery (II) (Fig. 6.27 B).

In most actinopterygial fishes, four pairs of aortic arches (III–VI) arise from the ventral aorta. They service the gills associated with five pharyngeal slits. In sturgeons and a few other species, the first pharyngeal slit persists as a small spiracle (slit), but in most fishes, this spiracle is absent in the adult. The first aortic arch is lost along with this first slit (Fig. 6.29A).

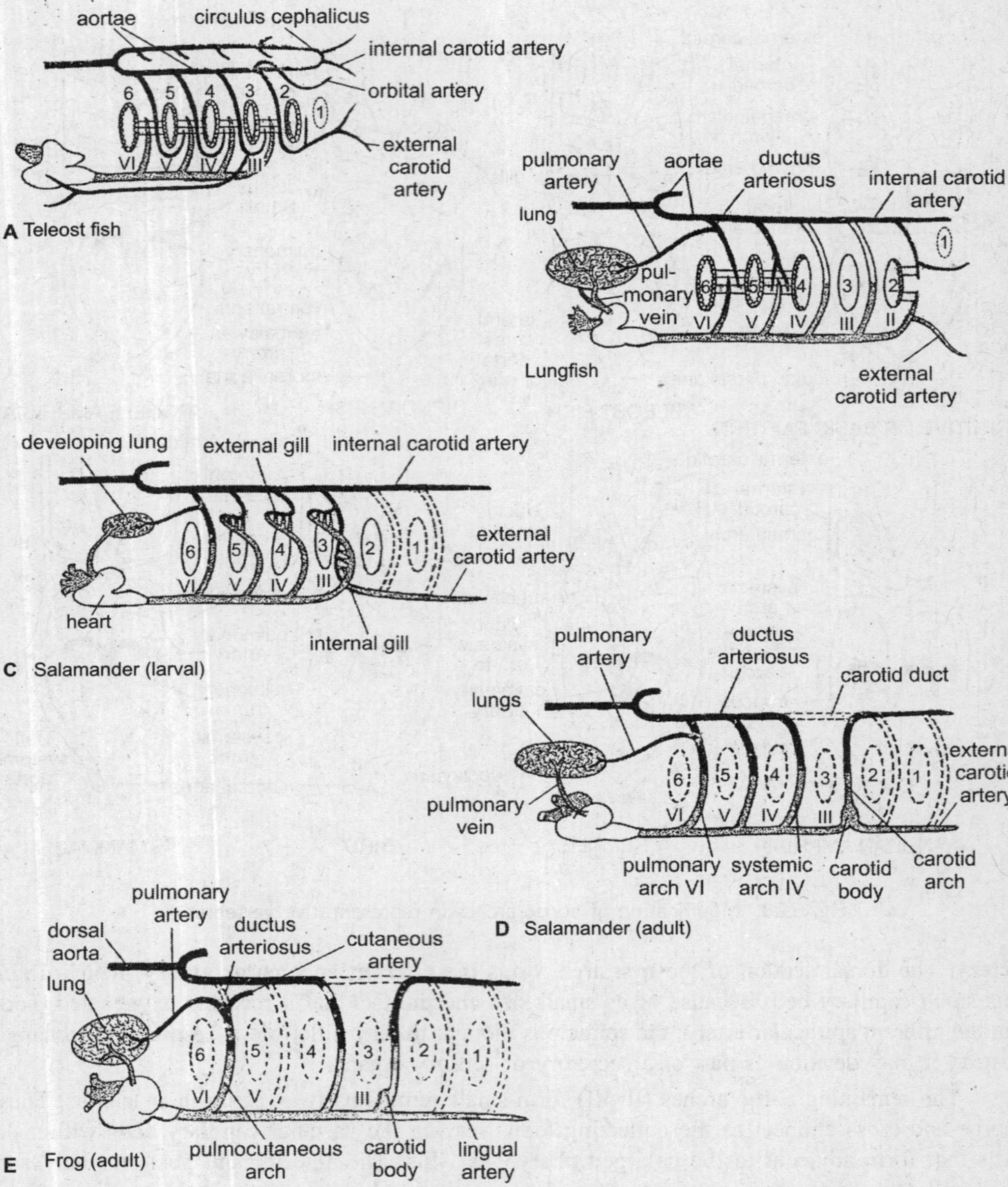

Fig. 6.29. Aortic arches of anamniotes and some derivatives. Diagrams of the basic six-arch pattern. A – Teleost fish. B – Lungfish (*Protopterus*). C – Neotenic and larval salamander, D – Adult salamander. E – Adult frog (after Kardong 2006).

In most fishes with supplementary air-breathing organs, oxygenated blood departing from these organs enters the general venous circulation, boosting the overall oxygen level of the blood returning to the heart. However, in lungfishes, blood leaving the highly vascularized lungs directly re-enters the heart via a separate vein (Fig. 6.29B). phylogenetically, separate left and right atria appear first in lungfishes, thereby establishing a separate pulmonary circuit to and from the lung.

In lungfishes and other bony fishes, the first pharyngeal slit is reduced to a spiracle that has no respiratory function. Its associated aortic arch (I) is reduced as well. In the Australian lungfish, *Neoceratodus*, the remaining five pharyngeal slits open to fully functional gills supplied by four aortic arches (III–VI). In the African lungfish, *Protopterus*, the functional gills are reduced further. The third and fourth gills are absent entirely, but their aortic arches (III–IV) persist (Fig. 6.29B). In all lungfishes, the efferent vessel of the most posterior aortic arch (VI) gives rise to the **pulmonary artery** but maintains its connection to the dorsal aorta via the short **ductus arteriosus**.

Were lungfishes imperfectly designed for the breathing? If one thinks carefully about the blood flow implied by above discussed pattern (Fig. 6.29B), he will understand the misconceptions it invites. For example, one can notice that if blood low in oxygen in the ventral aorta flows along its presumed route through arches II, V, and VI, it passes through the capillary beds of the gills, is replenished with oxygen, and enters the dorsal aorta as oxygened blood. But notice that in the African lungfish, blood low in oxygen in the ventral aorta seems to have an alternative route through arches III and IV, which lack gills. Theoretically, blood could reach the dorsal aorta unaltered, still lacking oxygen. If that happened, as the anatomy alone might suggest, then both oxygenated and deoxygenated blood would mix in the dorsal aorta, reduce the overall oxygen tension in blood flowing to the systemic tissues, and apparently defeat most advantages that air breathing lungs might bring.

Because of this, the aortic arch pattern of lungfishes seemed inefficient to early anatomists, but they ignored it, and perhaps even expected such a pattern, because they thought that lungfishes were neither complete water-breathers nor air-breathers. Lungfishes held a middle ground capable of a bit of both, but doing neither particularly well. This mistaken view that lungfishes were imperfectly designed compared to more advanced vertebrates, was tested when careful studies of their circulatory physiology were completed. In fact, very little mixing of oxygenated and deoxygenated blood actually occurs due to a partially divided heart (see **Kardong** 2006).

3. Aortic Arches in Amphibians. Like the lungfishes, the cardiovascular system of amphibians was thought to be imperfectly designed (**Kardong** 2006). The anatomical arrangement of amphibians' aortic arches suggests that some mixing occurs between oxygenated blood from the gills and deoxygenated blood returning from the body.

A. Salamanders. In amphibians, the first two aortic arches (I, II) disappear early in development, The pattern of the remaining arches differs between larvae and metamorphosed adults. In most **larval salamanders**, the next three aortic arches (III-V) carry external gills, and the last aortic arch (VI) sprouts the pulmonary artery to the developing lung (Fig. 6.29C,D). A notable exception is the neotenic salamander, *Necturus*, in which part of the sixth arch disappears and only its dorsal section persists, forming the base of the pulmonary artery. In most species of salamanders, the external gills are lost following the larva's transformation (*i.e*, metamorphosis) into the adult, but the aortic arches are retained as major systemic vessels.

The **carotid ductus** which is a short section of dorsal aorta between aortic arches III and IV, usually closes at metamorphosis (Fig. 6.29 D). This forces the carotids to fill with blood from a derivative of the ventral aorta. The section of ventral aorta between arches III and IV becomes the **common carotid artery**, which feeds the external carotid (from the ventral aorta) and the internal carotid (anterior section of dorsal aorta together with the third aortic arch). The **carotid body** is a small cluster of sensory cells associated with capillaries, usually located near the point at which the common carotids branch. The carotid body plays a role in sensing the gas content or pressure of the blood as well as having some endocrine function.

The next two arches (IV, V) constitute major systemic vessels that join the dorsal aorta. The final aortic arch (VI) also joins the dorsal aorta, its last short section forming the **ductus arteriosus**. Shortly before joining the dorsal aorta, the sixth aortic arch gives off the pulmonary artery, which itself divides into small branches to the floor of the mouth, pharynx, and esophagus before actually entering the lungs. In lungless salamanders, the pulmonary artery, if it persists, supplies the skin of the neck and back.

B. Frogs. In frogs, the larva usually has internal gills that reside on the last four arctic arches (III-VI), and the embryonic pulmonary artery buds from arch VI. At metamorphosis, these gills are lost together with the carotid duct and all of arch V. The aortic arches that persist (III, IV and VI) expand to supply blood to the head, body and pulmonary circuits, respectively. The III aortic arch and associated section of the anterior dorsal aorta become the internal carotid. The anterior extension of the ventral aorta is the external carotid. Internal and external carotids both branch from the common carotid, the section of ventral aorta between arches III and IV. A carotid body usually be found at the root of the internal carotid. The next enlarged aortic arch (IV) joins with the dorsal aorta, the major systemic artery supplying the body. The last arch (VI) loses its connection to the dorsal aorta because the ductus arteriosus closes and becomes the **pulmocutaneous artery**. One branch of the pulmocutaneous artery is now well - developed *pulmonary artery* that enters the lung. The other branch is the **cutaneous artery**, which delivers blood to the skin along the dorsal and lateral body wall.

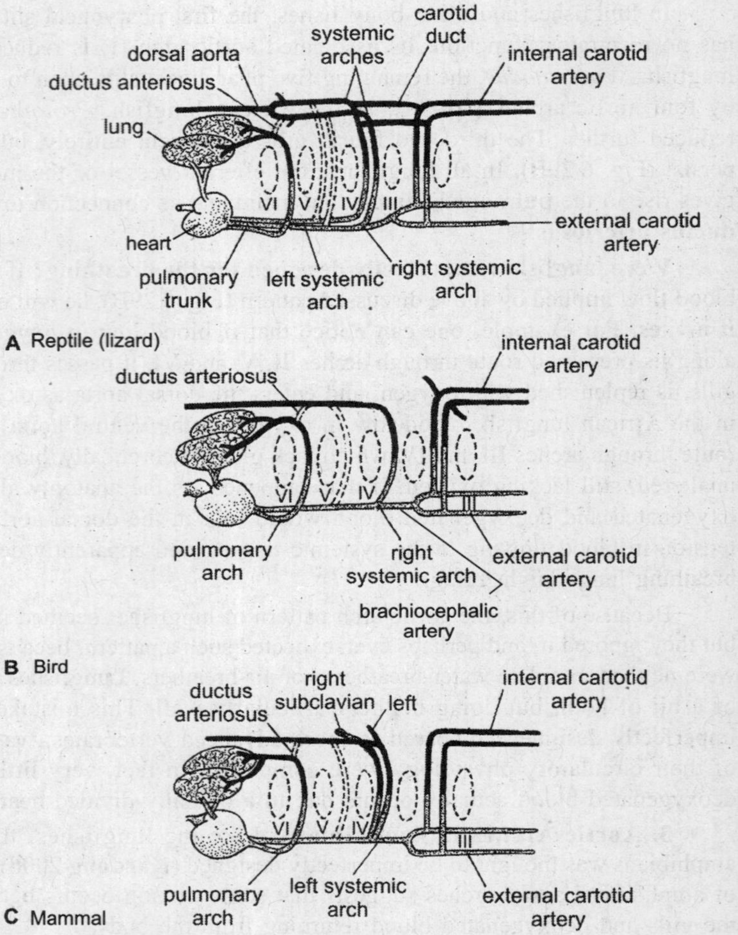

Fig. 6.30. Aortic arches of amniotes, Diagrams of derivatives of the basic six-arch pattern. A – Reptile. B – Bird. C – Mammal (after Kardong 2006).

Due to inefficiencies in amphibian blood flow, oxygenated blood returning to the heart was thought to mix with deoxygenated blood returning from systemic tissues. Suffice it to say, this does not happen. In fact, little mixing occurs (**Kardong** 2006).

4. Aortic Arches in Reptiles. Beginning in reptiles, but carried into birds and mammals, the symmetrical aortic arches of the embryo tend to become asymmetrical in the adult. Aortic arches III, IV and VI persist in reptiles, but most of the changes center on enhancement and modification of the fourth arch. Perhaps the most significant anatomical modification of the arterial system in reptiles is the subdivision of the ventral aorta. During embryonic development, the ventral aorta splits to form the bases of three separate arteries leaving the heart: the **left aortic arch**, the **right aortic arch**, and the **pulmonary trunk** (Fig. 6.30 A).

The pulmonary trunk incorporates the bases of the paired sixth arch and their branches as part of the **pulmonary arch** to the lungs. The base of left aortic arch, the left aortic arch (IV) itself, and the curved section of the left dorsal aorta into which it continues constitute the **left systemic arch**. The **right systemic arch** includes the same components on the right side of the body: the base of the right aortic arch, the right aortic arch itself, and the arched section of the right dorsal aorta. The two systemic arches unite behind the heart to form the **common dorsal aorta**. The right systemic arch tends to be the most prominent of the two, primarily because of the additional vessels that it supplies. For example, the **carotid arteries**, originating from the ventral aorta in more primitive vertebrates, arise in reptiles from the right systemic arch. Blood passing through the right systemic arch might flow to the body or enter the carotid arteries to supply the head. In most reptiles, the **subclavian arteries** branch from the dorsal aorta, but in some reptiles, they branch from the systemic arches. These modifications of the aortic arches in reptiles produce one pulmonary circuit and two systemic circuits, each of which arises independently from the heart.

5. Aortic Arches in Birds. In birds, the **right systemic arch** becomes predominant (Fig. 6.30 B). The bases of the aortic arch, the right aortic arch (IV), and the adjoining section of the right dorsal aorta form the right systemic arch during embryonic development. Its opposite member, the left systemic arch, never fully develops. The carotids arise generally from the same components of the aortic arches as in reptiles (aortic arch III and parts of the ventral and dorsal aortae), and they branch from the right systemic arch. However, the paired subclavians to the wings arise from the internal carotids and not from the dorsal aorta. The common carotids and subclavians supply the head and forelimbs, respectively. The common carotids can branch from the right systemic arch separately, or both can join to form a single carotid. A short but major vessel, the **brachiocephalic artery**, is present in few reptiles, especially turtles, but serves as the major anterior vessel in many birds. It also branches from the right systemic arch. Beyond this junction of the brachiocephalic artery, the systemic arch curves posteriorly to supply the rest of the body. In birds as in reptiles, the pulmonary arch forms from the bases of the paired sixth arch and their branches to supply both lungs.

6. Aortic Arches of Mammals. Up to six aortic arches arise in the mammalian embryo, but only three (*i.e.*, III, IV and VI) persist in adults as the major anterior arteries: the carotid arteries, the pulmonary arch and the systemic arch (Fig. 6.30C). The **carotid arteries** and **pulmonary arch** are assembled from the same arch components as those of reptiles. The mammalian carotids arise from the paired aortic arches (III) and parts of the ventral and dorsal aortae. The pulmonary arch forms from the bases or the paired sixth arch and its branches. The systemic arch arises embryologically from the left aortic arch (IV) and left member of the paired dorsal aorta, and therefore is a **left** systemic arch in mammals. The common carotids may share a brachiocephalic origin or branch independently from different points on the aortic arch. The other notable difference in mammals is in the formation of the subclavian arteries. The left subclavian departs from the left systemic arch in mammals. The right subclavian, however, includes the right aortic arch (IV), part of the adjoining right dorsal aorta, and the arteries that grows from these into the right limb.

Concluding Remark About Aortic Arches

In most fishes, the aortic arches deliver deoxygenated blood to the respiratory surfaces of the gills and then distribute oxygenated blood to tissues of the head (via the carotids) and remainder of the body (via the dorsal aortae). In lungfishes and tetrapods, the aortic arches form the pulmonary arch, the arterial circuit to the lungs, and the systemic arches, the arterial circuits to the rest of the body. The carotid arteries still bear the primary responsibility for supplying

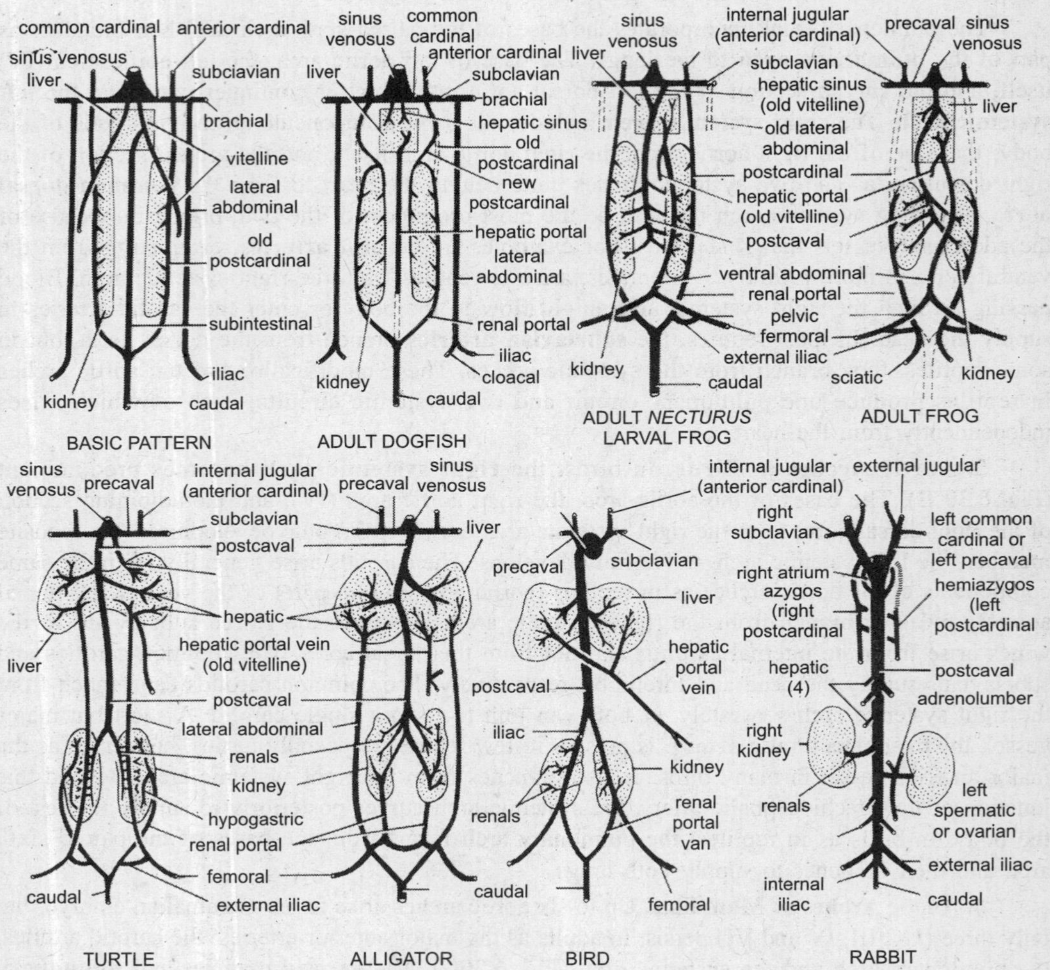

Fig. 6.31. Modifications of venous system in vertebrates. Broken line represent embryonic vessels that have disappeared in the adult.

blood to the head in tetrapods, but now they usually branch from one of the major systemic arches. The double systemic arches (left and right) present in amphibians and reptiles become reduced to a single systemic arch—the right in birds and the left in mammals. Although birds and mammals share many similarities, including endothermic, active lives, and diverse radiation, they arose out of different reptilian ancestors. Any similarities in their cardiovascular anatomies represent independent evolutionary innovations.

The basic six-arch pattern of aortic arches is a useful concept that allows us to track aortic arch derivaties and organize the diversity of anatomical modifications we encounter. Furthermore, the appearance of six-aortic arches during the embryonic development of living gnathostomes suggest that this is the ancestral pattern.

6.8 VENOUS CHANNELS AND THEIR MODIFICATIONS

A generalized venous system of vertebrates consists of the following major streams: 1-3. **Cardinals** (anterior, posterior and common cardinals); 4. **Renal portal**, 5. **Lateral abdominal**; 6. **Hepatic portal**; 7. **Hepatic sinuses**; and 8. **Coronary veins** (Fig. 6.31).

The hepatic portal system is derived from the embryonic subintestinal vein and the distal portion or the embryonic vitelline veins, and the hepatic sinuses are derived from the vitelline veins between liver and heart. Two additional streams develop in lungfishes and tetrapods: a **pulmonary stream** from the lungs and a postcava from the kidneys. These eight channels and their tributaries drain the entire body-head, trunk, tail and appendages. As development progresses, they are slowly modified by deletion of some vessels and addition of others. Modifications are few in stem craniates, more numerous as derived features within individual clade.

6.9 LYMPHATIC SYSTEM

Lymph and lymph channels are found in all craniates (Fig. 6.32). A lymphatic system consists of thin-walled **lymph channels**, **lymph** (a fluid in transit), **lymph hearts** as far up the phylogenetic scale as embryonic birds, **lymph nodes** in birds and mammals, and solitary or **aggregated masses** of **lymph nodules (Fig. 6.33**), the largest of which is the spleen (Fig. 6.35). In contrast to blood, lymph moves in one direction only: from the tissues toward the heart.

The lymph system begins in **lymph capillaries**, the walls of which are single-layered endothelial tubes, or in **lymph sinusoids**, which are branching and anastomosing tubes of

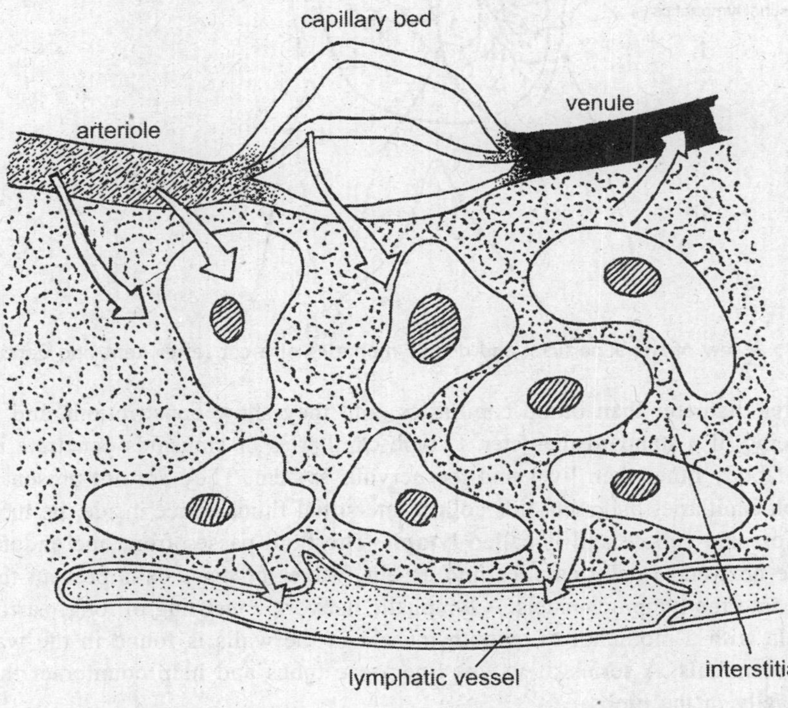

Fig. 6.32. Formation of lymph. The relatively high pressure in capillaries results in fluid from the blood leaking into surrounding tissues. Some of this interstitial fluid returns to the blood on the low-pressure venous side of circulation. Blind-ended channels called lymphatic vessels collect the excess fluid (lymph) and return it to the general circulation, usually through one of the large thoracic veins (arrows indicate movement of fluid) (after Kardong 2006).

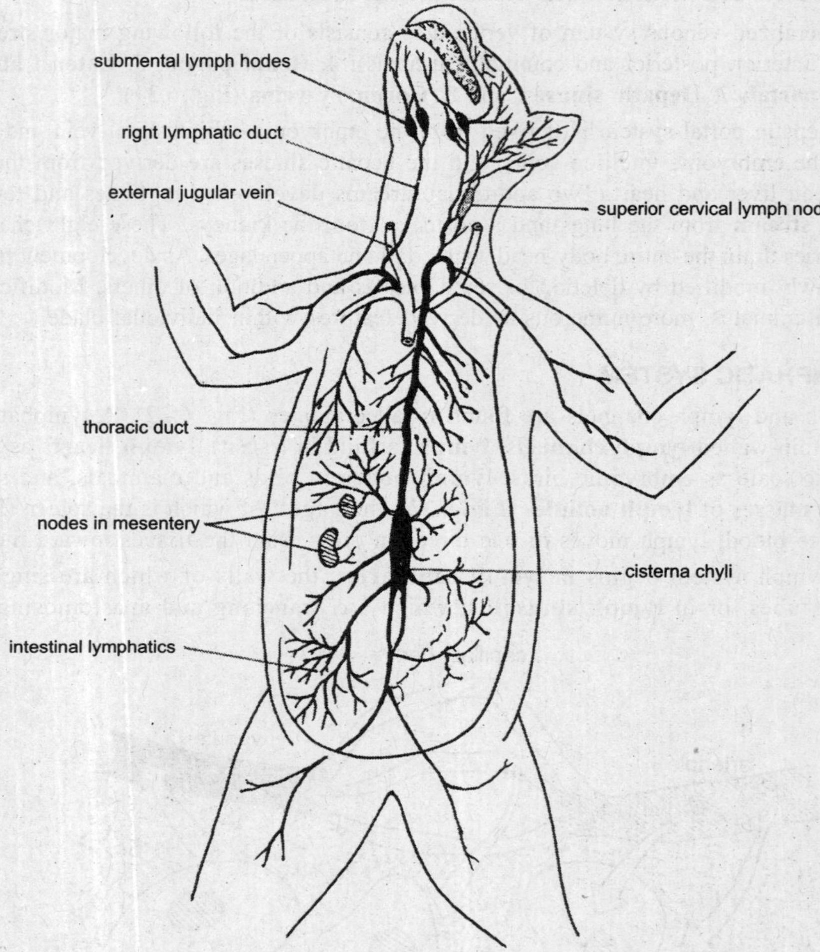

Fig. 6.33. A few of the superficial and deep lymphatics of a cat (after Kent and Carr 2002).

slightly greater diameter than blood capillaries, and they show constrictions and expansions rather than being of a standard diameter. Lymph capillaries or sinusoids penetrate most of the soft tissues of body other than liver and the nervous system. They are not present in skeletal tissues. Lymph capillaries and sinusoids collect interstitial fluids. Once inside the tube, the fluid is colourless or pale yellow and is called **lymph**. The fluid passes from one endothelial-lined channel to the next and finally empties into a vein. Valves of these exits prevent the influx of venous blood into the lymph channels. A lymphatic network consisting of long, narrow, discrete tubular vessels with a modicum of smooth muscle in the walls is found in the walls is only in birds and mammals. A series of valves line these tubes and help counteract the effect of gravity, especially in the limbs.

The collection of fats absorbed from the small intestine after a meal by the lymphatics in intestinal villi is derived feature within craniates (Fig. 6.34). If the meal was particularly fatty, the lymph in these vessels has a milky appearance. For this reason, these specific lymphatics are called **lacteals**. The lymph within them is called **chyle**. Certain lymphatics in living agnathans,

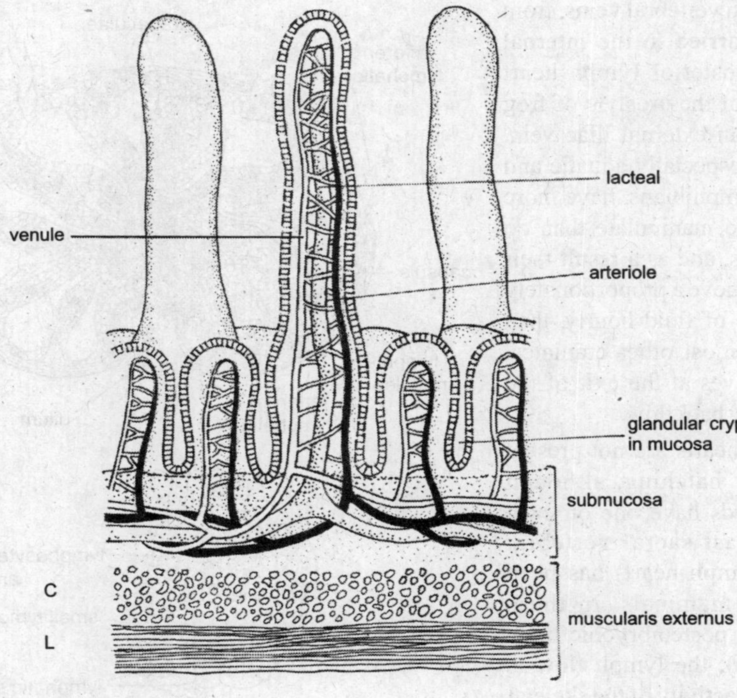

Fig. 6.34. Blood supply and lymph drainage of a villus. C, circular muscle; L, longitudinal muscle (after Kent and Carr 2001).

cartilaginous fishes, and even humans contain some red blood cells. The fluid in these vessels is called **hemolymph**.

Lymph channels that drain the body wall, limbs, and tail of craniates typically empty into nearby veins at the base of the tail, in the trunk, or in the neck. Those draining viscera are often paired in most craniates, but in mammals, a single **thoracic duct** commences in a large abdominal lymph sinus, the **cisterna chyli** (Fig. 6.33), and empties into the brachiocephalic or left subclavian vein, or into the external and internal veins (Fig. 6.33). The mammalian thoracic duct also receives lymphatics from the left side of the head and neck and from the left forelimb. One or more additional major lymphatics drain the right side of the body anteriorly.

The lymphatic system of anuran amphibians is notable because of the numerous **sinusoids** that take the place of capillaries at some locations. They especially prominent just under the skin over almost the entire body, where they form huge **lymph reservoirs** separated from each other by connective tissue septa that attach the skin to the underlying muscles at these locations only. The skin is very thin, and sinusoids may buffer the underlying muscles from the drying effects of air when a frog is out of water. Because of these subcutaneous lymph sinuses, the skin of frog is easily removed from the body.

The **flow of lymph** results from a number of factors, including lymph hearts situated at advantageous locations along lymph pathways in fishes, amphibians and reptiles (except post-embryonic birds). **Lymph hearts** are pulsating sinusoidal swellings on lymph pathways with thin walls containing striated muscle fibers of unknown embryonic origin. Frogs have two pairs of lymph hearts, urodeles have as many as 16 pairs, and caecilians as many as 100. The anterior pair of lymph hearts in frogs, located just behind the transverse processes of the third vertebra, pump lymph in vertebral veins, from which it is carried to the internal jugulars. A posterior lymph heart on each side of the urostyle of frog empties into an external iliac vein. Amphibians, especially aquatic and semiaquatic amphibians, have more tissue fluids to manipulate than do other craniates, and as a result their lymph hearts move a proportionately larger volume of fluid hourly, than the hearts of most other craniates. Semilunar valves at the exit of the hearts prevent backflow.

Lymph hearts are not present in birds after hatching, although embryonic birds have one on each side of the last sacral vertebra. None (*i.e.*, lymph heart) has been described in mammals. In these craniates (*i.e.*, postembryonic birds and mammals), the lymph flow is maintained by activity of the skeletal muscles as they contract and relax, by movements of the viscera, and by rhythmical changes in intrathoracic pressure that result from breathing.

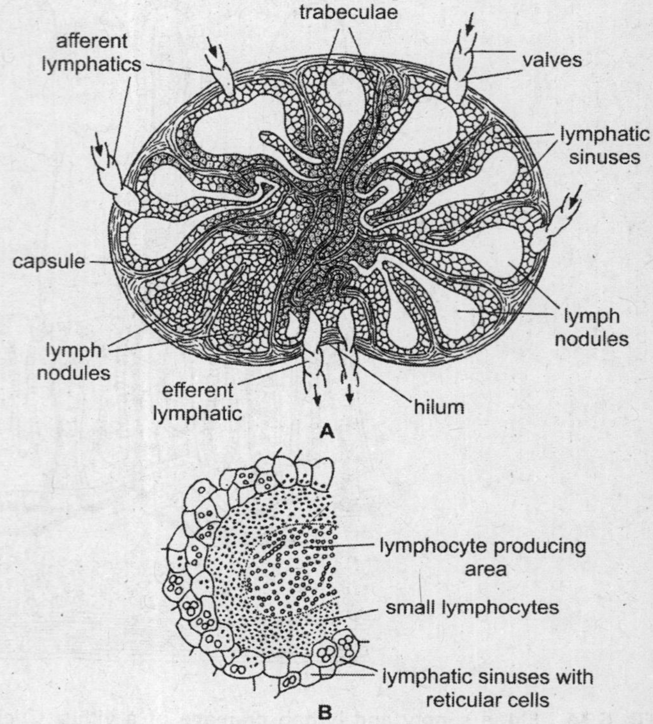

Fig. 6.35. Diagrammatic section. A – Lymph node in detail. B – A portion of nodule magnified.

Lymph nodes are masses of hemopoietic tissue interposed along the course of lymphatic channels of birds and mammals. They may be no larger than a pinhead, or they may be several centimeters in diameter. They are the "swollen glands" that one can feel in the neck, axilla, and groin of humans when there is inflammation in areas drained. Lymph enters a node via several **afferent lymphatics**, filters through the node, and leaves via a single large **efferent lymphatic**. The endothelium of their sinusoidal passageways includes many **phagocytes** that ingest bacteria and other foreign particles. Entangled in their connective tissue stroma are large numbers of small lymphocytes (Fig. 6.35). The lymph nodes are therefore the second line of defense against bacterial infections acquired through the skin, the first line of defense being granulocytes that assemble at the invaded area.

Other **lymphoid masses** develop at one site or another. The largest of these is the **spleen**, which is unique because it is interposed into the blood circulating stream. Other lymphoid masses include the **thymus** (absent in hagfishes), the **bursa of Fabricius** of very young birds, solitary nodules or aggregates of nodules (**Peyer's patches**) in the wall of the small intestine of amniotes, and mammalian **tonsils** and **adenoids**.

Spleen

In adult human beings, the spleen weighs about 150 gm. It is located in the left upper quadrant of the abdomen. It is supplied by a single **splenic artery**, which pierces the capsule at the hilum and divided into progressively smaller branches that remain surrounded by protective and supporting fibrous tuberculae. Small arterioles are surrounded by layers of lymphocytes called **periarteriolar lymphoid sheaths**. The lymphoid follicles, some of which contain germinal centers are attached to these sheaths. These dense lymphoid tissues constitute the **white pulp** of spleen. The arterioles ultimately end in vascular sinusoids. Among these a large number of macrophages, dendritic cells, sparse lymphocytes and plasma cells are scattered; these constitute the **red pulp**. The sinusoids form venules that drain into a **splenic vein.** This carries blood out of the spleen and into the portal circulation.

Functions of spleen are mostly shared by other organs and tissues, therefore it is *not essential for life.* Its function includes the following:

(1) **Formation of RBCs.** (i) during second trimester of uterine life; and (ii) in adults after destruction of red bone marrow.

(2) **Contain lymphoid tissues** which **form lymphocytes** and **plasma cells.**

(3) Being a part of tissue macrophage system, spleen **destroys** aged RBCs, platelets and WBCs.

(4) **Participates in defense** reaction against toxins (diphtereia, tetanus); bacteria and large parasites by the formation of antibodies and by phagocytosis. Therefore, in the absence of spleen, bacterial infections are more common and severe. Moreover malaria has a higher mortality rate, because the deformed RBCs that contain malaria parasite are not removed.

(5) **Reservoir of RBCs.** Contraction of spleen following injection of epinephrine hormone, increases RBCs release into peripheral blood.

QUESTIONS

Long Answer Questions

1. What are constituent parts of the circulatory system? Give a brief account of them.
2. Describe the evolution of heart in vertebrates.
3. Give a comparative account of hearts and aortic arches in reptiles, birds and mammals.
4. Describe aortic arches in fishes and amphibians.
5. Which taxa achieve a complete separation of oxygenated and deoxygenated blood.

Short Answer Questions

1. Describe different types of hearts in hagfishes.
2. Write short notes on the following;
   (*i*) Branchial heart;
   (*ii*) Trunchus arteriosus;
   (*iii*) Foramen of Pannizzale;
   (*iv*) Renal portal system;
   (*v*) Hepatic portal system;
   (*vi*) Spleen.

Very Short Answer Questions

1. What type of blood is present in cavum venosum of squamate heart?
2. How many chambers are present in the heart of lizard? Name them.
3. Why the wally of left ventricle of mammalian heart are thicker than the walls of right ventricle?
4. Why do all apertures of heart bear valves?

Fill in the Blanks

1. The conus arteriosus of heart of lamprey is muscular but lacks ………….. muscles.
2. In the heart of teleost fishes, the conus arteriosus is ………….. .
3. ………….. of bulbus arteriosus tend to protect delicate ………….. capillaries of fishes.
4. Lymph hearts are absent in postembryonic birds and ………….. .
5. Caecilians have upto ………….. lymph hearts.

Multiple Choice Questions

1. In urodele, conus (or truncus arteriosus) is replaced by
   (a) aorta (b) vena cava
   (c) foramen ovale (d) bulbus arteriosus
2. Smallest arteries are connected to smallest vein by
   (a) capillaries (b) arterioles
   (c) venules (d) muscles
3. The components of mammalian cardiac tissue having the highest propagation velocity is
   (a) Purkinje fiber (b) AV node
   (c) atrial muscle (d) ventricular muscle

ANSWERS

Very Short Answer Questions

1. Deoxygenated blood.
2. Six, i.e., two chambers in atrium, three chambers in ventricle and sinus venosus (see Kardong 2006).
3. The wall of left ventricle of mammalian heart are thicker than the walls of right ventricle because left ventricle has to pump blood to the whole body except lungs.
4. Heart valves do not permit the blood pumped back from the arteries.

Fill in the Blanks

1. Cardiac.
2. Absent.
3. Depulsation, gill.
4. Mammals.
5. 100.

Multiple Choice Questions

1. (d) 2. (a) 3. (a).

7

Urinogenital System

Evolutionary survival of animals depends on doing many things successfully: escaping from predators, procuring food, adjusting to the environment, and so on. All of this comes down to reproducing successfully, which is the primary biological role of the genital system. On the other hand, the urinary system is devoted to quite different functions: namely to the elimination of waste products, primarily ammonia, and to the regulation of water electrolyte balance. Although urinary and reproductive functions are quite different, anatomists treat both systems together as the **urogenital system** (or urinogenital system) because both share many of the same ducts (Box 7.1).

Box 7.1

Although the function of kidneys is different from that of gonads, the urinary and genital systems of gnathostomes are so interrelated both developmentally and anatomically, that neither can be discussed without frequent reference to the other. For this reason, urinary and genital organs are discussed in a single chapter. The term *urogenital* or *urinogenital* is a combination of the Greek combining form *ouro*-pertaining to urine, and an English derivative of the Latin word *genitalis,* which pertains to reproduction.

Anatomically, the urinary system includes the kidneys and the ducts that carry away their products, **urine**. The genital system includes the gonads and their ducts that carry away the products they form, **sperm** and **ova**. Embryologically, urinary and reproductive organs arise from the same or adjacent tissues and maintain close anatomical association throughout the organism's life.

7.1 URINARY SYSTEM

1. Evolutionary Origin of Kidneys in Vertebrates

Craniate life began in water, and the early stages of evolution of kidneys took place in that medium. Craniates that are submerged in freshwater unavoidably acquire water by absorbing it through the gills, oropharyngeal membranes, or skin and by unavoidably swallowing it with their food. Therefore, to maintain their tissue fluids in an osmotic state compatible with existence in this medium, it is essential that freshwater organisms **excrete water** and **conserve salt** (Box 7.2).

Box 7.2

The earliest role of kidneys was osmoregulation in the aquatic environment. Elimination of nitrogenous wastes is an added excretory role that has become significant in tetrapods kidneys and replaced gills in this function (**Kent** and **Carr** 2001).

Fishes that are submerged in salt water face a different osmotic problem. Instead of accumulating too much water, they are in danger of accumulating too much salt. Survival in salt water depends on **conserving water** and **excreting salt.** Craniates living in some environments on land face the possibility of becoming dehydrated. The major role of kidneys is **osmoregulation**, the maintenance of a ratio between the quantity of salt and water in the body fluids such that the organism can survive in its natural habitat, whether it be in water or on land. Osmoregulation is one aspect of a physiological function, the **homeostasis**, *i.e.*, the involuntary control of the chemical environment of cells.

Kidney maintain appropriate concentrations of water and salt in body tissues by eliminating any excess water, by preventing the escape of water when necessary, and by preventing the escape of water when necessary, and by regulating the excretion of certain salts. In the latter role, they have the assistance of extrarenal osmoregulatory organs (Box 7.3).

Box 7.3

Craniates that imbibe excessive quantities of salt (marine craniates, birds whose food comes from the sea, other reptiles that inhabit arid environments) excrete salt extrarenally via gills or rectal, nasal, orbital, oral, or integumentary **salt glands** (**Kent** and **Carr** 2001).

The production of nitrogenous wastes is a natural consequence of metabolism (Table 7.1). Mechanisms for the elimination of these byproducts vary among taxa. In fish, nitrogenous wastes, if soluble, can be eliminated by simple diffusion across any membrane exposed to the external environment (mostly by the gills and sometimes via the skin). Thus, in many fish, the kidney is not a primary organ in the elimination of nitrogenous wastes but serves homeostasis in maintaining water and salt balance. With the transition to a terrestrial habitat, the kidneys of tetrapods must take on the additional primary role of the elimination of nitrogenous wastes. The functional components of craniates kidneys and their adaptations is the next topic of discussion.

Whether the first craniates, hence craniates kidneys, evolved in fresh or salt water is not settled. Following two theories have been forwarded for the origin of kidneys in vertebrates:

Table 7.1 A comparison of the forms of nitrogenous wastes in craniates (source. Kent and Carr 2001).

Form of nitrogen	*Energy cost*[1]	*Toxicity*	*Solubility*	*Water requirement*[2]
1. Ammonia	Least costly	Highly toxic	Highly soluble	Greatest (300-500ml/1gN)
2. Urea	Intermediate	Minimal	Quite soluble	Intermediate (50ml/1gN)
3. Uric acid	Mostly costly	None	Slightly soluble, precipitates with pH change	Least (10ml/1gN)

1. Refers to the relative formation costs for nitrogenous wastes. Ammonia is a natural by product of metabolism, whereas urea and uric acid require energy to form.
2. This indicates the relative amount of water (in milliliters of water per gram of nitrogen) needed to excrete nitrogen in its various forms.

1. Freshwater origin of vertebrate kidneys. Advocates of this theory point to the design of the craniate kidneys in its ability to excrete excess water. As noted above, a consequence of life in freshwater is the accumulation of excess water and loss of salts.

2. Marine origin of vertebrate kidneys. Proponents of this theory point to fossil evidence. Many deposits formerly thought to be freshwater in origin are now known to be marginal marine. Recently described craniate fossils from the Lower Cambrian are found with a clearly marine invertebrate fauna (**Shu** et al., 1999). These new discoveries predate the occurrence of other known fossil craniates by 20 to 50 million years and are older than the famous Burgess shale in Canada.

Regardless of which view is correct, the functional components of craniate kidneys have been such that, when subjected to changing environments during geologic time, they were adaptable for life in salt water, in fresh water, on land, and even in a desert (Table 7.2).

Table 7.2 Environmental osmotic challenges and Evolutionary solutions in craniates (source: Kent and Carr 2001).

Environment	***Primary Osmotic Problem***	***Solution***	***Example Taxa***
Freshwater	• Environment is hypoosmotic ▪ Excessive water uptake • Must excrete excess water and conserve salts	• Waste excreted as ammonia ▪ Use excess water as solvent • Actively transport solutes (for retention)	• Freshwater teleosts, aquatic and semiaquative amphibians
Salt water	• Environment is hyperosmotic • Excessive water loss • Excessive uptake of salts (in food and drinking water) • Must retain water and eliminate salts	• Isoosmoticity • Lose glomerulus (↓ water loss) • Become hyperosmotic relative to environment ▪ Retain urea to achieve ▪ Increases water uptake ▪ Retain glomerulus to excrete water • Salt excretion glands: ▪ Rectal gland ▪ Salt glands on gills	• Hagfish • Marine teleosts • Elasmobranchs and *Latimeria* • Elasmobranchs • Teleosts
Terrestrial	• Environment is dry • Water and salts are limited. ▪ Must retain both water and salts • A secondary return to water ▪ Reintroduces above aquatic concerns	• N excreted in all three forms ▪ Balance costs vs. benefits ▪ Variable among taxa ▪ Variable in life histories • Reduce glomerulus • Solute recovery • Loop of Henle or equivalent • Salt glands (elimination)	• Nonavian reptiles and anurans inhabiting arid habitats • Birds and mammals • Marine taxa

2. Structure of Mammalian Kidney

The vertebrate kidneys are a pair of compact masses of tubules situated dorsal to the abdominal cavity. Urine produced by the tubules is ultimately released into the **cloaca** or its derivative, the **urinogenital sinus.**

A cutaway view of the mammalian kidney reveals the two regions: an outer **cortex** surrounding a deeper **medulla** (Fig. 7.1A). Urine produced by kidney enters the **minor calyx** and then the **major calyx**, which joins the **renal pelvis**, a common chamber leading to the **urinary bladder** via the **ureter**. Elimination of urine from the body occurs through the **urethra.** Within the kidney, the functional unit that forms the urine is the microscopic **uriniferous tubule** (Fig.7.1B; Fig. 7.2). The uriniferous tubule consists of two parts: the **nephron (nephric tubule)** and the **collecting tubule** into which the nephron empties. The number of uriniferous tubules varies from only a few hundred in the kidneys of cyclostomes to over a million per kidney in mammals, in whom the tubule of kidneys combined constitute over 120 km of tubing. The nephron forms urine. The collecting tubule affects the concentration of urine and conveys it to the minor calyx, the beginning of the excretory duct.

Glomeruli are tufts of microscopic capillary-like arterial loops, or retia mirabilis, on the pathway of an arteriole, where water, ions, metabolic waste products, and certain other

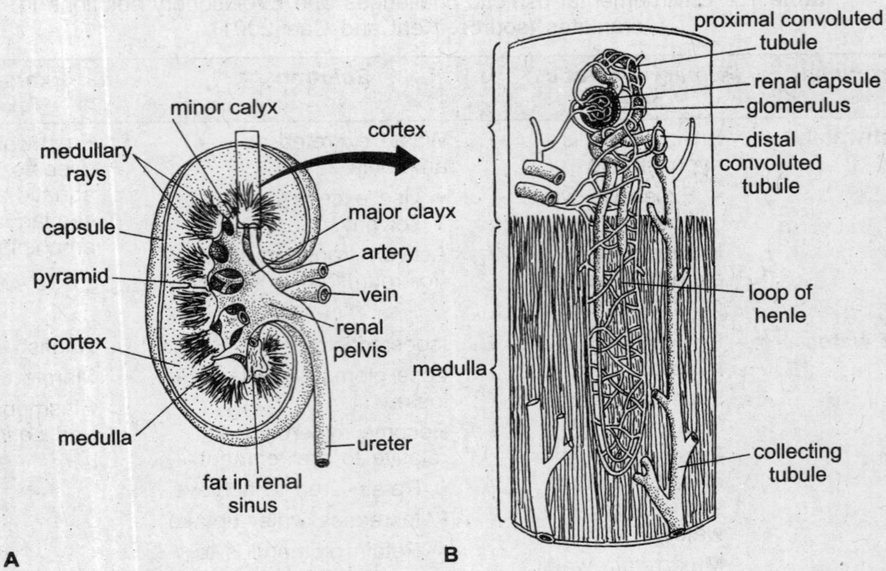

Fig. 7.1. Structure of the mammalian kidney. A – Section of kidney showing cortex, medulla, and departure of ureter. B – The uriniferous tubule begins in the cortex, loops through the medulla, and then returns to the cortex where it joins with the collecting tubule (after Kardong 2006).

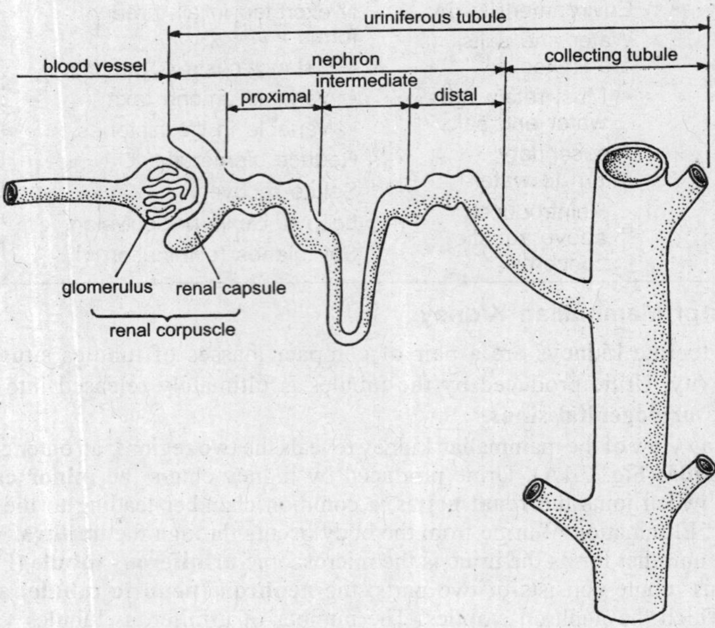

Fig. 7.2. Uriniferous tubule. The nephron (nephric tubule) and collecting tubule make up the uriniferous tubule. In turn, the nephron comprises the renal (Bowman's) capsule and the proximal, intermediate and distal tubules. The glomerulus is the capillary bed associated with the renal capsule. The excretory duct carries away waste products from several uriniferous tubules (after Kardong 2006).

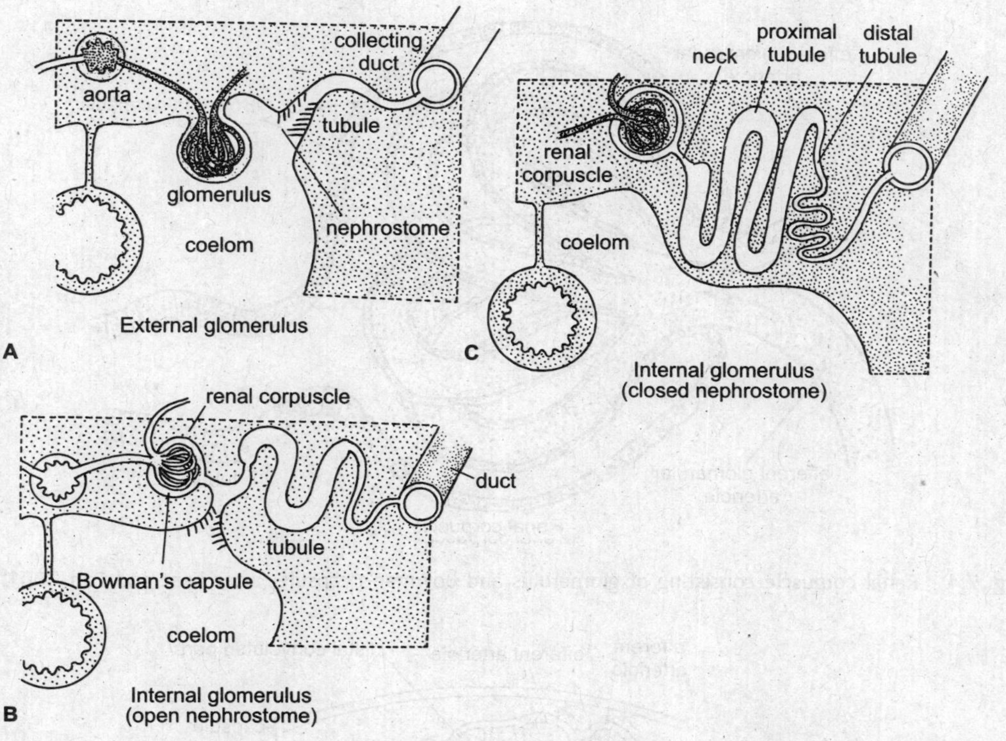

Fig. 7.3. Basic components of a craniate kidney. A – External glomerulus and tubule in the cephalic end of the segmental kidney of a larval hagfish. B – Tubule with internal glomerulus and open nephrostome, as in an apodan kidney. C – Generalized nonsegmental tubule of freshwater fishes, urodeles and anurans. The tubule of marine fishes often lack glomeruli (after Kent and Carr 2001).

constituents are removed from the bloodstream. In some fishes, they are large enough to be seen with the naked eye or a hand lens. In others, they are microscopic. The most primitive glomeruli are suspended in the coelom surrounded by peritoneum. They discharge their filterate into the coelomic fluid, which is then swept into a peritoneal funnel, or **nephrostome**, leading to a tubule (Fig. 7.3 A). These are **external glomeruli.** In today's craniates, external glomeruli are confined to embryos and larvae. Glomeruli in adults are embedded within the dorsal body wall (hence said to be retroperitoneal and are ensheathed by **Bowman's capsule**, a delicate double-walled outgrowth from a kidney tubule (Fig.7.3 B; Fig.7.4). Its inner wall adheres to the surfaces of the vascular loops. The capsular cavity collects the glomerular filterate, which then passes into a renal tubule. These are **internal glomeruli**. A glomerulus and the surrounding capsule constitute a **renal corpuscle** (Fig.7.4). A renal corpuscle, renal tubule, and the associate **peritubular** capillaries constitute a **nephron**, the functional unit of a gnathostome kidney (Fig. 7.1).

Supplying a glomerulus is an **afferent glomerular arteriole**, and emerging from the glomerulus is an **efferent glomerular arteriole** of *lesser diameter*. This has the effect of increasing the blood pressure within glomerulus. The afferent arterioles of agnathan are supplied directly from the dorsal aorta by segmental arteries. In gnathostomes, they are terminal branches of renal arteries. Efferent glomerular arterioles supply peritubular capillaries (Fig.7.5). Efferent glomerular

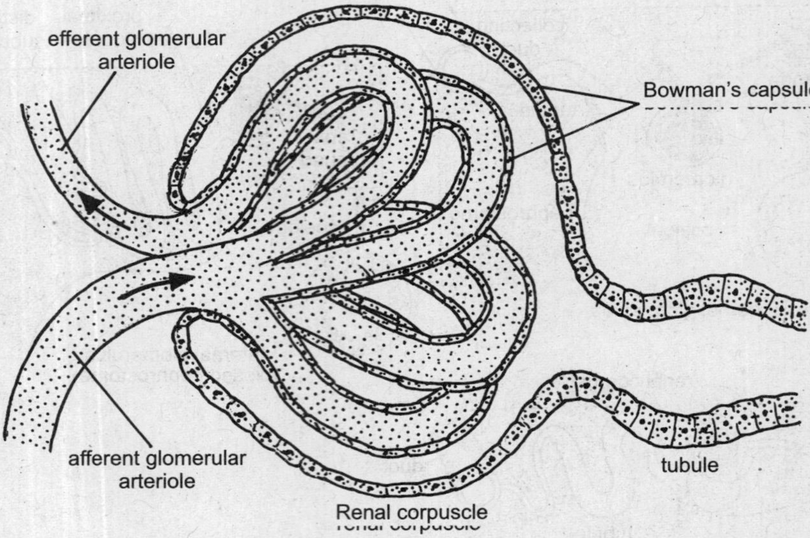

Fig. 7.4. Renal corpuscle consisting of glomerulus and Bowman's capsule (after Kent and Carr 2001).

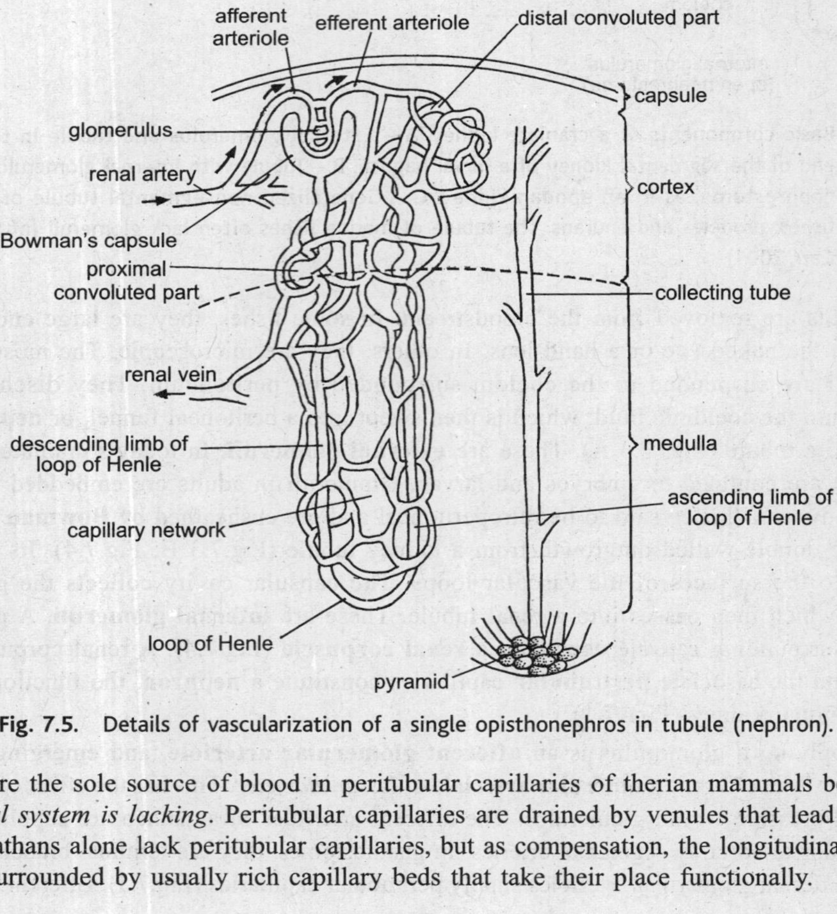

Fig. 7.5. Details of vascularization of a single opisthonephros in tubule (nephron).

arterioles are the sole source of blood in peritubular capillaries of therian mammals because a *renal portal system is lacking*. Peritubular capillaries are drained by venules that lead to renal veins. Agnathans alone lack peritubular capillaries, but as compensation, the longitudinal kidney ducts are surrounded by usually rich capillary beds that take their place functionally.

7.2 EMBRYONIC DEVELOPMENT OF KIDNEY AND TRIPARTITE CONCEPT OF KIDNEY ORGANIZATION (OR SUCCESSION OF KIDNEY AND DUCTS)

The kidneys form within the intermediate mesoderm located in the dorsal and posterior body wall of the embryo. In all vertebrates the kidney develop embryologically from **mesomere** tissue (of intermediate mesoderm), which lie in a retroperitoneal position along the dorsal edge of the developing coelom (Fig 7.6). Like the vertebral column and somatic muscles, the developing kidneys are segmented, like many other structures.

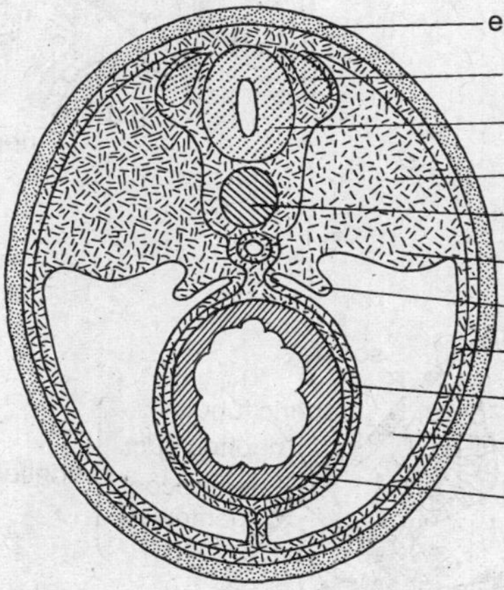

Fig. 7.6. Cross-section through a developing vertebrate to show the relationship of the nephrogenic intermediate mesoderm to other tissues.

At the onset of its differentiation, this posterior region of intermediate mesoderm expands, forming a **nephric ridge** that protrudes slightly from the dorsal wall of the body cavity (Fig.7.7A). In fact, the nephric ridge is a **nephrogenic region**, meaning that it is embryonic source of kidneys and their ducts.

The next structure to appear usually is paired **nephrotome**, the embryonic forerunner of the nephric tubule (Fig.7.7 B). The nephrotome is often segmental and contains the **nephrocoel**, a coelomic chamber that may open via a ciliated **peritoneal funnel** to the coelom. Next, the medial end of the nephrotome widens into a thin-walled renal capsule into which grows the glomerulus, a tuft of arterial capillaries. The lateral end of nephrotome grows outward. This outgrowth fuses with similar outgrowth from successive nephrotomes to form the common **nephric duct** (Fig.7.7 C). From this point in embryonic development, the modified nephrotome is more properly called a **nephric tubule.** It may retain its connection with the coelom via the persistent peritoneal funnel.

Tripartite Concept

Developmental and structural differences in the nephric tubules that arise within the nephric ridge inspired a view of kidney formation known as the **tripartite concept**. This concept envisions formation of nephric tubule in one of three locations within the nephric ridge. Subsequent loss, merger, or replacement of these tubules constitutes the developmental basis for the definitive adult kidneys. Specifically nephric tubules may arise within the anterior, middle, or posterior region of the nephric ridge, giving rise to a **pronephros**, **mesonephros** or **metanephros**, respectively (Fig.7.8; Fig.7.9 A–C; Fig. 7.12). In addition to positional differences, the three regions vary with respect to connections to the coelom. In the pronephros, the tubules retain their connections to the coelom through the peritoneal funnel; however, tubules arising within the middle or posterior regions are not connected to the coelom.

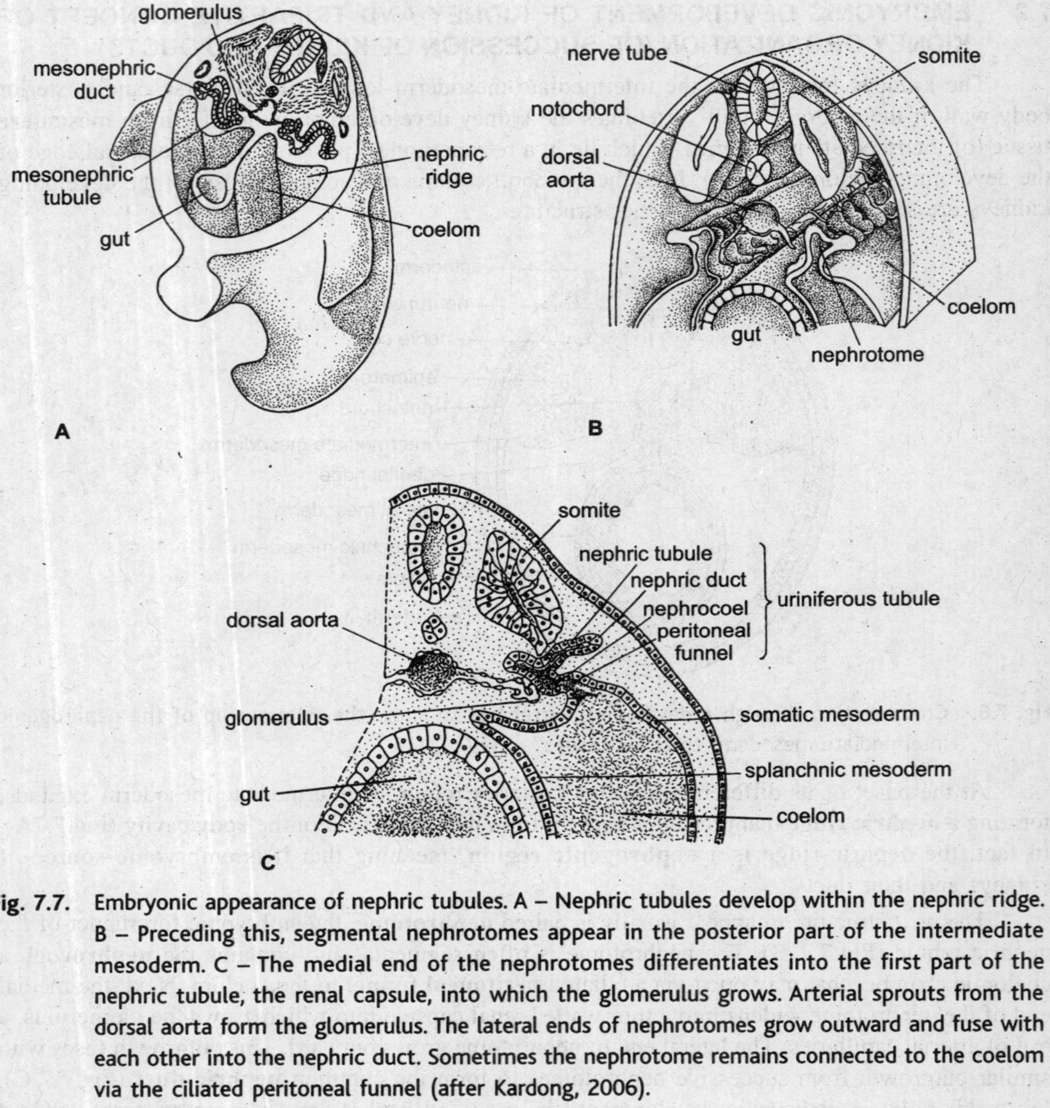

Fig. 7.7. Embryonic appearance of nephric tubules. A – Nephric tubules develop within the nephric ridge. B – Preceding this, segmental nephrotomes appear in the posterior part of the intermediate mesoderm. C – The medial end of the nephrotomes differentiates into the first part of the nephric tubule, the renal capsule, into which the glomerulus grows. Arterial sprouts from the dorsal aorta form the glomerulus. The lateral ends of nephrotomes grow outward and fuse with each other into the nephric duct. Sometimes the nephrotome remains connected to the coelom via the ciliated peritoneal funnel (after Kardong, 2006).

1. Holonephros. The kidney of the larval hagfish is structurally and functionally so generalized that it is thought to be very similar to a phylogenetically early kidney (Fig.7.11). It is called the **holonephros**. It lies in a retroperitoneal position, but extends the entire length of the body. Thus, it develops in a cephalocaudal sequence and has one kidney tubule per segment throughout its length. The tubules from the pronephros portion join and form the **Wolffian duct**, which runs the entire length of the nephrogenic tissue and terminates at the cloaca. As the mesonephros and metanephros develop, their tubules also hook on to the Wolffian duct.

Each holonephric tubule originates at the body cavity; its lumen is continuous with the coelomic fluid through the opening as its neck, called the **nephrotome**. The tubule cells in the nephrotome are **ciliated** and pump coelomic fluid into the tubule (Fig.7.11). Blood comes to each holonephric kidney tubules in one of the pair of renal arteries from the dorsal aorta. This renal artery terminates in a capillary knot called **external glomerulus** against the serosa (*i.e.*, the

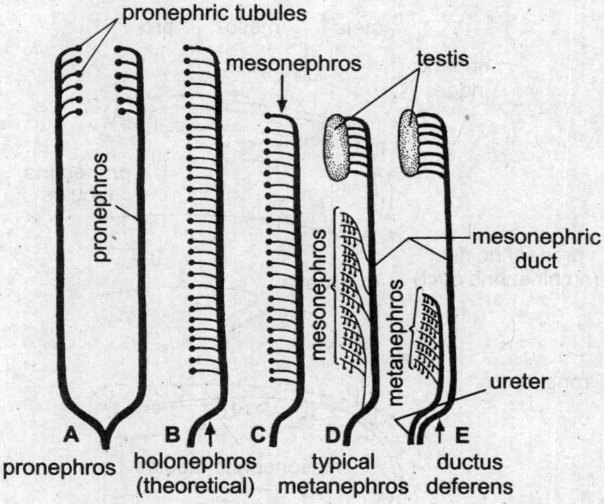

Fig. 7.8. Development of kidney in vertebrates. A – Pronephros; B – Holonephros; C – Mesonephros; D – and E – Metanephros.

peritoneal coat of a visceral organ) of the coelom. From the glomerulus, small vessels carry blood to the walls of tubule where another capillary network forms. A renal vein drains the tubules capillary blood into the posterior cardinal veins, there is no renal portal system.

2. Pronephros. It is an embryonic kidney of all vertebrates. Tubules that appear within the anterior part of nephric ridge are called **pronephric tubules**. These tubules join to form a common pronephric duct. This duct grows posteriorly in the nephric ridge eventually reaching and opening into the cloaca (Fig.7.9 A). Glomeruli may protrude into the roof of the body coelom, and fluid filters from them into the body cavity. Pronephric tubules then take up this coelomic fluid through ciliated peritoneal funnels, act on it, and eventually excrete the fluid as urine. However in most pronephric kidneys, glomeruli make direct contact with pronephric tubules.

Pronephric tubules become associated with glomeruli to form functional kidneys in larval cyclostomes, some adult fishes, and embryos of most lower vertebrates. Fluid filtered from the blood enters the tubules directly, and the peritoneal funnels may or may not remain open, depending on the species. In a few amniotes, usually only several pronephric tubules appear during embryonic development. They are not connected to the coelom and do not become functional. In most vertebrates, the embryonic pronephros regresses, and as it does, it is replaced by a second type of embryonic kidney, the mesonephros.

3. Mesonephros and Opisthonephros. Tubules of the mesonephric kidney arise in the middle of the nephric ridge. These **mesonephric tubules** do not produce a new duct but instead tap the pre-existing pronephric duct. To be consistent, the pronephric duct is now properly renamed the **mesonephric duct** (Fig.7.9 B).

The mesonephros usually becomes functional in the embryo, but if it persists into the adult, it is modified by incorporation of additional tubules arising within the posterior nephric edge. This extended mesonephric kidney with additional posterior tubules is termed the **opisthonephros** (Fig.7.9 D and Fig. 7.10). The opisthonephros is found in most adult fishes and amphibians. In amniotes, the mesonephros is replaced in later development by another advanced type of embryonic kidney, the metanephros.

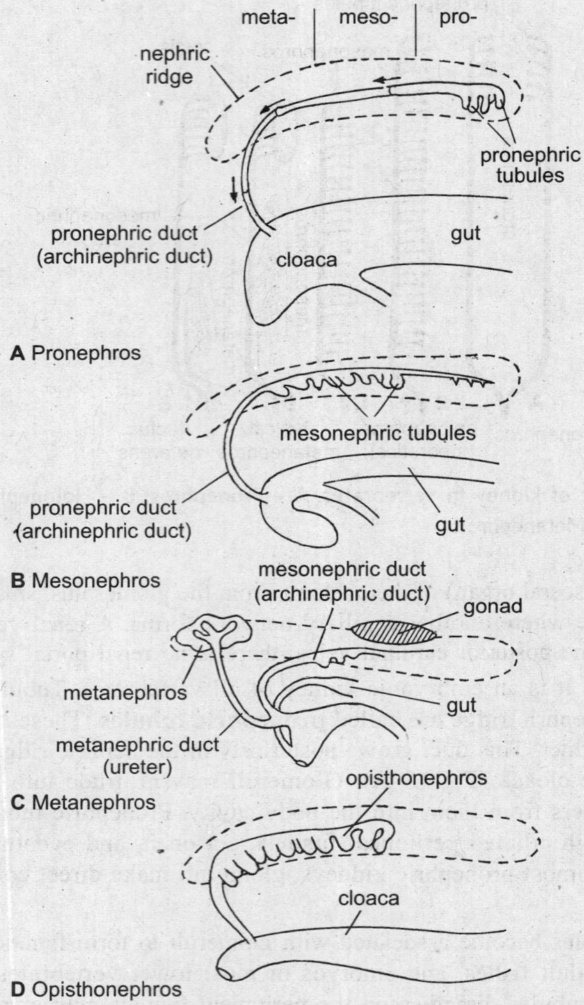

Fig. 7.9. Embryonic origin of the kidneys. Tubules forming the kidney arise in one of three regions of the nephric ridge: anterior (pro-), middle (meso-), or posterior (meta-). A – Pronephros. Tubules arise in the anterior part of the nephric ridge. They produce a pronephric duct that grows posteriorly in the nephric ridge and empties into the cloaca. Of the three types of kidneys, the pronephros is the first to arise during embryonic development. It becomes the adult kidney in a few fishes but is usually replaced during embryonic development by the mesonephros. B – Mesonephros. Tubules arise in the middle of the nephric ridge and tap into the existing pronephric duct, now appropriately renamed the mesonephric duct. The mesonephros is usually embryonic and transient. C – Metanephros. Sprouting from the mesonephric duct, the ureteric diverticulum (later the ureter) grows into the posterior section of the nephric ridge where it stimulates differentiation of tubules that form the metanephros. In males, the mesonephric duct usually takes over the task of sperm transport and is called the vas deferens. In the female, the mesonephric duct degenerates. D – Opisthonephros. Tubules arising from the middle and posterior nephric ridge form an extended kidney, the opisthonephros that may develop into the adult kidney of fishes and amphibians (after Kardong 2006).

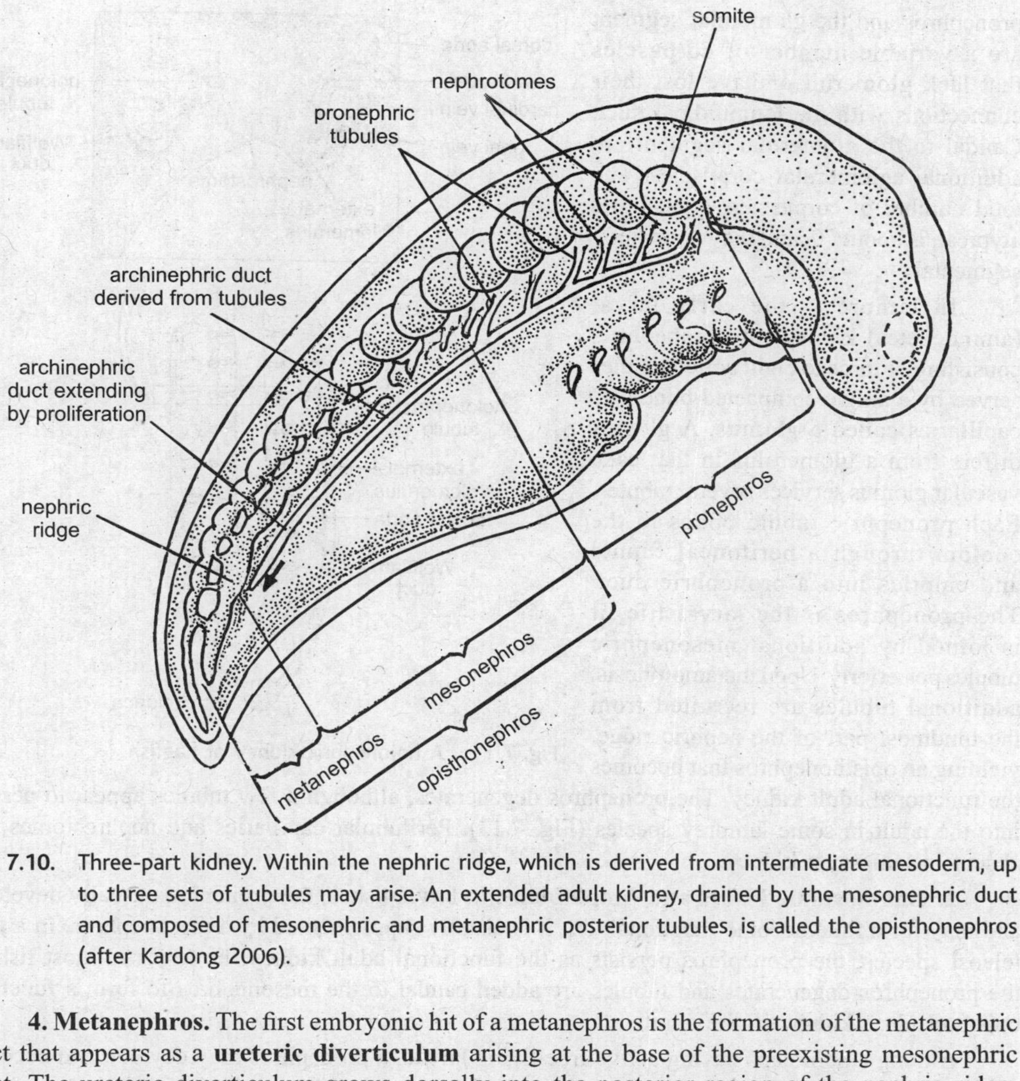

Fig. 7.10. Three-part kidney. Within the nephric ridge, which is derived from intermediate mesoderm, up to three sets of tubules may arise. An extended adult kidney, drained by the mesonephric duct and composed of mesonephric and metanephric posterior tubules, is called the opisthonephros (after Kardong 2006).

4. Metanephros. The first embryonic hit of a metanephros is the formation of the metanephric duct that appears as a **ureteric diverticulum** arising at the base of the preexisting mesonephric duct. The ureteric diverticulum grows dorsally into the posterior region of the nephric ridge. Here it enlarges and stimulates the growth of **metanephric tubules** that come to make up the metanephric kidney. The metanephros becomes the kidney of amniotes and metanephric duct is usually called the **ureter** (Fig.7.9 C).

The ontogenetic appearance of such kidneys (according to the tripartite concepts) seems to retrace their phylogenetic origins.

Phylogeny of Kidney

1. Agnathans. The kidney of the adult hagfish *Myxine* is an expression of the larval holonephros, with modifications (Fig. 7.11). All nephrostomes (or nephrotomes, see **Kardong, 2006)** have closed. The glomerular part occupies a 10 cm segment of the nephrogenic mesoderm commencing some distance anterior to the cloaca. This segment consists of 30 to 35 large renal corpuscles up to 1.5 mm in diameter, strictly segmental and connected to the longitudinal duct by very short tubules. There are no peritubular capillaries and no renal portal system. Between the regressed pronephros and the glomerular segment are a variable number of corpuscles that lack glomeruli or have lost their connections with the longitudinal duct. Caudal to the glomerular segment are additional aglomerular corpuscles. The total number of corpuscles, typical and atypical, is about 70, and they are strictly segmental.

In lampreys, the early larval (ammocoetes) kidneys are pronephric, consisting of three to eight coiled tubules served by a single compacted bundle of capillaries called a **glomus**. A glomus differs from a glomerulus in that each vascular glomus services several tubules. Each pronephric tubule opens to the coelom through a peritoneal funnel and empties into a pronephric duct. The pronephros is the larval life, it is joined by additional mesonephric tubules posteriorly. Upon metamorphosis, additional tubules are recruited from the hindmost part of the nephric ridge, yielding an opisthonephros that becomes the functional adult kidney. The pronephros degenerates, although a few tubules appear to persist into the adult in some lamprey species (Fig. 7.13). Peritubular capillaries and nephrostomes are absent in lamprey's kidney.

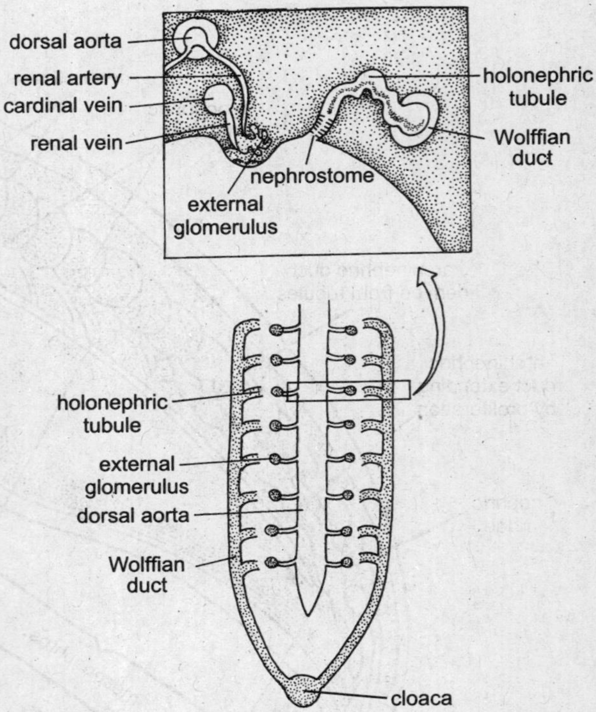

Fig. 7.11. A holonephric kidney of hagfish.

2. Gnathostome Fishes and Amphibians. In larval fishes, the pronephros usually develops and may for a time become functional, but it is usually supplemented by a mesonephros. In a few teleost species, the pronephros persists as the functional adult kidney; however in most fishes, the pronephros degenerates and tubules are added caudal to the mesonephros to form a function opisthonephric kidney in the adult.

Among amphibians having active free-living larvae, a pronephros may develop and become functional for a time. One or two pronephric tubules may contribute to the adult kidney as well. In caecilians, as many as a dozen pronephric tubules have been reported in the adult kidney. However, the early embryonic pronephros is usually succeeded by the larval mesonephros, which upon metamorphosis is replaced by an opisthonephros in most amphibians (Fig. 7.15). Nephrons within the opisthonephros tend to differentiate into proximal and distal regions before the joining the urinary ducts. In amphibians, as in many sharks and teleosts with opisthonephric kidneys, the anterior kidney tubules transport sperm, illustrating again the dual use of ducts that serve both genital and urinary system (Fig. 7.14).

3. Amniotes. In amniotes, the anterior end of the nephric ridge rarely produces pronephric tubules. When present, these are few in number and without excretory function. The predominant embryonic kidney is a mesonephros, but in all amniotes, it is supplemented in late development and completely replaced in the adult by the metanephros drained by a new urinary duct, the ureter (Fig.7.16). Metanephric tubules tend to be long with well-differentiated proximal, intermediate and distal regions. In mammals, in particular, the intermediate section of the tubules is specifically

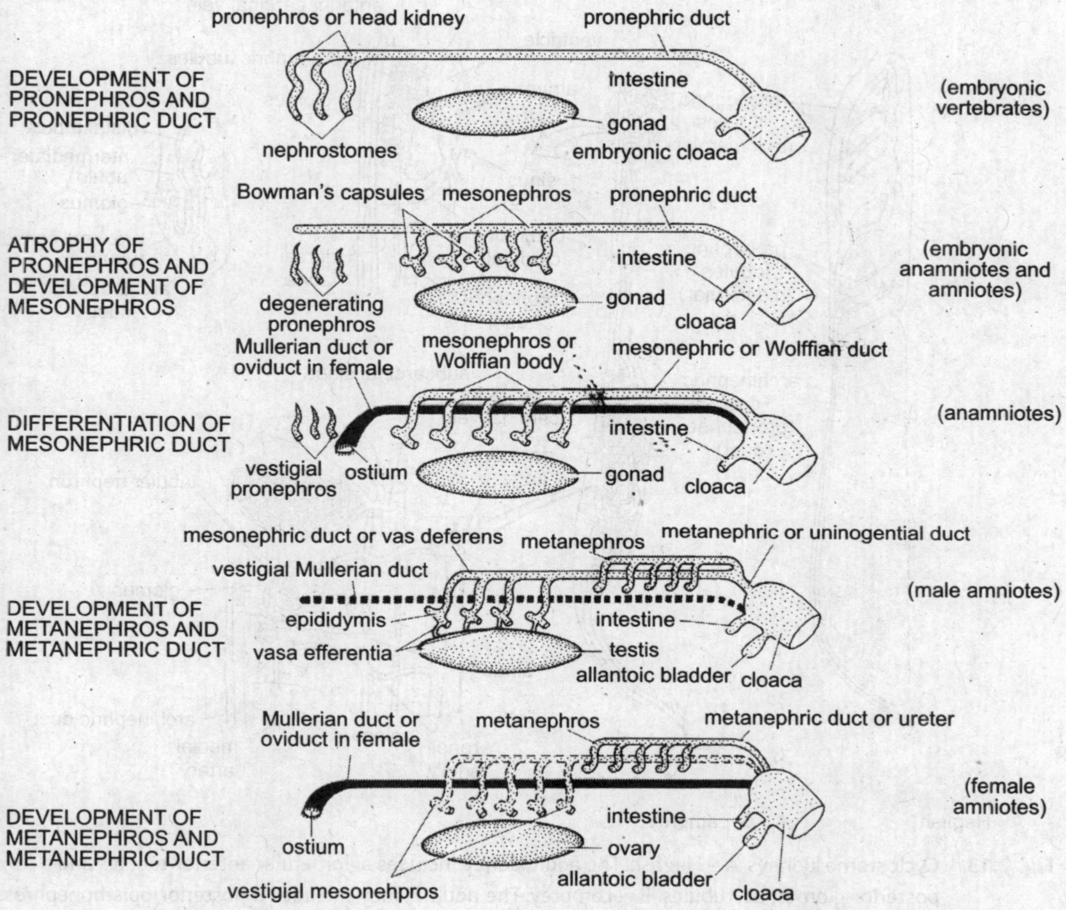

Fig. 7.12. Evolution of kidney in vertebrates.

elongated, consisting of the major part of the **loop of Henle**. This term refers to both a positional and a structural feature of the nephron. Positionally, the loop includes the part of the nephron that departs from the cortex and dips into the medulla (the descending limb), makes a sharp turn, and returns to the cortex (the ascending limb). Structurally, three regions contribute: the straight portion of the proximal tubule (Fig.7.1 B). One should note that the terms **descending** and **ascending limbs** refer to the parts of the loop that are departing or entering the cortex, respectively. The term **thick** and **thin** refer to the height of the epithelial cells forming loop. Cuboidal cells are thick and squamous cells are thin.

Loops of the nephron occur only in groups capable of producing **concentrated urine.** Among vertebrates, only the kidneys of mammals and some birds can produce urine in which solutes in the urine are more concentrated than in blood, and only these two groups possess nephrons with loops (Fig. 7.17). All mammalian nephrons have loops, specifically loops of Henle. Mammalian kidneys produce urine 2 to 25 times more concentrated than blood. Further, the ability to concentrate urine is correlated with loop length, and loop length is correlated with availability of water. The beaver has short loops of nephrons and excrete urine only about twice the osmotic concentration of its blood plasma, but some desert rodents have long loops of nephrons and can produce urine that is about 25 times as concentrated as their blood.

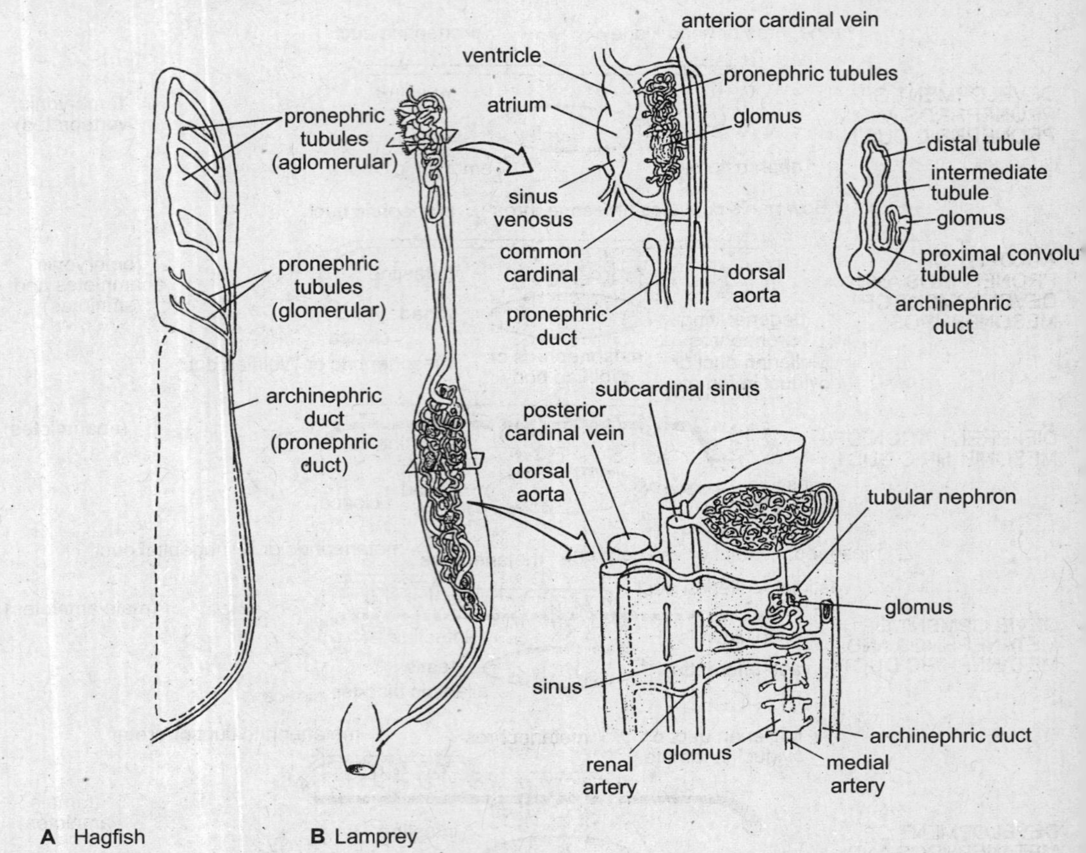

Fig. 7.13. Cyclostome kidneys. A – Hagfish. The adult kidney includes aglomerular anterior tubules and a few posterior glomerular tubules. B – Lamprey. The adult kidney includes a posterior opisthonephros. In some species, a few anterior pronephric tubules with peritoneal funnels may persist. Several pronephric tubules share a glomus, and each can be composed of proximal, intermediate, and distal sections (Kardong 2006).

In a few species of birds, the kidneys contain some nephrons with short, distinct loop of Henle in mammals, these short avian loops evolved independently. These avian kidneys exhibit a modest ability to produce concentrated urine. Their product is about two or four times more concentrated than their blood. However, the nephrons of most birds do not have loops. In the absence of a loop, the avian nephron is similar to the nephron of reptile.

Urinary Bladder

Before being excreted, urine is usually stored in specialized regions of the urinogenital system. In this way, the vertebrates can void urine at opportune times rather than continuously as it is formed. If water conservation is important, the urinary bladder sequesters the concentrated urine so that it does create osmotic pressure that draws water out of the tissues of the animal.

Urinary bladders are adaptations to life on land. They serve as reservoirs for water that may be needed later and therefore should not be wasted. Urinary bladders arise during embryonic development as an evagination from the ventral wall of the embryonic cloaca; in amphibians, turtles, basal lizards, and monotremes (the only metatherian tetrapods that have a urinary bladder

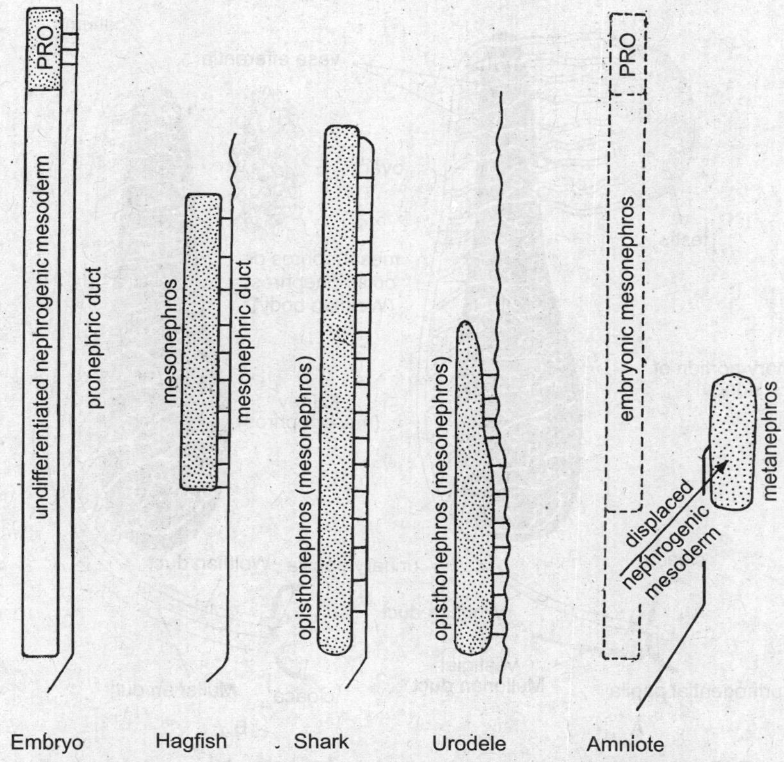

Fig. 7.14. Fate of the nephrogenic mesoderm in representative craniates. **PRo.** Pronephros. The pronephric duct persists in adult anamniotes to drain the adult kidney (after Kent and Carr 2001).

excepting a few retite birds), the bladder continues to open into the cloaca throughout life. Because the urethera also open into the cloaca in these tetrapods, urine backs up into the bladder from the cloaca unless the vent is open (see **Kent** and **Carr** 2001).

The urinary bladder develops from the extra- embryonic membrane and from the allantois also. However, most vertebrates have a urinary bladder. The urinary bladder is absent in cyclostomes and elasmobranchs among fish, snakes, crocodilians, some lizards and birds other than ostriches.

Types of Urinary Bladders. There are two general types of urinary bladders: tubal and cloacal. According to some workers, a third type of bladder, called **allantoic bladder** occurs in mammals and some reptiles as turtles and cetain lizards. Allantoic bladder is said to arise from the enlargement of the proximal or basal end of the embryonic allantoic stalk.

1. Tubal bladders. In fishes, urine is usually stored within the ends of the urinary ducts where they join the cloaca or open to outside. A urinary bladder of this type is mesodermal and noncloacal in origin. It is found among elasmobranchs, holocephalans, and most teleost fishes (Fig. 7.18 A; Fig. 7.19).

In many fish two independent bladders may form **vesica duplex** (Fig. 7.20A)- one near the end of each urinary duct, with the two ducts afterwards uniting into common passage way of exit; or the two may run together into a common bilobed bladder **vesica bicornis** (Fig.7.20 B) as in *Lepidosteus* and some other ganoids, or finally the two excretory ducts may first fuse and then expand into a single bladder, **vesica simplex** (Fig. 7.20 C) as for example in *Pike* (Fig.7.20 D) and *Esox*.

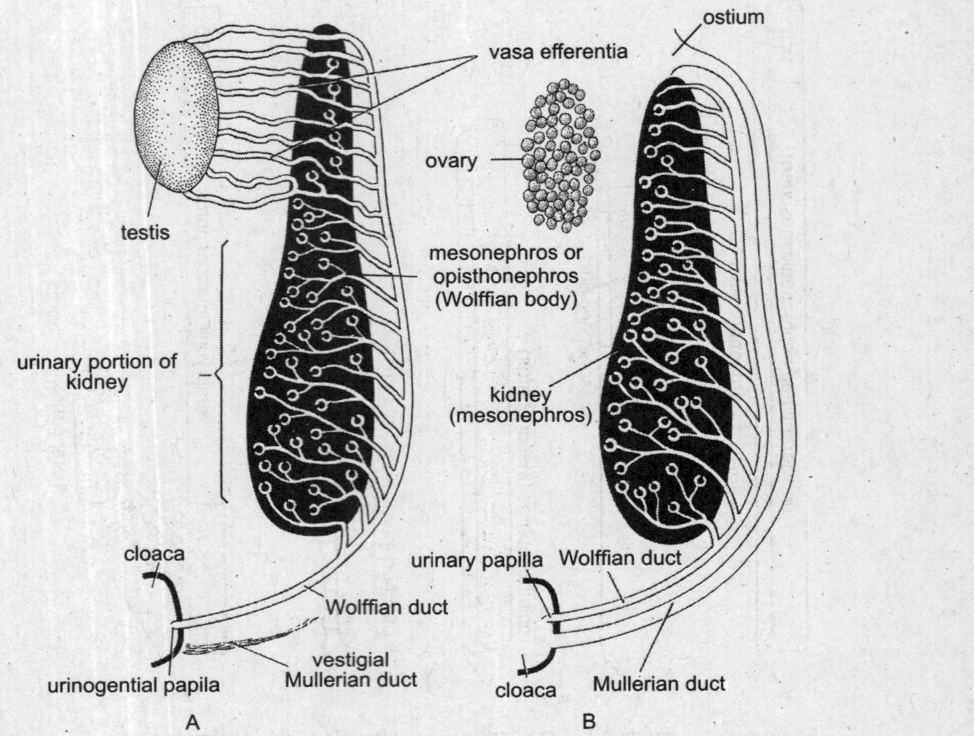

Fig. 7.15. Urogenital organs in an anamniote (*e.g.*, Amphibian). A – Male; B – Female.

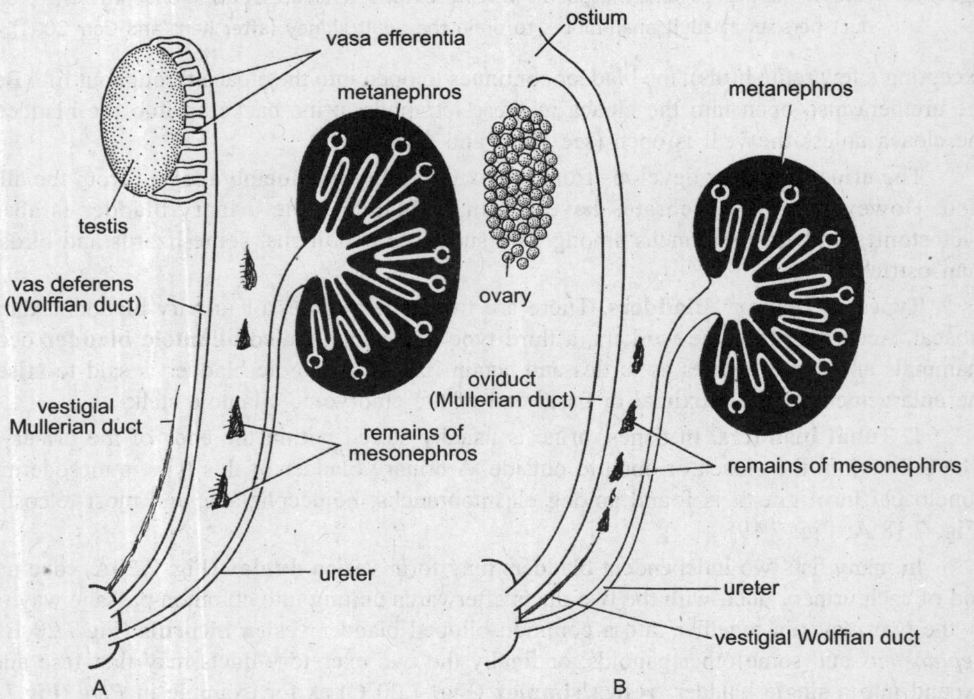

Fig. 7.16. Plan of the kidney and genital ducts in an amniote. A – Male; B – Female.

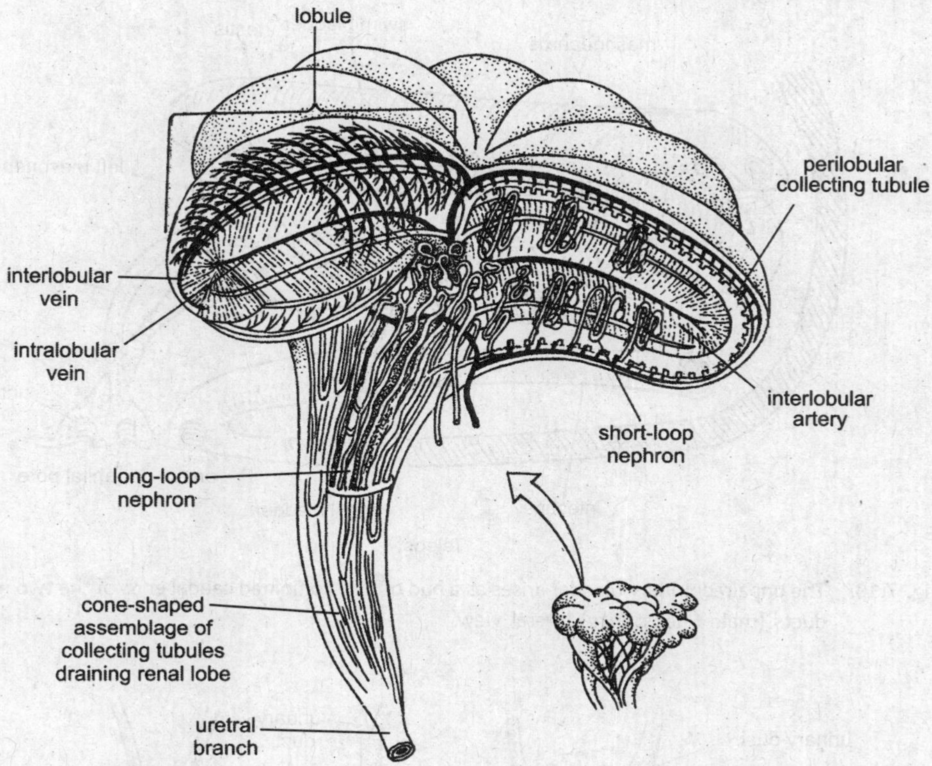

Fig. 7.17. Avian kidney. A section of the kidney is enlarged and cut away to reveal the arrangement of nephrons within and the blood supply to the nephrons (after Kardong 2006).

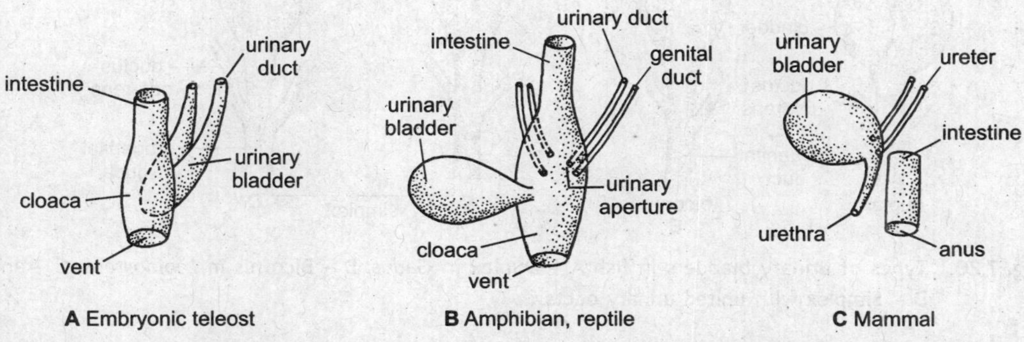

Fig. 7.18. Evolution of urinary bladder. A – In teleosts, the intestine and urinary ducts establish separate exits, anus and urinary pores, respectively. As a consequence the embryonic cloaca is lost in the adult. The teleost urinary bladder, when present, is formed from the expanded ends of the urinary ducts. B – and C – In tatrapods, the urinary bladder is an outgrowth of the cloaca. It empties into the cloaca in amphibians and reptiles (B) but exits in the urethera in mammals (C) (after Kardong 2006).

2. Cloacal bladder. In tetrapods, the urinary bladder arises as an outpocketing of the cloaca. Urine flowing from the urinary ducts usually empties into the cloaca first and then fills the baglike urinary bladder (Fig. 7.18 B). In mammals, the urinary ducts (ureters) empty directly into

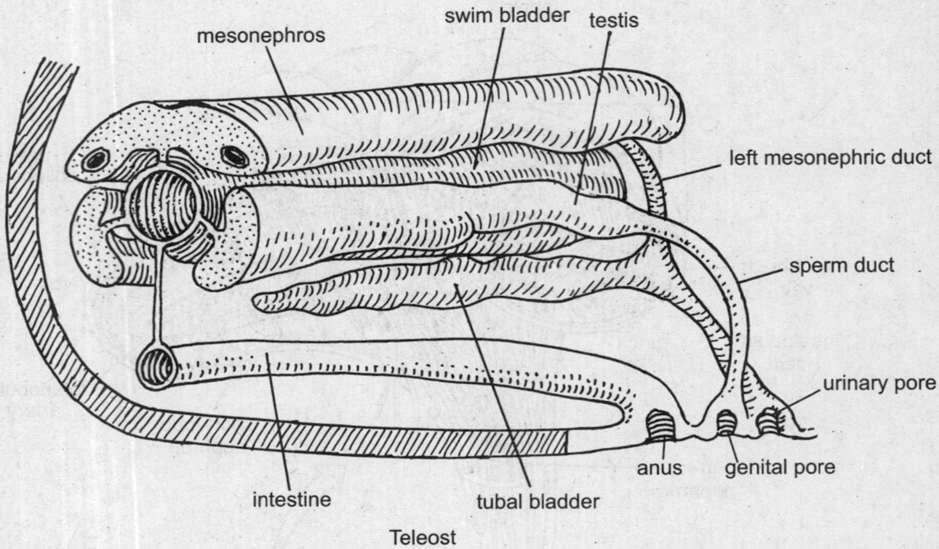

Fig. 7.19. The unpaired urinary bladder arises as a bud off the conjoined caudal ends of the two mesonephric ducts (male teleost *Pike*) lateral view.

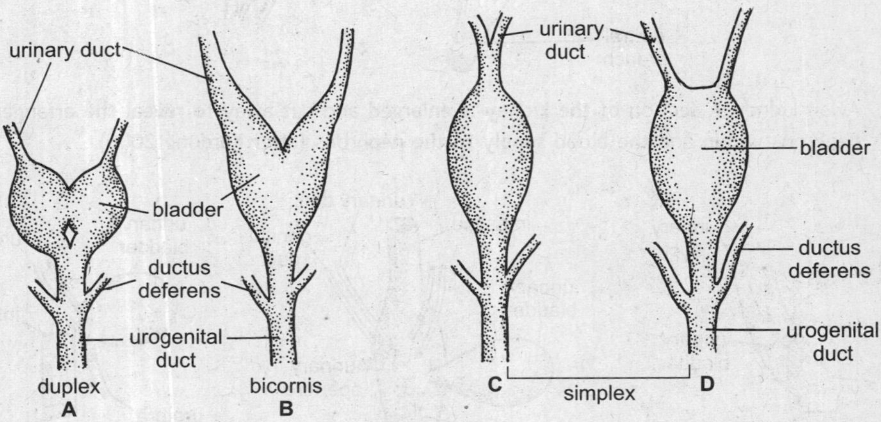

Fig. 7.20. Types of urinary bladders in fish A – Duplex in *Gadus*; B – Bicornis in *Lepidosteus*; C – and D – Simplex with united urinary ducts.

the urinary bladder (Fig. 7.18 C). The tetrapod urinary bladder appears first among amphibians and is present in *Sphenodon*, turtles, most lizards, ostriches among birds, and all mammals.

Cloacal bladder is located dorsally in lungfishes and ventrally in amphibians. In perennibranchiate amphibians it is considered elongated, but rounded and broadened in frogs and toads. Frequently, it is bilobed (Fig. 7.21) while in some urodeles (*e.g. Salamandra*, *Triton* and *Eurycea*, the lobes are prolonged into horn-like process.

Phylogeny of Bladder in Vertebrates

1. Fish. Most of the fish have the terminal enlargements or evaginations of the mesonephric ducts which are known as tubal bladders. The urinary sinus of male sharks and urinary sinus of female both of which have neither urogenital nor urinary papilla are the closest structure to a urinary bladder in shark. In dipnoans a small diverticulum of the dorsal wall of the cloaca is called urinary bladder.

2. Amphibians. In amphibians and all remaining tetrapods the bladders develop as evaginations from the ventral wall of the cloaca (Fig. 7.21). The urine gets collected in cloaca and then backs up into the bladder, where it is stored.

3. Reptiles. Turtles and some lizards have large bladders and some freshwater turtles have two accessory bladders. (Fig. 7.22) The urine back up into the bladder from the cloaca. In female turtles the water stored in bladders is used by female for softening and moistening the soil when building nest for the eggs.

4. Birds. There is no urinary bladder in birds. The lack of bladder reduces body weight and help in flight (*i.e.*, it is a flight adaptation). However ostrich (a flightless bird) has a urinary bladder.

5. Mammals. In mammals the kidney ducts empty directly into the bladder, and the bladder is drained by urethera (Fig .7.18C). In human beings urinary bladder lies in the pelvic cavity and is

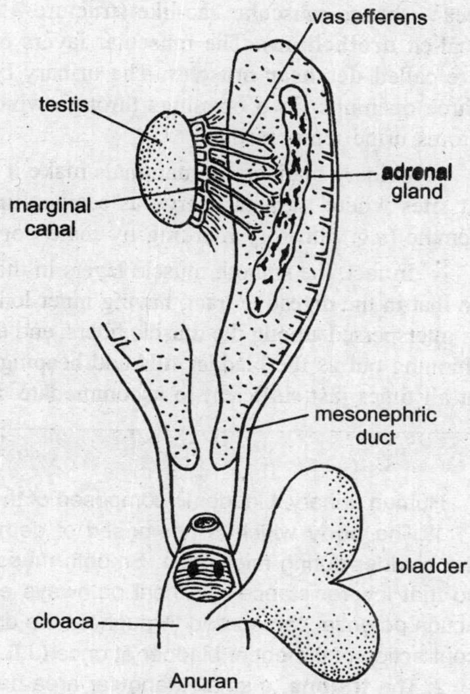

Fig. 7.21. Urogenital system and adrenal gland of a male frog, ventral view (after Kent and Carr 2001).

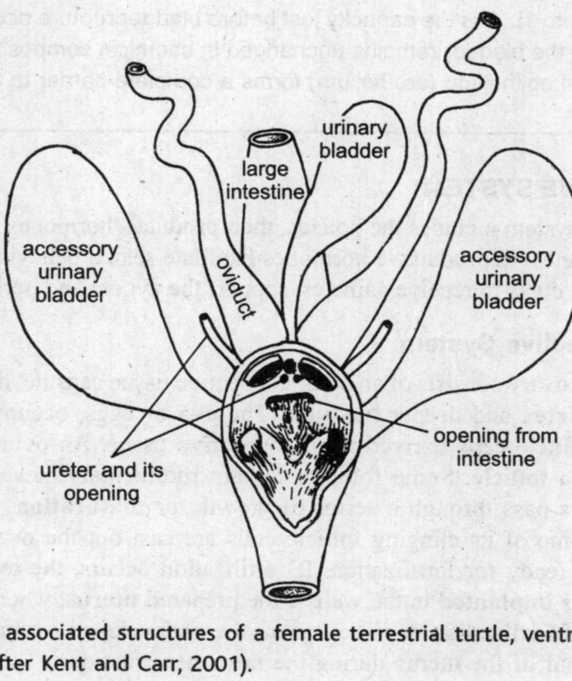

Fig. 7.22. Cloaca and associated structures of a female terrestrial turtle, ventral view. Clitoris has been removed (after Kent and Carr, 2001).

bean- shaped muscular sac-like structure. Its inner lining is composed of transitional epithelium (called **urothelium**). The muscular layers of urinary bladder comprise smooth muscle cells and are called **detrusor muscles.** The urinary bladder has a triangular area the **trigone** between the three openings. Two openings through which the ureter enters the bladder. The urinary bladder stores urine temporarily.

Urinary bladders in mammals make it possible for **pheromones** in the urine to be deposited at sites where they can serve as a signalling device to other members of the species, male or female (*e.g.,* territory marking by male dogs).

In fact, the smooth muscle layers in the urinary bladder wall are disposed in a reverse pattern to that in the digestive tract, having inner longitudinal and outer circular layers. Much elastic tissue is interspersed among the muscle fibers, and the cells of the urothelium lining slide over one another, thinning out as the bladder fills and heaping as it empties. Consequently, the size of the lumen is at all times just sufficient to accommodate the contained urine whatever the quantity (Box 7.4)

Box 7.4

Human urinary, bladder is composed of the following:

1. The **body** which is comprised of detrusor muscle. It is mainly responsible for emptying of the bladder during micturition. Smooth muscle cells of the detrusor muscle fuse with one another so that low-resistance electrical pathways exist from one muscle cell to the other. Therefore, as action potential can spread throughout the detrusor muscle, from one muscle cell to next, to cause contraction of the entire bladder at once (**J.E. Hall** 2011).
2. The **trigone**, a small triangular area near the mouth (neck) of urinary bladder, through which both ureters and urethera pass.
3. The **internal sphincter** to maintain tonic closure of the urethral opening and to prevent reflux of semen into bladder during ejaculation.
4. The **external sphincter** which is voluntary skeleton muscle.
5. The normal physiological capacity of bladder of adult human beings is 600 ml, however, its anatomical capacity is upto 1L. It is the capacity just before bladder rupture occurs (see **Jain** 2012).
6. The urine stored in the bladder remains unchanged in chemical composition since the luminal surface of the transitional epithelium (urothelium) forms a complete barrier to the passage of water and solutes.

7.3 REPRODUCTIVE SYSTEM

The reproductive system includes the gonads, their products, hormones and gametes, and the ducts that transport gametes. Reproductive hormones facilitate sexual behaviour and parental care, prepare the reproductive ducts to receive gametes, support the zygote and perform other functions.

Mammalian Reproductive System

In mammals, each ovary consists of an outer connective tissue capsule, the **tunica albuginea**, that encloses a thick **cortex** and deeper **medulla.** The **ova** or eggs, occupy the cortex and are wrapped in layers of **follicle cells** derived from connective tissue. An ovum plus its associated follicle cells is termed a **follicle**. Some follicles remain rudimentary, never change, and never release their ova. Others pass through a series of growth, or **maturation**, stages at the end of which the ovum and some of its clinging follicle cells are cast out the ovary in the process of **ovulation**, and become ready for fertilization. If fertilization occurs, the ovum continues down the **oviduct** and becomes **implanted** in the wall of the prepared **uterus**, where subsequent growth of the embryo occurs. If fertilization does not occur, the undeveloped ovum continues down the oviduct and is flushed out of the uterus during the next menstruation.

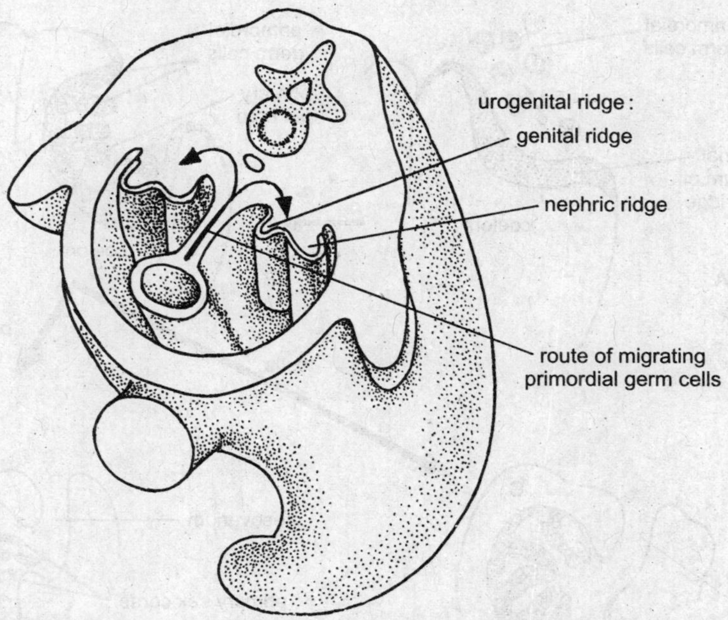

Fig. 7.23. Urogenital ridge. In the posterior part of the developing embryo, paired urogenital ridges arise within the roof of the coelom. The median ridges are the genital ridges and give rise to the gonads. The lateral nephric ridges give rise to the kidney and its ducts. Primordial germ cells that develop into eggs or sperms arise outside the gonads, migrate to them, and colonize the early gonad rudiments (after Kardong 2006).

Each mammalian testis also consists of an outer tunica albuginea, which encloses the **seminiferous tubules** that produce sperm. Within the walls of the seminiferous tubules, stem cells multiply and grow into sperm that eventually are released into the lumen. The coiled seminiferous tubules straighten forming **tubuli recti** just before they join the **rete testis**. Via the **efferent ductules** the rete testis joins the **epididymis** where sperm are temporarily stored. Upon ejaculation, sperm travel along the **vas deferens (ductus deferens)** into the urethera. Along the way, three accessory sex glands, the **seminal vesicle**, **prostate** and **bulbourethral (Cowper's) gland**, respectively, add their secretions as sperm move from the testes to the urethera. This fluid and the sperm constitute **seminal fluid**, or **semen**.

Embryological Development of Gonads

Gonads arise as a pair of embryonic **genital ridges**, which are thickenings of the coelomic mesothelium just medial to the mesonephroi (Fig. 7.23). The ridges are longer than the resulting mature gonads, which suggests that one time gonads may have extended the length of the pleuroperitoneal cavity, as they do in living agnathans. **Primordial germ cells** (progenitors of sperm and eggs) migrate into the genital ridges and establish themselves just internal to the peritoneum (Fig.7.24). This region of a gonadal primordium thereby becomes the **germinal epithelium.** Although primordial germ cells are first observed in the germinal epithelium, their progenitors are in the endoderm of the yolk sac and elsewhere before genital ridges have developed. From these locations they migrate to the genital ridges.

During early differentiation, the gonads have the potential to become either ovaries or testes. Under the influence of genes, hormones, and factors not fully indentified, these bisexual

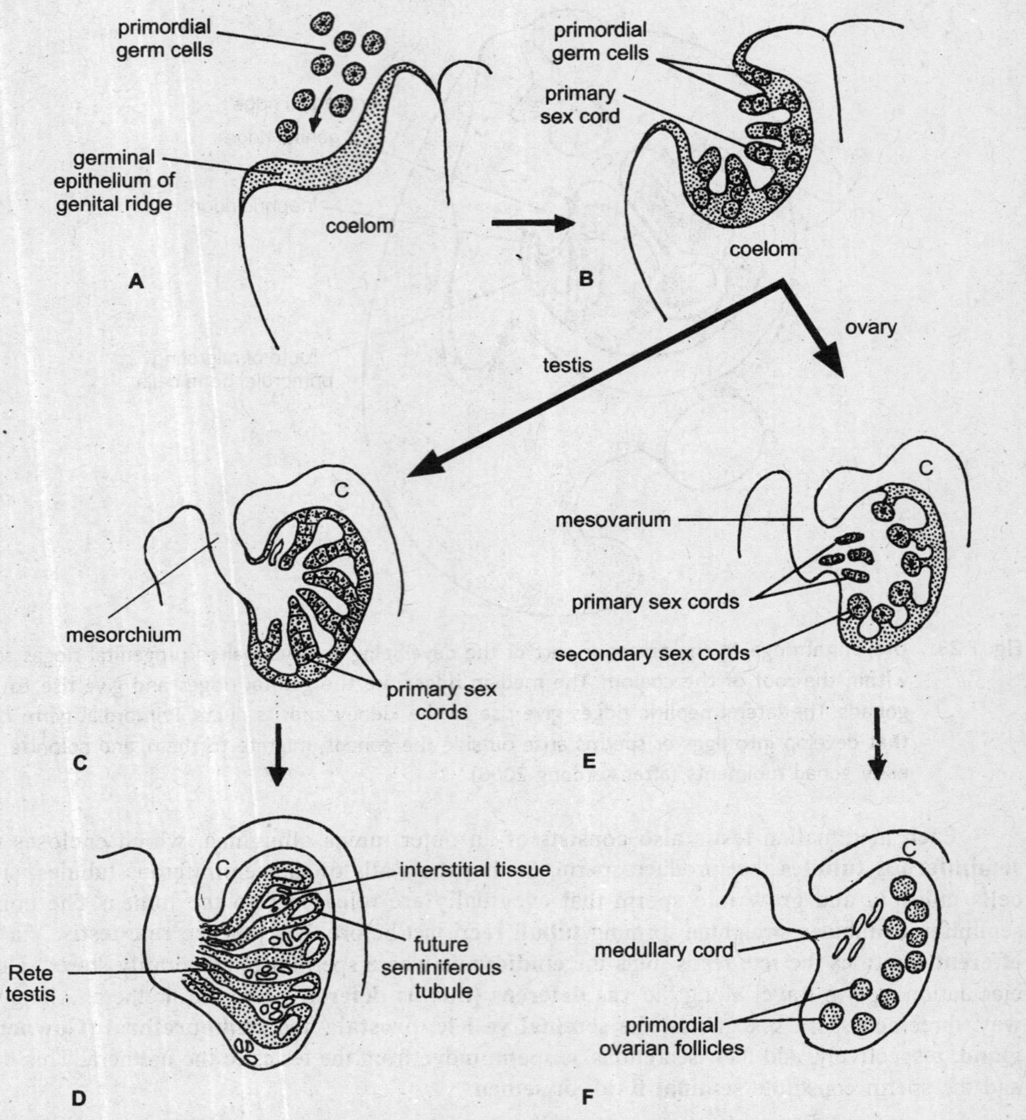

Fig. 7.24. Early development of mammalian gonads. A- Germinal epithelium on genital ridge of sexually indifferent primordium. B- Primary sex cords of bisexual stage. C- to F- Differentiating testis and ovary. C, coelom (after Kent and Carr 2001).

gonads develop into the gonads of a specific sex. As gonads enlarge, they generally acquire a dorsal mesentery, the **mesorchium** in males and the **mesovarium** in females.

The gonadal primordia that are becoming testes, **primary sex cords** invade the deeper region of the genital ridge and become **seminiferous tubules** which is lined with germinal epithelium (Fig.7.24). These tubules produce **sperm.** Primary sex cords form also in females, but they are not the source of **ova.** Oocytes form from **secondary sex cords** that also arise from the germinal epithelium (Fig.7.24 B, E,F). These cords do not grow deep into the medulla. Consequently, with the exception of some teleosts, eggs form just internal to the surfaces of the ovary and are shed to the exterior when mature. The primary sex cords of females regress, but remnants remain deep in the ovarian medulla, where they are called **medullary cords.**

In all craniates, *mature eggs are released into the coelom* or a compartment of it before being swept into an oviduct. *Sperm,* on the contrary, except in living agnathans, *are conducted in a closed system of vessels* that conveys them from the testis to the exterior without having entered the coelom. All living agnathan gametes are shed into the coelom.

Single-gonad vertebrates. A few adult craniates have a single gonad because of fusion of the two ridges across the midline, as in lampreys and a few teleosts, or because one of the juvenile gonads, most commonly an ovary, fails to differentiate. Among craniates with a single functional ovary are hagfishes, some viviparous elasmobranches, a few reptiles, and most species of birds. A few female mammals, including the platypus and some bats, have only one ovary.

Sex reversal. Examples of natural or experimental sex reversal have been demonstrated in many craniates. The embryonic testis of anurans consists of an anterior segment, called **Bidder's organ**, which in frogs usually disappears before sexual maturity, and a more caudal portion that becomes the functional testis. Bidder's organ persists in adult male toads (Fig.7.25) and contains large undifferentiated cells resembling immature ova. If the testes of a toad are removed experimentally and Bidder's organs are allowed to remain with a blood supply, the latter develop into functional ovaries, and the rudimentary female sex hormones produced by the new ovaries. Hens that have not been subjected to any treatment have been known to cease laying eggs, crow, and develop other male characteristics. This comes about when the left ovary atrophies and the right ovary, which is rudimentary in birds, enlarges and produces male hormones. However, these birds are not known to have produced sperm.

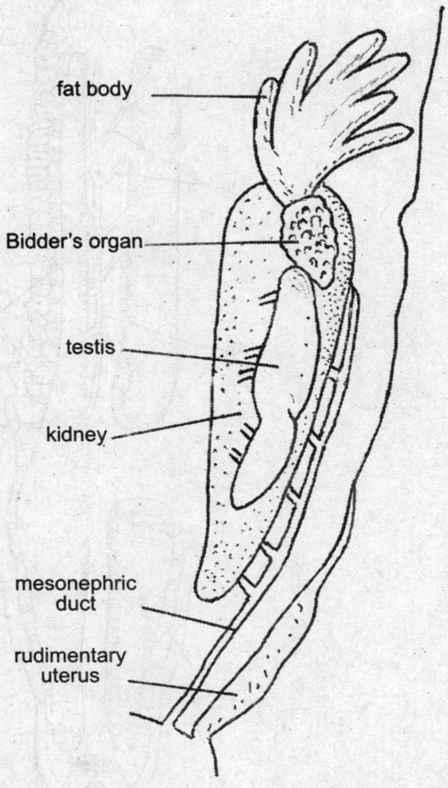

Fig. 7.25. Bidder's organ and the rudimentary female reproductive tract of a young male *Bufo*, ventral view. Only the left organs are illustrated. The mesonephric duct and rudimentary uterus empty into cloaca (after Kent and Carr 2001).

Testes and Male Genital Ducts

Craniate testes are essentially the same with few exceptions. Hagfishes, lampreys, some jawed fishes and urodeles lack seminiferous tubules. In these craniates, the germinal epithelium may be on the surface of the testis or deep within, and primordial germ cells migrate from the germinal epithelium into cystlike **seminiferous ampulae**, where the mature before being discharged. At the end of the spawning season, after hundreds of thousands of spermatozoa have been shed from each ampulla, the cysts collapse.

In some fishes and urodeles, there has been a tendency to form a new sperm duct to replace the mesonephric duct as a carrier of sperm (*e.g.,* marginal canal in Fig. 7.26). In teleosts, this has culminated in a mesonephric duct that carries no sperm whatsoever (Fig. 7.27). In all other

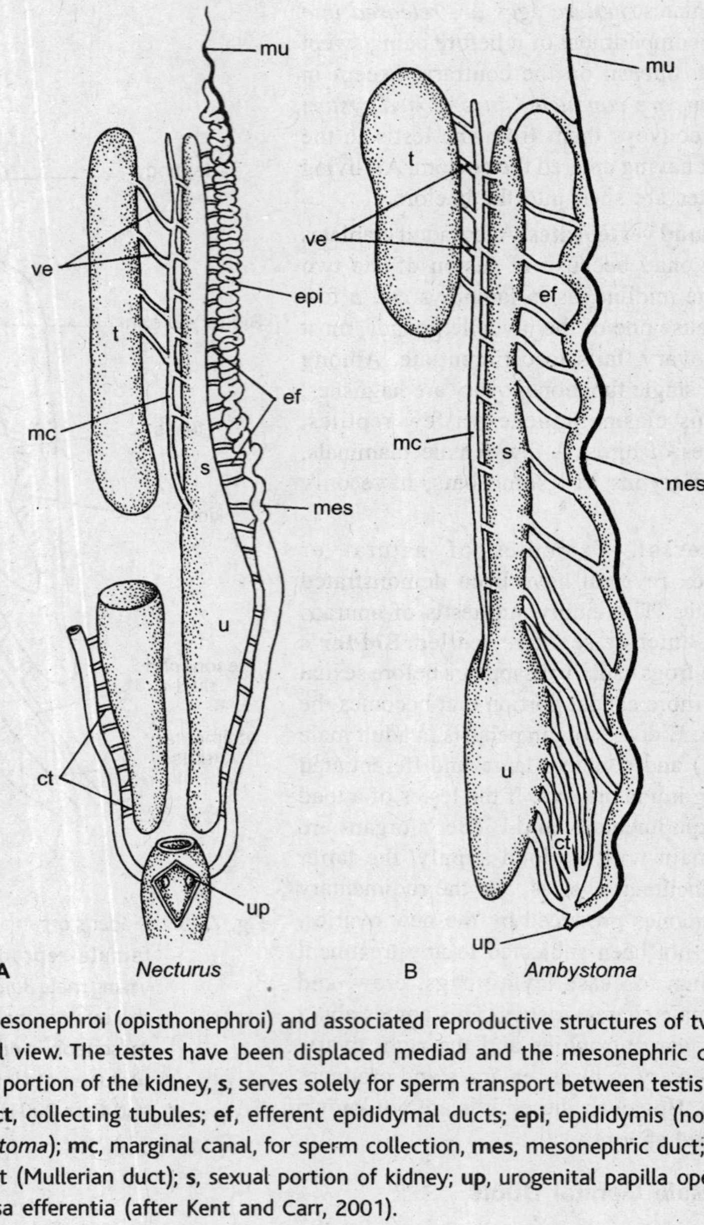

Fig. 7.26. Left mesonephroi (opisthonephroi) and associated reproductive structures of two male urodeles, ventral view. The testes have been displaced mediad and the mesonephric ducts laterad. The sexual portion of the kidney, **s**, serves solely for sperm transport between testis and mesonephric duct, **ct**, collecting tubules; **ef**, efferent epididymal ducts; **epi**, epididymis (not highly coiled in *Ambystoma*); **mc**, marginal canal, for sperm collection, **mes**, mesonephric duct; **mu**, rudimentary oviduct (Mullerian duct); **s**, sexual portion of kidney; **up**, urogenital papilla opening into cloaca; **ve**, vasa efferentia (after Kent and Carr, 2001).

fishes and amphibians, and from reptiles to humans, the embryonic mesonephric duct remains in adult males to carry sperm. A duct that carries only sperm is called **vas deferens** (ductus deferens (Box 7.4).

Box 7.4

The pronephric duct usually persists and drains the mesonephros or extended opisthonephros. It is renamed the **mesonephric duct** or **opisthonephric duct**, respectively. In some males, this duct transports sperm and is called the vas deferens. In females, it is known embryologically as wolffian duct. Because this duct plays different roles in different groups, the more general term **archinephric**

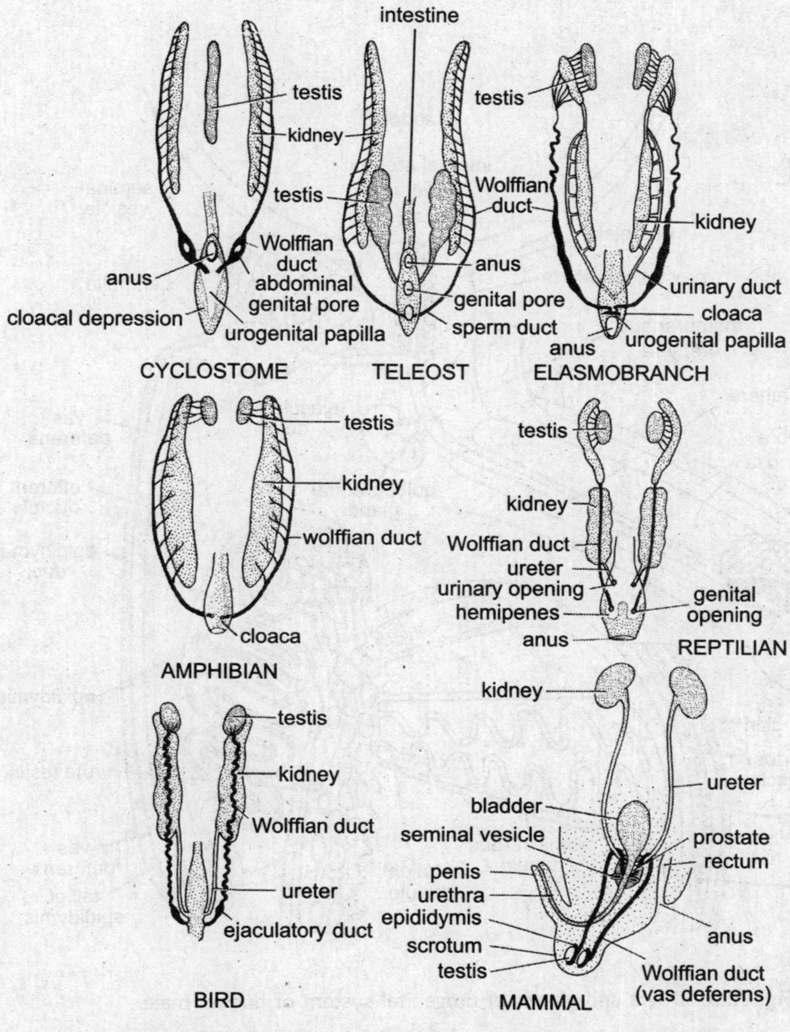

Fig. 7.27. Reproductive and urinary ducts of male vertebrates.

duct in preferred. The **metanephric duct** is commonly called the **ureter**. In some males, the kidney divides its services between reproductive and excretory roles. To recognize this, it is common to speak of the **reproductive kidney** and the **uriniferous kidney** (see **Kardong** 2006).

Spermatic duct generally empty into the cloaca or derivative thereof in craniate excluding therian mammals. In therian mammals, they empty into the urethera at the prostate gland (Fig. 7.27). This development is a result of the complete separation of the embryonic cloaca into a urogenital sinus and rectum.

As a result of caudal migration of the testes in mammals during the late fetal life, the spermatic ducts become "caught" or "hung up" on the ureters, so that thereafter they loop over the latter en route to the urethera.

A number of multicellular glands produce secretions that contribute to seminal fluid in amniotes. These are most abundant in mammals, in which one or more open into the genital tract at or near the junction of the spermatic ducts and urethera (Fig.7.28). Among these are

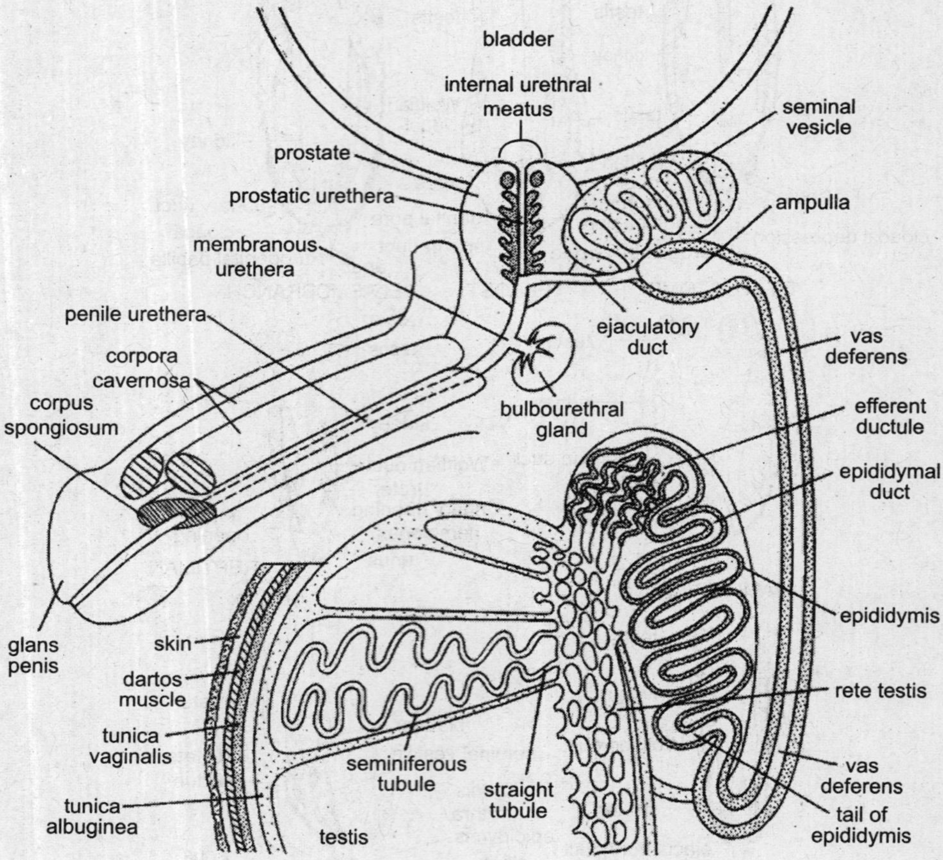

Fig. 7.28. Ducts and glands of urogenital system of human male.

ampullary glands, several median and paired **prostate glands, seminal vesicles, coagulating glands if bulbourethral (Cowper's) glands**. Note all are present in every mammalian order. Seminal vesicles have only a glandular role in mammals, although they serve as storage sites for sperm in other craniates. Coagulating glands cause semen to coagulate in the vagina. This results in a **copulation plug** that precludes another mating for several hours of days. Other glands produce mucus, nutritive secretions, or fluids that neutralize the slight acidity of the vaginal contents, there by providing a more suitable milieu for motile sperm.

Intromittant organs. In species in which fertilization is internal, males with few exceptions develop **intromittant or copulatory organs** for introducing sperm into the female reproductive tract. These are present in fishes with internal fertilization, in the anuran family Ascaphidae (frogs lacking eardrums), the reptiles, a few birds, and in all mammals. In urodeles with internal fertilization and in most birds, the cloaca is eversible, facilitating sperm transfer.

Male intromittant organs of fishes are modifications of pelvic fins (claspers) or anal fins (gonopodia). Snakes and lizards have paired hemipenes. Male turtles, crocodilians, some birds, and monotreme have an unpaired penis in the cloacal floor containing a corpus spongiosum (Fig.7.29). Therian mammals have an external penis with a corpus spongiosum and two corpora cavernosa (Fig.7.28).

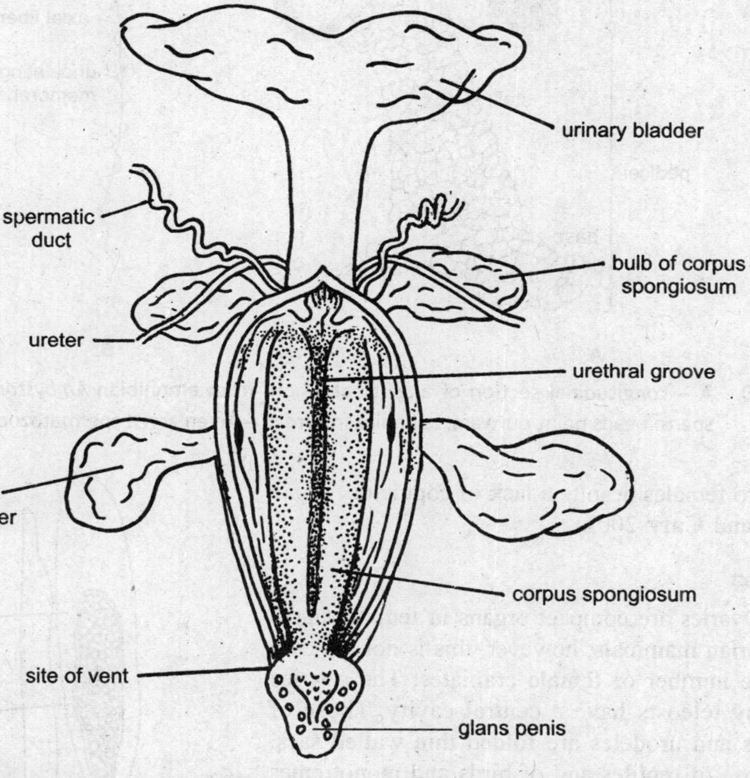

Fig. 7.29. Urogenital system of a male turtle. The cloacal floor viewed from above, showing the penis extruded. The rectum has been removed (after Kent and Carr, 2001).

Internal and External Fertilization

In viviparous craniates, fertilization takes place within the body of the female (called **internal fertilization).** Fertilization is also internal whenever eggs are covered by an impenetrable shell before being extruded, as in reptiles (also in some sharks).

In oviparous fishes, frogs and toads **external fertilization** is the rule despite an aquatic environment that dilutes the sperm. This is possible only because millions of sperm are shed over the eggs as the latter are being extruded. In apodans and urodeles, however the fertilization is usually internal even in oviparous species. Some male urodeles deposit a gelatinous packet of sperm called **spermatophore** (Fig.7.30), in the immediate vicinity of congregated females during the mating ritual. The spermatophore is generally picked up by the lips of a female's cloaca, although in some species it is placed in the cloaca by the male. The sperm eventually escape from the jelly and fertilize eggs, which are passing down the oviduct. In some species, the sperms are stored in crypts (called **spermathecae)** in the female tract and fertilize eggs that are ovulated at a later date when the environment is more suitable for embryonic development. The adaptation has a genetic basis. Packaging sperm in spermatophones make it possible to convey

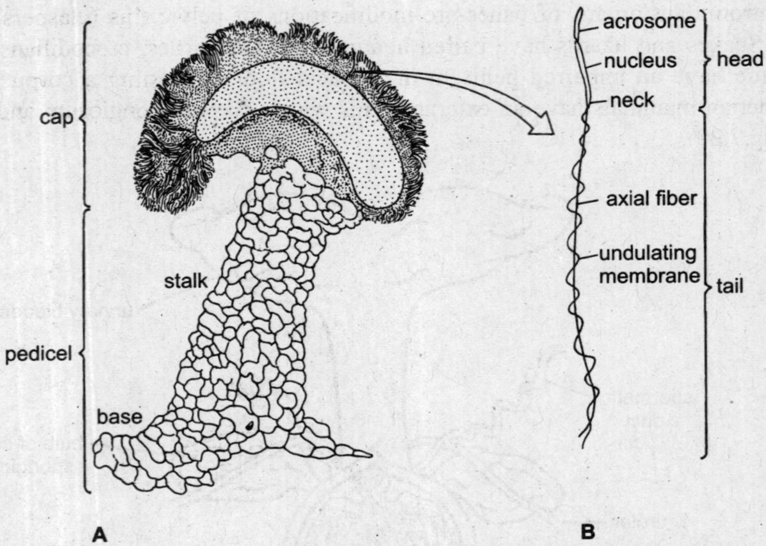

Fig. 7.30. A – Longitudinal section of a spermatophore from amphibian *Ambystoma texanum*. Generally, sperm heads point outward, tail point inward. B – An enlarged spermatozoon (after Kardong 2006).

sperm to females despite a lack of copulatory organ (**Kent** and **Carr** 2001).

Ovaries

Ovaries are compact organs in female sharks and therian mammals, however, this is not the case in large number of female craniates. The ovaries of many teleosts have a central cavity. Those of anurans and urodeles are folded thin walled sacs, and those of reptiles and of birds and monotremes exhibit numerous internal, irregular, fluid-filled lacunae. On the other hand, the ovaries of fishes other than teleosts are generally compact, as are those of turtles, crocodilians, and therian mammals.

The hollow ovary of teleosts results from entrapment of a small part of the coelomic cavity within the developing ovary. Because of this, the cavity (*i.e.,* an isolated pocket of coelom) is lined by the germinal epithelium; the small eggs, or in viviparous teleosts, the young, are discharged into that cavity. The latter may be directly continuous with an oviduct (Fig. 7.31) or partially surrounded by an oviducal funnel composed of peritoneal folds that have become so arranged as to form a tube. The ovary of some other teleosts is also hollow but for a different reason: it results from a secondary hollowing out of the interior of the ovary at each ovulation. However the eggs are discharged in

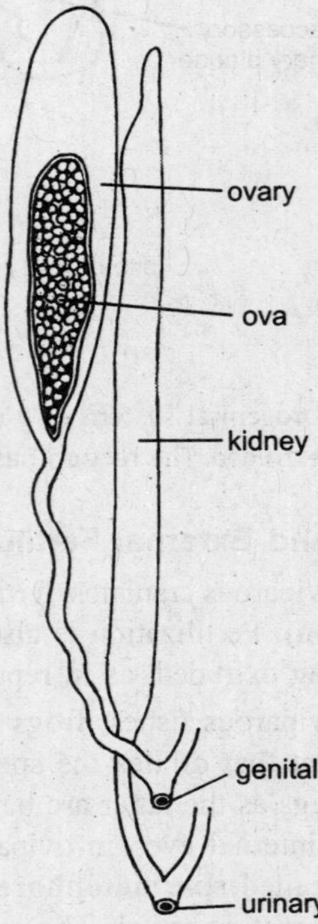

Fig. 7.31. Female reproductive system of a teleost. Ova are shed into the ovarian cavity. In some teleosts, the ovary reaches almost to the genital pore (after Kent and Carr 2001).

a similar manner. The germinal epithelium of most other craniate ovaries, whether compact, saccular, or lacunate, is in a cortex immediately internal to the visceral peritoneum, and the eggs are discharged into the coelom through eroded areas in the visceral peritoneum. At the end of each reproductive season, the ovaries of most craniates other than therian mammals regress to a state resembling those of juveniles.

The ovaries of therian mammals (Fig.7.32) are compact, the only cavitation being the fluid-filled **antra** within the egg follicle (called Graafian follicles) as ovulation approaches. The number of ovulated ova is not large in mammals — one to three, generally—although the number in some carnivores and small rodents is larger. The South African elephant shrew is atypical. It ovulates 50 or more eggs, although there is room for only two to implant and develop to term. The average number of mature eggs ovulated in most species is slightly larger than the average number that implants and the average number that implants exceeds the average number born. Thus, nature compensates for possible prenatal loss just as in teleost fishes, in which millions of eggs may be extruded, but far fewer are fertilized during external fertilization.

The fully mature mammalian Graafian follicle bulges on the surface of the ovary. The escape of ovum into the coelomic fluid is the process of **ovulation**. Following ovulation, the many cells that remain in the ovulated follicle are altered histologically and physiologically under the stimulus of luteinizing hormone from the pituitary, and they organize a **corpus luteum**

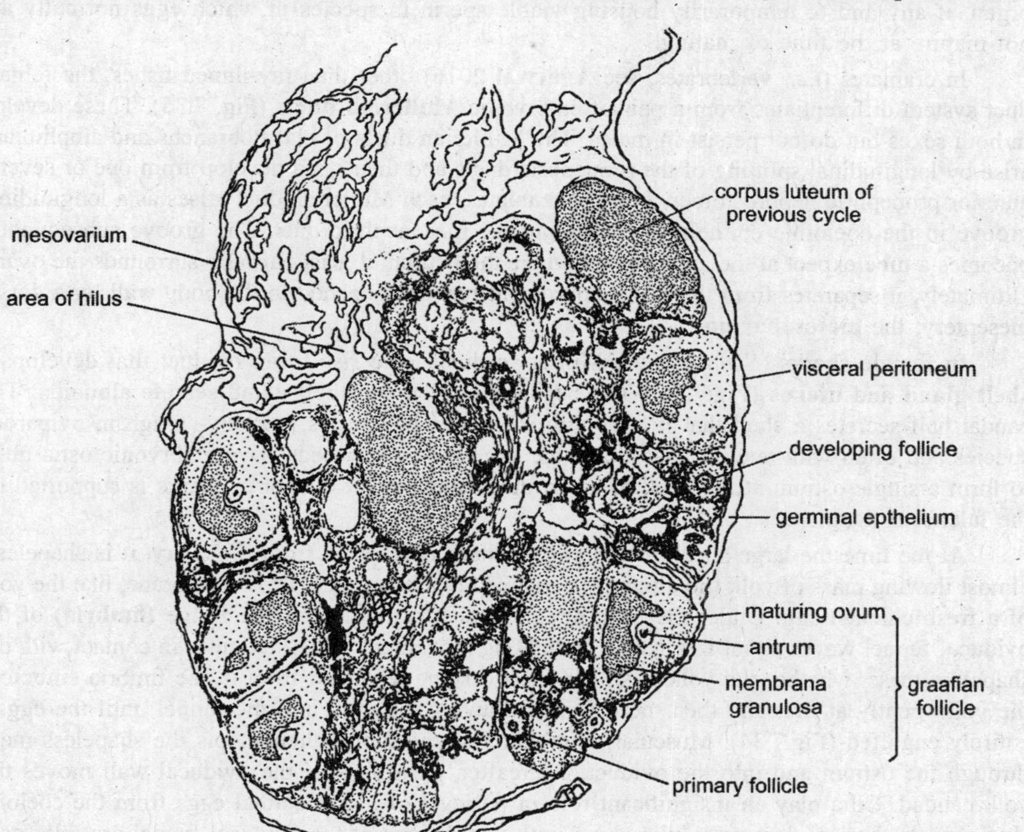

Fig. 7.32. Section of hamster ovary at approach of estrus (heat) and ovulation, which occurs every fourth day unless pregnancy occurs. The follicles are in the germinal epithelium. The hamster is a burrowing rodent with a short tail and large cheek pouches (Kent and Carr 2001).

(Fig.7.32). Therefore, the corpus luteum is major source of the **progesterone** (hormone) required for the maintenance of pregnancy. Before ovulation, **estrogen** (hormone) is the predominant secretion of these cells.

Corpora lutea are not confined to mammals. They have been identified in every craniate class having taxa that retain ovulated eggs or embryos in the female tract for a period time.

Female Genital Tracts in Craniates (Excluding Therian Mammals)

Agnathans have no genital ducts (Fig. 7.33). In ray-finned fishes, the genital ducts are aberrant (*i.e.,* not normal). However, the female duct systems of craniates other than therian mammals all reflect a generalized pattern of structure. It consists of paired muscular and glandular **oviducts** that begin at an **ostium**, an opening to the coelom, leading to the neck of an **oviducal funnel** (or infundibulum) (Fig. 7.34). Ciliated and fringed margins of ostium are the **fimbria** that embrace the ovary. Usually oviducts are not connected to ovaries directly. Oviducts end in the cloaca or a subdivision of it. Any differences along the length of the tracts are in the degree of regional specializations that support species. Specific functions such as 1. collecting eggs from the coelom; 2. coating them with protective or nutrient substances; 3. temporarily housing eggs, embryos, or larvae and maintaining them until the eggs are shed or the young are delivered; 4. expelling the eggs or the young; 5. receiving the male intromittent organ, if any and 6. temporarily housing viable sperm in species in which eggs normally are not mature at the time of mating.

In craniates (*i.e.,* vertebrates; see **Agarwal** 2016) other than ray-finned fishes, the female duct system differentiates from a pair of embryonic **Mullerian ducts** (Fig. 7.35). These develop in both sexes but do not persist in males. The Mullerian ducts of elasmobranchs and amphibians arise by longitudinal splitting of the pronephric ducts and their ostia develop from one or several anterior pronephric nephrostomes. In other craniates, each Mullerian duct arises as a longitudinal groove in the coelomic epithelium paralleling the mesonephric duct. The groove subsequently becomes a tube expect at the anterior end where the ostium at least partially surrounds the ovary. Ultimately, it separates from the mesonephros but remains attached to the body wall by a dorsal mesentery; the **mesotubarium**.

In female sharks (*Squalus*), the Mullerian ducts give rise to an oviduct that develops a **shell gland** and **uterus** (Fig.7.34). The cephalic half of the shell gland secrete albumen. The caudal half secretes a shell that is very thin is viviparous species, leathery- tough in oviparous species and often with tendrils that attach the egg case to seaweeds. Two embryonic ostia unite to form a single ostium at the neck of an oviducal funnel. The funnel in sharks is supported by the falciform ligament.

At the time the large-sized mature egg of shark is released from the ovary, it is shapeless, almost flowing mass of yolk (no albumin) contained in a nonrigid plasma membrane, like the yolk of a fresh chicken egg. Under the influence of ovulatory hormones, the fringe **(fimbria)** of the oviducal funnel waves about in a gentle undulating movement. When it comes in contact with the shapeless mass, whether the yolk is still in the ovary or separated from it, the fimbria embraces the yolk, gently at first and then more firmly, drawing the yolk into the funnel until the egg is entirely engulfed (Fig.7.34). Muscular contraction of the funnel then shoots the shapeless mass through the ostium and into the oviduct. Thereafter, peristalsis of the oviducal wall moves the yolk caudad. Cilia play an insignificant role in retrieval of macrolecithal eggs from the coelom. However, in therian mammals, cilia along with undulations of the oviducal funnel are important because the eggs are microlecithal.

In teleost, the oviducts are not homologous with those of other craniates. They arise as caudally directed folds of the coelomic peritoneum created by the genital ridges as the

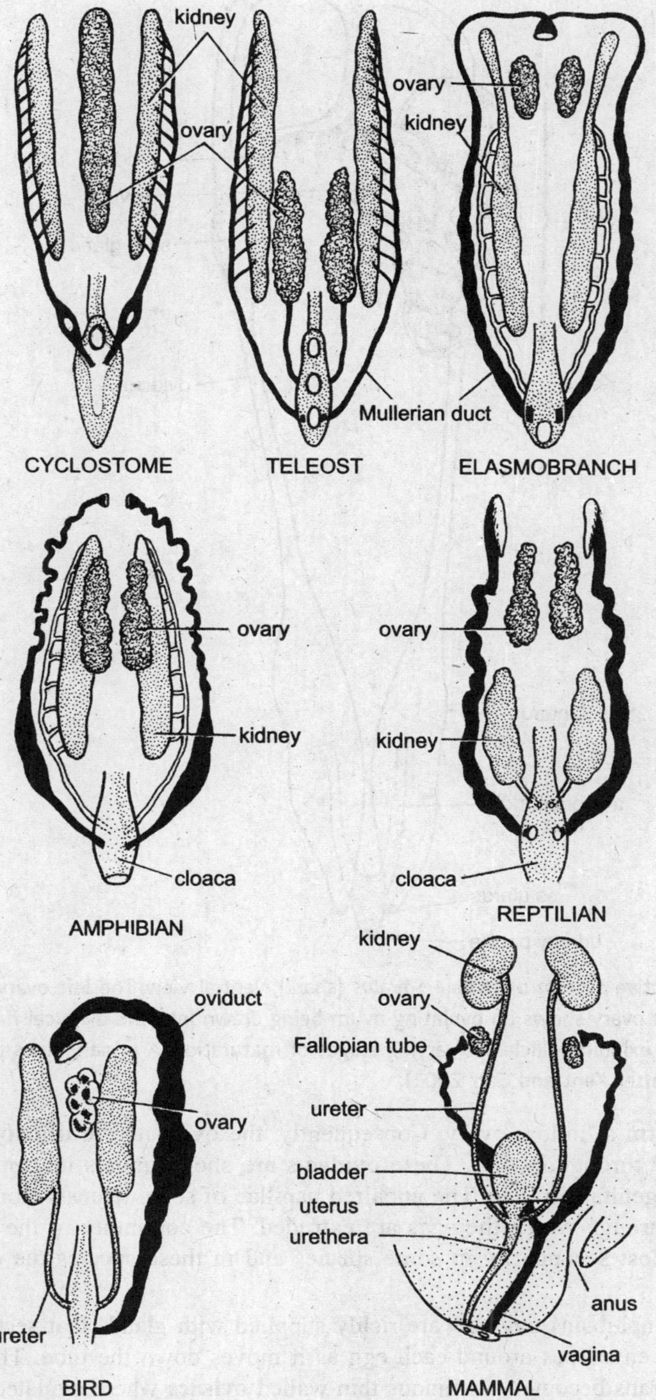

Fig. 7.33. Reproductive and urinary ducts of female vertebrates.

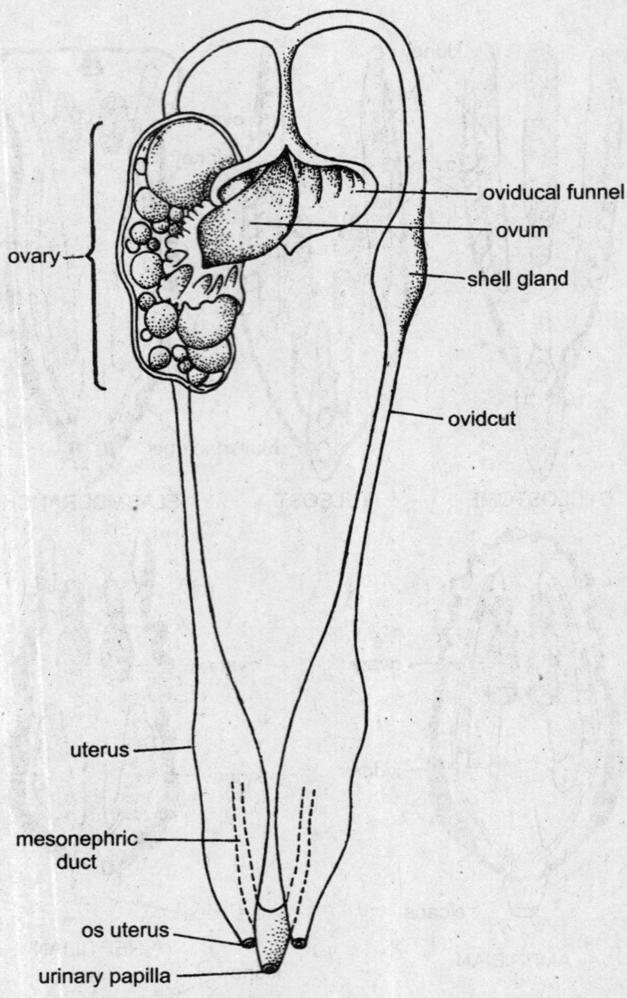

Fig. 7.34. Reproductive system of female *Squalus* (shark), ventral view. The left ovary has been removed. The right ovary shows an ovulating ovum being drawn into the oviducal funnel. An adult ovary typically exhibits follicles at varying stages of maturation. A urinary sinus precedes the urinary papilla (after Kent and Carr 2001).

latter entrap to form a hollow ovary. Consequently, the oviducts are usually continuous with the ovarian cavity throughout life. These oviducts are short funnels that open into paired or unpaired external genital papillae. The unpaired papillae of some teleosts elongate to form tube like **ovipositors** through which the eggs are extruded. The continuity of the oviducts with the ovarian cavity is lost secondarily in some species and in these species the eggs are extruded into the coelom.

In case of amphibians oviducts are richly supplied with glands that secrete several layers of protective jelly envelopes around each egg as it moves down the tube. The caudal portions of oviducts in anurans become voluminous thin-walled **ovisacs** where ovulated eggs accumulate until they shed. In terrestrial anurans, this occurs after a heavy rain has provided appropriate pools for egg-laying during amplexus (mating). Similar sacs in ovoviviparous urodeles shelter unborn young. The oviducts of dipnoans are similar to those of oviparous urodeles.

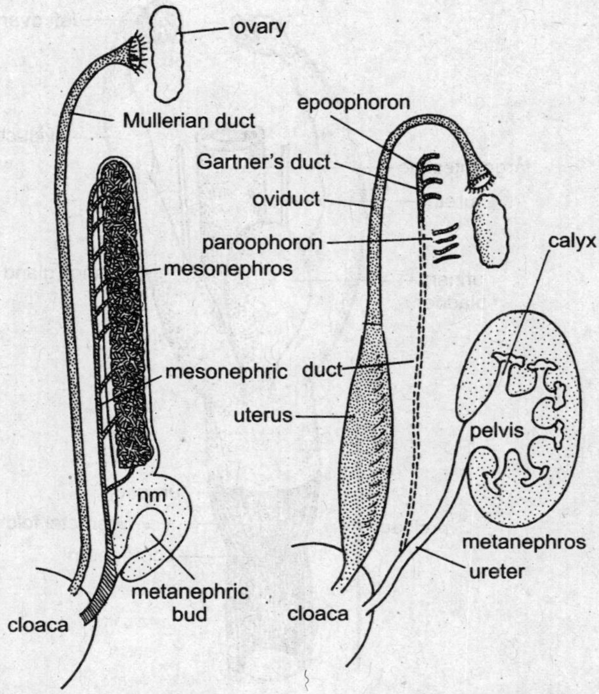

Fig. 7.35. Developmental changes in urogenital system of female mammals. In the early stage (A), the mesonephric kidney and ducts are present, and the metanephric bud has formed at the caudal end of the nephrogenic mesoderm (nm). A Mullerian duct (oviduct) parallels the mesonephric duct. In the later stage (B) the mesonephric kidney and duct have regressed except for remnants (epo-and paro-oproron), and the Mullerian duct has differentiated to form a female reproductive tract (after Kent and Carr 2001).

In reptiles and monotreme mammals the female tracts confirm closely to the basic pattern (Fig. 7.36). However, only one of the two embryonic Mullerian ducts may differentiate, as in crocodilians and most birds. The rudimentary right Mullerian duct of birds is seen in figure 7.37. Albumen glands line part of the oviduct, and there is usually a prominent shell gland just interior to the cloaca (Fig. 7.37 and Fig. 7.36). The shell remain leathery in species whose eggs develop in moist environments; shell becomes brittle in eggs that develop in dry air. The difference depends on the nature of the secretions. Ovoviviparous reptiles as well as oviparous ones cover their eggs with a shell.

Although the female tracts of birds are similar to those of reptiles, some of the anatomical terminology differs. The albumen-secreting area in oviduct of fowl is called the **magnum**, the shell gland is a **uterus**, and the short glandular and muscular segment at the entrance to the cloaca is a **vagina** (Fig.7.37). The vagina secretes mucus that seals the pores of the shell, preventing the loss of vital water vapour from within the shell without affecting the supply of oxygen to the chorio-allantoic membrane. The vagina than expels the egg to the exterior.

Spermathecae, which are dorsal diverticula of the female cloaca or tubular crypts in the oviducal lining, are present in many urodeles, lizards, snakes and domestic fowl. They store sperm at the time of next ovulation, which in some species, in weeks or mouths after mating (**Kent** and **Carr** 2001).

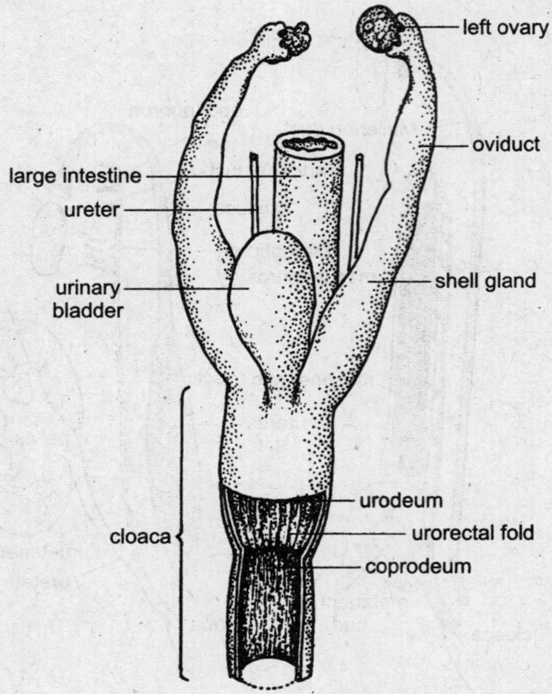

Fig. 7.36. Genital tract and cloaca of a female monotreme, ventral view. The right ovary is usually the smaller, or it is rudimentary. The cephalic half of the cloaca is partitioned by the urorectal fold into a urodeum and coprodeum (after Kent and Carr 2001).

Female Tracts of Therian Mammals

Paired Mullerian ducts in mammals, like the other craniates, give rise to the female genital tract beginning with the ostium of the oviduct. However, the Mullerian ducts of eutherian mammals unite at their caudal ends to one degree or another. The extent of the union of the caudal ends of oviducts varies with the mammalian order. The tract of eutherian, mammals, therefore, consists of two oviducts, a uterus with or without horns (cornu), and a vagina.

Oviducts

Oviducts (or **Fallopian tubes** as they are also called in mammals) are relatively short, small in diameter (accommoding microlecithal eggs), more or less convoluted, and lined with cilia. The oviducal funnel has a fringed border. In many mammals, a membranous fold of peritoneum, the **ovarian bursa**, envelops the ovary and the oviducal funnel, entrapping them and a small coelomic cul-de-sac (French for a street or passage closed at one end). The bursa increases the probability that ovulated eggs will enter the oviduct rather than be lost in coelomic fluid, perhaps to implant somewhere in the coelomic cavity. The ovarian bursa may be broadly open to the main coelom, as in cats and rabbits; it may communicate with the coelom by a mere slit, as in most carnivores and rats; or it may be completely closed off from the coelom, as in hamsters. There is no bursa in humans. **Ectopic pregnancies** with fetuses implanted in the ventral abdominal wall are occasionally found in cats, rabbits and humans. These fetuses, of course, cannot be delivered.

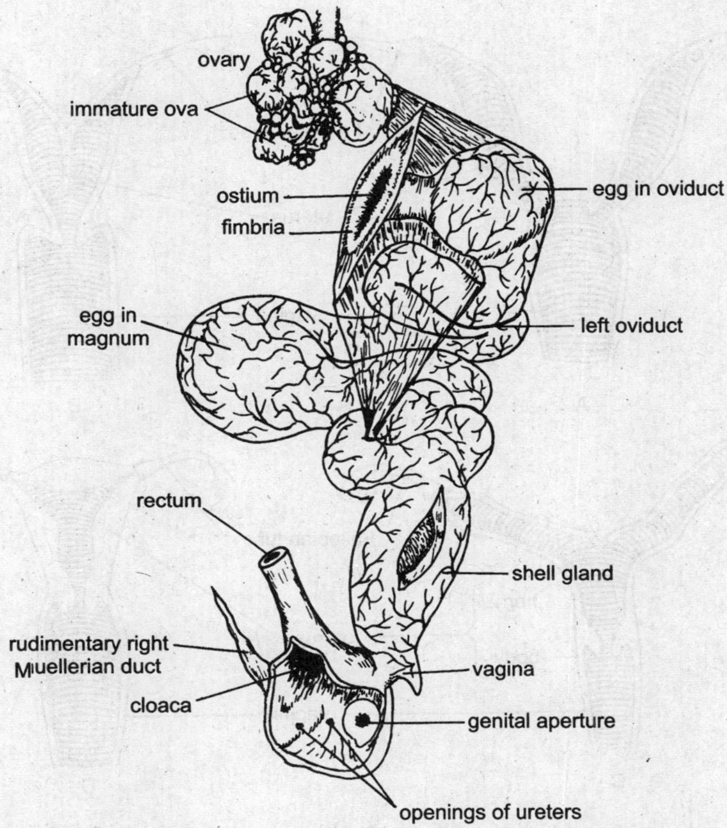

Fig. 7.37. Genital duct of a hen. The unusual presence of two eggs in the oviduct may be noted (after Kent and Carr 2001).

Uteri

The Mullerian ducts of marsupials are paired all the way to the urogenital sinus, a subdivision of the embryonic cloaca. Marsupials therefore have two completely separate uteri (a **duplex uterus)** and two vaginas (Fig.7.38). The uteri of eutherian mammals exhibit varying degrees of fusion distally, resulting most often in a **bicornuate uterus**—a uterus with two horns in which the young develop (Fig 7.39). A uterus with two horns may have two totally separate passageways within the body of the uterus although this is not recognizeable from external view (Fig. 7.38 B). When this is the case, the uterus is more precisely termed **bipartite** (*e.g.*, rabbit and hamster). In some mammals, one uterine horn is much larger than the horn in impalas; even though both ovaries produce viable eggs.

Maximal fusion of the Mullerian ducts occurs in armadillos, some bats and in monkeys, apes and humans, in which fusions begins at the ends of short oviducts and there are no uterine horns. This is **simplex uterus** (Fig. 7.38). Blastocysts implant in the body of the uterus, and there is usually only one fetus per pregnancy. Armadillos are exceptional; they always give birth to identical quadruplets.

The narrow neck, or **cervix**, of the uterus of therian mammals projects into the vagina as the **lips of the cervix.** These surround the opening, or **os uteri**, of the uterus into the **vagina**. Sperm deposited in the vagina pass between the lips of the cervix en route to the upper end of

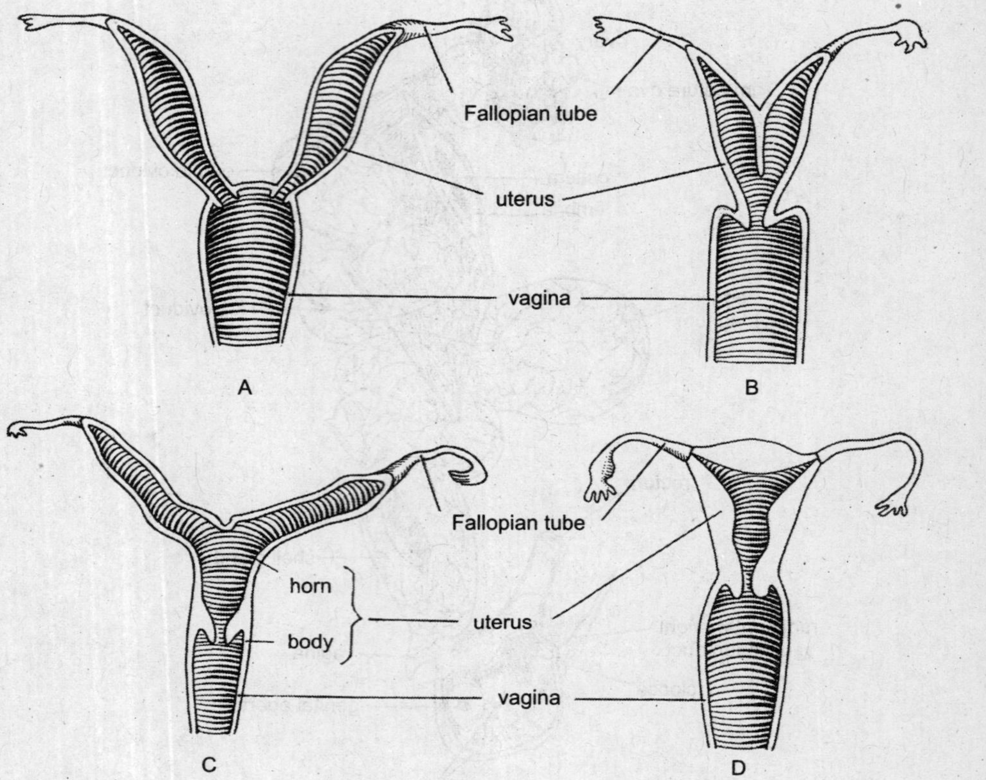

Fig. 7.38. Types of uteri found in mammals. A – Duplex type in marsupials: B – Bipartite type of some rodents (*e.g.,* rabbit, hamster); C – Bicorunate type of carnivores; D – Simplex type of primates.

the oviduct, where fertilization takes place. The uterine lining (**endometrium**) of therian mammals becomes highly vascular under the stimulus of hormones before implantation of blastocysts. The thick muscular layer of the uterine wall (**myometrium**) assists in ejection of the young at birth, provided it, too, has been hormonally prepared. Before **delivery**, the cervix must dilate to provide a passageway large enough for the young to pass through. Dilation takes place under the influence of the hormone relaxin from the ovary.

Vaginas

The vaginas of therian mamamals is the unpaired terminal portion of the embryonic Mullerian ducts (Fig.7.40). The cloaca contributes to it in some mammals (Fig.7.40 A, C grey part). Except in rodents and as a derived character in some primates, the vagina empties into the **urogenital sinus**, a derivative of the embryonic cloaca, *e.g.,* rabbit (Fig. 7.39; Fig.7.40). In many rodents, the vagina opens directly to the exterior (Fig. 7.41C). In apes and humans, vagina opens into a shallow **vestibule** of the **vulva**, which also receives the urethera (Fig. 7.38D). The vestibule is bordered by two fleshy folds the **labia minora**, derived from the genital folds (Fig. 7.41B). It is overlaid by two additional folds, the **labia majora**, derived from the genital swellings. The latter are homologues of the scrotal sacs of males. The vagina is lined by a cornified epithelium for reception of the penis.

In marsupials, two pouch-like components of the **median vagina**, lie against the urogenital sinus, separated from the latter by thin septa. At birth the fetus is usually forced through the septa

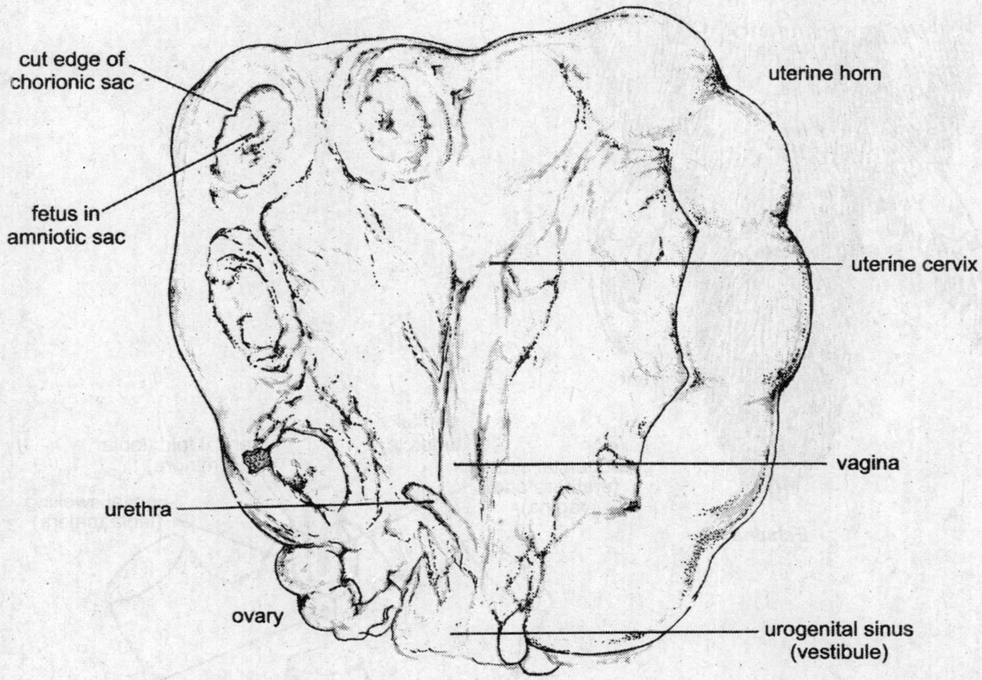

Fig. 7.39. Gravid bicornuate uterus of dog. The chorionic sacs have been partially cut away to expose the fetuses within the amniotic sacs. The body of the uterus is insignificant in the pregnant dog (after Kent and Carr 2001).

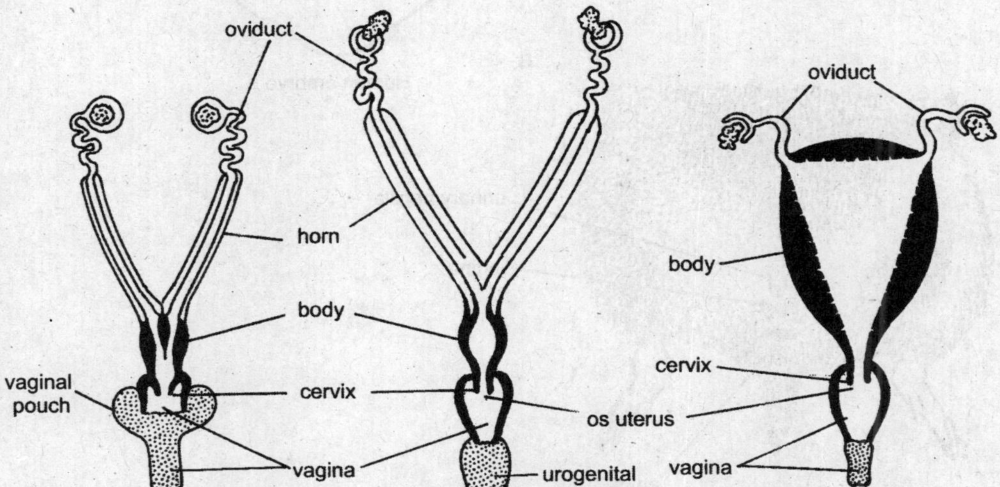

Fig. 7.40. Varieties of mammalian uteri. Blackened regions are the fused caudal ends of the Mullerian ducts; shaded part is a derivative of cloaca. Note the two lumens in the body of the bipartite uterus. The urogenital sinus in ungulates, as in female mammals, receives the vagina and urethera (after Kent and Carr 2001).

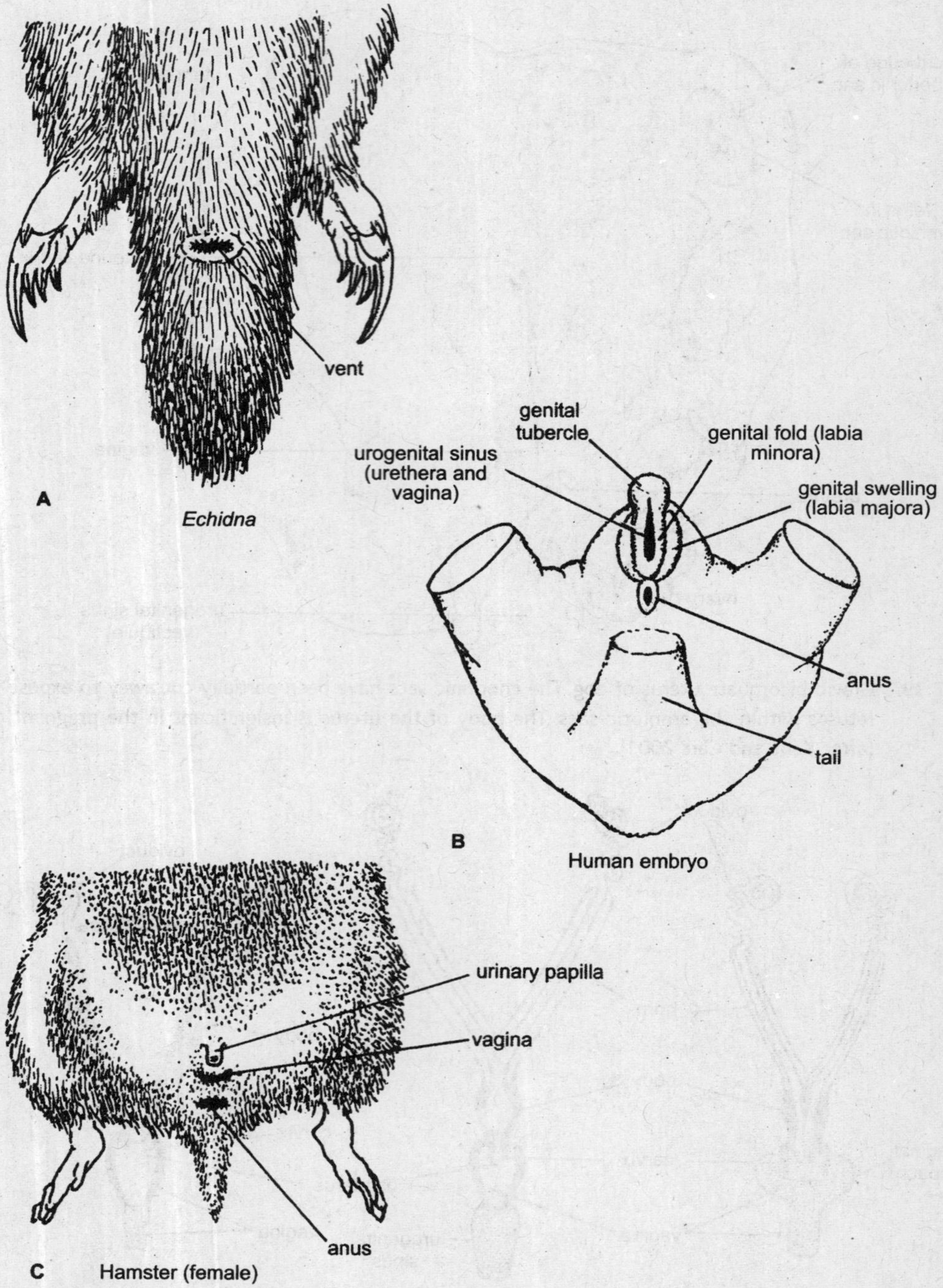

Fig. 7.41. External openings from mammalian cloaca or its derivatives. A – Cloacal vent of monotreme *Echidna*. B – External genitalia of bisexual stage of 75 mm human embryo (about 12 weeks of age). The cloaca at this stage has been divided into urogenital sinus and rectum. C – Perianal region of female hamster. At the tip of the urinary papilla is the urethral aperture (after Kent and Carr 2001).

directly into the urogenital sinus, and the passageway thus established may remain throughout life. As an adaptation to paired vagina, some male marsupials have a bifurcated penis.

Cloaca

The cloaca is a common chamber receiving products from kidneys, intestines and often gonads. It opens to the outside through a cloacal opening or **vent** (Latin *cloaca* means 'sewer'). The cloaca arises at some point during embryonic development in all vertebrates, but in many it becomes subdivided, lost or incorporated into other adult structures (Fig. 7.42). A well developed cloaca occurs in adult sharks and lungfishes (Fig. 7.42 B,D). But in teleosts, distinct urinary, anal and genital openings are present, replacing the cloaca (Fig. 7.42 F). Among tetrapods, a cloaca is present in amphibians, reptiles, birds and monotreme. A shallow cloaca persists even in marsupials.

Apparently a cloaca is a primitive vertebrate feature because it occurs in most primitive gnathostomes and persists in the embryos of almost all vertebrates. Its absence in chimaeras (Holocephali), ray- finned bony fishes (Actinopterygii), *Latimeria* (Coelacanth), and most placental mammals (Eutheria) may represent independent losses.

Embryologically, the cloaca arises from hindgut endoderm and proctodeal ectoderm. Structurally, three functions influence it: defecation, urination, and copulation. Each function tends to be associated with a compartment and each compartment is controlled by muscles that regulate the entry and departure of products from the intestines, kidneys and gonads. The most proximal (nearest) compartment is the **coprodeum** into which the intestine empties. The **urodeum** receives products from urinary and genital ducts. The most distal compartment is the **proctodeum**, which functions in copulation and in many amniotes develops a penis (Fig.7.42A). Many urogenital ducts, upon approach to the cloaca, inflate slightly to form an expanded **urogenital sinus.** These ducts often open into the cloaca via a small projection called the **urogenital papilla**.

Hans Gadow, a classical anatomist, suggested that each of the three cloacal compartments was separated from the other by folds in the mucosal wall: 1. the **rectocoprodeal fold** between intestine and coprodeum; 2. the **coprourodeal fold** between coprodeum and urodeum and proctodeum. Although such folds occur in many vertebrates, they are low or absent in some, making it difficult to delineate boundaries between compartments of the cloaca. Gadow's terminology describing compartments and folds was based on tetrapods, but it is now applied to fishes as well.

Evolution of cloaca

1. The cloaca in most amphibians is simple. Folds usually delineate the coprodeum and urodeum, but in the absence of an intromittent organ or a uroproctodeal fold, the proctodeum is not anatomically demarcated from the rest of cloaca (Fig.7.42B).

2. Among reptiles, the cloaca of *Sphenodon* is subdivided by folds into three compartments; the proctodeum is simplified and lacks a penis. The cloaca of snakes and lizards also has three compartments, but the proctodeum is usually reduced (Fig.7.42C to E). The internal subdivision of the cloaca is much less distinct in turtles (Fig.7.42F), and in crocodilians the coprodeum and urodeum and to a lesser extent, the proctodeum are more or less united into a single large chamber (Fig.7.42 G).

3. In birds, the cloacal folds are quite variable. The ostrich cloaca has a rectocoprodeal fold Fig. 7.42H), but this is apparently lacking in other groups (Fig.7.42 I). In birds, the proctodeum is associated with a **cloacal bursa** (also called **bursa of Fabricus**) that has an immune function.

4. The cloaca persists in *monotremes*, where distinct coprourodeal and uroproctodeal folds demarcate the urodeum from other compartments (Fig.7.42J). The ureter and vas deferens open to a urogenital sinus, but urine flows directly into the urodeum, and semen flows through a

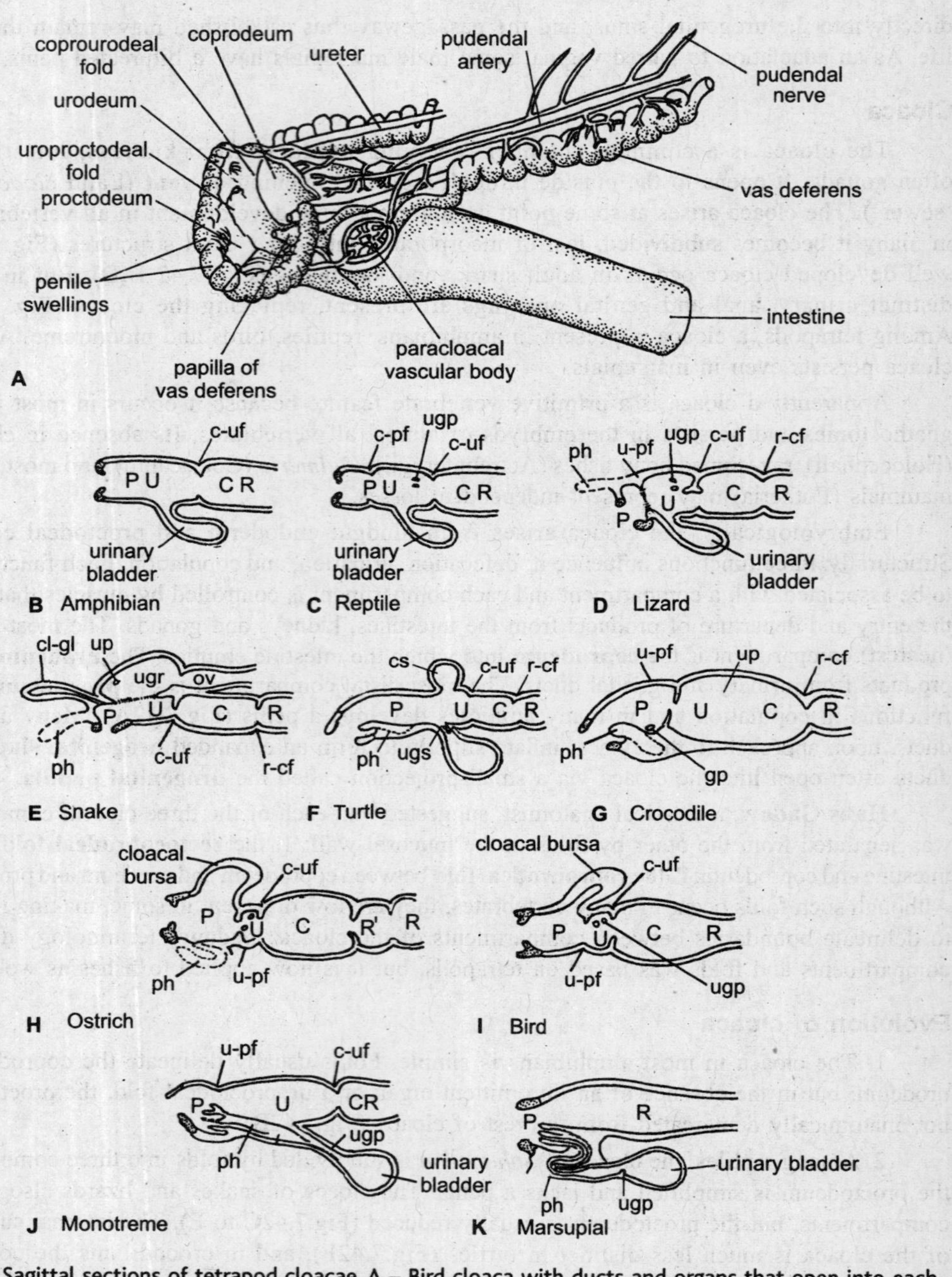

Fig. 7.42. Sagittal sections of tetrapod cloacae. A – Bird cloaca with ducts and organs that open into each of the three chambers; B – Amphibians; C – Reptile, *Sphenodon*; D – Lizard, *Lacerta*; E – Snake; *Torpidonotus*; F – Turtle, *Pseudemys*; G – Crocodile; H – Ostrich; I – Bird; J – Mammal, monotreme; K – Marsupial. Parts of the cloaca include the coprodeum (C), proctodeum (P), rectum (R), and urodeum (U). Other abbreviations: coprourodeal fold (c-uf), cloacal gland (cl-gl), cloacal sac (cs), genital pore (gp), oviduct (ov), penis (ph), rectocoprodeal fold (r-cf), urinary pore (up), urogenital pores (ugp), uroproctodeal fold (u-pf), urogenital sinus (ugs), urogenital reservoir (ugr) (after Kardong 2006).

sperm duct within the penis. *Marsupials* possess a reduced cloaca that is represented primarily by the proctodeum (Fig.7.42 K). The ectoderm part of the cloaca persists in *eutherian mammals*—some rodents and insectivores, however in all other placental mammals, the cloaca divides in the sexually indifferent stage and forms separate orifices from the coprodeum and urodeum. Generally, the coprodeum becomes the rectal region of the digestive tract with an anal opening. The urodeum yields separate structures, depending on the sex. In the male, the urogenital sinus becomes the urethera that transports sperm and urinary products. In the female, the urogenital sinus divides again to produce a urethral opening for the urinary system and a vaginal opening for the reproductive system.

QUESTIONS

Long Answer Questions

1. Why are structures of osmoregulation and reproduction united within a single system—the urogenital system?
2. Compare holonephric, mesonephric, opisthonephric and metanephric kidneys.
3. What are the functions of the pronephric, opisthonephric and metanephric kidneys?
4. Why do we not find urinary bladders in most fishes? What function do they serve?
5. Compare the passage of gametes in males of females.
6. What are the intromittent organs in sharks, reptiles and therian mammals.
7. What are the structures that are formed by the caudal fusion of oviducts in therian mammals.

Short Answer Questions

1. Write short note on the following:
   (i) Nephron;
   (ii) Renal corpuscle;
   (iii) Pronephros;
   (iv) Mullerian duct;
   (v) Archinephros;
   (vi) Mesonephros;
   (vii) Metanephros;
   (viii) Opisthonephros.

Very Short Answer Question

1. What is glomus?
2. What is cloaca?
3. Define ureter.
4. What is uriniferous tubule?
5. What is glomerulus?
6. Define the urinary bladder.

Fill in the Blanks

1. The nephron or nephric tubule of a ________ forms______.
2. The collecting tubules of ________ ________ affects the _________ of urine.

3. Renal corpuscle of nephron includes ______ and ______ capsule.
4. At the end of each reproductive season, the ovaries of most craniates other than ______ mammals regress.
5. The ovaries of therian mammals are ______.
6. Most preferred term for the mesonephric duct is ______.
7. The shell remains leathery in species whose eggs develop in ______ environment.

Multiple Choice Questions

1. Hypothetical primitive kidney is
   (a) pronephros (b) archinephros
   (c) metanephros (d) nephron
2. Functional unit of kidney is
   (a) neuron (b) nephron
   (c) glomeruli (d) renal corpuscle
3. In humans, each kidney has nearly
   (a) 10,000 nephrons (b) 100,000 nephrons
   (c) 50,000 nephrons. (d) 1 million nephrons
4. Archinephros is found in
   (a) fishes (b) amphibian
   (c) reptiles (d) larval cyclostomes
5. Wolffian body is
   (a) archinephros (b) pronephros
   (c) mesonephros (d) metanephros
6. Urinary bladder opens to exterior through
   (a) pelvis (b) tubule
   (c) urethra (d) ureter
7. In mammals testes are found in
   (a) scrotal canal (b) inguinal canal
   (c) Bidder's organ (d) Peyer's patches
8. Henle's loop and major portion of the collecting tubules are situated in the kidney's
   (a) cortex (b) pelvis
   (c) medulla (d) pyramid
9. Oviduct in vertebrates is modified
   (a) urinary duct (b) inguinal canal
   (c) Wolffian duct (d) Mullerian duct

ANSWERS

Very short Answer Questions

1. Glomus is a part of primitive kidney (*e.g.*, pronephros of 15 mm larval frog). It is a vascular glomerulus like part of primitive kidney which services several nephrons.
2. Cloaca means sewer. It is a common terminus for digestive and urinary tracts.

3. Ureter is a metanephric duct arising as a ureteric diverticulum and draining the metanephros.
4. Uriniferous tubule is functional unit of the kidney composed of the nephron and collecting tubule.
5. Glomerulus is a small bed of capillaries associated with the uriniferous tubule.
6. Urinary bladder is a evagination of the embryonic cloaca for urine storage.

Fill in the Blanks

1. Kidney, urine
2. Uriniferous tubule, concentration.
3. Glomerulus, renal.
4. Therian.
5. Compact.
6. Archinephric.
7. Moist.

Multiple Choice Questions

1. (*b*) **2.** (*b*) **3.** (*d*) **4.** (*d*) **5.** (*c*)
6. (*c*) **7.** (*a*) **8.** (*c*) **9.** (*d*).

8

Nervous System

8.1 COMPOSITION AND ACTIONS OF NERVOUS SYSTEM OF CHORDATES

The nervous system plays three basic roles. It makes aware the organism with its external environment and stimulates the organism to orient itself favourably in that environment; it participates in regulation of internal environment; and it serves as a storage site for information. These functions are accomplished by the nerves, spinal cord, and brain in association with receptors (sense organs) and effectors (chiefly muscles and glands (see **Kent** and **Carr** 2001).

The nervous system is divided into the **central nervous system (CNS)**, which includes the brain and spinal cord, and the **peripheral nervous system (PNS)**. A third division of nervous division is called **autonomic nervous system (ANS)**. ANS innervates smooth muscles and cardiac muscles and glands. ANS is made of two ganglionated sympathetic nerves, ganglia in the head and viscera, and their connecting nerves. The autonomic nervous system is often regarded as a part of the peripheral nervous system because the two are connected (see **Kardong** 2006).But all the three divisions of NS are connected both structurally and functionally. The nervous system *receives* stimuli from one or more **receptors** and *transmits* information to one or more **effectors** that responds to stimulation. Effectors include **mechanical effectors**, such as muscles, and **chemical effectors**, such as glands. Thus, responses of the nervous system involve the muscle contraction and glandular secretions. The nervous system regulates an animal's performance by integrating immediate incoming sensory information with stored information, the results of past experience, and by then translating past and present information into action through the effectors.

The nervous system includes millions of nerve cells, each of which establishes thousands of contacts with other nerve cells, so the total number of interconnections is astronomical. That is why analysis of the function of the nervous system is a difficult task.

Types of cells within the Nervous System

There are two types of cells within the nervous system: **neurons** and **neuroglial cells** or **glia**.

Neuroglia

Neuroglia ("nerve" and glue") cells do not transmit impulses. They support, nourish and insulate neurons. All neuroglia bind together nervous tissue, and they can be specialized (Fig. 8.1) **Astrocytes** pass nutrients between blood capillaries and neurons; **microglia** engulf foreign material and bacteria; **oligodendroglia** and **Schwann cells** insulate the axons of neurons; and **ependymal cells** line the central canal of brain and spinal cord (Fig. 8.2).

Neurons

Neurons are specialized for long-distance transmission of electrical stimuli throughout the body. The neuron is the structural and functional unit of the nervous system. It consists of the nerve cell body, the **perikaryon**, the **body** or **soma** of the neuron, and thin cell processes called **nerve fibers** (or processes or neurites) if they are long (Fig. 8.3). The processes are o

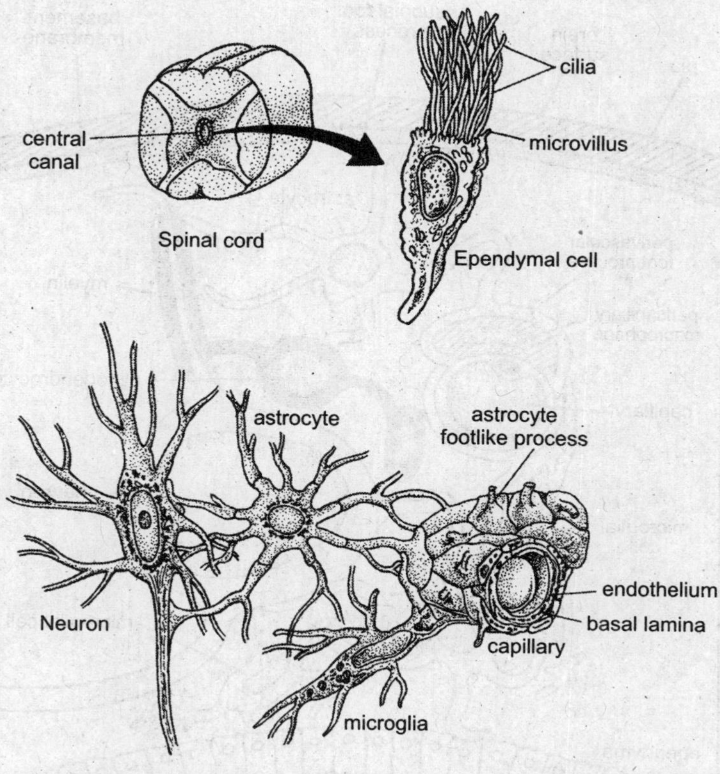

Fig. 8.1. Four types of neuroglia found within the central nervous system. Astrocytes form cytoplasmic connections to transport nutrients between blood capillaries and neurons. Phagocytic microglia engulf stray or foreign materials. Ependymal cells line the central canal of the central nervous system (after Kardong 2006).

two types. There usually is one **axon** per neuron and one or many **dendrites**. Dendrites transmit incoming electrical impulses toward the perikaryon. Neurons are grouped by number of their processes. **Unipolar neurons** have a single stem that divides into a dendrite and axon. **Bipolar neurons** have two processes, usually at opposite ends. **Multipolar neurons** have many processes associated with the cell body (Fig. 8.4).

Neurons and their processes are often known by different terms, depending on whether they occur in the CNS or the PNS. For example, collection of axons traveling together is a **nerve tract** in the CNS and a **nerve** in the PNS. A collection of nerve cell bodies is a **nucleus** in the CNS and a **ganglion** in the PNS. Neuroglial cells wrap some axons in a thick myelin sheath. Such fibers are called **myelinated nerves,** and those without sheaths are unmyelinated nerves (Fig. 8.5). A neuroglial cell that produces the myelin sheath is an oligodendroglial cell in the CNS and a **Schwann cell** in the PNS (Fig. 8.6). The nodes of Ranvier are indentations between adjacent neuroglial cells in the myelin sheath.

Regeneration and Addition of Neurons

Some peripheral nerves, if not damaged too severely, can regrow, sprouting a new axon from several axonal stump or from the perikaryon that slowly grows down the Schwann cell tube to reestablish innervations of the effector organ. Once formed, nerves of the central nervous

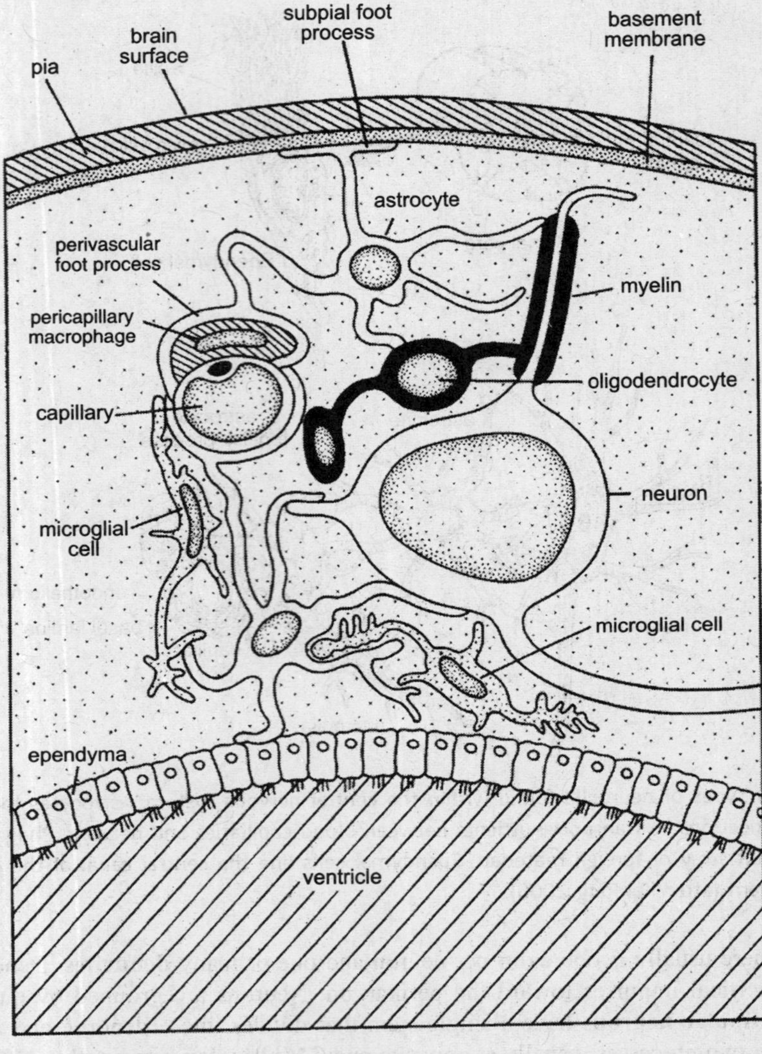

Fig. 8.2. Neuroglia in CNS. Oligodendrocyte insulate axons within the central nervous system.

system were thought to lack the ability to replace themselves. More recent evidence now suggest otherwise. In fact, across vertebrates so far studied, neurons of the CNS are replaced regularly (see **Kardong** 2006). Even in the adult mammalian brain, thousands of new neurons are added daily. Although such new neurons are a tiny proportion of the total population, over a life time of addition this can be considerable. Additional neurons in adults are especially evident in parts of the brain important for learning and memory. In birds, new neurons are added seasonally to areas concerned with courtship.

Transmission of information within nervous system: Nerve impulse, Graded potential and Action potential

Information traveling through the nervous system is transmitted in the form of **electrical** and **chemical signals**. Electrical signals are **nerve impulses** that travel within the plasma membrane of the neuron and are of two kinds: graded potentials and action potentials (**Kardong** 2006). A

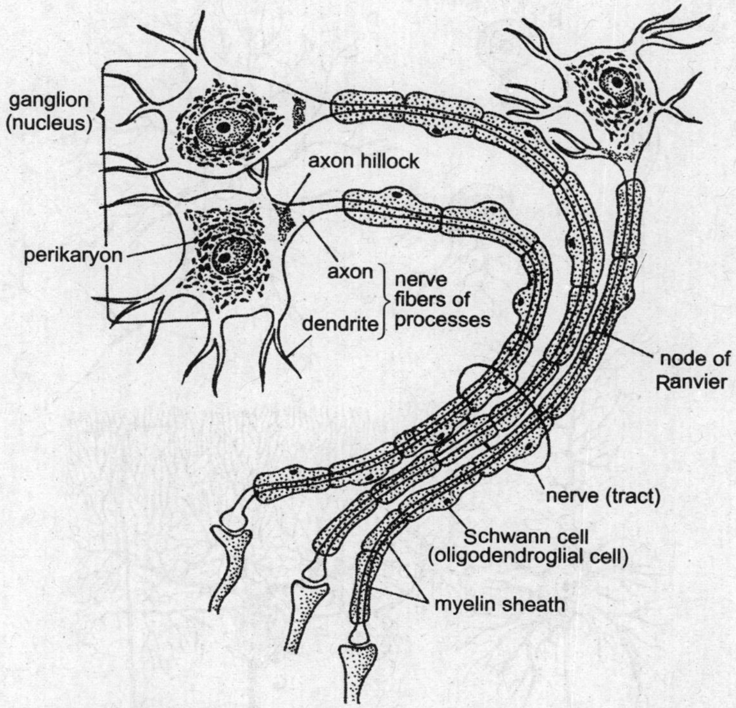

Fig. 8.3. Structure of a neuron. The cell nucleus and surrounding cytoplasm form the cell body of a neuron (perikaryon). Nerve fibers or processes are cytoplasmic extensions from the perikaryon. Axons carry impulses away from the perikaryon and dendrites carry impulses toward it. The same structures are given different names in the peripheral and the central nervous systems. Central nervous terms are given in parentheses (after Kardong 2006).

graded potential is a wave of electrical excitation proportional to the magnitude of the stimulus that triggers it. The graded potential declines in magnitude as it travels along a nerve fiber. An **action potential** is an all-or-none phenomenon. Once initiated, it propagates without decrement along a nerve fiber. Action potentials are often used for long – distance signaling in the nervous system. Within the dendrites and perikaryons, nerve impulses are usually graded potentials, but they become action potentials as they travel out the axon.

Chemical signals are generated at synapses which are gaps between the junctions of neurons (Fig. 8.7). These gaps occur between the processes of one neuron and the next, and between axons and perikarya. Upon arrival at the terminus of an axon, an electrical impulse stimulates the release of stored *neurotransmitters* into the tiny space between processes. Neurotransmitters diffuse across this synaptic junction and settle on the associated cellular process of the next neuron.

When collected in sufficient concentration, neurotransmitters initiate an electrical impulse in the next neuron. Excess and spent neurotransmitter molecules are rapidly inactivated to prevent prolonged effects. Neurotransmitters must quickly reach a threshold level to initiate an electrical impulse in the next neuron. Thus, passage of information through chain of connected neurons includes alternating events of electrical and chemical transmission involving nerve impulses and neurotransmitters, respectively (Fig. 8.8B).

Synapses (of neurons) introduce **control** into the *processing* of information transfer. If synapses were absent and neurons were in direct contact with each other, excitation in

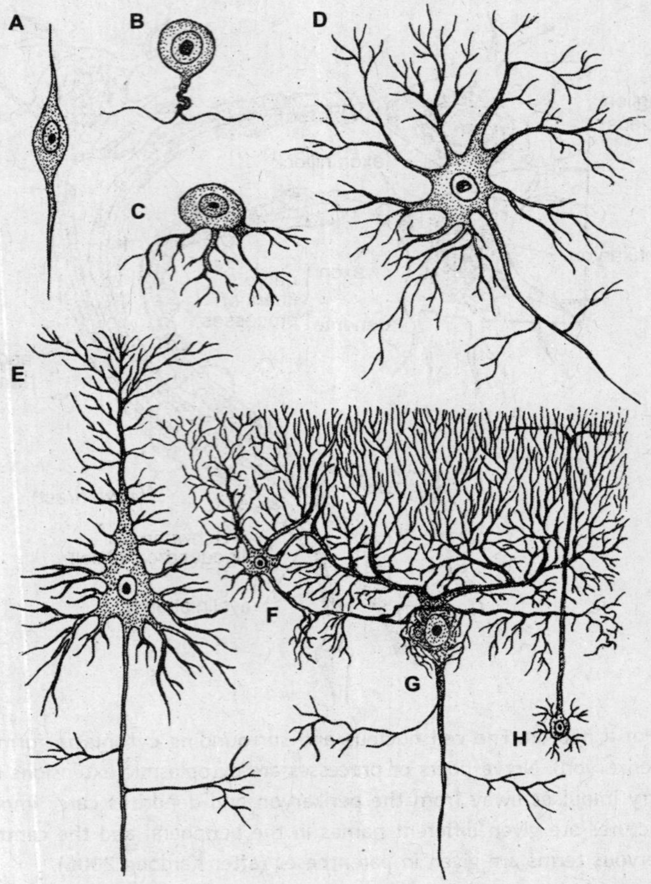

Fig. 8.4. Types of neurons. A – Bipolar neuron; B – Unipolar neuron; C – H – Multipolar neurons (after Kardong 2006).

one neuron might spread inevitably throughout an entire network of interconnected neurons like ripples on a pond, without any local control. Synapses break up a network of neurons into *information-processing units*. Whether an impulse is transmitted to the next neuron in a sequence depends on whether or not there is a sufficient concentration of neurotransmitters at the synapse. Where neurons converge, transmission of a single impulse from one neuron might be insufficient. Several impulses might need to arrive simultaneously in order to release enough neurotransmitter molecules to trigger an electrical impulse in the next neuron. Convergence promotes **summation** of information. Conversely, if a neuron sends branches to several circuits, information diverges and is distributed to appropriate areas. Branches of a single axon are called **collateral branches**. Inhibition also affects the flow of information by decreasing the responsiveness of neurons to incoming information. Convergence, divergence, and inhibition are modes of information processing that take advantage of the character of synapse (Fig. 8.7) furthermore, the structure of neurons at a synapse ensures that trans mission across the gap occurs in only one direction.

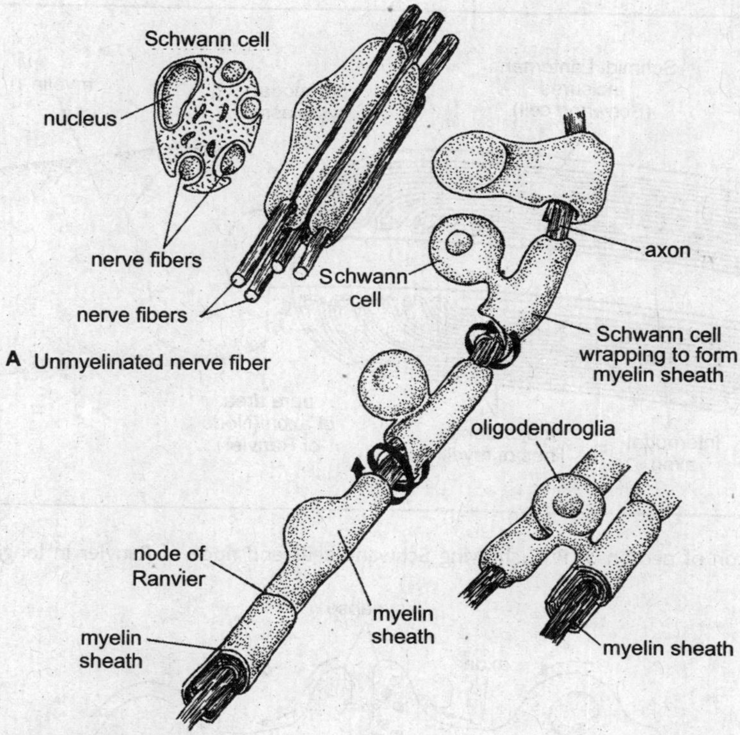

Fig. 8.5. Myelinated and unmyelinated nerve fibers. A – Despite their name, unmyelinated nerve fibers are associated with neuroglial cells. Usually there are several fibers per neuroglial cell, but these neuroglial cells are not wrapped repeatedly around the fibers as they are in myelinated nerves. B – The myelin sheath is formed by a neuroglial cell that is repeatedly wrapped around a section of the nerve fiber. In the peripheral nervous system, the neuroglial cell is a Schwann cell. In the central nervous system, it is an oligodendroglial cell. Successive neuroglial cells collectively form the myelin sheath. The boundaries between them are termed nodes of Ranvier (Kardong 2006).

Neurosecretory cells

Neurosecretory cells are specialized neurons. Most neurons release neurotransmitters at the end of their axons. Neurosecretory cells also release secretions at the ends of their axons, but these secretions are delivered into a blood capillary and transported to a target tissue. Newrosecretory cells are thus endocrine in function.

Functional Division of the Nervous system

The body functions of the nervous system are somatic or visceral .**Somatic functions** are those which are carried out by the voluntary muscles, skeleton, and the skin and its derivatives. **Visceral functions** are those which are performed by the digestive, circulatory, excretory,endocrine, and the urinogenital system. Both somatic and visceral functions are divided into two component each: (i) **afferent** or **sensory**; and (ii) **efferent** or **motor**. Thus the nervous system has four functional divisions:

1. Somatic sensory. The somatic sensory fibers carry impulses from the skin, sense organs, voluntary muscles and joints to the central nervous system.

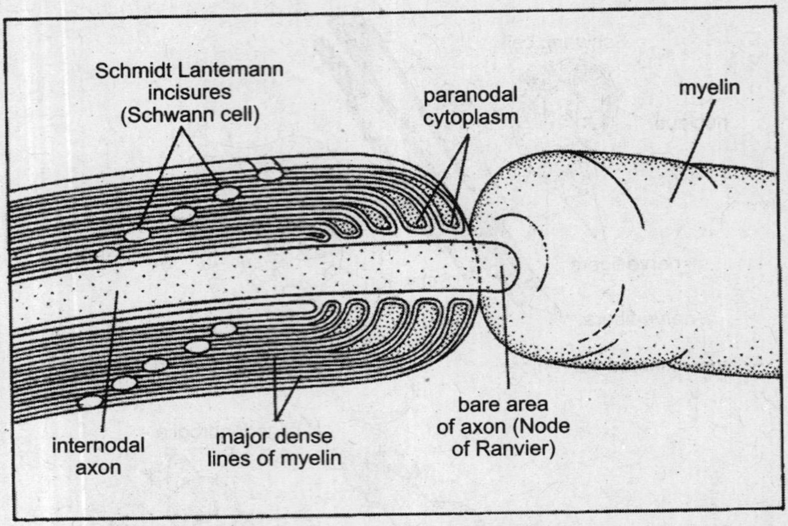

Fig. 8.6. An axon of neuron of PNS showing Schwann cells and node of Ranvier in longitudinal section.

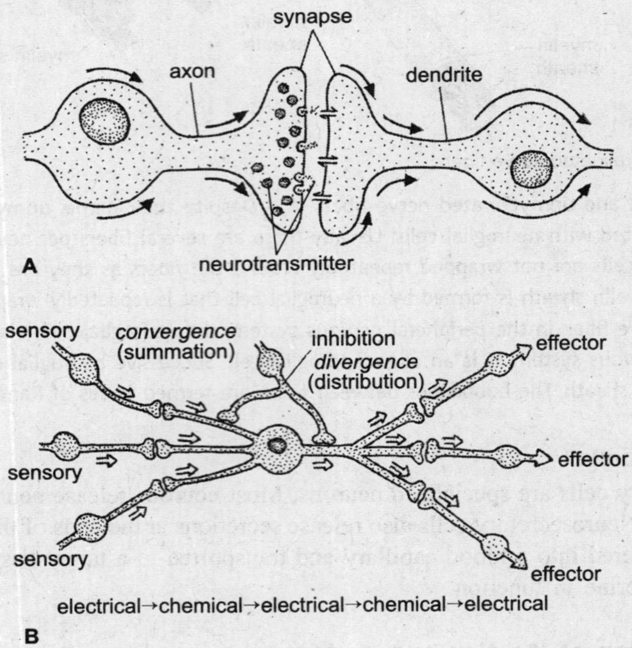

Fig. 8.7. Transmission of information within the nervous system. A – Neurons transmit and receive stimuli as electrical impulses along their fibers. Synapses are junctions between nerve cells. Axons release chemical messengers (*i.e.*, neurotransmitters) that diffuse across the synapse. When they arrive at the dendrite in sufficient concentration, the neurotransmitter molecules initiate electrical impulse in the next neuron. B – Synapses assist in processing information. Electrical input can converge or diverge. Inputs of some neurons inhibit or reduce the sensitivity of other neurons (after Kardong 2006).

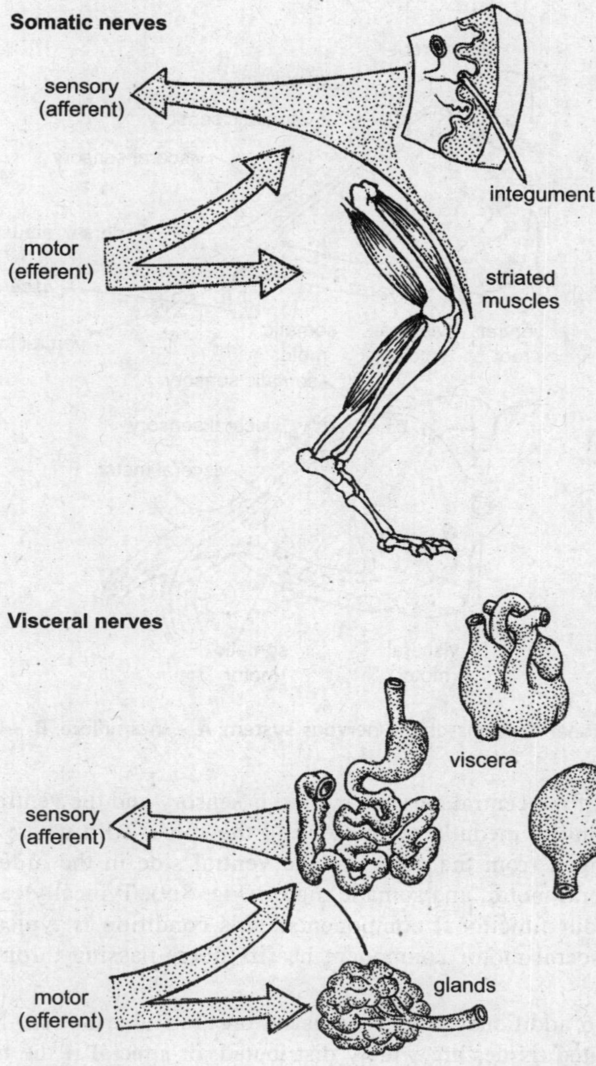

Fig. 8.8. Functional categories of neurons of the peripheral nervous systems. Some neurons supply somatic tissues, others visceral tissues. They can be sensory and respond to stimuli from these tissues, or they can be motor and deliver stimuli to these tissues (after Kardong, 2006).

2. Somatic motor. The somatic motor fibers carry impulses from the central nervous system to the voluntary muscles.

3. Visceral sensory. The visceral sensory component carries impulses from the viscera and organs of taste and smell to the central nervous system.

4. Visceral motor. The visceral motor component carries imputes from the central nervous system to the involuntary muscles, bronchial muscles and glands.

Association or adjustor neurons. These neurons are located in the brain and spinal cord. They link the afferent and efferent nerve fibers through synapses.

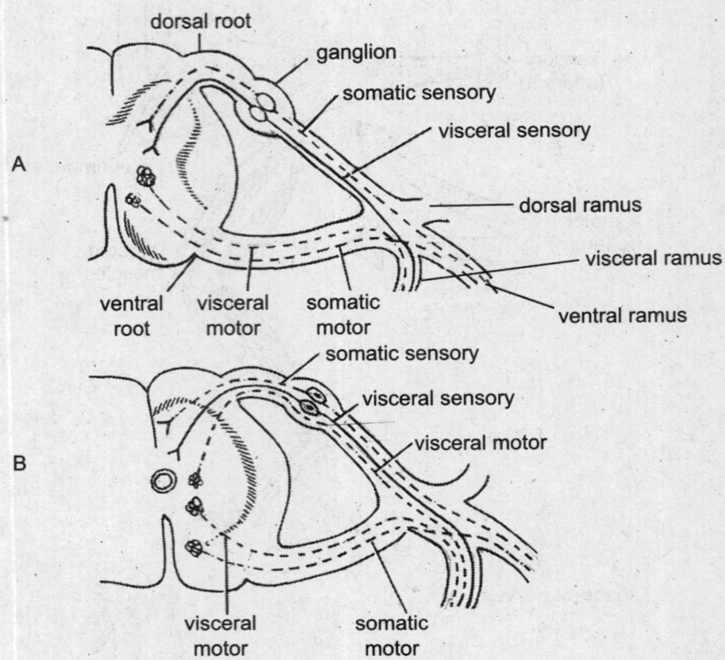

Fig. 8.9. Functional divisions of the nervous system: A – In amniota; B – In anamniota.

The dorsal half of the central nervous system is sensory and the ventral half is motor, this is especially marked in the medulla oblongata and the spinal cord where the four functional components are arranged from the dorsal to the ventral side in the order, somatic sensory, visceral sensory, visceral motor, and somatic motor (Fig. 8.9). Typically each peripheral nerve is also made of all four functional components. This condition is typical of amniotes, but in anamniotes the visceral motor component has its fibers passing through both dorsal and ventral roots.

Nerves have two additional properties based on their distribution, Neurons are termed **general** if the innervated tissues are widely distributed, or **special** if the tissues are restricted in location. Thus, **general somatic neurons** innervate sense organs or supply effectors to the integument and most striated muscles. **Special somatic neurons** are associated with somatic sense organs (*e.g.,* eyes).

Differentiation of the primary vesicles to form an adult brain is accomplished by (1) localized thickenings of the lateral walls and floor of some vesicles, and (2) dorsal, lateral, or ventral evaginations, median or paired, in others. The surface structures that develop in all craniates are readily seen to be homologues (Fig. 8.12).

The embryonic prosencephalon differs from the rest of the neural tube in not being divided into alar and basal plates. Early in development, except in ray-finned fishes, prosencephalon exhibits two pairs of evaginations, the **telencephalic vesicles**, the walls of which become the cerebral hemispheres, and the **optic vesicles**, which become the retina of eye balls (Fig. 8.10C). The optic vesicles become associated with the diencephalon. The mesencephalon does not undergo subdivision. The mesencephalon becomes the metencephalon and myelencephalon.

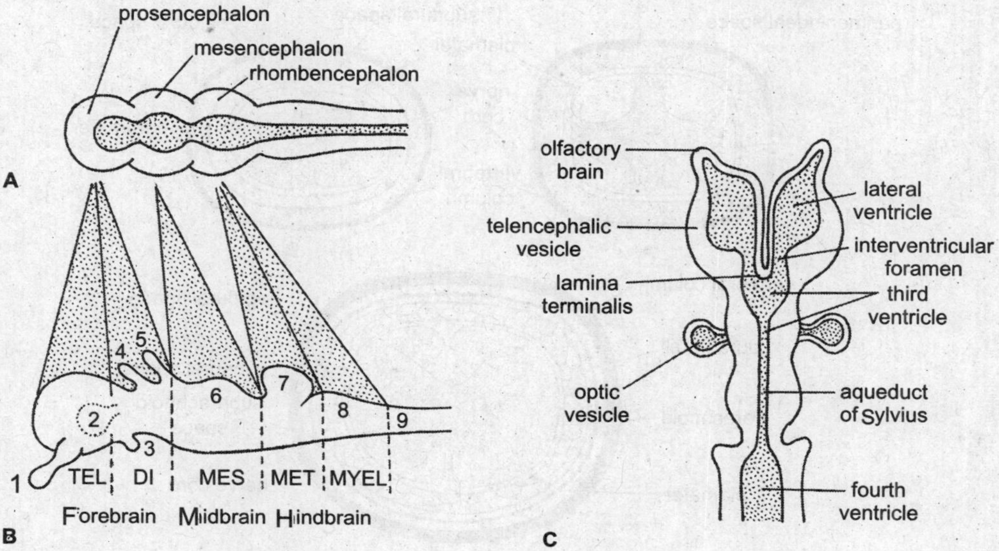

Fig. 8.10. Primary brain vesicles of vertebrates and initial stages of differentiation, generalized seheme. A – Three vesicle stage, sagittal view. B – Early differentiation; lateral view. 1, olfactory bulb; 2, future site of optic vesicle; 3, neural lobe of pituitary; 4, parapineal; 5, pineal; 6, optic lobe; 7, cerebellum; 8, medulla; 9, postcranial neural tube, C – Initial formation of cerebral hemispheres, except in actinopterygians (a type of bony fishes), frontal section (after Kent and Carr 2001).

Meninges

The adult brain is surrounded by the meninges (Fig. 8.11). Closely applied to the outer glial membrane of spinal cord and brain in most fishes is delicate connective tissue membrane, the **meninx primitiva**. A similar meninx appears during embryonic development in nonmammalian craniates (vertebrates), but it eventually differentiates into a vascular **leptomeninx** and a dense outer membrane, the **duramater**. Mammals have three meninges – an inner **piamater**, a filmy web-like **arachnoid**, and a tough fibrous **duramater** (outer membrane). The latter loosely surrounds the other meninges. The piamater adheres closely to the glial membrane of the cord, and it is attached to the arachnoid by a network of fibrous strands that traverse a sub-arachnoid space filled with cerebrospinal fluid. A similar subdural space separates the arachnoid from the duramater, and a peridural space filled with adipose tissue separates the cord and its meninges from the bony **vertebral canal**. The cerebrospinal fluid and the cushion of fat buffer the mammalian cord against mechanical trauma (physical injury). The same meninges surround the brain, but there is no peridural fat, the duramater having a somewhat different relationship to the periosteal lining of the skull.

Olfactory organs, inner ears or effectors (*e.g.*, branchiomeric muscles, ciliary eye muscles, extrinsic ocular muscles) that are limited in distribution. **General visceral_neurons** innervate sensory organs or supply effectors in glands or smooth muscles of the digestive tract, heart and other viscera. **Special visceral neurons** concerned with sensory input innervate the taste buds and olfactory epithelium.

8.2 COMPARATIVE ANATOMY OF BRAIN

Generally, the brain of each vertebrate species reflects the demands of information processing required by its habitat and mode of life (Fig. 8.43). Cavefish, for example, have reduced eyes and live in caves, a permanently dark subterranean environment. Correspondingly, the tectum of

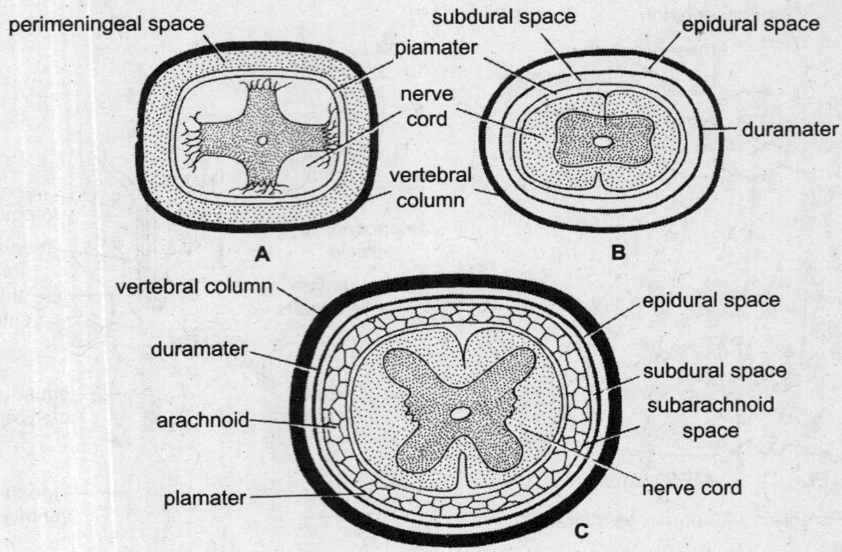

Fig. 8.11. T.S. or nerve cord showing meninges. A – Fish; B – Amphibian, reptile, and bird; C – Mammal.

the midbrain which normally receives visual input is reduced as well. On the other hand, when visual information constitutes a large part of the brain's sensory input, as in salmon, the tectum is enlarged. Thus, reduction or loss of sensory input from an exteroceptor or interoceptor results in corresponding reduction or loss of brain nuclei that receive and process this information, whereas increased sensory input leads to increased prominence of the appropriate association (Kardong 2006).

The vertebrate brain forms embryologically from the neural tube anterior to the spinal cord. In all craniates (vertebrates) from fishes to humans, brain exhibits three primary brain vesicles, a **prosencephalon** (future forebrain), **mesencephalon** (future midbrain), and **rhombencephalon** (future hindbrain, Fig, 8.10A). The rhombencephalon was named for its shape. The primary vesicles soon differentiate to form the anlagen of the five major subdivisions of the adult brain: **telencephalon** (TEL), **diencephalon** (D1), **mesencephalon** (MES), **metencephalon** (MET), and **myelencephalon** (MYEL; Fig. 8.10B).

Subdivisions of Brain and their Components

The major components of the brain are listed in Table 8.1.

Table 8.1 Subdivisions of the vertebrate brain and their major components.

Subdivisions		*Chief components*
1. Prosencephalon (forebrain)	1. Telencephalon	1. Cerebral hemispheres (I ventricle or rhinocoel) 2. Lateral ventricles (II ventricles or paracoels)
	2. Diencephalon	1. Epithalamus (roof) 2. Thalamus (sides) 3. Hypothalamus (floor) 4. Third (III) ventricle (diacoel)
2. Mesencephalon (midbrain)		1. Tectum 2. Tegumentum 3. Cerebral aqueduct (iter)

3.	Rhombencephalon (hindbrain)	Metencephalon	1. Cerebellum 2. Tegmentum 3. Fourth (IV) Venticle (metacoel)
		Myelencephalon	1. Medulla oblongata 2. Fourth (IV) ventricle

1. Hindbrain: Metencephalon and Myelencephalon

Myelencephalon is represented chiefly by the **medulla oblongata**, which merges imperceptibly with the spinal cord. The area of transition is characterized internally by gradual relocation of the fiber tracts, which are the white matter. As a result, whereas in the spinal cord the gray matter is centrally located, in the brain it is dispersed into small and large nuclear masses of gray matter interspersed among fiber tracts.

A conspicuous feature of the hindbrain is the **cerebellum**, a dorsal evagination of the metencephalon (Fig. 8.12 and Fig. 8.13). In mammals it is connected to the brain stem by three pairs of **cerebral peduncles** (stalks) composed of afferent and efferent fiber tracts (Fig. 8.14, 3-5).

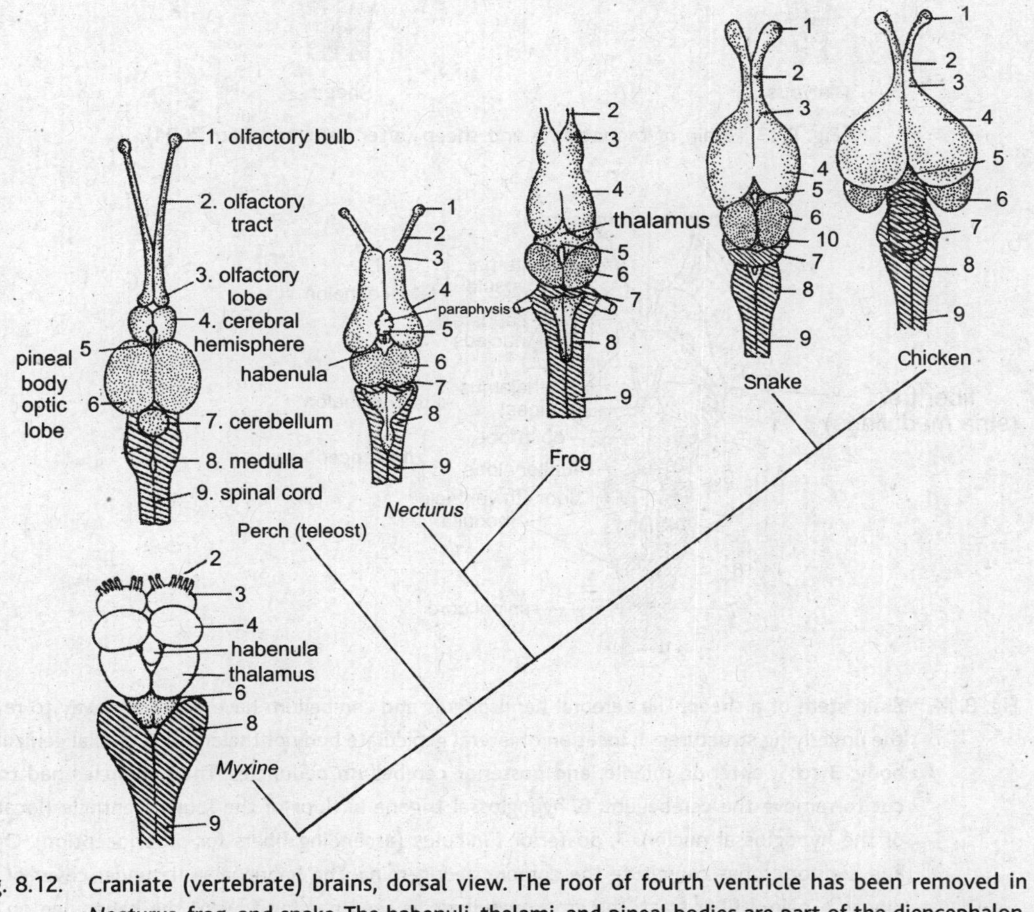

Fig. 8.12. Craniate (vertebrate) brains, dorsal view. The roof of fourth ventricle has been removed in *Necturus*, frog, and snake. The habenuli, thalami, and pineal bodies are part of the diencephalon. 1 to 9, see perch (teleost) brain for key; 10, auditory lobe of ammiotes. Mammalian brains are shown in Fig. 8.13 (after Kent and Carr 2001).

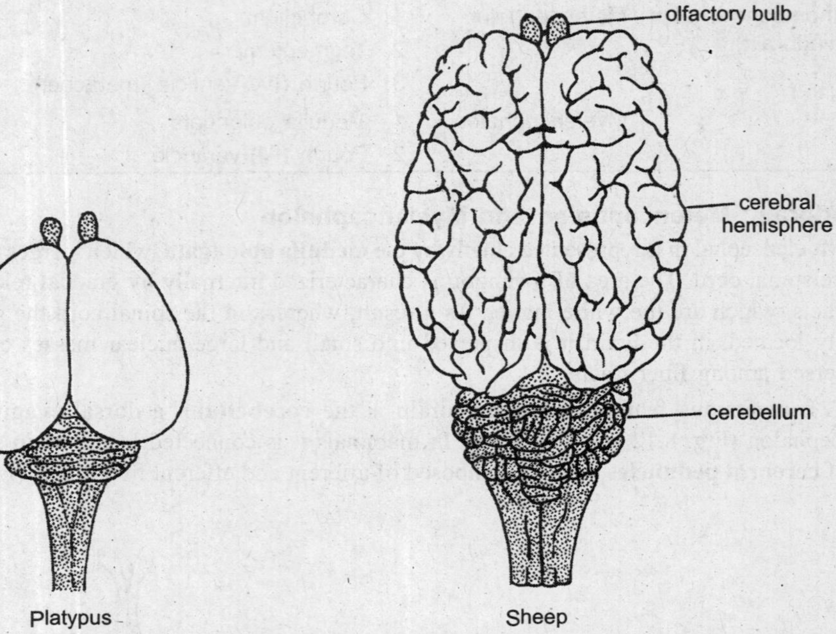

Fig. 8.13. Brain of monotreme and sheep (after Kent and Carr 2001).

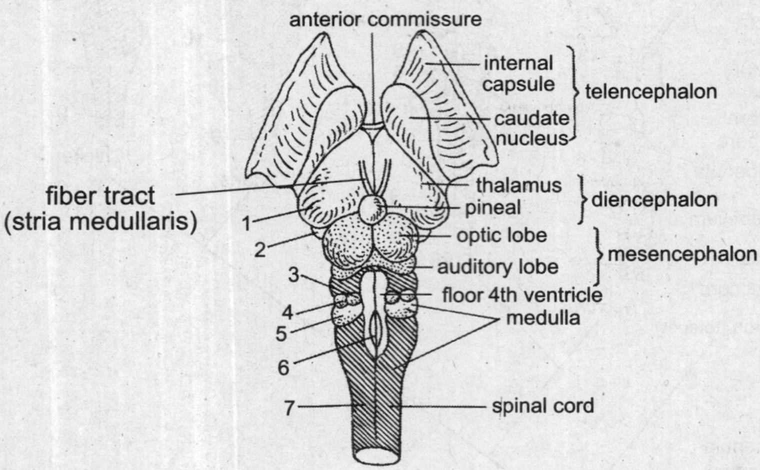

Fig. 8.14. Brain stem of a sheep. The cerebral hemispheres and cerebellum have been cut away to reveal the underlying structures. 1, location of lateral geniculate body of thalamus; 2, medial geniculate body; 3 to 5, anterior, middle, and posterior cerebellum peduncles. The peduncles had to be cut to remove the cerebellum. 6, hypoglossal trigone in floor of the fourth ventricle (location of the hypoglossal nuclei). 7, posterior funiculus (ascending fibers for proprioception). Optic and auditory lobes constitute the corpora qadrigemina. The habenualae lie under cover of the pineal. A paired fiber tract, the stria medullaris, is seen passing toward the habenulae on the surface of the thalamus (after Kent and Carr 2001).

The cerebellum coordinates the response of skeletal muscles to input from the membranous labyrinth, lateral-line canals of fishes, proprioceptors in muscles, joints, and tendons, and from reflex and voluntary motor centers in the brain stem and forebrain. The size of cerebellum is correlated with the complexity or the activities of striated musculature. It is larger in fishes than in amphibians because swimming, which involves schooling, vertical movements, adjusting to water currents, and keeping the dorsal part of the body form tipping over, requires more synergistic muscle activity than dragging the belly along the ground or squatting on a lily pad (*i.e.,* a leaf of water lily or lotus). Lacking a large cerebellum, aquatic urodeles rely to a large extent on spinal cord reflexes and primitive nuclei in the hindbrain for muscle coordination when swimming. The cerebellum is largest in birds and mammals, which require a large computer-like neural center to coordinate the muscles of the head, neck, trunk, and appendages in such diverse activities of flying, running, climbing, balancing, or, in case of humans, playing a piano or flute. The cell bodies (perikaryons) of the cerebellum are on the surface. In living agnathans, the cerebellum is not well developed and does not cause a bulge on the brain.

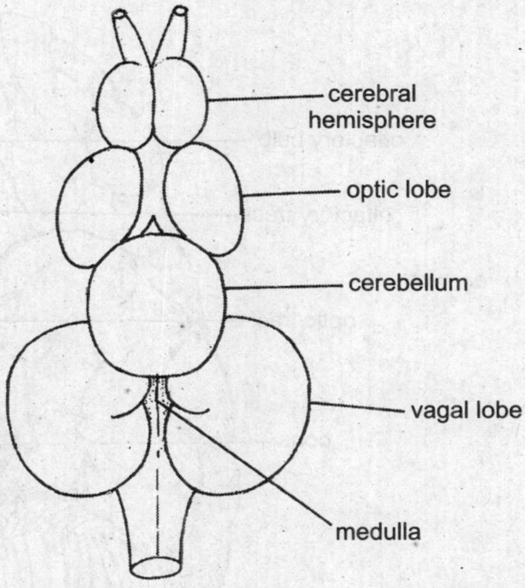

Fig. 8.15. Brain of the buffalo fish, *Carpiodes velifer*. Note unusual bulge (vagal lobe) on the alar plate of the medulla. Here terminate the many incoming taste fibers characteristic of bottom feeders (after Kent and Carr 2001).

Other topographical features of the hindbrain include various swellings caused by underlying nuclei, and elevated ridges or transverse bands from superficial fiber tracts. There topographical markings are most prominent in mammals. One sensor nucleus **(nucleus solitarius)** in the alar plate becomes enormous in fishes that have taste buds over the entire surface of the body. The resulting swelling, or **vagal lobe** (Fig. 8.15), is the termination of the many incoming sensory fibers for taste. The nucleus contains second-order sensory neurons whose fibers are projected to reflex and relay centers elsewhere in the brain. Among ventral ridges on the mammalian hindbrain are the **pyramids** (Fig. 8.16), which contain the corticospinal tracts that carry voluntary motor impulses from the cerebral cortex to the cord, and the **ventral decussation of the pons** in those mammals that have relatively large cerebellum (Fig. 8.17). The hindbrain also houses a reflex vasomotor center that maintain normal blood pressure and a reflex respiratory center that is vital in maintaining pulmonary respiration.

The cavity of the hindbrain is the **fourth (IV) ventricle**. The cerebellum is part of its roof. The remainder of the anterior and posterior of the cerebellum is a thin membrane that includes the ependyma of the neural tube and, posteriorly, a layer of highly vascular piamater from which vascular tufts hang into the cavity. The tufts constitute the **choroid plexus** of the fourth ventricle.

2. Midbrain: Mesencephalon

The **tectum**, or roof, of the mesencephalon displays a pair of prominent **optic lobes** in all craniates (Fig. 8.12 and Fig. 8.14). These bulging gray masses serve partly as reflex and relay centers that receive impulses originating in the retina. Optic lobes are especially large in birds, which have large eyes and depend on visual stimuli for much information about the environment. A pair of **auditory lobes** lies caudal to the optic lobes in the tectum commencing with amniote, and the four bodies (*i.e.,* optic and auditory lobes) constitute the **corpora quadrigemina** (Fig. 8.14). Fishes have

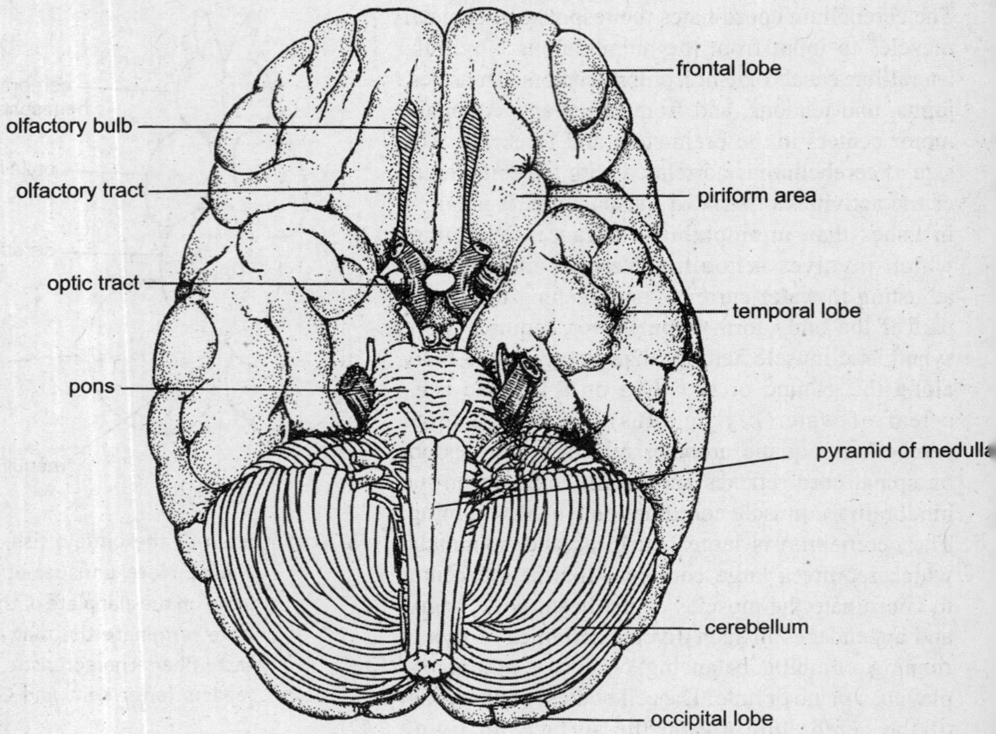

Fig. 8.16. Human brain, ventral view (after Kent and Carr 2001).

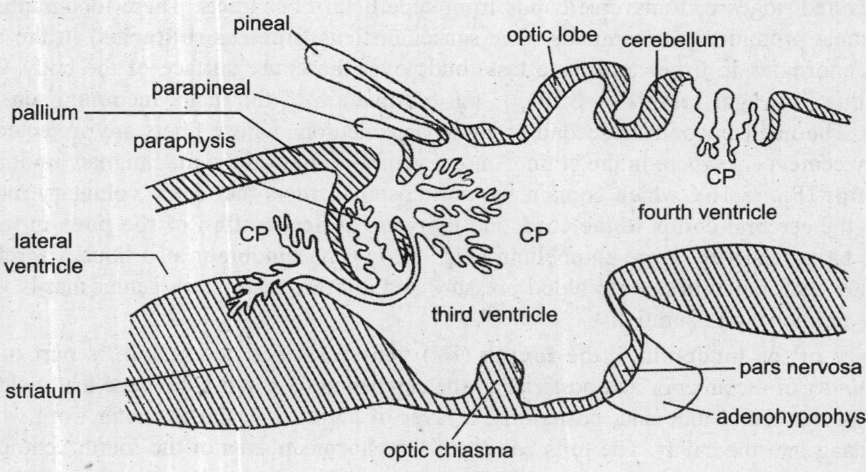

Fig. 8.17. Diencephalon and adjacent areas of a larval anuran brain, sagittal section, anterior end to left. **CP**, choroid plexus of lateral, third and fourth ventricles. The choroid plexus of the lateral ventricle enters the latter via an interventricular foramen (after Kent and Carr 2001).

auditory nuclei in this location, but they are not large enough to bulge on the surface. The auditory lobes receive input from the part of the membranous labyrinth that is sensitive to vibratory stimuli, and also from other sources. Phylogenetically, they enlarge as the receptors for sound expand. The

mesencephalic nucleus of the trigeminal nerve is a prominent gray mass in the alar plate. It contains the cell bodies of the proprioceptive fibers of most, if not all, of the cranial nerves that have them.

The basal plate of the mesencephalon, the **tegmentum**, is greatly thickened nuclei and by aggregates of fiber tracts that connect higher level of the brain with the hindbrain and spinal cord. In mammals, some of these nuclei and tracts are massive. One of these is the **red nucleus**, a large mass of gray matter that in fresh brains, has a pink tinge. Most of its afferent fibers come from the cerebellum, some fibers come from the cerebral cortex via a massive pair of fiber tracts, the **cerebral peduncles**. Efferent impulses from the red nucleus influence, directly or via relays, motor functions of the medulla and spinal cord, thus participating in control of the striated musculature.

The ventricle of the midbrain is quite large in fishes and amphibians and extends dorsad into the hollow optic lobes (Fig. 8.17). The derived condition for the optic lobes is that they are not hollow, and in the midbrain ventricle in this case is constricted to a narrow canal, the **cerebral aqueduct,** also known as the **aqueduct of Sylvius** and the iter (Fig. 8.10C).

3. Forebrain: 1. Diencephalon

The diencephalon consists of three major components: **epithalamus**, **thalamus** and **hypothalamus** (Fig. 8.18). The neurocoel of diencephalon is the **third (III) ventricle.**

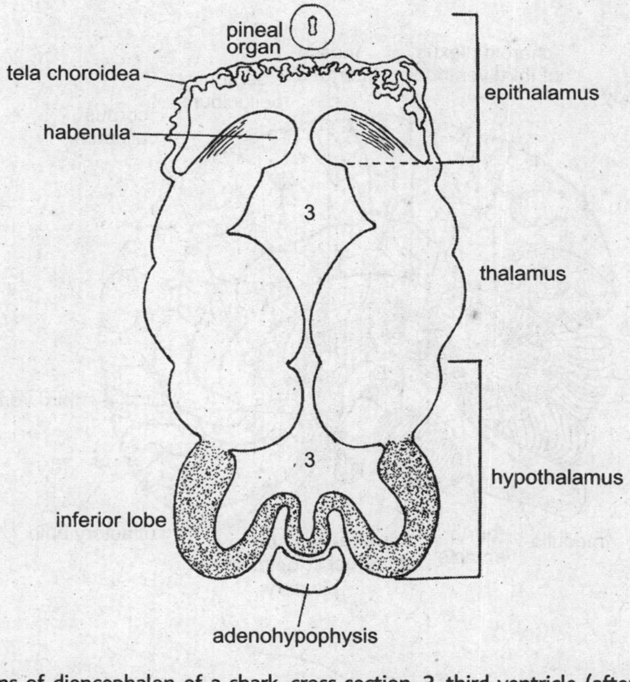

Fig. 8.18. Major regions of diencephalon of a shark, cross section. 3, third ventricle (after Kent and Carr 2001).

Epithalamus. The epithalamus is the dorsalmost component of the diencephalon and, as such, constitutes the roof of the third ventricle. It consisits of a **pineal** or **parapineal organ**, or both, a choroid plexus like that of the fourth ventricle, and a pair of elevated thickenings (median in *Myxine*), the **habenulae** (Fig. 8.18).

The **pineal** is a club-shaped or knob-like organ, sometimes threadlike or saccular, projecting above the diencephalon (Fig. 8.17). Pineal is connected to diencephalon by a stalk that frequently contains an extension of the third ventricle. In lampreys, the pineal is a **photoreceptor**. In gnathostomes, pineal functions as an **endocrine** organ that is stimulated in part by light that first enters the body through the retina. The pineal is vestigial or absent in a few craniates, including hagfishes, crocodilians, and a few permanently aquatic mammals. Pineal is relatively large in primates and sheep (Fig. 8.14).

A **median eye** was a constant feature of bony fishes and placoderms of the Devonian and of ancestral amphibians and amniotes. Living sharks and teleosts possess an endocrine pineal suggesting that the median eye (organ) of fossil fishes and early tetrapods also consisted a pineal. In contrast to lampreys, the parapineal in *Sphenodon* and lizards serve as a photosensitive organ – a **parietal eye** when functional as a photoreceptive organ, it lies under a patch of translucent skin.

The **habenulae** (Fig. 8.18) are elevations of a pair of underlying habenular nuclei that perform an ancient role associated with olfaction. They receive fibers from olfactory nuclei, and discharge impulses to the thalamus and midbrain, bring out reflex responses to olfactory stimuli. One incoming tract, the **stria medullaris,** is seen in Figure 8.14 emerging from the thalamus and passing caudad to disappear beneath in sheep. The habenulae are largest in species such as sharks and bloodhounds (hunting dogs having a very keen sense of smell, used in tracking), which depend heavily on olfaction for locating food. Habenlae are inconspicuous in most birds, many of which have a poorly developed olfactory sense. They are poorly developed in permanently aquatic mammals. The habenulae of *Myxine* is a median elevation (Fig. 8.12), but the underlying nuclei are paired.

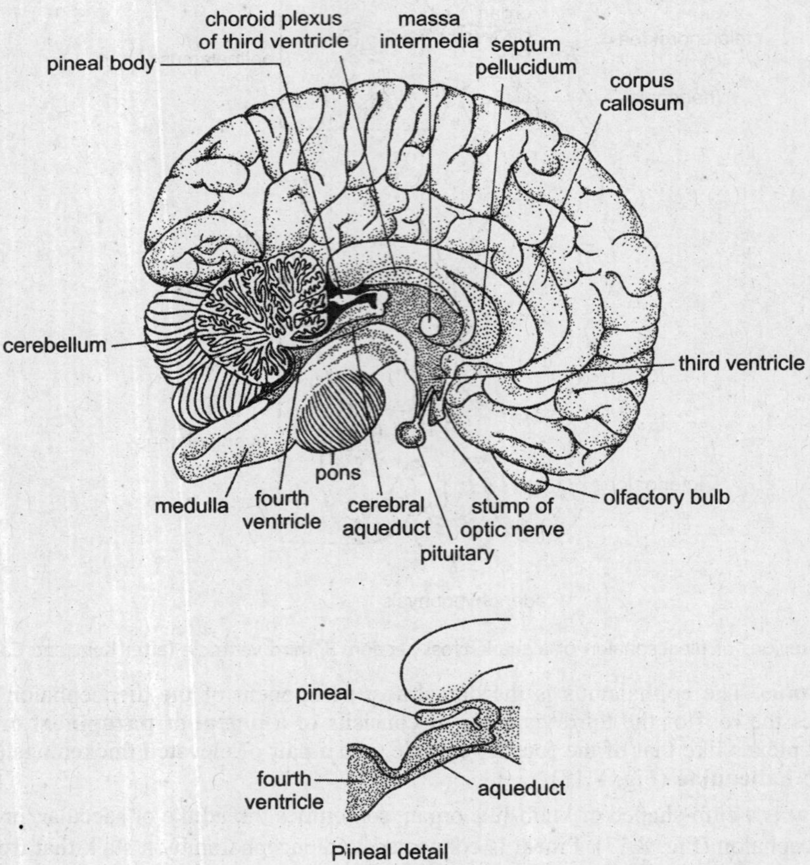

Fig. 8.19. Human brain, sagittal section (after Kent and Carr 2001).

Thalamus

The thalamus is the largest structure of the diencephalon. It is a paired mass of many nuclei in the latetral walls of the third ventricle, and it bulges dorsally just behind the cerebral hemispheres (Fig. 8.12; *Myxine* and frogs). In amniota, it is hidden by the caudal poles of the cerebral hemispheres and must be observed on a brain stem preparation (Fig. 8.14).

All *sensory pathways* ascending to the telencephalon from the spinal cord, hindbrain, or midbrain, synapse in a thalamic nucleus before continuing to the telencephalon. With development of the **neocortex** in mammals, the thalamic nuclei have become so numerous that the left and right thalami bulge into the third ventricle and meet at one site to form a oval bridge of gray matter, the **massa intermedia**, or false commissure (Fig. 8.19). The term **false commissure** reflects the fact that it does not consist of decussating fiber tracts as do true commissures.

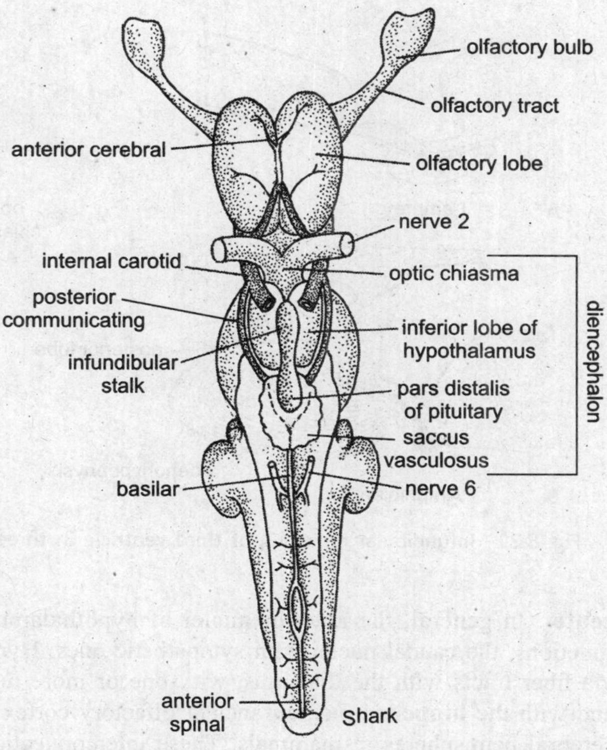

Fig. 8.20. Brain of shark, *Seqalus*, ventral view (after Kent and Carr 2001).

Hypothalamus and associated Structures

A ventral view of the brain reveals the optic chiasma, inferior lobes or other components of the hypothalamus, a saccus vasculosus in fishes only, and the pituitary (Fig. 8.16 and Fig. 8.20).

The **optic chiasma** is the cephalic boundary of the diencephalon, ventrally. This is where the optic nerves reach the brain, and where some or all of its fibers cross to the opposite side (Fig. 8.16). Caudal to the optic chiasma are the components of hypothalamus and at the caudal boundary of the diencephalon, the pituitary.

The floor of diencephalon exhibits a mid-ventral evagination within which is an **infundibular recess** (recess is a small space set into a surface) of the third ventricle (Fig. 8.21). An elongated evagination results in an infundibular stalk (Fig. 8.21C). The diencephalic floor at the end of the infundibular recess or stlak becomes the **posterior lobe (neural lobe, pars nervosa)** of the pituitary. Intimately associated with it is the adenohypophysis, which arises embryonically from the roof of the stomodaeum.

The hypothalamus is the floor and venrolateral walls of the third ventricle (Fig. 8.18 and Fig. 8.21c, hypothalamic nucei). It is prominent on the ventral surface of diencephalon in cartilaginous fishes because of the extraordinary size of the **inferior lobes** which have connections with the cerebellum (Fig. 8.20). In all craniates, the hypothalamus is an important neural center for homeostasis, a visceral function. Hypothalamic nuclei exert major reflex control over the autonomic nervous system, produce neurohormones that regulate the pituitary and gonads, and monitor the sodium chloride (NaCl) and glucose content of the blood. There is an **appetite regulating**

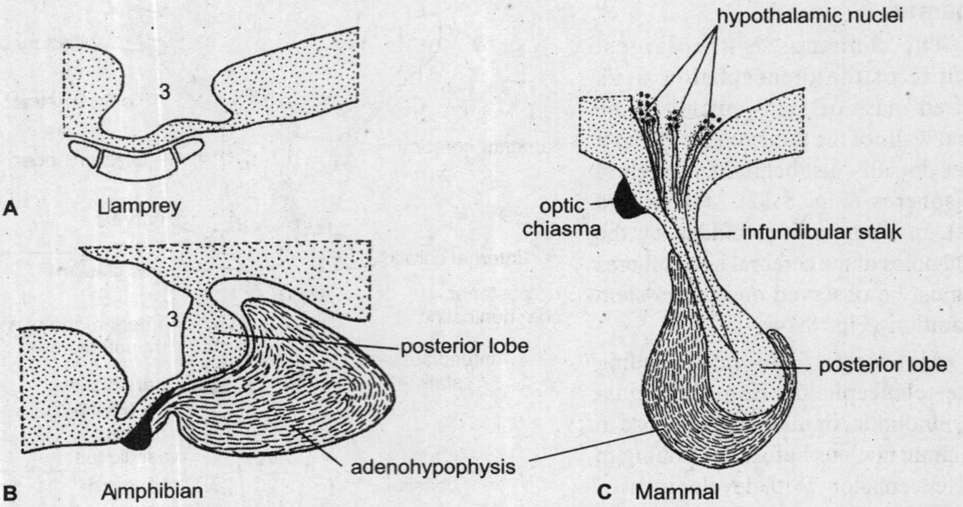

Fig. 8.21. Infundibular recess, 3, of third ventricle in three vertebrates (after Kent and Carr 2001).

center. In general, the anterior nuclei of hypothalamus are associated with parasympathetic functions, the caudal nuclei with sympathetic ones. Hypothalamic nuclei also have connections via fiber tracts with the thalamus, with one or more nuclei of the basal ganglia (Fig. 8.22 F), and with the **hippocampus**, an ancient olfactory cortex tucked under the temporal lobes of the cerebral hemispheres in mammals. These interconnections, the **limbic system**, are involved in evoking emotional responses in primates, an indication of the close relationship between visceral functioning and the emotions.

The **saccus vasculosus** is a highly vascularized, thin walled ventral evagination of the diencephalic floor of elasmobranch and ray-finned fishes, located just behind the pituitary

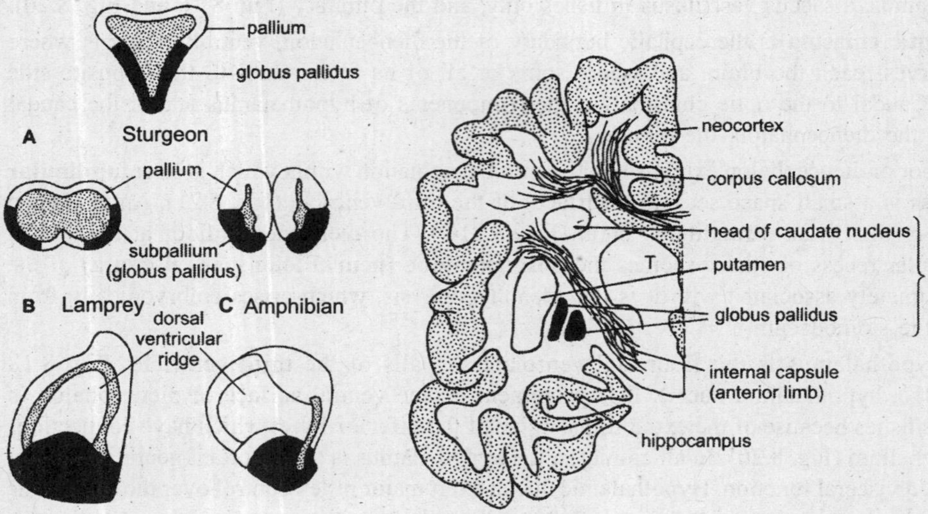

Fig. 8.22. The globbus pallidus from fishes to humans. D-F only the left cerebral hemispheres are shown. T, thalamus (after Kent and Carr 2001).

(Fig. 8.20). Within it is a fluid-filled recess of the third ventricle. The saccus vasculosus is a sense organ of jawed fishes other than dipnoans. The sac is largest in deep–sea fishes, not well developed in shallow fresh-water fishes, rudimentary in living agnathans and absent in lung-fishes. Sac is lined with hair cells, the cilia of which project into the cerebrospinal fluid. Its sensory nerve fibers pass to the hypothalamus and other brain centers. Probably this sense organ monitors the pressure of the cerebrospinal fluid (**Kent** and **Carr** 2001).

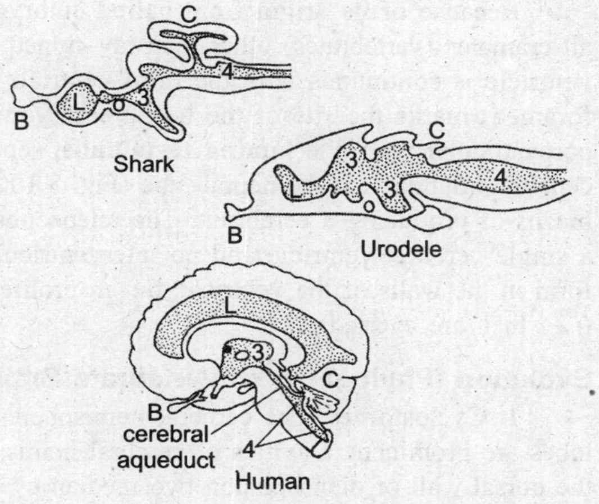

Fig. 8.23. **Sagittal brain section of chordates showing the ventricles (shaded areas.) B, olfactory bulb; C, cerebellum; L, 3 and 4, lateral, third and fourth ventricles. The black foramen in the third ventricle is the interventricular foramen. Immediately behind it (oval) is the massa intermedia (after Kent and Carr 2001).**

Third Ventricle

Third ventricle is part of forebrain, which is continuous caudad with the cerebral aqueduct of the midbrain and cranially with the ventricle in each cerebral hemisphere via a left and right **interventricular foramen** such as in humans (Fig. 8.10 and Fig. 8.23). The third ventricle becomes compressed laterally with expansion of the thalamus toward the midline. An optic recess extends toward the optic chiasma, the infundibular recess extends into the infundibular stalk, and a narrow canal extends into the stalk of the pineal. A choroid plexus is present in ventricular roof.

Telencephalon

The telencephalon (part of forebrain) in all craniates (vertebrates) other than ray – finned fishes is the portion of the brain that develops in the walls, floor and some regions of the roof of the paired embryonic telencephalic vesicles (Fig. 8.10C). The typical craniate telencephalon therefore consists of paired left and right hemispheres, each with a ventricle. These hemispheres, which anatomically are mirror images, are referred to collectively as the **cerebrum**. Extending forward from each hemisphere is an **olfactory tract** and **bulb** (Fig. 8.12). In mammals, these are hidden from dorsal view by the enormous outgrowth of the hemispheres, a result of millions of years of evolutionary hemispheric expansion (Fig. 8.19). The olfactory bulbs lie in contact with the ethmoid certilages or bones and receive axon-like processes from the olfactory cells in the olfactory epithelium.

The regions of the cerebrum devoted to olfaction – olfactory bulbs, tracts, lobes, and nuclei that process olfactory information – are prominent in those craniates that depend on olfactory lobes that are larger than the rest of the cerebrum (Fig. 8.20). On the other hand, olfactory components remain rudimentary or are vestigial in many marine mammals, and they are poorly developed in birds other than those that feed on carrion. Olfaction is also a relatively minor source of environmental information in bats and primates, including humans.

At the boundary between the diencephalon and telencephalon, the thin roof of the embryonic neurocoel evaginates dorsal in some members of every craniate class to form a wrinkled thin – walled sac, the **paraphysis** (Fig. 8.17). However, in amniotes other than *Sphenodon*, it is limited to embryos. Paraphysis resembles a chronic plexus, but seems to have a different role.

Because of its origin from paired embryonic vesicles, the neurocoel of the cerebrum in all craniates (vertebrates) other than ray–finned fishes consists of paired lateral ventricles. Each ventricle is continuous with the third ventricle via an interventricular foramen. This site (*i.e.,* foramen) marks the site of the initial embryonic evagination of the cerebrum. In mammals, a perpendicular wall, the **lamina terminalis**, separates the two foramina and marks the original cephalic boundary of the neural tube (Fig. 8.10C). Ray–finned fishes employ a slightly different means of producing a cerebrum. The telencephalic neurocoel remains unpaired; hence, there is a single cerebral ventricle and no interventricular septum. As in other craniates, paired nuclei form in the walls of the ventricle, but in proliferating, the dorsal and lateral walls turn outward (*i.e.,* they are everted).

Evolution (Phylogeny) of Vertebrate Brain

1. Cyclostomes. The cerebral hemispheres are barely indicated (Fig. 8.24). Two olfactory lobes are prominent. Cavities of cerebral hemispheres (lateral ventricles) are rudimentary. From the dorsal wall of diencephalon two median eye – like structures may arise. The anterior one is the parietal body and the posterior one is the pineal body. These two are vestigeal in *Eptatretus* and absent in *Myxine*. The ventral part of diencephalon gives rise to the infundibulum of the hypophysis as well as the posterior lobe. The optic lobes are not well differentiated. Cerebellum is poorly developed due to sluggish and slow moving habit. Medulla oblongata is very well developed in all vertebrates since it serves as a center for many vital body activities (respiration,

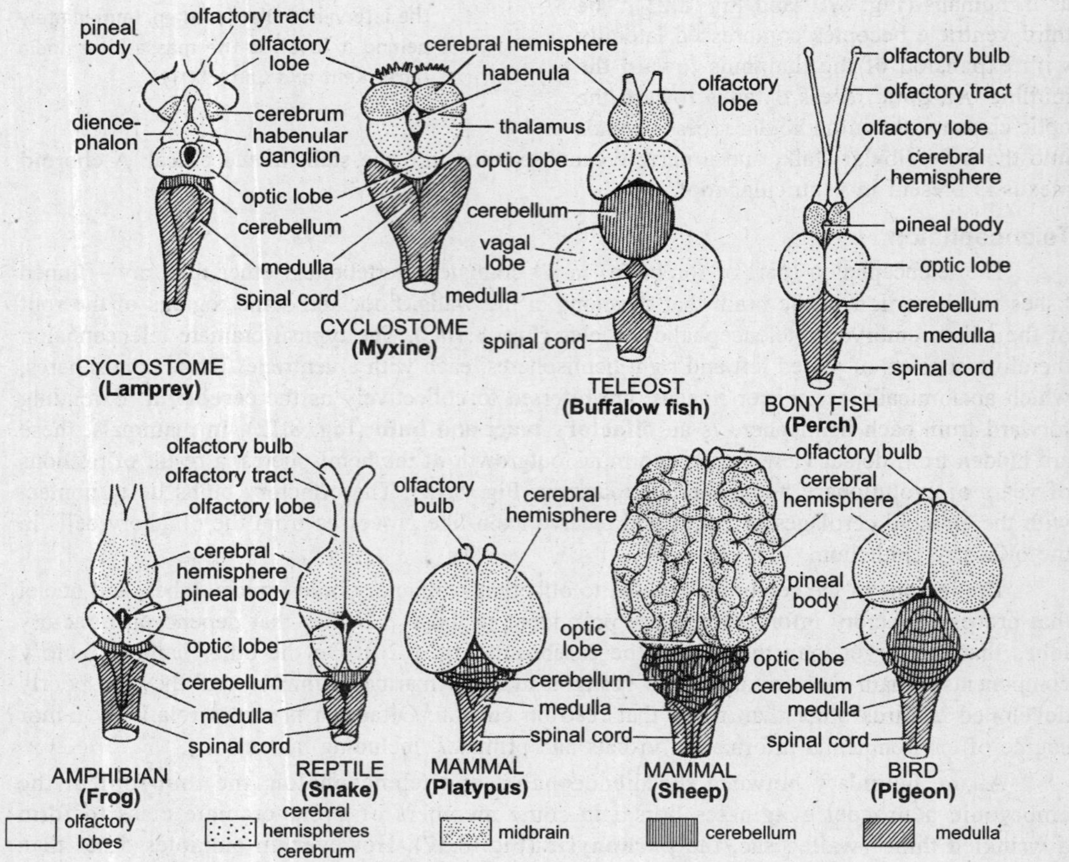

Fig. 8.24. Comparative anatomy of the vertebrate brains in dorsal view.

heart action, metabolism, etc). It is also serves as a relay center and in fishes is a center for lateral line system and inner ear.

2. Elasmobranchii. The brain of elasmobranchs is large and different from that of cyclostomes and bony fishes. In fishes, telencephalon is primarily olfactory in function and olfactory lobes are large in elasmobranchs. Each olfactory lobe has a rhinocoel cavity. The cerebral hemispheres (cerebrum) are better developed. Cerebrum in fishes, however, consists of a basal ganglionic mass (*i.e.,* nuclei) known as **corpus striatus** and a thin, dorsal, epithelial layer called the **pallium** composed essentially of nonnervous tissue and the center of brain activity is back in the mesencephalon. The **diencephalon** is narrow with a thin roof having an anterior choroid plexus, below which is the third ventricle. Its lateral walls form thick thalami which serves primarily as a relay center for olfactory and visual impulses and is associated with several sensory and motor (glandular) structures. Dorsally the diencephalon has a long tubular **pineal body** and the floor of the diencephalon give rise to the stalk or infundibulum of the hypophysis as well as the posterior lobe. A thin-walled vascular sensory organ, **succus vasculosus** is attached to hypophysis. This sensory organ is connected with the cerebellum by fiber tracts. This midbrain in fishes is the center of nervous coordination. Dorsally midbrain develops two prominences known as optic lobes concerned with visual reception. Cerebellum is a center of muscular coordination, it is larger than that of cyclostomes and dipnoans. Medulla is well developed and at its antero – lateral margins are present ruffle-like **restiform bodies**, which assist the cerebellum in the maintenance of equilibrium.

3. Osteichthyes (Bony fishes). The brain of bony fishes is more specialized than the elasmobranches. Olfactory lobes of the forebrain are large, having no olfactory peduncles and are closely applied in front of the cerebral hemispheres. Each of them contains a rhinocoel continuous with the lateral ventricles. The cerebral hemispheres are better developed. The roof or pallium of the cerebral hemispheres is thin and non-nervous, consequently it is not equivalent of the pallium of other animals. The corpus striatum is very thick. The lateral vertricle is undivided. Diencephalon is small hidden dorsally by the midbrain. Only the pineal body with stalk is present. Parapineal body is not found in modern bony fishes. Melatonin has been identified in the pineal gland of certain fishes. This neurohormone has an inhibitory effect on gonadial activity. In immature pacific salmon (genus *Oncorhynchus*) melatonin content is six times more than that of adults. On the ventral side, the diencephalon has as infundibulum to which pituitary gland is attached below. Optic lobes are very large (Fig. 8.24). Cerebellum is larger than that of elasmobranchs. In some fishes, **restiform bodies**, are also present. **Vagal lobes** are the bulgings from the antero-lateral sides of medulla and are found in bottom feeders, which have taste buds on their body surface.

4. Amphibians. The nervous system of amphibians is still basically like that of fishes. The pallium of cerebral hemispheres is invaded for the first time in vertebrates by nerve cells due to which the cerebral hemispheres becomes enlarged in accordance with more complex activities of locomotion, hibernation, breeding, etc. The olfactory lobes are fused together in the median plane. The corpora striata form a thick floor of the cerebrum. The diencephalon is short but optic fibers end in it. The pineal body is found in all amphibians, but only anurans possess a parietal body. The optic lobes are large and are the center of brain activity like fishes. The members of this class are notably slow and sluggish in their movement, so the cerebellum is very small, like a simple narrow transverse band. Medulla is also small, except in caecilians, flexures are lacking in amphibian brain.

5. Reptiles. The brain shows many features which are characteristic of the mammalian brain. In all anamniotes the midbrain is the center of brain activity, but in reptiles for the first time there is a shift in the nerve center to the cerebrum. This is correlated with a marked increase in the size of the cerebral hemisphere as a result of invasion of the pallium by many nerve cells so as to form **neopallium**. Two long olfactory lobes are larger than that of amphibians due to greater thickness and enlargement of corpora striata. Parietal eye is found in *Sphenodon* and some modern lizards, but absent or vestigial in some other reptiles. Behind the optic lobes are present a pair of **auditory lobes** which are not hollow. The third ventricle is reduced and form a narrow **cerebral aqueduct**. Cerebellum is relatively larger than that of amphibians. However, it does not attain the size of cerebellum of sharks, and not near the size of birds or mammals. This is correlated with relatively limited locomotive powers of most reptiles.

6. Aves. The brain of birds show a considerable advance over that of reptiles. In shape it is short broad and rounded. The olfactory lobes are extremely small and degenerate due to poor sense of smell. The **cerebral hemispheres** are very large and covers the diencephalon and optic lobes. Its big size is due to enlargement of the corpus striatum rather than the cerebral cortex. The cerebral cortex is smooth. The size of corpora striata makes the lateral ventricles small. The **diencephalon** is covered dorsally by the cerebral hemispheres and cerebellum. It has a small dorsal pineal body, an anterior choroid plexus and a narrow third ventricle. The **thalami** are highly developed and are connected by fibers to the corpora striata and spinal cord.

The **midbrain** of birds is highly developed with exceptionally large rounded optic lobes. This seems to be correlated with the remarkable powers of sight that birds possess.

The **cerebellum** of birds is larger than in reptiles and is deeply fissured, although it is not as large as in mammals. Ventral to cerebellum, the avian brain shows the beginning of the development of a pon. In bird's cerebellum masses of nerve cell bodies (perikaryons) form nuclei so that it has outer layers of gray matter and an inner mass of white matter, forming a branching pattern, called **arbor vitae**. A small projection of the fourth ventricle extends into the cerebellum. It controls equilibrium and movement in all places during flight. The medulla oblongata has a posterior dorsal choroid plexus and a fourth ventricle, which does not extend into the cerebellum.

7. Mammalia. The brain reaches its highest development in mammals with better integration and mastery over the environment. The cerebral hemispheres reaching the status of a dominant integrating part of the brain and acting as coordinating centers of the brain. The cerebral hemispheres in Prototheria are smaller and smooth like that of reptiles. In Metatheria, they are larger and smooth. In most eutherians, the cerebral hemispheres are very large, projecting forwards above the olfactory lobes and backwards above the diencephalon and midbrain and divided into lobes. They have convolutions on the surface so that there are ridges or **gyri** and depressions or **sulci**. The outer layer or cortex of the cerebrum is composed of gray matter. The right and left cerebral hemispheres are connected with one another by a broad white commissure, called the **corpus callosum**. The olfactory lobes are small compared with those of lower vertebrates.

The diencephalon consists of a dorsal **epithalamus**, a lateral **thalamus**, and a ventral **hypothalamus**. A pineal gland is present on the roof of the diencephalon, but it shows no eye-like structure. Thalamus is an important relay center. Hypothalamus is very important is mammals and consists of four parts. These are the **infundibulum** forming the stalk and posterior lobe of the pituitary, the **optic chiasma**, where right and left optic nerves cross en route to the brain, and the **mammillary bodies**, which integrate olfactory impulses. The hypothalamus controls many mammalian functions including blood pressure, sleep, water content, fat and carbohydrate metabolism, body temperature and possibly rhythmic activities such as moulting, mighration and pituitary secretion.

The mammal's midbrain is of less importance than in lower vertebrates, in which it is really the brain center. Midbrain is divided into four prominences, called the **corpora quadrigemina**. The two superior lobes are concerned with sight, whereas the two inferior lobes are probably concerned with hearing. The third (III) ventricle or **iter** of midbrain is laterally compressed vertical passage called the **cerebral aqueduct**.

The **cerebellum** which is the center of control of body movement, is most highly developed in mammals. Its surface is convoluted and it is divided into a number of lobes: central lobe or **vermis**, two lateral lobes, and two outer floccular lobes. Beneath the cerebellum is a typical mammalian structure, the **pons**. This relay center is a conspicuous feature of the ventral metencephalon. **Pons varolli** have crossing or decussating fibers connecting opposite sides of the cerebral cortex. These are the transverse nerve fibers. Besides pons, there is a short tract of transverse fibers, called **corpus trapezoideum** relaying impulses for sound.

The **medulla oblongata** lies ventrally and is much thickened. It has a dorsal posterior choroid plexus below which is fourth ventricle joined in front to the iter behind to the central canal. The medulla has centers for regulation of digestion, respiration and circulation.

Choroid Plexus and Cerebrospinal Fluid (CSF)

The cavities of the brain, the neural cannal of the cord, and the subarachnoid space, when present are filled with **cerebrospinal fluid (CSF)**. This is a clear, watery liquid of low specific gravity that is removed from the blood by diffusion and secretion. Most of it is secreted into the ventricles by the choroid plexuses; in mammals, at least, a smaller amount comes from the bloodstream directly. **Choroid plexuses** consist of tufts of a highly vascularized piamater and ependyma that dangle into the third and fourth ventricles. The plexus in the third ventricle extend forward into the lateral ventricles via the interventricular **foramina** (Fig. 8.17). Obstruction of these foramina results in accumulation of fluid in the lateral ventricles. When this condition is present in fetal or neonatal mammals, the fontanel of the immature skull allow the head to enlarge, a condition known as **hydrocephalus**. The plexus in the fourth ventricles is in the roof of the medulla.

Cerebrospinal fluid moves sluggishly caudad from the fourth ventricle into the neural canal of the cord (Fig. 8.25). In mammals, it also enters the subarachnoid spaces via a median and two lateral aperture in the roof of the fourth ventricle under cover of the cerebellum. These spaces surrounding the cord of mammals exhibit several cisterns where fluid normally accumulates. One of these in humans surrounds the cauda equina in the lumbosacral region, from which a **spinal tap** may aspirate some of the fluid for study of its contents or diagnosis.

From the subarachnoid spaces, the CSF travels along narrow perivascular spaces (Box 8.1) that accompany blood vessels entering or leaving the substance of the cord and brain, it follows the rootlets of the spinal and cranial nerves where they pierce the subarachnoid spaces and duramater to emerge from the cord (Fig. 8.26) and it seeps into the cavity surrounding membranous labyrinth where it contributes to the perilymph.

Box 8.1 Perivascular space of CNS

The pia is only loosely adherent to the vessels, so a space, the **perivascular space,** exists between it and each vessel. Therefore, perivascular spaces follow both the arteries and the veins into the brain as far as the arterioles and venules go. No true lymphatics are present in brain tissue, but perivascular space, in effect, are a specialized lymphatic system for the brain (**Hall**, 2011). In addition to transporting fluid and proteins, the perivascular spaces transport extraneous particulate matter out of the brain. For example, whenever infection occur in human brain, dead white blood cells and other infectious debris are carried away through the perivascular spaces.

Cerebrospinal fluid of the brain and cord of nonmammalian craniates is removed by diffusion into adjacent venous channels. In mammals, clusters of **arachnoid granulations,** which are tufts of the pia-arachnoid membrane, project into large subdural venous sinuses (a derived condition in many craniates), and the fluid diffuses out of the granulations and venous blood of the sinuses.

Functions of CSF. In addition to cushioning the brain and cord against mechanical trauma, cerebrospinal fluid selectively exchanges metabolites with the tissues it bathes.

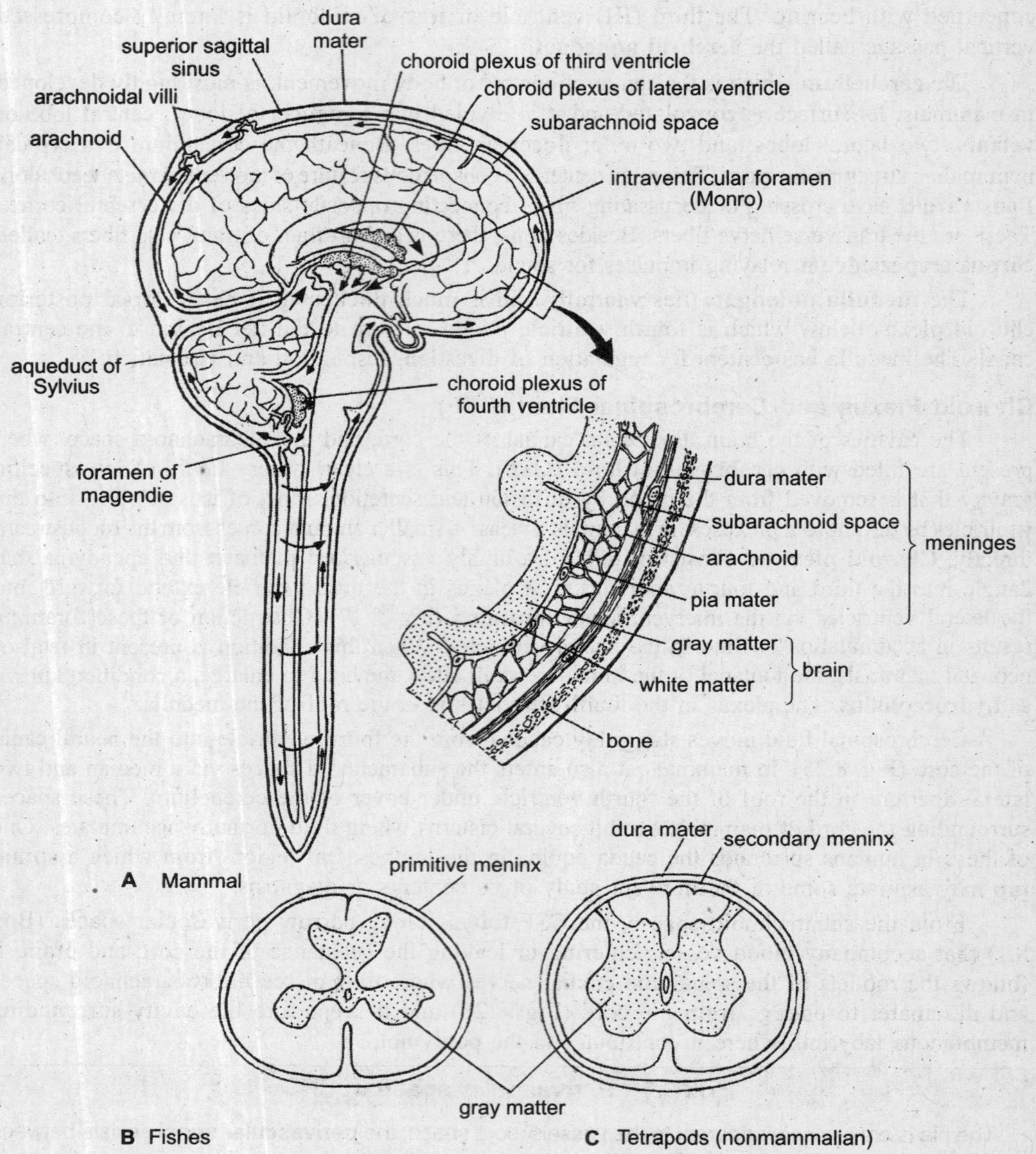

Fig. 8.25. Cerebrospinal fluid and meninges. A – Arrows trace the circulation of cerebrospinal fluid through the brain and the spinal cord of a mammal. The triple - layered meninges are enlarged to the right. B – The meninges of fishes consist of a single thin layer, the primitive meninx. C – In all tetrapods except mammals, the meninges are double layered and consist of a outer duramater and inner secondary meninx (after Kardong 2006).

8.3 AUTONOMIC NERVOUS SYSTEM (ANS)

Early anatomists noticed that visceral activity did not appear to be under voluntary control. The peripheral nerves and ganglia associated with visceral activity seemed to be autonomous, or independent of the rest of the nervous system. Collectively, they were considered to constitute the autonomic nervous system, of functional division of the peripheral nervous system that governs

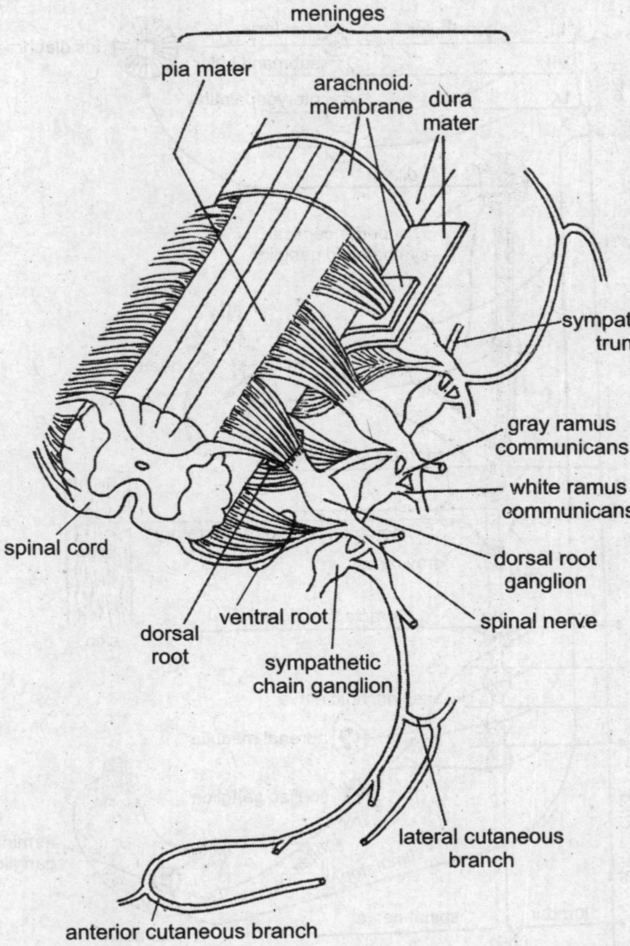

Fig. 8.26. Cutaway section of the spinal cord in a mammal illustrates the three meningeal layers: duramater, arachnoid and piamater. Branches of the spinal nerve are shown along with their connections to the sympathetic chain (after Kardong 2006).

visceral activity. Both sensory and motor fibers are included in autonomic nervous system. **Autonomic sensory fibers,** monitor an internal environment of the organism, *i.e.,* blood pressure, oxygen and carbon dioxide tension, core and skin temperature, and activity of the viscera. **Autonomic motor fibers** are general visceral motor neurons that innervate cardiac muscles, smooth muscle and glands. Thus, they control the digestive tract, blood vessels, respiratory tree, bladder, sex organs, and other general body viscera. Because the autonomic nervous system includes the general visceral motor circuits, pre- and postganglionic nervous in series characterize the motor innervation to each organ (Fig. 8.27).

Conscious centers also can affect visceral activity controlled by the autonomic nervous system. For example, through practiced meditation or through deliberate effort that bring a chilling thought to mind, it is possible to affect the heartbeat or the release of sweat. But, for the most part, the autonomic system operates subconsciously and is not under voluntary control. Reflexes control activities maintaining the internal environment. The neuronal circuitry of the autonomic nervous system basically includes four neurons linked in a reflex loop: a sensory neuron that synapses with an associated neuron that synapses with a preganglionic motor neuron in series with a postganglionic motor neuron.

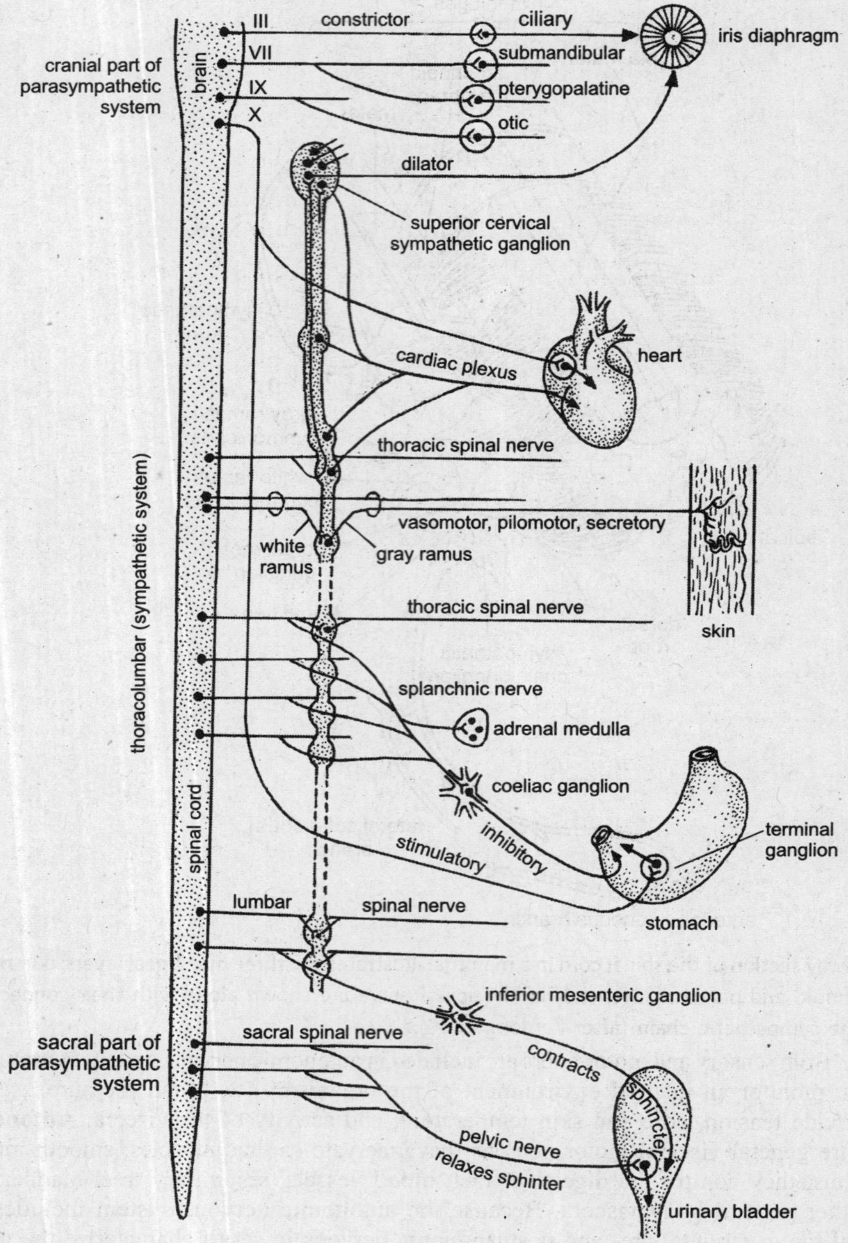

Fig. 8.27. Representative components of the mammalian autonomic nervous system. Autonomic fibers are not confined to the spinal nerves shown. Arrows show dual control of visceral effectors by craniosacral, and thoracolumbar systems (= out flows). The cell bodies (solid dots) of preganglionic fibers are in the brain or cord; those of postganglionic fibers are in the sympathetic, collateral or terminal (= visceral) gonglion (after Kent and Carr 2001).

Functional Division of Autonomic NS

In mammals, the autonomic nervous system is divided into *two* contrasting, antagonistic systems of control over visceral activity: the sympathetic system and the parasympathetic system.

1. Sympathetic nervous system. It prepares the body for strenuous action by increasing activity of the viscera, although it slows digestive processes. Stimulation of the sympathetic nervous system **inhibits** activity of the alimentary canal but **promotes** contraction of the spleen (causing it to release extra red blood cells into the general circulation), **increases** heart rate and blood pressure, **dilates** coronary blood vessels, and **mobilizes** glucose from glycogen storage in the liver. Thus, the sympathetic nervous system prepares the individual to fight or to flee (Table 8.2).

The general visceral motor nerves that participate in sympathetic activity depart from the thoracic and lumber regions of the mammalian spinal cord. This activity is referred to as the **thoracolumbar outflow.** The sympathetic preganglionic neuron is usually short and synapses in the sympathetic chain ganglion or in a ganglion located away from the vertebral column. The postganglionic fiber is usually long (Fig. 8.28).

2. Parasympathetic nervous system. It restores the body to a restful or vegetative state by lowering its activity level, although digestion is stimulated. The effects of the parasympathetic system are antagonistic to those of the sympathetic system. Parasympathetic NS enhances digestion, slow heart rate, drops blood pressure, constricts coronary vessels, and promotes glycogen formation.

Table 8.2 Functional divisions of the autonomic nervous system (Source: after Kardong 2006).

Organ/Activity	*Sympathetic stimulation*	*Parasympathetic stimulation*
1. Eye		
Ciliary muscles	Relaxation	Contraction
Pupil	Dilation	Constriction
2. Glands		
Salivary	Vasoconstriction Slight secretion	Vasodilation Copious secretion
Gastric	Inhibition of secretion	Stimulation of secretion
Pancreas	Inhibition of secretion	Stimulation of secretion
Lacrimal	None	Secretion
Sweat	Sweating	None
3. Digestive tract		
Sphincters	Increase tone	Decrease tone
Walls	Decrease motility	Increase motility
4. Liver	Glucose release	None
5. Gallbladder	Relaxation	Contraction
6. Bladder		
Smooth muscle	Relaxation	Contraction
Spincter	Contraction	Relaxation
7. Adrenal gland	Secretion (preganglionic neuron innervation)	None
8. Heart		
Muscle	Increase rate and force	Slowed rate
Coronary arteries	Dilation	Constriction
9. Lungs (bronchi)	Dilation	Constriction
10. Spleen	Contraction	Relaxation
11. Blood vessels		
Abdomen	Constriction	None
Skin	Constriction	None
12. Sex organs		
Penis	Ejaculation	Erection
Clitoris	?	Erection
13. Metabolism	Increased	None

Participating visceral motor neurons include cranial nerves vii, ix and x together with spinal nerves departing from the sacral region. This is referred to as **craniosacral outflow.** Parasympathetic preganglionic fibers are long and reach to the wall of the organ they innervate and synapse with very short postganglionic fibers (Fig. 8.28).

Adrenergic and cholinergic control. The sympathetic system is said to be adrenergic because the neurotransmitters released during stimulation are **adrenaline** and **noradrenaline** (also termed **epinephrine** and **norepinephrine**). The parasympathetic system is said to be cholinergic because the neurotransmitter released is **acetylcholine**. Acetylcholine is also released between pre- and postganglionic fibers in both systems and at junctions between nerves and skeletal muscles.

Almost every visceral organ in mammals has sympathetic and parasympathetic innervation (Fig. 8.29 and Table 8.2) Exceptions to this double innervation include the adrenal glands peripheral blood vessels, and sweat glands, all of which receive only sympathetic innervations. Cessation of sympathetic stimulation allows these organs to return to a resting state.

The adrenal gland is also exceptional in that it is innervated by the preganglionic fiber only, the postganglionic fiber is absent. Because epinephrine and norepinephrine serve both as adrenergic chemical signals of the sympathetic circuit and as hormones produced by the adrenal gland, there is a possibility for chemical confusion. But the preganglionic neuron releases acetylcholine rather than adrenaline or similar chemicals, so **direct innervation** of the adrenal gland by preganglionic fibers removes the possibility of chemical ambiguity between parasympathetic innervation and hormonal stimulation by the gland.

Anatomic Divisions of the Autonomic NS

The division of the autonomic nervous system into sympathetic and parasympathetic functional components holds reasonably well for mammals, however, in other vertebrates, the comparative anatomy of the autonomic nervous system is poorly understood. Most viscera receive

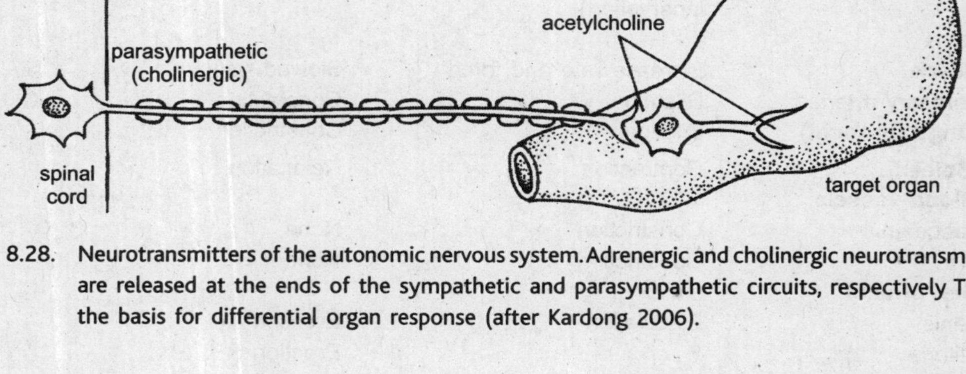

Fig. 8.28. Neurotransmitters of the autonomic nervous system. Adrenergic and cholinergic neurotransmitters are released at the ends of the sympathetic and parasympathetic circuits, respectively This is the basis for differential organ response (after Kardong 2006).

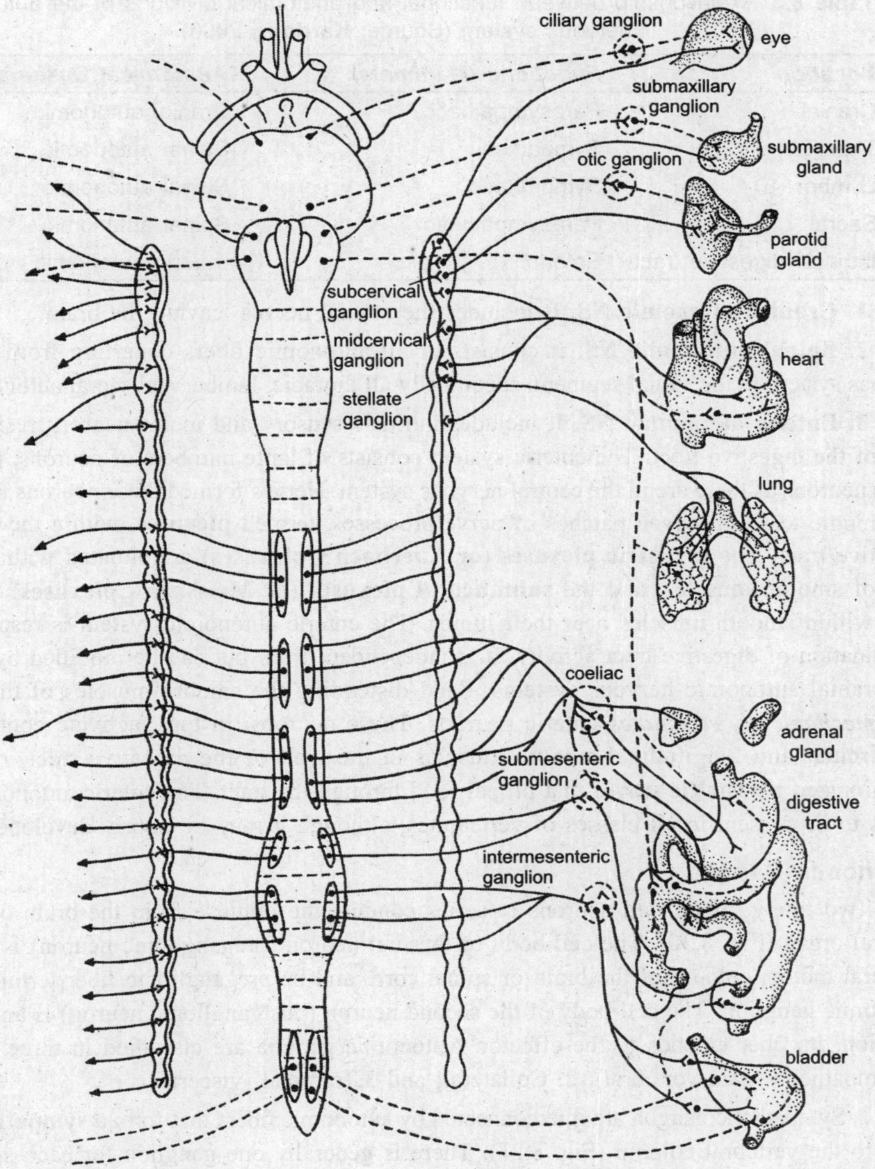

Fig. 8.29. Sympathetic (solid lines)and parasympathetic (dashed lines) systems in a mammal. Note the double innervation of most organs preganglionic and postganglionic fibers are indicated (after Kardong 2006).

contrasting sympathetic and parasympathetic innervation, but these functional divisions do not always correspond to thoracolumbar and craniosacral outflow, respectively, often in nonmammalian vertebrates, the autonomic nerves departing from their regions have a mixed function.

There are *three* anatomical divisions of the autonomic NS: cranial autonomic, spinal autonomic and enteric autonomic systems (Table 8.3).

Table 8.3 Relationship between functional and anatomical divisions of the autonomic nervous system (Source: **Kardong** 2006).

	Location	*Functions (Mammals)*	*Anatomical Designation*
1.	Cranial	Parasympathetic	Cranial autonomic
2.	Thoracic	Sympathetic	Spinal autonomic
3.	Lumber	Sympathetic	Spinal autonomic
4.	Sacral	Parasympathetic	Spinal autonomic
5.	Intrinsic digestive tract	Enteric	Enteric autonomic

1. Cranial autonomic NS. It includes the cranial nerves leaving the brain.

2. Spinal autonomic NS. It consists of all autonomic fibers departing from the central nervous system in the spinal segment, specifically all thoracic, lumber and sacral autonomic fibers.

3. Enteric autonomic NS. It includes intrinsic sensory and motor neurons residing in the wall of the digestive tract. The enteric system consists of large numbers of neurons, probably as many neurons as there are in the central nervous system. Nerves formed these neurons interconnect and mingle to from woven patches of nerve processes, termed **plexuses,** within the wall of the digestive tract. The **myentric plexuses (or Auerbach's plexuses)** are situated within the outer wall of smooth muscles, and the **submucosal plexuses (or Meissner's plexuses)** are located deep within smooth muscles near their lumen. The enteric autonomic system is responsible for coordination of digestive tract activity. It is independent from but can be modified by the spinal and cranial autonomic nervous systems. Food distending the smooth muscles of the digestive tract *mechanically stimulates* enteric neurons. These neurons, in turn, activate contractions of the circular and longitudinal smooth muscles in the wall of the digestive tract, resulting in synchronized **peristaltic waves** that propel food through the tract. The enteric autonomic system seems to be present in all classes of vertebrates, although it may be poorly developed in some.

Autonomic Ganglia

Two finely myelinated neurons in series conduct the impulse from the brain or cord to a visceral organ (Fig. 8.30). The cell body of the first neuron (preganglionic neuron) is in the VM (visceral motor) column of the brain or spinal cord, and its preganglionic fiber terminates in an autonomic ganglion. The cell body of the second neuron (postganglionic neuron) is an autonomic ganglion. Its fiber extends to the effector. Autonomic ganglia are classified in three categories: 1. Sympathetic (para vertebral); 2. Collateral; and 3.Terminal (visceral).

1. Sympathetic ganglia are interconnected by autonomic fibers that form a **sympathetic trunk** close to the vertebral column (Fig. 8.27). There is generally one ganglion for each spinal nerve except in the cervical region, where there are fewer, and in the sacrum, where there are none. The sympathetic ganglia in the trunk are connected to a spinal nerve by white and gray rami communicates (Fig. 8.27). White rami contain preganglionic fibers from the cord and afferent visceral fibers with cell bodies in the dorsal root ganglion. Gray rami constain postganglionic fibers to the skin.

In general, fibers with cell bodies in the cervical ganglion of the sympathetic trunk supply visceral organs in the head, neck and thorax. Fibers with cell bodies in other sympathetic ganglia supply vasomotor, pilomotor, and secretory fibers to the skin via gray rami and spinal nerves (Note. Vaso**motor** supply cutaneous arterioles, **pilomotor fibers** supply muscles that move hairs, and **secretory fibers** supply skin glands). In cold-blooded craniates, the cutaneous fibers also supply some types of chromatophores, and in birds they supply **plumomotor fibers** – muscles

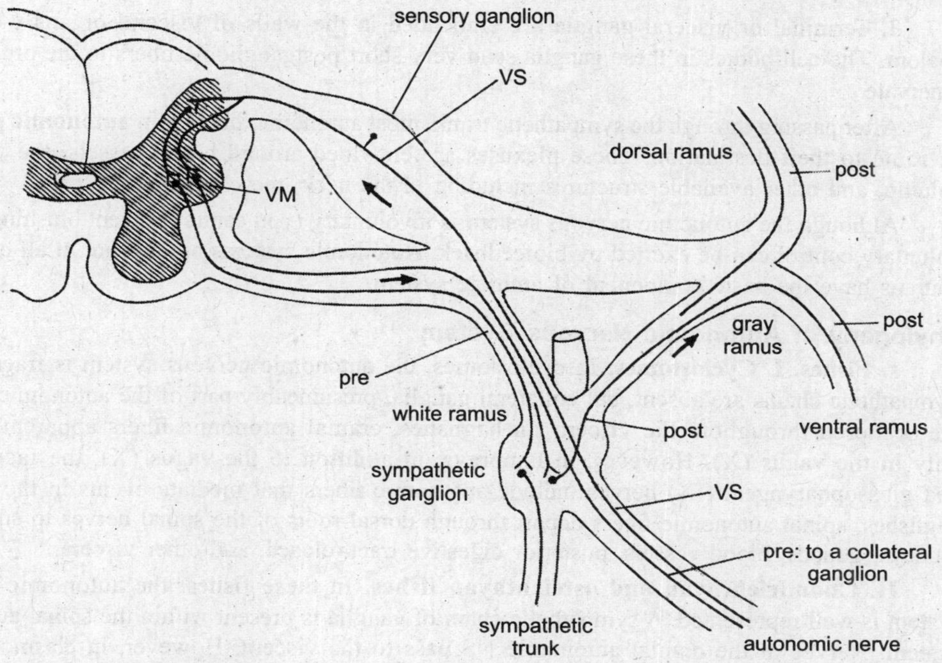

Fig. 8.30. Motor fibers of autonomic nervous system in a thoracic spinal nerve. Fibers in the dorsal and ventral rami are vasomotor, pilomotror or secretory. The motor fiber in the autonomic nerve distal to the ganglion will synapse in a collateral (see splanchinc nerve, Fig. 8.27). VS, visceral sensory fiber, VM, visceral motor cell body in lateral horn; **Post**, **Pre**, postganglionic and preganglionic fiber (after Kent and Carr 2001).

that move feathers. A considerable majority of preganglionic fibers that enter the sympathetic trunk pass through sympathetic ganglia without synapsing, enter an *autonomic nerve* such as the splanchnic, and terminate in the coeliac and other collateral ganglia in the abdominal cavity.

2. Postganglionic fibers with cell bodies in collateral ganglia in the head–**ciliary, submandibular, sphenopalatine (pterygopalatine)** and **otic**–innervate. Visceral effectors in the head. Their neural connections given in Table 8.4. Postganglionic fibers with cell bodies in collateral ganglia in the coelom, such as the coeliac and inferior mesenteric, innervate organs in the abdomen. The collateral ganglia in the abdomen are part of the autonomic plexuses on the major arterial branches of the abdominal aorta.

Table 8.4 Innervation and peripheral distribution of autonomic ganglia of the head of mammals (Source. **Kent** and **Carr** 2001).

	Ganglion	*Receives fibers from*	*Projects fibers to*
1.	Ciliary	Oculomotor nerve	1. Ciliary muscle of eye 2. Constrictor muscles of iris
2.	Submandibular	Facial nerve	1. Submandibular gland 2. Sublingual gland
3.	Sphenopalatine (pterygopalatine)	Facial nerve	1. Lacrimal gland 2. Glands of nose and pharyngeal mucosa
4.	Otic	Glossopharyngeal nerve	1. Parotid gland

3. Terminal or visceral ganglia are embedded in the walls of visceral organs within the coelom. The cell bodies in these ganglia send very short postganglionic fibers to the organs they innervate.

After passing through the sympathetic trunk, most autonomic fibers join **autonomic plexuses** en route to their destination. These plexuses are entwined around blood vessels, the vertebral column, and other available structures including in the neck, the esophagus.

Although the autonomic nervous system is involuntary (you cannot prevent blushing), some voluntary control can be exerted by biofeedback. Autonomic systems are present in all craniates; anurans have the basic component of amniote system.

Phylogeny of Autonomic Nervous System

I. Fishes. I. Cyclostomes. In cyclostomes, the autonomic nervous system is fragmentary. Sympathetic chains are absent, but collateral ganglia, presumeably part of the autonomic system, are scattered throughout the viscera. In hagfishes, cranial autonomic fibers apparently occur only in the vagus (X). However, in lampreys, in addition to the vagus (X), the facial (VII) and glossopharyngeal (IX) nerves include autonomic fibers that mediate events in the gills. In hagfishes, spinal autonomic fibers depart through dorsal roots of the spinal nerves to supply the kidneys, gonads, blood vessels, posterior digestive tract, cloaca, and other viscera.

II. Chondrichthyan and osteichthyan fishes. In these fishes, the autonomic nervous system is well represented. A sympathetic chain of ganglia is present within the spinal autonomic system. Nerves of the cranial autonomic NS pass to the viscera. However, in elasmobranchs, collateral ganglia are absent and sympathetic chains apparently do not contribute fibers to the cranial nerves. The vagus is well developed with branches to the stomach and heart, but the heart apparently lacks a sympathetic counterpart to the inhibitory vagal innervation (Fig. 8.31). Further, the sympathetic ganglia of elasmobranchs are associated with populations of chromaffin cells, which are neural crest derivatives that in teleosts and most tetrapods (except urodeles) become separated from the ganglia (Box 8.2). In most teleosts, collateral ganglia occur, and some spinal nerve fibers are shared with cranial nerves (Fig. 8.31B).

Box 8.2

Chromaffin cell is any cell of the suprarenal organs in lower vertebrates, of the adrenal medulla in mammals, of the paraganglia, or of the carotid bodies that stains with chromium salts.

Tetrapods. The autonomic system is well developed in tetrapods. The paired sympathetic chain is present, collateral ganglia are dispersed among the viscera, and cranial and spinal nerves are well delineated. Autonomic outflow in amphibian spinal nerves passes through the ventral roots, (Fig. 8.32A), but it is still unclear whether motor fibers also occur in the dorsal root (Fig. 8.32B, C).

Concluding remarks. With the exception of cyclostomes, the anatomical organization of the autonomic nervous system is similar in all vertebrate classes. The **spinal autonomic outflow** includes a paired sympathetic chain (except in cyclostomes and elasmobranchs), with some fibers contributing to cranial nerves. The **cranial autonomic outflow** includes the facial (VII), glossopharyngeal (IX), and vagus (X) cranial nerves. In fishes, however, the facial and glossopharyngeal can be reduced. In vertebrates with eyes, the oculomotor (III) cranial nerve may send fibers to the iris and ciliary muscles in the eyes.

In mammals (humans), the autonomic NS is better known, and the circuits for **motor outflow** can be traced with greater confidence. The human autonomic NS includes a sympathetic **thoracolumnber outflow** and a parasympathetic **craniosacral outflow.** The existence of a sacral

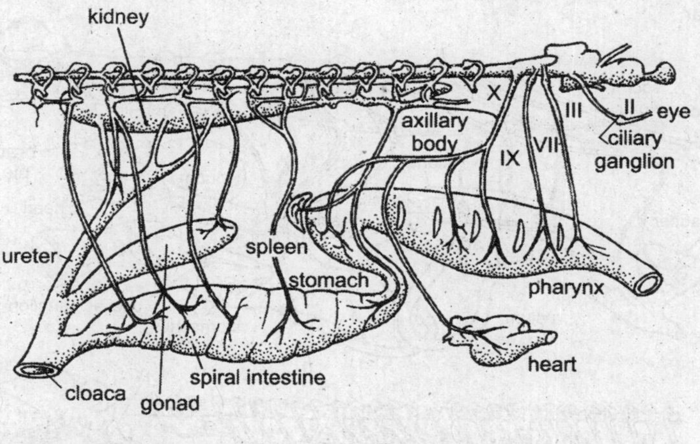

A Elasmobranch (Shark)

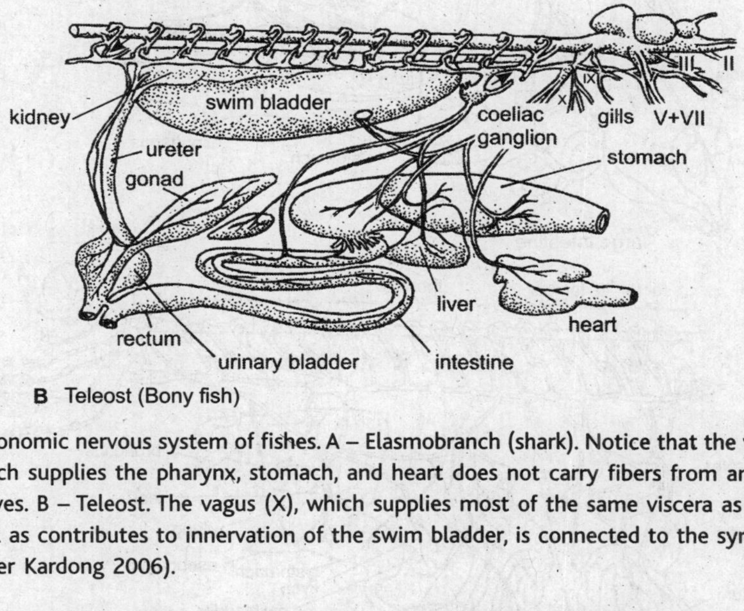

B Teleost (Bony fish)

Fig. 8.31. Autonomic nervous system of fishes. A – Elasmobranch (shark). Notice that the vagus nerve (X), which supplies the pharynx, stomach, and heart does not carry fibers from any of the spinal nerves. B – Teleost. The vagus (X), which supplies most of the same viscera as in the shark as well as contributes to innervation of the swim bladder, is connected to the sympathetic chain (after Kardong 2006).

parasympathetic system in other vertebrates is still uncertain. The pelvic nerves of amphibians arising from the posterior end of the spinal cord and supplying the urinary bladder and rectum have traditionally been regarded as parasympathetic fibers mingle with thoracolumbar sympathetic fibers in the pelvic plexus, making it difficult to trace the posterior circuitry of these two systems on their way to visceral effectors.

Somatic and visceral effectors receive motor information. Effectors and receptors are linked through the CNS. Control of much of the body's activity involve simple reflexes. The somatic reflex arch is primarily involved in controlling skeletal muscles. The visceral reflex arc is the basic component of autonomic NS, which is responsible for monitoring internal visceral action.

8.4 SPINAL CORD

The spinal cord occupies the bony vertebral canal formed by the centra of the vertebrae and the successive neural arches. It is flattened in agnathans but tends to be rounded or quadralateral

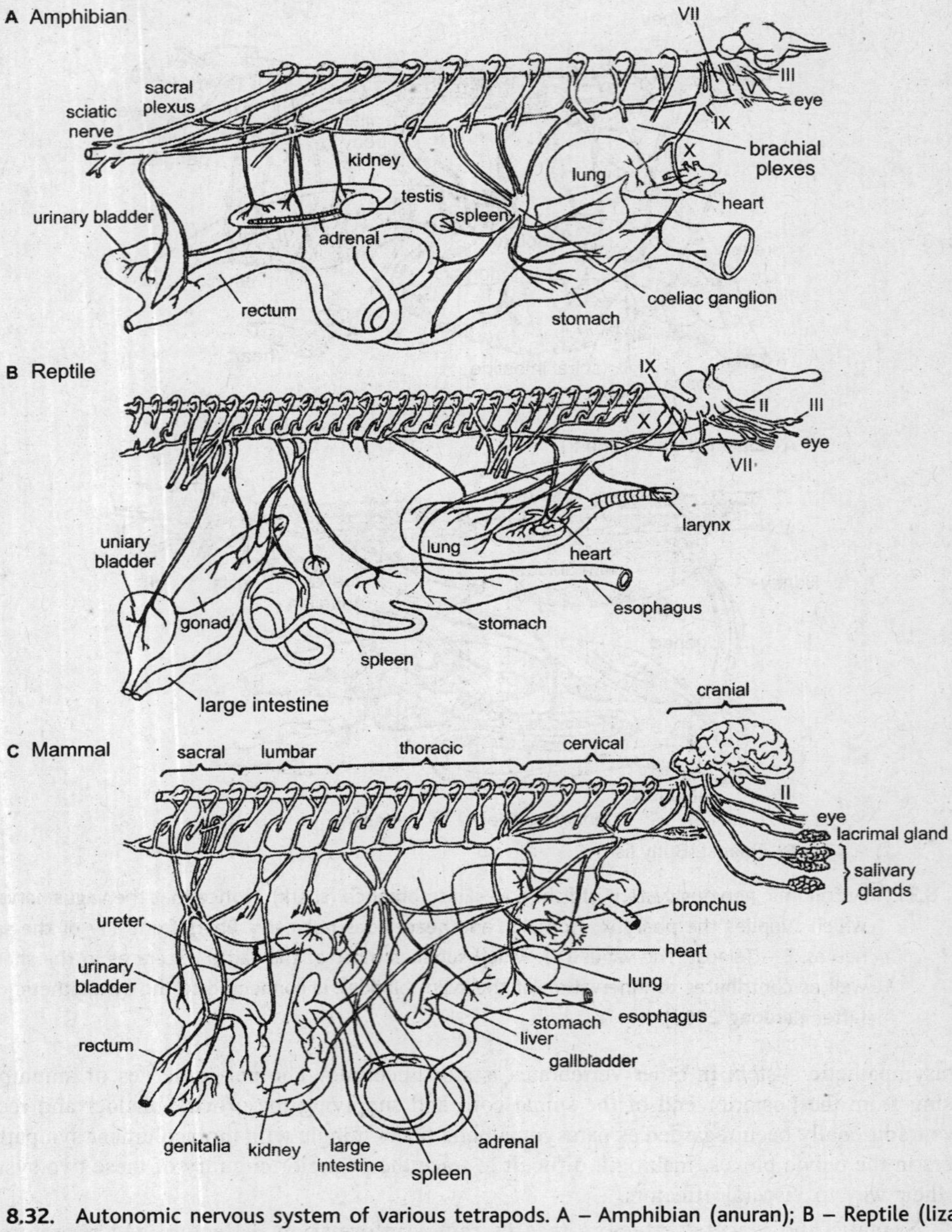

Fig. 8.32. Autonomic nervous system of various tetrapods. A – Amphibian (anuran); B – Reptile (lizard); C – Mammal (placental) (after Kardong 2006).

as a derived condition in some tetrapods. Generally the neurocoel is relatively large in craninate narrower or even obstructed at some levels in mammals.

The spinal cord begins at the foramen magnum, but there is no abrupt transition between cord and brain. Instead there is a gradual rearrangement of the gray matter in the transitional area. In the spinal cord, the gray matter is confined to the area surrounding the neural canal. In the brain stem (*i.e.,* brain without the cerebral hemispheres of forebrain and cerebellum of hind brain); it is dispersed into many separate gray masses, or nuclei, separated by fiber tracts. The transition from spinal cord to brain in mammals is completed over the length of about one body segment (Fig. 8.38). In other craniates the transitional region is longer, and the nerves in the transitional region (*e.g.,* occipitospinal nerves) are atypical.

The vertebrate spinal cord, like the brain, is organized into two regions and named because of their appearance in first preparations (Fig. 8.33). The **gray matter** of the spinal cord includes nerve cell bodies that lie within the core of the spinal cord. Dorsal and ventral extensions of the gray matter are the **dorsal horns** and **ventral horns**, respectively. The dorsal horns contain the bodies of neurons receiving incoming sensory information. The ventral horns contain the bodies of motor neurons (Fig. 8.34). The **white matter** of the spinal cord surrounds the gray matter. It is predominatly composed of nerve fibers linking different levels of the spinal cord with each other and with the brain. Many of these fibers are myelinated, creating their white colour.

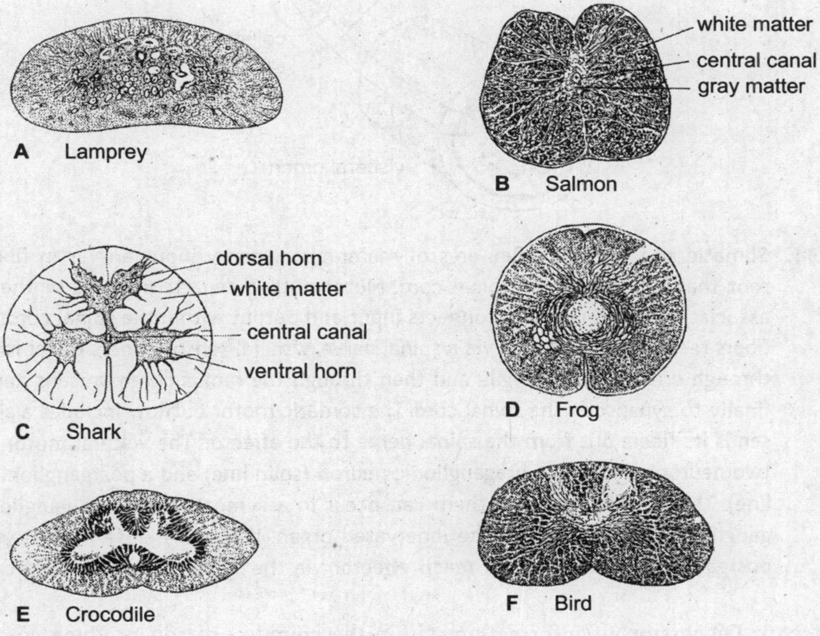

Fig. 8.33. Cross sections of vertebrate spinal cord (after Kardong 2006).

The spinal cord functions in two capacities. It establishes simple reflexes and contain pathways of diverging and converging information.

Spinal Reflexes

The spinal cord completes the reflex loop between sensory input and motor output. In doing this the spinal cord selects the effectors to be activated or inhibited. Although the spinal cord operates at the reflex level, it also contains circuits that coordinate different parts of the cord.

Arriving sensory fibers synapse in the dorsal horn of the gray matter with association neuron (Fig. 8.35). Association neurons carry the impulse to the ventral horn on the same side, to the opposite side, or to a different level of the spinal cord or brain. In the ventral horn, the association neuron synapses with a motor neuron whose axon travels out the ventral root to

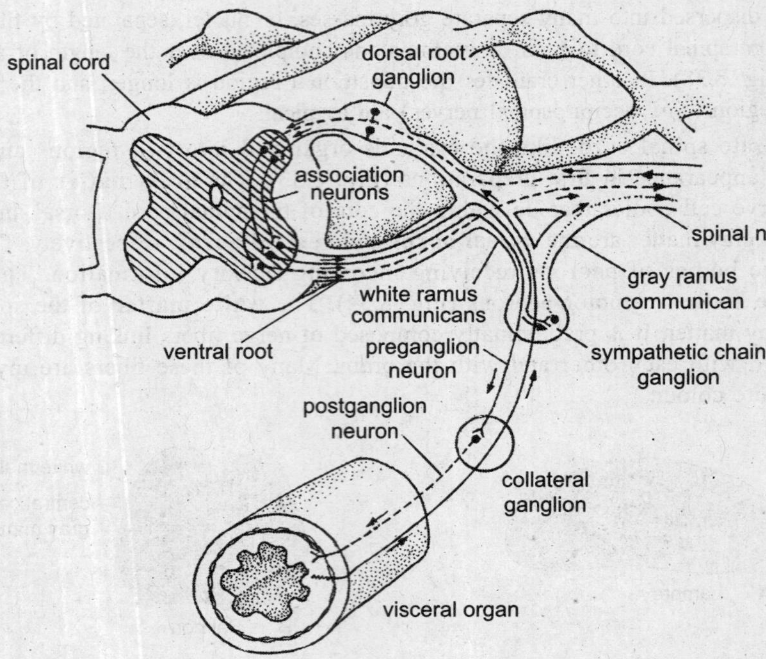

Fig. 8.34. Somatic and visceral reflex arcs of mammals. Sensory input arrives in fibers of the dorsal root that synapse in the spinal cord. Motor output departs in fibers of the ventral root. An association neuron usually connects input and output within the spinal cord. Somatic sensory fibers reach the dorsal root via a spinal nerve. Visceral sensory fibers travel from visceral organ through one or more ganglia and then through the ramus communicans and the dorsal root finally to synapse in the spinal cord. The somatic motor outflow includes a single neuron that sends its fibers out from the spinal nerve to the effector. The visceral motor outflow includes two neurons in series: a preganglionic neuron (solid line) and a postganglionic neuron (dotted line). The synapse between them can occur in a sympathetic chain ganglion, in a collateral ganglion, or in the wall of the innervated organ. If they synapse in the sympathetic chain, postganglionic fibers usually reach effector via the spinal nerve (after Kardong 2006).

the effector. Dispersion of information within the spinal cord can produce complex responses to stimuli without involving higher centers. For example, if an animal should inadvertently place its feet on a sharp object, the reflex to withdraw it could involve as few as **three** neurons (Fig. 8.36). The first, the *afferent sensory* **neuron,** carries the painful stimulus to the spinal cord, where it synapses with an *association neuron*. The association neuron transmits the stimulus to the ventral horn, where it synapses with a *motor neuron* whose axon conveys the impulse to appropriate retractor muscles that contract and withdraw the foot. Association neurons that connect to appropriate levels on the opposite side of the cord reach motor neurons innervating extensor muscles in the opposite leg. These muscles contract, extend the leg and prevent the animal's collapse when it lifts its other leg from the sharp object. The circuitry involved demands a connection between painful stimulus and appropriate effectors (retractor and extensor muscles). It needs not involve higher brain centers. Usually, association neurons also convey the painful stimulus to conscious centers of the brain, where it is perceived (Fig. 8.35B); however, by the time the higher centers become aware of the surprise trauma to the foot, the spinal reflex to retract it is already underway.

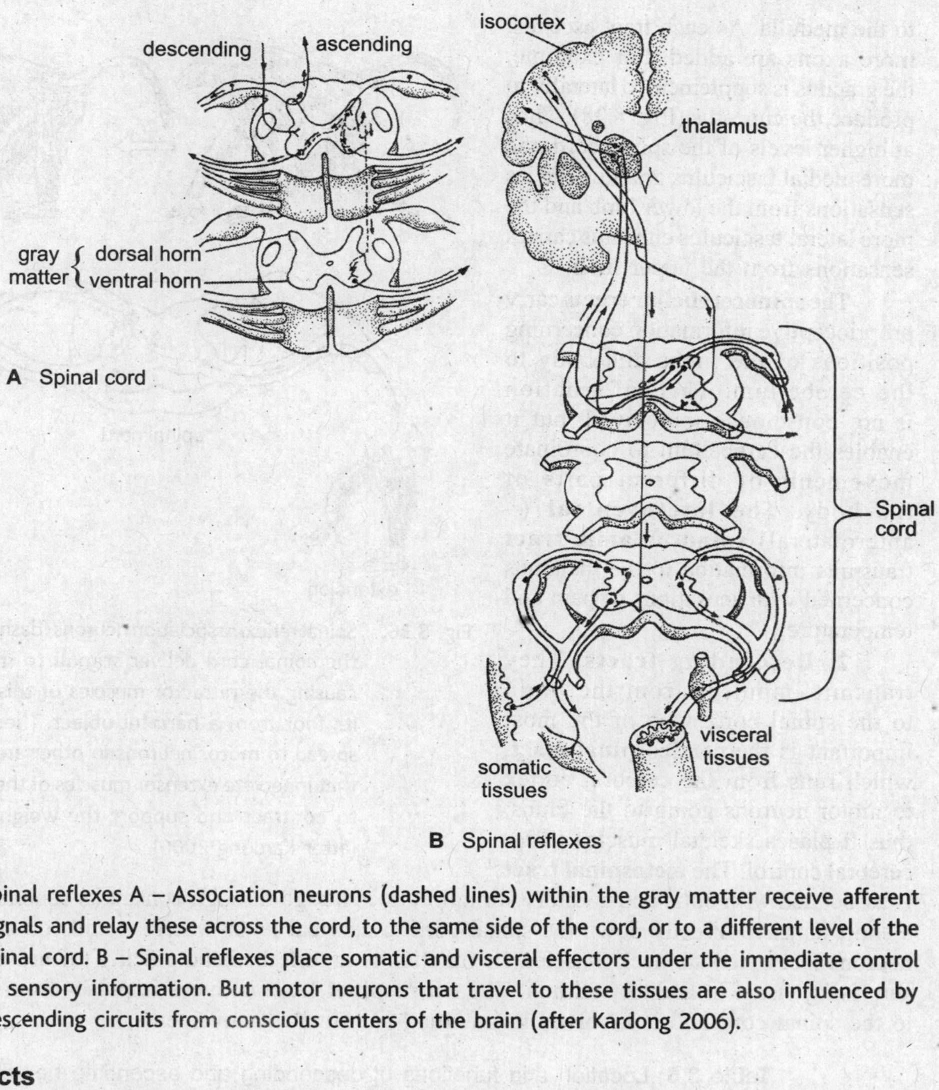

Fig. 8.35. Spinal reflexes A – Association neurons (dashed lines) within the gray matter receive afferent signals and relay these across the cord, to the same side of the cord, or to a different level of the spinal cord. B – Spinal reflexes place somatic and visceral effectors under the immediate control of sensory information. But motor neurons that travel to these tissues are also influenced by descending circuits from conscious centers of the brain (after Kardong 2006).

Spinal Tracts

Not all information is processed at the level of the spinal cord. Much, perhaps most, information is carried to higher levels of the nervous system for evaluation. The resulting decisions are carried down the spinal cord to appropriate effectors. Nerve fibers carrying similar information tend to travel together in **nerve tracts**, bundles of similar fibers that may be ascending or descending tracts, depending on whether they convey information up or down the cord, respectively (Fig. 8.37). They are usually named for their source and their destination. For example, the spinothalamic tract begins in the spinal cord and extends the thalamus (Table 8.5).

Wars, accidents, and diseases can lead to localized wounds of the spinal cord that sever the ascending or descending flow of information. In humans, such losses of function have been correlated with the specific region in which the wound occurred and used to map the positions of these nerve tracts.

1. Ascending tracts. They carry sensory impulses from the spinal cord to the brain. Among the most prominent are the **fasciculus gracilus** and **fasciculus cuneatus**, located in the dorsal region of the spinal cord. Both carry **proprioceptive stimuli** and sensations associated with **posture**

to the medulla. As each tract ascends, more axons are added. For example, the gracilus is supplemented laterally to produce the cuneatus (Fig. 8.38). Thus at higher levels of the spinal cord, the more medial fasciculus gracilus carries sensations from the lower limb and the more lateral fasciculus cuneatus carries sensations from the upper limb.

The **spinocerebellar tracts** carry proprioceptive information concerning positions of the limbs and body to the cerebellum. This information is not consciously perceived, but it enables the cerebellum to coordinate movements of different parts of the body. The **lateroventral** (= anterolateral) **spinothalamic tract** transmits information to the thalamus concerned with sensations of pain and temperature.

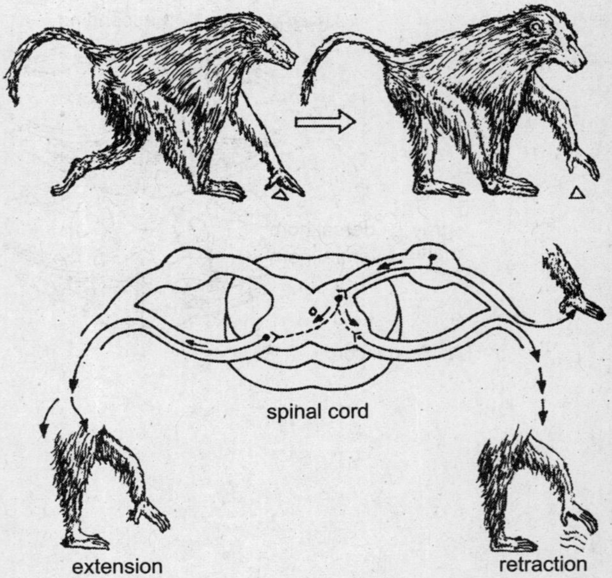

Fig. 8.36. Spinal reflex. Association neurons (dashed lines) within the spinal cord deliver stimuli to motor neurons, causing the retractor muscles of this animal to lift its foot from a harmful object. These stimuli also spread to motor neurons in other areas of the cord that innervate extensor muscles of the opposite limb to contract and support the weight of the body (after Kardong 2006).

2. Descending tracts. They transmit impulses from the brain to the spinal cord. One of the most important is the **corticospinal tract**, which runs from the cerebral cortex to motor neurons going to the limbs; thus it places skeletal muscles under cerebral control. The **tectospinal tract** is associated with optic and auditory stimuli. It does not go through conscious centers, but extends a short distance directly down the spinal cord to cervical levels where it ends on somatic motor neurons that innervate neck muscles. Its function is to turn the head quickly toward threatening or surprising visual or auditory stimuli. The **rubrospinal tract** conveys impulses from the midbrain to the spinal cord and is involved in initiating coordinated movements.

Table 8.5 Location and functions of descending and ascending nerve tracts of the spinal cord (Source: **Kardong** 2006)

	Tract	*Source*	*Destination*	*Function*
I.	**Descending tracts**			
1.	Lateral and ventral corticospinal tracts	Cerebral cortex	Spinal cord	Motor connections direct from cortex to primary motor neurons of arms and legs (*i.e.*, places motor neurons under direct voluntary cortical control)
2.	Rubro – spinal tract	Midbrain (*i.e.*, red nucleus of tegmentum)	Spinal cord	Motor connections in spinal cord
3.	Lateral and ventral reticulo – spinal tract	Medulla reticular formation	Spinal cord (dorsal horn)	Postural reflexes

	Tract	Source	Destination	Function
4.	Tectospinal tract	Midbrain (colliculus, roof)	Spinal cord	Visual and auditory stimuli to limbs and trunk
5.	Vestibulo – spinal tract	Medulla (vestibular nucleus)	Spinal cord	Postural reflexes accomplished by axial and limb musculature
II.	**Ascending tracts**			
1.	Fasciculus gracilis and fasciculus cuneatus	Spinal cord	Medulla	Fine tactile information, Golgi tendon organs, Pacinian corpuscles in joints
2.	Dorsal and ventral spinocerebral tracts	Spinal cord	Cerebellum via peduncle	Proprioceptive information from muscles to cerebellum, intrafusal muscle fibers
3.	Latero – ventral spinothalamic tract	Spinal cord	Thalamus	Pain and temperature sensations to thalamus
4.	Spinotectal tract	Spinal cord	Midbrain (tectum)	Proprioceptive information from neck and shoulders
5.	Spinoreticular tract	Spinal cord	Medulla (reticular formation)	Pain and sensations from internal organs

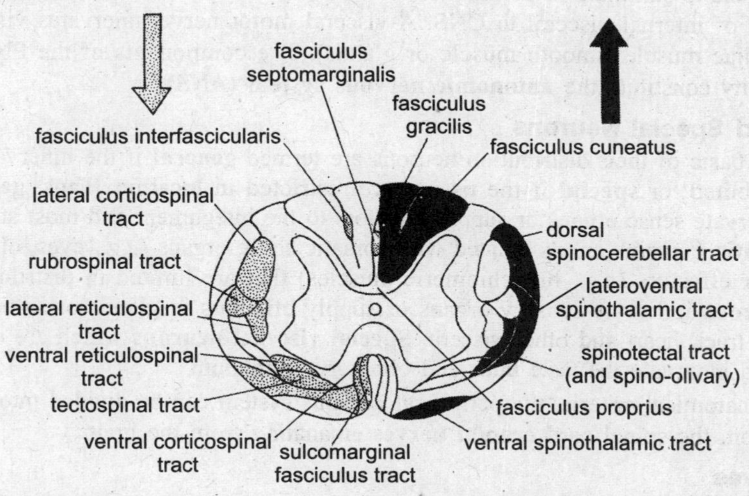

Fig. 8.37. Cross section of the human spinal cord showing approximate locations of ascending (right) and descending (left) nerve tract (after Kardong 2006).

8.5 PERIPHERAL NERVOUS SYSTEM

Peripheral nerves serve either somatic or visceral tissues and carry sensory or motor information (Fig. 8.8). **Somatic nerves** pass to or from somatic tissues such as skeletal muscles, skin and their derivatives. **Visceral nerves** pass to or from viscera such as involuntary muscles and glands. Nerves carrying information from tissues to the central nervous system are **afferent, or sensory, neurons**. Nerves carrying information away from the CNS to effectors are **efferent, or motor, neurons**. Thus, a *somatic sensory nerve* might carry information about touch, pain or

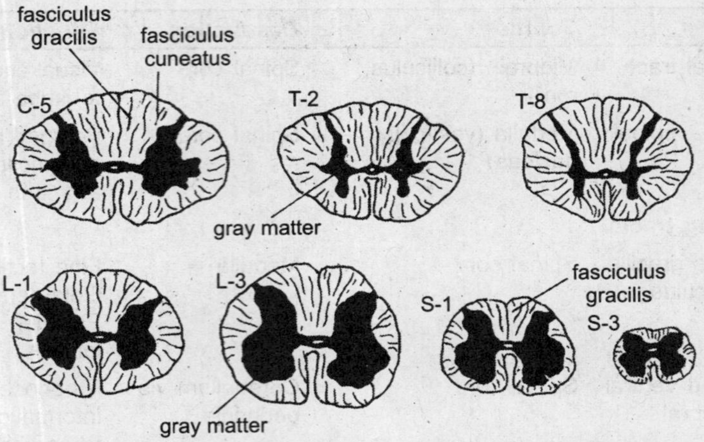

Fig. 8.38. Gray and white matter at various level of the human spinal cord. Sections are identified by their region – cervical (C), thoracic (T), lumber (L) and sacral (S) – and by specific vertebrae numbered (arabic number) within each of these regions from which they came. At the levels of the arms (C-5) and legs (L-3), additional sensory and motor fibers enter and leave the spinal cord. This is reflected in more extensive gray matter compared to other regions of the cord (e.g., T-2, T-8). Note the addition of the fasciculus cuneatus at the highest level on the cord. Primarily, this carries sensory information from the arms (after Kardong 2006).

temperature from the skin to the CNS. A *somatic motor nerve* carries impulses from the CNS to a striated muscle to stimulate its contraction. A *visceral sensory nerve* delivers information about the condition of internal visceral to CNS. A visceral motor nerve innervates visceral effectors (such as cardiac muscle, smooth muscle or glands). The components of the PNS that control visceral activity constitute the **autonomic nervous system (ANS)**.

General and Special Neurons

On the basic of their distribution neurons are termed **general** if the innervated tissues are widely distributed, or **special** if the tissues are restricted in location. Thus, **general somatic neurons** innervate sense organs or supply effectors to the integument and most striated muscles. **Special somatic neurons** are associated with somatic sense organs (*e.g.,* eyes, olfactory organs, inner ears) or effectors (*e.g.,* branchiomeric muscles) that are limited in distribution. **General visceral neurons** innervate sensory organs or supply effectors in glands or smooth muscles of the digestive tract, heart and other viscera. **Special visceral neurons** which are concerned with sensory input, innervate the taste buds and olfactory epithelium.

From anatomical criteria, the peripheral nervous system can be divided into **spinal nerves** emanating from the spinal cord **cranial nerves** emanating from the brain.

Spinal Nerves

Spinal nerves are sequentially arranged and numbered (C-1, T-1, L-1, S-1) according to their association with regions of the vertebral column (cervical, thoracic, lumber, sacral). Early anatomists recognized **dorsal** and **ventral roots** of each spinal nerve. Afferent fibers enter the spinal cord via the dorsal root, and efferent fibers leave by way of the ventral root. The dorsal root ganglion, a swelling in the dorsal root, is a collection of neuron bodies whose axons contribute the spinal nerves (Fig. 8.39).

Parallel to the spinal cord and attached to each spinal nerve through the **rammus communicans** is the **sympathetic chain** of ganglia (paravertebral ganglia), a paired series of linked ganglia adjacent to the vertebral column or notochord (Fig. 8.39A, B). Other peripheral ganglia form the **collateral ganglia** (pre-vertebral ganglia). The paired **cervical, coeliac** and **mesenteric ganglia** are examples

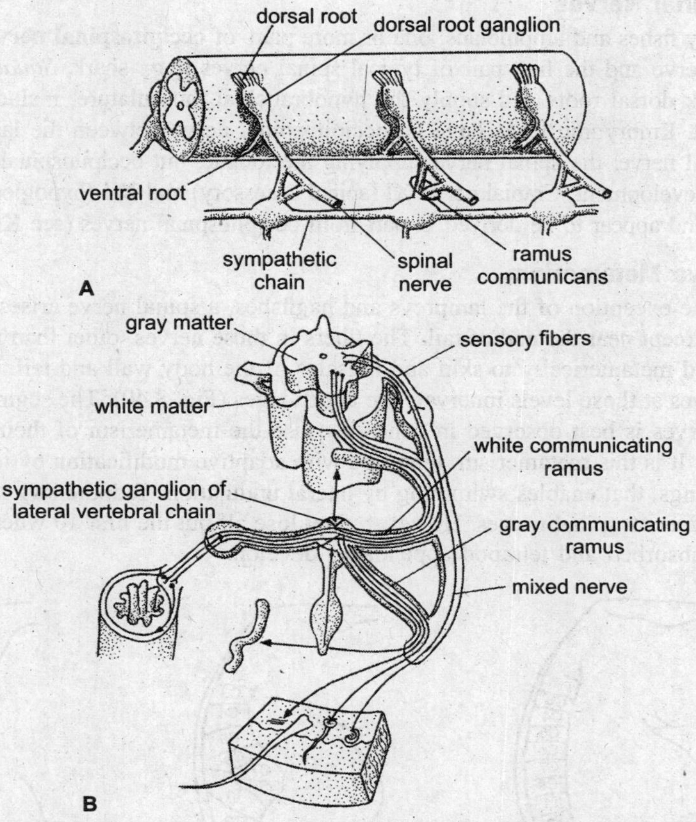

Fig. 8.39. Spinal nerve anatomy. A – Dorsal and ventral roots connect spinal nerves to the spinal cord. A dorsal root is enlarged into a dorsal root ganglion. Spinal nerves join with the sympathetic chain through communicating rami. B – Configuration of sensory and motor neuronal routes in an adult mammal (after Kardong 2006).

of the collateral ganglia. The **visceral ganglia** occur within the walls of visceral effector organs (Fig. 8.39B). Thus, there are three types of ganglia: sympathetic, collateral and visceral.

Peripheral nerves in the trunk arise during embryonic development from *two* sources. 1. One source is the neurons that differentiate within the spinal cord. Axonal processes sprout from these neurons and grow outward to the ganglia or to the effectors they supply. 2. The other source is the **neural crests** to specific locations and sprout processes that grow back to the central nervous system and out to the tissues they innervate. The *ventral root* arise from neurons in the spinal cord that send fibers out from the spinal cord. The *dorsal root* arises from cells of neural crest origin that send fibers into the spinal cord. In gnathostomes, the two roots usually merge to form the composite spinal nerve and attached sympathetic chain. In lampreys roots of spinal nerves do not unite. However, in hagfishes they unite in the trunk but not in the tail.

The fibers of each spinal nerve innervate restricted structures at the level of the cord. This is especially pronounced with spinal nerve innervation of somatic tissues. Each growing spinal nerve tends to accompany its adjacent embryonic **myotome**, the source of somatic muscles, and its **dermatome**, the source of dermal connective tissue and muscle, as they spread and differentiate during development. Once it is differentiated, a spinal nerve supplies the skeletal muscles derived from its adjacent myotome and receives somatic sensory input from the restricted area of the body surface differentiated from its dermatome.

Occipitospinal Nerves

In many fishes and amphibians, one or more pairs of **occipitospinal nerves** arise between the last cranial nerve and the first pair of typical spinal nerves (*e.g.,* shark, *Squalus acanthias*). They generally lack dorsal roots and supply the hypobranchial musculature, including tongue muscles when present. Embryonic frogs have an occipitospinal nerve between the last cranial nerve and the first spinal nerve; the spinal nerve supplying the tongue, but occipitospinal nerve is suppressed during later development. Cranial nerve XI (spinal accessory) and XII (hypoglossal) of amniote lack dorsal roots and appear to be derived in part from occipitospinal nerves (see **Kent** and **Carr** 2001).

Spinal Nerve Metamerism

With the exception of the lampreys and hagfishes, a spinal nerve arises from each segment of the cord except near the end of tail. The fibers in these nerves, other than fibers to the viscera, are distributed metamerically to skin and muscles of the body wall and tail. Where fins or limbs develop, nerves at those levels innervate the appendages (Fig. 8.40). The segmental distribution of the spinal nerves is best observed in fishes because the metamerism of their body wall muscles is so evident. It is this metamerism, inherited with adaptive modification by tetrapods from fishes to human beings, that enables swimming by lateral undulation. Tadpole swim like fishes and they have as many as 40 spinal nerves. However, they lose all but the first 10 when at metamorphosis, the tail is reabsorbed and tetrapod appendages develop.

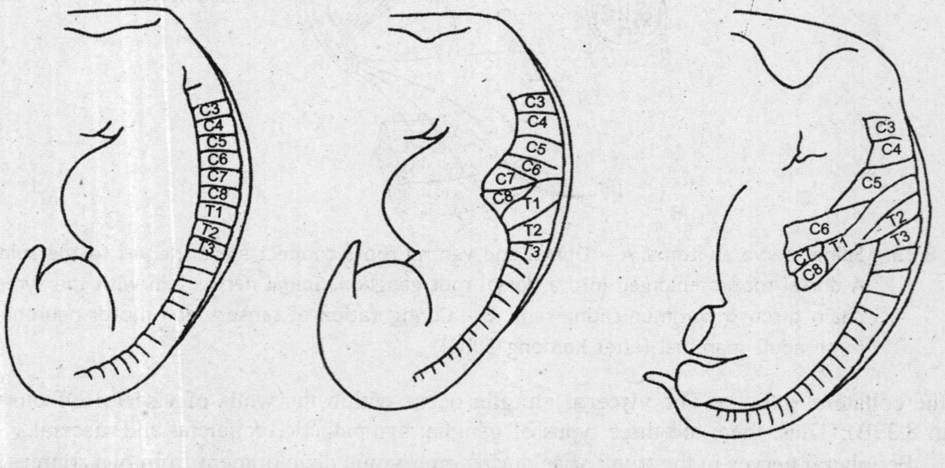

Fig. 8.40. Metameric innervation of the skin of a mammalian forelimb. C, cervical, and T, thoracic somites and the skin areas supplied by their associated spinal nerves (after Kent and Carr 2001).

Rami and Plexuses

Shortly after emerging from the vertebral canal, spinal nerves divide into two more branches (Fig. 8.41). A **dorsal ramus** supplies the epaxial muscles and skin to the midventral raphe. In the thoracic and lumber region of mammals, **white** and **gray rami communicates** (sing. ramus communicans) connect the spinal nerves with the ganglia of the sympathetic trunk (Fig. 8.41). These rami conduct impulses that have visceral function.

The ventral rami of successive spinal nerves of tetrapods unite to from **spinal nerve plexuses** which are sites where two or more nerves or branches of nerves converge and their fibers commingle in one or more common trunks to being distributed. These plexuses are relatively simple in fishes but become increasingly complex in tetrapods. The chief spinal nerve plexuses are the **brachial** and **lumber** (**cervicobrachial** and **lumbosacral** in tetrapods that have cervical, thoracic, lumber and sacral vertebrae). Nerves emerge from spinal nerve plexuses to supply the

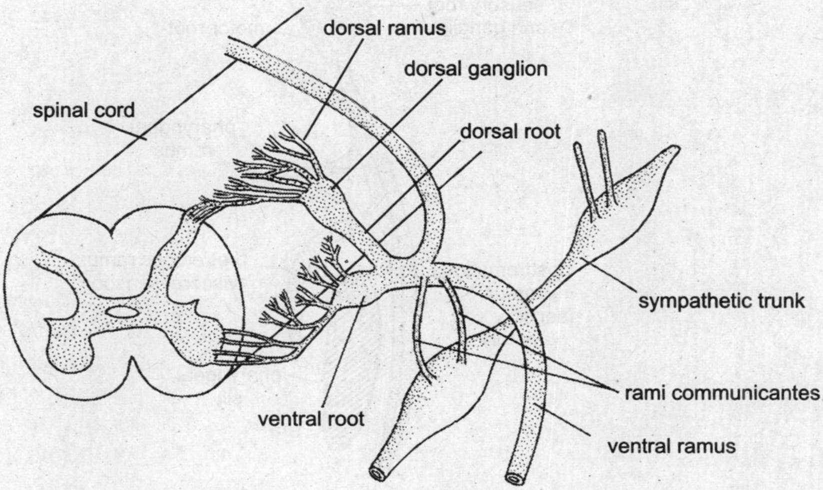

Fig. 8.41. Primary branches of thoracic and lumber spinal nerves in mammals (after Kent and Carr 2001).

fins and limbs. Highly complex spinal nerve plexuses with fibers that also supply pelvic viscera are present in the plexus of mammals including humans.

Cranial Nerves

Cranial nerves have roots enclosed in the braincase. Most cranial nerves are named and numbered by roman numerals from anterior to posterior (Box 8.3) The conventional system for numbering these nerves is sometime inconsistent. For example, most anamniotes are said to have ten numbered cranial nerves (I, II and so on) plus six pairs of unnumbered (*e.g.,*anterior lateral-line or ALL) lateral line cranial nerves. A few anamniotes and all amniotes are said to have 12 cranial nerves.

Box 8.3

Cranial nerves can be presented numerically, but this neglects the important developmental and functional aspects of these nerves. A full understanding of cranial nerves is achieved only with an appreciation of development. This provides information on the unique origin of some nerves and composite nature of other. Functional categories of nerves are **sensory**, **motor** and **mixed** cranial nerves.

Cranial nerves or their components are derived from three embryological tissues; neurectoderm from the brain, neural crest, and\or ectodermal placodes. Two nerves represent outgrowths of nervous system: the retina of the eye and the epiphyseal complex (pineal and parapineal). Ectodermal placodes contribute the ganglion cells of the terminal (o), olfactory (1), vomeronasal (VM), anterior and posterior lateral - line (*i.e.*, ALL and PLL) and octaval or vestibulo- cochlear (VIII) nerves and the taste bud components of cranial nerves VII (facial), IX (glossopharyngeal) and X (vagus). Neural crest forms the ganglion cells associated with cranial nerves V (trigeminal), VII (facial), IX (glossopharyngeal) and X (vagus) (see Kent and Carr 2001).

Like spinal nerves the cranial nerves supply somatic and visceral tissues and carry general sensory and motor information. Some cranial nerves consists of only sensory or only motor fibers; other nerves are mixed, containing both types. Cranial nerves concerned with localized senses (*e.g.,* sight, hearing, lateral line olfaction, taste) are called **special cranial nerves** to distinguish them from those concerned with the sensory or motor innervation of the more widely distributed viscera, called **general cranial nerves.**

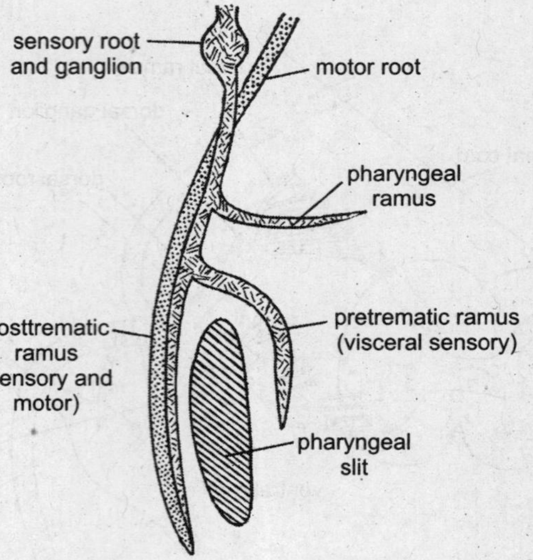

Fig. 8.42. Components of a cranial nerve in a fish. The pharyngeal ramus, to the lining of the pharynx, and the small pretrematic ramus to the front of the pharyngeal slit both carry visceral sensory fibers. The dorsal ramus from the skin is composed of somatic sensory fibers. The posttrematic ramus running down the back of pharyngeal slit includes both sensory and motor fibers. Rostral is to right of the figure.

Primitively all cranial nerves serving the branchial pouches are formed of three branches per pouch: **pretrematic, posttrematic** and **pharyngeal** (Fig. 8.42). In amniotes, these tend to be lost or their homologies become uncertain.

Cranial Nerves of Mammals and other Vertebrates

1. Nervous terminalis (O). The **terminal nerve** may be testimony to an ancient anterior head segment that has been lost. The terminal nerve is a nerve or perhaps a complex of nerves that arises from olfactory placodes. It is present in all classes of gnathostomes except birds (Fig. 8.43). It runs to blood vessels of the olfactory epithelium in the olfactory sac and carries *visceral sensory* and some *motor* fibers. A role in reproduction has also been suggested.

2. Olfactory nerve (I). The **olfactory nerve** is a **sensory nerve** concerned with the sense of smell. Olfactory cells lie in the mucous membrane of the olfactory sac. A short axon leads from each cell to the olfactory bulb (Fig. 8.44). Each axon constitutes an olfactory fiber. Collectively, the olfactory fibers form the short olfactory nerve, which is the only cranial nerve composed of axons of the receptor cells themselves.

3. Optic nerve (II). Strictly speaking the **optic nerve** is a nerve but a **sensory tract** (see **Kardong** 2006). That is, collection of fibers in the CNS. However, once it develops as an outpocketing of the brain. However, once it is differentiated it lies outside the brain. Its fibers synapse in the thalamus and midbrain.

4. Oculomotor nerve (III). The **oculomotor nerve**, primarily supply extrinsic eye muscles (superior rectus, medial rectus, inferior rectus, and inferior oblique muscles) derived from preotic myotomes. It is a **motor nerve** that also carries a few visceral motor fibers to iris and ciliary body of the eye. Fibers arise in the oculomotor nucleus in the floor of the midbrain.

5. Trochlear nerve (IV). The **trochlear nerve** is a **motor nerve** that supplies the extrinsic superior oblique eye muscle. Fibers arise in the trochlear nucleus of the midbrain.

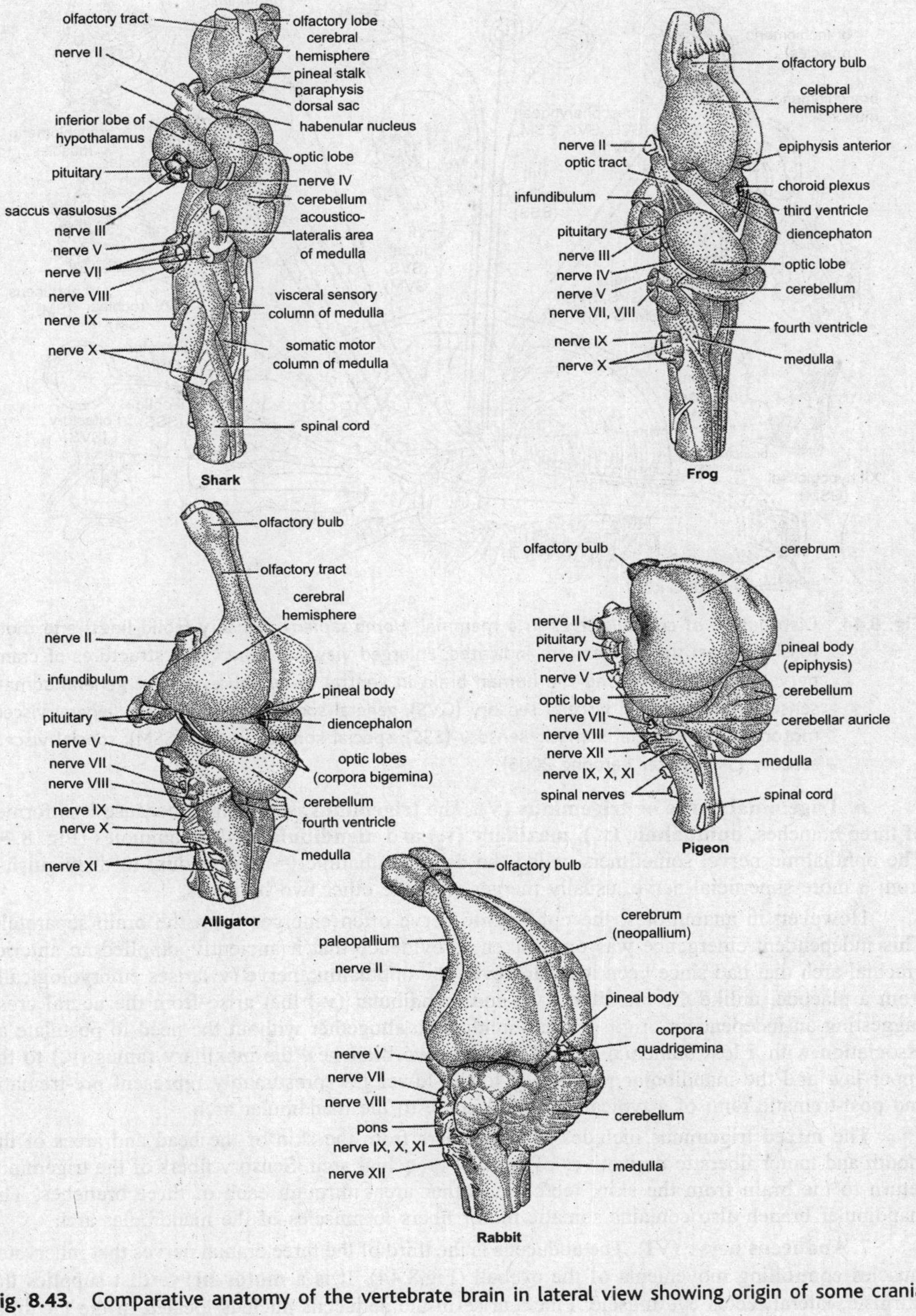

Fig. 8.43. Comparative anatomy of the vertebrate brain in lateral view showing origin of some cranial nerves.

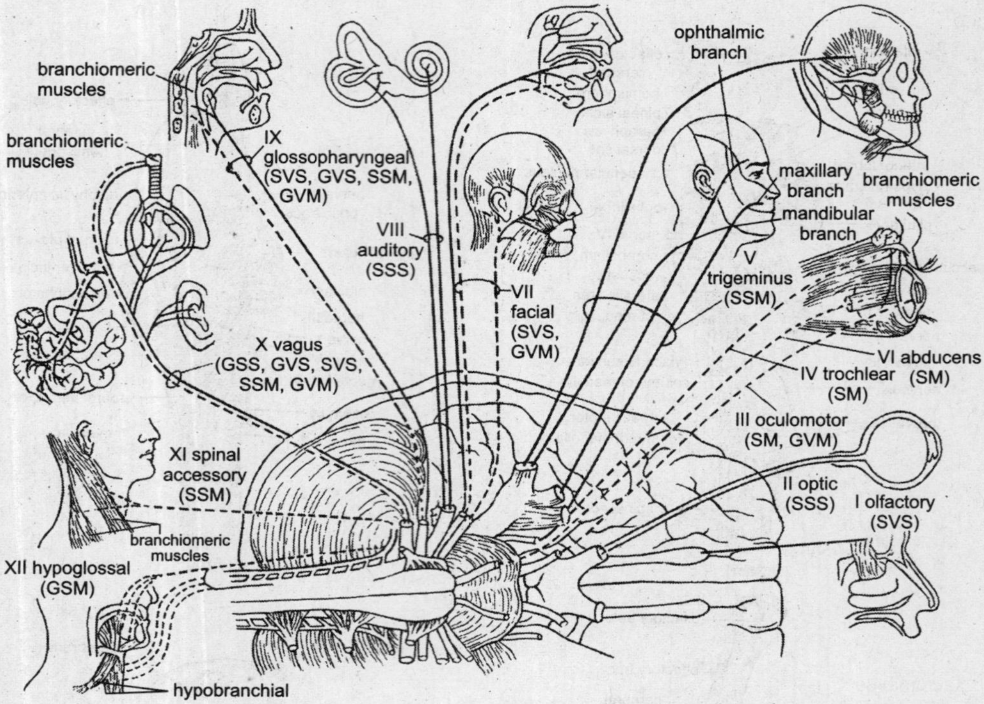

Fig. 8.44. Distribution of cranial nerves in a mammal, *Homo sapiens*. Sensory (solid lines) and motor (dashed lines) nerve fibers are indicated. Enlarged views of innervated structures of cranial nerves are shown around the human brain in ventral view. Abbreviations: general somatic sensory (GSS), general visceral sensory (GVS), general somatic motor (GSM), general visceral motor (GVM), special somatic sensory (SSS); special somatic motor (SSM), special visceral sensory (SVS) (after Kardong 2006).

6. Trigeminal nerve or trigeminus (V). The **trigeminus** is so named because it is formed of three branches: **ophthalmic (v_1)**, **maxillary (v_2)** and **mandibular (v_3)** in amniotes (Fig. 8.44) The ophthalmic nerve, sometimers called the deep ophthalmic (= profoundus) to distinguish it from a more superficial nerve, usually merges with the other two branches.

However, in anamniotes, the ophthalmic nerve often emerges from the brain separately. This independent emergence was once taken as evidence that it anciently supplied an anterior brachial arch that had since been lost. However, the ophthalmic nerve (v_1)arises embryologically from a placode, unlike the maxillary (v_2) and mandibular (v_3) that arise from the neural crest, suggesting an independent origin for the ophthalmic altogether without the need to postulate an association with a lost branchial arch. The other two branches, the maxillary ramus (v_2) to the upper jaw and the mandibular ramus (v_3) to the lower jaw presumably represent pre-trematic and post-trematic rami of a typical branchial nerve to the mandibular arch.

The **mixed** trigeminus includes sensory fibers from the skin of the head and areas of the mouth and motor fibers to derivatives of the first branchial arch. Sensory fibers of the trigeminus return to the brain from the skin, teeth, and other areas through each of three branches. The mandibular branch also contains somatic motor fibers to muscles of the mandibular arch.

7. Abducens nerve (VI). The abducens is the third of the three cranial nerves that innervates muscles controlling movements of the eyeball (Fig.8.44). It is a **motor nerve** that supplies the extrinsic, lateral rectus eye muscle. Fibers arise in the abducens nucleus located in the medulla.

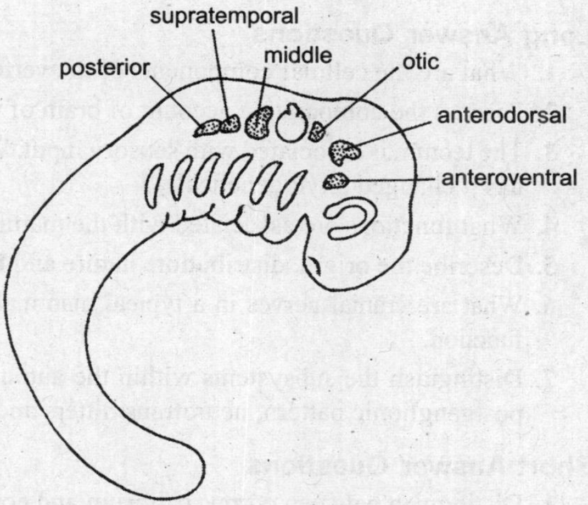

Fig. 8.45. Embryonic development of cranial nerves. The lateral line cranial nerves arise from the dorsolateral placodes (after Kardong 2006).

8. Facial nerve (VII). The **facial nerve** is a **mixed nerve;** it includes sensory fibers from taste buds as well as motor fibers to derivatives of the second (hyoid arch). This nerve also carries a substantial number of somatic sensory fibers to the skin. In fishes, the skin of entire opercular complex is innervated by the facial nerve.

9. Octavel (= auditory) nerve (VIII). The sensory **octavel nerve** (is also known as **acoustic vestibulocochlear** or **statoacoustic** nerve). It carries **sensory** fibers from the inner ear which is concerned with balance and hearing. The nerve synapses in several regions of the medulla.

10. Glossopharyngeal nerve (IX). The mixed **glossopharyngeal nerve** supplies the third branchial arch. It contains **sensory** fibers from branchial arch. It contains sensory fibers from the taste buds, the first gill pouch, and the adjacent pharyngeal lining. **Motor** fibers innervate muscles of the third branchial arch.

11. Vagus nerve (X). The term **vagus** is Latin for wandering and aptly applies to this mixed nerve. The vagus meanders widely, serving areas of the mouth, pharynx, and most of the viscera. It is formed by the union of several roots across several head segment. Occasionally additional nerves of the lateral line merge with the vagus.

12. Spinal accessory nerve (XI). In anamniotes, the **spinal accessory nerve** is probably composed of a branch of the vagus nerve and several occipitospinal nerves. In amniotes, especially in birds and mammals, it is a small but distinct **motor nerve** that supplies derivatives of the cucullaris muscle (cleidomastoid, sternomastoid, trapezius). A few of its fibers accompany the vagus nerve to supply part of the pharynx and larynx and perhaps the heart. Fibers arise from several nuclei within the medulla.

13. Hypoglossal nerve (XII). The hypoglossal nerve is a **motor nerve** that innervates hyoid and tongue muscles. Fibers originate in the hypoglossal nucleus within the medulla.

14. Lateral line nerves. In addition to formally numbered cranial nerves, fishes possess preotic and postotic **lateral line craninal nerves** that are rooted in medulla and supply the lateral line system (Fig. 8.45). They were thought to be components of facial (VII), glossopharyngeal (IX) and vagal (X) nerves but they are now recognized as independent cranial nerves, derivatives of the dorsolateral placodes (octavolateralis placodes). Unfortunately this late recognition as distinct cranial nerves has left them without any identifying roman numerals. In most jawed fishes and some amphibians, upto six pairs of lateral line nerves are present. Three are preotic (rostral to the otic vesicle, future inner ear): the anterodorsal, anteroventral, and one otic lateral line nerves; three postotic: the middle, supratemporal, and posterior lateral line nerves. Because the lateral line and the octaval nerves share organizational similarities they are often spoken of as the **octavolaterlis system** or **acousticolateralis**.

QUESTIONS

Long Answer Questions

1. What are the cellular components of the vertebrate nervous system? Describe them.
2. Discuss the comparative account of brain of vertebrates.
3. The tectum is associated with sensory input. What senses are processed in the tectum, and how has it changed phylogenetically?
4. What functions are associated with the mammalian neocortex?
5. Describe the origin, distribution, nature and function of cranial nerves in a vertebrate.
6. What are cranial nerves in a typical mammal? List them by abbreviation, name, and primary function.
7. Distinguish the subsystems within the autonomic nervous system. Include location, pre-and postganglionic pattern, neurotransmitters and function.

Short Answer Questions

1. Distinguish between corpus callosum and corpus striatum.
2. Draw a typical thoracic spinal nerve in a mammal. Include rootlets, roots, rami, sympathetics trunk, and trace in the four functional groups: VM, VS, SM, and SS.
3. Write short note on the following
   (i) Pineal eye
   (ii) Autonomic nervous system
   (iii) Neuron
   (iv) Synapse
   (v) Neuroglia
   (vi) Cerebellum
   (vii) Habenulae.

Very short Answer Question

1. Define the graded potential.
2. What is decussation of pyramids?
3. What are the pons?

Fill in the Blanks

1. Within the dendrites and perikaryons, nerve impulses are usually ________ potentials.
2. In the axon the nerve impulses are ________ potentials.
3. The habenulae are largest in species such sharks and ________.

Multiple choice Questions

1. Bipolar neurons are found in
   (*a*) cornea (*b*) conjunctiva
   (*c*) lens (*d*) retina
2. A mass of cell bodies (perikaryons) within the gray matter of brain or spinal cord is
   (*a*) neuroglia (*b*) ganglion
   (*c*) nucleus (*d*) cyton
3. Connective tissue of nervous system is
   (*a*) adipose (*b*) areolar
   (*c*) neuroglia (*d*) cartilage

4. what are arbor vitae?
   (*a*) grooves and ridges of cerebral hemispheres
   (*b*) floor of mesencephalon
   (*c*) four divisions of optic lobe
   (*d*) branching of gray matter in cerebral hemisphere
5. Central nervous system in vertebrate is derived from
   (*a*) basal plate (*b*) blastopore
   (*c*) neural plate (*d*) none
6. Cerebral hemispheres in eutherians connected internally by
   (*a*) corpus albicans (*b*) corpus callosum
   (*c*) corpus luteum (*d*) aqueduct of Sylvius
7. Function of restiform bodies in elasmobranchs
   (*a*) swimming (*b*) steering
   (*c*) maintenance of equilibrium (*d*) sound production
8. Connecting link between the receptors, central nervous system and effector is
   (*a*) central nervous system
   (*b*) peripheral nervous system
   (*c*) sympathetic nervous system
   (*d*) parasympathetic nervous system
9. Terminal nerves are numbered
   (*a*) 0 (*b*) 11
   (*c*) 12 (*d*) 13

ANSWERS

Very short Answer Questions

1. When the neuron is stimulated brief electrical disturbances are caused. These disturbances lead to changes in membrane potential, which are called **graded potential**. A single such potential is not enough to excite a neuron into doing anything but many graded potentials acting at the same time can change the voltage difference across the membrane enough to initiate an action potential.
2. Decussation of pyramids is crossing of the fibers of the corticospinal tracts from one side of the central nervous system to the other near the junction of the medulla and the spinal cord.
3. Pons is a broad mass of chiefly transverse nerve fibers conspicuous on the ventral surface of the mammalian brain at the anterior end of the medulla oblongata.

Fill in the Blanks.

1. Graded
2. Action
3. Bloodhounds (dogs).

Multiple choice Questions

1. (*d*) **2.** (*c*) **3.** (*c*) **4.** (*d*) **5.** (*c*)
6. (*b*) **7.** (*c*) **8.** (*c*) **9.** (*a*).

9

Sense Organs

9.1 INTRODUCTION

To survive, an animal must react to danger and take advantage of opportunity. Appropriate responses require information about the external environment, the body's internal physiology and previous experience. The results of previous experience are recorded within the nervous system as memory, but **sensory receptors** monitor the external and internal environment (Fig. 9.1). Sensory receptors are specialized organs that respond to selected

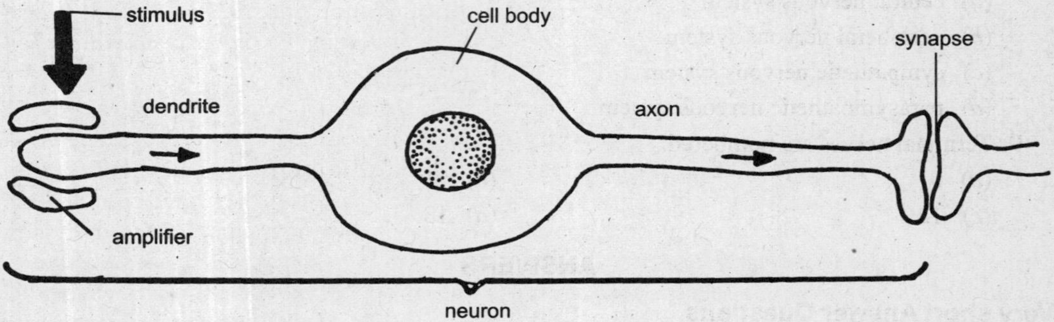

Fig. 9.1. Sensory receptor. A sensory receptor is usually composed of the dendrites of a neuron and can include tissue that amplifies the stimulus. The receptor is a transducer that transforms a stimulus into an electrical impulse or burst of electrical impulses that spread to the cell body and along its axon to other neurons, usually in the central nervous system (after Kardong 2006).

information. Sensory receptors code or translate environmental energies into nerve impulses that are transmitted to the central nervous system (CNS) via afferent fibers. These impulses may or may not be received at conscious levels of the brain.

Perception. Sensations that we as humans become conscious of are referred to as **perception**. Our view of the world is partially determined by the kinds of information our sensory receptors detect and by how that information is processed. Vertebrates differ in their ability to perceive stimuli. Bats and even dogs hear sounds at frequencies our ears cannot. Hawks can hunt high above the ground and detect tiny rodents scampering below. Rattlesnake hunt in light too dim for human to see in. Yet humans see the world in color, while most other mammals see it only in black and white. Our perception of the world is limited or enhanced by the availability and sensitivity of our sensory, receptors.

It should also be highlighted that conscious sensations are an organism's subjective (biased) interpretation of the environment. The environment contains chemicals and photons of light, but the senses of taste and colour are interpretations of these phenomena. Similarly, pain does not exist in the environment. You cannot measure pain as you can measure temperature or force or solar radiation. Taste, colour, and pain are perceptions arising out of events in the brain itself.

Nerve impulses carried by sensory nerves are electrical impulses. The optic nerve carries the same kind of electrical impulses as the auditory nerves, olfactory nerves, taste nerves and so on. *Different sensations result from different ways in which the nervous system interprets signal from different sensory receptors.* Because the impulses are the same, the central nervous system can be fooled. Artificial stimulation of the auditory nerve is perceived as a sound. Artificial stimulation of the optic nerve is perceived as light. Mechanical pressure on the eyeball stimulates the optic nerve to send electrical impulses to the brain. These impulses are interpreted as what they are expected to be, sensations of light, rather than as mechanical stimuli. This is why a mechanical blow of the eye can make one "see star".

While discussing perception, we should distinguish an environmental stimulus from the way in which it is interpreted, but this is seldom done. In common usage and even in scientific research, convention rules. We speak of the senses of seeing, hearing, tasting, smelling, and so on as if the stimulus and the perception of the stimulus were the same. The treatment of sensory receptors in this chapter attempts to separate the kind of environmental energy or stimulus monitored from the nervous system's interpretation of this stimulus (see **Kardong** 2006).

Components of Sensory Organs

Sensory neurons are nerve cells specialized for **detecting** and **transmitting** information about the external or internal environment. Each sensory neuron sends out slender processes, or **nerve fibers.** Sensory receptors usually contain **dendrites,** processes that are responsive to stimuli and carry impulses toward the body of the nerve cell. A sensory neuron usually has an **axon** as well, a nerve process that transmits impulses away from the cell body to other neuron.

The sensory receptor act as a **transducer**, a device that translates energy from one form into another. The microphone of a public address system, which translates sound waves into electrical energy, is another example of transducer. Most sensory receptors translate light, mechanical, or chemical stimuli into electrical impulses. Often the tip of sensory nerve fiber is associated with accessory tissues that amplify the stimulus and thereby increase the sensitivity of receptor. A sensory nerve fiber with its associated tissues is termed a **sensory organ.**

9.2 CLASSIFICATION OF SENSORY ORGANS

Sensory organs of vertebrates can be classified according to several criteria (Box 9.2):

1. **Somatic sensory organs** refer to those of the skin, body surface, and skeletal muscles.
2. **Visceral sensory organs** reside in the viscera.
3. **Exteroceptors** receive sensations from the environment.
4. **Interoceptors** respond to sensations from organs.
5. **Proprioceptor** is a sensory organ located in striated muscles, joints and tendons, is a special type of introceptor (Fig. 9.2).

According to **Kardong** (2006), all sensory organs of vertebrates are broadly divided into two main groups: **General Sensory Organs** which are widely distributed throughout the body and are concerned with sensations of touch, temperature and proprioception (Fig. 9.3).

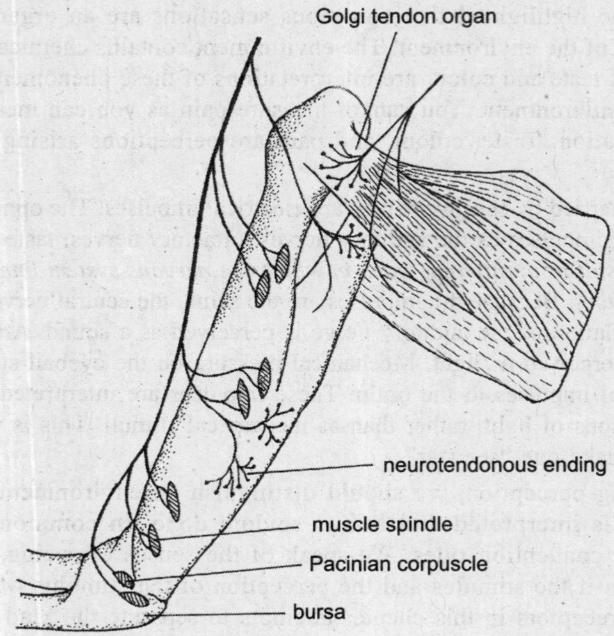

Fig. 9.2. Representative proprioceptor sites (exaggerated) occurring in muscle, tendons and bursa (*i.e.*, a sac or pouch) (after Kent and Carr 2001).

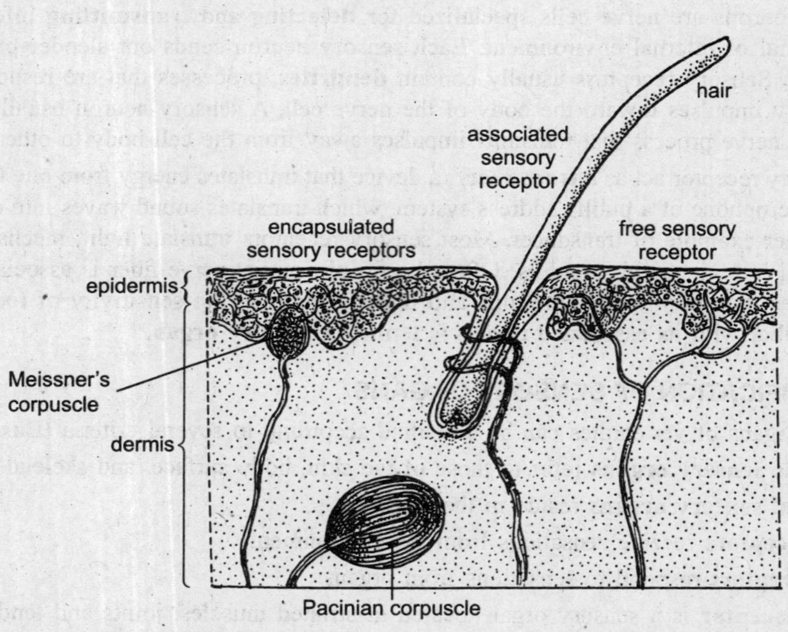

Fig. 9.3. General sensory receptors of mammals. A free nerve ending, encapsulated sensory receptor, and an associated sensory receptor of a hair follicle (after Kardong 2006).

Special Sensory Organs which are localized and often specialized (Box 9.1).

Box 9.1 Special senses and Motor Acts

The evolution of vertebrates shows a trend called **cephalization** in which special sensory organs develop in the head of animals, along with the corresponding development of the brain. These special sensory systems, which include the visual, auditory, vestibular, olfactory, and gustatory systems, detect and analyze light, sound and chemical signals in the environment, as well as signal the position and movement of the head. The stimuli detected and transduced by these systems are most familiar to us when they provide conscious awareness of our environment, but they are equally important as sensory input for reflexive and subconscious behavior.

Further, movements are the major way that vertebrate animals interact with the world. Most of human activities, whether running, reaching, eating, talking, writing, or reading, ultimately involve motor acts. Thus, motor control is a major task of the nervous system, and from an evolutionary prospective, it is probably the reason that nervous system is devoted to motor control which can be defined as the generation of signals to coordinate contraction of the musculature of the body and head either to maintain a posture or to make a movement – transition between two postures; **Koeppen** and **Stanton** 2010).

Box 9.2 Classification of Sensory Organs (Modern View)

1. **General Sensory Organs**
   (i) Free sensory receptors: Cutaneous reception
   (ii) Encapsulated sensory receptors
   - Meissner's capsule
   - Corpuscle of Ruffini
   - End bulbs of Krause
   - Pacinian corpuscles
   - Hair and nerve fibers
   (Hair units and field units)
   (iii) Associated Sensory Receptors
   (a) Proprioceptive receptors
   - Muscle spindles
   - Golgi tendon organs
2. **Special Sensory Organs**
   (i) Chemoreceptors
   - Olfaction (odors)
   - Vomerolofaction (vomodors)
   - Tastebuds (gustatory)
   (ii) Radiation receptors
   - Photoreceptors (Eye; pineal complex)
   (iii) Infrared receptors (Thermoreceptors)
   (Pit organs)
   (iv) Mechanoreceptors (Hair cells)
   - Pain and temperature sensation
   (Nociceptors which have ability to sense tissue damaging mechanical, thermal or chemical stimuli)
   - Lateral line system
   (Neuromast organs)
   - Vestibular apparatus (Equilibratory)
   - Auditory system (ear) (Auditory)
   (v) Electroreceptors
   (vi) Earth's magnetic field
   (*e.g.*, sea turtles)
   (vii) General visceral receptor

Source: Kardong; 2006; **Koeppen** and **Stanton** 2010)

9.3 VISUAL AND OTHER RADIATION RECEPTORS

Radiation travels in waves. Cosmic radiation has the shortest wavelength, and radio waves have the longest. Together with intermediate wavelengths, these constitute the **electromagnetic spectrum** (Fig. 9.4). From the standpoint of an organism radiation carries information about its intensity, wavelength, and direction.

No organism taps the full range of information available throughout the electromagnetic spectrum. Organisms can perceive only a limited range of wavelengths. Insects, such as honeybees and butterflies, can see ultraviolet (UV) radiation, as do many fishes, reptiles, and birds. Most mammals have lost the ability to see UV light, except perhaps for some rodents. The urine of these rodents reflects UV, making its detection by social rodents a useful signal. On the other side of the visual range (Fig. 9.4), some vertebrates, such as vampire bats, pythons, and pit viper snakes can detect infrared (IR) radiation using specialized surface receptors. However, most vertebrates can perceive only a narrow band of electromagnetic radiation between about 380 and 760 nm. This restricted band is called "visible" light, meaning that we can see it (Fig. 9.4). When we talk of "light", we are being quite narrow-minded because we really mean this very narrow range of electromagnetic radiation out of much wider spectrum. Similarly, when we speak of the sense of "vision", we are referring to the ability to perceive light within this narrow range.

Vertebrates have evolved a variety of sense organs that gather electromagnetic radiation. Different regions of the spectrum represent different energies (Fig. 9.4) and persent different levels of stimuli to sensory receptors.

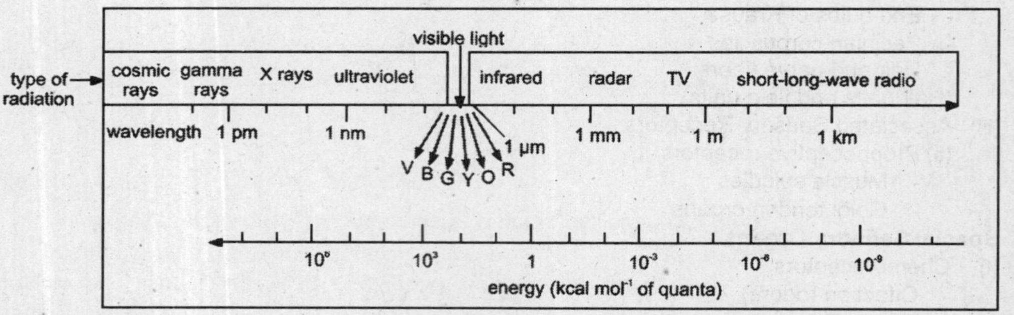

Fig. 9.4. The spectrum of electromagnetic radiation. Between extremely short cosmic band of "visible light", to which human eyes are normally sensitive. Wavelength increases to right. Energy within the electromagnetic radiation increases to left. Abbreviations: violet (V), blue (B), green (G), yellow (Y), orange (O), red (R) (after Kardong 2006).

Photoreceptors

Eye photoreceptors are sensitive to light. The most obvious and best understood photoreceptor is the eye. The vertebrate eye can focus light on photosensitive cells to form an image of the environment. The capacity to focus light on objects at different distances is called **visual accommodation**.

The nervous system takes advantage of the physical properties of light to interpret the images presented to it. Differences in light intensity are interpreted as **contrasts.** Within the visible spectrum, different wavelengths are interpreted as **different colours.**

Structure of Eye

The mammalian eye has three layers: sclera, uvea and retina (Fig. 9.5).

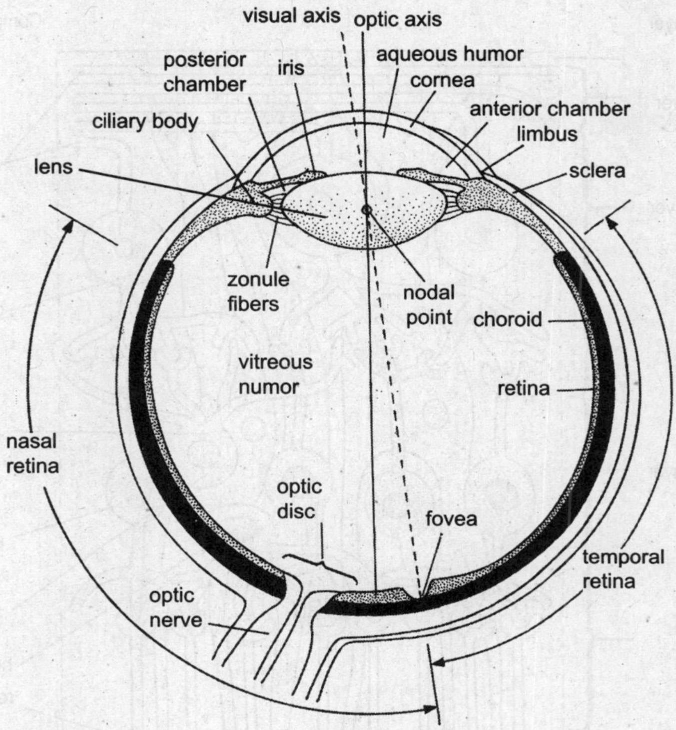

Fig. 9.5. View of a horizontal section of the right human eye (after Koeppen and Stanton 2010).

Sclera. The sclera is the outer layer of the eye. It forms the "white of eye" and consists of a tough capsule of connective tissue to which extrinsic ocular muscles attach. Contractions of these muscles rotate the eyeball in its orbit to direct the gaze toward an object of interest. The sclera helps define the shape of the eyeball. In birds, reptiles, and fishes, small plates of bone called **scleral ossicles** are often present to help hold the shape of the sclera. At the front of the eye, the sclera clears to become the transparent **cornea**.

Uvea. The middle layer of the eye is called **uvea** and it is composed of three regions. The **choroid** adjacent to the retina is the largest region. Because it is highly vascular, the choroid provides nutritional support to the ocular tissue. The choroid is pigmented. In some nocturnal vertebrates, choroid includes a special reflective material, called **tapetum lucidum**. Under the condition of dim light, this structure reflects the limited light to restimulate the light-sensitive cells in the retina. The tapetum lucidum produces the "eye shine" of mammals seen at light in car head-lights or in flashlight beams.

The second region of uvea, the **ciliary body** is a tiny circle of smooth muscle around the interior of the eye ball. The **ciliary muscle** controls visual accommodation. It is attached to the flexible **lens** through a circular **suspensory ligament**. Tension on the lens tends to deform it, whereas relaxation allows the lens to restore its shape elastically.

Uvea's third region is **iris** which is a thin continuation of the uvea across the front of the eyeball. The **pupil** is not a structure but an opening defined by the free edge of the iris. Tiny smooth muscles within the iris act like a diaphragm to reduce or enlarge the size of the pupil and regulate the amount of light that enters the eye.

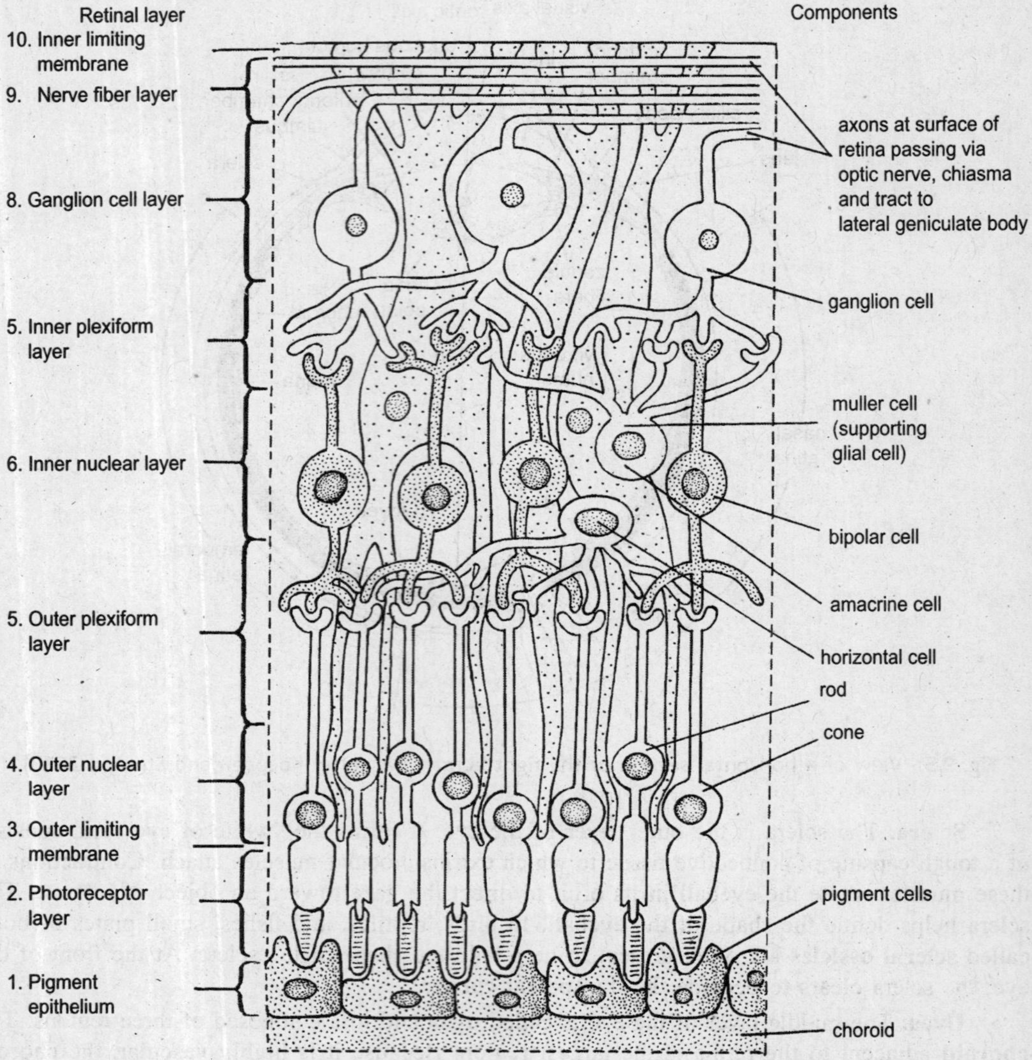

Fig. 9.6. Layers of retina. Neurons indirectly connect light-sensitive rods and cones to ganglion cells that form the optic nerve. Light impinging on retina comes from the top of the figure and passes through all the superficial layers to reach the rods and cones. Muller cells (*i.e.*, supporting glial cells) form outer and inner limiting membrane (after Koeppen and Stanton 2010).

Retina. Retina is the innermost layer of the eye which is photosensitive and composed of three layers. The deepest layer of cells within the retina contains the **photoreceptor cells**. These include **rods**, which are sensitive to low levels of illumination but not colours, and **cones**, which are sensitive to colours in bright light. The photoreceptor cells synapse with **horizontal cells** and **bipolar cells** (Fig. 9.6). Proximal to these are layered **amacrine cells** and **ganglion cells**. This arrangement means that light entering the eye and falling on the retina passes sequentially through ganglion, amacrine, bipolar, and horizontal cells before it reaches the photoreceptive rods and cones (Fig. 9.6). The functional significance, if any, of this arrangement is not known (Box 9.3).

Box 9.3

1. Pigmented epithelium. It contains **melanin** pigment which along with the pigmented choroid **absorbs** extra amount of light, thus preventing reflection of rays back through the retina. If this light gets refracted back, it would result in "blurring" of vision. This epithelium has the phagocytotic functions. Thus, cells of pigmented layer extend tentacle-like processes into photoreceptive layer and can (i) provide nutrients and remove waste from photoreceptors; (ii) phagocytose the ends of outer segments of the rods and cones which are continuously shed; (iii) reconvert metabolized photopigments into a from that can be reused after it is transported back to the photoreceptors.

2. Muller cells maintain internal geometry of the retina.

3. The selective responsiveness to motion or direction is a function of amacrine cell system. Their horizontal interconnections cause a sharpening of the edges of any stimulated field on the retina.

4. Inner plexiform layer is the site of major processing of the visual image.

5. Cells of ganglion cell layer transmit visual information to brain (see **Agarwal** 2016).

In a few vertebrates (*e.g.*, humans), the retina is indented, forming a **fovea** (Fig. 9.5). The fovea is the point at the back of the eyeball where light converges. Composed entirely of cones, fovea forms the point of sharpest focus. Although rods are absent from fovea, they increase peripherally.

In fact, raptorial birds (owl, hawks, eagles and other birds of prey) and some lizards have two foveas in each eye. The grand champion of foveas is probably that of hawks, which have a human-sized eyeball with approximately one million cones per square millimeter in each fovea. The few other birds with two foveas in each eye are diurnal birds that catch flying insects.

Towards the medial quadrant of the retina is a **blind spot**, called **optic disc**, the epithelium of which lacks both rods and cones. This is where the optic nerve fibers leave the retina and blood vessels enter and leave.

Within the eye are the three chambers. Two lie in front of the lens: the **anterior chamber** between the iris and the cornea, and the small **posterior chamber** between the iris and the lens. The third and largest **vitreal chamber** is located behind the lens. These chambers are filled with a transparent fluid that helps maintain the shape of the eyeball. The anterior and posterior chambers are filled with watery **aqueous humor**. The vitreal chamber contains a thick **vitreous humor** sometime called the **vitreous body** because it can be dissected from the eye as a single viscous plug.

Embryonic Development of Eye. Embryologically, the eye is a composite structure formed from surrounding mesenchyme and the optic placode, a thickened neuroectodermal tissue that gives rise to the optic vesicle. Development of the eye begins with the appearance of paired outgrowths, the **optic vesicles**, from the sides of the future telencephalon (Fig. 9.7 A). As the optic vesicles approach the overlying ectoderm it thickens to become the optic placode and invaginates to form the **lens primordium** (Fig. 9.7 B). The optic placode pinches off to settle into an indentation, the **optic cup**. Mesenchyme surrounding the developing eyeball condenses to produce the outer coats of the eye (Fig. 9.7 C, D). Thus, the ectoderm gives rise to the eyelid, cornea and lens. The mesenchyme forms the choroid and sclera, and the iris and retina develop from the optic cup. The optic stalk comes to carry the axons of ganglionic cells that project to optic areas in the diencephalon. Although this stalk is actually an extension of the brain and should therefore be termed a tract, in practice it is often termed the **optic "nerve"** and connect as the second cranial nerve.

Evolution (Phylogeny) of Vertebrate Eye

Cyclostomes

In lampreys, the striated **corneal muscle** of myotomal origin is attached to the spectacle, a clear area of skin over the cornea. Contraction of this muscle tends to draw the spectacle taut

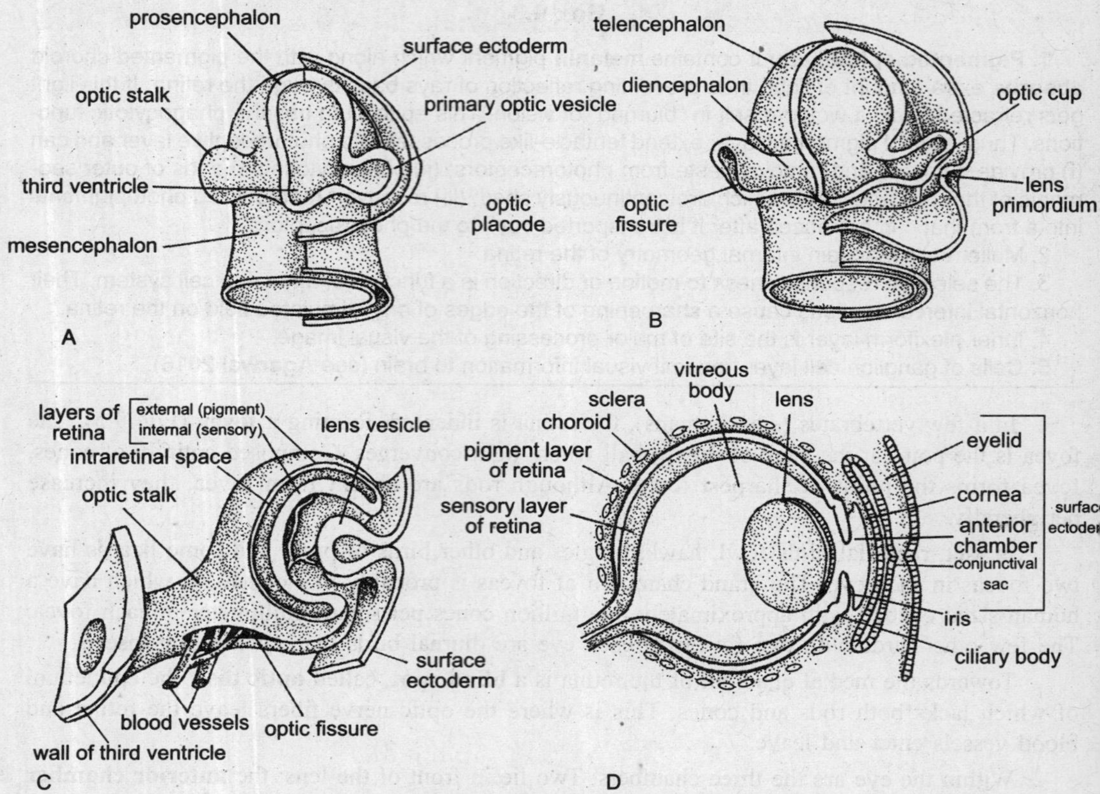

Fig. 9.7. Embryonic development of the vertebrate eye. A – Early embryo in whom outwardly growing optic vesicles eventually meet the thickened optic placode in the ectoderm. B – slightly later, interaction between optic vesicle and optic placode leads to initial differentiation of the lens and what becomes a double – layered secondary optic vesicle or optic cup. C – and D – successive stages in separation of the lens within the optic vesicle. The eyelids and cornea form, and the wall of the eyeball differentiates into distinct layers. After the lens is pinched off, openings appear within the surface epithelium to delineate the cornea and anterior chamber (after Kardong 2006).

and flatten the cornea. This in turn pushes the lens closer to the retina. Accommodation is thus accomplished by deformation of the eyeball from outside. Upon relaxation of the corneal muscle, elasticity of the cornea and vitreous humour return the lens to a resting position (Fig. 9.8).

Fishes. In cartilaginous and bony fish, the eyeball is often supported by scleral bones or cartilages. The lens is nearly round. It is held by the suspensory ligament and moved by the **retractor** or **protractor lentis muscle**, which is inserted directly on the lens. To focus the image, the retractor lentis pulls the lens forward in elasmobranchs and backward in teleosts.

Amphibians. The amphibian eye is sometimes reinforced by a cup or ring of scleral cartilage. The lens is nearly round and held by a circular suspensory ligament. The retractor lentis muscle of amphibians is inserted on the base of the suspensory ligament instead of directly on the lens. The lens is normally focused on distant objects. In order for amphibians to view near objects, the retractor lentis pulls the lens forward.

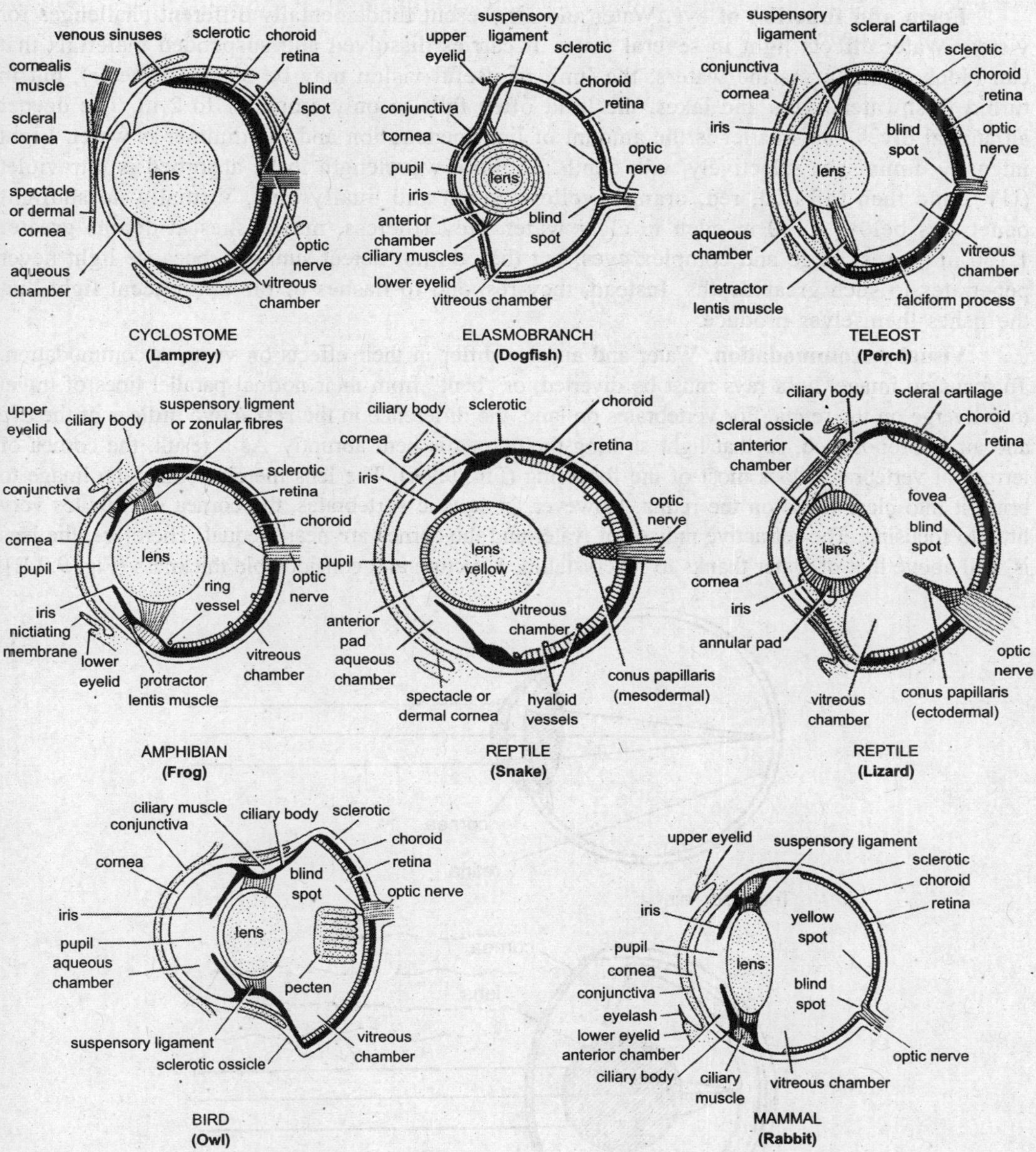

Fig. 9.8. Eyes of different vertebrates in sagittal section.

Amniota. Among amniotes, except snakes, the lens changes shape to accommodate the visual image. This usually involves contraction of the ciliary muscle, which may squeeze the lens to change its shape or act through the circular suspensory ligament to stretch the lens and make it flatter. Relaxation of the muscle allows the lens to return, thanks to its resilience, to a more rounded shape. In some reptiles and some birds, but not in mammals, scleral ossicles are present. They are particularly well developed in raptors (birds of prey such as kite) that reach high rates of aerial speed. A **papillary cone** in reptiles or a **pecten** in birds is projected in vitreous chamber from the back wall of the eye. These structures are thought to provide supplemental nutritional support for deep ocular tissues.

Form and function of eye. Water and air present fundamentally different challenges for vision. Water affects light in several ways. It carries dissolved and suspended materials that can block light. In marine waters, the limit of useful vision may be 30 m (= meter), but in turbid freshwater rivers and lakes, the limit often falls to only around 1 to 2 m. The deeper an animal dives, the smaller is the amount of light penetration and the dimmer is water. Light intensity diminishes selectively with depth. The first wavelength to be absorbed is ultraviolet (UV), and then infrared, red, orange, yellow, green and finally blue. Virtually no sunlight penetrates below 1,100 m even in clear water. Nevertheless, most fishes at depths greater 1,100 m possess large and complex eyes, but they do not detect sunlight, because light never penetrates to such great depths. Instead, they respond to flashes of bioluminescent light that the fishes themselves produce.

Visual Accommodation. Water and air also differ in their effects on visual accommodation. To focus an image, light rays must be diverted, or "bent" from their normal parallel lines of travel to converge on the retina. For vertebrates on land, the difference in the **refractive indices** of the eye and air is pronounced, so that light striking the cornea is bent abruptly. As a result, the cornea of terrestrial vertebrates does most of the focussing (Fig. 9.9A). The lens merely refines the image to bring it into clear focus on the retina. However, in aquatic vertebrates, the cornea contributes very little to focusing. The refractive indices of water and the cornea are nearly equal. Therefore, the lens is well above that of water thanks to its crystalline structure and considerable thickness (Fig. 9.9 B).

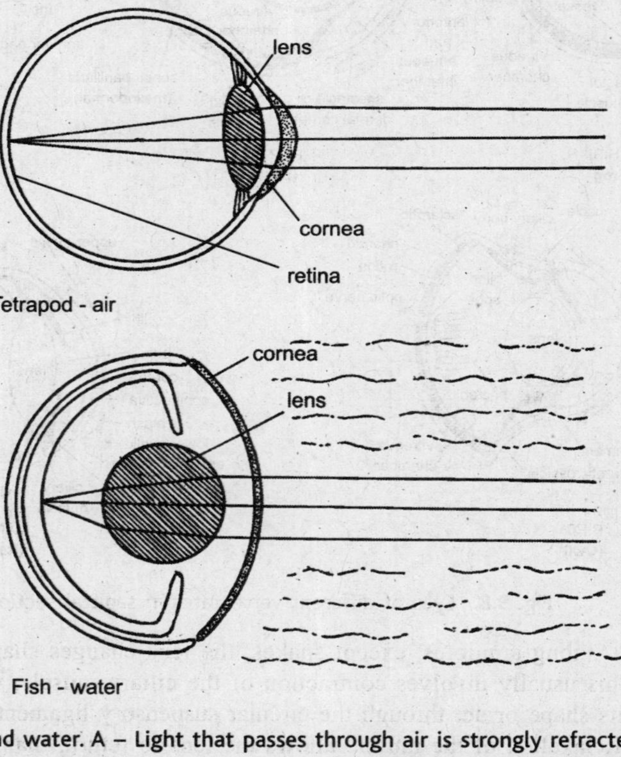

A Tetrapod - air

B Fish - water

Fig. 9.9. Vision in air and water. A – Light that passes through air is strongly refracted when it passes through the cornea; therefore cornea is primarily responsible for focusing light rays. The lens fine – tunes the focused image. B – Because the cornea has refractive properties similar to those of water, incoming light is affected very little when it first enters the eye, therefore, the large lens bears primary responsibility for bringing light rays into focus on the retina (after Kardong 2006).

Focussing in terrestrial vertebrates, except for snakes, depend on changes in **lens shape** (Fig. 9.10C, D). In contrast to this, focusing in fishes depends on changes in **lens position** (Fig. 9.10A, B). It is unclear whether the fish eye at rest is focused on close or distant objects. Different mechanisms of accommodation are employed by different groups of fishes. When electrically stimulated, the *retractor lentis* muscle move the lens posteriorly in some fishes, obliquely downward in others, and obliquely upward in still others.

In trout fish, the lens is round and the retina is ellipsoid, so the two are not concentric. This may mean that at rest the near field of view at the center and the far field at the periphery are in focus. During accommodation, the lens is pulled backward. This does not change the focus on peripheral objects much, but it focuses distant objects in the center of the field of view.

Photoreception

Most vertebrates have two overlapping visual systems within the retina. One is **scotopic**–responsible for visual ability in dim light, and the other **photopic**–responsible for colour vision in bright light. Rods provide night vision, sensitive to dim light, but lack colour perception. Cones

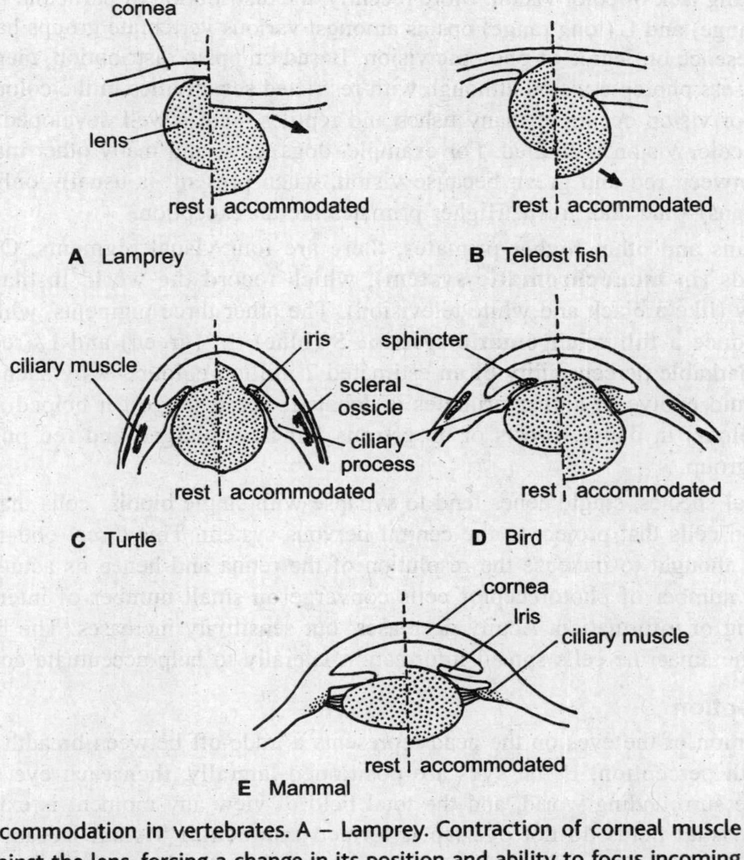

Fig. 9.10. Accommodation in vertebrates. A – Lamprey. Contraction of corneal muscle pulls the cornea against the lens, forcing a change in its position and ability to focus incoming light. B – Teleost. Accommodation in the teleost depends on a change in lens position. C – Turtle and D – Bird. Accommodation in both involves the iris sphincter muscle squeezing the lens. E – Mammal. Accommodation is brought about by relaxation of the suspensory ligaments of the eye (after Kardong 2006).

provide color vision and are made up of three classes, each with absorptive pigment that respond maximally to light with spectral sensitivities in the blue, green, and red ranges (actually in the violet, blue-green and yellow-green ranges). Alternatively, these are now more logically termed short (S)–, medium (M)–and long (L) range cone sensitivities, respectively. The absorptive pigments in rods and cones belong to a class of visual pigments termed **opsins,** proteins conjugated with a light–absorbing derivatives of vitamin A.

The proportions of rods and cones in retina vary considerably in different animals. In animals active under bright conditions, such as diurnal tetrapods and diurnal shallow-water fishes, both rods and cones are present. Most cones are concentrated near the fovea, and rods predominate at the periphery. Nocturnal animals or those that inhabit dimly lit waters are adapted to poor illumination. In these groups, cones are few or absent, and the retina is composed almost entirely of rods.

Conventionally, vertebrate scotopic and photopic vision was deduced from the relative abundance of rods and cones in the retina. The retinas of all vertebrates possess rods and hence, all vertebrates possess some degree of scotopic vision. However, many vertebrates lack retinal cones, suggesting lack of color vision. More recently, the distribution of particular S (short range), M (medium range) and L (long range) opsins amongst various vertebrate groups has been used to deduce the presence or degree of photopic vision. Based on opsin distribution, many, if not most, vertebrate possess photopic vision although with restricted sensitivities in the color range. Highly developed color vision occurs in many fishes and reptiles, and is well developed in most birds. In mammals, color vision is limited. For example, dogs, cats, and many other mammals cannot distinguish between red and green because vision, when present, is usually only **dichromatic** (two cone opsins)–blue and green. Higher primates are an exception.

In humans and other higher primates, there are four visual pigments. One pigment is present in rods (in **monochromatic system)**, which record the world in black, white and shades of gray (like a black and white television). The other three pigments, which are present in cones, produce a full **trichromatic system:** S (blue), M (green) and L (red). This gives humans a remarkable perceptibility of an estimated 2 million colours. Why such a color acuity and range would evolve in higher primates is debated. Some suggest it helped spot ripe fruits (change in colour) in dense forests or an estrous female (enlarged and red pubic swellings) amongst the group.

In diurnal species, single cones tend to synapse with single bipolar cells that synapse with single ganglion cells that project to the central nervous system. This direct one–to–one transfer of impulses is thought to increase the resolution of the retina and hence its acuity. In nocturnal species, large number of photoreceptor cells converge on small number of interneurons; thus, there is pooling of information. Acuity decreases, but sensitivity increases. The horizontal cells and perhaps the amacrine cells spread information laterally to help accentuate contrasts.

Depth Perception

The position of the eyes on the head represents a trade-off between breadth and the visual field and depth perception. If the eyes are positioned laterally, then each eye scans separate portions of the surrounding world, and the total field of view any moment is extensive. Vision in which the visual fields do not overlap is termed **monocular vision**. Because it allows an individual to see a large portion of its surroundings and to detect potential threats from most directions, it is common in prey animals. Strictly monocular vision in which the visual fields of the two eyes are wholly separate is relatively rare. It occurs in cyclostomes, some sharks, salamanders, penguins, and whales.

The visual field overlaps in animals with **binocular vision** (Fig. 9.11). Extensive overlap of visual fields characterize humans. We have as much as 140° of binocular vision, leaving 30° of monocular vision on each side. Binocular vision is important in most classes of vertebrates. Birds have up to 70° of overlap, reptiles up to 45°, and some fishes as much as 40°. Within the area of overlap, the two visual fields, merge into a single image, producing **stereoscopic vision.** The advantage of stereoscopic vision is that it gives a sense of depth perception. If a person with two functioning eyes.

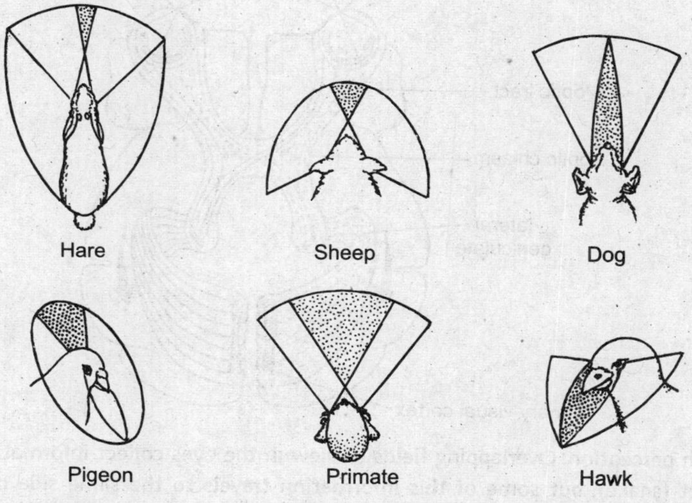

Fig. 9.11. Monocular and binocular vision in birds and mammals. The degree of overlap of visual fields (indicated by shading) varies considerably. Large panoramic fields of view characterize animals susceptible to attack.

Depth perception comes from the method of processing visual information. In binocular vision, the visual field seen by each eye is divided. Half the input goes to the same side and the other half crosses via the **optic chiasma** to the opposite side of the brain. The result is to bring information gathered by both eyes to the same side of the brain. The brain compares the parallax of the two images. **Parallax** is the slightly different view one gets of a distant object from two different points of view. Look at a distant lamppost from one position, and then step a few feet laterally and look at it again. Slightly more of one side can be seen and less of the opposite side. The position of the post relative to background reference points changes as well. This is parallax (**Kardong**, 2006). The nervous system takes advantage of parallax that results from differences in positions of the two eyes. Each eye registers a slightly different image due to the distance between the eyes. Although this difference is slight, it is enough for the nervous system to produce a sense of depth from differences in parallax (Fig. 9.12).

Visual accommodation also contributes to depth perception. Even with a single eye, the degree of accommodation required to bring the object into focus can be used to interpret its distance.

Integration of Visual Information. In lower vertebrates, the optic nerve tends to run directly into the midbrain. In amniotes, the axons in the optic tract travel to one of three regions: the **lateral geniculate nuclei** of the thalamus, the tectum of the midbrain, and the pretectal area in the tegmentum of the midbrain.

Most fibers of the optic tract travel to the paired lateral geniculate nuclei of the thalamus. From here, fibers relay information to the primary visual cortex of the cerebrum (Fig. 9.13).

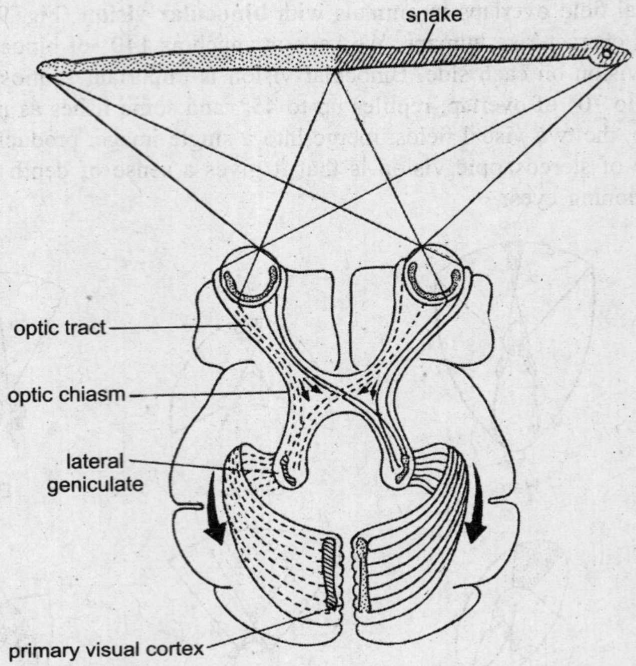

Fig. 9.12. Depth perception. Overlapping fields of view in the eyes collect information about the same object (snake), but some of this information travels to the same side of the brain via the optic chiasma, whereas other information travels to the visual areas on the opposite side of the brain. This permits similar information from the slightly differently placed eyes to be brought together and compared. Such comparisons are thought to be the basis for depth perception. Notice the specific organization of visual information. Information from the left visual field (tail of snake) is received by the right side of each eye. The right visual field (snake head) is received by the left side of each eye. Nerve fibers from the medial retina cross in the optic chiasma. Lateral fibers do not cross. Overall, information from visual field (tail) is processed in the primary visual cortex of the right cerebral hemisphere. Information from the right visual field (head) is processed in the left cerebral hemisphere (after Kardong 2006).

Thalamic cells and photoreceptor cells and some cells of the visual cortex are more specialized. Some of these respond to visual images in the shape of slits, bars, or edges. Others respond only to moving or to stationary visual edges. Such specialized cells add more information about the size, shape, and movement of the visual image. In other words, the retina responds to light intensity and wavelength, but in the visual cortex, the image gains contour, orientation, and motion. Image formation might be linked to pictures composed of dots of discrete information. As edges, shades, and shapes are sorted, the image progressively emerges. Furthermore, visual information is thought to reach a conscious level in the visual cortex. When it reaches the cortex the visual world is consciously perceived.

Before entering the thalamus, some ganglion cells of the optic tract send branches into the midbrain to contact the tectum. In addition to this visual input neurons of the **superior colliculum** receive information about sound, head position, and feedback from the visual cortex. The tectum, in turn, produces motor output to muscles that rotate the eyes, head, and even the trunk toward the visual stimuli. Humans have two eyes and two visual fields. Mentally, these merge into one field, in part because of the intricately synchronized movements of both eyes.

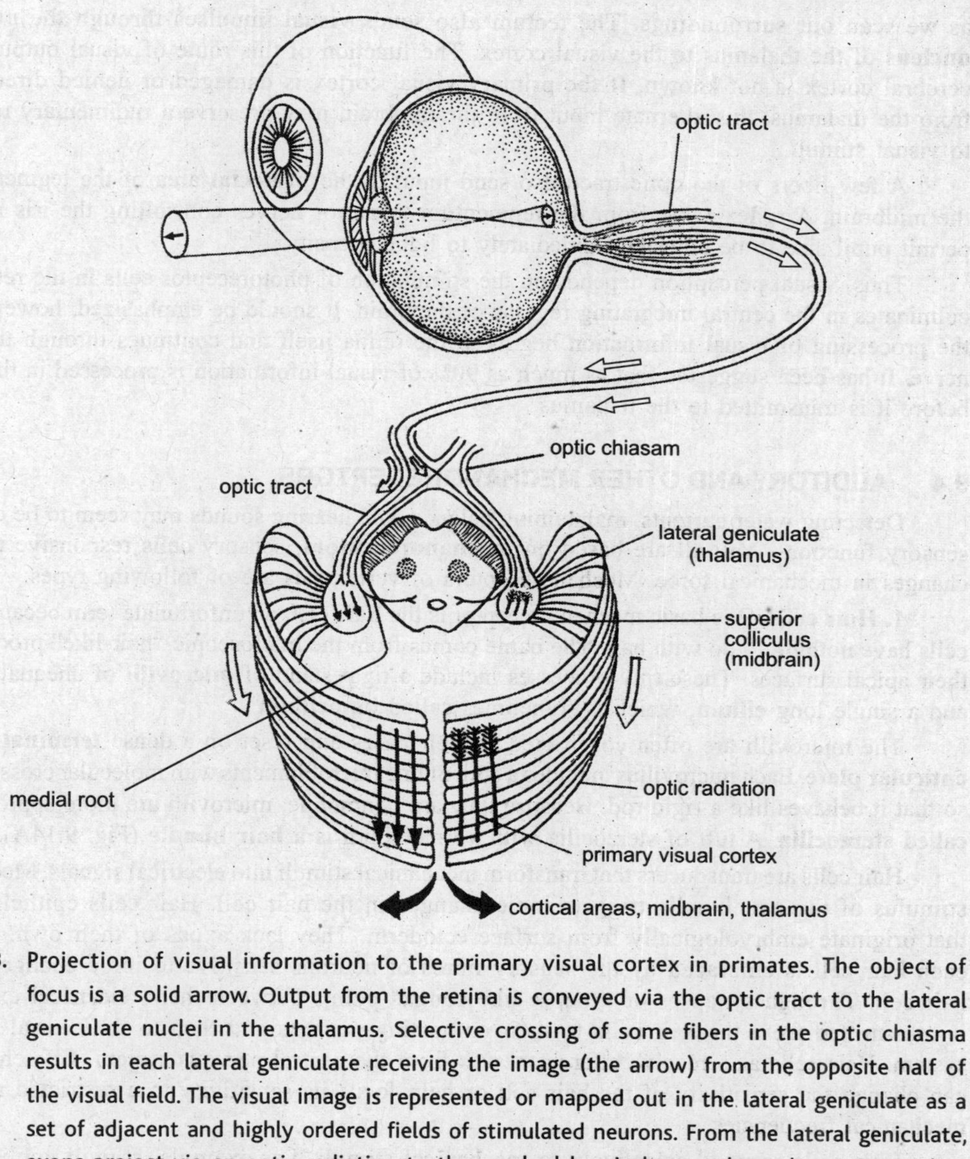

Fig. 9.13. Projection of visual information to the primary visual cortex in primates. The object of focus is a solid arrow. Output from the retina is conveyed via the optic tract to the lateral geniculate nuclei in the thalamus. Selective crossing of some fibers in the optic chiasma results in each lateral geniculate receiving the image (the arrow) from the opposite half of the visual field. The visual image is represented or mapped out in the lateral geniculate as a set of adjacent and highly ordered fields of stimulated neurons. From the lateral geniculate, axons project via an optic radiation to the cerebral hemispheres, where the corresponding parts of the image are now represented as a pile of superimposed fields in the primary visual cortex. This ordering of multiple representations of an image both in the thalamus (lateral geniculate) and in the cerebral hemispheres (primary visual cortex), although not entirely understood, is thought to contribute to neuroanalysis of the binocular image. Notice that the visual pathway does not end in the primary visual cortex but continue to the midbrain and even back to the thalamus. A small branch of the optic tract, the medial root, does not enter the thalamus but instead leads to the superior colliculus (optic tectum). The medial root synapses with motor neurons that control the visual orientation reflex involving eyeball movements, head turning, and rotation of the trunk (after Kardong 2006).

as we scan our surroundings. The tectum also sends visual impulses through the **pulvinar nucleus** of the thalamus to the visual cortex. The function of this route of visual output to the cerebral cortex is not known. If the primary visual cortex is damaged or denied direct input from the thalamus, this alternate input from the midbrain may preserve a rudimentary response to visual stimuli.

A few fibers of the optic tract also send input to the **pretectal** area of the tegmentum of the midbrain. A reflex relay from the tegmentum to motor nerves controlling the iris muscles permit pupil size to be adjusted immediately to light intensity.

Thus, visual perception depends on the stimulation of photoreceptor cells in the retina and culminates in the central integrating regions of the brain. It should be emphasized, however, that the processing of visual information begins in the retina itself and continues through the optic nerve. It has been suggested that as much as 90% of visual information is processed in the retina before it is transmitted to the thalamus.

9.4 AUDITORY AND OTHER MECHANORECEPTORS

Detecting water currents, maintaining balance, and hearing sounds may seem to be different sensory functions. Yet, all are based on **mechanoreceptors**, sensory cells responsive to small changes in mechanical force. Mechanoreceptors of vertebrates are of following types:

1. Hair cells. One basic mechano receptor is the **hair cell**, an unfortunate term because these cells have nothing to do with hair. The name comes from the microscopic "hair-like" processes at their apical surfaces. These tiny processes include a tight stand of **microvilli** of unequal lengths and a single long **cilium**, which is sometimes called **kinocilium**.

The microvilli are often constricted at their bases and reset on a dense **terminal web** or **cuticular plate**. Each microvillus includes a core of fine microfilaments with molecular cross-bridges so that it behaves like a rigid rod. Because they are nonmobile, microvilli are more appropriately called **stereocilia**. A tuft of stereocilia with a kinocilium is a **hair bundle** (Fig. 9.14A).

Hair cells are transducers that transform mechanical stimuli into electrical signals. Mechanical stimulus of the hair bundle triggers ionic changes in the hair cell. Hair cells epithelial cells that originate embryologically from surface ectoderm. They lack axons of their own. Instead, each hair cell is embraced by the sensory fibres of neurons sensitive to ionic changes in the hair cell. Through synapses or synapse-like contact points between them, electrical excitation is transmitted from the hair cells to their embracing neurons and then to the central nervous system. Hair cells also receive efferent nerves from the central nervous system. Efferent nerves can change the sensitivity of the hair cell or help focus its sensitivity to a restricted range of mechanical frequencies.

Hair cells respond selectively to mechanical stimuli. For example, stimuli applied from one direction will trigger a hair cell, whereas stimuli applied from the opposite direction will not (Fig. 9.14 C). Selectivity is thought to result from the asymmetry of the hair bundle itself, due to the different lengths of its stereocilia.

2. Neuromast organ. A neuromast organ is a small collection of hair cells, supporting cells, and sensory nerve fibres, composing the most common arrangement of a mechanoreceptor. The projecting hair bundles are usually embedded in a gelatinous cap called the **cupula** (Fig. 9.14B). The cupola most likely accentuates (*i.e.*, makes more noticeable) the mechanical stimulation of the hair cells, thereby increasing their sensitivity. The neuromast organ, or a modification of it, is the fundamental component of all three types of mechanoreceptive systems: the lateral line system, which detects water currents; the vestibular apparatus, which senses changes in equilibrium; and the auditory system, which responds to sound.

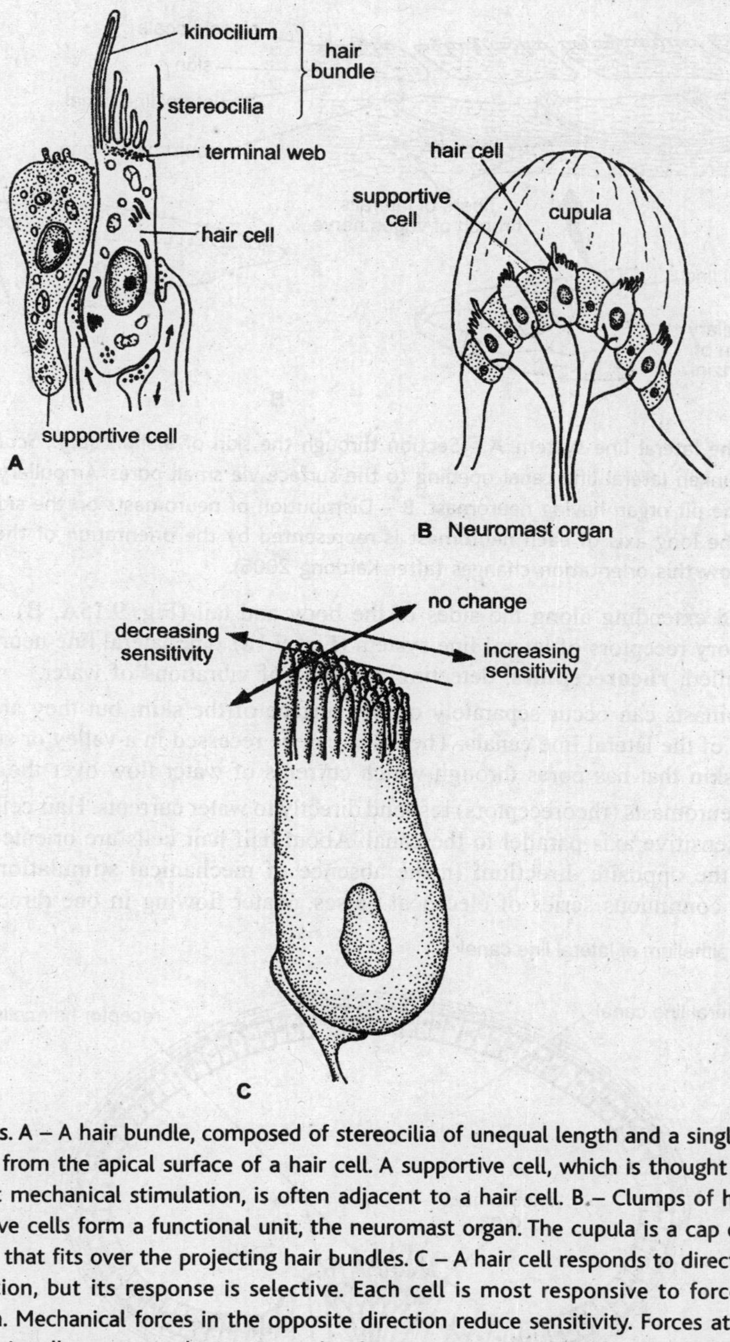

Fig. 9.14. Hair cells. A – A hair bundle, composed of stereocilia of unequal length and a single kinocilium, projects from the apical surface of a hair cell. A supportive cell, which is thought not to react to direct mechanical stimulation, is often adjacent to a hair cell. B.– Clumps of hair cells and supportive cells form a functional unit, the neuromast organ. The cupula is a cap of gelatinous material that fits over the projecting hair bundles. C – A hair cell responds to direct mechanical stimulation, but its response is selective. Each cell is most responsive to forces from one direction. Mechanical forces in the opposite direction reduce sensitivity. Forces at right angles to the hair cells cause no change. By such selective responses, hair cells indicate the directions of the mechanical forces that impinge upon them (after Kardong 2006).

3. Lateral line system. The **lateral line system** is present within the skin of most cyclostomes, fishes, and aquatic amphibians (Fig. 9.17), but it is unknown in terrestrial vertebrates, including aquatic birds and mammals. Lateral line system consists of long, recessed (recess is a small space set back in a wall or set into a surface) grooves, or **lateral line canals**, concentrated on

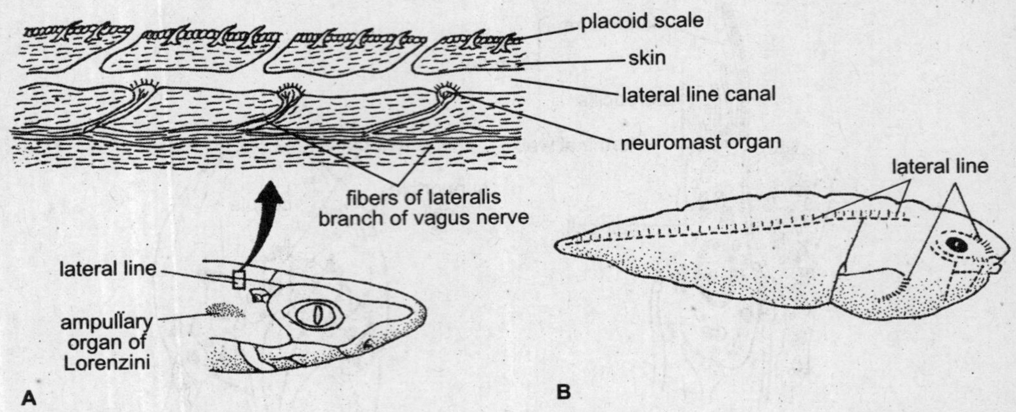

Fig. 9.15. The lateral line system. A – Section through the skin of a shark (*e.g., Scoliodon*) showing the sunken lateral line canal opening to the surface via small pores. Ampullary organ of Lorenzi is the pit organ having neuromast. B – Distribution of neuromasts on the side of a frog tadpole. The long axis of each neuromast is represented by the orientation of the short bars. Notice how this orientation changes (after Kardong 2006).

the head and extending along the sides of the body and tail (Fig. 9.15A, B). Neuromast organs are the sensory receptors of lateral line system (Fig. 9.16). The lateral line neuromasts are current receptors called, **rheoreceptors**, detecting any type of vibrations of water.

Neuromasts can occur separately on the surface of the skin, but they are usually found at the bottoms of the lateral line canals. The canal can be recessed in a valley or sunken and covered by surface skin that has pores through which currents of water flow over the neuromast organs.

The neuromasts (rheoreceptors) respond directly to water currents. Hair cells are oriented with their most sensitive axis parallel to the canal. About half hair cells are oriented in one direction, the rest in the opposite direction. In the absence of mechanical stimulation, each neuromast generates a continuous series of electrical pulses. Water flowing in one direction stimulates an

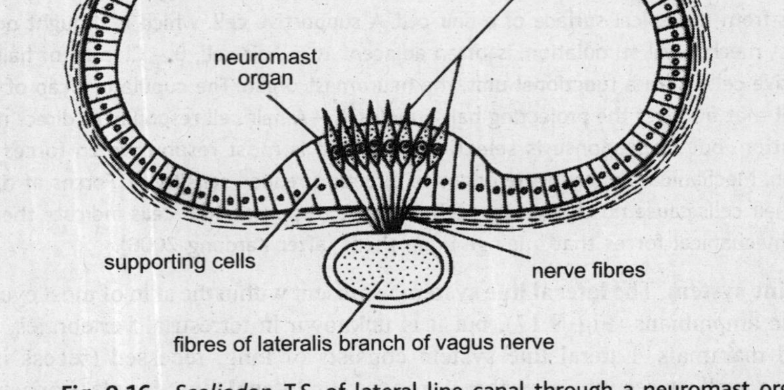

Fig. 9.16. *Scoliodon*. T.S. of lateral line canal through a neuromast organ.

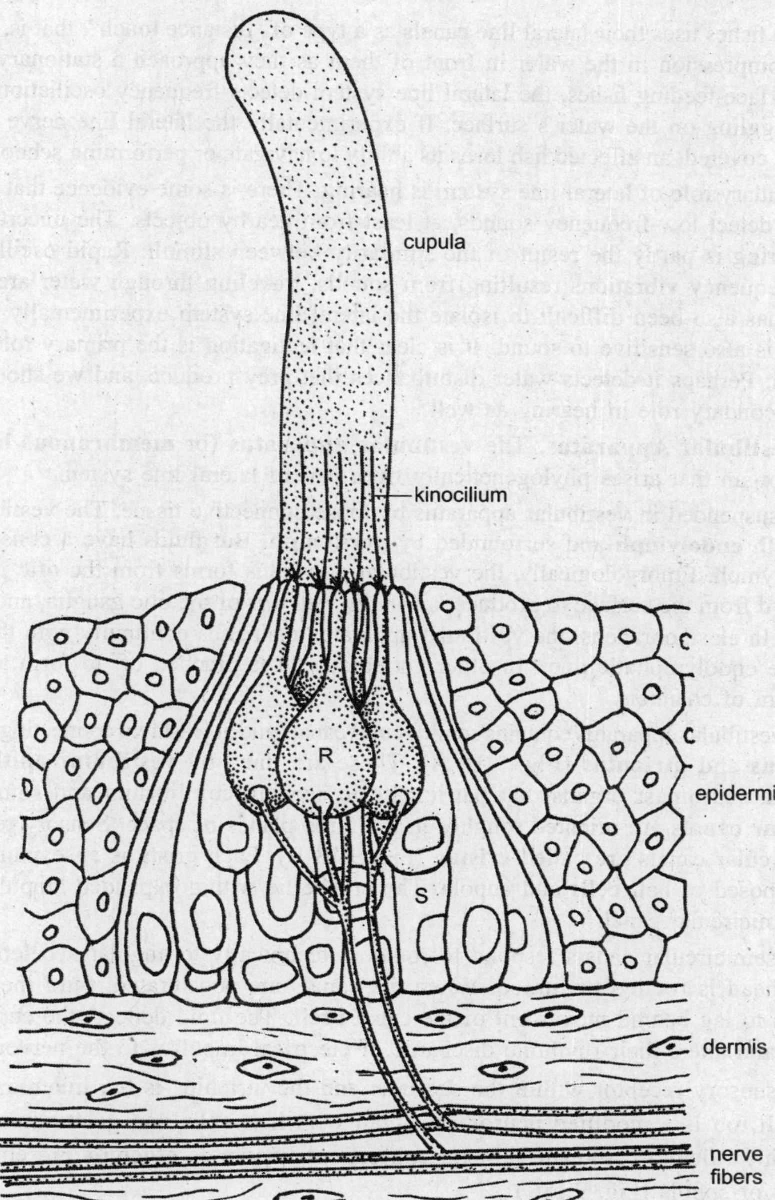

Fig. 9.17. A neuromast organ in the epidermis of *Necturus*. **R**, hair cell (receptor cell) with a single kinocilium and three or four short delicate stereocilia (at the base of kinocilium) extending into the cupula. Contacting the base of the hair cell is a heavily myelinated sensory nerve fiber; **S**, supporting cell (after Kent and Carr 2001).

increase in this discharge rate. If flow is in the opposite direction, the discharge rate falls below its resting rate. If the water passes at right angles to these neuromasts, the resting electrical charge is unaffected. This provides information about the animal's direction of movement and about disturbances in the water. Even cave fish that are blind can navigate around obstacles in their environment because of their lateral line system.

Some fishes uses their lateral line canals as a type of "distance touch", that is, it allows them to detect compression in the water in front of them as they approach a stationary object ahead. In some surface-feeding fishes, the lateral line system detects frequency oscillations produced by insects struggling on the water's surface. If experimentally the lateral line nerve is cut or if its grooves are covered, an affected fish loses its ability to navigate or performing schooling behaviour.

Secondary role of lateral line system is hearing. There is some evidence that the lateral line canals can detect low-frequency sounds, at least from nearby objects. The uncertainty about its role in hearing is partly the result of the similarity between stimuli. Rapid oscillation of water and low-frequency vibrations resulting from sounds, traveling through water are mechanically similar. It has also been difficult to isolate the lateral line system experimentally from the inner ear, which is also sensitive to sound. It is clear that navigation is the primary role of the lateral line system. Perhaps it detects water disturbances that prey produce, and we should allow for a possible secondary role in hearing as well.

4. Vestibular Apparatus. The **vestibular apparatus** (or **membranous labyrinth)** is a balancing organ that arises phylogenetically from part of lateral line system.

It is suspended in vestibular apparatus by loose connective tissue. The vestibular apparatus is filled with **endolymph** and surrounded by **perilymph**. But fluids have a consistency similar to that of lymph. Embryologically, the vestibular apparatus forms from the **otic placode**, which sinks inward from the surface to produce *hair cells*, *neurons* of the otic ganglia, and the *vestibular apparatus*. In elasmobranchs, the vestibular apparatus maintains continuity with the environment through the endolymphatic duct. In other vertebrates, it is pinched off to form a closed, fluid–filled system of channels.

The vestibular apparatus contains semicircular canals and at least two connecting compartments: the **sacculus** and **utriculus** (Fig. 9.18 A). These are lined by **vestibular epithelium** within which arise neuromast organs that participate in sensing equilibrium and sounds. The three **semicircular canals** are oriented roughly in the three planes of space. Sensory receptors within the semicircular canals are called **cristae** (Fig. 9.18 B). Each crista is an expanded neuromast organ composed of hair cells and cupola. The cristae lie within expanded ampullae at the base of each semicircular canal.

The semicircular canals respond to rotation, technically to angular acceleration produced when the head is rotated or turned. When the canals are accelerated, fluid inertia causes the endolymph to lag behind movement of the canal itself. The fluid deflects the cupola, stimulates hair cells, and alters their rhythmic discharge of electrical impulses to the nervous system.

The sensory receptor within the sacculus and the utriculus is the **macula**, or the **otolith receptor**. It too is a modified neuromast organ with hair cells and gelatinous cupola, but in addition, tiny calcium carbonate mineral concretions known as **otoconia** are embedded within the surface of cupola (Fig. 9.18C).

Maculae respond to changes in orientation within a gravitational field. Moving the head or changing position of the body tilts the maculare and changes their orientation with respect to gravity. Acceleration of the body also causes the maculae to respond. The otoconia are inertial masses that accentuate shearing displacements of the hair bundles in response to linear acceleration or changes in head orientation.

The vestibular apparatus keeps the central nervous system informed as to whether an animal is at rest or in motion and conveys information about its orientation. The maculae sense gravity and linear acceleration, and the cristore respond to angular acceleration. In fishes, the vestibular apparatus is also responsive to sound, although this function does not seem to be well developed. In fishes, reptiles, birds, and especially mammals, the vestibular apparatus produces a specialized region of sound reception, the **lagena** (Fig. 9.19).

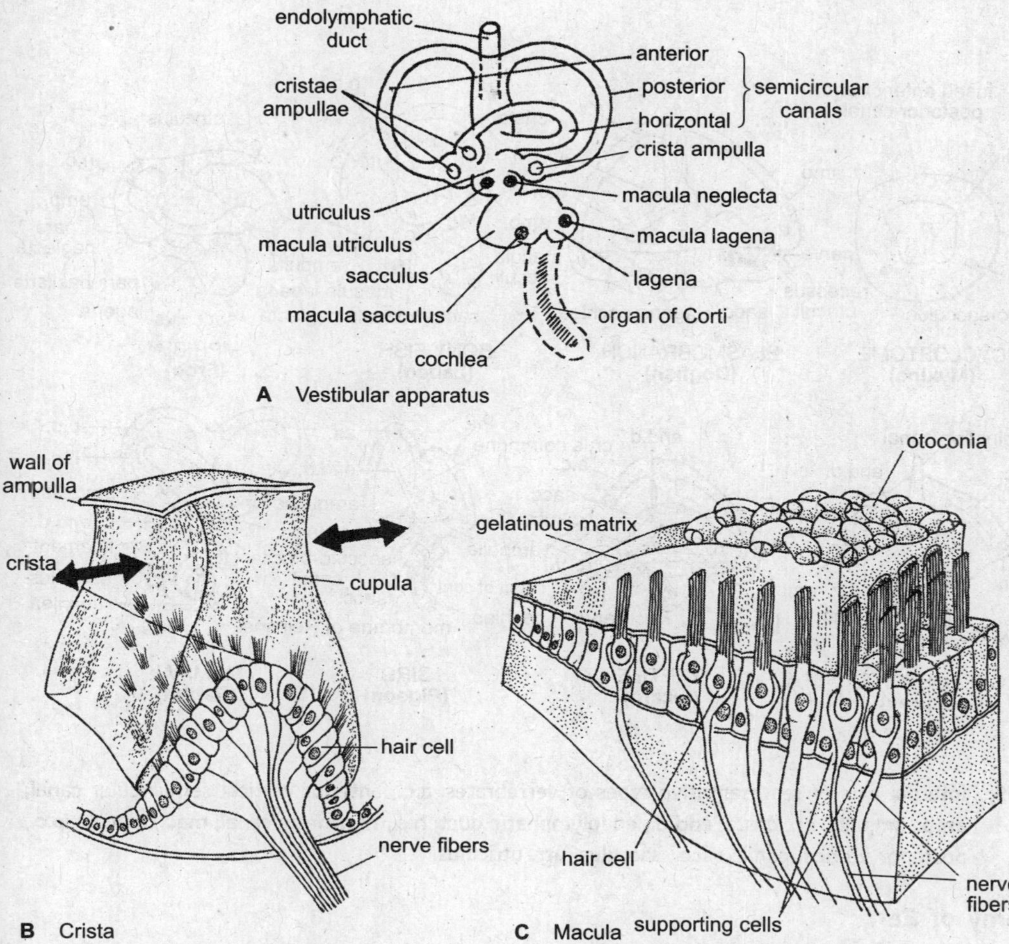

Fig. 9.18. Vestibular apparatus. A – Generalized vestibular apparatus showing the three semicircular canals and major compartments: utriculus, sacculus, and lagena. B – The crista is an expanded neuromast organ. One crista resides at the base of each semicircular canal in an enlarged region, the ampulla. The gelatinous cupola extends across the ampulla and is attached to the opposite wall. Acceleration of the head (arrows) produces a shearing force of endolymphatic fluid against the cupola, which bends and deforms hair cells embedded within it. C – The maculae form a neuromast platform containing otoconia, These maculae reside in the three compartments of the vestibular apparatus. They derive their names from these compartments. In some species a fourth· macula is present, the macula neglecta (after Kardong 2006).

5. Auditory System. The lagena is involved in hearing. It develops as an enlargement of the sacculus, so it is part of the vestibular apparatus. In terrestrial vertebrates, it tends to lengthen, and in most mammals it becomes coiled into **cochlea.** Within the lagena, or cochlea in mammals, lies the sensory receptor of sound, the **organ of Corti** which is a specialized strip of neuromasts connected to the brain via the auditory nerve. The lagena, like the rest of the vestibular apparatus, lies within bone or cartilage of the skull. The "ear" includes up to three adjoining compartments: the external, middle, and inner ears (Fig. 9.20).

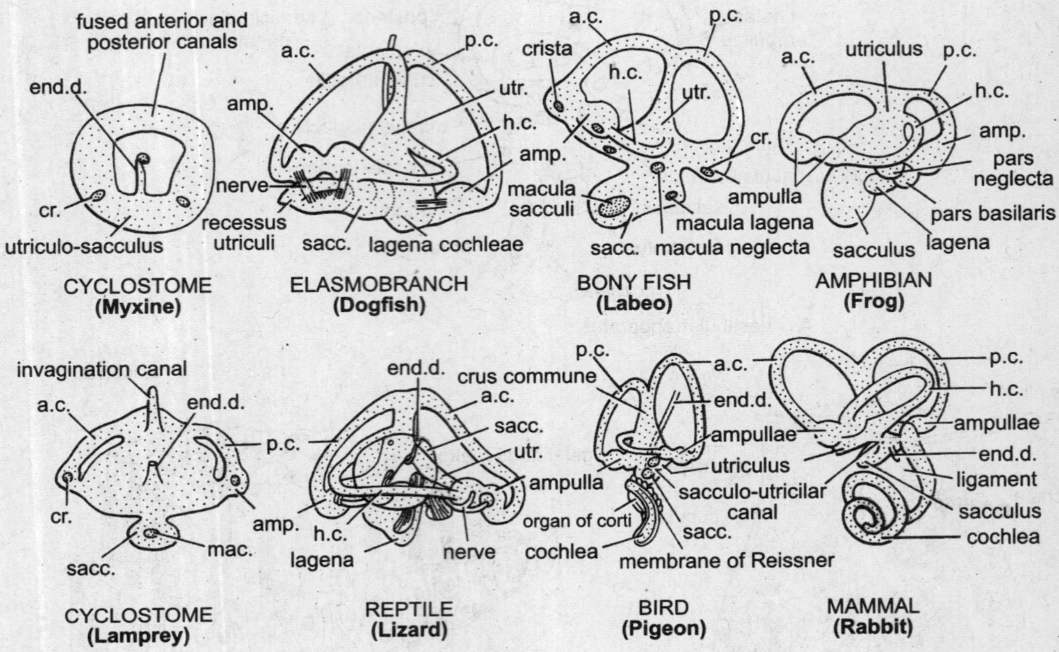

Fig. 9.19. Internal ears of representative types of vertebrates. **a.c.**, anterior vertical semicircular canal; **amp.**, ampulla; **cr.**, crista; **end d.**, endolymphatic duct; **h.c.**, horizontal canal; **mac.**, macula; **p.c.**, posterior vertical canal; **sacc.**, sacculus; **utr.** utriculus.

Anatomy of Ear

1. External ear. It is absent in fishes and amphibians, but present in lizards and crocodilians. External ear consists of a short, indented tube, the **external auditory meatus** which opens to the surface through the **external orifice.** In birds and mammals, the external auditory meatus is elongated. What most people call the "ear" is correctly termed **pinna**. This external cartilaginous flap surrounding the external orifice is present in most mammals. The irregular shape of the pinna helps to differentiate sounds approaching from different directions and channel them into the external auditory meatus. Just as two eyes with overlapping visual fields give stereoscopic vision, paired ears provide stereophonic hearing.

2. Middle ear. It consists of upto three parts: a tympanic membrane or **tympanum**, a middle ear cavity or meatus, and one to three tiny bones, or **middle ear ossicles.** The tympanum or **eardrum**, is a thin, taut membrane at the body surface in frogs and a few reptiles, but it is often recessed at the bottom of external auditory meatus in living reptiles as well as in most birds and mammals. Evidently, the tympanum has evolved multiple times in tetrapods.

The first pharyngeal pouch enlarges as the tubotympanic recess. Its expanded end forms the **middle ear cavity** (Fig. 9.22A-E). The remainder of the recess stays open to form the **eustachian tube**, which maintains continuity between the middle ear cavity and the pharynx.

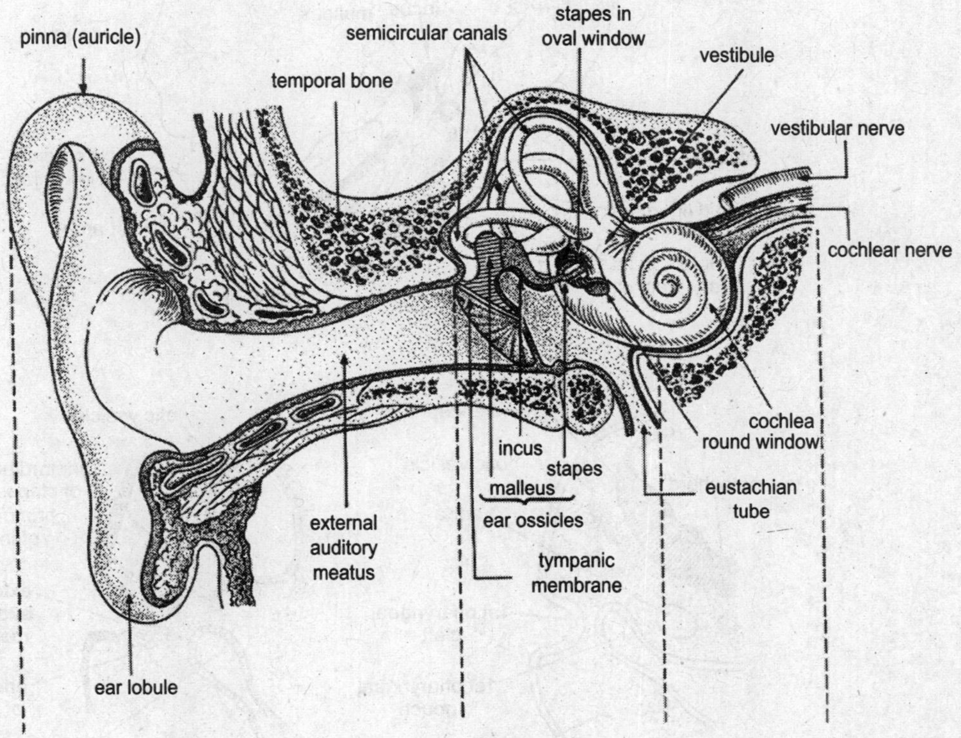

Fig. 9.20. Structure of human ear.

The first tetrapod middle ear ossicle to differentiate was the **stapes**, or columella, a derivative of the first **hyomandibula**, where it functions primarily in jaw suspension in fishes. At first it was a stout bone, lodged firmly between brain case and palatoquadrate, continuing in its primary role as a mechanical buttress (support). Its contact with the braincase placed it strategically so that it might secondarily transfer sound vibrations to the nearby inner ear. In subsequent tetrapods, the stapes abandoned its role in cranial support and became dedicated to sound transmission. The stapes is lightened, making it more responsive to vibrations. Independently, in various tetrapod lineages, stapes became associated with a tympanum, and enclosed within its own middle ear chamber, where it transfers sound form the tympanum to the inner ear. In some amphibians and in reptiles and birds, the stapes is tipped with a cartilaginous extension, the **extracolumella**. This cartilaginous structure, derived from the hyoid arch, rests on the undersurface of the tympanum.

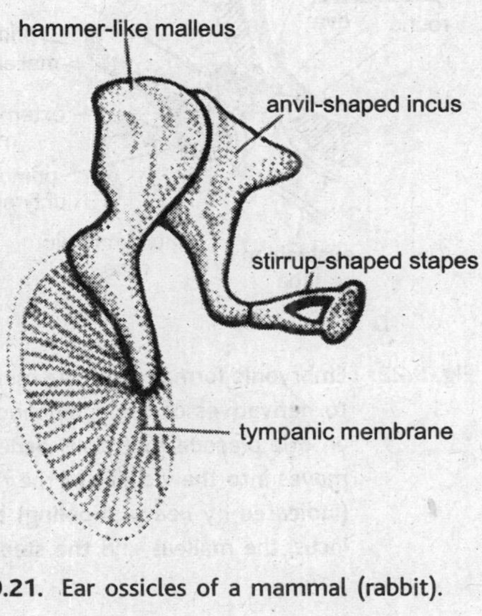

Fig. 9.21. Ear ossicles of a mammal (rabbit).

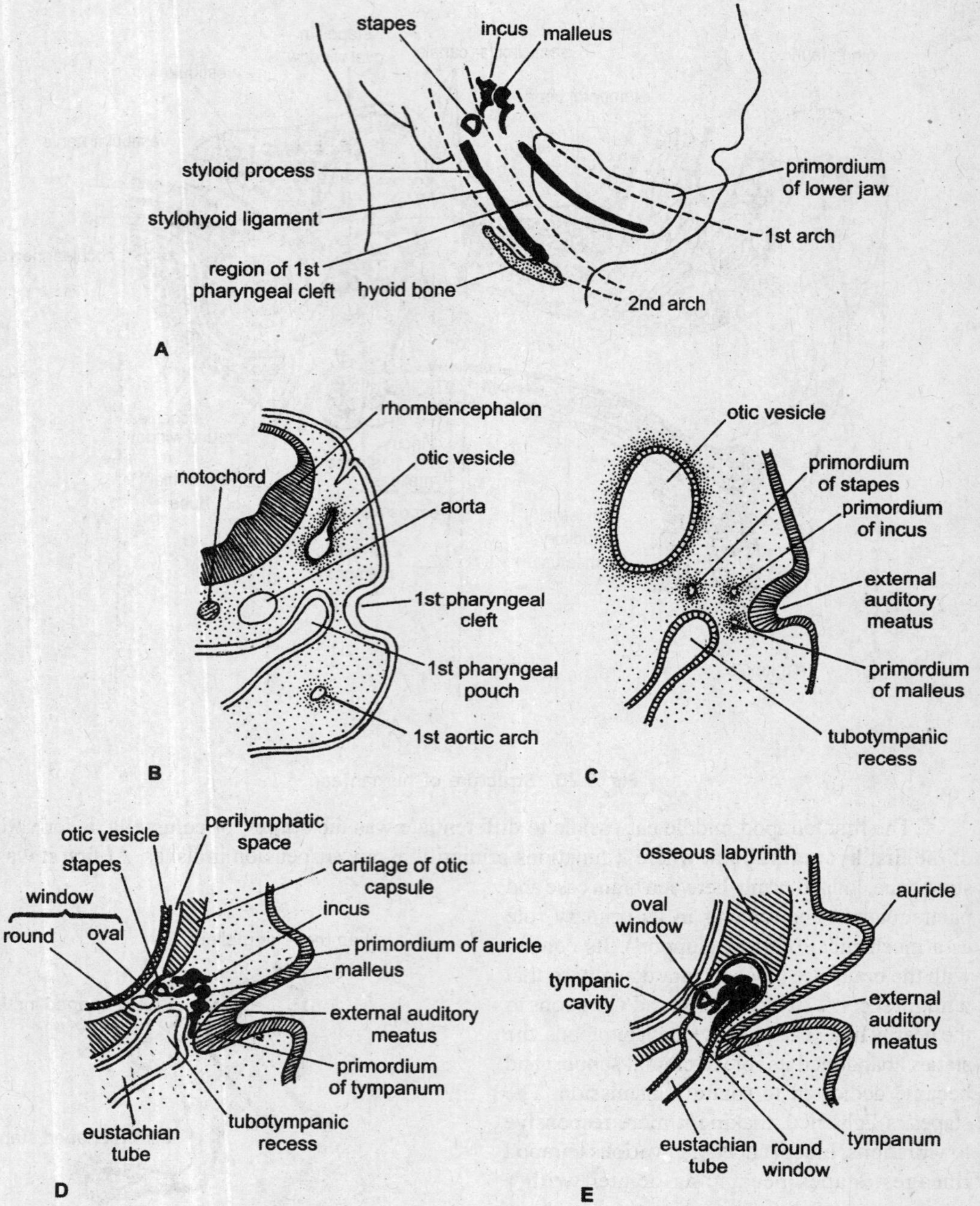

Fig. 9.22. Embryonic formation of the middle (human) ear. A – Location of the middle ear ossicles relative to derivatives of the splanchnocranium. B – The surface of the ectoderm thickness, forming an otic placode that sinks beneath the skin and gives rise to the otic vesicle. The otic vesicle moves into the vicinity of the first pharyngeal cleft and pharyngeal pouches. C – mesenchyme (indicated by heavy stippling) begins to condense and differentiate into the ear ossicles: the incus, the malleus and the stapes (D, E) (after Kardong 2006).

In mammals, there are three middle ear bones. The **stapes** is much reduced and lightened from the early amniote condition. The **incus** and **malleus** are derivatives of the **quadrate** and **articular** bones, respectively. The malleus, incus and stapes form a articulated chain that spans the middle ear cavity from the tympanum to the inner ear (Fig. 9.21).

3. **Inner ear.** The inner ear includes the vestibular apparatus and the surrounding perilymphatic spaces. As mentioned, in birds and mammals the auditory part of the vestibular apparatus is the tubular lagena. In mammals, the lagena forms a coiled cochlea (Fig. 9.23). The organ of Corti runs along a central channel suspended within the lagena. Two parallel perilymphatic channels run on either side. Thus the cochlea consists of three coiled fluid-filled channels. Two perilymphatic channels are the **scala vestibuli**, and the **scala tympani**, and the canal between them is usually called the **scala media** (cochlear duct).

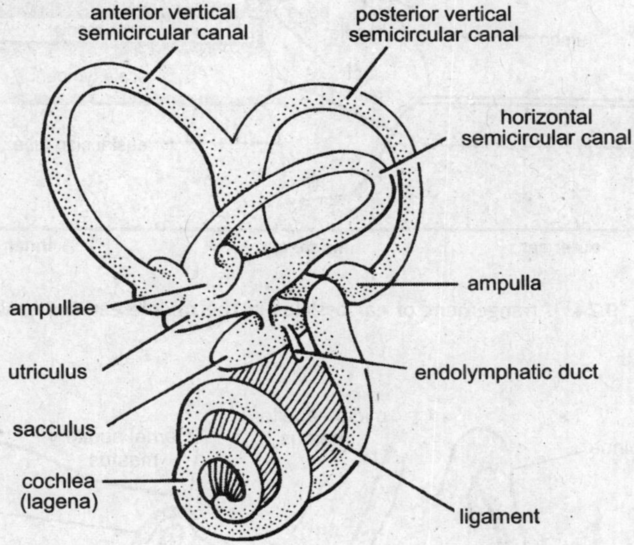

Fig. 9.23. Mammalian (rabbit) vestibular apparatus. Note that the lagena is lengthened and coiled to form the cochlea.

The **basilar membrane** separates scala tympani from the scala media, and the organ of Corti vibrates with the basilar membrane in response to sound waves. In many vertebrates, the hair bundles of the organ of Corti are embedded in a firm plate, the **tectorial membrane**. **Reissner's membrane** is located between the scala vestibuli and scala media (Fig. 9.25).

Sound enters inner ear through the **fenestra ovalis** (or fenestra vestibuli) or oval window (Fig. 9.24). One end of the thin stapes (or columella) expands into a pedicle, or **footplate**, that occupies this window so that sound waves pass from this ear ossicle to the fluid filling chambers of the inner ear. The **fenestra rotunda**, or round window, at the end of perilymphatic channel is sealed by a flexible membrane (Fig. 9.24). The auditory apparatus,containing sensitive hair cells, floats in fluid between the oval window and round window. In all vertebrates, sounds conveyed to this fluid vibrate and mechanically stimulate the auditory receptors.

Function of the Ear

In most mammals (human beings, etc.) the pinna deflects sounds into the external auditory meatus, where they vibrate the tympanum. All mammals have three ear ossicles that amplify these

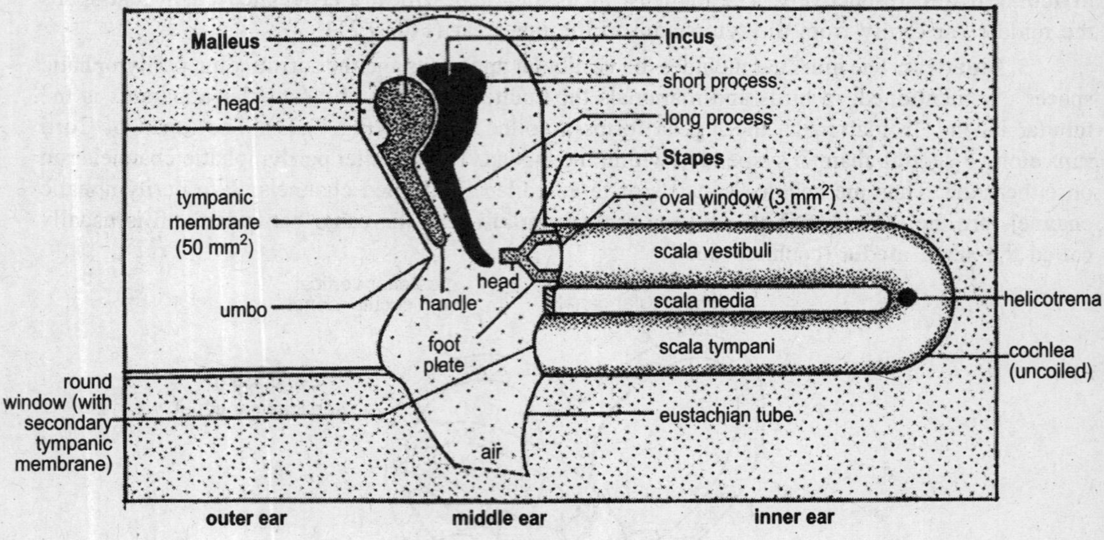

Fig. 9.24. Arrangement of ear ossicles in the middle ear and scalae in the cochlea.

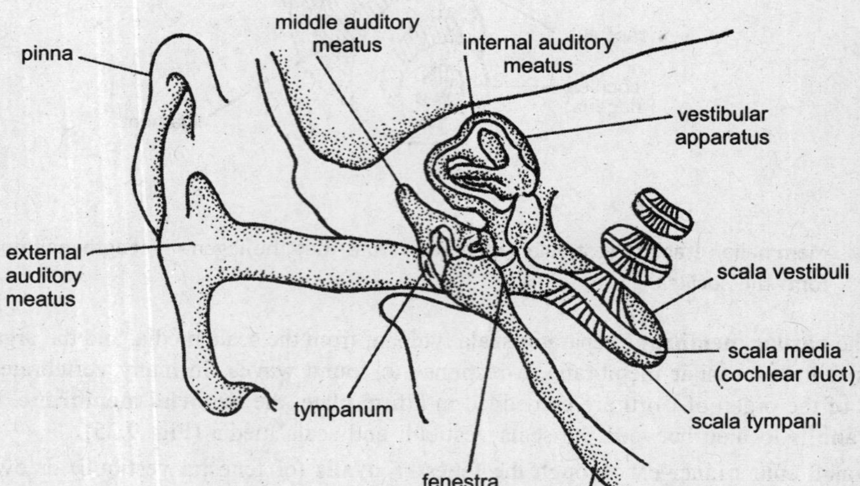

Fig. 9.25. Anatomy of human ear. External, middle and inner ears (after Kardong 2006).

vibrations and carry them to the fenestra ovalis (Fig. 9.20). From the fenestra ovalis, vibrations ripple through the fluid in the extensive cochlea. As mentioned, the cochlea is composed of three parallel channels. The middle channel (**cochlear duct**) includes the organ of Corti, which

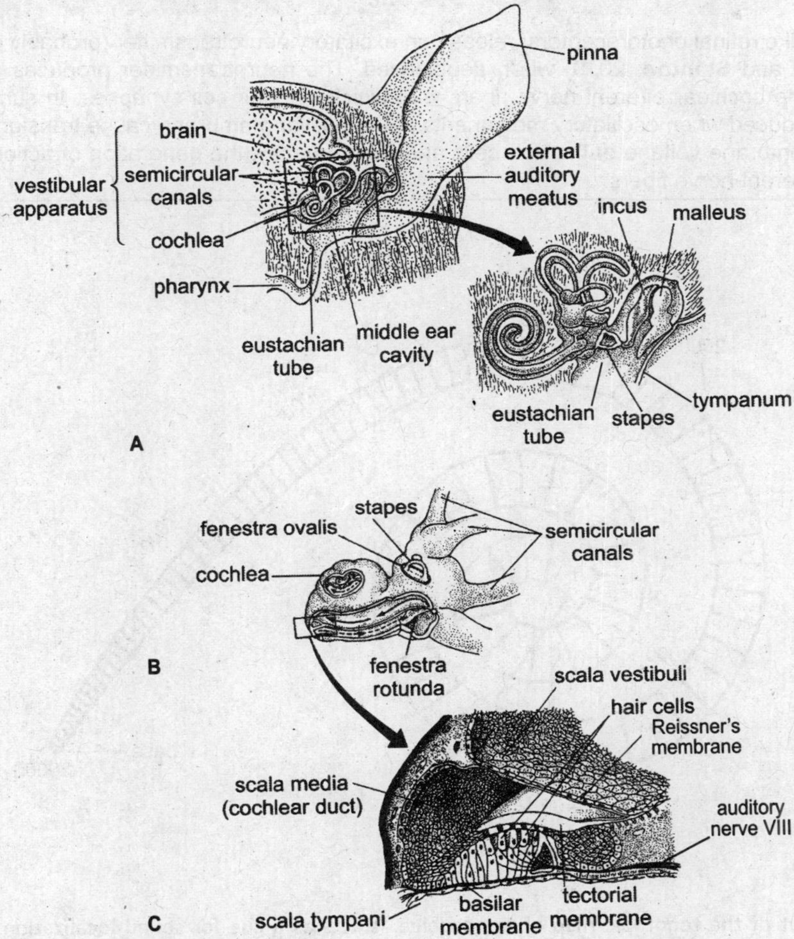

Fig. 9.26. Mammalian ear. A - Cross section through a mammalian skull. B - Internal structure of the cochlea. Section through the organ of Corti showing inner and outer rows of hair cells and the tectorial membrane in which hair bundles are embedded. Sound waves travel first in the scala vestibuli (solid arrow) before passing at the apex of the cochlea into the scala tympani (dashed arrow) (after Kardong 2006).

consists of an outer and inner row of hair cells. There may be 20,000 to 25,000 of these cells. Hair bundles are embedded in the tectorial memberane (Fig. 9.26).

Discrimination of Different Frequencies

Sound waves stimulate the neuromast organs (*e.g.*, auditory papilla, organ of Corti) the sites of sound sensation (Note. Auditory papilla occurs in the ear of reptiles). In some species, the hair bundles protrude directly into the surround fluid. When this fluid vibrates in response to sound waves, the bundles are bent by the moving fluid and stimulated. In most higher vertebrates, hair bundles are embedded in the tectorial membrane. Sound waves impart motion to the tectorial membrane, which then deflects the hair bundles. This action stimulates hair cells (Box 9.5).

Box 9.5.

Hair cells, like retinal photoreceptors, release an excitatory neurotransmitter (probably glutamate, see **Koeppen** and **Stanton,** 2010) when depolarized. The neurotransmitter produces **generator potential** in the cochlear efferent nerve fibers with which the hair cell synapses. In summary, the sound is transduced when oscillatory movements of the basilar membrane cause transient changes in the transmembrane voltage of the hair cells and consequently, the generation of action potential in cochlear afferent nerve fibers.

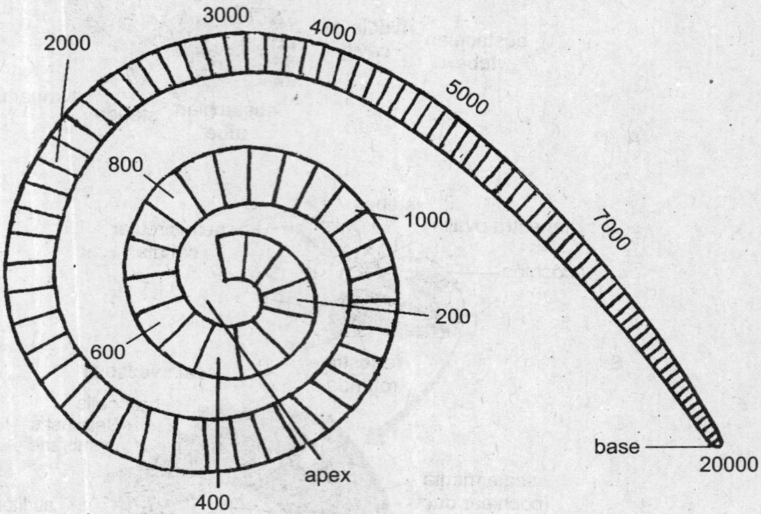

Fig. 9.27. Layout of the tonotopic map of the cochlea. It acts as a cue for sound localization. Locations along the organ of Corti showing how waves of different frequencies have their maximum effects at different places. The numbers refer to hertz (Note. Hertz is a SI unit of frequency, equal to one cycle per second).

In addition to being able to detect sound waves, auditory receptors can discriminate between different frequencies (tones) of sound. Thus, the inner ear is also a **resonance analyzer**. Hair cells are tuned into only a narrow range of frequencies. In mammals, differences in the orientation of hair cells in the inner and outer rows of the organs of Corti produce differences in sensitivity in different regions. Sequential grading of the tuned hair cells along the organ of Corti produces tone discrimination over a range of frequencies.

However, there may be more to tone detection than the tuning of hair cells. In mammals, the basilar membrane on which the organ of Corti resides, changes gradually in width as the cochlea twists (Fig. 9.27). The basilar membrane may be like a harp in that each section of the membrane may resonate only to a distinctive range of frequencies. If this is the case, tones that enter the cochlea impart the greatest motion to the section of the basilar membrane that corresponds to their frequency; thus, specific tones stimulate specific sections of hair cells. It has been suggested that such differential stimulation of hair cells could contribute to tone discrimination.

Place theory of hearing

The basilar membrane serves as a **frequency analyzer**; it distributes the stimulus along the organ of Corti so that different hair cells will respond differentially to particular frequencies of sound. This is basic of the **place theory of hearing**. In addition, hair cells located at different places along the organ of Corti are **tuned** to different frequencies because of difference in their stereocilia and biophysical properties. As a result of these factors, the basilar membrane and organ of Corti have a so-called **tonotopic map** (Fig. 9.27).

QUESTIONS

Long Answer Questions

1. Sense organs provide information about the external environment. What type of information is necessary for survival.
2. The neuromast appears to be the fundamental unit for several sensory modalities. What is it and how does it function in the different modalities?
3. Describe the structure of eye in vertebrates. Also explain the process of focussing mechanism in vertebrates.
4. Describe the structure and function of ear in mammals.

Short Answer Questions

1. What are receptors? Classify them.
2. Write a note on organ of Corti.
3. Write short notes on hair cells and neuromast organs.
4. Draw a detailed labelled diagram showing the structure of mammalian eye.
5. Draw a detailed labelled anatomical diagram of human/mammalian ear.
6. Write short notes on the following.
   (*i*) Cochlea;
   (*ii*) Mechanoreceptors;
   (*iii*) Proprioceptor.

Very Short Answer Questions

1. What is scotopic vision?
2. Define the photopic vision?
3. Name the visual pigment of rods and cones of retina.

Fill in the blanks

1. A crista of vestibular apparatus is an ________ ________organ.
2. _______ of ampulla of internal ear respond to ________ _________ during motion of vertebrates.
3. Trichromatic system of human being has three types of visual pigment in their cones: __________, ___________ and _______.
4. The retinas of all vertebrates have rods, hence, all of them possess some degree of_________ vision.

Multiple Choice Questions

1. What are the receptors of touch and pressure, vibrations and balance?
   (*a*) thermoreceptors (*b*) chemoreceptors
   (*c*) mechanoreaptors (*d*) photoreceptors
2. Muscles have which types of receptors?
   (*a*) mechanoreceptors (*b*) thermoreceptors
   (*c*) chemoreceptors (*d*) photoreceptors
3. Pecten is found in the eye of
   (*a*) fish (*b*) frog
   (*c*) lizard/snake (*d*) bird

ANSWERS

Very Short Answer Questions

1. Scotopic vision is sensitivity to dimlight (*i.e.*, black and white colours). It takes place by rods of retina.
2. Photopic vision is colour vision in bright light. It takes place by cones of retina.
3. Opsins.

Fill in the Blanks

1. Expanded, neuromast.
2. Cristae, angular, acceleration.
3. S (blue), M (green), L (red).
4. Scotopic.

Multiple Choice Questions

1. (*c*) 2. (*a*) 3. (*d*).

CC IX

Animal Physiology: Life Sustaining Systems

1

Definition, History and Branches of Physiology

1.1 DEFINITION

The study of the way in which the organs and organ systems in a living organism function is called **physiology (Indge**, 1997). Physiology is the study of function in living organisms. Based on knowledge of anatomy, physiology seeks to demonstrate the manner in which organs perform their tasks, and in which the body is organised and maintained in a state of **homeostasis**. The aim of physiology is to explain the physical and chemical factors that are responsible for the origin, development and advancement of life.

The term *physiology* is originated from two Greek words: *physis* = nature, and *logos* = study. It means *the study of function and vital processes of living organisms and their constituent parts*. Since organs of living organisms are composed of chemical substances, so the analysis of their functions involves the application of the laws of chemistry and physics. Since physiology is a dynamic science, its basic concepts are mostly supplemented with experimental evidence. The ultimate aim of physiology is to understand, in physical and chemical terms, the mechanisms that operate in living organisms at all levels, ranging from the subcellular to the integrated whole animal.

Unlike metaphysics, the science of physiology holds that there is nothing "occult" or supernatural about life. All the living processes of an organism can be explained, or better, should be explained, with the help of underlying structural peculiarities and the chemical and physical changes.

In an animal, the various body functions require the coordinated activity of a number of tissues and organs. The human brain, for example, cannot function without a constant supply of blood, carrying oxygen and glucose, provided by the pumping of the heart. The heart cannot survive more than minutes without oxygen supplied via the circulating blood by the lungs. The lungs cannot function without neural commands to the respiratory muscles from the brain.

1.2 HISTORY

The basic idea for the growth and development of physiology was the requirement of medicine. The ancient physicians had considerable interest in physiology as they realised that it is impossible to cure a human disease without knowing the structure of the human body and function of its organs (Box 1.1).

Box 1.1

Herophilus (320 B.C.). Alexandrian physician regarded as **Father of Scientific Anatomy** and one of the first dissectors. He distinguished nerves from tendons and partially recognised their role. His work survives only through Galen's writings.

Galen of Pergamum (A.D.130–200). Greek physician at the court of the Emperor Marcus Aurelius. His writing drew together the best of classical medicine and provided through the medieval period to Renaissance. Galen himself contributed many original and careful observation in anatomy and physiology.

William Harvey (1578–1657). British physician who discovered the circulation of the blood. He showed that the heart acts as a pump and that the blood circulates endlessly about the body; that there are valves in the heart and veins so that the blood can flow in one direction only; and that the necessary pressure come only from the lower left-hand side of the heart. Harvey's discoveries demolished the theories of Galen that blood was consumed at the body's periphery and that left and right sides of the heart were connected by pores. He also made important studies of the development of the embryo.

Marcello Malpighi (1628–1694). Italian physician and biologist. Malpighi is considered as the **"Father of Microscopic Anatomy"**. He was a discoverer of the capillaries and a pioneer in several fields of medicine and biology.

Jan Swammerdam (1637–1680). Dutch microscopist. His precision enabled him to make many discoveries, including red blood cells (1658).

The term '*physiology*' was first introduced by the French physician **Jean-Fernel** in 1542. **Andreas Vesalius** (1514–1564) dissected human body and described several organs for the first time in his book *On the structure of human body* (1543). Classic demonstration of blood circulation by **William Harvey** in 1628 was the key experiment that laid the foundation of modern physiology. This work threw an entirely new light on the subject of physiology in general and had great influence on its progress. In 1658 **Swammerdam** described the red blood corpuscles. In 1660, **Marcello Malpighi** demonstrated the capillaries in the lung of a frog and was able to complete the description of blood circulation discovered by **Harvey**. In 1738, **Daniel Bernoulli** (1700–1782; Swiss mathematician, anatomist and botanist) observed the movement of fluids in the body and laid down the foundation of haemodynamics. Blood pressure was first measured in arteries of animals by **Stephen Hales** (1677–1761). This discovery strengthened the idea that bodily processes could be measured quantitatively more than a century after Harvey's key experiment. In 1777, **Reamer** and **Spallanzani** (1729–1799) discovered the functions and properties of gastric juice secreted by the stomach cells or glands. In 1778, **Lavoisier** demonstrated that oxidation of organic compounds of the body is carried out by the oxygen present in the air (Box 1.2). Later **Lavoisier** and **Laplace** demonstrated that oxidative processes yield same amount of heat both during the oxidation of "carbonic combustions" in the organism and during their combustion outside the body of organism.

Box 1.2

Antoine Laurent Lavoisier (1743–1794). He was a French scientist who coined the term *Oxygen* for Priestley's "dephlogisticated air"; investigated the chemistry of respiration and demonstrated analogy of respiration with combustion.

Albrecht von Haller (1708–1777). He was a Swiss biologist and regarded as founder of experimental anatomy. He investigated the respiration, the blood circulation, the nervous system and irritability and sensibility of different types of body tissue, in all cases relying on experiment.

Growth of Physiology in 19th Century

In 1814, the catalytic action of enzymes was demonstrated by **Kirchhoff**. This was the beginning of extensive investigations that have been fruitful in understanding the metabolic activity of living protoplasm. In 1833, the concept of reflex action was discovered by **Hall**. In 1838, **Jones J. Berzelius** (1779–1848) coined the term **protein**. **Mulder** (1839) studied nature of proteins and regarded them of first or primary importance in the structure of living matter. It was **Verworn** who stated that living matter might be defined as the material in which protein metabolism occurs. In 1947, **H.L.M. von Helmholtz** (1821–1894; German physiologist) formulated the *law of conservation of energy*. He was also first to measure the speed of nerve impulses and invented the ophthalmoscope (1850). In 1851, **Heinrich Muller** (1820–1864; German anatomist) discovered **rhodopsin**. In 1856, with **Rudolph**

A. von Kolliker (1817–1905), Muller demonstrated that each contraction of a frog's heart produces an electrical current. **Stannius** (1852) experimented on the heart and his experiments started many investigations on the nature of the heart beat (Box 1.3). **Claude Bernard** (1859) pointed out that complex organisms live in two environments-an external environment and an internal environment. Internal environment surrounds the cells and tissues of body and is characteristic of the animal species and remains relatively constant. This internal environment serves as medium for the exchange of foods and wastes and for distribution of chemical messengers.

Box 1.3

Sir Charles Bell (1774–1842) was a Scottish anatomist. He was a pioneer investigator of the working of the nervous system. His most important discovery was to distinguish the function of sensory and motor nerves.

Johannes Peter Muller (1810–1858) was a German physiologist and anatomist who described the Mullerian duct in 1825. During 1834–40, he published a handbook of physiology that ranks as the first great text on the subject produced in the 19th century.

William Beaumont (1758–1853). He was an army physician noted for his researches into the human digestive system. While on an assignment in northern Michigan in 1822 he treated a trapper with a serious stomach wound; when the wound healed, an opening (or fistula) into the victim's stomach remained, through which Beaumont could extract gastric juice for analysis.

Claude Bernard (1813–1878) was a French physiologist. He followed the work of **Beaumont** and opened artificial fistulas in animals to study their digestive systems. He demonstrated the role of the *pancreas* in digestion. Bernard also discussed the presence and function of glycogen in liver (1856) and in 1851 he reported the existence of the vasomotor nerve. In 1859, Bernard recognised the importance of constancy of internal environment of animals.

Ludwig (1872) described the gas exchange of the blood. **Pasteur** (1876) discovered the *Pasteur effect* and showed that yeasts and mould cause the fermentation of sugar to alcohol and CO_2. **Kuhne** (1878) coined the term *enzyme* and discovered the nature of enzyme. In 1880, **Sidney Ringer** (1835–1910) showed the influence of blood ions on the heart contraction. He developed a balanced solution for keeping isolated organs functional for long period of time. **Metchinkoff** (1882) highlighted the role of phagocytosis in immunity (Box 1.4). The theory that microbes are ingested and destroyed by certain white blood cells (phagocytes) shares with the theory of chemical bodies (antibodies), the chief explanation of the body's natural immunity.

Box 1.4

Sir William Maddock Bayliss (1860–1924). English physiologist who, in collaboration with **Ernest Henry Starling** introduced the term *hormones* to describe substances that when secreted from one part of the body, have a specific effect on another. In 1902, **Bayliss** and **Starling** discovered *secretin*, one of the first hormone to be identified. Bayliss also introduced the saline injections for cases of surgical shock during World War I.

Ernest Henry Starling (1866–1927). British physiologist who worked with Bayliss on the study of hormones. He also did research into the heart, demonstrating the **Starling equilibrium** in 1896, *i.e.*, the balance between hydrostatic pressure, causing fluids to flow out of the capillary membrane and osmotic pressure, causing fluid to be absorbed from the tissue into the capillary.

Elie Metchnikoff (1845–1916). Russian biologist who shared with **Paul Ehrlich** the 1908 Nobel Prize for Physiology or Medicine for his discovery of phagocytes (in humans called **leucocytes**) and their role in defending the body from bacteria.

Paul Ehrlich (1854–1915). German bacteriologist and immunologist, the *Founder of chemotherapy* and an early pioneer of hematology. His discoveries include: a method of staining (1882) and hence identifying the tuberculosis bacillus; the reason for immunity in terms of the chemistry of antibodies and antigens (for which he was awarded with **Metchnikoff** the 1908 Nobel Prize for Physiology or Medicine); and the use of the drug *Salvarsan* to cure syphilis, the first drug to be used in treating the root cause of a disease (1911).

Albrech Kossel (1853–1927). German biochemist awarded the 1910 Nobel Prize for Physiology or Medicine for his work on *proteins* and *nucleic acids*, principally for showing that "nuclein" from cellular sources was not one substance but was composed of a protein and a non-protein (nucleic acid) component. He also demonstrated the presence of purines and pyrimidines in nucleic acids.

Our knowledge concerning the relation between the pancreas and carbohydrate metabolism began with the classical demonstration by **von Mering** and **Minkowski** (1889) that removal of the pancreas in the dog was followed by the development of symptoms of diabetes mellitus. In 1893, the anatomy and physiology of the auriculo-ventricular node and bundle was discovered by **Wilhelm His** (1863–1934; German physician), *i.e.*, Bundle of His of heart was investigated. In 1896, **Buchner** succeeded in extracting from yeast cell a substance (later on called **zymase**) that was active in fermentation as the original yeast. In 1897, **Sir C.S. Sherrington** (1857–1952; British neurophysiologist) discovered the concept of the synapse in the nervous system. He shared with **E.D. Adrian** (1889–1977; English physiologist who explained the function of neurons in nervous system) the 1932 Noble Prize for Physiology or Medicine. In 1898, mitochondria were discovered by **Benda**.

Growth of Physiology in 20th Century

In 1900, **Karl Landsteiner** (1868–1943; Austrian-born US pathologist) discovered the human blood group (*e.g.*, ABO system of blood typing) and awarded the 1930 Nobel Prize for Physiology or Medicine. This fundamental discovery made successful blood transfusions as well as initiated the great amount of work on the biochemistry and physiology of blood. **E.A. von Behring** (1854–1917; German bacteriologist) was awarded *first Nobel Prize* for Physiology or Medicine (1901) for his work in developing an **antitoxin** giving protection against diphtheria. **Bayliss** and **Starling** (1902) discovered the first hormone, called **secretin**, produced by the wall of alimentary canal (*i.e.*, intestine). In 1904, **Cannon** described the mechanics of digestion by the help of X-rays. In 1905, **Haldane** and **Priestley** discovered the role of CO_2 in the regulation of breathing. **Einthoven** (1906) discovered the mechanics of electrocardiogram. **I.P. Pavlov** (1849–1936), Russian physiologist and 1904 Nobel Laureate, studied physiology of digestive system. In 1906, **Robert Barany** (1876–1936; Austrian-born physiologist) devised two tests to measure vestibular function (of inner ear). He was awarded 1914 Nobel Prize for Physiology or Medicine for this discovery. In 1907, **Hopkins** showed the relationship of lactic acid to muscle contraction. **E.T. Kocher** (1841–1917; Swiss surgeon) was awarded the Nobel Prize for Physiology or Medicine for his discovery of the relationship between the thyroid gland and cretinism. In 1918, **E.H. Straling** (1866–1927; British physiologist) discovered **the law of heart**. According to this law, the more the ventricles are filled with incoming blood, the greater is the force of their contraction at systole. **S.A.S. Krogh** (1874–1949; Danish physiologist) awarded the 1920 Nobel Prize for Physiology or Medicine for his discovery that capillaries contract and expand so as to vary the amount of blood oxygen supplied to parts of the body in accordance with their individual requirements. **Langley** (1921) discovered the concept of autonomic nervous system. In 1922, **F. Banting** along with **C. Best** isolated the insulin hormone. **Kopec** (1922) showed that the brain was necessary for insect metamorphosis and thus demonstrated the role of hormonal factors in invertebrate physiology. **Houssay** (1924) discovered the role of pituitary gland in regulation of carbohydrate metabolism (Box 1.5). In 1925, **Barcroft** explained the function of spleen and **Mast** discovered the nature of amoeboid movement.

Box 1.5

William Einthoven (1860–1927; Dutch physiologist). He was awarded the 1924 Nobel Prize for Physiology or Medicine for his invention of technique of investigation of heart action with electrocardiograph.

Sir Frederick Gowland Hopkins (1861–1947; British biochemist). He shared with **Eijkmann** the 1929 Nobel Prize for Physiology or Medicine for his work showing the necessity of certain dietary elements (now identified as **vitamins**) for the maintenance of health.

Christian Eijkman (1858–1930; Dutch pathologist). Following a trip to Indonesia (1886) to investigate **beriberi** disease he was able to show that the disease resulted from dietary deficiency. This led to the discovery of vitamins. For his work he was awarded (with Sir F.G. Hopkins) the 1929 Nobel Prize for Physiology or Medicine.

Sir Frederick Grant Banting (1891–1941; Canadian physiologist). He along with **C.H. Best**, first of all isolated the hormone **insulin** from the pancreas of dog (1922). For his work, **Banting** shared the 1923 Nobel Prize for Physiology or Medicine with J.J.R. **Macleod**.

Bernardo Alberto Houssay (1887–1971; Argentinian physiologist). He was awarded with (**C. Cori** and **G. Cori**) the 1947 Nobel Prize for Physiology or Medicine for his discovery that certain hormones produced by pituitary gland were responsible for regulating the blood's sugar and insulin content.

In 1926, **A.V. Hill** measured the heat production in the nerves by applying the principle of thermocouple (Box 1.6). **Eggleton** (1927) discovered the role of phosphocreatine in muscle contraction. **A. Harden** studied enzyme action in fermentation of sugars and demonstrated the existence of **coenzymes** (Box 1.6). In 1929, American physiologist **Walter B. Cannon** coined the term **homeostasis** for Bernard's constancy of internal environment. In the same year, **Karl Lohmann** discovered ADP in muscles. **Lewis** (1931) gave the concept of **pinocytosis**. In 1932, **Krebs** and **Henseleit** discovered the reactions of conversion of ammonia into urea. In 1937, **Krebs** discovered the Krebs cycle (or citric acid cycle) which has been the main focal point in the study of cellular physiology (Box 1.6). **Richard Kuhn** (1900–1967; German chemist) studied chemistry of carotenoids and vitamins and was awarded the 1938 Nobel Prize for Chemistry. In 1940, **Landsteiner** and **Wiener** discovered the **Rh-blood factor**. In 1941, **Fritz Lipmann** (1845–1921) discovered formation of ATP from ADP. He is regarded as **"Father of ATP cycle"**. In the same year, **Szent-Gyorgyi** demonstrated the role of ATP in muscular contraction (Box 1.6). In 1945, **Cori** showed the hormone influence on enzyme activity (Box 1.6). **Coenzyme A** was discovered in 1947 by **Lipmann**. **Szent-Gyorgyi** (1947) forwarded the concept of the contractile substance, the **actomysin**. **Martin** and **Synge** (1944) developed a biochemical technique, called **paper chromatography** (Box 1.6). In 1952 **James** and **Martin** developed gas-liquid chromatography.

Box 1.6

Sir Archibald Vivian Hill (1886–1977). British physiologist who shared with **Otto Meyerhof** (1884–1951; German physiologist) the 1922 Nobel Prize for Physiology or Medicine for their independent work on the biochemistry of **muscle contraction** and **relaxation**.

Sir Arthur Harden (1865–1940). British biochemist awarded with **Euler-Chelpin** the 1929 Nobel Prize for Chemistry for their work on enzyme action in the fermentation of sugars. Harden demonstrated the existence of coenzymes, and Euler-Chelpin determined the structure of the first coenzyme, called **cozymase**.

Albert Szent-Gyorgyi (1893–1986). Hungarian-born US biochemist awarded the 1937 Nobel Prize for Physiology or Medicine for work on biological combustion processes, especially in relation to vitamin C.

Carl Ferdinand Cori (1896–1984). Czech-born US biochemist who shared the 1947 Nobel Prize for Physiology or Medicine with his wife **Gerty Theresa Radnitz Cori** (1896–1957), for their joint explanation of the processes by which glycogen is broken down and reformed in the body. **B.A. Houssay** also shared in 1947 Nobel Prize for Physiology or Medicine.

Sir Hans Adolf Krebs (1900–1981). German-born British biochemist. In 1932, Krebs discovered with the German biochemist **Kurt Henseleit** a series of chemical reactions occurring in mammalian tissue by which ammonia is converted into urea (urea cycle). In 1937 Krebs discovered an essential series of intermediate reactions in the oxidation of food-stuffs, the Krebs cycle. In 1953, Krebs was awarded the Nobel Prize (with **Lipmann**) for Physiology or Medicine for his research on metabolic processes.

In 1953, **Watson** and **Crick** proposed a model of DNA that explained the chemical structure of DNA. Both share 1962 Nobel Prize for Physiology or Medicine for their work establishing the "double helix" molecular model of DNA. In 1954, **Huxley** and **Hanson** proposed a **sliding hypothesis** for muscle contraction (Box 1.7). In 1955–1957, **Kornberg** and **Ochoa** synthesized nucleic acids. Both were awarded the 1959 Nobel Prize for Physiology or Medicine for discovering an enzyme, called **DNA polymerase** that could produce from a mixture of nucleotides exact replicas of DNA molecules. In 1956 **Borsook** and **Zamecnick** found that in rat liver the ribosomes are the site of protein synthesis. Later on, **Hoagland** and **Zamecnick** demonstrated the involvement of transfer RNA in the process of protein synthesis. In 1957, **R.W. Holley** established nucleotide structure of a nucleic acid. **Francis Crick** proposed that transfer RNA functioned as an adapter molecule in protein synthesis and he also recognised the coding sequence of mRNA.

Box 1.7

Sir Andrew Fielding Huxley (b.1917). British physiologist. He shared the 1963 Nobel Prize for Physiology or Medicine with **Sir A.L. Hodkin** (b.1914; British physiologist) and **Sir J. Eccles** (b.1903; Australian physiologist) for his work with Hodkin on the chemical basis of nerve impulse transmission (across synapses).

Robert William Holley (b.1922). US biochemist who shared with **Nirenberg** and **Khorana** the 1968 Nobel Prize for Physiology or Medicine for his work in establishing for the first time the nucleotide structure of a nucleic acid.

Marshall Warren Nirenberg (b.1927). US biochemist received 1968 Nobel Prize for his major contributions toward the decipherment of the genetic code.

Har Gobind Khorana (b.1922). Indian-born US biochemist. He got 1968 Nobel Prize for his major contributions toward deciphering the genetic code.

Andre Michael Lwoff (1902–1994). French microbiologist awarded with **F. Jacob** and **J. Monod** the 1965 Nobel Prize for Physiology or Medicine for his work on lysogeny, a process of genetic intersection between bacteriophage and bacteria.

Jacques Lucien Monod (1910–1976). French biochemist who received 1965 Nobel Prize for his work with Jacob on regulatory gene action in bacteria.

Francois Jacob (b.1920). French biologist who received 1965 Nobel Prize for his work on regulatory gene action in bacteria.

In 1961, **Jacob** and **Monad** recognised the role of messenger RNA in the genetic code and **Nirenberg** invented a method to form the code for amino acids (Box 1.7). **A.F.H. Huxley** worked on chemical basis of nerve impulse and got 1963 Nobel Prize for Physiology or Medicine. **H.K. Hartline** (1903–1983; US physiologist), 1967 Nobel Laureate, studied the function of retinal nerve cells. In 1966, **Khorana** demonstrated the proof of code assignments in the genetic code. Thus by alternating codon (CUG and UCU) in an artificial mRNA chain it was possible to synthesize a polypeptide of alternating amino acids (leucine and serine) for which these codons respectively stood. In 1970, **Khorana** synthesized the first artificial gene that contained 77 nucleotides. The first wholly **artificial gene** with the potential to function inside a living cell was synthesized by Khorana in 1973. This gene is a 126-unit DNA fragment that codes for the synthesis of tyrosine transfer RNA in *E.coli*. This one is said to be the first gene that has the potential for both functioning and having its product detected in a living cell (Box 1.7).

Box 1.8

Albert Claude (1899–1983; Belgian-US biologist) who shared the 1974 Nobel Prize for Physiology or Medicine with **Georg Emil Palade** and **Christian Rene de Duve** for demonstrating the usefulness of the electron microscope and the centrifuge in biological studies. Claude pioneered the study of the internal structure of cells.

Georg Emil Palade (b. 1912). Romanian-born US cell biologist, Nobel Laureate of 1974, discovered the nature of the granules found in cell cytoplasm that are now called **ribosomes**.

C.R. de Duve (b. 1917). British-born Belgian-US biochemist, 1974 Nobel Laureate, discovered the lysosomes.

Earl Wilbur Sutherland, 1971 Nobel Laureate, demonstrated the role of cyclic adenosine 3′, 5′-monophosphate (cAMP) in intracellular mediation of hormonal effects on various cellular processes such as lipid metabolism, membrane transport and cell proliferation. The 1972 Nobel Prize for Physiology or Medicine has been awarded to **G.M. Edelman** (b. 1929; US biochemist) and **R. Porter** (1917–1985; British biochemist) in recognition of their outstanding contributions in discovering the chemical structure of antibodies. This discovery has initiated great amount of research in immunobiology all over the world, producing results of applied value for clinical diagnosis and therapy.

The 1974 Nobel Prize for Physiology or Medicine has been awarded jointly to **Albert Claude, George Emil Palade** and **Christian de Duve** for their reforms that started the modern science of Cell Biology nearly 30 years ago and also for their major contributions to the understanding of inner working of living cells (Box 1.8).

The 1975 Nobel Prize for Physiology or Medicine has been awarded to **Renato Dulbecco** (b. 1914; Italian-born Anglo-American physiologist), **Howard Martin Temin** (b. 1934; US virologist) and **David Baltimore** (b. 1938; US virologist) for their discoveries concerning interaction between tumor virus and the genetic material of the cell.

The 1976 Nobel Prize for Physiology or Medicine was awarded to **Daniel Carleton Gajdusek** (b. 1923; US physiologist) for finding out the cause (a slow virus transmitted during cannibalistic rites) of a fatal degenerative disease called **Kuru** among the Fore tribes of New Guinea. This prize was shared by **Baruch Samuel Blumberg** (b. 1925; US physiologist) who identified a protein (called the "Australian antigen") in the blood of an Australian aborigine that proved to be closely related to the virus which causes debilitating liver disease, hepatitis B.

The 1977 Nobel Prize for Physiology or Medicine was shared by **Rosalyn Sussman Yalow** (b. 1921; US medical physicist; shared half of the Nobel Prize) for developing the technique of radioimmunoassay for estimation of hormones in blood and **Roger Guillemin** (b. 1924; French-born UK physiologist) and **Andrew Schally** (b. 1926; Polish-born US physiologist) for isolating, characterising and synthesizing three polypeptides involved in the control of the anterior pituitary gland in the brain.

The 1978 Nobel Prize for Physiology or Medicine has been awarded to **Daniel Nathanas** (b. 1928; US scientist) and **Hamilton O. Smith** (b. 1931; US molecular biologist) and **Warner Arber** (b. 1929; Swiss microbiologist) for their discovery of restriction enzymes and their application to the problems of molecular biology. The restriction enzymes provide the 'chemical knives' that split chromosomes (*i.e.*, DNA molecules) into their component genes.

Peter Dennish Mitchell (1920–1992) was a British biochemist who received 1978 Nobel Prize for Chemistry for his theory of how cells obtain energy from foodstuffs and sunlight (credited for chemiosmotic mechanism of oxidative phosphorylation).

The 1979 Nobel Prize for Physiology or Medicine has been awarded jointly to **Mr. Allan Mcleod Cormack** (b. 1924; US physicist) and **Sir Godfrey Newbold Hounsfield** (b. 1919; British engineer) for the development of computer-assisted **tomography**, a new method for X-ray diagnosis, particularly in neurosurgery.

The 1980 Nobel Prize for Physiology or Medicine has been awarded to **George D. Snell** (b. 1903; US geneticist), **Baruj Benacerraf** (b. 1920; US pathologist) and **Jean B.G. Dausset** (b. 1916; French geneticist) for their research work that established links between genetics and immunology. They worked on histocompatability system and discovered the so-called immune-response genes.

The 1981 Nobel Prize for Physiology or Medicine was awarded to **Roger W. Sperry** (1913–1994; US psychobiologist; received half of prize for his research into the organisation of nerve cells in the brain), **David Hunter Hubel** (b. 1926; US neurobiologist) and **Torsten Wiesel** (b. 1924; Swedish neurobiologist). **Hubel** and **Wiesel** shared half of Nobel Prize for their research on recording from single cells which revealed the distribution and structure of nerve cells in the brain.

The 1982 Nobel Prize for Physiology or Medicine was awarded to **Sir John Robert Vane** (b. 1927; British pharmacologist), **Sune K. Bergstrom** (b. 1916; Swedish biochemist) and **Bengt Ingemar Samuelsson** (b. 1934; Swedish biochemist) for their research on prostaglandins.

The 1983 Nobel Prize for Physiology or Medicine was awarded to **Barbara McClintock** (1902–1992; US geneticist) for her discovery of mobile genetic elements (transposons) in maize chromosme, *i.e.*, for her theory of gene regulation.

The 1984 Nobel Prize for Physiology or Medicine was awarded to **Niels Jerne** (1907–1973), **Georges Koehler** (b. 1946; German pharmacologist) and **Cesar Milstein** (b. 1927; Argentinian biochemist) for discovering the principle for the production of monoclonal antibodies.

Sir Alec Jeffrey (1984) discovered DNA fingerprinting; he is regarded as father of DNA fingerprinting.

The 1985 Nobel Prize for Physiology or Medicine was awarded to **Michael S. Brown** (b. 1941; US biochemist) and **Joseph L. Goldstein** (b. 1940; US biochemist) for their work on the metabolism of cholesterol. They studied that cholesterol penetrates into the cells and participates in the building up of membrane structures. The ultimate aim of their research was to know the causes of artherosclersois.

The 1986 Nobel Prize for Physiology or Medicine was awarded to **Stanley Cohen** (b. 1922; US biochemist) and **Rita Levimontalcini** (b. 1909; Italian-born US biochemist) for their discoveries of nerve and epidermal growth factors.

The 1987 Nobel Prize for Physiology or Medicine was awarded to **Susumu Tonegawa** (b. 1939; Japanse biologist based in US) for his discovery of the genetic principle for the generation of a huge number of antibodies in the human body.

The 1988 Nobel Prize for Physiology or Medicine was awarded to **J.W. Black** (b. 1929; Scottish pharmacologist), **G.B. Elion** (b. 1918; US biochemist) and **G. Hitchings** (b. 1905; US biochemist) for their discoveries of new drugs '**propranolol**' used in the treatment of heart disease and '**cimetidine**' for treatment of cancer (leukemia). The cancer drugs could block the nucleic acid synthesis in cancer cells without damaging normal cells.

The 1989 Nobel Prize for Physiology or Medicine was awarded to **John M. Bishop** (b. 1936; US virologist) and **Harold E. Varmus** for their contribution to cancer research.

The 1990 Nobel Prize for Physiology or Medicine was awarded to **Edward D. Thomas** (b. 1920; US surgeon) and **Joseph E. Murray** (b. 1919; US plastic surgeon) for their separate contributions to techniques for overcoming rejection by the immune system following transplantation of organs and tissues. **Thomas** performed first successful bone transplantation in 1956 and **Murray** do first kidney transplantation in 1962 in humans.

The 1991 Nobel Prize for Physiology or Medicine was awarded to **Bert Sakmann** (b. 1942; German neurophysiologist) and **Erwin Neher** (b. 1944; German biophysiologist) for their development of the patch-clamp technique, used to detect extremely small electrical currents through cell membrane. They explained cell-cell communication and mode of action of acetylcholine on interaction with its receptor.

The 1992 Nobel Prize for Physiology or Medicine was awarded to **Edwin G. Krebs** (b. 1918; US biochemist) and **Edmond H. Fischer** (b. 1920; US biochemist) for their discoveries on reversible protein phosphorylation, a mechanism governing the activity of cell proteins. Thus, they explained the role of protein kinases and phosphatases in phosphorylation.

The 1993 Nobel Prize for Physiology or Medicine was awarded to **Richard J. Roberts** (b. 1943; US molecular biologist) and **Philip A. Sharp** (b. 1944; US molecular biologist) for their discovery of "**split genes**". Both reported independently that a single adenovirus messenger RNA molecule corresponds to four encoding regions of DNA. Thus, they showed that the genes of higher animals are fragmented in structure consisting of coding units or "**exons**" and noncoding sequences called **junk DNA** or "**introns**".

The 1994 Nobel Prize for Physiology or Medicine was awarded to **Alfred G. Gilman** and **Martin Rodbell** (both US scientists) for their discovery of structure and function of GTP-binding proteins (G-proteins pass on the message received by the receptors at the cell surfaces to inside the cells).

The 1995 Nobel Prize for Physiology or Medicine was awarded to **E.B. Lewis, C. Nusslein Volhard** and **Eric Wieschaus** for their discovery of **homeotic genes**. They found that genes which determine body pattern in *Drosophila* are also at work in vertebrates, including mammals.

The 1996 Nobel Prize for Physiology or Medicine was awarded to **Peter Doherty** and **Rolf M. Zinkernagel** for expalining the way by which the body detects virus infected cells. (**Note.** They discovered that the cytotoxic T-cells recognised the viral antigens on the infected target cells by the latter's association with MHC protein, sphenomenon called MHC-restriction).

The 1997 Nobel Prize for Physiology or Medicine was awareded to **Stanley B. Prusiner** for the discovery of **prions** which were responsible for certain diseases of central nervous system in animals and humans.

The 1998 Nobel Prize for Physiology or Medicine was awarded to **Robert Furchgott, Louis J. Ignarro** and **Ferid Murad** for the discovery that cells in the body make nitric oxide. Body's own nitric oxide acts as a signal telling blood vessels to dilate that in turn lowers blood pressure. Their work has wide ranging applications for treatment to atherosclerosis, septic shock and cancer.

The 1999 Nobel Prize for Physiology or Medicine was awarded to **Gunter Blobel**. He proved that the signal sequences (*i.e.,* address codes) in the form of a chain of amino acids present as a short tail at N-terminus of the protein, guide the proteins to their correct destination. Due to **signal sequence,** newly synthesized polypetide moves into cisternal space of the rough endoplasmic reticulum (RER) through a protein-lined, aqueous channel in the ER membrane (**Karp** 2010).

The 2000 Nobel Prize for Physiology or Medicine was awarded to **Arvid Carisson, Paul Greengard** and **Eric Kandel** for discovery of signal transduction in the nervous system.

The 2001 Nobel Prize for Physiology or Medicine was awarded to **Leland HartWell, Timothy Hunt** and **Paul Nurse** for discovering key regulators of the cell cycle.

The 2003 Nobel Prize for Physiology or Medicine was awarded jointly to **Paul C. Lauterbur** (b. 1929; US scientist) and **Peter Mansfield** (b. 1933; British scientist) for their discoveries concerning **"magnetic resonance imaging or MRI"**.

Andrew Fire and **Craig Mello,** two American scientists, in 1998 discovered the phenomena of **RNA interference (RNAi)** in which small double-standed RNA molecules can cleave mRNA and thus stop production of corresponding proteins. Both **A. Fire** and **C.Mello** received the Nobel Prize in Physiology and Medicine in 2006 for their discovery.

Shinya Yamanka discovered induced pluripotent cells (iPSCs) in 2006. **Yamanka** was awarded with **John Curdon,** the Nobel Prize in 2012 for their contribution to stem cell research **(Sumit Deswal, 2015)**.

1.3 BRANCHES OF ANIMAL PHYSIOLOGY

The field of physiology is divided into following *six* branches or divisions:

1. **General physiology.** This branch of physiology deals with the study of vital phenomena concerning the living organisms as a whole. All living organisms have some common characteristics and therefore, many things are common in their physiology.
2. **Cellular physiology.** In fact all principles of general physiology are based on a sound knowledge of cellular physiology, because cell is a unit of structure and function in all living beings.
3. **Special physiology.** It deals with the study of the functions of a particular organ or tissue in detail. For example, **endocrinology** deals with hormone secreting tissue and functions of various hormones in the body. **Neurophysiology** is concerned with the structure and functions

of nervous system. **Muscle physiology** deals with functions of muscles. **Renal physiology** deals with functions of kidney.

4. **Comparative physiology.** This branch of physiology is concerned with the study of the functions of an organ in a wide range of animals of different groups. Comparative animal physiology integrates and coordinates functional relationships which surpass special group of animals. It is concerned with the ways in which diverse animals perform similar functions. Genetically different animals may have striking similarities in characteristics and response to the same environmental stimulus. In contrast, closely related animals frequently react very differently to their surroundings. While other branches of physiology use such variables as the light, temperature, oxygen tension and hormone balance, comparative physiology uses, in addition, species or animal types as a variable for each function.

   In fact, physiological homologies and analogies are as real as morphological ones, and their study leads to unique kinds of biological generalizations. A great deal of information may be obtained from comparative physiology of different animal groups to show evolution of homeostatic controls. The gradual failure of homeostatic mechanisms due to illness (pathology) and age (gerontology) can also be learned from comparative physiology.

5. **Physiology of special groups.** This deals with the functional character in particular kinds of animals. The basic animal physiology is **human** and **mammalian physiology** and both provide the rational basis for much of the medicine and animal husbandary. **Insect physiology**, **Fish physiology** and **Physiology of parasites**, etc., are some other rapidly advancing branches of physiology.

6. **Applied and experimental physiology.** In this branch of physiology are included various closely related disciplines which are more or less concerned with the study of physiology. These branches are **Enzymology**, **Biochemistry**, **Biophysics**, **Animal Behaviour**, etc. Biochemistry and Biophysics attempt to describe and interpret the chemical and physical processes of biological materials in terms of the principles of organic chemistry, physical chemistry and physics.

Biophysics includes Physical biophysics, Physiochemical biophysics, Physiological biophysics and Mathematical biophysics. **Classical physiological biophysics** includes the problems such as bioelectricity, brain and heart measurements, volume conduction, membrane potential, etc. **Modern physiological biophysics** is concerned with the effects of high energy radiation, effect of physical and thermal shocks (radiation therapy, modern space medicine), system control and bioenergetics.

1.4 SIGNIFICANCE OF STUDY OF PHYSIOLOGY

In its true sense, physiology comprises all aspects of biology and indeed physiological control systems (adaptation and feedback systems) are quite complex. The significance of the study of physiology is self evident. Without the study of physiology we cannot understand and explain any of our normal or abnormal activities or the working process of our bodily systems. Animal structure and function are inseparably related to each other, and it seems quite reasonable to say that the knowledge of a structure is quite incomplete without the knowledge of its function. A knowledge of the physiology of pathological bacteria and disease-producing parasites and the recent investigation in physiology are a helpful tool in medicine.

QUESTIONS

Long Answer Questions

1. Define the physiology. What is the aim of physiology?
2. Give an account of brief history of the animal physiology.
3. Trace growth of animal physiology in the 19th century.
4. Trace growth of animal physiology in 20th century.

Short Answer Questions

1. Give a brief account of branches of animal physiology.
2. Write about significance of physiology.

Very Short Answer Questions

1. Define physiology.
2. Who formulated law of conservation of energy?
3. Which Nobel Laureate discovered the concept of synapse in nervous system?
4. What is the contribution in physiology of Hopkins?
5. Who is father of ATP cycle?
6. What is main contribution of P.D. Mitchell in the field of physiology?
7. Write about the contribution in physiology of Gunter Blobel.
8. What is Cori cycle?
9. Who discovered the role of cyclic AMP in intracellular mediation of hormonal effects?

Fill in the Blanks

1. I. P. Pavlov studied physiology of .... system.
2. Insulin hormone was isolated from the .... of dog by .... and Best.
3. Coenzyme A was discovered by ....
4. The 1953 Nobel Prize in Physiology or Medicine was awarded to .... and Lipmann.
5. Chemiosmotic mechanism of oxidative phosphorylation was formulated by ....
6. Urea cycle was discovered by H.D. Krebs and ....
7. Biochemistry of muscle contraction and relaxation was explained by A.V. Hill and ....

Multiple Choice Questions

1. The term physiology was first introduced by
   (*a*) William Harvey (*b*) R.J. Roberts
   (*c*) Jean-Fernal (*d*) Michael Brown
2. Who coined the term homeostasis?
   (*a*) G.M. Cooper (*b*) R. Holliday
   (*c*) W.B. Cannon (*d*) Jack Lange
3. Physiology is regarded as a tripod science. Which of the following are the three legs of physiology?
   (*a*) Physics, Maths, Chemistry
   (*b*) Biophysics, Biotechnology, Molecular Biology
   (*c*) Biochemistry, Biophysics, Physical chemistry
   (*d*) Physics, Chemistry, Anatomy
4. Signal hypothesis about trafficking of cellular proteins was proposed by
   (*a*) P.D. Mitchell (*b*) Dobberstein and Gunter Blobel
   (*c*) Claude Bernard (*d*) S. Tonegawa

ANSWERS

Very Short Answer Questions

1. Physiology is the study of function and vital processes of living organisms and their constituent parts.
2. H.L. M. Von Helmoltz.
3. Sir C.S. Sherrington.
4. In 1907, Hopkins showed the relationship of lactic acid to muscle contraction.
5. Fritz Lipmann, in 1941 discovered the formation of ATP from ADP.
6. P.D. Mitchell is credited for chemiosmotic mechanism of oxidative phosphorylation.
7. Protein trafficking by signal transduction, *i.e.*, in 1975 Dobberstein and G. Blobel proposed the signal hypothesis.
8. Cori cycle is the cycle in carbohydrate metabolism consisting of the conversion of glycogen to lactic acid in muscle, diffusion of lactic acid into the blood stream which carries it to the liver where it is converted into glycogen, and the breakdown of liver glycogen to glucose which is transported to the muscle by the blood stream and reconverted into glycogen.
9. E.W. Sutherland (1915–1975).

Fill in the Blanks

1. Digestive 2. Pancreas 3. F. Lipmann 4. H.D. Krebs
5. P.D. Mitchell 6. Kurt Henseleit 7. O. Meyerhof.

Multiple Choice Questions

1. (*c*) 2. (*c*) 3. (*d*) 4. (*b*).

2

Physiology of Digestion

:.1 INTRODUCTION

ll living organisms need *matter* to build up the body and *energy* to perform the metabolic reactions ha: sustain life. The total sum of the processes by which the living organisms obtain food and ne:abolise it to the materials that build up tissues and provide energy is called **nutrition**. The materials ıat provide the matter and energy to the organisms to keep alive are termed **nutrients**. A substance hich is eaten to supply the necessary nutrients to the body is termed **food**. Food and drink regularly onsumed by an animal form its **diet**.

Food contains the common ingredients such as carbohydrates, proteins, lipids, vitamins, minerals nd water. Only a few components of the diet such as salts, certain vitamins, water and other water oluble molecules are directly absorbed by the digestive tract. The major components of the diet are elatively complex molecules such as carbohydrates, proteins and fats. These polymers can not be bsorbed unless they are broken down into simpler compounds.

Digestion is a physico-chemical process by means of which the insoluble and complex food ıatters are converted into soluble and simple ones by the activation of different types of enzymes ontaining digestive juices that are liberated from different types of digestive glands. These glands re located in the different parts of the gastrointestinal tract.

lolozoic Nutrition

Most animals are holozoic. The term **holozoic** means obtaining food in the manner of most animals, y ingesting complex organic matter. This term is applied mainly to free-living animals which have specialized digestive tract, the **alimentary canal**. The characteristic processes involved in holozoic utrition are as follows:

1. Ingestion. It is the taking in of food.

2. Digestion. It is the breakdown of large organic molecules into simpler soluble molecules. **Iechanical digestion** involves mechanical breakdown of food, for example, by teeth. **Chemical igestion** involves the activity of enzymes. The type of chemical process these enzymes catalyse uring digestion is **hydrolysis**.

Two kinds of digestion. Digestion is of following two types:

(*i*) Intracellular digestion. When digestion of food takes place inside the cells of an animal nd is carried out in **food vacuoles** by the help of lysosomal enzymes, then it is called **intracellular igestion**. Content of **digestive vacuole** (food vacuole plus lysosomes) become first acidic (pH 5.6) and en alkaline (pH 7.3). The digestive enzymes include carbohydrases, aminopeptidase, exopeptidase, ndopeptidase, esterase, collagenase and nuclease. They are secreted at different times so that their fects are separated by time rather than spatially as in higher organisms. Digested food products are sorbed by surrounding cytosol and undigested material is eliminated by **exocytosis** at any point the surface of organism or cell.

This mode of digestion is a characteristic of lower animals such as protozoans and sponges. In coelenterates digestion is partially extracellular but largely intracellular. Many platyhelminthes (*e.g.*, turbellarian *Dugesia*), molluscs, arthropods (*Limulus*) and cephalochordates are considered as examples of animals in which occurrence of intracellular digestion represents a primitive condition. Thus, in *Lamellidens* (bivalve mollusc) partial digestion of starch (polysaccharides) and lipids occur extracellularly in the stomach; protein digestion, however, occurs extracellularly in the digestive glands. In *Palaemon* (Arthropoda) some cells of hepatopancreas perform intracellular digestion. In *Branchiostoma* or *Amphioxus* (Chordata) both extracellular and intracellular digestion occur.

(*ii*) Extracellular digestion. When digestion of food is practically completed in the lumen of the digestive tract, it is called **intercellular** or **extracellular digestion**. Thus, the development of a digestive tract, allows a more complete extracellular digestion which is characteristic of higher animals such as earthworms, cephalopods, crustaceans, insects, tunicates and all vertebrates.

The digestive system comprises of **alimentary canal** and **digestive glands**. The food stuffs are started with *mechanical breakdown* by teeth and tongue in the buccal cavity. Then it is moved from region to region by *peristaltic waves* of contraction of the smooth muscle fibers, in the wall of gut tube. The smooth muscle layer also provides a churning action which mixed the food with digestive juices. The *chemical changes* are hydrolytic and are catalyzed by enzymes. The necessary water required for hydrolysis is provided by copious secretion from the lining epithelium and glands of the alimentary tract. Mucus assists movement by acting as a lubricant and mobilizer. The **digestive enzymes** are secreted by glands (technically called **exocrine glands**) either in the gut wall (*e.g.*, glands of stomach and small intestine) or outside the alimentary tract (*e.g.*, pancreas, liver) in which case the secretion is poured into the gut along ducts.

3. Absorption. It is the uptake of the soluble molecules from the digestive region, across a membrane and into the body tissue proper. In digestive tract, the products of digestion are absorbed through the **villi** of the small intestine. In absorption, the food may pass directly into cells or first pass into the bloodstream to be transported to other regions of body.

4. Assimilation. It is using the absorbed molecules to provide energy or material to be incorporated into the body.

5. Egestion. It is the elimination from the body of undigested waste food materials. It is called **defecation**.

Types of Digestive Tract

Animals have following two types of digestive tracts:

1. Incomplete digestive tract. Incomplete digestive tract is found in coelenterates, flatworms (*e.g.*, *Dugesia*, *Fasciola hepatica*) and brittle stars (ophiuroids). It has a single opening the mouth which is used for taking food (ingestion) as well as for expelling indigestible materials (egestion). In animals with incomplete digestive tract, food must be digested and the indigestible remains expelled before the next meal can be taken. This condition (*i.e.*, absence of anus) also makes impossible any specialization of digestive tract into regions for storage, digestion or absorption. The incomplete digestive tract distributes the products of digestion to the body cells, thus, acting as circulatory system also. Due to this reason, it is also termed as **gastrovascular tract** (*e.g.*, in *Hydra*, *Obelia*, etc.).

2. Complete digestive tract. Complete digestive tract is found in roundworms (nematodes), segmented worms (annelids), molluscs, arthropods, echinoderms (except ophiuroids) and chordates. It has two opening: **mouth** for ingestion of food and cloacal aperture or **anus** for egestion. Such an arrangement allows the intake of food while digestion goes on and lets uninterrupted passage of food through the digestive tract. The products of digestion and indigestible food have no chance of mixing up. Moreover, the complete digestive tract has specialized regions of mastication, storage, chemical digestion, absorption and defaecation.

Basic Mechanism of Digestion

Digestion involves mechanical as well as chemical changes in the food. Mechanical changes are brought about by teeth, grinding organs such as gizzard and muscular action of stomach and intestinal walls. Chemical changes, whether in protozoans or metazoans from sponges to humans, involve splitting of complex, insoluble macromolecules into simpler, soluble micromolecules or subunits by the action of enzymes. The absorbable foods require no such alterations, being already fit for absorption. The nonabsorbable foods undergo a great amount of changes to that they become fit for absorption. For instance, the **proteins** and **polysaccharides** such as starches, glycogen, cellulose, etc., are polymers formed from simple **monomers** by removal of water; **lipids** (fats) are esters of fatty acids and glycerol, also formed by removal of water. The chemical bonds between the subunits (monomers) of these macromolecules are called **anhydro bonds**, being formed by removal of water. The breakdown of these food compounds (macromolecules) into subunits (monomers) occurs by the addition of water, a process termed as **hydrolysis**. Hydrolysis is a spontaneous reaction and releases energy in the form of heat. Spontaneous hydrolysis of food compounds is an extremely slow process, but can be speeded up by *catalysts* (or biocatalysts). The catalysts secreted by living cells are called **enzymes**. The digestive enzymes that accelerate hydrolysis are termed **hydrolases**. There are, however, specific hydrolases for the splitting of carbohydrates, lipids, proteins and nucleic acids. For example, hydrolysis of starch to glucose can be considered as an example of digestion:

$$\underset{\text{Starch}}{(C_6H_{10}O_5)^n} + \underset{\text{Water}}{nH_2O} \xrightarrow{\text{Enzyme}} \underset{\text{Glucose}}{nC_6H_{12}O_6}$$

The digestive glands release their secretions (saliva, HCl, enzymes, etc.) at proper time and in proper amount. The secretion (*synthesis* and *release* by the cells of glands) of digestive enzymes is an energy consuming process. Therefore, it would be wasteful, even may be harmful, if the glands secreted the enzymes all the time. So the flow of digestive juice is co-ordinated with the presence of food partly by nervous stimuli and partly by hormones.

2.2 PHYSIOLOGY OF DIGESTION

In humans, digestion of food starts in the oral/buccal cavity, continues in the stomach and is completed in the small intestine. There are, thus, three main phases of digestion: salivary, gastric and intestinal. Some digestion also occurs by bacterial (*i.e.*, microbial or symbiotic) activity in the large intestine. The food undergoes no change in the pharynx and oesophagus. These digestive organs only transfer the food from the oral cavity into the stomach by a process called **swallowing** or **deglutition**.

A. Salivary Phase of Digestion

Saliva is secreted mainly by three pairs of salivary glands: *submandibular glands*, *sublingual glands* and *parotid glands*. All of them are located in the oral cavity.

Salvia is a clear, tasteless, odourless, viscous fluid. It is slightly acidic (pH 6–7). Saliva has water, salts (sodium chloride or NaCl; hydrogen carbonate or HCO_3^-), phosphates, carbonates, Ca^{2+}, K^+, Mg^{2+}, sulphocyanide), mucin (a carbohydrate rich in glycoprotein) and enzyme: **ptyalin** or **salivary amylase** (L. *amylum* = starch). Saliva also contains **antibodies** (*e.g.*, IgA), the first immunological defense against bacteria and viruses; **lysozyme** which attacks the wall of bacteria, **lactoferrin** which binds iron and is bacteriostatic. It also contains **proline-rich proteins** that protect tooth enamel and bind toxic tannins.

About 1 to 1.5 liters of saliva and mucus are secreted daily in the buccal cavity.

Salivary phase of digestion occurs in buccal or oral cavity. The buccal cavity is lined by a stratified squamous epithelium. It contains tongue and teeth. The buccal cavity receives an **amylase** enzyme containing **saliva** from the salivary glands and **mucus** from numerous mucous glands located in its mucosa. In the oral cavity, the food undergoes many other operations besides the digestion. Thus in buccal cavity food is:

1. tasted by the taste buds (Box 2.1);
2. moistened and softened with mucus and saliva for easy chewing;

3. masticated by the grinding teeth to smaller particles which get mixed with **ptyalin** (salivary amylase enzyme) (optimal pH 3.0 to 6.0) and provide greater surface for their action;

4. partly digested by salivary enzymes;

5. lysozyme of saliva keeps the oral cavity free of pathogenic microorganisms by catalyzing the breakdown of their cell walls;

6. bicarbonates, phosphates and mucin act as buffers;

7. made into a slimy ball or **bolus** by the working of the tongue against the palate; and

8. finally swallowed into the stomach.

Box 2.1 Tasting of Food

Taste buds are located on the tongue and respond to sweet, salt, sour and bitter materials. Taste buds also occur on the palate. This is why the persons with artificial teeth, which seal off the palate, complain that their food does not taste as good as it did when they had natural teeth.

The digestion of carbohydrates and lipids is started in the buccal cavity. Human food contains three types of carbohydrates: *polysaccharides* (starch, glycogen and cellulose), *disaccharides* (sucrose or table/cane sugar and lactose or milk sugar) and *monosaccharides* (glucose, fructose and galactose). The monosaccharides are absorbed without any digestive change. The polysaccharides and disaccharides need digestion.

Salivary amylase enzyme, or **ptyalin**, is activated by chloride ions present in the saliva itself (Box 2.2). Amylase splits starch and glycogen first to **dextrins** (oligosaccharides) and then into disaccharide sugars such as **maltose** $\underset{\text{Starch}}{(C_6H_{10}O_5)_n} + H_2O \xrightarrow{\text{Ptyalin}} \underset{\text{Maltose}}{C_{12}H_{22}O_{11}}$.

Maltotriose and small dextrins, called **limit "dextrins"** (*i.e.*, salivary digestion in buccal cavity is restricted, because of the short time the food is retained here). However, about 30% of starch present in the food is hydrolysed in the oral cavity. The starch is more easily hydrolysed if it has been first heated. This is why human beings cook starch foods (bread, potatoes, etc.) before eating. As pH optimum of salivary amylase is close to neutrality (6.7), its action is finally stopped by the acidic gastric juice in the stomach. In fact, salivary digestion continues in the stomach till food and gastric juice become thoroughly mixed. This takes about half an hour and by this time about 40% of the starch is split by the salivary amylase.

Box 2.2 Salivary Amylase in other Mammals

1. Carnivorous mammals, such as lions, tigers, cats, dogs, etc., and domestic herbivores, such as cow and buffalo, which rapidly gulp large chunks of food into the stomach, lack a digestive enzyme in their saliva. Pig has ptyalin in its saliva.

2. Salivary amylase activity is high in humans, apes and rodents. Further in human beings the amylase activity varies greatly depending on the dietary habit. In the Tswan tribe of Botswanaland feeding habitually on carbohydrate rich tubers, about 248 amylase units/mL of saliva have been recorded, whereas in Bushman of Nambia who feed mostly meat, there are on an average only 22 amylase units/mL saliva. One unit of amylase is the amount required to digest 5 mL of 1% soluble starch in 10 minutes at 38°C (**Banerjee** *et al.*, 1999).

Saliva has no protease or lipase enzymes, hence, it has no action on proteins and fats. Mucin present in the saliva helps bind the semisolid, partially digested food particles into a slimy ball or bolus with the help of tongue and it also lubricates the food to facilitate swallowing.

Swallowing involves the passage of food bolus from buccal cavity to the stomach. Food does not merely slide down the oesophagus due to gravity, it is actively pushed along by **peristalsis**. This is shown by the fact that we can swallow while upside down on head or even on the moon. In a grazing horse, food moves up the neck against gravity.

B. Gastric Phase of Digestion

The gastric phase of digestion occurs in the stomach. The stomach is a distensible, muscular sac. It temporarily stores and partially digests the food. When stomach is undistended, its inner surface shows folds, called the **rugae**, which disappear as the stomach expands. A fully distended stomach can hold nearly 5 dm^3 of food. The thick mucosa of the stomach contains numerous mucus-secreting epithelial cells and tubular glands. (*Note*. There are about 35 million gastric glands in the human stomach; **Dhami** and **Dhami** 2001). In the stomach, food undergoes both mechanical as well as chemical changes.

1. Mechanical changes. Wall of the stomach undergoes periodic muscular contractions (15–20 times per minute). This activity churns and breaks the food mechanically and mixes it thoroughly with the gastric juice for proper action of enzymes. The cardiac and pyloric sphincters remain closed during churning of food.

2. Chemical changes. In the stomach, the food mixes with acidic gastric juice secreted by the gastric glands. The gastric juice contains **mucin, hydrochloric acid** (HCl), two proenzymes: **pepsinogen** (propepsin) and **prorennin** and an enzyme **gastric lipase**. Gastric juice lacks carbohydrate-digesting enzyme *carbohydrase*. All these components of the gastric juice serve various functions.

***(i)* Mucus.** Mucus is secreted by *neck mucous cells* and *surface mucous cells* of gastric mucosa. It lubricates the food and protects the stomach lining from being damaged and digested by its own HCl and proteolytic enzymes.

***(ii)* Hydrochloric acid.** HCl of gastric juice is secreted by *parietal cells* of gastric glands. It serves the following functions:

Functions of HCl

(*i*) It disinfects the food by killing bacteria.
(*ii*) It stops the action of salivary enzyme which has digested a part of dietary carbohydrate.
(*iii*) It activates pepsinogen to pepsin and prorennin to rennin.
(*iv*) It produces a proper acidic medium required for protein digestion.
(*v*) It softens the food and breaks down the cement material between the cells of food.
(*vi*) It coagulates the proteins to give more time to enzymes for their proper action on the substrate.
(*vii*) It makes calcium and iron salts suitable for absorption in the intestine.
(*viii*) It initiates hydrolysis of sucrose to glucose and fructose.
(*ix*) It splits nucleoproteins into nucleic acids and proteins.

***(iii)* Intrinsic gastric factor.** It is also secreted by parietal cells of gastric glands. It is a protein (*viz.*, 49-kDa glycoprotein) that aids in the absorption of vitamin B_{12} or **cyanocobalamin** in the small intestine. In the absence of this factor, vitamin B_{12} is not absorbed. Due to deficiency of this vitamin, the bone marrow fails to form red blood corpuscles in adequate numbers. This results in a disease, called **pernicious anaemia**.

Lack of the intrinsic gastric factor is associated with concurrent absence of HCl, a condition called **achlorhydria**.

***(iv)* Pepsinogen.** It is secreted by *chief cells* of gastric glands. Pepsinogen is a proenzyme which is inactive. It is splitted by HCl into an active **pepsin** enzyme and an inactivating fragment (*viz.*, pepsinogen, with a moleculer weight of 42,500, is converted into pepsin, with a molecular weight of 34,500 by the high concentration of H^+ in the gastric juice and also by previously formed pepsin). Then the pepsin itself continues to activate additional pepsinogen. Such an activation is called **autocatalytic reaction**. Pepsin hydrolyses the large molecules of proteins into smaller subunits such as **proteoses** and **peptones** in an acidic medium (pH 2) (Box 2.3).

Box 2.3

1. **Proteoses.** Any of various protein derivatives that are formed by the partial hydrolysis of proteins by digestive enzymes, that are not coagulated by heat and that are soluble in water but are precipitated from solution by saturation with ammonium sulphate.

2. **Peptones.** Any of various protein derivatives that are formed by the partial hydrolysis of proteins by digestive enzymes, that are not coagulated by heat and that are soluble in water but unlike proteoses are not precipitated from solution by saturation with ammonium sulphate.

In fact, pepsin can digest collagen of the white fibres of connective tissue but not keratin of hair, horns and nails. The invertebrates generally lack a pepsin-like enzyme which could act in an acid medium.

(v) Prorennin. It is secreted by *chief cells* of gastric glands. This proenzyme is also splitted by HCl into an active enzyme **rennin** or **chymosin** and an inactivating fragment (Box 2.4). Rennin hydrolyses the soluble milk protein, **casein**, into **paracasein** and **whey** proteins. Paracasein is spontaneously precipitated in the presence of calcium as insoluble **calcium paracaseinate**, forming solid curd or coagulated milk. The curd stays in the stomach for a longer period for proper action of pepsin. Pepsin converts calcium paracaseinate into proteoses, peptones and polypeptides (Fig. 2.1).

(*i*) $\underset{\text{(inactive)}}{\text{Prorennin}} \xrightarrow{\text{HCl}} \underset{\text{(active)}}{\text{Rennin}}$

(*ii*) Rennin + Casein → Paracasein

(*iii*) Paracasein + Ca^{2+} → $\underset{\text{(curdles)}}{\text{Calcium paracaseinate}}$

(*iv*) Calcium paracaseinate + Pepsin → Proteoses, peptones and polypeptides

Fig. 2.1. Mode of action of rennin and pepsin on milk.

Box 2.4

1. Rennin enzyme occurs in the stomach of human infants. As a person matures, production of rennin decreases greatly or stops altogether (**Murray** *et al.*, 2000). Curdling, however, is adequately brought about by pepsin and HCl.

2. **Rennet tablets** contain rennin enzyme that is obtained from calf. They are commonly used for curdling milk.

(vi) Gastric lipase. The heat of stomach is important in liquidizing dietary lipids. Gastric lipase converts some lipids into **monoglycerides** and **fatty acids**. Since gastric lipase is destroyed by free HCl, so, it works only in the initial stages, when acidity is not high. However, lingual lipase remains active in the stomach and can digest as much as 30% dietary glyceride.

Stomach contains a thick acidic mixture, called **chyme** which has formed due to mixing of gastric juice and semidigested food. The pyloric sphincter opens from time to time to let the chyme pass into the duodenum.

Functions of stomach. The stomach serves the following *four* main functions:

1. temporary storage of food;
2. mechanical churning of food;
3. partial digestion of proteins; and
4. regulation of flow of partially digested food (chyme) into the small intestine.

Storage of food in the stomach enables the mammals (humans) to take their food at intervals as a meal instead of continuously.

Time taken in gastric digestion. In humans, gastric digestion of an average meal is completed in about 3 to 5 hours. After about half an hour, some of the food (chyme) is ready to pass into the duodenum through the pyloric valve. Different food items remain in the stomach for different time periods, *e.g.*, carbohydrates for only 1 or 2 hours, proteins for upto 3 hours, and lipids for 3 to 6 hours. Thus, a fat-rich food should be taken if the next meal is expected to be available after a long time and an athelete should take carbohydrate-rich diet before competition.

C. Intestinal Phase of Digestion

Stomach is followed by **intestine** which is divided into two parts: small intestine and large intestine. The small intestine consists of *three* regions: about 2.5 m long proximal **duodenum**, about 2.4 m long middle **jejunum** and about 3.6 m long distal **ielum** (Fig. 2.2). Mucosa of small intestine possesses numerous finger-like projections, called **villi** whose walls are richly supplied with blood capillaries and lymph vessels and contain smooth muscle. The individual cells on the surface of the villi possess tiny **microvilli** on their free surface forming the so-called **brush border**. Over 6 m length of human intestine allows enough time for the digestion of food. Fold, villi and microvilli of small intestine greatly increase the surface for the digestion and absorption of digestion products.

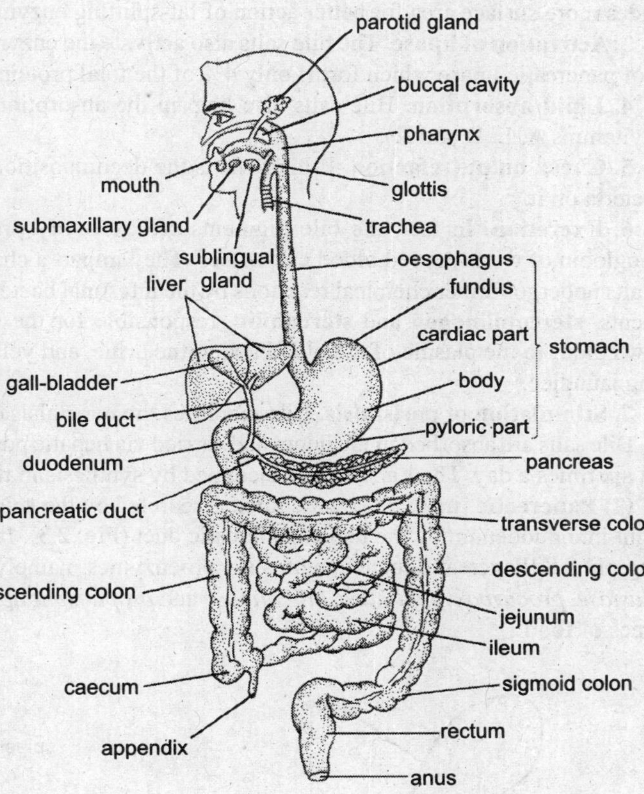

Fig. 2.2. Human digestive system.

***(a)* Digestion of food in duodenum.** Duodenum is the first, shortest and widest part of small intestine that extends from pylorus to jejunum. It receives the bile from liver and pancreatic juice from pancrease (Fig. 2.3). Villi of duodenum are leaf-shaped. Submucosal Brunner's glands of duodenum secrete **mucus** and alkaline watery fluid in first part of duodenum, so protecting its mucosa from acidic chyme entering from stomach. Action of pepsin is stopped in duodenum due to alkalinity of contents here. Chyme meets with following *two* juices:

(1) Bile. The bile is a yellowish brown or greenish alkaline fluid. It is bitter in taste. Bile is produced by hepatic cells of liver and is stored and concentrated in gall-bladder. Bile does not contain any digestive enzymes thus, has no chemical action on food. It contains water, mucin, lecithin, cholesterol, bile salts such as **sodium carbonate**, **sodium glycocholate** and **sodium taurocholate** and bile pigments, such as **biliverdin** (green) and **bilirubin** (yellow), fat and cell debris.

Functions of bile. Bile performs the following important functions:

1. Neutralisation of HCl. Bile's sodium carbonate neutralizes the acid of chyme. This is necessary as the enzymes of the pancreatic and intestinal juices act only in an alkaline medium (pH 8).

2. Emulsification. Bile's other salts, namely sodium glycocholate and sodium taurocholate, like the modern detergents, reduce the surface tension of lipids in the chyme, and emulsify them *i.e.*, they break large droplets of lipids (fats or oils) into smaller ones. The association of bile salts with the small fat droplets prevents the fusion of fat droplets and a stable emulsion is formed. This provides more surface area for better action of fat-splitting enzymes.

3. Activation of lipase. The bile salts also activate the enzyme lipase, *e.g.*, bile salts activate a type of pancreatic lipase which forms only 4% of the total protein in pancreatic juice.

4. Lipid absorption. Bile salts also help in the absorption of fats and fat-soluble vitamins (*i.e.*, vitamins A, E, D and K).

5. Check on putrefaction. Bile prevents the decomposition of food by checking the growth of bacteria on it.

6. Excretion. In fact, the bile pigments are excretory products formed by breakdown of haemoglobin of worn-out red blood corpuslces. They impart a characteristic colour to the bile. The bile salts undergo further chemical reactions by the intestinal bacteria and are converted to the brown pigments, **stercobilinogen** and **stercobilin**, responsible for the colour of faeces. Bile gives pale yellow colour to the plasma of the blood and to the urine, and yellow colour to the body of persons having jaundice.

7. Stimulation of peristalsis. Bile increases the peristalsis of the intestinal wall.

Bile salts are absorbed in the ileum and carried via hepatic portal vein to the liver for resecretion about six times a day. Their any loss is recouped by synthesis in the liver cell.

(2) Pancreatic juice. Pancreatic juice is secreted by the acini of the exocrine pancreas, and is brought into duodenum by the hepatopancreatic duct (Fig. 2.3). It is a clear, strongly alkaline fluid, having pH 8.8. Pancreatic juice contains six proenzymes, namely *trypsinogen*, *chymotrypsinogen*, *proelastase*, *procarboxypeptidase*, *procolipase* and *prophospholipase* and several enzymes to digest all types of food.

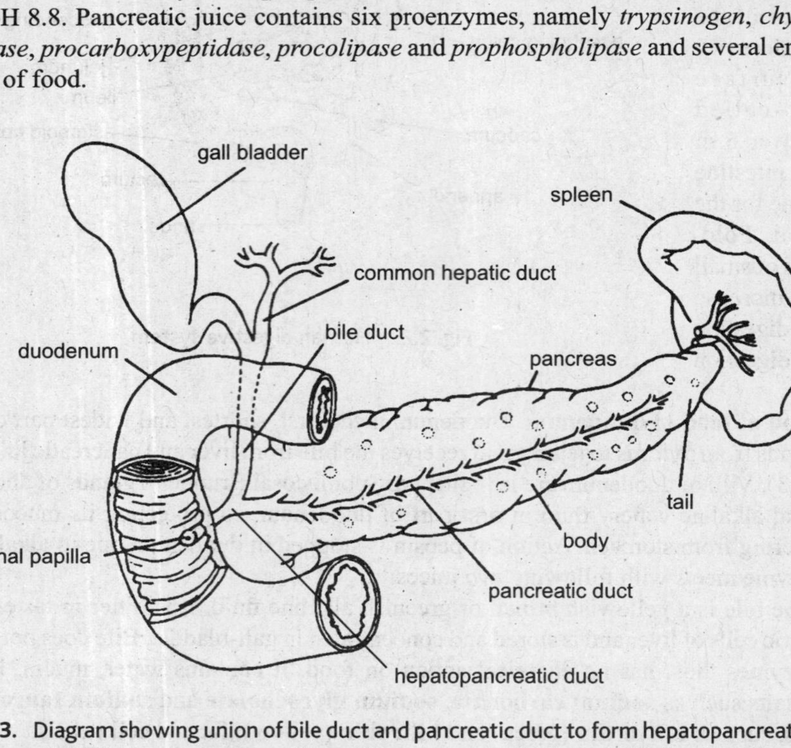

Fig. 2.3. Diagram showing union of bile duct and pancreatic duct to form hepatopancreatic duct. Sphincter of Oddi is located at the orifice of hepatopancreatic duct at duodenal papilla.

***(i)* Trypsinogen.** Its activation to an enzyme **trypsin** is started by a nondigestive enzyme, **enterokinase**, present in the intestinal juice or another trypsin molecule. The activation involves separation of an inactivating fragment from the proenzyme by enterokinase. Once formed, trypsin activates additional trypsinogen. This form an example of **autocatalysis**. (*Note*. A 249-residue polypeptide (trypsinogen) is inert until a 6-residue segment is cleaved from its NH_2 terminal end).

Trypsin breaks the proteins, specially basic proteins, into **peptides** (small groups of amino acids). Trypsin enzyme has a limited action on collagen and no action on casein and keratin. It coagulates the blood by hydrolyzing its fibrinogen into fibrin in blood feeding (sanguivorous) animals.

***(ii)* Chymotrypsinogen.** It is activated to an enzyme *Chymotrypsin* by trypsin. It also splits the proteins into peptides. Chymotrypsin clots milk in an alkaline medium, *i.e.*, casein of milk ultimately changes into calcium paracaseinate.

***(iii)* Proelastase.** Its activation to **elastase** is done by trypsin. It digests elastin protein.

***(iv)* Procarboxypeptidases.** These are activated to enzymes **carboxypeptidases** by trypsin. These enzymes separate individual amino acids adjacent to the free terminal carboxyl group (–COOH) of the peptides, reducing them into **dipeptides**. It, thus, produces amino acids and dipeptides.

***(v)* Procolipase.** It is activated by trypsin into **colipase** which helps in digestion of fat droplets.

***(vi)* Pancreatic lipase.** (Old name **steapsin**). It is the main lipid-digesting enzyme. Pancreatic lipase enzyme hydrolyses lipids in stages. In the first stage, it separates one fatty acid molecule from the substrate, changing the emulsified **triglyceride** into a **diglyceride**. In the second stage, the diglyceride is broken down into another fatty acids and a **monoglyceride**. In the third stage, the monoglyceride is hydrolysed to another fatty acid molecule and a glycerol molecule. Anyhow, normally *the digestion of lipids remains incomplete in the lumen of intestine*. Hydrolysis of lipids is a slow process and a few hours available for the action of lipase in the intestine are not enough for the completion of fat hydrolysis. This the end products of hydrolysis of the lipids are *fatty acids, glycerol, monoglycerides, diglycerides* and *triglycerides*. The fatty acids are insoluble but are kept in solution by the bile salts to affect their absorption.

Cholesterol ester hydrolase. The 4% of total protein in pancreatic juice is formed by **bile-salt activated lipase** or **cholesterol ester hydrolase**. It catalyses the hydrolysis of cholesterol esters, esters of fat-soluble vitamins, phospholipids and triglycerides.

***(vii)* Pancreatic α-amylase** (Old name **amylopsin**). It hydrolyses the remaining starch and glycogen first into **dextrins** (oligosaccharides) and then into disaccharides such as maltose, maltotriose and α-"limit" dextrins.

***(viii)* Nucleases.** Pancreatic juice contains following two nucleases: **ribonuclease** (or **RNAase**) and **deoxyribonuclease** (or DNAase). These break the nucleic acids (RNA and DNA) into the constituent nucleotides, namely, **ribonucleotides** and **deoxyribonucleotides**, respectively, by splitting the phosphate ester bonds between the adjacent nucleotides.

***(ix)* Prophospholipase A.** It is activated by trypsin into active phospholipase A which digests phospholipids into fatty acids and lysophospholipids (A phospholipid from which one fatty acid residue has been removed is called **lysophospholipid**).

Pancreatic juice, being alkaline, also serves to neutralize the acid of chyme and stops the action of pepsin enzyme like the bile. In duodenum, the food is converted into a very fine paste. It is called **chyle**.

***(b)* Digestion of food in post duodenal small intestine.** The intestinal glands, such as **Brunner's glands** and **crypts of Lieberkuhn**, secrete **intestinal juice** or **saccus entericus**. Like the pancreatic juice, the intestinal juice is also alkaline, having pH 8.3. It is a straw-coloured mixture of water and mucus and helps to increase the volume of fluid in gut. Entrocytes secrete the following enzymes which remain **membrane-bound** to microvilli of enterocytes of the small intestine (Fig. 2.4):

1. Enteropeptidase (Enterokinase). This proteolytic enzyme activates inactive trypsinogen to trypsin.

2. Aminopeptidases (Exopeptidases). These enzymes separate individual amino acids adjacent to the free terminal amino-group ($-NH_2$) of the peptides, reducing them to dipeptides.

3. Carboxypeptidases. These enzymes cleave C-terminal amino acid from the peptide (polypeptide).

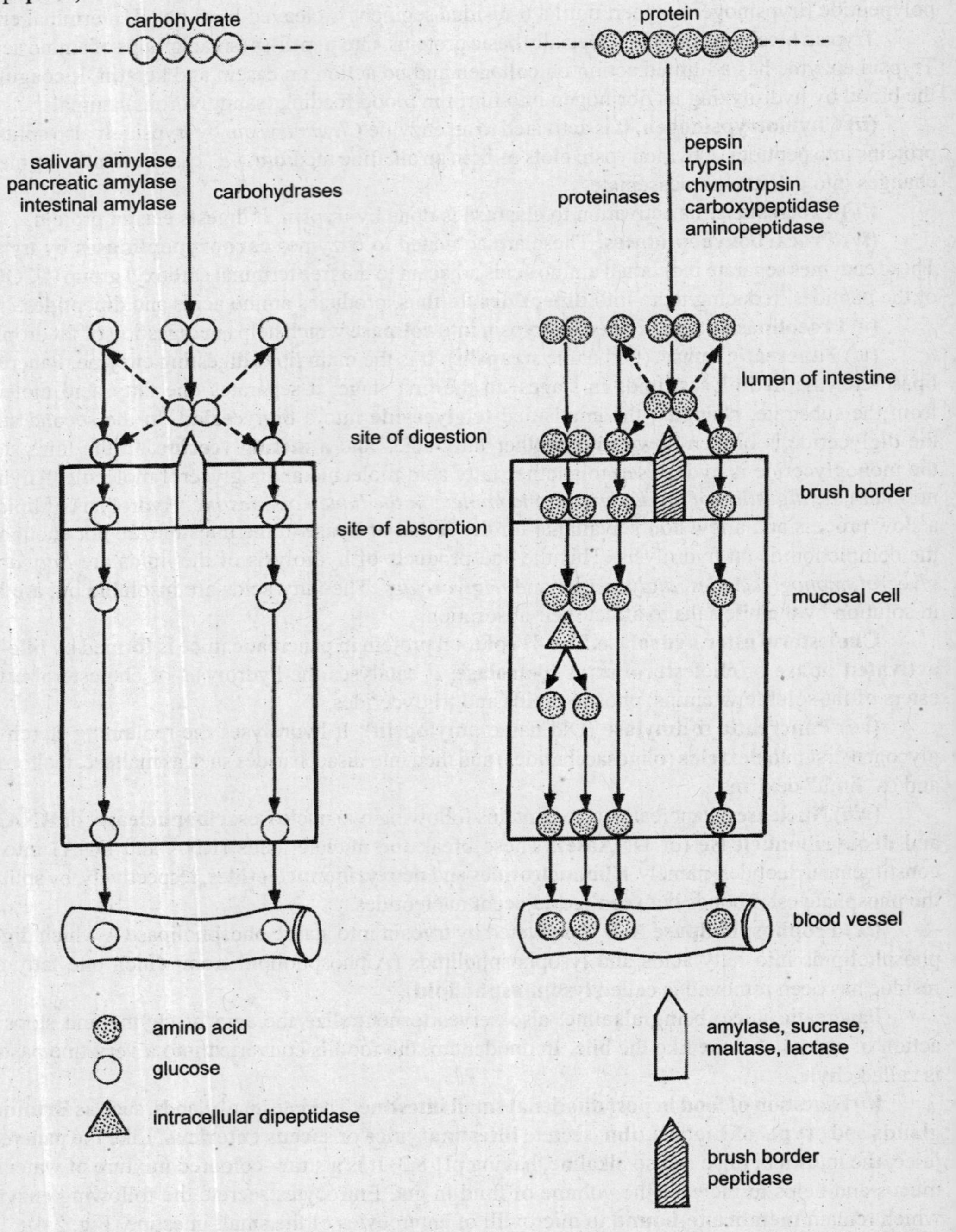

Fig. 2.4. Schematic diagram of two epithelial cells of intestinal villi, showing the final phase of digestion of disaccharides and peptides and absorption of digestion products.

4. Endopeptidases. These enzymes cleave between amino acid residues in midportion of the polypeptide.

5. Dipeptidases. These enzymes split the dipeptides into **amino acids**.

6. Maltase. It splits disaccharides such as maltose, maltotriose and α-dextrins into monosaccharide, the **glucose** (Box 2.5).

Box. 2.5

1. **α-glucosidase (or maltase)** changes maltose into glucose.
2. **Isomaltase (or 1:6 glucosidase)** hydrolyses isomaltose into glucose.
3. **"Limit dextrinase"** changes 'limit' dextrins (branched oligosaccharides) into glucose.
4. **α-dextrinase.** It converts α-dextrins, maltose and maltotriose into glucose.

7. Lactase. It converts lactose (milk sugar, a disaccharide) into **galactose** and **glucose** (Box 2.6).

8. Sucrase. It hydrolyses sucrose (table sugar, a disaccharide) into **fructose** and **glucose**.

9. Trehalase. It hydrolyses trehalose into glucose.

Maltase, lactase, sucrase and trehalase are collectively called **disaccharidases** as they hydrolyse disaccharides (double sugars) into monosaccharides (single sugars).

10. Polynucleotidase. it splits nucleic acids into nucleotides.

11. Nucleosidase. It hydrolyses nucleosides into nitrogen bases, pentose and phosphate.

12. Phosphatase. It removes phosphate from certain organic phosphates such as hexose phosphates, glycerophosphates and nucleotides derived from the diet and digestion of nucleic acids by nucleases.

13. Phospholipase. It attacks phospholipids to produce glycerol, fatty acids, phosphoric acid and bases such as choline.

Intestinal amylase of intestinal juice splits remaining starch and glycogen into disaccharides (double sugars) such as maltose, isomaltose and "limit" dextrins. In contrast to old contention, *there is no lipase enzyme* in the intestinal juice (**Ganong**, 1999; **Murray** *et al.*, 2000).

Box. 2.6

Among mammals, human being alone take milk even after becoming adult. In some humans, secretion of lactase enzyme decreases or ceases with age. This condition is called **lactose intolerance**. The lactose intolerant persons fail to digest lactose of milk. In their intestine, lactose is fermented, producing gases and acids. This causes **flatulence** (distention of stomach and intestine) **intestinal cramps** and **diarrhoea**. In yoghurt (*Dahi*), lactose is fermented into lactic acid and in case of cheese, lactose is left in whey.

Erepsin is a group of peptidase enzymes which hydrolyse polypeptides, tripeptides and dipeptides into amino acids. It has been recently found that enzymes secreted by enterocytes and complete the process of digestion in the small intestine are not released into the intestinal lumen but remain bound to the membrane of the microvilli (Fig. 2.6). Dipeptidases are located within these cells. At the level of brush border (=microvilli), the hydrolysis of disaccharides and dipeptides is effected. The end products of digestion, namely monosaccharides (*e.g.*, glucose) and amino acids are released into the intestinal lumen.

Mode of action of protein-digesting enzymes. Protein-splitting enzymes are of two types: endopeptidases and exopeptidases. The **endopeptidases** attack the polypeptide molecule at many points, breaking them into groups of amino acids. *Pepsin*, *trypsin* and *chymotrypsin* form the examples of endopeptidases. *Pepsin* breaks the bonds adjacent to tyrosine or phenylalanine, *trypsin* adjacent to lysine or arginine and *chymotrypsin* adjacent to tryosine, phenylalanine, tryptophan, methionine or leucine. The **exopeptidases** separate terminal amino acids from polypeptide chain, *e.g.*, carboxypeptidase and aminopeptidase. **Carboxypeptidase** separates the amino acids adjacent to the free terminal carboxyl group of the polypeptide chain. Aminopeptidase separates the amino acid adjacent to the free terminal amino group of the chain (Fig. 2.5).

H_2N—ala—gly—leu—yr—ala—phe—tys—val—glu—gly—COOH

↑ AP (ala—gly), ↑ C (leu—yr), ↑ P (ala—phe), ↑ T (tys—val), ↑ CP (glu—gly)

Amino end … Carboxyl end

Fig. 2.5. A polypeptide chain showing the points of attack of the enzymes pepsin (P), trypsin (T), chymotrypsin (C), aminopeptidase (AP) and carboxypeptidase (CP).

Proteases for digesting some animal fibrous proteins such as hair keratin, silk fibroin and wool protein, are not produced by vertebrates. Certain invertebrates, such as insects, secrete enzymes for digesting some fibrous proteins. Hence, they damage silk fabrics and wool garments.

Thus, small intestine is involved in completion of digestion and absorption of food.

(*c*) Digestion of food in large intestine. In the hindgut or large intestine, ions and water are absorbed specially by the **caecum** and **ascending colon**. The large intestine secretes a thick mucoid substance which acts as lubricant to facilitate the passage of faecal matter through the colon. The faecal matter is stored in the large intestine for about 36 hrs during which it is acted upon by a number of bacteria which impart the faeces their characteristic foul smell.

Colon is involved in the absorption of water, amino acids and vitamins formed by symbiotic bacteria and excretion of excess minerals. **Rectum** is involved in absorption of water and formation and storage of faeces.

2.3 ABSORPTION OF DIGESTED FOOD

Absorption is a physico-chemical process by means of which digestive end products are transferred from the intestinal lumen into the blood vessels via intestinal epithelium.

Mode of Absorption in Various Animals

In protozoans, food digested in the food vacuoles passes directly into the cytoplasm. In sponges, food digested in the food vacuoles in the **choanocytes** enters their cytoplasm and is distributed to other cells by diffusion and by amoebocytes. In coelenterates and flatworms, food digested partly or fully in the gastrovascular cavity get into the cells that line the cavity and reaches other tissues by diffusion. In nematodes (roundworms), digested food diffuses from the gut into the pseudocoelomic fluid and from there it is taken up by other tissues. In other invertebrates, and lower chordates digested food passes from the gut into the blood for distribution of the body. In vertebrates digested food enters from the alimentary canal into the blood and lymph, which carry it to the tissues. *Thus, the passage of digested food from the food vacuoles into the cytoplasm or from the digestive tract into the tissues, or into the blood and lymph is the absorption.*

Anatomical and Histological Basis of Absorption

The total quantity of fluid absorbed each day is approximately 9 liters. Out of this, small intestine absorbs 8 to 8.5 liters, rest is passed to colon for ultimate defaecation.

1. Stomach. Very little absorption of nutrients takes place in the stomach. Some water, salt, alcohol, and certain drugs, such as aspirin, are however, readily absorbed in the stomach.

2. Small intestine. It is chief place for absorption of digested food. For this purpose, the internal surface area of small intestine is greatly increased (upto 600 fold, Box 2.7) by (*a*) its enormous length (about 6.25 meter); (*b*) valvulae conniventes, plicae circulares or folds; (*c*) villi and (*d*) microvilli. The folds are close-set longitudinal or transverse projections of the mucosa into the lumen of the intestine. These occurs in bony fishes, amphibians and reptiles. The cartilaginous and bony fishes have a large **spiral valve** or **scroll valve** or **rolled valve** in short intestine. Some vertebrates have one or more blind pouches, the **caeca**, arising from the intestine. Spiral valves, rolled valves and caeca increase the absorptive surface of gut.

Box 2.7

Each intestinal epithelial cell (in humans) possess brush bordered surface that consist of about 600 microvilli, each 1 μ in length and 0.1 μ in diameter which increase the absorptive surface area another 20 fold. So combination of folds, villi and microvilli increases the absorptive area of the mucosa about 600 fold and makes a tremendous total area of about 250 square meter for the entire small intestine.

The villi are numerous (about 6 million in human's intestine), small, finger-shaped projections of mucosa. These are found in the birds and mammals. Each villus contains abundant blood capillaries, a centrally placed lymph vessel called the **lacteal**, and smooth muscles. Villi twist and shorten to quicken absorption by causing circulation of blood and lymph in them. Movements of villi are promoted by a hormone, **villikinin** secreted by intestinal mucosa in response to the presence of food in the intestine.

Mechanical mixing of food by muscular contractions of intestine helps absorption as this keeps renewing the fluid directly in contact with the absorptive surface (brush border of intestinal epithelium).

3. Large intestine. It can absorb more water and ions than absorbed by small intestine and almost no nutrients. Large intestine can absorb saline, glucose, alcohol and certain drugs to some extent.

Physiology of Absorption

Absorption through the gastrointestinal mucosa (*i.e.*, plasma membrane of each enterocyte cells) occurs by two types of processes: passive absorption and active absorption.

***(i)* Passive absorption.** It occurs down the concentration gradient. It takes place by physical processes such as simple diffusion, osmosis and facilitated diffusion.

***(a)* Simple diffusion.** It is a slow process. Small nutrient molecules dissolved in water diffuse into the blood across the mucous membrane till the concentration is higher in the intestinal contents than in the blood. When the concentration difference disappear, the process of diffusion stops. In this way, no nutrient can be totally absorbed by diffusion. Pentoses are absorbed by simple diffusion. About 98% of dietary lipids are absorbed by passive diffusion.

***(b)* Osmosis.** Water is absorbed by osmosis. The process of osmosis occurs so long as the solute concentration (and the resultant osmotic pressure) is higher in the blood than in the intestinal contents. Absorption of any solute raises the osmotic pressure of the blood, and brings about absorption of an equivalent amount of water. Water is absorbed partly in the small intestine and mostly in large intestine.

***(c)* Facilitated diffusion.** Fructose is absorbed by the process of facilitated diffusion. Like the diffusion, facilitated diffusion occurs only down the concentration gradient, it cannot absorb the entire amount of the nutrient in the intestinal contents. However, it is more rapid than the simple diffusion because the nutrient is carried across the cell membrane by the **carrier protein molecules**. These carriers do not use energy in the form of ATP molecules, so, there is no active participation of the cells.

***(ii)* Active absorption.** It occurs by biological processes such as active transport and endocytosis.

***(a)* Active transport.** This process is independent of concentration gradient. It can move nutrients (digestive end products) not only down the concentration but also against the concentration gradient. Active transport is brought about by the carrier protein molecules of plasma membrane of enterocytes. It utilizes energy derived from hydrolysis of ATP which are supplied by mitochondria. The process of active transport is far more rapid than diffusion. Active transport is adversely affected by low temperature, lack of oxygen, low concentration of Na^+ on mucosal surface and poisons such as cyanide. By active transport, nutrients can be absorbed completely from the intestinal contents. Substances of high nutritional importance, such as **sodium ions, glucose, galactose, amino acids, dipeptides, tripeptides, purine** and **pyrimidine bases of nucleic acids, water-soluble vitamins** (vitamin B_{12}, folate, thiamine, riboflavin, niacin, pyridoxine, pantothenate, biotin and ascorbic acid), etc., are absorbed by active transport mechanism (Box 2.8).

Box 2.8

1. Glucose/galactose and Na^+ share the same carrier protein, called **cotransporter** or **symport**, *e.g.*, **sodium-dependent glucose transporter** (**SGLT**) or Na^+-glucose co-transporter.

***(b)* Endocytosis.** Cells lining the small intestine take up small portions of chyle by endocytosis. Oligosaccharides, small proteins and cholesterol-containing LDL or low density lipoprotein are absorbed by the membrane of enterocyte by **receptor-mediated endocytosis**. In this type of endocytosis, a specific receptor on the surface of the plasma membrane "recognizes" an extracellular macromolecule and bind with it. The substane bound with the receptor is called the **ligand** (*e.g.*, cholestrol). The region of plasma membrane containing the receptor-ligand complex undergoes endocytosis.The whole process of receptor-mediated endocytosis, includes the following events: (*i*) interaction of ligands and cell surface receptors; (*ii*) formation of coated-pits and coated-vesicles (coat is made of clathrin protein); and (*iii*) formation of endocytic vesicles and their fusion with endosomes. An endosome has an acidic interior (pH 5–6) due to the presence of ATP-driven **proton** (H^+) **pumps** in its membrane and it contains lysosomal enzymes for digestion.

1. Absorption of Carbohydrates

The end products of carbohydrate digestion, monosaccharides are absorbed from the jejunum into the blood of the portal venous system. All the carbohydrates are absorbed in the form of monosaccharides, only a small fraction of a percent being absorbed as disaccharides and almost none as polysaccharides.

***(i)* Absorption of monosaccharides.**

***(a)* Pentose** is absorbed from the small intestine by **diffusion**.

***(b)* Galactose** is absorbed by duodenal and jejunal part of small intestine by **active transport** (by Na^+-glucose cotransporter protein).

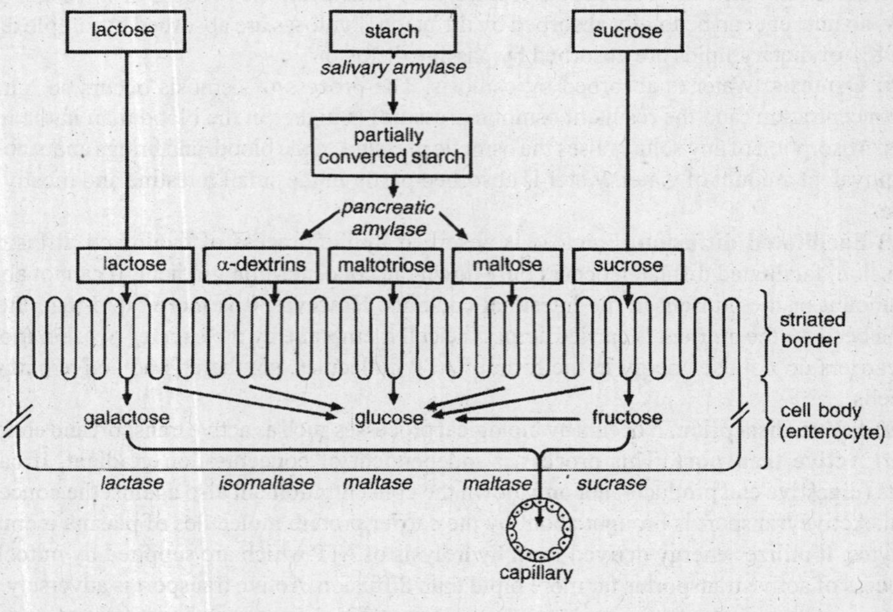

Fig. 2.6. Schematic diagram showing digestion and absorption of carbohydrates by an enterocyte. Enzymes superimposed over the striated border (*i.e.*, linked to glycocalyx) are bound to the plasma membrane of microvilli and act on their substrates in this location. Absorbed monosaccharides are then passed through the enterocyte and into the underlying capillaries that lead to the portal vein and via this vein to liver.

(*c*) **Mannose and fructose** are absorbed from the small intestine by **facilitated diffusion**.

(*d*) **Glucose** is absorbed from the duodenal and jejunal part of the small intestine by the process of **active transport** (by Na^+-glucose cotransporter protein).

(*ii*) **Absorption of disaccharides**

(*a*) **Maltose** is absorbed by receptor-mediated endocytosis. Some portion of maltose is absorbed from jejunum and it may also be absorbed from the first part of the ileum. Inside the cell, the one molecule of maltose is hydrolysed into two molecules of glucose.

(*b*) **Lactose** is absorbed from duodenum and may be absorbed by jejunum part of small intestine (Box 2.9).

(*c*) **Sucrose** is absorbed from the ileum and the distal part of the jejunum of the small intestine.

Box 2.9 Lactose Intolerance

Among the mammals, only human beings can digest the lactose found in the milk in adult stage. But with advancing age, the amount of lactase (enzyme) produced goes on decreasing so they cannot digest milk. In these persons lactose is fermented in the intestine producing gases (CO_2 and H_2) and acids. This causes flatuence, intestinal cramps and diarrhoea due to **lactose intolerance**. Such persons must take curd or yoghurt as the lactose is fermented to lactic acid in the yoghurt.

Further, during their intestinal absorption disaccharides (e.g., sucrose, maltose and lactose) first enter the enterocytes (**Millner** and **Crane** 1960). Sucrose is mainly absorbed in the jejunum in human beings. Inside the enterocytes of intestinal mucosa, these are hydrolysed by the disaccharidases (enzymes) such as sucrase, maltase and lactase respectively into their monosaccharides. These monosaccharides are then passed from the entrocytes into the capillary blood within the villus. So the earlier belief that dietary disaccharides are first broken into monosaccharides in intestinal lumen and then absorbed is not applicable. Out of these disaccharides, lactose is less taken by the enterocytes so lactose reaches the lower part of the intestine and promotes the growth of highly beneficial certain lactose - fermenting microorganisms.

2. Absorption of Fats

To recall, fat are finely emulsified in the small intestine by the detergent action of bile salts, lecithins and monosaccharides. The products of lipid digestion enter the body by a different process and a different route.

Micelles formation. The products of lipid digestion, namely, long chain fatty acids, monoglycerides, diglycerides, etc., are insoluble in water and cannot be directly absorbed from the intestinal contents. They combine with the bile salts and phospholipids to form small, spherical, water-soluble aggregates, called **micelles**. Micelles contain fatty acids, monoglycerides and cholesterol in their hydrophobic centres. Miceller formation solubilizes the fats and provides a mechanism for their transport to the enterocytes. Micelles move down their concentration gradient through the unstirred layer to the brush border of the mucosal cells. The lipids diffuse out of the micelles, and a saturated aqueous solution of the lipids is maintained in contact with the brush border of the mucosal cells.

Traditionally, the lipids were thought to enter the enterocytes by **passive diffusion**, but there is some evidence that carriers are involved. From the micelles, *fatty acids*, *glycerides*, *fat soluble vitamins* (*e.g.*, A, D, E and K vitamins) are actively absorbed in intestinal cells. Inside the cells the lipids are rapidly esterified, maintaining a favourable concentration gradient from the lumen into the cells. Unlike the mucosa of ileum, the rate of absorption of bile salts by the jejunal mucosa is low, and for the most part, the bile salts remain in the intestinal lumen, where they are available for the formation of new micelles. Thus, the bile salt micelles solubilize the lipids, transport them across the unstirred layer and keep a saturated solution of lipids in contact with the mucosal cells.

Fate of lipids inside enterocytes. The fate of the fatty acids in enterocytes depends on their size. Thus, fatty acids containing less than 10–12 carbon atoms pass from the mucosal cells directly into the portal blood, where they are transported as free (*i.e.*, unesterified) fatty acids. The fatty acids containing more than 10–12 carbon atom are re-esterified to triglycerides into mucous cells. In addition, some of the absorbed cholesterol is esterified. The triglycerides and cholesteryl esters are then coated with a layer of protein, cholesterol and phospholipid to form **chylomicrons**. Chylomicrons leave the intestinal cell and enter the **lacteals** (lymph vessels in the villi).

How chylomicrons are formed? In mucosal cells, most of the triglyceride is formed by **acylation** of the absorbed 2-monoglycerides, primarily in the **smooth endoplasmic reticulum**. (*Note*. Acylation is a process whereby acyl group is incorporated into a molecule by substitution. Acyl is a radical formed from organic acid by removal of a hydroxyl group (OH); its general formula is RCO). However, some of the triglyceride is formed from glycerophosphate, which in turn is a product of glucose catabolism. Glycerophosphate is converted into glycerolipids that participate in chylomicron formation (Fig. 2.7).

Acylation of glycerophosphate and formation of lipoproteins occur in **rough endoplasmic reticulum**. Carbohydrate moieties are added to the proteins in the **Golgi apparatus** and the finished chylomicrons are extruded by exocytosis from the basal or lateral aspects of the cell. Ultimately they reach lacteals of villi to be transported through the lymphatic system into the circulation. In fact, after a fatty meal, the lacteals are filled with a milky suspension of chylomicrons; it is from this milky appearance that the lacteals get their name. The chylomicrons is relatively stable in aqueous solution because of its hydrophilic protein covering.

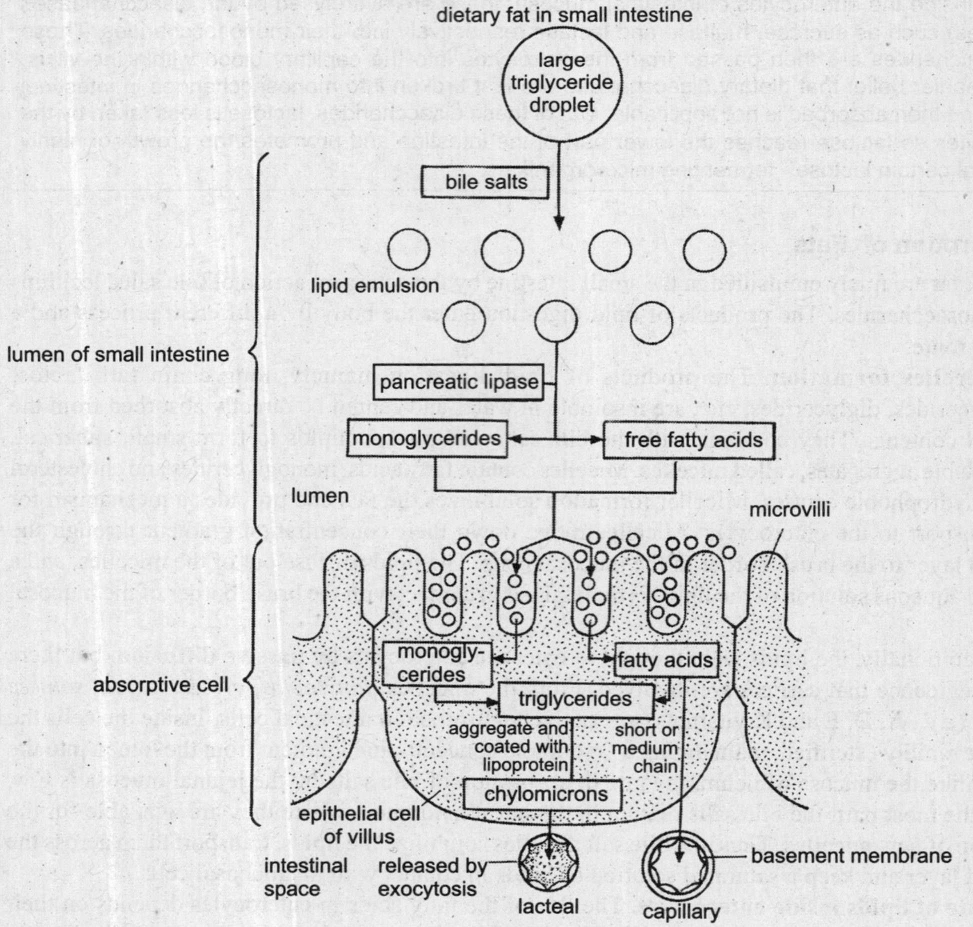

Fig. 2.7. Absorption of fats in intestine.

Absorption of long chain fatty acids is greatest in the upper parts of small intestine, but appreciable amounts are also absorbed in the ileum. On a moderate fat intake, 95% or more of the ingested fat is absorbed. **Short chain fatty acids** (SCFA) are produced by the action of bacteria on undigested complex carbohydrates from fruits and vegetables and major sites of their production and absorption is the colon. **Cholesterol** is readily absorbed from the small intestine if bile, fatty acids and pancreatic juice are present.

Glycerol enter the brush border of epithelial cells and are utilized. Glycerol is a sweet syrupy hygroscopic trihydroxy alcohol ($C_3H_8O_3$). Amino acids, monosaccharides and short chain fatty acids (SCFA) pass directly into the venous capillaries of the villus.

3. Absorption of Amino Acids

Amino acids are transported from the intestinal mucosa to serosa by active transport. The uptake of amino acids is closely coupled with the simultaneous uptake of Na^+ and it is the gradient of these ions acting across the plasma membrane of the mucosa cells that drive the amino acids pump.

Absorption of animo acids is rapid in the duodenum and jejunum but slow in the ileum. Approximately 50% of the digested protein comes from ingested food, 25% from proteins in digested juices, and 25% from desquamated mucosal cells. Only 2 to 5% of the protein in the small intestine escapes digestion and absorption. Some of the ingested protein enters the colon and is eventually digested by bacterial action (symbiotic digestion). Any protein present in the stool is not of dietary origin but comes from bacteria and cellular debris.

In infants, moderate amount of undigested proteins are also absorbed. The protein antibodies in maternal **colostrum** (*viz.*, milk secreted for a few days after parturition/ childbirth and characterized by high protein and antibody content) are mainly secretory immunoglobulins (IgAs) that cross mammary duct endothelial cells by transcytosis. They enter the circulation of the infant from the intestine and provide passive immunity against infections. Absorption of these proteins takes place by **endocytosis** and subsequent **exocytosis**.

Protein absorption declines with age but adult still absorb small quantities of proteins. Foreign proteins that enter the circulation elicit the formation of antibodies and the antigen-antibody reaction occurring upon subsequent entry of more of the same protein may provoke **allergic symptoms**. Thus, absorption of proteins from the intestine may explain the occurrence of allergic symptoms after eating certain foods. In fact, certain foods are more allergic than others. Crustaceans, molluscs and fish are common offenders and allergic responses to legumes, cow's milk and egg white are also relatively frequent.

Absorption in Large Intestine

The main function of large intestine (caecum and colon) is absorption of 90% of the fluid, it converts the 1000–2000 mL of isotonic chyle that enters it each day from the ileum to about 200-250 mL of semisolid faeces.

The absorption capacity of the mucosa of the large intestine is great. Na^+ is actively transported out of the colon, and water follows along the osmotic gradient thus generated. Normally there is net **secretion** of K^+ and HCO_3^- into the colon.

2.4 ASSIMILATION

Assimilation means the incorporation of the materials called **metabolites** after digestion of food into the tissue cells as their integral and homogeneous part. The various food materials reaching the cells of body are assimilated by the following methods:

1. Amino acids. Amino acids reach the liver by way of hepatic portal vein. Surplus amino acids get **deaminated** in the liver by separation of amino group (NH_2). The amino group combines with carbon dioxide to from urea, $CO(NH_2)_2$ (a process called **Krebs-Henseleit cycle** occurring in the mitochondria of liver cells). Urea is excreted by the kidneys in the form of urine. The remnants of amino acid molecules (*i.e.*, keto acids), after the deamination (*i.e.*, separation of amino groups) are converted into sugar or fat for use as sources of energy. Other amino acids are released by liver to the cells of the body. In body cells, they are transformed into the characteristic proteins of the human body by intracellular enzymes involved in protein synthesis. Some of these proteins (*e.g.*, histones, tubulin, actin, keratin, collagen, etc.) participate in the formation of the living materials (protoplasm), for repair and growth, and some act as enzymes and hormones to regulate various chemical reactions. Proteins are not stored in the body.

2. Monosaccharides. Molecules of glucose, fructose and galactose also reach the liver by way of hepatic portal vein. Here, fructose and galactose are converted into glucose. Excess of glucose is converted into **glycogen** in the liver cells (glycogenesis) and stored for supply at the time of need; and rest of the glucose is released to the body cells. Here, it is partly used as fuel and partly combined with other substances to form a basic part of cell contents. Some glucose is also stored in the muscle cells as glycogen. It is used as fuel in the muscle cells themselves and is not released to other cells. Some glucose is changed into fats (lipids) that are stored in the liver and muscles.

Glycogen stored in the liver and muscles would supply energy needs of the body for only 12 hours if they are used alone (during starvation). However, a human being of normal weight can usually survive without food for about 6 weeks, using fat reserves for energy. Anyway, thirst can lead to death within a few days.

3. Fats (Lipids). The absorbed lipids (fats) change into characteristic fats of the animal by rearrangement of fatty acids. Some fats are stored in the liver, bones, muscles and mesentries, around the heart and kidneys, and in the subcutaneous adipose tissue, whereas some enter into the composition of cell organelles, especially cell and nuclear membranes. Lipids are used as fuel also. In some cells, lipids are used in the synthesis of steroid hormones (*e.g.*, adrenal glands, ovaries, testes, etc.).

In fact, lipid molecules are constantly exchanged between the blood stream and the adipose tissue; every molecule of lipid in adipose tissue is replaced every three weeks.

Absorbed **minerals** are used in the biosynthesis of complex molecules such as proteins and hormones. They regulate transport mechanism across the plasma membrane and maintain ionic balance in the secretory products of the cell.

2.5 DIGESTION BY MICROBIAL ACTIVITY

Microorganisms, which are normal inhabitants of the alimentary canal, help in the digestion of food. Digestion of food by microbial activity is called **symbiotic digestion**.

1. Symbiotic Digestion in Humans

In humans, the flora of bacteria play a role in digestion in the large intestine. The chyle entering the large intestine (colon) consists of about 75 percent water, some fatty and proteinaceous substances, and excess enzymes. Water is absorbed into blood as the digestive waste materials move through the colon. The symbiotic bacteria, such as *Escherichia coli*, *Enterobacter aerogenes*, *Bacteroides fragilis*, *Streptococcus faecalis*, etc. (**Ganong**, 1999), residing in the colon feed on protein contents of the chyle, breaking them into amino acids which are absorbed into the blood. The colon bacteria also synthesize vitamin K, a number of B complex vitamins and folic acid, all of which are absorbed, Doctors recommend vitamins when prescribing oral antibiotic that kill bacteria of colon to avoid vitamin deficiency in the patient. The colon bacteria also convert the bile pigments into brown pigments that give colour to faeces.

Colon bacteria produce some of the foul smelling gases in the flatus. Organic acids formed from carbohydrates by bacteria are responsible for the slightly acidic reaction of the stool (pH 5.0–7.0). Intestinal bacteria also appear to play a role in **cholesterol metabolism**. A number of amines, including such potentially toxic substances such as **histamine** and **tyramine**, are formed in the colon by bacterial enzymes that decarboxylate amino acids. Other amines formed by the intestinal bacteria, especially **indole** and **skatole**, contribute to the odour of the faeces, as do **sulphides**. Ammonia is also produced in the colon and absorbed.

2. Symbiotic Digestion in Non-human Animals

Digestion by symbiotic microorganisms has a special significance for herbivorous animals because these animals lack enzymes for cellulose digestion.

***(i)* Symbiotic digestion in lagomorpha and coprophagy.** In rabbit and hare, the cellulose part of the food escapes the action of enzymes available in stomach and small intestine. It is partially digested in the **caecum** by bacterial activity. Bacteria secrete enzymes, **cellulases**, that break down the cellulose into simpler but still nonabsorbable, form. Also there is little absorption activity in the caecum. The simplified cellulose from the caecum forms a special kind of faeces at night; this faeces is soft, semifluid, mucus-covered and light-coloured. Such faeces are not dropped by the animal but are eaten directly from the anus (this act is called **caecotrophy**) and are passed through the gut once more. At this turn, the simplified cellulose is digested and the products of digestion are absorbed (in duodenum and ilieum). The vitamins formed in the large intestine by symbiotic bacteria are also absorbed from the faeces. The faeces passed out during the day are in the form of firm dark pellets, and are not eaten.

The habit of passing food twice through the gut ensures absorption of maximum amount of nutrients from it; and is called **coprophagy** (G. *copros* = excrement; *phagein* = to eat), **refection** or **pseudorumination**. The reingested faeces are not masticated and mixed with other food in the stomach, instead they are lodged separately in the **fundus** of the stomach (Fig. 2.8). The soft faeces are covered by a membrane and undergo fermentation in the stomach for many hours. Thus, the fundus of rabbit's stomach acts as a **fermentation chamber**, analogous to the rumen of cattle and sheep. One of the products of fermentation in rabbit is lactic acid.

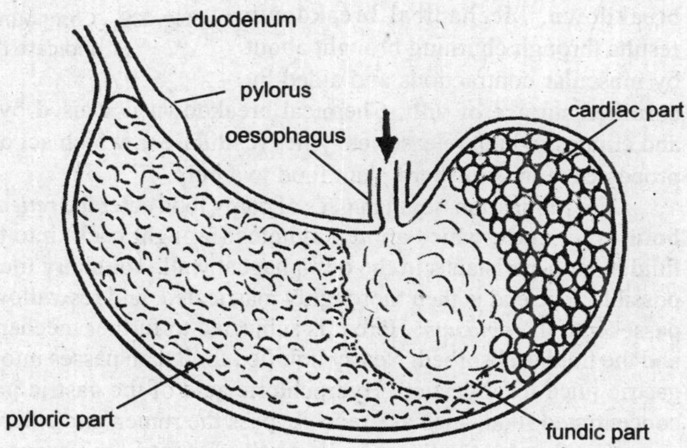

Fig. 2.8. Stomach of rabbit. Reingested faeces remain in the fundus while fermentation goes on. Ingested food is kept in the pyloric part which contains digestive glands.

In rabbit, prevention of coprophagy reduces digestive ability, protein utilization and nitrogen retention. Coprophagy also occurs in rats, shrews, marsupials and prosimian primates.

***(ii)* Symbiotic digestion in ruminants.** In ruminants (cud-chewing animals), such as cattle, buffalo, goats, sheep, camel, etc., cellulose part of the food is digested in a chamber, called **rumen**, of the stomach by symbiotic microorganisms, *viz.*, bacteria and ciliates. Many of these microorganisms produce enzymes (cellulase) that act on cellulose and simplify it. Fermentation produces certain short-chain fatty acids (such as acetic, butyric, propionic acids). These acids are neutralized by sodium bicarbonate of saliva, and are absorbed. Saliva also serves as proper fermentation medium for the microorganisms. Fermentation perhaps also produces certain proteins from inorganic nitrogen compounds, such as ammonium salt and also vitamin B complex. The proteins and vitamins are utilized by the microorganisms as well as the host. The microorganisms get energy from the chemical reactions of fermentation and an ideal temperature to live in return for a supply of cellulase enzyme. Fermentation in the rumen also produces CO_2 and methane (CH_4). These are allowed to escape by belching or **eructation** (voiding of gas *via* mouth).

The stomach of a ruminant mammal is called a **compound stomach** (Fig. 2.9). The ruminant stomach consists of four chambers: **rumen** (paunch), **reticulum** (honeycomb), **omasum** (psalterium) and **abomasum** (rennet). The first three chambers are lined with stratified squamous epithelium with cornified surface. The last chamber of stomach has the digestive glands. The ruminants take plant food in a large quantity. In order to reduce the duration of feeding, they swallow slightly chewed food mixed with copious (large in quantity) saliva. This food passes into the rumen. After feeding, they seek a quite, safe place, and sit down to ruminate leisurely.

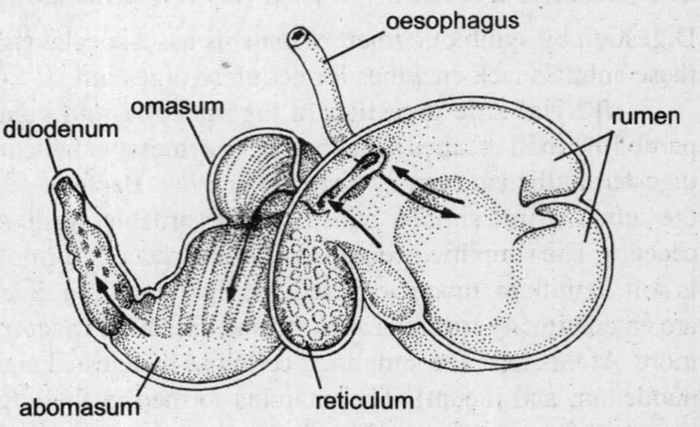

Fig. 2.9. Compound stomach of a ruminant animal (arrows indicate the passage of food through it).

In the rumen, the food undergoes mechanical as well as chemical breakdown. Mechanical breakdown results through churning brought about by muscular contractions and aided by cornified surface of villi. Chemical breakdown is caused by symbiotic microorganisms (bacteria and ciliates) which releases enzymes (cellulases), which act on cellulose and simplify it. These two processes reduce the hard plant food to a pulp.

For rumination, small masses of pulpy food enter the *reticulum*, where they are rolled into rounded **boluses**, or **cuds**, which are regurgitated (brought back) into the mouth along with a fair amount of fluid by anti-peristalsis in the oesophageal wall. Voluntary musculature of the oesphagus makes this possible. The cud is then thoroughly masticated, and reswallowed. Now the well chewed, pasty food passes into the *omasum*. Here it is subjected to further mechanical breakdown by the cornified folds and the fluid is absorbed. Nearly liquified food then passes into *abomasum*, where it is mixed with the gastric juice for digestion. Hydrochloric acid of the gastric juice destroys the microorganisms. The concentrated food, such as grains, bypass the rumen and enter the reticulum directly.

***(iii)* Symbiotic digestion in reptiles.** Many herbivorous reptiles also benefit from the symbiotic microorganisms. The marine green turtle (*Chelonia mydas*) feeds on sea weeds and digests cellulose with the aid of microorganisms. The common iguana (*Iguana iguana*) is herbivorous throughout life, and benefits from microbial fermentation in the hind gut.

***(iv)* Symbiotic digestion in termites.** In the intestine of termites, cellulose is digested by symbiotic flagellates, such as *Trichonympha campanula*. These protozoans enter young termites (nymphs) when they feed on fresh faeces of adult (this is made possible due to social behaviour of termites). The flagellates produce enzymes **β-glucosidases** which split cellulose into sugar. The latter is used by the protozoans as well as termites. It is interesting to note that the termites starve death without their intestinal flagellates, and the latter do not survive outside the termites.

The termites live virtually exclusively on wood. The cellulose from wood can provide enough energy but the wood has very low nitrogen content. It has been found that the nitrogen fixing bacteria also resides in the termite intestine, and they supply nitrogen for proteins.

Certain other invertebrates, namely garden snail (*Helix*), shipworm (*Teredo*) and silver fish (*Lepisma*), also can digest cellulose with the help of cellulase enzyme.

2.6 CONTROL ON ENZYME SECRETION

The digestive enzymes bring about chemical digestion of foodstuffs. In humans, these are contained in the **saliva**, **gastric juice**, **pancreatic juice**, **bile** and **succus entericus** secreted by salivary glands, gastric glands, pancreas, liver and intestine, respectively. These are secreted and released only when needed. Their secretion is stimulated by there types of stimuli, 1. *nervous*, 2. *mechanical*, and 3. *hormonal*.

1. Nervous stimuli. They arise from the stimulation of never ending by touch, sight and smell of food. The nerve impulses generated thereby are passed via central nervous system to trigger the activity of secretory glands. The secretion of digestive juices by nervous stimuli represent **reflex process** and reflex center lies in medulla.

2. Mechanical stimuli. These are produced by the stretching or contraction of the organ, caused by presence or absence of food. The stretch impulses cause stimulation of the sensory nerve endings and produce local reflexes.

3. Chemical stimuli. These stimulations are due to chemical substance (**hormones**). The hormones are absorbed (loaded) in the blood and carried to the appropriate glands which are stimulated to produce their secretion.

In the light of above classification, it has been found that—

(*i*) the salivary secretion is purely under the nervous control;

(*ii*) gastric and pancreatic secretion is controlled by the nervous and chemical stimuli;

(*iii*) bile secretion involves only chemical stimuli; and

(*iv*) intestinal secretion involves all the three stimuli.

1. Control of Secretion of Saliva

The secretion of salivary juice (salivation) is entirely under nervous control and is purely a **reflex phenomenon**. The salivary glands receive nerve supply from the **sympathetic** and **parasympathetic systems**. The **conditioned reflexes** are produced by the sight, smell, idea or talk of food. The **unconditioned reflexes** are produced by chewing and taste of food and by physical presence of food in buccal cavity. The stimulus produced is carried to the **salivary nuclei** present in the posterior part of pons of medulla. These supply the salivary glands by issuing parasympathetic nerve fiber (Fig. 2.10).

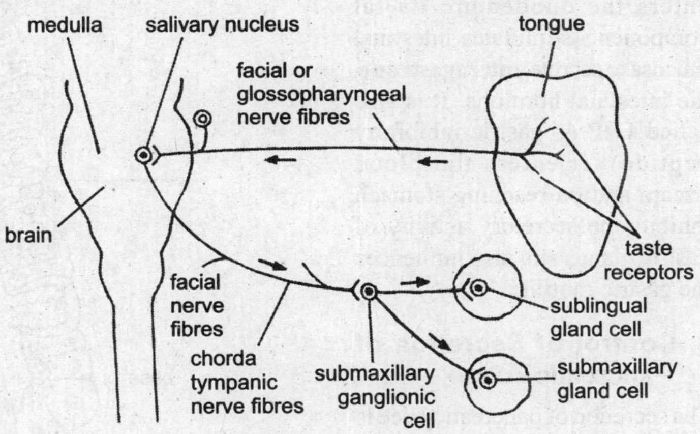

Fig. 2.10. Diagram to illustrate the reflex pathways for the secretion of saliva.

2. Control of Secretion of Gastric Juice

The secretion of gastric juice from the gastric glands into stomach is stimulated by **nervous** and **chemical excitation** and also by secretion of pancreatic juice and bile juice.

***(i)* Nervous excitation.** The **cephalic** or **psychic phase** of gastric secretion is under nervous control and is influenced both by unconditioned and conditioned reflexes. The reflex center for gastric secretion lies in the medulla. The taste of food (unconditioned reflex), the sight, smell or the thought of food (conditioned reflex) stimulate gastric glands to secrete gastric juice. The ending of vagus nerve release **acetylcholine** which stimulates gastric glands for secretion.

The gastric juice under nervous excitation is released before the food enters the stomach, and is called **appetite juice**.

***(ii)* Chemical excitation.** The chemical excitation occurs at the **gastric phase** and at the **intestinal phase** (of digestion).

***(a)* Gastric phase.** When food enters the stomach, its protein component and peptones provoke the mucosa of pyloric stomach for the formation of a hormone, called **gastrin**. Gastrin is a polypeptide hormone and is secreted by APUD cells of gastric glands (Fig. 2.11). The gastrin is released in blood stream and is brought back to the gastric glands, which are stimulated to secrete gastric juice. The food components which initiate gastric secretion are called **secretagogues**.

***(b)* Intestinal phase.** The gastric glands also get excited when product of gastric digestion enter the duodenum. The gastro-intestinal hormones responsible for this phase of secretion are **gastrin, secretin** and **vasoactive intestinal peptide (VIP)**.

When chyme from stomach enters the duodenum, its fat component stimulates intestinal mucosa to secrete **enterogastrone**, the intestinal hormone. It is also called **GIP** (= gastric inhibitory peptide). It enters the blood stream and on reaching stomach inhibits the secretory activity of gastric glands and also influences the gastric motility.

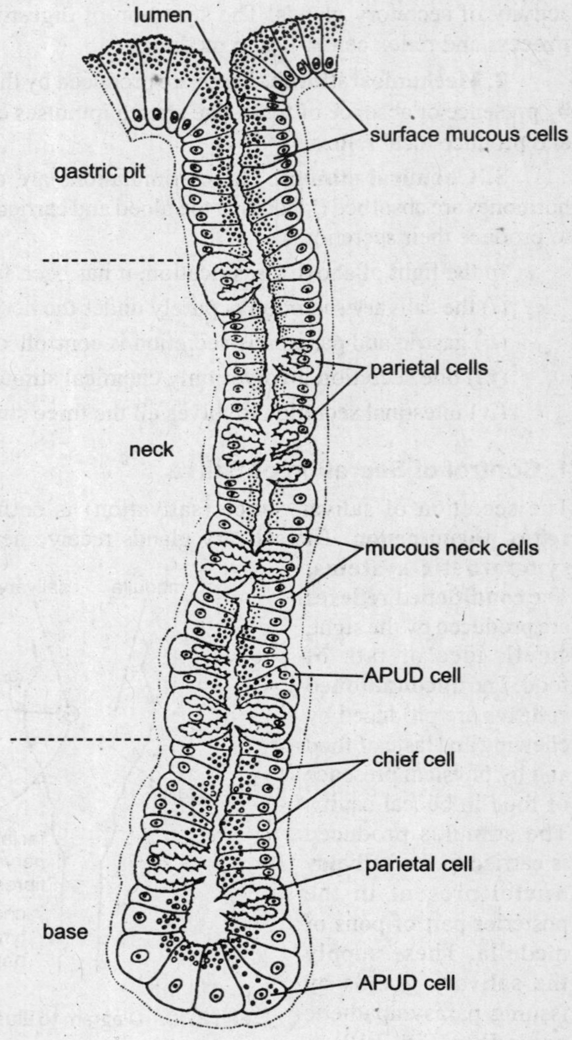

Fig. 2.11. Gastric (or fundic) gland of mammal showing positions of APUD cells. APUD = amine precursor uptake and decarboxylation (*i.e.*, enteroendocrine cells).

3. Control of Secretion of Pancreatic Juice

The secretion of pancreatic juice is both under nervous and hormonal control (Box 2.10).

***(i)* Nervous control.** It includes the reflexes caused by seeing or tasting food. The impulses are sent to the vagal nuclei in medulla and from there these are carried to the pancreatic secretory cells through vagal nerve fibers. The secretory cells are excited by the release of the acetylcholine at the nerve endings.

***(ii)* Hormonal control.** When acidic chyme from stomach enters duodenum it stimulates duodenal mucosa to liberate two peptide hormones **secretin** and **pancreozymin (PZ)**. Through hepatic portal system these reach pancreas and stimulate its secretion (Fig. 2.12).

Box 2.10

Pancreatic secretion involves three phases: **cephalic phase, gastric phase** and **intestinal phase.** During the cephalic phase, moderate amount of the pancreatic secretion is stimulated by acetylcholine released by the vagal nerve endings in the pancreas. During the gastric phase, moderate secretion of pancreatic enzymes is stimulated by nervous and gastric hormonal stimulation. During intestinal phase, large amount of pancreatic secretion is stimulated by secretin, pancreozymin and cholestokinin (CCK) hormones.

4. Control of Secretion of Bile Juice

Secretion of bile juice is exclusively under chemical control.

(*i*) **Bile salts.** The bile salts such as **sodium glycocholate** and **sodium taurocholate** are absorbed from the intestine and are carried to liver through hepatic portal circulation.

(*ii*) **Hormone.** When food enters duodenum it stimulates mucosa to secrete another hormone, the **cholecystokinin (CCK)**. It bring about contraction of gall bladder and relaxation of sphincter of Oddi (which is present in common bile duct). Vagus stimulation also causes the same effect and brings about expulsion of the bile into the duodenum.

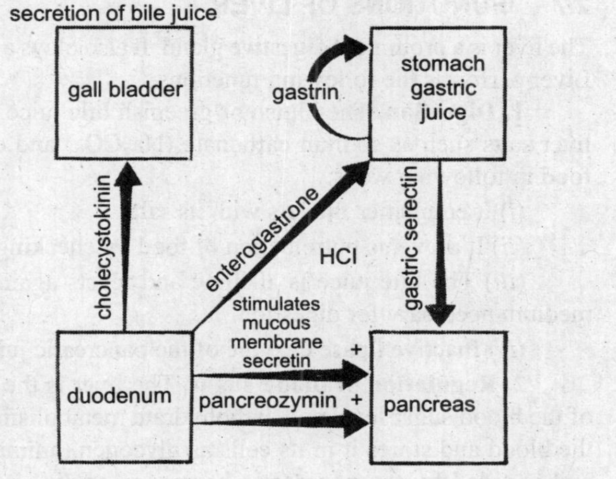

Fig. 2.12. Hormonal control of secretion of digestive juice from stomach, duodenum and pancreas.

5. Control of Secretion of Intestinal Juice (Succus Entericus)

The secretion of digestive juice from the intestinal mucosa is more or less continuous but it is considerably increased during the process of digestion. The stimulation is brought about both by mechanical as well as chemical stimuli (Fig. 2.13).

(*i*) **Mechanical stimuli.** The mechanical stimuli are produced by the presence of food in the intestine. The food induces the local sense organs (**Meissner's plexus**). These mechanical irritations stimulate the secretion of intestinal juice.

(*ii*) **Hormone.** The hormone **enterocrinin** secreted in intestinal mucosa induces the intestinal cells to secrete the digestive enzymes. **Secretin** hormone also plays some role in stimulation of intestinal mucosa.

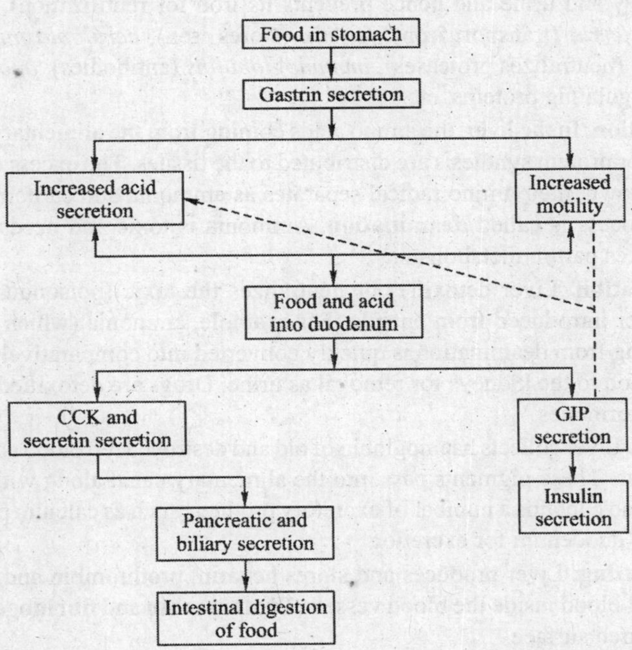

Fig. 2.13. Integrated action of gastrointestinal hormones in regulating digestion and utilization of absorbed materials. The dashed arrows show inhibition.

2.7 FUNCTIONS OF LIVER

The liver is a prominent digestive gland. It also plays a major role in the whole process of metabolism. Liver performs the following functions:

1. Digestion. The bluish or greenish bile juice is manufactured by the liver cells. It contains liver salts such as sodium carbonate (Na_2CO_3) and cholesterol. The bile helps in the digestion of food in following ways:

(*i*) It emulsifies the fats with its salts.

(*ii*) It prevents putrefaction of food by checking the growth of bacteria.

(*iii*) The bile juice is alkaline and reacts against the acidity of food to prepare the alkaline medium necessary for digestion.

(*iv*) Inactive lipase enzyme of the pancreatic juice is activated by the bile juice.

2. Regulation of blood sugar. The liver is the chief **homeostatic organ** in the maintenance of the blood sugar level and carbohydrate metabolism. The liver withdraw the excess of sugar from the blood and stores it in its cells as glycogen (animal starch). This process is called **glycogenesis** and is aided by the pancreatic hormone **insulin**. It takes place when the animal gets plenty of food.

In the condition of food shortage, the stored glycogen is changed into glucose and added to the blood stream for distribution to the body. This process is termed **glycogenolysis**. It is aided by another pancreatic hormone called **glucagon**.

3. Glyconeogenesis. Formation of glycogen from non-carbohydrate sources is called **glyconeogenesis**. Excess of amino acids/fats from very high protein/fat diet are changed into glycogen by the liver.

4. Lipogenesis. Formation of fat from glucose and amino acid is called **lipogenesis**. Liver converts excess glucose and amino acids into lipids or fats. Thus liver is vitally involved in the synthesis, storage and utilization of fats.

5. Protein synthesis. Liver synthesizes a variety (more than 20 types) of proteins of blood plasma such as *albumin*, *hepatoglobin* (which binds extracorpuscular haemoglobin, prevent its loss into the kidney and urine and hence prevents its iron for reutilization), *hemopexin* (binds to porphyrins), *transferrin* (transport iron), *ferritin* (stores iron), *ceruloplasmin* (transport copper), a_2*-macro-globulin* (neutralizes proteases), *immunoglobulins* (antibodies), *apolipoprotein B* (a lipid carrier), blood coagulating proteins, etc.

6. Deamination. In the liver, the amino acids coming from the alimentary canal are sorted out. Those necessary for protein synthesis are distributed to the tissues. The excess or unnecessary amino acids are broken down, their amino radical separates as ammonia and carbon chain changes into a keto acid. This process is called **deamination**. Ammonia is toxic and needs immediate removal, while keto acid takes part in metabolism.

7. Detoxification. Liver detoxifies or neutralizes the toxic (poisonous) substances formed in the body itself or introduced from outside. For example, ammonia (which is extremely toxic to body cells) resulting from deamination is quickly converted into comparatively harmless **urea**. The urea is sent via blood to the kidneys for removal as urine. Drugs are detoxified into liver. Liver also inactivates many hormones.

8. Excretion. Liver collects haemoglobin of old and destroyed red blood corpuscles and changes it into bile pigments. These pigments pass into the alimentary canal along with bile for elimination with faeces. Bile also contains a number of excretory products such as calcium phosphate, etc., which are discharges into duodenum for excretion.

9. Blood clotting. Liver produces and stores heparin, prothrombin and fibrinogen. **Heparin** prevents clotting of blood inside the blood vessels. **Prothrombin** and **fibrinogen** help in the clotting of blood at the injured surface.

10. Formation of RBCs. Liver produces red blood corpuscles in the embryo (fetus). The process of formation of red blood corpuscles (RBCs) in the embryo is called **erythropoiesis** (Box 2.11).

Box 2.11 Extramedullary Haematopoiesis

In the adult, red blood cells, many white blood cells and platelets are formed in the bone marrow. In the fetus (embryo), blood cells are also formed in the liver and spleen. This is called **extramedullary haematopoiesis**.

11. Destruction of red blood corpuscles. The liver destroys the old worn-out red blood corpuscles. Their haemoglobin is excreted as bile pigments.

12. Phagocytosis. Foreign matter, dead cells and bacteria are disposed of in the liver. These are engulfed and digested by large star-shaped cells called the **macrophages** or **Kupffer's cells**, that occur in the lining of liver sinusoids.

13. Synthesis of vitamin A. Liver synthesizes vitamin A from the β-carotene, an orange-yellow substance of carrots.

14. Storage. Beside glycogen , liver stores a variety of substances such as lipids (*e.g.*, fats, fatty acids and cholesterol), minerals such as coper and iron and vitamins namely vitamin A, B_{12}, D and E.

15. Production of heat. Liver generates heat by its active metabolism.

16. Osmoregulation. Liver produces a protein called **angiotensinogen** that helps the kidney in maintaining body fluid osmoregulation.

2.8 MOVEMENTS OF THE GASTRO-INTESTINAL TRACT

The function of different parts of alimentary canal are performed by a variety of movements peculiar to the goals of that region and the local structures. These movements are under the control of the nervous system and hormones (of endocrine system) which are produced by the gastro-intestinal tract.

1. Movements of Mouth and Pharynx

The mouth and pharynx perform two functions, *i.e.*, mastication and deglutition.

***(i)* Mastication.** Most animals engulf large pieces of food and are **macrophagus**. The food is first treated mechanically and masticated into slimy balls or **'bolus'** of food. The teeth on jaws, tongue, saliva, cheek pouches and jaw muscles help in mastication of food in a mammal.

In the act of mastication, the lower jaw is moved upwards, downwards and laterally. A combination of these movements allows the food particles to be finely ground by the teeth. The tongue and cheeks help in the act of mastication as these keep the food between the teeth. The tongue not only rolls the food from side to side during chewing but also tastes it and helps in selection and rejection of food. Mastication of food in the buccal cavity is made more efficient by the development of cheek-pouches with whose help a larger amount of food may be chewed at any one time.

During the process of mastication, saliva is poured in the mouth by salivary glands. The saliva helps to moisten and lump the food particles together to form a bolus. The bolus is then swallowed.

***(ii)* Deglutition or swallowing.** Deglutition is the process of swallowing which involves the passage of food bolus from buccal cavity to stomach. Swallowing in mammals has three stages: (*a*) **buccal**, (*b*) **pharyngeal** and (*c*) **oesophageal**. The buccal stage of swallowing is *voluntary*, but the pharyngeal and oesophageal stage of swallowing are *involuntary*.

Deglutition is started as a **voluntary action** by pressing the tongue against the hard palate. This pushes the food bolus into the pharynx. From here, swallowing continues by **involuntary action** or **reflex action**. The latter is started by stimulation of the sensory nerve ending in the pharynx. The swallowing reflex is controlled by the **medulla** of brain. As food reaches the pharynx, the soft palate and uvula (**Uvula** is the pendent fleshy lobe in the middle of the posterior border of the soft palate) rise to close the nasopharynx and the larynx moves upward to meet the epiglottis for closing the glottis. This stops the breathing for a moment and prevents the food from going the wrong way (Box 2.12).

Box 2.12

Swallowing normally is very quick and we are unaware of the momentary interruption of breathing. Breathing during swallowing would choke the respiratory passage. Entry of a piece of food into the glottis may pose threat to life. Fortunately, the **coughing reflex** quickly clears the air passage. This reflex is poor in an unconscious person. Hence, such a person should not be given the liquids.

Pharyngeal muscles contract, pushing the food bolus over the epiglottis into the oesophagus. The sphincter muscle at the beginning of the oesophagus relaxes for this purpose. The food bolus stretches the oesophagus and initiates another reflex action, causing the contraction of the circular muscles of the oesophageal wall above the bolus. As the food (bolus) is pushed down the oesophagus, it stimulates successive regions to contract by reflex action, producing a **peristaltic wave**. The cardiac sphincter of stomach opens as food bolus approaches to let it slip into the stomach. The sphincter then closes again to check food passing back into the oesophagus. The cardiac sphincter is much less effective in infants so that **regurgitation** is much more common in them than in the adults.

Peristalsis. Food is pushed through the gut by the muscles of the muscularis externa, the **outer longitudinal** and **inner circular layers** of muscle in the gut wall. Behind the bolus, the circular muscles contract, squeezing and contracting the gut. In front of the food, the longitudinal muscles contract, shortening this section of the gut and pulling it past the advancing bolus (Fig. 2.14).

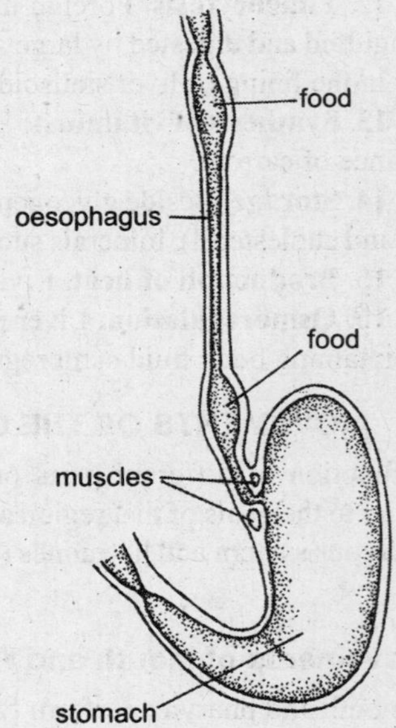

Fig. 2.14. Diagram of peristalsis in oesophagus.

Vomiting. If toxic materials or pathogenic (=disease-causing) organisms are swallowed inadvertently, **vomiting** occurs. Vomiting is a protective device as by this process harmful material from the alimentary canal is ejected out through the mouth. Vomiting involves relaxation of cardiac sphincter and a violent contraction of the diaphragm and abdominal muscles. These events are controlled by stimuli brought by vagus nerve from the medulla of brain. *Stomach plays a passive role during vomiting.*

Thus, vomiting is an important defence mechanism of body. Excessive vomiting is, however, harmful. It may upset the delicate pH balance of the body due to loss of acid along with stomach contents.

2. Gastric Movements

As food bolus enters and fills the stomach, the muscles stretch and volume is adjusted according to the contents. Wall of stomach undergoes muscular contractions (15–20 times per minute).This **churns** and breaks the food mechanically and mixes it thoroughly with the gastric juice for proper action of enzymes. For this role, the stomach has three types of smooth muscle layers: outer longitudinal, middle circular and inner oblique. The cardiac and pyloric sphincters remain closed during churning of food.

Peristalsis in stomach is started at the **incisura angularis** (a notch or bent in the lesser curvature of the stomach near its pyloric end). Later, the activity of the stomach increases; and the origin of the peristaltic progressively moves towards the fundus. The peristaltic movements propel the food (called chyme) into the duodenum.

The movements of the stomach beside being under the control of autonomic nervous system, are also regulated by certain hormones (*e.g.*, enterogasterone). During hunger, the stomach undergoes a type of sweeping peristalsis which are called **hunger contractions**. Although these contractions stimulate hunger, they have nothing to do with appetite as they quickly disappear when a small amount of food is taken.

3. Movements of Intestine

The intestine (or bowel) has following three types of movements:

(i) **Segmenting movements**, which are rhythmic contractions producing churning and mixing effects. These movements, however, do not move the bolus. The segmenting movements are not under the control of nervous system.

(ii) **Pendular movements**, which probably produce slight mixing effect and are not very important in humans.

(iii) **Peristaltic movements**, which can be initiated by mechanical, electrical or chemical stimuli. However, the most powerful stimulus is the presence of food in the intestine. When the food reaches the duodenum, regular peristaltic movements are initiated. At intervals powerful peristaltic contractions take place which extend through the entire length of the intestine. This is called **peristaltic rush** and it moves the food for a considerable distance in the intestine. The peristaltic rush causes the relaxation of the ilio-colic sphincter so that part of food (chyme) is pushed into colon. The frequency of the peristaltic rush is increased if an irritant poison or a purgative is taken.

The **villi** of intestine are also in a state of constant movement. Such movements are due to contractions of smooth muscle fibres within each villus. **Motilin** hormone secreted by mucosa of duodenum and initiates **interdigestive intestinal motility**.

4. Movement in Large Intestine

Large intestine is about 1.5 meter long and 7 cm thick. Large intestine includes *three* distinct regions: 1. Caecum; 2. Colon and 3. Rectum. Ileum opens into large intestine at junction of caecum and colon and its opening is regulated by **ileo-caecal valve**.

1. Caecum. It is a 6 cm long bag-like structure and contains a 9 cm long vermiform appendix. Lining of caecum contains many mucous glands. Solitary lymph follicles are present in caecum and appendix.

2. Colon. It is about 1 meter long and has diameter greater than that of the small intestine. Colon is divided into *four* parts: **ascending colon**, **transverse colon**, **descending colon** and **sigmoid colon** (Fig. 2.15). The fibres of colon's external muscular layer are collected into three longitudinal bands, called **teniae coli**. Because these bands are shorter than the rest of colon, the wall of the colon forms outpushing called **haustea** between teniae. *There are no villi on mucosa*. The colonic glands are short inward projections of the mucosa that secrete mucus.

The movements of the colon include *segmentation contractions* and *peristaltic waves* like those occurring in the small intestine. Segmentation contractions mix the contents of colon and expose more and more of them to the mucosa to facilitate absorption. Peristaltic waves propel the contents towards the rectum. A third type of contraction that occurs only in the colon is the **mass action contraction** which moves material from one portion of the colon to another.

5. Defaecation

Faeces can remain in the colon for about 36 hours, then moving into rectum, where they are stored briefly before discharge via the anus. This movement is spontaneous and involuntary. It typically occurs after a meal, when chyme from the small intestine passes into large intestine. Widening on the entry of faeces into it stimulates the nerves in its wall, and this produces an urge for defaecation. This urge can be resisted by constricting the outer anal sphincter. Delayed defaecation hardens the faeces by further absorption of water from it and leads to **constipation**.

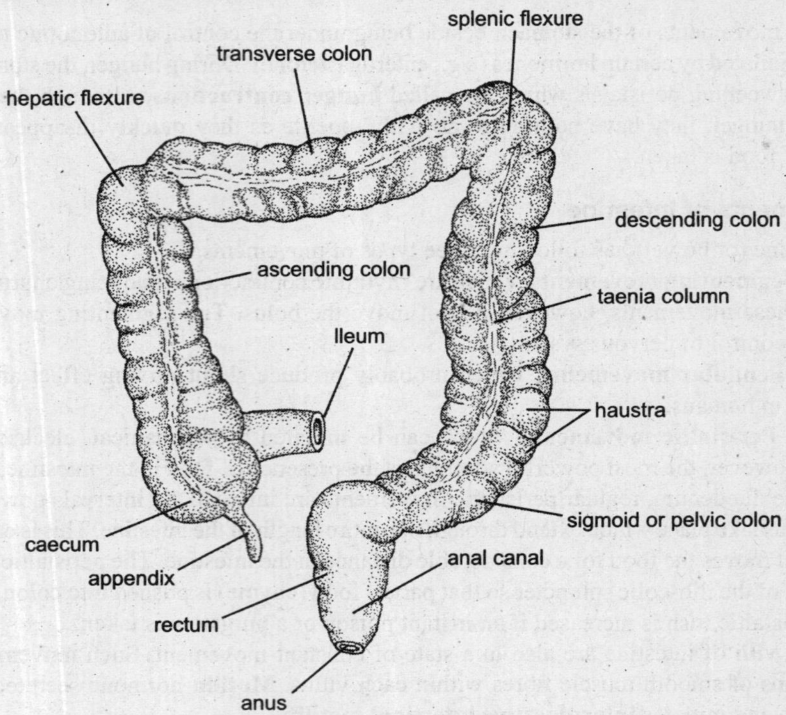

Fig. 2.15. The human colon.

Defaecation itself involves a coordinated action of muscles of large intestine, abdominal muscles, diaphragm and anal sphincter. The act is partly voluntary and partly involuntary. It increases abdominal pressure which makes the faeces to pass out.

QUESTIONS

Long Answer Questions

1. Describe the physiology of digestion in different parts of the alimentary canal of mammals.
2. What is digestion? Describe the digestion of carbohydrates, fats and proteins in mammals. *(CCSU Meerut 2006, 2008, 2011)*
3. Define digestion. Starting from mouth describe the changes undergone by food material in different parts of the alimentary canal. *(CCSU Meerut 2010)*
4. What is digestion? Describe the digestion and absorption of carbohydrates in mammals. *(CCSU Meerut 2009)*
5. Give an account of movements of gastrointestinal tract.
6. Explain enzymatic digestion of chyme in the small intestine. *(Punjab 1998)*
7. Describe the role of pancreas in the digestion of proteins and carbohydrates.
8. Explain composition, functions and regulation of hepatic and pancreatic juices. *(HPU 2007)*
9. Describe the mechanism of absorption of fats in alimentary canal of man in starch digestion. *(HPU 2006)*
10. Give an account of absorption of glucose, amino acids and products of fat digestion in the intestine.
11. What is assimilation? Discuss the process of assimilation in the body of the vertebrates.
12. Explain the digestion by microbial activity. *(Punjab 1999)*
13. Why the process of digestion is essential? Describe the hormonal control of digestion in mammals. *(CCSU Meerut 2013)*
14. Describe the structure and functions of digestive glands of mammal. *(CCSU Meerut 2014)*

Short Answer Question

1. Give the difference between the intracellular digestion and extracellular digestion.
2. Write down function of liver. (*Rohilkhand 1999; CCSU Meerut 2000; HPU 2009*)
3. Give the composition and functions of bile. (*CCSU Meerut 2011*)
4. Give an account of nervous and hormonal control of the digestive glands.
5. Write short notes on the following:
   (*i*) Saliva; (*Nagpur 2002*)
   (*ii*) Peristals
   (*iii*) Hormonal regulation of digestive enzymes;
   (*iv*) Defaecation;
   (*v*) Deglutition;
   (*vi*) Chylomicrons; (*Punjab 1999*)
   (*vii*) Coprophagy (*Punjab 1999*)
   (*viii*) Balanced diet (*HPU 2009*)
   (*ix*) Symbiotic digestion of cellulose (*HPU 2009*)
6. Describe biological importance of proteins. (*HPU 2009*)
7. What causes lactose intolerance?
8. Write a note on symbiotic digestion of cellulose. (*HPU 2009*)
9. Name the symbiotic bacteria of human colon.

Very Short Answer Questions

1. What is the role of hydrolases?
2. Which component of food is digested in the oral cavity?
3. How is ptyalin activated?
4. Which food component is not digested in the stomach?
5. Name the enzyme that acts within the mucosal cells of the small intestine.
6. Where are microvilli found?
7. Which hormones stimulate the secretion of pancreatic juice?
8. What is function of nondigestive enzyme of the intestinal juice?
9. What is coprophagy?
10. Where does digestion of fats occur in the alimentary canal?
11. What is function of chymotripsin?
12. Proteolytic or protein-digesting enzymes are always secreted in inactive form. Give reason.
13. Name the hormones that induce secretion of pancreatic juice.
14. What is the role of cholecystokinin?
15. How does bile juice assist in digestion of food?
16. Where is bilirubin found?
17. What is main function of bile?
18. What happens when gall bladder of humans is removed?
19. Which hormone inhibits gastric motility?
20. Give name of enzyme present in human saliva.
21. In which medium ptyalin of the saliva is secreted?
22. What is the functions of HCl in stomach?
23. Where the oxyntic cells are located?
24. What is function of trypsin?
25. Give the term for muscular contractions in the stomach.

26. What are peristaltic movements?
27. What is the role of cholecystokinin?

Fill in the Blanks

1. Feeding is controlled by two hypothalamic centers: feeding center and .... center.
2. The act of swallowing is also called ....
3. The enzyme .... of saliva disinfects the oral cavity.
4. Chyme leaves the stomach and .... leaves the small intestine.
5. Gastric digestion of an average meal is completed in .... to .... hours.
6. .... fats are absorbed in the lacteals.
7. Intestinal juice contains a nondigestive enzyme. It is named ....
8. Intestinal juice is also called ....
9. Premature passage of food into the intestine is prevented by ....
10. Intestinal bacteria produce vitamin ....
11. Fatty acids and glycerol are absorbed into ....
12. Trypsinogen is activated to trypsin by ....
13. Secretion of bile juice is regulated by .... hormone.

Multiple Choice Questions

1. Liver is the only organ which normally
   (*a*) synthesizes heparin (*b*) synthesizes urea
   (*c*) degrades haemoglobin (*d*) converts glycogen to glucose
   (*e*) all above
2. Chymotrypsinogen is activated into chymotrypsin by
   (*a*) trypsin (*b*) pepsin
   (*c*) bile salts (*d*) fatty acids
3. Bilrubin and biliviridin are found in
   (*a*) blood (*b*) bile
   (*c*) pancreatic juice (*d*) saliva
4. HCl is secreted by the
   (*a*) parietal cells (*b*) chief cells
   (*c*) goblet cells (*d*) paneth cells
5. Casein is milk
   (*a*) bacterium (*b*) sugar
   (*c*) protein (*d*) fat
6. Amino acids are absorbed in
   (*a*) blood capillaries of villi (*b*) wall of rectum
   (*c*) lacteals of villi (*d*) lacteals and blood capillaries of villi
7. Sphincter of Oddi is located in
   (*a*) gastrium (*b*) urinary bladder
   (*c*) biliary duct (*d*) central nervous system
8. Most potent stimulus for secretin is
   (*a*) dilation of intestine (*b*) acid chyme
   (*c*) protein (*d*) fat

9. Pepsinogen is secreted from
(*a*) parietal cells (*b*) chief cells
(*c*) G-cells (*d*) mucus cells

10. Which of the following is absorbed in the colon
(*a*) electrolytes (*b*) bile salts
(*c*) proteins (*d*) iron

11. Stomach is the site of digestion mainly of
(*a*) fats (*b*) carbohydrates
(*c*) proteins (*d*) all of these

12. All digestive enzymes are
(*a*) ligases (*b*) oxidases
(*c*) hydrolases (*d*) transferases

13. As HCl is to pepsinogen, so is enterokinase to
(*a*) rennin (*b*) trypsinogen
(*c*) pepsin (*d*) none of these

14. Iron is absorbed almost entirely in
(*a*) stomach (*b*) duodenum
(*c*) jejunum (*d*) ileum

15. Water absorption in mainly a function of
(*a*) appendix (*b*) stomach
(*c*) intestine (*d*) colon

16. Which of the following processes involved in nutrition occurs outside the digestive tract
(*a*) ingestion (*b*) digestion
(*c*) egestion (*d*) assimilation
(*e*) absorption

17. Liver produces
(*a*) digestive enzymes (*b*) hormones
(*c*) bile (*d*) HCl

18. Glycogen is formed from
(*a*) glucose (*b*) proteins
(*c*) fats (*d*) all of these

19. Which of the following gastrointestinal glands lie completely outside the wall of the GI system?
(*a*) mucosa glands (*b*) oxyntic glands
(*c*) salivary glands (*d*) pyloric glands

20. Which of the following surfaces of the gastrointestinal mucosa is poorest in absorption?
(*a*) brush border of microvilli (*b*) villi
(*c*) stomach lining (*d*) plicae circularis

21. Brunner glands are present in
(*a*) stomach (*b*) liver
(*c*) small intestine (*d*) large intestine

22. Secretion of which of the following is not under hormonal control?
(*a*) saliva (*b*) gastric juice
(*c*) pancreatic juice (*d*) bile

23. Select the strongest lipolytic enzyme
(*a*) chymotrypsin (*b*) steapsin
(*c*) intestinal lipase (*d*) phospholipase

ANSWERS

Very Short Answer Questions

1. Accelerates breakdown of large molecules into small ones.
2. Carbohydrates. (*e.g.*, starch).
3. By chloride ions of saliva.
4. Carbohydrates.
5. Dipeptidase.
6. Mucosal cells of small intestine.
7. Secretin, cholecystokinin.
8. Activation of trypsin.
9. Passing food twice via gut (eating its own faeces).
10. Digestion (or hydrolysis) of fats occurs mainly in duodenum and is completed in small intestine
11. It helps in digestion of proteins into peptones and polypeptides in duodenum.
12. Proteolytic enzymes cannot be released in the alimentary canal, nor they can be stored in the lysosomes because they will digest the cells that produce them.
13. (*i*) Hormone secretin and (*ii*) hormone, pancreozymin.
14. Cholecystokinin is a hormone, secreted by duodenal, mucosa. It stimulates contraction of gall bladder to release bile juice.
15. By emulsifying the fats and making the food alkaline.
16. Bile.
17. To emulsify fats for digestion.
18. Digestion of fats is not possible.
19. Bulbogastrone secreted by upper part of small intestine.
20. Ptyalin or salivary amylase.
21. Neutral medium.
22. To activate trypsinogen to trypsin.
23. In gastric epithelium.
24. It digests proteins in duodenum in an alkaline medium.
25. Peristalsis.
26. Peristaltic movements are rhythmic waves of contraction passing backward over the gut wall. These are created by contraction and relaxation of longitudinal and circular muscles of gut wall
27. The cholecystokinin (CCK) bring about contraction of gall bladder and relaxation of sphincter of Oddi for expulsion of bile into duodenum.

Fill in the Blanks

1. Satiety	**2.** Deglutition	**3.** Lysozyme	**4.** Chyle
5. 3, 5	**6.** Fats	**7.** Enterokinase	**8.** Saccus entericus
9. Pyloric sphincter	**10.** B_2 and K	**11.** Lacteal capillaries	**12.** Enterokinase
13. Cholecystokinin or CCK.			

Multiple Choice Questions

1. (*e*)	**2.** (*a*)	**3.** (*b*)	**4.** (*a*)	**5.** (*c*)	**6.** (*a*)
7. (*c*)	**8.** (*b*)	**9.** (*b*)	**10.** (*a*)	**11.** (*c*)	**12.** (*c*)
13. (*b*)	**14.** (*b*)	**15.** (*d*)	**16.** (*d*)	**17.** (*c*)	**18.** (*d*)
19. (*c*)	**20.** (*c*)	**21.** (*c*)	**22.** (*a*)	**23.** (*b*).	

3

Physiology of Respiration

3.1 INTRODUCTION

Body tissues or cells utilize oxygen (O_2) and produce carbon dioxide (CO_2) as a result of metabolism. The main function of the **respiratory system** is to extract O_2 from the environment (atmosphere or hydrosphere or air/water) to deliver it to the tissues or cells and take out CO_2 from the tissues or cells and discharge it into the environment (**Jain**, 2012). **Respiration** can be defined as the exchange of environmental oxygen with the carbon dioxide produced in the cells during oxidation at a moist surface. In other words, the process of respiration involves: 1. intake of molecular oxygen from the environment; 2. oxidation of food with the incoming oxygen; 3. elimination of carbon dioxide produced in oxidation; 4. release of energy in small doses during oxidation and 5. conservation of energy so released in biologically useful forms, such as ATP.

1. Respiratory Medium

Most of the Earth's oxygen occurs in air but some is dissolved in water. Thus, air or water may serve as the source of oxygen for the respiration of animals. The source of oxygen is called **respiratory medium**. The respiratory medium supplies oxygen to body's respiratory surface. The latter may be the general body surface or some specialized area such as gill or lung. Most of the animal's cell lie some distance from the respiratory surface and obtain oxygen from **tissue fluid**, which bathes all cells. Blood brings oxygen from the respiratory surface to the tissue fluid and carries carbon dioxide from the tissue fluid to the respiratory surface.

Exchange of gases across a respiratory membrane occurs by **diffusion**, but can distribute substances over a short distance only, about 0.5 mm. Therefore, animals having substances more distant than 0.5 mm from the respiratory surface require **circulatory system** to transport the gases between the respiratory surface and the tissues fluid present around the cells.

2. Respiratory Substrates

A foodstuff used in oxidation for producing energy is called a **respiratory substrate**. Any foodstuff—carbohydrate, fat (lipid) or protein—may be used in cellular respiration.

1. Carbohydrates. The carbohydrates are usually the first choice of most cells. The brain cells of mammals cannot use anything but glucose. Polysaccharides and disaccharides are hydrolyzed to monosaccharides before they enter the respiratory pathway:

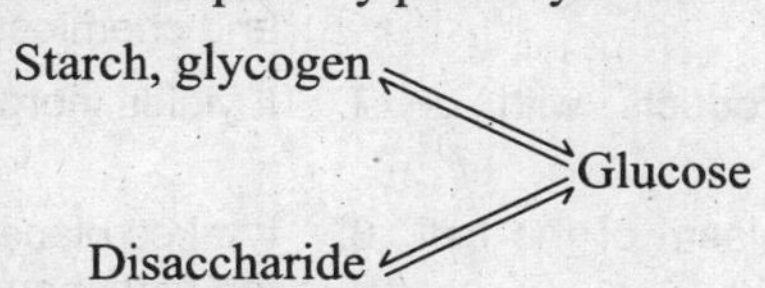

2. Fats (lipids). The fats or lipids form the 'first reserve' and are mainly used when carbohydrate reserves have been exhausted. However, in the skeletal muscle cells, if glucose and fatty acids are both available, then cells prefer the fatty acids for the oxidation.

3. Proteins. Proteins have other essential functions in the animal's body. Therefore, they are used only when all carbohydrate and fat reserves have been used up, as during long starvation. However, protein of a cell is gradually broken down and new one synthesized to take its place. The cell material is thus, never quite the same from time to time.

In cellular respiration, the initial reactions vary with different foodstuffs used (*e.g.*, glycolysis for glucose, β-oxidation for fatty acids and deamination for the amino acids), but eventually all foodstuffs join a common scheme of oxidation, the Krebs cycle.

Respiration which uses protein is called **protoplasmic respiration**, whereas that which uses carbohydrate or fat is termed **floating respiration**. Protoplasmic respiration cannot be carried on for long because it breaks down the cells own material and it produces the toxic ammonia.

3. Types and Phases of Respiration

Respiration is of two types: anaerobic and aerobic (Table 3.1).

1. Anaerobic respiration. In the anaerobic respiration, decarboxylation reactions can occur independently of the reactions in which molecular oxygen is utilised. For example, yeasts can degrade glucose to ethanol (ethyl alcohol) and carbon dioxde without utilising oxygen:

$$\underset{\text{Glucose}}{C_6H_{12}O_6} \rightarrow \underset{\text{Ethanol}}{2C_2H_5OH} + \underset{\text{Carbon dioxide}}{2CO_2} + \text{Energy}$$

In certain bacteria and parasitic worms [*Ascaris* (roundworm), tapeworm], glucose is metabolised to lactic acid without the use of oxygen and without the formation of carbon dioxide:

$$\underset{\text{Glucose}}{C_6H_{12}O_6} \rightarrow \underset{\text{Lactic acid}}{2CH_5COCOOH} + \text{Energy}$$

The process of releasing energy without the use of oxygen is called **anaerobic respiration**. It is also termed **fermentation**. Fermentation yields only about 5% of the energy available in glucose because the latter is incompletely oxidized. Anaerobic respiration is, thus, a wasteful process. The organisms which carry on anaerobic respiration are called **anaerobes**.

Table 3.1 Differences between anaerobic respiration and aerobic respiration.

	Anaerobic metabolism		*Aerobic metabolism*
1.	It occurs in a few oraganisms (yeast, some bacteria, parasitic worms).	1.	It occurs in the majority of organisms.
2.	It does not use molecular O_2.	2.	It uses molecular O_2.
3.	It may or may not release CO_2.	3.	It always releases CO_2.
4.	It does not produce H_2O.	4.	It produces H_2O.
5.	It produces less energy (only 5 per cent of that available in glucose).	5.	It produces much more energy (whole that is available in glucose).
6.	It consists of 2 steps: glycolysis and incomplete breakdown of pyruvate.	6.	It consists of 5 steps: glycolysis, pyruvate oxidation, TCA cycle (Krebs cycle), ETS and chemiosmotic ATP synthesis.
7.	It yields organic end products with or without inorganic one.	7.	It yields inorganic end products.
8.	It takes place in the cytoplasm of the cell only.	8.	It takes place in the cytoplasm and partly in the mitochondria of the cell

2. Aerobic respiration. In most animals and plants, the utilization of oxygen and the production of carbon dioxide occur together in respiration. Such a respiration is called **aerobic respiration**. It releases entire energy available in glucose as the latter is fully oxidized to carbon dioxide and water.

$$\underset{\text{Glucose}}{C_6H_{12}O_6} + 6O_2 \rightarrow 6CO_2 + 6H_2O + \text{Energy}$$

The organisms which carry on this type of respiration are termed **aerobes**.

Types of Aerobic Respiration

Aerobic respiration is of two main types: direct and indirect.

***(a)* Direct respiration.** It is the exchange of environmental oxygen with the carbon dioxide of the body cells without the aid of blood. It is found in bacteria, protists, plants, sponges, coelenterates, flatworms, roundworms, most arthropods.

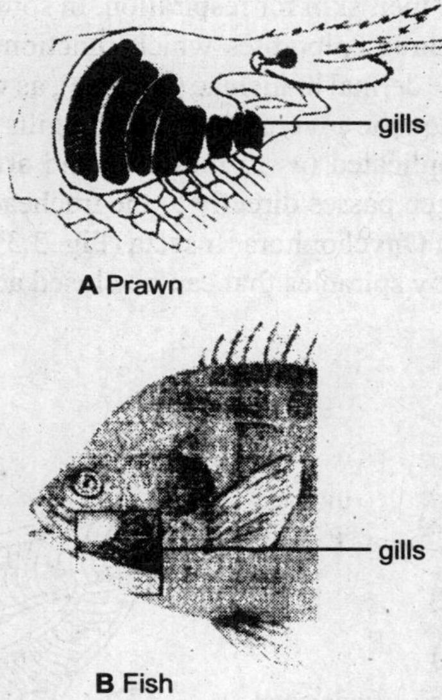

Fig. 3.1. Branchial respiration: Gills of prawn and fish for aquatic respiration.

***(b)* Indirect respiration.** It involves special respiratory organs, such as skin, buccopharyngeal lining, gills (Fig. 3.1) and lungs (Fig. 3.2), and needs the help of blood. The respiration in the skin, buccopharyngeal lining, gills and lungs is respectively called **cutaneous, buccopharyngeal, branchial** and **pulmonary respiration**. Cutaneous respiration takes place in annelids, some crustaceans and amphibians. It occurs both in water and in air. Buccopharyngeal respiration is found in certain amphibians such as frog and toad. It occurs in the air. Branchial respiration is found in many annelids, most crustaceans and molluscs, some

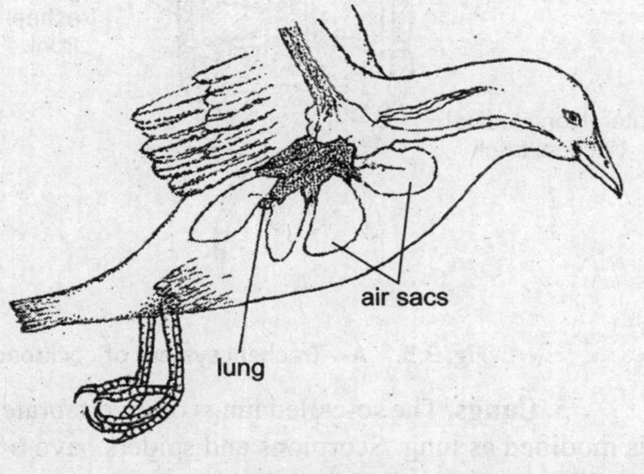

Fig. 3.2. Pulmonary respiration: Lungs and air sacs of birds for aerial mode of respiration.

insect larvae, echinoderms, all fishes and some amphibians. It occurs in water only. Pulmonary respiration is found in snails, some amphibians and in all reptiles, birds and mammals. Lungs get air in one of the two ways: **positive-pressure breathing** in which air is forced into the lungs as in amphibians, and **negative-pressure breathing** in which air is drawn into the lungs by lowering pressure around them as in other vertebrates.

Examples of direct and indirect respiration occur both in nonchordates and chordates.

4. Respiration in Nonchordates

Nonchordates depend upon the following modes of respiration:

1. Direct diffusion. In protists (*Amoeba proteus*), porifera, coelenterates (*Hydra, Obelia*) and flatworms (*Planaria*) oxygen is obtained by diffusion directly from the surrounding medium (water).

2. Integument. In most annelids, gaseous exchange is made through a moist skin provided with blood vessels or other derivatives of the integument such as **parapodia** which are highly vascular. A few crustaceans and molluscs also use their skin for respiration. In some molluscs (*e.g., Pila*) mantle is used and in *Palaemon*, there are three epipodites which function as respiratory organs. Some echinoderms make use of the finger like dermal branchiae (papulae), as well as tube feet, in respiration.

3. Gills. Gills are common in aquatic invertebrates. Some gills are simple in structure (as in marine annelids) and others are complicated (as in molluscs and arthropods). Numerous aquatic insects have gills through which oxygen passes directly to the tracheae.

4. Tracheae. These are found in Onychophora, Insecta (Fig. 3.3), Myriapoda, in some spiders and isopods. Tracheae usually opens by spiracles that can be closed according to the need.

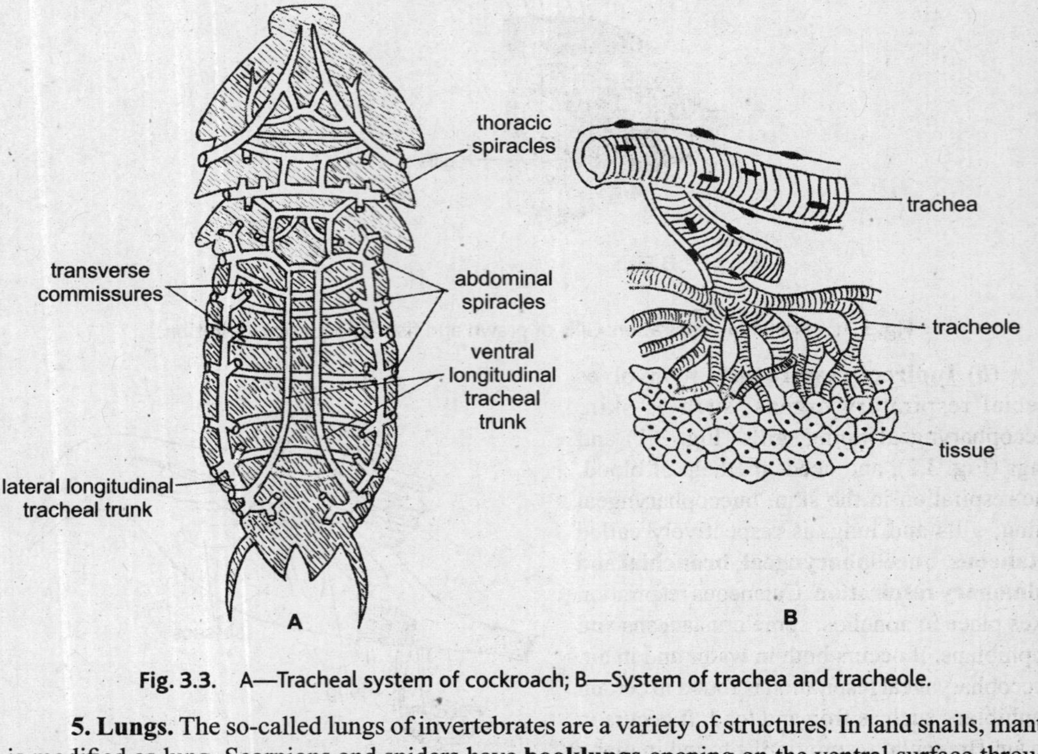

Fig. 3.3. A—Tracheal system of cockroach; B—System of trachea and tracheole.

5. Lungs. The so-called lungs of invertebrates are a variety of structures. In land snails, mantle is modified as lung. Scorpions and spiders have **booklungs** opening on the ventral surface through slit-like aperture called the **stigmata**.

6. Respiratory trees. Respiratory trees of holothurians (echinoderms) are filled with sea water by cloacal pumping and contractions of the trees themselves and gas exchanges with the coelomic fluid.

5. Respiration in Chordates

In chordates, the respiration is carried out by any of the following three devices:

(*i*) Direct surfaces such as the skin and pharyngeal regions. Many amphibians make use of their skin in respiration. Some salamanders depend on the skin and highly vascular pharyngeal region. Some fishes (and also eels) absorb oxygen through the skin.

(*ii*) Gills. Many aquatic chordates have gills, either external (protochordates and some larval amphibians) or internal (cyclostomes and fishes).

(*iii*) Lungs. Land vertebrates have the lungs for respiration. In amphibians, the lungs are simple, but in higher vertebrates there is a gradual increase in the internal respiratory surface of the lungs, thus making them more efficient organs of respiration.

Three Phases of Indirect Respiration

Indirect respiration has three phases:

1. External respiration (Ventilation or **Breathing)** *i.e.*, the absorption of O_2 and removal of CO_2 from the body as a whole.

2. Transport of gases (O_2 and CO_2) in the blood.

3. Internal respiration (Cellular respiration) *i.e.*, the utilization of O_2 and production of CO_2 by cells and the gaseous exchange between the cells and their fluid medium (Table 3.2).

Table 3.2 Differences between breathing and respiration.

	Breathing		*Cellular respiration*
1.	It is a preliminary step for respiration and is a physical process.	1.	It is a biochemical process involving exchange of gases and oxidation of food.
2.	It does not release any energy.	2.	It releases energy which is stored in ATP.
3.	It is an extracellular process.	3.	It is an intracellular process.
4.	It does not involve enzyme action.	4.	It involves a number of enzymes in oxidation.
5.	Breathing mechanism varies in different organisms.	5.	Respiratory mechanism is similar in all animals.
6.	It is confined to certain organs only.	6.	It occurs in all cells of the body.

3.2 RESPIRATION IN HUMAN BEINGS

Respiration provides oxygen to the tissues and removes carbon dioxide. The four major functions of respiration in humans are:

1. **Pulmonary ventilation**, which means the inflow and outflow of air between the atmosphere and the lungs alveoli.
2. **Diffusion of oxygen and carbon dioxide between the alveoli and the blood.**
3. **Transport of oxygen and carbon dioxide in the blood and body fluids** to and from the body's tissue cells; and
4. **Regulation of ventilation** and other facets of respiration.

Human Respiratory System

The human respiratory system may be divided into two major components: respiratory tract or conducting part and respiratory organs.

I. Respiratory Tract

It serves as a passageway for the fresh air to flow from outside to the lungs and for the foul air to return from the lungs to the exterior. Gas exchange does not occur here. The respiratory tract consists of following parts: 1. external nares; 2. nasal chambers; 3. internal nares; 4. pharynx; 5. larynx; 6. trachea; 7. brochi and 8. bronchioles (Fig. 3.4 and Fig. 3.5). Of these, some bronchi and all bronchioles lie within the lungs, other parts lie outside the lungs.

1. External nares or nostrils. These are a pair of slits at the lower end of the nose. Nostrils open into the nasal chambers.

2. Nasal chambers. The nasal chambers are a pair of passages in the head above the palate (Fig. 3.4). The two chambers are separated from each other by a median partition, the **nasal septum**. Each chamber has *three* regions: lower vestibular, middle respiratory and upper olfactory.

(*i*) Vestibular region. It lies just within the external naris. It is very short and lined with skin. It bears hairs and certain glands such as sweat gland and oil glands. The hair act as a sieve to check the entry of large dust particles.

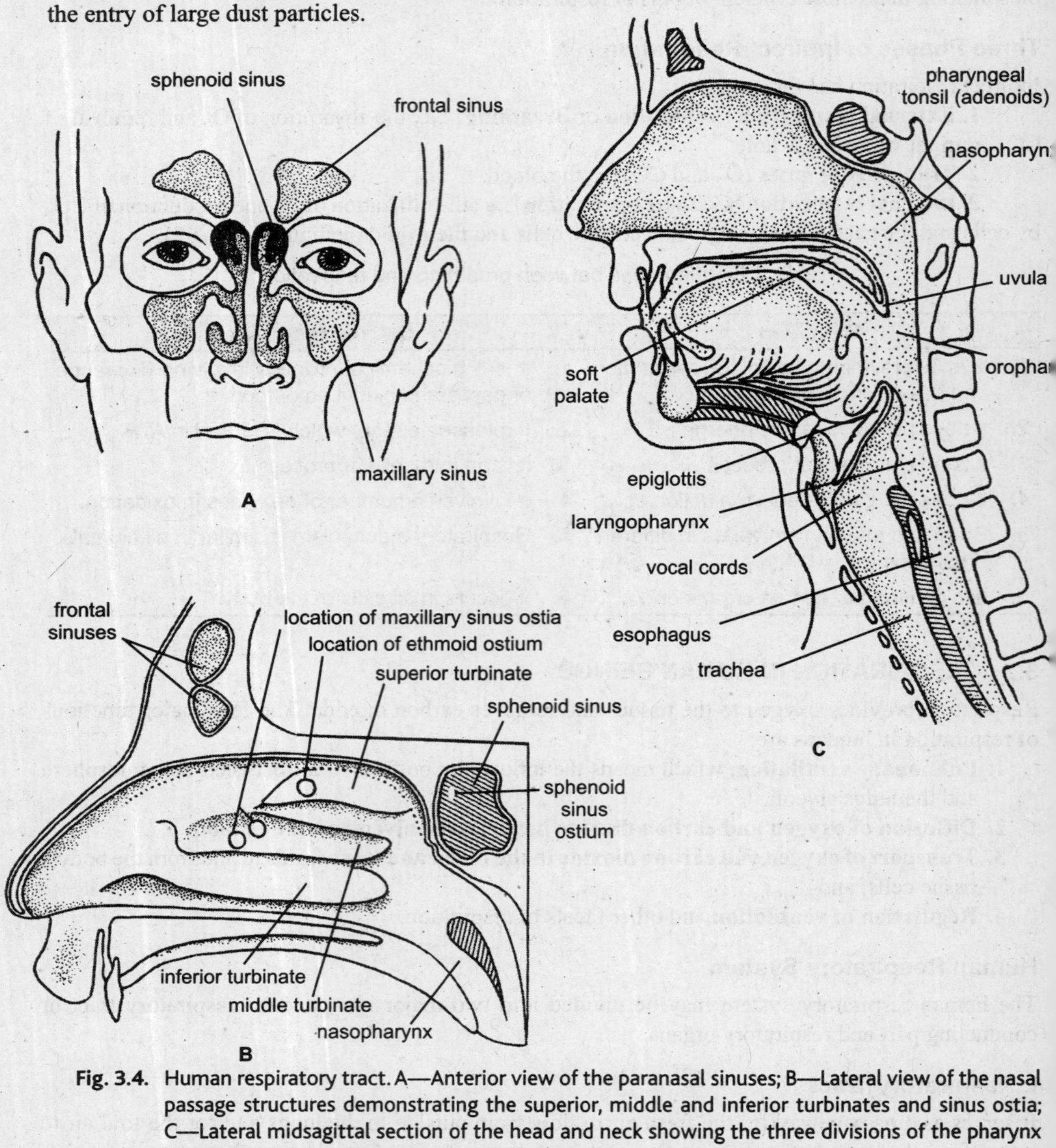

Fig. 3.4. Human respiratory tract. A—Anterior view of the paranasal sinuses; B—Lateral view of the nasal passage structures demonstrating the superior, middle and inferior turbinates and sinus ostia; C—Lateral midsagittal section of the head and neck showing the three divisions of the pharynx and surrounding upper airway structures (after Koeppen and Stanton 2010).

***(ii)* Respiratory region.** It is the middle region of the nasal chamber. It is lined with **respiratory epithelium**. This is a ciliated, pseudostratified columnar epithelium rich in **Goblet cells** and other gland cells such as **mucous cells** and **serous cells**. Goblet cells and mucous cells secrete a viscid fluid called **mucus**. The serous cells produce a watery fluid. The respiratory epithelium is highly vascular appears pink or reddish. The respiratory region acts as an air-conditioning and filtering unit in breathing.

***(iii)* Olfactory region**. It is the upper region of the nasal chamber. It is lined with **olfactory epithelium**. This epithelium is confined to the upper part of the nasal chamber and the superior **nasal concha**. It looks yellowish-brown. The olfactory region acts as an organ of smell. It detects the odour of the inspired air. If the odour is offensive or pungent, the air is not allowed to pass in. Thus, the organ of smell helps in selecting the air for respiration.

Nasal conchae. These are three shallow bony ridges which arise from the wall of each nasal chamber. These are individually named **superior**, **middle** and **inferior nasal conchae** (or **turbinates**) according to the position. The conchae or turbinates are covered with mucous membrane. They greatly increase the surface of the nasal chambers.

Paranasal sinuses. These are cavities in the skull bones such as frontal, maxillary, ethmoid and sphenoid (Fig. 3.4A, B). Goblet cells in their lining epithelium produce mucus that drains into the nasal chambers.

3. Internal nares or choanae. The nasal chambers open behind into the nasopharynx by internal nares.

4. Pharynx. The pharynx is a short, vertical, about 12 cm long tube behind the buccal cavity (Fig. 3.4B). The food and air passages cross here. Its upper part is called **nasopharynx**. The internal nares open into it. The middle part is called **oropharynx**. The oropharynx receives the buccal cavity. The lower part is called **laryngopharynx**. It leads into two tubes: the one at the front is the windpipe or **trachea**, and the one at the back is the foodpipe or **oesophagus**. The pharynx is lined with nonkeratinized stratified squamous epithelium which is ciliated in the nasopharynx.

Mouth serves as an alternate route for air when nasal chambers are blocked. It leads into the oropharynx through the buccal cavity.

5. Larynx. The larynx is often called the "Adam's apple" (the thyroid cartilage) and is prominent in human males than in females. Before puberty, the larynx is the upper part of the trachea. It is a short, tubular chamber supported by a cartilaginous framework. Larynx opens into the laryngopharynx by the slit-like aperture, the **glottis**. The glottis is always remains open except during swallowing. It bears a leaf like cartilaginous flap, the **epiglottis**, at its anterior margin. The epiglottis projects into pharynx opposite uvula. During swallowing, the larynx moves upward to meet the epiglottis. This closes the glottis to check the entry of food into it. Besides forming a part of the respiratory tract, it also serves as the **voice box** or **larynx**. Larynx has elastic vocal cords and is lined with ciliated epithelium, the cilia of which beat toward the pharynx. The larynx allows nothing but air to pass into the lungs. A laryngeal spasm (an involuntary muscular contraction), in the form of a cough, results if any substance enters the larynx. Entry of food into the respiratory tract can be fatal. The larynx is followed by trachea.

6. Trachea. The **wind pipe** or **trachea** is a thin-walled tube, about 11 cm long and 2.5 cm wide (Fig. 3.5). Trachea extends downward through the neck.

7. Primary bronchi. In the middle of the thorax, trachea divides into two tubes, the **major** or **primary bronchi** (Fig. 3.6). One primary bronchus enters the right lung, and the other, the left lung. The right primary bronchus further divides into three **lobar** or **secondary brochii** that likewise pass into two lobes of the left lung. The secondary bronchi subdivide into smaller **tertiary bronchi** which divide into still smaller **bronchioles**. The small **terminal bronchioles** give off **respiratory bronchioles** which is divided into **alveolar ducts**. The latter enter the **alveolar sacs**. The respiratory passageways within the lungs are referred to as the **bronchial tree**. Wall of the trachea, bronchi and bronchioles is composed of fibromuscular tissue and is lined by pseudostratified ciliated columnar epithelium rich in mucus-secreting cells. The mucus holds the dust and bacteria, which are swept by cilia toward the pharynx from where they are thrown out or swallowed. This ensures clean and bacteria-free air to the lungs. Terminal bronchioles are lined with simple ciliated columnar epithelium without mucous cells. The mucus, if present, may block these narrow tubules. A short distance down the respiratory bronchioles, epithelium changes to low nonciliated cuboidal type. The alveolar ducts also have thin, non-ciliated cuboidal epithelium. Cartilaginous rings, incomplete behind, support the walls of the trachea and the bronchi to prevent their collapsing. The trachea and bronchi, thus, remain permanently open even during negative pressure created by exhalation. This normally allows easy passage for air. The rings gradually become thinner, and finally disappear over the bronchioles. The bronchioles have smooth muscle controlled by automatic nervous system so that air flow is adjusted to suite metabolic needs.

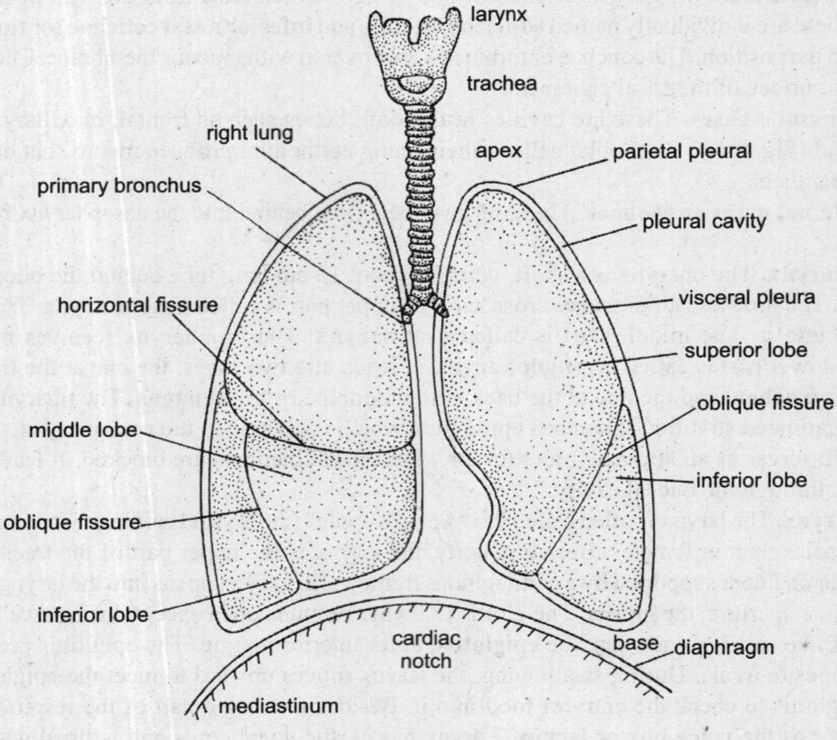

Fig. 3.5. Human trachea and lungs.

Mucociliary Clearance System

The respiratory system is lined with three very different and highly significant fluids: **periciliary fluid, mucus** and **surfactant**. Periciliary fluid and mucus are components of mucociliary clearance system and line the epithelium of the conducting airways from the trachea to the terminal bronchioles. Together they form the basis for the mucociliary clearance system which aids in the removal of particulates (*e.g.*, bacteria, viruses, toxins) from the lungs. Surfactant lines the epithelium of the alveolus and provides an **antistick function** that decrease surface tension in the alveolus.

The cells of respiratory epithelium maintain the level of periciliary fluid, a 5 μm layer of water and electrolytes in which cilia and the mucociliary transport system function. The depth of periciliary fluid is maintained by the movement of various ions, mainly **chloride secretion** and **sodium absorption**, across the epithelium. Mucus and inhaled particles are removed from the airways by the rhythmic beating of cilia on the top of pseudostratified columnar epithelium. Each epithelial cell contains about 250 cilia (**Koeppen** and **Stanton**, 2010).

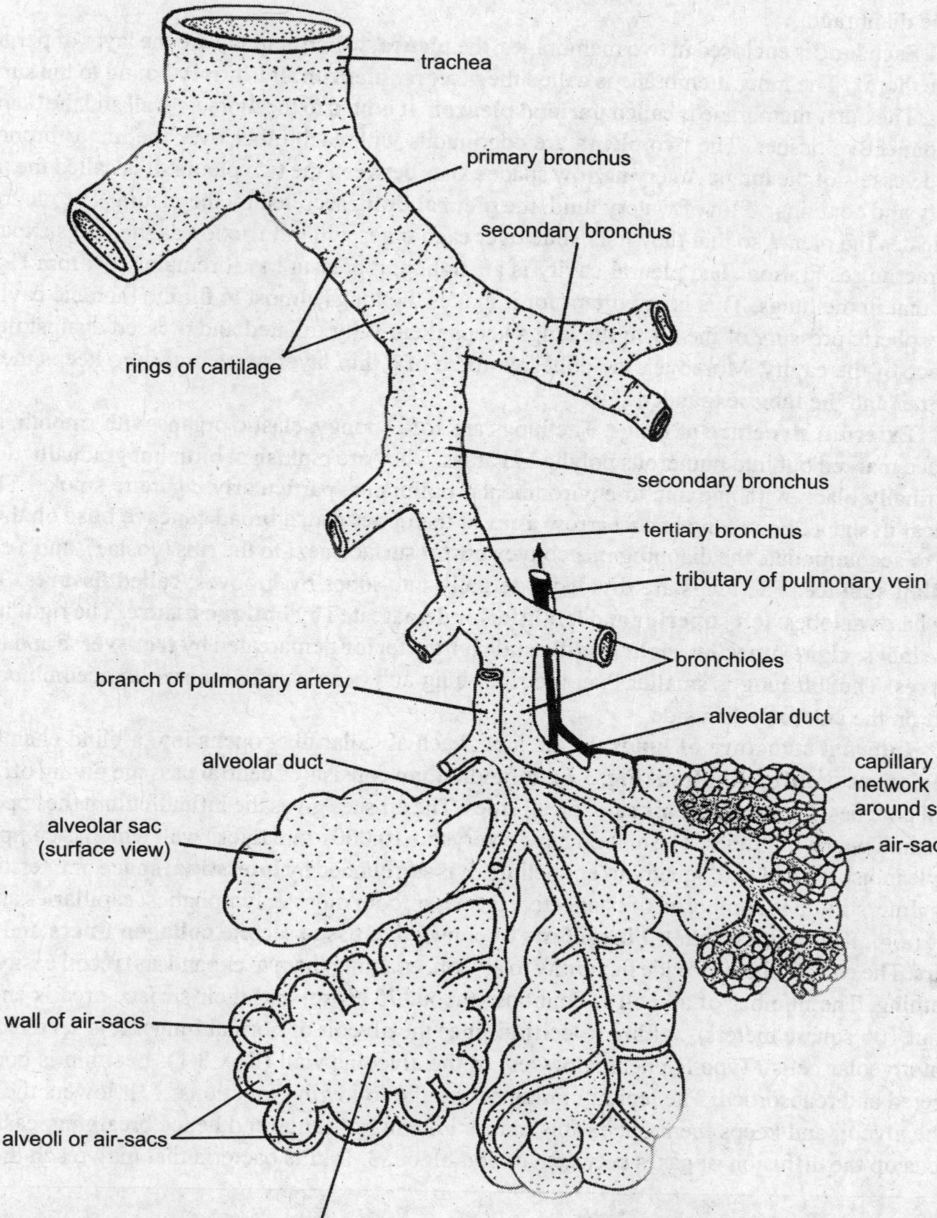

Fig. 3.6. Bronchi and their branches.

II. Respiratory Organs

The respiratory organs provide the surface for the exchange of gases. In human beings, the respiratory organs are a pair of lungs (Fig. 3.5). The lungs lie in the thoracic cavity on the sides of the heart. The thoracic cavity is an air-tight chamber enclosed behind by the thoracic vertebrae, laterally by the ribs embedded in intercostal muscles and in front by the sternum. Thoracic cavity is closed below by the diaphragm.

Each lung is enclosed in two membranes, the **pleura**, which actually are the layer of peritoneum of the thorax. The inner membrane is called the **visceral pleuron**. It is firmly bound to the surface of lungs. The outer membrane is called **parietal pleuron**. It is held to the thoracic wall and the diaphragm by connective tissues. The two pleura are continuous with each other over the major bronchi and blood vessels of the lungs. A very narrow space exists between the two pleura. It is called the **pleural cavity** and contains a film of watery fluid, the **pleural fluid**, secreted by the pleura. The pleural fluid lubricates the pleura so that they may slide over each other without friction as the lungs expand and contract in respiration. The pleural cavity is air tight and pressure in it remains 3–4 mm Hg lower than that in the lungs. This is important for it makes the lungs almost to fill the thoracic cavity. The atmospheric pressure of the air in the lungs keeps them fully inflated and pressed against the inner surface of the cavity. Moreover, the adhesive force of a thin layer of pleural fluid keeps the pleura together and the lungs expanded.

External structure of lungs. The lungs are soft, spongy, elastic organs with smooth, shining surface marked out into numerous polyhedral areas. They are pinkish at birth but gradually turn grey and finally black with age due to environmental pollutants, particularly cigarette smoke. They are conical in shape. Each lung has a narrow **apex** at the upper end, a broad concave **base** on the lower side to accommodate the diaphragm, a convex **costal** surface next to the ribs (costae), and a concave **median surface**. The lungs are divided externally into lobes by grooves, called **fissures**. The left lung has two lobes: **left superior** and **left inferior** demarcated by **oblique fissure**. The right lung has *three* lobes: **right superior**, **right middle** and **right inferior** demarcated by **transverse** and **oblique fissures**. The left lung is smaller than the right lung and has a **cardiac notch** to accommodate the heart on the anteromedian side.

Internal structure of lungs. In the lung, each alveolar duct opens into a blind chamber, the **alveolar sac** or **infundibulum** (Fig. 3.6). Infundibulum consists of central passage giving off several small pouches, the **alveoli** or **air sacs**, on all sides. The air sacs gives the infundibulum the appearance of a small bunch of grapes. The alveoli have a very thin (0.0001 mm thick) wall which is composed of simple moist, nonciliated, squamous epithelium. It is surrounded by **interstitial space** or interstitium of lung. Interstitium is composed of connective tissue, smooth muscle, lymphatics, capillaries and other cells (*e.g.*, Kulschitzky cells). **Fibroblasts** of connective tissue secrete **collagen fibers** and **elastin fibers**. The elastin fibers provide flexibility to alveoli, enabling them to expand and recoil easily during breathing. The number of alveoli is countless (about 75 crore) and their surface area is enormous (about 100 square meters). A film of **lecithin** lines the alveoli of normal lungs. It is secreted by the giant alveolar cells (Type II pneumatocytes) in the alveolus wall (Box 3.1). Lecithin is constantly secreted and reabsorbed. The lecithin lining serves the following functions: 1. It lowers the surface of the alveoli and keeps them open. This makes inflating of lungs and hence breathing, easier. 2. It speeds up the diffusion of gases between air and blood. 3. It kills bacteria that may reach the lungs.

Box 3.1

Pulmonary surfactant is a complex mixture of phospholipids (lecithin), neutral lipids (cholesterol and cholesteron esters), fatty acids and proteins. Surfactant is 85% to 90% lipids, predominantly phospholipids and 10% to 15% protein (*e.g.*, SP-A, SP-B, SP-C, SP-D). The major phospholipid is **phosphatidylcholine**, approximately 75% of which is present as **dipalmitoyl phosphatidylcholine** (**DPPC**) and decreses surface tension. The second most abundant phospholipid is **phosphatidylglycerol** (**PG**) which accounts for 1% to 10% of total surfactant. PG is important in spreading the surfactant over a large surface areas (**Koeppen** and **Stanton**, 2010).

Respiratory membrane. The alveolar wall is very thin (0.0001 mm thick). It is closely surrounded by a network of pulmonary capillaries which arise from the pulmonary artery and rejoin to form the pulmonary vein. The blood in the capillaries is separated from the inspired air in the alveoli by only two layers of cells, *i.e.*, capillary endothelium and alveolar epithelium, just 0.3 μm thick. Due to this, the alveolar wall seems to be a sheet of flowing blood and is called **respiratory membrane** (Fig. 3.7). It consists of five components: 1. alveolar epithelium; 2. epithelium basement membrane; 3. a thin interstitial space; 4. capillary basement membrane and 5. capillary endothelial membrane. The alveo-capillary barrier is still very thin (about 0.3 μm thick) and as a result, the diffusion of gases between the blood and alveolar air can occur easily and quickly.

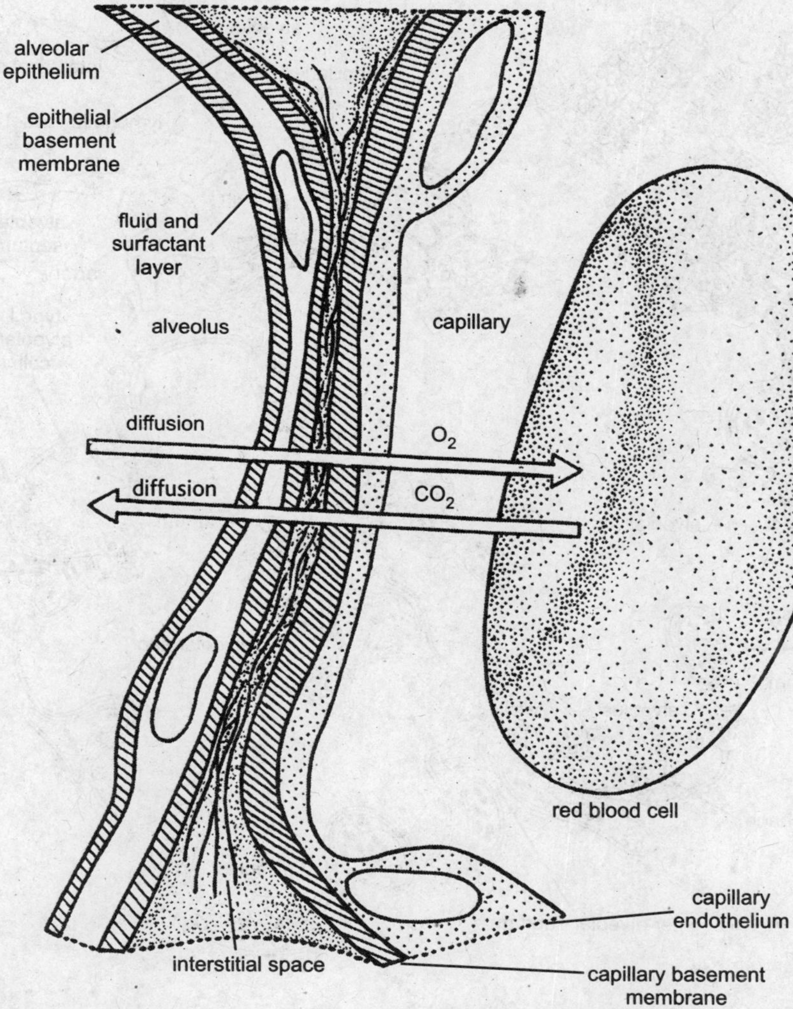

Fig. 3.7. Ultrastructure of the alveolar respiratory membrane.

Histology of lung. Alveolar epithelium is composed of type I and II alveolar cells or **pneumocytes**, occasional brush cells and septal connective tissue which contains the alveolar **macrophages** or **dust cells** which petrol and scavenge the surface to remove RBCs and pathogens (*e.g.*, T.B. bacteria).

1. Type I alveolar cells. These are extremely thin (flat) **squamous cells** that line most of the surface of the alveoli, *i.e.*, about 95 per cent. They form 40 per cent of alveolar epithelium.

2. Type II alveolar cells. These are **cuboidal** secretory cells interspersed among the type I cells. They represent 60 per cent of alveolar cell population numerically, but occupy very little (5 to 10 per cent) of alveolar surface. Type II pneumocytes tend to bulge into the air space. Their apical cytoplasm is filled with granules, called lamellar **bodies** which contain stacks of parallel membrane lamellae. The lamellae are rich in phospholipids, including a surface active agent called **surfactant** (Box 3.1). The lamellar bodies are released into the alveolar space by exocytosis and surfactant forms a monomolecular layer over the alveolar epithelium, thus, reducing the surface tension at the air-epithelium interface.

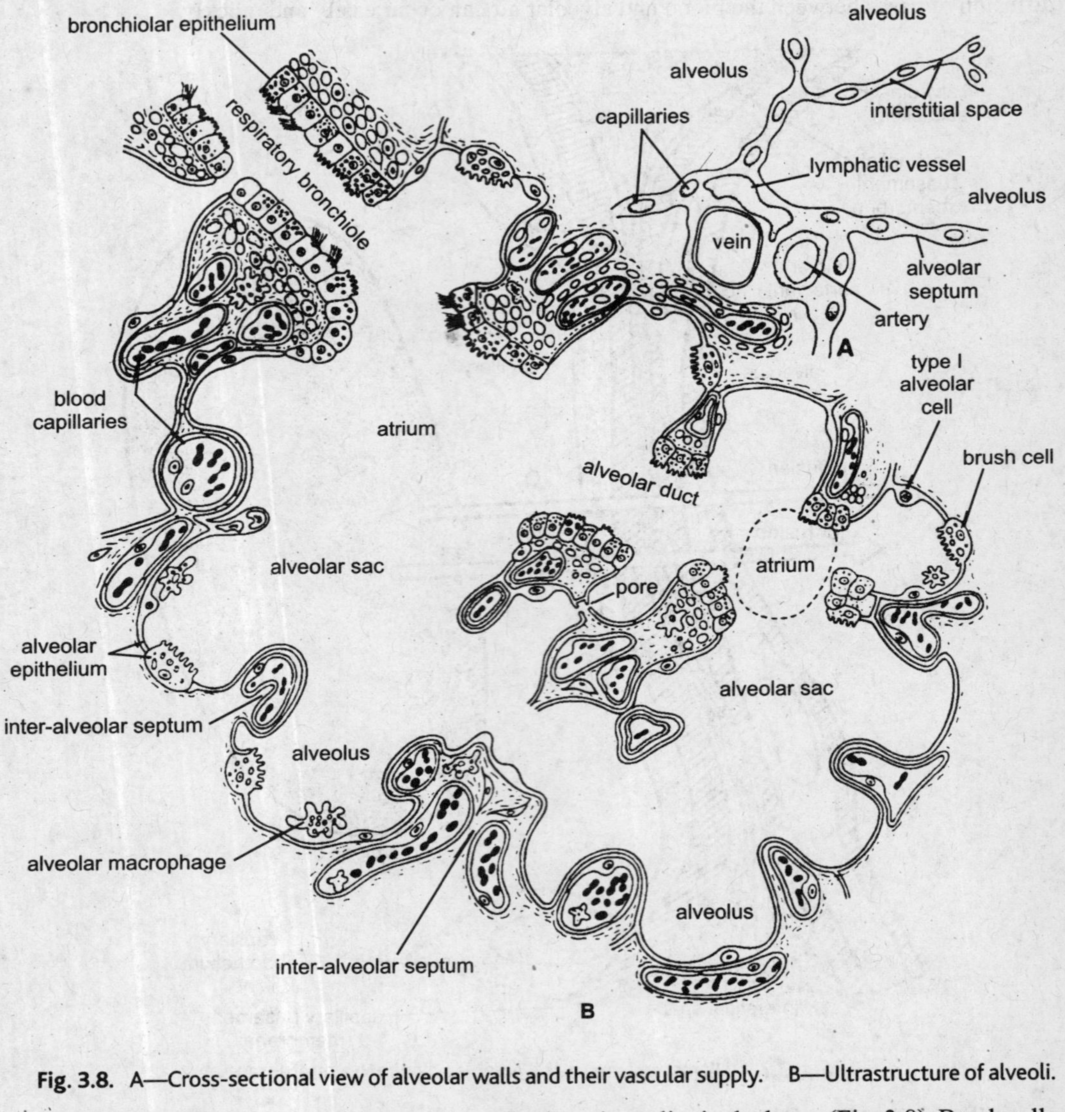

Fig. 3.8. A—Cross-sectional view of alveolar walls and their vascular supply. B—Ultrastructure of alveoli.

3. Brush cells. They are receptor cells to monitor air quality in the lungs (Fig. 3.8). Brush cells are less in number and are columnar cells that bear microvilli. The basal surface of the brush cells is in synaptic contact with an efferent nerve ending.

Breathing Mechanism (Mechanics of Pulmonary Respiration)

Lungs have little musculature and cannot expand and contract of their own. Breathing is brought about by alternate expansion and contraction of the thoracic cavity wherein the lungs lie (Fig. 3.9). This leads to intake of fresh air called **inspiration** (or inhalation or breathing in) and elimination of foul air called **expiration** (or exhalation or breathing out) respectively. Inspiration and expiration are together referred to as **respiratory movement**.

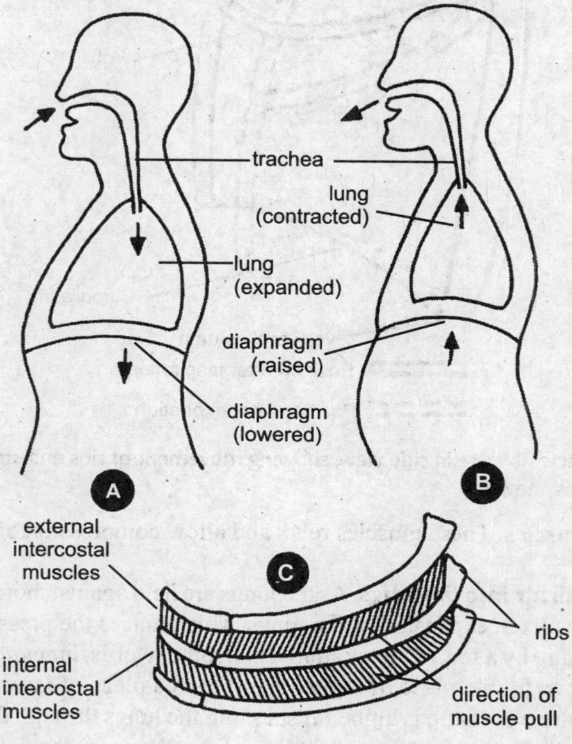

Fig. 3.9. Mechanism of breathing. A—Inspiration (chest cavity enlarged); B—Expiration (chest cavity reduced); C—Diagrammatic representation of intercostal muscles.

Mammalian (or human) lungs can be expanded or contracted in two ways: (i) By downward and upward movement of the diaphragm to expand or contract the chest cavity. (ii) By raising or lowering of the ribs to increase or decrease the diameter of the chest cavity (Box 3.2).

Box 3.2

In large and heavy animals (*e.g.*, elephant) movement of diaphragm plays a greater role than the movement of ribs in respiration.

A. Inspiration. Inspiration is an active process. It is brought about by diaphragm muscles and external intercostal muscles. These muscles are, therefore, called **inspiratory muscles**. The abdominal muscles play a passive role in the process.

(*a*) Diaphragm. The diaphragm is convex upward and has a peripheral muscle attached to the ribs and vertebral column. This muscle contracts and lowers the diaphragm, making it flat. This pushes the abdominal viscera downward and enlarges the thoracic cavity vertically.

(*b*) External intercostal muscles. These muscles slant forward and downward between the ribs (Fig. 3.9C). Their contraction pulls the ribs and sternum upward and outward. This enlarges the thoracic cavity from front-to-back as well as from side-to-side (Fig. 3.10).

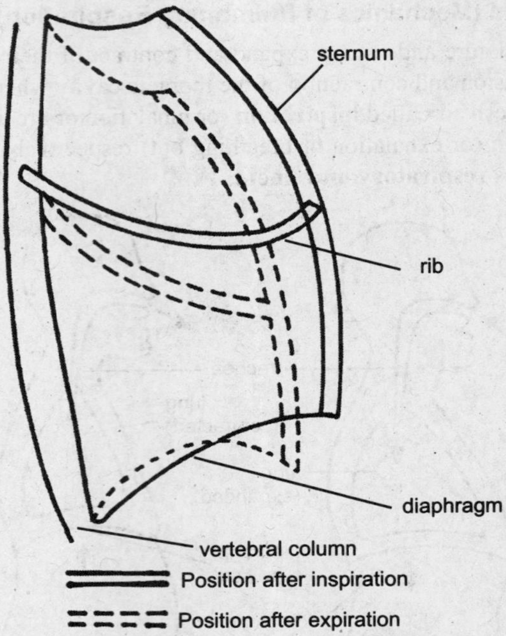

Fig 3.10. Human thoracic rib cage in side view, showing movement of ribs and sternum during breathing. Only one rib is shown.

(c) Abdominal muscles. These muscles relax and allow compression of abdominal organs by the diaphragm.

Movement of fresh air into the lungs. As the lungs are held against thoracic wall, enlargement of the thoracic cavity results in expansion of the lungs. This reduces the pressure of air in the lungs below atmospheric pressure by a few (–2 to –6) mm Hg. In other words, lungs come to have a negative pressure. Air always move from a place of higher pressure to a place of lower pressure. Hence, the fresh air from outside, where pressure is higher, rushes into the lungs through the respiratory passage till the pressure of air in the lungs becomes equal to the atmospheric pressure. On reaching the lungs, the fresh air is distributed by the bronchi, bronchioles and alveolar ducts to the alveoli. The fresh air follows the following route:

External nares → Nasal chamber → Internal nares → Pharynx → Glottis → Larynx → Trachea → Bronchi → Bronchioles → Alveolar ducts → Alveoli

With the expansion of the thoracic cavity and the lungs, additional blood is drawn into the blood vessels of the lungs and the thorax.

Conditioning and filtration of air. The air passing through the nasal chamber is subjected to four important processes:

1. Warming. Blood vessels of the conchae radiate heat like hot water pipes, and warm the air passing over them to the body temperature. Very cold winter air, if inhaled as such, may freeze the lung tissues.

2. Moistening. Fluid secreted by nasal mucous membrane is slowly evaporated and the vapours are added to the passing air, which gets almost saturated with moisture. Dry air would kill the lung cells.

3. Sterilization. Mucus is sticky and antiseptic. It holds and kills the bacteria of the air.

4. Cleaning. Hair present in the vestibular region entrap the coarse dust particles from the ingoing air. A large inhaled dust particle stimulates sensory cell in the nasal chamber. This cell signals the brain to cause a sneeze that forcefully throws out the particle. Mucus also holds dust particles of the passing air. The cilia of nasal chambers by their ceaseless beating "sweep" the dust and germs trapped in mucus to the exterior via external nares.

B. Expiration. The expelling of air or exhalation (*i.e.*, expiration) from the lungs, takes places when the size and the pressure of the thoracic cavity are reduced. The action involves the following movement.

(*i*) The ribs take their original position as a result of contraction of the internal intercostal muscles and because of their own weight.

(*ii*) The diaphragm relaxes and rises to resume its original position.

(*iii*) The compressed abdominal organs push up against the diaphragm.

As a result of these movements, the size of the thoracic cavity is reduced, the lungs become expressed, the pressure in them rises (become higher than the atmosphere) and air rushes out through the air passages.

A normal person breaths about 16–20 times/minute at rest. The breathing rate, however remains higher in children and at the time of muscular exercise.

3.3 LUNG VOLUMES AND CAPACITIES

These are of two main types:

A. Static lung volume and capacities (Time factor is not involved, therefore, expressed in mL or L).

B. Dynamic lung volumes and capacities (Time dependent therefore, expressed in mL/min or L/min).

A. Static Pulmonary or Lung Volumes and Capacities

I. Pulmonary Volumes.

1. Tidal Volume (TV). It is the volume of air breathed in or out of lungs, during quiet respiration.

Normal: 500 mL (of air).

(*i*) *Decreases* due to less contraction of respiratory muscles; *causes*: weakness of respiratory muscles or depression of respiratory center.

(*ii*) *Increases* in muscular exercise.

2. Inspiratory Reserve Volume (IRV). It is the maximal volume of air which can be inspired after completing a normal tidal inspiration, *i.e.*, inspired from the end-inspiratory position.

Normal: 2000–3200 mL (of air).

3. Expiratory Reserve Volume (ERV). It is the maximal volume of air which can be expired after a normal tidal expiration, *i.e.*, expired from the end-expiratory position.

Normal: 750–1000 mL (of air).

4. Residual Volume (RV). It is the volume of air which remains in lungs after a maximal expiration.

Normal: 1200 mL.

5. Closing Volume (CV). It is the lung volume above residual volume at which airways in the lower dependent parts of the lungs begin to close off because of the lesser **transmural pressure**, *i.e.*, pressure difference between intra-pulmonary and intra-pleural pressure, in the areas (**Jain**, 2012).

II. Pulmonary Capacities (Fig. 3.11).

1. Inspiratory Capacity (IC). It is the maximal volume of air, which can be inspired after completing tidal expiration, *i.e.*, from the end-expiratory position. It can be computed as: TV + IRV.

Normal: 2500–3700 mL.

2. Expiratory Capacity (EC). It is the maximal volume of air which can be expired after completing tidal inspiration, *i.e.*, from the end-inspiratory position. It can be computed as TV + ERV.

Normal: 1250–1500 mL.

3. Vital Capacity (VC). It is an important measure of pulmonary capacity (Box 3.3). It is the maximum amount of air a person can expel from the lungs after first filling the lungs to their maximum extent. It can be computed as: TV + IRV + ERV.

Normal: 4.8 liters in males and 3.2 liters in females.

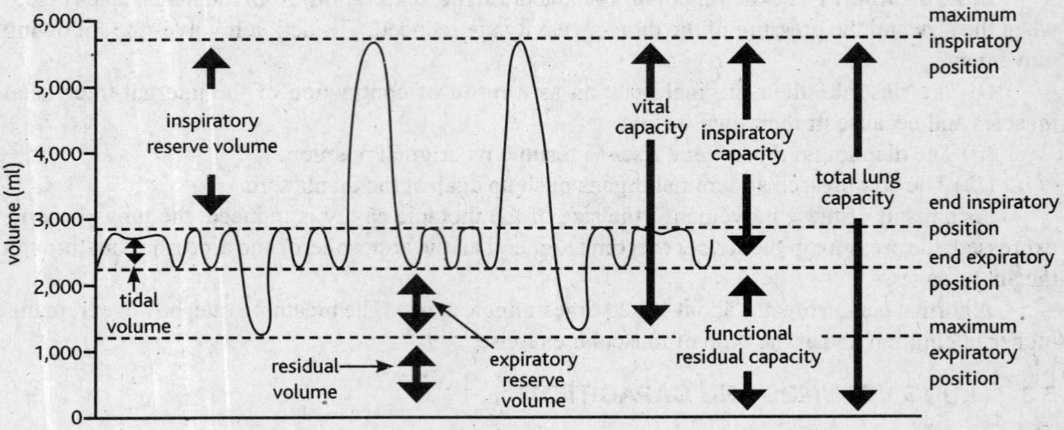

Fig. 3.11. Lung (pulmonary) volumes and capacities.

Box 3.3

Two stage VC. Vital capacity is measured in two stages:

Stage 1. Subject is asked to breath in maximally after a normal expiration; then he takes a few normal breaths.

Stage 2. Subject is asked to breath out maximally after a normal expiration. Sum of stage 1 and 2 volumes give the 'two stage VC'. It is slightly larger than 'one stage VC'.

Advantages of VC

1. It provides useful information about the strength of the respiratory muscles, therefore, maximum inspiratory and expiratory efforts can be assessed.
2. It gives the useful information about other aspects of pulmonary function through FEV (Forced Expiratory Volume in seconds).

Factors Affecting VC

A. Physiological

(*i*) *Physical dimension, i.e.*, size and development of the subject.

VC in male is 2.6 L/m^2 BSA, and females, 2.1 L/m^2 BSA.

VC is more in males because of (*a*) large chest sizes, (*b*) more muscle power; and (*c*) more BSA (body surface area).

(*ii*) **Age.** VC decreases in old age due to loss of elasticity of lungs (with aging, elastic fibers get replaced by fibrous tissue).

(*iii*) **Strength of respiratory muscles.** VC increases in swimmers and divers.

(*iv*) **Posture.** In standing position VC is more as compared to sitting or lying positions, because in standing position, (*a*) decrease in venous return decreases the pulmonary blood flow and (*b*) diaphragm descends down thus increasing inspiration.

(*v*) **Pregnancy.** VC decrease.

B. Pathological

(*i*) VC decreases in diseases of the respiratotry apparatus. For example , poliomyelitis, pulmonary fibrosis, respiratory obstruction, emphysema, pleural effusion, pulmonary oedema, pneumothorax, etc.

(*ii*) **Ascites,** *i.e.*, accumulation of fluid in the abdominal cavity causes VC to decrease.

4. Functional Residual Capacity (FRC). It is the volume of air which is contained in the lungs at end-expiratory position, *i.e.*, after completion of tidal expiration. It can be computed as: RV + BRV.

Normal. 2.5 liters.

5. Total Lung Capacity (TLC). It is the volume of air contained in the lungs after a maximal inspiration. It can be computed as: VC + RV.

Normal. 6 liters.

With the exception of FRC, RV and closing volume, all other volumes and capacities can be measured with the help of **simple spirometer**.

B. Dynamic Lung Volume and Capacity

1. Timed Vital Capacity (TVC) or Forced Vital Capacity (FVC)

If vital capacity (VC) is recorded on a karyograph (spirograph) at the known speed, volume of air expelled can be timed. Thus, FVC is the maximum volume of air which can be breathed out as 'forcefully' and 'rapidly' as possible following a maximum inspiration. Thus, TVC is exactly similar to VC except that there is a special stress on "rapid", forcible and complete exhalation.

Clinical Significance of FVC

FVC is used to distinguish between 'restrictive' and 'obstructive' lung disorders.

In *restrictive disorders* [*e.g.*, emphysema (Box 3.4)] chest expansion is restricted. Therefore, VC decreases while FEV, (Forced Expiratory Volume in 1 second) is normal.

Box 3.4

Emphysema is a degenerative disorder and is characterized by a loss of lung elasticity and replacement of alveoli with large air sacs. Thus loss of lung elasticity prevents full expansion of lungs (*i.e.*, airway restriction). Most common cause of emphysema is heavy cigarette smoking.

In **obstructive disorders** (*e.g.*, bronchial asthma), inspiration is normal but expiration is obstructed therefore, VC is normal while, FEV decreases.

2. Pulmonary Ventilation (PV)

This is the volume of air expired or inspired by the lungs in one minute, therefore,

$$\begin{aligned} PV &= TV \times RR \text{ per min.} \\ &= 500 \times 12 \\ &= 6 \text{ L/min., normally} \end{aligned}$$

3. Peak Expiratory Flow Rate (PEFR)

This is the expiratory flow rate during the peak of FVC (Box 3.5).

Normal. 400–450 L/min.

Box 3.5

It is a simple test of ventilatory function which is widely used in clinical practice, and is recorded with the help of **Wright's peak flow meter**. Normal values are dependent on age, sex and body build. There may be a marked fall in its value in cases of obstructive airway disease.

4. Maximum Breathing Capacity (MBC)

It is also called **Maximum Voluntary Ventilation or Maximum Ventilation Volume (MVV).** It is largest volume of air that can be moved into and out of the lungs in one minute by maximum voluntary effort.

Normal. 90–170 L/min.

5. Pulmonary Reserve (PR) or Breathing Reserve (BR)

It is maximum amount of air above the pulmonary ventilation, which can be breathed in and out of lungs in one minute. It can be computed as: MVV – PV.

PR is usually expressed as percentage of MVV, *i.e.*, (MVV – PV) × 100/MVV; and called as percentage pulmonary reserve or Dyspnoeic Index (DI).

Normal % PR (DI) = 60 – 70% (usually 90%). If < 60%, Dysponea is usually present.

3.4 MECHANISM OF GAS EXCHANGE

The exchange of respiratory gases (O_2 and CO_2) in the respiratory organs and the tissues takes place according to the laws of gases, and is, thus, a physical rather than a biological process. Gas exchange occurs partly by **diffusion** and partly by **facilitated diffusion**. The facilitated diffusion uses a carrier protein, namely, a cytochrome, which speeds up diffusion, allowing the blood to take up oxygen faster. **Diffusion** is the net flow of a substance from a region of higher concentration to a region of lower concentration. The diffusion is directly proportional to:

1. partial pressure gradient (difference) of gases on the two sides of a membrane between them;
2. thinness of the membrane;
3. surface area of the membrane; and
4. permeability of the membrane.

Moisture (in the respiratory surface) is also necessary because O_2 and CO_2 are more easily exchanged in cells when in a liquid medium.

Partial pressure. Partial pressure of a gas is the pressure it exerts in a mixture of gases, and is equal to the total pressure of the mixture divided by percentage of that gas in the mixture. For example, if the pressure of atmospheric air at sea level is 760 mm Hg and oxygen forms 21% of the air, the partial pressure of the oxygen will of 21% of 760 or 159 mm Hg. In other words, the partial pressure of a gas is proportional to its concentration in the mixture.

1. Gas exchange in respiratory organs. The partial pressure (or concentration) of oxygen (P_{O_2}) is higher in the water/air drawn to the gills/lungs than in venous blood flowing in the capillaries of the respiratory organs. Hence, oxygen diffuses from water/air into the venous blood (Fig. 3.12 and Fig. 3.13). In the case of lungs, oxygen of air dissolves in the mucus that lines the alveoli before diffusing into the blood. The partial pressure (or concentration) of carbon dioxide (P_{CO_2}) is higher in the blood flowing in the capillaries of the respiratory organs than in the water/air drawn to the gills/lungs with the result, carbon dioxide diffuses from the blood into the water/air (Fig. 3.12 and Fig. 3.14). Diffusion of both the gases occurs simultaneously.

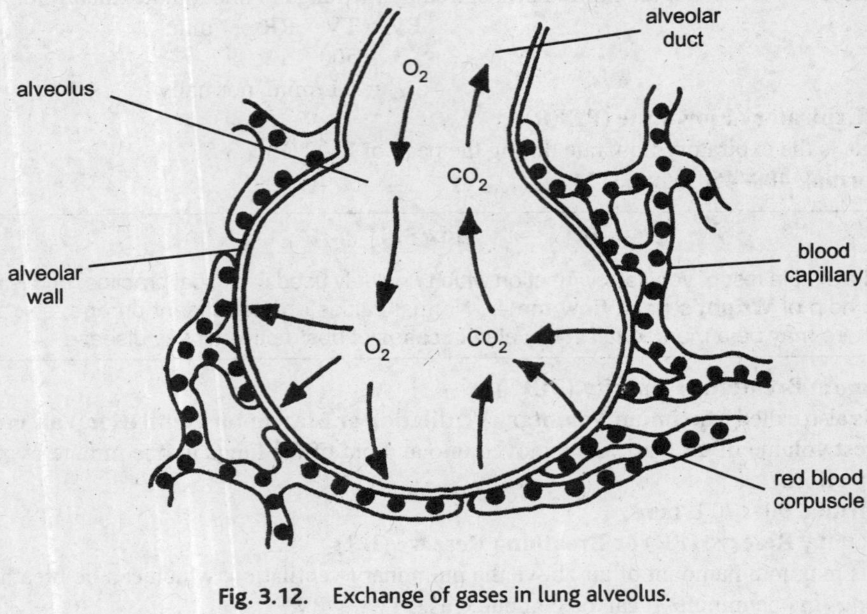

Fig. 3.12. Exchange of gases in lung alveolus.

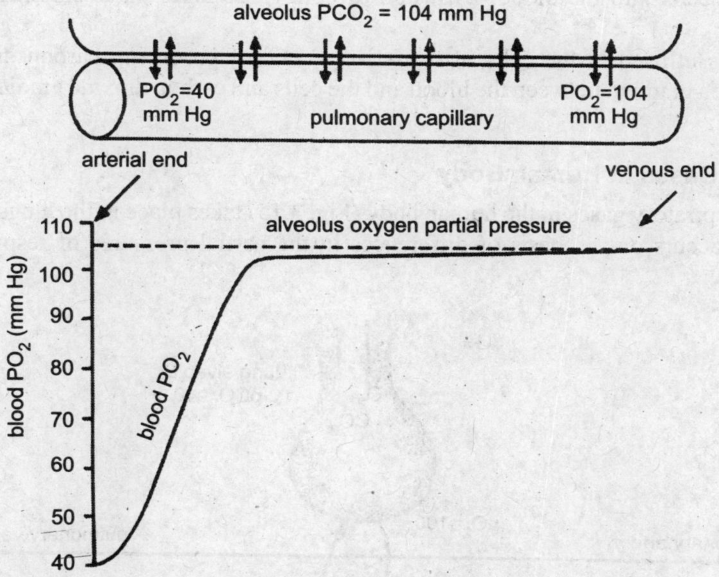

Fig. 3.13. Uptake of oxygen between alveolus and blood of pulmonary capillary.

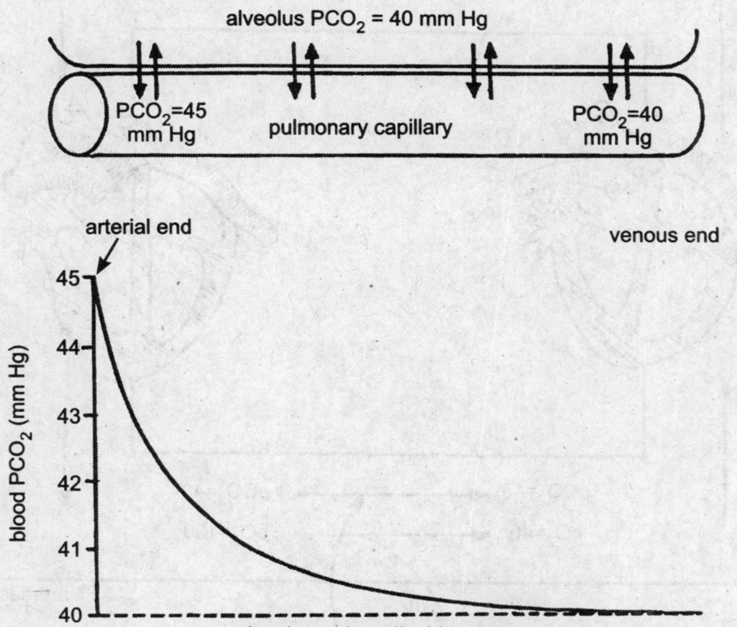

Fig. 3.14. Diffusion of CO_2 from the pulmonary blood into the alveolus.

2. Gas exchange in tissues. The gases behave in a similar manner in their exchange between the blood and the tissue cells. Here, the concentration of oxygen is higher in the blood than in the tissue cells, and the concentration of carbon dioxide is greater in the tissue cells than in the blood (Fig. 3.15). Hence, oxygen diffuses from the blood into the tissue cells, and carbon dioxide diffuses from the tissue cells into the blood. Diffusion of both gases takes place simultaneously and via tissue fluid.

Oxygen is utilised in the cells in oxidative reactions, producing carbon dioxide, and this maintains oxygen gradient between the blood and the cells and carbon dioxide gradient in the reverse direction.

Exchange of Gases in Human Body

Exchange of respiratory gases in the human body (Fig. 3.15) takes place in the lungs and the tissues. The exchange occurs on the basis of differences in the partial pressures of respiratory gases as explained below:

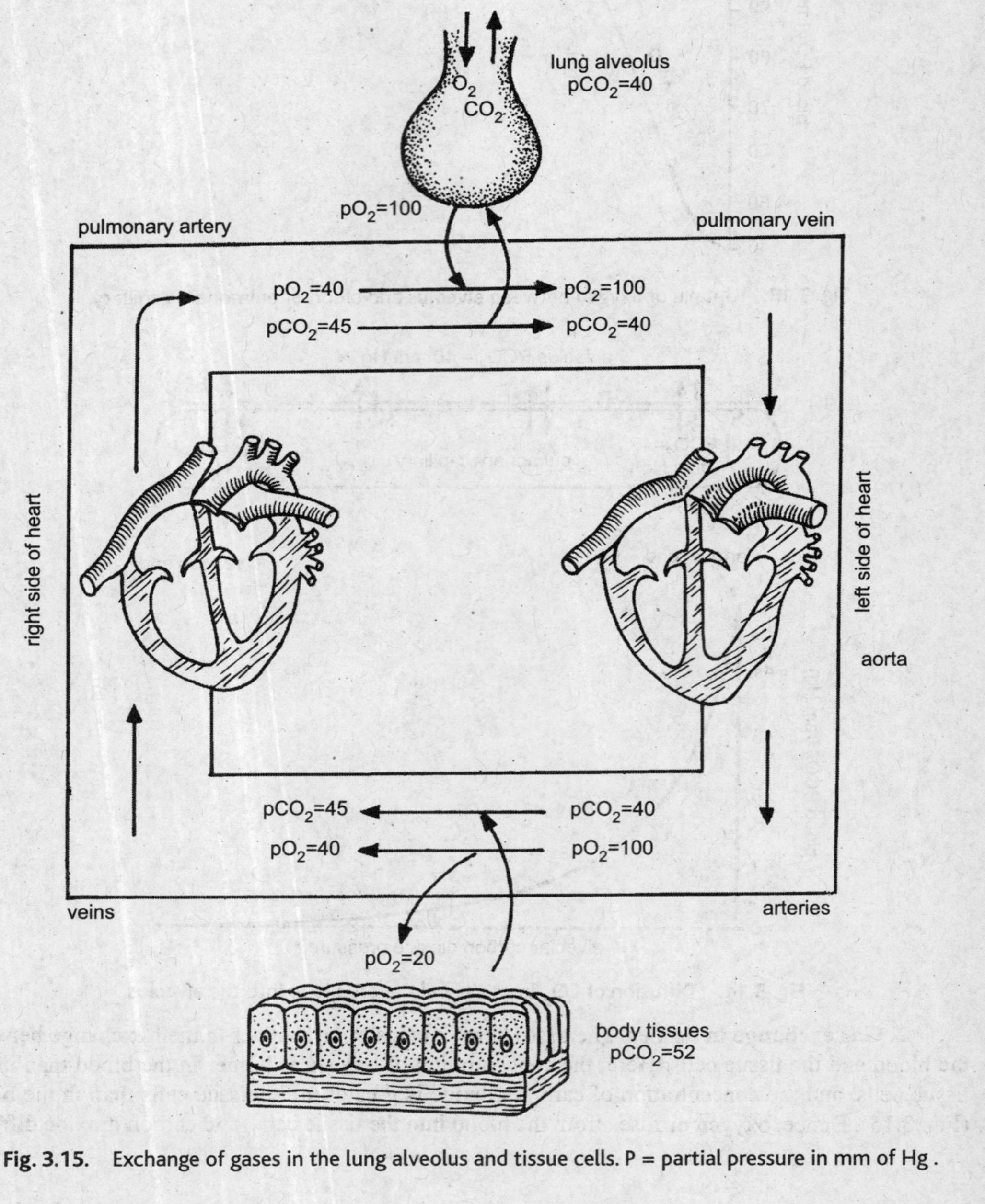

Fig. 3.15. Exchange of gases in the lung alveolus and tissue cells. P = partial pressure in mm of Hg.

A. Lungs. In the human lungs, exchange of gases occurs between the air in the alveoli and the blood in the capillaries around the alveoli (Fig. 3.15). In this exchange, the blood takes up O_2 from and releases CO_2 to the alveolar air.

***(i)* Uptake of O_2 by the blood.** The P_{O_2} (partial pressure of oxygen) of the alveolar air is 100 mm Hg, and P_{O_2} of blood in alveolar capillaries is 40 mm Hg. Due to the P_{O_2} difference of 60 mm Hg between air and blood, oxygen diffuses rapidly from the alveolar air into the blood of alveolar capillaries, raising the P_{O_2} of blood about 95 mm Hg.

***(ii)* Release of CO_2 by blood.** The P_{CO_2} (partial pressure of carbon dioxide) of blood reaching the alveolar capillaries in 45 mm Hg and P_{CO_2} of alveolar air is 40 mm Hg. Therefore, CO_2 diffuses from the blood of alveolar capillaries into the alveolar air till the P_{CO_2} of blood falls to 40 mm Hg.

The exchange of gases in the lung alveoli that raises the P_{O_2} of blood and lowers its P_{CO_2} is the *external respiration*. The blood oxygenated by this respiration is returned from the lungs by the pulmonary veins to the left side of the heart. The heart supplies the oxygenated blood to the body tissues.

B. Tissues. In the tissues, exchange of gases occur between the blood and the tissue cells (Fig. 3.15). This exchange occurs *via* tissue fluid that bathes the tissue cells. The blood reaching the tissue capillaries has P_{O_2} of 95 mm Hg and P_{CO_2} of 40 mm Hg. The tissue cells constantly use oxygen in oxidation which produces carbon dioxide. They have P_{O_2} of 20 mm Hg only and P_{CO_2} of 52 mm Hg. Because of P_{O_2} and P_{CO_2} differences between the blood and the tissue cells, oxygen separates from the oxyhaemoglobin and diffuses from the blood into the tissue fluid and thence into the tissue cells; and carbon dioxide diffuses from the tissue cells into the tissue fluid and thence into the blood in the tissue capillaries. Gases mostly diffuse through the tissue fluid as such, only small amounts dissolve in it.

Exchange of gases in the tissues that lowers the P_{O_2} of the blood and raises its P_{CO_2} is called **internal respiration**. The blood deoxygenated by this respiration returns to the right side of the heart which sends it to the lungs for reoxygenation.

The partial pressure of nitrogen (P_{N_2}) is the same (573 mm Hg) in the alveolar air and in the blood. This uniformity persists because nitrogen is an inert gas and it is not used in tissue cells.

Respiratory Pigments

Oxygen moves within the circulatory system on carrier proteins, called **respiratory pigments** that bind dissolved molecules of oxygen. A respiratory pigment is, therefore, a substance which binds free oxygen by forming temporary compound with oxygen in the region of high concentration and decomposes to liberate it on reaching the regions of low concentrations. Six types of respiratory pigments are known (Table 3.3).

Table 3.3 Some characteristics of respiratory pigments.

	Pigment	*Source*	*Location*	*Atoms of metal per mole*	*Molecular weight*
1.	Haemoglobin	Human being	Intracellular	4 Fe	6.5×10^4
		Frog	Intracellular	4 Fe	6.5×10^4
		Gastrophilus larva	Tracheal system	2 Fe	3.4×10^4
2.	Myoglobin	Mammalian muscle	Intracellular	1 Fe	1.7×10^4
3.	Erythrocruorin	Annelids	Plasma	96 Fe	3×10^6
		Molluscs	Plasma	96 Fe	1.5×10^4
4.	Chlorocruorin	Polychaete worms	Plasma	96 Fe	3.4×10^6
5.	Haemerythrin	Sipunculus	Intracellular	16 Fe	10.5×10^4
6.	Haemocyanin	Squid	Plasma	20 Cu	3.75×10^6
		Snail	Plasma	200 Cu	6.75×10^6

Functions of respiratory pigments

1. These are carriers of oxygen. In the absence of respiratory pigments, the blood would carry the oxygen only in solution and the amount would be very low.
2. These pigments have great affinity towards respiratory gases.
3. The respiratory pigments exhibit reversibility in their action as follows:

$$\text{Haemoglobin} + \text{Oxygen} \underset{\text{Respiratory organs}}{\rightleftharpoons} \text{Oxyhaemoglobin}$$

Example of Respiratory Pigments (Table 3.4)

1. Haemoglobin. Haemoglobin occurs in majority of vertebrates and invertebrates. In invertebrates it is found dissolved in plasma, whereas in vertebrates it is contained in the special cells, the **red blood cells** (in mammals, the **red blood corpuscles**). It is an iron-containing compound and consists of an **iron-porphyrin** (**heme**) coupled with a protein **globulin**. Haemoglobin is a macromolecule and consists of variable number of unit molecules. Each unit consists of one heme and its associated protein. Vertebrate haemoglobin has four unit molecules. The molecular weight of each unit is $17 - 18 \times 10^3$; therefore, the molecular weight of haemoglobin comes to $68 - 72 \times 10^3$.

The iron in haemoglobin is in **ferrous state** (Fe^{2+}). It may, therefore, be assumed that it has two primary valencies for holding the rings or porphyrin nucleus and secondary or residual valencies, out of which only two are utilised for nuclear attachment, so that the residual valencies remain free. Due to these free valencies haemoglobin has great affinity for oxygen and also for carbon monoxide.

Table 3.4 Properties of respiratory pigments in oxygenated and deoxygenated states.

Pigment	*Metal present*	*Colour*		*Occurrence*
		Oxygenated	*Deoxygenated*	
1. Haemoglobin	Iron	Red	Red	Vertebrates and most of the invertebrates
2. Haemocyanin	Copper	Blue	Colourless	Most molluscs and arthropods
3. Haemerythrin	Iron	Red	Colourless	*Sipunculus*, *Dendrostomum* (sipunculus worms), *Lingula* (brachiopoda), and *Magelona* (polychaete)
4. Chlorocruorin	Iron	Green	Green	Some polychaete worms

Haemoglobin is usually symbolised as Hb. Its heme resembles chlorophyll in structure, but it has iron, whereas the metal present in chlorophyll is magnesium. Heme constitutes the active and functionally significant portion of haemoglobin whereas globin serves simply as a carrier of heme.

Functions. Haemoglobin carries oxygen and supply it to the various tissue cells of the body. This function of haemoglobin is due to its property of forming a temporary compound **oxyhaemoglobin** with oxygen under high oxygen tension and this oxyhaemoglobin dissociates into oxygen and haemoglobin at places where oxygen tension is low. Each iron atom of haemoglobin combines with one molecule of oxygen and when all its iron have associated with oxygen, the haemoglobin is described saturated (Box 3.6). There is a definite ratio between the amount of haemoglobin and oxyhaemoglobin. This is represented by **dissociation curve**, which represents the relationship between oxygen tension in the medium and percent saturation of oxygen by the haemoglobin.

$$Hb + O_2 \rightarrow HbO_2$$

Box 3.6

Haemoglobin uncombined with O_2 and containing heme with ferrous iron (Fe^{2+}) is called **ferrohaemoglobin, deoxyhaemoglobin** or **reduced haemoglobin** is composed of four polypeptide chains, two α-chains and two β-chains, and its subunit structure is designated $\alpha_2A\beta_2B$.

Each unit of haemoglobin can carry an O_2 molecule attached to the remaining coordination bond. Since each of the four hemes contains an iron that serves as the point of loose physical attachment for a single molecule of O_2, a mammalian haemoglobin can carry a maximum of four molecules of molecular oxygen. When haemoglobin (Hb) is fully oxygenated, it is called **oxyhaemoglobin (HbO_2)** which contains 4 molecules of O_2 per molecule of haemoglobin

$$Hb_4 + 4O_2 \rightleftharpoons \underset{\text{Oxyhaemoglobin}}{Hb_4(O_2)_4}$$

This reaction proceeds to the right when O_2 is present in excess, as in the lungs.

When oxyhaemoglobin reaches the sites of low concentration, the oxyhaemoglobin dissociates liberating all its oxygen

$$HbO_2 \rightarrow Hb + O_2$$

Oxygen is carried by haemoglobin in blood largely in the form of HbO_2. When it reaches the tissues which are oxygen-poor in relation to blood. HbO_2 decomposes liberating free oxygen which is then available to the tissue for respiration. Haemoglobin is red when oxygenated and purple when deoxygenated.

Haemoglobin also functions as a carrier of small quantity of CO_2, conveying it as a waste product from the tissues to the lungs for excretion. However, the CO_2 is not transported by the haemoglobin's iron, but instead is carried in part in combination with the protein of the haemoglobin molecule.

Haemoglobin can also combine with 4 molecules of carbon monoxide (CO) to give carbon monoxide haemoglobin or carboxyhaemoglobin (HbCO), which is photosensitive and dissociates in light to release CO. The affinity of human Hb for CO is more than 200 times greater than its affinity for O_2. The brilliant cherry-red colour of HbCO is quite distinctive and is manifested in the skin and tissues of victim of carbon monoxide poisoning.

Four molecules of NO also combine with one molecule of Hb to give the very stable HbNO. Many agents, *e.g.*, peroxides, ferricyanide, can oxidise the Fe^{2+} in Hb to Fe^{3+} to give **methaemoglobin (met Hb)** which binds neither O_2 nor CO. **Cyanomethaemoglobin** is formed on reaction of met Hb with CN^-.

The properties of heme pigment vary in different animal species, and are generally related to the amount of carbon dioxide present. This is often low in water breathing animals and high in air breathing ones. This shows that the evolution of air breathing animals from water breathing ones involved changes not only in the morphology of respiratory organs but also in the chemical properties of the heme pigment of the blood.

2. Myoglobin. Myoglobin (Mb) of vertebrate muscle is very similar to Hb. While haemoglobin is a **oxygen carrier protein**, myoglobin is an **oxygen storage protein**. It stores oxygen (O_2) in resting skeletal muscle as **oxymyoglobin (MbO_2)** and on exercise releases O_2. Myoglobins have only about 30 residues homologous with either the α or β chains of vertebrate haemoglobins, but their tertiary structures are strikingly like those of Hb chains, with the same kinds of helical and nonhelical secondary structures. The most important structural differences between Hb and Mb, however, are the distributions of R groups on the surfaces of the molecules.

In muscles, myoglobin binds O_2 reversibly. However, it shows no Bohr effect (which is exhibited by Hb) with its O_2 dissociation curve is hyperbolic (which is sigmoid for human and frog haemoglobin). During muscle contraction, when the demand for O_2 is highest, O_2 dissociates from myoglobin and is available for oxidations. In humans, myoglobin is abundant only in cardiac muscle. In diving mammals, the myoglobin content of muscles is very high and facilitates submerging for long periods. The flight muscles of birds are also rich in myoglobin. At a venous PO_2 of 40 mm Hg, at which Hb is 66% saturated, myoglobin is still 94% saturated. At an O_2 tension of 10 mm Hg, Hb is 10% saturated, whereas myoglobin is 80% saturated. Myoglobin can accept O_2 from Hb and store it in the muscle cells for release to cytochrome oxidase when the O_2 supply is limited.

3. Erythrocruorin. Large haemoglobins, frequently called erythrocruorins, occur in the circulating plasma of many annelid worms and molluscs. The giant molecule of erythrocruorin from

Limnodrilus (MW = 3×10^6), consists of 108 subunits (MW = 2.8×10^4), each bearing one heme, arranged as a regular hexagonal cylinder having the dimensions 220 by 160 Å. This cylinder, in turn, consists of six lesser cylinders oriented in the same axis as the major cylinder axis. Thus, there are three layers of 36 protein subunits each in the overall packaged molecule. Since there are four different NH_2-terminal amino acids, it is assumed that there are four different subunit polypeptide chains in the molecule.

4. Chlorocruorin. Certain annelids (Sabellid and Serpulid worms) such as *Spirographis*, have iron-containing green blood pigment, the chlorocruorin (MW = 3.4×10^6). In this pigment, the porphyrin differs from protopophyrin in that the 2 vinyl group is replaced by a formyl group. It has close similarity with haemoglobin and an atom of iron combines with two atoms of oxygen. When such respiratory pigments are present in the coelomic fluid rather than in an organised circulation, they serve to store rather than transport O_2.

5. Haemerythrin. Haemerythrin is present in the blood of all sipunculid worms, a few polychaete forms, and the brachiopod *Lingula*. It is a rare, reddish-violet (or violet to pink, **Brusca** and **Brusca**, 2002), iron-containing respiratory pigment. The most carefully studied haemerythrin from *Sipunculus* has a molecular weight of 105,000 and contains 16 atoms of Fe^{2+}. Haemerythrin binds one molecule of oxygen with three iron atoms.

6. Haemocyanin. The blood plasma of many molluscs and arthropods (chelicerates, few myriapods and higher crustaceans) contain **haemocyanin**, blue pigment containing copper (instead of iron) but no haeme (or heme). Haemocyanins are always found in solution, never in cells, a characteristic probably related to the necessity for rapid oxygen unloading (**Brusca** and **Brusca**, 2002). The molecular weights of haemocyanins range from 0.5×10^6 to 10^7. Squid haemocyanin is a decamer (MW = 3.75×10^6), which on dilution or at high salt concentration dissociates into 5 dimers and then into 10 monomers. Each monomer contains two atoms of Cu^+ and both in monomeric and polymeric forms, bind one oxygen molecule (O_2) per two Cu^+ atoms. There is some evidence that O_2 binding and release involves oxidation and reduction of the copper. Haemocyanin is colourless when deoxygenated.

Transport of Oxygen

Oxygen is carried in the blood in two forms, *i.e.*, either in the dissolved form or in combination with haemoglobin. Some of the oxygen which enters the blood from the alveoli goes into simple solution in the plasma. This accounts for only a small per cent and would never be adequate to sustain the high metabolic activity of cells of higher animals. A loose combination of O_2 with haemoglobin (**oxygenation**) allows for the transportation of 50–100 times as much gas as could be transported in simple solution. In addition, the retention of haemoglobin within the red blood corpuscles adds immensely to the efficiency of this transportation system, because a non-cellular solution of haemoglobin rapidly escapes into tissues and is destroyed or eliminated.

The maximum amount of O_2 which the normal human blood can absorb is 20 ml per 100 ml of blood, *i.e.*, in human only a small amount (5%) of oxygen gas is transported from lungs to body tissues, while dissolved in blood plasma. The remaining 95% is bound to haemoglobin within red blood corpuscles. When O_2 passes from the lung alveoli into the lung capillaries, it diffuses into the blood. Each atom of iron in the heme portion of haemoglobin is combined with a molecule of O_2 to form oxyhaemoglobin which carries O_2 in the molecular rather than in ionic form:

$$Hb_4 + 4O_2 \rightleftharpoons Hb_4O_8 \text{ or } Hb_4(O_2)_4$$

Dissociation Curve

The O_2 pressure at which respiratory pigment (haemoglobin) becomes saturated vary greatly. In other words, the proportion of oxyhaemoglobin to haemoglobin present in the blood at any time is dependent upon the tension of O_2 in the blood. This is known as **dissociation curve** of oxyhaemoglobin (Fig. 3.16). This curve is S-shaped or sigmoid. This curve tells us what percentage of haemoglobin is saturated within oxygen at a given partial pressure of gas. It is relatively flat between P_{O_2} 100 mm

Hg to P_{O_2} 60 mm Hg and shows about 97% oxygen saturation of haemoglobin, ensuring adequate oxygenation of blood. The curve bends sharply between P_{O_2} 60 mm Hg and 40 mm Hg and oxygen saturation of haemoglobin falls to about 70%, thereby facilitating rapid dissociation of oxyhaemoglobin. The S-shaped curve is interpreted as under:

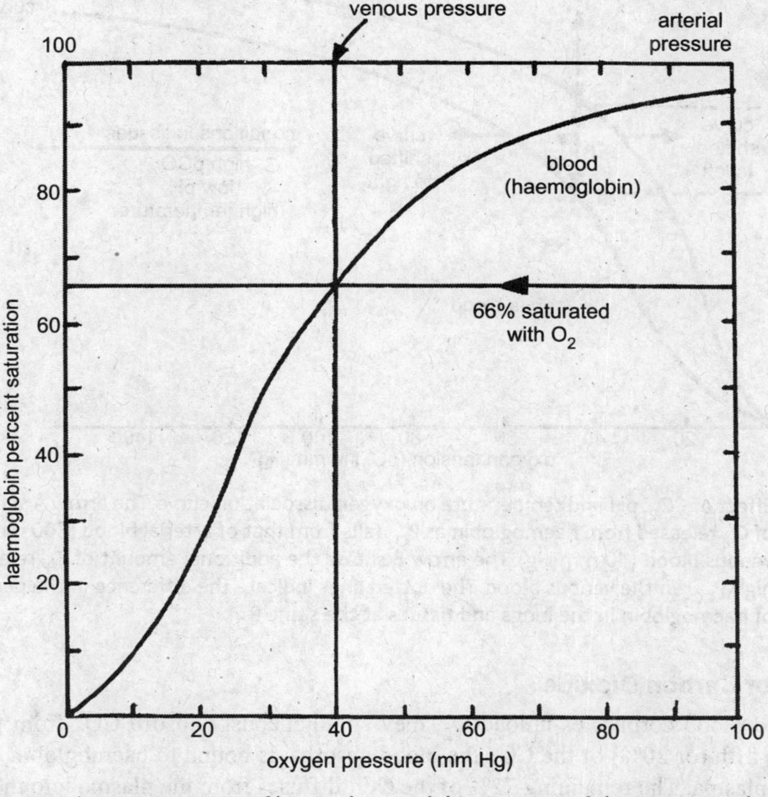

Fig. 3.16. Oxygen dissociation curve of human haemoglobin. Because of the interaction between the four heme groups in a haemoglobin molecule, the binding of O_2 to haemoglobin follows S-shape.

At low partial pressure, oxygen does not join haemoglobin extensively. At intermediate pressure, oxygen readily joins haemoglobin, small changes in P_{O_2} causing large changes in saturation of haemoglobin with oxygen. This indicates that haemoglobin gains oxygen more quickly after some oxygen is already linked to it. There is an interaction between the four heme units of haemoglobin molecule in oxygen binding, the second and third molecules of oxygen join more readily than the first. At high P_{O_2}, the interaction between haemoglobin units decreases, haemoglobin becomes almost fully saturated and very little change then occurs.

Bohr effect. Oxygen dissociation from oxyhaemoglobin is affected by carbon dioxide, pH and temperature. Relatively low partial pressure of carbon dioxide (P_{CO_2}), high pH and low temperature in lung capillaries increase the affinity of haemoglobin for oxygen so that the oxygen dissociation curve shifts to the left (Fig. 3.17). Relatively high P_{CO_2}, low pH and high temperature in the tissues due to active metabolism decrease the affinity of haemoglobin for oxygen and enhance the dissociation of oxyhaemoglobin, so that the oxygen dissociation curve shifts to the right (Fig. 3.17). The effect of increased acidity on haemoglobin is called **Bohr effect**. The Bohr effect facilitates the release of oxygen from oxyhaemoglobin in the tissues where production of CO_2 lowers the pH. In other words, it ensures adequate supply of O_2 to the tissues where it is needed for active metabolism. Bohr effect is attributed to conformational changes in haemoglobin molecules.

An increase in CO_2 tension over 40 mm Hg will increase the acidity and enhance the dissociation of oxygen from oxyhaemoglobin.

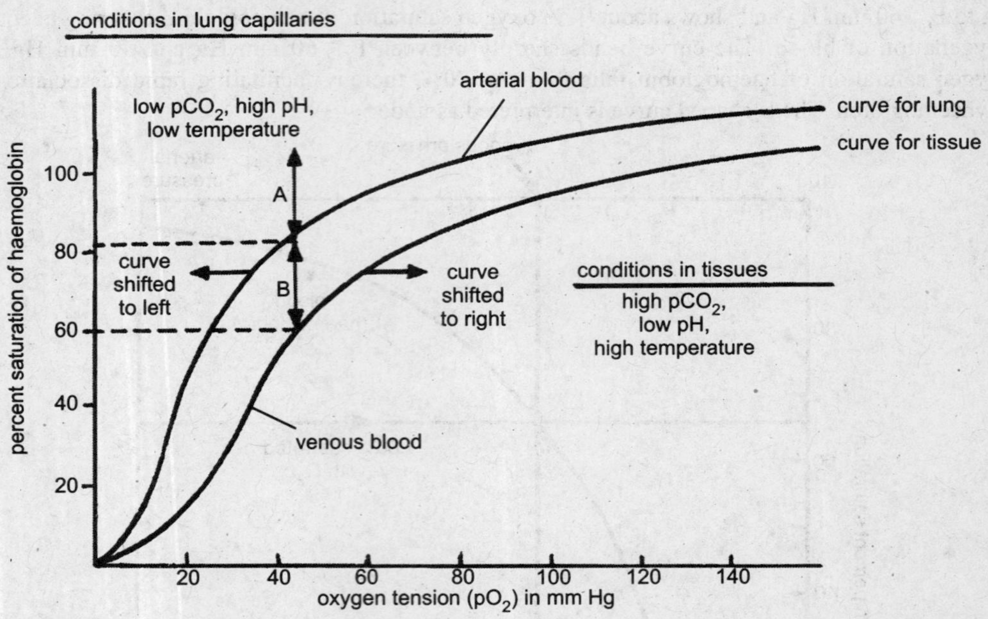

Fig. 3.17. Effect of CO_2, pH and temperature on oxygen dissociation curve. The arrow A shows the amount of O_2 released from haemoglobin as P_{O_2} falls from that of arterial blood (100 mm Hg) to that of venous blood (40 mm Hg). The arrow B shows the additional amount of O_2 released due to the high P_{CO_2} in the venous blood. The dotted lines indicate the difference in the percent saturation of haemoglobin in the lungs and tissues at the same P_{O_2}.

Transport of Carbon Dioxide

When the red blood corpuscles unload O_2, they simultaneously absorb CO_2 from the tissues. As much as one fifth (or 20%) of the CO_2 the blood absorbs, is bound to haemoglobin, another 8% is dissolved in plasma. The remaining 72% of the CO_2 diffuses from the plasma into the cytoplasm of the red blood corpuscles (RBCs).

Transport of Carbon Dioxide and the Chloride Shift

CO_2 is evolved in the body as a result of various activities of the cells and diffuses into blood. The total amount of CO_2 in the venous blood is about 60 ml per 100 ml blood and the arterial blood contains about 50 ml total CO_2 per 100 ml. Thus, a relatively small amount of CO_2 is given off in the lungs.

As a result of metabolism, tissues put out CO_2. It can be transported as follows:

1. As carbonates. More than 60% of the total output of CO_2 passes out into RBCs and is combined with water to form **carbonic acid** in the presence of an enzyme, **carbonic anhydrase**. The carbonic acid dissociates into **bicarbonate (HCO_3)** and hydrogen (H) ions.

This chemical process removes large amounts of CO_2 from the blood plasma, facilitating the diffusion of more CO_2 into it from the surrounding tissue. Such a facilitation is critical to CO_2 removal, since the difference in CO_2 concentration between blood and tissue is not large (only 5%).

$$CO_2 + H_2O \rightleftharpoons \underset{\text{Carbonic acid}}{H_2CO_3}$$

HCO_3 (bicarbonate) hooks up with K^+ present in RBCs and diffuses out into plasma. Thus CO_2 actually travels in the plasma as bicarbonate, *i.e.*, HCO_3. Plasma membrane of RBCs is almost impermeable to positively charged H^+ ions. Free H^+ ions in RBCs are buffered by the haemoglobin itself, which forms weak **haemoglobinic acid**. To maintain electroneutrality in the RBCs, the positively charged haemoglobin must be countered by chloride (Cl^-) from plasma.

$$Na^+ + HCO_3^- \rightarrow NaKCO_3$$
$$K^+ + HCO_3^- \rightarrow KHCO_3$$

The red blood corpuscles carry their cargo of bicarbonate ions back to the lungs. The lower CO_2 concentration in the air inside the lungs causes the carbonic anhydrase reaction to proceed in the reverse direction, releasing gaseous CO_2 which diffuses outward from the blood into the alveoli. With the next inhalation, this CO_2 leaves the body. The fifth of the CO_2 bound to haemoglobin also leaves because haemoglobin has a greater affinity for oxygen than for CO_2 at low CO_2 concentration. The diffusion of CO_2 outward from the red blood corpuscles causes the haemoglobin within these cells to release its bound CO_2 and take up oxygen instead. The red blood corpuscles, with their newly bound oxygen, then start the next respiratory journey.

In the dissociation of carbonic acid in RBCs, it is important to note that, for every bicarbonate ion produced, a hydrogen ion is also produced. Hydrogen ions cause lowering of pH, we might therefore expect the blood plasma to become highly acidic after absorbing CO_2 at the body tissues. But this never happens, the blood pH remains constant at pH 7.3 and a person will die if the pH of blood ever drops below 7.0. The maintenance of pH is possible because the blood contains certain inorganic ions (*e.g.*, HPO_4). Na^+ is in plasma and K^+ in RBCs and proteins which act as buffers. Much buffering is done by potassium in the blood.

$$HPO_4 + H^+ \rightleftharpoons H_2PO_4$$

Buffers take up most of the free hydrogen ions, *e.g.*, in RBCs buffering is provided by haemoglobin forming a weak haemoglobinic acid.

2. As carbamino compounds. Remaining about 35% of total output of CO_2, combines with the NH_2 group of haemoglobin and other proteins to form carbamino compounds; carbamino-haemoglobin is the chief product.

$$CO_2 + Hb.NH_2 \rightleftharpoons \underset{\text{Carbamino haemoglobin}}{Hb.NH.COOH}$$

At the lung surface and in its capillaries, carbamino haemoglobin breaks up, liberating its CO_2 to plasma and then to alveolar air. As the CO_2 from haemoglobin of red blood corpuscles passes out, K^+ and the bicarbonate ions of plasma re-enter the red blood corpuscles where H^+ of RBCs combines with HCO_3 forming H_2CO_3. Carbonic anhydrase converts H_2CO_3 back to CO_2 and H_2O. This CO_2 is now free to diffuse out of the plasma and into the alveoli of lungs. The reactions are briefly shown here

(*i*) $H_2CO_3 \xrightarrow{\text{Carbonic anhydrase}} H_2O + CO_2$

(*ii*) $NaHCO_3 \longrightarrow Na^+HCO_3^-$

(*iii*) $2NaHCO_3 \longrightarrow Na_2CO_3 + H_2O + CO_2$

3. Chloride shift. Another problem in CO_2 transport is that, when bicarbonate ions leave the red blood corpuscles, the hydrogen ions remain and even though picked up by buffers, they tend to cause the red corpuscles to lose their electrical neutrality and become positively charged. This problem is solved by what is known as **chloride shift**. The plasma always contains some free chloride ions which have a negative charge. Each time a carbonate ions leaves the red blood corpuscles, a chloride cation enters the cell, thus maintains the electrostatic neutrality of the cell (Box 3.7). Reverse reaction takes place at the lung surface, where while the bicarbonate ion re-enters the red blood corpuscles, the chloride cation shifts back to plasma (Fig. 3.18).

Box 3.7

During oxygenation of blood in the capillaries of respiratory surface, bicarbonate ions pass into RBCs for decomposition by the action of carbonic anhydrase. According to **Donnan effect**, if some ions move in, others must move out to maintain the balance. The membrane of the RBCs is relatively impermeable to Na^+ and K^+ ions, so the chloride ions pass into the plasma. This restores the anion-cation balance in the corpuscles.

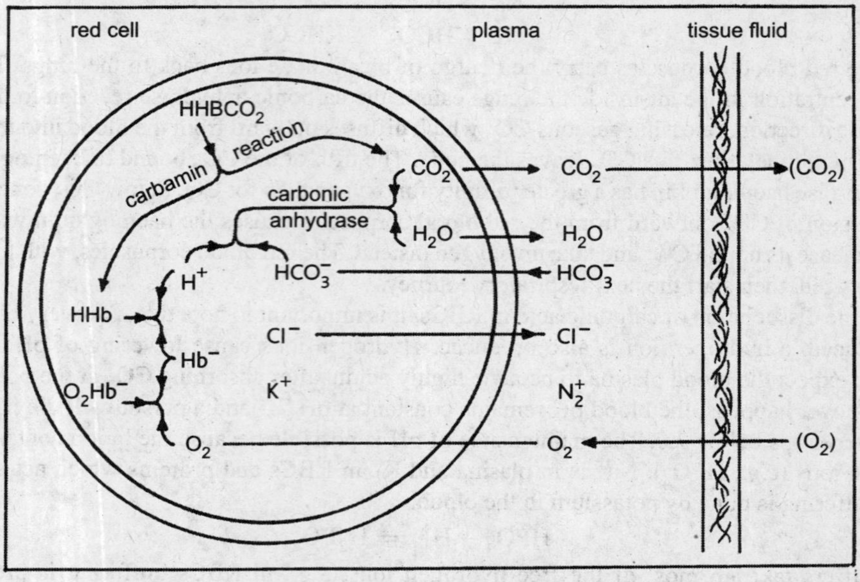

Fig. 3.18. Uptake of oxygen by RBC. Schematic summary of the processes that occur when haemoglobin takes up O_2 in the lungs.

It is also significant that H^+ ions are buffered by haemoglobin changing it to haemoglobinic acid. Acid haemoglobin encourages unloading of O_2 from oxyhaemoglobin in the tissues. Reverse process goes on in the alveolar capillary, where return of acid haemoglobin to neutral encourages loading of haemoglobin with O_2 to form oxyhaemoglobin.

The tidal flow of oxygen towards the tissues and of the carbon dioxide towards the respiratory surface is maintained by differential amounts of these gases at the two places. In the working tissue, carbon dioxide is at a higher concentration, which stimulates greater supply of the tissues.

Birds and other vertebrates laying shelled egg get rid of some of their CO_2 by converting it into carbonate ions, which are used up in the formation of calcium carbonate of the egg shell.

Carbon monoxide poisoning is due to the fact that haemoglobin combines even more readily with carbon monoxide than with oxygen. The result is carboxyhaemoglobin. Carbon monoxide molecules couples with haem group, thus leaving no space for oxygen molecules to take their rightful positions.

4. The carbon dioxide dissociation curve (Bohr effect). Winton and **Bayliss** (1955) constructed a CO_2 dissociation curve by plotting on the ordinate the total amount of CO_2 in the blood and on the abscissa the various CO_2 partial pressures. It has been seen that the curves for oxygenated and deoxygenated blood somewhat differ. This is because oxyhaemoglobin is slightly more acidic than haemoglobin. The oxygenated blood will bind slightly less CO_2. This phenomenon is closely related to the **Bohr effect** (**Bohr**, 1904) which shows the other side of the coin, *i.e.*, an increase in the CO_2 shifts the oxygen dissociation curve to the right giving more oxygen from the blood to the tissue. When CO_2 is added to the blood, it pushes the equilibrium between oxyhaemoglobin and haemoglobin in the direction of the weaker acid (haemoglobin). Because of this, more oxygen is released (Fig. 3.19).

Haldane Effect

The reciprocal effect of oxygenation on acid strength of haemoglobin, the so-called **Haldane effect**, accounts for the CO_2 exchange. In the lungs, chloride ions move out of the red blood corpuscles and bicarbonate ions move back in. The enzyme carbonic anhydrase then promotes the rapid reformation of free CO_2, and this gas diffuses from the blood into the lung alveoli.

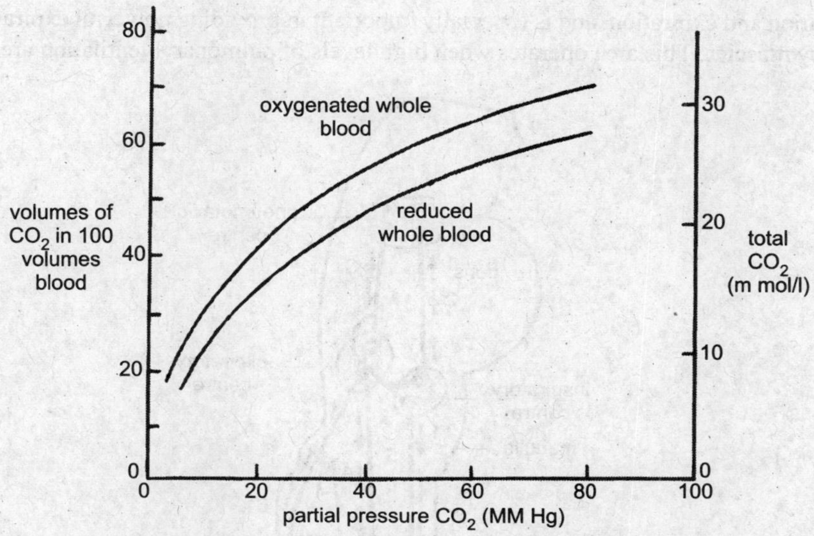

Fig. 3.19. Carbon dioxide dissociation curve for oxygenated and deoxygenated blood.

3.5 CONTROL OF BREATHING

Breathing or ventilation is controlled by the following (Fig. 3.20):

1. Dorsal respiratory group of neurons (or Dorsal inspiratory center). Most of these neurons are located within the nucleus of tractus solitarius of medulla of brain. The basic rhythm of respiration is generated by these neurons. Neurons emit repetitive bursts of inspiratory action potentials. From these neurons, nerve signals are transmitted to inspiratory muscles such as diaphragm and external intercostal muscles.

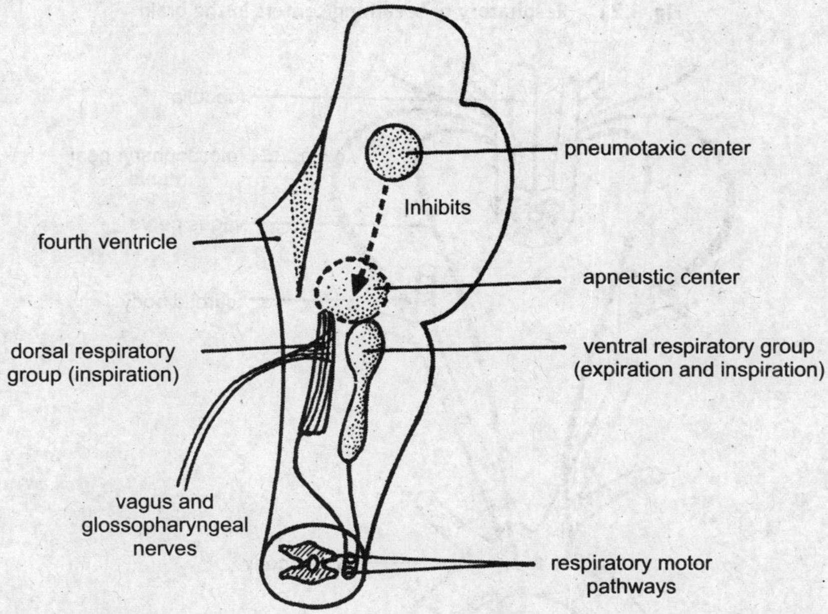

Fig. 3.20. Organization of the respiratory center.

2. Ventral respiratory group of neurons (or Ventral expiratory center). This group is located in nucleus ambiguus and nucleus retroambiguus of medulla. This ventral area contributes both inspiration and expiration, and is especially important in providing powerful expiratory signals to expiratory muscles. This area operates when high levels of pulmonary ventilation are required.

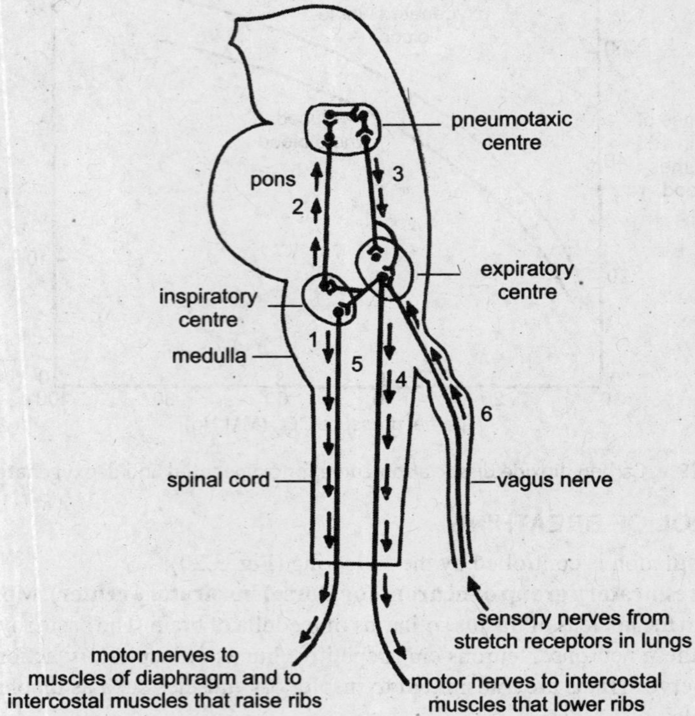

Fig. 3.21. Respiratory or breathing centers in the brain.

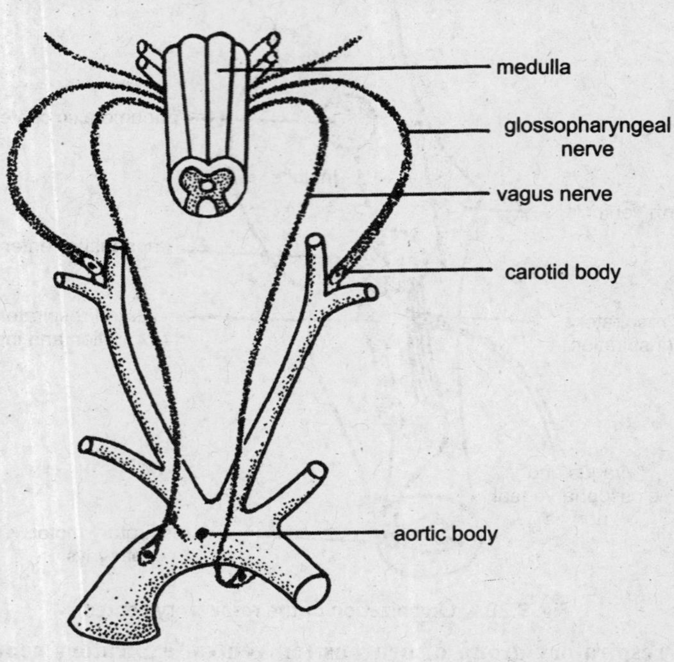

Fig. 3.22. Respiratory control by peripheral chemoreceptors in the carotid and aortic arches.

3. Apneustic center. It is located bilaterally in lower part of pons. It sends signals to dorsal respiratory group of neurons that prevent the "switch-off" of the inspiratory ramp signal. This leads to prolonged inspiration which is termed **apneusis**. It operates in association with the pneumotaxic center to control the depth of inspiration.

4. Pneumotaxic center. It is located bilaterally in nucleus parabrachialis of the upper pons (Fig. 3.21). The primary function of this center is to control the "switch off" point of inspiratory ramp, thus controlling the duration of inspiration.

5. Peripheral chemoreceptors present in the aortic arch and carotid sinus and respectively called **aortic bodies** and **carotid bodies** (Fig. 3.22; Fig. 3.23).

6. Chemoreceptors in brain.

7. Mechanoreceptors or stretch receptors. They are situated in the lung alveoli and bronchial tree.

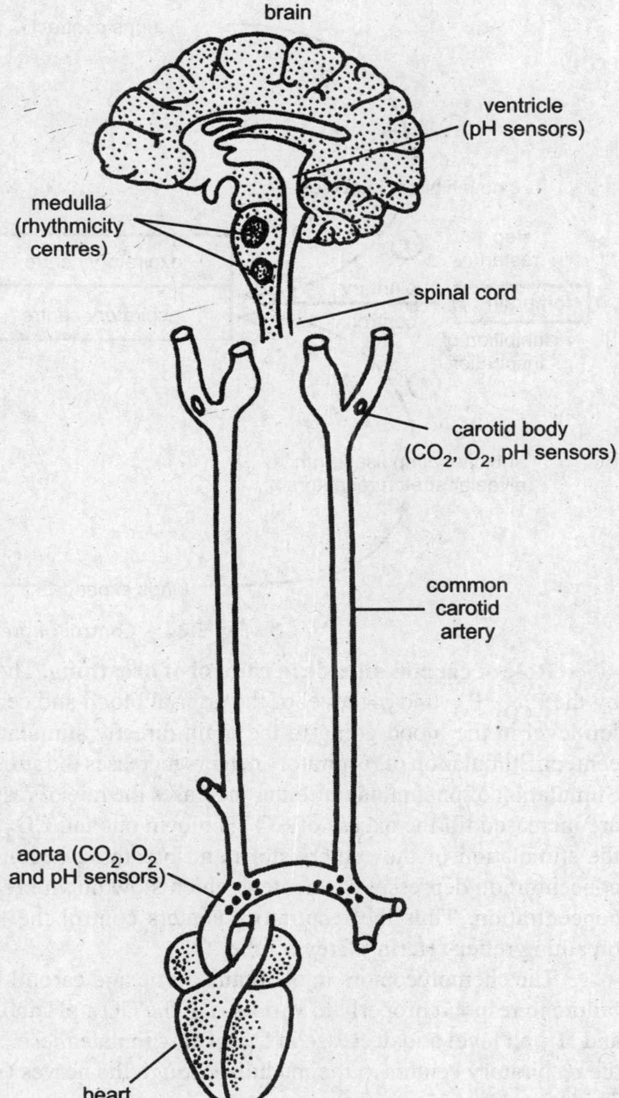

Fig. 3.23. Control centers for breathing.

The inspiratory center promotes inspiration (Fig. 3.24). The expiratory center stops inspiratory activity and promotes expiration. The breathing centers communicate with the diaphragm via **phrenic** and **thoracic nerves**. The bronchial tubes and alveoli are innervated by branches of the **vagus nerve**. Nerve impulses from the inspiratory activity and promotes expiration. The breathing centers communicate with the diaphragm via **phrenic** and **thoracic nerves**. The bronchial tubes and alveoli are innervated by branches of the vagus nerve. Nerve impulses from the inspiratory center stimulate the muscles of the diaphragm to flatten the latter and the intercostal muscles to raise the ribs. This brings about inspiration. The inspiratory center also sends impulses to the pneumotaxic center which relays them to the expiratory center. The expiratory center is excited and sends impulses to the intercostal muscles that lower the ribs and cause expiration. Other impulses pass from the expiratory center to the inspiratory center to inhibit it momentarily. When the impulses from the pneumotaxic center fade away, inspiration restarts and the cycle is repeated.

The stretch receptors in the lungs are stimulated by their expansion during inspiration and send impulses via the vagus nerve to stimulate the expiratory center and inhibit the inspiratory center. The respiratory muscles relax and expiration takes place. After expiration has occurred, the alveoli are no longer stimulated. Therefore, the expiratory center becomes inactive, and inspiration begins again. The whole cycle is repeated rhythmically throughout the life of the organism. The reflex from the stretch receptors serves as a second feedback mechanism for the regulation of the respiratory cycle.

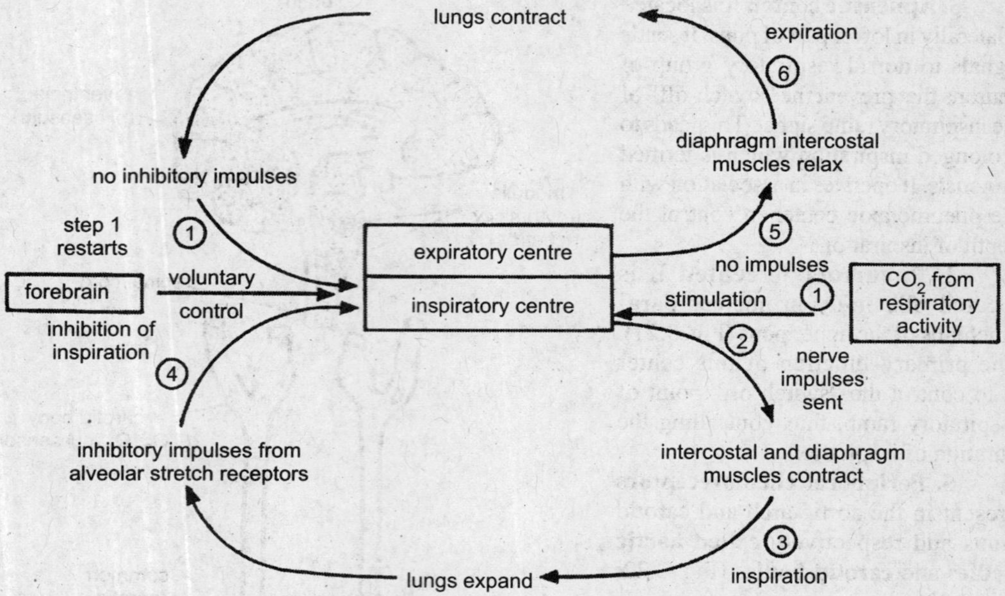

Fig. 3.24. Control of breathing.

Role of carbon dioxide in control of breathing. The rate of respiration is primarily controlled by the P_{CO_2}, P_{O_2} and pH level of the arterial blood and cerebrospinal fluid. Increase in CO_2 and H^+ ion level in the blood going to the brain directly stimulates both the respiratory and pneumotaxic centers. Stimulation of respiratory centers increases the strength of contraction of respiratory muscles. Stimulation of pneumotaxic center increases the rate of respiration. The depth and rate of respiration are increased till the excess of CO_2 is blown out and CO_2 and H^+ ion level returns to normal. Now the stimulation of the centers stops and breathing become normal. Decrease in CO_2 and H^+ ion concentration depresses the centers which slow down breathing to conserve CO_2 and raise H^+ ion concentration. Thus the respiratory centers control the lung ventilation by modifying the basic breathing reflex (Hering-Brever reflex).

The chemoreceptors in the brain, aortic and carotid bodies, act as a protection against body's failure to respond properly to variation in the CO_2, pH and O_2 content of the blood. Increase in CO_2 and H^+ ion level and decrease in O_2 level stimulate these chemoreceptors to send nerve impulses to the respiratory centers in the medulla through the nerves (*i.e.*, glossopharyngeal and vagus nerves). This stimulates the respiratory centers.

Other brain centers may modify the basic breathing pattern. We can have deep inspiration control, breathing for speech, pant like a dog, sigh and hold breath to check inhaling smoke or irritating gases. However, breathing is essentially an involuntary process, and we cannot kill ourselves by holding breath. Accumulation of carbon dioxide in the blood forces us to breathe.

QUESTIONS

Long Answer Questions

1. Differentiate between breathing and respiration. Describe the physiology of pulmonary respiration in a mammal. (*CCSU Meerut 2008*)
2. What is respiration? How carbon dioxide is transported in the blood of mammals? (*CCSU Meerut 2013*)
3. What do you know about tissue respiration? Describe the physiology of respiration in mammals. (*CCSU Meerut 2010*)

4. Explain the chemical control on respiration.
5. Describe various factors affecting the oxygen dissociation curve.

Short Answer Questions

1. Describe transportation of carbon dioxide during respiration. (*CCSU Meerut 2009*)
2. Enumerate the oxygen dissociation curve of haemoglobin. (*HPU 2009*)
3. Describe the chloride-shift. Give its significance. (*HPU 2007, 2009*)
4. Describe mechanism of breathing in man. (*HPU 2006*)

Very Short Answer Questions

1. Name two animals which carry on anaerobic respiration.
2. Where does exchange of gases occur in human body?
3. What is floating respiration?
4. Make a list of the respiratory pigments.
5. What is Haldane effect? (*CCSU Meerut 2013*)
6. State P_{O_2} and P_{CO_2} levels of pulmonary capillary blood.
7. How is Haldane effect beneficial?
8. What is chloride shift?
9. Enumerate various forms in which CO_2 is transported.
10. What is P_{50}?
11. What is Bohr's effect?
12. Why is O_2 dissociation curve sigmoid shape?
13. What is function of pneumotaxic centre? (*CCSU Meerut 2014*)
14. Name the forms in which O_2 is transported in blood.
15. What is pulmonary ventilation?
16. Define tidal volume (TV).
17. What is residual volume (RV)?
18. What is functional residual capacity (FRC)?
19. What is vital capacity (VC)?
20. What is methaemoglobin?
21. Why carbon monoxide is poisonous?
22. What is hypoxia?
23. What is the percentage of oxygen in inhaled air? (*CCSU Meerut 2010*)
24. What are the functions of paranasal sinuses?
25. What are secretions of respiratory epithelium of the nasal chamber?

Fill in the Blanks

1. Floating respiration uses .... or fats.
2. Respiration which uses proteins is termed .... respiration.
3. Air left in the lungs after normal expiration is called ....
4. Volume of air normally inspired or expired in one breath is called ....
5. Amount of air inhaled and exhaled with maximum effort is referred to as the .... of the lungs.
6. The volume of air left in the lungs after maximum expiration is called .... while the volume of air breathed out during a normal resting respiration is called ....
7. Alveolar P_{O_2} is .... than the venous P_{O_2} while atrial P_{O_2} is .... than the alveolar P_{O_2}.

Multiple Choice Questions

1. Exchange of gases between the alveolar air and blood of alveolar capillaries occurs by
   (*a*) osmosis (*b*) active transport
   (*c*) diffusion (*d*) pinocytosis
2. Most oxygen (O_2) in the blood is carried in the form of ________
   (*a*) carbonic acid (*b*) bicarbonate ion
   (*c*) oxyhaemoglobin (*d*) haemoglobin ion
3. The reaction of CO_2 with H_2O to form carbonic acid occurs
   (*a*) mainly in the tissue fluid (*b*) mainly in the blood plasma
   (*c*) mainly in the red blood corpuscles (*d*) equally well is (*a*), (*b*) and (*c*)
4. Which of the following has the highest oxygen-binding capacity?
   (*a*) human maternal haemoglobin (*b*) human foetal haemoglobin
   (*c*) llams haemoglobin (*d*) myoglobin
5. The oxygen concentration of expired air is
   (*a*) 4% (*b*) 10%
   (*c*) 16% (*d*) 20%
6. Carbonic anhydrase is active in
   (*a*) WBC (*b*) RBC
   (*c*) plasma (*d*) platelets
7. The volume of air taken in and given out at each normal inspiration and expiration is
   (*a*) vital capacity (*b*) residual volume
   (*c*) tidal capacity (*d*) tidal volume
8. Which of the following prevents collapsing of the trachea?
   (*a*) diaphragm (*b*) cartilaginous rings
   (*c*) ribs (*d*) muscles
9. In the process of transport of CO_2 which phenomenon occurs between RBC and plasma?
   (*a*) adsorption (*b*) absorption
   (*c*) chloride shift (*d*) osmosis
10. Oxygen dissociation curve of haemoglobin is
    (*a*) sigmoid (*b*) linear
    (*c*) hyperbolic (*d*) parabolic
11. Pneumatic and inhibitory centers are associated with
    (*a*) respiration (*b*) breathing
    (*c*) inspiration (*d*) expiration
12. Exchange of oxygen and carbon dioxide at the respiratory surface occurs through
    (*a*) active transport (*b*) passive transport
    (*c*) osmosis (*d*) diffusion
13. Hamburger phenomenon is also known as
    (*a*) bicarbonate shift (*b*) sodium shift
    (*c*) chloride shift (*d*) H^+ shift
14. Oxygen carrying shift of blood is
    (*a*) 20% (*b*) 30%
    (*c*) 40% (*d*) 50%
15. Medullary inspiratory center is under the control of
    (*a*) nervous (*b*) physical
    (*c*) chemical (*d*) electrical
16. At high carbon tension, oxygen dissociation curve of the haemoglobin shift to
    (*a*) left (*b*) right
    (*c*) straight (*d*) no change

17. The greatest volume of air that can be expired after a maximum inspiratory effort is its
(*a*) residual volume (*b*) vital capacity
(*c*) tidal volume (*d*) lung volume

18. The affinity between CO and haemoglobin is greater than CO_2 by
(*a*) 2 times (*b*) 20 times
(*c*) 250 times (*d*) 1000 times

19. Volume of air remaining in lungs after maximum respiratory effort is
(*a*) vital capacity (*b*) tidal volume
(*c*) residual volume (*d*) lung capacity

20. Chloride shift is required for the transport of
(*a*) oxygen (*b*) carbon dioxide
(*c*) nitrogen (*d*) both (*a*) and (*b*)

21. The tidal volume of human lung is
(*a*) 500 ml (*b*) 1 liter
(*c*) 1.5 liter (*d*) 4.5 liter

22. Respiratory mechanism is under the control of
(*a*) central nervous system (*b*) sympathetic nervous system
(*c*) autonomic nervous system (*d*) parasympathetic nervous system

23. The residual volume of human lung is
(*a*) 500 ml (*b*) 1 liter
(*c*) 1.5 liter (*d*) 3–4.5 liter

24. The control of breathing is centered in which of the following areas of the brain?
(*a*) anterior pituitary (*b*) corpus callosum
(*c*) cerebellum (*d*) medulla oblongata

25. CO_2 dissociates from carbaminohaemoglobin when
(*a*) P_{CO_2} is high and P_{O_2} is low (*b*) P_{O_2} is high and P_{CO_2} is low
(*c*) P_{CO_2} and P_{O_2} are equal (*d*) none of these

ANSWERS

Very Short Answer Questions

1. Tapeworm, *Ascaris*.

2. Lung alveoli and tissue cells.

3. That uses carbohydrates or fats.

4. Haemoglobin, haemocyanin, haemoerythrin, chlorocruorin.

5. Deoxyhaemoglobin (reduced haemoglobin) combines with more CO_2 as compared to oxyhaemoglobin. This is known as Haldane's effect.

6. P_{O_2} = 40 mm Hg; P_{CO_2} = 45 mm Hg.

7. Haldane's effect is beneficial because it almost doubles the quantity of CO_2 to be carried from the tissues to lungs and also doubles the amount of CO_2 to be excreted by lungs.

8. At the tissues when CO_2 enters the blood, it reacts with water to form H_2CO_3 which dissociates into H^+ and HCO_3^- ions. H^+ ions formed in the reaction are buffered by proteins (plasma proteins and haemoglobin). This leads to increased HCO_3^- inside the red cells as compared to that in the plasma. Therefore, HCO_3^- ions diffuse from RBC to plasma. As HCO_3^- is a negatively charged ion and the plasma membrane of RBC is relatively impermeable to cations, inside of the cells become less negatively charged. In order to neutralise the effect, negatively charged chloride ions (Cl^-) diffuse from plasma into the red blood corpuscles. This is called **chloride shift**.

9. CO_2 is transported in the following three forms: 1. dissolved state in plasma (7%); 2. in bicarbonate form (70%); 3. in the form of carbamino compound (23%).

10. Partial pressure of O_2 at which haemoglobin 50% saturated is termed P_{50}.

11. Shifting of O_2 dissociation curve due to changes in P_{CO_2} is known as Bohr's effect.

12. Sigmoid curve is due to: 1. Presence of haemoglobin in red blood corpuscles. 2. Haemoglobin combines with four molecules of O_2. When first three molecules of O_2 combine with haemoglobin, its affinity for O_2 is greatly increased and the reaction occurs very rapidly.

13. Controls the duration of inspiration.

14. O_2 is transported by the blood in two form: 1. Dissolved in plasma (3%). 2. In combination with haemoglobin (97%).

15. Pulmonary ventilation is a cycle process by which fresh air enters lungs and equal volume of air is exhaled. It is volume of air taken in or given out during normal tidal respiration.

16. Tidal volume is the volume of air that is taken in or given out during normal quite respiration. It is 500 ml in a normal adult male.

17. The residual volume (RV) is the volume of air that remains in the lungs after forceful expiration. It is 1200 ml in a normal adult male.

18. Functional residual capacity (FRC) is the volume of air remaining in the lungs after normal tidal inspiration. It is equal to addition of RV and ERV (= expiratory reserve volume). It is 2300 ml in a normal adult male.

19. Vital capacity is equal to the maximum volume of air that can be expired after deepest possible respiration. It is the addition of respiratory reserve volume (IRV), tidal volume (TV) and ERV. It is 4600 ml in a normal male adult.

20. When ion in haemoglobin, which is in ferrous (Fe^{2+}) state, is oxidised to ferric (Fe^{3+}) state, methaemoglobin is formed which is incapable of carrying O_2. It is formed in the red blood corpuscles at a slow rate and is reconverted to functional haemoglobin by reducing compounds in red cells during metabolism.

21. Carbon monoxide (CO) gas combines with haemoglobin at the same site where O_2 combines. As haemoglobin has 250 times greater affinity for CO as compared to that for O_2, even, with carbon monoxide pressure 0.4 mm Hg, it competes equally with O_2 for combination with haemoglobin. Over half of haemoglobin binds with CO when its pressure is more than 0.4 mm Hg. This can be lethal as it causes hypoxia.

22. Hypoxia is inadequate oxygen supply to the tissue.

23. 21%.

24. The paranasal sinuses have two major functions: they lighten the skull, which makes up right posture easier, and they offer resonance to the voice. They may also protect the brain during frontal trauma.

25. The secretory cells of respiratory epithelium of nasal chamber produce mucus, important immunoglobulins, inflammatory mediators and interferons which are first line in the host defense.

Fill in the Blanks

1. Carbohydrate **2.** Protoplasmic **3.** Functional residual capacity
4. Tidal volume **5.** Vital capacity **6.** Residual volume, tidal volume
7. Higher, lower.

Multiple Choice Questions

1. (*c*)	**2.** (*c*)	**3.** (*c*)	**4.** (*d*)	**5.** (*c*)	**6.** (*b*)
7. (*d*)	**8.** (*b*)	**9.** (*c*)	**10.** (*a*)	**11.** (*b*)	**12.** (*d*)
13. (*c*)	**14.** (*a*)	**15.** (*c*)	**16.** (*b*)	**17.** (*b*)	**18.** (*c*)
19. (*c*)	**20.** (*b*)	**21.** (*a*)	**22.** (*a*)	**23.** (*c*)	**24.** (*d*)
25. (*b*).					

4

Renal Physiology (Excretion)

4.1 INTRODUCTION

Throwing out of waste product is known as **excretion**. Metabolism (*i.e.*, anabolism and catabolism) produces a wide variety of substances in addition to energy. Some of these matabolic byproducts and endproducts are reutilised for metabolic activities. However, certain byproducts and endproducts of metabolism prove to be toxic and act as poison for the body tissues if retained. Hence, such substances are eliminated from the body through excretion. Thus, *excretion is the separation and elimination process of the metabolic wastes from the body.* The wastes eliminated are called **excretory products**. The organs concerned with excretion are called **excretory organs**.

Excretory Organs in the Animal Kingdom

The excretory organs of various groups of animals have been summarised in Table 4.1.

Table 4.1 Main nitrogens wastes and excretory organs of various animal groups.

	Groups of animal	*Main nitrogenous and other wastes*	*Excretory organs*
1.	Protozoa	Ammonia	General body surface
2.	Platyhelminthes	Ammonia, fatty acids, iron	Protonephridia-flame cells
3.	Polychaetes	Ammonia	Protonephridia-solenocytes (Protonephromixium, see Brusca and Brusca 2002)
4.	Annelids	Ammonia, urea	Metanephridium and chloragogen cells
5.	Insects, arachnids	Uric acid, guanine, xanthine, ammonia (in aquatic forms)	Malpighian tubules
6.	Crustacea	Ammonia	Green glands
7.	Mollusca	Ammonia	Metanephridia or kidney (organ of Bojanus and Kaper's gland)
8.	Cyclostomata (*Myxine*)	Ammonia, urea	Pronephric kidney
9.	Cyclostomata (*Petromyzon*), Fishes and Amphibians	Ammonia, urea, TMO in marine teleosts, creatine and uric acid in freshwater fish	Mesonephric kidney
10.	Reptiles, Birds and Mammals	Urea, uric acid	Metanephric kidney

Various Excretory Products

Wide variety of excretory products excreted by different animals are the following:

1. Urea. It is derived from ammonia through ornithine cycle. It is excreted by semiterrestrial animals which are exposed to scarcity of water, but is not necessarily surrounded by large quantities of water, *e.g.*, earthworms, shore gastropods, amphibians, semiterrestrial turtles and tortoises, elasmobranch fishes and mammals.

2. Uric acids. It is excreted by true terrestrial vertebrates, *e.g.*, insects, terrestrial gastropods, terrestrial reptiles and birds.

3. Amino acids. Amino acids are the end products of protein digestion. Excess of amino acids are excreted as such, *e.g.*, Molluscs (*Unio*, *Limnaea*) and Echinoderms (*Asterias*, *Paracentrotus*).

4. Ammonia. It is formed by the deamination of amino acids. Ammonia (NH_3) is a main excretory product in most of the aquatic animals such as protozoa, polychaetes, crustaceans, molluscs, teleost fishes, aquatic larva of insects, amphibian tadpoles, tortoises and turtles.

5. Trimethylamine oxide (TMO). It is excreted by marine teleosts.

6. Hippuric acids. It is an excretory product of some mammals. This chemical is derived from benzoic acid. When benzoic acid is present in the diet, it combines with glycine to form hippuric acid.

7. Guanine. It is excreted by spiders.

8. Ornithic acids. It is excreted by few birds when benzoic acid is present in the diet.

9. Creatine. Few mammals excrete creatine.

10. Creatinine. Some freshwater teleosts and mammals excrete creatinine. It is derived from creatine.

Besides the above mentioned substances, other excretory products are allantoinic acid, purines and pyrimidines, etc.

4.2 CLASSIFICATION OF ANIMALS ACCORDING TO NITROGENOUS WASTES

In animals the major nitrogen-containing substances (*e.g.*, proteins and nucleic acids) are never completely metabolized to yield free nitrogen. But nitrogen-containing (*i.e.*, nitrogenous) end products such as **ammonia, urea** and **uric acid** are formed (Fig. 4.1). Depending on the chief nitrogenous excretory products, the animals may be divided into three groups, namely ammonotelic, ureotelic and uricotelic animals.

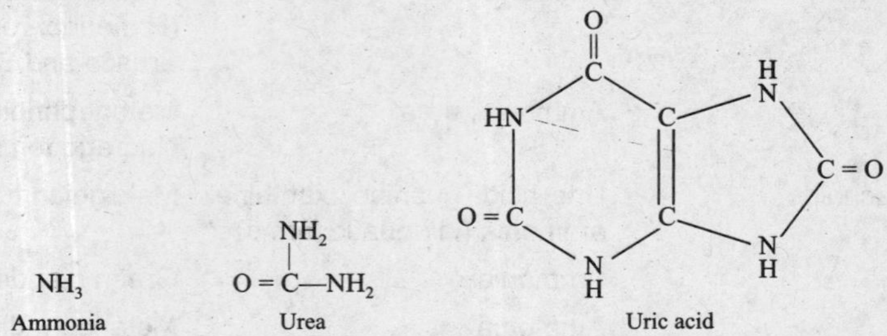

Fig. 4.1. Nitrogen-containing end products of protein metabolism.

1. Ammonotelic animals. The animals that excrete nitrogen in the form of ammonia are called **ammonotelic**. The chief chemical state in which metabolic nitrogenous waste products are excreted by different animals depends upon the given species and the evolutionary adaptations which it has made to the availability of water in its environment. Most animals living in water are ammonotelic, because they can excrete ammonia as such into the surrounding water without converting it to the less toxic form such as urea or uric acid. Thus, most aquatic invertebrates and freshwater fishes are, in general, are ammonotelic as they have an unlimited supply of water at their disposal into which ammonia is rapidly diffused from their bodies. In protozoa, the quantity of ammonia excreted varies with diet, stage in culture cycle and species. Echinoderms are ammonotelic but they also give off considerable amount of amino acids. Polychaete worms are ammonotelic. However, urea and traces of uric aids are reported in *Arenicola*'s coelomic fluid. The cephalopds, *Sepia* and *Octopus*, excrete most of their nitrogen as NH_3 (ammonia) and the pelecypods and some gastropods are also ammonotelic.

Among chordates freshwater fish are ammonotelic, although they also excrete some urea. Many of the marine teleosts are ammonotelic, but also excrete considerable amounts of urea and other products such as **trimethylamine oxide (TMO)**. TMO is a soluble nontoxic substance. However, that ammonotelism is definitely related to the habitat and the mode of life, can be illustrated by citing the example of the African lung fish, *Protopterus*. It produces urea. Thus ammonotelism is very strictly determined by the supply of water.

2. Ureotelic animals. The animals that excrete nitrogen in the form of urea is called **ureotelic**. Terrestrial animals have limited supply of water and cannot get rid of ammonia unless it is converted into more complex and less toxic compounds. In ureotelic animals (*e.g.*, elasmobranchs, amphibians and mammals), urea is formed in liver from the nitrogen of amino acids (by Ornithine cycle) and then excreted by the kidney. Ureotelic animals still require water but in small amounts to dissolve urea so that it may be excreted with the urine. Thus a change from ammonotelic to ureotelic appears to be closely correlated with the supply of water. When water is plentiful, the tadpole of frog is ammonotelic. But as it metamorphoses and leaves water for land, it becomes ureotelic.

Mammals are ureotelic. Their embryonic life is more or less aquatic and the embryonic nitrogenous wastes (*i.e.*, urea) diffuse rapidly into the maternal circulation. However, there is need for conservation of water in adult mammals, and urea may be concentrated by as such as a hundred times by the reabsorption of water in the kidney tubules. According to **Schmit-Nielsen** (1952), in desert animals such as the kangaroo rat (*Dipodomys*) and possibly in other mammals depending on high protein and low water intake, there may be active tubular secretion of urea. Ruminants can use urea as an ultimate protein source, and on a low protein diet a camel or sheep can reduce its urea excretion twenty fold.

3. Uricotelic animals. The animals that excrete nitrogen in the form of uric acid are called **uricotelic**. The excretion of uric acid by uricotelic animals is the last step in the course of the evolution of nitrogen excretion. Since uric acid is almost insoluble in water, its excretion is definitely advantageous over the ureotelism. Terrestrial reptiles and birds are, therefore, uricotelic. In these animals uric acid is excreted by the kidney tubules directly into tubular urine. It precipitates out of solution, leaving the water free to be reabsorbed and is finally released as a white paste. The urine formed in the malpighian tubules of insects is rich in uric acid and there may be a mass of crystals left after absorption of water from the secreted solution. Terrestrial gastropods convert their ammonia into uric acid by their metanephridia or kidney.

All animals that produce uric acid from amino acids deposit eggs outside the maternal organism. Since the embryo must develop on the limited supply of water present in the egg, it would be better if the excretory products are deposited as insoluble substance.

Thus, it is quite clear from the above description of ammonotelic, ureotelic and uricotelic animals, that the nitrogen excretion is a labile character. It has changed during evolution of several groups according to their habitat and *relationship exists between nitrogen excretion and evolutionary adaptations of animals*. In some animals the excretory pattern is changed with the life cycle. In others, changes in nitrogen excretion may be induced by the diet, availability of water, and osmotic stress.

Besides proteins many other compounds contain nitrogen. Among these are the nucleic acids, which in humans give rise to the uric acid that is present in a small amount in the urine. In other mammals, the uric acid is transformed by an enzyme to **allantoin**. The allantoin is further broken down to urea in many amphibians and fishes. However, in aquatic invertebrates ammonia is frequently the end product of the nucleic acid metabolism.

Urea Formation by Ornithine Cycle

During catabolism of proteins, amino acids undergo deamination (in hepatic cells) and ammonia is liberated. Free ammonia is toxic to the body cells and needs to be eliminated from the body. In vertebrates toxic ammonia in converted into nontoxic urea by the liver cells. The mechanism involved in the formation of urea constitutes **urea cycle**, **ornithine cycle** or **Krebs** and **Henseliet cycle** (Fig. 4.2). The urea is formed principally in the liver, then it enters the bloodstream and is eventually excreted by the kidney. The ornithine cycle was described by **Krebs** and **Henseleit** (in 1932).

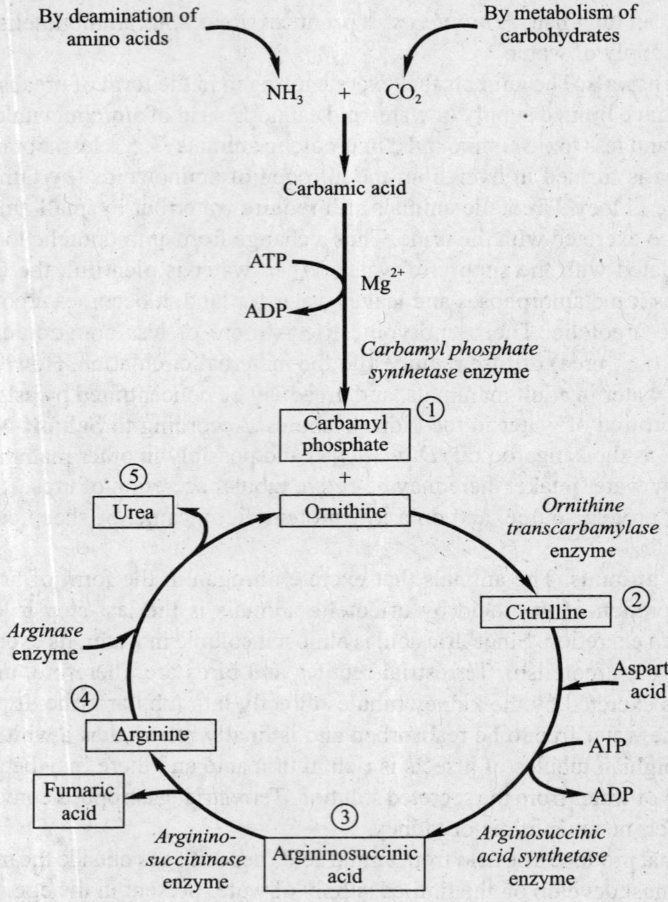

Fig. 4.2. Ornithine cycle.

The first step of the ornithine cycle involves the enzymatic reaction whereby CO_2, NH_3 and ATP, in the presence of a certain derivative of glutamic acids, are converted to the compound **carbamyl phosphate**, ADP and H_3PO_4 (Equation 4.1, Fig. 4.3). The compound in turn reacts with the amino acid **ornithine** in the presence of enzyme *ornithine transcarbamylase* to form the amino acid **citrulline** (Equation 4.3 and 4.4) lead to the formation of the amino acid **arginine** whereby citrulline and aspartic acid react in the presence of ATP first form **arginino-succinic acid** which is then converted to arginine and fumaric acid. Thus, aspartic acid provides one of its amino group to citrulline to form arginine. Some of the arginine may subsequently be used for protein synthesis, whereas some may be broken down by the hydrolyzing enzyme *arginase* to yield urea and one of the starting materials, ornithine, thus completing the urea cycle. In this way two moles of NH_3 (one via aspartic acid and the other via carbamyl phosphate), one CO_2 and three moles of ATP used in the cycle to form one molecule of urea (Box 4.1).

Equation 4.1

$$\text{Ammonia} + CO_2 + 2ATP \xrightarrow[\text{Cofactor (N-acetyl glutamic acid)}]{\textit{Carbamyl phosphate synthetase}} \text{Carbamyl phosphate} + 2ADP + H_3PO_4$$

Equation 4.2

$$\text{Carbamyl phosphate} + \text{Ornithine} \xrightarrow{\textit{Ornithine transcarbamylase}} \text{Citrulline} + H_3PO_4$$

Equation 4.3

$$\text{Citrulline} + \text{Aspartic acid} \xrightarrow{\textit{Arginino succinic synthetase} + Mg^{2+}} \text{Arginino succinic acid} + AMP + PP$$

Equation 4.4

$$\text{Arginino succinic acid} \xrightarrow{\textit{Argino succinic cleavage enzyme}} \text{Arginine} + \text{Fumaric acid}$$

Equation 4.5

$$\text{Arginine} + H_2O \xrightarrow[\text{(in liver)}]{\textit{Arginase}} \text{Ornithine} + \text{Urea}$$

Fig. 4.3. Various chemical reactions of ornithine cycle.

Box 4.1 Consumption of ATP Molecules in Ornithine Cycle

In ornithine cycle three ATP molecules are consumed:

(*i*)	In the formation of active CO_2	= 1ATP
(*ii*)	In the synthesis of carbamyl phosphate	= 1ATP
(*iii*)	In the formation of arginine- succinate	= 1ATP
	Total	= 3ATP

The ornithine cycle operates in the liver of mammals, frogs and turtles. It is absent in telesot livers but it functions in elasmobranchs. Adult uricotelic birds and reptiles too lack in ornithine cycle.

Uric Acid Formation

Hydrolysis of nucleic acids yields purines and pyrimidines. Pyimidines are catabolised to ammonia, bicarbonate and either acetyl CoA or succinyl CoA (**Horton** *et al.*, 2006). Purines are bases which are excreted as such in several animals. But reptiles, birds, insects and other animals which possess the three enzymes, *adenase*, *guanase* (used to convert xanthine into guanine) and *xanthine oxidase* or *xanthine dehydrogenase* are able to covert the purine bases into **hypoxanthine** and **xanthine** and oxidise the latter into uric acid:

$$\text{Hypoxanthine} \xrightarrow[\textit{or Xanthine dehydrogenase}]{\textit{Xanthine oxidase}} \text{Xanthine} \xrightarrow[\textit{or Xanthine dehydrogenase}]{\textit{Xanthine oxidase}} \text{Uric acid}$$

In reptiles, birds and insects, there is no uricase and a large amount of nitrogenous waste is sent out as uric acid (Box 4.2).

Box 4.2

1. Birds, some reptiles and primates (including humans) convert purine nucleotides to uric acid, which is excreted.
2. In birds and reptiles, amino acid catabolism also leads to uric acid.
3. In mammals, surplus nitrogen from amino acid catabolism is disposed of in the form of urea.
4. Birds and reptiles cannot further catabolise uric acid, but many organisms degrade uric acid to other products such as **allantoin** (mammals, turtles, some insects, gastropods), **allantoate** (some bony fishes), **urea** (most fishes, amphibians, freshwater molluscs) and CO_2 plus **ammonia** (crustaceans) (**Horton** *et al.*, 2006).

4.3 FUNCTIONAL ANATOMY OF KIDNEYS

Structure and function are closely linked in the human kidney. Human kidneys are paired organs that lie on the posterior wall of abdomen behind the peritoneum on either side of the vertebral column. In an adult human, each kidney weights between 115 and 170 g and is approximately 11 cm long, 6 cm wide and 3 cm thick.

Outer surface of each kidney is convex and inner concave which faces the vertebral column. A depression called **hilus** is present on the concave side, from where **ureter** takes its origin and joins the urinary bladder backwardly. The renal artery and renal vein pass in and out at the hilus.

Internally each kidney is distinguished into two zones—an outer dark red zone, **cortex** and inner pale red zone, the **medulla**. Ureter entering through the hilus expands, forming a wide funnel-shaped structure, called **pelvis**, which has at its free end a number of cup-like cavities called **calyces** and each cavity is known as **calyx.** Medulla is distinguished into cone-like structures, called **renal pyramids** (Fig. 4.4).

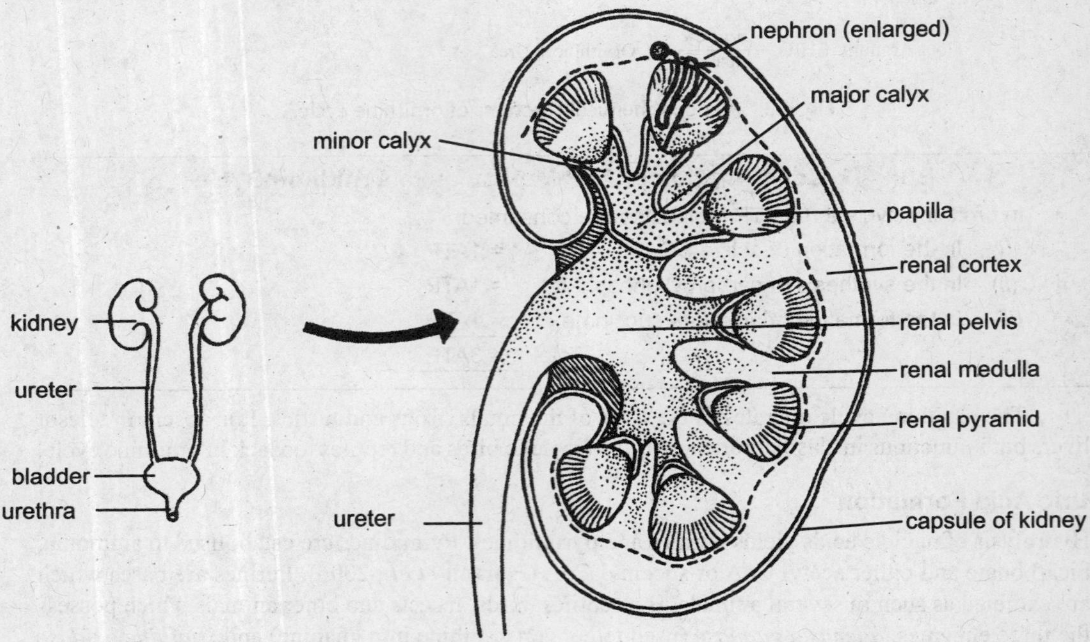

Fig. 4.4. General organisation of the kidneys and the urinary system (after Guyton and Hall 2011).

Ultrastructure of Nephron

The functional unit of the kidney is the **nephron**. Each human kidney contains approximately 1.2 million nephrons, which are hollow tubes composed of single cell layer. Nephrons are of two types: **1. Cortical nephrons.** Glomeruli are present near the surface of kidney. These nephrons constitute about 86% of total nephrons. **2. Juxtamedullary nephrons.** Glomeruli lie at the junction of cortex and medulla of the kidney. These constitute 14% of the total nephrons of a kidney (Table 4.2).

Table 4.2 Differences between cortical nephron and Juxta - medullary nephron.

	Character	Cortical nephron	Juxta - medullary nephron
1.	Position of glomerulus	In the upper part of renal cortex.	In the deeper part of renal cortex, near the cortico-medullary junction

	Character	Cortical nephron	Juxta - medullary nephron
2.	Size of arterioles	Afferent arteriole is broader than efferent arteriole.	Afferent and efferent arterioles have same diameter.
3	Loop of Henele	Descending limb is thin walled while ascending limb is thick walled. It does not extend upto renal papilla.	Both the limbs of loop of Henle are thin walled. It extends upon the tip of renal papilla.
4.	Blood capillaries	Efferent arteriole forms a network of peritubular capillaries.	Efferent arteriole forms loop-like blood capillaries called vasa rectae.
5.	Function	These are involved in urine formation.	These are involved in sodium reabsorption and urine concentration.

The nephron consists of a **renal corpuscle, proximal convoluted tubule (PTC), loop of Henle, distal convoluted tubule (DCT)** and **collecting duct system** (Fig. 4.5). The renal corpuscle consists of glomerular capillaries (or glomerulus) and **Bowman's capsule**. The proximal convoluted tubule (PCT) initially forms several coils, followed by a straight piece that descends toward medulla of kidney. The next segment is the loop of Henle, which is composed of the straight part of the proximal tubule the *descending thin limb* (only in nephrons with long loops of Henle) and the *thick ascending limb*. Near the end of thick ascending limb, the nephron passes between the afferent and efferent arterioles of the same nephron. This short segment of the thick ascending limb is called **macula densa**. The distal tubule begins a short distance beyond the macula densa and extends to the point in the cortex where two or more nephrons join to form a cortical collecting duct. The **cortical collecting duct** enters the medulla and becomes the outer medullary collecting duct.

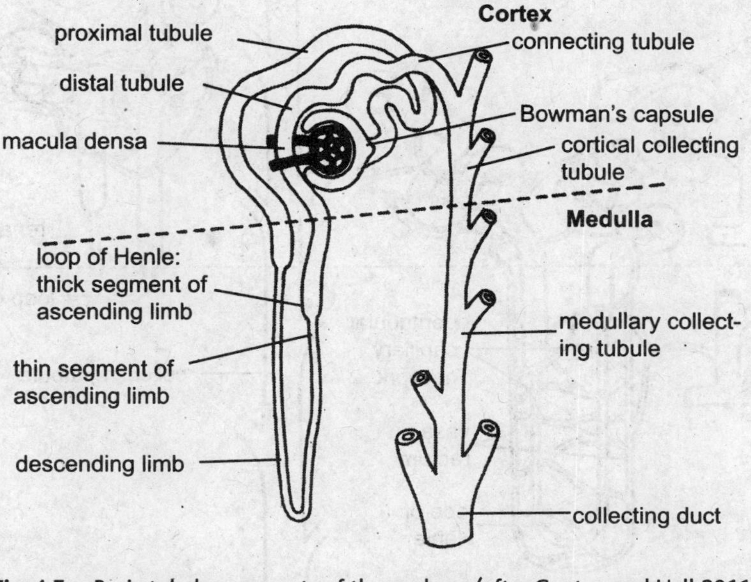

Fig. 4.5. Basic tubular segments of the nephron (after Guyton and Hall 2011).

Blood supply to nephron. Each nephron is supplied with blood by an **afferent arteriole,** a branch of the renal artery that subdivides to form the **glomerulus** (Fig. 4.6). The capillaries converge as they leave the capsule to from an **efferent arteriole**, but then subdivides again into a second network of capillaries, the **peritubular capillaries**. These capillaries are intermingled with the proximal and distal convoluted tubules of nephron.

Additional capillaries extend downward to form the **vasa recta**, the capillary system that serves to loop of Henle. The vasa recta is also a loop, with a descending vessel and an ascending vessel conveying blood in opposite directions.

Although the renal tubule and its surrounding capillaries are closely associated, they do not exchange materials directly across their walls; the tubules and capillaries are immersed in **interstitial fluid**, through which various substances pass back and forth between the plasma within capillaries and the filtrate within the nephron tubules.

Ultrastructure of the renal corpuscle. The first step in the urine formation starts with passive movement of a plasma ultrafiltrate from the glomerular capillaries (*i.e.*, glomerulus) into Bowman's space. The term *ultrafiltration* refers to the passive movement of an essentially protein-free fluid from the glomerular capillaries into Bowman's space. To appreciate the process of ultrafiltration one must understand the anatomy of the renal corpuscle. The glomerulus consists of a network of capillaries supplied by the afferent arteriole and drained by the efferent arteriole (Fig. 4.7). During embryological development, the glomerular capillaries press into the closed end of the proximal tubule to from the Bowman's capsule of renal corpuscle. The capillaries are covered by epithelial cells, called **podocyte** that form the visceral layer of Bowman's capsule. The visceral cells face outward at the vascular pole (*i.e.*, where the afferent and efferent arterioles enter and exit Bowman's capsule) to form the parietal layer of Bowman's capsule. The space between the visceral layer and the parietal layer is **urinary** or **Bowman's space**, which at the urinary pole (*i.e.*, where the proximal tubule joins Bowman's capsule) of the glomerulus becomes the lumen of the proximal tubule.

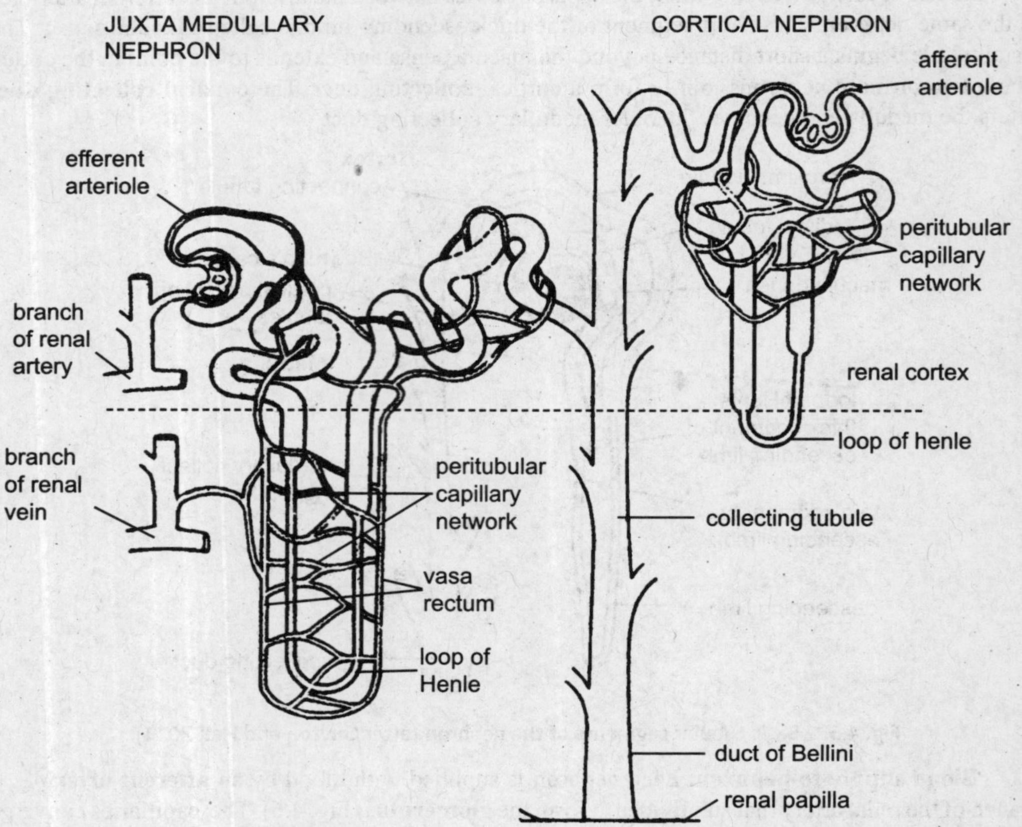

Fig. 4.6. Schematic of relations between blood vessels and tubular structures and juxtamedullary and cortical nephrons.

The endothelial cells of glomerular capillaries are covered by a basement membrane that is surrounded by podocyte (Fig. 4.8). The capillary endothelium, basement membrane and foot processes of podocytes form the so called **filtration barriers** (Fig. 4.9). The endothelium is fenestrated (*i.e.*, contains 700 A° holes, where 1A° = 10^{-10} m) and freely permeable to water, small solutes (such as Na+, urea and glucose), and most proteins but is not permeable to red blood corpuscles, white blood cells or platelets. Because endothelial cells express negatively charged glycoproteins on their surface, they may retard the filtration of very large anionic proteins into Bowman's space. In addition to their role as a barrier to filtration, the endothelial cells synthesize a number of vasoactive substances (*e.g.*, **nitric oxide** (**NO**), a **vasodilator,** and **endothelin I** (**ET-1**), **a vasoconstrictor**) that are important in controlling renal plasma flow (RPF).

The basement membrane, which is a porous matrix of negatively charged proteins, including type **IV collagen**, **laminin**, the proteoglycans **agrin** and **perlecan**, and **fibronectin**, is an important filtration barrier to plasma proteins. The basement membrane is thought to function primarily as a charge-selective filter in which ability of proteins to cross the filter is based on charge.

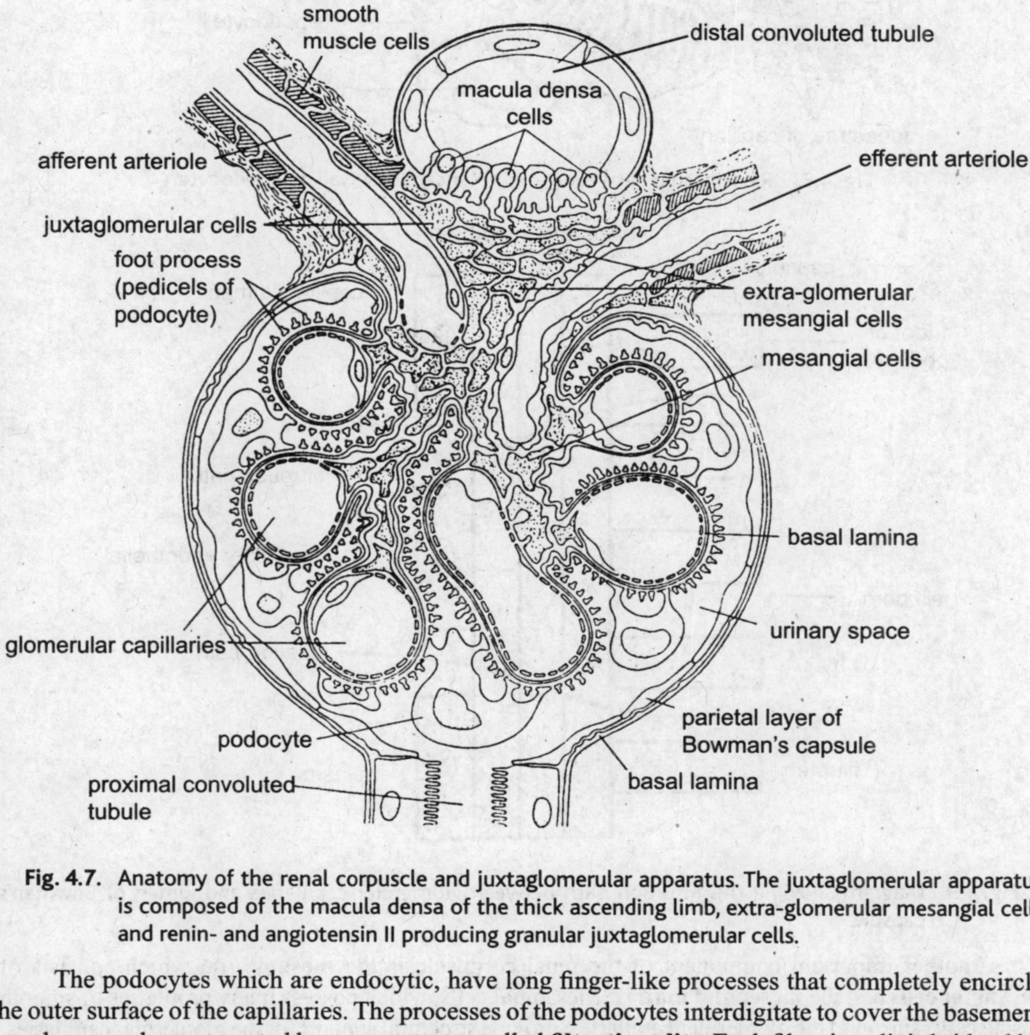

Fig. 4.7. Anatomy of the renal corpuscle and juxtaglomerular apparatus. The juxtaglomerular apparatus is composed of the macula densa of the thick ascending limb, extra-glomerular mesangial cells and renin- and angiotensin II producing granular juxtaglomerular cells.

The podocytes which are endocytic, have long finger-like processes that completely encircle the outer surface of the capillaries. The processes of the podocytes interdigitate to cover the basement membrane and are separated by apparent gaps called **filtration slits**. Each filtration slit is bridged by a thin diaphragm that contains pores with a dimension of 40 × 140A°. The **filtration slit diaphragm,** which appears as a continuous structure when viewed by electron microscopy, is composed of several proteins, including **nephrin (NPHS1), NEPH-1, podosin (NPHS2), α-actinin4 (ACTN4)** and **CD2-AP** (Fig. 4.9; Fig. 4.10). Filtration slits, which function primarily as a size-selective filter, keep the proteins and macromolecules that cross the basement membrane from entering Bowman's space.

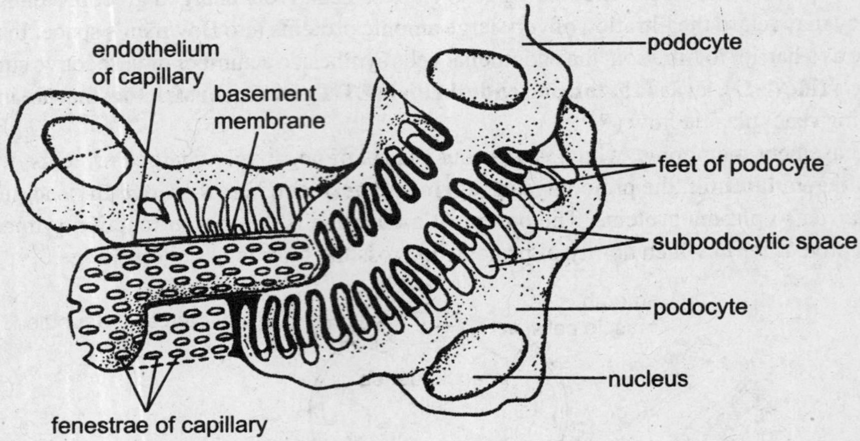

Fig. 4.8. Relationship between glomerular capillary wall and podocyte.

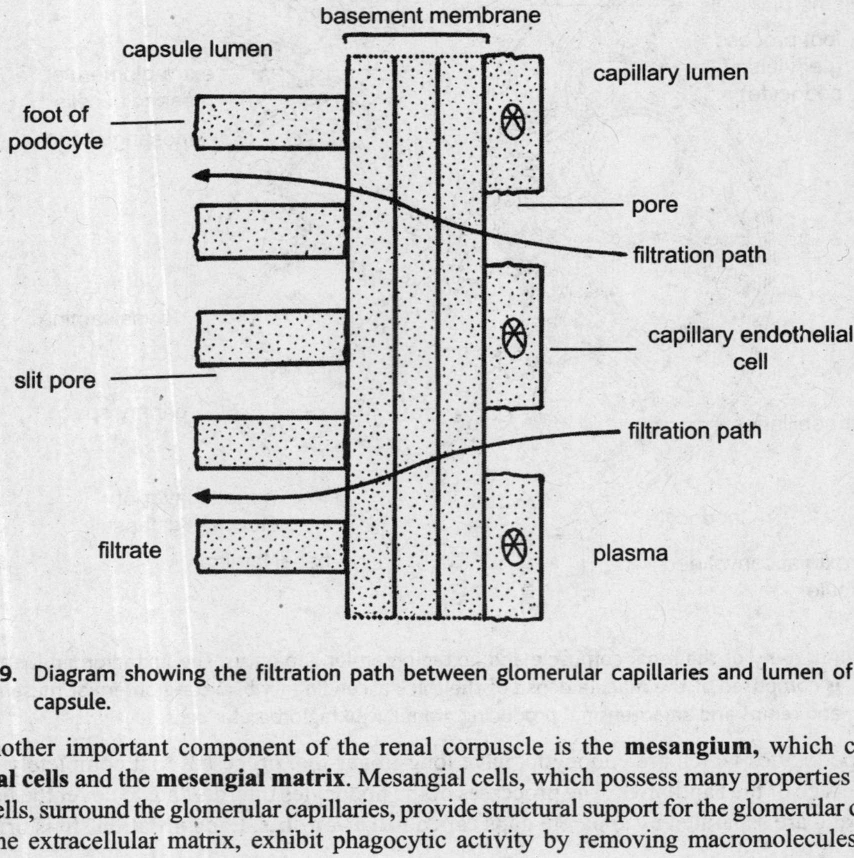

Fig. 4.9. Diagram showing the filtration path between glomerular capillaries and lumen of Bowman's capsule.

Another important component of the renal corpuscle is the **mesangium,** which consists of **mesangial cells** and the **mesengial matrix**. Mesangial cells, which possess many properties of smooth muscle cells, surround the glomerular capillaries, provide structural support for the glomerular capillaries, secrete the extracellular matrix, exhibit phagocytic activity by removing macromolecules from the mesangium, and secrete prostaglandins and proinflammatory cytokines. Because they also contract and are adjacent to glomerular capillaries, mesangial cells may influence the GFR by regulating blood flow through the glomerular capillaries or by altering the capillary surface area. Mesangial cells located outside the glomerulus (between the afferent and efferent arterioles) are called **extraglomerular mesangial cells**.

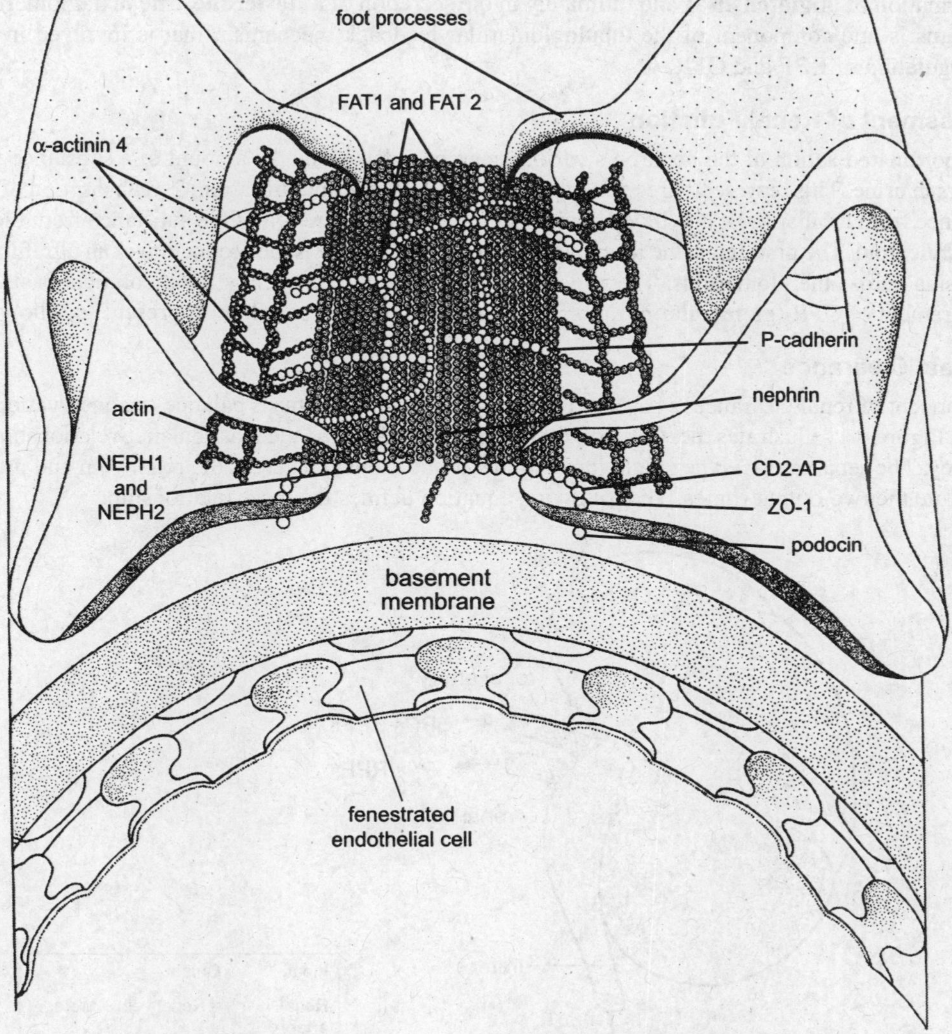

Fig. 4.10. Overview of the major proteins that form the slit diaphragm. Nephrins form opposite foot processes interdigitate in the center of the slit. In the slit, nephrin interacts with NEPH1 and NEPH2, FAT1 and FAT2 and P-cadherin. The intracellular domains of nephrin, NEPH1 and NEPH2 interact with podocin and CD2-AP, which connect these slit diaphragm proteins with ZO-1, α-actinin 4 and actin (after Koeppen and Stanton 2010).

Ultrastructure of the juxtaglomerular apparatus. The juxtaglomerular apparatus is one component of an important feedback mechanism, the tubuloglomerular feedback mechanism. Structures that make up the juxtaglomerular apparatus include the following (Fig. 4.6): 1. The **macula**

densa of the thick ascending limb; 2. Extraglomerular mesangial cells and 3. Renin-and angiotensin II- producing **granular cells** of the afferent arteriole.

The cells of the macula densa represent a morphologically distinct region of the thick ascending limb. This region passes through the angle formed by the afferent and efferent arterioles of the same nephron. The cells of the macula densa contact the extraglomerular mesangial cells and the granular cells of the afferent arterioles are derived from metanephric mesenchymal cells. They contain smooth muscle myofilaments and importantly, they manufacture, store and release **renin**. Renin is involved in the formation of **angiotensin II** and ultimately in the secretion of **aldosterone**. The juxtaglomerular apparatus is one component of the tubuloglomerular feedback mechanism that is involved in the autoregulation of RBF and GFR.

Assessment of Renal Function

The coordinated action of the nephron's various segments determine the amount of a substance that appears in urine. This represent three general processes: 1. glomerular filtration; 2. reabsorption of the substance from tubular back into blood, and 3. (in some cases) secretion of the substance from blood into tubule fluid. The first step in the formation of urine by the kidney is the production of an ultrafiltrate of plasma across the glomerulus. The concept of renal clearance, which is the theoretical basis for measurement of GFR (glomerular filtration rate) and RBF (renal blood flow) is present as follows:

1. Renal Clearance

The concept of renal clearance is based on the **Fick principle** (*i.e.*, mass balance or conservation of mass). Figure 4.11 illustrates the various factors required to describe the mass balance relationship of a kidney. The renal artery is the single input source to the kidney, whereas the renal vein and ureter constitute the two output routes. The following equation defines the mass relationship:

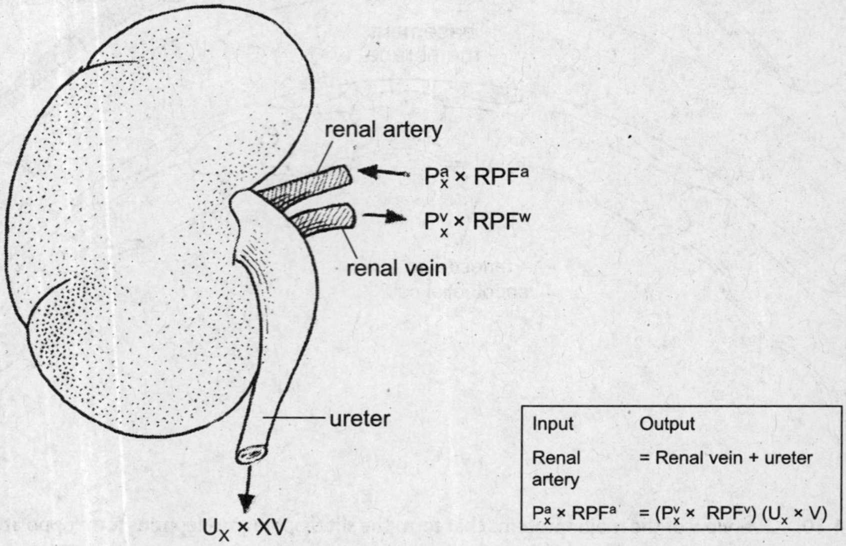

Fig. 4.11. Mass balance relationship for the kidney. See text for defination of symbols.

Equation 4.6

$$P^a_x \times RPF^a = (P^v_x \times RPF^v) + (RPF^v) + (U_x \times V)$$

Where

P^a_x and P^v_x are the concentration of substance X in the renal artery and renal plasma, respectively.

RPF^a and RPF^v are **renal plasma flow** rates in the artery and vein, respectively.

U_x is the concentration of substance X in urine. V is the urine flow.

Glomerular Filtration Rate (GFR)

The GFR is equal to the sum of the filtration rates of all functioning nephrons. Thus, it is an index of kidney function. A fall in GFR generally means that the kidney disease is progressing, whereas recovery generally suggests recuperation.

Creatinine is a byproduct of skeletal muscle creatinine metabolism, and it can be used to measure GFR. Creatinine is freely filtered across the glomerulus into Bowman's space, and to a first approximation, it is not reabsorbed, secreted, or metabolized by the cells of the nephron. Accordingly, the amount of creatinine excreted in urine per minute equals the amount of creatine filtered at the glomerulus each minute.

Equation 4.7

Amount filtered = Amount excreted

$$GFR + Pcr = Ucr \times V$$

Where

Pcr = plasma concentration of creatinine

Ucr = urine concentration of creatinine

V = urine flow

Renal Blood Flow (RBF)

Blood flow through the kidneys serves several important functions, including the following:

1. Indirectly determines the GFR.
2. Modifies the rate of solute and water reabsorption by the proximal tubule.
3. Participates in the concentration and dilution of urine.
4. Delivers O_2, nutrients and hormones to the cells of the nephron and returns CO_2 and reabsorbed fluid and solutes to the general circulation.
5. Deliver substrates for excretion in urine.

Blood flow through any organ may be represented by the following equation:

Equation 4.8

$$Q = \frac{\Delta P}{R}$$

Where

Q = blood flow

ΔP = mean arterial pressure minus venous pressure for that organ

R = resistance to flow through that organ

Accordingly, RBF is equal to the pressure difference between the renal artery and the renal vein divided by renal vascular resistance.

$$RBF = \frac{\text{Aortic pressure} - \text{Renal venous pressure}}{\text{Renal vascular resistance}}$$

The afferent arteriole, efferent arteriole and interlobular artery are the major resistance vessels in the kidneys and thereby determine renal vascular resistance.

Functions of Kidney

The kidney/nephron has the following functions:

1. Excretion of nitrogenous wastes. Kidney/nephron removes urea and uric acid from the blood.

2. Maintenance of water balance (osmoregulation). Kidney/nephron regulates fluid balance in the body by eliminating excess of water.

3. Regulation of pH balance. Kidney/nephron selectively filters out the acidic substances from the blood to maintain correct pH balance (about 7.4).

4. Regulation of salt contents. Kidney/nephron regulates the amount of mineral salts (sodium and potassium ions) in the body. A constant ionic balance is essential for the normal functioning of nerves and muscles, as well as for normal osmotic interactions between cells and interstitial fluid.

5. Elimination of other substances. Kidney/nephron also removes pigments, drugs, hormones (*e.g.*, HCG), poisons and excess vitamins from the blood.

6. Fluid homoeostasis. By the elimination of nitrogenous wastes, excess water, acidic substances and other unwanted materials from the blood, kidney/nephron maintains the volume, composition, pH, and osmotic pressure of the body fluids. In other words, it maintains fluid homoeostasis in the body. Thus, the composition of our body fluids is determined not by what the mouth takes, but by what the kidney/nephron keep.

7. Regulation of blood pressure. The kidney/nephron play a dominant role in regulation of arterial pressure by excreting variable amount of water and sodium. They also contribute to short term arterial pressure regulation by secreting vasoactive factors, such as **renin**, that lead to formation of vasoactive products.

8. Endocrine function of the kidney. Kidney also acts as the endocrine gland. It produces a glycoprotein hormone, called **erythropoietin** which stimulates the formation of red blood corpuscles. Kidney also produces **prostaglandins** which have a variety of effects on renal functions.

Kidney secretes **renin**, an acid protease (enzyme), into the **blood** stream. This enzyme acts on a plasma globulin (angiotensinogen) to form angiotensin I which, by the action of angiotensin-converting enzyme (ACE), is converted to angiotensin II (Fig. 4.12). Angiotensin II acts on adrenal cortex of adrenal gland and causes increased secretion of aldosterone (Fig 4.12). Aldosterone causes increased reabsorption of Na^+ from nephric filtrate to increase the blood volume. This forms a **RAAS axis** for regulation of osmoregulation. Angiotensin II also acts on the brain to increase blood pressure, water intake and secretion of vasopressin and ACTH.

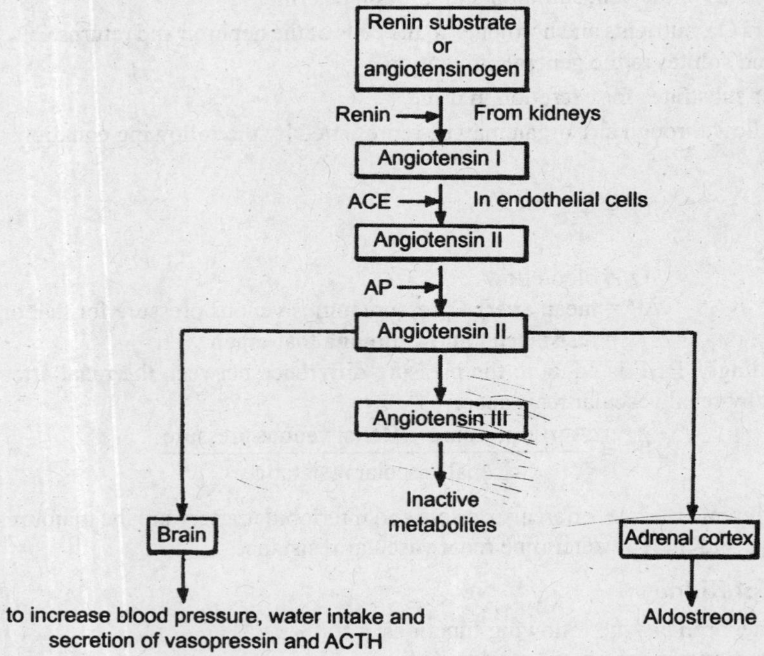

Fig. 4.12. The renin-angiotensin system. AP, aminopeptidase; ACE, angiotensin-converting enzyme.

The kidney converts the vitamin D into its more active form 1,25- dihydroxycholecalciferol which performs important regulatory actions on phosphate and calcium metabolism.

Several **kinins, bradykinin, basylbradykinin** or **kalidin** which are potent vasodilator peptides are also produced in the kidney. They have several important effects on renal function such as on renal circulation as well as on renal sodium and water excretion.

4.4 FORMATION OF URINE

I. Chemical Composition of Urine

The urine is an aqueous solution of organic nitrogenous wastes, inorganic salts and certain other metabolic substances. The various constituents of urine can be classified into following four categories:

1. Nitrogen-containing constituents

1. Urea; 2. Creatinine; 3. Ammonia; 4. Uric acid; 5. Hippuric acid; 6. Allantonin; 7. Indican, indoxyl potassium sulphate; 8. Undertermined nitrogen in various forms.

2. Nitrogen-free organic compounds

1. Carbohydrates and related compounds; 2. Phenolic compounds, *e.g.*, p-hydroxyphenylacetate; 3. Oxalates as oxalic acid; 4. Lactates as lactic acid; 5. Ketone substances and acetone, etc., 6. Glucuronates.

3. Inorganic constituents

1. Chlorides as NaCl; 2. Phosphates as PO_4; 3. Sulphates as SO_4; 4. Potassium; 5. Calcium as CaO; 6. Magnesium as MgO.

4. Water

In addition to the above mentioned substances, a few others are found in traces in the human urine. Among these are the purine bases, guanine, adenine, xanthine and hypoxanthine closely related to uric acid. Amino acids and proteins occur in traces. Several sulphur compounds such as chondroitin, sulphuric acid, thiocyanates, taurine derivatives and ethereal sulphate, etc., also are found in the urine. Under pathological conditions many other substances such as proteins of blood, glucose, acetone, acetoacetic acid, are also present in the urine.

Urine is pale lemon - yellow coloured fluid due to presence of **urochrome** pigment formed from the haemoglobin of dead RBCs in the liver cells. Other pigments found in urine are **urobilin** (gives brown colour) and **uroerythrin**.

The specific gravity of urine varies from 1.003 to 1.030. The pH of normal urine is 6.0. The characteristic odour of the urine is due to the presence of **urinod**. A little amount of ammonia (NH_3) present in the urine is also responsible for the odour of the urine. The chemical composition of urine shows it to be quite different from blood plasma as shown in Table 4.3.

Table 4.3 Showing relative concentration of some components of urine and blood (plasma).

	Component	Concentration	
		Urine	Plasma
1.	Water	97%	91.5%
2.	Protein	0% (normally)	8.0%
3.	Glucose	0% (normally)	0.1%
4.	Sodium chloride and other salts (*i.e.*, PO_4, SO_4, Ca, K, Mg)	0.6	0.85%
5.	Urea	2.0%	0.03%
6.	Creatinine	0.1%	0.001%
7.	Uric acid	0.05%	0.004%

II. Physiology of Urine Formation

The formation of urine involves three processes: 1. Filtration; 2. Reabsorption and 3. Tubular secretion. All these steps take place before the urine reaches the form in which it is eliminated from the body. Briefly, this means that certain fluid containing wastes is separated from the blood into the kidney tubules, that certain substances from this fluid are then returned to the blood by selective absorption and that finally other substances are added to the fluid before it is actually called the urine. The whole process of urine formation (Fig. 4.13) may be expressed mathematically as follows:

$$\text{Urinary excretion rate} = \text{Filtration rate} - \text{Resorption rate} + \text{Secretion}$$

(*i*) **Glomerular filtration.** The first step in the formation of urine is the separation of a cell and protein-free fluid by a simple mechanical process of **pressure filtration** of the circulating blood through the capillary walls of the glomerulus. The filtered substance is called the **capsular** or **glomerular filtrate** and in addition to water, it consists of salts, sugars, urea, uric acid and some other substances. This process of filtration occurs by the force of excess blood pressure in the glomerular capillaries over the osmotic pressure of the blood colloids. The hydrostatic pressure in the glomerular capillary is estimated to be about 70 mm Hg, which is considerably higher than in capillaries elsewhere. The osmotic pressure of the blood is about 28 mm Hg. There is a back pressure, *i.e.*, a hydrostatic pressure, in the capsule and tubule of about 10 mm Hg. Thus, the effective *glomerular filtration pressure* is equal to about 32 mm Hg [70 – (28 + 10)].

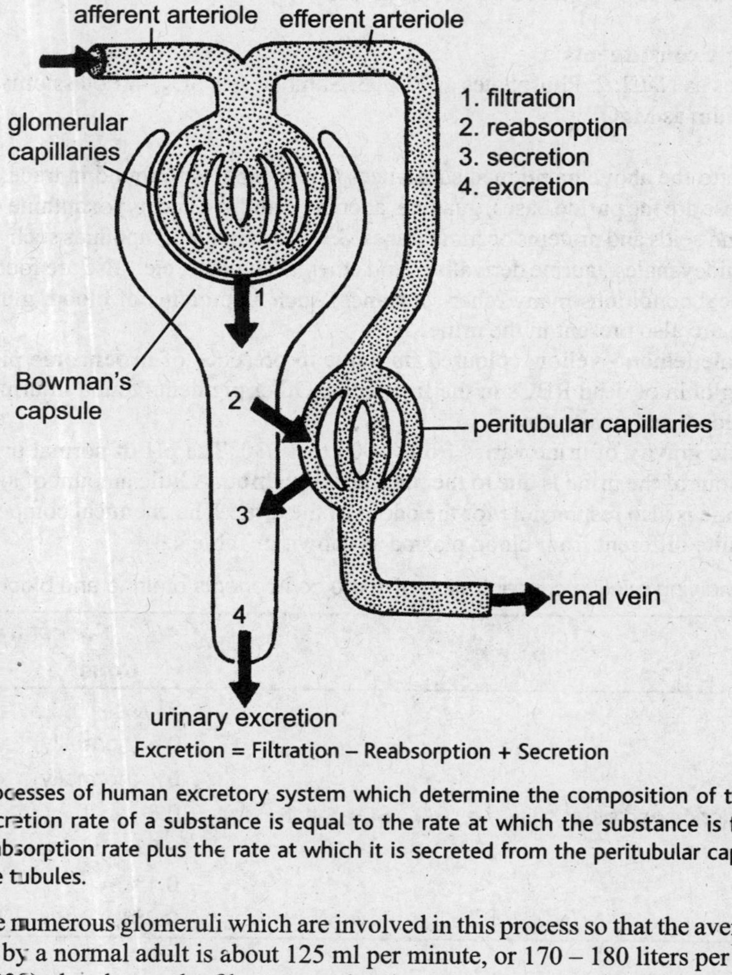

Fig. 4.13. Processes of human excretory system which determine the composition of the urine. Urinary excretion rate of a substance is equal to the rate at which the substance is filtered minus its reabsorption rate plus the rate at which it is secreted from the peritubular capillary blood into the tubules.

There are numerous glomeruli which are involved in this process so that the average glomerular filtrate formed by a normal adult is about 125 ml per minute, or 170 – 180 liters per day. According to **Richard** (1938), the glomerular filtrate contains the same composition of dissolved substances as plasma except for the absence of proteins.

Quantification of the *glomerular filtration rate* (*GFR*) helps in understanding how various substances are handled by the kidney. The GFR in normal human being is about 125 ml/min. Thus, substances whose excretory rates are less than their filtration rates must undergo net tubular reabsorption, whereas substance whose excretory rates are more than their filtration rates must undergo net secretion. The term **clearance** is generally used to quantify the removal of a substance from the blood during the passage through the kidneys. Clerance is defined as the equivalent volume of blood plasma that contains all of a peritubular substance excreted in the urine in

1 minute. It is expressed as $C = UV/P$ where C is clearance, and P is plasma concentration. The clearance of a substance that has a concentration in plasma identical with that in the glomerular filtrate and is neither reabsorbed nor secreted by the tubular epithelium is a measure of the rate of glomerular filtration. The substance that most nearly meets these requirements in **inulin**, a polymer of fructose found in the roots of certain plants. Other materials that have been used are **creatinine** and radioisotopes of **vitamin B_{12}.** All of these yield clearance values of about 100 to 125 per minute per 1.73 m^2 of surface area in human males and somewhat lower value in females.

Any substance that is completely filtered, has a clearance less than that of inulin, and is not bound to plasma protein must be **reabsorbed** as the glomerular filtrate flows through the tubules. Such substances are primarily Na^+, K^+, Cl^-, water, glucose, urea, amino acids and uric acid (Fig. 4.14). In contrast, any substance whose clearance is more than that of inulin must be **secreted** into the urine by tubular cells. NH_4^+, H^+ and N-methyl-nicotinamide behave in this manner as do some other substances such as penicillin, p- amino hippurate (PAH) and phenolsulphonphthalein. The ratio of inulin to PAH clearance is termed as the **filtration fraction** and is approximately 18% in normal individuals.

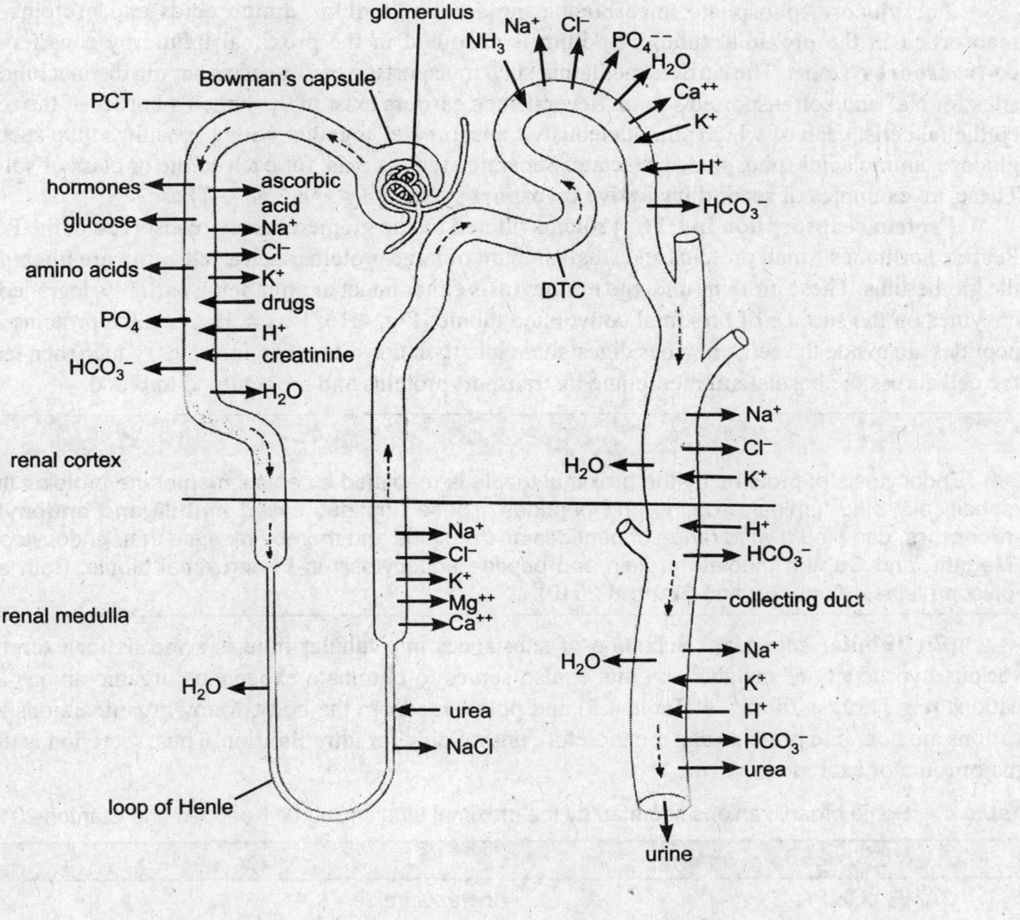

Fig. 4.14. Diagram showing tubular reabsorption and secretion.

Another parameter of renal excretory function is the **transport maximum (Tm)** which is *the maximum ability of the kidneys either to reabsorb or to secrete a given material*. For example, since no glucose is excreted when the plasma glucose concentration is 100 mg/dl, the kidneys must absorb 125 mg of glucose per minute in the proximal tubules. However, at an artificially elevated plasma glucose concentration of 400 mg/dl, that is, a glomerular filtration rate for glucose of 500 mg/minute, 200 mg/minute of glucose may be excreted. The difference (300 mg/minute) is the maximum rate at which the kidney can reabsorb glucose and is denoted as *Tm* for glucose.

***(ii)* Tubular reabsorption.** As the glomerular filtrate passes along the renal tubule from Bowman's capsule, most of its water and some of the dissolved substances are continuously reabsorbed through the walls of the tubule into the blood of the surrounding capillaries. Thus, in passing along the proximal convoluted tubule (PCT) alone, about 85% of water, Na^+, Cl^-, HCO_3^- and glucose of the glomerular filtrate are reabsorbed into the blood. As a result of further **selective reabsorption** in loop of Henle and the distal convoluted tubule (DCT) only about 1% of the original filtrate is left.

The substances that are absorbed in large amount (*e.g.*, water and glucose) are called **high-threshold substances**. Other, such as urea, uric acid and certain salts, that are partly absorbed and partly excreted are called **low-threshold substances**, and still others that are not absorbed at all are known as **non-threshold substances**. An active transport mechanism is found to operate in the reabsorption process.

The glucose, phosphate, bicarbonate, potassium, sodium, amino acids and proteins are reabsorbed in the proximal tubule. Sodium is absorbed in the proximal tubule by a variety of **co-transport systems.** The carriers mediating such transport are envisaged as having distinct binding sites for Na^+ and cotransported solute. Several such carriers exist in the lumen's border of the renal epithelial cells, each of which simultaneously transport Na^+ together with a specific solute such as glucose, amino acids, phosphate or lactate. Separate systems exist for each solute or class of solute. These are examples of **secondary active transport system,** *e.g.*, Na^+, K^+-ATPase.

Protein reabsorption in PTC. Proteins filtered by the glomerulus are reabsorbed in the PTC. Peptide hormones, small proteins and small amount of large proteins such as **albumin** are filtered by the glomerulus. These proteins undergo **endocytosis** either intact or after being partially degraded by enzymes on the surface of proximal convoluted tubule (Fig. 4.13; Box 4.3). Once the proteins and peptides are inside the cell, enzymes digest them into their constituent amino acids, which then leave the cell across the basolateral membrane by transport proteins and are returned to blood.

Box 4.3

Endocytosis of proteins by the proximal tubule is mediated by apical membrane proteins that specifically bind luminal proteins and peptides. These peptides called **multiligand endocytic receptors**, can bind a wide range of peptides and proteins and thereby mediate their endocytosis. **Megalin** and **Cubilin** mediate protein and peptide endocytosis in the proximal tubule. Both are glycoproteins. (**Koeppen** and **Stanton** 2010).

***(iii)* Tubular secretion.** Secretion of substances into tubular fluid is a means for excreting various byproducts of metabolism, and it also serves to eliminate exogenous organic anions and cations (*e.g.*, drugs; Table 4.4; Table 4.5) and pollutants from the body. Many organic anions and cations are bound to proteins and are therefore unavailable for ultrafiltration. Thus, secretion is their major route of excretion in urine.

Table 4.4 Some organic anions secreted by the proximal tubule (Source Koeppen and Stanton 2010).

	Endogenous anions		*Drugs*
1.	CAMP, cGMP	1.	Acetazolamide
2.	Bile salts	2.	Chlorothiazide
3.	Hippurates	3.	Furosemide

	Endogenous anions		*Drugs*
4.	Oxalate	4.	Penicillin
5.	Prostaglandins	5.	Probenecid
6.	Urate	6.	Salicylate (aspirin)
7.	Vitamins: Ascorbate, folate	7.	Hydrochlorothiazide
		8.	Bumetanide
		9.	Nosteroidal anti-inflammatory drugs (NSAIDS)
		10.	Indomethacin

Table 4.5 Some organic cations secreted by the proximal tubule (Source Koeppen and Stanton 2010).

	Endogenous		*Drugs*
1.	Creatinine	1.	Atropine
2.	Dopamine	2.	Isoproterenol
3.	Epinephrine	3.	Cimetidine
4.	Nor-epinephrine	4.	Morphine
		5.	Quinine
		6.	Amiloride
		7.	Procainamide

(*iv*) Loops of Henle and counter-current mechanism. The loop of Henle is a U-shaped segment of the nephron located in the renal medulla. It consists of two limbs: a **descending limb** that starts from the PCT and extends into the medulla; and an **ascending limb** which reenters the cortex and joins the DCT. Each limb has a **thick region** (30 µm) toward the cortex and a **thin region** (12 µm) toward the apices (papillae) of the pyramids. The thick regions are walled by columnar cells and the thin regions by flat cells. The loop of Henle are long in the birds and mammals which pass out **hypertonic** urine (urine more concentrated than blood plasma). The following functions occur in the Henle's loop:

1. The short, wide first part of the descending limb is impermeable to ions, urea and water. It simply transfers the nearly isotonic glomerular filtrate from the PCT to the long, thin second part of the descending limb.

2. The second part of the descending limb is freely permeable to water. Here water is reabsorbed exclusively via water channels called **AQP_1** (Box 4.4). The interstitial fluid around it has a high osmotic pressure due to high concentration of sodium chloride (NaCl) and urea in it. Therefore, water is drawn out of the filtrate by osmosis and carried away by the vasa recta.

Box 4.4

Protein pores called **water channels** or **aquaporins (AQPs)** mediate the transcellular reabsorption of water across many nephron segments. Aquaporins have a narrow pore that permits water molecules to diffuse through the membrane in single file. The pore is too narrow to permit passage of any hydrated ions (**Guyton** and **Hall** 2011).

In 2003, **Dr. Peter Agre** received the Nobel Prize in Chemistry for his discovery that AQPs regulate and facilitate **water transport** across cell membranes, a process essential to all living organisms. To date, 13 aquaporins have been identified. The AQP family is divide into two groups based on their permeability-characteristics. One group (*aquaporins*) is permeable to water, *e.g.*, **AQP0, AQP1, AQP2, AQP4, AQP5, AQP6** and **AQP8**. The other group (*aquaglyceroporins*) is permeable to water and small solutes, especially glycerol, *e.g.*, **AQP3, AQP7, AQP9** and **AQP10**. Aquaporins form tetramers in the plasma membrane of cell, with each subunit forming a water channel in the kidneys. **AQP1** is expressed in the apical and basolateral membranes of the proximal tubule and descending thin limb of Henle's loop (**Koeppen** and **Stanton** 2010). Aquaporins (*e.g.*, aquaporin-2) are not static in position but are altered in different physiological conditions.

3. Exit of water reduces the filtrate volume by 5%, and makes it more concentrated as it reaches the turn of the loop to enter the ascending loop.

4. The ascending limb is impermeable to water along its entire length. Its short, narrow first part is permeable to sodium chloride and urea. NaCl leaves the filtrate by diffusion as it passes away from the more concentrated region of the medulla toward the cortex and the urea enters the filtrate by diffusion. With rudimentary brush-border (– microvilli) and abundant mitochondria. These cells pass the sodium (Na^+) and chloride (Cl^-) ions out of the filtrate by active transport (*i.e.*, by Na^+, K^+-ATPase) into the medulla which becomes concentrated.

5. The filtrate becomes **hypotonic** to plasma due to loss of Na^+ and Cl^- ions. The ascending limb of Henle's loop is called the **diluting segment** (**Koeppen** and **Stanton** 2010).

Countercurrent Mechanism of Urine Concentration

Higher vertebrates (birds and mammals including humans) have evolved a *countercurrent mechanism* to excrete **hypertonic urine** (*i.e.*, urine more concentrated than blood) for conserving body water, so necessary for land life. Henle's loop and capillary loop (*i.e.*, vas rectum) play an important role in this mechanism. Glomerular fluid (filtrate) in the Henle's loop and blood in the vas rectum flow in opposite directions in the two limbs of the loop, toward the renal medulla in the descending limb (Fig. 4.14). These two **counter current systems** help in concentrating the urine. The idea of counter current system was first given by **Kuhn** and **Ryffel** in 1942.

Countercurrent multiplier system of loop of Henle. The flow of filtrate in the descending and the ascending limbs of Henle's loop is in opposite directions, *i.e.*, it is a countercurrent flow. The concentration difference at any given level of the loop between the fluids in the ascending and descending limbs is moderate, but along the length of the loop the difference is additive, *i.e.*, there is a multiplication effect (Fig. 4.15). Consequently, the final concentration of the filtrate at the base of the loop is much greater than at either end. The entire mechanism operating in the loop of Henle is therefore, called a **counter-current multiplier system**.

The exit of sodium chloride (NaCl) from the ascending limb increases the concentration of the interstitial fluid in the medulla of the kidney. The increase concentration of interstitial fluid draws out water by osmosis from the descending limb. This water quickly enters the vasa recta and is carried away. This maintains high concentration of solutes in the interstitial fluid and also in the filtrate. The filtrate flows round the loop and the countercurrent exchange continues. The constant cycling of solutes, both passively and actively, ensures a permanently high concentration of NaCl in the medulla.

Vasa Recta-A countercurrent exchange system. The vasa recta are capillary loops arising from the peritubular capillaries and extending close and parallel to the loops of Henle and collecting ducts into the renal medulla. A vas rectum consists of a narrow **descending capillary** and wider **ascending capillary** that run parallel to each other and give off branched loop at different levels along their length. The cells of the vasa recta are freely permeable to ions, water and urea. As the blood flows in descending capillary toward the renal medulla, water is drawn from the plasma by osmosis due to the progressive increasing concentration of the interstitial fluid, and sodium chloride (NaCl) and urea enters the plasma by diffusion. As the blood flows in the ascending capillary toward the renal cortex, the reverse occurs, *i.e.*, water reenters the plasma and sodium chloride and urea leaves it by diffusion due to progressive decrease in the concentration of the interstitial fluid. The vasa recta, thus, act as a **countercurrent-exchange system.**

The countercurrent-exchange system is advantageous as (*i*) it helps maintain concentration gradient in the renal medulla, along with the loop of Henle; (*ii*) it allows the osmotic concentration of plasma leaving the kidney to remain at a steady state; (*iii*) it, being a passive process, makes no metabolic demand on the kidneys.

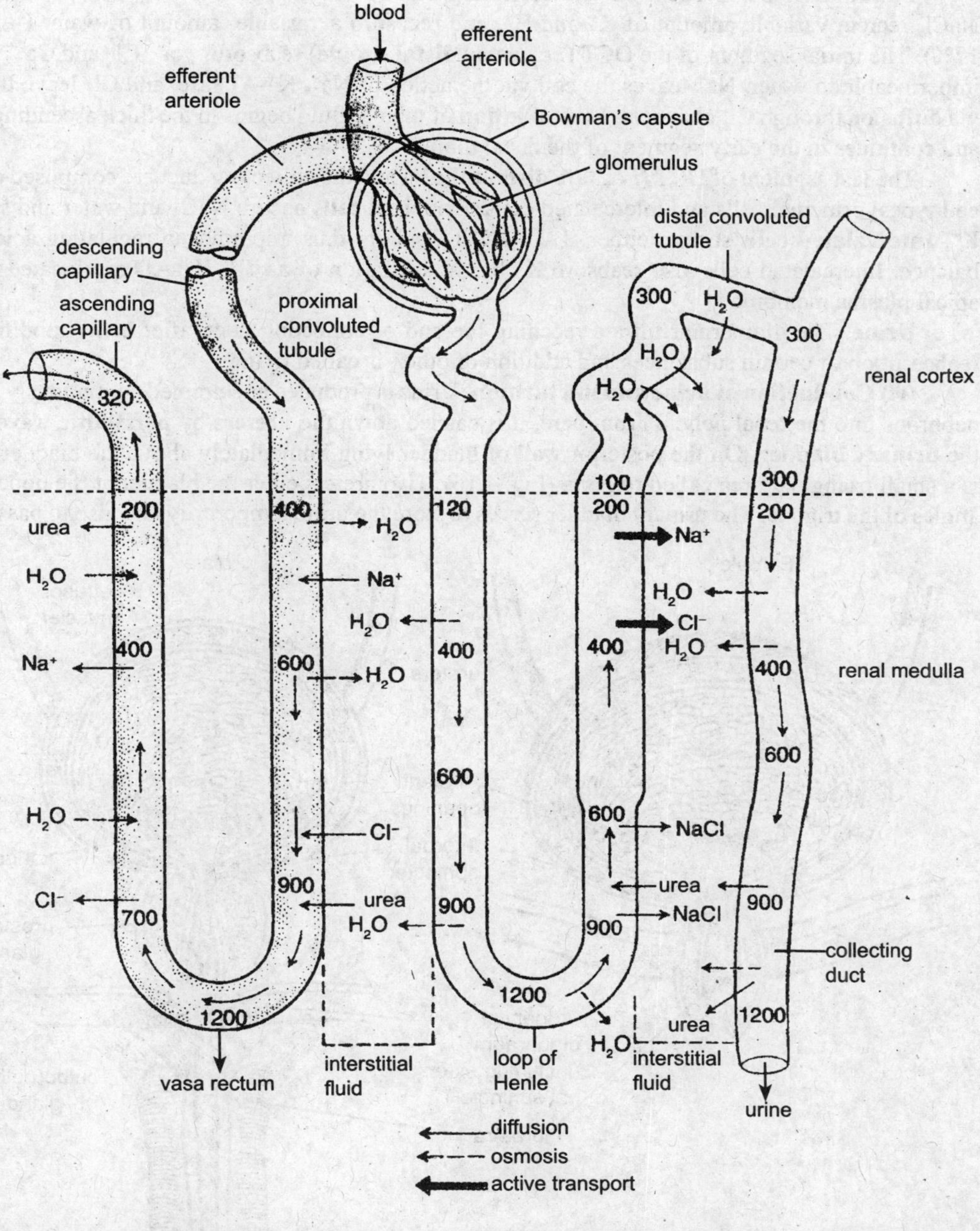

Fig. 4.15. The countercurrent mechanism operates in the loop of Henle and the vasa recta. The numerals refer to the concentration of solutes in million moles per liter.

(*v*) Distal convoluted tubule (DCT) and collecting duct. Like the PCT, the DCT is also greatly twisted and lies in the renal cortex. A short terminal, relatively straight part of the DCT is called **collecting tubule**. The DCT opens into the collecting duct.

The **collecting duct** is a large tube and receives the collecting tubules of many nephrons. It passes into the renal medulla and joins with a few such ducts, forming larger **ducts of Bellini**. The latter run through the medullary pyramids and open into the calyces, which leads into pelvis.

Function of DCT. The DCT and collecting duct reabsorb approximately 8% of the filtered NaCl, recrete variable amount of K^+ and H^+, and reabsorb a variable amount of water (≈ 8% to 17%). The initial segment of the DCT (*i.e.*, **early distal tubule**) reabsorbs Na^+, Cl^- and Ca^{2+} and is impermeable to water. Na^+ leaves the cell via the action of Na^+, K^+-ATPase, and Cl^- leave the cell via diffusion through Cl^- channels. Thus, **dilution** of tubular fluid begins in the thick ascending limb and continues in the early segment of the distal tubule.

The last segment of DCT (*i.e.*, **late distal tubule**) and the collecting duct are composed of two cell types: principal cells and intercalated cells. **Principal cells** absorb NaCl and water and secrete K^+. **Intercalated cells** secrete either H^+ or HCO_3^- and are thus important in regulating acid-base balance. Intercalated cells also reabsorb K^+ by the operation of an H^+, K^+-ATPase located in the apical plasma membrane.

Urine. The glomerular filtrate reaching the end of collecting duct, after being modified by reabsorption of certain substances and addition of other, is called **urine**.

(*vi*) Conduction of urine and micturition. Urine is produced and drained continuously by the nephrons into the renal pelvis. From here, it is carried down the **ureters** by **peristaltic** waves into the **urinary bladder**. On the posterior wall of bladder lying immediately above the bladder neck, is a small triangular area called **trigone** (Fig. 4.16). Two ureters enter the bladder at the uppermost angles of the trigone. The urinary bladder serves to store the urine temporarily and also to pass it out

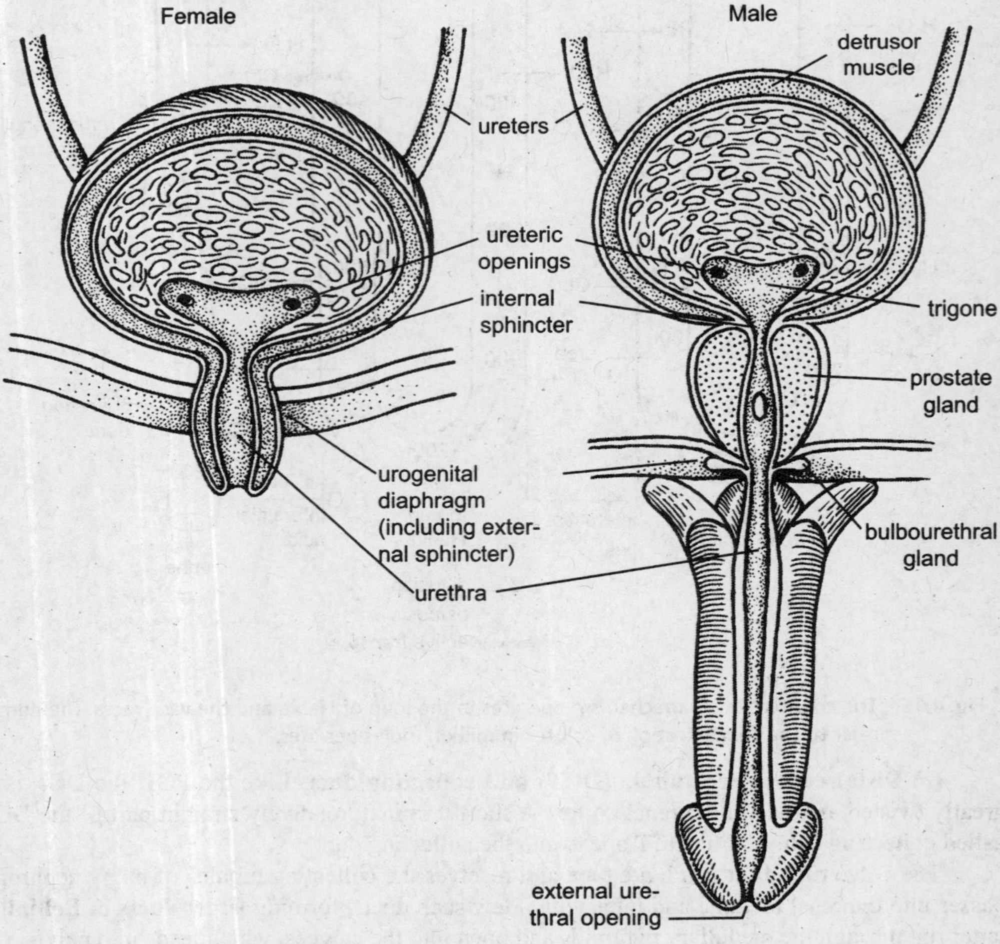

Fig. 4.16. Anatomy of the urinary bladder in human males and females.

at suitable intervals. The process of passing out urine is called **micturition** (L. *micturio* = to desire to make water). As urine collects , the muscular walls of the bladder distend to accommodate it. *Distension* of walls of urinary bladder (which is due to special smooth muscle called **detrusor muscle**; **Guyton** and **Hall** 2010) stimulates the sensory nerve ending in the bladder wall and this set up reflexes, which cause an urge to pass out urine. During the discharge of the urine, the bladder and urethral **sphincters** relax and the smooth muscles of the bladder wall gradually contract. This slowly drives the urine from the bladder through the urethra to the exterior. Reflex of the urine into the ureters is prevented because the terminal part of the ureters pass obliquely through the bladder wall and are consequently closed when the bladder wall contracts around them.

4.5 HORMONAL CONTROL OF URINE FORMATION

The hormones that control urine formation and volume of urine are the following:

1. Aldosterone or mineralocorticoid. This hormone influences and modifies tubular reabsorption of sodium (Na^+) ions. It increases reabsorption of sodium (Na^+) and chloride (Cl^-) but eliminates potassium (K^+) ions. When due to decrease of Na^+ ion concentration, the osmotic concentration of plasma tends to fall, adrenal cortex secretes more of aldosterone which increases reabsorption of sodium from the tubules into the blood. Once osmotic concentration of blood plasma returns to normal, the secretion of aldosterone comes down.

The secretion of aldosterone is under the control of ACTH of pituitary, osmotic concentration of blood plasma and renin-angiotensin mechanism.

2. Renin-angiotensin mechanism. It is activated when podocyte cells of nephron secrete the enzyme renin. Renin catalyses conversion of angiotensin protein of plasma into **angiotensin-II** in presence of converting enzyme. Angiotensin-II increases production of aldosterone and also causes vasoconstriction of systemic arterioles (Fig. 4.12).

3. Vasopressin or antidiuretic hormone (ADH). Permeability of the walls of the distal convoluted tubule (DCT) and collecting duct by ADH from the posterior lobe of pituitary. ADH is released when shortfall of water occurs in the plasma. It increases permeability of the DCT and collecting duct so water is passed from the filtrate into the cortex and medulla and a reduced amount of concentrated urine is produced.

ADH also increases the permeability of the collecting duct to urea which diffuses from the urine into the interstitial fluid in the medulla. Here it increases the osmotic concentration which draws additional water from the thin region of the descending limb of the Henle's loop.

Release of ADH is inhibited when the plasma has enough water in it. Wall of the DCT and collecting duct become impermeable to water, less water is reabsorbed, and abundant dilute urine is produced.

QUESTIONS

Long Answer Question

1. What is excretion? What are important nitrogenous waste products in vertebrates? Where and how is urea formed in mammals. *(Rohilkhand 2000, 2001)*
2. What are ammonotelic, ureotelic and uricotelic animals? Explain giving examples.
3. What is ornithine cycle? Explain how it help in urea formation? *(CCSU, Meerut 2001; Agra 2002)*
4. Review in detail the process of urine formation in a mammal. *(Kanpur 2003)*
5. Write an essay on the physiology of excretion in mammals.
6. Describe the structure of uriniferous tubule. Discuss how glomerular ultrafiltration is brought about. *(CCSU Meerut 2003, 2010)*
7. Describe the process of urine formation in mammal. How do hormones control urine formation. *(Rohilkhand 2002; CCSU Meerut 2013)*

8. Describe the function of mammalian kidney and explain the process of urine formation. *(Kanpur 2003)*

9. Give an account of functions of mammalian kidney. *(CCSU Meerut 2009)*

10. Describe in brief the control of urine formation. *(Allahabad 2001; Bundelkhand 2002; Poorvanchal 2003)*

11. Describe the composition of urine and explain the method of its formation. *(Rohilkhand 2003)*

12. What are the main components of urine? Explain the process of glomerular filtration.

13. Describe urine formation in a vertebrate. How do hormones regulate excretion? *(CCSU Meerut 2006; 2011)*

14. Describe the structural details of mammalian kidney and explain the process of urine formation. *(Osmania 2005)*

15. Describe the process of urine formation in mammals. How do hormones control urine formation? *(CCSU Meerut 2013)*

16. Draw a labelled diagram of a mammalian nephron and discuss the mechanism of urine formation. *(CCSU Meerut 2014)*

17. Describe the role of renin-angiotensinogen system, ANF and ADH in urine formation. *(HPU 2007)*

Short Answer Questions

1. Discuss ultrafiltration. *(Avadh 2004)*

2. Discuss role of glomerulus in excretion.

3. Give countercurrent theory of urine concentration in mammals. *(Avadh 2003; HPU 2006)*

4. Differentiate between ammonotelic, ureotelic and uricotelic animals.

5. Name various nitrogenous wastes produced in animals. Give one example of each.

6. Describe the role of ADH on glomerular filtrate formation. *(Avadh 2002)*

7. Write short notes on the following:

(*i*) Bowman's capsule;

(*ii*) Factors regulating excretion;

(*iii*) Glomerulus;

(*iv*) Ornithine cycle, urea cycle or Krebs-Hensleit cycle; *(Garhwal 2000)*

(*v*) Ultrafiltration;

(*vi*) Ureotelic animals; *(Poorvanchal 2003)*

(*vii*) Formation of urea; *(Kanpur 2004)*

(*viii*) Juxtamedullary nephrons;

(*ix*) Uricotelic animals; *(Agra 2000; Rohilkhand 2003; Purvanchal 2003)*

8. Give an account of function of mammalian kidneys. *(CCSU Meerut 2009)*

9. Describe various types of nitrogenous waste materials. *(MSU Tirunelveli 2007)*

10. Draw a neat and labelled diagram of internal structure of kidney of a mammal. *(CCSU Meerut 2011)*

11. Discuss mechanism of urine formation in man. *(MSU Tirunelveli 2007)*

12. What is selective reabsorption? How and where does it occur? Explain. *(HPU 2009)*

13. Differentiate between cortical nephron and juxta-medullary nephron. *(HPU 2010)*

Very Short Answer Questions

1. Where do the nephrons discharge the urine?

2. Which part of the nephron adds materials to the filtrate?

3. Name the enzyme produced by the kidney to convert angiotensinogen to angiotensin. From where this enzyme comes?
4. Name the accessory excretory organs of vertebrates.
5. What is micturition?
6. What is guano?
7. List the nitrogenous wastes of animals.
8. Name the three groups of animals with regard to their main nitrogenous waste matter.
9. What processes are associated with excretion and osmoregulation?
10. Which specialised cells are associated with ultrafiltration?
11. Where are the slit-pores are found in a kidney?
12. Name the substances which resist ultrafiltration.
13. Give an example of ureotelic animal. (*CCSU Meerut 2010*)
14. Differentiate between ureotelic and uricotelic animals. (*CCSU Meerut 2013*)
15. Give an example of an ammonotelic animal. (*CCSU Meerut 2007*)
16. Give an example of uricotelic animal. (*CCSU Meerut 2011*)
17. What is role of renin produced by kidney cortex? (*HPU 2009*)

Fill in the Blanks

1. Bowman's capsule encloses a bunch of capillaries called ....
2. A unit of structure and function in a kidney is a ....
3. Nephrons are of two types: .... and ....
4. The main nitrogenous waste materials include .... , urea and ....
5. Uric acid is formed by degradation of .... in mammals.
6. Ammonia is converted into urea in ....
7. The hormone ADH increases the absorption of .... from the collecting duct.
8. .... excretion is found in reptiles and birds.
9. Kidney are concerned with the function of .... and ....
10. Blood enters the glomerulus through .... arteriole and leaves glomerulus through its .... arterioles.
11. Two counter-current systems are formed in the kidneys by the .... and the ....
12. Urine formation involves three processes: ultrafiltration, ... and ....
13. A daily output of ultrafiltrate and urine in humans is .... and .... respectively.
14. Wall of human's urinary bladder contains special smooth muscles called ....
15. Two ureters posteriorly opens into a triangular structure called ....
16. Process of passing out urine is called ....
17. Cholesterol is excreted in the .... and waxes in the ....
18. Reabsorption of sodium and water in DCT is controlled by the hormone .... and ....
19. During micturition, the urinary bladder .... and the urethral sphincters ....
20. In the presence of ADH (= antidiuretic hormone) the .... and collecting duct becomes .... to water.
21. Yellow colour of the urine is due to a pigment named ....
22. Ultrafiltration in a nephron is brought about by .... (*HPU 2007*)

Multiple Choice Questions

(*HPU 2006*)

1. In mammalian kidneys, Bowman's capsule (or Malpighian corpuscles) occur in
   (*a*) cortex (*b*) medulla
   (*c*) pelvis (*d*) all of these
2. The vital morphological and physiological units of a mammalian kidneys are
   (*a*) ureters (*b*) nephridia
   (*c*) seminiferous tubules (*d*) uriniferous tubules or nephrons
3. In a mammalian kidney, loop of Henle are situated in
   (*a*) pelvis (*b*) cortex
   (*c*) medulla (*d*) renal columns of Bertini
4. Uricotelism is found in
   (*a*) frogs and toads (*b*) mammals and birds
   (*c*) fishes and fresh water protozoans (*d*) insects, birds and reptiles
5. In glycosuria, urine contains
   (*a*) amino acids (*b*) glucose
   (*c*) inorganic ions (*d*) epithelial cells
6. The major source of ammonia (NH_3) produced by the kidneys is
   (*a*) leucine (*b*) glycine
   (*c*) alanine (*d*) glutamine
7. Ultrafiltration in a nephron is brought about by
   (*a*) active transport (*b*) passive diffusion
   (*c*) blood pressure (*d*) osmotic gradient
8. Active reabsorption of glucose occurs in the
   (*a*) PCT (*b*) loop of Henle
   (*c*) DCT (*d*) collecting duct
9. The normal glomerular filtration rate is about
   (*a*) 25 ml/min (*b*) 50 ml/min
   (*c*) 120 ml/min (*d*) 300 ml/min
10. Most of the glomerular filtrate is reabsorbed in the
    (*a*) proximal convoluted tubule (PCT) (*b*) distal convoluted tubule (DCT)
    (*c*) thick region of loop of Henle (*d*) thin region of loop of Henle
11. Angiotensin stimulates adrenal cortex to secrete aldosterone which increases reabsorption of
    (*a*) sodium (*b*) potassium
    (*c*) calcium (*d*) magnesium
12. The effect of ADH is
    (*a*) reduced water loss by lungs (*b*) increased perspiration
    (*c*) decreased reabsorption of water by kidneys (*d*) increased reabsorption of water by kidneys
13. The moving fluid from the interstitial spaces into the capillaries is due to the
    (*a*) capillary colloidal osmotic pressure (*b*) capillary blood pressure
    (*c*) interstitial fluid hydrostatic pressure (*d*) interstitial fluid colloidal osmotic pressure
14. Effective filtration pressure (EFP) in the glomerulus of kidney of humans is about
    (*a*) 25 mm Hg (*b*) 50 mm Hg
    (*c*) 75 mm Hg (*d*) 80 mm Hg
15. Volume is regulated by
    (*a*) aldosterone (*b*) aldosterone and ADH
    (*c*) aldosterone, testosterone and ADH (*d*) ADH only
16. In humans, the urinary bladder opens into
    (*a*) uterus (*b*) urethra
    (*c*) vestibule (*d*) ureter

17. Excretory product of mammals is (*MSU Tirunelveli 2007*)
(*a*) ammonia (*b*) uric acid
(*c*) amino acid (*d*) urea

18. Excretory product as urea is formed by (*MSU Tirunelveli 2007*)
(*a*) uric acid cycle (*b*) Krebs cycle
(*c*) ornithine cycle (*d*) none of the above

ANSWERS

Very Short Answer Questions

1. Collecting duct.
2. Distal convoluted tubule (DCT).
3. Renin, liver.
4. Skin, lungs, gills, liver, large intestine.
5. Act of passing out urine from urinary bladder.
6. Faecal matter (containing solid excretory waste) of sea birds. Guano is a commercial source of uric acid.
7. Ammonia, urea, uric acid, trimethylamine oxide (TMO), purines, allantoin, creatine, creatinine, hippuric acid.
8. Ammonitelic, ureotelic and uricotelic.
9. Ultrafiltration, tubular reabsorption and tubular secretion.
10. Podocyte.
11. Between the feet of podocytes found in the visceral wall of Bowman's capsule.
12. Blood cells (RBCs, WBCs), platelets and proteins (albumen globulins).
13. Mammals (*e.g.*, rabbit).
14. *Ureotelic animals* are terrestrial animals (mammals) which excrete their metabolic nitrogen-containing wastes in the form of **urea**. Urea requires sufficient amount of water for its excretion. *Uricotelic* animals (*e.g.*, insects, birds and reptiles) are desert animals which have to conserve water. They excrete their nitrogenous waste in the form of crystals of *uric acid* which is insoluble in water.
15. Fresh water fish-*Lebeo*.
16. *Pigeon*.
17. Renin is a proteolytic enzyme of the blood that is produced and secreted by the juxtaglomerular cells of the kidney and hydrolyzes angiotensinogen to angiotensin I.

Fill in the Blanks

1. Glomerulus	**2.** Nephron	**3.** Cortical, juxtaglomerular
4. Ammonia, uric acid	**5.** Nucleic acid	**6.** Liver.
7. Water	**8.** Uricotelic	**9.** Excretion, osmoregulation
10. afferent, efferent	**11.** Henle's loop vasa recta	**12.** Tubular reabsorption, tubular secretion
13. 172 liters, 1 to 1.8 liters	**14.** Detrusor muscle	**15.** trigone
16. Micturition	**17.** Bile, sebum	**18.** Aldosterone, antidiuretic
19. contracts, relax.	**20.** DCT, permeable	**21.** Urochrome
22. Glomerulus.		

Multiple Choice Questions

1. (*a*)	**2.** (*d*)	**3.** (*c*)	**4.** (*d*)	**5.** (*b*)	**6.** (*d*)
7. (*c*)	**8.** (*a*)	**9.** (*c*)	**10.** (*a*)	**11.** (*a*)	**12.** (*c*)
13. (*a*)	**14.** (*a*)	**15.** (*b*)	**16.** (*b*)	**17.** (*d*)	**18.** (*c*).

5

Osmoregulation and Regulation of Acid-Base Balance

Regulation of the homeostasis of water and salt contents of the animal body is called **osmotic regulation** or **osmoregulation**. **Hober** (1902) first coined the term osmoregulation to the collective activity of the variety of mechanisms used by organisms to control water movement and water volumes. This implies the maintenance of internal osmotic concentration which is different from the surrounding medium. However, it is not sufficient if the *water volume* is regulated. The *internal ionic concentration* is also to be regulated as large differences exist in the external ionic concentrations as well as the ionic concentration in the different portions of the body of the animals. The various processes by which the internal ionic concentration of the body are maintained fall under **ionic regulation**. In most animals the osmoregulatory mechanism includes both of them because the regulation of one (water) always affect the other (salt).

Study of osmoregulation is a wide field. Various overall mechanisms regulating the internal water and solute levels, the excretory organs, cellular activities basic to these processes (osmosis, diffusion, active transport, etc.) and the molecular basis for these mechanisms are all included in the phenomenon of osmoregulation. Osmoregulation is also closely related to the other homeostatic functions such as pH and temperature regulation in which body water and ionic concentrations are involved.

If the water content of the body is to be maintained at a constant level, the amount of water entering the body must equal the water which leaves the body. Water enters the animal body in a number of ways, such as, by drinking along with food, by oxidative reactions in the body resulting in metabolic water or by osmosis from the surrounding medium. The **osmosis** is caused due to concentration gradients across the semipermeable membrane of the animal. The exit of water also occurs through several routes such as through urine, respiration, sweat, through the skin by osmosis and by exocytosis.

The process of osmoregulation is not a passive mechanical equilibrium. It results due to active and negative feedback mechanisms which control ionic and water movements.

5.1 OSMOTIC PRESSURE AND TONICITY

Movement of water across cell membranes occurs by the process of **osmosis**. The movement of water is passive, with the driving force for this movement being the osmotic pressure difference across the cell (plasma) membrane. **Osmotic pressure** is determined solely by the number of solute molecules in that solution. It is not dependent on such factors as the size of the molecules, their mass, or their chemical nature (*e.g.*, valence). Osmotic pressure (π), measured in atmosphere (atm) is calculated by **Vant Hoff's Law** as:

$$\pi = nCRT \qquad \textbf{Equation 5.1}$$

where, n = number of dissociable particles per molecule; C = total solute concentration; R = gas constant; T = temperature in degrees Kelvin.

For a molecule that does not dissociate in water, such as glucose or urea, a solution containing 1 m mol/L of these solutes at 37°C can exert an osmotic pressure of 2.54×10^{-2} atm as calculated by Equation 5.1 with the following values:

$$n = 1; C = 0.001 \text{ mol/L}; R = 0.082 \text{ atmL/mol °K}; T = 310°K$$

Because 1 atm equals 760 mm Hg at sea level, π for this solution can also be expressed as 19.3 mm Hg. Alternatively, osmotic pressure can be expressed in terms of osmolarity. Thus, regardless of the type of molecules, a solution containing 1 m mol/L of solute exerts an osmotic pressure of 1 m Osm/L.

For molecules that dissociate in a solution, n in Equation 5.1 will have a value other than 1. For example, a 150 m mol/L solution of NaCl has an osmolarity of approximately 300 m Osm/L because each molecule of NaCl dissociate into an Na^+ and Cl^- ion (*i.e.*, n = 2). (*Note.* NaCl does not completely dissociate in water. The value for n is 1.88 rather than 2. However, for simplicity the value of 2 is most often used). If dissociation of a molecule into its component ions is not complete, n will not be integer. Accordingly, the osmolarity of any solution can be calculated as:

$$\text{Osmolarity} = \text{Concentration} \times \text{Number of dissociable particles}$$

$$\text{m Osm/L} = \text{m mol/L} \times \text{number of particles/molecules}$$

Osmolarity versus Osmolality

Osmolarity refers to the osmotic pressure generated by the dissolved solute molecules in 1L of solvent, where as **osmolality** is the osmotic pressure generated by the solute molecules dissolved in 1kg of solvent. For dilute solutions, the difference between osmolarity and osmolality is insignificant. Measurements of osmolarity are dependent on temperature because the volume of solvent varies with temperture (*i.e.*, the volume is larger at higher temperature). In contrast, osmolality which is based on the mass of the solvent, osmolality is expressed as millionsmoles per kilogram water (m Oam/kg H_2O) (**Koeppen** and **Stanton** 2010).

Isotonic, Hypotonic and Hypertonic Animals

Animals whose body fluids have the same concentration as that of their surrounding medium (**isotonic** or **isosmotic animals**) never face the problem of osmoregulation as long as they live in such a medium. For example, most marine animals are isotonic, a they have their body fluids isotonic with sea water.

Animals which live in a medium of lower salt concentration (**hypotonic** or **hyposmotic animals**) have to face the difficulty of hydration and hence they have evolved special mechanisms to get rid of the excess water that usually enters their body and dilutes their body fluids. For example, freshwater protozoans and crustaceans are hypotonic and eliminates the excess water by their contractile vacuoles and excretory organs respectively.

In a similar way animals which live in a medium of highest salt concentration (**hypertonic** or **hyperosmotic animals**) have to evolve regulatory mechanisms by which excessive loss of water from the body is prevented. Since these animals (*e.g.*, bony fishes of sea) are in constant danger of dehydration, they compensate for the water loss by drinking the surrounding water or by other means.

5.2 MECHANISM OF OSMOREGULATION

Osmoregulation is an energy consuming process. As long as an animal remains alive, osmoregulation goes on. Once animals dies, osmoregulation stops.

Animals live in different types of environment; and they may live in freshwater, sea water or on land. There are various mechanisms in different animals for osmoregulation. They carry out one or more of the following mechanisms for osmoregulation.

1. Removal of Excess of Salts

When the marine animals drink sea water (rich in salts) to compensate their water loss, the salt content of the body increases. The excess of salt is removed from the body by any one of the following methods:

(*i*) The marine forms, *i.e.*, the fishes have **chloride secretory cells** in their gills. They secrete out the excess of salts present in the body fluid.

(*ii*) In the case of marine turtles and marine birds **salt glands** present in the head secrete out excess of salts.

2. Removal of Excess Water

In the case of freshwater animals, the body fluid is hypertonic (having more number of solutes) and the freshwater is hypotonic (having less number of solutes). Hence endosmosis takes place. The freshwater animals have the following devices to deal with excess of water entering the body of the animal.

(*i*) *Amoeba* removes the excess of water by contractile vacuole.

(*ii*) Crustaceans remove excess of water by the green glands or antennal glands (*e.g.*, *Astacus*).

(*iii*) Platyhelminthes (*e.g.*, *Gunda ulva*) stores the water in special kind of cells called **palisade parenchyma**.

(*iv*) Freshwater fishes removes excess of water by **glomerular kidney**.

3. Compensation of Salt Loss

When freshwater forms expel the excess of water, some amount of salt from the body is also lost. This lost salt is recovered by the following methods:

(*a*) Crustaceans and freshwater fishes have special kind of cells on the gills, called **chloride cells**. They absorb salts from the freshwater and add it to the body fluid.

(*b*) The kidney of the freshwater fishes is capable of reabsorbing some salts from the urine.

4. Compensation of Water Loss

In the case of marine animals, the body fluid is actually **hypotonic** and the sea water is **hypertonic**. Hence exosmosis occurs in the marine animals. As a result the marine animals are dehydrated. The loss of water is compensated by drinking sea water and by getting water from marine food. This is the reason that is why marine animals drink sea water.

The land animals are dehydrated by evaporation. They compensate the water loss by drinking water and by getting water from food and metabolism.

5.3 CATEGORIES OF ANIMALS ON THE BASIS OF OSMOREGULATION

There are two extreme patterns of response to osmotic conditions of the environment. Animals may be osmotically labile (dependent) and their body fluid concentration may change with the medium. These animals are called **osmoconformers**. Other animals are osmotically stable (independent) and when the medium changes, the internal concentration of the body fluid remains constant or unchanged. These animals are called **osmoregulators**. The term *poikilosmotic* and *homoisometic* are often applied respectively to osmoconforming and osmoregulating animals.

Euryhaline and Stenohaline Animals

In poikilosmotic animals, osmotic concentration of body fluid varies according to the osmotic concentration of the surrounding medium. They are of following two types:

1. Euryhaline animals. Animals which can tolerate a wide range of osmotic concentration (salinity), are called **euryhaline animals**. For example, *Mytilus* (mollusc) when left in sea water diluted to a great extent, can adjust the salinity of its haemolymph to exactly the same concentration and lives unaffected. Similarly *Aplysia* left in 75% sea water increases in weight by 20% in 2 hours, but loses salts and water in next 4 hours. In this way its original volume is regained and it is able to live with this diluted blood in the changed water content.

2. Stenohaline animals. Animals which can not tolerate wide range in salinity are known as **stenohaline animals**. Virtually all truly marine invertebrates are stenohaline and their body fluids are nearly isosmotic with their surrounding (Table 5.1). For example, a marine crustacean, spider-crab (*Maia*), if left in 80% sea water increases in its weight up to a certain extent due to the quick entry of water into its body. The excretory organs of the animal activate to get rid of excess water in the form of dilute urine. However, along with the excreted water some salts also escape out, making the body fluid more dilute and as a consequence the animal dies in 2 to 4 hours.

Examples of Osmoregulations

1. Osmoregulation in freshwater animals. Freshwater has salt concentration from 0.01 to 0.5 ppt, *i.e.*, usually less than 1% of sea water. Osmotic concentrations of some freshwater animals have been shown in Table 5.2.

Table 5.1 Osmotic concentration of body fluids of marine animals (after Willmer *et al.*, 2000).

	Genus (Phylum)	Osmotic Concentration (mOsm)
1.	*Aurelia* (Cnidaria)	1050
2.	*Mytilus* (Mollusca)	1148
3.	*Sepia* (Mollusca)	1160
4.	*Neanthes* (Annelida)	1108
5.	*Arenicola* (Annelida)	1120
6.	*Carcinus* (Crustacea)	1100
7.	*Echinus* (Echinodermata)	1065
8.	*Scyllium* (Elasmobranch fish)	1075
9.	*Squalus* (Elasmobranch fish)	1096
10.	*Gadus* (Teleost fish)	308
11.	*Paralichthys* (Teleost fish)	337
12.	Sea water	1000–1050

Table 5.2 Osmotic concentration of extracellular fluids of some freshwater animals (after Willmer *et al.*, 2000).

	Genus (animals group)	Osmotic concentration (mOsm)
1.	*Spongilla* (Sponges)	55
2.	*Chlorohydra* (Cnidarian)	45
3.	*Asplanchna* (Rotifer)	80
4.	*Anodonta* (Mollusc)	66
5.	*Pomacea* (Mollusc)	139
6.	*Pheretima* (Annelid)	152
7.	*Hirudo* (Annelid)	200
8.	*Lumbricus* (Annelid)	300
9.	*Astacus* (Crustacean)	477
10.	*Potamon* (Crustacean)	522
11.	*Aedes* (Insecta)	266
12.	*Ephemera* (Insecta)	237
13.	*Rana* (Chordata)	210
14.	*Carassius* (Chordata)	280
15.	*Alligator* (Chordata)	278
16.	*Anas* (Chordata)	294

All freshwater invertebrates, fishes and amphibians are *hypertonic* to the surrounding water. Since they are never completely impermeable, they have steady inflow of water and some loss of salt (Fig. 5.1). They eliminate excess amount of water as urine and by elaborating copious urine of very low concentration (hypotonic), they keep salt loss in the kidney to a minimum. However, some salt is lost and must be made up by food or by active absorption from the very dilute water. Numerous freshwater invertebrates can absorb salt actively and probably all of them have some salt absorbing mechanism.

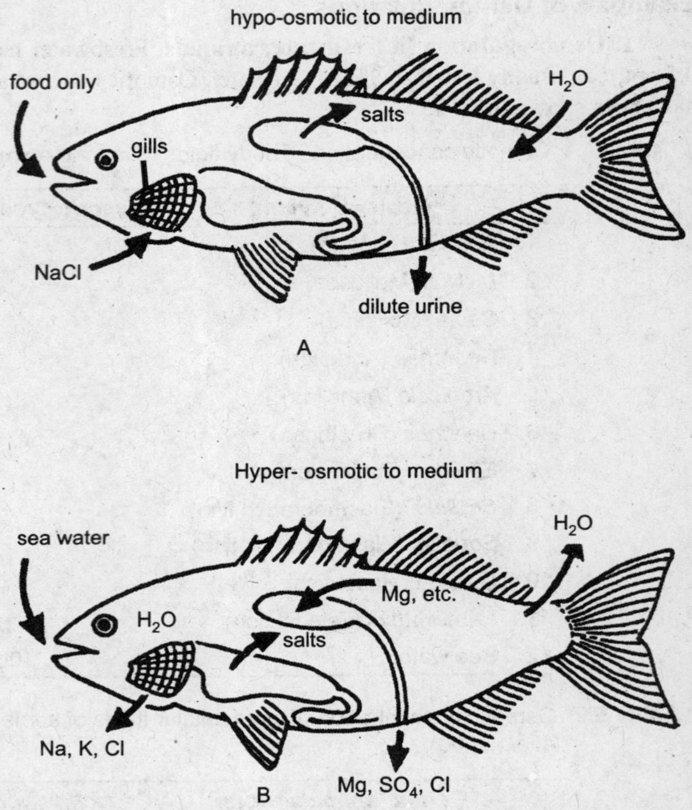

Fig. 5.1. The pathways of ions and water movement in osmoregulation. A—Freshwater fishes; B—Marine fishes.

Freshwater protozoa (*Amoeba*, *Paramecium*) living in *hypotonic* medium, eliminate excess water by their *contractile vacuoles*. The pulsation rate of the vacuole increases when more water enter into the body. In freshwater crustaceans the excess water is eliminated by their excretory organs, the *antennary* or *green glands*. The small amount of salt lost with the dilute urine is restored by active absorption of salts by the gills from the freshwater. Freshwater molluscs, *e.g.*, *Unio* and *Anodonta* also excrete a large amount of dilute urine and absorb salts to make up the loss. Similarly, in freshwater fishes, the water that enters the body is eliminated by *kidney*. The urine is quite low in salt and is highly *hypotonic*, as it is in most freshwater animals. The salts loss is balanced by active absorption (uptake) of salt in the gills.

Aquatic insects vary considerably in the permeability of the cuticle to water. Gills are more permeable than body surfaces, and larvae and nymphs tend to be more permeable than adults. Aquatic insects in air tend to loss more water by evaporation than do terrestrial insects. Larvae of aquatic insects conserve to extract salts not with their gills but with their guts. Either the rectal cuticle or special **anal papillae** developed by some larvae, such as *Aedes aegypti*, can absorb salts and the gut wall is histologically differentiated in these areas. In *A. aegypti*, the anal papillae are responsible for about 90% of salt conservation.

The freshwater crayfish, *Astacus*, secretes urine equal to about 4% of its body weight in 24 hr., osmotic concentration of which is about one-fifth that of its blood. The antennary organ of *Astacus* has a long canal between labyrinth and bladder (Fig. 5.2). The canal functions as a water-secreting segment. Salt loss is compensated for by the uptake of salts from the medium, even when the external salt concentration is very low, Cl^-, Na^+ and K^+ all being absorbed by the gills. *Astacus* can extract Na^+ from water containing as little as 0.02 to 0.09 mM/liter, and Cl^- from a medium which is hardly less dilute.

Amphibians are very similar to freshwater fish (*i.e.*, the body fluid is hypertonic and the surrounding freshwater is hypotenic), although the adult animals have no gills and the absorption of salts takes place from the whole slimy skin surface. The common frog can extract Cl^- from a medium containing only 0.01 mM/liter. In amphibians, resorption of salts takes place from urine by uriniferous tubules. As they have a low salt concentration, they have not been able to adapt to a marine life, for in salt water they would become dehydrated and salt would enter the body.

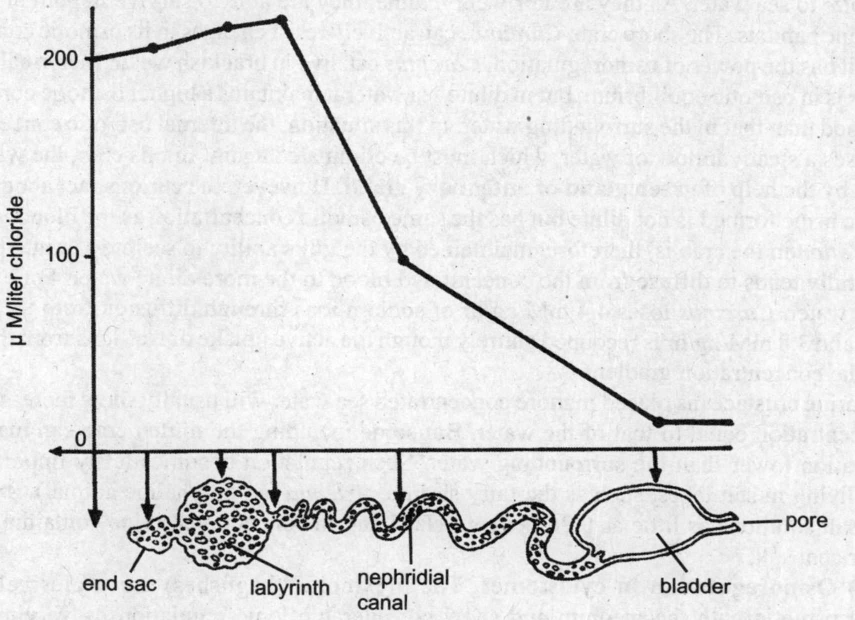

Fig. 5.2. Antennal gland of *Astacus (a decapod crustacean)*. Graph represents chloride concentration in different segments. Salt concentration decreases drastically towards the distal portion of tubule and bladder showing reabsorption of salts.

2. Osmoregulation in brackishwater animals. Brackishwater is generally defined as dilute sea water that ranges in concentration from 1.5% to 90% of that of pure sea water. Most brackishwater contain both marine animals which can tolerate low salinity, and true brackishwater animals which are not found in the sea. Typical marine animals in brackhishwater include the shore crab *Carcinus maenas* and the mussel *Mytilus edulis*. These marine forms tend to disappear from the brackishwater fauna when the water becomes equivalent to about 15% sea water concentration. An example of a true brackish water animal is the prawn *Palaeomonetes varians*, which can live in water ranging from 0.5% to 100% sea water, but does not in fact live in genuinely marine habitat.

In the planarian *Procerodes ulvae* which lives in *tidal estuaries* and areas of the shore in which freshwater enters the sea, a different pattern of osmoregulation is exhibited. If *Procerodus* is placed in sea water that is progressively diluted, the animal swells as it is permeable to water. But eventually, it adjusts its body volume to its previous level. By chemical analysis of the animal, the adjustments of its body volume can be shown to be achieved by simply secreting ions outwards. In freshwater, *Procerodus* has an internal salt concentration that is only one-tenth of that in sea water. The limitations of such a method are clear and the fact, that the blood concentration reached in fresh water is dangerously low is illustrated by the observation that *Procerodus* cannot live indefinitely in fresh water. Such sort of adjustments is only suitable for an animal living in tidal conditions.

3. Osmoregulation in marine animals. Most invertebrates living in the sea have an osmotic concentration in their body fluids which equals that of the surrounding sea water but they can regulate their ionic composition. When a marine invertebrate is placed in dilute sea water, the water enters the animal due to the higher osmotic concentration in the animal. But after some time a state of osmotic equilibrium is attained with the new environment. Most animals that are exclusively marine can tolerate only a limited change in the concentration of their medium. If the medium becomes too dilute, they would die. *Octopus* is *hyperosmotic* to sea water at about 1300 m Osm, with a continuous need to produce a slightly dilute urine and to absorb salt from the imbibed sea water.

(*i*) Osmoregulation in crustaceans. Some palaemonid shrimps and grapsid crabs are always *hyposmotic* to sea water. As they are fairly euryhaline, they are able to survive in both brackish and hypersaline habitats. The shore crab, *Carcinus*, can actively resist changes in its osmotic concentration because it has the power of osmoregulation. *Carcinus* can live in brackish water. In normal sea water, *Carcinus* is in osmotic equilibrium, but in dilute sea water it maintains a higher osmotic concentration in the blood than that in the surrounding water. In this situation, the internal osmotic concentration in crab causes a steady inflow of water, which must be eliminated again. In this crab, the water can be excreted by the help of **green gland** or **antennary gland**. However, the curious fact about *Carcinus* is that the urine formed is not dilute but has the same osmotic concentration as the blood. A *high salt concentration* in the crab is, therefore, maintained by the gill's ability to secrete salt into blood. The salt naturally tends to diffuse from the concentrated blood to the more dilute water. For example, in 40% sea water, *Carcinus* loses 4.4 mM/kg/hr of sodium ions through diffusion from its permeable surfaces and 3.8 mM/kg/hr is recouped entirely though the active uptake of Na^+ ions from the medium against the concentration gradient.

Marine crustaceans placed in more concentrated sea water will usually show increased blood's salt concentration equal to that of the water. But some including the mitten crab can maintain salt concentration lower than the surrounding water. Such regulation becomes fairly important in the animals living in salt lakes, such as the fairy shrimp, *Artemia* (an euryhaline animal). *Artemia* can exist in salt solutions as little as 0.26 per cent NaCl and in concentrated brine containing as much as 30 per cent salt.

(*ii*) Osmoregulation in cylostomes. The myxinoids (hagfishes) are exclusively marine. They are osmotic with the medium and show considerable ionic regulation by varying the total osmoconcentration largely by changes in amino acids. Lampreys show both hyposmotic and hyperosmotic regulation. When marine lampreys (*Petromyzon marinus*) enter freshwater they stop drinking, but if returned to brackishwater they drink and extrarenal extraction of Na^+ and Cl^- ions is activated. In freshwater, they can actively absorb chloride.

(*iii*) Osmoregulation in marine bony fishes. The body fluid of marine teleosts (bony fishes) is **hypotonic** and the sea water is **hypertonic**. Since they are impermeable, they loss water to the more concentrated sea water and are infact in constant danger of dehydration. Bony fishes compensate for the water loss by drinking the sea water and by absorbing water and salt (Na^+ and Cl^-) from the intestine (Fig. 5.1B). The fish kidney is unable to produce a concentrated urine to help eliminate the salt. Instead, the gills secrete the excess salt ions (Na^+, K^+ and Cl^-) into the sea water. Since such a secretion takes place from a lower concentration in the blood to a higher one outside, it is an active transport and needs energy. Some ions are lost in the faeces and urine (Mg, SO_4, Ca^{2+}, K^+ and PO_4).

(*iv*) Osmoregulation in elasmobranchs (cartilaginous fishes). The cartilaginous fishes have solved their osmotic problem with the help of their kidneys. Their blood contains sufficient amount of urea and trimethylamine oxide (TMO), thereby maintaining an osmotic concentration equal or more than that of sea water. Thus, for osmoregulation, their kidneys possess a special mechanism for pumping urea back into the blood as fast as it passes through the glomerular filter. Even the young elasmobranch is provided with its store of urea. This is obtained either directly from the parent as in the viviparous dogfish *Squalus*, or by its being placed in an egg case impermeable to urea. By the time they hatch or are born, the young elasmobranchs have developed their own **urea conserving** mechanism.

Rectal glands. As the osmotic concentration of the blood of the marine elasmobranchs is often 10 to 50 m Osm/kg more than that of the sea water, they can, therefore, use mechanisms similar to those of freshwater fish for the excretion of water. There is a tendency for osmotic inflow and excess water is removed by the kidneys. However, there is also inward diffusion of salt which must be eliminated. Excess salt is excreted not by the gills but by specialised organs, the **rectal glands**. For example, the dogfish *Squalus acanthians* secretes about 2 ml/kg/hr of solution through its rectal glands; this is volume about equal to the urine output of the animal. The rectal glands tend to eliminate excess *monovalent ions*, whereas the kidney eliminates *divalent ions*. The gut plays little or no role in water or solute balance in cartilaginous fishes. Some urea which is diffuses through the gut wall is eliminated in faeces.

(*v*) Osmoregulation in marine reptiles. Marine turtles, marine iguanas and sea snakes all have plasma osmotic and ionic concentration similar to those of bony fishes. Their integuments are highly impermeable and they also lack the exposed permeable surfaces of gills because they are air breathers. Their major problem is a continuing water loss because they must produce sufficient urine to eliminate metabolic wastes and they must lose some water in respiration. However, marine reptiles are not known to drink water and the physiological mechanisms used to control water are not known. The urine is produced at a low rate, and the kidneys of these animals have a low glomerular filtration rate. The salt intake occurs, in part at least through the food.

(*vi*) Osmoregulation in marine birds and mammals. In marine birds such as sea gulls, cormorants and pelicans that drink sea water, the excess salt is eliminated as NaCl by **nasal glands** of their heads. Marine mammals (*e.g.*, whales) do not drink water to any great extent and rely on their food as the sole source of water. Whales have a slightly higher blood osmotic concentration than terrestrial mammals. Most marine mammals feed on bony fishes rather than marine non-chordates. This is an advantage since teleosts have already performed the osmotic work of eliminating excess salts from their tissues.

(*vii*) Osmoregulation in migratory fishes. When an eel (*Anguilla*) reaches the freshwater habitat on its upstream migration, it faces problem of salt depletion and overhydration as compared with the tendency towards dehydration and salt excess in the ocean. A young salmon (*Onychorhynchus*) migrating downstream to its ocean habitat comes up against these conditions in reverse. Thus, both **anadromous** (marine fishes that migrate from sea to river for spawning, *e.g.*, salmon) and **catadromous** (freshwater fishes that migrate to sea for spawning, *e.g.*, *Anguilla*) fishes must be versatile in their osmotic adjustments. Salmon eliminates excess water by glomerular kidneys and salt loss via urine is compensated by chloride cells of gills.

Osmoregulation in terrestrial animals. All terrestrial animals lose water to the environment and therefore preserve water for avoiding danger of dehydration. The only truly successful land animals are higher arthropods and higher vertebrates. All other animals depend more on a moist environment. There are several ways by which water is gained or lost in land animals. The water gain includes drinking water, water in food and the oxidation water formed inside the body in the combustion of food materials, while water loss occurs through evaporation from the lungs and skin, urine formation and faeces.

To prevent the loss of water from the body, the land arthropods (arachnids, myriapods and insects) have a chitinous coat. The insects have been most successful in minimizing surface evaporation. They have a thin waxy cuticle of such unique properties that it not only prevent evaporation but can also take up water from a moist environment. Many terrestrial insects excrete nitrogenous wastes in solid or semisolid form and reabsorbs water in the posterior part of digestive tract. In a hydrated, *Periplaneta*, the rectal fluid (urine) is hyposmotic to haemolymph.

In higher vertebrates, protection against surface drying is afforded by horny scales (reptiles), feathers (birds) and hair (mammals) of the skin. Another means of water conservation in birds and mammals is that their kidney excretes a concentrated urine. This ability of their kidney depends upon the **loop of Henle**, the function of which is the resorption of further water. Those animals (*e.g.*, desert rats) which can produce the most concentrated urine have the longest loops of Henle.

5.4 HORMONAL CONTROL OF OSMOREGULATION

Endocrine secretions control osmotic and ionic balance in both invertebrates (= nonchordates) and vertebrates (= chordates). Hormones regulate by acting on gills, skin and excretory organs and on tissues in relation to extracellular ions. Following examples show hormonal regulation of osmotic and ionic balance in animals.

1. Neurohypophyseal hormones. Neurohypophysis of vertebrates produce following two hormones for water and salt balances in them. **1. Vasopressins** are known for their marked antidiuretic (ADH) or water conserving activity. **2. Oxytocins** have weak ADH activity.

2. Prolactin. In bony fishes, prolactin from the anterior pituitary enhances production of a dilute urine. This hormone tends to enhance fresh water osmoregulation in fish by both gills and kidneys.

3. Adrenal cortical hormones. Several hormones of the **adrenal cortex** also function in maintaining salt balance. **Aldosterone** increases Na^+ reabsorption by the bladder of a toad *Bufo marinus*. In mammals, aldosterone favours Na^+ retention and K^+ excretion. **Cortisone** enhances K^+ excretion and decreases Na^+ excretion, in salivary glands and in kidneys. In birds **glucocorticoids** from the adrenal cortex stimulate salt output by the salt glands.

4. Renin. In mammals, elevated blood Na^+ causes the release from juxtaglomerular cells of the kidneys a protein, called **renin**. Renin enzymatically converts angiotensin precursor (*i.e.*, angiotensinogen) in the blood to angiotensin I; this is converted to angiotensin II. Angiotensin II acts on the adrenal cortex to increase the secretion of **aldosterone**. The aldosterone stimulates resorption of Na^+ in the kidney, which in turn favours water retention.

5. Thyrocalcitonin. The hormone thyrocalcitonin is released from the **thyroid** in the presence of high serum Ca^{2+}. It lowers plasma Ca^{2+} by decreasing its release from the bones. In fish, cells of **ultimobranchial bodies** release a polypeptide which lowers blood Ca^{2+} level of eel and catfish (**Copp** 1969).

6. Neuro-endocrine regulation. Neuroendocrine regulation of salt and water balance has been observed in crustaceans and insects. In crayfish, ligation of the eyestalks increases water influx and urine production when in freshwater, and increases blood osmoconcentration when in 50% sea water; a brain extract counteracts these effects (**Kamemoto** and **Ono** 1969).

7. Role of atrial natriuretic factor (ANF) in osmoregulation. Atrial natriuretic factor (ANF) is a polypeptide hormone secreted by the smooth muscle cells of wall of atria of heart. It is secreted in response to increased blood volume and blood pressure. It is antagonistic of RAAS (renin-angiotensinogen aldosterone system) and inhibits the release of renin from the juxtaglomerular cells so reduces the secretion of aldosterone and reabsorption of NaCl of nephric filtrate into the interstitial fluid.

So ADH, RAAS and ANF regulate the functioning of kidney tubules by **feedback mechanisms** to regulate the salt concentration, blood pressure and blood volume, called **homeostasis**.

5.5 REGULATION OF ACID-BASE BALANCE

Renal regulation of acid-base balance

1. The kidneys are responsible for clearing the body of metabolically produced non-carbolic acids (Box 5.1). Therefore, normal urine reaction is acidic in nature.

Box 5.1

The human body produces large amounts of acids in two forms:

1. **Carbonic acid (H_2CO_3).** It is called **volatile acid** because CO_2 can be formed from H_2CO_3 and in turn can be eliminated by the lungs, and
2. **Non-carbonic acids.** These acids cannot be converted into CO_2 and therefore are called non-**volatile** or **fixed acids**, *e.g.*, H_2SO_4 (a product of protein catabolism), HCl, phosphoric acid (H_3PO_4, a product of phospholipid metabolism), keto acids (aceto-acid and β-hydroxy-butyric acid) and lactic acid.

2. Sources of non-carbonic acid or the sources of H^+

(*i*) As an end product of metabolism-the greatest source of hydrogen ion (H^+) is CO_2 produced as end product of cellular metabolism, the amount of CO_2 produced during a day in a normal individual is capable of forming 20–40 mEq of H^+.

(*ii*) Diets which are high in protein, generate between 50-100 mEq of H^+ per day.
(*iii*) Renal tubular generation of H^+.
(*iv*) Other sources of H^+ are the same as that of sources of non-carbonic acid.

Other sources of non-carbonic acid are the following:

I. Cellular metabolism produces CO_2, a large part of which is expelled from the lungs and remaining gets hydrated with body fluids to form carbonic acid (H_2CO_3).

II. High protein diets contain large amounts of phosphorus and sulphur; oxidation of these complexes forms anions of phosphate (PO_4^{3-}) and sulphate (SO_4^{2+}) which leads to formation of non-carbonic acid, e.g., H_2SO_4 and H_3PO_4.

III. Waste products of metabolism form acid phosphate ($H_2PO_4^-$).

IV. Strenuous exercise leads to formation of lactic acid from skeletal muscles.

V. Under some pathological conditions such as **starvation** or **uncontrolled diabetes mellitus** accumulation of organic acids such as acetoacetic acid and β-hydroxybutyric acid occurs.

VI. Ingestion of acidifying salts, e.g., NH_4Cl and $CaCl_2$ which in effect add HCl to the body.

VII. Renal tubular generation of H^+. Therefore, in renal failure, failure of excrete normal acid load leads to acidosis.

3. Body fluids are maintained in an alkaline state (PH = 7.4). In all body fluids, electrical neutrality must be maintained, anions such as phosphate or sulphate must be 'covered' by an equal amount of cations. This can only be achieved by daily excretion of the excess of anions in the urine. If these anions which are excreted in the urine are fully 'covered' by an equivalent amount of cations (mainly Na^+), this will produced serious consequences. However, this situation is prevented due to manufacture of two important cations, H^+ and NH_4^+ by the kidneys which 'cover' the excreted anions.

Role of kidney in Acid-Base Balance

1. The kidneys perform two major functions for maintaining the acid-base balance:

(*a*) The kidneys stabilize the standard HCO_3^- pool by obligatory reabsorption (mainly by PCT or proximal convoluted tubule) and by controlled reabsorption of filtered HCO_3^- (by the DCT or distal convoluted tubule and CT or collecting tubule).

(*b*) The kidneys excrete a daily load of 50–100 mEq (= milliequivalent) of metabolically produced non-carbonic acid, which represents a H^+ excretion of 1 mEq/kg of body weight/day.

2. The major sites of urine acidification are the DCT and CT since most of secreted H^+ in the PCT form CO_2 and H_2O from H_2CO_3. Essentially, all of the H^+ within the tubular lumen is from the tubular secretion of H^+ generated by metabolism.

3. The amount of acid secreted upon the subsequent events in the tubular urine. The maximal H^+ gradient against which the transport mechanism can secrete H^+ (in humans) corresponds to a urine pH of about 4.4.

4. If there were no buffers that 'tied up' in the urine, this pH would be reacted rapidly and H^+ (*i.e,* hydrogen ion) secretion stops. However, three important reactions (Buffer systems) in the tabular fluid remove free H^+, permitting more acid to be secreted.

5. The major buffer systems in kidneys are the following: 1. Bicarbonate system; 2. Dibasic phosphate system; and 3. Ammonium system.

The ammonium system (NH_3/NH_4^+) assists the conservation of Na^+ and HCO_3^- by the body.

Titratable acidity

1. Hydrogen ion (H^+) secreted into the lumen that reacted with HCO_3^- is not excreted, whereas H^+ secreted into the lumen that reacts with a non-bicarbonate buffer remains in the tubular fluid and is excreted.

2. The renal contribution of newly synthesized HCO_3^- is accompanied by the excretion of an equivalent amount of acid in the urine in the form of **titratable acid**, NH^+ or both. Therefore, each H^+ that reacts with buffer organic anions other than HCO_3^- contributes to the urinary titratable acidity. It is measured by determining the amount of alkali that must be added to the urine to return its pH of 7.4, i.e., the normal pH of the glomerular filtrate etc. PCT is the major nephron site where titratable acid is formed.

QUESTIONS

Long Answer Questions

1. How do marine and freshwater fish maintain osmotic homeostasis?
2. Explain why marine fish drink sea water and survive but humans cannot?
3. What is osmoregulation? Write an account of the regulatory processes in different habitats. (*MSU Tirunelveli 2007*)
4. Write an essay on osmoregulation in fishes.
5. Outline the processes involved in the secretion of H^+ into the renal tubules.
6. Name the process by which acid is removed from the body.

Short Answer Questions

1. Discuss the osmoregulation in migratory fishes.
2. Differentiate between the following terms:
   (*a*) Euryhaline and stenohaline animals;
   (*b*) Osmotic conformers and osmoregulators.
3. Describe osmoregulation in marine animals. (*Osmania 2005*)
4. Differentiate between volatile and nonvolatile acids.

Very Short Answer Questions

1. Define osmoregulation.
2. Who coined the term osmoregulation?
3. Differentiate between terms osmolarity and osmolality.
4. Give examples of the euryhaline animals.
5. Give an example of stenohaline animal.
6. How does ANF help in osmoregulation?
7. Enumerate the major buffer systems in kidney.

Fill in the Blanks

1. Annelids have .... and insects possess .... for excretion and osmoregulation.
2. The .... cells of gills of crustaceans and freshwater fish absorbs .... from freshwater and add it to the body fluid.
3. Dogfish, *Squalus acanthians* has .... glands for osmoregulation.
4. The .... hormone tends to enhance osmoregulation in freshwater bony fishes.
5. Osmotically labile or dependent animals are called .... , *i.e.*, their body fluid concentration may change with medium.
6. Osmotically stable or independent animals are called .... , *i.e.*, when the medium changes, the body fluid concentration remains constant or unchanged.
7. .... is the major site where litratable acid is formed.

Multiple Choice Questions

1. Animals in which the concentration of the body fluids changes according to the medium are called as
   (*a*) osmoregulators (*b*) osmotic conformers
   (*c*) limited volume regulators (*d*) none of these
2. Animals whose internal body fluid concentration remains constant and does not change according to the medium are
   (*a*) osmoregulators (*b*) osmotic conformers
   (*c*) limited volume regulators (*d*) none of these

3. Animals which tolerate a narrow range of variation in salinity are
(*a*) euryhaline (*b*) stenohaline
(*c*) isosmotic (*d*) poikilosmotic
4. Some animals such as estuarine crabs can survive in water of varying salinity. They are known as
(*a*) euryhaline (*b*) stenohaline
(*c*) hyperosmotic (*d*) hyposmotic
5. A poikilosmotic animal
(*a*) has a constant internal body fluid concentration
(*b*) varies internal body fluid concentration according to the medium
(*c*) swells in freshwater
(*d*) shrinks in salt water
6. Gills in fresh water fishes have special cell which
(*a*) excrete salt (*b*) absorb salt
(*c*) excrete urea (*d*) perform all these functions
7. The function of contractile vacuole in protozoans is
(*a*) excretion (*b*) osmoregulation
(*c*) nutrition (*d*) all of these
8. Fresh water fishes osmoregulate by producing
(*a*) dilute urine and absorb salts through gills
(*b*) concentrated urine and excreting salt through gills
(*c*) dilute urine and excreting salts through gills
(*d*) concentrated urine and absorb salts through gills
9. Mammals which produce highly concentrated urine have
(*a*) long loops of Henle (*b*) short loops of Henle
(*c*) no loops of Henle (*d*) aglomerular nephrons
10. Which of the following animals is euryhaline?
(*a*) freshwater tadpole (*b*) *Anguilla*
(*c*) shore crab, *Carcinus* (*d*) fairy shrimp, *Artemia*
11. In fish, which of the following hormones is involved in osmoregulation?
(*a*) adrenaline (*b*) prolactin
(*c*) thyrotropin (*d*) thyroxine
12. Birds that feed on marine animals ingest a lot of salt. They excrete most of it by means of
(*a*) rectal salt glands (*b*) nasal salt glands
(*c*) green glands (*d*) hypertonic urine
13. Which vertebrate organ is mainly responsible for osmoregulation
(*a*) skin (*b*) lungs
(*c*) kidney (*d*) hypothalamus
14. Which of the following substances are retained by marine bony fishes for maintaining the osmolarity of body fluids
(*a*) sodium (*b*) magnesium
(*c*) calcium (*d*) urea
15. Chloride secreting cells are found in
(*a*) elasmobranchs (*b*) fresh water bony fishes
(*c*) marine bony fishes (*d*) crustaceans
16. Greatest source of H^+ in the body is
(*a*) end product of cellular metabolism (*b*) high protein diet
(*c*) synthesized by renal tubular cells (*d*) waste product of metabolism

17. Cations manufactured by the kidney to cover the excreted anions in urine are
   (*a*) H^+ and Na^+ (*b*) NH_4^+ and H^+
   (*c*) Na^+ and H^+ (*d*) K^+ and H^+
18. The maximal H^+ gradient against which transport mechanism can secrete H^+ corresponds to a urine pH of
   (*a*) 4.4 (*b*) 4.8
   (*c*) 5.2 (*d*) 6.0
19. Major source of ammonia production in renal tubular cell is by
   (*a*) deamination of glutamine
   (*b*) deamination of glutamic acid
   (*c*) deamination of other amino acids: glycine, asparagine, alanine
   (*d*) comes directly from arterial blood
20. Body fluid pH depends primarily on the renal control of
   (*a*) reabsorption of sodium in exchange of potassium
   (*b*) secretion of H^+ in exchange for ammonia
   (*c*) secretion of H^+ in exchange for potassium
   (*d*) secretion of H^+ in exchange for sodium
21. The buffer system in the kidney to excrete H^+ is all except
   (*a*) bicarbonate (*b*) dibasic phosphate
   (*c*) ammonia (*d*) urate

ANSWERS

Very Short Answer Questions

1. Regulation of homeostasis of water and salt contents of the animal body.
2. Hober in 1902.
3. *Osmolarity* refers to the osmotic pressure generated by dissolved solute molecules in one liter solvent. *Osmolality* is the osmotic pressure generated by the solute molecules dissolved in one kilogram of solvent.
4. *Mytilus*, *Aplysia* and fairy shrimp-*Artemia*.
5. Spider crab-*Maia*.
6. ANF or atrial natriuretic factor is a peptide hormone secreted by the cardiac atria that is pharmacological doses lowers blood pressure and promotes salt excretion.
7. In kidney following three major buffer systems are present: (1) bicarbonate system; (2) dibasic phosphate system; and (3) ammonia system.

Fill in the Blanks

1. Metanephridia, Malpighian tubules
2. Chloride, salts
3. Rectal
4. Prolactin, osmoregulation
5. Osmoconformers
6. Osmoregulators
7. PCT or proximal convoluted tubule.

Multiple Choice Questions

1. (*b*)	2. (*a*)	3. (*b*)	4. (*c*)	5. (*b*)	6. (*b*)
7. (*b*)	8. (*a*)	9. (*a*)	10. (*d*)	11. (*b*)	12. (*b*)
13. (*c*)	14. (*a*)	15. (*c*)	16. (*a*)	17. (*b*)	18. (*a*)
19. (*a*)	20. (*d*)	21. (*d*).			

6

BLOOD

6.1 BLOOD AND ITS FUNCTIONS

Blood and lymph are fluid or liquid connective tissues. Their extracellular matrix (called **plasma)** is fluid and devoid of fibres. The volume of matrix, however, exceeds that of the cells. The cells of the fluid tissues (*i.e.*, erythrocytes and leucocytes in blood and only leucocytes in lymph) are freely suspended in the plasma (= matrix).

A. Blood

Blood is a fluid connective tissue that circulates through the cardiovascular system. It is red-coloured and constitutes about 8 per cent of the body weight, *i.e.,* 5.6 litres in 70 kg human being.

Composition of blood. The blood is composed of a faintly yellow transparent fluid, the plasma and floating in this fluid are numerous cells or corpuscles of different kinds. The plasma constitutes approximately 55 per cent and the corpuscles about 45 per cent of the volume of blood (Table 6.1).

Table 6.1 Composition of blood plasma.

S. No.	Component	Percentage
1.	Water	91–92
2.	Proteins (fibrinogen, globulin, albumin)	7–8
3.	Other solutes	1–2
	(*i*) Electrolytes (Na^+, K^+, Ca^{2+}, Mg^{2+}, Cl^-, HCO_3^-)	
	(*ii*) Non-protein nitrogen substances (urea, uric acid, creatine, creatinine, ammonium salts)	
	(*iii*) Nutrients (glucose, lipids, amino acids)	
	(*iv*) Regulatory substances (hormones, enzymes)	
	(*v*) Respiratory gases (oxygen, carbon dioxide)	

A. Formed elements of blood. Since the cells of blood are originated as stem cells in bone marrow, they are called **formed elements of blood.** They include the following elements:

1. **Red blood cells/corpuscles** which are also called erythrocytes (corpuscles = a small round body).
2. **White blood cells** which are also called **leucocytes.**
3. **Platelets** which are non-nucleated fragments of megakaryocytes.

Hematocrit. Plasma is liquid intercellular material or matrix that imparts to the blood its fluid properties. The relative volume of cells and plasma is about 45 per cent and 55 per cent, respectively. This value is called a **hematocrit.** It is obtained by centrifuging a blood sample with anticoagulants added. The hematocrit reading is then obtained by measuring the per cent volume in the centrifuge tube occupied by the red cells compared with the whole blood volume. A normal reading is about 45; thus, 45 per cent of blood consists of red blood cells.

The cell fraction consists mainly of packed erythrocytes (Fig. 6.1). At the upper part of the packed cells is a narrow layer, called the **buff coat.** It constitutes about 1 per cent of the entire sample, but contain all the **leucocytes** and **platelets.** There are far more red blood cells (5×10^6 / mm^3 of blood) than white blood cells ($6 - 9 \times 10^3$ / mm^3 of blood).

B. Plasma

It is non-living homogeneous fluid portion of the blood. It is colourless but appears grey to yellow when in large amounts. Plasma's composition has been summarized in Table 6.1. Its composition varies in different parts of body but normally contains about 90 per cent of water and varying percentage of proteins, salts, sugars and fats.

Plasma Proteins

1. Fibrinogens. These are largest blood proteins which are made in the liver and function in blood clotting.

2. Albumins. These are smallest proteins of blood. They are made in the liver and are responsible for exerting the major osmotic pressure on the blood vessel wall (called **colloidal osmotic pressure**).

3. Globulins. These include **immunoglobulins (antibodies)** and glycoproteins such as fibronectins.

Functions of Blood Plasma

1. Transport of digested food. Plasma transports digested food to different organs and tissues of the body.

2. Removal of excretory substances. Waste substances from the tissues are removed and carried to the kidneys for elimination.

3. Disposal of CO_2. Major portion of CO_2 from the tissues is removed by plasma, where it remains dissolved in the form of bicarbonate.

4. Transport of oxygen (O_2). Certain percentage of O_2 is also transported by plasma. Even the oxygen is bounded by the red blood corpuscles is first dissolved in the plasma before reaching the cell.

5. Distribution of hormones. The secretions of various endocrine glands are transported by plasma.

6. Distribution of vitamins. Plasma transports and distributes vitamins absorbed along with digested food to various body cells.

7. Regulation of water balance. Plasma supplies water to different tissues and removes excess of water formed during metabolic processes.

8. Osmotic balance. Plasma proteins and minerals in the plasma exert osmotic pressure and maintain blood pressure.

9. Protection against disease. Immunoglobulins of plasma act as antibodies and neutralise the harmful effects of foreign agents.

10. Regulation of body temperature. Blood plasma regulates the body temperature by transporting heat from the deeply seated heat producing organs to peripheral parts of the body.

11. Blood clotting. The plasma proteins fibrinogens, prothrombin and some other blood-clotting factors are present in the plasma which help in the clotting of blood in case of injury.

12. Lymph formation. The lymph is filtered plasma.

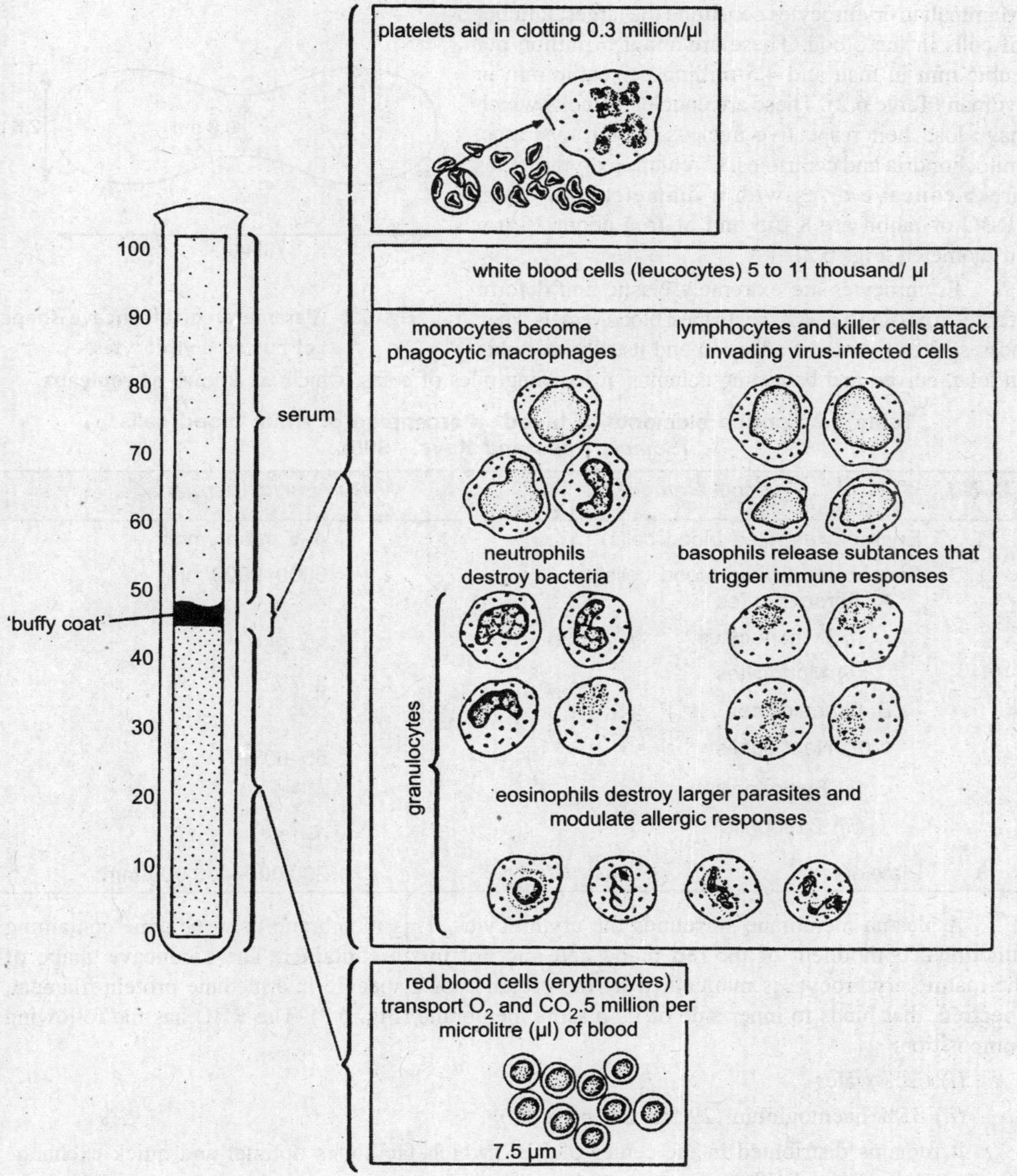

Fig. 6.1. The composition of human blood. The figure at left shows blood from a normal person after removing the clotting proteins and centrifuging it to make the cell settles. A little over half the volume is the fluid serum. Approximately 45% (a little less in women) is red blood cells. Between serum and red blood cells is a "buffy coat" containing platelets and white blood cells.

C. Blood Cells

I. Erythrocytes or red blood corpuscles (RBCs).

Mammalian erythrocytes constitute the largest number of cells in the blood. These are about 5 million per cubic mm in man and 4.5 million per cubic mm in woman (Table 6.2). These are enucleated cells which have lost their respective nucleus, Golgi apparatus, mitochondria and centrioles. Circulating erythrocytes are **biconcave** discs with a diameter of 7.8 µm (RBC of rabbit are 8 µm and of frog about 25 µm in diameter; Fig. 6.2).

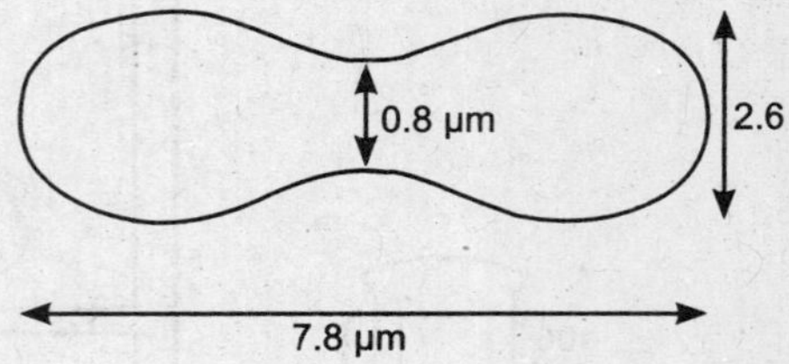

Fig. 6.2. Parameter of biconcave shape of human erythrocytes.

Erythrocytes are extremely elastic and deform readily in passing through the smallest blood vessels. They possess great power of adhesion and usually assemble in long, curved and branching columns, resembling piles of coins, which are known as **rouleaux.**

Table 6.2 Formed elements of blood. *Percentage of white blood cells. *(Source:* **Ross and Kaye, 1995).**

S. No.	*Formed element*	*Number or percentage*
1.	Erythrocytes (Red blood cells)	**4–5** million/mm^3
2.	Leucocytes (White blood cells)	6000–9000/mm^3
	A. Agranulocytes	
	(*i*) Lymphocytes	30–35%*
	(*ii*) Monocytes	3–7%*
	B. Granulocytes	
	(*i*) Neutrophils	55–60%*
	(*ii*) Eosinophils	2–5%*
	(*iii*) Basophils	0–1%*
3.	Platelets	200,000–400,000/mm^3

A plasma membrane surrounds the erythrocytes. This membrane is an enzyme-containing functional component of the red blood cell and not just a container. The biconcave shape of the mature erythrocyte is maintained by the presence of a specific intermediate protein filament, **spectrin,** that binds to inner side of its plasma membrane (Fig. 6.3). The RBC has the following composition:

(*i*) 62% water

(*ii*) 35% haemoglobin (29.5 ± 2.5 pg/RBC).

It remains distributed in the center of RBC which facilitates optimal and quick exchange of gases (*e.g.*, O_2 and CO_2).

(*iii*) 2.5

(*a*) Sugar-glucose

(*b*) Lipids-cephalin, cholesterol and lecithin

(*c*) Protein. **Glutathiones,** albumin like insoluble protein, acts as reducing agent, thus prevents damage of haemoglobin.

(*d*) Enzymes of glycolytic system; carbonic anhydrase and catalase.

(*e*) Vitamin derivatives, and

(*f*) Ions - Na^+, K^+, Ca^+, PO_4^{3-} and SO_4^{2-}

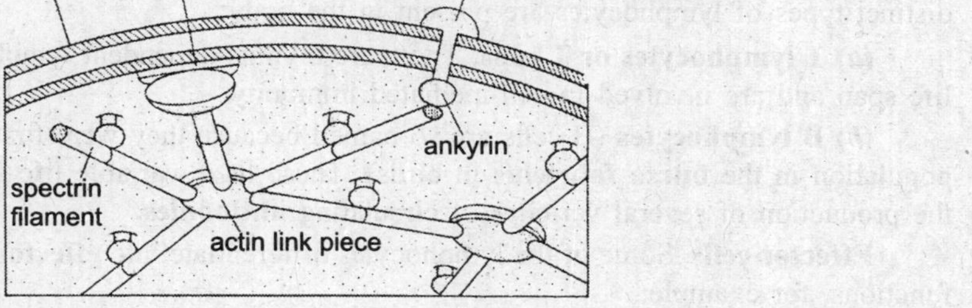

Fig. 6.3. Diagram shows red blood cell cytoskeleton, which maintains its distinct shape. A filamentous skeleton of the protein spectrin is anchored to the cell membrane by three main proteins (band 3 protein, ankyrin and band 4 protein) with short actin pieces, about 15 monomers long, linking spectrin to band 4 protein.

Erythrocytes are pale or greenish yellow in colour. But in group these appear red. The colour is due to a pigmented protein **haemoglobin** that fills the red blood cells; about 26.5 crore molecules of haemoglobin are packed in each RBC. Oxygen and carbon dioxide are transported bound to a haemoglobin. The disc shape of the erythrocyte facilitates gas exchange because more haemoglobin molecules are closer to the plasma membrane than they would be in a spherical cell. Thus gases have less distance to diffuse within the cell to reach a binding site on the haemoglobin.

Haemoglobin consists of four polypeptide chains complexed to iron-containing **haeme** groups. There is variation in the structure of the polypeptide present, haemoglobin is designated as HbA, HbA_2 and HbF. Adult haemoglobin is about 96% HbA, 2% HbA_2 and 2% HbF. HbF is the principal form of haemoglobin in fetus and its persistence in high percentage in the adult is indicative of certain form of anaemia in human beings. Haemoglobin is similar in composition and structure to **myoglobin,** the oxygen-binding protein found in striated muscle.

Functions. Erythrocytes perform their functions only within the bloodstream. Their role is restricted to binding oxygen for delivery to the tissues and in exchange, binding carbon dioxide for removal from the tissue.

2. Leucocytes or white blood cells. These are nucleated amoeboid cells which exhibit active amoeboid movement and are able to pass through the wall of blood vessels (diapedesis; Box 6.1). Their number varies from 6000 to 9000 per cubic mm but may be greatly increased under pathological conditions. WBCs have a defensive role, destroying infecting organisms such as bacteria and viruses, as well as assisting in the removal of dead or damaged tissues. These are of following two types:

A. Agranulocytes. These are agranular cells whose cytoplasm is homogeneous without granules or with azurophilic granules (*i.e.*, lysosomes). They are mononucleate having a single large slightly indented nucleus. These are of following types:

Box 6.1 WBCs Form Transient Component of Blood

In contrast to RBCs, WBCs leave the blood through the walls of capillaries and venules to enter connective tissues, lymphatic tissues and bone marrow (this phenomenon is called **diapedesis**), where they perform their specific functions. Therefore, whereas red blood cells must be considered essential components for the normal functions of blood, white blood cells must be regarded as transients within blood. That is, WBCs use the bloodstream as a vehicle for transport to specific sites within the body.

***(i)* Lymphocytes**. They are small blood cells (6–9 μm diameter) with relatively large ovoid or kidney-shaped **nuclei** surrounded by a thin layer of homogeneous and basophilic cytoplasm. These are second most numerous white blood cells in the blood; these constitute 30–35 per cent of total leucocytes, *i.e.*, about 1500 to 2500 per cubic mm of blood. Following two functionally distinct types of lymphocytes are present in the body:

***(a)* T lymphocytes or T cells.** These are thymus-dependent lymphocytes. They have a long life span and are involved in cell-mediated immunity.

***(b)* B lymphocytes** (B cells are so named because they were first recognized as a separate population in the **bursa fabricius** in birds). These have variable life spans and are involved in the production of several varieties of circulating **antibodies.**

Effector cells. Some of the lymphocytes differentiate into **effector cells** which have specific functions, for example:

1. B lymphocytes may divide several times, producing more B lymphocytes and large clone of cells that differentiate into **plasma cells** involved in production of antibodies.

2. T lymphocytes may undergo several rounds of cell division, producing cells that differentiate into **cytotoxic T lymphocytes** such as **killer T lymphocytes, helper T lymphocytes** and **suppressor T lymphocytes.**

3. Some B and T cells do not undergo differentiation into effector cells but serve as long-lived **memory cells,** which are circulating lymphocytes that are primed to respond more rapidly and to greater extent to subsequent exposure to their specific antigens.

***(ii)* Monocytes.** They are large-sized agranulocytes with small, oval or indented (*i.e.*, kidney-shaped or horse-shoe shaped) nucleus and large amount of cytoplasm. Although classified as agranular, there are often numerous small, dense azurophilic granules (lysosomes) in the cytoplasm of the monocyte. Monocytes constitute 3–7 per cent of leucocytes; *i.e.* 200 to 700 per cubic mm and measure 12–20 μm in diameter.

In the peripheral blood, monocytes are in transit from the bone marrow to the body tissues, where they will differentiate into the various phagocytes of the **mononuclear phagocytic system,** *i.e.*, the connective tissue macrophages (histiocytes), osteoclasts, alveolar macrophages, perisinusoidal macrophages in liver (Kupffer cells) and macrophages of lymph nodes, spleen and bone marrow, among others. Monocytes remain in the blood only for about three days.

During inflammation, the monocyte leaves the blood vessel at the site of inflammation, transforms into a tissue macrophage, and participates in the phagocytosis of bacteria, other cells and tissue debris.

B. Granulocytes. These white blood cells contain lobulated nuclei and different types of granules in cytoplasm. Granulocytes may be of following types:

***(i)* Neutrophils.** These are spherical cells and are most numerous of WBCs and most common of granulocytes. These are 55–60 per cent of total leucocytes, *i.e.*, 4000 to 5000 per cubic mm. The nucleus is highly **polymorphic** having 3 to 5 lobes of nuclear material joined by thinner nuclear strands (Box 6.2). The cytoplasm of neutrophils is rich in **glycogen** (for anaerobic respiration in tissues) and is differentiated into nongranular **peripheral** and granular layer of **inner mass**. There are two types of granules:

Box 6.2

1. Neutrophils are named for their lack of characteristic cytoplasmic staining, but they are also identified by the multilobed shape of their nucleus. They are, thus, also called **polymorphonuclear neutrophils** and **polymorphs.**

2. In neutrophils chromatin too has characteristic arrangement. Wide regions of heterochromatin are located chiefly at the periphery of nucleus, in contact with the nuclear envelops.

3. In human female, about 3% of nuclei exhibit a small condensed nuclear appendage (drumstick chromosome), which represent the quiescent X chromosome or **Barr body.**

(1) Specific granules. These are small and most numerous. They contain bacteriostatic and bacteriocidal agents, such as **lysozyme, alkaline phosphatases** (marker enzymes), etc. Alkaline phosphatase containing granules are called **phosphasomes.**

(2) Azurophilic granules. These are larger and less numerous. They are lysosomes, having **peroxidases** (*e.g.,* myeloperoxidase) and lysosomal enzymes.

Neutrophils are highly motile and phagocytic cells. Their primary function is to digest and destroy invading microorganisms. They function extravascularly to cause **acute inflammation** which is the local tissue response at the near site of injury. Acute inflammation is traditionally characterized by **heat, redness, swelling, pain** and **loss of function.** The inflammatory response occurs in the connective tissue into which plasma and formed elements of the blood leak from blood vessels damaged by injury. The neutrophils are the most numerous of the first wave of cells to enter the inflammatory site except, of course, for extravasated RBCs. They engage in **active phagocytosis** of bacteria and other foreign organisms and **passive phagocytosis** of damaged connective tissue cells, red blood cells and fibrin.

When a bacterium is phagocytosed, the specific granules fuse with the membrane of the phagosome within 30-60 seconds, and their bacteriocidal content is emptied into the phagosome. Azurophilic granules fuse with the phagosome-specific granule complex somewhat later; the hydrolytic enzymes of the lysosomal azurophilic granules digest the microorganism. Many neutrophils die in this process and release their lysosomal enzymes into extracellular space to cause liquefaction of adjacent tissues. The accumulation of dead bacteria, dead neutrophils and tissue fluid form the thick yellowish exudate called **pus.**

***(ii)* Basophils.** These are most rare white blood cells forming 0.5–1 per cent of the total leucocytes. A basophil is about the same size as a neutrophil and is so designated because the numerous large granules in its cytoplasm stain with basic dyes. The nucleus of a basophil is a S-shaped having two or more constrictions. These granules contain a variety of substances namely hydrolytic enzymes, heparin (also heparan sulphate), chondroitin sulphate, histamine, leucotriene and slow-reacting substance (SRS) or anaphylaxis. **Anaphylaxis** means hypersensitivity (as to foreign proteins or drugs) resulting from sensitization following prior contact with the causative agent.

1. Histamine and **SRS** or **anaphylaxis** are vasoactive agents that among other actions, brings out the dilation of small blood vessels.

2. Leucotriene-3 participates in allergic responses such as bronchoconstriction in asthma.

3. Heparan sulphate is a sulphated glycosaminoglycan that is closely related to the heparin found in granules of tissue mast cells.

***(iii)* Acidophils or eosinophils.** These are spherical cells having **bilobed** nuclei. These comprise 2–5 per cent of total leucocytes, *i.e.,* 70 to 300 per cubic mm of blood. The cytoplasm of eosinophils contains numerous large elongated granules measuring 1 μm or more in their largest dimensions. They are retractile and have affinity for acidic dyes (Box 6.3).

Box 6.3

The centre of the granules of eosinophils contains a **crystalloid body** that is readily seen with the electron microscope. The large granules with crystalloid inclusions are the lysosomes; it is the crystalloid that is responsible for the refractility of granules in the LM (= light microscope). It contains an arginine rich protein called **major basic protein** that probably accounts for the intense acidophilia of the granules. The granules contain peroxidase, histaminase, arylsulphatase and other hydrolytic enzymes. Histaminase neutralizes the activity of histamine and arylsulphatase neutralizes the action of SRS (= slow-reacting substance) or anaphylaxis.

Eosinophils play an important role in immunity, allergy and hypersensitivity. Eosinophils are known to release arylsulphatase and histaminase at sites of allergic reaction, thereby moderating the potentially deleterious effects of the inflammatory vasoactive agents. Eosinophils are engaged in phagocytosis of antigen-antibody complexes. Thus, the count of eosinophils in blood samples of individuals suffering from allergies and parasitic infections is usually high.

Platelets

Platelets are small disc-shaped anucleate cells (Fig. 6.4) and are formed by the cytoplasmic fragments of huge precursor cells called **megakaryocytes** in the bone marrow. Human platelets have a life span of about 10 days; they are seen in a peripheral blood smear either singly or in clusters. They are about 2 μm in diameter and exhibit a deeply stained core, the **granulomere** and a lightly stained periphery, the **hyalomere**. Microtubules and microfilaments are the major components of the hyalomere. The prominent circumferential band of microtubules in the hyalomere is probably responsible for maintaining the biconvex (lens like) shape of platelet.

Platelets contain mitochondria, microtubules, glycogen granules, occasional Golgi elements and ribosomes, as well as enzyme system, for aerobic and anaerobic respiration. Their most conspicuous organelles, however, are following four types of granules:

1. Alpha (= α) granules. These are variable in shape and size and contain three important groups of proteins (platelet exclusive proteins, coagulation factors and other proteins).

2. Dense or delta (= δ) granules. These are electron-dense granules containing serotonin.

3. Lysosomes. These are membrane-bound vesicles containing acid hydrolases.

4. Peroxisomes. These are few in number and have peroxidase enzyme.

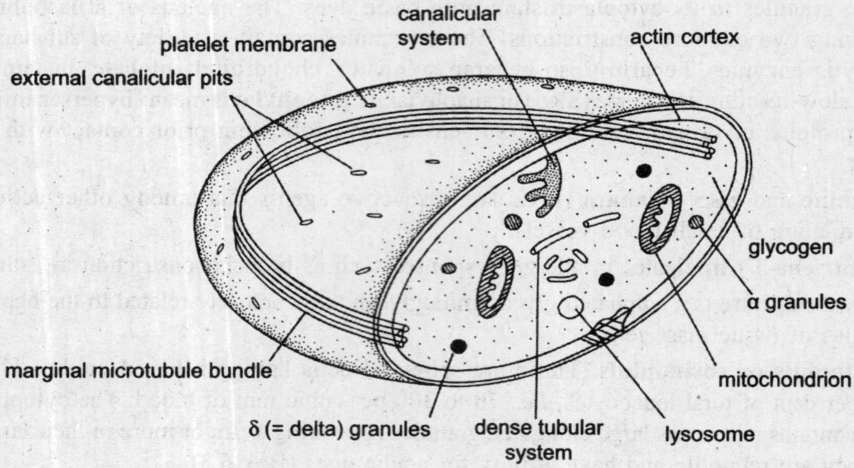

Fig. 6.4. Structure of a blood platelet.

D. Lymph

The circulatory system of humans is not closed in the strict sense. At the tissue level, leakage of leucocytes and plasma having respiratory gases (oxygen), nutrients glucose, amino acids, lipids and salts) and hormones takes place from the blood capillaries (Fig. 6.5). The tissue cells, therefore, get bathed with this intercellular fluid, called **tissue fluid.** Some of the tissue fluid returns back to the capillaries towards the venous end. However, rest of the tissue fluid drains into blindly ending lymphatic capillaries, and once inside these, the fluid is termed **lymph.**

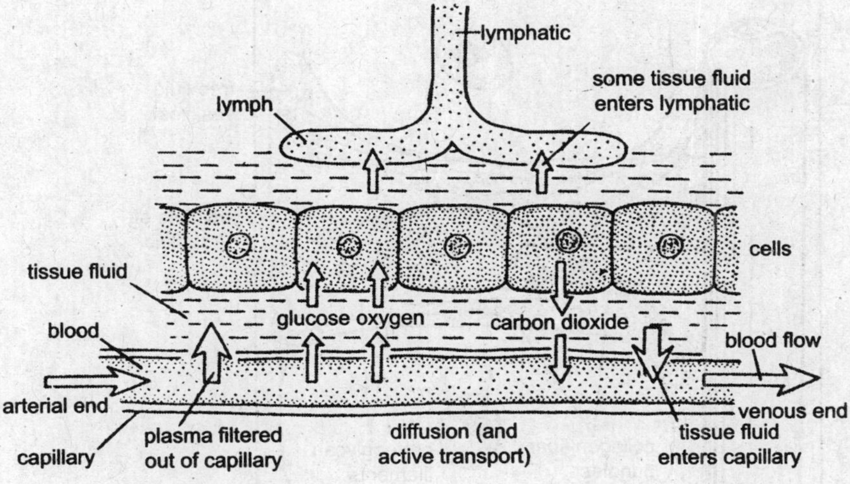

Fig. 6.5. Relationship between blood, tissue fluid and lymph.

Thus, the lymph is a filtered fluid that lies between blood and tissue fluid. It is a colourless fluid. Lymph contains lymphocytes (leucocytes) and glucose, salts, amino acids, vitamins, urea, CO_2, oxygen and nutrients in the same concentration as present in plasma, but proteins are present in comparatively low concentration.

Functions of lymph. Lymph is a connecting medium between tissue fluid and blood. It serves following functions:

1. Lymph brings back into the blood the cell debris, colloid particles and part of blood plasma that has diffused out from arterial capillaries.

2. It brings CO_2 and nitrogenous wastes from tissue fluid to blood.

3. In the villi of small intestine, lymph capillaries are called **lacteals** which help in the absorption of fats.

4. Leucocytes of lymph too perform the function of defence and tissue repair.

Interstitium and Interstitial Fluid

About one sixth of the total volume of the body consists of spaces between cells, which collectively are called the **interstitium.** The fluid in these spaces is the **interstitial fluid** (Fig. 6.6). Interstitium contains two major types of solid structures: 1. **collagen fiber bundles** and **2. proteoglycan filaments.** The collagen fiber bundles extend long distances in the interstitium. They are extremely strong and therefore provide most of the tensional strength of the tissues. The proteoglycan filaments, however, are extremely thin coiled or twisted molecules composed of 98 per cent **hyaluronic acid** and 2 per cent protein. These molecules are very thin and form a mat of very fine reticular filaments forming the so-called **brush pile.**

The fluid in the interstitium is derived by filtration and diffusion from the capillaries. It contains almost the same constituents as plasma except for much lower concentration of proteins because proteins do not easily pass outward through the pores of the capillaries.

The interstitial fluid is entrapped mainly in the minute spaces among the proteoglycan filaments. This combination of proteoglycan filaments and fluid entrapped within them has the characteristics of gel and therefore is called **tissue gel.**

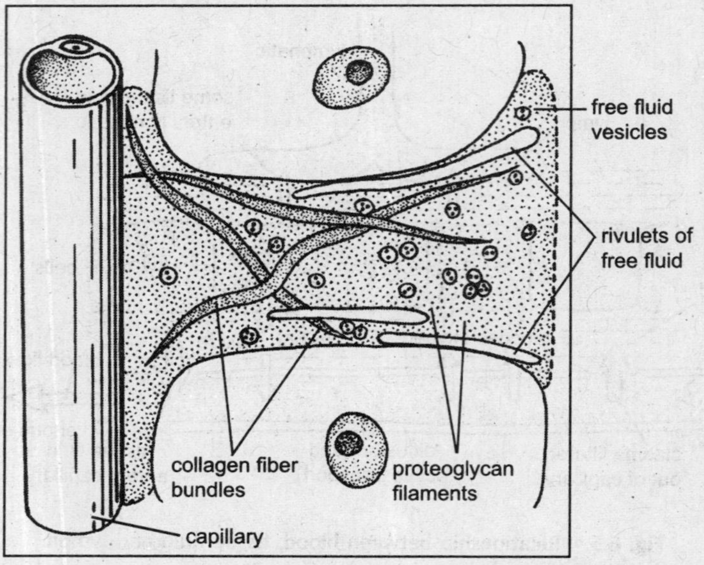

Fig. 6.6. Structure of the interstitium. Proteoglycan filaments are everywhere in the spaces between the collagen fiber bundles. Free fluid vesicles and small amounts of free fluid in the form of rivulets occasionally also occur (after Guyton and Hall 2011).

Because of the large numbr of proteoglycan filaments, it is difficult for fluid to flow easily through the tissue gel. Instead, fluid mainly diffuses through the gel; that is, it moves molecules by molecules from one place to another by kinetic, thermal motion rather than by large numbers of molecules moving together.

Diffusion through the gel occurs about 95 to 99 per cent as rapidly as it does through free fluid. For the short distances between the capillaries and the tissue cells, this diffusion allows rapid transport through the interstitium not only of water molecules but also of electrolytes, small molecular weight nutrients, cellular excreta (ammonia), oxygen, carbon dioxide, and so forth.

"Free" fluid in the interstitium. Although almost all the fluid in the iterstitium normally is entrapped within the tissue gel, occasionally small **rivulets of "free fluid"** and **small free fluid vesicles** are also present. It means fluid that is free of the proteoglycan molecules, can flow freely.

6.2 STRUCTURE AND FUNCTIONS OF HAEMOGLOBIN

The red, oxygen carrying pigment in the erythrocytes (red blood cells or RBCs) of vertebrates is **haemoglobin.** Haemoglobin, the respiratory pigment, performs two important biological functions: 1. Delivery of oxygen (O_2) from the lungs to the tissues. 2. Transport of carbon dioxide (CO_2) and protons from tissues to lungs for excretion.

Haemoglobin consists of the protein **globin** (polypeptide) united with the red iron pigment **haem** (or heme).

Structure

Haemoglobin is a conjugated protein, cantaining **globin**- the apoprotein part and the **haeme**- the non-protein part (prosthetic group). Haemoglobin is a tetrameric allosteric protein (Fig. 6.7).

1. Haeme. It is an iron-containing porphyrin, called **iron-protoporphyrin IX**. The porphyrin nucleus is **tetrapyrrole**, *i.e.*, it consists of 4 'pyrrole rings' joined together by 4 methine (=CH–) bridges. The pyrrole rings are numbered I, II, III and IV; the carbon atoms of the methine bridges are labelled α, β, γ, δ; the position on pyrrole rings to which side chains are attached numbered 1 to 8. The side chains at 1,3,5 and 8 position are methyl (– CH_3); 2 and 4 are vinyl (– CH = CH_2); 6 and 7 are propionic acid (–$CH_2 \cdot CH_2 \cdot COOH$) or propionyl (Fig. 6.8).

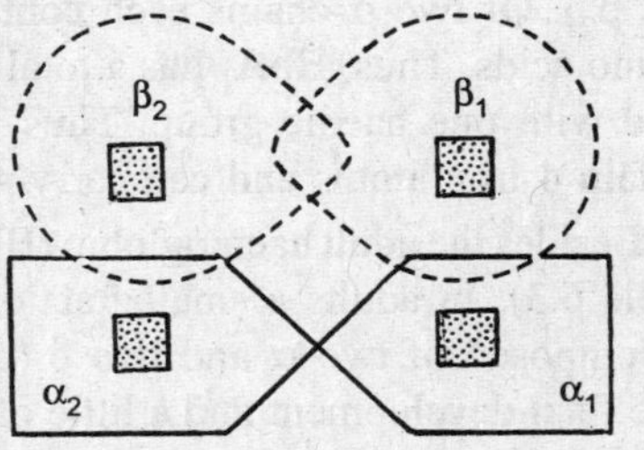

Fig. 6.7. Diagrammatic representation of haemoglobin with 2α and 2β chains (shaded blocks represent haeme).

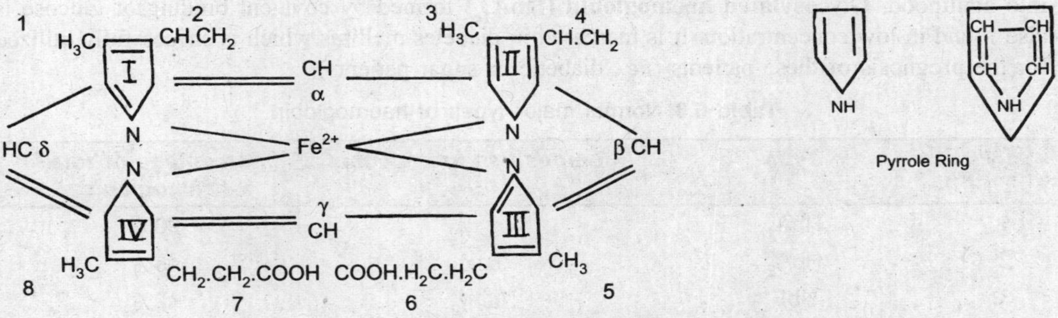

Fig. 6.8. Structure of haemoglobin molecule.

Haeme is common prosthetic group present in cytochromes, in certain enzymes such as catalase, tryptophan pyrolase, and chlorophyll (Mg^{2+}). In case of cytochromes, oxidation and reduction of iron ($Fe^{2+} \rightleftharpoons Fe^{3+}$) is essential for electron transport chain.

2. Iron. The iron in haeme is in **ferrous** (Fe^{2+}) form (Fig. 6.9). The iron is attached to the 'N' of each pyrrole ring. Each Fe^{2+} combines loosely and reversibly with one molecule of oxygen. Combination of haeme with oxygen is called **oxygenation** and not oxidation, because, after combination with oxygen, iron in the haeme stays in Fe^{2+} state. Therefore, the oxygen does not become ionic oxygen but is carried as molecular oxygen.

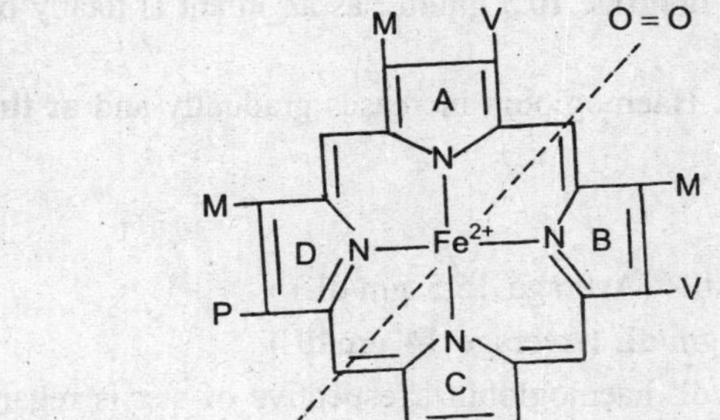

Fig. 6.9. Structure of haem (**M**, methyl (–CH_3); **V**, vinyl (– CH_2 – CH_2–); **P**, propionyl (– CH_2 – CH_2 – CH_2).

3. **Globin.** Globin is a protein built from four polypeptide chains of different primary structures (monomeric units), two α and two β chains. Therefore, the normal adults haemoglobin (HbA_1) is written as HbA_1 (α_2 β_2). Of two α-chains each contains 141 amino acids and of two β-chains each contains 146 amino acids. Thus, HbA_1 has a total of 574 amino acid residues. Each polypeptide chain is associated with one haeme group. Thus, there are four haeme to the one molecule of haemoglobin, contain 4 iron atoms and can carry 4 molecules (8 atoms) of oxygen.

Other forms of Haemoglobin. Besides the adult haemoglobin (Hb A_1), other types of haemoglobins are also found in humans (Table 6.3). In adults a small fraction (<5%) of haemoglobin, known as **HbA_2** is present. HbA_2 is composed of two α and two δ (delta) chains. **Fetal haemoglobin (HbF)** is synthesized during the fetal development and a little of it may be present even in adults.

Hereditary persistence of fetal haemoglobin (HPFH)

HPFH is a condition in which fetal haemoglobin synthesis is not terminated at birth but continues into adulthood. **Glycosylated haemoglobin (HbA_{1c})** formed by covalent binding of glucose is also found in low concentration. It is increased in diabetes mellitus which is successfully utilized for the prognosis of these patients (*i.e.*, diabetic or sugar patients).

Table 6.3 Normal major types of haemoglobin.

S. No.	*Type*	*Composition and symbol*	*Percentage of total haemoglobin*
1.	HbA_1	α_2 β_2	90%
2.	HbA_2	α_2 δ_2	<5%
3.	HbF	α_2 γ_2	<2%
4.	HbA_{1c}	α_2 β_2-glucose	<5%

Myoglobin. Myoglobin (Mb) is monomeric oxygen binding haemoprotein found in heart and skeletal muscle. It has a single polypeptide (153 amino acid) chain with a single haeme moiety. Myogloblin (mol. wt. 17000) structurally resembles the individual subunits of haemoglobin molecule. Myoglobin functions as a **reservoir for oxygen.** It further serves as oxygen carrier that promotes the transport of oxygen rapidly respiring muscle cells.

Normal Values of Haemoglobin

1. **At birth.** 23 gm/dL, because RBC count is more.

2. **At the end of 3 months.** 10.5 gm/dL, as an infant is totally on milk feed which is devoid of iron.

3. **After 3 months.** Haemoglobin increases gradually and **at the end of 1 year** it becomes 12.5 gm/dL.

4. Adults

Males. 14–18 gm/dL (Average 15.5 gm/dL)

Females. 12–15.5 gm/dL (Average 14 gm/dL)

Clinically 14.8 gm/dL haemoglobin irrespective of sex is regarded as 100% haemoglobin.

When blood is equilibrated with 100% oxygen (pO_2 = 760 mm Hg), the normal haemoglobin becomes 100% saturated.

1 gm/dL haemoglobin when fully saturated combines with 1.34 mL oxygen, therefore, haemoglobin concentration is an index of **oxygen-carrying capacity** of blood. Normal values— males 21 mL/dL, females, 18 mL/dL.

Functions of Haemoglobin

Haemoglobin is largely responsible for the **transport of O_2** from lungs to tissues. It also helps to **transport CO_2** from tissues to the lungs.

1. Binding of O_2 to haemoglobin (*i.e.*, **oxyhaemoglobin**). One molecule of haemoglobin (with four haemes) can bind with four molecules of O_2. This is in contrast to myoglobin (with one haeme) which can bind with only one molecule of oxygen. In other words, each haeme moiety can bind with one O_2.

Oxygen dissociation curve. The **binding** ability of *haemoglobin with O_2* at different partial pressures of oxygen (pO_2) can be measured by a graphic representation known as oxygen dissociation curve (Fig. 6.10).

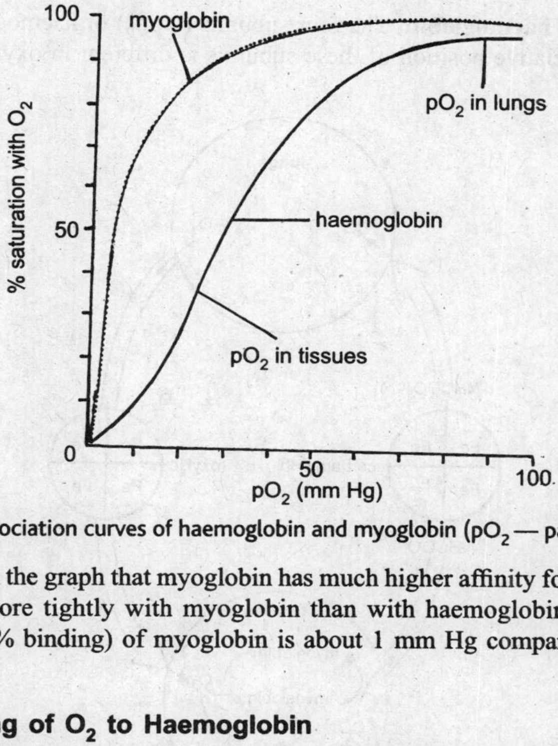

Fig. 6.10. Oxygen dissociation curves of haemoglobin and myoglobin (pO_2 — partial pressure of oxygen).

It is evident from the graph that myoglobin has much higher affinity for O_2 than haemoglobin. Hence O_2 is bound more tightly with myoglobin than with haemoglobin. Further, pO_2 needed for half saturation (50% binding) of myoglobin is about 1 mm Hg compared to about 26 nm Hg for haemoglobin.

Cooperative Binding of O_2 to Haemoglobin

The oxygen dissociation curve for haemoglobin is sigmoid (S – shaped) in shape (Fig. 6.10). This indicates that the binding of oxygen to one haeme increases the binding of oxygen to other haemes. Thus the affinity of Hb for the last O_2 is about 100 times greater than the binding of the first O_2 to Hb. This phenomenon is referred to as **cooperative binding of oxygen (O_2) to haemoglobin (Hb)** or simply **haeme-haeme interaction** (Fig. 6.11). On the other hand, release of O_2 from one haeme facilitates the release of O_2 from others. In short, there is a communication among haeme groups in the haemoglobin function.

Transport of oxygen to the tissues. In the lungs, where the concentration of O_2 is high (hence high pO_2), the haemoglobin gets fully saturated (loaded) with O_2. Conversely, at the tissue level, where the O_2 concentration is low (hence low pO_2), the oxyhaemoglobin releases (unloaded) the O_2 for cellular respiration. This is often mediated by binding O_2 to myoglobin which serves as the immediate reservoir and supplier of O_2 to the tissues (Fig. 6.12).

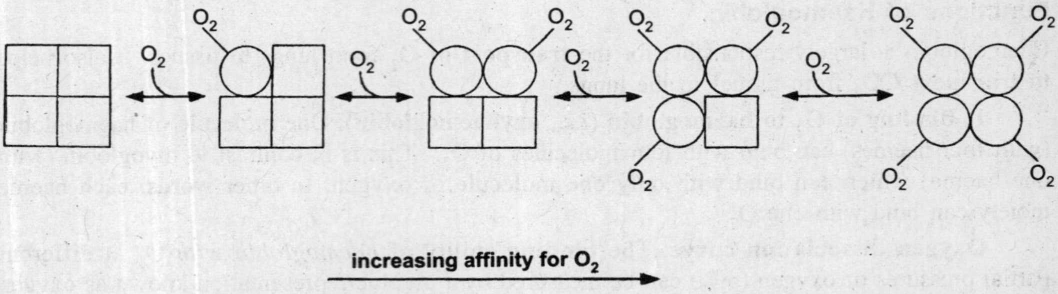

Fig. 6.11. Cooperative binding of oxygen (O_2) to haemoglobin. (□, T or taut form; O, R or relaxed form).

T or R forms of haemoglobin. The four subunits ($\alpha_2\ \beta_2$) of haemoglobin are held together by weak forces. The relative position of these subunits is different in oxyhaemoglobin compared to deoxyhaemoglobin.

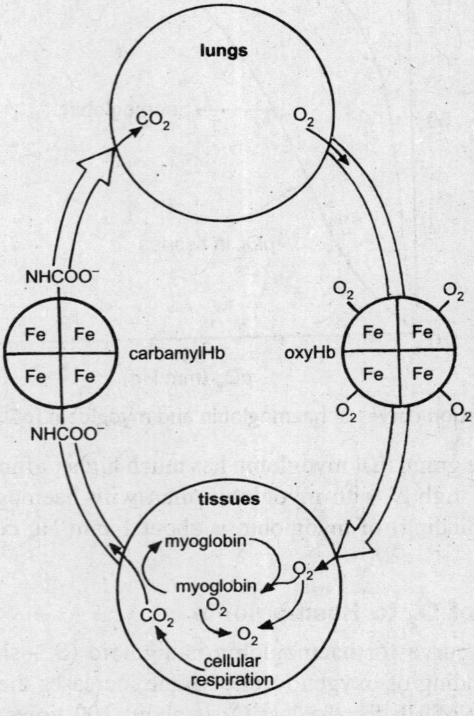

Fig. 6.12. Diagrammatic representation of transport of O_2 and CO_2 by haemoglobin.

T-form of Hb. The deoxy form of haemoglobin (Hb) exists in a T or taut (tense) form. The hydrogen and ionic bonds limit the movement of monomers. Therefore, the **T-form of Hb has low oxygen affinity**.

R-form of Hb. The binding of O_2 destabilizes some of the hydrogen and ionic bonds particularly between α β dimers. This results in a relaxed form or R- form of Hb. Wherein the subunits move a little freely. Therefore, the **R-form of Hb has high oxygen affinity**.

The existence of haemoglobin in two forms (T and R) suitably explains the allosteric behaviour of haemoglobin (Fig. 6.11).

2. Transport of CO_2 by haemoglobin. In aerobic metabolism, for every molecule of O_2 utilized, one molecule of CO_2 is liberated. Haemoglobin actively participate in the transport of CO_2 from the tissues to the lungs. About 15% of CO_2 carried in blood directly binds with Hb. The rest of the tissue CO_2 is transported as bicarbonate (HCO_3^-).

Carbon dioxide molecules are bound to the uncharged α- amino acids of haemoglobin to form carbomyl haemoglobin as show below

$$Hb-NH+CO_2 \rightleftharpoons Hb-NH-COO^- + H^+$$

The oxyhaemoglobin (oxyHb) can bind 0.15 moles CO_2/mole haeme, where deoxyHb can bind 0.40 moles CO_2/mole haeme. The binding of CO_2 stabilized the T (taut) form of haemoglobin structure, resulting in decreased O_2 affinity for haemoglobin.

Haemoglobin also helps in the transport of CO_2 as bicarbonate (Fig. 6.13). As the CO_2 enters the blood from tissues, the enzyme **carbonic anhydrase** present in erythrocytes catalyzes the formation of carbonic acid (H_2CO_3). Bicarbonate (HCO_3^-) and proton (H^+) are released on dissociation of carboinic acid. Haemoglobin acts as a buffer and immediately binds with protons. It is estimated that every two protons bound to Hb, 4 oxygen moleculer are released to the tissues. In the lungs, binding of O_2 to Hb result in the release of protons. The bicarbonate and protons combine to form carbonic acid. The carbonic acid is acted upon by carbonic anhydrase to release CO_2, which is exhaled.

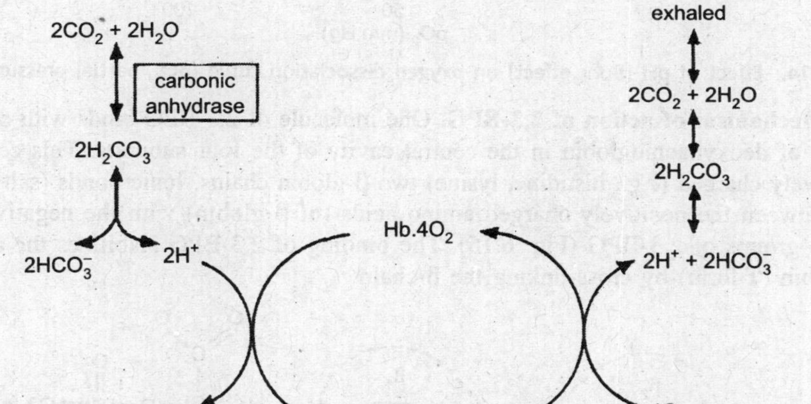

Fig. 6.13. Transport of CO_2 through the mediation of haemoglobin.

Effects of 2,3-Bisphosphoglycerate (2,3- BPG) on O_2 Affinity of Hb

2,3-BPG or 2,3-bisphosphoglycerate (formerly called 2,3-diphosphoglycerate) is the most abundant organic phosphate is the erythrocytes. Its molar concentration is approximately equivalent to that of haemoglobin. 2,3-BPG is produced in the erythrocytes from an intermediate (1, 3-bisphoshoglycerate) of glycolysis. This short pathway, referred to as **Rapaport-Leubering cycle**, is part of carbohydrate metabolism and includes following roles:

1. Binding of 2,3-BPG to deoxyhaemoglobin. 2,3-BPG regulates the binding of O_2 to haemoglobin. It specifically binds to deoxyhaemoglobin (and not to oxyhaemoglobin) and decreases the O_2 affinity to Hb. The effect of 2,3-BGP on Hb may be summarized as follows:

$$\underset{\text{OxyHb}}{HbO_2} + 2,3-BPG \longrightarrow \underset{\substack{\text{DeoxyHb bound to}\\ \text{2,3-BPG}}}{Hb-2,3-BPG} + O_2$$

The reduced affinity of O_2 to Hb facilitates the **release of O_2** at the partial pressure found in the tissues. The 2,3-BPG shifts the oxygen dissociation curve to the right (Fig. 6.14).

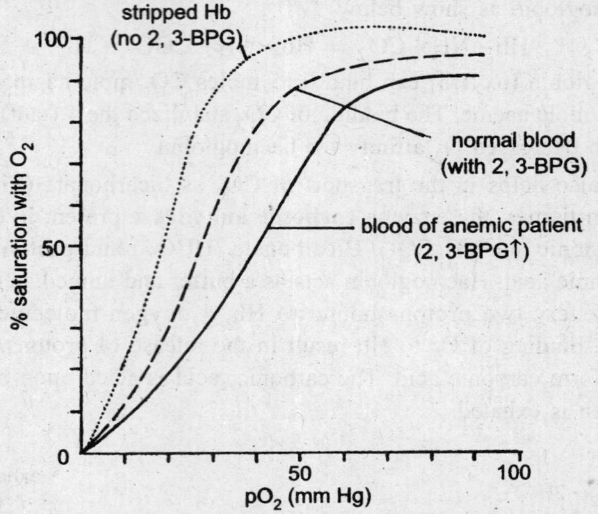

Fig. 6.14. Effect of pH (Bohr effect) on oxygen dissociation curve (pO_2, partial pressure of O_2).

2. Mechanism of action of 2,3-BPG. One molecule of 2,3-BPG binds with one molecule (tetramer) of deoxyhaemoglobin in the central cavity of the four subunits. This central pocket has positively charged (*e.g.*, histidine, lysine) two β-globin chains. Ionic bonds (salt bridges) are formed between the positively charged amino acids (of β globin) with the negatively charged phosphate groups of 2,3-BPG (Fig. 6.15). The binding of 2,3-BPG stabilizes the doxygenated haemoglobin (T-form) by cross-linking the β-chain.

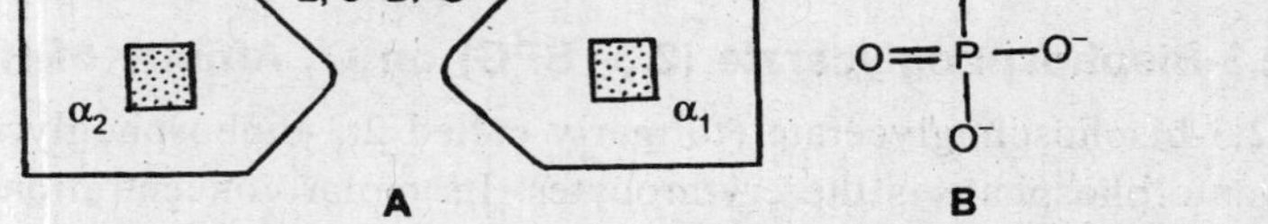

Fig. 6.15. A – Diagrammatic representation of binding of 2, 3- BPG to deoxy-haemoglobin; B – Structure of 2,3- BPG.

On oxygenation of haemoglobin, 2,3-BPG is expelled from the pocket and the oxyhaemoglobin attains the R- form of structure.

Clinical significance of 2,3-BPG. Since the binding of 2,3-BPG with haemoglobin is primarily associated with the release of O_2 to the tissues, this small molecule assumes a lot of biomedical significance. The erythrocyte levels of 2,3-BPG are related to tissue demands of oxygen supply.

1. In hypoxia. The concentration of 2,3-BPG in erythrocytes is **elevated** in O_2 supply (*e.g.*, adaptation to high altitude).

2. In anemia. 2, 3-BPG levels are **increased** in severe anemia in order to cope up with the oxygen demands of the body. This is an adaptation to supply as much as O_2 as possible to the tissue, despite the low haemoglobin level.

3. In blood transfusion. Storage of blood in acid citrate-dextrose medium results in the **decreased concentration of 2,3-BPG.** Such blood when transfused fails to supply O_2 to the tissues immediately. Addition of **inosine** (hypoxanthine-ribose) to the stored blood prevents the decreased of 2,3-BPG.

4. Fetal haemoglobin (HbF). The binding of 2,3-BGP to fetal haemoglobin is very weak. Therefore, HbF has **higher affinity for O_2** compared to adult haemoglobin (HbA) This may be needed for the transfer of oxygen from the maternal blood to the fetus.

Haemoglobin Derivatives

Haemoglobin (specifically Haeme) combines with different ligands and forms haemoglobin derivateives. The normal blood contains oxyHb and deoxyHb. Besides these, **methaemoglobin (metHb)** and **carboxyhaemoglobin (COHb)** are the important derivatives.

Methaemoglobin. For the biological function of haemoglobin (*i.e.*, to carry oxygen), the iron should remain in the ferrous (Fe^{2+}) state. Haemoglobin (Fe^{2+}) can be oxidized to methaemoglobin (Fe^{3+}, *i.e.*, iron is in ferric state). In normal, circumstances, however, molecular oxygen does not oxidise Hb, it only loosely binds to form oxyhaemoglobin.

The oxidation of haemoglobin to methaemoglobin (metHb) may be caused in the living system by **H_2O_2, free radicals,** and **drugs**. The methaemoglobin (with Fe^{3+}) is unable to bind to O_2. Instead a water molecule occupies the oxygen site in the haeme of metHb.

In normal circumstances, the occasional oxidation of haemoglobin is corrected by the enzyme **methaemoglobin reductase** present in the erythrocytes (Fig. 6.16).

Carboxyhaemoglobin (COHb). Carbon monoxide (CO) is a toxic compound (an air pollutant or industrial pollutant) that can bind with Hb in the same manner as O_2 binds. CO forms carboxyhaemoglobin (COHb). However, CO has about 200 times more affinity than O_2 for binding with Hb.

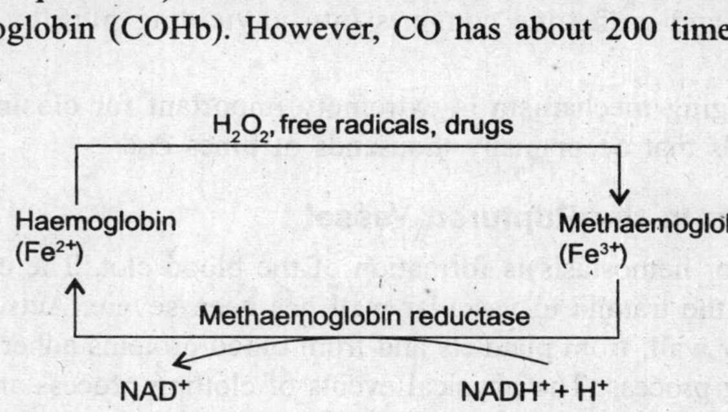

Fig. 6.16. Conversion of haemoglobin to methaemoglobin and vice versa.

Clinical manifestation of *carbon monoxide (CO) toxicity* observed when the COHb concentration exceeds 20%. The symptoms of carbon monoxide toxicity include headache, nausea, breathlessness, vomiting and irritability. Administration of O_2 through oxygen masks will help to reverse the manifestation CO toxicity.

6.3 HEMOSTASIS AND BLOOD COAGULATION

The term **hemostasis** means prevention of blood loss. Whenever a blood vessel is severed or ruptured, hemostasis is achieved by several mechanisms: 1. vascular constriction, 2. formation of a platelet plug, 3. formation of a blood clot as a result of blood coagulation, and 4. eventual growth of fibrous tissue into the blood clot to close the hole in the vessel permanently.

1. Vascular Constriction

Immediately after a blood vessel has been cut or ruptured, the trauma (Box 6.4) to the vessel wall causes the smooth muscle in the wall to contract, this immediately reduces the flow of blood from the ruptured vessel. The contraction results from (i) local myogenic spasm. 2. local autacoid factors and 3. nervous reflexes. The nervous reflexes are initiated by pain nerve impulses or other sensory impulses that originate from the traumatized vessel or nearby tissue. In ruptured small vessels, blood platelets tend to release a vasoconstrictor called **thromboxane A (Guyton** and **Hall,** 2011).

Box 6.4

Trauma. An injury or wound to living tissue caused by extrinsic agent, accident, surgery, etc.
Autacoid factor. An autacoid factor is physiologically active substance such as serotonin, bradykinin or angiotensin, produced by and acting within the body.

2. Formation of the Platelet Plug

If the cut in the blood vessel is very small, it is often sealed by a **platelet plug** rather than by a blood clot. When blood platelets come in contact with a damaged vascular surface, especially with **collagen fibers** in the vascular wall, the platelets immediately change their own characteristics drastically. They start to swell; they assume irregular forms with numerous irradiating pseudopods protruding from their surfaces; their contractile proteins contract forcefully and cause the release of granules that contain multiple active factors; they become sticky so that they adhere to collagen in the tissues and to a protein called **von Willebrand factor** that leaks into the traumatized tissue from the plasma; they secrete large quantities of ADP; and their enzymes form **thromboxane A**. The ADP and thromboxane in turn act on nearby platelets to activate them as well, and the stickiness of these additional platelets causes them to adhere to the original activated platelets. Such a newly formed **platelet plug** is initially loose plug and with time converts into unyielding plug by deposition of **fibrin threads** of blood clot.

The platelet-plugging mechanism is extremely important for closing minute ruptures in very small blood vessels that occur many thousands of times daily.

3. Blood Coagulation in the Ruptured Vessel

The third mechanism for hemostasis is formation of the blood clot. The clot begins to develop in 15 to 20 seconds if the trauma to vascular wall has been severe. Activator substances from the traumatised vascular wall, from platelets and from blood proteins adhering to the traumatised wall initiate the clotting process. The physical events of clotting process are shown in Fig. 6.17. The table 6.4 list the most important factors of blood coagulation.

Blood Coagulation Factors

An International Commission on Blood Coagulation in 1954 suggested a system of nomenclature of blood coagulation factors, which are designated by Roman numerals according to the recommendation of the numerical order, which indicates the historical sequence of their discovery.

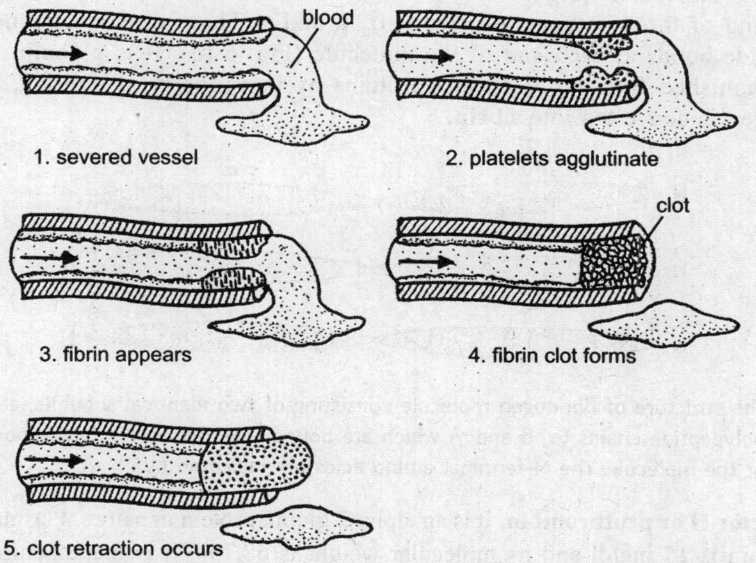

Fig. 6.17. Clotting process in a traumatised blood vessel (after Guyton and Hall 2011).

Table 6.4 Clotting factors in blood and their synonyms.

S. No.	Clotting factor	Synonyms
1.	Fibrinogen	Factor I
2.	Prothrombin	Factor II
3.	Tissue factor	Factor III; tissue thromboplastin
4.	Calcium ions	Factor IV
5.	Factor V	Proaccelerin; labile factor; accelerator-globulin (Ac-G)
6.	Factor VII	Serum prothrombin conversion accelerator (SPCA); proconvertin; stable factor
7.	Factor VIII	Antihemophilic factor (AHF); antihemophilic factor A; platelet cofactor I; antihemophilic globulin (AHG)
8.	Factor IX	Plasma thromboplastin component (PTC); Christmas factor; antihemophilic factor B
9.	Factor X	Stuart factor; Stuart-Prower factor; prothrombinase
10.	Factor XI	Plasma thromboplastin antecedent (PTA); antihemophilic factor C
11.	Factor XII	Hageman factor; glass factor
12.	Factor XIII	Fibrin stabilizing factor; fibrinase
13.	Prekallikrein	Fletcher factor
14.	High-molecular weight kininogen (HMWK)	Fitzgerald factor
15.	Platelets	

Note. There is no Factor VI.

1. Factor I or fibrinogen. Plasma concentration of fibrinogen is 100 to 700 mg/dl and its molecular weight is 340,000. The molecule has two identical subunits, each consisting of three polypeptide chains α, β and γ. The chains are bound together by disulphide bonds at each end of the molecule (Fig. 6.18). It is globulin in nature and is distinguished from other plasma proteins by its property of clotting, during which fibrinogen is converted into **fibrin.**

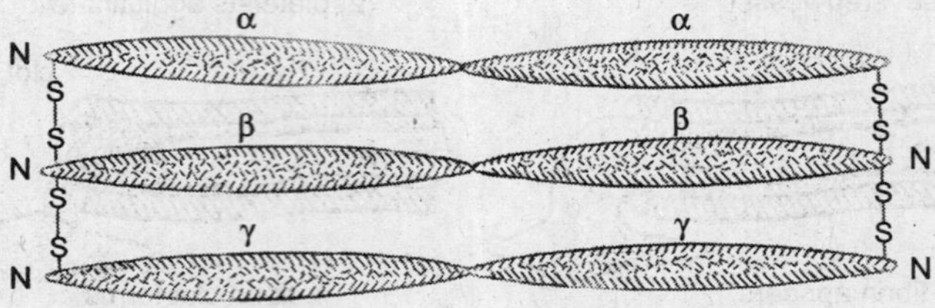

Fig. 6.18. The structure of fibrinogen molecule consisting of two identical subunits, each having three polypeptide chains (α, β and γ) which are bound together by disulphide bonds at each end of the molecule. The N-terminal amino acids are indicated by N.

***(ii)* Factor II or prothrombin.** It is an alpha-2 globin protein in nature. Plasma concentration of prothrombin is 15 mg/dl and its molecular weight is 68,700. It is manufactured in the liver. However, vitamin K is essential for the formation of prothrombin. During clotting, prothrombin is converted into thrombin by means of prothrombin activator, which is also called **thromboplastin** or **thrombokinase.**

***(iii)* Factor III or thromboplastin or tissue factor.** It is a protein having a molecular weight 43000. It is derived from two sources: *intrinsic* in the plasma and *extrinsic* in the tissues. **Intrinsic thromboplastin** is formed in the plasma due to interaction between different factors, *e.g.,* **factor XII** (Hageman factor), **factor XI** (AHF), **calcium ions** (factor IV), **factor** V (AcG) and **factor X** (prothrombinase). **Extrinsic** or **tissue thromboplastin** is formed from different tissues, *e.g.,* extracts of lungs, placenta, brain, etc., as a result of injury. So far it was known that prothrombin was converted into thrombin with the help of calcium ions and thromboplastin liberated from the damaged tissues. Later on, it was discovered that various plasma factors, *e.g.,* **factor VII** or **proconvertin,** are required for such conversion and the process is called **extrinsic thromboplastin formation.**

***(iv)* Factor IV or calcium.** The plasma concentration of Ca^{2+} ions is 0.5 mg/dl. Ionic calcium acts as a **cafactor** in coagulation. It is essential for the formation of both intrinsic and extrinsic thromboplastins and also in the conversion of prothrombin into thrombin.

***(v)* Fator** V **or proaccelerin or labile factor or accelerator globulin (AC-G). It is a** heat-labile protein which is present in the plasma. The plasma concentration **of** factor V is 1 mg/dl and its molecular weight is 330,000. It is necessary for the complete conversion of prothrobin into thrombin by the extrinsic or intrinsic thromboplastin.

***(vi)* Factor VI or accelerin or Va.** This factor is a hypothetical activation product of factor V.

***(vii)* Factor VII or SPCA or proconvertin or stable factor.** This factor is present in the plasma and is not used up during clotting. It is a protein associated with prothrombin. The plasma concentration of proconvertin is 0.05 mg/dl and its molecular weight is 45,500. It accelerates the formation of tissue thromboplastin, being activated by extract released from the damaged tissue. During blood clotting proconvertin is changed into **convertin.**

(*viii*) Factor VIII or antihemophilic factor (AHF) or antihaemophilic globulin (AHG) or platelet cofactor. The plasma concentration of antihaemophilic factor or von Willbrand factor is 1.6 mg/dl and its molecular weight is 1.2×10^6, having 15% carbohydrate content. Factor VIII helps in the formation of intrinsic thromboplastin and intrinsic prothrombin conversion. It is present in the plasma and disappears when the blood clots. AHG is a protein closely associated to fibrinogen. This factor is antihaemophilic.

(*ix*) Factor IX or Christmas factor or PTC or platelet cofactor II or antihaemophilic factor B. The plasma concentration of Christmas factor is 0.3 mg/dl and its molecular weight is 57,000. It has 26% carbohydrate content. This factor is necessary for intrinsic thromboplastin formation. The absence of this factor leads to a disease known as **Haemophilia B.** This type of bleeder's disease was first detected in a patient named **Christmas,** hence the name Christmas factor. This factor is not used up during clotting.

(*x*) Factor X or Stuart-Prower factor. The plasma concentration of this factor is 1.5 mg/dl and its molecular weight is 59,000. It has 15% carbohydrate content. It is a circulating plasma zymogen, which on activation by factor IXa serves to activate prothrombin to thrombin.

(*xi*) Factor XI or PTA or antihaemophilic factor C. The plasma concentration of fator XI is 0.6 mg/dl and its molecular weight is 160,000. It has 12% carbohydrate content. This is activated by the active Hageman factor (Xlla) and ultimately leads to formation of thrombin.

(*xii*) Factor XII or Hagemen factor. (Note. Hageman was the first patient in whom a deficiency of factor XII was observed in 1963). The plasma concentration of factor XII is 2.9 mg/dl and its molecular weight is 76,000. It has 15% carbohydrate content. This is a protein in nature. Inactive form is activated on surface contact. This is also known as the clot-promoting factor or glass factor.

(*xiii*) Factor XIII or fibrin-stabilising or Laki-Lorand factor (LLF) or fibrinase. The active form along with calcium ions converts soft fibrin clot to a solid, fibrous state. The plasma concentration of factor XIII is 2.0 mg/dl and its molecular weight is 320,000. It has 2.5% carbohydrate content.

Other Factors for Clotting

1. High-molecular-weight kininogen (HMW-k) or **Fitzgerald factor.** This factor catalyses the intial reaction in which inactive factor XII is converted to active factor XII (*i.e.,* Xlla).

2. Pre-Kallikrein (Pre-k) or Fletcher factor. Pre-Kallikrein or inactive plasma kallikrein is activated by pre-Kallikrein activators, which are proteolytic fragments of the active form of clotting factor XII (Hageman factor).

3. Kallikrein. Plasma kallikrein also catalyses the activation of factor XII.

4. Platelet phospholipid (PL). Phospholipids from aggregated platelets (PL) and Ca^{2+} ions are necessary for full activation of factor X. This also catalyses the conversion of prothrombin to thrombin.

5. Protein C. Although not listed as clotting factor, protein C is often referred to as factor XIV. Its molecular weight is 62,000. It is a circulating zymogen of protease which on activation by thrombin specifically inactivates factor V and factor VIII by proteolysis. Congenital absence of protein C leads to uncontrolled intravascular clotting and, in general, death in infancy.

6. Role of thrombin. Thrombin is a glycoprotein. It autocatalyses its own formation by converting plasma Ac-globulin into active serum Ac-globulin and by labilising platelets to accelerate thromboplastin generation and liberates vasoconstrictors. By acting as a proteinase, it splits only four peptide bonds in conversion of fibrogen to fibrin. These bonds exist between arginine and glycine residues.

Role of phospholipid. Phospholipid **kephalin** (cephalin) helps in the formation of **prothrombinase**. In the intrinsic system it is in the platelet factor 3, a phospholipid which participates in the conversion of prothrombin to thrombin by factor X and V, and in the extrinsic system in tissue thromboplastin.

Role of vitamin K. Adequate dietary intake of vitamin K (coagulation vitamin) is requried for normal production in the liver of prothrombin as well as factor VII, IX and X. Vitamin K is required to form the binding sites for Ca^{2+} in prothrombin as well as in factors VII, IX and X.

The Ca^{2+} ion binding sites are formed by clusters of carboxyglutamate residues in the amino terminal regions of each factor.

Mechanism of Blood Clotting or Blood Coagulation

Basic concept of blood coagulation. More than 50 important substances that cause or affect blood coagulation have been found in the blood and in the tissues (**Guyton** and **Hall** 2011). Some of these substances *promote* coagulation, these are called **procoagulants.** Remaining of these substances *inhibit* coagulation, these are called **anticoagulants.** Whether blood will coagulate depends on the balance between these two groups of substances. In the blood stream, the anticoagulants normally predominant, so the blood does not coagulate while it is circulating in the blood vessels. But when a vessel is ruptured, procoagulants from the area of tissue damage become "activated" and override the anticoagulants, and then a clot does develop.

Three Steps of Blood Coagulation

The process of clotting of blood is initiated by platelets in mammals and by thrombocytes in other vertebrates. The injured cells release substances that attract the platelets/thrombocytes. They gather at and stick to the injured surface of the blood vessel. Their mass alone may physically plug a cut in a very small vessel. The contact of platelets (or thrombocytes) with the collagen fibers exposed by the injury causes them to disintegrate and release three substances: serotonin, thromboxane A_2 (**Guyton** and **Hall** 2011) and thromboplastin (or thrombokinase). Thromboplastin is also released by damaged tissue cells. It is a lipoprotein. Serotonin and thromboplastin minimize loss of blood from the injury in following two ways:

1. Serotonin is produced by argentaffin cells located in the crypts of Lieberkuhn in the intestinal mucosa. It is carried by blood platelets and causes the contraction of blood vessels (vasoconstriction) at the site of bleeding. This reduces the blood loss, and also makes it less likely for the clot formed to plug the injury to be swept out by flow of blood. Serotonin also lowers the blood pressure and this reduces the amount of bleeding to some extent.

2. Thromboplastin initiates the process of clot formation. It is not present in the circulating blood. It is relesed from the breakdown of platelets and damaged tissue cells at the site of injury. The blood escaping from the wound comes in contact with air and mixes with thromboplastin. This starts the clotting process. Clot formation occurs in three steps (**Morowitz** 1904).

(*i*) Thromboplastin, along with the clotting factors VII and X (plasma enzymes), catalyzes the conversion of the inactive plasma protein **prothrombin** to its active form **thrombin.** The change requires calcium (Ca^{2+}) ions. Prothrombin is kept inactive in the plasma by a substance called **antiprothrombin.**

(*ii*) Thrombin acts as a proteolytic enzyme to separate two peptides (fibrinopeptidases A and B) from the large, soluble, globular plasma protein **fibrogen** molecule to from insoluble, filamentous proetin **fibrin** monomer.

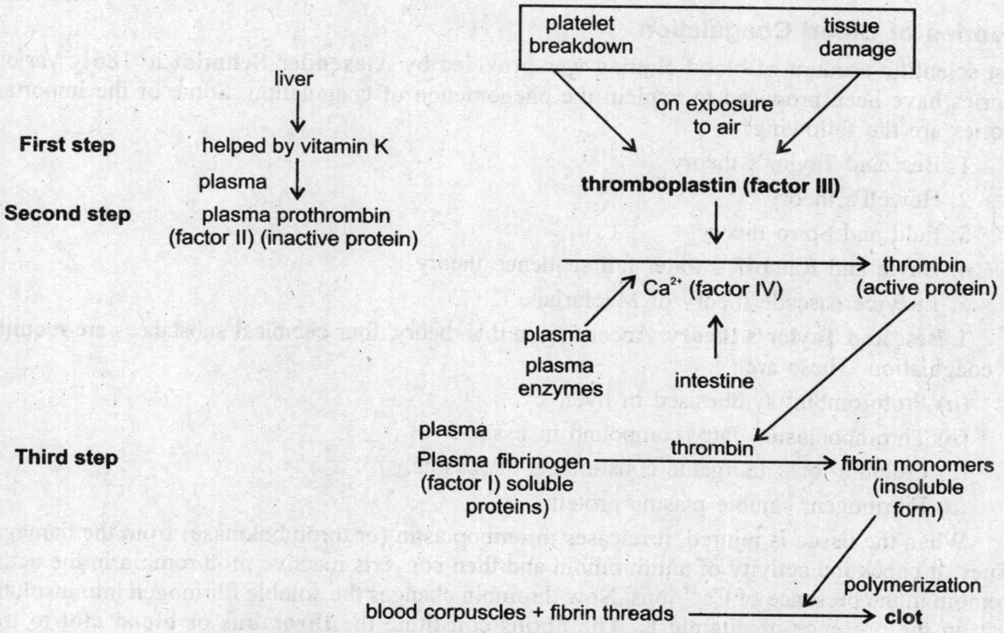

Fig. 6.19. Mechanism of blood clothing or coagulation.

Thrombin also causes further aggregation and breakdown of the platelets at the injured site of the blood vessel for more thromboplastin. This is called **positive feedback mechanism.**

(*iii*) The fibrin monomers polymerise to form long, sticky fibers. The fibrin threads form a fine network over the wound. The threads being sticky, adhere to the surrounding tissues. The fibrin threads are so closely interlaced that even the erythrocytes cannot pass through the interstices. The net traps blood corpuscles (RBCs, WBCs, platelets) to from a crust, the **clot** (Fig. 6.19). The clot seal the minor wound and stops bleeding and also checks entry of pathogens. A fibrin thread consists of seversal filaments having alternating light and dark bands. The corresponding bands of adjacent filaments are in alignment. The platelets entrapped in the clot release more thromboplastin, which keeps the clotting cycle going. Clot platelets also polymerise actin and myosin into a contractile apparatus. The latter contracts, and the initially soft, jelly-like clot becomes relatively firm and soild. The part of blood plasma in the clotted area is squeezed out as a clear fluid called **blood serum**. The serum is plasma except that it lacks fibrinogen and corpucles. It, is therefore, unable to clot.

Role of vitamin K in blood clotting. Vitamin K is necessary for the synthesis of prothrombin in the liver. If vitamin K is inadequate in the diet or is not absorbed in the intestine, blood clotting becomes inefficient.

Natural anticoagulants. A substance that checks the clotting of blood is called **anticoagulant**. Blood does not clot in uninjured vessels because their smooth lining does not protmote the breakdown of platelets, and the blood in them contains a strong natural anticoagulant, **heparin**. Heparin is a mucopolysaccharide produced in the liver. It checks the change of prothrombin into thrombin and also of fibrinogen into fibrin in conjunction with a serum protein cofactor (heparin cofactor).

Haemophilia. Lack or low concentration of any of the essential clotting factors would cause excessive bleeding if injury occurs. This condition is called **bleeder's disease** or **haemophilia**.

It is due to an inheritable, sex-linked, recessive gene mutation on a X charomosome. Hence, haemophilia occurs only in males (X^hY), the females are carriers (XX^h). It is transmitted from the mother to the son (*i.e.*, this is possible by the homozygous haemophilic mother X^hX^h).

Theories of Blood Coagulation

First scientific account of blood clotting was provided by **Alexender Schmidt** in 1861. Various theories have been proposed to explain the phenomenon of coagulation. Some of the important theories are the following:

1. Best and Taylor's theory
2. Howell's theory
3. Fuld and Spiro theory
4. Davie and Ratnoff's water fall sequence theory
5. Enzyme cascade theory of Macfarlane

1. Best and Taylor's theory. According to this theory, four chemical substances are required for coagulation. These are

(*a*) Prothrombin: synthesised in liver.

(*b*) Thromboplastin: fatty compound in tissues.

(*c*) Calcium ions: inorganic constituents of plasma.

(*d*) Fibrinogen: soluble plasma protein.

When the tissue is injured, it releases thromboplastin (or thrombokinase) from the damaged tissues. It check the activity of antithrombin and then converts inactive prothrombin in the active thrombin in the presence of Ca^{2+} ions. Now thrombin changes the soluble fibrinogen into insoluble fibrin in the presence of vitamin K. The fibrins constitute the **thrombus** or **blood clot** to trap the blood corpuscles to block the injured part of blood vessel.

2. Howell's theory. According to this theory, thrombokinase enzyme does not play any role in conversion of prothrombin into thrombin. This conversion is done in the presence of Ca^{2+} ions alone. It was also suggested that an enzyme **tryptase** is released by the injured tissues and it helps in conversion of prothrombin into thrombin in the presence of calcium ions (Fig. 6.20).

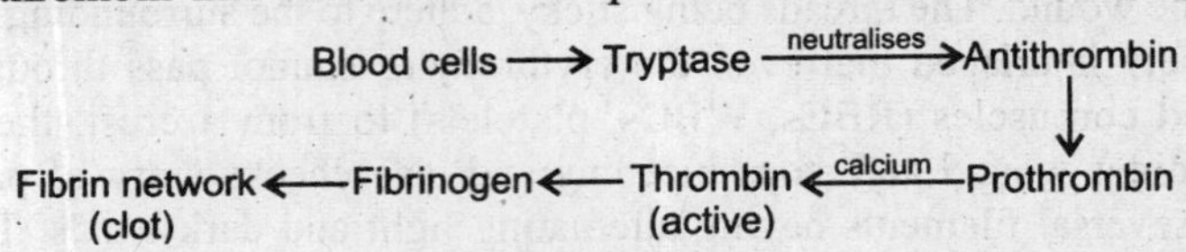

Fig. 6.20. Steps of process of blood clotting according to Howell's theory.

3. Fuld and Spiro theory. According to this theory, injured blood platelets release thromboplastin which converts prothrombin into thrombin in the presence of calcium ions. Thrombin converts soluble fibrinogen into insoluble fibrin thread to form **clot.**

4. Davie and Ratnoff's Water fall sequence theory. Davie and **Ratnoff** (1965) have proposed the so-called **waterfall sequence theory** to explain the succession of events taking place in the coagulation of blood. The sequence of events leading to blood coagulation depends upon whether **thromboplastin** is present or not. It is the active substance in aqueous extracts of all tissues. Thus, a wound that results in bleeding also disrupts tissue causing the liberation of thromboplastin. This substance then initiates the sequence shown in Fig. 6.21. Because a substance is required that is not a normal constituent of blood, this sequence is called **extrinsic system of coagulation.** On the other hand when all the substances involved are normally found in the blood, this sequence is called **intrinsic system of coagulation.** In both the systems each coagulation factor exists in the plasma in an inactive (proenzyme) form and is activated sequentially until thrombin is formed, which converts fibrinogen into fibrin (Fig. 6.21).

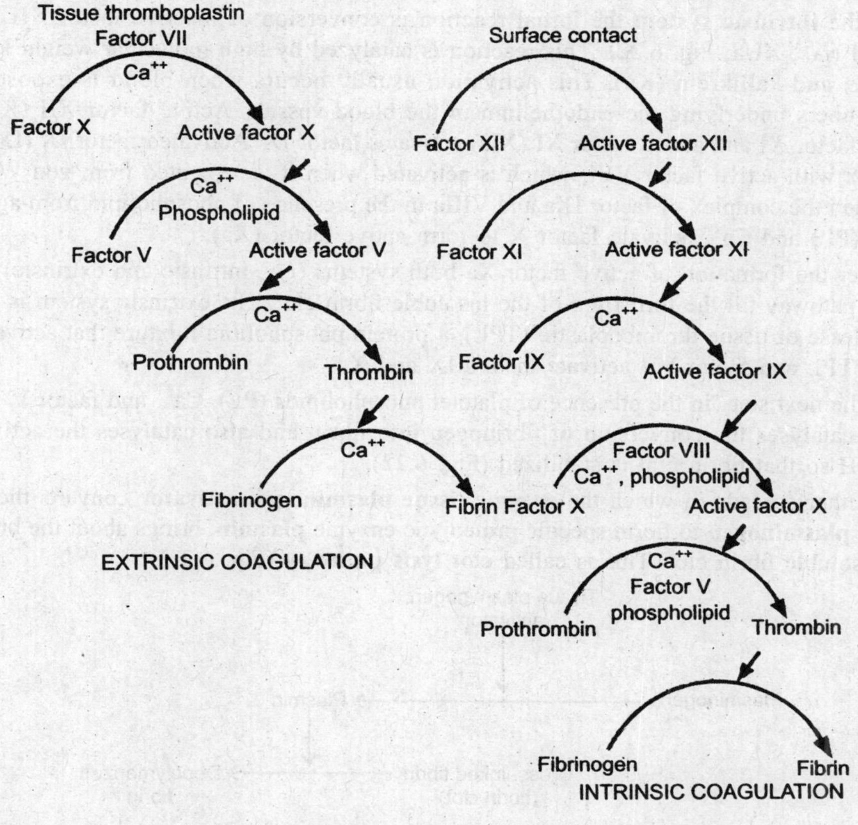

Fig. 6.21. The waterfall scheme of extrinsic and intrinsic blood coagulation.

5. Enzyme cascade theory of Macfarlane. This theory was proposed by **Macfarlane** (1967) and it has suggested that blood coagulation mechanism consists of a large number of enzymatic reactions. The term **cascade** expresses the relation-ship whereby an enzyme acts on a precursor, converting it to an enzyme that then act on another precursor, converting it to an active enzyme and so on.

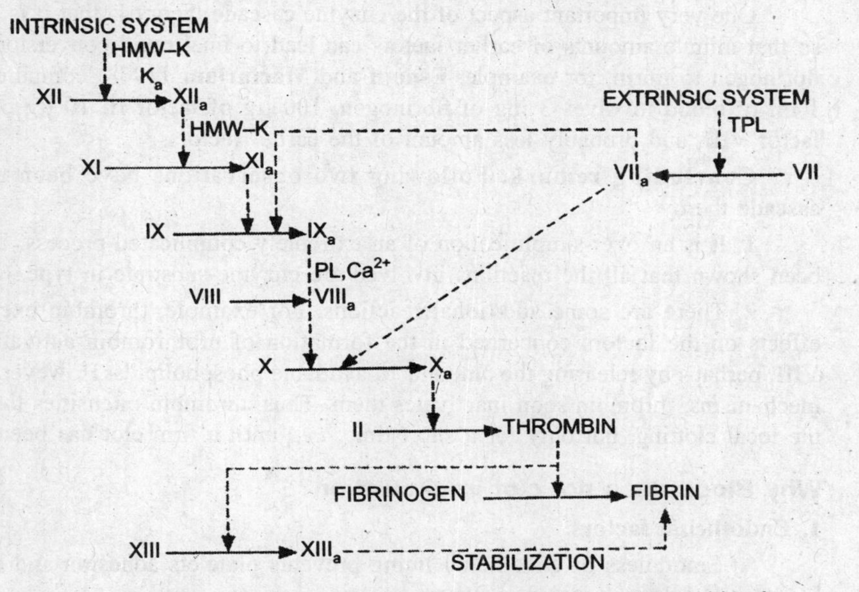

Fig. 6.22. The scheme of cascade reactions leading to blood coagulation by activation of intrinsic and extrinsic factors.

In the **intrinsic system** the initial reaction is conversion of inactive factor XII to active factor XII (*i.e.*, XUa; Fig. 6.22). This reaction is catalyzed by high molecular weight kininogen (HMW-K) and kallikrein (Ka). This activation usually occurs when blood is exposed to the collagen fibers underlying the endothelium in the blood vessels. Active factor XII (XIIa) then activates factor XI and active factor XI (XIa) activates factor IX. Activated factor IX (IXa) forms a complex with active factor VIII, which is activated when it is separated from von Willebrand factor. Then the complex of factor IXa and VIIIa in the presence of phospholipid from aggregated platelets (PL) and Ca^{2+} activate factor X to form active factor (Xa).

After the formation of active factor Xa both systems (*i.e.*, intrinsic and extrinsic) follow a common pathway till the formation of the insoluble fibrin clot. The extrinsic system is triggered by the release of tissue thromboplastin (TPL), a protein phospholipid mixture that activate factor VII. The TPL and factor VII activate factors IX and X.

In the next step, in the presence of platelet phospholipids (PL), Ca^{2+} and factor V, activated factor X catalyses the conversion of fibrinogen into fibrin and also catalyses the activation of factor XIII so that fibrin clot is stabilized (Fig. 6.22).

Another cascade in which the enzyme **tissue plasminogen activator** converts the inactive precursor **plasminogen** to fibrin specific proteolytic enzyme **plasmin**. brings about the breakdown of the insoluble fibrin clot. This is called **clot lysis pathway** (Fig. 6.23).

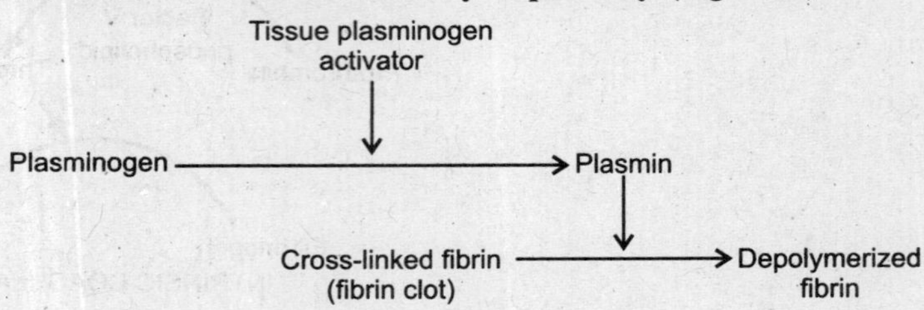

Fig. 6.23. Clot lysis pathway.

One very important aspect of the enzyme cascade theory is that it is an amplifying system so that minute amounts of earlier factors can lead to final rapid conversion of large amounts of fibrinogen to fibrin, for example, **Esnouf** and **Macfarlane** (1968) computed that the clotting of 1 ml of blood involves 3 mg of fibrinogen, 100 μg of factor II, 10 μg of factor X, 0.1 μg of factor VIII, and probably less amount of the earlier factors.

Concluding remark. Following two observations have been made about enzyme cascade theroy:

1. It is an over-simplification of an extremely complicated process. Because it has not yet been shown that all the reactions involved are enzyme-substrate in type.

2. There are some additional reactions. For example, thrombin exerts positive feedback effects on the factors concerned in the formation of prothrombin activator. It activates factor VIII, perhaps by releasing the amounts of available phospholipids. However, after activating these mechanisms, thrombin soon inactivates them. Thus, thrombin intensifies the reactions promoting the local clotting, but only for a short time. *i.e.,* until a firm clot has been formed.

Why Blood does not clot in circulation?

1. Endothelial factors.

(*i*) Smothness of endothelial lining prevents platelets adhesion and extension of clot into blood vessel.

(*ii*) Negatively charged particles (such as glycocalyx, a mucopolysaccharide) present over endothelial lining, repel the clotting factors, *i.e.*, protein anions, thereby prevent clotting.

2. Velocity of circulation if increases, it leads to clotting.

3. Presence of natural anticoagulants in the blood, *e.g.*, heparin and protein C.

All endothelial cells except those in the cerebral microcirculation produce **thrombomodulin,** a thrombin binding protein that converts thrombin into **protein C activator**. This activates **protein C** (a naturally occurring anti-coagulant protein) along with its cofactor **proteins,** inactivates factor V and VIII and also inactivates an inhibitor of "tissue plasminogen activator", increasing the formation of **plasminogen (fibrinolysin)** (Fig. 6.24).

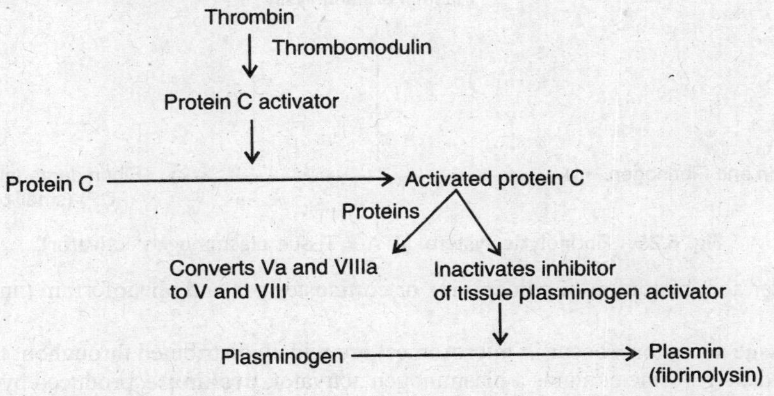

Fig. 6.24. Steps involved in the formation of plasmin (fibrinolysin).

4. Simultaneous activation of **fibrinolytic system** along with clotting mechanism.

5. In the event of spontaneous clot formation, liver removes the activated clotting factors from the circulation.

Box 6.5 Thrombosis

Formation of clots inside blood vessels is called **thrombosis**. The bit of thrombus when break off and travel in the blood stream, is called **emboli**.

Causes of Thrombosis

***(i)* Sluggish blood flow**. It allows activated clotting factors to accumulate instead to be washed away. This is commonly seen in leg veins after delivery or surgery.

***(ii)* Damange to tunica intima by atherosclerotic plaques**. They are particularly prone to occur in the coronary and cerebral vessels.

***(iii)* Mutation** in the gene for factor V, thus it fails to get activated by protein C.

***(iv)* Mutation** in antithrombin III and proteins S.

Complications due to thrombosis. Thrombi/emboli occlude the blood supply to the organ in which they form and damage it.

4. Anticoagulant Mechanism–Fibrolytic System

Factors that initiate clotting mechanism also stimulate the dissolution of the blood clot, called **fibrinolysis** (Fig. 6.25). Fibrinolysis is due to the action of proteolytic enzyme **fibrinolysin** or **plasmin**. It is present in the circulation as inactive **plasminogen** (or **profibrinolysin**) which get converted to plasmin by the action of thrombin and *tissue plasminogen activator* (*TPA*) which is released by tissue damage. Plasmin lyses fibrin and fibrinogen, with the production of **fibrinogen degradation products (FDP)** that inhibit thrombin. However, this process of fibrinolysis is much slower than clotting process.

Factors Affecting Fibrinolytic System

A. Promoting Factors.

(*i*) Stress and strains (physical or mental), *e.g.*, violent exercise, surgical operation; that is why in violent sudden death, the blood is in fluid state and incoagulable as a result of fibrinolysis.

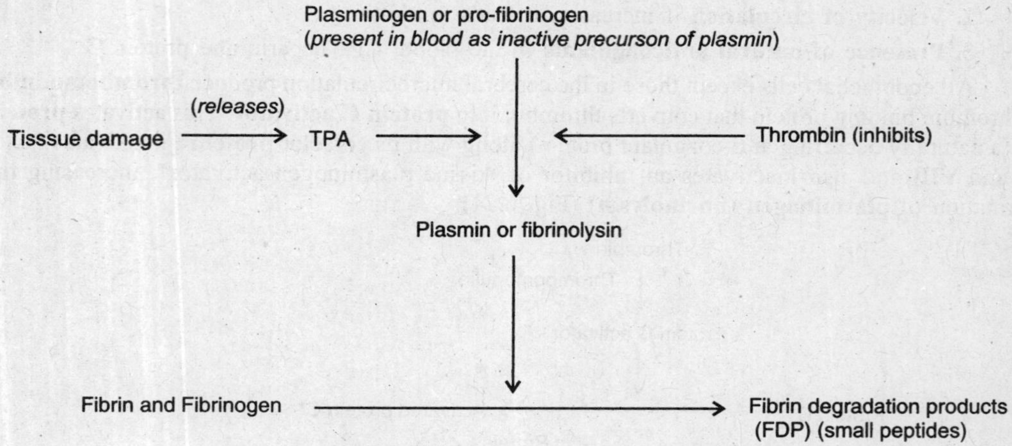

Fig. 6.25. Fibrinolytic system. (TPA = Tissue plasminogen activator).

(*ii*) After administration of epinephrine or corticosteriods and phenoformin (an anti-diabetic agent).

(*iii*) Tissue activators (occur in microsomes) are widely distributed throughout the body cells and body fluids, *e.g.*, urine contains a plasminogen activator, **urokinase**, produced by kidney cells.

B. Inhibiting factors. Antiplasmin prevents activation of plasminogen, *e.g.*, E-amino caproic acid (EACA) and oprotinin or trasylol (tryspin inhibitor).

Physiological Significance of Fibrinolytic System

1. In physiological conditions the clotting system of plasma is continually forming small amounts of fibrin which is deposited to form a thin layer on vascular endothelium, and that the fibrinolytic system is constantly in action to prevent excessive fibrin formation, therefore (i) if clotting system predominates it leads to intravascular thrombosis; and (ii) if fibrinolytic system predominates it leads to tendency of bleeding. However both of these actions remain balanced.

2. Plasmin can form "kinins" (*e.g.*, bradykinin, kallidin).

3. It also plays a role in cell movement and in ovulation producing defects in growth and fertility.

Anticoagulants

Several substances, known as **anticoagulants**, are in use to inhibit the blood clotting. Calcium is essentially required for certain reactions of blood coagulation. The substances which bind with Ca^{2+} are very effective as anticoagulants. These include oxalate, fluoride, EDTA, citrate, etc.

Types of Anticoagulants

Natural Anticoagulants

1. Heparin. It is a powerful anticoagulant which was first isolated from liver (hence its name). Heparin is also present in many other organs such as lungs. It is a polysaccharide containing many sulphate groups with MV between 15000-18000. Heparin facilitates the action of antithrombin III, thereby inhibiting the active forms of clotting factors IX, X, XI and XII. This is responsible for its anticoagulation property.

Origin of heparin. Heparin is secreted by (i) granules of circulating basophils; (ii) granules of **mast cells**. Its characteristic features are: (a) Mast cells are granulated wandering cells and are found in large numbers in tissues that are rich in connective tissue; and are arranged in close proximity to the walls of small blood vessels. (b) Mast cells contain immunoglobulin IgE (reagin) receptors on their surfaces and discharge the content of their granules when IgE-coated antigen binds to the receptors, thus mediate alergic and inflammatory reactions. (c) Mast cells also contain

heparin, **histamine** and **proteases**. (d) Mast cell may aid in defence against parasitic infestation and response to natural immunity. Heparin is destroyed by an enzyme **heparinase** in the liver.

Functions of heparin. Heparin is responsible for fluidity of blood. Post-operatively heparin prevents spread of intravascular thrombosis.

2. Antithrombin or Heparin Cofactor II, which inhibits thrombin.

3. Protein C which is activated by thrombin. Active protein C hydrolyses and inactivates clotting factors V and VIII.

Synthetic Anticoagulants

These are effective orally. These include

(*i*) Coumarin derivatives, *e.g.*, Dicoumarol;

(*ii*) Warfarin;

(*iii*) Phenindione; and

(*iv*) Nicoumalone, etc.

Mode of action. 1. Vitamin K antagonists act by substrate competitive inhibition of vitamin K in liver, *i.e.*, they occupy vitamin K receptor sites in the liver and prevent vitamin K to carry out its normal physiological function. Vitamin K deficiency thus produced results in deficiency of prothrombin factor VII, IX and X, proteins and protein C and hence decrease in blood coagulability.

2. Clotting can be prevented *in vitro* if Ca^{2+} is removed from the blood:

(*i*) by addition of substances which form insoluble salts with Ca^{2+} such as **sodium citrate** or **sodium oxalate** or **sodium edetate** (EDTA = ehtylene diamine tetra acetic acid).

(*ii*) by **chelating agents** which bind Ca^{2+}. However, *in view*, a plasma Ca^{2+} level low enough to interfere with blood clotting is incompatible with life.

3. Snake Venom of Malayan (Malaysian) pit Viper. (*i*) *In vitro*, it has a direct anticoagulent effect on fibrinogen by forming imperfect fibrin polymer.

(*ii*) *In vivo*, (a) by *defibrination* (destruction of fibrinogen), and (b) by stimulating conversion of plasminogen to plasmin, *i.e.*, stimulation of fibrinolytic system.

4. Arvin (Ancord). It is also a type of snake venom. Arvin is a glycoprotein, used therapeutically; anticoagulation is due to defibrination which produces **febrinogenopenia**.

5. Cold. Keeping blood at 5–10°C, postpones coagulation but does not absolutely prevents its occurrence. Ice when applied to the surface of the body for arresting haemorrhage, prevents bleeding by inducing reflex vasoconstriction.

Humoral Control of the Circulation

Humoral control of the circulation means controls by substances secreted or absorbed into the body fluids, such as **hormones** and **locally produced factors**. Some of these substances are formed by special glands and transported in the blood throughout the entire body. Others are formed in local tissue areas and cause only local circulatory effects. Humoral factors that affect circulatory function are of following two types:

1. Vasoconstrictor Agents

Norepinephrine and Epinephrine. *Norepinephrine* is an especially powerful vasoconstictor hormone; *epinephrine* is less so and in some tissues even causes mild vasodilation (A special example of vasodilation caused by epinephrine occurs to dilate the coronary arteries during increased heart activity). **Angiotensin II**. It is another powerful vasoconstrictor substance. Angiotensin II increases arterial pressure.

Vasopressin. It is also called **antidiuretic hormone (ADH)**. It is even more powerful than angiotensin II as a vasoconstrictor. Only minute amounts of vasopressin is secreted but even then it is known to increase arterial pressure (**Guyton** and **Hall**, 2011).

2. Vasodilator agents

Kallikrein-Kinninogen system. The circulating vasodilator substances (or agents) are the peptides which bring about the dilation of blood vessels throughout the human body. For example:

(*i*) **Bradykinin**: found in plasma (nonpeptide);

(*ii*) **Lysyl bradykinin**: found in body tissues or kalidin (decapeptide);

(*iii*) **Methionyllysyl**: found in urine.

Kinins are formed from the circulating **alpha 2 globulins**, the **kininogens** by the action of proteolytic enzymes, **kallikreins**. Kinins released is inhibited by glucocorticoids (Fig. 6.26).

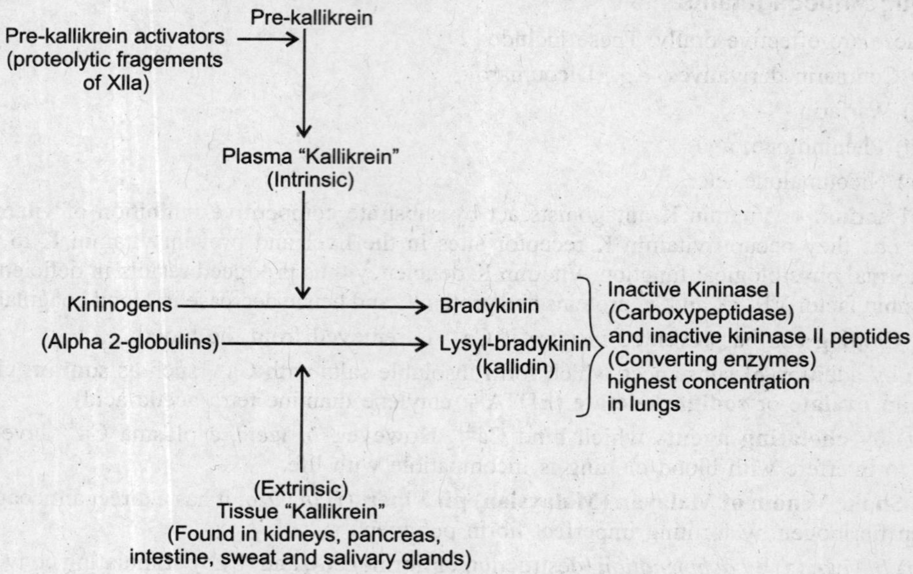

Fig. 6.26. Formation and metabolism of kinins.

Actions of Kinins

The actions of kinins resemble those of histamine. (*i*) Kinins cause contraction of most types of smooth muscles, *e.g.*, ileum, uterus, bronchiole. They also cause vasodilation of vascular smooth muscles via nitric oxide, thus producing: (a) decrease in systemic blood pressure (BP); and (b) increase blood flow in actively secreting glands specially salivary, sweat and exocrine pancreas.

(*ii*) Increase vascular permeability by an action on venules which causes escape of plasma proteins.

(*iii*) Excite sensory nerve endings producing pain (which is enhanced by 5-HT (5, hydroxy-tryptamine; Box 6.6).

Box 6.6 5-hydroxytryptamine (5HT)

5HT is a **local hormone** (*i.e.*, it acts at the sites of its synthesis and release). It is also called **serotonin** because when acting directly, it is a cardiac stimulant and a vasoconstictor. 5-HT is also called **enteramine** as GIT (or gastrointestinal tract) contain 80–90% of all the 5HT in the body. Blood platelets obain their 5HT while passing through the GIT (Note. Platelets cannot synthesize 5HT). Normal blood serum contains 0.1–1.0 mg/L of 5HT level which is liberated from the platelets during clotting.

Serotonin (*i.e.*, 5HT) is produced from tryptophan (an essential amino acid) and is metabolized by monoamine oxidase. Generally, 5HT has two types of effects: (a) *excitatory effect* on motor pathways; and (b) an *inhibitory effect* on the sensory pathways).

(*iv*) The high concentration promote migration of leucocytes from blood to tissues.

The actions (*ii*), (*iii*) and (*iv*) are characteristically seen during inflammation, therefore kinins are responsible for **inflammation**.

(*v*) Kinins are formed by antigen-antibody interactions and may thus help to produce bronchioconstriction in bronchial asthma.

(*vi*) Probably, play a role in thermo-regulatory vascular adjustments.

Histamine. Histamine is released in essentially every tissue of the body if the tissue becomes damaged or inflamed or is the subject of an allergic reaction. Most of the histamine is derived from **mast cells** in the damaged tissues and from **basophils** in the blood. Histamine has a powerful vasodilator effect on the arterioles and like bradykinin, has the ability to increase greatly capillary porosity, allowing leakage of both fluid and plasma protein into the tissues (causing oedema).

6.4 COMPLEMENT SYSTEM

(Enzymatic proteins that kill invading microbes)

The vertebrate blood employs a very effective chemical defense, the **complement system,** that complements its cells cellular defenses. The complement system consists of a battery of more than a dozen different proteins that normally circulate in the blood plasma in an inactive state. Its defensive activity is triggered by the cell walls of bacteria and fungi (and also by binding of antibodies to invading microbes). On detection of bacterial cells wall, these proteins interact to form a membrane attack complex (MAC) like that of produce by natural killer cells. The MAC inserts itself into the pathogen's cell membrane, forming a hole. Like a dagger through the heart, this wound is fatal to the invading cell, water is pulled in by osmosis, causes it to swell, and burst. The complement proteins can lyse, antibody-coated eukaryotic cells and bacteria (**cytolysis**). Complement system can mediate inflammation and activate phagocytic cells. In general, complement proteins amplify the effect of antibodies and is the major effector of humoral branch of immune system.

Complement system consists of many different proteins (enzymes). Some of these are serine esterases (proteases). There are at least 120 different interacting proteins in the complement system. The complement proteins are designated by the letter C and by numbers C1, C2, C3, and so on upto C9. The complement C1 is actually an assembly of subunits designated as C1q, C1r and C1s (Therefore there are 11 proteins in the complement system) (Box 6.7). These subunits are known to be held together by bonds other than the usual covalent bonds and calcium ions (Ca^{++}) is needed to keep the assembly intact. If the calcium ion (Ca^{2+}) is removed by chelating agents, the subunits of C1 protein disaasemble. The entire assembly is believed to consists of one molecule of subunit C1q two of C1r and four of C1s.

Box 6.7

The eleven proteins involved in antigen-antibody reactions is known as **complement system**. It was described by a Belgian **Jules Bordet** in 1895. Complement works with antibodies to cause opsonization, chemotaxis and lysis of bacterial cells.

The pathway for activation of the complement set into motion by the interaction of antigen and antibody molecules is known as **classical pathway**.

Louis Pillemes (1955) proposed an *altenative pathway* or *properdin system* for activation of complement system.

The complement system acts in a cascade fashion. Collectively the complement proteins make up much of the globulin fraction of serum. Within the plasma and other body fluids, complement proteins are in an inactive state. They usually are activated after the binding of antibodies to antigens and are specifically directed against the target molecules identified by the antibodies.

Functions of Complement Proteins

The complement system is a powerful mediator of inflammation and mediates protection against microbes. The complement proteins help antibody responses and have a major role in innate defense mechanism. Their main functions are the following:

1. Attraction of neutrophils to the site of microbial attack (chemotaxis).

2. Opsonisation, *i.e.*, enhancement of attachment of the microbe to the phagocyte.

3. Initiation of acute **inflammation** as a result of degranulation of mast cells.

4. Killing of microbes by MAC (membrane attack complex) which produces pores in the target cell or plasma membrane thereby killing it.

5. Clearance of immune system complexes from the circulation and **deposition** in the spleen and liver.

The details of mechanism of various proteins of complement system leading to **haemolysis** has been summarized in Table 6.5

Table 6.5 Functions of complement system.

S. No.	*Reaction*	*Biochemical events*
1.	E + A = EA	Erythrocyte reacts with anti-erythrocyte antibody.
2.	EA + C1 = EAClq	C1 attaches to antibody at a site on C1q converts C1r to active form by cleavage.
3.	C1r → C1s*	Bound C1q converts C1r to active form by cleavage of C1r.
4.	C1s → $C1s^-$	C1r activates C1s by cleavage of C1s of C1r.
5.	C4 → C4a + $C4b^-$	C1s cleaves C4 into C4a and C4b* and C2 into C2a and C2b*.
6.	C2 → C2a + $C2b^-$	Has anaphylatoxin activity.
7.	C4b* + C2a* → $C4b2a^-$	$C4b^-$ and C2a combine to form C3 convertase.
8.	C3 → C3a + C3b*	C4b2b cleaves C3a and C3b; C3a casues degranulation of mast cells (anaphylatoxin).
9.	C3b* + C4b2a = C4b2a3b	C3b* binds to activated biomolecular complex of C4b2a to form a trimolecular complex that is a specific enzyme for C5, C5 convertase. The macrophages have receptors for C3b, so that C3b acts as opsonin. The C3b component of this complex binds C5 and alters its conformation so that the C4b2a component can cleave C5 into C5a which diffuses away and C5b which attaches to C6.
10.	C5 = C5a + C5b*	C5 is cleaved into C5a and C5b* convertase; C5a has chaemotactic activity for polymorphonuclear neutrophils.
11.	C5b* + C6789 + C5b9	C5b* reacts with other complement components to produce a macromolecular complex that has the ability to alter permeability; C8 is most likely the active component with Ca increasing efficiency of C8 and producing maximal cell lysis.

Activation of the Complement

Two different but convergent pathways can stimulate (*i.e.*, activate) the complement cascade. These are 1. immunoglobulin dependent pathway (or classical pathway) and 2. immunoglobulin independent pathway (or Alternative pathway (Fig. 6.27; Fig. 6.31; Fig. 6.32).

I. Antibody dependent pathway or Classical pathway. Activation of the classical pathway requires initiation by the interaction of antibodies with an antigen that is usually cell bound (Fig. 6.29). However, some microbial products (lipid A of endotoxin and staphylococcal protein A) or plasmin (a proteolytic enzyme that dissolves the fibrin of blood clot) may activate C1 directly without antibody participation. Once a pair of antibodies recognizes and attaches to antigen, the complement protein C1 (which actually consists of three subunits) bind to two or more antibodies, thus, activating C1. In the presence of calcium ions (*i.e.*, Ca^{2+}), a trimolecular complex ($C1q\ r_2\ s_2$ Ag Ab) that has esterase activity is repaidly formed. The activated C1s subcomponent attacks and cleaves its natural substrate in serum (C2b and C4). This leads to binding of a portion of each molecule (C2b and C4b) to the antigen-complement complex with the release of small C4a and C2a fragments. The released C2 fragments is generally called **C2a**.

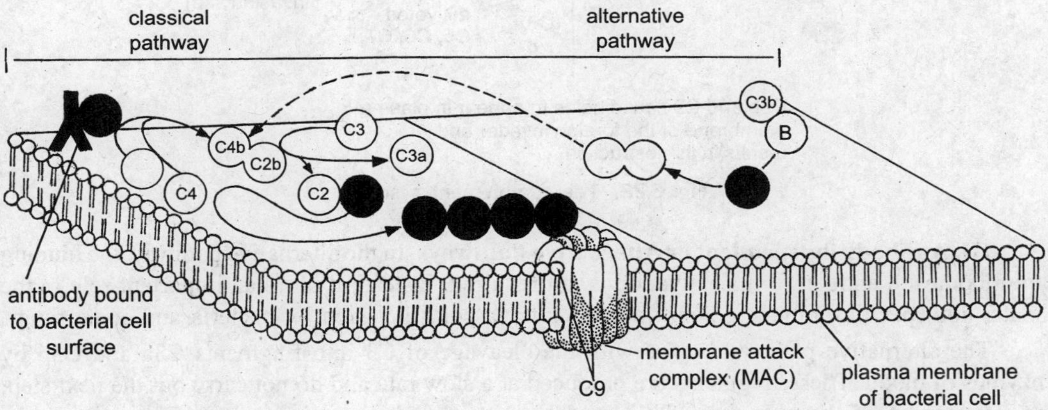

Fig. 6.27. Diagrammatic representation of the classical and alternative pathways of complement activation.

This complex forms a macromolecular structure called MAC (*i.e.*, Membrane attack complex). The MAC forms a large transmembrane channel that disrupts the target cells/plasma membrane (Fig. 6.30) resulting in leakage of cell's contents and enable ions and small molecules to move through it.

II. The Lectin Pathway. This pathway is similar to classical pathway and does not depend on antibody for activation. After initiation it proceeds through the action of C4 and C2, to produce a C5 convertase. Since the pathway is activated by binding of mannose-binding lectic (MBL) to mannose residues on carbohydrates on the surface of microorganisms, it is called the **lectin pathway**. MBL is an acute phase protein and functions like C1q of classifical pathway. MBL-associated serine protease (MASP) binds to MBL bound to the surface of microorganisms. This causes cleavage and activation of C4. THe structure of MASP is similar to C1r and C1s and resembles their activities. Hence lactin pathway also represents an important innate defence mechanism (Fig. 6.28).

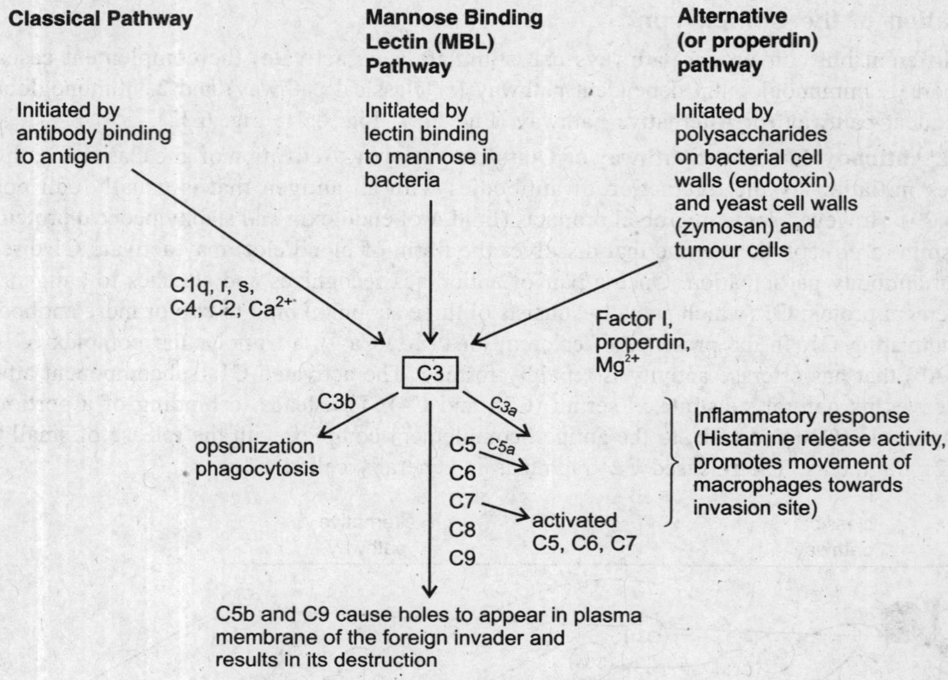

Fig. 6.28. The complement system.

III. Antibody independent or Alternative Pathways. In the alternative pathway, the binding of antibodies to antigen for its activation is not required. Instead, it plays an important role in the innate, non-specific immune defense against intravascular invasion by bacteria and some fungi.

The alternative pathway begins with the cleavage of C3 into fragments C3a and C3b by enzymes of blood. These fragments are produced at a slow rate and do not carry out the next step, the cleavage of C5, because free C3b is rapidly cleaved and inactive fragments. However, when C3b binds to lipopolysaccharide (LPS) of outer membrane of these bacterial cells to aggregates of immunoglobulins - IgA or IgE, or to some endotoxins, it becomes stable. A protein in blood termed **factor B** adsorbs to bound C3b and is cleaved into two fragments by factor D, leading to the formation of active enzyme C3bBb (this complex is called **C3 convertase** because it cleaves more C3 to C3a and C3b). The enzyme C3bBb is further stabilized by a second blood protein, called **properdin**, and is changed into the C5 convertase (C3Bb3b). The convertase enzyme then cleaves C5 to C5a and C5b. The steps that follow in the alternative pathway are identical with the last steps of the classical pathway (Fig. 6.27).

The complement activation, thus, provides a basis for consideration of the function of complement as an integrated system during an organisms defensive efforts. Gram-negative bacteria arriving at a local tissue site will interact with components of the alternative pathway. This results in the generation of biologically active fragments, opsonization of the bacteria and initiation of the lytic sequence. If the bacteria persists or if they invade the animal a second time, antibody response will also activate the classical pathway. The classical pathway is much more rapid and efficient in mediating opsonization and complement fragment generation.

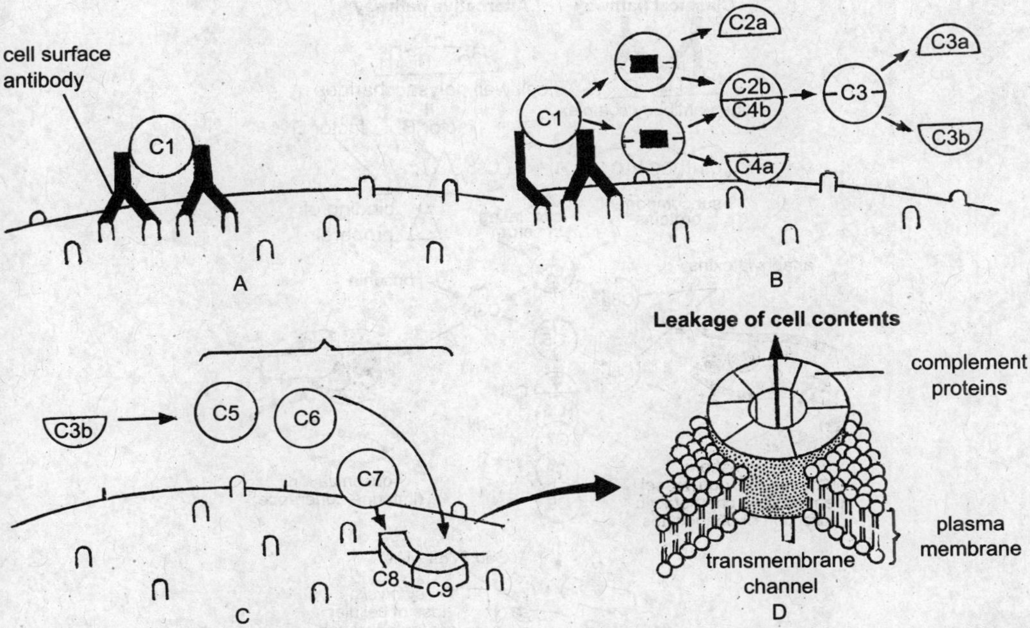

Fig. 6.29. Cytolysis caused by complement via the classical pathway. A – As soon as the antibodies recognize and attach to the antigen, the protein binds to adjacent antibodies. B – C1 acts as an enzyme and splits C2a and C4 proteins into fragments. The active fragment is called C3b. C – C3b initiates series of reactions involving C5-C9. These are together called the membrane attack complex (MAC). This complex forms circular transmembrane channel in the antigenic plasma membrane and C9 proteins play a key role. As a result, there is a leakage of cell contents, *i.e.*, cytolysis. D – Transmembrane channel.

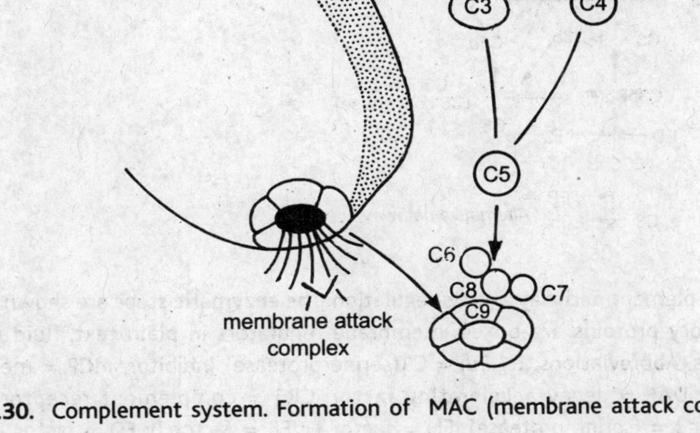

Fig. 6.30. Complement system. Formation of MAC (membrane attack complex).

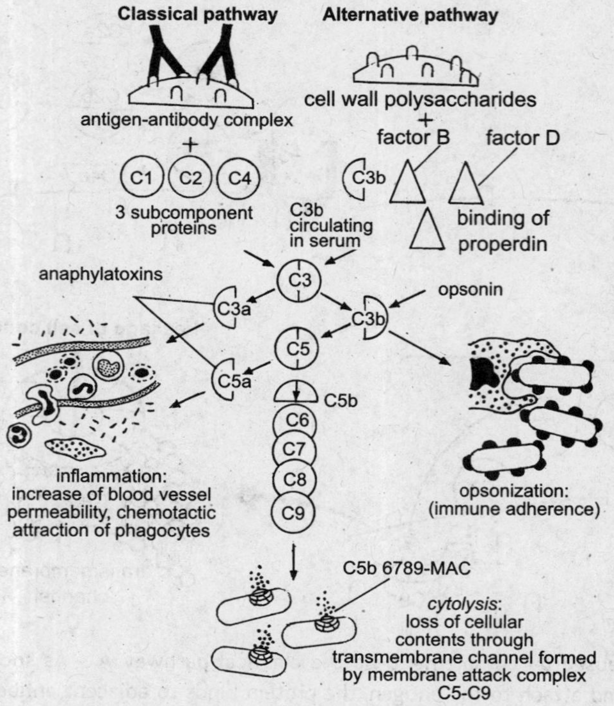

Fig. 6.31. Two complement systems: Classical and alternative pathways.

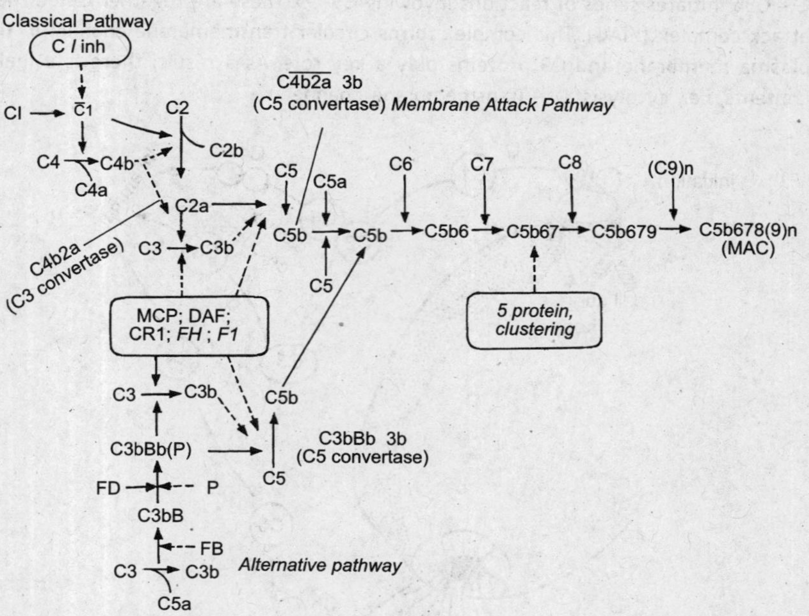

Fig. 6.32. The complement pathway and its regulation. The enzymatic steps are shown by heavy arrows. Regulatory proteins are boxed, membrane regulators in plain text, fluid phase regulators in italics. Abbreviations: *C1 inh* = C1(serine protease) inhibitor; MCP = membrane cofactor protein; DAF = decay accelerating factor; CR1 = complement receptor1; FI = factor I; PI-factor I = (serine protease), FH = factor H; FB = factor B; FD = factor D; P = properdin; MAC = membrane attack complex.

Table 6.6 Proteins of the complement cascade in serum.

Complement proteins		*Mode of Action*
Recognition Unit		**Function**
C1	q	Binds to the Fc portion of antigen-antibody complex.
	r	Subunit of C1; activates C1s.
	s	Cleaves C4 and C2 due to its enzymatic activity.
Activation Unit		
C2		Cause viral neutralization.
C3	a	Anaphylatoxin, immunoregulatory.
	b	Key component of the alternative pathway and major opsonins in serum.
C4	a	Anaphylatoxin.
	b	Cause viral neutralization
Membrane Attack Unit		
C5	a	Anaphylatoxin; principal chemotactic factor in serum: induces neutrophil attachment to blood vessel walls.
	b	Initiates membrane attack.
C6, C7, C8, C9		Participate with C5b in the formation of membrane attack complex that lyses targeted cells.
Alternative pathway.		
Factor B		Causes macrophage spreading on surfaces: precursor of C3 convertase.
Factor D		Cleaves factor B to form active C3b in alternative pathway.
Properdin		Stabililizes alternative pathway convertase.
Regulatory proteins		
Factor H		Promotes C3b1 breakdown and regulates alternative payhway.
Factor 1		Degrades C3b and regulates alternative pathway.
C4b binding protein		Inhibits assembly and accelerates decay of C4bC2a.
C1/NH-complex		Binds to and dissociates C1r and c1s from C1.
S protein		Binds fluid-phase C5b67 prevents memberane attachment.

Biological Effects of Products of Complement

The generation of small complement fragments C3a, C4a and C5a leads to several important inflammatory effects. They bind to receptors on mast cells and blood basophils which release **histamine** and pharmacologically active mediators. The blood supply to the area increases markedly (**hyperemia**) as blood vessels dilate due to released histamine. These fragments called as **anaphylatoxins** also cause the release of the phagocytic cells from the bone-marrow into the circulation. Phagoctyes then make their way to the site of hyperemia, where in the presence of C5a, they attach to the endothelium and leave the blood vessels. The C5a induces a directed, chemotactic migration of *neutrophils* to the site of complement activation. Macrophages in the area can synthesize even more complement components to interact with the bacteria. The ingestion and ultimate destruction of the bacteria by the neutrophils and macrophages is promoted by all these defensive events (Fig. 6.33).

The biology of C3 is quite complex (Table 6.6). C3b is, in fact, covalently bound to cell surfaces. Thus, once bound to a cell surface, C3b remains attached until the cell is phagocytosed.

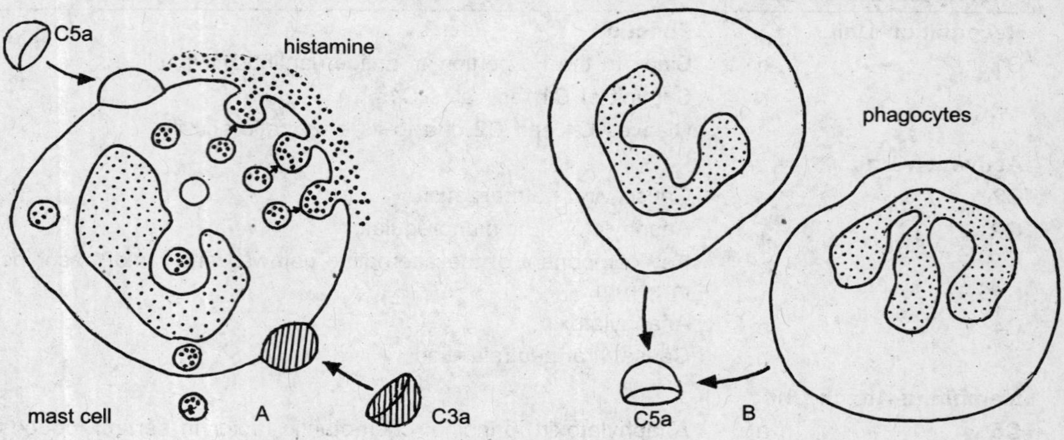

Fig. 6.33. Inflammation stimulated by complement. A – C3a and C5a bind to mast cells, basophils and platelets to trigger the release of histamine, which increases blood vessel permeability. B – C5a functions as a chemotactic factor that attracts phagocytes to the site of complement activation.

There are three receptors for C3b and these receptors are distributed on different cell types. Finally, the fragmentation of C3 generates several separate peptides and from C3b (the longest peptide) four more peptides are produced under various conditions, each having distinct biological properties. The emphasis here is that there are many complement systems which collectively exert a wide range of relatively non-specific antibacterial effects. This also suggests that complement is probably an ancient protective mechanism that preceded the evolution of an immune system. As the immune system evolved, however, it united with the pre-existing protective mechanisms in the armaments of resistance against infectious diseases.

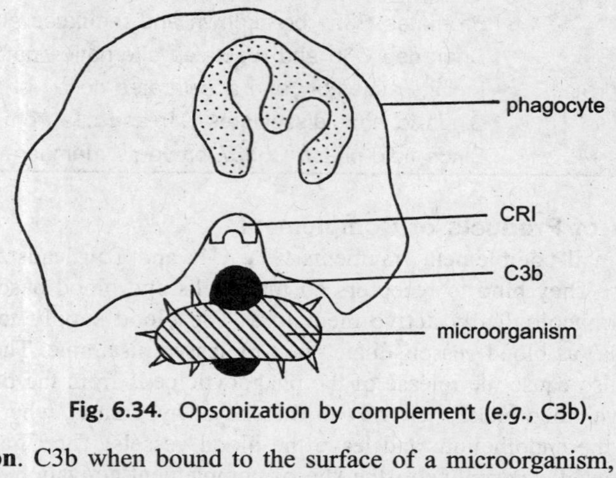

Fig. 6.34. Opsonization by complement (*e.g.*, C3b).

Opsonization. C3b when bound to the surface of a microorganism, the latter can interact with special receptors on phagocytes to promote phagocytosis. This is called **opsonization** ('make tasty) as C3b functions as an opsonin by coating the microorganism and promoting attachment of phagocyte to the microbe. Once complement is activated, its destructive capabilities usually ceases very quickly in order to minimize destruction of the host cells. This is achieved by various regulatory proteins present in host's blood and on certain cells, *e.g.*, blood cells (Fig. 6.34).

Toxin neutralization. Immunity to a disease such as diphtheria depends on the production of specific antibodies that inactivate the toxin produced by the bacteria. The process is termed as **toxin neutralization**. Once neutralized, the toxin-antibody complex is either unable to attach to receptor sites on target cells of the host or unable to enter the cell. For example, diphtheria toxin inhibits protein synthesis after binding to the cell surface by the B fragment and subsequent passage of the active A fragment into the cytosol of the target cell. Thus, antibody blocks the toxic effect by inhibiting the entry of the A fragment or the binding of the B fragment. Antiserum containing neutralizing antibody against a toxin is called **antitoxin**.

Complement as a protector. In the presence of C5b, the complement components C6, C7, C8 and C9 assemble and cause "hole" in cell or plasma membrane of the microorganism. Although this process has been studied most extensively using erythrocytes, some strains of bacteria can be lysed by complement (via the immunoglobulin dependent pathway). However, individuals with congenital deficiencies is one of the C5-C9 components generally are more susceptible to bacterial infections or other forms of observable disease. Thus, the complement effector function of creating "holes in plasma membranes" is apparently of marginal selective value. The rarity of these complement deficiencies does, however, indicate the existence of a significant selection pressure that maintains the functional phenotype.

It was earlier thought that one "hole" in any erythrocyte membrane is sufficient for the lysis of cell. It has, however, recently been shown to affect lysis, a minimum threshold number of "holes" is required.

The complement components C5b-C8 assemble in equimolar amounts as a complex embedded in the plasma membrane, but as such this is insufficient to cause formation of a hole. Upto five more molecules of C9 must be incorporated before osmotic lysis takes place. It is quite possible that the additional molecules of C9 serve to enlarge the hole to allow rapid ion flux and osmotic lysis. C3b is generated as a key intermediate by both the immunoglobulin dependent and immunoglobulin independent pathways of complement activation. The alternative pathway is strictly non-immune and probably represents a more evolutionarily ancient protective mechanism. It is activated only by large number of bacteria which provide the protective surface necessary for stabilizing C3b. The immunoglobulin dependent or classical pathway, however, does not generate large amounts of C3b. However, if the C3b that is produced remains bound to a suitable protective surface, it can initiate an autocatalytic cycle of immunoglobulin independent (alternative) pathway and produces very large amounts of C3b.

Significance of the Complement System

The importance of the complement system can be determined by the study of consequences of inherited deficiencies and by evaluating the diseases they produce or permit. Individuals with C4, C2 and C1 deficiencies are not at a higher risk for bacterial diseases. As C2 is necessary in solubilizing antigen immunoglobulin complexes in serum, its deficiency is a serious disease state. Complexes of immunoglobulin and antigen are filtered out in the renal glomeruli, where they form deposits that can lead to **kidney failure**. A few examples of C5, C6 and C8 deficiency have been reported, but no effective disease susceptibility is found (except for perhaps meningococcal or gonococcal infections).

The deficiency of C3 causes serious bacterial infections. The sera of such individuals manifest very weak opsonic activity for bacterial phagocytosis. A disease known as hereditary **angioneurotic oedema**, which is characterized by recurrent swelling in various parts of the body is caused due to a deficiency in CI-INH inhibitor. Deficiencies of C1, C2 and C4 cause collagen vascular disorders that result in **hypersensitivity**. The defects in C5 through C9 result in increased susceptibility to *Neissera meningitidis* and *N. gonorrhoeae* infections.

6.4 HAEMOPOIESIS

In humans, to keep the normal number of active erythrocytes circulating in the blood vessels, new corpuscles are being constantly manufactured at the rate of 1.5 million per second and the old and weak ones are destroyed. The process by which new corpuscles are produced is called **haemopoiesis** and the tissues involved in the haemopoiesis are called **haemopoietic tissues**.

Haemopoiesis is the development of blood cells, *i.e.*, RBCs, WBCs and platelets. The term is haemopoiesis includes the following three parts:

1. **Erythropoiesis**, *i.e.*, development of erythrocytes or red blood cells (RBCs);
2. **Leucopoiesis**, *i.e.*, development of leucocytes or white blood cells (WBCs); and
3. **Megakaryocytopoiesis**, *i.e.*, development of megakaryocytes which produce platelets.

There is a regular sequence of organs and tissues taking up haemopoiesis from the early embryonic stage to the adult. First in the series is the **yolk sac**, at least in embryos where it is present. In others, it is body **mesenchyme**. With the formation of **liver**, haemopoiesis shifts to this organ and in the adult fishes and amphibians, liver, also the **spleen**, continue to remain haemopoetic. In some fish and amphibian larvae, the opisthonephric **kidneys** become haemopoietic, at least for sometime. Besides liver and spleen, the **thymus**, **lymph glands** and part of **alimentary canal** produce blood cells in the adult fishes and amphibians. **Bone marrow** becomes haemopoietic for the first time in the tailless amphibians, especially during the peak periods of their active lives. In all other vertebrates and especially mammals, **red bone marrow** and lymph glands are the only haemopoietic tissues in the adult. Red bone marrow tissue is found at the ends of the long bones, in the vertebral centra, ribs and the sternum.

Theories of Haemopoiesis

1. Monophyletic theory. According to this most acceptable theory, different types of blood cells arise from a single **ancestral cell**, called **pluripotent stem cell** present in the bone marrow (Box 6.8; Fig. 6.35).

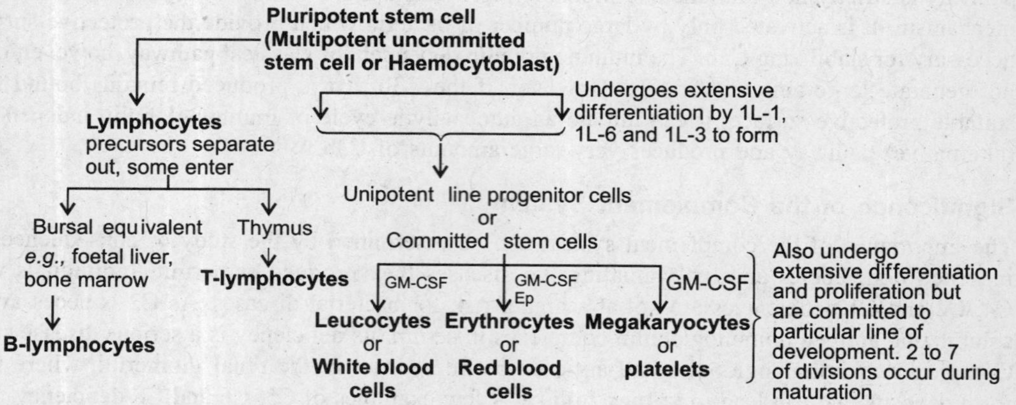

Fig. 6.35. Monophyletic theory of haemopoiesis. (*pluri*, many; *potent*, capable of producing; *IL*, interleukins; *EP*, erythropoietin; *GM-CSF*, granulocyte macrophage colony stimulating factors). *Note*. Differentiated mature cells in peripheral blood lose the power of further division, exception is lymphocyte.

Box 6.8

75% of the cells in the red bone marrow belong to the WBC producing **myeloid** series (*i.e.*, WBCs and their precursors) and only 25% are of **erythroid series** (*i.e.*, RBSs and their precursors).

2. Polyphyletic theory. There are separate stem cells present in the red bone marrow for each main variety of blood cell, *i.e.*, granulocytes, monocytes, lymphocytes, erythrocytes and megakaryocytes (platelets).

Process of Monophyletic Haemopoiesis

The primary cells of the red bone marrow from which all kinds of blood cells can be derived are called **haemocytoblasts**, which are irregularly shaped with a large deeply staining nucleus and a clear cytoplasm (Fig. 6.36), destined to give rise to erythrocytes are termed **haemocytoerythroblasts** which, after repeated divisions, give rise to the precursors of red blood corpuscles, the **pro-erythroblasts**. These cells after further differentiation, such as taking up the characteristic shape and acquiring haemoglobin in the cytoplasm, become transformed into erythroblasts. In mammals, differentiation of pro-erythroblast first begins to shrink and becomes rounded, intense synthetic activity in the nucleus can be visualised in stained preparations, the pro-erythroblast, is now called a **normoblast**. Cytoplasm of the normoblast, acquires haemoglobin and the nucleus shrinks to a small dense mass, finally the nucleus is expelled out and the enucleated cell is now a mature **erythrocyte** ready to be launched in the blood circulation. In a newly formed erythrocyte, the cytoplasm shows a network of fine threads forming a reticulum, hence, the erythrocyte for some time is called a **reticulocyte**. Blood contains a substance that causes the reticulocyte to ripen into mature RBCs. Various tissues produce this ripening substance, *e.g.*, pyloric mucosa in pig and in human beings the mucosa of lower part of the stomach.

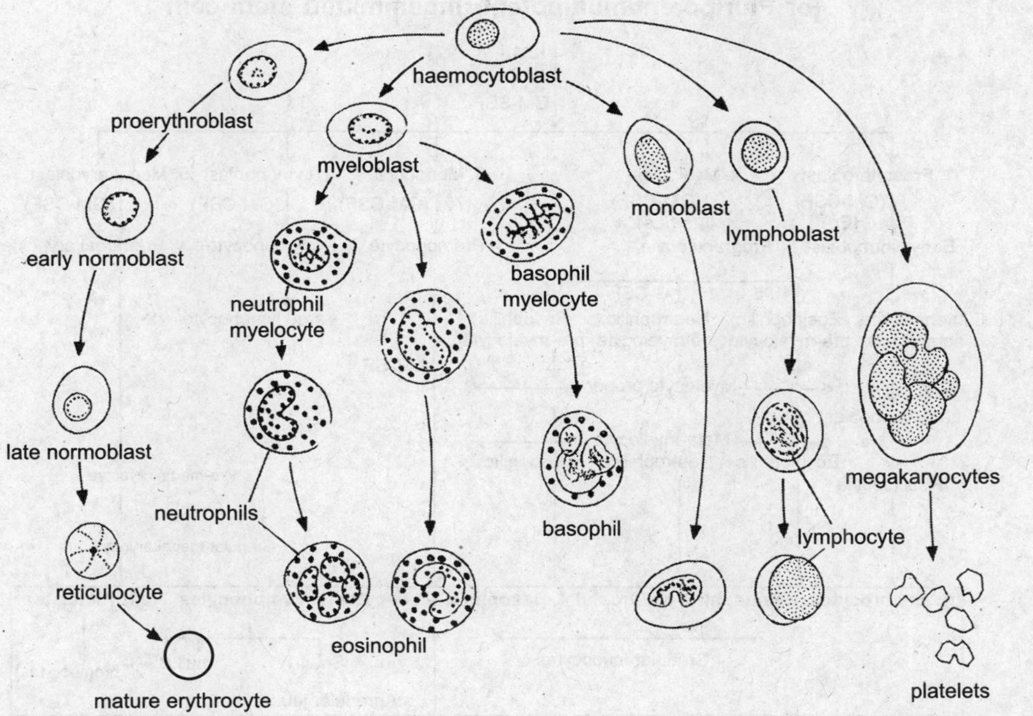

Fig. 6.36. Haemopoiesis in bone marrow. Formation of different types of cells from haemocytoblast.

Box 6.9

The rate of production of RBCs is subject to adjustments and requirements of the body, for instance, the greater demand of oxygen induces the red bone marrow to speed up production of RBCs. In human beings, about 10 million RBCs are formed every second and an almost equal number of old, weak RBCs are destroyed in the spleen. The products of breakdown of RBCs are conveyed to the liver, where they elaborate into bile pigments, the iron part is returned to erythroid tissue.

In a healthy person, number of all types of cells in the peripheral blood is kept constant within narrow limits by "feedback control mechanisms". This shows that the rate of formation, release and destruction of blood cell is balanced.

Regulation of Monophyletic Haemopoiesis

In a healthy person, the proliferation and maturation of blood cells that enter the blood from bone marrow is regulated with great precision by (a) Interleukins (ILs), and (b) Colony-stimulating factors (CSFs) (Box 6.9).

Interleukins (ILs). They are hormone-like chemical messengers with known amino acid sequence. They are produced by lymphocytes, macrophages and other body cells, *e.g.*, endotherlial cells, neurons, glial cells activated T-cells, fibroblasts, etc. ILs are the same as **cytokines**, but in cytokines amino acid sequence of a factor is not known. Initially, the messengers secreted by lymphocytes were often called **lymphokines**. However, since they are produced by other cells as well, therefore, more appropriate term **cytokines** is used now. Cytokines play an important role in regulation of the immune system (Fig. 6.37).

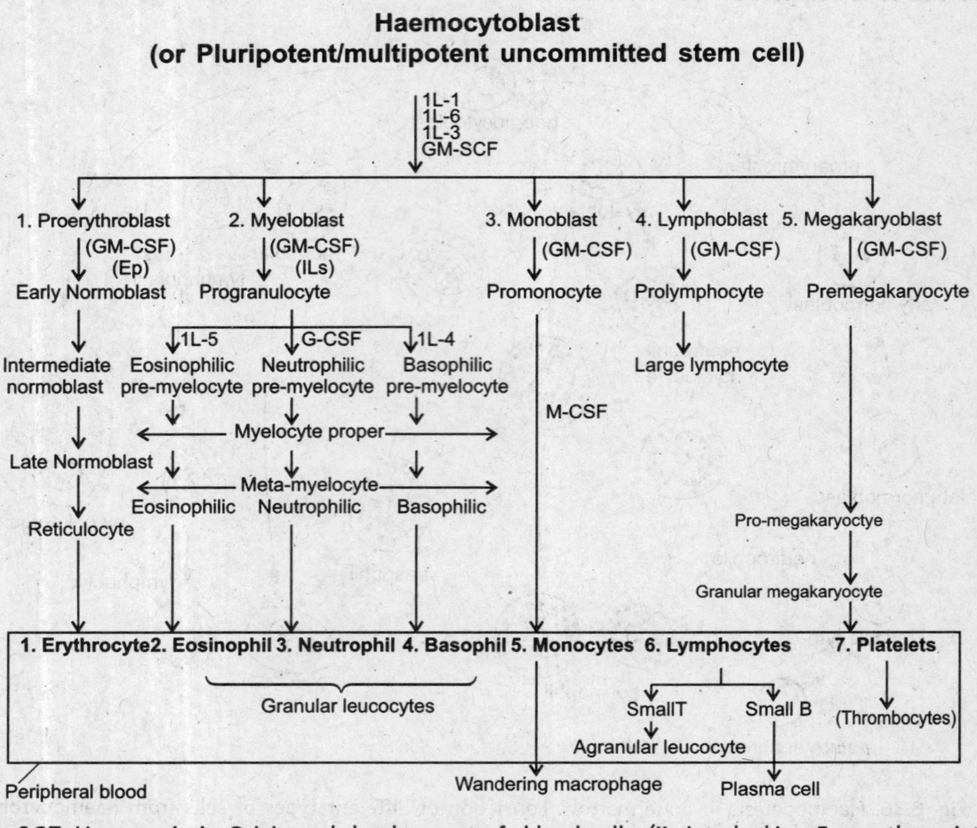

Fig. 6.37. Haemopoiesis: Origin and development of blood cells. (**IL**, interleukins; **Ep**, erythropoietin; **GM-CSF**, granulocyte macrophage colony stimulating factor).

Colony stimulating factors (CSFs). These are growth factors or hormones. They cause appropriate single stem cell to proliferate in culture medium, *e.g.*, soft agar, forming colonies of daughter cells.

IL-1, IL-6 and IL-3 act as sequence to convert multipotent uncommitted stem cells to committed stem cells (Fig. 6.37; Table 6.7). CSF and IL have overlapping action. Furthermore, they activate and sustain mature blood cells.

Table 6.7 Principal Interleukins (ILs) and Colony Stimulating Factors (CSFs) of human beings.

S. No.	*Type*	*Functions*
1.	IL-1, IL-3 (or multi-CSF) and IL-6	Increase secretion of CSF and count of all blood cells except lymphocytes
2.	IL-4	Increases basophil production
3.	IL-5	Increases eosinophil production
4.	IL-11	Increases production of granulocytes, RBCs and platelets
5.	G-CSF (Granulocyte-CSF)	Increases production of granulocytes
6.	M-CSF (Macrophage-CSF)	Increases production of monocyte
7.	GM-CSF (Granulocyte-macrophage-CSF)	Increases production of neutrophil, monocyte
8.	Ep (Erythropoietin)	Increases production of RBCs (or erythrocytes)
9.	Thrombopoietin	Increases production of platelets

A. Erythropoiesis

I. During intrauterine life. It includes the following three stages:

1. Mesoblastic stage. In early human embryo upto 3 months of foetal life, erythrocytes (RBCs) are formed from mesoderm of yolk sac or **area vasculosa**. Since erythropoiesis occurs within the blood vessels, therefore this stage is also called **intravascular erythropoiesis** (Fig. 6.38).

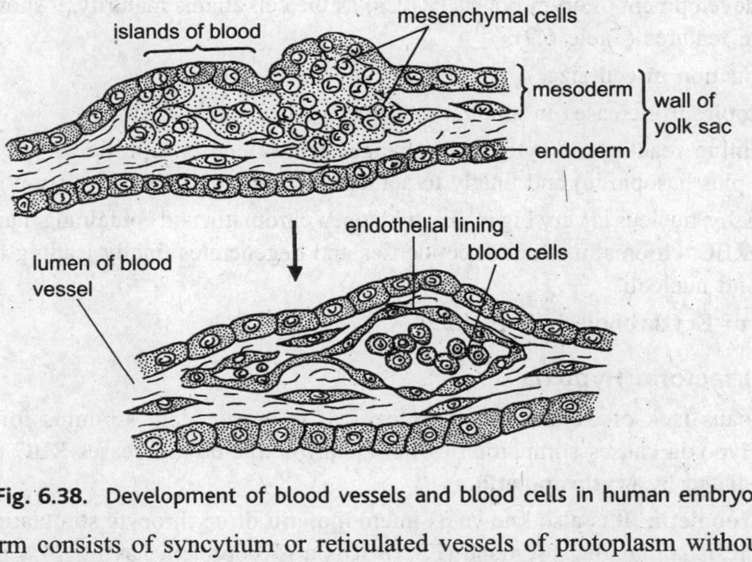

Fig. 6.38. Development of blood vessels and blood cells in human embryo.

Mesoderm consists of syncytium or reticulated vessels of protoplasm without distinct cell outlines which gives rise to network of capillary vessels lined by endothelium and containing plasma (formed by liquefaction of cytoplasm). Endothelial cells proliferate and differentiate to form masses of nucleated haemoglobin bearing cells; these cells get detached from capillaries and finally lose their nuclei to give rise to non-nucleated discs (RBCs).

2. Hepatic stage. After 3 months of foetal life, liver and spleen are the site of blood formation. Nucleated RBCs develop from the mesenchyme between the blood vessels and the tissue cells.

3. Myeloid stage. From the middle of foetal life erythropoiesis occurs in bone marrow.

Hepatic and myeloid stages are called **extravascular erythropoiesis**.

II. Erythropoisis in children. In children erythropoiesis occurs in (i) All bones with red bone marrow (mainly); (ii) Liver, and (iii) Spleen.

III. Erythropoiesis in adults. After 18–20 years of age, erythropoiesis takes place in red bone marrow. Red bone marrow occurs in (i) ends of long bones such as humerus and femur, because shaft is converted to yellow bone marrow (Table 6.8); (ii) skull; (iii) vertebrae; (iv) ribs; (v) sternum, and (vi) pelvis. Bone marrow cells, called **haemocytoblasts**, are few in number and can be stimulated to form any cell in the body. They are also capable of completely replacing the bone marrow when injected into a host whose one bone marrow has been completely destroyed.

Table 6.8 Differences between red bone marrow and yellow bone marrow.

S. No.	Red bone marrow		Yellow bone marrow
1.	It consists of numerous blood cells of all kinds and their precursors (erythroid and myeloid).	1.	It consists of fat cells, blood vessels and a minimal framework of reticulum cells and fibers.
2.	It is also called active marrow.	2.	It is also called inactive marrow.
3.	It occurs at the ends of long bones and other sites.	3.	It occurs in shafts of bones.

Stages in Erythropoiesis

During the development of erythrocytes (RBCs) as the cell attains maturity, it shows the following characteristic features (Table 6.9):

1. Reduction in cell size.

2. Cytoplasm increases in amount and nucleus decreases in size.

3. Staining reaction of cytoplasm changes from deep basophilic or polychromatophilic (acidophilic plus basophilic) and finally to acidophilic; due to gradual reduction in RNA amount.

4. Initially, nucleus is very big in size with open chromatin and containing many nucleoli; with maturity of RBC, chromatin material condenses and degenerates finally leading to disappearance of nucleus and nucleoli.

Regulation of Erythropoiesis (Fig. 6.39)

A. General factors: Hypoxia

Hypoxia means lack of oxygen at tissue level. It is most potent stimulus for the production of RBCs. Hypoxia causes stimulation of bone marrow thereby increases RBC production. This effect is mediated by **erythropoietin**.

Erythropoietin. It is also known as **haemopoietin** or erythrocyte stimulating factors (ESF) or erythropoiesis stimulating hormone (ESH). It is a glycoprotein having 74% protein and 26% carbohydrate. This protein contains 165 amino acids (MW 46000) and half life of 5 hours.

Table 6.9 Stages of erythropoiesis.

S. No.	Name of the cell	Cell size	Nucleus	Cytoplasm		Mitosis
				Staining	Haemoglobin	
1.	Haemocytoblast	19–23 µm	Very big–occupies almost whole of the cell with open chromatin, containing 4–5 nucleoli; deep basophilic	Rim all around the nucleoli; deep basophillic	Absent	Active mitosis (++)
2.	Proerythroblast	15–20 µm	Occupies 3/4 of cell volume; 2–3 nucleoli; chromatin open	Slightly more in amount; deep basophilic	Absent	Active mitosis (++)
3.	Early normoblast	14–16 µm	Size decreases; no nucleoli; chromatin condenses	Further increase in amount; less basophilic	Absent	-do-
4.	Intermediate normoblast	10–14 µm	Nucleus size further decreases; chromatin further condenses	Further increase in amount; less basophilic	-do-	-do-
5.	Late normoblast					
	(*i*) Early	8–10 µm	Nucleus very small with chromatin dot cart-wheel appearance	Increases markedly	Further increases in amount	Stops here
	(*ii*) Late	7–8 µm	Nucleus degenerates, becomes uniformly deeply stained pycnotic	Further increases; more acidic, less basophilic	-do-	Absent
6.	Reticulocytes	7–8 µm	No nucleus, remnants of RNA present	Acidophilic	-do-	-do-
7.	Erythrocyte	7.2–7.4 µm	Nil	-do-	-do-	-do-

Note. Normal (*i.e.*, normoblastic) bone marrow contains 30% proerythroblasts and early normoblast cells and 70% of intermediate and late normoblasts. Stages 1 to 5 are completed within 7 days; stages 6 and 7 take 2 days for the completion.

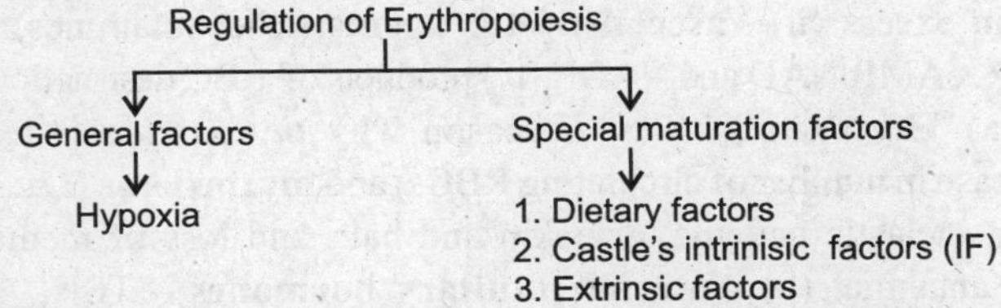

Fig. 6.39. Factors involved in regulation of erythropoiesis.

Sources and metabolism of erythropoietin.

1. Erythropoietin is mainly (85%) secreted by kidneys probably by interstitial cells (or cells of the endothelium) of the peritubular capillaries (*i.e.*, network of capillaries surrounding the renal tubules).

2. 15% erythropoietin is produced by extra renal sources such as liver parenchymal cells and cells of tissue macrophage system. The hypoxia stimulus has to be much greater in normal persons to trigger extra renal erythropoietin production. Erythropoietin is inactivated in the liver and kidneys and its main site of excretion is urine.

Formation and release of erythropoietin. Hypoxia of kidneys causes release of **renal erythropoietic factor (REF)** or **erythrogenin** which acts on a plasma substrate, called **erythropoietinogen**, which is an α-globulin to form erythropoietin (Fig. 6.40). Hypoxia also increases α-globulin formation from the liver. Erythroietin production reaches its peak within 24 hours of hyoxic stimulus.

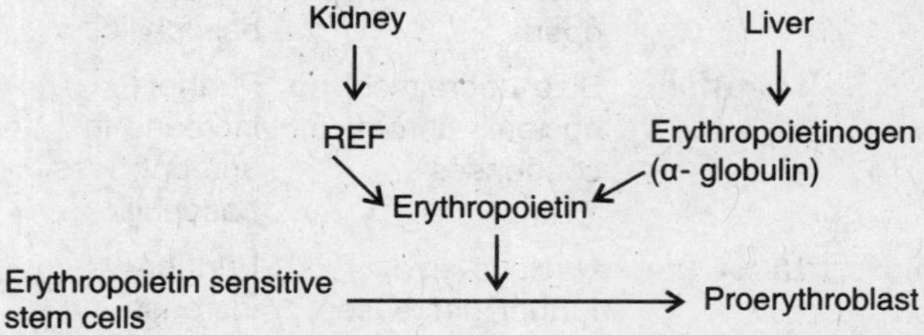

Fig. 6.40. Summary of steps of formation, release and action of erythropoietin.

Mode of action of erythropoietin

1. Erythropoietin causes early differentiation of **erythropoietin sensitive stem cells** in the bone marrow into proerythroblasts and subsequently to mature RBCs. Thus, increasing RBC count in 2–3 days.

2. It increases release of reticulocytes from the bone marrow.

3. It increases synthesis of RNA, DNA, and ferritin which increases haem (or heme) synthesis, thus increasing haemoglobin synthesis in already existing normoblasts (Box 6.10).

Box 6.10

1. Erythropoietin receptors have tyrosine kinase activity, that acts by inhibiting apoptosis of RBCs resulting in their increased growth and development.

2. Erythropoietin promotes erythropoiesis not only in adult bone marrow but also foetal yolk sac, liver and spleen.

Factors affecting erythropoietic production

1. Increase in production of erythropoietin.

(i) *Hypoxia* due to – (a) haemorrhage; (b) high altitude; (c) cardio-respiratory disturbances; (d) methaemoglobin excess (ii) **Vasoconstrictor agents** (catecholamines) due to renal hypoxia. (iii) *Nucleotides, e.g.*, cAMP, NAD and NADP. (iv) products of RBC destruction, called **haemolysates**. (v) *Hormones* – (a) **Thyroxine**, *i.e.*, it is reason why *thyrotoxicosis* leads to **polycythemia** (*i.e.*, abnormal increase in number of circulating RBCs) and **myxoedema** (*i.e.*, severe hypothyroidism chacterized by firm inelastic oedema, dry skin and hair, and loss of mental and physical vigor) is associated with anaemia. (b) **Anterior pituitary hormones** – TSH, ACTH; GH; LH; FSH and prolactin. Therefore, pituitary disturbances produce anaemia. (c) **Androgens**. They stimulate erythropoiesis by increasing REF production and also potentiate the action of erythropoietin. (vi) Other factors, *e.g.*, cobalt salts.

2. Decrease in production of erythropoietin.

(i) Oestrogen: (a) decrease hepatic synthesis of globulin; (b) depresses the erythropoietic response to hypoxia. It is due to this reason, *in females, the RBC count is less as compared to males*. (ii) Chronic renal diseases; (iii) Protein deficiency; (iv) Cirrhosis of the liver; (iv) Chronic inflammatory diseases.

B. Special Maturing Factors (Factors Responsible for Final Maturation of RBCs).

1. Dietary factors. (i) Proteins help in globin formation. (ii) Iron, manganese, copper, cobalt, nickel help in haem formation. (iii) Calcium increases iron absorption from GIT. (iv) VItamine C, B_{12} and folic acid help in synthesis of nucleic acids.

2. Castle's intrinsic factors (IF). It is a glycoprotein having M.W. 50,000 to 60,000 and produced by the parietal (oxyntic) cells of the stomach, therefore, it decreases due to: (i) big gastric ulcer, (ii) malignancy of stomach and (iii) gastrectomy (removal of stomach).

IF helps in absorption of vitamin B_{12} from ileum. One molecule of IF binds with one molecule of vitamin B_{12} to form **IFB_{12} complex** which then binds to specific receptor in the ileal mucosa. IF is split off from this complex and vitamin B_{12} is released into the portal blood. This process does not require energy but is calcium dependent and is optimal at pH 7.0. IF-B_{12} complex is resistant to GIT proteolytic enzymes.

3. Extrinsic factors (EF). These are present in certain foods and are essentially vitamin B_{12} and folic acid (Table 6.10). IF with extrinsic factor form **haematinic principle** which helps in maturation of erythrocytes (RBCs) (*i.e.*, for conversion of pro-erythroblast to mature

Table 6.10 Comparison of vitamin B_{12} and folic acids (the two extrinsic factors for erthyropoiesis).

S. No.	*Features*	*Vitamin B_{12}*	*Folic acid (Pteryolglutamic acid)*
1.	Source	Absent in plants (unless contaminated with bacteria) but present in all animal tissues. It is also synthesized by bacteria in water, soil or GIT.	Green leafy vegetables; yeast; liver, etc. Food folate is rapidly destroyed by heating
2.	Daily requirements	2 μgm	100–200 μgm
3.	Amount present in daily normal mixed diet	5–30 μgm	500–800 μgm
4.	Normal plasma concentration	300–400 pg/ml	3–20 ng/mL
5.	Site of absorption	Only from ileum in the presence of I.F.	
6.	Storage	In liver it is approximately 1000 times its amount which is present in normal diet taken daily; therefore many months of defective vitamin B_{12} absorption will produce its deficiency.	Liver, RBCs and WBCs (total body stores 2–3 mg)

S. No.	Features	Vitamin B_{12}	Folic acid (Pteryolglutamic acid)
7.	Excretion	30–40% of daily intake in faeces, also excreted in urine	Urine
8.	Function	(a) Promotes the maturation and normalization of RBCs (plays an essential role with folic acid in synthesis of nucleic acid - DNA) (b) Increase WBC and platelet count through its action on bone marrow. (c) Miantain normal activity of nervous system. (d) Helps in myelination of nerve fibres.	(a) As of vitamin B_{12} It is reduced to tetrafolic acid and then acts as a conezyme in different biological systems.

B. Leucopoiesis

Development of leucocytes or white blood cells (WBCs) occurs in following two steps: 1. Granulopoiesis and 2. Agranulopoiesis.

1. Granulopoiesis. Granulocytes develop mainly and exclusively in the red bone marrow. It is wholly **extravascular** process and is completed in 3 days (Table 6.11). Between the fat cells of red bone marrow are present branched **reticulum cells** (*i.e.*, irregular in outline, free from granules or mitochondria with faint basophilic cytoplasm). Reticulum cells multiply by mitosis forming primitive WBCs (stem cells) → myeloblast → premyelocyte → myelocyte proper → metamyelocyte → mature WBCs.

Table 6.11 Stages of granulopoiesis.

S.No.	Stage	Cell Size	Nucleus	Cytoplasm	Others
1.	Primitive WBCs	18–23 μm	Large, spherical, many nucleoli; open chromatin	Basophilic; thin rim of cytoplasm all around the nucleus	Mitosis+++
2.	Myeloblast	16–20 μm	Large, pale or purple blue; many nucleoli; fine stippled chromatin	Amount increases; narrow blue rim without granules	Mitosis+++ show active protein synthesis; highly developed endoplasmic reticulum
3.	Myelocyte-A (premyelocyte)	14–18 μm	Size decreases, nucleoli disappears, chromatin condenses	Amount increase further	Mitosis++

S.No.	Stage	Cell Size	Nucleus	Cytoplasm	Others
4.	Myelocyte-B (Myelocyte proper), *i.e.*, Neutrophil	12–16 μm	Round, nucleus size further decreases	Amount increases, less basophilic **granules appear** with special staining reaction; maximum granules in neutrophils, scanty in eosinophils and basophils	Multiplication is maximum
5.	Myelocyte-C (Meta-myelocyte)	10–14 μm	Nucleus size decreases; becomes indented and lobed	Amount further increases and becomes more liquid; **granules show amoeboid** movement maximum in neutrophils)	Mitosis stops
6.	Mature WBC (Neutrophil, Eosinophil, Basophil)	–	–	–	–

2. Agranulopoiesis

Agranulocytes also develop to a slight extent in the bone marrow but their main site of origin is in the **lymphoid tissues**, such as thymus, spleen, lymph nodes, Peyer's patches, tonsils, etc., from stem cells that originally come from the bone marrow. In humans, 3.5×10^{10} lymphocytes are in the peripheral blood and rest are in the lymphoid organs (Box 6.11).

Box 6.11

In the fetal stages the thymus gland is the chief source of lymphocyte production. In the early stages of fetal development lymphocyte producing cells are "blue-printed" from the thymus and transferred to lymph tissues (Thymus secretes the humoral factors such as **thymosin** which checks development of lymphocytes). The thymus gradually loses its lymphopoietic activity so that the major source of lymphocyte production in the adult is the lymph tissue.

Lymphocytes of thymic origin are concerned with **cellular immunity**, whereas lymphocytes "processed" in the lymph tissue of alimentary canal, tonsils, lymph nodes and appendix are concerned with **humoral immunity**. When exposed to foreign **antigens**, these lymphocytes are activated (called **immunoblasts**) and differentiate into **plasma cells** or **immunocytes**. These cells secrete the **immune globulins** or antibodies.

Regulation of granulopoiesis. For every circulating granulocytes, 50–100 mature cells are held in bone marrow reserve. The relative constancy of blood granulocyte count is maintained by an efficient 'feedback' mechanism by circulating factors (Fig. 6.41; Fig. 6.42).

Senile Leucocytes

Old leucocytes or WBCs are characterized by:

1. Loss of mobility.
2. Nucleus seen lying free; nuclear lobulation increases.
3. Poorly stained granules.
4. Cells break up readily while making blood smear.

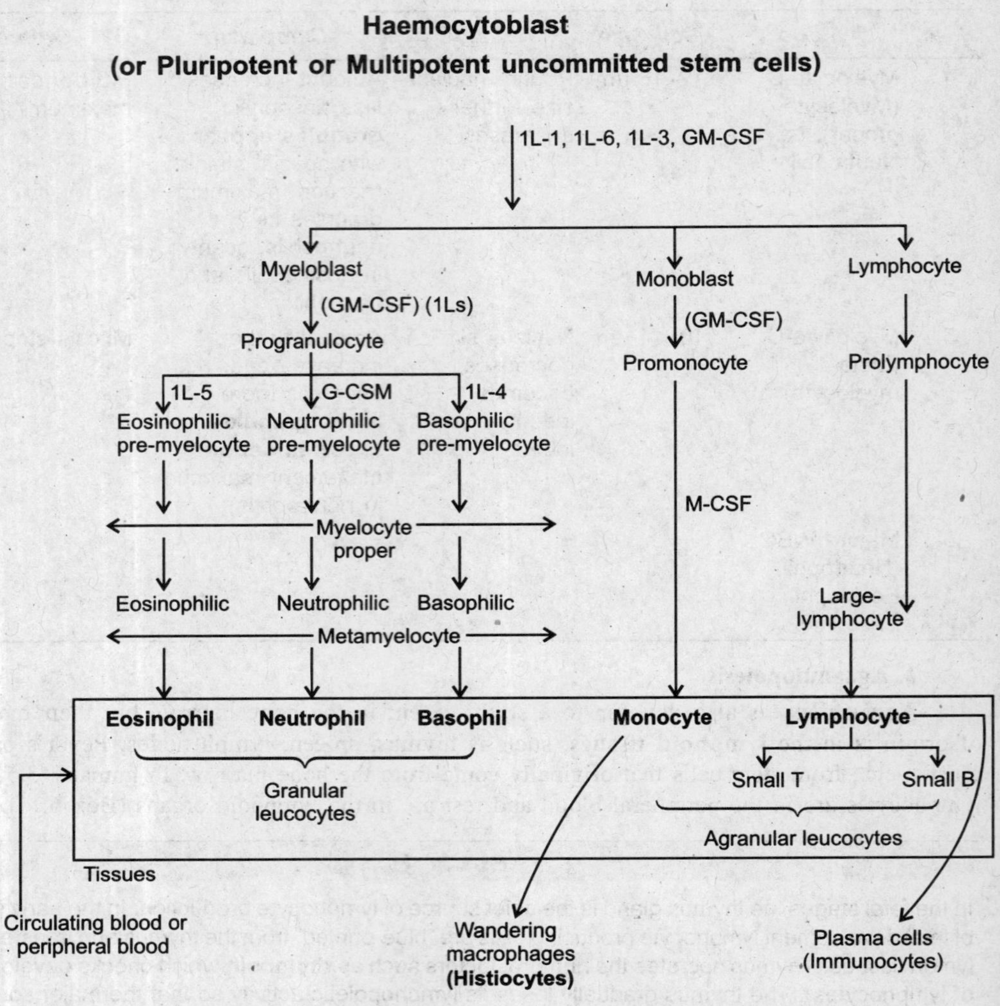

Fig. 6.41. Leucopoiesis–origin and development of leucocytes or white blood cells (WBCs). **(IL**, interleukins; **Ep**, Erythropoietin; **GM-CSF**, Granulocyte macrophage colony stimulating factor).

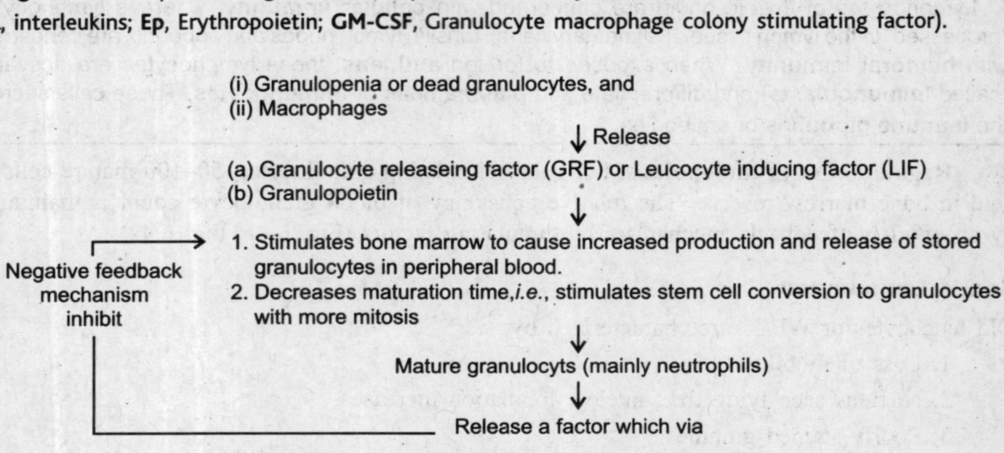

Fig. 6.42. Schematic diagram of granulopoiesis.

3. Thrombopoiesis or Megakaryocytopoiesis

Site of origin of megakaryocytes is red bone marrow. Megakaryocytopoiesis occurs in following steps:

Pluripotent stem cell
↓
Committed stem cell
(polyploid precursor cells)
↓
Megakaryoblast (stage I)
↓
Pro-megakaryocyte (stage II)
↓
Granular megakaryocyte (stage III)
↓
Platelets

Megakaryocytes are giant cells with 35–160 μm diameter, multinucleated (contain irregular ring of lobed nuclei) with dense granular cytoplasm. Platelet formation begins with the formation of micro-vesicles which join to form a demarcation membrane for platelets. Platelets are formed within the cytoplasm of granular megakaryocyte by pinching off bits of cytoplasm and are released into the circulation when the cell dies. One megakaryocyte forms 2000–4000 platelets by this process.

Platelets production is precisely regulated by feedback mechanism by **colony stimulating factor and thrombopoietic stimulating factor (TSF)** or **thrombopoetin**, produced by the liver or kidneys. It is probably released in circulation following excessive destruction of platelets or when the thrombocytopenia develops. It promotes the production of megakaryocytes from committed stem cells.

6.5 BLOOD GROUPS

The human blood is a fluid connective tissue containing two principal components: **cells** (red, white and platelets) and liquid **(plasma).** The plasma is composed of clotting protein fibrinogen and serum. When, blood of one individual is mixed with the serum of other person, then in certain cases clumping of red blood cells takes place but, in other cases do not. The clumping or **agglutination** of red blood cells takes place due to the **antigen-antibody reaction** which is first of all discovered by **Landstenier** in 1900.

Immunity has been defined by **Sir Macfarlane Burnet** as "the capacity to recognize the intrusion of material foreign to the body and to mobilise the cells and cell products to help remove that particular sort of foreign material with greater speed and effectiveness". **Immune response** forms second line of defence after the phagocytes (of blood) which according the definition above, are also part of the immune system. The immune response is the production of antibodies in response to antigens. Each antigen is recognized by a specific antibody (Fig. 6.43).

Antibodies and Antigens

An **antibody** is a molecule that is synthesized by an animal in response to the presence of foreign substances known as **antigens.** Each antibody is a protein molecule called **immunoglobulin** or **agglutinin.** Its structure consists of two **heavy chains** (H-chains) and two **light chains** (L-chains) (Fig. 6.44). The antibody has a **constant** and **variable part,** the variable part acting something like a key which specifically fits into lock (Fig. 6.45). The human body can produce an estimated 100 million different antibodies, recognizing all kinds of foreign substances, including many body has never met. It does this by suffling different sections of the gene, which produce the variable regions (like building different models from a basic set of shapes). There are five different types of of antibodies :**I_gM, I_gD, I_gG, I_gE** and **I_gA** (Box 6.14). Each class of antibodies may be either bound to the lymphocyte cell surface or it may be present in a free-floating state in the circulatory system.

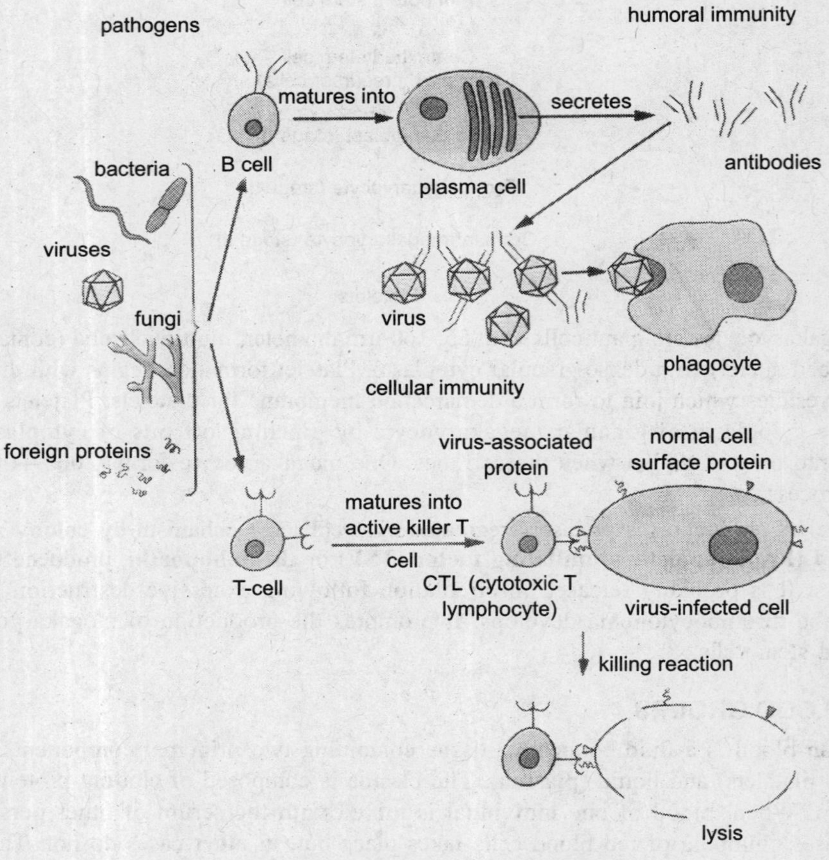

Fig. 6.43. Immunity: humoral and cellular.

An antigen (also called **agglutinogen**) is a molecule which can cause antibody formation. All cells possess antigens in their cell surface membranes which act as markers, enabling cells to recognise each other. Antigens are usually proteins, glycoproteins or glycolipids; although any complex molecule can be antigenic. The body can distinguish its own antigens (*'self'*) from foreign antigens (*'non-self'*) and normally only makes antibodies against non-self antigens. Microorganisms carry antigens on their surfaces.

Two systems of immunity have been developed by mammals, a **cellular immunity** or **cell-mediated immune response** and a **humoral immunity** or **humoral immune response** (Fig. 6.46). The two systems involve the development of two types of lymphocytes, the T and B cell. Both types of lymphocytes arise from precursor cells in the bone marrow. The influence of the thymus gland is essential in the development of the T cells, hence the name **T cells. B cells** are named after the bursa, a branch of gut in birds in which their B cell develop. The bursa is absent in mammals, although equivalent tissue exists. Each type of cell has an enormous capacity for recognizing the millions of different antigens that exist. When an antibody-antigen reaction occurs it prevents, in a variety of ways, the antigen or antigen possessing organism from acting upon the body in a harmful way.

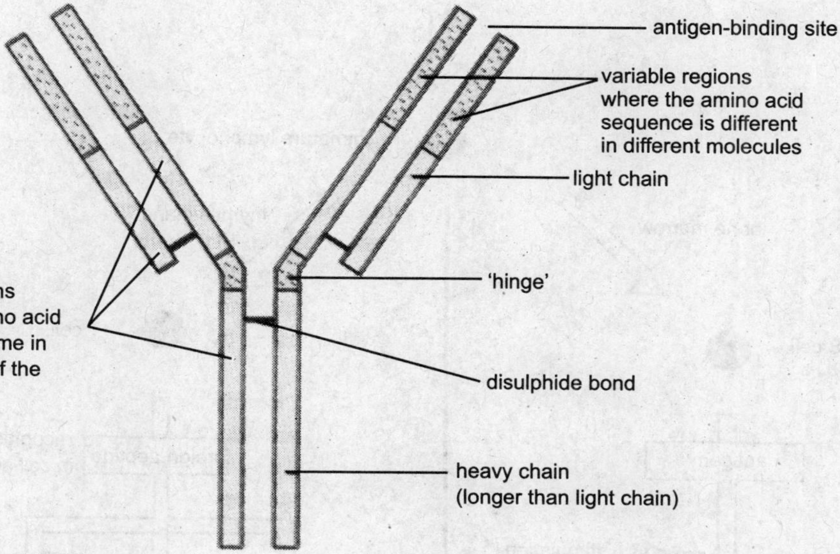

Fig. 6.44. Antibody molecule. Diagram of the basic structure of an antibody molecule. The molecule is made up of four polypeptide chains, two heavy and two light chains. Antigens are bound between the light and heavy chains of the variable regions (after Taylor *et al.*, 2002).

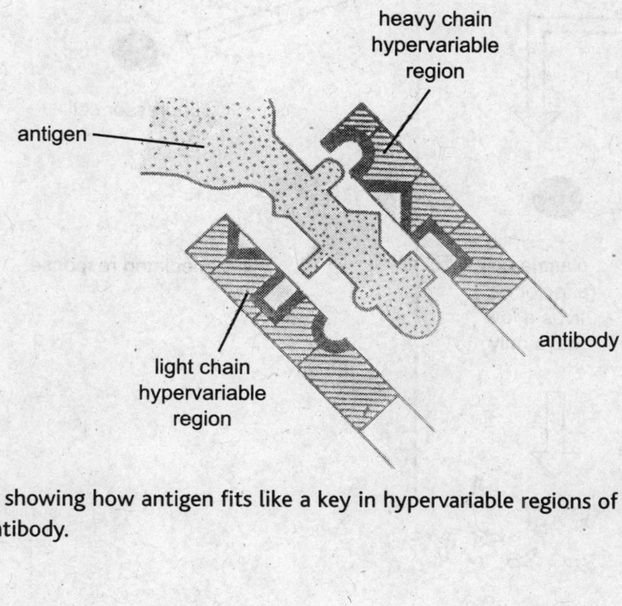

Fig. 6.45. Diagram showing how antigen fits like a key in hypervariable regions of heavy and light chains of an antibody.

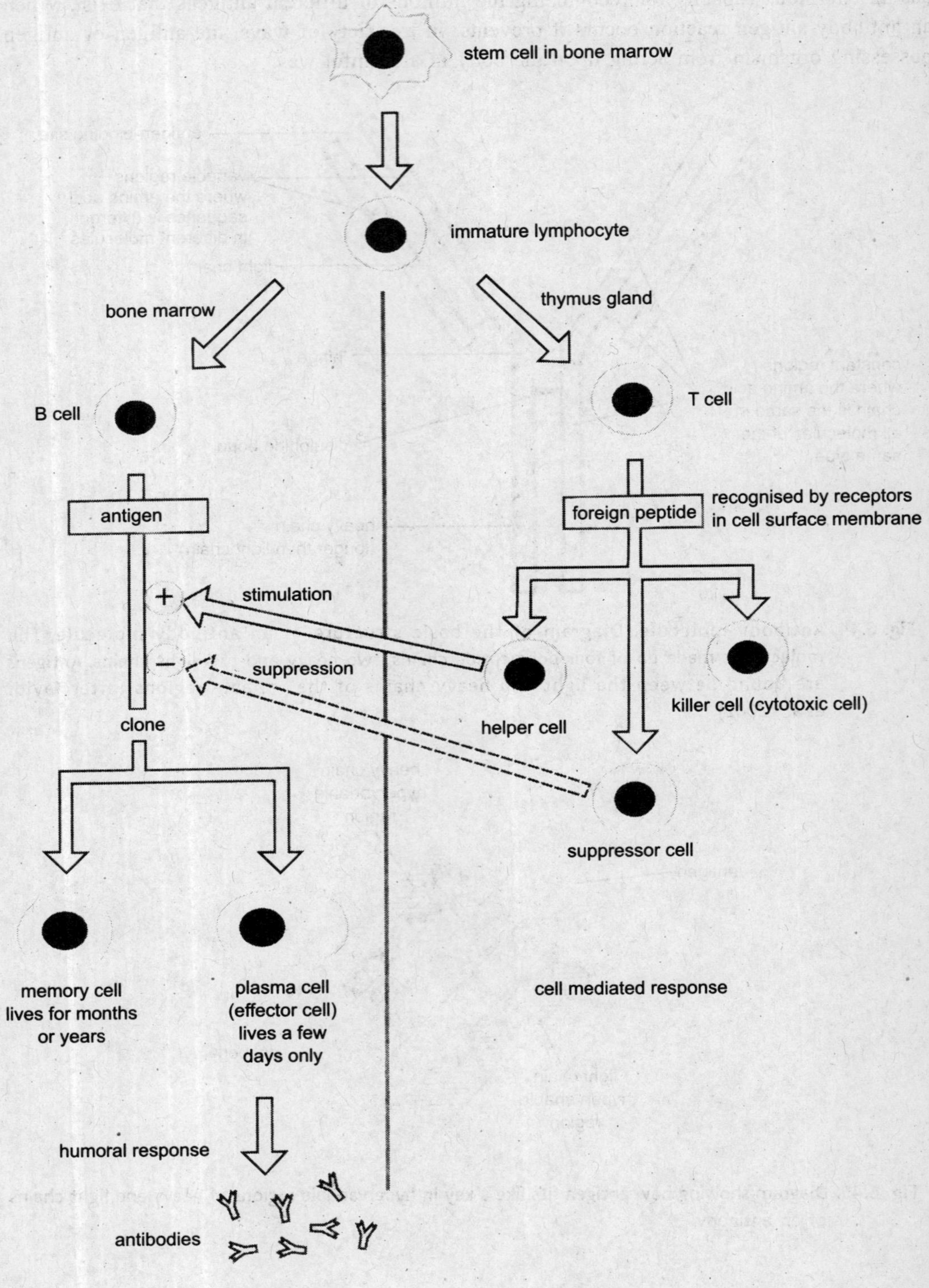

Fig. 6.46. Interaction between cell-mediated and humoral immune response (after Taylor *et al.*, 2002).

2. TYPES OF IMMUNITY

Immunity may be described as active or passive.

(*i*) **Natural active immunity.** This kind of immunity is obtained as a result of an infection. The body manufactures its own antibodies when exposed to an infectious agent. Because memory cells (B lymphocytes), produced on exposure to the first infection, are able to stimulate the production of massive quantities of antibody when exposed to the same antigen again, this type of immunity is of massive quantities of antibody when exposed to the same antigen again, this type of immunity is most effective and generally persists for a long time, sometimes even for life.

(*ii*) **Natural passive immunity.** In **passive immunity** antibodies from one individual are passed into another individual. They give immediate protection, unlike active immunity which takes a few days or week, to build up. However, passive immunity provides protection against infection for a few weeks, because the antibodies are broken down by the body's natural processes, so their numbers are slowly fall and protection is lost.

Passive immunity may be gained naturally. For example, antibodies from a mother can cross the placenta and enter her fetus. In this way they provide protection for the body until its own immune system is fully functional. Passive immunity may also be provided by **colostrum,** the first secretion of mammary glands. The baby absorbs the antibodies through its gut.

1. ABO Blood Types

Blood differs chemically from person to person, and when two different (incompatible) blood types are mixed agglutination (clumping together) of erythrocytes (red blood cells) results. The basis of these chemical differences is naturally occurring antigens on the membranes of red blood cells and epithelial cells. The best known of these is the ABO blood groups (Fig. 6.47; Box 6.12).

The ABO Blood System

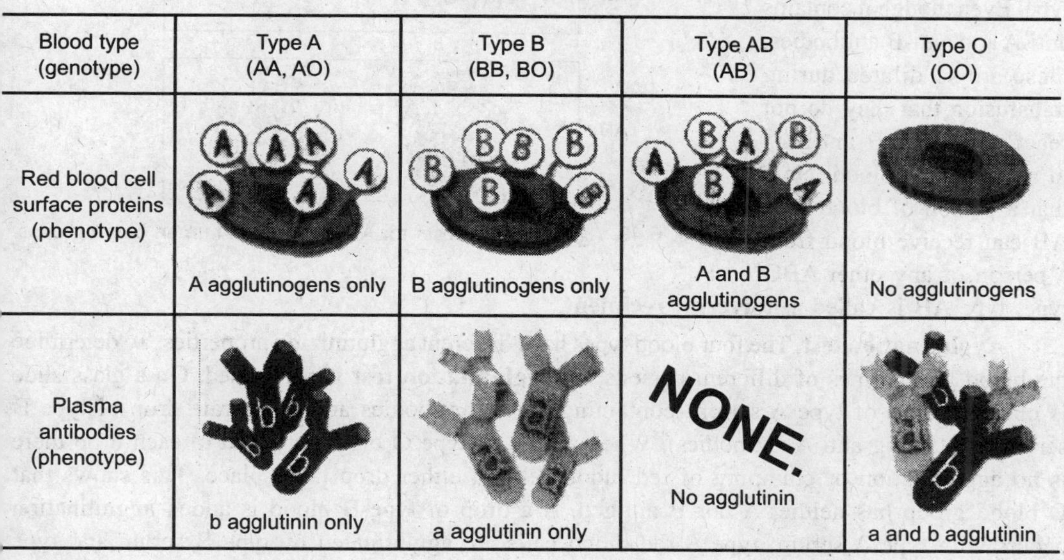

Fig. 6.47. The ABO blood type of human beings.

Box 6.12

Landsteiner in 1900 and 1902 discovered two kinds of **agglutinogens** or **antigens,** called **A** and **B antigens** from the surface of red blood cells of human blood. He found that out of A and B antigens, a person may contain either one (*i.e.,* A or B antigen) or neither of them. Accordingly, he recognized three kinds of **blood types** or **blood groups: type A, type B,** and **type O.** The fourth and most rare, the **AB blood group** or type, was discovered in 1902, by two of **Landsteiner's** students, **Von Decastello** and **Sturli.**

There is an odd feature about the ABO system. Normally we would expect that a type A individual would develop antibodies against type B blood only if B cells were introduced into the body. In fact, *type A persons always have anti-B antibodies in their blood, even without the prior exposure to type B blood.* Similarly, type B individuals carry anti-A antibodies. Type AB blood has neither anti-A nor anti-B antibodies (since if it did, it would destroy its own blood cells), and type O blood has both anti-A and anti-B antibodies.

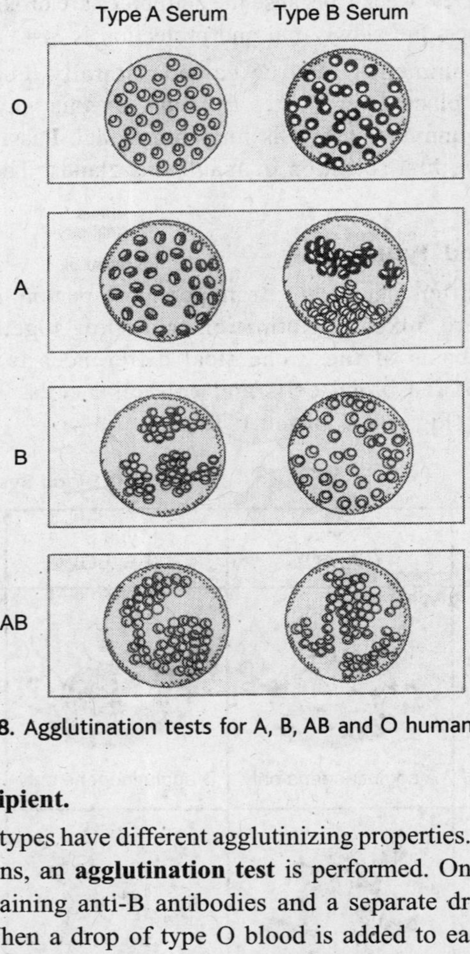

Fig. 6.48. Agglutination tests for A, B, AB and O human blood groups.

People with type O blood are called **universal donors,** because their erythrocytes lack antigens and their blood can be infused into a person with any blood type. Even though it contains anti-A and anti-B antibodies, these are so diluted during transfusion that they do not react with A or B antigens in a recipient's blood. Note that a person of blood type AB can receive blood from a person of any other ABO type; type AB is called a **universal recipient.**

Agglutination test. The four blood types have different agglutinizing properties. To determine the blood group types of different persons, an **agglutination test** is performed. On a glass slide is placed a drop of type A serum (containing anti-B antibodies and a separate drop of type B serum (containing anti-A antibodies). When a drop of type O blood is added to each drop there is no agglutination or clumping of red blood cells in either drop takes place. This shows that O blood group has neither A nor B antigen. If a drop of type B blood is added agglutination occurs with type A serum; type A red blood cells are agglutinated by type B serum and type A B red blood cells are agglutinated by both sera. The agglutination tests for four types of human blood has been illustrated in Figure 6.48. and Figure 6.49.

Table 6.12 Genetic control of the human ABO blood groups (Source: D.L. Hartl and E.W. Jones, 1999).

S. No.	Geno-type	Antigen present on red blood cells	ABO blood group phenotype	Antibodies present in blood fluid	Blood type that can be tolerated in transfusion	Blood types that can accept blood for transfusion
1.	I^A/I^A	A	Type A	Anti-B	A and O	A and AB
2.	I^A/I^O	A	Type A	Anti-B	A and O	A and AB
3.	I^B/I^B	B	Type B	Anti-A	B and O	B and AB
4.	I^B/I^O	B	Type B	Anti-A	B and O	B and AB
5.	I^A/I^B	A and B	Type AB	Neither anti-A nor anti-B	A, B, AB and O	AB only
6.	I^O/I^O	Neither A nor B	Type O	anti-A and anti-B	O only	A, B, AB and O

Biochemical Basis of ABO system

A biochemical basis of the ABO blood type system has now been carefully worked out. The A and B antigens are actually carbohydrate groups (sugars) that are bound to lipid molecules (fatty acids) protruding from the membranes of the erythrocyte. The specificity of A and B antigens is based on the terminal sugar of carbohydrate group.

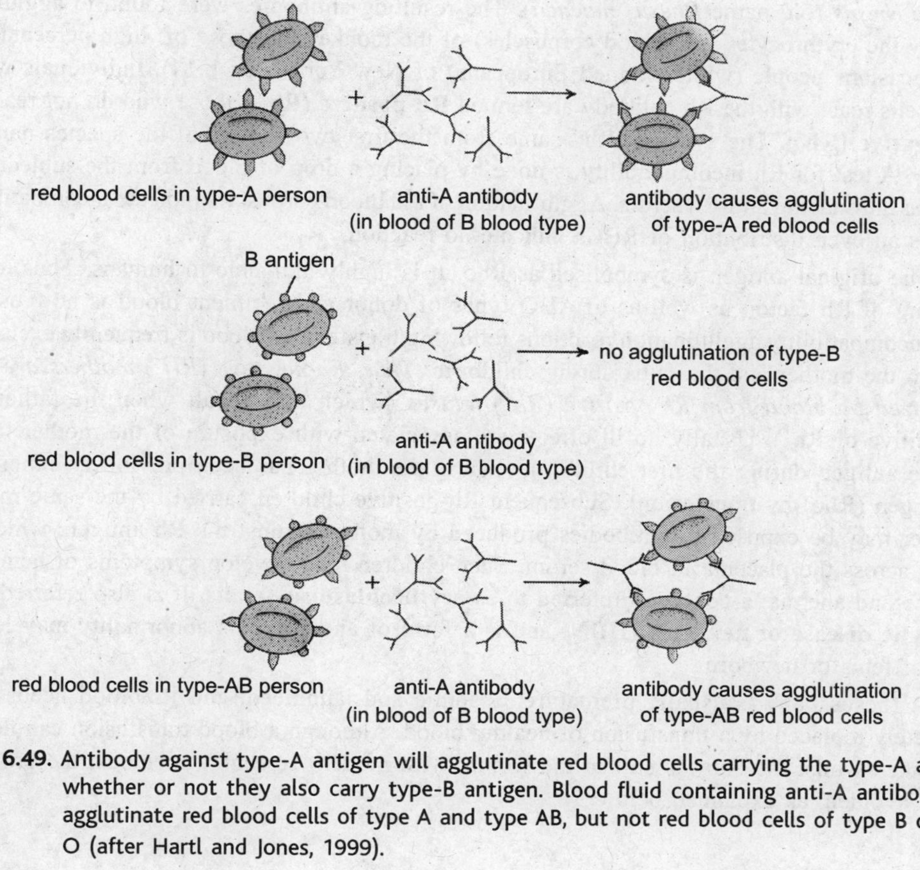

Fig. 6.49. Antibody against type-A antigen will agglutinate red blood cells carrying the type-A antigen whether or not they also carry type-B antigen. Blood fluid containing anti-A antibody will agglutinate red blood cells of type A and type AB, but not red blood cells of type B or type O (after Hartl and Jones, 1999).

Clinical Significance of ABO Blood Groups

The clinical significance of the ABO blood groups is that transfusion of blood containing A or B red cell antigens into persons who make antibodies against these antigens results in an agglutination reaction in which the donor red blood cells are clumped (Box 6.13). In this reaction, the anti-A antibody will agglutinate red blood cells of either blood type A or blood type AB, because both carry the A antigen. Similarly, anti-B antibody will agglutinate red blood cells of either blood type B or blood type AB. When the blood cells agglutinate, many blood vessels are blocked, and the recipient of the transfusion goes into shock and may die. Incompatibility in the other direction, in which the donor blood contains antibodies against the recipient's red blood cells is usually acceptable because the donor's antibodies are diluted so rapidly that clumping is avoided. The type of compatible blood transfusions are shown in last two columns of Table 6.12.

Box 6.13. Blood Typing

Blood typing is important for the prevention of transfusion reaction. **Blood groups** reflect cell surface antigens coded by polymorphic genes that are usually inherited in a simple Mendelian fashion. These antigens are most often identified by **serologic tests.**

3. Rh Factor (Antigen) Alleles in Humans

The **Rh factor** or **Rh antigen** found on the surface of erythrocytes, was discovered in 1940 by **K. Landsteiner** and **A.S. Wiener** from rabbits immunized with the blood of the rhesus monkey *Macaca rhesus* (old name *Rhesus macaca*). The resulting antibodies were found to agglutinate not only the erythrocytes (red blood corpuscles) of the monkey but those of high percentage of the Caucasians people (white skinned Europeans) of New York (Box 6.14). Individuals whose blood cells react with the Rh antibody are termed **Rh-positive (Rh^+),** those who do not react are **Rh-negative (Rh^-).** The symbol "Rh" came from the first two letters of the species name of monkey. A test for Rh incompatibility is done by placing a drop of blood from the subject on a slide and introducing anti-Rh serum. Agglutination of red blood corpuscles indicate incompatibility, whereas an even distribution of RBCs indicate no reaction.

The original antigen is symbolised as Rho. It is highly antigenic to humans. Thus, cross-matching of Rh factor, as well as of ABO types of donor and recipient blood is now used to avoid incompatibility agglutination reactions following transfusion. Blood is frequently exchanged between the mother and the fetus during childbirth. *Thus, Rh-negative (RH^-) mothers are often immunized by blood from Rh-positive (Rh^+) fetuses* (which may result when the fathers are Rh-positive or Rh^+). Usually no ill effects are associated with exposure of the mother to Rh-positive antigen during the first childbirth (unless the mother has been previously exposed to Rh antigen (Rho) by transfusion). Subsequent Rh-positive children carried by the same mother however may be exposed to antibodies produced by mother against the Rh antigen, which are carried across the placenta in blood serum. Such children may develop symptoms of hemolytic jaundice and anemia, a condition referred to as **erythroblastosis fetalis.** It is also referred to as **hemolytic disease of newborn (HDN)** and is a form of anemia. This abnormality may lead to death of fetus or newborn.

A rhesus baby is usually premature, anaemic and jaundiced, and its blood needs to be completely replaced by a transfusion of healthy blood. Although a blood transfusion can now be undertaken whilst the baby is still in the womb, with modern screening methods, the problems can be avoided, as explained below:

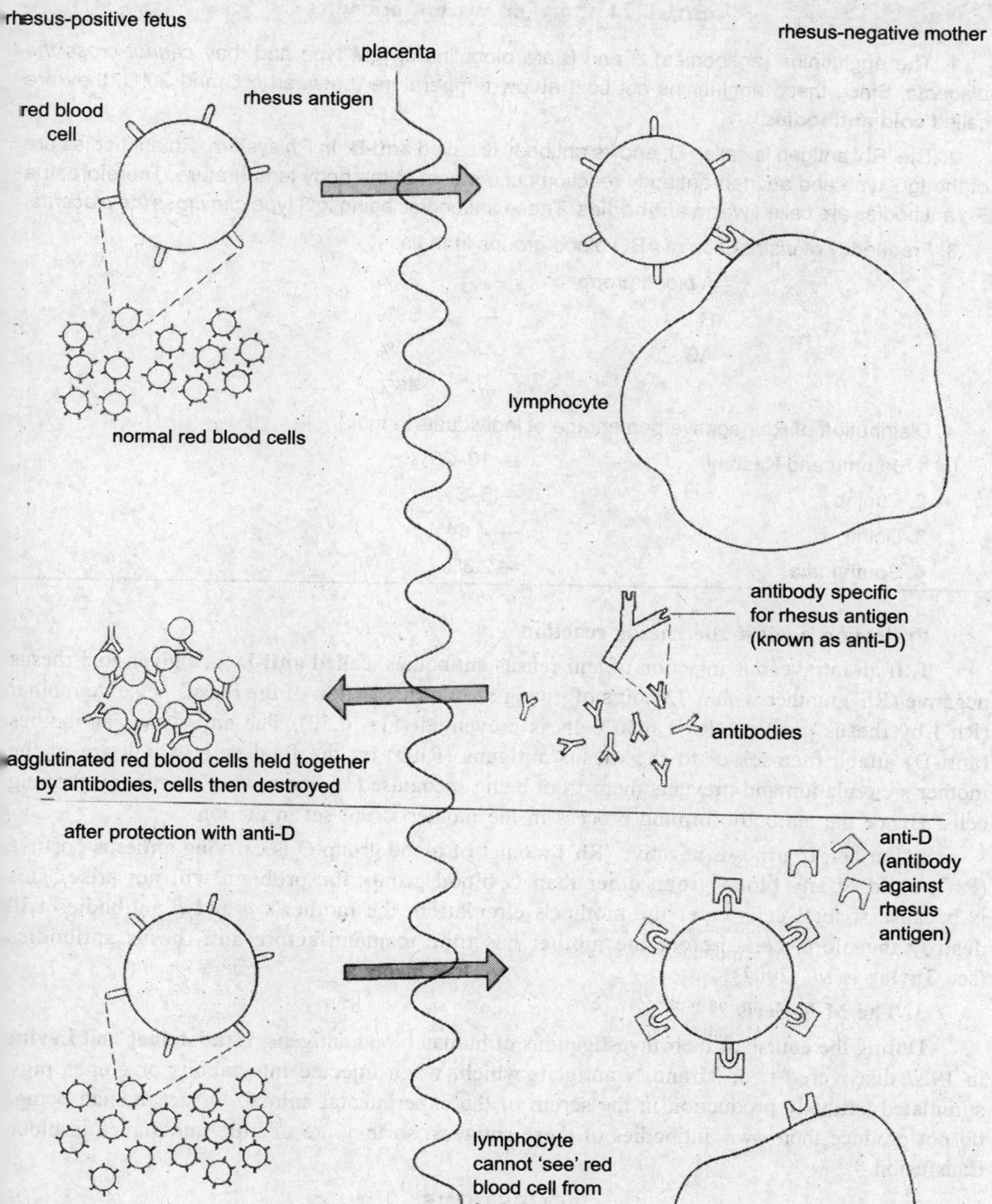

Fig. 6.50. Rhesus disease and its prevention (after Taylor *et al.*, 2002).

Box 6.14 Cold and Warm Antibodies

1. The agglutinins (antibodies) α and β are globulins of IgM type and they *cannot cross the placenta*. Since these agglutinins act best at low temperature (between 5°C and 20°C, they are called **cold antibodies**.

2. The 'Rh' antigen is called D, and its antibody is called **anti-D**. In Rh system, Rh antibodies are of the IgG type and antigen-antibody reaction occurs best at the body temperature. Therefore, the Rh antibodies are called **warm antibodies**. These antibodies being IgG type *can cross the placenta*.

3. Frequency of distribution of ABO blood groups in India.

A blood group	—	21%
B " "	—	39%
AB " "	—	9%
O " "	—	46%

4. Distribution of Rh negative percentage of individuals in India.

1. Jammu and Kashmir	— 10–20%
2. Punjab	— 5–6%
3. Delhi	— 4.84%
4. South India	— 2.3%

Protection against the rhesus reaction

1. If an intravenous injection of anti-rehsus antibodies, called **anti-D**, is a given to a rhesus negative (Rh^-) mother within 72 hours of giving birth, sensitisation of the rhesus negative mother (Rh^-) by rhesus positive (Rh^+) fetal cells is prevented (Fig. 6.50). The anti-rhesus antibodies (anti-D) attach themselves to the rhesus antigens (RhO) on the fetal cells which are in the mother's circulation and prevents them from being recognised by the mother's antibody forming cells. Hence the antibody forming process in the mother is not set in motion.

2. Further, if a rhesus negative (Rh^-) mother of blood group O is carrying a rhesus positive (Rh^+) child of any blood group other than O blood group, the problem will not arise. This is because if fetal cells enter the mother's circulation, the mother's *a* and *b* antibodies will destroy the blood cells before the mother has time to manufacture anti-rhesus antibodies (see **Taylor** *et al.*, 2002).

3. **The M-N Seris**

During the course of their investigations of human blood antigens, **Landsteiner** and **Levine** in 1927 discovered two, **M** and **N** antigens which, when injected into rabbits or guinea pigs stimulated antibody production in the serum of the experimental animal. In fact, human beings do not produce their own antibodies of these antigens, so they are of little importance in blood transfusion.

QUESTIONS

Long Answer Questions

1. Give an account of blood constituents and their functions. *(CCSU Meerut 2010)*

2. Give an account of the composition of blood plasma. Write down the functions of plasma.

3. Give an account of erythrocytes and leucocytes of blood.

4. Explain the mechanism of blood clotting.

5. Explain different theories regarding mechanism of blood coagulation. (*CCSU Meerut 2014*)
6. Describe the process of haemopoiesis. (*HPU 2007*)
7. Describe the complement systems of blood.

Short Answer Questions

1. What are the functions of haemoglobin?
2. Describe the mechanism of blood coagulation. (*HPU 2006*)
3. Differentiate five types of leucocytes of human blood.
4. Differentiate granulocytes and agranulocytes.
5. Explain haemostasis. (*CCSU Meerut 2013*)
6. Blood clotting—modern view. (*Osmania 2005*)
7. Describe the process of coagulation of blood. Enumerate the important factors involved in coagulation. (*CCSU Meerut 2006, 2008*)
8. What is the role of platelets in blood clotting?
9. Describe the process of erythropoiesis.
10. Why the blood does not clot inside the blood vessels? Explain.
11. Write short notes on the following:
    (*i*) Diapedesis;
    (*ii*) Haemostasis;
    (*iii*) Complement cascade;
    (*iv*) Fibrolytic system;
    (*v*) Kinins;
    (*vi*) Erythroblastosis fetalis.

Very Short Answer Questions

1. Name the largest type of blood corpuscles.
2. Write about anticoagulants. (*CCSU Meerut 2013*)
3. Agranulocytes in the blood. (*Osmania 2005*)
4. What is glutathiones?
5. Give an example of vasoconstrictor agent.
6. What is kinin?
7. Define the kininogen.
8. Define the complement.
9. What is blood group?
10. Define the Rh factor.

Fill in the Blanks

1. Life span of mammalian erythrocyte is .... (*HPU 2016*)
2. Largest leucocytes are .... (*HPU 2016*)
3. Process of formation of blood cells is called ....
4. Interstitium contains two types of protein solid structures: .... fiber bundles and .... filaments.

5. The term .... means prevention of blood loss.
6. Thromboxane A is a ....
7. Factor II for blood coagulation is ....
8. Ions of calcium serve as .... in blood coagulation.
9. The so-called Christmas factor for blood coagulation is ....
10. AHF is also called .... and its absence in blood results in .... disease.

Multiple Choice Questions

1. Which of the following in not primarily a function of blood plasma?
   (*a*) transport of antibodies (*b*) transport of hormones
   (*c*) transport of chylomicrons (*d*) transport of oxygen
2. Normally which of the following show highest percentage in blood
   (*a*) neutrophils (*b*) eosinophils
   (*c*) basophils (*d*) lymphocytes
3. In a normal case which of the following show the least percentage?
   (*a*) eosinophils (*b*) lymphocytes
   (*c*) basophils (*d*) monocytes
4. The deficiency of platelets in blood is called
   (*a*) polythemia (*b*) thrombocytopenia
   (*c*) thalassimia (*d*) none of these
5. Packed cells volume is the percentage of the blood volume made by
   (*a*) red blood cells (*b*) white blood cells
   (c) platelets (*d*) all the cells
6. Which of the following stimulates stem cells in the bone marrow to produce red blood cells?
   (*a*) erythropoietin (*b*) fibrinogen
   (*c*) plasminogen (*d*) platelets
7. The enzyme cascade theory of blood coagulation was suggested by
   (*a*) Davie and Ratnoff (*b*) E.O. Wilson
   (*c*) Macfarlane (*d*) M. Pearl
8. Among the following which is the Human Blood Coagulation Factor I.
   (*a*) prothrombin (*b*) fibrinogen
   (*c*) thrombin (*d*) thromboplastin
9. The haemophilia is due to the absence of
   (*a*) platelets (*b*) factor VIII
   (*c*) calcium (*d*) factor XIII
10. The prevention of blood loss is called
   (*a*) homeostasis (*b*) haemostasis
   (*c*) cybernetics (*d*) none of these

11. The volume of blood present in adult human is __________ liters.

(*a*) 10 (*b*) 5

(*c*) 2 (*d*) 1

12. The ion required for coagulation of blood is

(*a*) Na (*b*) Mg

(*c*) Ca (*d*) K

13. The vitamin necessary for blood clotting is

(*a*) A (*b*) B

(*c*) C (*d*) K

14. The formation of blood clot in blood vessel is known as

(*a*) coagulation (*b*) agglutination

(*c*) thrombosis (*d*) precipitation

15. Largest corpuscles present in mammalian blood are

(*a*) erythrocytes (*b*) lymphocytes

(*c*) monocytes (*d*) neutrophils

16. Which one engulfs pathogens rapidly?

(*a*) acidophils (*b*) basophils

(*c*) monocytes (*d*) neutrophils

17. Prothrombin present in normal circulating blood is formed by

(*a*) pancreas (*b*) stomach

(*c*) spleen (*d*) liver

18. Heparin present in normal circulating blood is formed by

(*a*) granulocytes (*b*) mast cells

(*c*) monocytes (*d*) platelets

19. Glutathiones within the RBCs

(*a*) acts as an enzyme of glycolytic system

(*b*) prevents damage of haemoglobin.

(*c*) required in the formation of NADPH

(*d*) maintains the shape and flexibility of RBCs

20. Normal reticulocyte count at birth is

(*a*) 1–2% (*b*) 2–6%

(*c*) 6–10% (*d*) 30–40%

21. Interleukins

(*a*) are hormone like chemical messengers

(*b*) amino acid sequence is not known

(*c*) are produced by macrophages only

(*d*) are also called lymphokines

22. In a normal individual, erythropoiesis is completed in .... days.

(*a*) 3 (*b*) 7

(*c*) 14 (*d*) 20

23. The commonest site of haemopoiesis in a 3 month old foetus is

(*a*) liver (*b*) spleen

(*c*) bone marrow (*d*) gut

24. Which bone does not contain the red marrow?

(*a*) vertebrae (*b*) ribs

(*c*) sternum (*d*) clavicle

25. Total leucocyte count at birth is

(*a*) 4000–11000/μL (*b*) 11000–15000/μL

(*c*) 15000–18000/μL (*d*) 2000/μL

26. The life span of platelets is

(*a*) 4 days (*b*) 8–12 days

(*c*) 20–30 days (*d*) 90 days

27. Thrombocytopenia commonly occurs in

(*a*) splenectomy (*b*) after surgery

(*c*) dengue fever (*d*) stress conditions

ANSWERS

Very Short Answer Questions

1. Monocytes.
2. Substances which inhibit the blood clotting are called anticoagulants, *e.g.*, heparin (a natural anticoagulant) and sodium citrate (a synthetic anticoagulant).
3. Agranulocytes of blood are lymphocytes and monocytes.
4. Glutathione is albumin-like insoluble protein which acts as a reducing agent, and prevents damage of haemoglobin.
5. Norepinephrine or Angiotensin II or ADH (antidiuretic hormone or vasopressin).
6. Kinin is any of various polypeptide hormones that are formed locally in the tissues and cause dilation of blood vessels and contraction of smooth muscle.
7. The kininogen is an inactive precursor of a kinin.
8. Complement is the thermolabile group of proteins in normal blood serum and plasma that in combination with antibodies causes destruction especially of particulate antigens (as bacteria and foreign blood corpuscles).
9. Blood group is one of the classes (as A, B, AB or O) into which individual vertebrates and especially human beings or their blood can be separated on the basis of the presence or absence of specific antigens in the blood (*i.e.*, on the RBCs).
10. The Rh factor is Rh antigen which is found on the surface of erythrocytes of rhesus monkeys and human beings. It was discovered by **K. Landsteiner** and **A.S. Wiener** in 1940.

Fill in the Blanks

1. 120 days.

2. Monocytes.

3. Haemopoiesis.

4. Collagen, proteoglycan.

5. Haemeostasis.

6. Vasoconstrictor.

7. Prothrombin.

8. Factor IV.

9. Factor IX.

10. Factor VIII, haemophilia.

Multiple Choice Questions

1. (*d*)	**2.** (*a*)	**3.** (*c*)	**4.** (*b*)
5. (*d*)	**6.** (*a*)	**7.** (*c*)	**8.** (*b*)
9. (*b*)	**10.** (*b*)	**11.** (*b*)	**12.** (*c*)
13. (*d*)	**14.** (*c*)	**15.** (*c*)	**16.** (*d*)
17. (*d*)	**18.** (*b*)	**19.** (*b*)	**20.** (*b*)
21. (*a*)	**22.** (*b*)	**23.** (*a*)	**24.** (*d*)
25. (*d*)	**26.** (*b*)	**27.** (*c*).	

7

Physiology of Heart

7.1 OVERVIEW OF CIRCULATION

The **circulatory system** transports and distributes essential substances to tissues and removes metabolic byproducts. This system also participates in homeostatic mechanisms such as regulation of body temperature (or thermoregulation), maintenance of fluid balance, and adjustments of O_2 and nutrient supply under various physiological states. The cardiovascular system that accomplishes these tasks is composed of a pump (the **heart**), a series of distributing and collecting tubes **(blood vessels)**, and an extensive system of thin vessels **(capillaries)** that permit rapid exchange between the tissues and vascular channels. Blood vessels throughout the body are filled with a heterogeneous fluid (**blood**) that is essential for the transport processes performed by the heart and blood vessels.

1. Heart

The heart consists of two pumps in series: One pump propels blood through the lungs for exchange of O_2 and CO_2 (the **pulmonary circulation)** and the other pump propels blood to all other tissues of the body (the **systemic circulation**). Flow of blood through the heart is one way (unidirectional). Unidirectional flow through the heart is achieved by the appropriate arrangement of **flap valves**. Although cardiac output is intermittent (*i.e.*, stop-go), continuous flow to body tissues (periphery) occurs by distention of the **aorta** and its branches during ventricular contraction (**systole**) and by elastic recoil of the walls of the large arteries with forward propulsion of the blood during ventricular relaxation **(diastole).**

2. Cardiovascular Circuit

In the normal intact circulation (*i.e.*, closed circulatory system) the total volume of blood is constant, and an increase in the volume of blood in one area must be accompanied by a decrease in another. However, the distribution of blood circulating to the different regions of the body is determined by the output of the **left-ventricle** and by contractile state of the resistance vessels (arterioles) of these regions. The circulatory system is composed of conduits arranged in series and in parallel (Fig. 7.1). This arrangement has important implications in terms of resistance, flow, and pressure in blood vessels.

Blood entering the **right ventricle** *via* the right atrium is pumped through the pulmonary arterial system at a mean pressure about one seventh that is in the systemic arteries. The blood then passes through the lung capillaries, where CO, in the blood is released and O_2 is taken up. The O_2-rich blood returns via the pulmonary veins to the left atrium, where it is pumped from the ventricle to the periphery, thus completing the cycle.

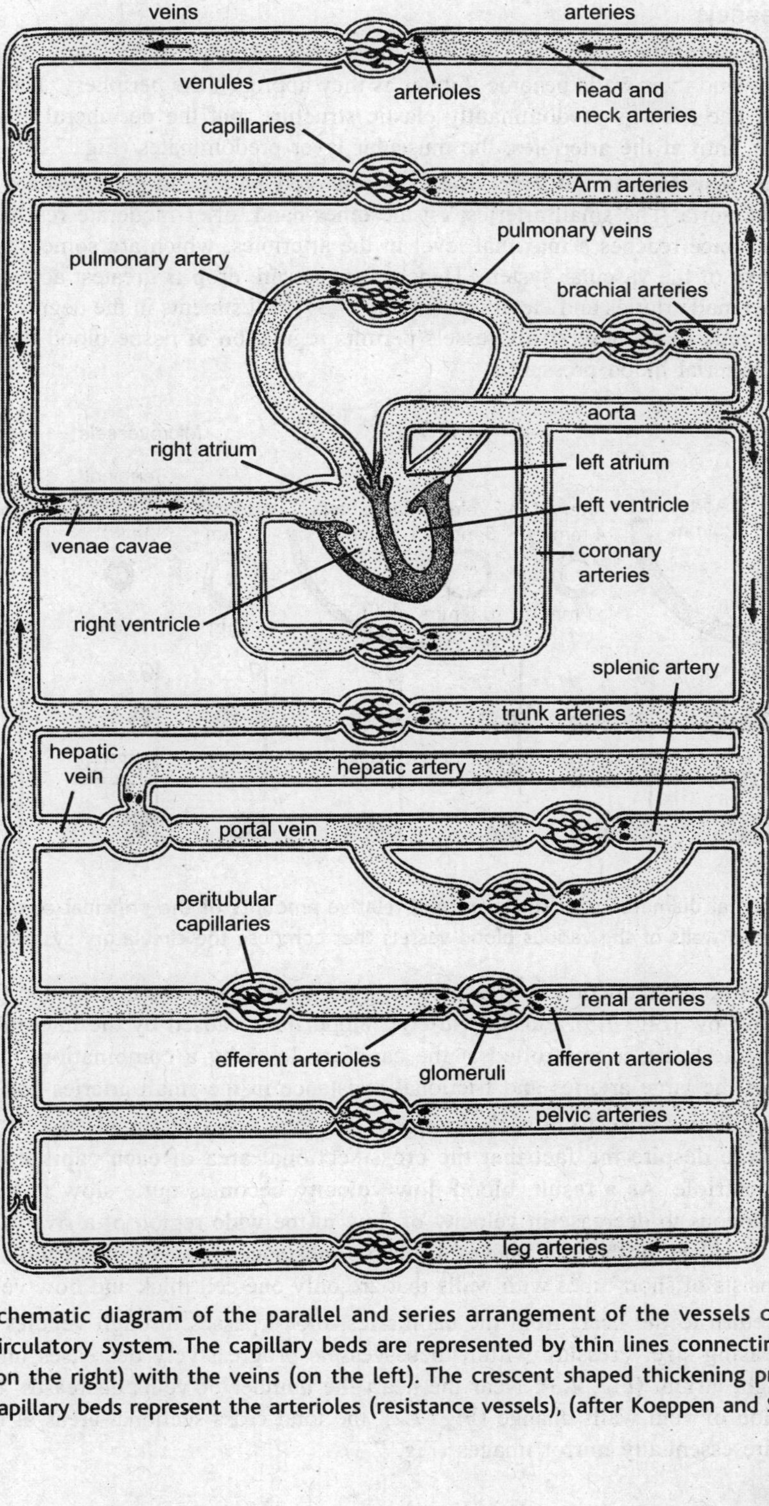

Fig. 7.1. Schematic diagram of the parallel and series arrangement of the vessels composing the circulatory system. The capillary beds are represented by thin lines connecting the arteries (on the right) with the veins (on the left). The crescent shaped thickening proximal to the capillary beds represent the arterioles (resistance vessels), (after Koeppen and Stanton 2010).

3. Blood Vessels

Blood moves rapidly thorough the aorta and its arterial branches (arterioles). These branches become narrow and their walls become thinner as they approach the periphery. They also change histologically. The aorta is predominantly elastic structure, but the peripheral arteries become more muscular until at the arterioles, the muscular layer predominates (Fig. 7.2).

Frictional resistance in the large arteries is relatively small and pressures are only slightly less than in the aorta. The small arteries, on the other hand, offer moderate resistance to blood flow. This resistance reaches a maximal level in the arterioles, which are sometimes referred to as the **stopcocks** of the vascular system. Hence, the pressure drop is greatest across the terminal segment of the small arteries and the arterioles (Fig. 7.3). Adjustments in the degree of contraction of the circular muscle of these small vessels permits regulation of tissue blood flow and aids in the control of arterial blood pressure.

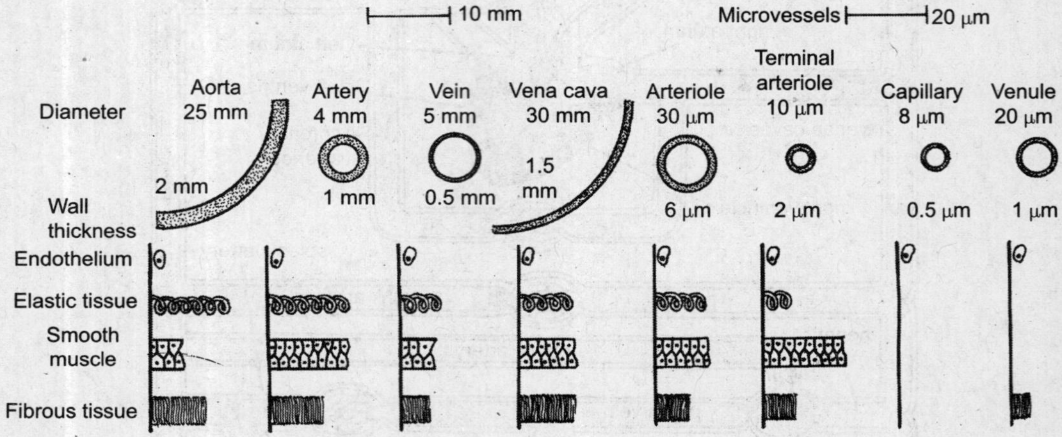

Fig. 7.2. Internal diameter, wall thickness and relative amounts of the principal components of the vessel walls of the various blood vessels that compose the circulatory system.

In addition to the reduction in pressure along the arterioles, there is a change from pulsatile to steady blood flow (Fig. 7.3). Pulsatile arterial blood flow, caused by the intermittent ejection of blood from the heart, is controlled at the capillary level by a combination of two factors: distensibility of the large arteries and frictional resistance in the small arteries and arterioles.

Many capillaries arise from each arteriole. The total cross-sectional area of the capillary bed is very large despite the fact that the cross-sectional area of each capillary is less than that of each centriole. As a result, blood flow velocity becomes quite slow in the capillaries (Fig. 7.3), analogous to decrease in velocity of flow in the wide region of a river. Conditions in the capillaries are ideal for the exchange of diffusible substances between blood and tissue because capillaries consists of short tubes with walls that are only one cell thick and flow velocity is low.

On its return to the heart from the capillaries, blood passes through venules and through veins of increasing size. Pressure within these vessels progressively decreases until the blood reaches the right atrium (Fig. 7.3). Near the heart the number of veins decreases, the thickness and composition of vein walls change (Fig. 7.2), the total cross-sectional areas at each level of the vascular are essentially mirror images (Fig. 7.3).

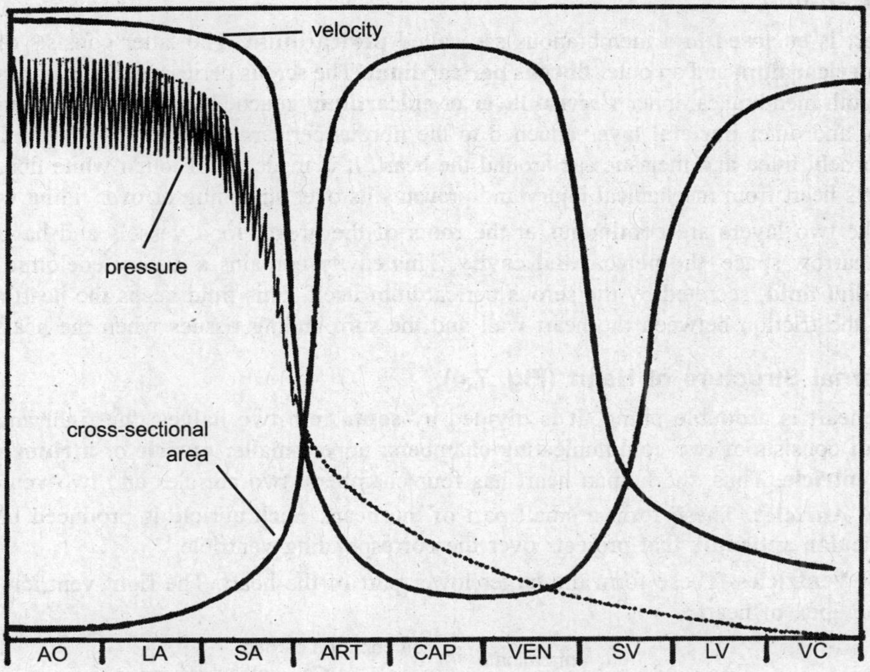

Fig. 7.3. Phasic pressure, velocity of flow, and cross-sectional area of the systemic circulation. The important features are the inverse relationship between velocity and cross-sectional area, the major pressure drop across the small arteries and arterioles, and the maximal cross-sectional area and minimal flow rate in the capillaries. AO, aorta; ART, arteriole; CAP, capillaries; LA, large arteria; LV, large veins; SA, small arteries; SV, small veins; VC, venae cavae; VEN, venules.

7.2 HEART

A knowledge of morphology and histology of human heart will help understand its physiology. The study of the heart, its structure, working and abnormalities is called **cardiology** (Greek, *kardia* = heart; *logous* = study).

1. Appearance and Position

The heart of human beings is a hollow, muscular organ, roughly of the size of one's fist, being 12 cm from apex to base, 8 cm frcm left to right, and 6 cm from front to back. It is reddish-brown in colour and somewhat conical in shape. Heart is situated almost in the middle line in the thoracic cavity between the lungs. Its broad **base** faces upwards and backwards. Its narrow **apex** is directed downward, forward and slightly to the left. Heart is enclosed in the bony thoracic cage formed by backbone, ribs and breast bones and diaphragm. Two grooves or **sulci** are seen on the surface of heart. These are:

(*i*) Coronary sulcus or Atrio-ventricular groove. It is a transverse groove which divides heart into an anterior and smaller **auricular part** and posterior longer and thicker **ventricular part.**

(*ii*) Interventricular groove. It is an oblique longitudinal groove on the ventral side of ventricular part of the heart. Apex of the heart is formed of only left ventricle.

2. Pericardium

The heart is enclosed in a membranous sac called **pericardium**. The latter consists of an inner **serous pericardium** and an outer **fibrous pericardium**. The serous pericardium is composed of two thin smooth membranes: inner **visceral layer** or **epicardium** attached to the heart forming its outer covering and outer **parietal layer** attached to the fibrous pericardium. Fibrous pericardium is the tough loosely fitted and inelastic sac around the heart. It is made up of tough white fibrous tissue. It protects heart from mechanical injury and prevents its over stretching or over filling with blood.

The two layers are continuous at the roots of the great blood vessels and have between them a narrow space, the **pericardial cavity**. This cavity contains a watery coelomic fluid, the **pericardial fluid**, secreted by the serous pericardium itself. This fluid keeps the heart moist and reduces the friction between the heart wall and the surrounding tissues when the heart beats.

3. External Structure of Heart (Fig. 7.4)

Human heart is a double pump. It is divided by **septa** into two halves: the right and the left. Each half consists of two communicating chambers: upper smaller **auricle** or **atrium** and lower larger **ventricle**. Thus, the human heart has four chambers: two auricles and two ventricles.

(*a*) Auricles. These form a small part of the heart. Each auricle is produced into a flap, the **auricular appendix** that projects over the corresponding ventricle.

(*b*) Ventricles. These form the larger lower part of the heart. The right ventricle does not reach the apex of heart.

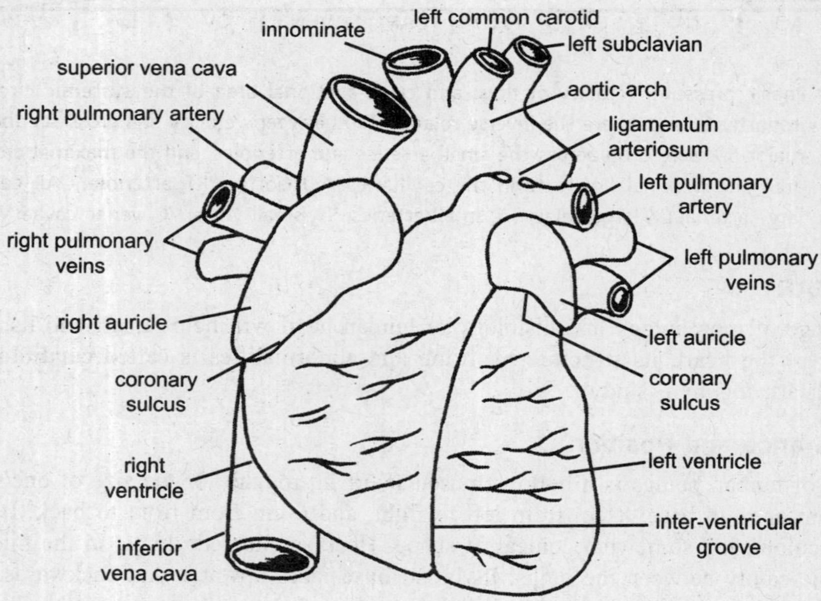

Fig. 7.4. External view of human heart from the ventral side.

4. Internal Structure of Heart (Fig. 7.5 and Fig. 7.6)

(*i*) Auricles. The auricles have thin walls, because they have to force the blood into the ventricles, that lie just below them. The inner surface of the auricles is smooth. The two auricles are separated from each other by a partition, the **interauricular** or **inter atrial septum** (Fig. 7.6). The septum has an oval, thin area called the **fossa ovalis**. It marks the position of an opening, the **foramen ovale**, between the two auricles in the foetus.

***(ii)* Ventricles.** The ventricles have thicker walls than the auricles have and wall of left ventricle is about three times as thick as that of the right ventricle. This is because the left ventricle has to pump blood to the farthest end of the body, whereas the right ventricle has to send the blood to the lungs, which lie nearby. Correspondingly, the blood entering the aortic arch from the left ventricle is at a much higher blood pressure than the blood entering the pulmonary arch.

The inner surface of the ventricles is raised into a network of a low, muscular ridges, called the **columnae carneae** or **tuberculae carneae**, and also has a few large, conical, muscular elevations known as the **musculi papillares** or **papillary muscles**. The two ventricles are separated from each other by a thick, curved partition, the **interventricular septum**. The cavity of the right ventricle is smaller and arc-like, whereas that of left ventricle is larger and nearly circular.

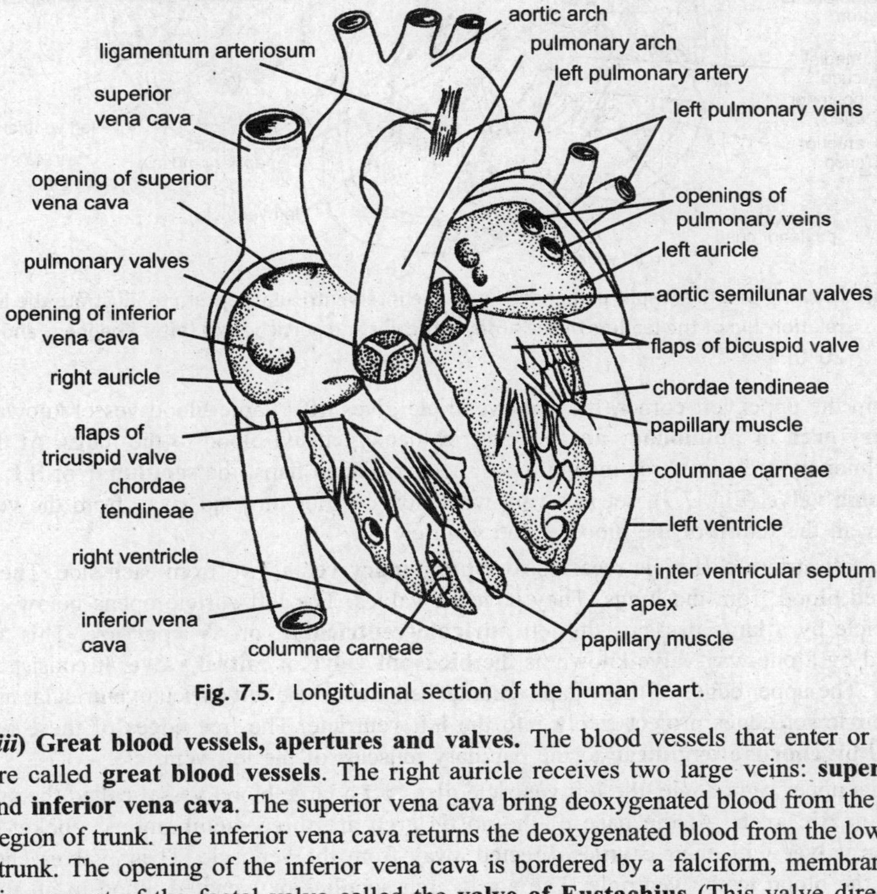

Fig. 7.5. Longitudinal section of the human heart.

***(iii)* Great blood vessels, apertures and valves.** The blood vessels that enter or leave the heart are called **great blood vessels**. The right auricle receives two large veins: **superior vena cava** and **inferior vena cava**. The superior vena cava bring deoxygenated blood from the head and upper region of trunk. The inferior vena cava returns the deoxygenated blood from the lower region of the trunk. The opening of the inferior vena cava is bordered by a falciform, membranous fold, which is remnant of the foetal valve called the **valve of Eustachius** (This valve directs blood from the inferior vena cava into the left auricle via foramen ovale during foetal life). The right auricle also receives a small **coronary sinus** (not shown in Fig. 7.6). It returns deoxygenated blood from the heart wall. Its opening is bordered by a small fold, the **valve of the besius**. The right auricle opens into the right ventricle through a wide passage, the **right auriculoventricular**, or **A-V aperture**. This aperture is guarded by a one-way valve called the **tricuspid valve** (Fig. 7.7). The valve consists of three membranous flaps, which are attached to the margin of the right auriculoventricular aperture above, but project freely into the ventricle below. Their free lower edges are fixed to the papillary muscles of the right ventricle by a number of tough white cords, the **chordae tendineae**. The chordae tendineae checks the pushing of the flaps into the auricles during the contraction of the ventricle.

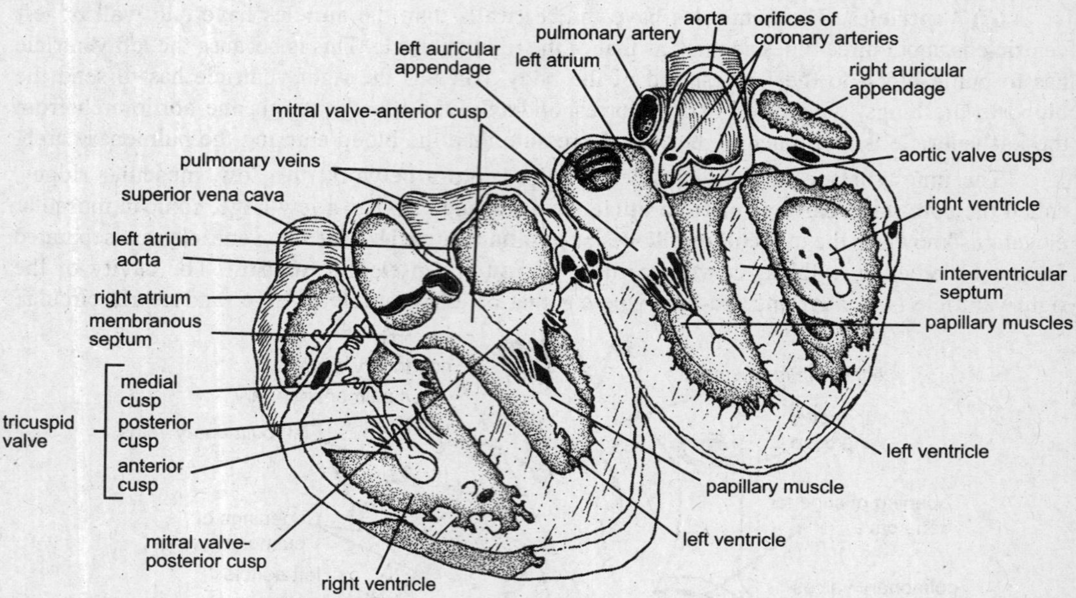

Fig. 7.6. Drawing of a heart split perpendicular to the interventricular septum to illustrate the anatomic relationship of the leaflets of the atrioventricular and aortic valves (after Koeppens and Stanton 2010).

From the upper left corner, the right ventricle gives off a large blood vessel known as the **pulmonary arch** or **pulmonary aorta**. It carries deoxygenated blood to the lungs. At the base of the pulmonary arch are three membranous, pocket-shaped flaps, the **semilunar** or **S.L. valves** or **pulmonic valve** (Fig. 7.7), set in a ring with their cavities directed away from the ventricle. They prevent the return of the blood to the ventricle.

The left atrium or auricle receives four **pulmonary veins**, two from each side. They bring oxygenated blood from the lungs. They have no valves. The left auricle opens below into the left ventricle by a large passage, the left **auriculo-ventricular**, or **AV aperture**. This aperture is guarded by a one-way valve known as the **bicuspid valve** or **mitral valve**. It consists of two flaps only. The upper edges of these flaps are attached around the left auriculoventricular aperture, while their lower edges project freely into the left ventricle. The free edges of these flaps are connected by **chordae tendineae** to the papillary muscles of the left ventricle.

At its upper right angle, the left ventricle gives off a large blood vessel called the **systemic aorta** or **aortic arch**. At the base of the aortic arch are three membranous, pocket-shaped **semilunar valves** with their cavities directed away from the ventricle. These valves check the return of the blood to the ventricle. The aortic arch supplies oxygenated blood to all the parts of the body, except the lungs.

Histology of Heart

Heart is a highly modified blood vessel. Its wall is composed of following three concentric tissues: 1. Epicardium, 2. Myocardium and 3. Endocardium.

1. Epicardium. It is the outer epithelium of thin cells. It is in fact the visceral layer of serous pericardium.

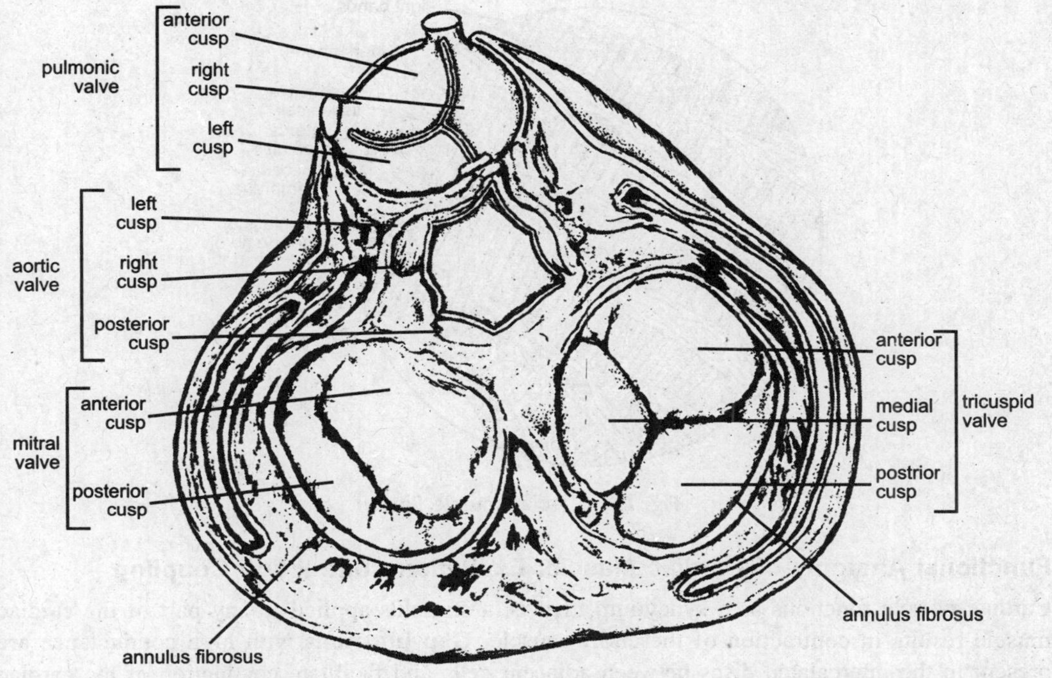

Fig. 7.7. Four cardiac valves as viewed from the base of the heart. Note how the leaflets overlap in the closed valves (after Koeppens and Stanton 2010).

2. Myocardium. It is the middle thick layer of **cardiac muscle** and connective tissue. The cardiac muscle consists of short, cylindrical fibers that branch and anastomose so as to from a network (Fig. 7.8). This arrangement permits a quick spread of contraction waves and accounts for the rhythmic contractions characteristic of heart muscle. The fibers are uninucleated, the nucleus lying at the center. They contain sarcolemma, sarcoplasmic reticulum, T-tubules and numerous large mitochondria (see Fig. 7.8). They contain striated contractile fibers (or myofilaments). At their ends, cardiac muscle fibers have specialized regions of plasma membrane, forming the **intercalated discs** between adjacent fibers. These discs function as boosters of the wave of muscle contraction and permit transmission of the wave from one fiber to the next. No neurons are found in the wall of the heart.

The cardiac muscle has a very rich blood supply and contracts quickly without getting fatigued. Its contraction is involuntary.

Function of Heart

The human heart contracts 72 times per minute at rest and about 2.5 billion times in life time of an individual. It pumps over 7,000 liter of blood daily, *i.e.,* 5 to 6 liters per minute per ventricle.

During each heart beat the heart chambers contract and relax in a specific sequence. The contraction and relaxation of a cardiac chamber are known as **systole** and **diastole** respectively. The movements of cardiac chambers are repeated in a cyclic manner during each heart beat. These events in each **heart beat** constitute a **cardiac cycle**.

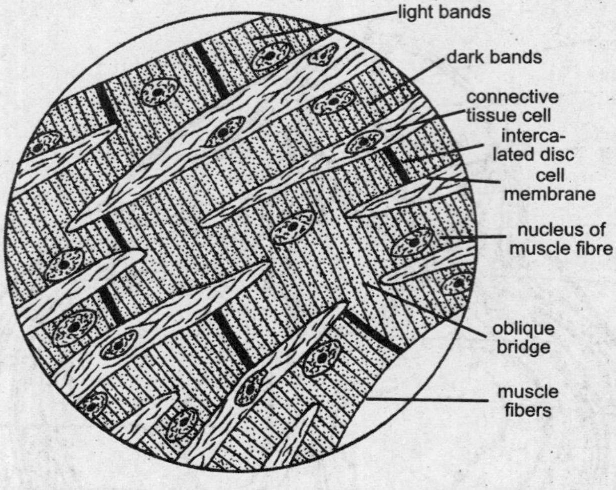

Fig. 7.8. Cardiac muscle fibers.

Functional Anatomy of Cardiac Muscle: Excitation-Contraction Coupling

Cardiac muscle functions as a syncytium; that is, a stimulus applied to any part of the cardiac muscle results in contraction of the entire muscle. **Gap junctions** with high conductance are present in the intercalated discs between adjacent cells and facilitate conduction of the cardiac impulse from one cell to the next.

Cardiac muscle must contract repetitively for a life-time, and hence it requires a continuous supply of O_2. Cardiac muscle is very rich in mitochondria. The large number of mitochondria which have the enzymes necessary for oxidative phosphorylation, permits rapid oxidation of substrates and synthesis of ATP and thus sustains the myocardial energy requirements.

To provide oxygen (O_2) and substrate for its metabolic machinery, the myocardium is also endowed with a rich capillary supply, about one capillary per muscle fiber. Thus, diffusion distances are short and O_2, CO_2, substrates, and waste material can move rapidly between the myocardial cell and capillary. The **transverse (T) tubular system** within myocardial cells participate in this exchange of substances between capillary blood and myocardial cells. T tubule system also plays a key role in excitation-contraction coupling.

Excitation Contraction Coupling

The earliest studies on isolated hearts indicated that optimal concentrations of Na^+, K^+ and Ca^{2+} in extracellular fluid are necessary for contraction of cardiac muscle. Without Na^+, the heart is not excitable and will not beat.

On excitation, voltage-gated Ca^{2+} channels open to admit extracellular Ca^{2+} into the cell. The influx of Ca^{2+} triggers the release of Ca^{2+} from the sarcoplasmic reticulum (Fig. 7.9). The elevated Ca^{2+}, elicits contraction of the myofilaments. Relaxation is accomplished via resting cytosolic Ca^{2+} by pumping Ca^{2+} back into the sarcoplasmic reticulum and exchanging it for extracellular Na^+ across the sarcolemma. Velocity and force of contraction are functions of Ca^{2+}. Force and velocity are inversely related, so with no load, velocity is maximal. In an isovolumic contraction, no external shortening occurs.

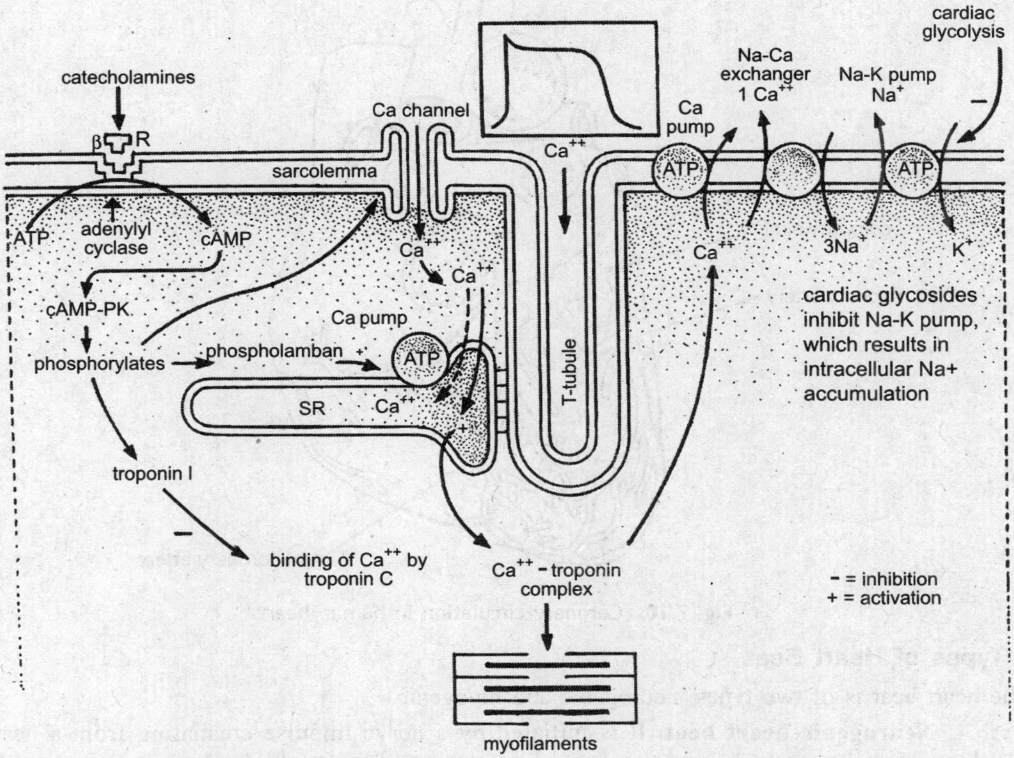

Fig. 7.9. Schematic diagram of the movement of calcium ions (Ca^{2+}) in excitation-contraction coupling in cardiac muscle. Influx of Ca^{2+} from interstitial fluid during excitation triggers release of Ca^{2+} from the sarcoplasmic reticulum (SR). The free cytosolic Ca^{2+} activates contraction of the myofilaments (systole). Relaxation (diastole) occurs as a result of uptake of Ca^{2+} by the SR, by extrusion of intracellular Ca^{2+} by the $3Na^+ - 1\ Ca^{2+}$ antiporter, and to a limited degree by the Ca^{2+}-ATPase pump. βR, β-adrenergic receptor; cAMP-PK, cAMP-dependent protein kinase (Koeppen and Stanton 2010).

Nourishment of Heart (Coronary Circulation)

A pair of **coronary arteries** *(i.e.,* left and right coronary arteries) arise from the systemic aorta just beyond its semilunar valves and break up into capillaries to vascularise the heart wall. Blood is returned from the heart wall by a number of **coronary veins** that join to from a **coronary sinus**. This sinus opens into the right auricle. The arteries and veins of the heart wall form the **coronary circulation** (Fig. 7.10). The blood filling the heart chambers is quite far from most of the tissues forming the thick heart wall. Moreover, the food and oxygen requirement of the heart wall are high in view of its constant beating. Any damage to the coronary vessels will disrupt supply of nutrients and oxygen to the heart tissue. It may prove very serious. (Note. Only inner 75 to 100 um thick endocardial surface can obtain significant amount of nutrients from the blood present in heart-chambers during heart beat; see **Joshi** and **Joshi Mendhur** 2010).

7.3 HEART BEAT

The spontaneous and rhythmic contraction and relaxation of the heart to pump out and receive blood to and from the body is called **heart beat**.

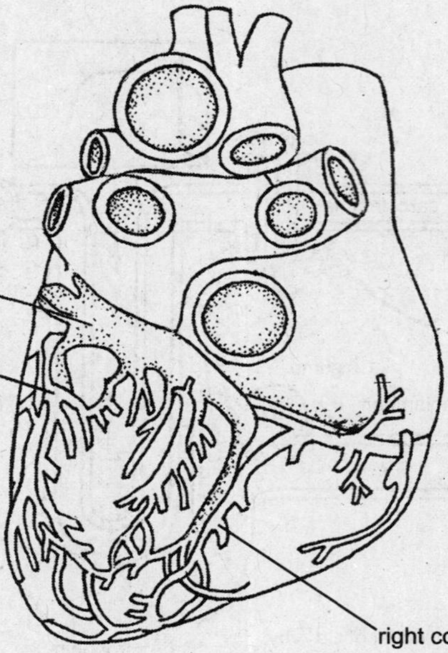

Fig. 7.10. Coronary circulation in human heart.

I. Types of Heart Beat

The heart beat is of two types: neurogenic and myogenic.

1. Neurogenic heart beat. It is initiated by a nerve impulse emanating from a nerve ganglion situated near the heart. It is shown by the heart of most arthropods and some annelids.

2. Myogenic heart beat. It is initiated by a patch of modified heart muscle itself without requiring an external stimulation. It is exhibited by the hearts of molluscs and vertebrates. The heart in these animals has an inherent capacity to contract rhythmically without any external stimulus. A heart removed from frog's body continues to beat for a long time if immersed in a well oxygenated **Ringer's solution** (Box 7.1) that supplies essential elements and nutrients. This shows that the heart beat has its mechanism in the heart itself, and is independent of nervous and endocrine systems. These systems, however, control the rate of heart beat. The properties of **automaticity** (the ability to initiate its own beat) and **rhythmicity** (the regularity of pacemaking activity) allow a perfused heart to beat even when it is completely removed from the body.

Box 7.1

Ringer's solution contains 8.6 gm of NaCl, 0.3g of KCl and 0.33g of $CaCl_2$ in each 1000 ml of distilled water. It has been suggested by **Sydney Ringer**, a British physilologist (1835–1910).

II. Mechanism of Heart Beat

The heart beat results from a wave of electrical depolarization or action potential, called **cardiac impulse**, that is originated in and is conducted along the tracts of specialized cardiac muscle fibers called **Purkinje fibers**. These tracts are collectively referred to as **pacemaker** or **nodal tissue** (Fig. 7.11). Impulses are conducted about 10 times faster in the Purkinje fibers than in the ordinary cardiac muscle fibers.

Structure of nodal tissue. The nodal tissue consists of following five types of components: 1. **Sinoatrial** or **sinuauricular node** (or **SA node**); 2. **Auriculo-ventricular node** (or **AV node**); 3. **His-Tawara bundle** or **bundle of His** (Named after the German physiologist

Wilhelm His, 1868–1913) or **Auriculoventricular bundle** (AV bundle), 4. **Right** and **left bundles,** and 5. **Purkinje fibers.** The *SA node* is located in the wall of right auricle below the opening of the superior vena cava. This is the place where the primitive sinus venosus is incorporated in the heart of the embryo. The SA node is, thus, a vestige of the sinus venosus. It consists of thin fibers, about one-third the size of other cardiac muscle fibers. Its fiber has few myofibrils and a few nerve endings from the autonomic nervous system. The *AV node* lies at the base of right auricle near interauricular septum. Like SA node, it comprises muscle fibers having fewer myofibrils. The *His-Tawara bundle* originates from the A-V node, passes downward in the interventricular septum and divides into two branches, the **left** and **right bundles**, one going into the wall of each ventricle. In the musculature of the ventricles, the two branches break up into a large number of *Purkinje fibers* that are distributed to all part of the ventricle.

The Purkinje fibers have the following general characteristics:

1. They are wider (50–70 μ) than the ordinary cardiac muscle fibers (15 μ).
2. They have relatively more sarcoplasm.
3. They contain a large amount of glycogen.
4. Their myofibrils are arranged peripherally round the stock of glycogen.

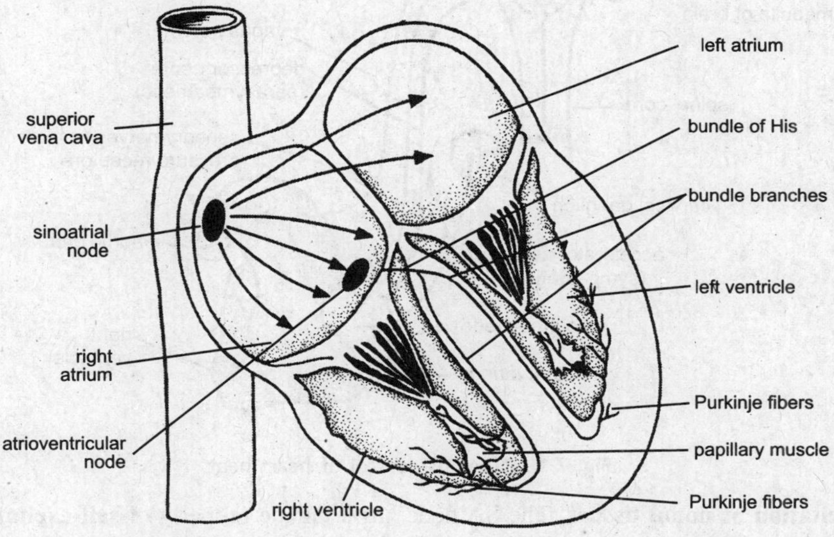

Fig. 7.11. The cardiac conduction system. Arrows show spread of waves of excitation.

Working of nodal tissue. The cells of the SA node maintain a differential **ionic concentration** across their membranes of –90 mV due to continual diffusion of Na^+ ions into them. This produces a depolarization which propagates an **action potential** in the cells adjacent to the SA node. As this wave of excitation passes across the muscle fibers of the heart, it causes them to contract. The SA node is called the **pacemaker** of the heart because each wave of excitation begins here and acts as the stimulus for the next wave of excitation. The wave of contraction spreads over both the auricles more or less simultaneously at a rate of about 1 meter per second along the muscle fibers that fan out from the pacemaker (Fig. 7.11). The auricular contraction cannot pass over the ventricles because the musculatures of the auricles and ventricles are completely separated by an atrioventricular septum, the **annular pad**, of fibrous connective tissue. The auricular contraction, however, stimulates the AV node which generates a fresh wave of contraction that passes over both the ventricles simultaneously at a rate of 5 meter per second along the bundle of His and its ramifications. The ventricular contraction being at the *apex* of the heart and passes quickly toward the origin of the pulmonary and systemic arches. There is a gap of approximately 0.15 second in conduction from the SA node to the AV node. This permits atrial systole to be completed before the ventricular systole begins (Box 7.2).

Box 7.2

No nerves are involved in the spread of a contraction wave through cardiac muscle. This is indicated by the fact that in the embryo the heart beats before the nerves develop and that heart muscle cells grow in tissue culture also contract rhythmically without any external stimulus.

It takes only 0.8 second from the time the S-A node produces its electric stimulation until the ventricles contract. The speed of the excitation wave over the heart is made possible by tight connections of the junction, called **intercalated discs,** between the cardiac muscle cells.

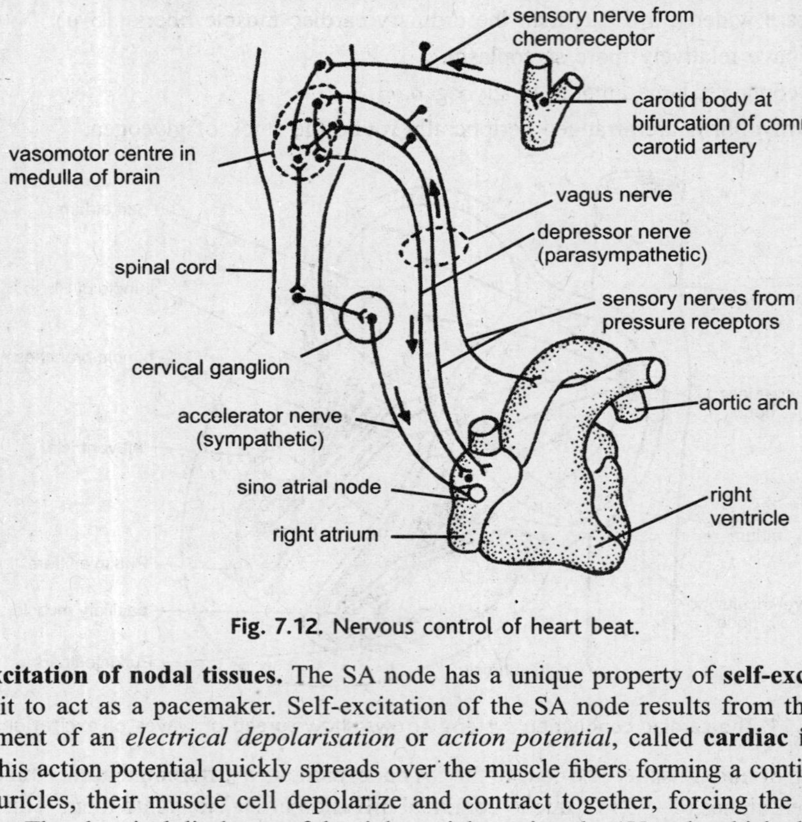

Fig. 7.12. Nervous control of heart beat.

Excitation of nodal tissues. The SA node has a unique property of **self-excitation,** which enables it to act as a pacemaker. Self-excitation of the SA node results from the spontaneous development of an *electrical depolarisation* or *action potential*, called **cardiac impulse**, in its fibers. This action potential quickly spreads over the muscle fibers forming a continuous network in the auricles, their muscle cell depolarize and contract together, forcing the blood into the ventricles. The electrical discharge of the right auricle excites the AV node which also depolarizes. The wave of depolarisation then quickly spreads over the muscle fibers of the ventricles through the bundle of His. This results in the contraction of the ventricles from apex upward, forcing the blood into the pulmonary and systemic arches.

Depolarisation of the muscle cells of SA node occurs again after depolarisation and takes place 72 times per minute. The SA node not only act as a pacemaker but also establishes by its inherent rhythmicity the basic rate at which the heart beats.

Factors Controlling Heart Beat

Though the heart beat is automatic, the rate of beat and the volume of blood pumped per beat is regulated by some extrinsic factors to adjust the heart output to body requirement. These factors include nervous regulation, hormonal regulation, oxygen supply, temperature, blood pressure, H^+ concentration, inorganic ions and cardiotonic factors.

1. Nervous regulation. The heart is innervated by the nerve fibers from vagus nerve (*i.e.*, 10 cranial nerve), *i.e.*, parasympathetic fibers and the sympathetic nerve fibers from spinal cord (Fig. 7.12). In myogenic heart of vertebrates, the **parasympathetic fibers** of **vagus nerve** cause slowing the heart beat rate. The stimulated nerve releases a chemical **acetylcholine** near sinoauricular (SA) node. Acetylcholine is inactivated by an enzyme *acetylcholinesterase* after its action is over.

The **sympathetic nerve fibers** of vagus accelerate the heart beat. The nerve endings of sympathetic nerve fibers produce **norepinephrine**, which increases the heart rate. An enzyme *monoamine oxidase* inactivates norepinephrine to prevent continuous stimulation of heart wall.

The **cardio-inhibitory center** present in the medulla sends parasympathetic impulses to the SA node through parasympathetic nerve fibers of vagus nerve.

The **cardio-inhibitory center** of medulla sends sympathetic impulses through sympathetic nerve fibers to the heart.

The **baroreceptors** (chemoreceptors) located in the carotid sinus and in aorta are sensitive to changes in pressure and send afferent nerve fibers to the cardio-inhibitory and cardio-accelator centers of medulla. The stimulation of baroreceptors by increase in aortic blood pressure stimulates cardio-inhibitory center of medulla which produces inhibitory cardiac reflex. A decrease in aortic or carotid blood pressure initiates reflex for acceleration of heart beat.

2. Hormonal control. The important hormones which control the rate of heart beat are adrenaline, noradrenaline and thyroxine.

Adrenaline or **epinephrine** secreted by the adrenal gland increases the rate of heart beat and a rise in blood pressure by causing constrictions of blood vessels.

Noradrenaline or **norepinephrine** also increases blood pressure, because of its constricting effect on the blood vessels.

Both of these hormones accelerate the heart beat, the norepinephrine under normal conditions and the epinephrine at the time of emergency such as vigorous exercise and emotions such as fear and anger.

Thyroxine is secreted by the thyroid gland. Its higher concentration in blood increases the heart rate, whereas its low concentration in blood slow down the rate of heart beat.

Other Factors

1. Inadequate **oxygen supply** causes weakness and irregularity of heart beat. Sudden anoxia of heart causes **cardiac pain.**

2. Lowering of **body temperature** slows down heart rate, where as the warming speeds up the cardiac contraction.

3. Coronary circulation is greatly affected by **blood pressure**, which in turn affects the general upkeep of heart, *i.e.*, supply of food (glucose) and oxygen to the heart muscles.

4. Slight alkalinity is essential for optimum cardiac output. Change in pH (*i.e.*, H^+-ion concentration) either way disturbs the cardiac efficiency. Rise of H^+ ion concentration reduces the strength of contraction and prolongs relaxation. Increased alkalinity causes opposite effects.

5. Presence of inorganic ions (such as Na^+, Ca^{2+} and K^+) in appropriate proportion in the surrounding medium is essential for the maintenance of normal activity of heart. If concentration of K^+ ions is increased, the force of heart contraction is gradually decreased. Increase in Ca^{2+} ions increases the heart contraction and further increase of calcium ions stops the heart in systole stage, *i.e.*, heart becomes contracted (contracture). This condition is called **calcium rigour**.

6. Phospholipid (β-palmitoyl-lyso-lecithin) and three plasma globulins, the **cardioglobulin, A, B** and **C** present in blood act as **cardiotonic factors**. A cardiotonic substance tends to increase the tonus of the heart muscles.

7.4 CARDIAC CYCLE

Heart contracts rhythmically and ceaselessly throughout human life. The sequence of events which takes place during the completion of one heart beat is known as **cardiac cycle**. The time period consumed by different events of one cardiac cycle is the **cardiac cycle time**. The human heart beats at a rate of 72 times per minute at normal conditions.

During the cardiac cycle, blood flows through the cardiac chambers in a specific manner and direction. The events that take place during the completion of one heart beat are: 1. Joint diastole; 2. Atrial systole; 3. Ventricular systole; and 4. Ventricular diastole.

1. Joint diastole. During this phase, all the chambers of the heart are *relaxed* and blood flows into the atria (auricles) through the great veins (*i.e.,* superior and inferior vena cavae). Some blood also flows slowly from the atria to the respective ventricles through AV valves. But no blood flows from the ventricles to the great arteries and the semilunar valves remain closed. The ventricles may be half-filled with blood by the end of joint diastole. At the end of this phase, the next heart beat starts with the auricular (or atrial) systole.

2. Atrial systole. The atrial contraction or systole begins when the atrial chambers are full of blood. A wave of contraction is initiated at the sinoatrial (SA) node which spreads to all parts of atria, forces blood into the ventricles which are still in diastole. This is accompanied by opening of the atrio-ventricular valves. The atrial systole lasts for about 0.15 sec. When this wave of contraction reaches the atrio-ventricular (AV) node it is also excited. It takes some time and accounts for a brief pause after the atrial systole, before ventricular systole begins.

3. Ventricular systole. Once the AV node is excited, the wave of contraction spreads to all the parts of the ventricles through the bundle of His. Because of this, both the ventricles and their parts contracts simultaneously. Due to ventricular systole, the pressure inside the ventricular cavities increases and the tricuspid and bicuspid valves close. The closure of these valves produces **first heart sound** of "**lubb**". When the pressure inside the ventricles exceeds that in the arteries, the semilunar valves (of pulmonary artery and of aorta) open and blood spurts into the pulmonary artery and aorta from right ventricle and left-ventricle respectively. Ventricular systole lasts for about 0.3 second.

As the blood is ejected from the ventricular cavities into arteries their pressure begins to fall. When it becomes lower than the arteries, the semilunar valves close and this causes **second heart sound** or **"dup".** This sound is short pitched and lasts longer than 'lubb' sound. The second heart sound indeed includes sounds produce by closure of pulmonic valve plus closure of aortic valve; **(Koeppen** and **Stanton** 2010). Dup sound indicates end of ventricular systole and sound beginning of ventricular diastole.

During ventricular diastole the atrio-ventricular valves remain closed because the pressure within the ventricles is still greater than that within the atria. During this stage the atria remain in a relaxed state **(atrial diastole)** and blood flows from the veins into the relaxed atrium.

4. Ventricular diastole. After ventricular systole, the ventricles undergo relaxation and increase in size. As they relax, the pressure within the ventricles decreases. Once it falls sufficiently, atrio-ventricular valves open and blood rushes from atria into the ventricles. At the ends of ventricular systole, ventricles strat relaxing, *i.e.*, when ventricular systole starts, the atria are still in their diastolic phase. It means at one time in cardiac cycle, auricles and ventricles both are in diastole for a brief period. This is called **joint diastole.**

Cardiac Output

Cardiac output or heart output is the amount of blood pumped by the heart into the aorta in one minute. Heart beats 70–72 times per minute and pumps out 70 ml blood during each beat.

Cardiac output = Stroke volume × 70/72 times per minute
= about 5 liters

The amount is only a little less than a total volume of blood (about 6.8 liters) present in the body. Mild exercise such as walking raises the cardiac output to about 11 liters. During vigorous exercise, heart output increases to 25 liters and 40 liters in the case of a trained athlete.

Cardiac output changes with age, size, sex, temperature, etc. It is directly proportional to the size of the organism, amount of blood entering the heart, metabolic rate, etc. It is inversely proportional to age and peripheral resistance, *i.e.,* thickness of arterial wall.

Heart Sounds

Each heart beat is accompanied by two heart sounds, **first heart sound** and **second heart sound**. The first heart sound marks the beginning of ventricular systole and is due to closure of atrio-ventricular valves. This is followed by second heart sound which marks the end of ventricular systole and begining of its diastole. It is produced by the closure of semilunar valves.

With a stethoscope, these heart sounds are heard as lubb and dup.

1. The **lubb sound (first heart sound)** is lower and lasts longer. It is produced by the contraction of ventricular muscles and the vibrations set up by the **closure of tricuspid** and **bicuspid valves**. It is also called **systolic sound**.

2. The **dup sound (second heart sound)** is caused by the closure of **semilunar valves** (pulmonic valve and aortic valve) of the pulmonary trunk and aorta. It is also called **diastolic sound**.

3. **Heart murmur.** It is abnormal sound produced either by incomplete closing of valves (called **valvular insufficiency**) or by their narrowing **(stenosis)**.

7.5 REGULATION OF HEART PUMPING

When a person is at rest, the heart pumps only 4 to 6 liters of blood each minute. During severe exercise, the heart may be required to pump four to seven times this amount. The basic means by which the volume of blood pumped by the heart is regulated are (i) intrinsic caridac regulation of pumping in response to changes in volume of blood flowing into the heart and (ii) control of heart rate and strength of heart pumping by the autonomic nervous system.

1. Intrinsic regulation of heart pumping. Under most conditions, the amount of blood pumped by the heart each minute is normally determined almost entirely by the rate of blood flow into the heart from the veins, which is called **venous return**. That is, each peripheral tissue of the body controls its own local blood flow, and all the local tissue flows combine and return by way of the veins to the right atrium. The heart, in turn, automatically pumps this incoming blood into the arteries so that it can flow around the circuit again.

The intrinsic ability of the heart to adapt to increasing volumes of inflowing blood is called the **Frank-Starling mechanism of the heart**, in honor of **Otto Frank** or **Ernest starling** (1910), two great physiologists of a century ago (Box 7.3). Basically, the Frank-Starling mechanism means that the greater the heart muscle is stretched during filling, the greater is the force of contraction and the greater the quantity of blood pumped into the aorta. Or stated in another way: *within physiological limits, the heart pumps all the blood that returns to it by the way of the veins.*

Box 7.3

Frank, Otto (1865–1944). German physiologist Frank served as professor at several German universities, his longest tenure being at Munich. He pioneered in the development of exact physiological recording techniques. In 1904, he published the first study of direct monitoring of heart sounds. He devised a great number of recording instruments. He published the fruits of his research in a handbook of physiological methodology. Areas of his research included the origin of the arterial pulse, a determination of the volume of blood in the heart, and the functioning of the aorta. Frank also made original contribution to the physiology of ear (see **R.W. Pearse, Jr.**, 1996).

Starling, Ernest Henry (1866–1927). British physiologist. Starling was one of outstanding physiologist of his time. He served for many years as professor of physiology at university college in London. His foremost contributions to physiology include an understanding of the maintenance of fluid balance in body tissues, the regulatory role of endocrine secretions, and mechanical controls on heart function. Much of his work was in collaboration with the physiologist **Willian Bayliss** (1860–1924). In 1902 they isolated a substance that is released into blood from the epithelial cells of duodenum, which in turn stimulates secretion into the intestine of pancreatic digestive juice. Two years later (in 1904), **Starling** coined the term *hormone* as the name for such substances. In 1918 he formulated a law concerning the heart (*i.e.*, Frank-Starling law of heart), stating that (other conditions being equal) the greater filling of the heart during diastole, the greatest following systole. Thus the heart is able to maintain a constant beat despite considerable variations in blood flow. This law of heart led to the concept upon which the law of muscle contraction is based (Merriam Webster's Medical Desk Dictionary 1996).

Frank-Starling Mechanism. When an extra amount of blood flows into the ventricles, the cardiac muscle itself is stretched to greater length. This in turn causes the muscle to contract with increased force because the actin and myosin filaments are brought to a more nearly optimal degree of overlap for force generation. Therefore, the ventricle, because of is increased pumping, automatically pumps the extra blood into the arteries. This ability of stretched muscle, upto an optimal length, to contract with increased work output is characteristic of all striated muscle and is not simply a characteristic of cardiac muscle. In addition to the important effect of lengthening the heart muscle, still an other factor increases the heart pumping when its volume is increased. Stretch of the right atrial wall directly increases the heart rate by 10 to 20 percent; this too, helps increase the amount of blood pumped each minute, although its contribution is much less than that of Frank-Starling mechanism (**Guyton** and **Hall** 2011).

7.6 ELECTROCARDIOGRAM

The changes in **electrical potential** associated with the initiation and spread of cardiac impulse can be recorded from the surface of the body by an appropriate instrument. The instrument is usually an extremely sensitive galvanometer known as the **electrocardiograph**. The graphic recording, made on a moving strip of paper, are known as **electrocardiogram (ECG or EKG)** (Box. 7.4).

Box 7.4

The science of electrocardiography is about 100 years old. The techniques of recording electrical activity from the heart were originally developed in 1906 by a Dutch physiologist, **William Einthoven** (1860–1927), called **Father of Electrocardiography**. Most modern clinical electrocardiograph used is the **Pen recorder** in which a pen records the ECG directly on a moving sheet of 'mm' graph paper at a speed of 25 mm/sec. In this apparatus, three electrographic bipolar limb leads are used. Electrocardiograph may be connected to an **oscilloscope** which shows ECG on a TV-type screen (*i.e.*, monitor). Einthoven proposed a law, which states if the electrical potentials of any two limb leads are known, third electric potential can be determined by summing the first two.

It is already mentioned that the muscle cells of the heart are connected with each other through tight junctions or nexuses and as such the impulse travels very rapidly over entire section (atria and ventricles) of the heart so that these behave as a single unit. The amount of current generated when the excitation spreads over the atria and ventricles is quite high; it creates a field of electricity which traverses over the entire body surface. The strength of the potential in the field gets feebler as the electric field moves away from the heart. The tissue between the heart and the body surface acts as a **volume conductor**, for example, like a solution of electrolytes. The current generated by the heart is so much that the other action potentials of the body produce very slight distortions in the recordings of the electrical potential of the heart.

The actual measurements (in unipolar recording) are carried out by a number of **electrodes**. One of these is always the **recording** or **exploring electrode** and the other an **indifferent electrode** which is kept at zero potential. The position of the electrodes is designated by the letter V (v = voltage). There are four standard leads – VR (right arm), VL (left arm), VF (left foot) and VC (chest) Various combination of these leads are used.

Figure 7.13 shows a typical electrocardiogram taken from the right and left arm. ECG is formed of a series of ridges and furrows. ECG taken when patient is lying down is called **resting ECG** but ECG taken when patient is exercising is called **stress ECG**.

In a ECG (Fig. 7.13), there are three essential components. These are P, QRS complex and T to which a fourth component U is sometimes added. The letters P, QRS and T were selected since they have not been previously included in the physiology terminology, and thus would not lead to confusion with other systems. In the intact heart, the **P wave** corresponds the **atria activation**, **QRS complex (fast wave)** with the activity of the ventricle during the transmission of the excitation through the Purkinje fibers and **T wave (slow wave)** with the regression of the impulse from the ventricle. The **U wave** frequently occurs in enlarged hearts of athletes.

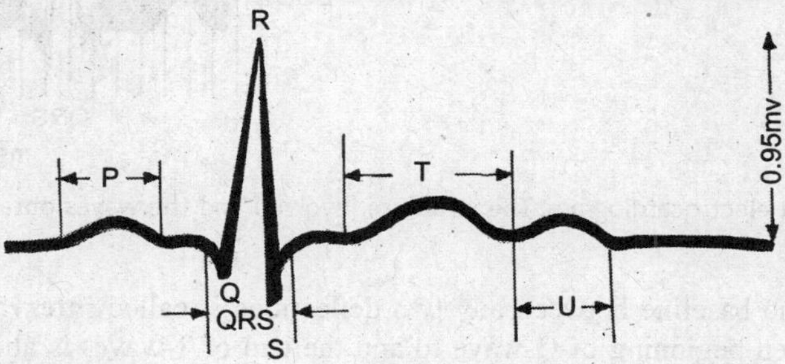

Fig. 7.13. A typical human electrocardiogram (ECG) recorded from the two arms.

The duration of each electrical event is as follows: The P wave lasts 60–110 msec; 120-200 msec for the PR interval (time for atrioventricular conduction), about 60–110 msec for QRS complex; and 200 msec for the duration of T (Fig. 7.14). The amplitude of the electrical changes recorded by electrocardiograph in a normal person are as follows: 0.1 mV for P, 0.95 mV for R, and 0.30 mV for T. These values, however, must not be considered to be absolute. There are important variations in cardiac potential recorded at different distances from the heart.

In ECG, P-wave indicates the impulse of contraction generated by S-A node. It indicates the electric potentials generated as the atria depolarize prior to contraction. QRS-wave indicates the spread of impulse of contraction from A-V node to the wall of ventricles and is caused by electric potentials generated prior to contraction of ventricles. Both P-wave and QRS-wave are collectively called **depolarization waves**. The T-wave represents the relaxation of ventricles or recovery from the state of depolarization, so called **repolarization wave**. It occurs about 0.25 to

0.35 seconds after the ventricular contraction. So ECG is formed of both depolarization and repolarization waves. During depolarization, a cardiac muscle fiber has reverse polarity (positively charged inside and negatively charged outside), while it is opposite in depolarization. The atria repolarize about 0.15 to 0.20 seconds after the P-wave but atrial T-wave is not observed because it is dominated by large QRS-wave of ventricular depolarization.

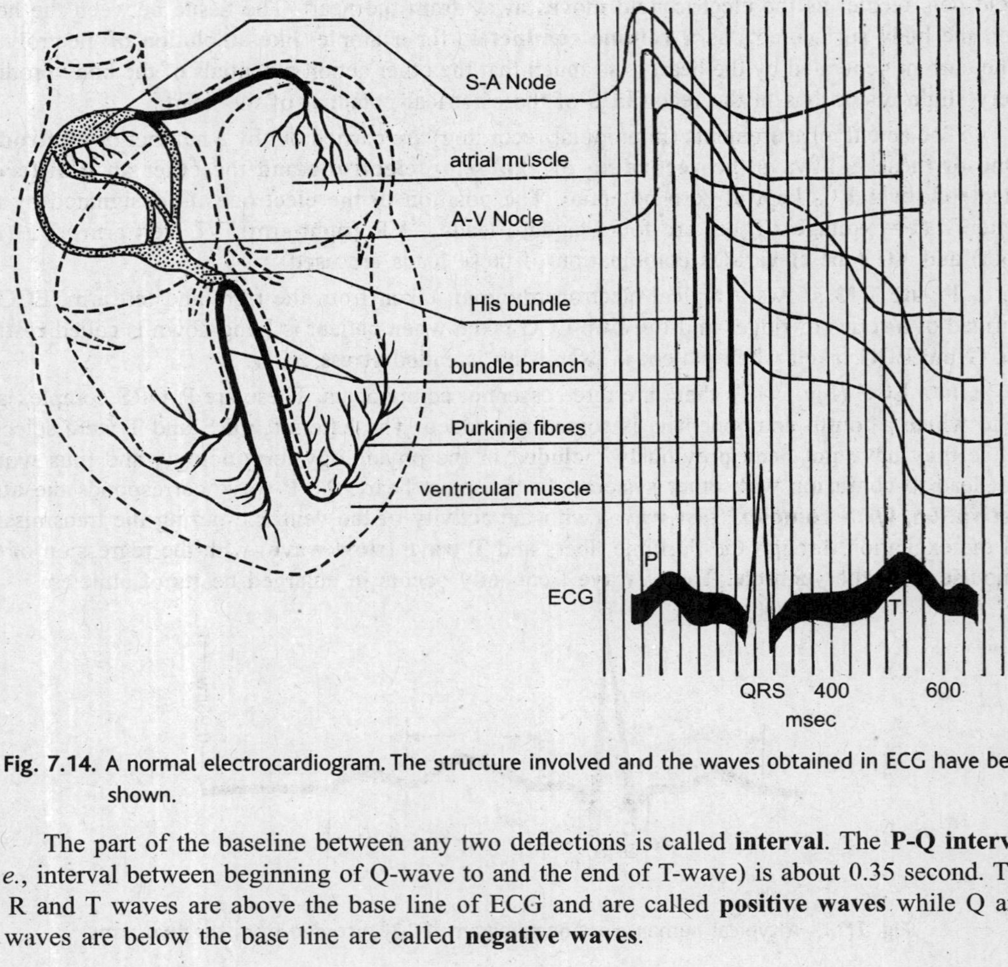

Fig. 7.14. A normal electrocardiogram. The structure involved and the waves obtained in ECG have been shown.

The part of the baseline between any two deflections is called **interval**. The **P-Q interval** (*i.e.*, interval between beginning of Q-wave to and the end of T-wave) is about 0.35 second. The P, R and T waves are above the base line of ECG and are called **positive waves** while Q and S waves are below the base line are called **negative waves**.

Significance of ECG

Because electrocardiogram (ECG) reveals facts about the origin and conduction of action potential of the heart, electrocardiograph is a tool which detects abnormality of the heart. Thus, ECG provides a method of evaluating excitation events, arrhythmias (*i.e.*, an alteration in rhythm of the heart beat either in time or force), tissue damage, the presence of ischemia (*i.e.*, localized tissue anaemia due to obstruction of the inflow of arterial blood) or necrosis and hypertrophy of the heart. The ECG does not provide any information on the mechanical performance of heart and at times even abnormalities may not be recorded.

7.7 BLOOD PRESSURE AND ITS REGULATION

The force of blood against the wall of the blood vessel is known as **blood pressure**. It is produced by the discharge of blood from the heart. Blood pressure differes a quite a bit in the different parts of the circulatory system. It is high in the arteries, gradually falls in the arterioles and capillaries and is very low in the veins. The maximum pressure is encountered in the aorta. It is normally 140 mm of mercury (Hg). The marked gradient of blood pressure allows the blood flowing from the arteries to the veins through the capillaries (Fig. 7.15).

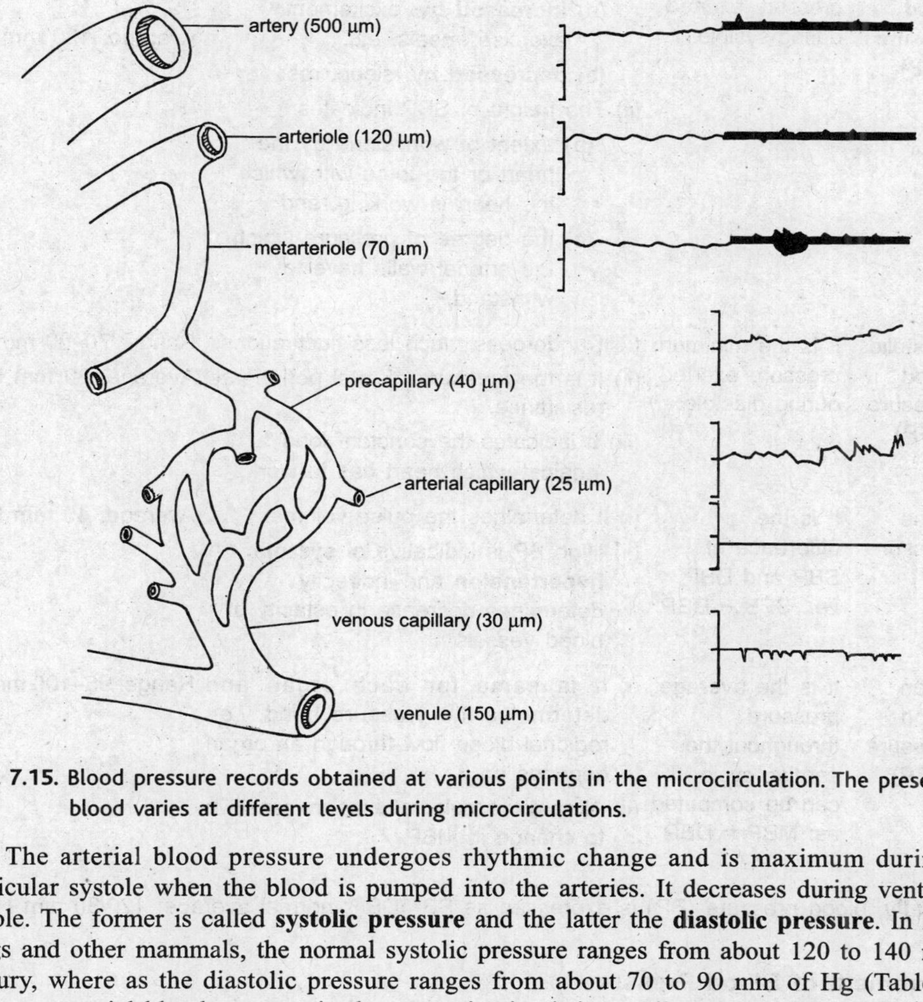

Fig. 7.15. Blood pressure records obtained at various points in the microcirculation. The presence of blood varies at different levels during microcirculations.

The arterial blood pressure undergoes rhythmic change and is maximum during the ventricular systole when the blood is pumped into the arteries. It decreases during ventricular diastole. The former is called **systolic pressure** and the latter the **diastolic pressure**. In human beings and other mammals, the normal systolic pressure ranges from about 120 to 140 mm of mercury, where as the diastolic pressure ranges from about 70 to 90 mm of Hg (Table 7.1). In human arterial blood pressure in the systemic circuit is usually measured in the upper arm (brachial artery), where systolic value of 120 mm Hg and diastolic value of about 80 mm Hg are normal in young adult male at rest. Note that these pressures apply to upper arm only, the values would not be same for the lower arm, the leg, or any other part of the body. In general the blood pressure decreases continuously as the blood moves farther away from the heart. An exception is found in the arteries of human feet, where the average pressure is considerably higher (by 80 mm Hg) than the heart. The average blood pressure in the head is considerably lower (by 40 mm Hg) than the pressure in the heart due to effect of gravity to get blood to the brain. It is because of this that the pressure necessary to supply blood to the brain of giraffe is very high as the head may be the meters above the heart. The average blood pressure of the giraff measures about 260 mm Hg, more than twice as high as human. The giraff's arteries are much thicker than others, an adaptation to withstand high blood pressure.

Table 7.1 Components of systemic arterial blood pressure

Component	*Definition*	*Characteristic features*	*Normal values*
1. Systolic blood pressure (SBP)	It is the maximum pressure exerted during systole.	(i) It undergoes considerable fluctuation: (a) **increased by**: excitement, exercise, meals, etc. (b) **decreased by**: sleep, rest (ii) The height of SBP indicates: (a) extent of work done by the heart or the force with which the heart is working, and (b) the degree of pressure which the arterial walls have to withstand.	Range: 100–130 mm Hg Average; 120 mm Hg
2. Diastolic blood pressure (DBP)	It is the minimum pressure exerted during diastole.	(i) It undergoes much less fluctuations. (ii) It is the measure of total peripheral resistance. (iii) It indicates the constant load against which heart has to work.	Range: 70–80 mm Hg Average: 80 mm Hg
3. Pulse pressure (PP)	It is the difference of SBP and DBP, *i.e.*, SPB – DBP.	(i) It determines the pulse volume. (ii) High BP is indicative of **systolic hypertension** and indirectly determines decrease in estacity of blood vessels.	Average: 40 mm Hg
4. Mean blood pressure (MBP)	It is the average pressure throughout the cardiac cycle. It can be computed as: MBP = DBP + 1/3 PP	(i) It is same for each organ and determines the pressure head, *i.e.*, regional blood flow through an organ depends on it. (ii) All cardiovascular reflexes are sensitive to change in MBP.	Range: 95–100 mm Hg

Clinically, blood pressure (BP) is expressed as SBP/DBP; normal average: 120/80 mm Hg

Measurement of Blood Pressure

Blood pressure is measured with the aid of an apparatus known as **sphygmomanometer** which makes it possible to measure the amount of air pressure equal to the blood pressure in the artery. The measurement is made in terms of how many millimeters high the air pressure raises a column of mercury in glass tube. The sphygmomanometer consists of a rubber cuff and column of mercury that is marked off in millimeters. The cuff is wrapped around the arm over the brachial artery (the chief artery of the upper arm) and air is pumped into the cuff by means of the bulb (Fig. 7.16). In this way, air pressure is exerted against the outside of an artery. Air is pumped in until the air pressure exceeds the blood pressure within the artery or, in other words, until it compresses the artery. At this point, no pulse can be heard through the stethoscope placed over the brachial artery at the bend of the elbow along the inner margin of the biceps muscle.

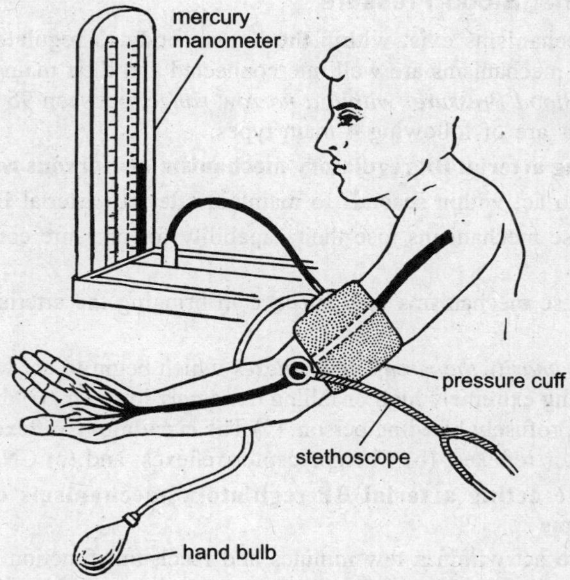

Fig. 7.16. Blood pressure measurement by sphygmomanometer.

By slowly releasing the air in the cuff, the air pressure is decreased, until it approximately equals the blood pressure within the artery. At this time the artery opens slightly and a small gush of blood comes through, producing the first-tap sound. This is followed by increasingly louder sounds which suddenly change, becoming lower, then disappear altogether. The doctor or nurse must train himself/herself to hear these different sounds and simultaneously to read the column of mercury since the first-tap like sound represent the systolic pressure. **Systolic blood pressure** is the force with which the blood is pushing against the wall of artery when the ventricles are contracting. The lowest point at which the sounds can be heard, just before they disappear is approximately equal to the **diastolic blood pressure**, or the force of the blood when the ventricles are relaxed. Systolic blood pressure gives valuable informations about the force of the left ventricle and diastolic pressure gives the information about the resistance of the blood vessels. Medically, diastolic pressure is more important than systolic pressure, because it indicates the strain or pressure to which the walls of the blood vessels are constantly subjected.

Blood of the arteries of the average adult human being exerts pressure equal to that required to raise a column of mercury about 120 mm high in a glass tube during systole of the ventricles and 80 mm high during diastole. This is expected as BP of 120/80 mm Hg. It should be clear now that the blood pressure fluctuates considerably during each heart beat. The difference between systolic and diastolic pressure is called **pulse pressure**.

The average blood pressure rises with age due to the decrease in the elasticity of arteries. As a rule BP is lower in females than in males.

The blood pressure, however, is affected by a number of factors such as age, sex, heredity and other physical and emotional states. The following factors also affect the human blood pressure:

1. Blood volume;

2. Blood vessel space;

3. Stretch reflexes;

4. Carbon dioxide (CO_2).

Regulation of Arterial Blood Pressure

The various mechanisms exist within the human body to regulate the systemic arterial blood pressure. These mechanisms are well interconnected and their main aim is *to maintain the normal MBP (Mean Blood Pressure) within a narrow range*, between 95 to 100 mm Hg. These regulatory mechanisms are of following 4 main types:

1. Rapidly acting arterial BP regulatory mechanism or Nervous regulatory mechanisms

(i) They begin to act within seconds to maintain after the arterial BP becomes normal.

(ii) Most of these mechanisms lose their capability for pressure control after a few hours or a few days.

(iii) None of these mechanisms ever succeed in bringing the arterial BP all the way back to normal.

(iv) These are *primarily the circulatory reflexes* which begin to act within seconds and help to control BP from rising extremely high or falling extremely low, *e.g.*, (i) during sudden change in body posture; and (ii) profusely bleeding person. (v) The **circulatory reflexes** of these mechanisms include (a) Baroreceptor reflexes; (b) Chemoreceptor reflexes; and (c) CNS ischaemic response.

2. Intermediate acting arterial BP regulatory mechanisms or Intrinsic physical regulatory mechanisms

(i) They begin to act within a few minutes and reach full function within a few hours.

(ii) These mechanisms remain functional from few days to a month only.

(iii) They primarily correct any alteration in BP by altering the blood volume.

The mechanisms include (i) capillary fluid shift mechanisms (Fig. 7.17), and (ii) stress relaxation and reverse stress relaxation mechanism.

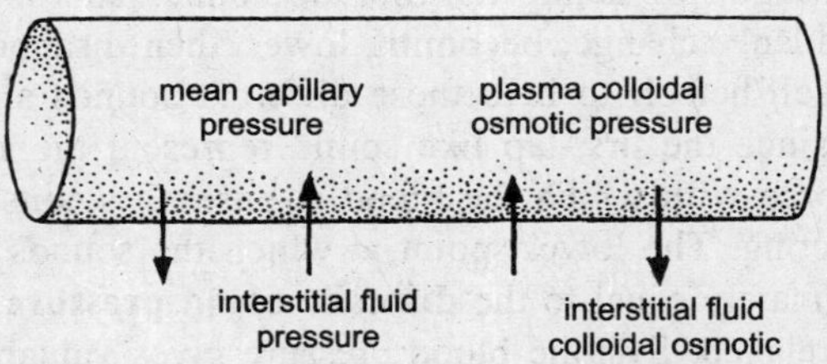

Fig. 7.17. Capillary fluid shift mechanism. Since mean capillary pressure is directly proportional to arterial BP, therefore, rise in arterial BP increases hydrostatic pressure at arterial end and fluid shifts out of capillaries to the interstitial fluid compartments. Thus, blood volume decreases to restore the BP (blood pressure).

3. Long term arterial BP regulatory mechanisms or Autoregulation of BP by kidneys

(i) These mechanism are slow to begin acting. Generally it takes 3–10 days for these mechanisms to come to complete equilibrium.

(ii) Their function is to control the arterial BP over a period of days to years.

(iii) The effectiveness of most of these mechanisms becomes steadily greater with time.

(iv) These mechanisms have the ability to bring the arterial BP all the way back to normal.

The mechanisms are of two types:

(a) Direct mechanisms which involve control of blood volume by kidneys directly. They are called **renal fluid mechanisms**.

(b) Indirect mechanisms which involve control of kidney functions indirectly via hormonal systems. For example (a) Aldosterone system, and (b) Renin-angiotensin system.

Further, degree of sympathetic stimulation of kidneys can change renal function tremendously. Likewise, ADH plays an important role in the regulation of arterial BP.

QUESTIONS

Long Answer Questions

1. Describe structure and function of mammal's heart. *(CCSU Meerut 2007)*
2. What is heart beat? Explain the physiology of heart beat in mammalian heart in detail. *(Rohikhand 2002; Bundelkhand 2002; CCSU Meerut 2006, 2011)*
3. What is myogenic heart? Describe the factors which regulate the rate of heart beat.
4. What do you understand by myogenic heart and neurogenic heart? Describe the origin and transmission of a heart beat in mammals. *(Avadh 2004)*
5. What are pacemakers? Explain their role in heart beat. *(Avadh 2002)*
6. What is cardiac cycle? Describe pressure changes, sound changes, and electric changes during cardiac cycle.
7. Describe the origin and conduction of heart beat. *(HPU, 2007, 2009)*
8. Explain various aspects of cardiac output.
9. Define blood pressure (BP) and its various components. Give the characteristic features of each component of BP.

Short Answer Questions

1. Write short notes on the following:
   (*a*) Blood pressure; (*b*) Cardiac cycle; (*c*) Diastole;
   (*d*) Heart beat; (*e*) Myogenic heart; (*f*) Pace maker;
   (*g*) Systole; (*h*) Cardiac output.
2. Draw neat and labelled diagram of following
   (*i*) V.S of mammalian heart (showing contractile centers). *(CCSU Meerut 2009)*
   (*ii*) Pace maker.
   (*iii*) Exitation-contraction coupling.
   (*iv*) Mechanism of mammalian heart contraction. *(Osmania 2005)*
   (*v*) BP regulatory mechanisms.
3. Write a note on electrocardiogram. Give its significance. *(HPU 2009)*
4. Describe the factors regulating the cardiac output.

Very Short Answer Questions

1. Name two circulatory systems found in man. *(CCSU Meerut 2009)*
2. Name the artery that supply blood to the heart. *(CCSU Meerut 2010)*
3. Why the walls of left ventricle of heart are thicker than the wall s of right ventricle?
4. Give the terms used for the contraction and relaxation of heart chambers.
5. Give one advantage of the closed circulatory system over the open circulatory system.
6. One aortic arch is better than two aortic arches. Explain.
7. What is electrocardiograph?
8. What are vasa vasora?
9. Which is shorter, systole or diastole?
10. Give reason why the systolic pressure of heart is more than diastolic pressure?
11. During cardiac cycle where do the lubb and dup sounds originate?
12. Where do the cardiac impulses orginate?

13. Which part of the heart contains pure or oxygenated blood?
14. What are great blood vessels?
15. What is the size and weight of human heart?
16. What do you mean by contraction and relaxation of heart chambers?
17. How the cardiac waves is conducted from atria to the ventricles?
18. Which nerve slows down the heart beat?
19. Where is SA node (or sinoatrial node) is located.
20. What is the function of sinoatrial node?
21. Write whether the circulatory system of human being is open type or closed type.
22. Why is sinu-auricular (S-A) node called pace maker?
23. What is Frank-Starling's law of heart?

Fill in the Blanks

1. Arthropods have an .... circulatory system.
2. The fact that the blood circulates in the body was discovered by ....
3. .... vessels vascularise the heart wall.
4. Human heart beats about .... times per minute.
5. Contraction of heart starts at the .... node, which is called ....
6. A wave for contraction initiated by .... node is spread over the ventricle by ....
7. Amount of blood pumped by heart per minute is called ....
8. Cardiac output of a person at rest is about .... liters.
9. Heart sound lubb is produced by the closure of .... and .... valves.
10. Closure of the .... valves produces dup sound.
11. For a healthy adult person at rest the systolic and diastolic pressures are .... and .... respectively.
12. Heart sounds are heard with an apparatus named ....
13. The .... valves close shortly after the ventricular diastole begins.
14. Heart beat is of two types: myogenic found in .... and .... in arthropods.
15. .... is the study of blood vessels.
16. .... have valves to check backflow of blood, ... do not.
17. The graphic recording made on a moving sheet of 'mm' graphic paper is called ....
18. In humans, ... in the systemic circuit is measured in the upper arm (of .... artery).
19. DBP (..................) ranges from .... to .... mm Hg.
20. SBP (..................) ranges from .... to .... mm Hg.

Multiple Choice Questions

1. Thickest muscle layer in human heart is found in
   (*a*) right auricle (*b*) right ventricle
   (*c*) left auricle (*d*) left ventricle

2. The pace maker of the heart is
(*a*) bundle of His (*b*) Purkinje fibers
(*c*) AV node (*d*) SA node

3. Cardiac impulse originates from
(*a*) AV node (*b*) SV node
(*c*) SA node (*d*) nerves

4. All arteries carry oxygenated blood except
(*a*) hepatic artery (*b*) renal artery
(*c*) pulmonary artery (*d*) coronary artery

5. Normal BP of a healthy person is
(*a*) 120/80 (*b*) 80/120
(*c*) 140/60 (*d*) 200/130

6. The valves which allow blood flow from the ventricle into arteries and not in opposite direction
(*a*) semilunar valve and tricuspid valve
(*b*) aortic valve and Mitral valve
(*c*) atrio-ventricular valve and semilunar valve
(*d*) bicuspid valve and tricuspid valve

7. Main aim of various BP regulatory mechanisms is to maintain the normal
(*a*) systolic BP (*b*) diastolic BP
(*c*) mean BP (*d*) pulse pressure

8. Which component of systemic arterial BP undergoes much less fluctuations?
(*a*) systolic BP (*b*) diastolic BP
(*c*) mean BP (*d*) pulse pressure

9. Which mechanism is most effective in returning BP precisely back to normal?
(*a*) baroreceptor reflexes (*b*) chemoreceptor reflexes
(*c*) capillary fluid shift mechanism (*d*) long-term BP regulatory mechanism

10. Diastolic BP is mainly determined by
(*a*) capillary permeability (*b*) venous capacitance
(*c*) venous elasticity (*d*) peripheral resistance

11. Intercalated discs
(*a*) provide strong union between the myocardial fibers
(*b*) provide low resistance bridges for the spread to excitation from one cardiac muscle to another
(*c*) permit cardiac muscle to function as a syncytium
(*d*) are also known as gap junctions

12. Under basal conditions caloric need of heart is largely met by
(*a*) fat (*b*) protein
(*c*) carbohyrates (*d*) ketones

ANSWERS

Very Short Answer Questions

1. Pulmonary and systemic systems.
2. Coronary artery.
3. The walls of left ventricle are thicker than the walls of right ventricle because left ventricle has to pump blood to the whole body except lungs.
4. Systole, diastole.
5. Blood flows more rapidly and with higher pressure in the closed circulatory system than in open one.
6. Blood passing through two aortic arches loses pressure faster than when it flows via a single aortic arch.
7. Apparatus used to record electric events in the heart.
8. Vessels that vascularise the wall of the arteries and veins that are 1 mm or more in diameter.
9. Systole.
10. During systole phase blood from the heart circulates in the whole body whereas in diastole, blood returns back to the heart.
11. Lubb in auricles and dup in ventricles.
12. On SA node.
13. The left half of heart.
14. These are aorta and vena cavae.
15. About the size of fist, weighs about 300g.
16. Systole and diastole.
17. AV node (or Aurio-ventricular node).
18. Vagus nerve.
19. On the right wall of right auricle.
20. It serves as pace maker.
21. It is closed type in humans.
22. SA node is called pacemaker of human heart because it controls the rate of beat of the heart.
23. Within physiological limits, the force of contraction of the ventricular muscle fibers is directly proportional to its initial length, *i.e.*, larger the initial length of the cardiac muscle fibers, greater will be the force of contraction of the ventricles.

Fill in the Blanks

1. Open.**2.** William Harvey. **3.** Coronary. **4.** 72. **5.** Sinoatrial or sinuauricular, pace maker. **6.** Atrioventircular (AV), bundle of His. **7.** Cardiac output. **8.** 5. **9.** Bicuspid, tricuspid. **10.** Semilunar. **11.** 120, 80 mm Hg. **12.** Stethoscope. **13.** Semilunar. **14.** Human, neurogenic. **15.** Angiology. **16.** Veins, arteries. **17.** Electrocardiogram (ECG). **18.** Arterial blood pressure, brachial. **19.** Diastolic blood pressure, 70, 80. **20.** Systolic blood pressure, 100, 130.

Multiple Choice Questions

1. (*d*)	**2.** (*d*)	**3.** (*c*)	**4.** (*c*)
5. (*a*)	**6.** (*c*)	**7.** (*c*)	**8.** (*b*)
9. (*d*)	**10.** (*d*)	**11.** (*a*)	**12.** (*a*).

CC X

Biochemistry of Metabolic Processes

1

Overview of Metabolism

Hundreds of reactions simultaneously take place in a living cell, in a well-organized and integrated manner. The entire spectrum of chemical reactions occurring in the living systems are collectively referred to as **metabolism.** A **metabolic pathway** constitutes a series of enzymatic reactions to produce specific products. The term **metabolite** is applied to a substrate or an intermediate or a product of metabolic reactions.

1.1 CATABOLISM AND ANABOLISM

The fate of dietary components after digestion and absorption constitute **intermediary metabolism**. It encompasses a wide field that not only seeks to describe the metabolic pathways taken by individual molecules but also attempts to understand their interrelationships and the mechanisms that regulate the flow of metabolites through the pathways. Metabolic pathways are of three types:

1. Anabolic pathways. These are those metabolic pathways which are involved in the synthesis of the compounds constituting the body's structure and machinery. Protein synthesis is such a pathway. The free energy required for these processes comes from the next category.

2. Catabolic pathways. These pathways involve oxidative processes that release free energy, usually in the form of high-energy phosphate or reducing equivalents, *e.g.,* the respiratory chain and oxidative phosphorylation.

3. Amphibolic pathways. These have more than one function and occur at the "crossroads" of metabolism, acting as links between the anabolic and catabolic pathways, *e.g.,* citric acid cycle.

Stages of catabolism. The very purpose of catabolism is to trap the energy of the biomolecules in the form of ATP and to generate the substances (precursors) required for the synthesis of complex molecules. Catabolism occurs in three stages (Fig. 1.2).

1. Conversion of complex molecules into their building blocks. Polysaccharides are broken down to monosaccharides, lipids to free fatty acids and glycerol, and proteins to amino acids.

2. Formation of simple intermediates. The building blocks produced in stage 1 are degraded to simple intermediates such as pyruvate and acetyl CoA. These intermediates are not readily identifiable as carbohydrates, lipids and proteins. A small quantity of energy (as ATP) is captured in stage 2.

3. Final oxidation of acetyl CoA. Acetyl CoA completely oxidized to CO_2, liberating NADH and $FADH_2$ that finally get oxidized to release large quantity of energy (as ATP). Krebs cycle (or citric acid cycle) is the common metabolic pathway involved in the final oxidation of all energy-rich molecules. This pathway accepts all carbon compounds (pyruvate, succinate, etc.) derived from carbohydrates, lipids and proteins.

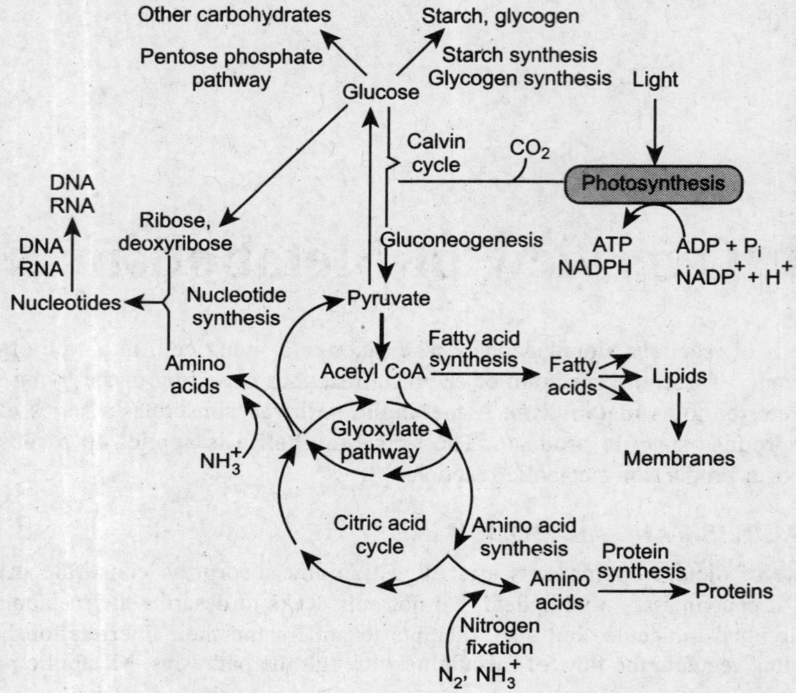

Fig. 1.1. Overview of anabolic metabolic pathways. Large molecules are synthesized from smaller ones by adding carbon (usually derived from CO_2) and nitrogen (usually as NH_3^+). The main pathways include the citric acid cycle, which supplies the intermediates in amino acid biosynthesis, and gluconeogenesis which results in the production of glucose. The energy for biosynthetic pathways is supplied by light in photosynthetic organisms or by breakdown of inorganic molecules in other autotrophs (after Horton *et al.*, 2006).

Stages of anabolism. For the synthesis of large variety of complex molecules, the starting materials are relatively few. These include pyruvate, acetyl CoA and the intermediates of citric acid cycle (Fig. 1.1). Besides the availability of precursors, the anabolic reactions are dependent on the **supply of energy** (as ATP or GTP) and reducing equivalents (as NADPH + H^+).

The anabolic and catabolic pathways are not reversible and operate independently. As such the metabolic pathways occur in specific cellular locations (mitochondria, endoplasmic reticulum, etc.) and are controlled by different regulatory signals.

Types of Metabolic Reactions

The biochemical reactions are mainly of four types:

1. Oxidation-reduction.
2. Group transfer.
3. Rearrangement and isomerization.
4. Make and break of carbons-carbon bonds.

These reactions are catalyzed by specific enzymes-more than 2,000 known enzymes so far.

1.2 COMPARTMENTALIZATION OF METABOLIC PATHWAYS

Some metabolic pathways are localised to particular regions within a cell. For example, the pathways of membrane-associated electron transport coupled to **ATP synthesis** takes place within

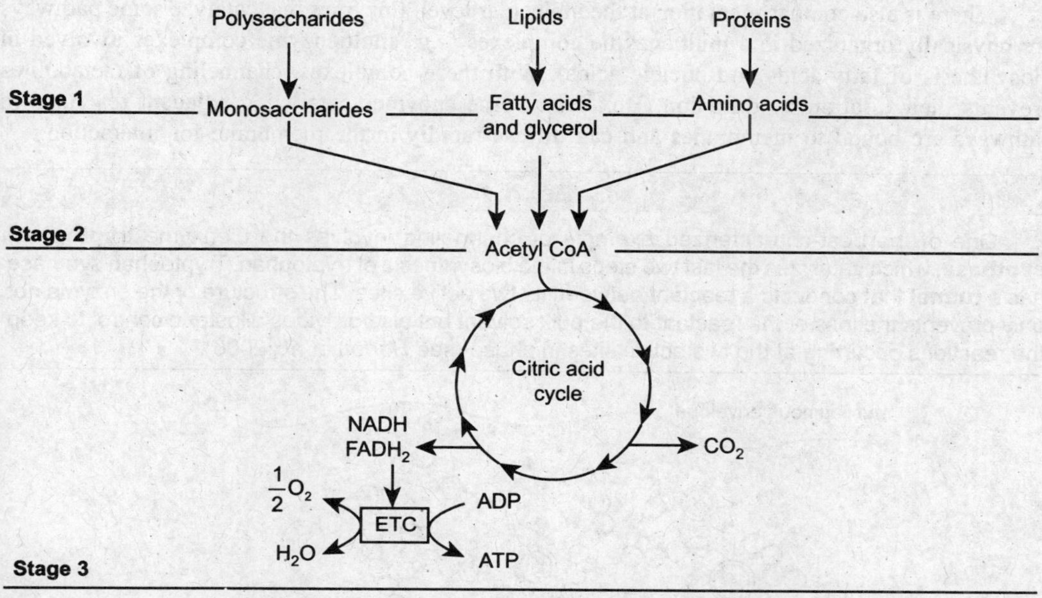

Fig. 1.2. The three stages of catabolism. ETC-Electron Transport Chain.

the membrane. In bacteria this pathway is located in the plasma membrane and in eukaryotes it is found in the mitochondrial membrane. **Photosynthesis** is another example of a membrane associated pathway in bacteria and eukaryotes.

In eukaryotes, metabolic pathways are localized within several membrane-bound compartments (Fig. 1.3). For example, the enzymes that catalyze fatty acid synthesis are located inside mitochondria. One consequence of compartmentation is that separate pools of metabolite can be found within a cell. This arrangement permits simultaneous operation of opposing metabolic pathways. Compartmentation can also offer the advantage of high local concentrations of metabolites and coordinated regulation of enzymes. Some of the enzymes that catalyze reactions in mitochondria (which have evolved from a symbiotic prokaryote) are encoded by mitochondrial genes, this origin explains their compartmentation.

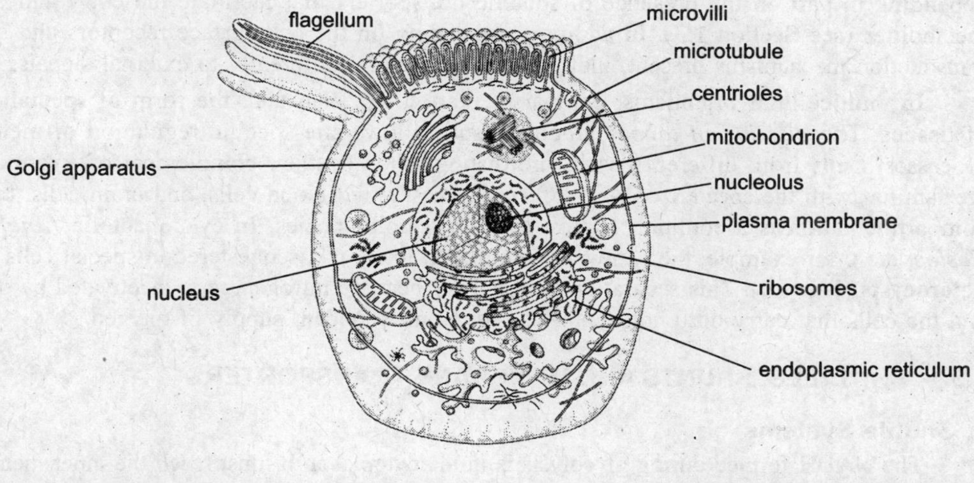

Fig. 1.3. Compartmentation of metabolic processes within a eukaryotic cell (Not all organelles shown).

There is also compartmentation at the molecular level. Enzymes that catalyze some pathways are physically organized into multienzyme complexes (*e.g.*, multienzyme complexes involved in biosynthesis of fatty acids and nucleic acids). With these complexes, channeling of metabolites prevents their dilution by diffusion (Box 1.1). Some enzymes catalyzing adjacent reactions in pathways are bound to membranes and can diffuse rapidly in the membrane for interaction.

Box 1.1

One of the best-characterized example of channeling involves the bienzyme **tryptophan synthase,** which catalyzes the last two steps in the biosynthesis of tryptophan. Tryptophan synthase has a **tunnel** that conducts a reactant between its two active sites. The structure of the enzyme not only prevents the loss of the reactant to the bulk solvent but also provides allosteric control to keep the reactions occurring at the two active sites in phase (see *Horton et al.*, 2006).

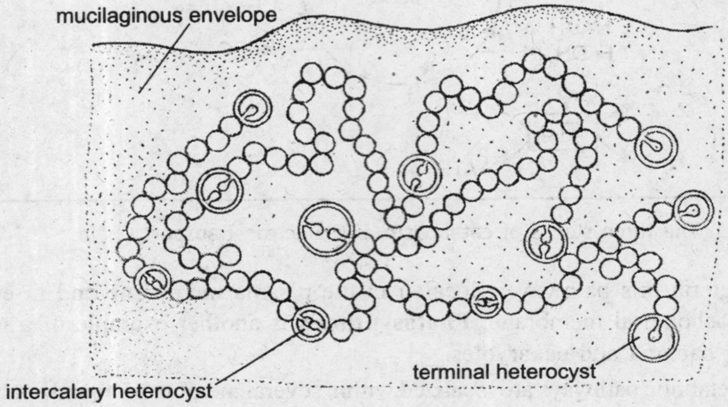

Fig. 1.4. *Nostoc sp.* Many species of cyanobacteria form long multicellular filaments. Some specialized cells (heterocysts) have adapted to carry out nitrogen fixation. These heterocysts are of two types: intercalary and terminal; they have become rounded and are surrounded by a thickened cell wall. Heterocysts are connected to adjacent cells by internal pores. The formation of heterocysts is an example of the compartmentation of metabolic pathways.

Individual cells of multicellular organisms maintain different concentrations of metabolites, depending in part on the presence of specific transporters that facilitate the entry and exit of metabolites (see Section 1.3). In addition, depending on the cell-surface receptors and signal-transduction mechanisms present, individual cells respond differently to external signals.

In multicellular organisms, compartmentation can also take the form of specialization of tissues. The *division of labour* among tissues allows site-specific regulation of metabolic processes. Cells from different tissues are distinguished by their complement of enzymes. We are familiar with the specialized roles of muscle tissue, red blood cells, and brain cells, but cell compartmentation is a common feature even in simple species. In cyanobacteria *(Anabaena, Nostoc,* etc.), for example, the pathway for nitrogen fixation is sequestered in special cells called **heterocysts** (Fig. 1.4). This separation is necessary because nitrogenase is inactivated by oxygen and the cells that carry out photosynthesis require an abundant supply of oxygen.

1.3 SHUTTLES, SHUNTS AND MEMBRANE TRANSPORTERS

1. Shuttle Systems

The NADH formed during glycolysis is in the cytoplasm. It must reach the inner membrane of mitochondria where ETS (respiratory chain) is located so that ATP may be produced. The inner mitochondrial membrane is impermeable to NADH. A special electron carrier system, called the **shuttle system**, located in the inner mitochondrial membrane, picks up the electrons from the hydrogen of NADH present in the cytoplasm, transfers them across the mitochondrial membranes and delivers them to the electron carriers inside the mitochondria (Box 1.2). There are two distinct shuttle systems: the less efficient glycerol-phosphate shuttle and the more efficient malate-aspartate shuttle (Fig. 1.5).

Box 1.2

Shuttle means cause to move backwards and forwards (*i.e.,* to and fro movement). **Shunt** means to send (railway wagons, coaches, etc.) from one track to another, especially to keep a track clear for important traffic.

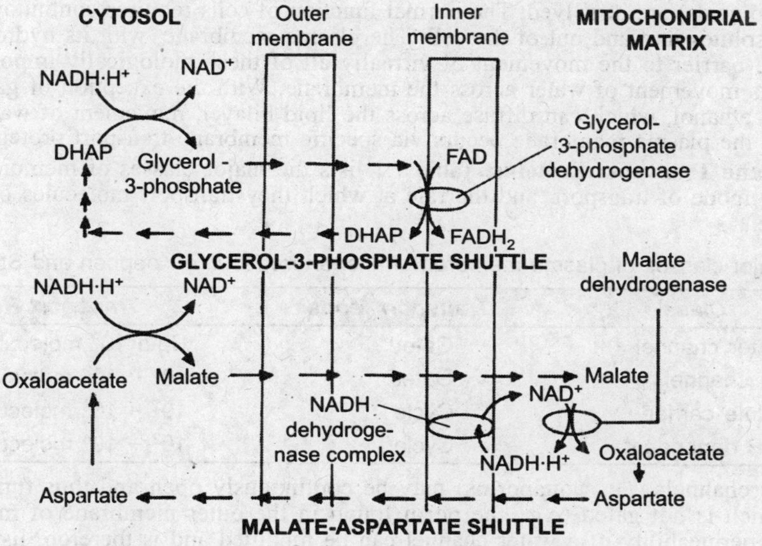

Fig. 1.5. The two electron transport shuttles.

(1) Glycerol-3-phosphate shuttle. In this case, the electrons from a cytosolic NADH molecule are transferred to a molecule of dihydroxyacetone phosphate (DHAP) in a reaction mediated by glycerol phosphate dehydrogenase enzyme. DHAP is reduced to **glycerol 3-phosphate** which diffuses through mitochondrial membranes and directly feeds its electrons into the succinate-dehydrogenase complex of ETS. These electron pairs yield only 2ATP molecules per pair. This shuttle operates in cells of skeletal muscle and brain.

(2) Malate-aspartate shuttle. This reaction is mediated by malate dehydrogenase enzyme and the electrons of a cytosolic NADH molecule are transferred to a molecule of oxaloacetate (Fig. 1.5). Consequently, NADH is reoxidizes to NAD^+ for reuse in glycolysis, whereas oxaloacetate is reduced to **malate.** The malate easily diffuses into mitochondria and transfers its electrons to a NAD^+ molecule of mitochondrial matrix in a reaction mediated by mitochondrial malate dehydrogenase. Consequently, malate is reoxidized to oxaloacetate and mitochondrial NAD^+ is reduce to NADH (Box 1.3). The inner mitochondrial membrane is impermeable to oxaloacetate also. Hence, the latter is transmitted to **aspartate** which easily diffuses out of the mitochondria. In cytosol, aspartate is transaminated back to oxaloacetate so that the shuttle reactions may be repeated. The NADH molecules, thus formed in mitochondrial matrix, transfer their electron pairs to the **NADH-dehydrogenase** complex of ETS, so that each electron pair yields three ATP molecules. This shuttle operates in cells of liver, kidney and the heart.

Box 1.3

NADH (in mitochondria) can be used directly in biosynthesis reactions such as gluconeogenesis (where glyceraldehyde 3-phosphate dehydrogenase operates in the reverse direction), fatty acid synthesis, amino acid synthesis, and nucleotide synthesis (**Horton** *et al.*, 2006).

2. Hexose Monophosphate (HMP) Shunt

It is the direct oxidative pathway of glucose. HMP shunt is anabolic in nature, since it is concerned with the biosynthesis of NADPH (used in biosynthesis of fatty acids and steroids) and pentoses (used in synthesis of nucleic acids). The enzymes of HMP shunt are located in the cytosol. The tissues such as liver, adipose tissue, adrenal gland, erythrocytes, testes and lactating mammary glands are highly active in HMP shunt (see **Satyanarayana** and **Chakrapani** 2013).

3. Membrane Transporters

Intracellular and extracellular fluid is composed primarily of H_2O in which solutes (*e.g.*, ions, glucose, amino acids) are dissolved. The normal function of cells requires continuous movement of water and solutes into and out of the cell. The plasma membrane, with its hydrophobic core, is an effective barrier to the movement of virtually all of these biologically important solutes. It also restricts movement of water across the membrane. With the exception of gases (*e.g.*, O_2 and CO_2) and ethanol, which can diffuse across the lipid bilayer, movement of water and other solutes across the plasma membrane occurs via specific membrane transport proteins (Fig. 1.6).

Membrane Transport Proteins. Table 1.1 lists the major classes of membrane transport proteins, their mode of transport, and the rate at which they transport molecules or ions across the membrane.

Table 1.1. Major classes of plasma membrane transporters (source Koeppen and Stanton, 2006).

	Class	*Transport Mode*	*Transport Rate*
1.	Water channel	Gated*	Upto 10^9 molecules/sec
2.	Ion channel	Gated	10^6 – 10^8 molecules/sec
3.	Solute carrier	Cycle	10^2 – 10^4 molecules/sec
4.	ATP dependent	Cycle	10^2 – 10^4 molecules/sec

* Water channels (*i.e.*, aquaporins) may be continuously open and thus function similar to a pore, which is not gated (*e.g.*, the porin found in the outer membrane of mitochondria). However, the permeability of a water channel can be modified and is therefore listed as gated.

1. Water channels. Water channels, or **aquaporins (AQPs),** are the main route for water movement into and out of the cell. They are widely distributed throughout the body, although different isoforms are found in different cell types. To date, 12 AQPs have been identified. The amount of H_2O that can enter or leave the cell via AQPs can be regulated by altering the number of AQPs in the membrane or by changing their permeability (*i.e.*, gating). Changes in pH have been identified as one factor that can modulate the permeability of AQPs.

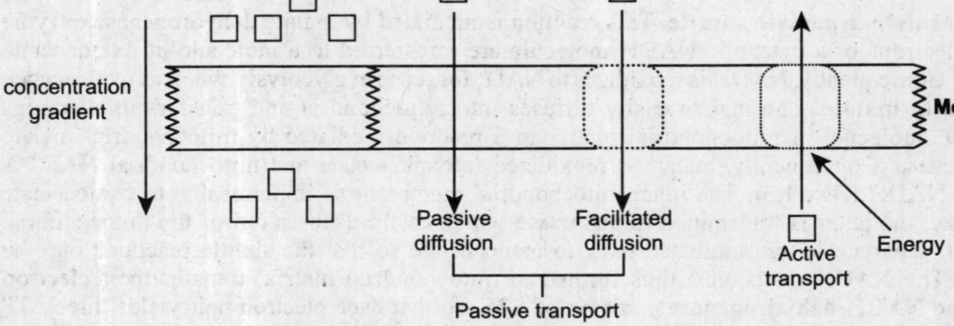

Fig. 1.6. Mechanism of transport across biological membranes.

2. Ion channels. Ion channels are found in all cells and are especially important for the function of excitable cells (*e.g.,* neurons and muscle cells). Ion channels are classified by their selectivity (*i.e.,* the ions that pass through the channel, Table 1.2). At one extreme, they can be highly selective by allowing only a specific ion through. At the other extreme, they may be nonselective and allowed all or a group of cations or anion through. Channels are also characterized by their conductance, which is typically expressed in picosiemens (ps). The range in conductance is considerable, with some channels having a conductance of only 1 to 2 ps and others having a conductance of greater than 100 ps. For some channels, conductance varies depending on which direction the ion is moving. For example, if the channel has greater conductance when ions are moving into the cell versus out of the cell, the channel is said to be an **inward rectifier.** Finally ion channels can be classified by their mechanism of gating. Ion channels fluctuate between an open state or a closed state, a process called **gating**. Factors that can control the gating include membrane **voltage,** extra-cellular **agonists** or **antagonists** (*e.g.,* acetylcholine) is an extracellular agonist that controls the gating of a cation-selective channel in the motor end plate of skeletal muscle cells, intracellular messengers (*e.g.,* Ca^{2+}, ATP, cGMP), and mechanical stretch of the plasma membrane. Transmembrane ion flux can be regulated by changing the number of channels in the membrane or by gating of the channels.

3. Solute carriers. Solute carriers represent a large family of membrane transporters, with more than 40 different types (> 300 specific transporters) already identified. These carriers are divided into following three major functional groups:

1. Uniporters. They transport a single molecule across the membrane. The transporter that brings glucose into the cell (GLUT2) is an important member of this group.

2. Symporters. They couple the movement of two or more molecules/ions across the membrane. As the name implies, the molecules are transported in the same direction.

Table 1.2. Examples of plasma membrane transporters.

1. Water channels

Aquaporin (AQP–multiple isoforms).

2. Ion channels

(i) Na^+	Multiple channels exist for each ion listed. They are distinguished by their selectivity, conductance, and mode of regulation (*i.e.,* gating).
(ii) K^+	
(iii) Ca^{2+}	
(iv) Cl^-	
(v) Anion	
(vi) Cation	

3. Solute carriers (Fig. 1.7)

A. Uniport

(i) Glucose (GLUT2)

(ii) Fructose (GLUT5)

(iii) Urea (UT-A1)

(iv) Fe^{3+} (ferroprotin/REG-1)

B. Symport

(i) $1Na^+$– glucose (SGLT-2)

(ii) $2Na^+$– glucose (SGLT-1)

(iii) Na^+– amino acid (multiple transporters)

(iv) Na^+– Cl^- (NCC/TSC)

(v) $1Na^+$, $1K^+$, $2Cl^-$ (NKCC2)

(vi) Na^+– $3HCO_3^-$ (NBC1)

(vii) $3Na^+$– P_1 (type IIa phosphate transporter)

(viii) $2Na^+$– 11^- (NIS)

(ix) Na^+– bile acid (NTCP - multiple isoforms)

(x) $3Na^+$– dicarboxylate (SDCT - multiple isoforms)

(xi) H^+– Fe^{3+} (DCT–1)

(xii) K^+– Cl^- (KCC–multiple isoforms)

C. Antiport

(i) Na^+– H^+ (NHE–multiforms)

(ii) Cl^-– HCO_3 (AE–1/band three and pendrin)

(iii) $3Na^+$– Ca^+ (NCX - multiple isoforms)

(iv) Organic anions (OAT - multiple transporter for different anions)

(v) Organic cations (OCT and OCTN - multiple isoforms)

D. Transport ATPases

(1) P-type

(i) Na^+, K^+ – ATPase

(ii) H^+,K+ – ATPase

(iii) H^+, Ca^{2+} – ATPase (PMCA)

(2) V-type

(i) H^+ – ATPase

E. ABC Transporters

(i) Cystic fibrosis transmembrane regulator (CFTR)

(ii) Multidrug resistance protein (MRP–1)

(iii) Organic anion (MRP–2)

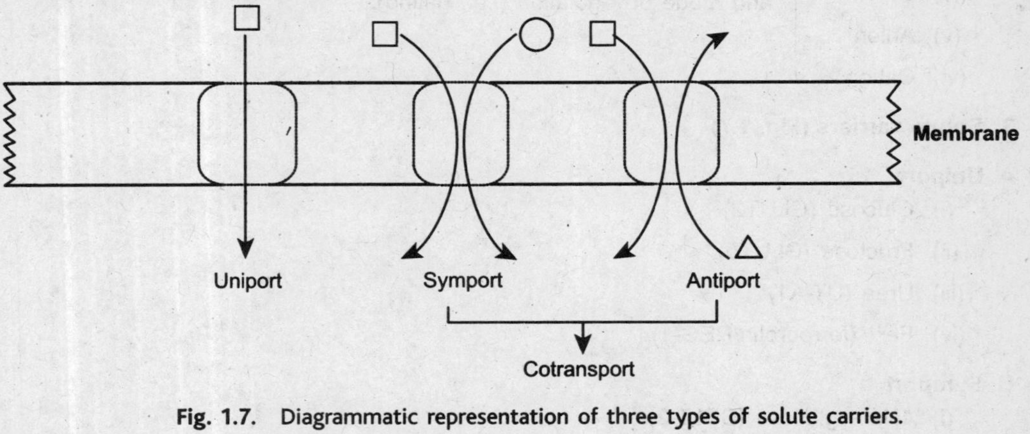

Fig. 1.7. Diagrammatic representation of three types of solute carriers.

Cotransport is another term used to describe this group of solute carriers. The $1Na^+$, $1K^+$, $2Cl^-$ symporter found in the kidney (NKCC2), which is critically important for diluting and concentrating urine, is an example of a member of this group.

3. Antiporters. They also couples the movement of two or more molecules/ions across the membrane. However, in this case the molecules/ions are transported in opposite directions. The term **exchangers** and **counter transporters** are also used to describe this group of solute carriers. The $Na^+–H^+$ antiporter is an example of this group of solute carriers. One isoform (NHE–1) of this antiporter is found in all cells and plays an important role in regulating intracellular pH.

4. ATP-dependent transporters. ATP-dependent transporters, as their name implies, use the energy in ATP to drive the movement of molecules/ions across the membrane. There are two groups of ATP-dependent transporters:

***(i)* ATPase ion transporters.** They are also two types : P-type and V-type ATPases. (*Note.* F-type ATPase are found in mitochondria and are responsible for ATP synthesis).

(a) **P-type ATPases.** They are also phosphorylated during the transport cycle. Na^+, K^+–ATPase is an important example of a P-type ATPase (Fig. 1.8). With the hydrolysis of each ATP molecule, $3Na^+$ ions are transported out of the cell and $2K^+$ ions into the cell. Na^+ –K^+ –ATPase (or Na^+–K^+ pump) is present in all cells and plays a critical role in establishing cellular ions and electrical gradients, as well as maintaining cell volume.

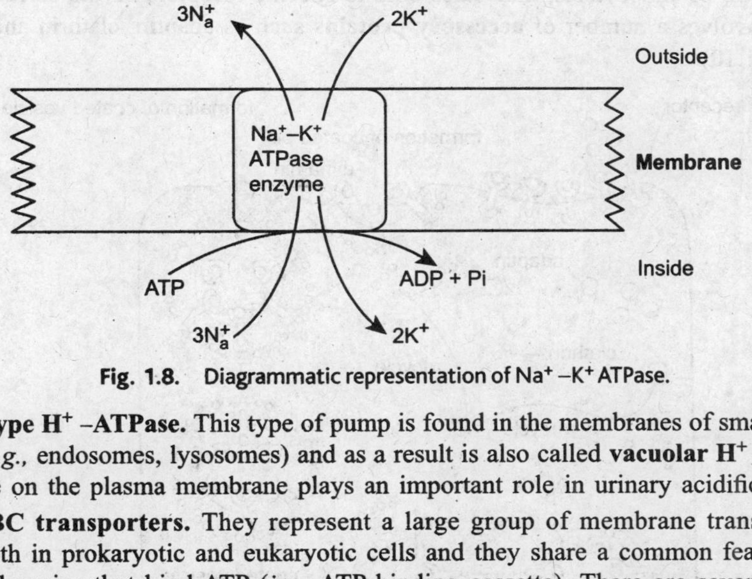

Fig. 1.8. Diagrammatic representation of Na^+ –K^+ ATPase.

(b) **V-type H^+ –ATPase.** This type of pump is found in the membranes of small intracellular organelles (*e.g.*, endosomes, lysosomes) and as a result is also called **vacuolar H^+ –ATPase**. The H^+ – ATPase on the plasma membrane plays an important role in urinary acidification.

***(ii)* ABC transporters.** They represent a large group of membrane transporters. They are found both in prokaryotic and eukaryotic cells and they share a common feature of having amino acid domains that bind ATP (*i.e.*, ATP-binding cassette). There are seven subgroups of ABC transporters in humans and more than 40 specific ABC transporters have been identified. They transport a diverse group of molecules/ions, including Cl^-, cholesterol, bile acids, drugs, iron and organic anions.

5. Vesicular Transport

Solute and water can be brought into the cell by the process of **endocytosis** and released from the cell by the process of **exocytosis** (Fig. 1.9). In both processes the integrity of the plasma membrane is maintained, and the vesicles that are formed allow transfer of the contents between cellular compartments. In some cells (*e.g.*, the epithelial cells lining the gastrointestinal tract), endocytosis across one membrane of the cell is followed by exocytosis across the opposite membrane. This allows the transport of substances across the epithelium, a process termed **transcytosis**.

Type of endocytosis. Endocytosis can be subdivided into the following three mechanisms:

(a) Pinocytosis. It consists of the nonspecific uptake of small molecules and water into the cell. Pinocytosis is a prominent feature of the endothelial cells that line capillaries and is responsible for a portion of the fluid exchange that occurs across blood vessels.

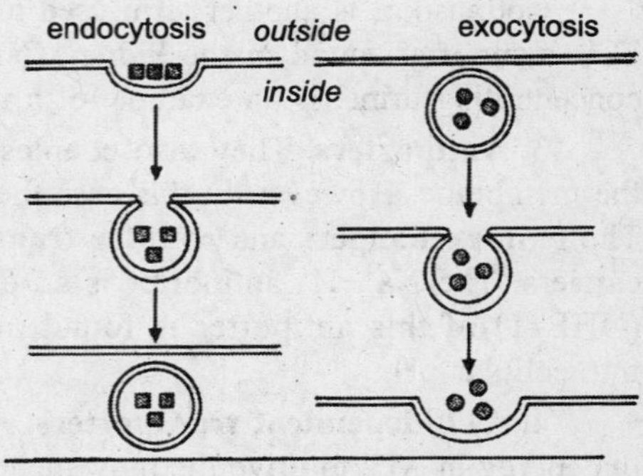

Fig. 1.9. Diagrammatic representation of endocytosis and exocytosis.

2. Phagocytosis. It allows the internalization of large particles (*e.g.,* bacteria, cell debris), Phagocytosis is an important characteristic of cells in the immune system (*e.g.*, neutrophils and macrophages). Often but not always, phagocytosis is a *receptor-mediated process*. For example macrophages have receptors on their surface that bind the Fc portion of immunoglobulins. When bacteria invade the body, they are often coated with antibody, a process called **opsonization.** These bacteria then attach to the membrane of macrophages via the Fc portion of the immunoglobulin and are phagocytosed and destroyed inside the cell.

3. Receptor-mediated endocytosis. It allows the uptake of specific molecules into the cell. In this form of endocytosis, molecules bind to specific receptors on the surface of the cell. Endocytosis involves a number of **accessory proteins** such as adaptin, clathrin and the GTPase dyamin (Fig. 1.10).

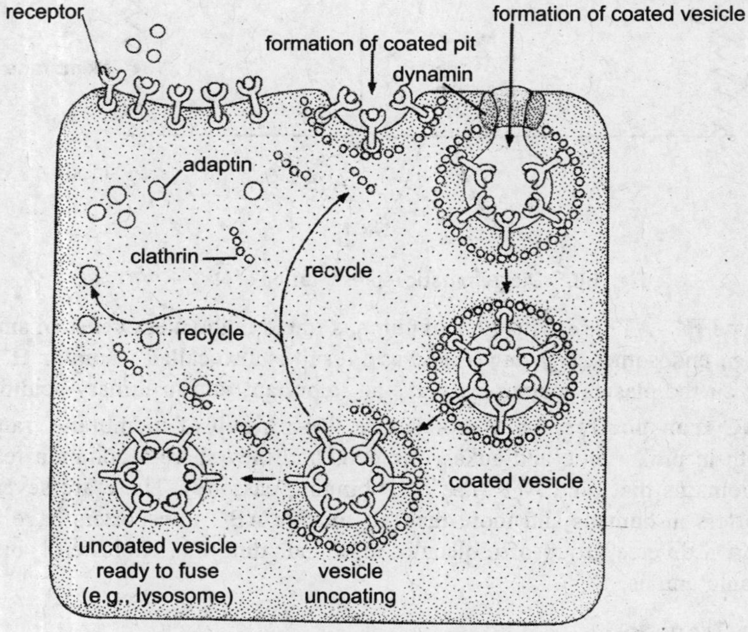

Fig. 1.10. Receptor-mediated endocytosis. A receptor on the surface of the cell binds the ligand. A clathrin-coated pit is formed, with adaptin linking the receptor molecules to the clathrin. Dynamin, a GTPase, assists in separation of the endocytic vesicle from the membrane. Once inside the cell, the clathrin and adaptin molecules dissociate and are recycled. The uncoated vesicle is then ready to fuse with other organelles in the cell (*e.g.*, lysosome) (after Koeppen and Stanton 2006).

B. Types of exocytosis. Exocytosis can be either **constitutive** or **regulated. Constitutive secretion** is seen, for example, in plasma cells that are secreting immunoglobulin or in fibroblasts secreting collagen. **Regulated secretion** occurs in endocrine cells, neurons and exocrine glandular cells (pancreatic acinar cells). In these cells the secretory product (*e.g.*, hormone, neurotransmitter, or digestive enzyme), after synthesis and processing in rough ER and Golgi apparatus, is stored in cytoplasm in secretory granules until an appropriate signal for secretion is received. These signals may be hormonal or neural. Once the cell receives the appropriate stimulus, the secretory vesicle fuses with the plasma membrane releases its contents into the extracellular fluid. Fusion of the vesicle with the plasma membrane releases its contents into the extracellular fluid. Fusion of the vesicle with the membrane is mediated by a number of accessory proteins. One important group is the SNAREs. These membrane proteins help target the secretory vesicle to the plasma membrane. The process of **secretion** is usually triggered by an increase in intracellular (Ca^{2+}). However, two notable exceptions to this general rule exist: **renin secretion** by juxtaglomerular cells of the kidney is triggered by a decrease in intracellular Ca^{2+} as in the secretion of parathyroid hormone (PTH) by the parathyroid gland.

1.4 ATP AS ENERGY CURRENCY OF CELL AND COUPLED REACTIONS

A number of nucleoside triphosphates are coenzymes (Box 1.3) Adenosine triphosphate (ATP) is by far is most abundant. Other common examples are GTP, S-adenosyl–methionine (SAM) and nucleotide sugars such as uridine diphosphate glucose (UDP-glucose). ATP is a versatile reactant that can donate its phosphoryl, pyrophosphoryl, adenylyl (AMP), or adenosyl groups in group transfer reactions (Fig. 1.11).

$$\text{Methionine} + \text{ATP} \rightarrow \text{S-Adenosylmethione} + P_i + PP_i$$

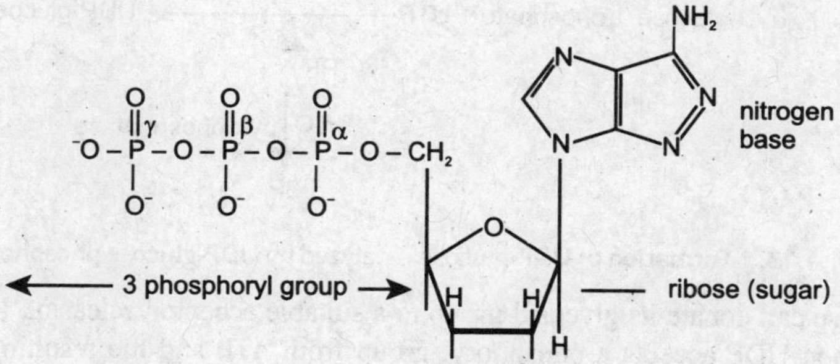

Fig. 1.11. ATP (Adenosine triphosphate).

Box 1.3

Cofactor. An inorganic ion or organic molecule required by an apoenzyme to convert it to a holoenzyme. There are two types of cofactors: essential ions and coenzymes.

Coenzymes. An organic molecule required by an enzyme for full activity. Coenzymes can be further classified as cosubstrates or prosthetic groups.

Nucleophile. An electron-rich species that is negatively charged or contain unshared electron pairs and is attracted to chemical species that are positively charged or electron-deficient (electrophiles).

Nucleoside. A purine or pyrimidine N-glycoside of ribose or deoxyribose.

The most common reaction involving ATP is phosphoryl group transfer. In reactions catalyzed by kinase enzymes, for example, the γ-phosphoryl group of ATP is transferred to a nucleophile (Box 1.3), leaving ADP. The second most common reaction is nucleotidyl-group transfer (transfer of the AMP moiety), leaving pyrophosphate (PPi). ATP plays a central role in metabolism.

ATP is also the source of several other metabolic coenzymes. One, **S-adenosyl methionine** or **SAM** (Fig. 1.12), is synthesized by the reaction of methionine with ATP:

$$\text{Methionine} + \text{ATP} \rightarrow \text{S-Adenosylmethionine} + P_i + PP_i$$

Unlike the thiomethyl group of methionine, the positively charged sulphonium of S-adenosylmethionine reacts readily with nucleophilic acceptors and is the donor of virtually all the methyl groups used in biosynthetic reactions. For example, it is required for the conversion of the hormone norepinephrine to epinephrine. Methylation reactions that require S-adenosylmethionine (SAM) include methylation of phospholipids, proteins, DNA and RNA.

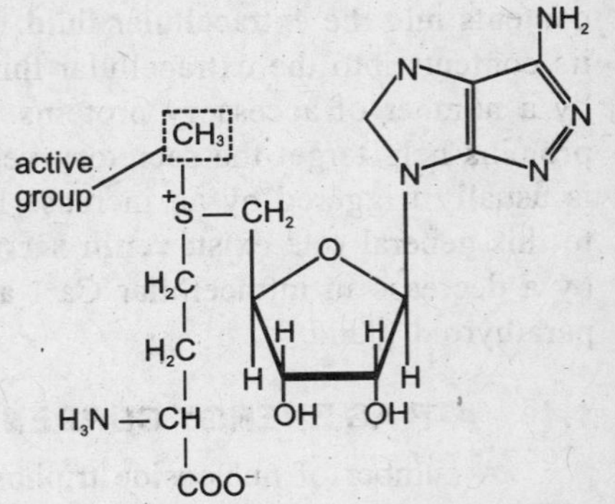

Fig. 1.12. S-Adenosylmethionine (SAM).

Nucleotide-sugar coenzymes are involved in carbohydrate metabolism. The most common nucleotide sugar, uridine diphosphate glucose (UDP–glucose), is formed by the reaction of glucose 1-phosphate with uridine triphosphate (UTP), a nucleoside triphosphate in which the adenine of ATP is replaced by uracil moiety (Fig. 1.13).

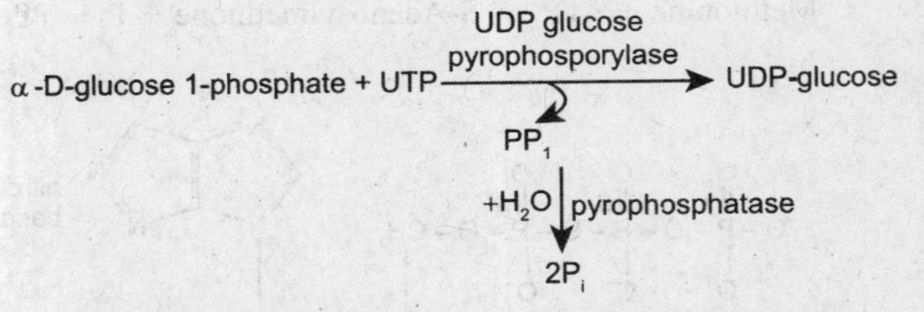

Fig. 1.13. Formation of UDP-glucose catalyzed by UDP-glucose phosphorylase.

UDP-glucose can donate its glycosyl group to a suitable acceptor, releasing UDP. UDP-glucose is regenerated when UDP accepts a phosphoryl group from ATP and the resulting UTP reacts with another molecule of glucose 1 phosphate.

Both the sugar and the nucleoside of nucleotide-sugar coenzymes may vary. For example, derivatives of the nucleotides cytidine diphosphate (CDP), which has cytosine as its nitrogenous base, will be encountered in lipid metabolism (see Chapter 3).

The production of ATP is one of the most important reactions in metabolism. The synthesis of most ATP is coupled to membrane-associated electron passport (Chapter 5). In electron transport the energy of reduced coenzymes, such as NADH, is used to generate an electrochemical gradient of protons across a cell membrane. The potential energy of this gradient is harnessed to derive the phosphorylation of ADP to ATP.

$$\text{ADP} + \text{Pi} \rightarrow \text{ATP} + H_2O$$

The reactions of membrane associated electron transport and coupled ATP synthesis are similar in many ways to the reactions that capture light energy during photosynthesis (See **Horton** *et al.*, 2006).

The Free Energy of ATP

ATP contains one phosphate ester formed by linkage of the α-phosphoryl group to the 5'-oxygen of ribose and two phosphoanhydrides formed by the α, β and β, γ linkages between phosphoryl groups (Fig. 1.14). ATP is donor of several metabolic groups, usually a phosphoryl group, leaving ADP, or a nucleotidyl (AMP) group, laving inorganic pyrophosphate PP_i. Both reactions require the cleavage of a phosphoanhydride linkage. The various groups of ATP are usually transferred to acceptors other than water. Nevertheless hydrolytic reactions provide useful estimates of the Gibbs free energy changes involved. Table 1.3. lists the standard Gibbs free of hydrolysis (AG°, hydrolysis) for ATP and AMP, and Fig. 1.14) shows the hydrolytic cleavage of each of the phosphoanhydrides of ATP.

Table 1.3. Standard Gibbs free energies of hydrolysis for ATP, AMP and pyrophosphate (Source: Horton *et al.*, 2006).

	Reactants and products	*ΔG° hydrolysis (kJ mol⁻¹)*
1.	$ATP + H_2O \rightarrow ADP + P_i + H^+$	–32
2.	$ADP + H_2O \rightarrow AMP + PP_i + H^+$	–45
3.	$AMP + H_2O \rightarrow$ Adenosine $+ P_i$	–13
4.	$PP_i + H_2O \rightarrow 2P_i$	–29

P_i (inorganic phosphate) = HPO_4^{2-}

PP_i (pyrophosphate) = $HP_2O_7^{3-}$

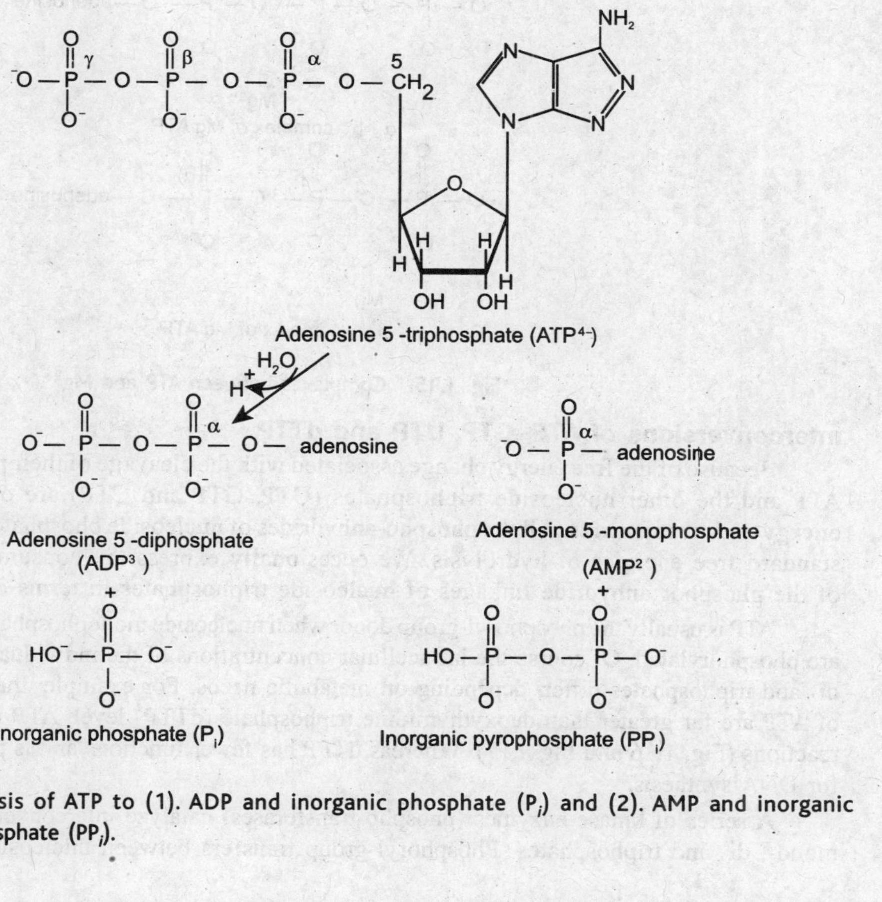

Fig. 1.14. Hydrolysis of ATP to (1). ADP and inorganic phosphate (P_i) and (2). AMP and inorganic pyrophosphate (PP_i).

Note from Table 1.3 that whereas cleavage of the phosphoester in AMP releases only 13kJ mol^{-1} under standard conditions, cleavage of either of the phosphoanhydrides in ADP or ATP releases at least 30 kJ mol^{-1} under standard conditions. Table 1.3 also gives the standard Gibbs free energy change for the hydrolysis of pyrophosphate. All cells contain an enzyme called **pyrophosphatase** that catalyzes this reaction. The cellular concentration of pyrophosphate is maintained at a very low concentration as a consequence of this highly favourable reaction. This means that the hydrolysis of ATP to AMP + pyrophosphate will always be associated with a negative Gibbs free energy change, even when the AMP concentration is significant.

Complexes of ATP and Mg (Magnesium) or Cations

Nucleoside diphosphates and triphosphates in both aqueous solution and at the active sites of enzymes are usually present as complexes with magnesium (or sometimes manganese) ions. These cations coordinate with the oxygen atoms of the phosphate groups, forming six-membered rings (Fig. 1.15). A magnesium ion can form several different complexes with ATP; the complexes involving the α and β and the β and γ groups are shown in Fig. 1.15. Formation of the β and γ complex is favoured in aqueous solution. The nucleic acids are also usually complexed counterions such as Mg^{2+} or cationic proteins. For convenience, we usually refer to the nucleoside triphosphates and adenosite triphosphates (ATP), guanosine triphosphate (GTP), cytidine triphosphate (CTP) and uridine triphosphate (UTP), however, these molecules actually exist as complexes with Mg^{2+} in cells.

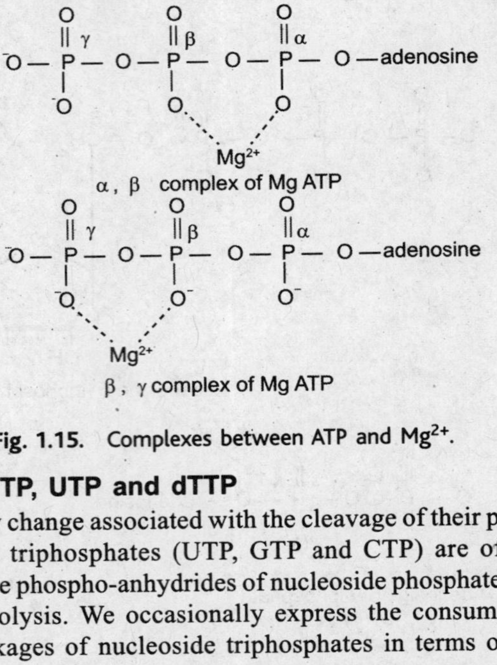

Fig. 1.15. Complexes between ATP and Mg^{2+}.

Interconversions of ATP, GTP, UTP and dTTP

Because of the free energy change associated with the cleavage of their phosphoanhydrides, ATP and the other nucleoside triphosphates (UTP, GTP and CTP) are often referred to as **energy-rich compounds.** All the phospho-anhydrides of nucleoside phosphates have nearly equal standard free energies of hydrolysis. We occasionally express the consumption or formation of the phosphor anhydride linkages of nucleoside triphosphates in terms of ATP equivalents.

ATP is usually the phosphoryl-group donor when nucleoside monophosphates and diphosphates are phosphorylated. Of course the intracellular concentrations of the individual nucleoside mono-, di- and triphosphates differ, depending on metabolic needs. For example, the intracellular levels of ATP are far greater than deoxythymidine triphosphate (dTTP) level. ATP is involved in many reactions (Fig. 1.16 and Fig. 1.17). Whereas dTPP has fewer functions and is primarily a substrate for DNA synthesis.

A series of kinase enzymes (phospho-transferases) catalyze interconversions of nucleoside mono-, di- and triphosphates. Phosphoryl-group transfers between nucleoside phosphates have

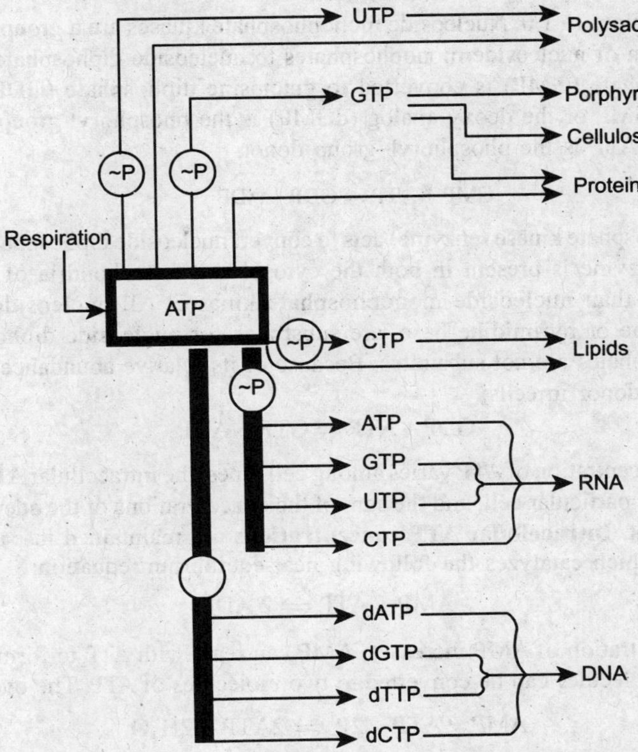

Fig. 1.16. Channelling of phosphate bond energy by ATP into specific biosynthetic routes of the cell (after De Robertis *et al.*, 1970).

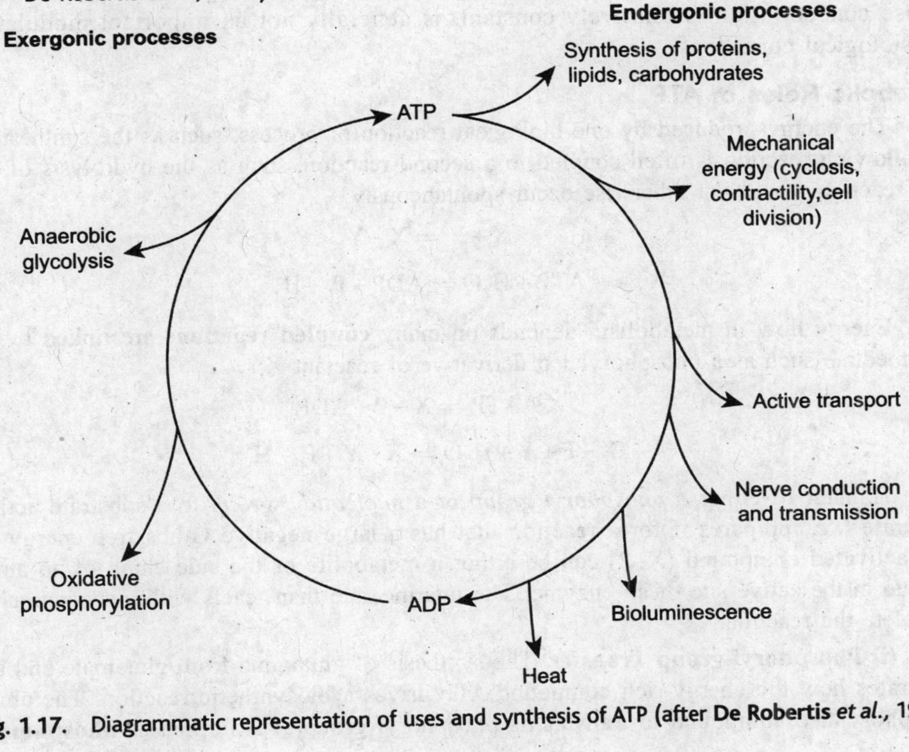

Fig. 1.17. Diagrammatic representation of uses and synthesis of ATP (after **De Robertis** *et al.*, 1970).

equilibrium constants close to 1.0. Nucleoside monophosphate kinases are a group of enzymes that catalyze the conversion of nucleoside monophosphates to nucleoside diphosphates. For example, guanosine monophosphate (GMP) is converted to guanosine diphosphate (GDP) by the action of guanylate kinase. GMP or the deoxy analog (dGMP) is the phosphoryl group acceptor in the reaction and ATP or dATP is the phosphoryl–group donor.

$$\text{GMP} + \text{ATP} \rightleftharpoons \text{GDP} + \text{ADP}$$

Nucleoside diphosphate kinase (enzyme) acts to convert nucleoside diphosphates to nucleoside triphosphates. This enzyme is present in both the cytosol and mitochondria of eukaryotes and is much less specific than nucleoside monophosphate kinases. All nucleoside diphosphates, regardless of the purine or pyrimidine base, are substrates for nucleoside diphosphates kinase. Nucleoside monophosphates are not substrates. Because of its relative abundance, ATP is usually the phosphoryl group donor in cells:

$$\text{GDP} + \text{ATP} \rightleftharpoons \text{GTP} + \text{ADP}$$

Although the concentration of ATP varies among cell types, the intracellular ATP concentration fluctuates little within a particular cell, and the sum of the concentrations of the adenine nucleotides remains nearly constant. **Intracellular ATP concentrations** are maintained in part by the action of adenylate kinase, which catalyzes the following near-equilibrium equation:

$$\text{AMP} + \text{ATP} \rightleftharpoons 2\ \text{ADP}$$

When the concentration of AMP increases, AMP can react with ATP to form two molecules of ADP. These ADP molecules can be converted to two molecules of ATP. The overall process is:

$$\text{AMP} + \text{ATP} + 2\text{P}_i \rightleftharpoons 2\text{ATP} + 2\text{H}_2\text{O}$$

ATP concentrations in cells are greater than ADP or AMP concentrations. ADP and AMP are often effective allosteric modulators of some energy-yielding metabolic processes. ATP, whose concentration is relatively constant, is generally not an important modulator under physiological conditions.

Metabolic Roles of ATP

The energy produced by one biological reaction or process, such as the synthesis of X–Y in following reaction is often coupled to a second reaction, such as the hydrolysis of ATP. The first reaction would not otherwise occur spontaneously.

$$\text{X} + \text{Y} \rightleftharpoons \text{X} - \text{Y}$$

$$\text{ATP} + \text{H}_2\text{O} \rightleftharpoons \text{ADP} + \text{P}_i + \text{H}^+$$

Energy flow in metabolism depends on many **coupled reactions** are linked by a shared intermediate such as a phosphorylated derivative of reactant X:

$$\text{X} + \text{ATP} \rightleftharpoons \text{X} - \text{P} + \text{ADP}$$

$$\text{X} - \text{P} + \text{Y} + \text{H}_2\text{O} \rightleftharpoons \text{X} - \text{Y} + \text{P}_i + \text{H}^+$$

Transfer of either *a phosphoryl group* or a *nucleotidyl group* to a substrate activate that substrate (*i.e.,* prepares if for a reaction that has a large negative Gibbs free energy change). The activated compound (X–P) can be either a metabolite or the side chain of an amino acid residue in the active site of an enzyme. The intermediate than reacts with a second substrate to complete the reaction.

(i) Phosphoryl-group Transfer. The synthesis of glutamine from glutamate and ammonia illustrates how the energy-rich compound ATP drives a biosynthetic reaction. The ability of a phosphorylated compound to transfer its phosphoryl group(s) is termed as **phosphoryl group**

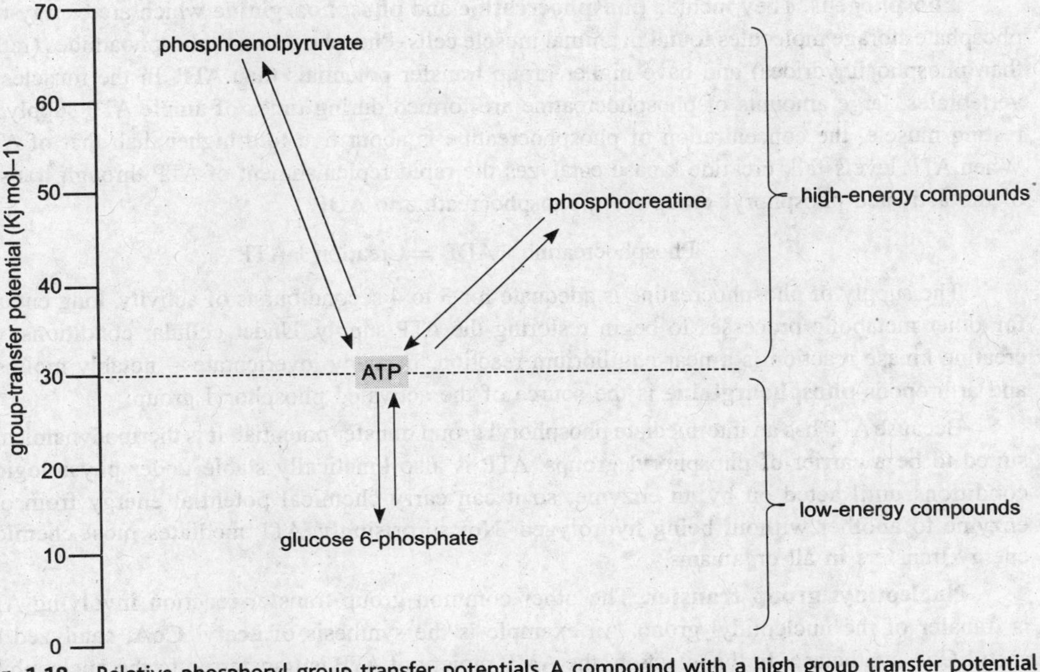

Fig. 1.18. Relative phosphoryl group transfer potentials. A compound with a high group transfer potential (*i.e.*, a large negative value of ΔG' hydrolysis) can donate its phosphoryl group to a compound that is less energy-rich. The reaction arrows indicate the predominant direction of phosphoryl - group transfer under standard conditions (after Horton *et al.*, 2006).

transfer potential (Fig. 1.18), or simply group-transfer potential. Some compounds such as phosphoanhydrides, are excellent phosphoryl group donors. In fact, the group-transfer potential is a measure of the free energy required for formation of the phosphorylated compound. Often one kinase catalyzes transfer of a phosphoryl group from an excellent donor to ADP to form ATP, which is then acts as a donor for a different kinase-catalyzed reaction.

(*ii*) Production of ATP by Phosphoryl group Transfer. Metabolites with high phosphoryl group-transfer potentials, including carboxylic acid phosphate anhydrides, can donate a phosphoryl group to ADP to form ATP. Some of these energy-rich compounds are intermediates in catabolic pathways; others are energy storage compounds.

For example, **phosphoenolpyruvate,** an intermediate in the glycolytic pathway, has the highest phosphoryl group transfer potential known. The standard free energy of phosphoenol pyruvate hydrolysis is –62 kJ mol^{-1}. The free energy of hydrolysis for phosphoenol pyruvate can be understood by picturing the molecule as an enol whose structure is locked by attachment of the phosphoryl group when the phosphoryl group is removed, the molecule spontaneously form the much more stable keto tautomer (Fig. 1.19). Transfer of the phosphoryl group from phosphoenol pyruvate to ADP is catalyzed by the enzyme pyruvate kinase. Because the ΔG° for the reaction is about –30 kJ mol^{-1}, the equilibrium for this reaction under standard conditions lie far in the direction of transfer of the phosphoryl group from phosphoenol pyruvate the ADP. In cells, this metabolically irreversible reaction is an important source of ATP.

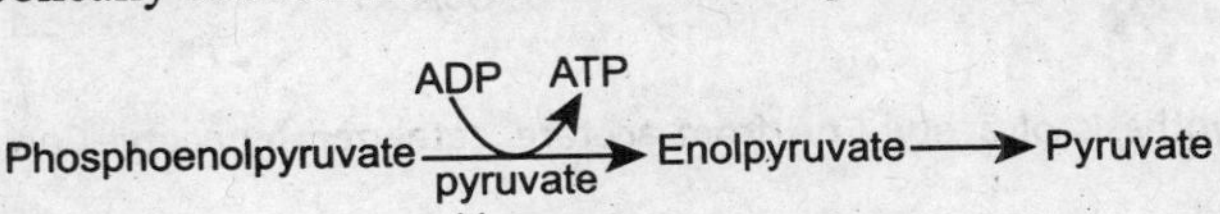

Fig. 1.19. Transfer of the phosphoryl group from phosphoenol pyruvate to ADP.

Phosphogens. They include **phosphocreatine** and **phosphoarginine** which are energy-rich phosphate storage molecules found in animal muscle cells. Phosphogens are phosphoamides (rather than phosphoanhydrides) and have higher group transfer potentials than ATP. In the muscles of vertebrates, large amounts of phosphocreatine are formed during times of ample ATP supply. In resting muscle, the concentration of phosphocreatine is about five fold higher than that of ATP. When ATP levels fall, creatine kinase catalyzes the rapid replenishment of ATP through transfer of the activated phosphoryl group from phosphocreatine to ADP.

$$\text{Phosphocreatine} + \text{ADP} \rightleftharpoons \text{Creation} + \text{ATP}$$

The supply of phosphocreatine is adequate for 3 to 4 second bursts of activity, long enough **for** other metabolic processes to begin restoring the ATP supply. Under cellular conditions, the creatine kinase reaction is a near equilibrium reaction. In many invertebrates—notably molluscs and arthropods-**phosphoarginine** is the source of the activated phosphoryl group.

Because ATP has an intermediate phosphoryl group transfer potential, it is thermodynamically suited to be a carrier of phosphoryl groups. ATP is also kinetically stable under physiological conditions until acted on by an enzyme, so it can carry chemical potential energy from one enzyme to another without being hydrolyzed. Not surprisingly, ATP mediates most chemical energy transfers in all organisms.

Nucleotidyl group transfer. The other common group-transfer reaction involving ATP is transfer of the nucleotidyl group. An example is the synthesis of **acetyl CoA,** catalyzed by acetyl CoA synthetase. In this reaction, the AMP moiety of ATP is transferred to the nucleophilic carboxylate group of acetate to form acetyl-adenylate intermediate (Fig. 1.20). Pyrophosphate

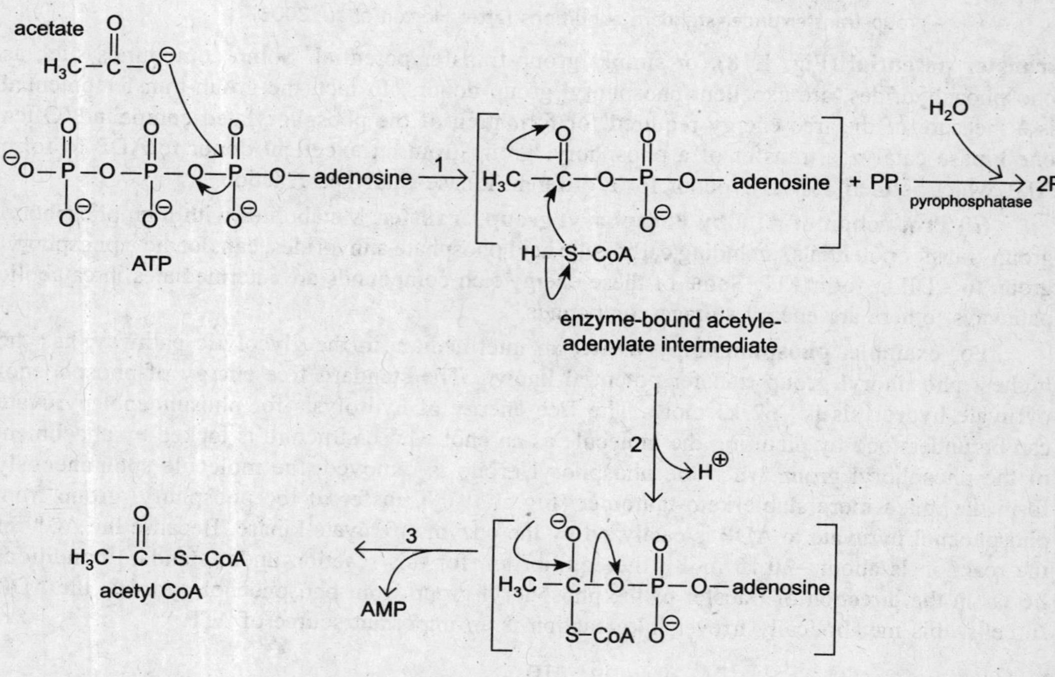

Fig. 1.20. Synthesis of acetyl CoA from acetate, catalyzed by acetyl-CoA synthetase.

(PP_i) is released in this step. The reactive intermediate is shielded from nonenzymatic hydrolysis by tight binding within the active site of the enzyme. The reaction is completed by transfer of the acetyl group to the nucleophilic sulphur atom of coenzyme A, leading to the formation of acetyl CoA and AMP.

The synthesis of acetyl CoA also illustrates how the removal of a product can cause a metabolic reaction toward completion.

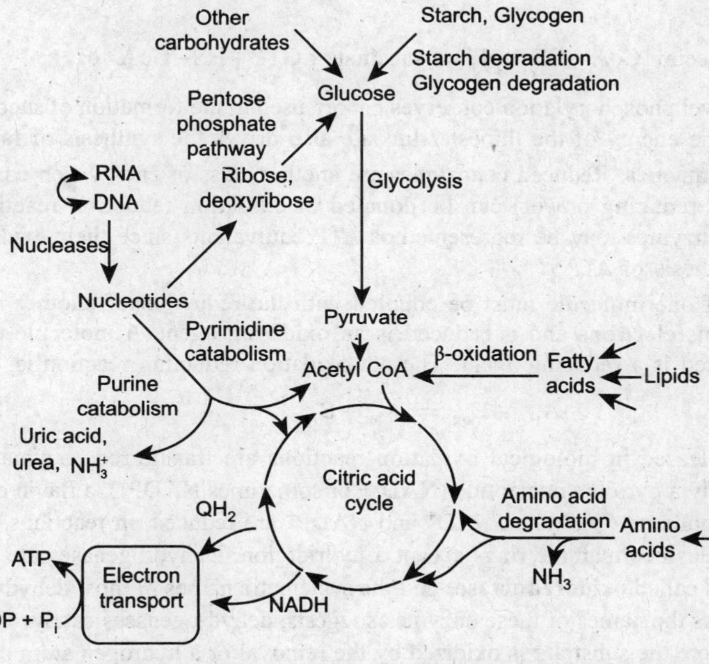

Fig. 1.21. Overview of catabolic pathways. Amino acids, nucleotides, monosaccharides and fatty acids are formed by enzymatic hydrolysis of their respective polymers. They are then degraded in oxidative reactions and energy is conserved in ATP and reduced coenzymes (mostly NADH) (after Horton *et al.*, 2006).

1.5 REDUCING EQUIVALENTS AND COFACTORS

1. Thioesters. Thioesters are another class of energy-rich compounds. **Acetyl CoA,** which occupies a central position in metabolism is one example of thioesters (Fig. 1.21). The high energy of thioesters can be used in generating ATP equivalents or in transferring the acyl groups to acceptor molecules. Acyl groups are attached to coenzyme A (or acyl carrier protein) via a thioester linkage:

$$R-\overset{\overset{\displaystyle O}{\|}}{C}-S-\text{Coenzyme A}$$

Thioesters resemble oxygen-acid anhydrides in reactivity. Sulphur is in the same group of the periodic table as oxygen; however, thioesters are less stable than oxygen esters because the unshared electrons of the sulphur atom are not as effectively delocalized in a thioester as the unshared electrons in an oxygen ester. The energy associated with the thioester linkage is similar to the energy of each of the phosphoanhydride linkages of ATP. The standard Gibbs free energy change for hydrolysis of acetyl CoA is –31 kJ mol^{-1}.

$$\text{Acetyl-CoA} \xrightarrow[\quad]{H_2O \quad \text{HS-CoA}} \text{Acetone}^- + H^+$$

Despite its high energy of hydrolysis, a CoA thioester resists nonenzymatic hydrolysis at neutral pH values. In other words, it is kinetically stable in the absence of appropriate catalysts.

The high energy of a CoA thioester is used in the fifth step of the citric acid cycle (or Krebs cycle), when the thioester succinyl CoA reacts with GDP (sometimes ADP) and P_i to form GTP (or ATP).

$$\text{Succinyl CoA} + \text{GDP} + P_i \rightleftharpoons \text{succinate} + \text{GTP} + \text{HS} - \text{CoA}$$

This substrate-level phosphorylation conserves energy used in the formation of succinyl CoA as ATP equivalents. The energy of the thioester linkage also drives the synthesis of fatty acids.

2. Reduced Coenzymes. Reduced coenzymes are another class of energy-rich compounds. Their high energy (or reducing power) can be donated in oxidation-reduction reactions. The energy of reduced coenzymes may be represented as ATP equivalents since their oxidation can be coupled to the synthesis of ATP.

The oxidation of one molecule must be coupled with the reduction of another molecule. A molecule that accepts electrons and is reduced is an oxidizing agent. A molecule that loses electrons and is oxidized is a reducing agent. The net oxidation reduction reaction is

$$A_{red} + B_{ox} \rightleftharpoons A_{ox} + B_{red}$$

The electrons released in biological oxidation reactions are transferred enzymatically to oxidizing agents, usually a pyridine nucleotide (NAD + or sometimes $NADP^+$), a flavin coenzyme (FMN or FAD) or ubiquinone (Q). When NAD^+ and $NADP^+$ are reduced, in reactions catalyzed by dehydrogenases, their nicotinamide rings accept a hydride ion. Dehydrogenases are members of the class of enzymes called **oxidoreductases** and the systematic names of most dehydrogenases refer to oxireductase. As the names of these enzymes suggests, dehydrogenases catalyze oxidation reduction reactions where the substrate is oxidized by the removal or a hydrogen atom of hydride ion. One electron is lost when a hydrogen atom (composed of one proton and one electron) is removed and two electrons are lost when a hydride ions (composed of one proton and two electrons) is removed. Remember that oxidation is loss of electrons.

NADH and NADPH, along with QH_2 supply reducing power. $FMNH_2$ and $FADH_2$ are reduced enzyme-bound intermediates in some oxidation reactions. The reduced coenzymes NADH and NADPH have similar structures and play similar roles in oxidation-reduction reactions.

1.6 REGULATORY MECHANISMS OF INTERMEDIATE MECHANISMS

Metabolic pathways are **regulated** by rapid mechanisms affecting the activity of existing enzymes, *e.g.,* allosteric and covalent modification. The latter is often initiated by the action of hormones. Hormones also regulate by longer term mechanisms, by promotion or inhibition of the synthesis of enzyme protein through changes in gene expression.

1.7 BASAL METABOLISM

Even when a person seems completely inactive, as in sleep, energy consuming vital functions, such as breathing, heart beat, peristalsis, secretion, etc., go on. However, the rate of oxidation is greatly reduced because the need for energy is much less due to little muscular work and greatly reduced mental activity. The amount of energy expended by an awake person just to maintain the vital functions at room temperature when at complete physical and mental rest and no food is being digested (about 12 hours after a meal) is called **basal metabolism.**

The energy requirements of the body are expressed quantitatively in terms of calories. It is determined by measuring oxygen consumption of a fasting person at complete rest and at 20°C over a short period of time. The basal metabolism of an average adult human male needs about 1600 kilo-calories per day (kcal/day). This is called the **basal metabolic rate (BMR)** of a person. This means that the food which can provide 1600 kilocalories of available chemical energy should be taken daily just to maintain the vital functions of body. The BMR is slightly (about 5%) less for an adult human female. The chemical reactions occur more rapidly at higher temperature, and the basal metabolic rate (BMR) increases about 5% for each degree the body temperature rises. This is one reason why the weight of body is lost during fever. Another reason is that an unwell person takes less food.

QUESTIONS

Long Answer Questions

1. What are important functional groups? Give the characteristic chemical reactions.
2. Give an account of membrane transport proteins.

Short Answer Questions

1. Give an account of shuttle system for NADH transport in mitochondria.
2. Write short notes on the following:
   (i) Aquaporins;
   (ii) Ion channels;
   (iii) ATP dependent transporters;
   (iv) ABC transporters;
   (v) ATP as energy currency.

Very short Answer Questions

1. What is metabolism?
2. Define intermediary metabolism.
3. Give an example of an anabolic metabolic reaction.
4. Give an example of catabolic metabolic reaction.
5. Give an example of amphibolic metabolic pathway.
6. What is the structural feature common to ATP, FAD, NAD^+ and CoA?
7. Distinguish between anabolic and catabolic reactions.
8. How is the energy generated during metabolic processes usually stored for later use?
9. Almost all cell contain the enzyme inorganic pyrophosphatase, which catalyzes the hydrolysis of PPi to Pi. What effect does the presence of this enzyme has on the synthesis of acetyl CoA?
10. Name two electron shuttles of inner mitochondrial membrane.
11. What is SAM?
12. Give an example of nucleotide sugar?
13. What is function of P-type ATPase?
14. What is phosphogen?
15. Define the coenzyme.
16. What is cofactor?

Fill in the blanks

1. ________ mitochondrial membrane has ______ shuttles for transporting ______ of cytoplasm produced during glycolysis.
2. ________ shuttle operates in cells of skeletal muscle and brain.
3. ______ shuttle operates in cells of liver, kidney and heart.
4. There occur about 12 types of _______ of aquaporins (AQPs).
5. Example of uniporter solute carrier is __________.
6. __________ transport two or more molecule in the same direction.
7. Antiporters are called _________ or ________.
8. V type _______ occurs in membrane of endosomes and lysosomes.
9. ________ (or exocytosis) occurs in plasma ______ secreting immunoglobulins.
10. ________ (or exocytosis) occurs in neurons, endocrine cells and pancreatic acinar cells.

Multiple choice Questions

1. Oxidation and reduction
   (a) are defined as loss of electrons
   (b) are both endergonic reactions
   (c) always occur together
   (d) proceed only under aerobic conditions
2. ATP synthesis occurs in
   (a) chloroplasts
   (b) mitochondria
   (c) chloroplasts and mitochria
   (d) all cell organelles
3. The role played by ATP in biochemical reaction is that of
   (a) a reducing agent
   (b) a coenzyme
   (c) an energy-donar substance
   (d) an energy-donar substance or a coenzyme

ANSWERS

Very Short Answer Questions

1. Metabolism is the entire spectrum of chemical reactions occurring in the living organism.
2. The fate of dietary components after digestion and absorption forms intermediary metabolism *i.e.* entire range of catabolic anabolic reactions in animals. It does not include metabolism of nucleic acids.
3. Protein synthesis.
4. Digestion and respiration (glycolysis, glycogenolysis, etc).
5. Krebs cycle or citric acid cycle.
6. An ADP unit.
7. An anabolic reaction consumes ATP (*i.e.*, it is an endergonic reaction); a catabolic reaction generates ATP (*i.e.*, it is an exergonic reaction).

8. In ATP and other compounds containing high-energy phosphate, and in glycogen.
9. Inorganic pyrophosphatase enzyme catalyzes the hydrolysis of pyrophosphate and drives the net reaction toward the synthesis of acetyl-CoA.
10. Malate-aspartate shuttle and glycerol- phosphate shuttle.
11. SAM stands for S-adenosyl methionine and it is a coenzyme. It causes methylation of phospholipids, proteins, DNA and RNA.
12. Uridine diphosphate glucose (UDP-glucose).
13. With the hydrolysis of each ATP by P-type ATP-ase 3 Na^+ ions are transported out of the cell and $2K^+$ ions are transported into the cell.
14. Phosphagen is a high-energy phosphate-storage molecule found in animal muscle cells. Phosphagens are phosphoamides and have a higher phosphoryl-group-transfer potential than ATP.
15. Coenzyme is organic molecule required by an enzyme for full activity. Coenzymes can be further classified as cosubstrates and prosthatic groups.
16. Cofactor is an inorganic ion or organic molecule required by an apoenzyme to convert it to a holoenzyme. There are two types of cofactors: essential ions and coenzymes.

Fill in the Blanks

1. Inner, two, NADH.
2. Glycerol-phoshate.
3. Malate-aspartate.
4. Isoforms.
5. GLUT2.
6. Symports.
7. Exchanger, counter transporter.
8. H^+ ATPase.
9. Constitutive secretion.
10. Regulated secretion.

Multiple Choice Questions

1. (c) 2. (c) 3. (d).

2

Carbohydrate Metabolism

2.1 DEGRADATION OF GLUCOSE

Glucose is degraded to provide energy (ATP) in five major phases: 1. Glycolysis; 2. Oxidation of pyruvate; 3. Krebs cycle; 4. Transfer of electrons over the ETS (= Electron Transport Chain) and 5. Chemiosmotic ATP synthesis. Glycolysis occurs in the cytoplasm and is an anaerobic process; whereas krebs cycle and other processes take place in the mitochondria and are aerobic see (Table 2.1). The product of one process forms the fuel for the next process.

Glycolysis or Embden-Meyerhof Pathways

In the process of glycolysis (Gr. *glykys-sweet, lysis*-destroy) the six carbon chain glucose or molecule is hydrolysed into two molecules of 3 carbon chain pyruvic acid molecule (Box 2.1). The hydrolysis of glucose takes place in various steps and each step is catalyzed by a specific enzyme. These steps have been worked out in 1930s by two German biochemists, **Gustav Embden** and **Otto Meyerhof** and therefore, known as **Embden-Meyerhof pathways** after the name of its discoverers. The Embden-Meyerhof pathways takes place in the cytoplasm and it does not require any amount of oxygen.

Box 2.1

At cellular pH, the organic acids, *e.g.*, pyruvic acid, lactic acid, etc., dissociate, losing hydrogen ion from carboxyl group, forming pyruvate ion, lactate ion, etc., there useful form in the cell.

Glycolysis involves a single oxidation and releases only a small amount of chemical energy to the cell in the form of ATP. Certain bacteria and some invertebrate parasites depend entirely on glycolysis for their energy requirement. The skeletal muscle also depend heavily on glycolysis for energy during excessive activity when oxygen supply becomes insufficient for aerobic respiration.

Glycolysis consists of following nine steps (Fig. 2.l): In first three steps of glycolysis two molecules of ATP are used to provide glucose energy so that the activation energy barrier is lowered.

1. Phosphorylation. The third phosphate group separates from adenosine triphosphate (ATP) molecule converting the latter in adenosine diphosphate (ADP) and releasing energy. With the help of this energy, the phosphate group combines with glucose to form glucose 6-phosphate. The reaction is catalyzed by the enzyme, *hexokinase.* Thus, a molecule of ATP is consumed in this step of glycolysis.

2. Rearrangement (Isomerisation). Glucose 6-phosphate is changed into its isomer fructose 6-phosphate by rearrangement. The rearrangement is catalyzed by an enzyme, *phosphoglucoisomerase.*

3. Phosphorylation. Fructose 6-phosphate combines with another phosphate group from another ATP molecule, yielding fructose 1, 6-diphosphate and ADP. This reaction is catalyzed by an enzyme *phosphofructokinase* and consumes another molecule of ATP. Phosphorylation reactions activate the sugar and prevents its escape from the cell.

4. Splitting and Interconversion. Fructose 1, 6-diphosphate now splits into two 3-carbon phosphorylated sugars (triose phosphates): dihydroxyacetone phosphate (DHAP) and 3-phosphoglyceraldehyde (3-PGAL) or glyceraldehyde 3-phosphate. The reaction is catalyzed by an enzyme *aldolase.* DHAP is converted to PGAL with the help of an enzymes *triose phosphate isomerase.* The fate of each 3 PGAL molecule is the same as discussed below.

5. Phosphorylation and Dehydrogenation. In phosphorylation, 3-phosphoglyceraldehyde combines with a phosphate group derived from inorganic phosphoric acid (H_3PO_4) found in the cytosol, not from ATP, forming 1, 3-diphosphoglycerate (or diphosphoglyceric acid). The reaction is catalyzed by a specific enzyme.

(*a*) In *dehydrogenation,* a pair of hydrogen atoms separate from 3-phosphoglyceraldehyde, Their separation releases a large amount of energy. A part of this energy is stored in newly formed phosphate bond of 1, 3-diphosphoglycerate, making it a high energy bond. Separation of hydrogen is catalyzed by an enzyme, *phosphoglyceraldehyde dehydrogenase.*

(*b*) As stated above, two hydrogen (H) atoms (*i.e.,* 2 protons and 2 electrons) separate from 3-phosphoglyceraldehyde. Of these, one complete hydrogen atom (proton and electron) and one additional electron are picked up by NAD^+ which gets reduced to NADH. The remaining one hydrogen proton or ion (H^+) remains free in the cytosol.

$$2H^+ + 2e^- + NAD^+ \rightarrow NADH + H^+$$

NADH is a high energy compound, carrying the rest of the energy released by the separation of hydrogen atoms from 3-PGAL. The energy is actually released by transfer of electrons from 3-PGAL to NAD. The NADH provides energy to convert ADP to ATP by passing electrons over the electron transmitter system (ETS) if oxygen is available. The overall reaction is:

$$\text{3-PGAL} + NAD^+ + P_i^{2+} \rightarrow \text{1,3-diphosphoglycerate} + NADH + H^+$$

6. Dephosphorylation. High-energy phosphate group on carbon 1 of 1,3-diphosphoglycerate is transferred to a molecule of ADP, converting it into an ATP molecule. The 1,3-diphosphoglycerate changes to 3-phosphoglycerate due to loss of a phosphate group. The reaction is catalyzed by an enzyme, *diphosphoglycerokinase.*

$$\text{1,3-diphosphoglycerate} + ADP \rightarrow \text{3-phosphoglycerate} + ATP$$

7. Rearrangement (Intramolecular phosphoryl-group transfer). The phosphate group on the third carbon of 3-phosphoglycerate shifts to the second carbon. This change is aided by the enzyme *phosphoglyceromutase.* Metabolic product of this reaction is **2-phosphoglycerate.**

8. Dehydration. The 2-phosphoglycerate loses water in the presence of an enzyme, *enolase,* and changes into phosphoenol pyruvate. The latter undergoes molecular rearrangement that transforms its phosphate group into a high-energy phosphate bond.

9. Dephosphorylation. High-energy phosphate group of phosphoenol pyruvate is transferred to a molecule of ADP with the help of an enzyme *pyruvate kinase.* This produces the pyruvate (or pyruvic acid) and a molecule of ATP.

All the enzymes, reactants, intermediates and the products of glycolysis are dissolved in the cytosol. Their interaction depends on random collusion brought about by kinetic movements.

The net energy yield of chain reactions of glycolysis is the production of two ATP molecules from one molecule of glucose. Under aerobic conditions, the end products of glycolysis are pyruvic acid and reduced coenzyme NAD (*i.e.,* NADH). NADH carries two electrons, taken from glyceraldehyde-3-phosphate and contains little energy. However, under anaerobic conditions, pyruvic acid remains in the cytosol (cytoplasmic matrix) and is used as a hydrogen acceptor and converted into lactic acid:

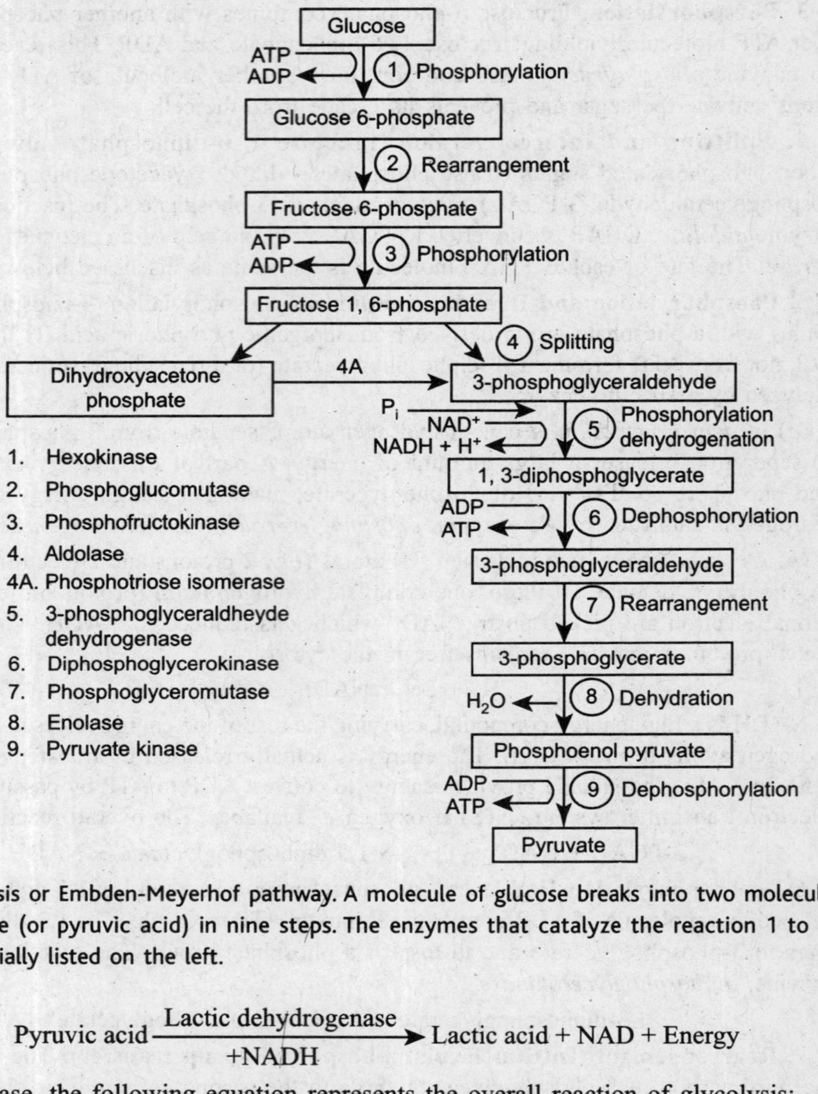

Fig. 2.1. Glycolysis or Embden-Meyerhof pathway. A molecule of glucose breaks into two molecules of pyruvate (or pyruvic acid) in nine steps. The enzymes that catalyze the reaction 1 to 9 are sequentially listed on the left.

$$\text{Pyruvic acid} \xrightarrow[\text{+NADH}]{\text{Lactic dehydrogenase}} \text{Lactic acid + NAD + Energy}$$

In above case, the following equation represents the overall reaction of glycolysis:

$$\underset{\text{Glucose}}{C_6H_{12}O_6} + P_i + 2ADP \rightarrow \underset{\text{Lactic acid}}{2C_3H_6O_3} + 2ATP + H_2O$$

Oxidative Decarboxylation

In aerobic organisms, since pyruvic acid still contains a larger amount of energy, it must undergo further degradation, but this time inside the mitochondria. This is done in three consecutive steps: **oxidative decarboxylation** (removal of carboxyl or—COOH group), **Krebs cycle** and **oxidative phosphorylation**. Pyruvic acid directly enters the mitochondrial matrix (Box 2.2) and is converted into **acetyl-CoA** by the help of a huge enzyme, called **pyruvic acid dehydrogenase** (Box 2.3; Fig. 2.2). The two NADH molecules (which are generated during glycolysis) cannot penetrate directly into the mitochondria, so their electrons are transferred to **dihydroxyacetone phosphate**, which shuttles them into the mitochondria. This process utilizes, one ATP molecule for each NADH; in all two ATP molecules are consumed for two NADH molecules. When both of these NADH pass through ETS, they tend to generate 6 ATP molecules.

Box 2.2

Pyruvate passes through outer mitochondrial membrane via aqueous channels formed by transmembrane proteins, called **porins**. However, in order to pass through inner membrane of mitochondria a specific transport protein in required for most metabolites. **Pyruvate translocase** specifically transports pyruvate in **symport** with H^+. Once inside the mitochondria, pyruvate can be converted to acetyl CoA and CO_2.

Box 2.3

The size of pyruvate dehydrogenase complex is enormous. It is several times bigger than that a ribosome. It is found in the matrix of mitochondria and chloroplasts.

The pyruvate dehydrogenase complex is a multienzyme complex containing multiple copies of three distinct enzymatic activities: 1. pyruvate lipoamide 2-oxidoreductase (E_1); 2. acetyl CoA: dihydrolipoamide 5-acetyltransferase (E_2) and 3. dihydrolipoamide: NAD^+ oxidoreductase (E3) (see **Horton** *et al.*, 2006).

Pyruvic acid dehydrogenase and its action. Sometimes two enzymes that catalyze sequential reactions form an **enzyme complex** and the product of the first enzyme does not have to diffuse through the cytoplasm to encounter the second enzyme. The second reaction begins as soon as the first is over. Some large enzyme (multienzyme) aggregates carry out whole series of reactions without losing contact with the substrate. For example, the conversion of pyruvic acid to acetyl CoA proceeds in *three* chemical steps, all of which take place on the same large multienzyme complex *(i.e.,* pyruvate dehydrogenase). Pyruvate dehydrogenase occurs in the mitochondrial matrix and is larger than a ribosome in size. It contains multiple copies of three enzymes namely **pyruvic acid dehydrogenase** or **E_1, dihydrolipoyl transacetylase** (or E_2) and **dihydrolipoyl dehydrogenase** (or E_3). It also contains five coenzymes (*e.g.,* NAD, coenzyme A, etc.) and **two regulatory proteins** (*e.g.,* protein kinase and protein phosphatase; both regulating the activity of pyruvic acid dehydrogenase, turning it off whenever ATP levels are high).

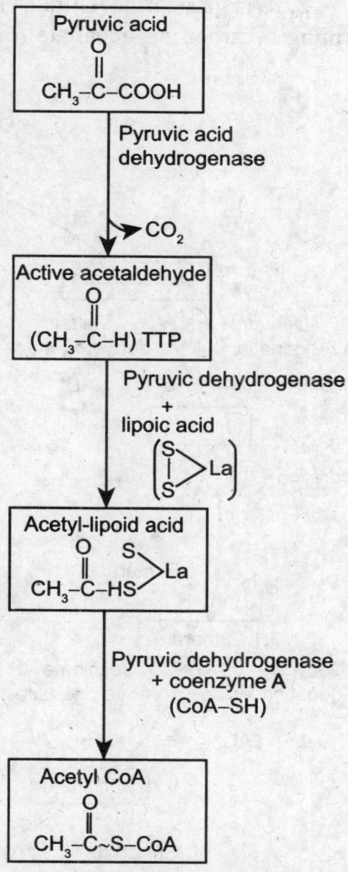

Fig. 2.2. The oxidative decarboxylation of pyrurvic acid to acetyl coenzyme A.

Thus, during oxidative decarboxylation of one molecule of pyruvic acid, one mole of CO_2 is produced and one NAD is reduced to NADH. The end product of this reaction is a 2-carbon compound, the **acetyl group** which is attached to coenzyme A to produce the carrier molecule, called **acetyl CoA.**

Acetyl-CoA is often called **active acetate** because it is a central intermediate in metabolism. It may be used in the synthesis of fatty acids or some amino acids. It can also be oxidized further in the cell's energy metabolism.

Krebs Cycle or Citric Acid Cycle

This cycle has been named after the German biochemist in England **Sir Hans Adolph Krebs** (1900–1981) who discovered it in 1937. He won Nobel Prize for this work in 1953. Krebs cycle is also called **citric acid cycle** after one of the participating compounds.

Krebs cycle takes place in the mitochondrial matrix. It produces a high-energy phosphate GTP, and forms reduced coenzymes NADH and FADH. The reduced coenzymes yield more high-energy phosphate ATP in ETS (Electron transport/transmitter system). Krebs cycle also receives the intermediates of fat and protein oxidation. Some intermediates of this cycle are used by the cell for synthesizing amino acids or lipids, when required. Krebs cycle is regarded as the central pathway of the cell, because the catabolic pathway converge upon it and the anabolic pathways diverge from it.

Eight Cyclic Steps of Krebs Cycle (Fig. 2.3)

1. Condensation. Acetyl-CoA reacts in the presence of water with the 4 carbon oxaloacetate normally present in the cell, forming 6 carbon citrate (citric acid) and freeing coenzyme A (Or CoA) for reuse in the pyruvate oxidation. The high energy bond of acetyl coenzyme A provides the energy for this condensation reaction. The reaction is catalyzed by the *citrate synthetase/ synthase* enzyme. The citrate has three carboxyl groups. Hence, the Krebs cycle is also called **tricarboxylic acid cycle** or **TCA cycle.**

2. Reorganization. Citrate undergoes reorganization in the presence of an enzyme, *aconitase,* forming 6 carbon cis-aconitate (or cis-aconitic acid) and releasing water.

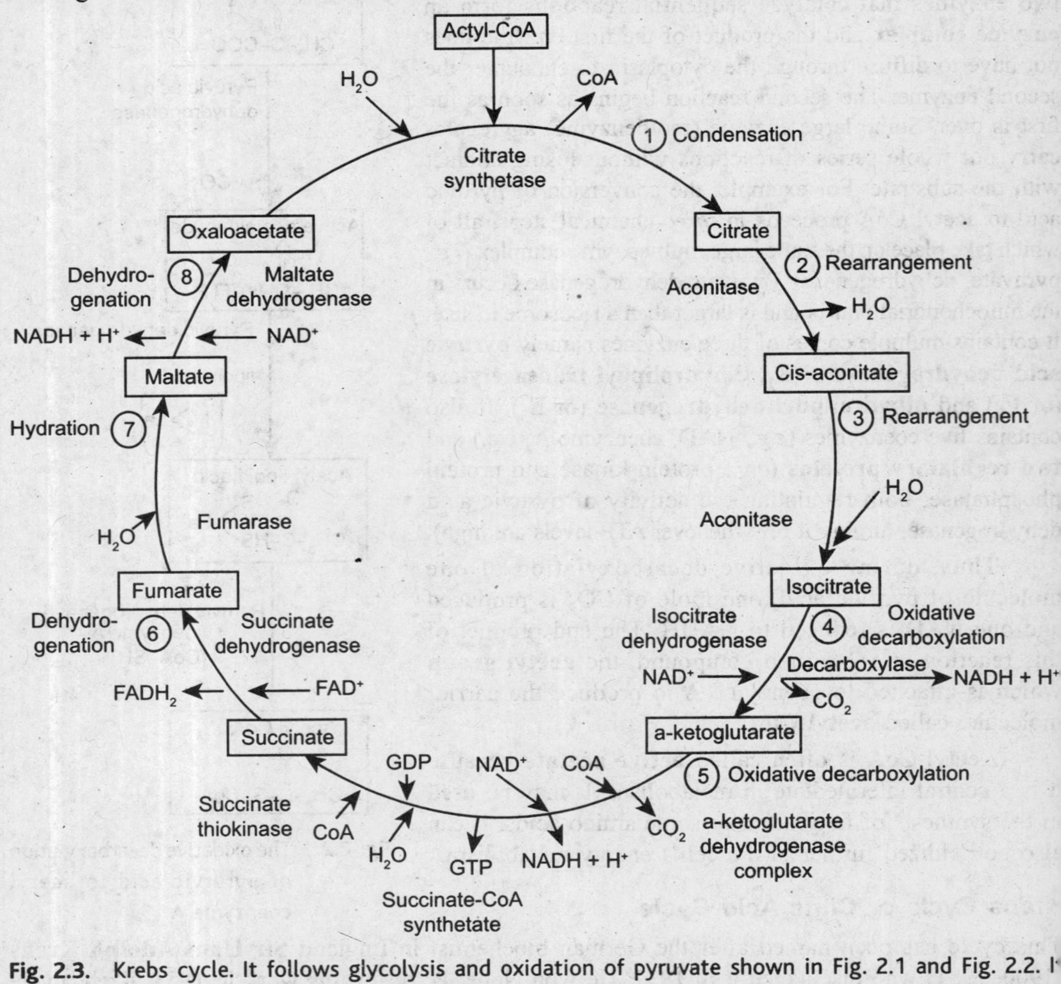

Fig. 2.3. Krebs cycle. It follows glycolysis and oxidation of pyruvate shown in Fig. 2.1 and Fig. 2.2. It involves two processes: (*i*) removal of hydrogen and (*ii*) breaking of carbon dioxide units one by one

3. Reorganization. Cis-aconitate is further reorganized into 6 carbon isocitrate (or iso-citric acid) by the enzyme, *aconitase* with the addition of water.

4. First Oxidative Decarboxylation. Isocitrate gives off a pair of hydrogen atoms (oxidation) and a molecule of CO_2 (decarboxylation) and becomes 5 carbon α-ketoglutarate (or α-ketoglutaric acid). Oxygen to oxidise a carbon atom to carbon dioxide is taken from a water molecule. The pair of hydrogen atoms gives two electrons and one H^+ to NAD', forming NADH + H^+. The enzyme *isocitrate dehydrogenase* catalyzes a reaction. NADH generates ATP by transferring its electrons over the ETS.

The α-ketoglutarate is an important Krebs cycle intermediate. This molecule may remain in this cycle or enter biosynthetic pathway forming amino acids.

5. Second Oxidative Decarboxylation. This is two stage process:

(a) In **stage l,** coenzyme A reacts with α-ketoglutarate, forming 4 carbon succinyl-coenzyme A and releasing carbon dioxide and a pair of hydrogen atoms. The reaction is catalyzed by a *α-ketoglutarate dehydrogenase enzyme* (Box 2.4). Oxygen to oxidize a carbon atom to CO_2 is taken from a molecule of water. The pair of hydrogen atoms pass two electrons and one H^+ to NAD^+, forming NADH + H^+ (Fig. 2.3).

Box 2.4

The α-*ketoglutarate dehydrogenase* enzyme is a large complex that resembles *pyruvate dehydrogenase* in both structure and function. The same coenzymes are involved and reaction mechanism is the same. The *three* component enzymes of the α-ketoglutarate dehydrogenase complex are 1. α-*ketoglutarate dehydrogenase* (E_1, containing TPP or thiamine pyrophosphate which is a coenzyme); 2. *dihydrolipoamide succinyl transferase* (E_2, containing a lipomide swinging arm) and 3. *dihydrolipoamide dehydrogenase* (E_3, NAD^+), (see **Horton** *et al.*, 2006).

(b) In the **stage 2**, succinyl-coenzyme A splits into 4 carbon succinate and coenzyme A with the addition of water. The coenzyme A transfers its high energy (located at sulphur bond) to a phosphate group that joins GDP (guanosine diphosphate) forming GTP (guanosine triphosphate). The latter is an energy carrier like ATP. It is used in certain biosynthetic reactions. GTP is the only high energy phosphate produced in the Krebs cycle. The stage 2 reaction is catalyzed by *succinyl-CoA synthase* enzyme.

6. **Dehydrogenation.** This process converts succinate (or succinic acid) into 4 carbon fumarate with the help of an enzyme *succinate dehydrogenase* and liberates a pair of hydrogen atoms. The latter pass to FAD^+ (flavin adenine dinucleotide) forming $FADH_2$. Hydrogen is carried by FAD in the form of whole atoms.

Succinate and fumarate are important intermediates. They may be converted into some amino acids or used in the synthesis of certain hormones and other materials.

7. **Hydration.** This process restores oxaloacetate (or oxaloacetic acids) by removing a pair of hydrogen atoms from malate with the help of an enzyme, *malate* (or *malic) dehydrogenase.* The pair of hydrogen atoms pass two electrons and one H^+ to NAD^+, forming NADH + H^+.

Oxaloacetate combines with acetyl-coenzyme A to form citrate (or citric acid) and so the Krebs cycle continues.

All the enzymes, reactants, intermediates and products of the TCA cycle (or Krebs cycle) also are found in the aqueous solution in the matrix of mitochondria, except the enzymes *α-ketoglutarate dehydrogenase* and *succinate dehydrogenase* which are located in the inner mitochondrial membrane.

The entire Krebs cycle may be represented in the following equation:

$$2CH_3CO \sim S - CoA + 8H_2O + 6NAD^+ + 2FAD^+ + 2FAD^+ + 2GDP + 2H_3PO_4^{2+} \rightarrow$$

$$4CO_2 + 2H_2O + 6NADH + 2\ FADH_2 + GTP + 6H^+$$

Table 2.1 Differences between glycolysis and Krebs cycle.

Glycolysis	*Krebs Cycle*
1. It takes place in the cytoplasm.	1. It takes place in the matrix of mitochondria.
2. It occurs in aerobic as well as anaerobic respiration.	2. It occurs in aerobic respiration only.
3. It consists of 9 steps.	3. It consists of 8 steps.
4. It is linear pathway.	4. It is a cyclic pathway.
5. It oxidizes glucose partly, producing pyruvic acid (or pyruvate).	5. It oxidizes acetyl coenzyme A fully.
6. It consumes 2ATP molecules.	6. It does not consume ATP.
7. It generates 2ATP molecule net from one glucose molecule.	7. It generates 2GTP molecules from 2 acetyl coenzyme molecule.
8. It yields 2NADH per glucose molecule.	8. It yields 8NADH molecules and 2FADH molecules.
9. It does not produce CO_2.	9. It produces CO_2.
10. All the enzymes catalyzing glycolytic reactions are dissolved in cytosol.	10. Two enzymes of Krebs cycle reactions (i.e. α ketoglutarate dehydrogenase and succinate dehydrogenase) are located in the inner mitochondrial membrane, all other are dissolved in the matrix.

2.2 HEXOSE MONOPHOSPHATE (HMP) SHUNT OR PENTOSE PHOSPHATE PATHWAY

Whole of the glucose is not oxidized by way of glycolysis and citric acid cycle (or Krebs cycle). A good part of it is oxidized by way of pentose phosphate pathway or **Warburg-Dickens pathway.** This pathway occurs mainly in the liver, adipose tissue, lactating mammary glands, adrenal cortex, gonads and erythrocytes. Pentose phosphate pathway takes place in the cytoplasm (Fig. 2.4).

The complete pentose phosphate pathway has two stages: an **oxidative stage** and a **nonoxidative stage.** In the oxidative stage, NADPH is produced along with ribulose 5- phosphate. The nonoxidative stage disposes of the pentose phosphate formed in oxidative steps by providing a route to glycolysis.

Oxidative Stage of HMP Shunt. It includes following 5 steps:

1. Dehydrogenation. Glucose 6-phosphate undergoes dehydrogenation at carbon 1 and changes into **6-phosphogluconolactone.** The reaction is catalyzed by enzyme *glucose 6-phosphate dehydrogenase*.

The coenzyme $NADP^+$ acts as a hydrogen acceptor and changes to NADPH + H^+. Divalent cations such as calcium (Ca^{2+}) or magnesium (Mg^{2+}) are needed as activators.

2. Hydration. The 6-phosphogluconolactone is changed to **6-phosphogluconic acid** with the addition of a water molecule by enzyme *glucolactone hydrolase.*

3. Dehydrogenation. The 6-phosphogluconic acid now undergoes dehydrogenation at carbon 3 and changes to **3-keto-6-phosphogluconic acid.** The reaction is catalyzed by enzyme *phosphogluconate dehydrogenase.* Another molecule of coenzyme $NADP^+$ accepts a pair of hydrogen atoms and becomes reduced to NADPH + H^+. Calcium (Ca^{2+}), magnesium (Mg^{2+}) or manganese (Mn^{2+}) ions act as activators.

4. Decarboxylation. 3-keto-6-phosphogluconic acid loses first carbon as carbon dioxide and changes to **ribulose 5-phosphate.** The reaction is catalyzed by enzyme *decarboxylase.*

5. Isomerization. The ribulose 5-phosphate is isomerized to **ribose 5-phosphate** by the enzyme *phosphopentose epimerase.*

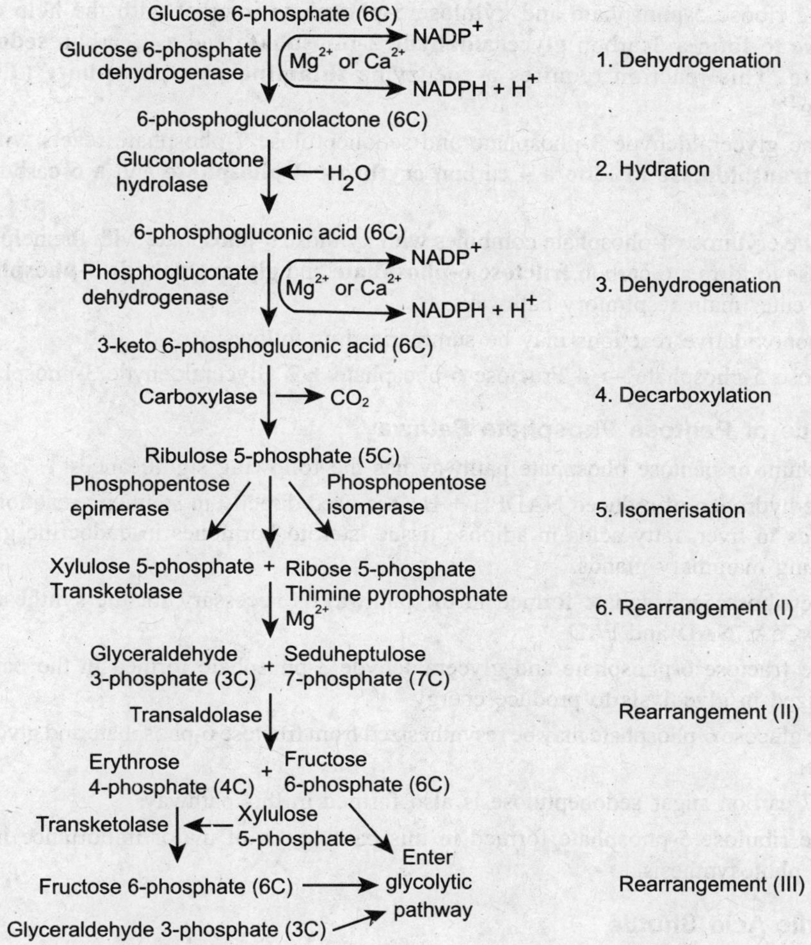

Fig. 2.4. Hexose monophosphate shut or pentose phosphate pathway.

The ribulose 5-phosphate may also be isomerized to **xylulose 5-phosphate** by the enzyme *phosphopentose epimerase.*

The reactions of oxidative stage of pentose phosphate pathway may be summed up as follows:

$$\text{6-Glucose 6-phosphate} + 12\text{NADP} + 6H_2O \rightarrow$$
$$\text{6-Ribose 5-phosphate} + 12\text{NADPH}^+ + 12H^+ + 6CO_2$$

Energetics of HMP Shunt. NADPH + H^+ formed in cytoplasm in HMP shunt reacts with NAD present in the mitochondria across the mitochondrial membrane, forming NADP and NADH + H^+. Twelve NADH molecules result from six glucose 6-phosphates. Twelve hydrogen pairs produce 36 (*i.e.*, 12 × 3) ATP molecules on transfer to oxygen over ETS. Six molecules of glucose produce via HMP shunt as much energy as produce by one glucose molecules via glycolysis and Krebs cycle.

Nonoxidative Stage of HMP Shunt–Splitting and Rearrangement

Nonoxidative reactions by splitting and rearranging the structure of pentose Phosphate produce sugars with 3, 4, 5, 6 or 7 carbon atoms without forming ATP or releasing CO_2. These reactions include the following:

(*i*) The ribose 5-phosphate and xylulose 5-phosphate interact with the help of enzyme *transketolase* to form a 3carbon **glyceraldehyde 3-phosphate** and a 7 carbon **sedoheptulose 7-phosphate.** This reaction requires a coenzyme **thiamine pyrophosphate (TPP)** and a cofactor Mg^{2+}.

(*ii*) The glyceraldehyde 3-phosphate and sedoheptulose 7-phosphate reacts with the help of enzyme **transaldolase** to form a 4 carbon **erythrose 4-phosphate** and a 6 carbon **fructose 6-phosphate.**

(*iii*) The erythrose 4-phosphate combines with xylulose 5-phosphate with the help of enzyme **transketolase** to form a 6 carbon **fructose 6-phosphate** and **glyceraldehyde 3-phosphate.** These compounds enter main respiratory pathways.

The nonoxidative reactions may be summarized as follows:

6 Ribose 5-phosphate → 4 Fructose 6-phosphate + 2 Glyceraldehyde 3-phosphate

Significance of Pentose Phosphate Pathway

The HMP shunt or pentose phosphate pathway has the following significances:

1. The hydrogen of reduced NADPH + H+ are used directly in synthetic reactions, forming phospholipids in liver, fatty acids in adipose tissue, steroid hormones in endocrine glands, milk fat in lactating mammary glands.

2. The ribose 5-phosphate formed in the pathway is necessary for the synthesis of RNA, DNA, ATP, CoA, NAD and FAD.

3. The fructose 6-phosphate and glyceraldehyde 3-phosphate formed in the pathway may be metabolized in glycolysis to produce energy.

4. The glucose 6-phosphate may be resynthesized from fructose 6-phosphate and glyceraldehyde 3-phosphate.

5. A 7 carbon sugar sedoheptulose is also formed in this pathway.

6. The ribulose 5-phosphate formed in this pathway is of major importance in reactions involved in photosynthesis.

Dicarboxylic Acid Shuttle

The dicarboxylic acid shuttle is a mechanism for reversing not last glycolytic reaction which is otherwise not reversible. It converts *pyruvic acid* to *phosphoenol pyruvic acid* having two carboxyl groups in the following manner:

1. Carboxylation. The pyruvic acid is first changed to **oxaloacetic acid** in the mitochondria by adding a CO_2 molecule in the presence of the enzyme **pyruvic carboxylase.** The reaction is energized by ATP:

$$\text{Pyruvic acid} \xrightarrow[\text{ATP} \quad \text{ADP} + P_i]{\text{Pyruvic carboxylase}} \text{Oxaloacetic acid}$$

All amino acids which are intermediates in Krebs cycle can give rise to oxaloacetic acid by running through citric acid cycle. Other amino acids which can form pyruvic acid can also produce oxaloacetic acid via citric acid cycle (Fig. 1.1). Aspartic acid directly forms oxaloacetic acid cycle.

2. Phosphorylation and Decarboxylation. Oxaloacetic acid is reduced to malic acid which diffuses out into the cytoplasm. Here it is reoxidized to oxaloacetic acid. The latter is then changed to **phosphoenol pyruvic acid** by the enzyme *phosphoenol pyruvic carboxvkinase* in the presence of GTP:

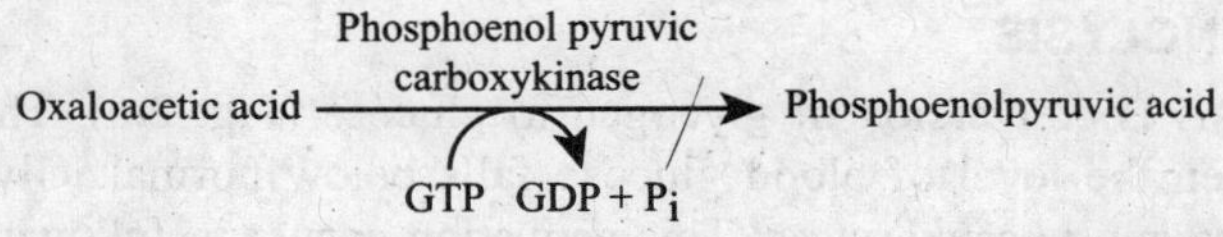

Significance. The dicarboxylic acid shuttle helps in gluconeogenesis. It brings amino acids into glycolytic pathway, which can form glucose by reversal of reactions.

2.3 GLUCONEOGENESIS

Gluconeogenesis is the formation of glucose from non-carbohydrate sources such as glucogenic amino acids, glycerol and lactic acid. It occurs when diet is deficient in carbohydrates, during starvation when dietary glucose is not available, and in diabetes mellitus when body excretes glucose in urine. It mainly takes place in the liver and to some extent in the cortex of kidneys. The starting point of the gluconeogenic pathway is pyruvic acid and the main points of entry into the pathway are pyruvic acid, oxaloacetic acid a 3-phosphoglyceraldehyde (Fig. 2.5).

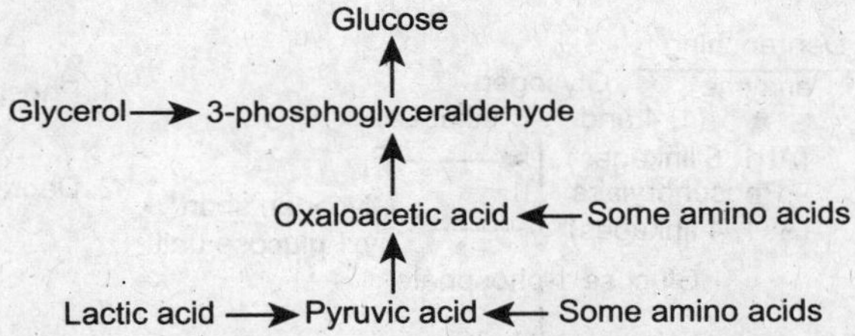

Fig. 2.5. Pathway of gluconeogenesis.

***(i)* Gluconeogenesis from Amino Acids.** The conversion of amino acids into phosphoenol pyruvic acid has been described in dicarboxylic acid shuttle. Phophoenol pyruvic acid is then processed by glycolytic reactions in reverse order upto the formation of fructose 1,6-diphosphate. The latter is changed to fructose 6-phosphate by an *enzyme fructose 1,6-diphosphatase.* This enzyme is present in liver, kidneys and striated muscle cells, but not in adipose tissue. Therefore, gluconeogenesis does not occur in adipose tissue. Fructose 6-phosphate is changed to glucose 6-phosphate by reversed action of the enzyme *phosphogluco-isomerase.* Glucose 6-phosphate is converted to glucose by an enzyme *glucose 6-phosphatase.* This enzyme is present in the liver and kidneys which add glucose to blood. It is absent in the muscle cells which cannot release glucose into the blood.

***(ii)* Gluconeogenesis from Glycerol.** Glycerol is formed by hydrolysis of fat. It is changed to dihydroxyacetone phosphate by phosphorylation and dehydrogenation with the help of an enzyme *glycerokinase* in the presence of ATP and NAD. Dihydroxyacetone phosphate is converted to 3-phosphoglyceraldehyde by an enzyme *phosphotriose isomerase.* The 3-phosphoglyceraldehyde is a normal intermediate in glycolytic pathway. It can be converted into glucose by some of the glucogenic reactions already mentioned for amino acids and by some reversed glycolytic reactions.

***(iii)* Gluconeogenesis from Lactic Acid.** Lactic acid formed in anaerobic respiration (fermentation) is changed to pyruvic acid with the help of an enzyme *lactate dehydrogenase.* Coenzyme NAD acts as hydrogen acceptor and becomes reduced to NAD + H^+. Pyruvic acid can then be processed for change into glucose as described above. It can also hydrolyze ATP (into ADP and phosphate) to pump protons in the opposite direction, if the electrochemical proton gradient is reduced.

Significance of Gluconeognesis. Gluconeognesis provides glucose to the body during hypoglycemia (decreased blood sugar) also clears the blood of metabolic products of tissues such as lactic acid formed in muscles and glycerol formed in adipose tissue.

2.4 GLYCOGENOLYSIS

Glycogenolysis is conversion of glycogen to glucose (Fig. 2.6). This biochemical event occurs in liver when the level of blood glucose falls below normal. Glycogenolysis requires activation of the enzyme *phosphorylase.* This activation occurs in following three steps:

(*i*) Adrenaline hormone activates the enzyme *adenyl cyclase.*

(ii) $\text{ATP} \xrightarrow{\text{Adenyl cyclase}} \text{Cyclic} - \text{AMP} + \text{PP}_i$

(*iii*) $\text{Inactive phosphorylase} \xrightarrow{\text{Cyclic-AMP}} \text{Active phosphorylase}$

Glycogenolysis then proceeds in following *four* steps:

1. Phosphorylation. The active *phosphorylase* enzyme in the presence of inorganic phosphate, breaks 1,4-glycosidic bond in glycogen molecule and removes glucose unit which is phosphorylated to **glucose 1-phosphate.** This makes glycogen molecule shorter by one glucose molecule.

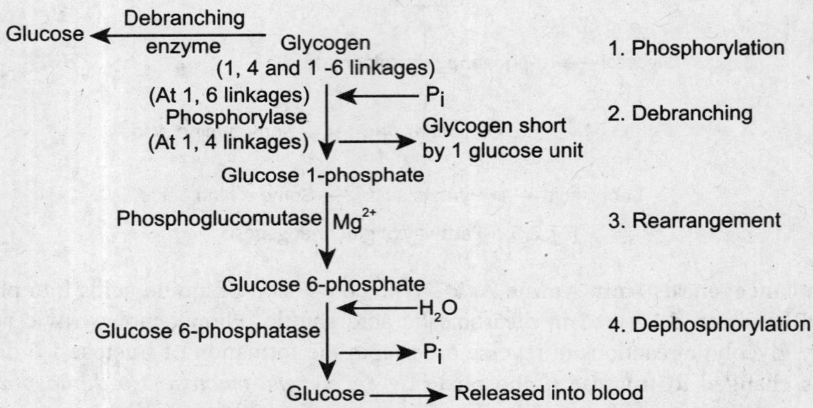

Fig. 2.6. Figure showing four vital steps of glycogenolysis in liver.

2. Debranching (Hydrolysis). A debranching enzyme *(amylo-1,6-glucosidase)* hydrolytically splits the glucose molecule from 1,6-linkage (branching point) and liberates **free glucose.**

3. Rearrangement. Glucose 1-phosphate is changed to **glucose 6-phosphate** by an enzyme *phosphoglucomutase* in the presence of cofactor Mg^{2+}.

4. Dephosphorylation. Glucose 6-phosphate is changed to **glucose** by an enzyme *glucose 6-phosphatase* in the liver with the release of phosphate. The glucose formed is released into the blood to raise its level. Hepatic glycogenolysis is stimulated by the hormones **epinephrine** and **glucagon.**

Muscle Glycogenolysis. In the skeletal muscle glycogenolysis stops at the formation of glucose 6-phosphate because there is no *glucose 6-phosphatase* enzyme in muscle cells. Therefore, the skeletal muscle cells do not release glucose into the blood. They can, however, indirectly utilise the glucose 6-phosphate, forming blood lactic acid which can then be converted to glucose in the liver (Cori cycle). Epinephrine alone stimulates muscle glycogenolysis.

2.5 GLYCOGENESIS

Glycogenesis is the formation of glycogen from glucose. It is a anabolic activity. It takes place mainly in the liver and muscle cells, but occurs in most other tissues also. Glycogen is a much-branched polymer of glucose units. Its synthesis requires the presence of a glycogen primer (*e.g.,* glycogenin protein; see Satyanaryana and Chakrapani, 2013) and activation of glucose,

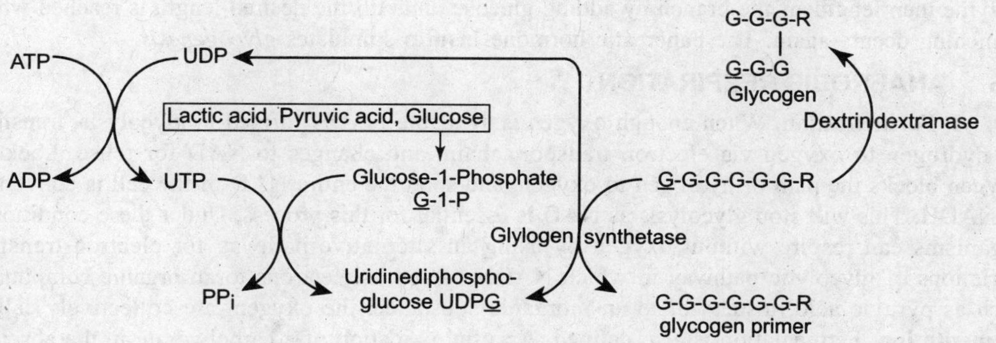

Fig. 2.7. Lengthening and branching of glycogen molecule.

i.e., formation of glucose-1-phosphate. The glucose units are added to the end of the chain as well as to glucose residues within the chain to initiate branch chains (Fig. 2.7).

Glycogenesis is accomplished in 5 steps (Fig. 2.8).

1. Phosphorylation. It takes place in presence of an enzyme **glucokinase**, a cofactor Mg^{2+} and energy-carrier ATP and produces **glucose 6-phosphate.**

Formation of glucose 6-phosphate is a **"locking device"** to keep glucose within the cells because the cell (plasma) membrane is permeable to glucose and not to glucose 6-phosphate.

2. Rearrangement. Glucose 6-phosphate is changed to **glucose 1-phosphate** with the help of an enzyme **phosphoglucomutase** and cofactor Mg^{2+}.

3. Dephosphorylation. Glucose 1-phosphate reacts with uridine triphosphate (UTP), forming **uridine diphosphate glucose (UDPG)** complex and pyrophosphate (PP_i). The reaction is catalyzed by an enzyme, *UDPG pyrophosphorylase.*

4. Condensation. Glucose of UDPG is transferred to a preformed glycogen molecule (a primer) by an enzyme *glycogen synthetase* (also called UDPG-glycogen glucosidase) with the release of uridine diphosphate. Carbon 1 of glucose molecule forms a glycosidic bond with the carbon 4 of the terminal glucose unit of the primer glycogen molecule, which thus becomes larger by one glucose unit.

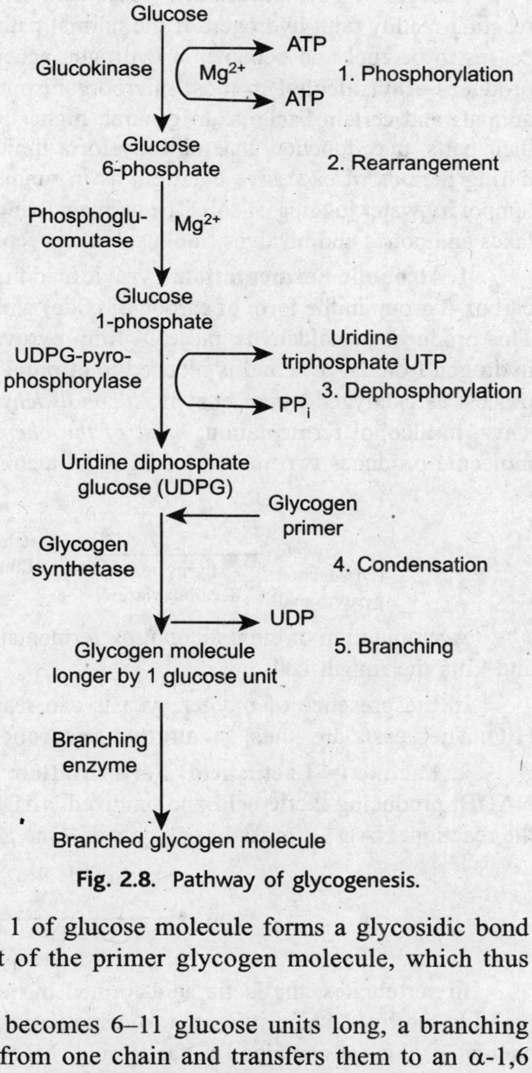

Fig. 2.8. Pathway of glycogenesis.

5. Branching. When glycogen molecule becomes 6–11 glucose units long, a branching enzyme *dextrin dextranase* cleaves fragments from one chain and transfers them to an α-1,6 linkage in another, thereby starting a branching point in the glycogen molecule. The *synthetase* enzyme then lengthens the branch by adding glucose units till the desired length is reached when branching occurs again. The pancreatic hormone **insulin** stimulates *glycogenesis.*

2.6 ANAEROBIC RESPIRATION

I. Fermentation. When enough oxygen is available NADH formed in glycolysis, transfers its hydrogen to oxygen via electron transport chain, and changes to NAD for reuse. Lack of oxygen blocks the path of hydrogen to oxygen, and soon the entire NAD of the cell is converted to NADH. This will stop glycolysis as NAD is essential for this process. Under these conditions, organisms can respire without oxygen by using an alternative pathway for electron transfer. Variations of glycolytic pathway in which NADH passes its electrons to an organic compound, such as pyruvic acid, instead of to an inorganic substance, the oxygen, are collectively called **fermentation.** Fermentation is also defined as partial oxidation of a carbohydrate in the absence of oxygen or simply as an **anaerobic respiration.**

Process of Fermentation. Pyruvic acid formed at the end of glycolysis has the property of reacting readily with hydrogen. If the normal path of hydrogen to oxygen is blocked, pyruvic acid ceases to be fuel and becomes a hydrogen acceptor. Use of pyruvic acid as hydrogen acceptor produces **ethyl alcohol** in most microorganisms, such as yeasts and forms **lactic acid** in some animals and certain bacteria. In general, higher plants and animals are aerobes, although they or their parts, may function anaerobically for a limited period of time. Animals respire anaerobically during periods of excessive exertion, as in running, and plant roots survive anaerobically during temporary water logging of soil. The animals living as intestinal parasites, in the oxygen-free mud of lakes and ponds and bivalves (molluscs) that keep their shell closed for long periods are anaerobes.

1. Alcoholic Fermentation. Pyruvic acid first undergoes decarboxylation (*i.e.,* removal of the carboxyl group in the form of carbon dioxide) with the help of an enzyme *pyruvate decarboxylase.* This produces acetaldehyde and CO_2 from pyruvate (or pyruvic acid). Acetaldehyde then accepts hydrogen from NADH and is reduced to **ethanol** or **ethyl alcohol,** producing oxidized NAD. The process is catalyzed by an enzyme *alcohol dehydrogenase.* The alcohol is excreted by the cell. Thus, in alcohol fermentation, *most of the chemical energy of glucose is wasted.* One glucose molecule produces two molecules of ethyl alcohol.

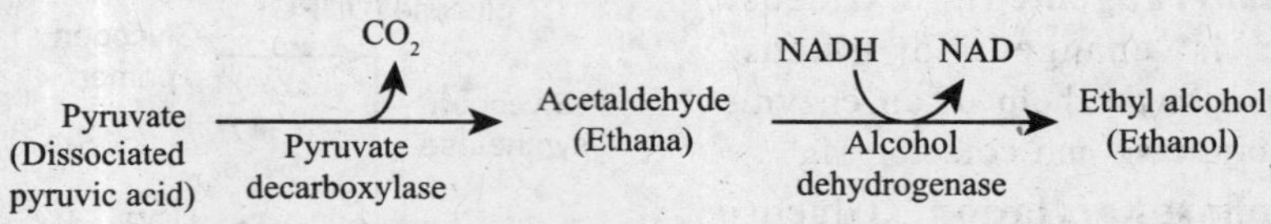

Accumulation of ethyl alcohol by fermentation in a culture of yeast may stop multiplication and kills the fungal cells.

In the presence of oxygen, yeasts can respire aerobically, oxidizing glucose to CO_2 and H_2O. The yeasts are, thus, **facultative anaerobes**.

2. Lactate (= Lactic acid) Fermentation. Pyruvate (Pyruvic acid) receives electrons from NADH, producing **lactic acid** and oxidized NAD^+. The enzyme **lactose dehydrogenase** catalyzes the reaction. NAD^+ is reused in glycolysis. One glucose molecule produces two lactate molecules.

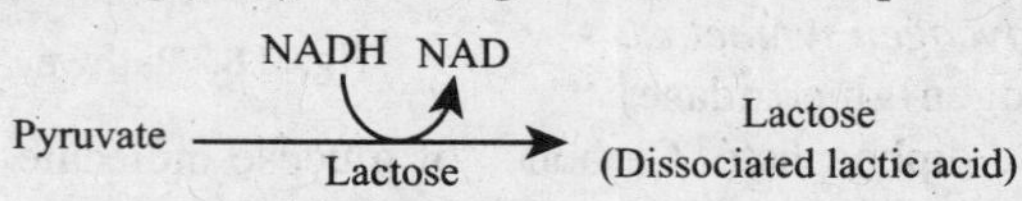

In vertebrates, the lactic acid formed in muscles in anaerobic respiration is carried by the blood to the liver where it is used to resynthesize glucose. Thus, these animals save the chemical energy of glucose contained in lactic acid. Microorganisms, which carry on lactic fermentation, excrete lactic acid. They, thus lose the energy of glucose contained in lactic acid.

Energy from Fermentation. Alcohol fermentation and lactic acid fermentation produce 57000 calories and 47000 respectively per mole of glucose as compared to complete oxidation of glucose that yields 690,000 calories. In both fermentation processes, only two ATP molecules are formed per glucose molecule partly oxidized. This is in contrast to 36 ATP molecules generated in complete oxidation of glucose molecule under aerobic condition. Thus, in terms of ATP, fermentation yields only about 5% of the energy obtainable by aerobic respiration.

QUESTIONS

Long Answer Questions

1. Explain the mechanism of glycolysis.
2. Give an account of oxidative decarboxylation.
3. Enumerate the mechanism of Krebs cycle. What is its significance.
4. Describe the various modes of gluconeogenesis.
5. Describe the hexose monophosphate pathway or shunt. Also write about its significance.

Short Answer Questions.

1. Write short notes on the following:
   (i) Dicarboxylic acid shuttle;
   (ii) Glycogenesis;
   (iii) Anaerobic respiration or fermentation;
   (iv) Conversion of pyruvate to acetyl CoA;
   (v) TPP in carbohydrate metabolism;
   (vi) UDPG.

Very Short Answer Questions

1. Why muscle cells do not release glucose in blood during their glycogen metabolism?
2. In which tissues occur the gluconeogenesis from amino acids?
3. Give full forms of TPP.
4. Give full form of UDPG.
5. Which enzyme is involved in lactic acid formation in vertebrate muscle cell.

Fill in the blanks

1. Gluconeogenesis does not occur in adipose tissues, since they lack the __________ enzyme.
2. Glycogenesis is an __________ activity.
3. Glycogenolysis in liver is stimulated by two hormones: __________ and __________.
4. During glycogenesis, the step of branching involves two enzymes __________ (called __________) and __________.
5. The name of the protein that has been identified to serve as a primer for glycogen synthesis is __________.
6. Substrate level phosphorylation in Krebs cycle is catalysed by the enzyme __________.

Multiple Choices Questions

1. Glycosis
   (a) produces no ATP
   (b) has no connection with the respiratory chain
   (c) reduces two molecules of NAD for every glucose molecule processed
   (d) takes place in mitochondria
2. Most enzymes of Krebs cycle are present in
   (a) cytoplasm
   (b) intermembrane space of mitochondria
   (c) mitochondrial matrix
   (d) inner mitochondrial membrane
3. Which of these conversions in Krebs cycle indicates substrate level phosphorylation?
   (a) succinic acid to α-ketoglutamic acid
   (b) succinic acid to maleic acid
   (c) succinyl CoA to succinic acid
   (d) malic acid to oxaloacetic acid
4. Gluconeogenesis is
   (a) a series of reactions through which glucose is converted into glycogen
   (b) the formation of glucose or glycogen from non-carbohydrate sources
   (c) reactions through which glycogen is converted into glucose
   (d) also called hexose monophosphate shunt

ANSWER

Very Short Answer Questions

1. The enzyme glucose 6 - phosphatase is absent in the muscle cells, so they do not release glucose (in glycogen metabolism) to blood.
2. In liver, kidneys and striated muscle cells.
3. Thiamine pyrophosphate (a coenzyme).
4. Uridine diphosphate glucose (a by product of glycogensis).
5. Lactose dehydrogenase converts pyruvate into lactate during anaerobic respiration.

Fill in the blanks

1. Fructose 1, 6 - diphosphate.
2. Anabolic.
3. Epinephrine, glucagon.
4. Dextrin dextranase, branching, synthetase.
5. Glycogenin.
6. Succinate thiokinase.

Multiple Choice Questions

1. (c) 2. (c) 3. (b) 4. (b)

3

Lipid Metabolism

Fats or lipids are indispensable for cell structure and function. Due to their hydrophobic and nonpolar nature, lipids differ from rest of body compounds (biomolecules) and are unique in their action.

Fats are the major constituents of dietary lipids. Digestion of fats mainly occurs in the intestine where enzyme **pancreatic lipase** breaks down fats into fatty acids and glycogen. Digestion of fats also occur in the mouth and stomach but is very small in amount. Fats are water insoluble compounds and enzyme lipases are water soluble.

In the intestine, first bile salts break the large fat particles into smaller fine particles. The process is known as **emulsification**. Emulsification increases the surface area of the fats for the action of enzyme lipase. Enzyme lipase breaks down fats (*i.e.*, triacyl glycerols) into fatty acids-1, 2 diacyl glycerols and 1 acylglycerol.

3.1 LIPIDS AND FATTY ACIDS IN BODY

Lipids constitute about 15-20% of the body weight in humans. Triacylglycerols (formerly triglycerides) are the most abundant lipids comprising 85-90% of the body fluids. Most of the triacylglycerols (TG; also called **neutral fat** or **depot fat)** are stored in the **adipose tissue** and serve as energy reserve of the body. This is contrast to carbohydrates and proteins which cannot be stored to a significant extent for energy purposes. Fats also act as an insulating material for maintaining the body temperature of animals.

1. Why should fat be the fuel reserve of the body? Triacylglycerols are the most predominant storage form of energy. There are two main reasons for fat being the fuel reserve of body.

1. Triacylglycerols (TG) are **highly concentrated form** of energy yielding 9 cal/g, in contrast to carbohydrates and proteins that produce only 4 cal/g. This is because fatty acids found in TG are in the reduced form.
2. The triacylglycerols (TG) are non-polar and hydrophobic in nature, hence stored in pure form without any association with water (anhydrous form). On the other hand, glycogen and proteins are polar. One gram of glycogen combines with 2g of water for storage.

For the two reasons stated above, one gram of fat stored in the body yields nearly six times as much as energy as one gram of (hydrated) glycogen. In a healthy, adult individual (weighing 70 kg), about 10-11 kg of fat is stored (mostly in adipose tissue) which corresponds to a fuel reserve of 100,000 Cals. If this much of energy were to be stored as glycogen (instead of fat), then the weight of the person would be 70 + 55 = 122 kg). This data explains why fat has been chosen as a fuel reserve during evolution.

Long chain fatty acids (of fat) are the ideal storage fuel reserves of the body. Fats can support the body's energy needs for long periods of food deprivation. In extreme cases, humans can fast and survive for 60-90 days, and the obese persons can survive even longer (6 months to one year) without food.

Hibernating animals offer good example for utilizing fat reserve as fuel. For instance, bears go on hibernation for about 7 months and during this entire period, the energy is derived from the degradation of fat stores. During its migration, the ruby-throated humming birds fly non-stop between New England and West Indies (about 2400 km) at a speed of 40 km/hr for 60 hours. This is possible only due to the stored fat.

2. Other Vital Lipids of Body. Phospholipids, glycolipids and cholesterol are major components of cell membranes. Cholesterol is also a precursor for bile acids and steroid hormones. Arachidonic acid (an unsaturated fatty acid) is the substrate for the synthesis of certain intercellular regulators such as prostaglandins, thromboxanes (Txs; a class of molecules involved in platelet aggregation during blood cloting), prostacyclins (*i.e.,* eicosanoids), etc.

3. Transport of Lipids. The insoluble lipids are solubilized in association with proteins to form lipoproteins in which form lipids are transported in the blood stream.

4. Plasma Lipids. The various fractions of lipids in the plasma can be estimated by different methods after extracting them with lipid solvents. The plasma levels of lipids (Table 3.1) are often useful for asserting the health of individuals.

Table 3.1 The plasma concentration of lipids (*i.e.*, lipid profile in humans).

	Lipid fraction	*Reference value* (mg/dl)
1.	Total lipids	400–600
2.	Total cholesterol	150–200
	(i) LDL–cholesterol	80–150
	(ii) HDL–cholesterol	30–60
	(iii) VLDL–cholesterol	20–40
3.	Triglycerides	75–150
4.	Phospholipids	150–200
5.	Free fatty acids	5–15

5. Dynamic State of Body Lipids. Previously it was believed that the lipids are inert storage compounds and are less significant metabolically. However, later experiments with isotope studies have proved that the body lipids are continuously being degraded and synthesized. So fat stored in the adipose tissue is the fuel reserve of the body. This is in a dynamic state.

The triacylglycerols transported from intestine as chylomicrons; Box 3.1 and liver (as VLDL) are stored in the adipose tissue. Besides, they are also utilized by muscle, liver, heart, etc., as per the needs of the body (Fig. 3.1).

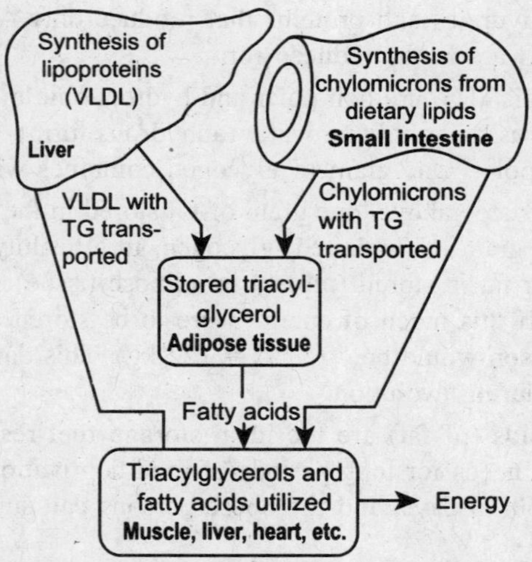

Fig. 3.1. Overview of fat metabolism; TG = triacylglycerol.

Box 3.1

After their absorption, products of lipid digestion-fatty acids, diacylglycerols and monoacylglycerols get resynthesized into triglycerols. These triglycerols are packaged in the form of lipoprotein particle known as **chylomicrons**. These chylomicrons are passed through the circulatory system to other tissues. On reaching the specific tissues, *i.e.*, liver and adipose tissues, the triglycerols of chylomicrons are hydrolyzed by the enzyme lipoprotein lipase into monoglycerol and fatty acids which are absorbed and stored in the tissues in the form of triacylglycerols.

6. Mobilization of Fat from Adipose Tissue. Triacylglycerol (TG) is the stored fat in the adipose tissue. The enzyme, namely hormone-sensitive *triacylglycerol lipase,* removes the fatty acid either from carbon 1 or 3 of the triacylglycerol to form diacylglycerol. The other two fatty acids of TG are cleaved by additional lipases specific for diacylglycerol and monoacylglycerol. The complete degradation of triacylglycerol to glycerol and free acids is known as **lipolysis** (Fig. 3.2).

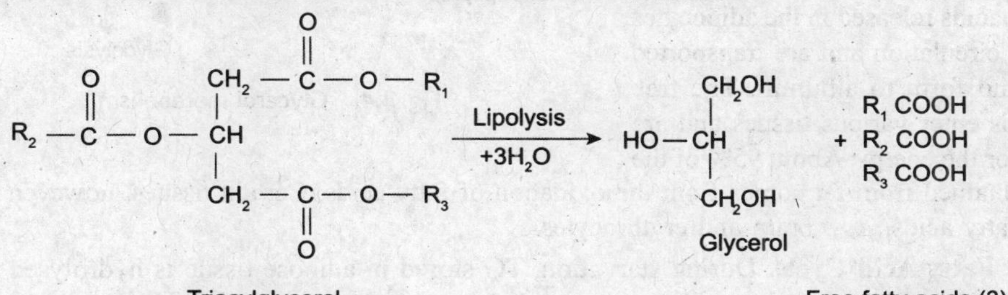

Fig. 3.2. Complete hydrolysis (or lipolysis) of triacylglycerol.

7. Regulation of hormone-sensitive TG-lipase. Hormone sensitive TG-lipase is so named because its activity is mostly controlled by hormones. Lipase is present in an active form 'b' and is activated (*i.e.,* phosphorylated) by cAMP (cyclic AMP) dependent protein kinase to lipase 'a'. Several hormones such as **epinephrine** (most effective), **norepinephrine, thyroxine, ACTH, etc.,** enhance the activity of adenylate cyclase and thus increase lipid hydrolysis or lipolysis. On the other hand, insulin decreases cAMP levels and thereby inactivates lipase. **Caffeine** promotes lipolysis by increasing cAMP levels through its inhibition on phosphodiesterase activity. The control of cAMP mediated lipolysis is illustrated in Fig. 3.3.

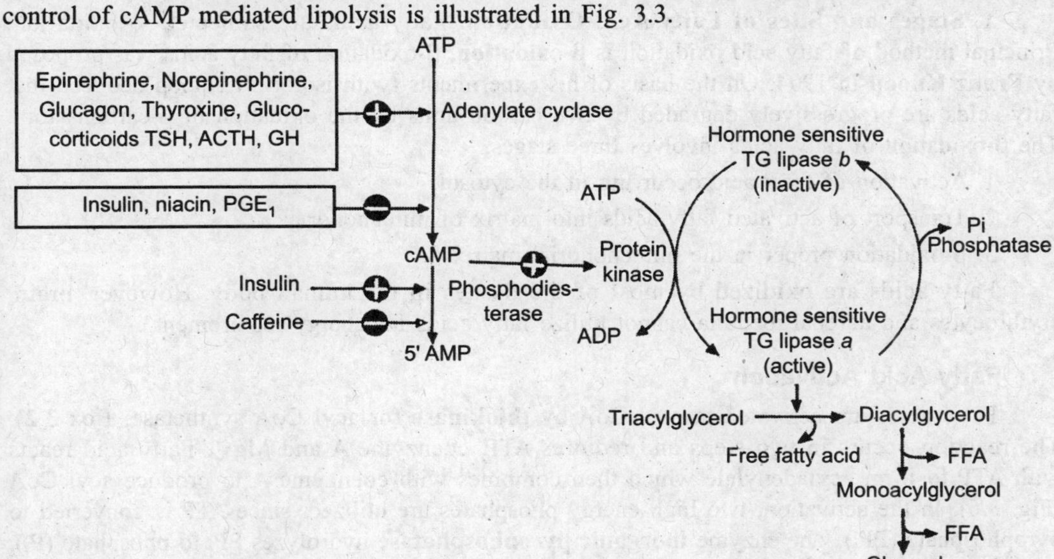

Fig. 3.3. Control of lipolysis in adipose tissue through cyclic AMP (⊕ – promoting and ⊖ – inhibiting effect; TSH = Thyroid stimulating hormone; ACTH = Adrenocorticotrophic hormone; GH = Growth hormone; PGE = Prostaglandin E1; TG = Triacylglycerol; FFA = Free fatty acid).

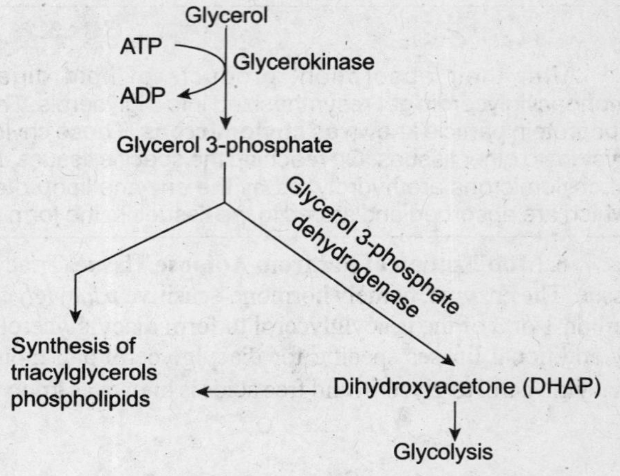

Fig. 3.4. Glycerol metabolism.

8. Fate of Glycerol. The adipose tissue lacks the enzyme *glycerol kinase,* hence glycerol produced in lipolysis, cannot be phosphorylated here. It is transported to liver where it is activated to glycerol 3-phosphate. The latter may be used for the synthesis of triglycerols and phospholipids. Glycerol 3-phosphate may also enter glycolysis by getting converted to dihydroxyacetone phosphate (DHAP; Fig. 3.4).

9. Fate of Free Fatty Acids (FFA). The fatty acids released in the adipocytes enter the circulation and are transported in a bound form to albumin. The free fatty acids enter various tissues and are utilized for the energy. About 95% of the energy obtained from fat comes from the oxidation of fatty acids. Certain tissues, however, cannot oxidize fatty acids, *e.g.,* brain and erythrocytes.

10. Fatty Acid Cycle. During starvation, TG stored in adipose tissue is hydrolysed to free fatty acids (for oxidation) to provide energy to skeletal and cardiac muscles. However, about 65% of these FFA are converted to TG, and sent back to adipose tissue for deposition. This process of lipolysis of TG and reesterification of FFA to TG is called **fatty acid cycle**.

3.2 FATTY ACID OXIDATION

The fatty acids in the body are mostly oxidized by **β-oxidation** which occurs inside the mitochondria. β-oxidation may be defined as the **oxidation of fatty acids on the β-carbon atom.** This results in the sequential removal of a two carbon fragment, **acetyl CoA**.

1. Stages and Sites of Fatty Acid Oxidation. Fatty acids are subjected to oxidation. Principal method of fatty acid oxidation is **β-oxidation.** β-oxidation of fatty acids was proposed by **Franz Knoop** in 1904. On the basis of his experiments (with isotopes) Knoop deduced that fatty acids are progressively degraded by two carbon units by the oxidation of β-carbon atom. The β-oxidation of fatty acids involves three stages.

1. Activation of fatty acid occurring in the cytosol;
2. Transport of activated fatty acids into matrix of mitochondria;
3. β-oxidation proper in the mitochondrial matrix.

Fatty acids are oxidized by most of the tissues in the human body. However, brain, erythrocytes and adrenal medulla cannot utilize fatty acids for energy requirement.

Fatty Acid Activation

Fatty acids are activated to acyl CoA by **thiokinase** (or acyl CoA synthetase; Box 3.2). The reaction occurs in two steps and requires ATP, coenzyme A and Mg^{2+}. Fatty acid reacts with ATP to form acyladenylate which then combines with coenzyme A to produce acyl CoA (Fig. 3.6). In the activation, two high energy phosphates are utilized, since ATP is converted to pyrophosphate (PP_i). The enzyme **inorganic pyrophosphatase** hydrolyzes PP_i to phosphate (P_i). The immediate elimination of PP_i makes this reaction totally irreversible.

Box 3.2

Acyl-CoA synthetases (enzymes) are found in the endoplasmic reticulum, peroxisomes and inside and on the outer membrane of mitochondria (Fig. 3.5). Several acyl-CoA synthetases have been described, each specific for fatty acids of different chain length.

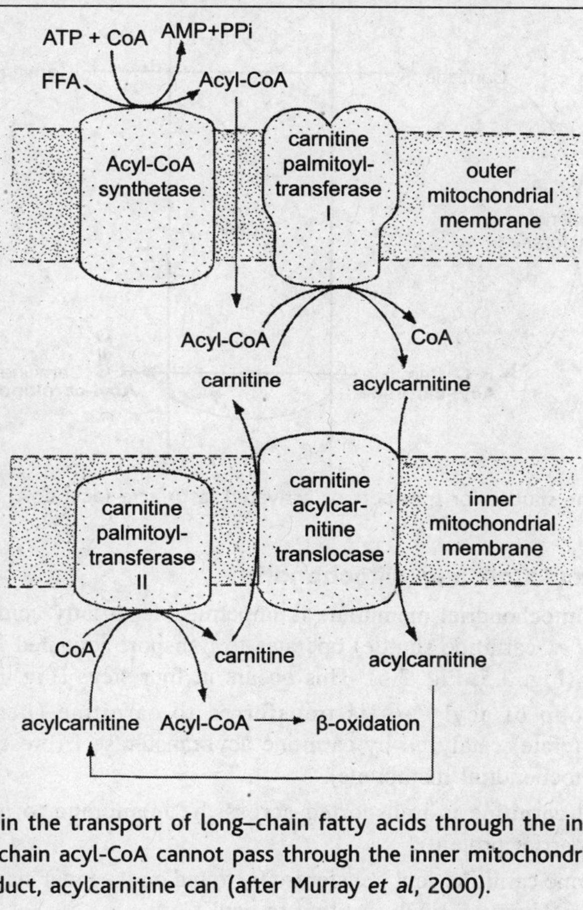

Fig. 3.5. Role of carnitine in the transport of long–chain fatty acids through the inner mitochondrial membrane. Long–chain acyl-CoA cannot pass through the inner mitochondrial membrane, but its metabolic product, acylcarnitine can (after Murray *et al*, 2000).

Three different thiokinases, to activate long chain (10-20 carbon), medium chain (4-12 carbon) and short chain (<>4 carbon) fatty acids have been identified.

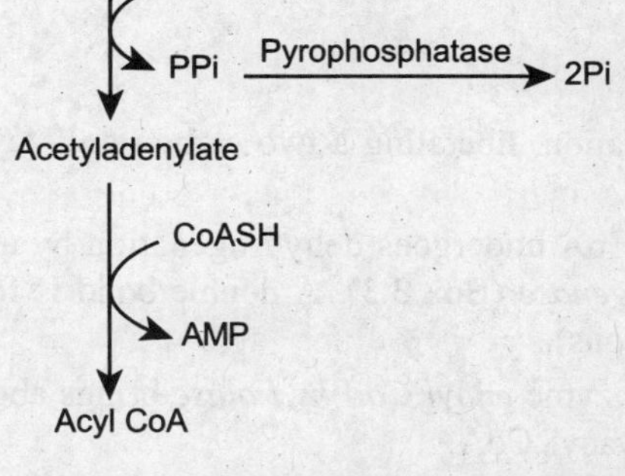

Fig. 3.6. Activation of fatty acid to acyl CoA by the enzyme thiokinase (acyl-CoA synthetase).

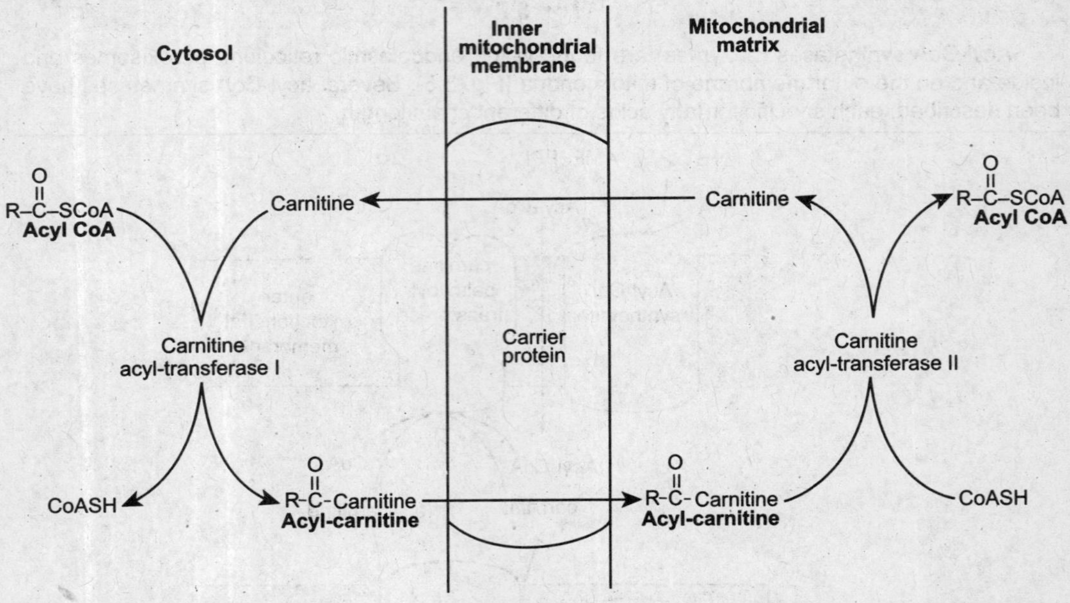

Fig. 3.7. Carnitine shuttle for transport of activated fatty *acid* (acyl CoA) into the mitochondrion.

Transport of Acyl CoA into Mitochondria

The inner mitochondrial membrane is impermeable to fatty acids. A specialized **carnitine carrier system** (*i.e.,* carnitine shuttle) operates to transport activated fatty acids from cytosol to the mitochondria (Fig. 3.5; Fig. 3.6). This occurs in four steps (Fig. 3.7).

1. Acyl group of acyl CoA is transferred to carnitine (*i.e.,* β-hydroxy γ-trimethyl aminobutyrate) catalyzed by carnitine acyltransferase I (present on the outer surface of inner mitochondrial membrane).
2. The acyl-carnitine is transported across the membrane to mitochondrial matrix by a specific carrier protein.
3. The enzyme carnitine acyl transferase II (found on the inner surface of inner mitochondrial membrane) converts acyl-carnitine to acyl CoA.
4. The carnitine released return to cytosol for reuse.

It should be noted that the coenzyme A used for activation is different from the one that finally combines with fatty acid in the mitochondria to form acyl CoA. Thus, the cell has two separate pools (cytosolic and mitochondrial) of coenzyme A.

β-Oxidation Proper

Each cycle of β-oxidation, liberating a two carbon unit-acyl-CoA, occurs in a sequence of four reactions (Fig. 3.8).

1. **Oxidation.** Acyl CoA undergoes dehydrogenation by an FAD-dependent flavoenzyme, *acyl CoA dehydrogenase* (Box 3.3). A double bond is formed between α and β carbons (*i.e.,* 2 and 3 carbons).
2. **Hydration.** The enzyme *enoyl CoA hydrolase* brings about the hydration of double bond to form β-hydroxyacyl CoA.

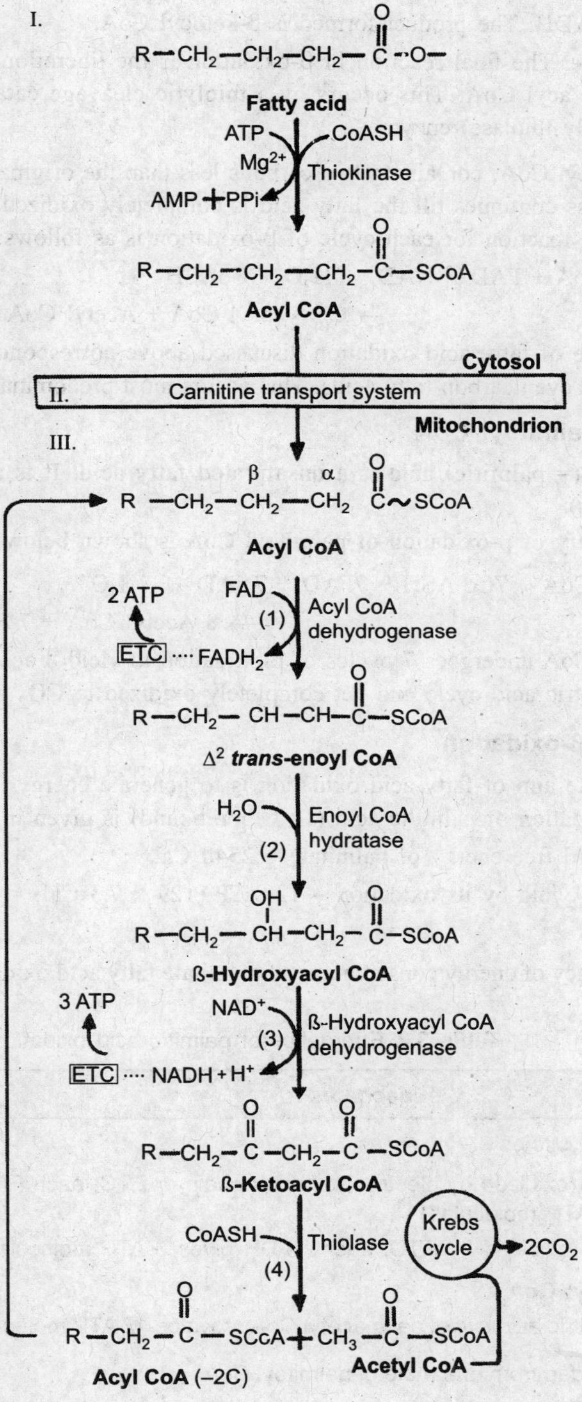

Fig. 3.8. β-oxidation of fatty acids, Palmitoyl CoA (16 carbon fatty acid) undergoes seven cycles to yield 8 acetyl CoA. [I = activation, II = transport, III = β-oxidation proper: 1. Oxidation; 2. Hydration; 3. Oxidation and 4. Cleavage].

3. **Oxidation.** The enzyme β-*hydroxyacyl CoA dehydrogenase* catalyzes the second oxidation and generates NADH. The product formed is β-ketocyl CoA.

4. **Cleavage.** The final reaction in β-oxidation is the liberation of a 2 carbon, fragment, acetyl CoA from acyl CoA. This occurs by a thiolytic cleavage catalysed by β-*ketoacyl CoA thiolase* (or simply thiolase) enzyme.

The new acyl CoA, containing two carbons less than the original, reenters the β-oxidation cycle. The process continues till the fatty acid is completely oxidized.

The overall reaction for each cycle of β-oxidation is as follows:

$$\text{Cn Acyl CoA} + \text{FAD} + \text{NAD}^+ + H_2O + \text{CoASH}$$
$$\rightarrow C_{(n-2)} \text{ acyl CoA} + \text{Acetyl CoA} + FADH_2 + \text{NADH} + H^+$$

The scheme of fatty acid oxidation discussed above corresponds to saturated (having no double bond) and even carbon fatty acids. This occurs most predominantly in biological systems.

Oxidation of Palmitoyl CoA

Palmitolic (= palmitic) acid is a unsaturated fatty acid. It is a even number fatty acid (having 16 atoms).

The summary of β-oxidation of palmitoyl CoA is shown below:

$$\text{Palmitoyl CoA} + \text{7Co ASH} + \text{7FAD} + \text{7NAD}^+ + 7H_2O$$
$$\rightarrow \text{8 Acetyl CoA} + 7\ FADH_2 + \text{7 NADH} + 7H^+$$

Palmitoyl CoA undergoes 7 cycles of β-oxidation to yield 8 acetyl CoA (Fig. 3.8). Acetyl CoA can enter citric acid cycle and get completely oxidized to CO_2 and H_2O.

Energetics of β-oxidation

The ultimate aim of fatty acid oxidation is to generate energy. The energy obtained from the complete oxidation of palmitic acid (16 C compound) is given in Table 3.2 and Fig. 3.9.

The standard free energy of palmitate = 2340 Cal

The energy yield by its oxidation – 129ATP (129 × 7.3 Cal) = 940 Cal

$$\text{The efficiency of energy conservation by palmitate fatty acid oxidation} = \frac{940}{2340} \times 100 = 40\%$$

Table 3.2 Energetics of palmitic acid oxidation.

Mechanism	*ATP yield*
I. β-oxidation 7 cycles	
(i) 7 $FADH_2$ (oxidized by electron transport chain or ETC; each $FADH_2$ gives 2ATP molecules)	14
(ii) 7 NADH (oxidized by ETC, each NADH gives 3 ATP molccules)	21
II. From 8 acetyl CoA	
Oxidized by citric acid cycle, each acetyl CoA provides 12 ATP molecules)	96
Total energy yielded from one mole of palmitoyl CoA	131
Minus energy utilized for activation (i.e., formation of palmitoyl CoA)	–2
Net yield for one molecule of palmitoyl CoA	129

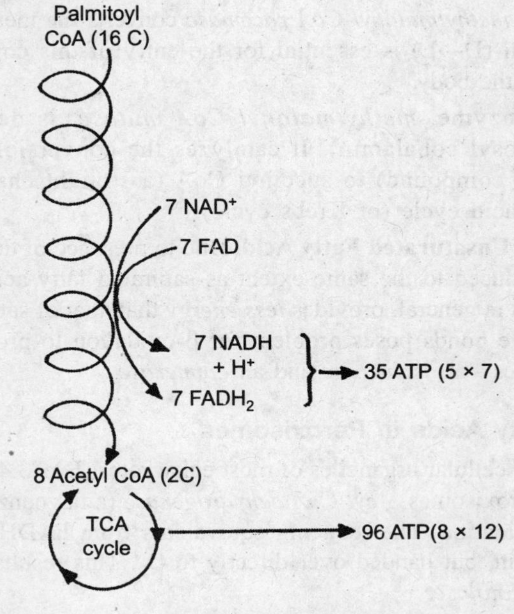

Fig. 3.9. An overview of oxidation of palmitic acid.

Box 3.3 SIDS-A disorder due to blockade in β-oxidation

The **sudden infant death syndrome (SIDS)** is unexpected death of healthy infants, usually overnight. The real cause of SIDS is not known. It is now estimated that at least 10% of SIDS is due to *deficiency of medium chain acyl CoA dehydrogenase*. The enzyme defect has a frequency of 1 in 10,000 births and is, in fact, more prevalent than *phenylketonuria* (a genetic disease). The occurrence of SIDS is explained as follows.

Glucose is the principal source of energy, soon after eating or feeding babies. After a few hours, the glucose level and its utilization decrease and the rate of fatty acid oxidation must simultaneously increase to meet the energy needs. The sudden death in infants is due to a **blockade in β-oxidation** caused by a deficiency in the enzyme medium chain acyl CoA dehydrogenase (MCAD).

2. Oxidation of Odd Carbon Chain Fatty Acids. The β-oxidation of saturated fatty acids containing odd number of carbon atoms proceeds in the same manner as described above for even carbon atoms. The only difference is that in the last and final β-oxidation cycle, a three-carbon fragment is left behind (in place of 2 carbon unit for saturated fatty acids). This compound is **propionyl CoA** which is converted to succinyl CoA as follows (Fig. 3.10).

1. Propionyl CoA (3C compound) is carboxylated in the presence of ATP, CO_2 and vitamin **biotin** to D-methylmalonyl CoA.

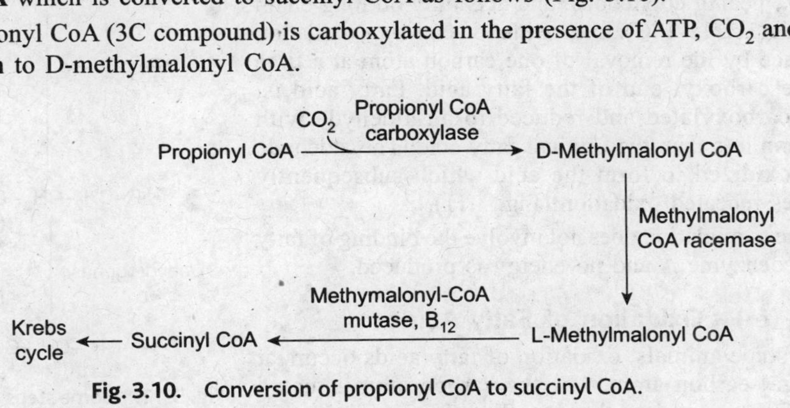

Fig. 3.10. Conversion of propionyl CoA to succinyl CoA.

2. The enzyme *methylmalonyl CoA racemase* converts the methylmalonyl CoA to L-form. This reaction (D→L) is essential for the entry of this compound into the metabolic reactions of the body.
3. The next enzyme, *methylmalonyl CoA mutase,* is dependent on **vitamin B_{12}** (deoxyadenosyl cobalamin). It catalyzes the conversion of methyl-malonyl CoA (a branched compound) to succinyl CoA (a straight chain compound), which can enter citric acid cycle (or Krebs cycle).

3. Oxidation of Unsaturated Fatty Acid. Due to presence of double bonds, the unsaturated fatty acids are not reduced to the same extent as saturated fatty acids. Therefore, oxidation of unsaturated fatty acids in general, provides *less energy* than that of saturated fatty acids. However, the presence of double bonds poses problem for β-oxidation to proceed. This is overcome by two additional enzymes—an *isomerase* and an *epimerase.*

β-Oxidation of Fatty Acids in Peroxisomes

Peroxisomes are cellular organelles of most eukaryotes (Box 3.4). The β-oxidation occurs in a modified form in peroxisomes. *Acyl CoA dehydrogenase* (a flavoenzyme) leads to the formation of $FADH_2$ as in β-oxidation. The reducing equivalents from $FADH_2$ are not transferred to the electron transport chain, but handed over directly to O_2. This results in the formation of H_2O_2, which is cleaved by *catalase.*

$$E\text{–}FADH_2 + H_2O \rightarrow E\text{-}FAD + H_2O_2$$

$$H_2O_2 \xrightarrow{\text{Catalase}} H_2O + {}^1/_2\, O_2$$

There is no ATP synthesized in peroxisomal β-oxidation of fatty acids, since the reducing equivalents do not pass through ETC. However, heat is liberated.

Peroximal oxidation is induced by high fat diet and administration of hypolipidemic drugs (*e.g.,* clofibrate).

Box 3.4 Cerebrohepatorenal or Zellweger Syndrome

This rare disorder (also called PBD, or peroxisome biogenesis disorder) is characterized due to absence of peroxisomes in almost all tissues. As a result **long chain fatty acids (C_{26}–C_{38})** are not oxidized. They accumulate in tissues, particularly in brain, liver and kidney and cause this fatal syndrome (infant dies within one year after birth).

Alpha (α) - Oxidation of Fatty acids

It was reported by **P.K. Stumpf.** It occurs in plants only, *e.g.,* peanut cotyledons, and acts only on long chain fatty acids (*i.e.,* 13-18C). In α-oxidation pathway, oxidation takes place by the removal of one carbon atom at a time from the carboxyl end of the fatty acid. Fatty acid is, first, decarboxylated and reduced to an aldehyde with one carbon less than the original fatty acid. The aldehyde is then oxidized to form the acid which subsequently undergoes repeated oxidation (Fig. 3.11).

The α-oxidation does not involve the binding of fatty acid to coenzyme A and no energy is produced.

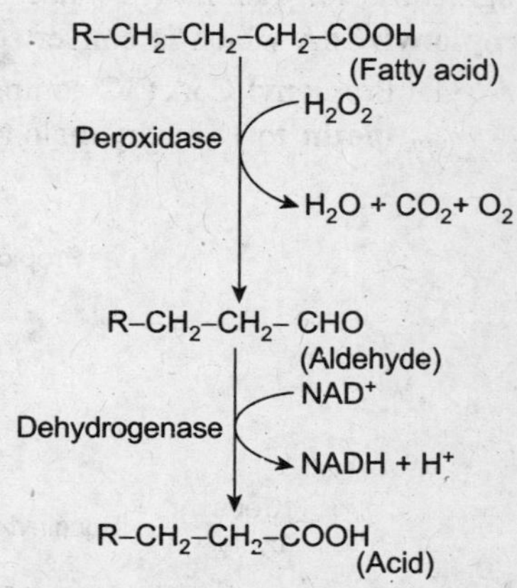

Fig. 3.11. Some steps of alpha oxidation.

Omega (ω) - Oxidation of Fatty Acids

In some animals, oxidation of fatty acids occurs at ω (omega) carbon atom, *i.e.,* the carbon atom situated farthest from the carboxyl group. Site of omega oxidation is the endoplasmic reticulum of many tissues. It involves the **hydroxylation** of fatty acids on the methyl carbon (omega carbon) at the other end of the molecule from the carboxyl group. Usually medium chain length fatty acids are involved. Omega (ω) oxidation results in the production of dicarboxylic acid. Dicarboxylic acid undergoes β oxidation and forms successive smaller dicarboxylic acids (Fig. 3.12).

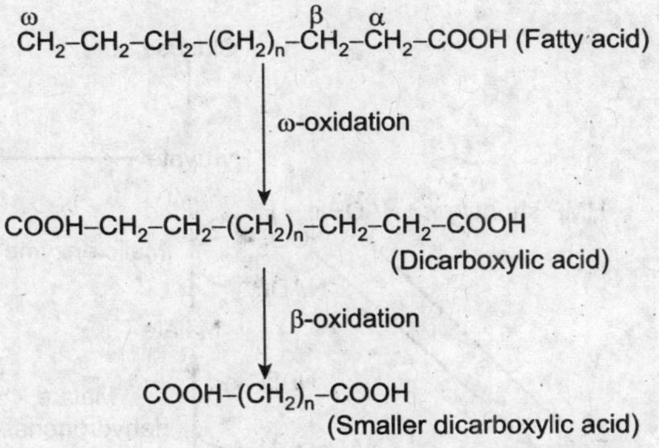

Fig. 3.12. Some steps of omega (ω) oxidation.

Significance of Omega Oxidation

1. It produces the fatty acids with even number of carbon atoms.

2. Aldehydes formed in omega oxidation may be used to form long chain alcohols.

Metabolic Water is Obtained from Oxidation of Fatty Acids. Fatty acid oxidation is accompanied by the production of water referred to **metabolic water**. For instance, when one molecule of palmitic acid is oxidized, it releases 16 molecules of water. This metabolic water has great significance in some animals. For example, camel can store lipids in its hump which is good source of water, besides energy supply. For this reason, camel can travel in deserts for long periods even without food and water supply.

3.3 BIOSYNTHESIS OF FATTY ACIDS

The dietary carbohydrates and amino acids, when consumed in excess, can be converted to fatty acids and stored as triacylglycerols. *De novo* (new) synthesis of fatty acids occurs predominantly in liver, brain, lungs, kidney, adipose tissue and lactating mammary gland. The enzyme machinery for fatty acid production is located in the cytosomal fraction of the cell. Acetyl CoA is the source of carbon atoms while NADPH provides the reducing equivalents and ATP supplies energy for fatty acid formation.

Three Steps of de novo Synthesis of Palmitic Acid

Palmitic Acid. As mentioned earlier, **palmitic acid** is a straight chain 16-carbon saturated fatty acid (CH_3–$(CH_2)_{14}$–COOH) and is the "stem" fatty acid as it is first to be synthesized. All other fatty acids are formed by modification of palmitic acid. Acetyl CoA is the **primer** and the chain growth starts at the carboxyl group of acetyl CoA and elongates by successive addition of acetyl residues at the carboxyl end of the growing chain.

The process of *de novo* biosynthesis of palmitic acid involves the following *three* steps:

1. Production of acetyl CoA and NADPH;
2. Conversion of acetyl CoA to malonyl CoA;
3. Reactions of fatty acid synthase complex.

1. Production of acetyl CoA and NADH. Acetyl CoA and NADPH are the prerequisite for fatty acid synthesis. Acetyl CoA is produced in the mitochondria by the oxidation of pyruvate and fatty acids, degradation of carbon skeleton of certain amino acids, and from ketone bodies. Mitochondria, however, are not permeable to acetyl CoA. A alternate arrangement is made for the transfer of acetyl CoA to cytosol. Acetyl CoA condenses with oxaloacetate in mitochondria to form citrate. Citrate is freely transported to cytosol where it is cleaved by **citrate lyase** to liberate acetyl CoA and oxaloacetate. Oxaloacetate in the cytosol is converted to malate (Fig. 3.13).

The enzyme *malate dehydrogenase* converts malate to pyruvate. NADPH and CO_2 are generated in this reaction. Both of them are utilized in fatty acid synthesis.

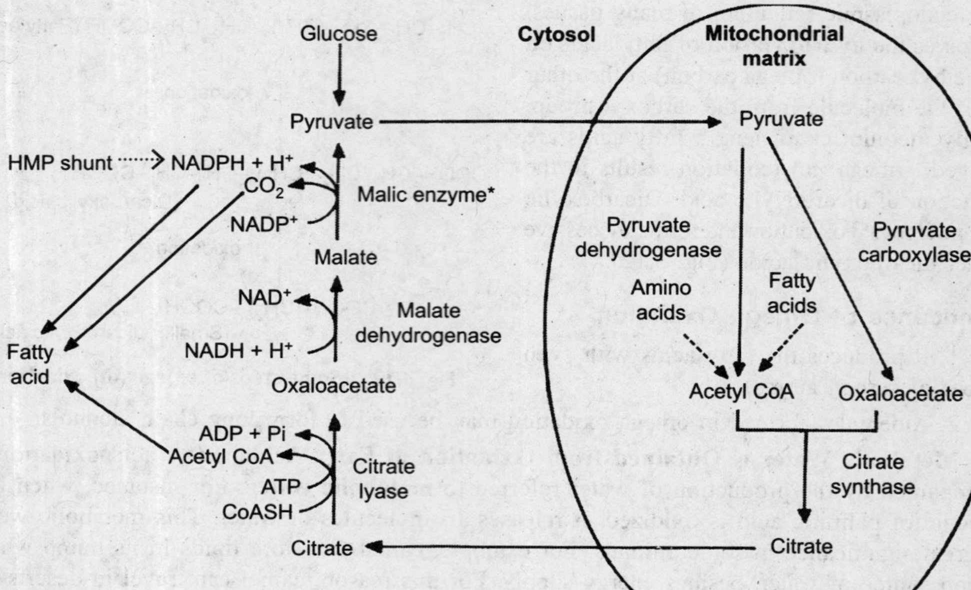

Fig. 3.13. Transfer of acetyl CoA from mitochondria to cytosol (HMP shunt = Hexose monophosphate shunt: * = also known as malate dehydrogenase).

The **transport** of acetyl CoA from mitochondria to cytosol is coupled with the cytosomal production of NADPH and CO_2 which is highly advantageous to the cell for optimum synthesis of fatty acids.

2. Formation of Malonyl CoA. Acetyl CoA is carboxylated to malonyl CoA by the enzyme *acetyl CoA carboxylase* (Fig. 3.14). This is an ATP–dependent reaction and requires *biotin* for CO_2 fixation. The mechanism of acetyl CoA carboxylase is similar to that of pyruvate carboxylase. Acetyl CoA carboxylase is a regulatory enzyme in fatty acid synthesis (Box 3.5).

Box 3.5

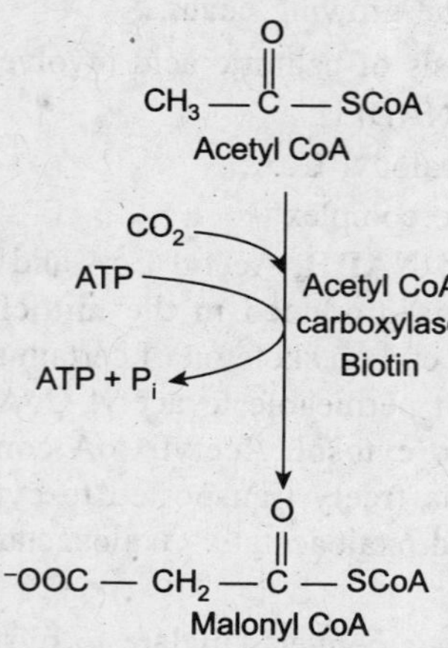

Fig. 3.14. Conversion of acetyl CoA to malonyl CoA.

The enzyme Acetyl CoA carboxylase is a multipurpose complex containing biotin; biotin carboxyl carrier; transcarboxylase and an allosteric (regulatory site). It is activated by citrate and inhibited by long chain fatty acyl CoA. The biotin acts as an intermediate carrier of CO_2 molecule. It is energy dependent step for which energy is provided by ATP. The CO_2 donor is HCO_3 (Fig. 3.14). Malonyl provides the necessary acetyl residues which get added to primer called acetyl CoA.

Acyl Carrier Protein (ACP; Fig. 3.15). Acyl carrier protein (ACP) has **thiol group** (–SH); it is a polypeptide chain of 77 amino acids (Mol. Wt = 10,000). Its serine moiety is in combination with 4'-phospho-pantetheine prosthetic group which is analogous to that coenzyme A. ACP serves as an anchor to which the acyl intermediates are esterified during the reaction. The acyl group of acetyl CoA and malonyl CoA are transferred to the thiol group of ACP in the presence of acetyl **transacylase** and **malonyl transacylase.**

Acetyl-CoA + ACP-SP ⇌ Acetyl-ACP + Co.ASH

Malonyl-CoA + ACP-SP ⇌ Malonyl-ACP + Co.ASH

Acetyl–ACP and malonyl – ACP now reacts in a specific manner to form **acetoacyl–ACP** and a CO_2 molecule is released.

Acetyl-ACP + Malonyl-ACP ⇌ Acetoacyl-ACP + CO_2 + ACP-SH

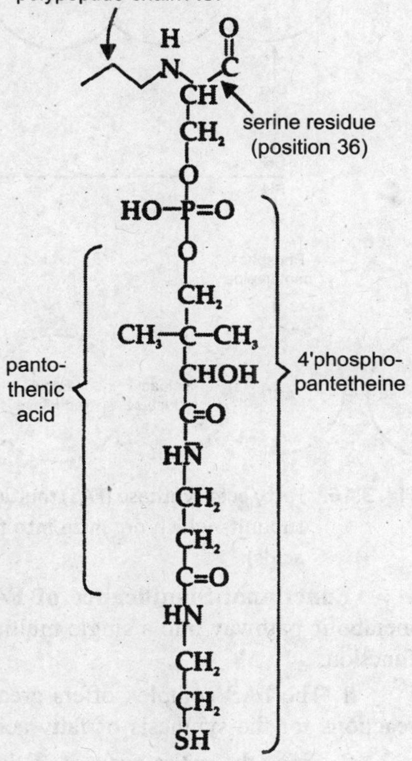

Fig. 3.15. ACP or acyl carrier protein.

Reactions of Fatty Acid Synthetase Complex

The remaining reactions of fatty acid synthesis are catalyzed by **FAS complex.**

(*i*) FAS Complex (Fatty Acid Synthase (= Synthetase) Complex). The diagrammatic representation of the model for fatty acid synthase (FAS) **multienzyme complex** is depicted in Fig. 3.16. This model is provisional and is mainly based on the work of **Wakil** (Box 3.6).

Box 3.6

The fatty acid-synthesizing systems from three sources have been investigated in some detail: that from yeast by **Lynen** (1952), with a molecular weight of 2.3×10^6; that from pigeon liver by **Wakil** (1961), with a molecular weight of 4.5×10^{15}, and that from *Escherichia coli* by **Vagelos** (1964). Of these systems, that from *E. coli* is perhaps the best understood, at present **(Jain** *et al.*, 2014).

Fatty acid synthase (FAS) is a **dimmer** composed of two identical subunits (monomers), each with a molecular weight of 240,000 daltons. Each subunits contains the activities of 7 enzymes of FAS and an ACP (Acyl carrier protein) with 4'-phosphopantetheine–SH group. The two subunits lie in antiparallel (head to tail) orientation. The –SH group of phosphopantetheine of one subunit is in close proximity to the –SH of cysteine residue (for the enzyme ketoacyl synthase) of the other subunit.

Each monomer of FAS contains all the enzyme activities of fatty acid synthesis. But only the dimer is functionally active. This is because the functional unit consists of half of each subunit interacting with the complementary half of the other. Thus, the FAS structure has both functional division and subunit division (Fig. 3.16). The two functional subunits of FAS independently operate and synthesize two fatty acids simultaneously.

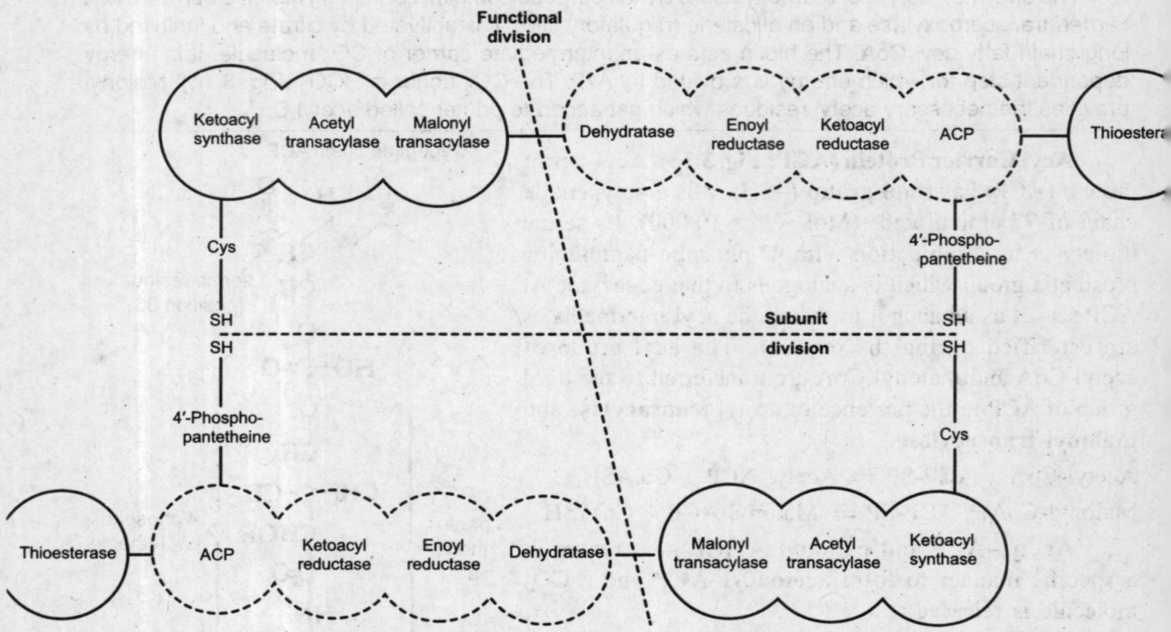

Fig. 3.16. Fatty acid synthase (FAS) multienzyme complex (ACP = Acyl carrier protein: FAS has two identical subunits which organize into two functional subunits to simultaneously synthesize two fatty acids).

Functional Significance of FAS Complex. The organization of different enzymes of metabolic pathway into a single multienzyme functional unit has distinct advantages for cellular function.

1. The FAS complex offers great efficiency that is free from interference of other cellular reactions for the synthesis of fatty acids.

2. Since the entire process of the metabolic pathway is confined to the complex, there are no permeability barriers for the various intermediates.

3. The multienzyme polypeptide complex is coded by a single gene. Thus, there is a **good coordination** in the synthesis of all enzymes of the FAS complex.

***(ii)* Formation of butyryl CoA.** The acetoacyl-ACP undergoes reduction with NADPH + H^+ to form **β-hydroxybutyryl-ACP** (or acyl-ACP) in the presence of enzyme *acetoacetyl CoA reductase* (Fig. 3.17).

The D-stereoisom erase of β-hydroxybutyryl-ACP formed is then dehydrated in the presence of enzyme *β-hydroxybutyryl CoA dehydratase* to form α, **β-unsaturated acyl-ACP complex** which is reduced to butyryl - ACP complex in the presence of enzyme *crotonyl - ACP reductase* and a reduced coenzyme NADH + H^+ with the formation of butyryl - ACP complex. First cycle is completed in which a molecule of malonyl –ACP has entered at the carboxyl end of growing fatty acyl chain with the displacement of the ACP molecule from the carboxyl group. Seven such cycles occur in the synthesis of palmitic acid. **Palmityl-ACP** is the ultimate end product which in the presence of a hydrolytic deacylase enzyme, decomposes and releases palmitic acid.

The human body can synthesize all the fatty acids except the essential, polysaturated fatty acids. These syntheses involve a variety of enzyme systems by which the palmitic acid is modified by 3 processes: elongation, desaturation and hydroxylation.

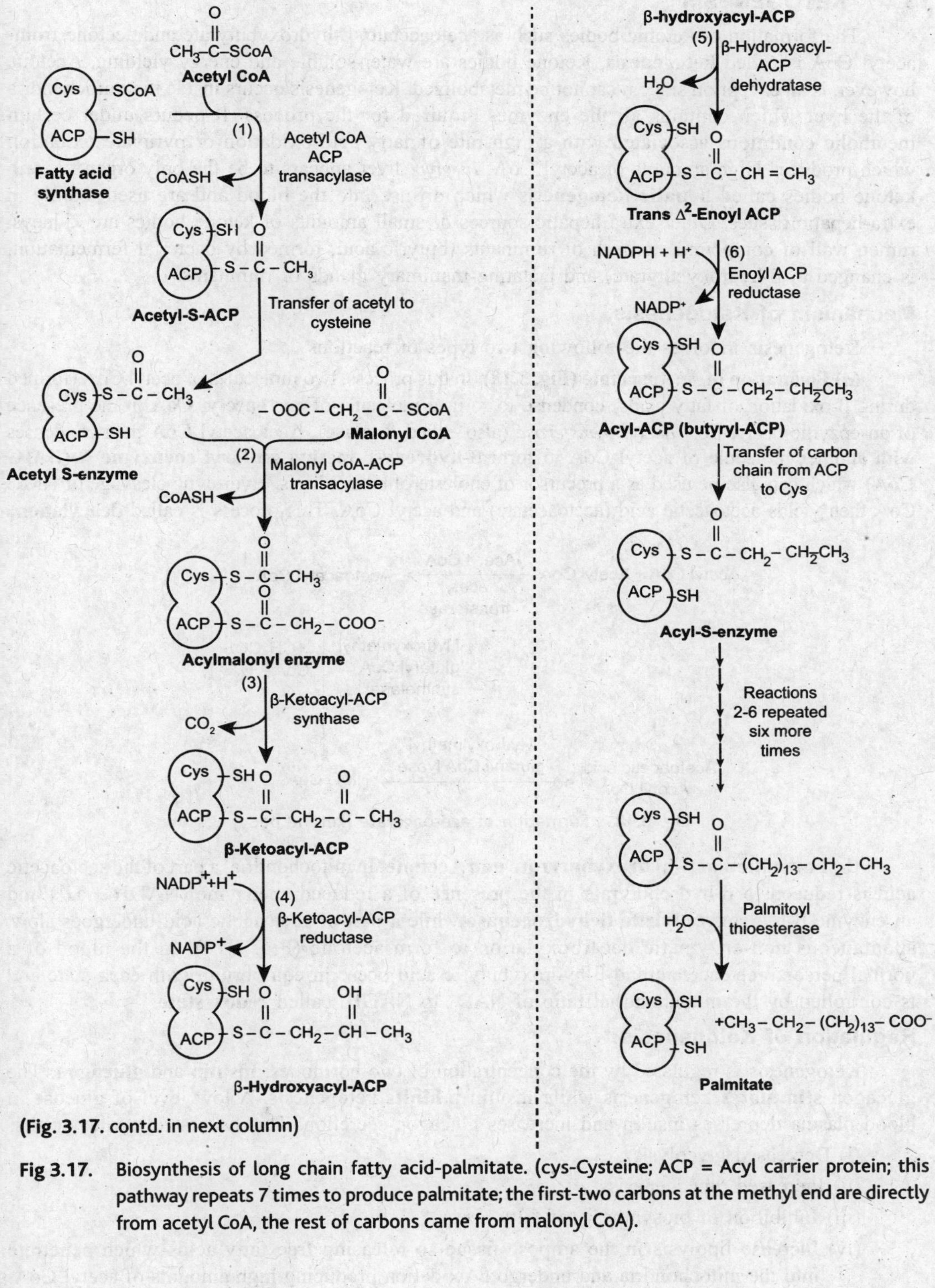

(Fig. 3.17. contd. in next column)

Fig 3.17. Biosynthesis of long chain fatty acid-palmitate. (cys-Cysteine; ACP = Acyl carrier protein; this pathway repeats 7 times to produce palmitate; the first-two carbons at the methyl end are directly from acetyl CoA, the rest of carbons came from malonyl CoA).

3.4 KETOGENESIS

The formation of ketonic bodies such as acetoacetate, β-hydroxybutyrate and acetone from-acetyl CoA is called **ketogenesis**. Ketone bodies are water-soluble and energy yielding. Acetone however, is an exception since it cannot be metabolized. Ketogenesis occurs inside the mitochondria of the liver which contains all the enzymes required for the process. It occurs under certain metabolic conditions associated with a high rate of fatty acid oxidation or pyruvate oxidation which produces large amounts of acetyl CoA. *In vivo,* liver appears to be the only organ to form ketone bodies called **hepatic ketogenesis** which diffuse into the blood and are used mainly in extra-hepatic tissues. Other extra hepatic sources of small amounts of ketone bodies are kidneys, rumen wall of compound stomach of ruminants (butyric acid, formed by bacterial fermentation, is changed to β-hydroxybutyrate) and lactating mammary glands of ruminants.

Mechanism of Ketogenesis

Ketogenesis involves the following two types of reactions.

***(a)* Formation of Acetoacetate (Fig. 3.18).** In this process, two molecules of acetyl CoA (formed during β-oxidation of fatty acids) condense to form a molecule of **acetoacetyl CoA** in the presence of an enzyme *acetyl CoA acetyl transferase* (also called **thiolase**). Acetoacetyl CoA then condenses with another molecule of acetyl CoA to form **β-hydroxy-β-methyl glutaryl coenzyme A (HMG CoA)** which can also be used as a precursor of cholesterol biosynthesis. Hydrolytic cleavage of HMG CoA then yields acetoacetic acid (acetoacetate) and acetyl CoA. This process is called **deacylation.**

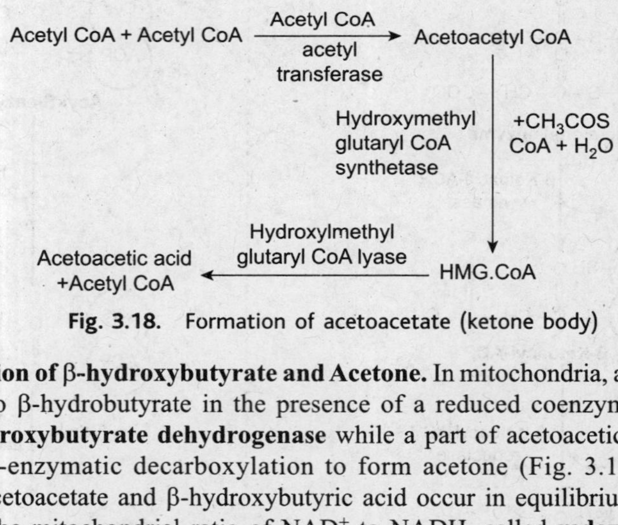

Fig. 3.18. Formation of acetoacetate (ketone body)

(b) Formation of β-hydroxybutyrate and Acetone. In mitochondria, a part of the acetoacetic acid is reduced to β-hydrobutyrate in the presence of a reduced coenzyme (NADH + H^+) and an enzyme **β-hydroxybutyrate dehydrogenase** while a part of acetoacetic acid undergoes slow, spontaneous non-enzymatic decarboxylation to form acetone (Fig. 3.19). In the blood of a normal person, acetoacetate and β-hydroxybutyric acid occur in equilibrium with each state and is controlled by the mitochondrial ratio of NAD^+ to NADH, called **redox state**.

Regulation of Ketogenesis

Ketogenesis is regulated by the concentration of two hormones: **insulin** and **glucagon**. The glucagon **stimulates** ketogenesis while insulin **inhibits** ketogenesis. A low level of glucose in blood plasma depresses insulin and increases glucagon secretion. This causes following events:

(i) Decreased glycolysis.

(ii) Increased glucogenesis.

(iii) Inhibition of biosynthesis of fatty acids.

(iv) Increase lipolysis in the adipose tissue so releasing free fatty acids which penetrate into the mitochondria and undergo β-oxidation producing high amounts of acetyl CoA.

(v) Increased ketogenesis from acetyl CoA.

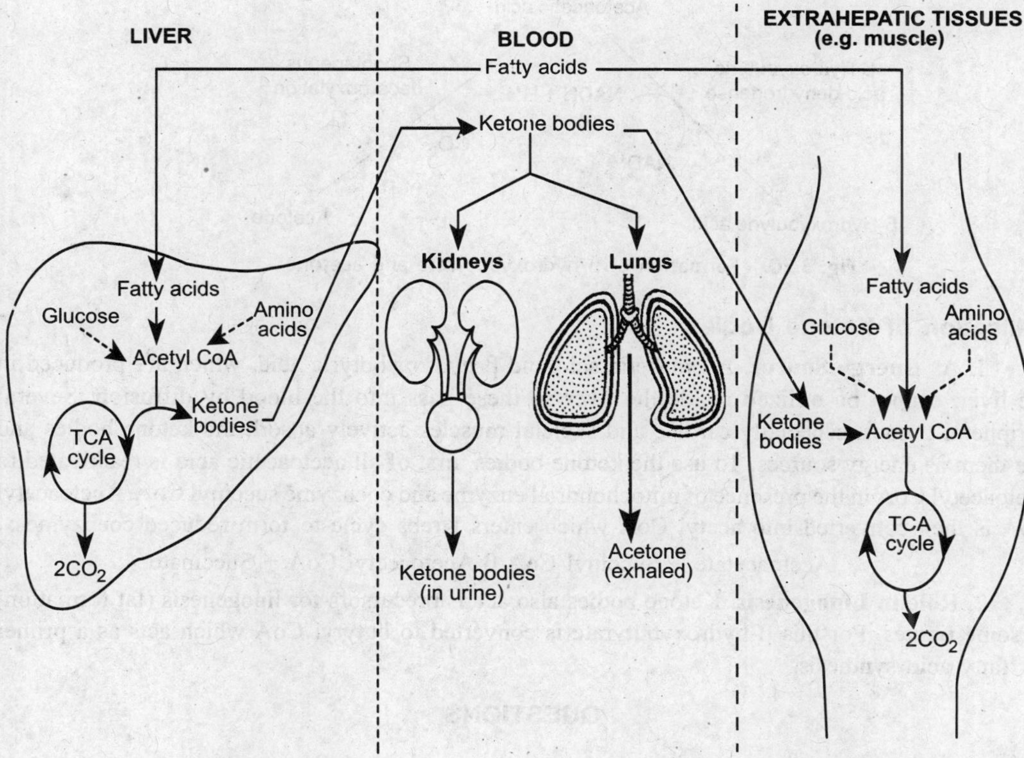

Fig. 3.19. Summary of ketone body synthesis, utilization and excertion.

Blood Levels of Ketone Bodies

The concentration of total ketone bodies in the blood of well fed mammals does not normally exceed 1 mg/dl and normal loss of ketone bodies via the urine is less than 1 mg/24 hours in humans. But the rapid formation and accumulation of ketone bodies in the blood plasma is called **ketonaemia** or **hyperketonaemia,** while the presence of higher amount of ketone bodies in the urine is called **ketonuria.** Such conditions generally occurs in a number of pathological conditions such as **diabetes mellitus**. Such condition is called **ketosis**.

Starvation. Starvation is accompanied by *increased degradation of fatty acids* (from the fuel reserve triacylglycerol) to meet the energy needs of the body. This causes an over production of acetyl CoA which cannot be fully handled by citric acid cycle (Krebs cycle). Further more citric acid cycle is impaired due to deficiency of oxaloacetate, since most of it is diverted for glucose synthesis to meet the essential requirements for tissues such as brain. The result is an accumulation of acetyl CoA and its diversion for overproduction of ketone bodies.

Acetoacetic acid and β-hydroxybutyric acid are moderately strong acids. When these acids accumulate in the blood plasma, they lead to **keto acidosis** (Fig. 3.20). If uncontrolled, it leads to rapid deterioration and death. It normally occurs when glucose available is markedly decreased and there is increased mobilization of trielvcerides stored in the adipose tissue.

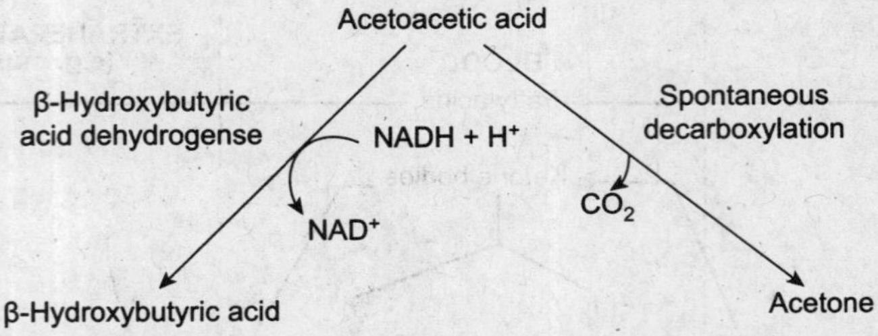

Fig. 3.20. Formation of β-hydroxybutyrate and acetone

Utilisation of Ketone Bodies

1. As Energy Source. Acetoacetic acid and β-hydroxybutyric acid, which are produced in the liver, cannot be utilized inside the liver so these pass into the blood by diffusion. Several peripheral tissues, especially cardiac and skeletal muscles actively absorb the ketone bodies and use them as energy sources. To use the ketone bodies, first of all acetoacetic acid is reactivated to acetoacetyl CoA in the presence of mitochondrial enzyme and coenzyme succinyl CoA. Acetoacetvl CoA is then converted into acetyl CoA which enters Krebs cycle to form reduced coenzymes:

Acetoacetate + Succinyl CoA β Acetoacetyl CoA + Succinate

2. Role in Lipogenesis. Ketone bodies also act as precursors for lipogenesis (fat formation) in some tissues. For this β-hydroxybutyrate is converted to butyryl CoA which acts as a primer for fatty acid synthesis.

QUESTIONS

Long Answer Questions

1. Give an account of fatty acid oxidation.
2. Describe the process of biosynthesis of fatty acids.
3. Give an account of ketogenesis.
4. Write about lipids and fatty acids in our body. What is their significance in human life?

Short Answer Questions

1. Describe β-oxidation of palmitolic acid (*i.e.*, even number fatty acid).
2. Describe the formation and utilization of ketone bodies.
3. Give an account of alpha oxidation of fatty acids.
4. Write short notes on the following:
   (*i*) Fatty acid synthase (or FAS) complex;
   (*ii*) Ketogenesis;
   (*iii*) Ketone bodies.

Very Short Answer Questions

1. What is ketosis?
2. What is ketogenesis?
3. What is ketone body?
4. In which body tissue triacylglycerols are stored?
5. What is full from of VLDL?

Fill in the Blanks

1. Triacylglycerols are highly concentrated form of __________.
2. Triacylglycerols are transported from intestine as ____________ to liver, muscles and heart.
3. ____________, ________, _______ operates to transport activated fatty acids from cytosol to the mitochondria.
4. Alpha oxidation of fatty acids occur in __________ only.
5. PBD or peroxisome biogenesis disorder causes death of human infants within one year after their births. PBD causes ________ syndrome.
6. Acyl carrier protein (ACP) has __________ amino acids.
7. FAS complex stands for ___________ ___________ ____________ complex.
8. Triacylglycerols are ____________ and ____________ in nature.
9. Presence of ketonic bodies in urine is called ____________.

Multiple Choice Questions

1. Oxidizing machinery of β-oxidation of fatty acids is located in
   (a) mitochondria (b) cytoplasm
   (c) Golgi body (d) endoplasmic reticulum
2. Hormone sensitive lipase activity is inhibited by the hormone
   (a) epinephrine (b) insulin
   (c) thyroxine (d) glucocorticoids
3. The two final products in the β-oxidation of odd chain fatty acids are
   (a) acetyl CoA and malonyl CoA (b) acetyl CoA and acetyl CoA
   (c) acetyl CoA and propionyl CoA (d) acetyl CoA and succinyl CoA

ANSWER

Very Short Answer Questions

1. A high concentration of ketone bodies in the blood and tissues is called ketosis.
2. The formation of ketonic bodies such as acetoacetate, β-hydroxybutyrate and acetone from acety CoA is called ketogenesis.
3. Acetoacetate, β-hydroxybutyrate or acetone.
4. Adipose tissue.
5. Very low density lipoprotein.

Fill in the blanks

1. Energy.
2. Chylomicrons.
3. Carnitine carrier system.
4. Plants.
5. Zellweger.
6. 77.
7. Fatty acid synthase.
8. Non-polar, hydrophobic.
9. Ketonuria.

Multiple Choice Questions

1. (a) 2. (b) 3. (c).

4

Protein Metabolism

Proteins are the most abundant organic compounds and constitute a major part of the body dry weight (10–12 kg in adult). They perform a wide variety of *static* (structural) and *dynamic* (enzymes, hormones, clotting factors, receptors, motor proteins (*e.g.,* myosin, dynein and kinesin; **Verma** and **Agarwal**, 2016). About half of the body protein (predominantly collagen) is present in the supportive tissue (skeleton and connective tissue) while the other half is intracellular.

Proteins are nitrogen containing macromolecules (polymers) consisting of L-α-amino acids as the repeating units. Of the 20 amino acids found in proteins, half can be synthesized by the body (called **non-essential amino acids**) while the rest have to be provided in the diet (called **essential amino acids**).

Individual amino acids are released when the proteins are degraded (or protolysed). Amino acids are not just the structural components of proteins. Each one of the 20 naturally occurring amino acids undergoes its own metabolism and performs specific functions. Some of the amino acids also serve as precursors for the synthesis of many biologically important compounds (*e.g.,* melanin, serotinin, creatine etc).Certain amino acids may directly act as neurotransmitters (*e.g.,* glycinc, aspartate, glutamate). Protein metabolism is more appropriately learnt as metabolism of amino acids.

4.1 CATABOLISM OF AMINO ACIDS

Metabolism of amino acids includes both biosynthesis of amino acids (*i.e., anabolism*) and *catabolism* (*i.e.,* catabolic breakdown) of amino acids. Different organisms vary considerably in their ability to synthesize amino acids, *e.g.,* human beings and albino rats can synthesize 10 of the 20 types of amino acids while the higher animals can synthesize all the non-essential amino acids only from ammonium ions but not from nitrites and nitrates. The twenty different amino acids are synthesized by different multienzyme systems. Amino acid metabolism includes **protein synthesis, formation of amphibolic intermediates, urea synthesis** and **synthesis of a variety** of physiologically active compounds such as **serotonin**.

Conversely, catabolism of amino acids includes a number of processes such as follows:

1. Transamination.
2. Deamination.
3. Decarboxylation.

In mammals (including human beings) the catabolism of amino acids mainly occurs in the liver and to some extent in the kidneys also (Fig. 4.1).

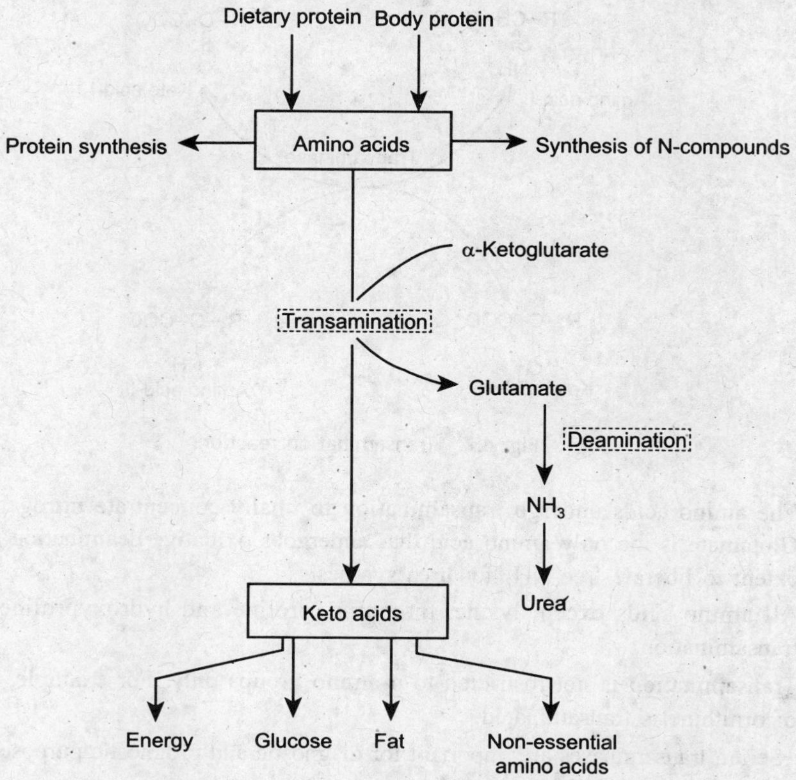

Fig. 4.1. An overview of amino acid metabolism.

4.2 TRANSAMINATION

Transamination means transfer of an amino ($-NH_2$) group from an amino acid to keto acid. This process involves the inter-conversion of a pair of amino acids and a pair of keto acids, catalyzed by a group of enzymes called *transaminases* (or *aminotransferases*). The enzymes are commonly found in the liver, kidneys and brain. Transaminase enzymes are found both in the mitochondria and in the cytosol of eukaryotic cells. Transamination was first reported by **Braunstein** and **Bychkou** (1939).

1. Characteristics of Transamination

1. All transaminases require *pyridoxal phosphate* (PLP), a coenzyme derived from vitamin B_6 or pyridoxine.
2. Specific transaminases exist for each pair of amino and keto acids. However, only two-namely, *aspartate transaminase* and *alanine transaminase*—make a significant contribution for transamination.
3. There is no free NH_3 (ammonia) liberated, only the transfer of amino group occurs.
4. Transamination is reversible (Fig. 4.2).
5. Transamination is very important for the redistribution of amino groups and production of non-essential amino acids, as per the requirement of the cell. It involves both catabolism (degradation) and anabolism (synthesis) of amino acids.
6. Transamination diverts the excess amino acids toward energy generation.

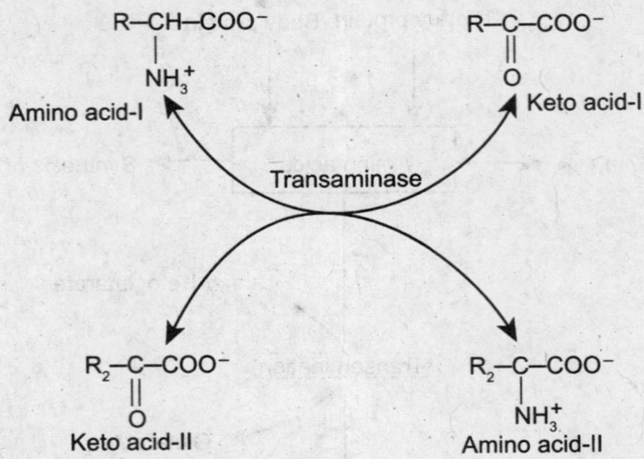

Fig. 4.2. Transamination reaction.

7. The amino acids undergo transamination to finally concentrate nitrogen in glutamate. Glutamate is the only amino acid that undergoes oxidative deamination to a significant extent to liberate free NH_3 for urea synthesis.
8. All amino acids except lysine, threonine, proline and hydroxyproline participate in transamination.
9. Transamination is not restricted to α-amino groups only. For example, δ-amino group of ornithine is transaminated.
10. Serum transaminases are important for diagnostic and prognostic purposes. For example, elevated levels of *alanine transaminase (ALT) or serum glutamate pyrovate transaminase (SGPT)* indicates liver diseases (*i.e.,* acute hepatitis of viral or toxic origin, jaundice and cirrhosis of liver. Likewise, elevated level of serum enzyme-*aspartate transaminase (AST) of serum glutamase transaminase* (SGOT) indicates heart attacks (or myocardial infarction).

2. Mechanism of Transamination

Transamination occurs in two stages (Fig. 4.3).

1. Transfer of the amino group to the coenzyme pyridoxal phosphate (PLP), which is bound to the coenzyme to form pyridoxamine phosphate.

2. The amino group of pyridoxamine phosphate is then transferred to a keto acid to produce a new amino acid and the enzyme with PLP is regenerated.

All the transaminases require a coenzyme, called *PLP* or *pyridoxal phosphate,* a derivative of vitamin B_6. The aldehyde group of **PLP** is linked with ε–amino group of lysine residue, at the active site of the enzyme forming a ***Schiff base*** (imine linkage). When an amino acid (substrate comes in contact with the enzyme, it displaces lysine and a new Schiff base linkage is formed. The amino acid-PLP-Schiff base tightly binds with the enzyme by non-covalent forces. **Snell** and **Braustein** proposed a **Ping-Pong Bi-Bi** mechanism involving a series of intermediates (aldimines and Ketimines) in transamination reaction.

4.3 DEAMINATION

Deamination means the removal of amino group from the amino acids as ammonia (NH_3). This liberated ammonia is used in **urea synthesis**. The carbon skeleton of amino acids is converted to keto acids. Deamination may be either oxidative or non-oxidative.

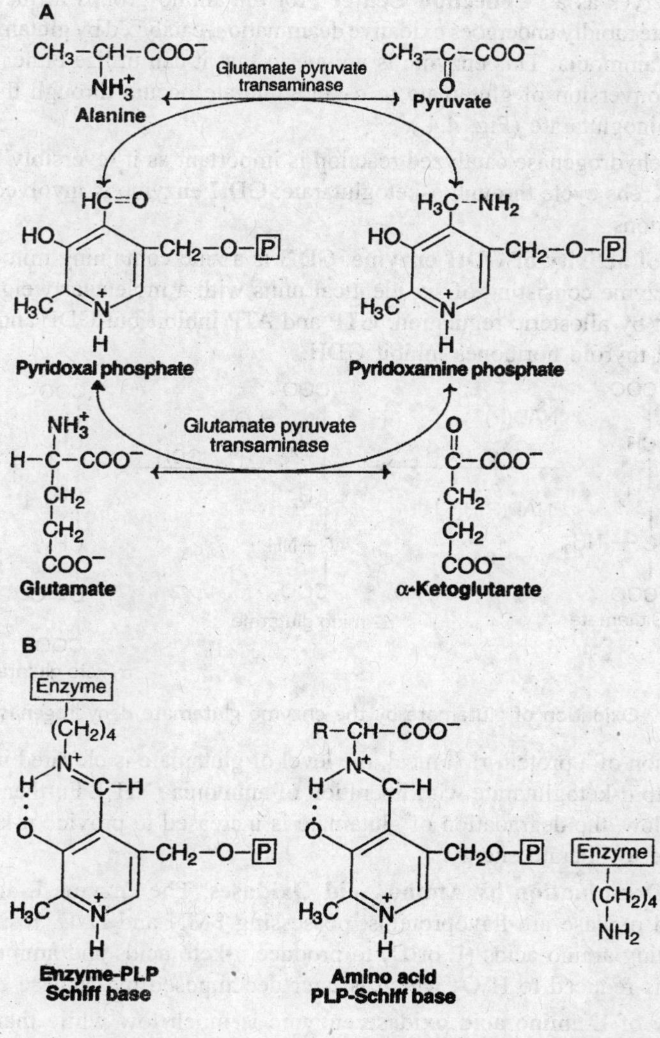

Fig. 4.3. Mechanism of transamination. A–Involvement of pyridoxal phosphate (PLP) in the transfer of amino group. B–Formation of enzyme-PLP-Schiff basa and amino acid-PLP-Schiff base. Note that when the amino acid binds, enzyme separates.

Box 4.1 *Transdeamination*

Generally, both transamination and deamination occur simultaneously, often involving glutamate as the central molecule. For this reason, some authors use the term transdeamination while describing the reactions of transamination and deamination, particularly involving glutamate.

A. Oxidative Deamination. Oxidative deamination is the liberation of free ammonia (NH_3) from the amino group of amino acids coupled with oxidation. This reaction occurs mostly in liver and kidney. The purpose of oxidative deamination is to provide NH_3 for urea synthesis and α-keto acids for a variety of reactions, including energy generation.

Role of GDH (or Glutamate Dehydrogenase). In the process of transamination, the amino groups of most amino acids are transferred to α-ketoglutarate to produce glutamate.

Thus, glutamate serves as a **"Collection Center"** for the amino groups in the biological system (Box 4.1). Glutamate rapidly undergoes oxidative deamination, catalyzed by glutamate dehydrogenase (GDH) to liberate ammonia. This enzyme is unique in that it can utilize either NAD^+ or $NADP^+$ as a coenzyme. Conversion of glutamate to α-ketoglutarate occurs through the formation of an intermediate, α-iminoglutarate (Fig. 4.4.).

Glutamate dehydrogenase catalyzed reaction is important as it reversibly links up glutamate metabolism with Krebs cycle through α-ketoglutarate. GDH enzyme is involved in both catabolic and anabolic reactions.

Regulation of activity of GDH enzyme. GDH is a zinc containing mitochondrial enzyme. It is a complex enzyme consisting of six identical units with a molecular weight of 56,000 each. GDH is controlled by allosteric regulation. GTP and ATP inhibit but GDP and ADP activate the GDH. Steroid and thyroid hormones inhibit GDH.

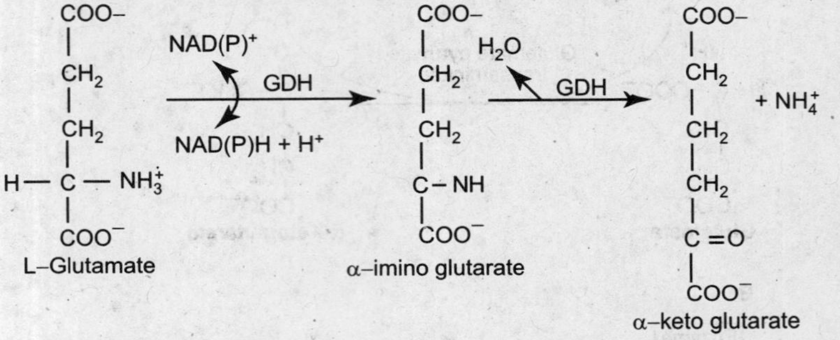

Fig. 4.4. Oxidation of glutamate by the enzyme glutamate dehydrogenase (GDH).

After ingestion of a protein-rich meal, the level of glutamate is elevated in liver. This amino acid is converted to α-ketoglutarate with liberation of ammonia (NH_3). Further, when the cellular, energy levels are low, the degradation of glutamate is increased to provide α-ketoglutarate which enters Krebs cycle to liberate energy.

Oxidative Deamination by Amino Acid Oxidases. The enzyme L-amino acid oxidase and D-amino acid oxidase are flavoproteins, possessing FMN and FAD, respectively. They act on the corresponding amino acids (L or D) to produce α-keto acids and ammonia (NH_3). In this reaction, oxygen is reduced to H_2O_2 which is later decomposed by catalase enzyme (Fig. 4.5).

The activity of L-amino acid oxidase enzyme is much low while that of D-amino acid oxidase is high in tissues (mostly liver and kidney). L-amino acid oxidase does not act on glycine and dicarboxylic acids. This enzyme, due to its very low activity, does not appear to play any significant role in the amino acid metabolism.

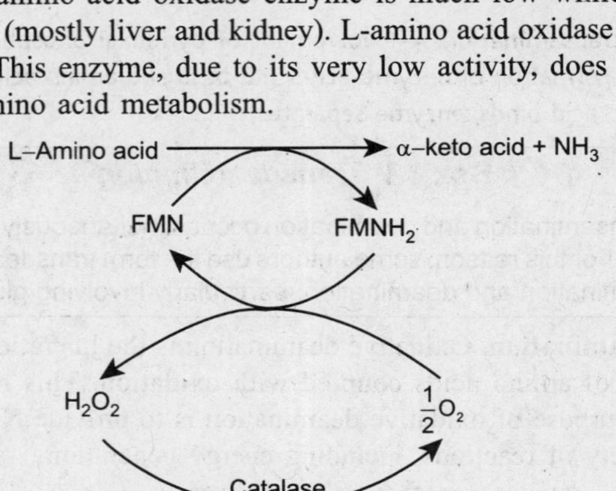

Fig. 4.5. Oxidative deamination of amino acid.

Fate of D-amino acids. D-amino acids are found in plants and microorganisms. They are not present in the mammalian proteins. However, D-amino acids are regularly taken in the diet and metabolized by the body. The enzyme D-amino acid oxidase converts them to the respective α-keto acids by oxidative deamination. These α-keto acids undergo transamination to be converted to L-amino acids which take part in various metabolic steps. Keto acids may be oxidized to generate energy or serve as precursors for glucose and fat synthesis. Thus, the enzyme D-amino acid oxidase is important as it initiates the first step for the conversion of D-amino acids to L-amino acids in the body (Fig. 4.6).

Non-oxidative Deamination. Some of the amino acids can be deaminated to liberate NH_3 without undergoing oxidation.

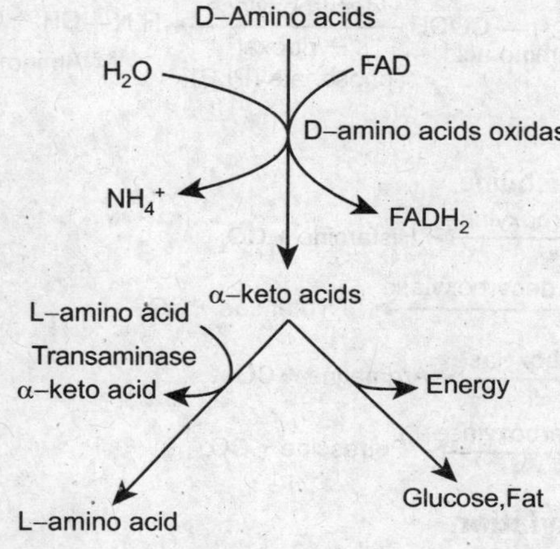

Fig. 4.6. Metabolic fates of D-amino acids.

***(i)* Role of Amino Acid Dehydrases Enzymes.** Serine, threonine and homoserine are the hydroxy amino acids. They undergo non-oxidative deamination catalysed by PLP (pyri-doxal phosphate) dependent dehydrases (dehydratase enzymes).

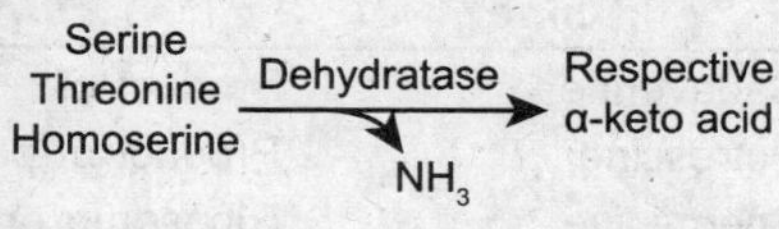

***(ii)* Amino Acid Desulphydrases.** The sulphur containing amino acids such as cysteine and homocysteine undergo deamination coupled with desulphydration to give keto acids.

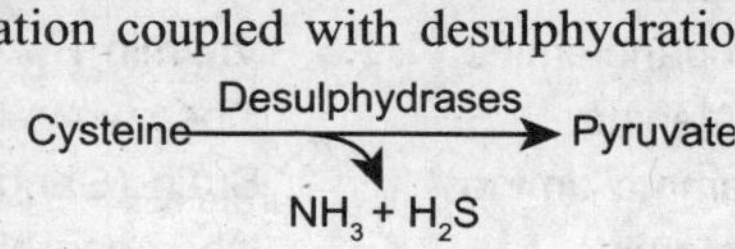

***(iii)* Deamination of Histidine.** The enzyme histidase acts on histidine to liberate NH_3 by a non-oxidative deamination process.

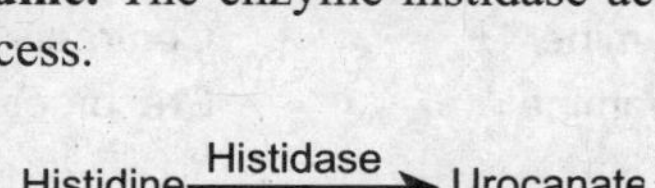

4.4 DECARBOXYLATION

Decarboxylation involves the removal of CO_2 from the carboxyl group of amino acids. Thus, it is a catabolic pathway.

Characteristics of Decarboxylation

1. Decarboxylation of amino acid is catalyzed by enzymes called **amino acid decarboxylases.**
2. Coenzyme required is the PLP (or pyridoxal phosphate), which is a derivative of vitamin B_6.
3. Amino acids are changed to corresponding *amines*, called **biogenic amines** (Table 4.1).
4. The cells containing such enzymes are liver and microorganisms found in intestinal tract.
5. Decarboxylation is involved in the **putrefaction** in intestine and produces **toxic amines** such as **tyramine, trypamine** and **cadaverine** from tyrosine, tryptophan and lysine.
6. These are mainly **irreversible reactions** except the decarboxylation of glutamic acid in brain tissue.

$$\underset{\alpha\text{-amino acid}}{H_2N-\overset{\overset{\displaystyle R}{|}}{C}H-COOH} \xrightarrow[\text{Pyrldoxal phosphate (PLP)}]{\text{Amino acid decarboxylase}} \underset{\text{Amino}}{H_2N-\overset{\overset{\displaystyle R}{|}}{C}H_2} + CO_2$$

Examples

1. Decarboxylation of histidine.

$$\text{Histidine} \xrightarrow{\text{Histidine decarboxylase}} \text{Histamine} + CO_2$$

2. $\text{Tryptophan} \xrightarrow{\text{Tryptophan decarboxylase}} \text{Tryptamine} + CO_2$

3. $\text{Arginine} \xrightarrow{\text{Arginine decarboxylase}} \text{Argmatine} + CO_2$

4. $\text{Ornithine} \xrightarrow{\text{Ornithine decarboxylase}} \text{Putrescine} + CO_2$

Significance of Decarboxylation

1. Many biogenic amines have important physiological and pharmacological effects.
2. Some act as important precursors of steroid hormones.
3. Some act as coenzymes.

Table 4.1 Some biologically important amines formed by decarboxylation.

	Amino acid	*Amine*	*Occurrence and significance*
1.	Lysine	Cadaverine	Product of putrefaction.
2.	Ornithine	Putrescine	Product of putrefaction.
3.	Methionine	Spermidine	Ribosome and sperm.
4.	Arginine	Argmatine	Putrefaction.
5.	Threonine	Propanolamine	Vitamin B_{12}.
6.	Aspartic acid	β-Alanine	Coenzyme A and pantothenic acids.
7.	Glutamic acid	Gamma amino	Brain (Ganglion inhabitor).
8.	Histidine	Histamine	Decrease blood pressure as acts as a vasodilator.
9.	Tyrosine	Tyramine	Contracts uterus.
10.	3, 4-Dihydroxy-phenyl alanine	Dopamine	Precursor of adrenalin and nor-adrenalin.

4.5 UREA CYCLE

Metabolism of Ammonia

Ammonia is constantly being liberated in the metabolism of amino acids (mostly) and other nitrogenous compounds. At the physiological pH, ammonia exists as a ***ammonium*** (NH_4^+) ***ion.***

***(i)* Formation of Ammonia.** The production of ammonia (NH_3) occurs from the amino acids (*i.e*, transamination and deamination), biogenic amines, amino groups of purines and pyrimidines and by the action of intestinal bacteria (urease) or urea.

***(ii)* Transport and Storage of Ammonia.** Due to regular and constant production of ammonia (NH_3) from various tissues, its concentration in the circulation (blood) is surprisingly low (normal plasma 10.20 mg/dl). This is mostly because the body has an efficient mechanism for ammonia transport and its immediate utilization for **urea synthesis**. The transport of ammonia between various tissues and the liver mostly occurs in the form of **glutamine** or **alanine** and not as free ammonia. Alanine is important for NH_3 (ammonia) transport from muscle to liver by *glucose-alanine cycle.*

***(iii)* Role of Glutamine is Ammonia Management.** Glutamine is a **storehouse of ammonia.** It is present at the highest concentration (*i.e.*, 8 mg/dl in adults) in blood among the amino acids. The amino acid glutamine serves as a storage and transport form of ammonia. Its synthesis mostly occurs in liver, brain and muscle. Ammonia is removed from the brain predominantly as glutamine. Glutamine is **freely diffusible** in tissues, hence easily transported.

A mitochondrial enzyme, called **glutamine synthetase** is responsible for the synthesis of glutamine from glutamate and ammonia (Fig. 4.7). This reaction is unidirectional requires ATP and Mg^+ ions.

Glutamate is deaminated by hydrolysis to release ammonia (as ammonium, NH_4^+, ions) by glutaminase enzyme. This enzyme is mostly found in kidney and intestinal cells.

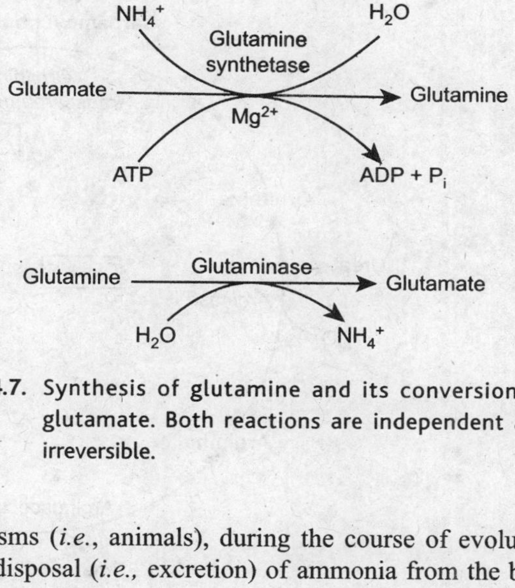

Fig. 4.7. Synthesis of glutamine and its conversion to glutamate. Both reactions are independent and irreversible.

***(iv)* Functions of Ammonia.** Ammonia is not just a waste product of nitrogen metabolism. It is involved (directly or via glutamine) for the **synthesis** of many compounds in the body. These include *non-essential amino acids, purines, pyrimidines, amino sugars, asparagine,* etc. Ammonium ions (NH_4^+) are very important to maintain **acid-base balance** of the body.

***(v)* Disposal of Ammonia.** The organisms (*i.e.*, animals), during the course of evolution, have developed different mechanism for the disposal (*i.e.*, excretion) of ammonia from the body. The animals in this regard are of three different types:

(*a*) **Ammonotelic Animals.** The aquatic animals dispose of ammonia (NH_3) into the surrounding water.

(*b*) **Uricotelic Animals.** In terrestrial animals such as reptiles and birds, ammonia is converted mostly of **uric acid** and is disposed off in that state.

(*c*) **Ureotelic Animals.** In mammals, the ammonia is converted into **urea.** Urea is a nontoxic and soluble compound, hence easily excreted as urine.

***(v)* Toxicity of Ammonia.** Even a marginal elevation in the blood ammonia concentration is harmful to the brain. Ammonia, when it accumulates in the human body, results in slurring of speech and blurring of the vision. It causes tremors. This may lead to coma and finally, death, if not corrected.

Urea Cycle

Urea is the end product of protein metabolism (or amino acid metabolism). The nitrogen of amino acids, converted to ammonia, is toxic to the body. It is converted to urea and detoxified. As such urea accounts for 80-90% of the nitrogen containing substances excreted in urine.

Urea is synthesized in liver and transported to kidneys for excretion in urine. Urea cycle or **ornithine cycle** is the *first metabolic cycle* that was explained by **Hans Krebs** and **Kurt Henseleit** (1932), hence, it is known as *Krebs-Henseleit cycle.* The individual reactions of urea cycle, however, were described in more detail later by **Ratner** and **Cohen.**

Urea has two amino ($-NH_2$) groups, one derived from NH_3, and the other from aspartate. Carbon atom of the urea is supplied by CO_2. Urea synthesis is a five-step cyclic process, with five distinct enzymes. The first two enzymes of urea cycle are present in mitochondria, while the rest are localized in cytosol (Fig. 4.8).

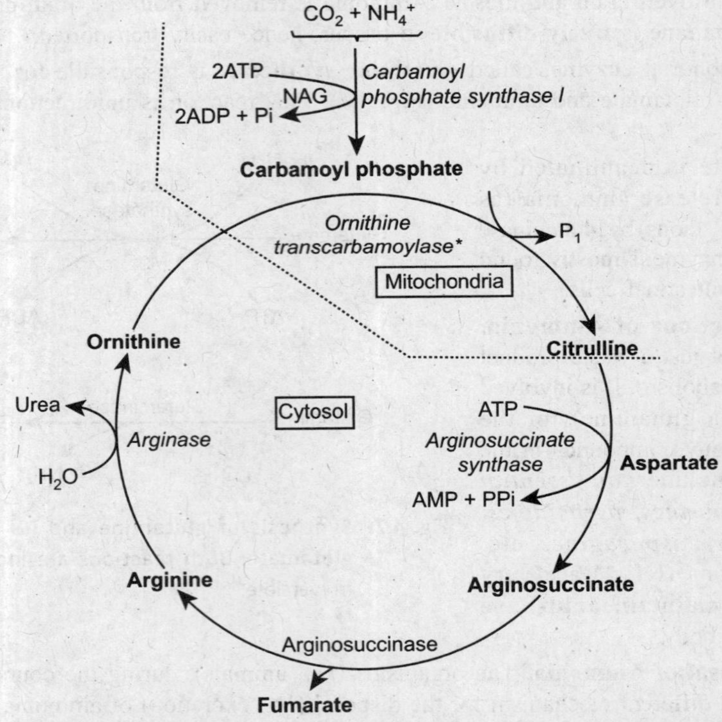

Fig. 4.8. Reactions of urea cycle, (NAG = N - Acetyl glutamate; * = mitochondrial enzymes; the rest of the enzymes are cytosomal).

1. Synthesis of Carbamoyl Phosphate. The enzyme *carbamoyl phosphate synthase I* (*CPSI*) of mitochondria catalyses the condensation of ammonium ion (NH_4^+) with CO_2 to form **carbamoyl phosphate**. This step consumes two ATP molecules and is irreversible, and rate limiting. CPSI (enzyme) requires N-acetyl glutamate (NAG) for its activity. Another enzyme, *carbamoyl phosphate synthase II* (*CPS II*)- involved in pyrimidine synthesis-is present in cytosol. It accepts amino group from glutamine and does not require NAG (N-acetyl glutamate) for its activity.

2. Formation of Citrulline. Citrulline is synthesized from carbamoyl phosphate and ornithine by *ornithine transcarbamoylase.* Ornithine is regenerated and used in urea cycle. Therefore, its role is comparable to that of oxaloacetate in citric acid cycle. *Ornithine* and citrulline are basic amino acids. (*Note.* Ornithine and citrulline are never found in protein structure due to lack of codons of DNA/mRNA). Citrulline produced in this reaction is transported to cytosol by a transporter system.

3. Synthesis of Arginosuccinate. The enzyme *arginosuccinate synthase* condenses citrulline with aspartate to produce arginosuccinate. The second amino group of urea is incorporated in this reaction. This step of urea cycle requires ATP which is cleaved to AMP and pyrophosphate (PP_i). The pyrophosphate is immediately broken down to inorganic phosphate (P_i).

4. Cleavage of Arginosuccinate. The enzyme arginosuccinase cleaves arginosuccinate to give arginine and fumarate. Arginine is the immediate precursor for urea. Fumarate liberated here provides a connecting link with Krebs cycle, gluconeogenesis, etc.

5. Formation of Urea. *Arginase* is the fifth and final enzyme that cleaves arginine to yield urea and ornithine. Ornithine, so regenerated, enters mitochondria for its reuse in the urea cycle. Arginase enzyme is activated by Co^{2+} and Mn^{2+}. Ornithine and lysine compete with arginine (competitive inhibition). Arginase is mostly found in the liver, while the rest of the enzymes of urea cycle (which are four in number) are also present in other tissues. For this reason **arginine synthesis** may occur to varying degrees in many tissues. But only the liver can ultimately produce urea.

Energetics of Urea Cycle

Urea cycle is irreversible and consumes ATP. Two ATP molecules are utilised for the synthesis of carbamoyl phosphate. One ATP is converted to AMP and PP_i to produce arginosuccinate which equals to 2ATP. Hence 4ATP are actually consumed.

$$NH_4^+ + CO_2 + \text{Aspartate} + 3ATP \rightarrow \text{Urea} + \text{Fumarate} + 2ADP + 2P_i + AMP + PP_i$$

Regulation of Urea Cycle

The first reaction of urea cycle is catalysed by the enzyme *carbamoyl phosphate synthase I* (CPSI) is **rate-limiting** reaction or committed step in urea synthesis. CPSI is allosterically activated by N-acetylglutamate (NAG). NAG is synthesized from glutamate and acetyl CoA by synthase and degraded by a hydrolase (Fig. 4.9).

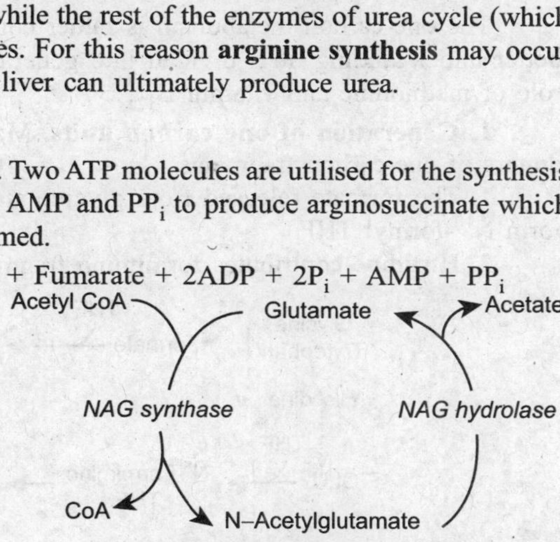

Fig. 4.9. Formation and degradation of N-acetylglutamate.

The rate of urea synthesis in liver is correlated with the concentration of N-acetylglutamate. High concentrations of arginine increase NAG. The consumption of a protein-rich meal increases the level of NAG in liver, leading to enhanced urea synthesis.

The enzymes carbamoyl phosphate synthase I and glutamate dehydrogenase are localized in the mitochondria. They coordinate with each other in the formation of NH_3 (ammonia), its utilization for the synthesis of **carbamoyl phosphate.** The remaining four enzymes of urea cycle are mostly regulated, by the concentration of their respective substrate.

Disposal of Urea

Urea produced in the liver freely diffuses and is transported in blood to **kidneys** and **excreted.** A small amount of urea enters the intestine where it is broken down to CO_2 and NH_3 by the bacterial enzyme urease. This ammonia is either lost in the feces or absorbed into the blood. In **renal failure**, the blood urea level is elevated (called **uremia**), resulting in diffusion of more urea into intestine and its breakdown to NH_3. ***Hyperammonemia*** (increased blood NH_3) is commonly seen in patients of kidney failure. For these patients, oral administration of antibiotics (*e.g.*, neomycin) to kill intestinal bacteria is advised.

4.6 ONE-CARBON METABOLISM

Amino acid metabolism is particularly important for the transfer or exchange of **one-carbon units.** The following one-carbon fragments are encountered in the biological reactions, which constitute one-carbon pool.

1. Methyl ($-CH_3$)
2. Hydroxy methyl ($-CH_2OH$)
3. Methylene ($-CH_2$)
4. Methenyl ($-CH =$)
5. Formyl ($-CH = O$)
6. Forminino ($-CH = NH$)

***Tetrahydrofolate* (*THF or FH4*)** is a versatile coenzyme that actively participates in **one-carbon metabolism**. With regard to the transfer of methyl groups for S-adenosylmethionine vitamin B_{12} is also involved besides THF.

The one-carbon unit covalently binds with THF at position N^5 or N^{10} or on both N^5 and N^{10} of pteroyl structure of folate.

The one-carbon metabolism is rather complex, involving many reactions. For the sake of better understanding, it is divided into generation and utilization of one-carbon units, and the role of methionine and vitamin B_{12}.

1. Generation of one carbon units. Many compounds (particularly amino acids) act as donors of one-carbon fragments.

1. The formate released from glycine and tryptophan metabolism combines with THF to form N^{10}-formyl THF.

2. Histidine contributes formimino fragment to produce N^5-formimino THF.

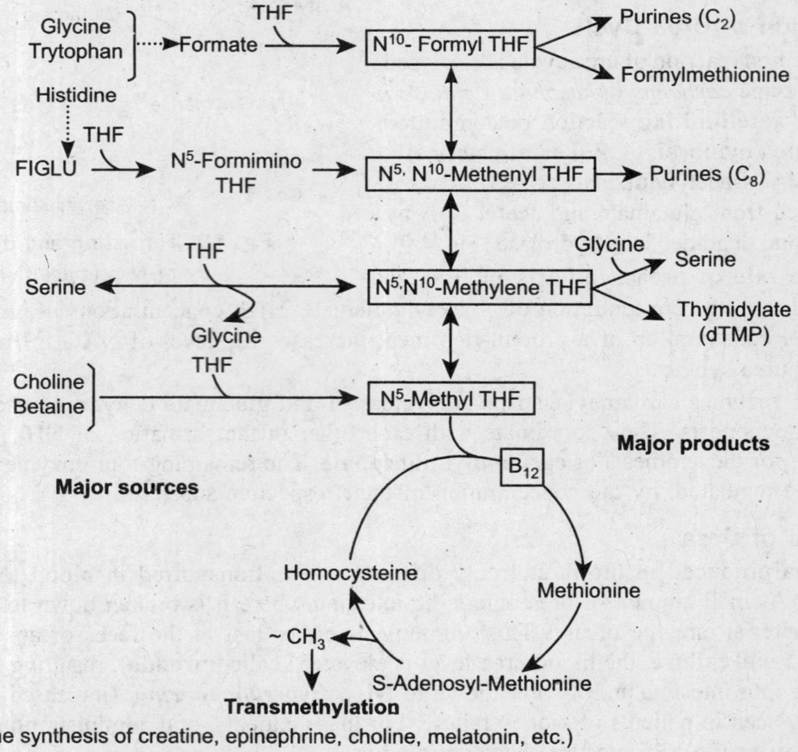

Fig. 4.10. Summary of one-carbon metabolism (THF = Tetrahydrofolate; ~ CH_3 = active methyl group).

3. When serine is converted to glycine, N^5, N^{10}-methylene THF is formed. This is the most predominant entry of one carbon units into one carbon pool.

4. Choline and betaine contribute to the formation of N^5-methyl THF.

The different derivatives of THF carrying one-carbon units are interconvertible and this is metabolically significant for the continuity of one-carbon pool (Fig. 4.10).

2. Utilization of one-carbon moieties. One-carbon fragments from THF are used for the synthesis of wide variety of compounds. These include purines, formyl-methionine tRNA (required for initation of protein synthesis), glycine, pyrimidine nucleotide (thymidylate), etc.

3. Role of methionine and B12 in one-carbon metabolism. Methyl ($-CH_3$) group is an important one-carbon unit. The role of active methionine as methyl donor in transmethylation reactions is already described. After the release of methyl group, methionine is converted to homocysteine. For the regeneration of meteionine, free homocysteine and N_5-methyl THF are required and this reaction is dependent on methylcobalamine (vitamin B_{12}). The one-carbon pool, under the control of THF, is linked with methionine metabolism (transmethylation) through vitamin B_{12}. Thus vitamin B_{12} is also involved in one-carbon metabolism.

4.7 FATE OF CARBON SKELETON OF AMINO ACIDS

After the removal of amino groups, the carbon skeleton of amino acids is converted to intermediates of Krebs cycle (or TCA cycle) or their precursors. The carbon skeleton finally has one or more of the following fates:

1. Oxidation via Krebs cycle to produce energy (about 10-15% of body needs).
2. Synthesis of glucose.
3. Formation of lipids-fatty acids and ketone bodies.
4. Synthesis of non-essential amino acids.

The carbon skeletons of amino acids of proteins are degraded to one of following seven products–**pyruvate, α-ketoglutarate, succinyl CoA, fumarate, oxaloaeetate, acetyl CoA and acetoacetate.** These products are also called **amphibolic (*Gr.*,** amphibole = uncertain) **intermediates** due to their multiple functions.

The amino acids are classified into two groups, based on the nature of the metabolic end products or carbon skeleton.

1. Glucogenic (or glycogenic) amino acids. 2. Ketogenic amino acids.

1. Glucogenic amino acids. These are the amino acids whose carbon skeleton is finally degraded to pyruvate or one of the intermediates of Krebs cycle (*e.g.*, Ketoglutarate succinyl CoA, fumarate and oxaloaeetate). These intermediates serve as good substrates for glucogenesis leading to formation of glucose or glycogen.

2. Ketogenic amino acids. The amino acids whose carbon skeleton is metabolized to acetyl CoA or acetoacetate can be converted to fat (*i.e.,* fatty acids or ketone bodies). Acetoacetate is a ketone body (besides β-hydroxybutyrate and acetone).

Some of the amino acids are both glycogenic and ketogenic since they serve as precursors for glucose as well as fat.

The classification of amino acids (glucogenic, ketogenic, or both) is given in Table 4.2. The various products obtained from the carbon skeleton of amino acids and their connection with the Krebs cycle is depicted in Fig. 4.11.

Amphibolic intermediates obtained from the degradation of amino acids are summarised here under.

1. Pyruvate, alanine, cysteine, glycine, hydroxyproline, serine and threonine.
2. **Ketoglutrate**. Glutamine, glutamate, arginine, histidine and proline.
3. **Succinyl CoA**. Isoleucine, methionine, threonine and valine.
4. **Fumarate**. Phenylalanine and tyrosine.
5. **Oxaloacetate**. Asparagine and aspartate.
6. **Acetyl CoA and Acetoacetate.** Phenylalanine, tyrosine, tryptophan, isoleucine. leucine and lysine.

Table 4.2. Classification of amino acids based on the fate of carbon skeleton.

Glycogenic (or Glucogenic)	*Glycogenic and Ketogenic*	*Ketogenic*
1. Alanine	1. Phenylalanine*	1. Leucine*
2. Arginine*	2. Isoleucine*	2. Lysine*
3. Aspartate	3. Tyrosine	
4. Cysteine	4. Tryptophan*	
5. Glutamine		
6. Glutamate		
7. Glycine		
8. Histidine*		
9. Hydroxyproline		
10. Methionine*		
11. Proline		
12. Serine		
13. Threonine*		
14. Valine*		

*Essential amino acids.

Leucine and lysine are only ketogenic, since they produce acetoacetate or acetyl CoA.

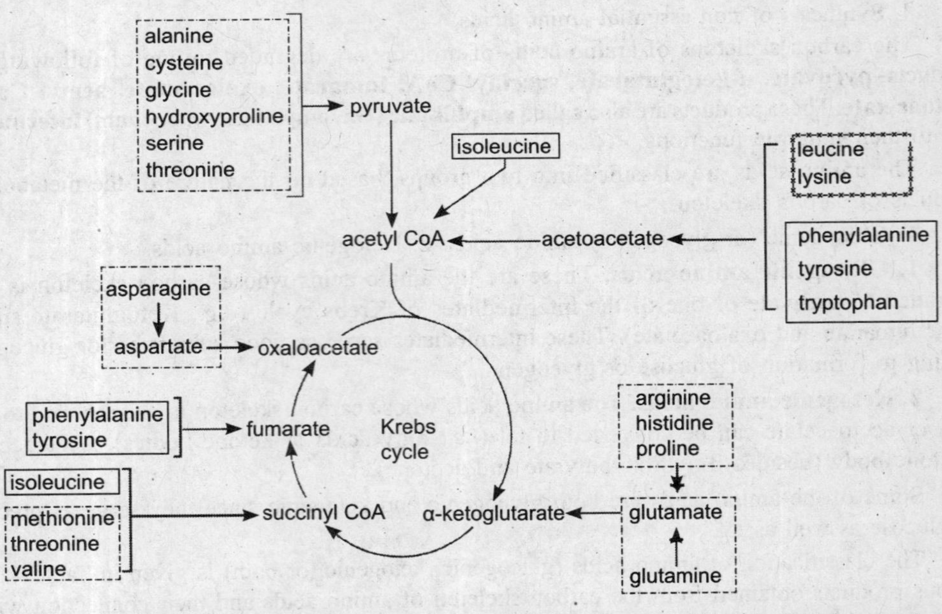

Fig. 4.11. Summary of the products formed from carbon skeleton of amino acids. (dashed box: glucogenic, solid box: glucogenic and ketogenic, crossed box: ketogenic).

Biosynthesis of Nonessential Amino Acids

Of the 20 amino acids, about half of them are non-essential in the diet, as they can be synthesized in human body. This is carried out by the biosynthesis of carbon skeleton, followed by the addition of amino group via transamination. In Table 4.3, sources of carbon skeleton for the synthesis of nonessential amino acids are given.

Table 4.3. The sources of carbon skeleton for the biosynthesis of non-essential amino acids.

Amino acids	*Sources of carbon Skeleton*
1. Glycine	Serine
2. Alanine	Pyruvate
3. Serine	Phosphoglycerate
4. Aspartic acid	Intermediates of Krebs cycle (4–9)
5. Asparagine	
6. Glutamic acid	
7. Glutamine	
8. Proline	
9. Cysteine	
10. Tyrosine	Serine (sulphur donated by methionine)
	Phenylalanine

QUESTIONS

Long Answer Questions

1. Describe various methods of catabolism of amino acids.
2. Describe the reactions in the synthesis of urea.
3. Explain urea cycle or ornithine cycle and its significance.
4. Describe the fate of carbon skeleton of amino acids.

Short Answer Questions

1. Describe various methods of oxidative deamination of amino acids.
2. Describe the mechanism of transamination taking the example of glutamate-pyruvate transaminase.
3. Describe the mechanism of oxidative decarboxylation of amino acids.
4. Write a note an urea cycle.
5. Write short note on the following:
   (i) Transamination; (ii) Deamination;
   (iii) One-carbon metabolism; (iv) Ammonia toxicity.

Very Short Answer Questions

1. Write an example of transamination.
2. Name two amino acid oxidases and their action.
3. Where does urea cycle operate?
4. Who explained urea cycle?
5. Name two non-proteinous amino acids which participate in urea cycle.
6. Give an example of amphibolic intermediate of protein metabolism.

Fill in the Blanks

1. Oxidative deamination of glutamic acid produces ___________.
2. Transamination of glutamic acid to pyruvic acid produces __________ and ___________.
3. Pyridoxal phosphate enzyme is derivative of vitamin ___________.
4. Urea cycle involves _________ and _________ amino acids.

5. The coenzyme that participates in transamination reaction is ____________.
6. The most important enzyme involved in oxidative deaminatin is __________.
7. N-acetylglutamate (NAG) is required for the activation of enzyme ________.
8. During oxidative deamination, amino acid glutamate serves as_____ ______for the amino groups.
9. _______ is the storehouse of ammonia in human body.

Multiple Choice Questions

1. The amino acid that does not participate in transamination
   (a) lysine (b) glutamate
   (c) alanine (d) tryptophan
2 The amino acids are said to be ketogenic when the carbon skeleton is finally degraded to
   (a) succinyl CoA (b) fumarate
   (c) acetyl CoA (d) pyruvate
3. The synthesis of urea occurs in
   (a) kidney (b) liver
   (c) muscle (d) brain

ANSWERS

Very Short Answer Questions

1. Alanine and glutamate.
2. The enzyms L-amino acid oxidase and D-amino oxidase are flavoproteins, possessing FMN and FAD, respectively. They act on the corresponding amino acids (L or D) to produce α-keto acids and ammonia (NH_3; Fig.4.5).
3. Liver.
4. Hans Krebs and Kurt Henseleit in 1932.
5. Ornithine and Citrulline.
6. Acetyl CoA.

Fill in the Blanks

1. α-ketoglutaric acid;
2. α-ketoglutarate, 2-alanine;
3. B6;
4. Ornithine, citrulline;
5. Pyridoxals phosphate (PLP);
6. Glutamate dehydrogenase;
7. Carbamoyl phosphate synthase I;
8. Collection center;
9. Glutamine.

Multiple Choice Questions

1. (a) 2. (c) 3. (b)

5

Oxidative Phosphorylation

A living organism is an **open system** and exchanges both matter and energy with its surroundings. Oganisms derive energy from their surroundings in two ways : (*i*) uptake of chemical fuels (*e.g.*, sugars, fats) as nutrients and extract energy by oxidizing them; and (*ii*) absorb energy from sunlight (*e.g.*, by photosynthesis in photoautotrophs). Most of the organisms completely rely an oxygen for their metabolism upon the cellular respiration (also called **oxidative phosphorylation**) in which cells derive energy by an enzymatically controlled stepped process in which a controlled reaction between H^+ and O_2 and end product is water. Though some organisms (*e.g.*, anaerobes) derive energy from their chemical fuels without the involvement of oxygen (**fermentation**). Both aerobic and anaerobic respiratory pathways are regulated by a specific sequence of enzymes, especially the **dehydrogenases**. A part of energy released is stored in high energy bonds of energy-rich molecules especially ATP and ADP.

Bioenergetics or **biochemical thermodynamics** deals with the study of energy changes (transfer and utilization) in biochemical reactions. The reactions are broadly classified as **exergonic** (energy releasing) and **endergonic** (energy consuming). Bioenergetics is concerned with the initial and final states of energy component of the reactant and not the mechanism of chemical reactions.

Free energy. The energy actually **available to do work** (utilizable) is known as **free energy. Changes** in the free (ΔG) are valuable in predicting the feasibility of chemical reactions. The reactions can occur spontaneously if they are accompanied by decrease in energy in free energy.

During a chemical reaction, heat may be released or absorbed. **Enthalpy** (ΔH) is a measure of the change in heat content of the reactants, composed of products.

Entropy (ΔS) represents a change in the randomness or **disorder of reactants and products**. Entropy attains a maximum as the reaction approaches equilibrium. The reactions of biological systems involve a temporary decrease in entropy.

The relation between the changes of free energy (ΔG) , enthalpy (ΔH) and entropy (ΔS) is expressed as

$$\Delta G \;=\; \Delta H - T\Delta S$$

T represents the absolute temperature in Kelvin ($K = 273^+$ °C).

The term **standard free energy** represented by ΔG^o (note the superscript °) is often used. It indicates the **free energy change** when the reactants or products are at a concentration of 1 mol/l at pH 7.0.

Negative and positive Δ**G**. If free energy change (ΔG) is represented by a negative sign, there is loss of free energy. The reaction is said to be *exergonic* and proceeds spontaneously. On the other hand, a positive ΔG indicates that energy must be supplied to the reactants. Such a reaction cannot proceed spontaneously and is *endergonic* in character.

The *hydrolysis* of ATP is a classical example of exergonic reaction :

$$ATP + H_2O \rightarrow ADP + P_i \ (\Delta G^o = -\ 7.3 \text{ Cal/mol})$$

The reverse of this reaction (*i.e.*, ADP + P_i = ATP) is *endergonic* and occurs only when there is a supply of energy of at least 7.3 Cal/mol (*i.e.*, ΔG^o is positive).

The free energy change becomes zero ($\Delta G^o = 0$) when a reaction is at equilibrium.

At a constant temperature and pressure, ΔG is dependent on the actual concentration of reactants and products. For the conversion of reactants A to product B (A → B), the following mathematical relation can be derived.

$$\Delta G = \Delta G^o + RT \text{ in } \frac{(B)}{(A)}$$

where ΔG^o = Standard free energy change

R = Gas constant (1.987 Cal/mol)

T = Absolute temperature (273^+ °C)

in = Natural logarithm

(B) = Concentration of product

(A) = Concentration of reactant.

Certain compounds are encountered in the biological system which on hydrolysis yield energy. The term high-energy compounds or **energy rich compounds** is usually applied to substances which possess sufficient free energy to liberate at least 7 Cal/mol at pH 7.0.

5.1 REDOX SYSTEMS

Chemically, **oxidation** is defined as the removal of electrons and **reduction** as the gain of electrons as illustrated by the oxidation of ferrous (Fe^{2+}) to ferric (Fe^{3+}) ion.

$$\underset{\text{Ferrous ion}}{Fe^{2+}} \xrightarrow{\nearrow e^- \text{ (electron)}} \underset{\text{Ferric ion}}{Fe^{3+}}$$

It follows that oxidation is always accompanied by reduction of electron acceptor. This principle of oxidation - reduction applies equally to biochemical systems and is an important concept underlying understanding of the nature of biologic oxidation. It will be appreciated that many biologic oxidations can take place without the participation of molecular oxygen, *e.g.*, dehydrogenation.

Biologic oxidation involves transfer of electrons from one compound (which undergoes oxidation) to another compound (which undergoes reduction) so involves the oxidation - reduction or **redox reactions** which is coupled with release of free energy.

Redox Reactions

All the metabolic reactions which involve flow of electrons are oxidation - reduction reactions or redox reactions as in these electron-donor is oxidized while electron acceptor is reduced. Both the electron donor (also called **reducing agent**) and the electron acceptor (also called **oxidizing agent**) together make a **conjugate redox pair**. A redox reaction represents as half reaction, *e.g.*,

$$\underset{\text{(Reducing agent)}}{Fe^{2+}} + \underset{\text{(Oxidizing agent)}}{Cu^{2+}} \rightleftharpoons \underset{\text{(Oxidized)}}{Fe^{3+}} + \underset{\text{(Reduced)}}{Cu^{+}}$$

In redox reactions, the electrons move from a compound having less **redox potential** (*i.e.*, measure of its affinity for electrons) to a compound having more redox potential. The difference in electron affinity between electron donor and electron acceptor develop an **electromotive force (emf)**. The electrons flow in proportion to emf. In ETC, the electrons flow is exergonic because O_2 has a higher electron affinity than intermediate electron carriers.

Transfer of electrons takes place in one of the four possible ways and in biological metabolic reactions, it is always regulated by specific enzymes such as **oxidases** (which use oxygen as electron acceptor) or **dehydrogenases** (which removes electrons/hydrogen atoms from the substrate) or **hydroperoxidases** (which uses H_2O_2 as a substrate) or **oxygenases** (which catalyze direct transfer of O_2 into the substrate). Out of these oxidases and dehydrogenases are involved in respiration, hydroperoxidases neutralise the free radicals, while oxygenases are involved in biotransformations.

1. Oxidases - dependent energy - releasing biological processes. Oxidases catalyze those biological reactions in which oxygen acts as a hydrogen acceptor from the reduced substrates. General mechanism catalyzed by oxidases can be shown by following reaction:

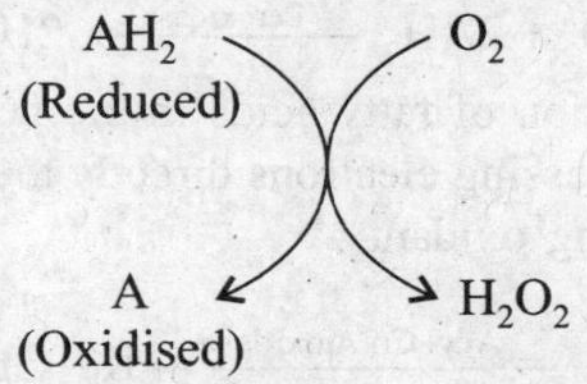

Oxidases exist in two different forms:

(*i*) Some oxidases are copper-containing *e.g.*, **cytochrome oxidase**. It is complex-IV of the ETC and carries the electrons from cytochrome C to molecular oxygen which is reduced to water.

(*ii*) Some oxidases are **flavoproteins** (have flavin-derived prosthetic groups), *e.g.*, **(a) D-amino acid oxidase**. It is present at high levels in the kidney and is involved in the detoxification of ingested D-amino acids derived from the bacterial cell walls and from grilled food stuffs, *e.g.*, D-glycine oxidase:

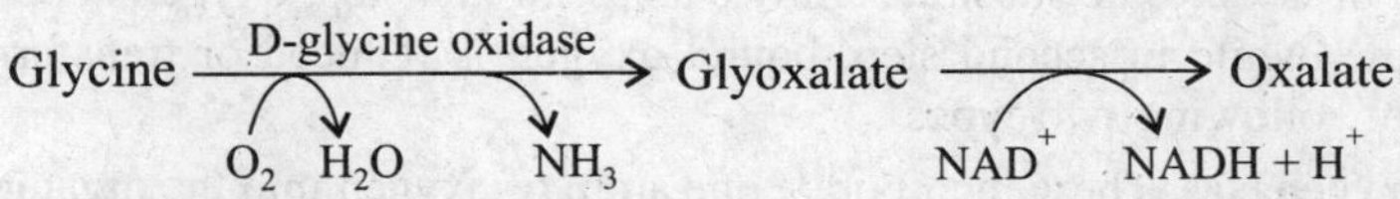

Glycine is an inhibitory neurotransmitter.

(b) Xanthine oxidase. It is a flavoprotein with an atom of molybednum and four sulphur centers in its prosthetic group. This enzyme oxidises hypoxanthine to xanthine and then to uric acid for excretion.

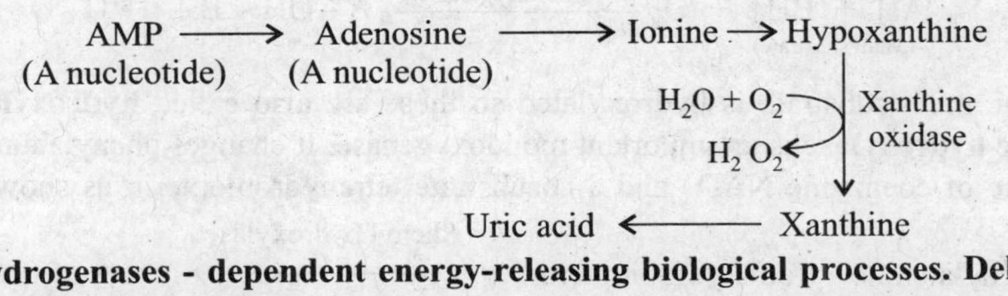

2. Dehydrogenases - dependent energy-releasing biological processes. Dehydrogenases regulate those metabolic reactions in which 2H-atoms are transferred from one compound to another compound in the presence of specific coenzymes which act as hydrogen carriers. So these enzymes control redox reactions and are important component of ETC. Coenzymes for dehydrogenases are derived either from nicotinamide (*e.g.*, NAD^+ or $NADP^+$) or riboflavin (*e.g.*, FMN or FAD). General mechanism catalyzed by dehydrogenases is shown in following reaction:

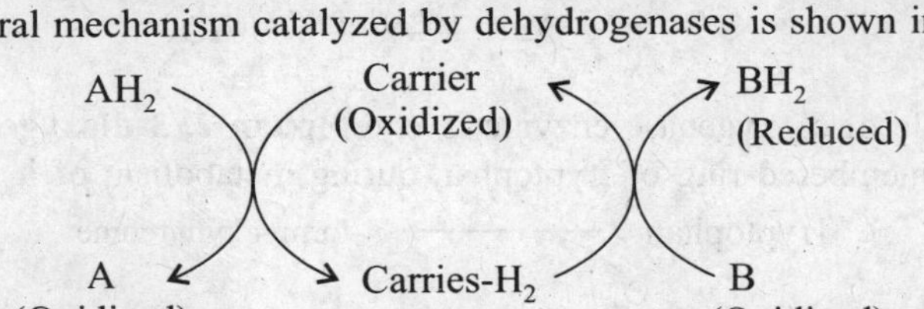

Examples:

1. Biological functions catalyzed by NAD^+ - linked dehydrogenases.
2. Biological functions catalyzed by $NADP^+$ - linked dehydrogenases.
3. Biological functions catalyzed by Flavin - linked dehydrogenases.

3. Hydroperoxidases - dependent energy - releasing biological processes. Hydroperoxidases are abundantly present in the peroxisomes and catalyze the metabolism of H_2O_2. These include two sets of enzymes: **peroxidases** and **catalases**. Peroxidases reduce H_2O_2 (hydrogen peroxide) at the expense of some other substrates and their general mechanism is as under

$$H_2O_2 + 2AH_2 \xrightarrow{\text{Peroxidase}} 2H_2O + A$$

For example, during β-oxidation of fatty acids in the peroxisomes, the enzyme peroxidase desaturates the fatty acyl Co.A by passing electrons directly to oxygen and producing H_2O_2 which is a strong and potentially damaging oxidant :

$$\underset{\text{Saturated fatty acyl Co.A}}{R-CH_2-CH_2-C(=O)S\sim Co.A} \xrightarrow[\text{(in peroxisome)}]{\text{Acyl Co.A oxidase}} \underset{\text{Unsaturated fatty acyl Co.A}}{R-CH=CH-C(=O)S\sim Co.A} + H_2O$$

Catalase enzyme which is also located in peroxisome, cleaves H_2O_2 immediately to H_2O and O_2.

$$2H_2O_2 \xrightarrow{\text{Catalase}} 2H_2O + O_2$$

4. Oxigenases - dependent energy - releasing biological processes. Oxigenases catalyze the incorporation of O_2 into the substrates in two steps. In first step, oxygen is bound to the active site of the enzyme, while in second step, bound oxygen is reduced or transferred to substrate. Oxigenases are of following two types :

***(i)* Monooxygenases.** These incorporate one atom of oxygen into the organic substrate while other oxygen atom is reduced to water. These enzymes are more abundant and more complex in action. These require two substrates to serve as reductants of two atoms of oxygen. Main substrate accepts one oxygen atom while cosubstrate provides two hydrogen atoms to oxygen atom.

$$\underset{\text{(Main substrate)}}{AH} + BH_2 + O_2 \xrightarrow{\text{Monooxygenase}} A-OH + B + H_2O$$

Here, the main substrate is hydroxylated so these are also called **hydroxylases**. Enzyme **phenylalanine hydroxylase** is an important monooxygenase. It changes phenylalanine to tyrosine in the presence of coenzyme NAD^+ and a cosubstrate tetrahydrobiopterin as shown below:

$$\text{Phenylalanine} + \text{Tetrahydrobiopterin} \xrightarrow[O_2 \quad NAD \quad NAD + H^+ \quad H_2O]{\text{Phenyl hydroxylase}} \longrightarrow \text{Tyrosine}$$

***(ii)* Dioxygenases**. These enzymes incorporate both atoms of oxygen into the organic substrate as under :

$$A + O_2 \xrightarrow{\text{Dioxygenase}} AO_2$$

A good example of dioxygenase enzyme is **tryptophan 2, 3-dioxygenase** which catalyzes the opening of five-membered ring of tryptophan during metabolism of it as shown below :

$$\text{Tryptophan} \xrightarrow[O_2]{} \text{N-formyl kynurenine}$$

Types of Energy System

As explained in biological oxidation, basic source of electrons is the flow of electrons due to which a number of reduced coenzymes such as $NADH_2$ and $FADH_2$ are formed especially during oxidative phosphorylation (oxidation of glucose in cell respiration). During the flow of electrons in various electron carrier complexes of ETC, one mole of glucose produces 36 or 38 energy rich ATP molecules. ATP is the primary source of energy which is fundamental to biological processes of growth, development and metabolism. It is the property of living organisms that they can harness energy from a variety of metabolic pathways. It is so as the life is dependent upon energy transformations. In various energy-rich biomolecules, the energy is bound in their chemical bonds. Usable energy is available for mechanical work (*e.g.*, muscle contraction) or anabolic/biosynthetic processes, when weaker bonds are broken and stronger bonds are made.

There are three types of energy-systems in the living organisms:

1. ATP-PC or phosphagen system.
2. Anaerobic or Lactic acid system.
3. Aerobic system or oxidative phosphorylation.

5.2 RESPIRATORY CHAIN AND OXIDATIVE PHOSPHORYLATION

Two molecules of $FADH_2$ and six molecules of NADH produced in Krebs cycle (from two molecules of acetyl-CoA) are oxidized by molecular O_2 in a **respiratory chain** or **electron transport system** or **ETS** involving a series of enzymes and coenzymes.

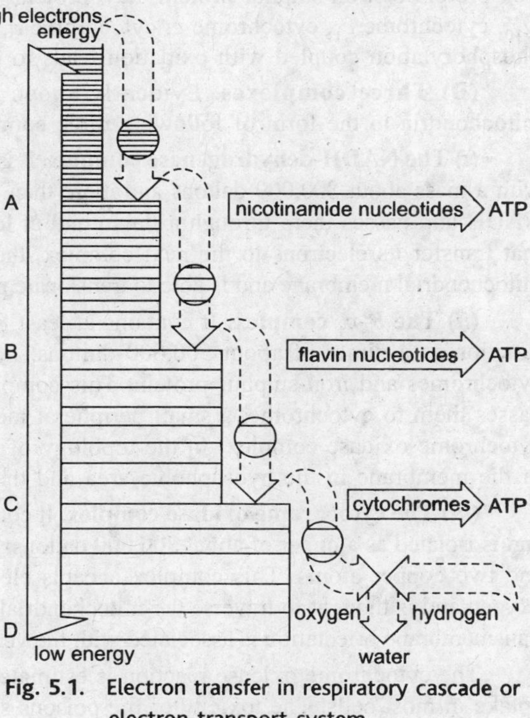

Fig. 5.1. Electron transfer in respiratory cascade or electron transport system.

In the electron transport system, the successive electron acceptors are at lower and lower energy levels. With each transfer to a lower energy level, the electrons release some of their potential energy. That is why this series is called an **electron cascade**, like a cascade of falling water (Fig. 5.1). At each stage, the released energy is used to form ATP. Since electron transport involves oxidation as well as phosphorylation (*i.e.*, ADP + P = ATP) this process by which cell system traps chemical energy is called **oxidative phosphorylation**. The passage of electrons from NAD to oxygen generates 3 ATP molecules, whereas the passage of electrons from FAD to oxygen generates only 2 ATP molecules.

(A) Compounds that occur in ETS. Following five types of compounds are associated with electron transport system of inner mitochondrial membrane:

***(i)* Pyridine-linked dehydrogenases** require as their coenzyme either NAD^+ or $NADP^+$ both of which can accept two electrons at a time. There are about 200 dehydrogenases for which NAD^+–linked compounds such as pyruvic acid dehydrogenase, are involved in ETS.

***(ii)* Flavin-linked dehydrogenases** (often called **flavoproteins** or **FPs**) require either FAD or FMN. Both are prosthetic groups whose isoalloxazine ring can accept two hydrogen atoms.

Flavin-linked enzymes are commonly involved in a number of enzyme systems such as fatty acid oxidation, amino acid oxidation and Krebs cycle activity (*e.g.*, succinic dehydrogenase or SDH).

***(iii)* Ubiquinones** were so named because of their occurrence in so many different organisms and their chemical resemblance to quinone. They are found in several different forms including the **plastoquinones** of chloroplasts. The form of ubiquinones present in mitochondria is often called **coenzyme Q_{10}** (CoQ_{10} or Q). CoQ_{10} is lipid soluble and accepts two hydrogen atoms (or two protons and two electrons) at a time.

***(iv)* Cytochromes** are proteins containing iron-porphyrin (haem) groups. There are a large number of cytochromes in cells; most are found in mitochondria, although some also function in the ER and in chloroplasts. Mitochondria have five types of cytochromes which are arranged in the following order in an inner membrane: cyt. *b*, cyt. c_1, cyt. *c*, cyt. *a* and cyt. a_3. All of them transfer electrons by reversible valence changes of the iron atom (trivalent ferric or $Fe^{3+} \rightleftharpoons$ bivalent ferrous or Fe^{2+}).

***(v)* Iron-sulphur proteins** (Fe_2S_2 and Fe_4S_4) are electron carriers of mitochondria containing iron and sulphur in equal amounts. The iron is reversibly oxidized during the electron transfer. Iron-sulphur proteins transfer one electron at a time.

All these components of ETS are arranged in the inner mitochondrial membrane in the following sequence: NAD-linked succinic dehydrogenase (SDH), flavoprotein (FAD), non-haem iron protein or iron-sulphur protein, flavoprotein (FAD), cytochrome b, ubiquinone or coenzyme Q_{10}, cytochrome c_1, cytochrome *c*, cytochrome *a*, cytochrome a_3 and three coupling sites, where phosphorylation coupled with oxidation leads to production of ATP.

(B) Three complexes. Evidently above described components of ETS occur in the mitochondria in the form of following three complexes (Fig. 5.2):

***(i)* The NADH-dehydrogenase complex.** It is the largest of the respiratory enzyme complexes, with a mass about 800,000 daltons and more than 22 polypeptide chains. It accepts electrons from NADH and passes them through a flavin and at least five iron-sulphur centres to ubiquinone (Q) that transfer its electrons to the next complex, the b–c_1, complex. This complex spans the inner mitochondrial membrane and is able to translocate protons across it from M side to C side (Fig. 5.2).

***(ii)* The b–c_1 complex.** It contains at least 8 different polypeptide chains and is thought to function as a dimer of about 500,000 daltons. Each monomer contains three haemes bound to cytochromes and iron-sulphur protein. This complex accepts electrons from ubiquinone (Q) and passes them to cytochrome, a small peripheral membrane protein that carries its electrons to the cytochrome oxidase complex. In the topology of this complex, the Q-site may be in the middle of the membrane in the hydrophobic area and the cytochrome c-site on the C side.

***(iii)* The cytochrome oxidase complex.** It comprises at least eight different polypeptide chains and is isolated as a dimer of about 300,000 daltons; each monomer contains two cytochromes (*a*, a_3) and two copper atoms. This complex accepts electrons from cytochrome *c* and passes them to oxygen and is thought to traverse the mitochondrial membrane, protruding on both surfaces. Such a transmembrane orientation is associated with the vectorial transport of protons across the membrane.

The cytochrome oxidase reaction is estimated to account for 90 percent of the total oxygen uptake in most cells. The toxicity of the poisons such as cyanide and azide is due to their ability to bind tightly to this complex and thereby block all electron transport.

(C) F_0–F_1 complex or coupling factors. One of the main proteins in the inner mitochondrial membrane is the multisubunit **coupling factor** (Fig. 5.3), the enzyme that actually synthesizes ATP and simultaneously acting as a proton pump (Box 5.1). A quite similar enzyme complex is located in the thylakoid membranes of chloroplasts and in the plasma membrane of bacterial cell. The coupling factor has two principal components:

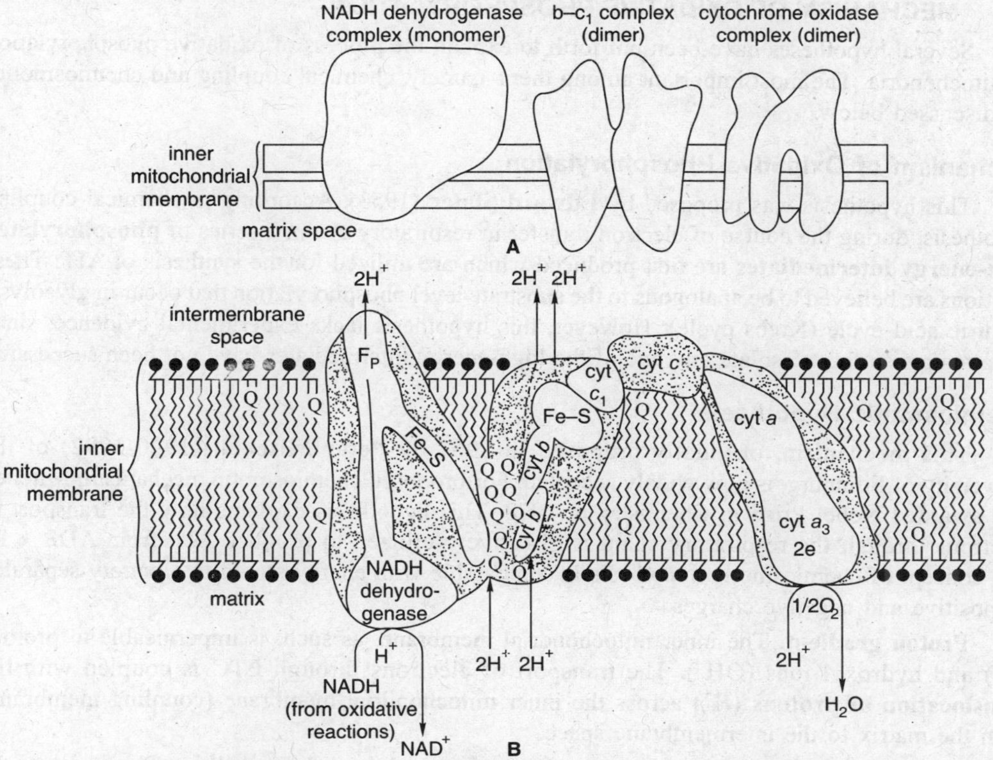

Fig. 5.2. Respiratory enzyme complexes. A—The relative sizes and shapes of the three respiratory enzyme complexes; B—Mode of transfer of H^+ through the three respiratory enzyme complexes of inner mitochondrial membrane.

***(a)* F_0-complex.** It is an integral membrane complex, composed of vary hydrophobic proteins—3 or 4 distinct polypeptides and one proteolipid—which together span the mitochondrial membrane. F_0-complex possesses the proton translocating mechanism. F_0-complex can be extracted only with strong detergents.

***(b)* F_1-particle.** Attached to the F_0 complex is F_1 particle, a complex of five distinct polypeptides: alpha (α), beta (β), gamma (γ), delta (δ) and epsilon (ε), with the probable composition of $\alpha_3\ \beta_3\gamma\delta\varepsilon$. F_1 forms the knob or 'tadpole' that protrudes on the matrix side of the inner mitochondrial membrane. F_1 particle can be detached from the membrane by mechanical agitation and is water soluble.

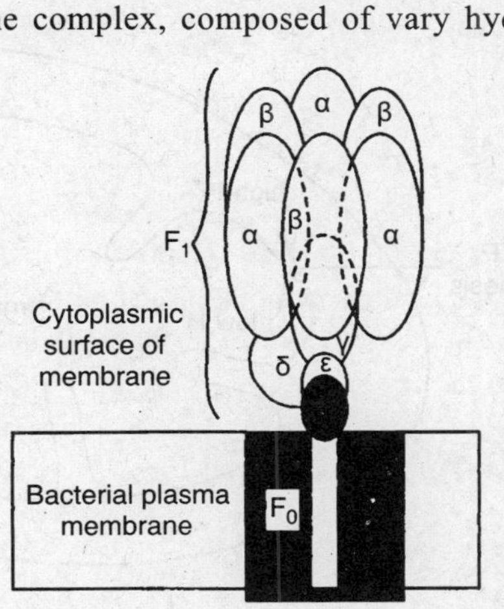

Fig. 5.3. Model of molecular structure of F_0 and F_1 particles in the bacterial plasma membrane. The core of F_1 particle is an $\alpha_3\beta_3$ complex. The δ polypeptide of F_1 and probably the γ are involved in binding F_1 to the membrane-embedded F_0-particle.

5.3 MECHANISM OF OXIDATIVE PHOSPHORYLATION

Several hypotheses have been put forth to explain the process of oxidative phosphorylation in mitochondria. The most important among them–namely, chemical coupling and chemiosmotic-are discussed below.

Mechanism of Oxidative Phosphorylation

This hypothesis was proposed by **Edward Slater** (1953). According to chemical coupling hypothesis, during the course of electron transfer in respiratory chain, a series of **phosphorylated high-energy intermediates** are first produced which are utilized for the synthesis of ATP. These reactions are believed to be analogous to the substrate level phosphorylation that occur in glycolysis or citric acid cycle (Krebs cycle). However, this hypothesis lacks experimental evidence, since all attempts so far, to isolate any one of the high-energy intermediates have not been successful.

Chemiosmotic Hypothesis

This mechanism, originally proposed in 1961 by **Peter Mitchell** (1920-1992) of the University of Edinburg, is now widely accepted. For proposed chemiosmotic mechanism, Mitchell was awarded Nobel Prize in Chemistry in 1978. This hypothesis explains how the transport of electrons through the respiratory chain is effectively utilized to produce ATP from ADP + Pi. The concept of chemiosmotic hypothesis is comparable with energy stored in a battery separated by positive and negative charges.

Proton gradient. The inner mitochondrial membrane, as such, is impermeable to protons (H^+) and hydroxyl ions (OH^-). The transport of electrons through ETC is coupled with the **translocation of protons (H^+)** across the inner mitochondrial membrane (coupling membrane) from the matrix to the intermembrane space.

The pumping of protons results in an **electrochemical** or **proton gradient**. This is due to the accumulation of more H^+ ions (low pH) on the outer side of the inner mitochondrial membrane than the inner side (Fig. 5.4). The proton gradient developed due to the electron flow in the respiratory chain is sufficient to result in the synthesis of ATP from ADP and Pi.

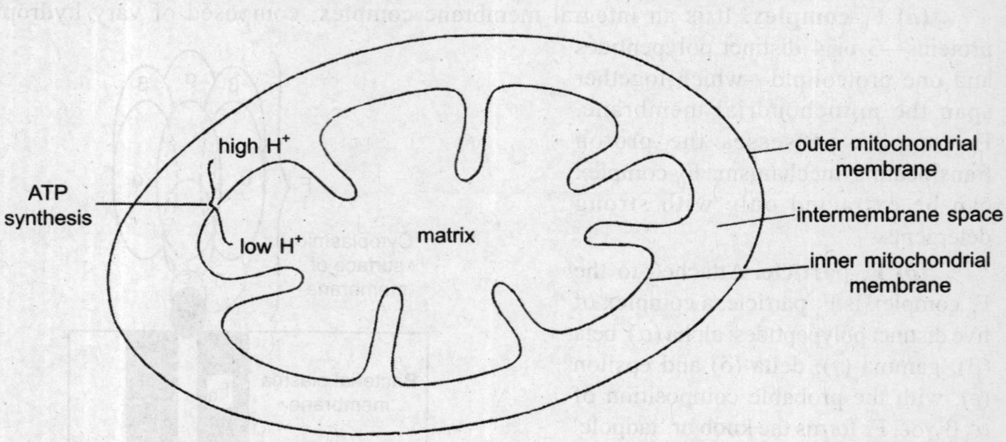

Fig. 5.4. Outline of chemiosmotic hypothesis for oxidative phosphorylation.

Enzyme system for ATP synthesis. ATP synthase enzyme present in complex V, utilise the proton gradient for the synthesis of ATP. The enzyme is also known as **ATPase** since it can hydrolyse ATP to ADP and Pi. ATP synthase is a complex enzyme and consists of two functional subunits, namely F_1 and F_0 (Fig. 5.5). Its structure is comparable with "lollipop".

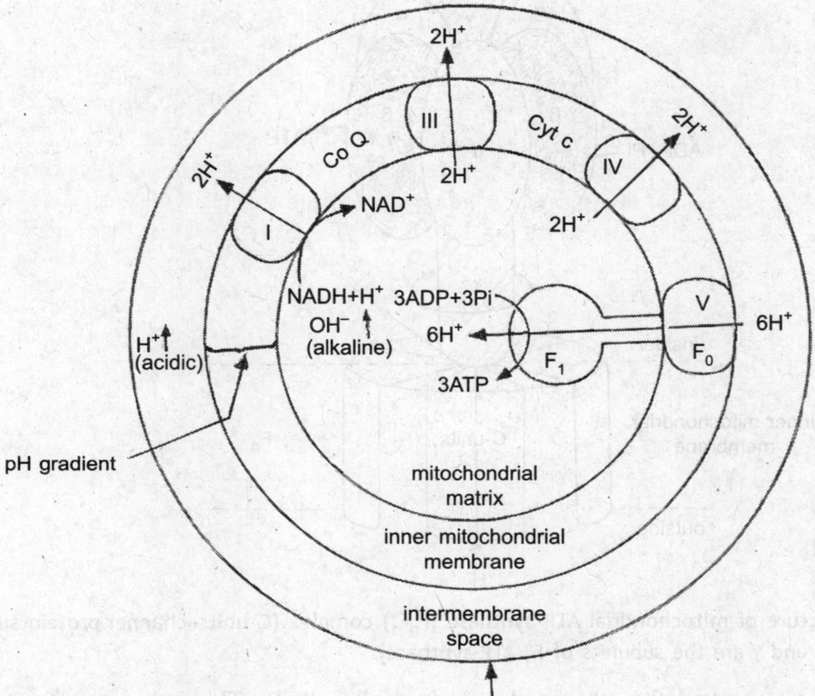

Fig. 5.5. Diagrammatic representation of chemiosmotic hypothesis for oxidative phosphorylation (I, III, IV and V–respiratory chain complexes; F_0, F_1–protein subunits for phosphorylation).

Box 5.1

The **subscript of Fo** is not zero but the letter O which denotes that it is the portion of the ATP synthase which binds the toxic antibiotic, oligomycin. Oligomycin is a potent inhibitor of this enzyme and thus of oxidative phosphorylation.

The protons that accumulate on the intermembrane space re-enter the mitochondrial matrix leading to the synthesis of ATP.

Rotary Motor Model for ATP Generation

Paul Boyer of UCLA in 1964 proposed a innovative hypothesis (For it he got Nobel Prize in Chemistry in 1997) that a conformational change in the mitochondrial membrane proteins lead to the synthesis of ATP. The original Boyer hypothesis, now considered as **rotary motor/engine driving model**, is widely accepted for the generation of ATP.

The enzyme ATP synthase is F_0F_1 complex (of complex V). The F_0 subcomplex is composed of channel protein 'C' subunits to which F_1–ATP synthase is attached (Fig. 5.6). F_1–ATP synthase consists of a central γ subunit surrounded by alternating α and β subunits (α_3 and β_3).

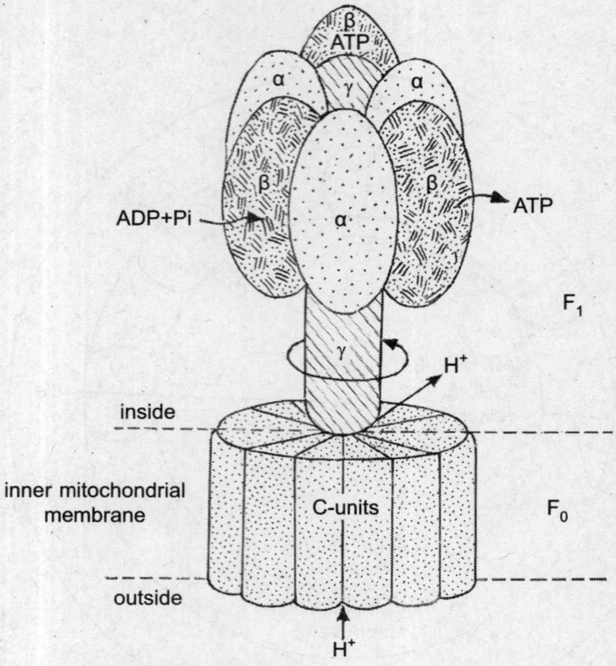

Fig. 5.6. Structure of mitochondrial ATP synthase (F_0F_1) complex (C units–channel protein subunits; α, β and γ are the subunits of F_1 ATP-synthase).

In response to proton flux, the γ–subunit physically rotates. This induces conformational changes in the β_3 subunits that finally leads to the release of ATP.

According to the **binding change mechanism** the three β subunits of F_1–ATP synthase adopt different conformations. One subunit has **open (O)** conformation, the second has **loose (L)** conformation while third one has **tight (T)** conformation.

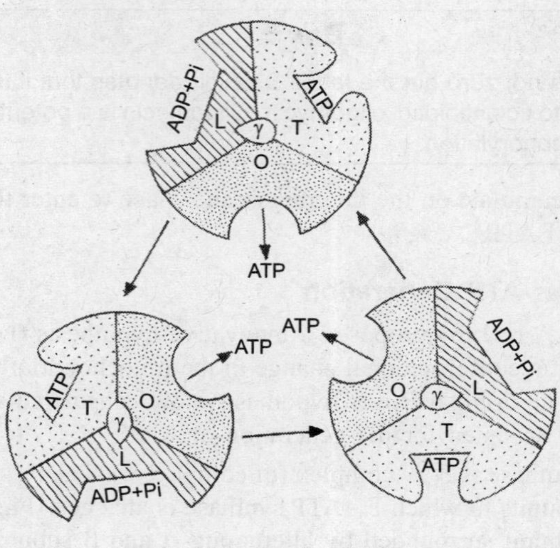

Fig. 5.7. The binding change model (rotary motor/engine driving model) for ATP synthesis by F_1–ATP synthase.

By an unknown mechanism, protons induce the rotation of γ subunits, which in turn induces conformation changes in β subunits. The substrates ADP and Pi bind to β subunit in L-confirmation. The L site changes to T confirmation and this leads to synthesis of ATP. The O site changes to L confirmation which binds to ADP and Pi. The T site changes to O conformation, and releases ATP. This cycle of conformation changes of β subunits is repeated. And three ATP molecules are generated for each revolution (Fig. 5.7).

It may be noted that the ATP release in O conformation is energy dependent (and not ATP synthesis) and very crucial rotary motor model for ATP generation.

The enzyme ATP synthase acts as a proton-driving motor, and is an example of **rotary catalysis**. Thus, ATP synthase is the world's smallest **molecular motor**.

Demonstration of rotational catalysis by single F_1 complex molecule. Direct evidence for the rotation of γ gamma subunit relative to the $\alpha\beta$ subunits has been obtained in a variety of experiments (**Noji** *et al.*, 1993, 1997). If "seeing is believing", then the most convincing demonstration came in 1993 from the work of **Masasuka Yoshida** and colleagues at Tokyo Institute of Technology and Keio University in Japan. These researchers devised an ingenious system to watch individual ATP synthase enzyme molecules catalyze the reverse reaction from that normally operating in the cell. First, they prepared a genetically engineered version of F_1 portion of the ATP synthase, consisting of three α subunits three β subunits, and a γ subunit ($\alpha_3\beta_3\gamma$) (Fig. 5.8). This polypeptide is then fixed to a glass coverslip by its head, and a short, fluorescently labeled actin filament was attached to the end of the γ subunit that projected into the medium (Fig. 5.8). When the preparation was incubated with ATP and observed under the microscope, the fluorescently labeled actin filament was seen to rotate like a microscopic propeller, powered by the energy released as ATP molecules were bound and hydrolyzed by the catalytic sites of the β subunits.

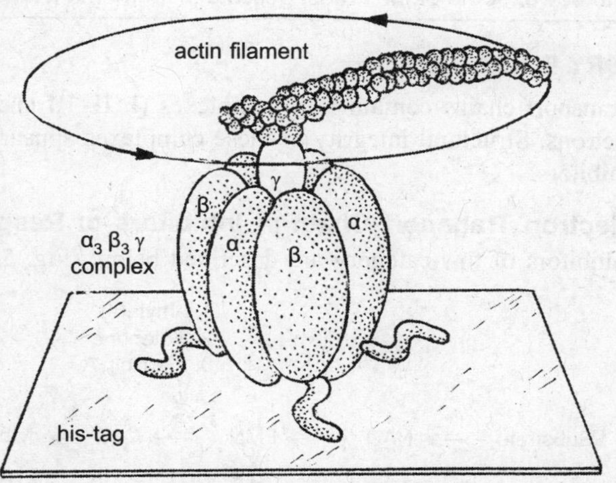

Fig. 5.8. Direct observation of rotational catalysis by bioengineering experiment of **Noji, Yasuda, Yoshida** and **Kinosita** (1997). To carry out the experiment, a modified version of a portion of the ATP synthase consisting of $\alpha_3\beta_3\gamma$ subunits was prepared. Each β subunit was modified to contain 10 histidine (His) residues (*i.e.*, amino acids) at its N-terminus, a site located on the outer (matrix) face of the F_1 head. The side chains of histidine have a high affinity for a substance (Ni-NTA), which was used to coat the coverslip. The γ subunit was modified replacing one of the serine residues near the end of the stalk with a cysteine residue, which provided a means to attach the fluorescently labeled actin filament. In the presence of ATP, the actin filament was observed to rotate counter clockwise (when viewed from the membrane side). At low ATP concentration, the actin filaments could be seen to rotate in steps of 120° (after Karp 2010).

More recently, a similar type of F_1 complex was "forced" to perform the more challenging activity that is normally carries out within the cell–the synthesis of ATP. In this study, a microscopic magnetic bead was attached to γ subunit of a single F_1 molecule and they caused to rotate to clockwise direction by subjecting the preparation to a revolving magnetic field. When one of these F_1 molecules was placed within a tiny transparent microchamber in the presence of ADP and Pi, the forced rotation of the γ subunit led to the synthesis of ATP, which built upto μM concentrations. As would be expected from the model, three molecules of ATP were synthesized with each 360° turn. When the magnetic field was switched off, the γ subunit rotated spontaneously in the reverse direction, driven by the hydrolysis of the recently synthesized ATP. Taken together, these innovative **bioengineering experiments** demonstrated evidently that ATP synthase operates as a rotary motor (Box 5.2).

Box 5.2 Biological Rotary "Nanomachines"

Rotary machines are common in our industrialized society; we use rotary turbines, rotary drills, rotary wheels and propellers, to name just a few. But rotary devices are extremely rare in living organisms. There are, for example, no known rotary organelles in eukaryotic cells, rotary joints in animals, or rotary feeding structures in the biological world. In fact, only two types of biological structures are known that contain rotating parts: ATP synthases (and related proteins that act as ionic pumps) and bacterial flagella, both of which can be described as **rotary nanomachines** because their size is measured in nanometers. Engineers have begun to invent nanoscale devices made of inorganic materials that may one day carry out various types of sub-microscopic mechanical activities. The construction of nanosized motors pose a particular challenge, and attempts have already been made to use ATP synthase to power simple, inorganic devices. Someday humans may be using ATP instead of electricity to power some of their most delicate instruments (**Karp** 2010).

5.4 RESPIRATORY INHIBITORS

The electron transport chains contain four complexes (I, II, III and IV) which participate in the transfer of electrons. Structural integrity of these complexes appears essential for its interaction with most inhibitors.

1. Inhibitors of Electron Transport Chain (= Inhibitors of Respiratory Chain)

The various inhibitors of this category are described below (Fig. 5.9).

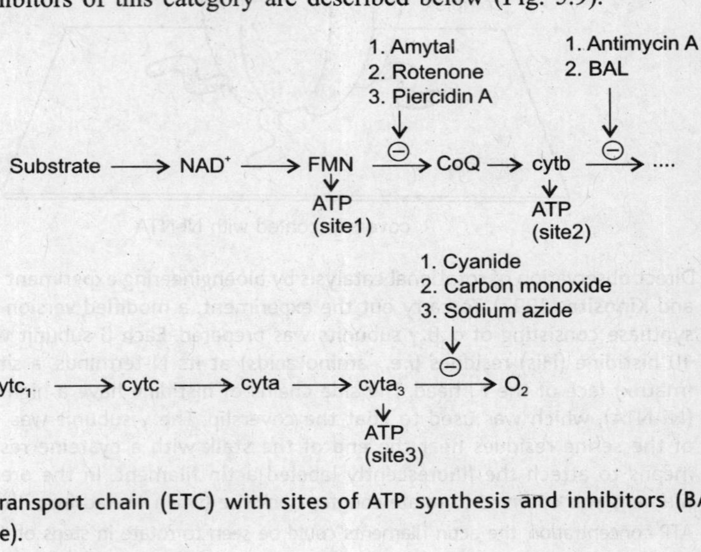

Fig. 5.9. Electron transport chain (ETC) with sites of ATP synthesis and inhibitors (BAL = British antilewisite).

***(i)* Rotenone.** Rotenone is a compound extracted from the roots of tropical plants such as *Derris elliptica* and *Lonchoncarpus nicou*. It complexes avidly with NADH dehydrogenase and acts between the Fe–5 proteins and ubiquinone. Only 30 nanomoles per gram of mitochondrial protein are effective for blocking the site. Rotenone is relatively nontoxic to mammals because it is absorbed poorly. However, exposure of the lungs to its dust is little more dangerous. Rotenone is intensely toxic to the fishes and insects as it readily passes into their gills and breathing tubes respectively. The Amazonian Indians, hence, use rotenone to poison fishes and kill the insects.

***(ii)* Piericidin A.** It is an antibiotic, produce by species of fungus *Streptomyces*. It has an action similar to that of rotenone.

***(iii)* Amytal (Barbiturates** and **Seconal)**. These also block NADH dehydrogenase, but are required in much higher concentrations for this purpose.

***(iv)* Antimycins.** These are also antibiotics and also produced by *Streptomyces*. They inhibit the respiratory chain at or around the site II and block electron flow between cytochrome b and C_1, which prevents ATP synthesis coupled to generation of a proton gradient at site II.

Antimycin A is an antibiotic; British antilewisite (BAL) is an antidote used against war-gas, both are inhibitors of the sites between cytochrome b and c_1.

***(v)* Dimercaprol.** It is identical in action to the antimycins.

***(vi)* Cyanides.** Inhalation of hydrogen cyanide (HCN) gas or ingestion of potassium cyanide (KCN) causes a rapid and extensive inhibition of the mitochondrial electron transport chain at the cytochrome oxidase step. *Cyanide is one of the most potent and rapidly acting poison known*. It binds to the Fe^{3+} (ferric form) in the haeme a_3 of cytochrome c oxidase that catalyzes the terminal step in the electron transport chain. Cyanide thus prevent the binding of oxygen as the final electron acceptor. Mitochondrial respiration and energy production cease, and cell death occur rapidly.

***(vii)* Azide.** Azide (N^{3-}) also blocks the electron flow between the cytochrome oxidase complex and oxygen. It reacts with the ferric form (Fe^{3+}) of this carrier.

***(viii)* Hydrogen sulphide (H_2S).** It is a lethal menace in all drilling operations, so on oil-drilling platforms at sea. *In vitro* tests reveal that 0.1 μM sulphide inhibits cytochrome oxidase more than does 0.3 μM cyanide.

***(ix)* Carbon monoxide (CO).** It also attacks between cytochrome oxidase and O_2 but, unlike cyanide and azide, CO inhibits the ferrous form (Fe^{2+}) of the electron carrier.

Phosphorylation coupled to generation of a proton gradient at site III does not occur in the presence of these inhibitors (*i.e.*, cyanides, azide, H_2S and CO) because electron flow is blocked.

2. Inhibitors of Oxidative Phosphorylation

These compounds inhibit both electron transport and oxidative phosphorylation.

***(i)* Oligomycins.** These polypeptide antibiotics are obtained from various species of *Streptomyces*. They inhibit transfer of high-energy phosphate to ADP and, therefore, also inhibit electron transfers coupled to phosphorylation.

***(ii)* Rutamycin.** This antibiotic also inhibits both electron transport and oxidative phosphorylation.

***(iii)* Atractylate (= Atractyloside).** This toxic glycoside is extracted from the rhizome of *Atractylis gummifera*, a plant native to Italy. It blocks oxidative phosphorylation by competing with ATP and ADP for a site on the ADP-ATP antiport of the inner mitochondrial membrane. Hence, it checks renewal of ATP supply in the cytosol.

(*iv*) Bongkrekate. It is a toxin formed by a bacteria (*Pseudomonas*) in a coconut preparation (called "bongkrek") from Java. It also blocks ADP-ATP antiport.

3. Uncouplers of Oxidative Phosphorylation

Uncoupling agents are compounds which dissociate (or uncouple) the synthesis of ATP from the transport of electrons through the cytochrome system. This means that the electron transport continue to function, leading to oxygen consumption but phosphorylation of ADP is inhibited. In the intact mitochondria, these two processes are intimately associated. When they are uncoupled, the transport of electrons speed up, there by pointing out that the phosphorylation of ADP has been a rate limiting process. In the presence of uncouplers, the free energy released by electron transport appears as heat, rather than as newly-made ATP. Uncoupling agents greatly enhance the permeability of the inner membrane to H^+. They are lipophilic and bind H^+ from one side of the membrane and carry it through the membrane toward the side with lower H^+ concentration. In Mitchell's hypothesis uncouplers are agents that are capable of destroying the vectorial, anisotropic structure of the membrane leading to elimination of the pH gradient. As the uncouplers bind and carry protons, they are also called **protonophores.** Uncoupling can be distinguished from inhibition. Uncoupling causes an increased oxygen consumption in the absence of increased utilization of ATP, whereas inhibition of phosphorylation (or inhibition of ADP-ATP antiport) diminishes consumption in normal coupled mitochondria. Examples of some common uncouplers are the following.

(*i*) 2, 4-dinitrophenol (DNP). Introduced by **Loomis** and **Lipmann**, dinitrophenol (Fig. 5.10) is one of the most effective agent for uncoupling respiratory-chain phosphorylation. The uncoupling action of DNP is due to the fact that both the phenol and the corresponding phenolate ion are significantly soluble in the lipid core of inner mitochondrial membrane. The phenol diffuses through the core toward the matrix, where it loses proton; the phenolate ion then diffuses back toward the cytosol side, where it picks up a proton to repeat the process (Fig. 5.11). DNP is also stimulates the activity of the enzyme ATPase, which is normally inactive as a hydrolytic enzyme in mitochondria.

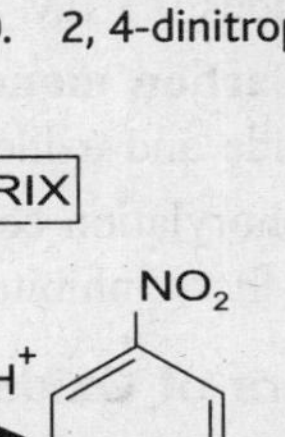

Fig. 5.10. 2, 4-dinitrophenol, DNP.

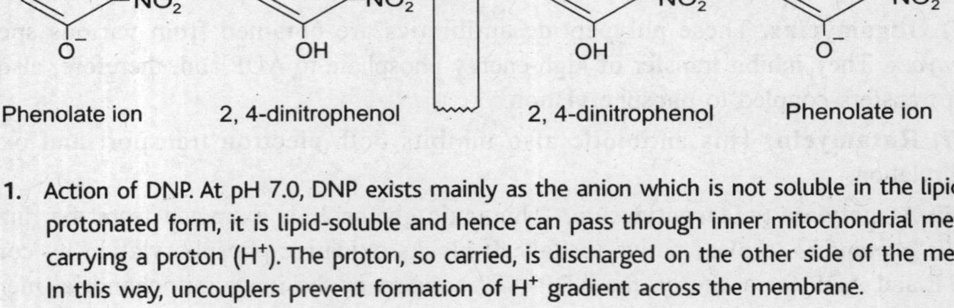

Fig. 5.11. Action of DNP. At pH 7.0, DNP exists mainly as the anion which is not soluble in the lipids. In its protonated form, it is lipid-soluble and hence can pass through inner mitochondrial membrane, carrying a proton (H^+). The proton, so carried, is discharged on the other side of the membrane. In this way, uncouplers prevent formation of H^+ gradient across the membrane.

***(ii)* Dicoumarol.** Dicoumarol or 3, 3′–methylene-bishydroxycoumarin, arises from the action of microorganisms on coumarin, a natural constituent of sweat clover, *Melilotus indica*. It has action identical to that of DNP. Dicoumarol is also an antagonist of vitamin K function.

***(iii)* m-Chlocarbonyl cyanide phenylhydrozone (CCCP).** Its action is also similar to that of DNP but it is about 100 times active than the latter.

4. Ionophores of Oxidative Phosphorylation

Ionophores ("ion carriers") are lipophilic substances, capable of binding and carrying specific cations through the biological membranes. Ionophores differ from the uncouplers in that they promote the transport of cations other than H^+ through the membrane. Some common types of ionophores are the following:

***(i)* Valinomycin.** This toxic antibiotic is a peptide hormone that binds potassium ion, K^+ and is synthesized by *Streptomyces*. Valinomycin is a repeating macrocyclic molecule made up of four kinds of residues (L–lactate, L–valine, D–hydroxyisovalerate and D–valine) taken 3 times. The four residues are alternately joined by ester and peptide bonds. It contains 6 peptide and 6 ester bonds with side chains consisting of hydrophobic alkyl radicals. The antibiotic forms a lipid-soluble complex with K^+ which readily passes through the inner mitochondrial membrane, whereas K^+ alone in the absence of valinomycin penetrates only very slowly. Valinomycin has a high degree of selectivity for K^+ as compared to Na^+. In fact, it binds K^+ about a thousand times as strongly as Na^+ bacause water has less attraction for K^+ than for Na^+. And it is energetically more costly to pull Na^+ away from water. Thus, valinomycin interferes with oxidative phosphorylation in mitochondria by making them permeable to K^+. The result is that mitochondria use the energy generated by electron transport to accumulate potassium ions (K^+) rather than to make ATP.

***(ii)* Gramicidin A.** Gramicidin A is a linear polypeptide consisting of 15 amino acid residues. Two noteworthy features of the molecule are that *(a)* the L-and D-amino acids alternate and that *(b)* both the N- and C-terminals of this polypeptide are modified (Fig. 5.12). Gramicidin A promotes penetration not only of potassium ions (K^+) but also of sodium ions (Na^+) and several other monovalent cations through the inner membrane. Unlike valinomycin, gramicidins do not complex cations. Rather they induce ion permeability of membranes at concentrations as low as 10^{-10} M by forming dimers which in effect 'provide' tubes or channels which span the membrane and through which the cations passes (Fig. 5.13). They are therefore better called as **ion channels** in contrast to valinomycin and allied substances which are classed as "ion carriers".

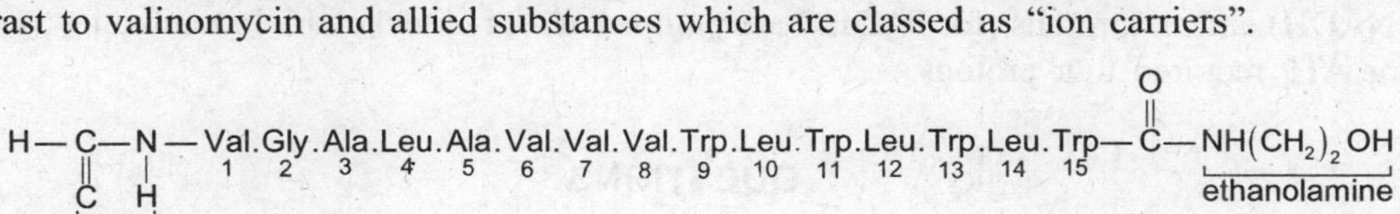

Fig. 5.12. Gramicidin A.

***(iii)* Nigericin.** Like valinomycin, it acts as an ionophore for K^+ but in exchange for H^+. It, therefore, abolishes the pH gradient across the membrane. In the presence of both valinomycin and nigericin, both the membrane potential and the pH gradient are eliminated and phosphorylation is therefore, completely inhibited.

Lastly, certain interlocking and regulatory mechanisms control glycolysis so that pyruvate is produced at a rate at which it is required by the citric acid cycle, a process which donates electrons for the oxidative phosphorylation.

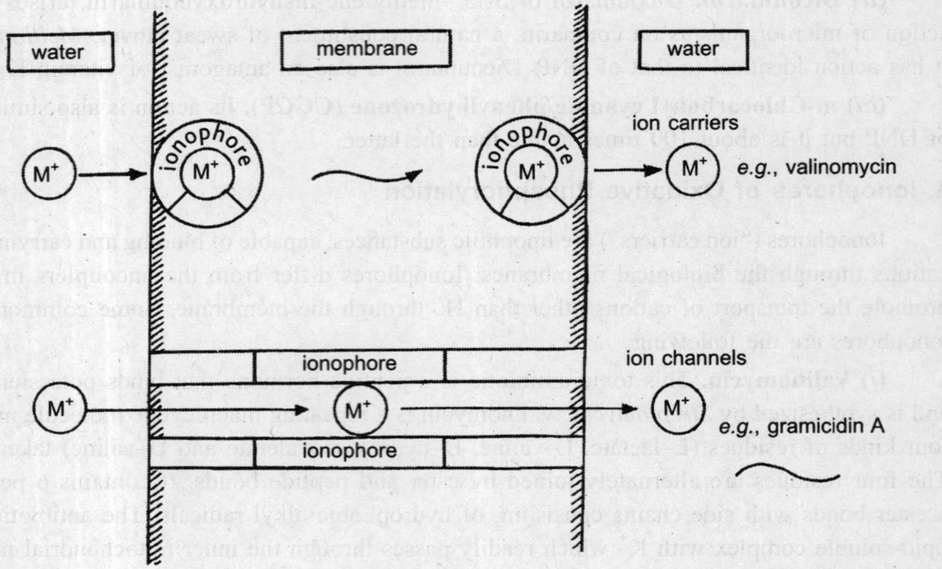

Fig. 5.13. Schematic representation of the action of ionophores on membranes; M^+ = metal ions.

5.5 P:O RATIO

The P:O ratio refers to the number of **inorganic phosphate (Pi)** molecules utilized (*i.e.*, organic phosphate) generation for every atom of **oxygen (1/2 O_2)** consumed. In other words, P:O ratio represents the number of molecules of ATP synthesized per pair of electrons carried through ETC.

The mitochondrial oxidation of NADH with a classical P:O ratio of 3 can be represented by the following equation:

$$NADH + H^+ + \frac{1}{2}\ O_2 + 3\ ADP + 3Pi \rightarrow NAD^+ + 3ATP + 4\ H_2O$$

Further, a P:O ratio of 2 has been assigned to the oxidation of $FADH_2$. There is a strong evidence now to suggest a P:O ratio of 2.5 for NADH, and 1.5 for $FADH_2$, since 10 protons (for NADH) and six protons (for $FADH_2$) are pumped across mitochondrial membrane. Synthesis of one ATP required four protons.

QUESTIONS

Long Answer Questions

1. Describe the components of electron transport chain and discuss the oxidation of NADH.
2. Define oxidative phosphorylation. Discuss chemiosmotic hypothesis in detail.
3. Discuss about the inhibitors of ETC and oxidative phosphorylation.

Short Answer Questions

1. Write short notes on the following:
   (*i*) High-energy bonds

(*ii*) Sites of oxidative phosphorylation

(*iii*) Uncouplers

(*iv*) Redox potential

(*v*) P:O ratio

(*vi*) ATP synthase

Very Short Answer Questions

1. What is the effect of each of following inhibitors on electron transport and ATP formation by the respiratory chain?

(*a*) Azide
(*b*) Atractyloside
(*c*) Rotenone
(*d*) DNP
(*e*) Carbon monoxide
(*f*) Antimycin A

Fill in the Blanks

1. The storage form of high-energy compound in vertebrates is ....

2. A more negative redox potential represents a greater tendency to lose ....

3. The electron transport chain is located in ....

4. The component of electron transport chain which can directly react with molecular oxygen ....

5. The site of ETC inhibited by cyanide ....

6. The bonds responsible for a majority of high-energy compounds are .... bonds.

Multiple Choice Questions

1. Name the compound with the greatest standard free energy

(*a*) ATP
(*b*) phosphocreatine
(*c*) cyclic AMP
(*d*) phosphoenolpyruvate

2. One of the following components of ETC possesses isoprenoid units

(*a*) coenzyme a
(*b*) cytochrome c
(*c*) cytochrome b
(*d*) non-heme iron

3. The P : O ratio for the oxidation of $FADH_2$ is

(*a*) 1
(*b*) 2
(*c*) 3
(*d*) 4

4. Inner mitochondrial membrane is impermeable to

(*a*) H^+
(*b*) K^+
(*c*) OH^-
(*d*) All of them

5. ATP synthase activity is associated with the mitochondrial enzyme complex

(*a*) V
(*b*) III
(*c*) IV
(*d*) I

6. Iron-containing compounds that act as hydrogen acceptors in the respiratory chain are

(*a*) flavoproteins
(*b*) dehydrogenase
(*c*) cytochrome
(*d*) oxidases

ANSWERS

Very Short Answer Questions

1. (*a*) Azide blocks electron transport and proton pumping at complex III.

(*b*) Atractyloside blocks electron transport and ATP synthesis by inhibiting the exchange of ATP and ADP across the inner mitochondrial membrane.

(*c*) DNP blocks ATP synthesis without inhibiting electron transport by dissipating the proton gradient.

(*d*) Carbon monoxide blocks electron transport and proton pumping at complex III.

(*e*) Antimycin A blocks electron transport and proton pumping at complex II.

Fill in the Blanks

1. Phosphoarginine.
2. Electrons.
3. Inner mitochondrial membrane.
4. Cytochrome oxidase (cyt a + a_3).
5. Cytochrome a + a_3.
6. Phosphoanhydride bonds.

Multiple Choice Questions

1. (*d*) **2.** (*a*) **3.** (*b*) **4.** (*d*) **5.** (*a*) **6.** (*c*).

INDEX

D

F

H

I

N

S

W

X

Y

Z